CHEMISTRY FOR COMPETITIONS

KINDLY NOTE:

The answers (A), (B), (C), (D) and (E) in the case of <u>Assertion and Reasoning questions</u> are based on:

(A) If both the assertion and reason are true, and the R is the correct explanation of A.

(B) If both the A and R are true, but the R is not a correct explanation of the A.

(C) If the A is true, but the R is false.

(D) If both the A and R are false.

(E) If the A is false, but the R is true.

CHEMISTRY FOR COMPETITIONS

T.S.S. RAO
M.Sc., Ph.D., Dr. Sc. (Germany)

T. SAMPURNA
M.Sc., Ph.D., Dr. rer. nat. (Germany)

CBS

CBS PUBLISHERS & DISTRIBUTORS
NEW DELHI • BANGALORE

ISBN : 81-239-1137-8

First Edition : 2004

Publishing Director : Vinod K. Jain

Published by :
Satish Kumar Jain for CBS Publishers & Distributors,
4596/1-A, 11 Darya Ganj, New Delhi - 110 002 (India)
E-mail : cbspubs@del3.vsnl.net.in
Website : http://www.cbspd.com

Branch Office :
Seema House, 2975, 17th Cross, K.R. Road,
Bansankari 2nd Stage, Bangalore - 560070
Fax : 080-6771680 • E-mail : cbsbng@vsnl.net

Printed at :
J.S. Offset Printers, Delhi - 110 051

PREFACE

The authors are of firm opinion that a thorough and sincere study of the concepts and practice of the objectives is the only surest way to selection for professional courses at the institutions of higher learning. An effort to guide the readers to relevant concepts in brief has been made. This book is planned for use as a tool for facing the Medical and Engineering entrance examinations at National level. The objectives included are of the levels required for success at IIT-Screening, CBSE-PMT, CBSE-AIEEE, AIIMS, JIPMER, BHU, AFMC, CPMT, EAMCET, M.P.-PMT and PET levels. 150 thought provoking questions with hints helpful for IIT-JEE and CBSE-PMT mains are incorporated as per the latest syllabus. The book is an invaluable source of chemistry useful to candidates appearing in PSC, UPSC, UGC and CSIR (NET/GATE) tests.

We shall welcome constructive criticism and suggestions with a view to improve our effort. We shall be highly grateful to the readers if the errors, obscurities and omissions detected are brought to our notice. The repetition of objective questions is not avoided as they were asked in different important exams and as they are useful to the students to know the level of each exam.

The authors are thankful to **Shri Praveen Kumar Rathore** B.Sc., M.A., PGDCA and **Smt. Vanisha Rathore** B.Sc., M.Lib.I.Sc., PGDCA for setting the manuscipt in the present attractive form. We thank **Shri Vipin Jain** M.Com for the application of Chem. Sketch and ISIS Draw -2.2 throughout this book for structural and reaction representations.

We are extremely thankful to .... for prompt publication of this book.

C-99, University Campus
SAGAR (M.P.) 470 003
e-mail: *drtssrao@yahoo.co.in*

T. S. S. Rao
T. Sampurna

Regarding the book

- **Each chapter** contains the following parts in sequence

    1. Concepts/fundamentals of the topics as per the latest CBSE syllabus.
    2. Things to remember (physical chemistry section), stressing on the main points often asked in exams and the formulae required for solving the problems.
    3. Solved examples (physical chemistry section) to stimulate the thought process for solving problems.
    4. Compilation of multiple choice type questions and fill in the blank type questions asked in various national and state level medical and engineering entrance examinations with thoroughly prepared answers.

- **Thought Provoking Questions** with hints which will be of help in preparing for short answer type questions.
- **Nobel Laureates in Chemistry** upto 2003 with a mention of their invention/discovery/contribution.

CONTENTS

PERIODIC TABLE

The Earlier Classifications were based on

1. Metals and non metals: B, Si, Ge, As, Te are metalloids. Hg is a metal but liquid at room temperature. Iodine is a non-metal but has metallic lusture (Lusture: Shiny; malleable: able to be beaten into shapes; ductile: able to be drawn into wires).

2. Valency: H, F, Cu, Ag, etc. are monovalent but differ vastly.

After Dalton's atomic theory the scientists tried to seek relationship between the properties of the elements and their atomic masses.

3. Dobereiner's Triads: e.g. Li^{7}, Na^{23}, K^{39}; all the known elements can't be grouped into triads. Fe Co, Ni have nearly same atomic masses.

4. Newland's Law of Octaves: When arranged in order of their increasing atomic masses, the properties of the eighth element were the repetition of the first. Noble gases were not known; classification broke down after Ca.

5. Lother Mayer: emphasized physical characteristics Atomic volume = atomic mass / density.

6. Mendeleev (1869) Periodic Law: The properties of the elements are a periodic function of their atomic mass(the properties of the elements vary in relation to their atomic masses). The existence of noble gases was not predicted. The position of isotopes can't be explained. Only 63 elements were known at that time.

Isotopes: Cl^{35}_{17}, Cl^{37}_{17}; C^{12}_{6}, C^{13}_{6} & C^{14}_{6} (Same atomic number)

Isobars: Ca^{40}_{20}, K^{40}_{19}, Ar^{40}_{18}; Xe^{130}_{54}, Ba^{130}_{56} (Same atomic mass)

Isotones: Si^{30}_{14}, P^{31}_{15}, S^{32}_{16}; K^{39}_{19}, Ca^{40}_{20} (Same number of neutrons)

Nuclear isomers: differ in radioactive property (U_x $t_{0.5}$, 1.4 min; U_z $t_{0.5}$, 6.7 hours)

Modern Periodic Table: Periodic law is based on atomic number. According to IUPAC (1984) system there are 18 groups and 7 periods.

Bohr's Classification: The basis is electronic configuration.

1. Noble gases: outer shell has 8 electrons (He: 2 only) $n\,s^2\,p^6$ configuration.

2. Representative elements: outer shell incomplete $ns^{1 \text{ or } 2}$; $n\,s^2\,n\,p^{1-6}$

3. Transition elements: outer shell as well as penultimate shell incomplete $(n-1)\,d^{1-10}\,n\,s^{1 \text{ or } 2}$

4. Inner transition elements: outer shell, penultimate and prepenultimate shell (d and f orbitals) incomplete $(n-2)\,f^{1-14}\,(n-1)\,d^{0 \text{ or } 1}\,n\,s^2$

Description of Periods

1st period ---- 2 elements; H and He- non metals

2nd period ---- 8 elements

3rd period ---- 8 elements

4th period ---- 18 elements Sc^{21} -- Zn^{30} (3 d)

5th period ---- 18 elements Y^{39} -- Cd^{48} (4 d)

6th period ---- 32 elements La^{57}, Hf^{72} -- Hg^{80} (5 d); Ce^{58} -- Lu^{71} (14 Lanthanides)

7th period ---- 26 elements Ac^{89}, Rf^{104}, Db^{105}, Sg^{106}, Ns^{107}, Hs^{108}, Mt^{109}, Dt^{110} (Darmstadtium) (6d); Th^{90} - Lr^{103} (14 Actinides)

Cu -- 2, 8, 8, $4s^1$, $3d^{10}$.

Cr -- 2, 8, 8, $4s^1$, $3d^5$,

Fe -- 2, 8, 8, $4s^2$, $3d^6$

The copper family: Cu, Ag, Au; Zinc family: Zn, Cd, Hg are actually representative elements but since they share many properties with the transition elements they are generally classified into latter.

Atomic Properties

1. Atomic volume: It is the volume occupied by one mole of atoms of the element when it is a solid.

Atomic radius: Distance between the centre of the nucleus and outermost shell of electrons.

Covalent radius: Half of the distance between the two nuclei of two covalent bonded atoms of the same element in a molecule.

van der Waal's radius : Distance between two non-bonded atoms of two adjacent molecules.

In noble gases the atomic radii are only the van der Waal's radii which are naturally higher than the covalent radii of other elements.

Generally the covalent radii are smaller than the others. It is expressed in nm or A^0

Ionic radius: Cations - decreases; anions- more than neutral atoms. $A^+ < A < A^-$ or $A^- > A > A^+$

$Mn^{2+} > Fe^{2+} > Ni^{2+} > Cu^{2+} > Zn^{2+}$; $Fe^{+2} > Fe^{+3}$

Iso- electronic ions: They have same number of electrons.

$Na^+ > Mg^{+2} > Al^{+3} > Si^{+4}$; $N^{3-} > O^{2-} > F^- > Na^+ > Mg^{+2}$;

$O^{2-} > F^- > Ne > Na^+$; $Ne > N^{3-} > O^{2-} > F^{-1}$

Lanthanide Contraction

2. Ionisation energy: Energy or potential requried to remove the loosely bound electron from the atom. Cations are formed.

In groups - decreases; periods-increases. Measured in electron volts.

$I.E_1$: F>N >O >C ; F> N> O> H> C> Be> B

Cl > P> S > Si; $I.E_2$: O> F > N > C

-I.E. depends upon the following factors:

(a) Atomic size (b) Increase in nuclear charge

(c) Screening or shielding effect: The screening effect constant is represented by σ. More the number of inner orbital electrons higher is the screening effect. The magnitude of σ is determined by Slater's rules.

(d) Penetration effect: The s-electrons experience more attraction from the nucleus than p, d or f electrons of the same principal shell, s > p > d > f (due to the shape of the s-orbital, electron penetrates nearer to the nucleus).

(e) Electronic arrangements: The order of stability:

completely filled orbitals > half filled orbitals > half of the available orbitals half filled. Stability: $d^5 < P^3 < d^{10} < P^6$. In the former two cases it is due to symmetrical distribution of electrons.

3. Electron affinity: Energy released when an electron is taken up by an atom or ion. Anions are formed. Electron affinities are calculated from Born - Haber cycle. E.A. of Cl is the highest. E.A. of metals - Lowest; E.A. of non metals - very high. Order of E.A. : Cl > F> Br>I ;
S > Se > O ; N > P > As

E.A. of F is unexpectedly low. It is probably due to the small size of the atom. The addition of an extra electron produces a high electron density in a relatively compact 2p- subshell and therefore a strong electron - electron repulsion.

The E.A. of Be, Ca, N, P are practically zero. This is due to completely filled sub-shells or half filled orbitals (Stable configurations). EA_1 is exoergic; EA_2 is always endoergic e.g. EA_1 for oxygen =+140.9 k J mol^{-1}; EA_2 for oxygen = -770 k J / mol.

4. Electronegativity: Tendency of attracting electron by a neutral atom. The concept was introduced by Pauling (1932).

-Factors: (a) Size of the atom (b) Type of ion (cation, anion) (c) Hybridisation: with lower energy, s-orbitals show greater electron attracting power. Hence hybrid orbitals with greater s-character are found to be more electronegative.Higher the I.E. greater the EA & higher the E.N.

EN: F>O>N> C; Cl > S> P> Si; F> O> N> Cl > C

(a) R. S. Mulliken: EN is proportional to the average of ionization energy and the electron affinity EN =IP+EA/2

(b) Pauling: Based on the difference in the bond energies of diatomic molecules. Electronegativity of flourine is assigned arbitrarily 4.0. The E.N. of other elements have been calculated by the formula.

$X_A - X_B = 0.208 [E_{A-B} - (E_{A-A} \times E_{B-B})^{1/2}]^{1/2}$

X_A and X_B = electronegativities of A and B;

E_{A-B}, E_{A-A}, E_{B-B} = Bond energies of A-B, A_2 and B_2 molecules respectively.

Nature of the bond between two atoms

If the difference $X_A - X_B = 0$ --> purely covalent bond. $X_A - X_B$ = Small--> polar covalent bond. $X_A - X_B = 1.7$--> 50% covalent, 50 % ionic; $X_A - X_B$ = very high-->more ionic.

Greater the value of difference ($X_A - X_B$), more stable will be the bond.

H-F (60% ionic character) C-F (43 %) ; C-Cl (11%); C-Br (3%); C-I (1%)

5.Valency: Valency with respect to hydrogen $1 \rightarrow 4 \rightarrow 0$; with respect to oxygen: $1 \rightarrow 7$. e.g. Li_2O, BeO, B_2O_3, CO_2, N_2O_5, SO_3, Cl_2O_7.

6. Metallic Character: In a period-decreases; in a group-increases. Reducing power increases down the group.

(7) Diagonal relationship: Explained by Fajan's theory. The diagonal relationship is due to their similar polarizing power i.e. Ionic charge/ (Ionic radius)^2

Li	Be	B	C	N	O	F
Na	Mg	Al	Si	P	S	Cl

Non-metallic character & nuclear charge increases

Metallic character increases

There is increase in both metallic & non-metallic character (compensates each other)

8. Density: In a period it first increases to the central members and decreases afterwards. In solids Osmium has the highest density; in liquids Hg has the highest.

9. m. p. & b.p.: The elements with low atomic volumes have high m.p. Among metals Tungsten has the highest m.p. (3410°C); carbon among non-metals (3727°C). Helium has the lowest m.p. (-270°C). m.p. and b.p. are directly proportional to the strength of the metallic bond.

10. Nature of oxides: In a period the nature changes from basic to acidic. In a group basicity increases. Oxides of metals are generally basic and the oxides of non metals are acidic. The oxides of metalloids are amphoteric. The oxides of Al, Zn, Sn, As and Sb are amphoteric. Generally, the acidic character increases with increase in oxidation number.

HYDROGEN

Hydrogen has similarities to alkali metals and halogens. Lightest element. m.p. 14 K. No neutrons are present.

Similarities with alkali metals

1. Electronic configuration
2. Valency (univalent positive element + 1)
3. Affinity for non metals: NaCl - HCl ; Na_2O - H_2O
4. Positive ion formation H_3O^+
5. Reducing nature: $Cu_2O + 2Na \rightarrow Cu + Na_2O$; $CuO + H_2 \rightarrow Cu + H_2O$

Similarities with halogens

1. Electronic configuration
2. Ionisation potential: F > H > Cl> Br > I
3. Atomicity: Diatomic, Cp / Cv = 1.40 (specific heat at constant pressure / volume)

Believe it or not!!!

The surface temperature of the Sun has been estimated to be 6000°C. The temperature of the Sun increases constantly from the outer surface towards its centre. Scientists have estimated the temperature at the centre of the sun to be around 14000000°C. At this temperature, nuclear fusion reactions are going on inside the sun. In these reactions, 4 H atoms combine to make 1 He atom, the mass of which is less than that of the 4 H atoms.

4. Non metallic character: bad conductor of heat & electricity

5. Negative ion formation: Na H $\longrightarrow$ $Na^+ + H^-$

6. Reaction with metals $2Na + H_2 \longrightarrow 2NaH$

7. Formation of covalent compounds e.g. CH_4, SiH_4 etc.

8. Replacement of hydrogen in alkanes by halogens.

Preparation of hydrogen

1. With cold water: $2Na(Hg) + 2H_2O \longrightarrow 2NaOH + H_2$
2. Hot water or Steam: $Zn + H_2O \longrightarrow ZnO + H_2$
3. Amphoteric metals (Zn, Al, Sn, Pb, Si) react with boiling NaOH or KOH e.g. $2Al + 2NaOH + 2H_2O \longrightarrow 2NaAlO_2$ (Sodium meta aluminate) + $3H_2$
4. Very pure hydrogen: $2Al + 2KOH + H_2O \longrightarrow 2KAlO_2 + 3H_2$ (militiary purposes, Uyeno's method)
5. Bosch's process: $C + H_2O \xrightarrow{1000^\circ C} \underset{\text{water gas}}{CO + H_2}$

Purification: purified by passing through $AgNO_3$ (PH_3, AsH_3), $Pb(CH_3COO)_2$ (H_2S) and KOH (CO_2,SO_2absorbed) solutions successively. Collected and dried over P_2O_5 or anhydrous $Ca\,Cl_2$.

Industrial application:(1) $3H_2 + N_2 \underset{}{\overset{Fe,Mo}{\rightleftharpoons}} 2NH_3$;

(2) $CO + 2H_2 \xrightarrow[Cr_2O_3, 300^\circ C]{ZnO\ /} CH_3OH$

- Adsorbed hydrogen is called occluded hydrogen.

Ortho and para hydrogen

ortho para
proton spin

Ordinary hydrogen is 75 % Ortho + 25 % Para

Isotopes of Hydrogen :

Protium H^1;

Deuterium H^2 or D^2 (reacts slowly);

Tritium : H^3 or T^3 (only in nuclear transformations).

-Deuterium is used as a tracer & projectile

-D_2O is used as a moderator for slowing down the velocity of neutrons in nuclear reactors.

-Water: Temporary hardness (due to bicarbonates of Ca and Mg) is removed by boiling.

Permanent hardness : The degree of hardness is the number of parts of calcium carbonate or equivalent to various Ca and Mg salts present in water. Removed by (i) Washing Soda (ii) Permutit (iii) Calgon (iv) Ion exchange resins.

H_2O_2 :

(1) $Na_2O_2 + H_2SO_4 \longrightarrow Na_2SO_4 + H_2O_2$

(2) $BaO_2 + H_2SO_4 \longrightarrow BaSO_4 + H_2O_2$

Merk's Process :

(3) $BaO_2 + CO_2 \longrightarrow BaCO_4 \xrightarrow{H_2O} BaCO_3 + H_2O_2$

Manufacture : (1) By electrolysis of cold 50% H_2SO_4 at high current density using platinum electrode.

$H_2SO_4 \rightleftharpoons H^+ + HSO_4^-$

$2HSO_4^- \longrightarrow H_2S_2O_8 + 2e^-$ at anode

$2H^+ + 2e^- \longrightarrow H_2$ at cathode

$\underset{\text{Marshall's acid}}{H_2S_2O_8} + 2H_2O \longrightarrow 2H_2SO_4 + H_2O_2$

(2) Autoxidation of 2-butyl anthraquinone

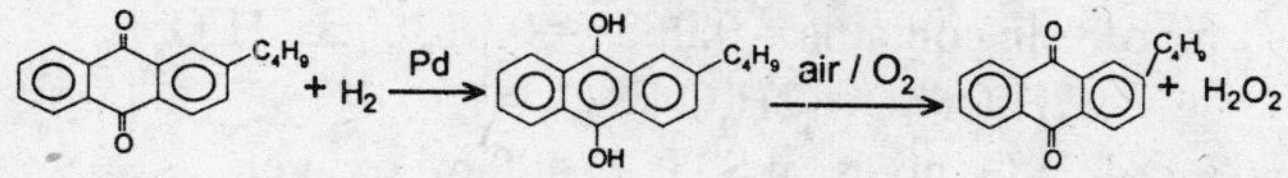

(3) Oxidation of 2-propanol :

$CH_3CHOH + O_2 \longrightarrow CH_3COCH_3 + H_2O_2$

- H_2O_2 is stored in pure aluminium or in stone jars in dark. Stored in polythene bags also. Stabilizers like orthophosphoric acid, acetanilide or sodium stannate are used. Concentrated by distillation under reduced pressure.

- H_2O_2 is an oxidising, reducing and bleaching agent (silk, wool, hair etc. can't be bleached by Cl_2).It is used as antichlor.

- A solution of chromic acid in sulphuric acid (acidified potassium dichromate solution) is oxidised to blue peroxide of chromium CrO_5.

$K_2Cr_2O_7 + H_2SO_4 + 4H_2O_2 \longrightarrow 2CrO_5 + K_2SO_4 + 5H_2O$

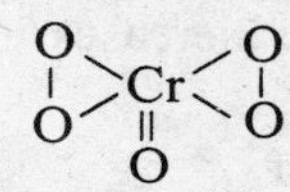

Butter fly structure

- The aqueous solution is very weakly acidic due to hydrolysis

$H_2O_2 + H_2O \rightleftharpoons H_3O^+ + HO_2^-$

The strength of H_2O_2 : It is expressed in volumes. This is an indication of the volume of oxygen measured at NTP, which can be obtained by decomposing one volume of H_2O_2.

A 10 volume of H_2O_2 means 1 ml of 10 vol. H_2O_2 gives 10 ml of O_2 at NTP.

1.Calculate the amount of H_2O_2 in g. in 10 ml. 10 volume of H_2O_2.

Sol. 1ml. of 10 vol. H_2O_2 will give 10 ml. O_2 at NTP

10 ml. of 10 vol. H_2O_2 gives 100 ml. O_2 at NTP

$2H_2O_2 \longrightarrow 2H_2O + O_2$

$2(2+32) \longrightarrow 22.4$ lit of O_2

22,400 ml.of O_2 is obtained from 68g.of H_2O_2

100 ml " " " $\frac{68 \times 100}{22,400} = 0.3035g.\ H_2O_2$

Sometimes the concentration of H_2O_2 is expressed as percentage of H_2O_2 in soln i.e. g / 100 ml.

Here 3.035 % or 30.35 g / lit

$$\text{Normality} = \frac{\text{conc. or strength per lit}}{\text{equivalent weight}} = \frac{30.35}{17} = 1.785N$$

2. A 5.0 cm³ soln. of H_2O_2 liberates 0.508 g. of iodine from acidified KI solution. Calculate the strength of H_2O_2 solution in terms of volume strength at STP.

Sol.: $H_2O_2 + H_2SO_4 + 2KI \longrightarrow K_2SO_4 + I_2 + 2H_2O$ 34 g
2 x 127 = 254g

254g. of I_2 is obtained from 34 g of H_2O_2
0.508 g of I_2 " " H_2O_2 = (34/254) x 0.508
= 0.068 g.

5 ml of soln contains 0.068 g of H_2O_2
1 ml of soln contains = 0.068/5=0.0136 g of H_2O_2

$$2H_2O_2 \longrightarrow 2H_2O + O_2$$

68 g of H_2O_2 gives at STP, 22400 ml oxygen
0.0136 g of " " = (22400/68) x 0.0136
= 4.48 ml.

i.e. 1 ml of H_2O_2 soln. gives 4.48 ml of oxygen.
so, strength of H_2O_2 soln. = 4.48 volume.

ALKALI METALS -1st GROUP

Six alkali metals. Na(7th) and K(8th) abundant in nature. The hydroxides of these metals are very strong alkalies. Fr obtained only as a result of radioactivity decay.

1. **Electronic Configuration :** ns^1
2. **Atomic & Ionic size :** Increases
3. **Reactivity :** Increases from Li to Cs as the trend of the loss of electron increases.
4. **E.N., I.P. :** Decreases
5. **Metallic character:** Increases with increase in atomic number.
6. **Density:** Na > K > Li

This is due to unusual increase in atomic size of K.

7. **Conductivity :** Good conductors because the loosely held valence electrons are free to move throughout the metal structure.
8. **m.p. and b.ps :** Decreases with increase in size (loosely packed metal lattices)
9. **Hydration of Ions :** The smaller the size of the ion the greater is the degree of hydration and hydration energy. As a result of degree of hydration, ionic radii in water decrease from Li^+ to Cs^+ i.e. Li^+ ion which should have been smallest is actually largest in size in water. Thus the ionic radii in water decrease in the order $Li^+ > Na^+ > K^+ > Rb^+ > Cs^+$.

Smaller the ion greater the velocity and hence greater the conductance.

Ionic conductance : $Cs^+ > Rb^+ > K^+ > Na^+ > Li^+$

10. **Reducing power:** Alkali metals are strong reducing agents. Li is the strongest because of its greater heat of hydration. Lithium has maximum E^0 value. The tendency of an element to loose an electron in solution is measured by its standard oxidation potential value (E^0).

$M(s) \longrightarrow M(g)$

$M(g) \longrightarrow M^+(g) + e^-$ Ionisation energy

$M^+(g) + H_2O \longrightarrow M^+(aq)$ Hydration energy

11. **Flame colours :** When alkali metal or its compound is heated in a bunsen flame the electrons get excited. When the electrons return to their original level the excitation energy absorbed is released in the form of light in the visible region of the spectrum. Li (crimson), Na (golden yellow), K (violet), Rb & Cs (violet red).
12. **Lattice energy :** The energy released during the formation of one mole of a crystal from the constituent ions present in the gaseous state. The lattice energy of salts of alkali metals having a given anion decreases with increase in atomic number of the alkali metal.
13. **Reactivity with H_2O and Alcohol:** increases down the group.
14. **Action with air :**

$$M \xrightarrow{O_2} M_2O \xrightarrow{H_2O} MOH \xrightarrow{CO_2} M_2CO_3$$
$$M_2O \xrightarrow{O_2} M_2O_2$$

15. **Solubility in liquid ammonia:** Alkali metals give deep blue colour due to solvated electrons or ammoniated electrons. When metals are dissolved in ammonia, the ammonia molecules solvate the cation as well as the electron

$$M + NH_3 \longrightarrow M^+(NH_3)_x + e^-(NH_3)_y$$

16. **Oxides :**normal oxides(K_2O) contain O^{2-} ion; peroxides(K_2O_2)O_2^{2-} ions; superoxides (KO_2)O_2^- ion.
17. **Hydroxides :** Basicity increases

CsOH > RbOH > KOH > NaOH > LiOH

(Cs^+ more soluble in water, therefore more basic)

18. **Carbonates & Bicarbonates:** Stability increases down the group.

$LiHCO_3$ doesn't exist as solid. Carbonates are more stable than bicarbonates. Carbonate solutions are alkaline due to hydrolysis.

$$Na_2CO_3 + 2H_2O \longrightarrow 2Na^+ + 2OH^- + H_2CO_3$$

Carbonates of Ist group are more stable than 2nd group carbonates.

19. **Organometallic Compounds :**

$$C_2H_5Cl + 2Li \longrightarrow C_2H_5Li + LiCl$$
$$C_6H_5Cl + 2Li \longrightarrow C_6H_5Li + LiCl$$

Specific Characters of Li :

(1) Much harder like Mg
(2) m.p. & b.p. much higher
(3) LiO_2 and MgO sparingly soluble in water
(4) Halides : covalent character
(5) Formation of nitrides : $Li + N_2 \longrightarrow Li_3N$; Mg_3N_2
The nitrides are ionic compounds.
(6) Bicarbonates of Li and Mg more soluble than carbonates.
(7) LiCl like $MgCl_2$ undergoes hydrolysis in hot water.
-LiCl : in air conditioning plants to regulate humidity.

-NaOH - Caustic soda; KOH - Caustic potash
-$Na_2CO_3.10H_2O$ - washing soda; $Na_2S_2O_3.5H_2O$ - hypo
-$NaHCO_3$ - baking soda;
$Na(NH_4)HPO_4.4H_2O$ - Microcosmic salt
-$Na_2SO_4.10H_2O$ -Glauber's salt ; $NaNO_3$-Chile saltpetre
-Euchlorine-ClO_2
$2KClO_3 + 4HCl \longrightarrow 2KCl+Cl_2+2ClO_2+2H_2O$
-Oxone : Na_2O_2 : for purification of air in submarines, aeroplanes, hospitals, theatres etc.

Cs, Rb, K: used in photoelectric cells since the valence electron is loosely held in the alkali metals, they emit electrons under the influence of light (photoelectric effect).

Caustic soda: NaOH :Prepared by Gossage process; Lowig process; Nelson cell process; Castner-Kellner cell.

Washing soda: $Na_2CO_3.10H_2O$: Leblanc process; Solvay ammonia soda process.

- KCl & $MgCl_2$: deliquescent (liquifying in air) and hygroscopic

(1) Hydrogen peroxide reacts with
(i) PbS; (ii) $K_4[Fe(CN)_6]$; (iii) O_3 and (iii) KI as
(i) $4H_2O_2 + PbS \longrightarrow PbSO_4 + 4H_2O$
(ii) $H_2O_2+2K_4[Fe(CN)_6]+H_2SO_4 \longrightarrow 2K_3[Fe(CN)_6]+K_2SO_4+2H_2O$
(iii) $H_2O_2 + O_3 \longrightarrow H_2O + 2O_2$
(iv) $H_2O_2 + 2KI+H_2SO_4 \longrightarrow K_2SO_4 + 2H_2O+I_2$

(2) Excess of caustic soda solution is added to
(i) $ZnSO_4$ (or $ZnCl_2$); (ii) $Al_2(SO_4)_3$ (or $Al(NO_3)_3$)
(i) $ZnCl_2+2NaOH \longrightarrow Zn(OH)_2 + 2NaCl$
$Zn(OH)_2+2NaOH \longrightarrow Na_2ZnO_2 + 2H_2O$
Sodium zincate
(ii) $Al_2(SO_4)_3+6NaOH \longrightarrow 2Al(OH)_3+ 3Na_2SO_4$
$Al(OH)_3 + NaOH \longrightarrow NaAlO_2 + 2H_2O$
Sodium meta aluminate

(3) KI reacts with (i) $CuSO_4$ (ii) $AgCl_2$ (iii) I_2 as
(i) $CuSO_4 + 2KI \longrightarrow K_2SO_4 + CuI_2$; $2CuI_2 \longrightarrow Cu_2I_2 + I_2$
(ii) $2KI + HgCl_2 \longrightarrow HgI_2 +2K Cl$
$HgI_2+ 2KI \longrightarrow K_2HgI_4$
Nessler's reagent
(iii) $KI + I_2 \longrightarrow KI_3$(Iodine dissolves in KI sol.)

(4) $6NaOH + 12S \longrightarrow Na_2S_2O_3+2Na_2S_5+3H_2O$
Hot & Conc. Sod. pentasulphide

(5) $6NaOH+3I_2 \longrightarrow 5NaI+NaIO_3+3H_2O$
Sod. Iodate

(6) $6NaOH+3Cl_2 \longrightarrow 5NaCl+NaClO_3+3H_2O$
Sod. Chlorate

(7) $4P + 3KOH+3H_2O \xrightarrow{\Delta} PH_3+3KH_2PO_2$
Yellow

(8) $4P + 3NaOH+3H_2O \xrightarrow{\Delta} PH_3+3NaH_2PO_2$
White

ALKALINE EARTH METALS - 2nd GROUP

All are alkaline earth metals except Be.

Alkaline Earth Metals : All chemists referred to any non-metallic substance insoluble in H_2O and unchanged by fire as an "earth". The earths of this group i.e. CaO and MgO give alkaline reactions.

1.**Electronic Configuration :** ns^2

2.Atoms of II group are smaller than 1st group due to the increased nuclear charge. This difference is responsible for higher densities, higher ionisation potential, higher m.p. & b.p.s, increased hardness. Smaller atomic size and comparatively high nuclear charge reduce the chemical reactivity of Be. Stronger bonding and closer packing in the metallic structure is due to contribution of two electrons per atom to metallic bonding (therefore, harder, heavier, have higher m.p.s).

Density : Ca < Mg (In Mg 'L' shell is completely filled whereas in Ca 'M' shell can accommodate 18 electrons, although 8 are present. Here the volume increases compared to weight (due to protons) and therefore density is less.

3. **E.N.:** decreases
4. **Metallic nature :** Increases
5. **Reactivity of metals :** Increases
6. **Reducing nature :** Increases
7. **Electropositive character :** Increases
8. **Magnetic property :** Diamagnetic
9. **Flame colour :** Ca-brick red; Sr-crimson red; Ba-apple green. In Be and Mg, the electrons are tightly held. Hence electron excitation becomes unlikely.
10.**Complex Compound Formation :**Require small size, high nuclear charge and vacant orbitals. Be, Mg and Ca forms complexes.

Ca and Mg form complexes with EDTA and Mg with 8-hydroxyquinoline.

11. **Hydrides :** BeH_2, MgH_2 - covalent; others ionic.
12. **Oxides :** Basicity: BeO< MgO<CaO<SrO< BaO
13.**Hydroxides:**Basicity and solubility order:
$Ba(OH)_2>Sr(OH)_2>Ca(OH)_2 > Mg(OH)_2 > Be(OH)_2$
14.**Sulphates :** solubility in water
$BeSO_4>MgSO_4>CaSO_4> SrSO_4 > BaSO_4$
As lattice energy of $BaSO_4$ is low (Ba^{2+} size is more) and can't overcome hydration energy, it is insoluble in H_2O.
15.**Bicarbonates:** They are stable only in solution. Their decomposition to carbonates is responsible for the formation of stalactites & stalagnites.
.16.**Carbonates:** stability
$BaCO_3>SrCO_3> CaCO_3 > MgCO_3 > BeCO_3$
17.**Halides:** BeF_2 & $BeCl_2$ covalent. Others electrovalent.
18.**Sulphides:** The sulphides of this group are not precipitated because of hydrolysis
$2CaS + 2H_2O \longrightarrow Ca(OH)_2 + Ca(HS)_2$
19.**Double salt formation:**
$KCl.MgCl_2.6H_2O$ - Carnalite
20.**Organometallic Compounds:** R Mg X

21. **Carbides :** $Be_2C + 2H_2O \longrightarrow CH_4 + 2BeO$

$Al_4C_3 + 6H_2O \longrightarrow 3CH_4 + 2Al_2O_3$

$CaC_2 + 2H_2O \longrightarrow Ca(OH)_2 + C_2H_2$

- $MgSO_4.7H_2O$ - Epsom salt
- $CaSO_4.2H_2O$ - Gypsum, alabaster or selenite
- $CaSO_4.\frac{1}{2}H_2O$ or $2CaSO_4.H_2O$ - Plaster of paris

$$\underset{\text{monoclinic}}{-CaSO_4.2H_2O} \xrightarrow{\Delta} \underset{\text{Orthorhombic}}{CaSO_4.2H_2O} \xrightarrow[H_2O]{120°c} \underset{\text{plaster of paris}}{2CaSO_4.H_2O} \xrightarrow{200°c} \underset{\text{anhydrous}}{CaSO_4}$$

- $BaSO_4 + ZnS$: Lithopone (white pigment)
- Sorel's cement : $MgCl_2.5MgO.x\ H_2O$
- Milk of Magnesia : $Mg(OH)_2$
- Quick lime or burnt lime : CaO
- Slaked lime : $Ca(OH)_2$
- Portland Cement : CaO (50-60%) SiO_2(20-25%) Al_2O_3 (5-10%) MgO (2-3%) Fe_2O_3 (1-2%) SO_3 (1-2%)
- Magnalium : Al + Mg (airships)
- Electron metal : Mg + Zn (air crafts)
- Duralumin : Al + Mg + Mn + Cu (air ships)
- CaH_2 - hydrolith
- Magnesia alba: $3MgCO_3.Mg(OH)_2.3H_20$ (used as a dental abrassive in tooth pastes)
- Chlorophyll is an important complex of Mg.

BORON FAMILY - 13th GROUP

Al, most abundant among metals and third most abundant of all the elements. Tl is Duckbill Platypus (as it shows similarities with many other elements).

1. **Electronic Configuration:** ns^2p^1
2. **MetallicCharacter:** Increases; Al is more + ve than other elements. Small size and high nuclear charge provide B with non metallic nature.
3. **Density :** Al < B (aluminium exceptionally low)
4. **Reducing power:** Al strong reducing agent due to high heat of energy.
5. **Oxidation state:** B is trivalent. Tl +1 oxidation state is more stable than +3 oxidation state. This is due to inert pair effect. In metal borides e.g., Mg_3B_2, B is in -3 state. B, forms generally covalent compounds while the others form electrovalent compounds also.
6. **Lewis acidity:**

$BF_3 < BCl_3 < BBr_3 < BI_3$

This can be explained in terms of back donation of electrons from the halogen atom to B atom.

BX_3 compounds are covalent compounds formed by sp^2 hybridisation. Three sp^2 hybridized orbitals overlap with the three singly filled p-orbitals of F atom giving BF_3 molecule.

Lateral overlap of the vacant 2p orbitals of B with one completely filled 2p orbitals of F occurs leading to the formation of p_π-p_π bonds between B-F. In this process B-F bond acquires some double bond character. The tendency to form P_π-P_π bond is maximum in the case of BF_3 and falls sharply as we move to BCl_3, BBr_3 and BI_3, because of appreciable difference in their energy levels. B>Al>Ga>In.

7. **Formation of double salts** :

$K_2SO_4.Al_2(SO_4)_3.24H_2O$; $Tl_2SO_4.Al_2(SO_4)_3.24H_2O$

8. **Hydroxides:** $B(OH)_3$ feebly acidic; Al_2O_3, Ga_2O_3 amphoteric; In_2O_3 - basic
9. **Hydrides :**

$$4\ LiH + AlCl_3 \longrightarrow LiAlH_4 + 3LiCl$$

B_2H_6, diborane is electron deficient compound; banana bond bridging three atoms is present.

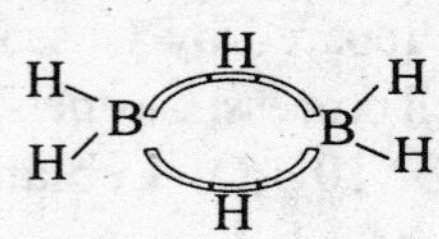

10. **Oxides:** B_2O_3 is an anhydride of ortho-boric acid (H_3BO_3)

$\alpha-Al_2O_3$-corrundum; Υ-Al_2O_3-activated alumina is used in column chromatography as an adsorbant.

11. **Trihalides :** BX_3 are covalent and gaseous. Fluorides of other metals ionic. BF_3, $AlCl_3$ are used as Friedel-Craft catalysts.
12. **Borax bead test :**

Borax is used for borax bead test.

$$Na_2B_4O_7.10H_2O \xrightarrow{\Delta} \underset{\text{anhydrous}}{Na_2B_4O_7} \xrightarrow{740°c} \underset{\text{Sodium metaborate}}{NaBO_2 + B_2O_3}$$

$$CuSO_4 + B_2O_3 \longrightarrow CuO.B_2O_3 + SO_3$$

$$CuO.B_2O_3 \downarrow Cu(BO_2)_2$$

Colours of metaborates :

Cu	Fe	Co	Cr	Ni
Blue	Green	Blue	Green	Brown

-Borazine or Borazole ($B_3N_3H_6$): Inorganic benzene

$$B_2H_6 + 2NH_3 \xrightarrow{\Delta} B_2H_6{\cdot}2NH_3 \xrightarrow{200°C} B_3N_3H_6$$

-**Permutit :** $NaAlSiO_4.3H_2O$ - used in softening of water.

-**Borax:** $Na_2B_4O_7.10H_2O$ (unrefined borax is Tincal Suhaga (55% Borax)

-**Goldschmidt alumino thermic process:** The heat evolved in the oxidation of aluminium is used in thermite welding of metals

$$Oxide + Al \longrightarrow Al_2O_3 + metal + heat\ energy$$

Al acts as a reducing agent. This principle is used in the extraction of metals and non metals e.g. Cr, Mo, Mn, B, Si etc.

Borate test : $3C_2H_5OH + H_3BO_3 \longrightarrow \underset{\text{green edge flame}}{B(OC_2H_5)_3 + 3H_2O}$

-Boron, Boron steel or Boron carbide rods are used in atomic reactors to control the nuclear reactions.

$$-\underset{\text{Solution}}{2Al + 2NaOH + 2H_2O} \longrightarrow \underset{\text{Sod. meta aluminate}}{2NaAlO_2 + 3H_2}$$

$$2Al + 6NaOH \xrightarrow{\text{fused}} \underset{\text{Sod. aluminate}}{2NaAlO_3 + 3H_2}$$

CARBON FAMILY- 14th GROUP

C,Si- non metals;Ge-metalloid; Sn,Pb-metals.

C - is an essential constituent of all organic matter.

Si - is an essential constituent of all inorganic matter.

1. **Electronic Configuration** : ns^2p^2

2. **m.p and b.p** of carbon exceedingly high (m.p. 4100°C) Due to allotropism, carbon shows different m.ps.

3.**I.E.** : Carbon quite high due to its small size.

4.**Oxidation state** : All elements have oxidation state+ 4 in common. +2 state becomes increasingly important for Ge, Sn, Pb. Since inert pair effect increases from Ge to Pb, the stability of M^{4+} ions decreases.Due to this Pb^{2+} compounds are more stable than Pb^{4+}. C_2^{2-} ions are present in $Ca^{2+}C_2^{2-}$ (Calcium Carbide).

Stability : $Ge^{2+} < Sn^{2+} < Pb^{2+}$; $Ge^{4+} > Sn^{4+} > Pb^{4+}$; $Pb^{2+} > Pb^{4+}$. Ge and Sn :+4 state more stable.

5. **Inert pair effect :** Inert pair effect is based on the decrease in the covalent bond forming ability of heavy elements due to :

(i) There is much repulsion between the inner electrons of the bonded atom and therefore, the bond energy is quite low.

(ii) As the size of the atom increases, the valence electrons get farther from the nucleus. The large 5-s and 6-s orbitals belonging to the heavier elements are more diffused and their ability to form an effective overlap is small.

Greater the inert pair effect, greater is the stability of the ion or compound.

6. **Catenation** (Tendency to form chains)**:**The tendency of an element to form chains depends upon the strength of the element - element bond which is maximum in the case of carbon. (C-C 348 k.J/mole; Si-Si 222 k.J./mole.; Ge-Ge 167 k.J./mole; Sn-Sn 155k.J./mole). Also in most of the carbon compounds, other bonds formed with H or O etc. are also strong.

The atom with 4 outer electrons can't for stereochemical reasons satisfy all its 4 valencies in a discrete molecule and accordingly forms non-volatile giant structural units of the diamond type. On the otherhand the 5 electron outer arrangements in 15th group are conductive to the formation of discrete molecules which are comparatively free of attraction for each other.

7.**Allotropy:**Except Pb all other elements exhibit the allotropy.

Carbon ⟶ Amorphous forms (microcrystalline) coal, charcoal, lamp black, coke etc.

↓

Crytalline (a) Diamond (purest, hardest) sp^3;

(b) Graphite sp^2, lubricant in machines, good conductor.

Buckminster fullerene - C_{60} (1985); C_{70} recent one.

8.**Formation of complexes:** carbon can't form complexes as d-orbitals are not present (no vacant orbitals). Carbon shows 4 covalency only. Other elements can form complexes e.g. SiF_6 etc.

9. **Hydrides :**

Thermal stability: $CH_4 > SiH_4 > GeH_4 > SnH_4 > PbH_4$

Reducing Nature: $CH_4 < SiH_4 < GeH_4 < SnH_4 < PbH_4$

Silanes, SiH_4 are more reactive than alkanes due to -

(i) less E.N. (ii) large size (iii) presence of d-orbitals in Si.

10.**Oxides:** CO neutral, sp; CO_2 weakly acidic, sp; SiO_2 acidic; GeO_2, SnO_2 and PbO_2 amphoteric.

11.**Halides:** CCl_4 doesnot undergo hydrolysis. Other halides are largely hydrolysed by water due to the utilisation of d-orbitals to which water molecules can get attached.

$PbBr_4$ and PbI_4 do not exist: It can be explained on the basis of strong oxidising nature of Pb^{4+}. The ions Br^- and I^- are reducing agents i.e., they reduce Pb^{4+} to Pb^{2+} ions.

12.**Carbides:** The carbides of metals like Ti, Zr, Hf, V, Nb, Ta, Mo, W etc. are interstitial compounds.

- Animal / bone charcoal have adsorption capability.
- CO combines with Haemoglobin to form stable Carboxy Haemoglobin.
- Dry ice is solid CO_2 and is used as refrigerant under the name drikold.
- Producer gas:$CO+N_2$; water gas: $CO+H_2$+a little of CO_2.
- water glass: Na_2SiO_3.
- Pink salt : ammonium chlorostannate $(NH_4)_2SnCl_6$.
- Mosaic gold : SnS_2; type metal : Pb+Sb+Sn
- Silicon carbide : SiC carborendum (is used to grind stones; for knife sharpner).
- PbO : Litharge or Massicoat; Crooke's Glass : doesn't allow UV rays to pass, contains CeO_2.
- Pyrex glass: Zn, barium boro-silicates and aluminium silicate.
- Natural gas:CH_4 mainly ; -Biogas : CH_4mainly
- LPG : butane mainly ; -Gobar gas : CH_4 mainly;
- Oil gas : $CH_4 + C_2H_4 + C_2H_2$
- Coal gas : H_2 mainly
- Plumbo Solvency effect : Lead is sparingly soluble in aerated water to form lead hydroxide.

$$2Pb + O_2 + 2H_2O \longrightarrow 2\,Pb(OH)_2$$

Silicates:

Orthosilicates contain SiO^{4-} ion

Pyrosilicates contain $Si_2O_7^{6-}$ ion

Sheet silicates contain SiO_4^{4-} ion e.g. talc, kaolin, quartz.

1. Washing soda has the formula [CBSE-90]
(A) $Na_2CO_3 . 7H_2O$ (B) $Na_2CO_3 . 10H_2O$
(C) $Na_2CO_3 . H_2O$ (D) Na_2CO_3

2. Poisonous gas present in the exhaust fumes of a jet plane is [CBSE-90]
(A) fluorocarbon (B) carbon monoxide
(C) methane (D) C_2H_2

3. Gypsum is [CBSE-90]
(A) $CaSO_4$ (B) $CaSO_4 . 1/2H_2O$
(C) $CaSO_4 . 2H_2O$ (D) $CaSO_4 . 3H_2O$

4. Alkali metals lose electrons in [CBSE-90]
(A) s-orbitals (B) p-orbitals (C) d-orbitals (D) f-orbitals

5. In a potassium atom, electronic energy levels are in the following order [CBSE-91]
(A) 4s > 3d (B) 4s > 4p (C) 4s < 3d (D) 4s<3p

6. The correct order of second ionisation potential of C,N,O and F is [CBSE-91]
(A) C > N > O > F (B) O > N > F > C
(C) O > F > N > C (D) F > O > N > C

7. Which of following has the lowest ionisation potential? [CBSE-91]
(A) $3d^2$ (B) $4s^1$ (C) $3p^6$ (D) $2p^6$

8. Glass is a [CBSE-91]
(A) Liquid (B) transparent organic polymer
(C) solid (D) Super-cooled liquid

9. Basicity of orthophosphoric acid is [CBSE-91]
(A) 2 (B) 3 (C) 4 (D) 5

10. Which is more electronegative ? [CBSE-91]
(A) F (B) Cl (C) I (D) None of these

11. Ce^{58} is a member of [CBSE-91]
(A) s-block (B) p-block (C) d-block (D) f-block

12. If an isotope of hydrogen has two neutrons in its atom, its atomic number and atomic mass number will respectively be. [CBSE-92]
(A) 2 and 1 (B) 3 and 1 (C) 1 and 1 (D) 1 and 3

13. A sudden large jump between the values of second and third ionisation energies of an element would be associated with the electronic configuration. [CBSE-92]
(A) $1s^2 2s^2 2p^6 3s^1$ (B) $1s^2 2s^2 2p^6 3s^2 3p^1$
(C) $1s^2 2s^2 2p^6 3s^2 3p^2$ (D) $1s^2 2s^2 2p^6 3s^2$

14. The lustre of a metal is on account of [CBSE-92]
(A) high density of metals (B) high polish of metals
(C) chemical inertness of metals
(D) reflection of light due to presence of free electrons

15. Electronic configuration of calcium atom may be written as [CBSE-92]
(A) Ne, $4p^2$ (B) Ar, $4s^2$ (C) Ne, $4s^2$ (D) Kr, $4p^2$

16. H_3PO_2 is the formula for one of the phosphorus acids. Its name and basicity are respectively. [CBSE-92]
(A) phosphorus acid and two
(B) hypophosphorus acid and two
(C) hypophosphoric acid and one
(D) hypophsphoric acid and two

17. Water gas is produced by [CBSE-92]
(A) passing steam through a red hot coke bed
(B) saturating hydrogen with moisture
(C) mixing oxygen and hydrogen in the ratio of 1:2
(D) heating a mixture of CO_2 and CH_4 in petroleum refineries

18. Which one of the following pairs is isoelectronic ? [CBSE-92]
(A) Ar and Cl (B) Na^+ and Ne
(C) Na^+ and Mg (D) Mg and Ne

19. In a group of the periodic table the ionisation potentials of elements decrease from the top to the bottom because of [CBSE-92]
(A) increase of densities
(B) decrease in chemical reactivities
(C) decrease in electronegativities
(D) increase in atomic size

20. Which one of the following substances is used in the laboratory for a fast drying of neutral gases [CBSE-92]
(A) Phoshorus pentoxide (B) Active charcoal
(C) Anhydrous calcium chloride (D) Na_3PO_4

21. Select from amongst the following the correct statement : [CBSE-92]
(A) Sodium metal is stored under kerosene oil
(B) One of the oxides of carbon is a basic oxide
(C) Metals can form only basic oxides
(D) To prevent combustion of white phosphorus with oxygen it is kept in kerosene oil.

22. On electrolysis of dilute sulphuric acid using platinum electrodes, the product obtained at the anode will be
(A) Hydrogen (B) oxygen [CBSE-92]
(C) Hydrogen sulphide (D) Sulphur dioxide

23. Which one of the following types of forces bind together the carbon atoms in diamond ? [CBSE-92]
(A) Ionic (B) Covalent
(C) Dipolar (D) van der Waals

24. In graphite, electrons are [CBSE-93]
(A) localised on every third C- atom
(B) present in anti-bonding orbitals
(C) localised on each C-atom
(D) spread out between the structure

25. Which compound will show the highest lattice energy ? [CBSE-93]
(A) RbF (B) CsF (C) NaF (D) KF

26. Which of the following has the greatest electron affinity ? [CBSE-93]
(A) Fluorine (B) Bromine (C) Chlorine (D) Iodine

27. Impossible configuration is of [CBSE-93]
(A) $1s^2, 2s^2, 2p^2, 3s^1$ (B) $1s^2, 2s^2, 2p^6$
(C) $1s^2, 2s^2$ (D) None

28. Which of the following has largest size ? [CBSE-93]
(A) Na (B) Na+ (C) Na^- (D) None of the above

29. Which one of the following pairs of atoms or ions will have identical ground state electronic configuration ? [CBSE-94]
(A) Li^+ and He^- (B) Cl^- and Ar
(C) Na and K (D) F^+ and Ne

30. The formula for Calcium chlorite is [CBSE-94]
(A) $Ca(ClO_4)_2$ (B) $Ca(ClO_3)_2$
(C) $CaClO_2$ (D) $Ca(ClO_2)_2$

31. Which of the following ions is not isoelctronic with O^{2-} ? [CBSE-94]
(A) N^{3-} (B) F^- (C) Tl^{++} (D) Na^+

32. A solid with high electrical and thermal conductivity from the following is [CBSE-94]
(A) silicon (B) lithium (C) Sodium chloride (D) ice

33. A 10 volume solution of H_2O_2 means [CBSE-94]
(A) 10 grams of H_2O_2 solution liberate 10 mL of O_2 at STP
(B) 10 gram equivalents of H_2O_2 liberates 10 mL of O_2 at STP
(C) 10 litres of H_2O_2 liberates 10 mL of O_2 at STP
(D) 1 mL of solution of H_2O_2 liberates 10 mL of O_2 at STP.

34. Which of the following is a false statement about boric acid H_3BO_3 ? [CBSE-94]
(A) It is a strong tribasic acid.
(B) It does not act as proton donor but acts as a Lewis acid by accepting hydroxyl ion.
(C) It is prepared by acidifying an aqueous solution of borax.
(D) It has a layer structure in which planar BO_3 units are joined by hydrogen bonds.

35. Carbon and silicon belong to Gr.IV. The maximum coordination number of carbon in commonly occurring compounds is four whereas that of silicon is six. This is due to [CBSE-94]
(A) larger size of silicon
(B) availability of low lying d-orbitals in silicon.
(C) silicon being vulnerable to attack by nucleophilic reagents
(D) more electropositive nature of silicon.

36. Which one of the following oxides of nitrogen is paramagnetic. [CBSE-94]
(A) N_2O (B) N_2O (C) NO_2 (D) N_2O_2

37. All the following substances react with water. The pair that yields the same gaseous product is. [CBSE-94]
(A) K and KO_2 (B) Ca and CaH_2
(C) Na and Na_2O_2 (D) Ba and BaO_2

38. Which of the products is formed when $SiCl_4$ vapour is passed over hot Mg ? [CBSE-94]
(A) $SiCl_2 + MgCl_2$ (B) $Si + MgCl_2$
(C) $Mg_2Si + Cl_2$ (D) $MgSiCl_6$

39. Which of the following statements is false ? [CBSE-94]
(A) Zinc disolves in hot NaOH solution.
(B) Tin (IV) chloride is made by dissolving tin in conc. HCl
(C) Carbon monoxide reduces iron (III) oxide to iron
(D) Mercury (II) iodide dissolves in excess of KI solution.

40. Which one of the following metal ions plays an important role in muscle contraction ?
(A) K^+ (B) Na^+ (C) Mg^{2+} (D) Ca^{2+}

41. The electronic configuration of an element is $1s^2\ 2s^2\ 2p^6\ 3s^2\ 3p^3$. What is the atomic number of the element which is just below the above element in the periodic table? [CBSE-95]
(A) 33 (B) 34 (C) 31 (D) 49

42. BCl_3 is a planar molecule whereas NCl_3 is pyramidal because [CBSE-95]
(A) BCl_3 has no lone pair of electrons but NCl_3 has a lone pair of electrons.
(B) B-Cl bond is more polar than N-Cl bond
(C) nitrogen atom is smaller than boron atom.
(D) N-Cl bond is more covalent than B-Cl bond.

43. Identify the correct statement: [CBSE-95]
(A) Gypsum contains a lower percentage of calcium than plaster of Paris
(B) Gypsum is obtained by heating plaster of Paris.
(C) Plaster of Paris can be obtained by hydration of gypsum.
(D) Plaster of Paris is obtained by partial oxidation of gypsum.

44. The solubility in water of sulphates down the Be group is Be>Mg>Ca>Sr>Ba. This is due to [CBSE-95]
(A) high heat of solvation for smaller ions like Be^{2+}
(B) increasing molecular weight.
(C) decreasing lattice energy
(D) increase in melting points

45. Aluminium (III) chloride forms a dimer because [CBSE-95]
(A) higher coordination number can be achieved by aluminium.
(B) aluminium has high ionization energy
(C) aluminium belong to 3rd group
(D) it cannot form a trimer.

46. Mercury is the only metal which is liquid at 0°C. This is due to its [CBSE-95]
(A) very high ionisation energy and weak metallic bond.
(B) low ionisation potential
(C) high atomic weight (D) high vapour pressure.

47. Which of the following has the smallest size? [CBSE-96]
(A) Na^+ (B) Mg^{+2} (C) Cl^- (D) F^-

48. $KMnO_4$ reacts with oxalic acid according to the equation: [CBSE-96]
$2MnO_4^- + 5C_2O_4^{2-} + 16H^+ \dashrightarrow 2Mn^{2+} + 10CO_2 + 8H_2O$
Here 20 ml of 0.1 M $KMnO_4$ is equivalent to
(A) 20ml of 0.5 M $C_2H_2O_4$ (B) 50ml of 0.1 M $C_2H_2O_4$
(C) 50 ml of 0.5 M $C_2H_2O_4$ (D) 20 ml if 0.1 M $C_2H_2O_4$

49. Which of the following is largest ? [CBSE-96]
(A) Cl^- (B) S^{2-} (C) Na^+ (D) F^-

50. The structural formula of hypophosphorus acid is [CBSE-97]

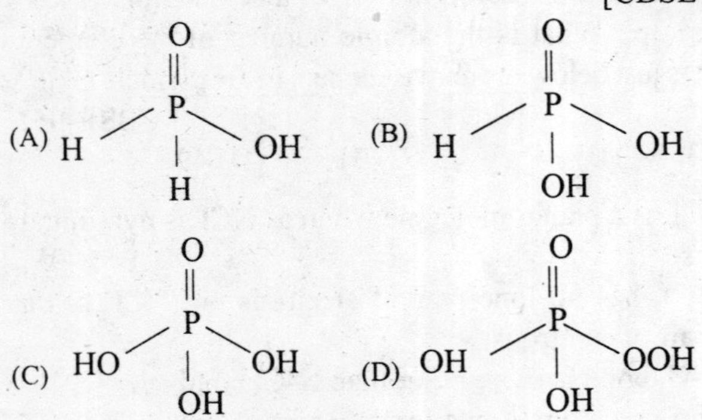

51. Which one of the following is correct order of the size of iodine species ? [CBSE-97]
(A) $I > I^- > I^+$ (B) $I > I^+ > I^-$
(C) $I^+ > I^- > I$ (D) $I^- > I > I^+$

52. The volume strength of 1.5 N H_2O_2 solution is [CBSE-97]
(A) 4.8 (B) 5.2 (C) 8.8 (D) 8.4

53. Calcium is obtained by the [CBSE-97]
(A) roasting of limestone
(B) electrolysis of solution of calcium chloride in H_2O
(C) reduction of calcium chloride with carbon
(D) electrolysis of molten anhydrous calcium chloride

54. The hydride ion H-is stronger base than its hydroxide ion OH^-. Which of the following reaction will occur if sodium hydride (NaH) is dissolved in water ? [CBSE-97]
(A) $H^-(aq) + H_2O \rightarrow H_2O$
(B) $H^-(aq) + H_2O \rightarrow OH^- + H_2$
(C) $H^- + H_2O \rightarrow$ No reaction (D) None of these .

55. In graphite, electrons are [CBSE-97]
(A) localised on every third C-atom
(B) present in anti-bonding orbital
(C) localised on each C-atom
(D) spread out between the structure

56. When a substance A reacts with water it produces a combustible gas B and a solution of substance C in water. When another substance D react with this solution of C, it also produces the same gas B on warming but D can produce B on reaction with dilute sulphuric acid at room temp. A imparts a deep golden yellow colour to a smokeless flame of Bunsen burner. A,B, C and D are respectively [CBSE-98]
(A) K, H_2, KOH, Al (B) Na, H_2, NaOH, Zn
(C) CaC_2, C_2H_2, $Ca(OH)_2$, Fe (D) Ca, H_2, $Ca(OH)_2$,Sn

57. The first ionization potential (eV) of Be and B respectively are. [CBSE-98]
(A) 8.29, 9.32 (B) 9.32, 8.29
(C) 9.32, 9.32 (D) 8.29, 8.29

58. Which one of the following pairs of substances on reaction will not evolve H_2 gas ? [CBSE-98]
(A) Iron and H_2SO_4 (B) Copper and HCl (aqueous)
(C) Sodium and Ethyl alcohol
(D) Iron and steam

59. Which one of the following elements constitutes a major impurity in pig iron ? [CBSE-98]
(A) Oxygen (B) Silicon (C) Graphite (D) sulphur

60. The structure of H_2O_2 is [CBSE-99]
(A) Planar (B) Non-planar
(C) Spherical (D) Linear

61. Which of the following element has the maximum electron affinity ? [CBSE-99]
(A) Cl (B) Br (C) I (D) F

62. The type of hybridisation of boron in diborane is [CBSE-99]
(A) sp-hybridisation (B) sp^2-hybridisation
(C) sp^3-hybridisation (D) sp^3d^2-hybridisation

63. Which of the following shows electrical conduction ? [CBSE-99]
(A) Potassium (B) Graphite (C) Diamond (D) Sodium

64. Cassiterite is an ore of [CBSE-99]
(A) Mn (B) Ni (C) Sb (D) Sn

65. Percentage of lead in lead pencil is [CBSE-99]
(A) zero (B) 20 (C) 80 (D) 70

66. Set containing iso-electric species [CBSE-2000]
(A) C_2^{2-}, NO^+, CN^-, O_2^{+2} (B) CO, NO, O_2, CN
(C) CO_2, NO_2, O_2, N_2O_5 (D) CO, CO_2, NO, NO_2

67. Which of the following is incorrect with respect to property indicated ? [CBSE-2000]
(A) EN $\rightarrow F_2 > Cl_2 > Br_2$ (B) EA$\rightarrow Cl_2 > F_2 > Br_2$
(C) Oxidising power $\rightarrow F_2 > Cl_2 > Br_2$
(D) Bond energy $\rightarrow F_2 > Cl_2 > Br_2$

68. $PbO_2 \rightarrow PbO$ $\Delta G298 < 0$

$SnO_2 \rightarrow SnO$ $\Delta G298 > 0$

Most probable oxidation states of Pb and Sn are- [CBSE-2001]

(A) Pb^{4+}, Sn^{4+} (B) Pb^{4+}, Sn^{2+} (C) Pb^{2+}, Sn^{2+}(D) Pb^{2+}, Sn^{4+}

69. In which of the following bond angle is maximum ? [CBSE-2001]

(A) NH_3 (B) NH_4^+ (C) PCl_5 (D) SCl_2

70. Which of the following statements is not correct ? [CBSE-2001]

(A) $La(OH)_3$ is less basic than $Lu(OH)_3$

(B) In lanthanide series ionic radius of Ln^{+3} ions decrease.

(C) La is actually an element of transition series rather lanthanide.

(D) Atomic radius of Zr and Hf are same because of lanthanide contraction.

71. Correct order of Ist IP among following elements Be, B, C, N, O is - [CBSE-2001]

(A) B<Be<C<O<N (B) B<Be<C<N<O

(C) Be<B<C<N<O (D) Be<B<C<O<N

72. In X - H ....Y, X and Y both are electronegative elements - [CBSE-2001]

(A) Electron density on X will increase and on H will decrease

(B) On both X and H electron density will increase

(C) On both X and H electron density will decrease

(D) On X electron density will decrease and on H it increases

73. Main axis of a diatomic molecule is Z. Atomic orbitals p_X and p_Y overlap to form- [CBSE-2001]

(A) π Molecular orbital (B) σ Molecular orbital

(C) δ Molecular orbital

(D) No Molecular orbital will be formed.

74. Which of the following will exhibit maximum molar conducitivity ? [CBSE-2001]

(A) $K_4[Fe(CN)_6]$ (B) $[Co(NH_3)_6]Cl_3$

(C) $[Cu(NH_3)_4]Cl_2$ (D) $[Ni(CO)_4]$

75. Which of the following order is wrong [CBSE-2002]

(A) $NH_3 < PH_3 < AsH_3$ - Acidic

(B) Li < Be < B < C - I[st] IP

(C) $Al_2O_3 < MgO < Na_2O < K_2O$ - Basic

(D) $Li^+ < Na^+ < K^+ < Cs^+$ - Ionic radius

76. In borax bead test which compound is formed [CBSE-2002]

(A) Ortho borate (B) Meta borate

(C) Double oxide (D) Tetra borate

77. The ions O^{2-}, F^-, Mg^+, Mg^{2+} and Al^{3+} are isoelectronic. Their ionic radii show [CBSE-2003]

(A) a decrease from O^{2-} to F^- and then increase from Na^+ to Al^{3+}

(B) a significant increase from O^{2-} to Al^{3+}

(C) a significant decrease from O^{2-} to Al^{3+}

(D) an increase from O^{2-} to F^- and then decrease from Na^+ to Al^{3+}

78. Which of the following isoelectronic species has more electrons than neutrons ? [AIIMS-93]

(A) O^{2-} (B) F^- (C) Na^+ (D) Mg^{2+}

79. The fact that BCl_3 is a planar molecule, while NCl_3 is pyramidal can be rationalized in many ways. Of the following statements the best rationalisation is [AIIMS-93]

(A) Nitrogen is more electronegative than boron

(B) The nitrogen atom in NCl_3 has a lone pair of electrons while the boron atom in BCl_3 does not have

(C) The nitrogen atom is smaller than the boron atom

(D) The boron atom in BCl_3 is sp^3 hybridised while nitrogen atom in NCl_3 is sp^2 hybridised.

80. The following diagram illustrates a possible electronic configuration of which of the following species ? [AIIMS-93]

↑↓		↑↓		↑↓	↑↓	↑↓		↑↓		↑↓	↑	↑

(A) Cl^+ (B) Cl^- (C) Ar^{2-} (D) P^{3-}

81. In which of the following compounds, the ratio of anion size to the cation size is the lowest ? [AIIMS-93]

(A) CsI (B) LiI (C) LiF (D) CsF

82. Which of the following is isoelectronic with CO ? [AIIMS-93]

(A) CN^- (B) O^{2+} (C) O^{2-} (D) N^{2+}

83. Which of the following elements will have the highest electron affinity ? [AIIMS-94]

(A) Nitrogen (B) Chlorine (C) Fluorine (D) Phosphorus

84. Which of the following glasses is the most heat resistant ? [AIIMS-94]

(A) Flint glass (B) Pyrex glass

(C) Bottle glass (D) Water glass

85. The number of electrons in $[_{19}K^{40}]^{-1}$ is [AIIMS-94]

(A) 19 (B) 20 (C) 28 (D) 40

86. Which of the following is Alum ? [AIIMS-94]

(A) $NaAlO_2$ (B) $KCl.MgCl_2.6H_2O$

(C) $FeSO_4(NH_4)_2SO_4.6H_2O$

(D) $Na_2SO_4.Al_2(SO_4)_3.24H_2O$

87. Which of the following metal reacts with water [AIIMS-94]

(A) Nickel (B) Silver (C) Copper (D) Sodium

88. Permanent hardness of water can be removed by adding [AIIMS-94]

(A) K (B) Cl_2 (C) Na_2CO_3 (D) $Ca(OCl)Cl$

89. $MgCl_2.6H_2O$ when heated gives

(A) Magnesium oxide (B) Magnesium chloride
(C) Magnesium dichloride (D) Magnesium oxychloride

90. Which of the following show bonds in silica [AIIMS-94]
(A) Si–Si–Si–Si (B) Si–C–Si–O–Si
(C) Si–C–Si–C–Si (D) –Si–O–Si–O–Si–

91. Assertion : Helium and beryllium have similar outer electronic configuration of type ns^2. Reason : Both are chemically inert. [AIIMS-94]
(A) A (B) B (C) C (D) D

92. Assertion : The first ionization energy of aluminium is lower than that of magnesium Reason : The ionic radius of aluminium is smaller than that of magnesium [AIIMS-94]
(A) A (B) B (C) C (D) D

93. The number of water molecules in gypsum and plaster of pairs is respectively [AIIMS-94]
(A) 2 and 1/2 (B) 2 and 1 (C) 5 and 2 (D) 7 and 2

94. The long form of Periodic Table is based on [AIIMS-94]
(A) Atomic number (B) Electronegativity
(C) Mass of the atom (D) Shape of the atom

95. Which of the following is a strong reducing agent ? [AIIMS-94]
(A) Cl (B) Ca (C) Cr (D) Na

96. Which of the following is the least stable and its existence is doubtful ? [AIIMS-95]
(A) Cl_4 (B) GeI_4 (C) SnI_4 (D) PbI_4

97. Which of the following alkaline earth metal has highest ionic mobility in aqueous solution ? [AIIMS-95]
(A) Ca^{2+} (B) Mg^{2+} (C) Be^{2+} (D) Ba^{2+}

98. Lithopone, a white pigment, consists of [AIIMS-95]
(A) PbS and MgO (B) $BaSO_4$ and $PbSO_4$
(C) ZnS and $BaSO_4$ (D) Al_2O_3 and $CaCO_3$

99. Euchlorine is produced by heating a mixture of [AIIMS-95]
(A) KCl + conc. HCl (B) $KClO_3$ + conc. HCl
(C) KCl + conc. H_2SO_4 (D) K_2ClO_3 + conc.H_2SO_4

100. Which of the following is the electronic configuration of Cu^{2+} (Z = 29) [AIIMS-95]
(A) [Ar] $4s^1\ 3d^8$ (B) [Ar] $4s^1\ 3d^{10}$
(C) [Ar] $3d^9$ (D) [Ar] $4s^2 3d^{10} 4p^1$

101. Assertion : Alkali metals impart colour to the flame. Reason : Their ionization energies are low. [AIIMS-95]
(A) A (B) B (C) C (D) D

102. Maximum covalency of an element of atomic number 7 is [AIIMS-95]
(A) 2 (B) 3 (C) 4 (D) 5

103. If an atom is reduced, its oxidation number [AIIMS-95]
(A) Increases (B) Slightly decreases
(C) Does not change (D) Sharply decreases

104. Which of the following has the maximum electonegativity [AIIMS-95]
(A) F (B) O (C) C (D) N

105. In the P^{3-}, S^{2-} and Cl^- ions, the increasing order of size is [AIIMS-96]
(A) $Cl^- > S^{2-} > P^{3-}$ (B) $P^{3-} < S^{2-} < Cl^-$
(C) $S^{2-} < Cl^- < P^{3-}$ (D) $S^{2-} < P^{3-} < Cl^-$

106. The volume of 3.0 N-H_2O_2 strength is [AIIMS-96]
(A) 4.2 litres (B) 8.4 litres
(C) 16.8 litres (D) 33.6 litres

107. Silicon is an important constituent of [AIIMS-96]
(A) Rocks (B) Animals (C) Alloys (D) Vegetables

108. Thermite is a mixture of [AIIMS-96]
(A) Fe+Al (B) Cu + Mg (C) Zn + Mg (D) Fe_2O_3 + Al

109. Assertion : Nitrogen is unreactive at room temperature but becomes reactive at elevated temperatures (on heating) or in the presence of catalysts. Reason : In nitrogen molecules, there is extensive delocalization of electrons. [AIIMS-96]
(A) A (B) B (C) C (D) D

110. If a neutral atom is converted into a cation, then its [AIIMS-96]
(A) Size increases (B) Size decreases
(C) Atomic weight increases (D) Atomic weight decreases

111. Which of the following is the correct sequence of atomic weights of the given elements ? [AIIMS-96]
(A) Fe > Co > Ni (B) Co > Ni > Fe
(C) Ni > Co > Fe (D) Co > Fe > Ni

112. Which of the following does not react with AgCl ? [AIIMS-97]
(A) $Na_2S_2O_3$ (B) NH_4OH (C) $NaNO_3$ (D) Na_2CO_3

113. Which of the following compounds on reaction with NaOH and Na_2O_2 gives yellow colour ? [AIIMS-97]
(A) Cr $(OH)_3$ (B) Zn $(OH)_2$
(C) Al $(OH)_3$ (D) None of these

114. Assertion : Pb^{4+} can be reduced easily to Pb^{2+} Reason : Pb^{2+} is paramagnetic. [AIIMS-97]
(A) A (B) B (C) C (D) D

115. Assertion : Na_2SO_4 is solubie in water while barium sulphate is insoluble. Reason : Lattice energy of $BaSO_4$ exceeds its hydration energy. [AIIMS-97]
(A) A (B) B (C) C (D) D

116. Assertion : Electron affinity of oxygen is less than

fluorine but greater than nitrogen. Reason : Ionisation potential is as follows N > O < F. [AIIMS-97]
(A) A (B) B (C) C (D) D

117. Assertion : Ionisation potential of Be (at.no.4) is less than B (atomic no.5)Reason : The first electron released from Be is of p-orbital but that from B is of s-orbital. [AIIMS-97]
(A) A (B) B (C) C (D) D

118. Assertion : Li resembles Mg. Reason : Li^+ has same size as Mg^{2+} [AIIMS-97]
(A) A (B) B (C) C (D) D

119. Assertion : Tungsten has very high melting point. Reason : Tungsten is a covalent compound. [AIIMS-97]
(A) A (B) B (C) C (D) D

120. Which of the following elements are analogous to the lanthanides ? [AIIMS-98]
(A) Actinides (B) Borides (C) Carbides (D) Hydrides

121. What is the oxidation number of sulphur in $Na_2S_4O_6$? [AIIMS-98]
(A) $\frac{2}{3}$ (B) $\frac{3}{2}$ (C) $\frac{3}{5}$ (D) $\frac{5}{2}$

122. Which of the following is the most electropositive element ? [AIIMS-98]
(A) Aluminium (B) Magnesium
(C) Phosphorus (D) Sulphur

123. Which of the following substances is used, in the laboratory, for fast drying of neutral gases ? [AIIMS-98]
(A) Sodium phosphate (B) Phosphorus pentoxide
(C) Sodium sulphate (D) Anhydrous calcium chloride

124. Assertion : Alkali metals impart colour to the flame. Reason : Their ionization energies are low. [AIIMS-98]
(A) A (B) B (C) C (D) D

125. The correct order of increasing ionic character is [AIIMS-99]
(A) $BeCl_2 < MgCl_2 < CaCl_2 < BaCl_2$
(B) $BeCl_2 < MgCl_2 < BaCl_2 < CaCl_2$
(C) $BeCl_2 < BaCl_2 < MgCl_2 < CaCl_2$
(D) $BeCl_2 < CaCl_2 < MgCl_2 < BaCl_2$

126. Oxidation state of osmium (Os) in OsO_4 is [AIIMS-99]
(A) +4 (B) +6 (C) +7 (D) +8

127. Which of the following shows electrical conduction? [AIIMS-99]
(A) Sodium (B) Graphite (C) Diamond (D) Potassium

128. The radii of F, F^-, O and O^{2-} are in the order of [AIIMS-99]
(A) $O^{2-} > F^- > O > F$ (B) $O^{2-} > F^- > F > O$
(C) $F^- > O^{2-} > F > O$ (D) $O^{2-} > O > F^- > F$

129. Which of the following has maximum ionization potential [AIIMS-99]
(A) Na (B) K (C) Be (D) Mg

130.Solubility in water of sulphates down the Be group is Be > Mg > Ca > Sr > Ba.This is due to [AIIMS-99]
(A) High ionization energy (B) Increase in melting point
(C) Higher co-ordination number (D) All of these

131. The formula of microcosmic salt and the product obtained by heating it is [AIIMS-2000]
(A) $Na(NH_4)PO_4$; $NaPO_3$ (B) $Na(NH_4)_2HPO_4$; $NaPO_3$
(C) $Na(NH_4)HPO_4$; $NaPO_3$ (D) $Na(NH_4)HPO_4$; $NaPO_3$

132. 12 g of an alkaline earth metal gives 14.8 gm of its nitride. Atomic weight of the metal is [AIIMS-2000]
(A) 12 (B) 20 (C) 40 (D) 14.8

133. Which one of the following has the largest size [AIIMS-2000]
(A) Br^- (B) I^- (C) I (D) I^+

134. How many unpaired electrons are present in an atom of configuration $1s^2\ 2s^2\ 2p^4$? [AIIMS-2000]
(A) 1 (B) 2 (C) 5 (D) 7

135. Which is explosive ? [AIIMS-2001]
(A) Ba_3N_2 (B) NaN_3 (C) KN_3 (D) AgN_3

136. Which is hardest ? [AIIMS-2001]
(A) Boron nitirde (B) Boron carbide
(C) Silicon boride (D) Aluminium boride

137. Amongst these compounds which is least soluble ? [AIIMS-2001]
(A) MgF_2 (B) CaF_2 (C) SrF_2 (D) BaF_2

138. In diborane formation of which intermediate is not possible : [AIIMS-2001]
(A) $B_2H(CH_3)_5$ (B) $B_2H_2(CH_3)_4$
(C) $B_2H_3(CH_3)_3$ (D) $B_2H_4(CH_3)_2$

139. Which of the following hydroxide is insoluble in water ? [AIIMS-2001]
(A) $Be(OH)_2$ (B) $Mg(OH)_2$ (C) $Ca(OH)_2$ (D) $Ba(OH)_2$

140. Which of the following is used as a moderator in nuclear reactor ? [AIIMS-2001]
(A) D_2O (B) N_2O (C) H_2O (D) NaOH

141. The hardness of water is measured by ? [AIIMS-2001]
(A) E.D.T.A. method (B) Distribution method
(C) Conductivity method (D) All of these

142. Assertion : Carbon tetrachloride is resistant towards hydrolysis, whereas silicon tetrachloride is readily hydrolyzed. Reason : Silicon has bigger size and can accommodate larger number of water molecules around itself. [AIIMS-2001]
(A) A (B) B (C) C (D) D

143. Assertion : Carbon dioxide (CO_2) is not soluble in water. Reason : Carbon monoxide (CO) is not good reducing agent. [AIIMS-2001]
(A) A (B) B (C) C (D) D

144. Which of the following reactions produces hydrogen? [AIIMS-2002]
(A) $Mg + H_2O$ (B) $BaO_2 + HCl$
(C) $H_2S_4O_8 + H_2O$ (D) $Na_2O_2 + 2HCl$

145. Which of the following does not have valence electron in 3d-subshell ? [AIIMS-2002]
(A) Fe (III) (B) Mn (II) (C) Cr (I) (D) P (0)

146. Assertion : Diamond is a bad conductor. Reason : Graphite is a good conductor. [AIIMS-2002]
(A) A (B) B (C) C (D) D (E) E

147. Assertion (A) : Potassium and cesium are used in photoelectric cells. Reason (R) : Potassium and cesium emit electrons on exposure to light. [AIIMS-2002]
(A) A (B) B (C) C (D) D (E) E

148. The paramagnetic species is- [AIIMS-2003]
(A) KO_2 (B) SiO_2 (C) TiO_2 (D) BaO_2

149. The reagent commonly used to determine hardness of water titrimetrically is - [AIIMS-2003]
(A) Oxalic acid (B) Disodium salt of EDTA
(C) Sodium citrate (D) Sodium thiosulphate

150. On dissolving moderate amount of sodium metal in liquid NH_3 at low temperature, which one of the following does not occur ? [AIIMS-2003]
(A) Blue coloured solution is obtained
(B) Na+ ions are formed in the solution
(C) Liquid NH_3 becomes good conductor of electricity
(D) Liquid ammonia remains diamagnetic

151. (A) Barium is not required for normal biological function in human. [AIIMS-2003]
(R) Barium does not show variable oxidation state.
(A) (B) (C) (D)

152. (A) The O–Obond length in H_2O_2 is shorter than that of O_2F_2. [AIIMS-2003]
(R) H_2O_2 is an ionic compound.
(A) (B) (C) (D)

153. Moderate electrical conductivity is shown by : [JIPMER-90]
(A) Silica (B) Graphite
(C) Diamond (D) Carborundum

154. The ion that is isoelectronic with CO is : [JIPMER-90]
(A) CN^- (B) O_2^+ (C) O_2^- (D) N_2^+

155. The element with highest second ionisation potential is : [JIPMER-90]
(A) C (B) N (C) O (D) F

156. The element with highest first ionisation potential is [JIPMER-90]
(A) B (B) N (C) O (D) C

157. Which has the maximum electronegative character : [JIPMER-90]
(A) Oxygen (B) Nitrogen (C) Fluorine (D) Astatine

158. The oxide that gives H_2O_2 on treatment with dilute acid is : [JIPMER-90]
(A) PbO_2 (B) MnO_2 (C) Na_2O_2 (C) TiO_2

159. Colour is imparted to glass by mixing [JIPMER-90]
(A) Synthetic dye (B) Metal oxide
(C) Oxide of non-metal (D) Coloured salt

160. The most electro negative element is :[JIPMER-90]
(A) Fluorine (B) Oxygen (C) Nitrogen (D) Sulphur

161. Which is soluble in acetone [JIPMER-90]
(A) KI (B) NaI (C) LiI (D) NaCl

162. The total number of electrons in one molecule of CO_2 is : [JIPMER-90]
(A) 22 (B) 44 (C) 66 (D) 88

163. If three electrons are lost by a metal ion M^{3+}, its final oxidation number would be : [JIPMER-90]
(A) 0 (B) +2 (C) +4 (D) +6

164. Which has the minimum atomic radius[JIPMER-90]
(A) N (B) Na (C) K (D) F

165. The species that does not contain peroxide ions is [JIPMER-91]
(A) PbO_2 (B) H_2O_2 (C) SrO_2 (D) BaO_2

166. Which one of the following is an acidic salt : [JIPMER-91]
(A) Na_2S (B) NaHS (C) Na_2SO_3 (D) Na_2SO_4

167. Lunar caustic is : [JIPMER-91]
(A) AgCl (B) $AgNO_3$ (C) NaOH (D) KNO_3

168. Which of the following has the maximum ionization potential : [JIPMER-92]
(A) F (B) C (C) N (D) Ne

169. Which electronic configuration represents alkali metals : [JIPMER-92]
(A) ns^1 (B) $(n-1)d^0\ ns^1$
(C) $(n-1)a^5ns^1$ (D) $(n-1)d^{10}\ ns^2$

170. The valency shell of calcium contains [JIPMER-92]
(A) 2 electrons (B) 4 electrons
(C) 6 electrons (D) 8 electrons

171. Red lead is : [JIPMER-93]
(A) Pb_3O_4 (B) PbO (C) $PbCrO_4.PbO$ (D) PbO_2

172. The main factor for small B–F bonds in BF_3 is :

[JIPMER-93]

(A) Large electronegativity of fluorine
(B) Three centered two electron bonds in BF_3
(C) pπ–pπ back bonding (D) pπ–dπ back bonding

173. Which of the following configurations corresponds to the element with maximum electron affinity ? [JIPMER-93]
(A) [He] $2s^2 2p^5$ (B) [Ar] $4s^2 3s^{10} 4p^5$
(C) [Ne] $3s^2 3p^5$ (D) [Kr] $5s^2 4d^{10} 5p^5$

174. The electronegativity difference between two atoms X and Y in a bond is 2.0, the percentage covalent character in the bond is about. [JIPMER-93]
(A) 50% (B) 54% (C) 46% (D) 8%

175. Which of the following alkali metal ion has the smallest ionic mobility : [JIPMER-93]
(A) Li^+ (B) Cs^+ (C) K^+ (D) Na^+

176. Which of the following ion has smallest size: [JIPMER-93]
(A) Ca^{2+} (B) Cl^- (C) S^{2-} (D) K^+

177. The most basic among the following is : [JIPMER-93]
(A) La $(OH)_3$ (B) Lu $(OH)_3$ (C) $Ce(OH)_3$ (D) Al $(OH)_3$

178. A gas which burns with a blue flame is [JIPMER-94]
(A) CO_2 (B) HCl (C) CO (D) NO_2

179. Heavy water is [JIPMER-94]
(A) H_2O (B) H_2O_2 (C) D_2O (D) $(H_2O)_n$

180. When the 3d orbital is complete the new electron will enter the [JIPMER-94]
(A) 4p orbital (B) 4f orbital
(C) 4s orbital (D) 4d orbital

181. Pb_3O_4 is a [JIPMER-94]
(A) Double oxide (B) Complex oxide
(C) Normal oxide (D) Mixed oxide

182. Pick a saline hydride from the following [JIPMER-94]
(A) CaH_2 (B) HF (C) H_2O (D) None

183. Lunar caustic is [JIPMER-94]
(A) $NaNO_2$ (B) $AgNO_3$ (C) $Cu(NO_3)_2$ (D) $Au(NO_3)_2$

184. Which one of the following metal is found free in nature [JIPMER-94]
(A) Au (B) Ba (C) Na (D) Ca

185. Rubies and sapphires contain [JIPMER-94]
(A) Al_2O_3 (B) $Al_2(SO_4)_3$
(C) $Al_2(CO_3)_3$ (D) Al_2O_3 and aluminates

186. Which of the following metals shows allotropy [JIPMER-94]
(A) Sn (B) Ca (C) K (D) Pb

187. Which of the following contains lead? [JIPMER-94]
(A) Soda glass (B) Potash glass
(C) Flint glass (D) Bottle glass

188. Which of the following gaseous atoms has the highest value of IE ? [JIPMER-95]
(A) Mg (B) P (C) Si (D) Al

199. The second ionization potential of an element M is the energy required to [JIPMER-95]
(A) Remove one mole of electron from one mole of gaseous cation of the element.
(B) Remove one mole of electron from one mole of gaseous anion
(C) Remove one mole of electron from one mole of monovalent gaseous cations of the element
(D) Remove 2 moles of electron from one mole of gaseous atoms.

190. Which of the following compounds does not exist ? [JIPMER-95]
(A) $AsCl_5$ (B) $SbCl_3$ (C) $BiCl_5$ (D) $SbCl_5$

191. All elements in third period have [JIPMER-95]
(A)3 complete sub-shells (B) an atomic number 3
(C) 3 valence electrons
(D) 3 electrons less than the octet

192. Which of the following carbonate will not decompose on heating [JIPMER-96]
(A) $CaCO_3$ (B) $BaCO_3$ (C) Na_2CO_3 (D) $CdCO_3$

193. Which of the following cannot be hydrolysed ? [JIPMER-96]
(A) $GeCl_4$ (B) $SnCl_4$ (C) $SiCl_4$ (D) CCl_4

194. Be exhibits the diagonal relationship with [JIPMER-96]
(A) Mg (B) Li (C) Si (D) Al

195. Which of the following hydroxide is a strongest base ? [JIPMER-96]
(A) Li OH (B) Na OH (C) KOH (D) $Ca(OH)_2$

196. The most electronegative element of 1st period is : [JIPMER-96]
(A) He (B) H (C) Li (D) Unpredictable

197. Which one is the electron deficient compound ? [JIPMER-96]
(A) NH_3 (B) PCl_3 (C) BCl_3 (D) ClF_3

198. Which of the following isoelectronic species has the smallest size ? [JIPMER-97]
(A) S^{2-} (B) Ar (C) Ca^{2+} (D) Ti^{4+}

199. Which of the following is not true ? [JIPMER-97]
An element with high electronegativity is generally
(A) Having a high ionization potential
(B) Having a high electron affinity

(C) A non metal (D) A metal

200. The value of IE_1, IE_2, IE_3, IE_4, and IE_5 of an element are 7.1, 14.3, 34.5, 46.8 and 162.2 eV respectively. The element is likely to be. [JIPMER-97]
(A) Na (B) Si (C) F (D) Ca

201. The element with the highest first ionization potential is. [JIPMER-97]
(A) Boron (B) Carbon (C) Nitrogen (D) Oxygen

202. Heavy water is obtained by [JIPMER-97]
(A) Dissolving heavy salts in water
(B) Distillation of water
(C) Removing Ca and Mg salts from water
(D) Prolonged electrolysis of water

203. Which of the following tests is used to identify H_2O_2? [JIPMER-97]
(A) Tailing of mercury (B) Perchromic acid
(C) Dichromate paper (D) Lead acetate paper

204. Which of the following molecules and ions cannot exist ? [JIPMER-97]
(A) H_2^{2+} (B) H_2^+ (C) H_2 (D) H_2^-

205. Lead pipes are not used as water pipes owing to [JIPMER-97]
(A) Plumbosolvency (B) Toxicity
(C) Action of CO_2 (D) Precipitation of $Pb(OH)_2$

206. Cement is [JIPMER-97]
(A) Calcium silicates + calcium aluminates
(B) Calcium silicates + aluminium silicates + gypsum
(C) Calcium silicates + magnesium silicates
(D) Calcium silicates + magnesium silicates + gypsum

207. Which of the following is not a periodic property ? [JIPMER-97]
(A) Density (B) Atomic volume
(C) Mass number (D) Ionization energy

208. Which of the following gases is colourless and odourless ? [JIPMER-97]
(A) CO (B) H_2S (C) HCl (D) NH_3

209. The superior type of laboratory apparatus are made of [JIPMER-97]
(A) Soda-lime glass (B) Hard glass
(C) Flint glass (D) Pyrex glass

210. Ortho and Para hydrogen [JIPMER-97]
(A) Have different masses
(B) Have different atomic number
(C) Differ in the spin of their nuclei
(D) Do not differ in the spin of their electrons

211. Which of the following has minimum ionization energy ? [JIPMER-98]
(A) Ge (B) Se (C) As (D) Br

212. The smallest among the following ions is [JIPMER-98]
(A) Na^+ (B) Mg^{+2} (C) Ba^{+2} (D) Al^{3+}

213. Lithopone is [JIPMER-98]
(A) $ZnS + BaSO_4$ (B) $BaS + ZnSO_4$
(C) $ZnS + CaSO_4$ (D) $ZnSO_4 + CaS$

214. Hydrogen has a tendency to gain one electron to acquire helium configuration. In this respect it resembles [JIPMER-98]
(A) Halogens (B) Actinides
(C) Transition elements (D) Alkali metals

215. In the isotopes of an element [JIPMER-98]
(A) The atomic number remains the same
(B) The mass number remains the same
(C) Both atomic number and mass number are conserved
(D) Neither the atomic number nor mass number is conserved.

216. The fundamental basis of the present-day Periodic table is that elements are [JIPMER-98]
(A) Arranged in the order of increasing atomic weights.
(B) Grouped according to chemical properties
(C) Arranged in the order of increasing number of neutrons in the atomic nucleus
(D) Arranged in the order of increasing number of protons in the nucleus.

217. The type of glass used in making lenses and prisms is [JIPMER-98]
(A) A flint glass (B) Jena glass
(C) Pyrex glass (D) Quartz glass

218. The compound which does not possess a peroxide linkage is. [JIPMER-98]
(A) Na_2O_2 (B) CrO_5 (C) H_2SO_5 (D) PbO_2

219. The oxide which is not reduced by hydrogen is. [JIPMER-98]
(A) Ag_2O (B) K_2O (C) Fe_2O_3 (D) P_4O_{10}

220. The correct order of electron affinity of F, Cl, Br, and I is. [JIPMER-99]
(A) F > Cl > Br > I (B) I > Br > Cl > F
(C) Cl > F > Br > I (D) Br > Cl > F > I

221. In P^{3-}, S^{2-}, Cl^- ions the increasing order of size is [JIPMER-99]
(A) Cl^-, S^{2-}, P^{3-} (B) P^{3-}, S^{2-}, Cl^-
(C) S^{2-}, Cl^{-1}, P^3 (D) S^{2-}, P^{3-}, Cl^-

222. The only oxide formed by an element X has the formula X_2O_3, while element Y forms a hydride YH_2. On the basis of this information the most likely formula for the stable compound formed between X and Y is. [JIPMER-99]
(A) X_3Y_2 (B) X_2Y_3 (C) XY (D) X_2Y

223. Phosphorous has the oxidation state of +3 in. [JIPMER-99]
(A) meta-phosphoric acid (B) Ortho-phosphoric acid
(C)Ortho-phosphorousacid (D)Hypo-phosphorous acid

224. Which of the following would lose water when exposed to the atomophere ? [JIPMER-99]
(A) Anhydrous sodium carbonate
(B) Caustic soda (C) Concentrated H_2SO_4
(D) A saturated solution of washing soda

225. Pick the ion with the smallest radius
[JIPMER-99] (A) Mg^{2+} (B) P^{3-} (C) Si^{4+}
(D) Al^{3+}

226. Which of the following ions has the highest polarizing power ? [JIPMER-99]
(A) Na^+ (B) Ca^{2+} (C) Mg^{2+} (D) Al^{3+}

227. The approximate composition of ordinary glass is [JIPMER-99]
(A) $Na_2O.CaO.6SiO_2$ (B) $Na_2O.CaO.SiO_2$
(C) $K_2O.CaO.6SiO_2$ (D) $Na_2O.CaO.SiO_2$

228. Which set has the same number of unpaired electrons in their ground state [JIPMER-2000]
(A) Cl^-, Fe^{3+}, Cr^{3+} (B) Na^+, Mg^{2+}, Al
(C) Na, P, Cl (D) N, P, V

229. Which species does not exist [JIPMER-2000]
(A) $(SiCl_6)^{2-}$ (B) $(CCl_6)^{2-}$ (C) $(GeCl^6)^{2-}$ (D) $(SnCl_6)^{2-}$

230. The velency shell of calcium contains [JIPMER-2000]
(A) 8 electrons (B) 6 electrons
(C) 4 electrons (D) 2 electrons

231. Although CO is neutral, it shows acidic nature on reaction with at high P and T [JIPMER-2000]
(A) LiOH (B) NaOH (C) $Ca(OH)_2$ (D) $Mg(OH)_2$

232. The decomposition of H_2O_2 can be slowed down by the addition of small amount of phosphoric acid which act as [JIPMER-2000]
(A) Promoter (B) Inhibitor (C) Detainer (D) Stopper

233. The most dangerous method of preparing hydrogen would be by the action of HCl on [JIPMER-2000]
(A) Al (B) K (C) Fe (D) Zn

234. The ionic carbide is [JIPMER-2000]
(A) ZnC (B) TiC (C) SiC (D) CaC_2

235. Lead pipes are not suitable for drinking water because [JIPMER-2000]
(A) A layer of lead dioxide is deposited over pipes
(B) Lead reacts with air to form litharge
(C) Lead reacts with water containing air to form $Pb(OH)_2$
(D) Lead forms basic lead carbonate.

236. PbO_2 is [JIPMER-2000]
(A) Basic (B) Acidic (C) Neutral (D) Amphoteric

237. N_2 combines with metal to from [JIPMER-2000]
(A) Nitride (B) Nitrate
(C) Nitrite (D) Nitrosyl chloride

238. Which one is least basic [JIPMER-2000]
(A) BI_3 (B) BBr_3 (C) BCl_3 (D) BF_3

239. An element X with the electronic configuration $1s^22s^22p^63s^2$ would be expected to from the chloride with the formula [JIPMER-2000]
(A) XCl_3 (B) XCl_2 (C) XCl (D) X_2Cl

240. Which pair of atoms or ions will have same configuration [JIPMER-2001]
(A) F^+ and Ne (B) Li^+ and He
(C) Cl^- and Ar (D) Na and K

241. The reaction $H_2S + H_2O_2 \rightarrow 2H_2O + S$ shows [JIPMER-2001]
(A) Oxidizing action of H_2O_2
(B) Reducing action of H_2O_2
(C) Alkaline nature of H_2O_2 (D) Acidic nature of H_2O_2

242. To obtain chromium from chromic oxide (Cr_2O_3), the method used is [JIPMER-2001]
(A) Alumino-thermic process
(B) Electrolytic reduction
(C) Carbon reduction (D) Carbon monoxide reduction.

243. Deuterium resembles hydrogen in chemical properties but reacts [JIPMER-2001]
(A) More vigorously than hydrogen
(B) Faster than hydrogen
(C) Slower than hydrogen (D) Just as hydrogen.

244. Diborane reacts with water to form [JIPMER-2001]
(A) $H_3BO_3 + H_2$ (B) H_2 (C) HBO_2 (D) H_3BO_3

245. Litharge is [JIPMER-2001]
(A) $Pb(CH_3COO)_2$ (B) Pb_3O_4 (C) PbO_2 (D) PbO

246. Concentrated nitric acid reacts with iodine to give [JIPMER-2001]
(A) $HOIO_3$ (B) $HOIO_2$ (C) HOI (D) HI

247. Red lead is an example of a/an . . . oxide [JIPMER-2001]
(A) Basic (B) Super (C) Mixed (D) Amphoteric

248. Which of the following has the smallest size? [JIPMER-2001]
(A) Mg^{2+} (B) Na^+ (C) Al^{3+} (D) Si^{4+}

249. The compounds of alkaline earth metals have the following magnetic nature [JIPMER-2002]
(A) Diamagnetic (B) Paramagnetic
(C) Ferromagnetic (D) Antiferromagnetic

250. Which of the following has largest size ? [JIPMER-2002]

(A) Al (B) Al^{+} (C) Al^{2+} (D) Al^{3+}

251. Producer gas is [JIPMER-2002]

(A) $CO + N_2$ (B) $CH_4 + H_2 + CO + CO_2$
(C) $CO + H_2 + H_2$ (D) $CH_4 + CO + H_2$

252. Which of the following is in the increasing order of the ionic character. [JIPMER-2002]

(A) $PbCl_4 < PbCl_2 < CaCl_2 < NaCl$
(B) $PbCl_2 < PbCl_4 < CaCl_2 < NaCl$
(C) $PbCl_2 < PbCl_4 < NaCl < CaCl_2$
(D) $PbCl_4 < PbCl_2 < NaCl_2 < CaCl_2$

253. The elements of group 14 have valance electrons, out of these the element having maximum tendency to exhibit +2 oxidation state is [JIPMER-2002]

(A) Sn (B) C (C) Ge (D) Pb

254. 20 volume H_2O_2 solution has a strength of about [JIPMER-2002]

(A) 30% (B) 10% (C) 6% (D) 3%

255. The principal constituent of pyrex glass is [JIPMER-2002]

(A) Cl (B) Pb (C) B (D) Zn

256. Semi water gas is [JIPMER-2002]

(A) $CO + H_2 + N_2$ (B) $CO + H_2$ (C) $CO + N_2$ (D) None

257. The formula of plaster of paris is [JIPMER-2002]

(A) $CaSO_4 + H_2O$ (B) $CaSO_4$
(C) $CaSO_4 + 2H_2$ (D) $2CaSO_4 + H_2O$

258. Which of the following is not a drying and dehydrating agent ? [JIPMER-2002]

(A) P_2O_5 (B) Silica gel
(C) Hydrated $CaCl_2$ (D) Conc. H_2SO_4

259.The chemical formula of plaster of Paris is[BHU-95]

(A) $CaSO_4.2H_2O$ (B) $CaSO_4.H_2O$
(C) $CaSO_4.\frac{1}{2}H_2O$ (D) $CaSO_4.\frac{1}{4}H_2O$

260. $PbCl_4$ exists but $PbBr_4$ and PbI_4 do not, because of [BHU-95]

(A) Inability of bromine and iodine to oxidize Pb^{+2} to Pb^{+4}
(B) Br^- and I^- ions are bigger in size
(C) More electropositive nature of bromine and iodine
(D) Chlorine is a gas

261. Which of the following is a microcosmic salt ? [BHU-95]

(A) $Na(NH_4).H_2O$ (B) $Na(NH_4)HPO_4.4H_2O$
(C) $Na(NH_3)HPO_4.4H_2O$ (D) $K(NH_4)HPO_3.2H_2O$

262. The chemical formula of lithopone is [BHU-95]

(A) $BaS + ZnSO_4$ (B) $ZnS + BaSO_4$
(C) $BaSO_4 + ZnCO_3$ (D) $BaCO_3 + ZnSO_4$

263. Which of the following is called Baeyer's regent ? [BHU-95]

(A) Saturated $KMnO_4$ solution
(B) Neutral $KMnO_4$ solution
(C) Alkaline $KMnO_4$ solution
(D) Acidic $KMnO_4$ solution

264. Which of the following represents noble gas configuration? [BHU-95]

(A) $1s^22s^22p^62s^23p^63d^{10}4s^24p^64d^{10}5s^25p^6$
(B) $1s^22s^22p^63s^23p^63d^{10}4s^24f^{14}5s^25p^65p^1$
(C) $1s^22s^22p^63s^23p^63d^{10}4s^24p^64d^{10}5s^25p^65d^16s^2$
(D) $1s^22s^22p^63s^23p^63d^{10}4s^24p^64d^{10}5s^25p^65d^56s^2$

265. Silver chloride is soluble in methylamine due to the formation of [BHU-95]

(A) AgOH (B) $Ag(CH_3NH_2)Cl$
(C) $[Ag(CH_3NH_2)_2]Cl$ (D) $Ag + CH_3Cl + NH_4Cl$

266. Which of the following statement is correct ? [BHU-95]

(A) $SnCl_4$ can not be reduced
(B) Mosaic gold has formula SnS_2
(C) Xenon has configuration (outer) ns_2p_6
(D) None of these

267. In which of the following reactions, hydrogen peroxide is a reducing agent ? [BHU-95]

(A) $2FeCl_2 + 2HCl + H_2O_2 \rightarrow 2FeCl_3 + 2H_2O$
(B) $Cl_2 + H_2O_2 \rightarrow 2HCl + O_2$
(C) $2H_1 + H_2O_2 \rightarrow 2H_2O + I_2$
(D) $H_2SO_3 + H_2O_2 \rightarrow H_2SO_4 + H_2O$

268. The absorption of hydrogen by palladium is called [BHU-96]

(A) Hydration (B) Reduction
(C) Occlusion (D) Hydrogenation

269. Which of the following is metalloid ? [BHU-96]

(A) Pb (B) Zn (C) As (D) Na

270. The amount of energy, which is released due to the addition of extra electron to the outermost orbit of gaseous atom, is called [BHU-96]

(A) Electron capacity (B) Electron affinity
(C) Ionisation potential (D) Electronegativity

271. Which of the following is the most electronegative element ? [BHU-96]

(A) Fluorine (B) Chlorine (C) Bromine (D) Iodine

272. Element boron is obtained by the reduction of oxides of [BHU-96]

(A) Non-metal (B) Noble gas
(C) Electronegative metal (D) Electropositive metal

273. Which of the following elements has the maximum electron affinity ? [BHU-97]

(A) Cl (B) F (C) Br (D) I

274. D_2O is used more in [BHU-97]
(A) Chemical industry (B) Nuclear reactor
(C) Pharmaceutical preparations
(D) Insecticide preparation

275. What is lye ? [BHU-97]
(A) 10% solution of NaCl (B) 10% solution of KOH
(C) 10% solution of NaOH (D) 10% solution of Na_2CO_3

276. Carborundum is [BHU-98]
(A) BN (B) SiO_2 (C) SiC (D) CS_2

277. Which metal is protected by a layer of its own oxide? [BHU-98]
(A) Aluminium (B) Silver (C) Gold (D) Iron

278. Which of the following oxides reacts with HCl and NaOH ? [BHU-98]
(A) CaO (B) ZnO (C) N_2O_5 (D) CO_2

279. Electronic configuration of calcium atom may be written as [BHU-99]
(A) [Ne] $2s^2$ (B) [Ne] $3s^2$ (C) [Ar] $4s^2$ (D) [Xe] sp^2

280. Which one of the following statements is wrong for gases ? [BHU-99]
(A) Gases do not have a definite shape and volume
(B) Volume of the gas is equal to volume of container confining the gas
(C) Confined gas exerts uniform pressure on the walls of its container in all directions
(D) Mass of the gas can not be determined by weighing a container in which gas is enclosed

281. If HNO_3 changes into N_2O, the oxidation number is changed by [BHU-99]
(A) 2 (B) 6 (C) 0 (D) + 4

282. In which of the following reactions, H_2O_2 is acting as a reducing agent ? [BHU-99]
(A) $H_2O_2 + SO_2 \rightarrow H_2SO_4$ (B) $2KI + H_2O_2 \rightarrow 2KOH + I_2$
(C) $PbS + 4H_2O_2 \rightarrow PbSO_4 + 4H_2O$
(D) $Ag_2O + H_2O_2 \rightarrow 2Ag + H_2O + O_2$

283. The metal that forms a self protecting film of oxide, to prevent corrosion, is [BHU-99]
(A) Cu (B) Al (C) Na (D) Au

284. Which of the following ion is the largest in size [BHU-99]
(A) K+ (B) S^{2-} (C) Cl^- (D) Ca^{2+}

285. The type of hybridisation of boron in diborane is [BHU-99]
(A) sp-hybridisation (B) sp^2-hybridisation
(C) sp^3-hybridisation (D) sp^2d^2-hybridisation

286. Which ion is detected by Nessler's reagent ? [BHU-99]
(A) NH_4^+ (B) PO_4^{3-} (C) MnO_4 (D) CrO_4^{2-}

287. Correct formula of borax is [BHU-2000]
(A) $Na_2[B_4O_5(OH)_4]$ $8H_2O$ (B) $Na_2B_4O_7.4H_2O$
(C) $Na_2[B_4O_5(OH)_4]$ $10H_2O$ (D) $Na_2B_4O_7.8H_2O$

288. Electronic configuration of most electronegative element is [BHU-2000]
(A) $1s^22s^22p^43s^1$ (B) $1s^22s^22p^63s^23p^5$
(C) $1s^22s^22p^5$ (D) $1s^22s^22p^63s^23p^6$

289. Optical glass has high percentage of [BHU-2000]
(A) MgO (B) Al_2O_3 (C) B_2O_3 (D) PbO

290. Epsom salt is [BHU-2000]
(A) $MgSO_4$ (B) $MgSO_4.2H_2O$
(C) $MgSO_4.7H_2O$ (D) $MgSO_4.\frac{1}{2}H_2O$

291. The material used in warfares for making smoke screens for military purposes is [BHU-2000]
(A) PH_3 (B) CCl_4 (C) $SiCl_4$ (D) PCl_5

292. Lithopone is [BHU-2000]
(A) $ZnSO_4 + PbS$ (B) $ZnS + BaSO_4$
(C) $ZnS + Ba(OH)_2$ (D) $ZnSO_4 + BaSO_4$

293. Which of the following represents noble gas configuration ? [BHU-2000]
(A) $1s^2,2s^22p^6,3s^23p^63d^{10}, 4s^24p^64d^{10}, 5s^2, 5p^6$
(B) $1s^2,2s^22p^6,3s^23p^63d^{10},4s^24f^{14},5s^2, 5p^65d^1$
(C) $1s^2,2s^22p^6,3s^23p^63d^{10},4s^24p^64d^{10},5s^25p^65d^1,6s^2$
(D) $1s^2,2s^22p^6,3s^23p^63d^{10},4s^24p^64d^{10},5s^25p^65d^5,6s^2$

294. Electrolysis of potassium sulphate yields [BHU-2000]
(A) $K_2S_2O_8$ (B) K_2SO_8 (C) KS_2O_4 (D) KSO_4

295. What is Bayer's Reagent ? [BHU-2000]
(A) Acidified $KMnO_4$ (B) Alkaline $KMnO_4$
(C) $ZnSO_4$ (D) $CuSO_4.2H_2O$

296. Isotopes are elements which have same [BHU-2001]
(A) Number of neutrons (B) Mass number
(C) Atomic number (D) Number of atoms

297. Electronic configuration of chromium is [BHU-2001]
(A) $[Ne]3s^23p^63d^44s^2$ (B) $[Ne]3s^23p^63d^54s^1$
(C) $[Ne]3s^23p^63d^14s^2$ (D) $[Ne]3s^23p^64s^24d^4$

298. Among the alkaline earth metals the element forming predominantly covalent compound is [BHU-2001]
(A) Be (B) Mg (C) Sr (D) Ca

299. In which of the following reaction H_2O_2 acts as reducing agent ? [BHU-2001]
(A) $2KI + H_2O \rightarrow 2KOH + I_2$
(B) $Na_2AsO_3 + H_2O_2 \rightarrow Na_3AsO_4 + H_2O$
(C) $K_2Cr_2O_7 + H_2SO_4 + 4H_2O_2 \rightarrow 2CrO_5 + K_2SO_4 + 5H_2O$
(D) $PbO_2 + H_2O_2 \rightarrow PbO + H_2O + O_2$

300. In a periodic table there are [BHU-2001]
(A) 7 periods and 9 groups (B) 18 periods and 9 groups
(C) 7 periods and 18 groups (D) 7 periods and 10 groups

301. Which of the following is not hydrolysed ? [BHU-2001]
(A) $SiCl_4$ (B) $PbCl_2$ (C) NaCl (D) CCl_4

302. The expected outermost electronic configuration of Pt(78) is [BHU-2002]
(A) $[Xe]6s^2\ 4f^{14}5d^8$ (B) $[Xe]6s^2\ 4f^{14}5d^7$
(C) $[Xe]6s^1\ 4f^{14}5d^9$ (D) $[Xe]6s^1\ 4f^{14}5d^{10}$

303. Magnetic moment of sodium in ground state is [BHU-2002]
(A) $\sqrt{6}$ (B) 2 (C) $\sqrt{3}$ (D) None of these

304. Volume strength of H_2O_2 in a 3.4% (w/v) solution of H_2O_2 is [BHU-2002]
(A) 11.2 ml (B) 5.6 ml (C) 2.8 ml (D) 22.4 ml

305. The temporary hardness of water due to calcium bicarbonate can be removed by adding [BHU-2002]
(A) $CaCO_3$ (B) $Ca(OH)_2$ (C) $CaCl_2$ (D) HCl

306. Which of the following has the largest anionic size ? [BHU-2002]
(A) F^- (B) Cl^- (C) O_2^- (D) I^-

307. Elements having maximum no. of stable isotopes is [BHU-2002]
(A) Sn (B) Fe (C) Pb (D) Os

308. Which element of 2nd period forms most acidic oxide ? [BHU-2002]
(A) F (B) N (C) B (D) C

309. The chemical process in production of steel from haematite involves [BHU-2002]
(A) oxidation (B) reduction (C) A & B (D) none

310. The hydrolysis of trimethyl chlorosilane $(CH_3)_3SiCl$ yields [BHU-2002]
(A) $[Si(CH_3)_3]_2O$ (B) $(CH_3)_3SiOH$
(C) $(CH_3)_3Si\ H$ (D) Does not hydrolyse

311. The % of H_2 in H_2O_2 is (H = 1, O = 16)[AFMC-91]
(A) 55.88 (B) 5.8 (C) 6.25 (D) 55.92

312. Electron is an ore of [AFMC-91]
(A) Mg (B) Ca (C) Sr (D) K

313. The halide that is not hydrolysed is [AFMC-91]
(A) $SiCl_4$ (B) CCl_4 (C) SiF_4 (D) $PbCl_4$

314. Lithopone is a combination of ZnS and [AFMC-92]
(A) $PbSO_4$ (B) $CaSO_4$ (C) $SrSO_4$ (D) $BaSO_4$

315.Which has the highest electro-negativity?[AFMC-92]
(A) Cl (B) Br (C) F (D) I

316. Decomposition of H_2O_2 is favoured by [AFMC-92]
(A) Traces of acid (B) Alcohol
(C) Acetanilide (D) MnO_2

317. In the preparation of sodium carbonate (Na_2CO_3) which of the following is used ? [AFMC-92]
(A) Slaked lime (B) Quick lime
(C) Limestone (D) Sodium hydroxide

318. Arrange in the increasing order of ionic radii of following elements O, C, F, Cl, Br [AFMC-94]
(A) F, O, C, Cl, Br (B) F, C, O, Cl, Br
(C) F, Cl, Br, O, C, (D) C, O, P, Cl, Br

319. Melting point is highest for [AFMC-94]
(A) B (B) Al (C) Ga (D) In

320. Melting point is lowest for [AFMC-94]
(A) Be (B) Mg (C) Ca (D) Sr

321. The lightest metal is [AFMC-94]
(A) Li (B) Fe (C) Cu (D) Ag

322. Hydrogen can not be obtained by [AFMC-94]
(A) Zn and dil. H_2SO_4 (B) Zn + dil. HNO_3
(C) Mg + $HgSO_4$ (D) Mg and H_2O

323. Pure nascent hydrogen is best obtained by [AFMC-94]
(A) Na and C_2H_5OH (B) Zn and dil. H_2SO_4
(C) Al and NaOH (D) All of these

324. Which gas is formed when CaC_2 is reacted with H_2O ? [AFMC-94]
(A) Methane (B) Ethane (C) Acetylene (D) Ethylene

325. CO_2 is obtained by heating [AFMC-94]
(A) Na_2CO_3 (B) K_2CO_3
(C) $NaHCO_3$ (D) None of the above

326. Which of the following is electron deficient ? [AFMC-94]
(A) BCl_3 (B) PCl_3 (C) PCl_5 (D) NH_3

327. Which of the following has unchanged valency ? [AFMC-94]
(A) H (B) Na (C) Fe (D) Oxygen

328. Which of the following gives brick red colour to the flame test ? [AFMC-94]
(A) Calcium (B) Barium
(C) Strontium (D) None of these

329. Living in the atmosphere of CO is dangerous because it [AFMC-94]
(A) Reduces organic matter of tissues
(B) Dries up the blood
(C) Combines with O_2 present inside to form CO_2
(D) Combines with hemoglobin and makes it incapable to absorb oxygen

330. The most electron affinity elements is [AFMC-95]
(A) F (B) Cl (C) Br (D) I

331. Arrange the following in increasing order of their atomic radius : Na, K, Mg, Rb [AFMC-95]
(A) Mg < K < Na < Rb (B) Mg < Na < K < Rb

(C) Mg < Na < K < Rb (D) Na < K < Rb < Mg

332. The hydride ions are iso-electronic with [AFMC-95]
(A) H (B) He^+ (C) He (D) Be

333. Fusion mixture is [AFMC-95]
(A) $K_2CO_3 + Na_2CO_3$ (B) $KHSO_4 + NaHSO_4$
(C) $K_2CO_3 + NaHSO_4$ (D) $KHSO_4 + Na_2SO_3$

334. Last molecule of H_2O is evolved from H_2O_2 by [AFMC-95]
(A) Crystallisation (B) Evaporation
(C) Distillation under reduced pressure (D) Electrolysis

335. Which of the following does not give Borax bead test ? [AFMC-95]
(A) Chromium (B) Ferrous salt (C) Sodium (D) Cobalt

336. Which of the following reacts with water with high rate ? [AFMC-95]
(A) Li (B) K (C) Na (D) Rb

337. $AlCl_3$ is [AFMC-95]
(A) Anhydrous and covalent (B) Anhydrous and ionic
(C) Covalent and basic (D) Co-ordinate and acidic

338. Which of the following is correct sequence for ionic radius ? [AFMC-96]
(A) Na < Al < Cl < Ar (B) Na > Al > Cl > Ar
(C) Na < Al > Cl < Ar (D) Na < Al < Cl > Ar

339. Diamond is used in glass cutting due to[AFMC-96]
(A) Hard substance (B) High R.I.
(C) High M.B. (D) High metallic bonding

340. Which of the following is metalloid ? [AFMC-96]
(A) Te (B) S (C) Se (D) At

341. Which of the following is used as moderator ? [AFMC-96]
(A) D_2O (B) Alum (C) H_2O (D) None of these

342. Potash Alum is used as [AFMC-96]
(A) Water softner (B) Disinfectant
(C) Mordant (D) Coolant

343. Which of the following statements, if any, regarding hydrogen peroxide is false ? [AFMC-96]
(A) It is decomposed by MnO_2
(B) It is more stable in basic solution
(C) It behaves as a reducing agent towards acidified $KMnO_4$ (D) It is strong oxidizing agent

344. Formula of Plaster of Paris [AFMC-96]
(A) $CaSO_4.H_2O$ (B) $CaSO_4.\ 2H_2O$
(C) $CaSO_4\ .½H_2O$ (D) $2CaSO_4.\ ½\ H_2O$

345. Which of the following set belongs to same period ? [AFMC-96]
(A) Li, Na, K (B) Li, Mg, Ca
(C) Cu, Ni, Zn (D) F, Cl, Br

346. Tl show variable valency due to [AFMC-96]
(A) Lone pair effect (B) Inert pair effect
(C) High M.P. (D) High B.P.

347. $NaHCO_3$ is prepared by [AFMC-96]
(A) Solvay process(B) Bosch process
(C) Down process (D) None of these

348. Arrange the elements in increasing order of atomic radius - Na, Rb, K, Mg. [AFMC-97]
(A) Na, K, Mg, Rb (B) K, Na, Mg, Rb
(C) Na, Mg, K, Rb (D) Rb, K, Mg,. Na

349. Which is the largest stable atom ? [AFMC-97]
(A) Bi (B) Al (C) U (D) Pb

350. Which of the following statements is correct ? [AFMC-97]
(A) Hydrogen has same ionization potential as alkali metals
(B) H-has same electronegativity as halògens
(C) H has oxidation no. -1 and +1
(D) It will not be liberated at anode

351. The electronic configuration [AFMC-97]
$1s^2 2s^2 2p_X^1 2p_Y^1 2p_Z^1$ is of
(A) Oxygen (B) Nitrogen (C) Hydrogen (D) Fluorine

352. What is heavy water ? [AFMC-97]
(A) $H_2\ ^{18}O$ (B) $H_2\ ^{16}O$ (C) H_2O_3 (D) D_2O

353. How much quick lime can be obtained from 25 gm of $CaCO_3$? [AFMC-97]
(A) 28 gm (B) 14 gm (C) 56 gm (D) None of these

354. Which of the following is not efflorescent ? [AFMC-97]
(A) $CuSO_4$ (B) Hydrated $CuSO_4$
(C) NaOH (D) All of these

355. Electronic formula of chromium is [AFMC-97]
(A) $3d^6 4s^1$ (B) $3d^5 4s^1$ (C) $3d^4 4s^1$ (D) $3d^4 4s^2$

356. Which of the following is correct for Hydrogen ? [AFMC-97]
(A) It can form bonds in +1 as well as -1 oxidation state
(B) It is always collected at anode
(C) It has very high ionization potential
(D) None of these

357. Which of the following is true for diamond ? [AFMC-97]
(A) It is a good conductor of electricity (B) It is soft
(C) It is a bad conductor of heat
(D) It is made up of C,H and O

358. A sudden large jump between the values of second and third ionization energies of an element would be associated with which the following electronic configuration ? [AFMC-98]
(A) $1s^2 2s^2 p^6 3s^1$ (B) $1s^2 2s^2 p^6 3s^2 3p^1$

(C) $1s^2 2s^2 p^6 3s^2 3p^2$ (D) $1s^2 2s^2 p^6 3s^2$

359. The lusture of a metal is due to [AFMC-98]
(A) Its high density (B) Its high polishing
(C) Its chemical inertness (D) Presence of free electrons

360. Anhydrous $AlCl_3$ is prepared from [AFMC-98]
(A) Conc. HCl and aluminium metal
(B) Aluminium and Cl_2
(C) Dry HCl gas + heated aluminium metal
(D) Dilute HCl and aluminium metal

361. Which of the following will not displace hydrogen ? [AFMC-98]
(A) Ba (B) Pb (C) Hg (D) Sn

362. Ag can be obtained from $AgNO_3$ by reacting it with [AFMC-98]
(A) PH_3 (B) AsH_3 (C) Na_2CO_3 (D) NH_3

363. The correct order of the increasing ionic character is [AFMC-98]
(A) $BeCl_2 < MgCl_2 < CaCl_2 < BaCl_2$
(B) $BeCl_2 < MgCl_2 < BaCl_2 < CaCl_2$
(C) $BeCl_2 < BaCl_2 < MgCl_2 < CaCl_2$
(D) $BaCl_2 < CaCl_2 < MgCl_2 < BeCl_2$

364. Which of the following is obtained when N_2 reacts with calcium carbide ? [AFMC-98]
(A) Calcium cyanate (B) Calcium acetate
(C) Calcium cyanamide (D) Calcium carbonate

365. Which of the following substances is used in laboratory for fast drying of neutral gases ? [AFMC-99]
(A) Active charcoal (B) Sodium phosphate
(C) Phosphorus pentoxide
(D) Anhydrous calcium chloride

366. The number of electrons in $[_{19}K^{40}]^{-1}$ is [AFMC-99]
(A) 20 (B) 40 (C) 18 (D) 19

367. Carnalite has which of the following composition ? [AFMC-99]
(A) Na_2AlF_5 (B) $KCl . MgCl_2 .6H_2O$
(C) Fe_3O_4 (D) Na_3AlF_6

368. Nessler's reagent is represented by [AFMC-99]
(A) $K_2 [HgI_2]$ (B) HgI_2 (C) $K_2 [HgI_4]$ (D) $HgOHgI_2$

369. Which of the following will not displace hydrogen ? [AFMC-99]
(A) Sn (B) Hg (C) Ba (D) Pb

370. Emery consists of [AFMC-99]
(A) Impure carborundum (B) Impure graphite
(C) Purest form of iron (D) Impure corundum

371. Which of the following is the most stable oxidation state of aluminium ? [AFMC-99]
(A) -1 (B) + 3 (C) +2 (D) +1

372. Which of the following has least electron affinity ? [AFMC-2000]
(A) Oxygen (B) Argon (C) Nitrogen (D) Boron

373. Which of the following have same number of electrons in the outermost orbit ? [AFMC-2000]
(A) P, Sb (B) N, O (C) As, Bi (D) P, Ca

374. During electrolysis of water, hydrogen and oxygen are liberated in the ratio of [AFMC-2000]
(A) 1: 1 (B) 2 :1 (C) 1 :2 (D) 1 : 18

375. Which of the following pairs is isoelectronic ? [AFMC-2000]
(A) Ar and Cl^- (B) Na^+ and Ne
(C) Na^+ and Mg (D) Mg and Ne

376. Which of the following does not give a Borax bead test ? [AFMC-2000]
(A) Chromium (B) Ferrous salt
(C) Sodium (D) Cobalt

377. Acidified solution of potassium dichromate on treatment with hydrogen peroxide yields [AFMC-2000]
(A) $CrO_3 + H_2O$ (B) $CrO_5 + H_2O$
(C) $H_2Cr_2O_7 + H_2O$ (D) $Cr_2O_3 + H_2O + O_2$

378. The element with highest electron affinity among halogens is [AFMC-2001]
(A) F (B) Cl (C) Br (D) I

879. Metal always found in free state is [AFMC-2001]
(A) Gold (B) Silver (C) Copper (D) Sodium

380. Hypo is used in photography because of its [AFMC-2001]
(A) Oxidising behaviour (B) Reducing behaviour
(C) Complex forming behaviour
(D) Reaction with light

381. Which of the following set of atomic numbers belong to that of alkali metals ? [AFMC-2001]
(A) 1, 12, 30, 4, 62 (B) 37, 19, 3, 55
(C) 9, 17, 35, 53 (D) 12, 20, 56, 88

382. Modern Periodic Law states that properties of elements are periodic functions of [AFMC-2001]
(A) Ionisation potential (B) Atomic number
(C) Atomic mass (D) Atomic volume

383. Electronic configuration of calcium atom may be written as [AFMC-2002]
(A) $[Ne] 4p^2$ (B) $[Ar] 4s^2$ (C) $[Ne] 4s^2$ (D) $[Kr] 4p^2$

384. Nitrolim is [AFMC-2002]
(A) $CaC_2 + N_2$ (B) $Ca(CN)_2 + Ca(NO_3)_2$
(C) $CaCN_2 + C$ (D) None of these

385. The element with atomic number 51 exists in block [AFMC-2002]
(A) s (B) d (C) p (D) f

386. Which of the following is most electronegative? [AFMC-2002]
(A) C (B) Si (C) Pb (D) Sn

387. Which of the order for ionization energy is correct ? [AFMC-2002]
(A) Be > b > C > N > O (B) B < Be < C < O < N
(C) B < Be < C < N < N (D) B < Be < N < C < O

388. Amongst the following elements (whose electronic configurations are given below), the one having the highest ionization energy is : [IIT-90]
(A) [Ne] $3s^2$ $3p^1$ (B) [Ne] $3s^2$ $3p^3$
(C) [Ne] $3s^2$ $3p^2$ (D) [Ar] $3d^{10}$ $4s^2$ $4p^3$

389. When zeolite, which is hydrated sodium aluminium silicate is treated with hard water the sodium ions are exchanged with. [IIT-90]
(A) H^+ ions (B) Ca^{2+} ions (C)SO_4^{2-} ions
(D) Mg^{2+} ions (E) OH^- ions.

390. The volume strength of 1.5 N H_2O_2 solution is : [IIT-91]
(A) 4.8 (B) 8.4 (C) 3.0 (D) 8.0

391. The oxidation states of the most electronegative elements in the products of the reaction, BaO_2 with dil. H_2SO_4 are : [IIT-91]
(A) 0 and −1 (B) −1 and −2
(C) −2 and 0 (D) −2 and +1.

392. The statement that is not correct for the periodic classification of elements is : [IIT-92]
(A) The properties of elements are the periodic functions of their atomic numbers
(B) Non-metallic elements are lesser in number than metallic elements.
(C) The first ionisation energies along a period do not vary in a regular manner with increase in atomic number.
(D) For transition elements the d -subshells are filled with electrons monotonically with increase in atomic number.

393. The species that do not contain peroxide ions are [IIT-92]
(A)PbO_2 (B)H_2O_2 (C) SrO_2 (D) BaO_2

394. The material used in solar cells contains [IIT-93]
(A) Cs (B) Si (C) Sn (D) Ti

395. Pick out the isoelectronic structure from the following. [IIT-93]
I : CH_3^+ II : H_3O^+ III : NH_3^+ IV : CH_3^-
(A) I and II (B) III and IV
(C) I and III (D) II,III and IV

396. Ca^{2+} has smaller ionic radius than K^+ because it has .... [IIT-93]

397. The hydrolysis of trialkylchlorosilane, R_3SiCl, yields .... [IIT-94]

398. The outermost electronic configuration of Cr is .... [IIT-94]

399. The two types of bonds present in B_2H_6 are covalent and .... [IIT-94]

400. The following compounds have been arranged in order of their increasing thermal stabilities. Identify the correct order [IIT-96]
K_2CO_3(I); $MgCO_3$(II); $CaCO_3$(III); $BeCO_3$(IV)
(A) I < II < III < IV (B) IV< II< III < I
(C) IV < II < I < III (D) II < IV < III < I

401. Sodium thiosulphate is prepared by [IIT-96]
(A) reducing Na_2SO_4 solution with H_2S
(B) boiling Na_2SO_3 solution and S in alkaline medium
(C) neutralising $H_2S_2O_3$ with NaOH
(D) boiling Na_2SO_3 solution with S in acidic medium.

402. Which of the following halides is least stable and has doubtful existence ? [IIT-96]
(A) Cl_4 (B) GeI_4 (C) SnI_4 (D) PbI_4

403. Which of the following oxides is neutral [IIT-96]
(A) CO (B) SnO_2 (C) ZnO (D) SiO_2

404. The incorrect statement among the following is : [IIT-97]
(A) the first ionization potential of Al is less than the first ionization potential of Mg.
(B) the second ionization potential of Mg is greater than the second ionization potential of Na
(C) the first ionization potential of Na is less than the first ionization potential of Mg.
(D) the third ionization potential of Mg is greater than the third ionization potential of Al.

405. Among KO_2, AlO_2^-, BaO_2 and NO_2^+, unpaired electron is present in [IIT-97]
(A) NO_2^+ and BaO_2 (B) KO_2 and AlO_2^-
(C) KO_2 only (D) BaO_2 only

406. Property of the alkaline earth metals that increases with their atomic number is [IIT-97]
(A) ionisation energy
(B) solubility of their hydroxides
(C) solubility of their sulphates (D) electronegativity

407. Ammonium dichromate is used in some fireworks. The green coloured powder blown in air is. [IIT-97]
(A) CrO_3 (B) Cr_2O_3 (C) Cr (D) CrO (O_2)

408. On heating , $Rb(ICl_2)$ will decompose giving ....and .... [IIT-97]

409. Compounds that formally contain Pb^{4+} are easily reduced to Pb^{2+}.The stability of lower oxidation state is due to .... [IIT-97]

410. Sodium nitrate decomposes above ~800°C to give,

[IIT-98]

(A) N_2 (B) O_2 (C) NO_2 (D) Na_2O

411. Highly pure dilute solution of sodium in liquid ammonia [IIT-98]
(A) shows blue colour
(B) exhibits electrical conductivity
(C) produces sodium amide (D) produces hydrogen gas

412. Ionic radii of [IIT-99]
(A) $Ti^{4+} < Mn^{7+}$ (B) $^{35}Cl^- < ^{37}Cl^-$
(C) $K^+ > Cl^-$ (D) $P^{3+} > P^{5+}$

413. The correct order of radii is [IIT-2000]
(A) N < Be < B (B) $F^- < O^{2-} < N^{3-}$
(C) Na < Li < K (D) $Fe^{3+} < Fe^{2} < Fe^{4-}$

414. The electronic configuration of a element is $1s^22s^22p^63s^23p^63d^54s^1$. This represents its [IIT-2000]
(A) excited state (B) ground state
(C) cationic form (D) anionic form

415. A : The first ionization energy of Be is greater than that of B R : 2p orbital is lower in energy than 2s
(A) A (B) B (C) C (D) D [IIT-2000]

416. The set representing the correct order of first ionization potential. [IIT-2001]
(A) K > Na > Li (B) Be > Mg > Ca
(C) B > C > N (D) Ge > Si > C

417. A : Between $SiCl_4$ and CCl_4, only $SiCl_4$ reacts with water. R : $SiCl_4$ is ionic and CCl_4 is covalent. [IIT-2001]
(A) A (B) B (C) C (D) D

418. An aqueous solution of a substance gives a white precipitate on treatment with dilute hydrochloric acid, which dissolves on heating. When hydrogen sulfide is passed through the hot acidic solution, a black precipitate is obtained. The substance is a [IIT-2002]
(A) Hg_2^{2+} salt (B) Cu^{2+} salt (C) Ag^+ salt (D) Pb^{2+} salt

419. Identify the correct order of acidic strengths of CO_2, CuO, CaO, H_2O [IIT-2002]
(A) CaO < CuO < H_2O < CO_2
(B) H_2O < CuO < CaO < CO_2
(C) CaO < H_2O < CuO < CO_2
(D) H_2O < CO_2 < CaO < CuO

420. Identity the least stable ion amongst the following [IIT-2002]
(A) Li^- (B) Be^- (C) B^- (D) C^-

421. When MnO_2 fused with KOH in presence of air, then compound formed and its colour is [IIT-2003]
(A) K_2MnO_4 and deep blues
(B) K_2MnO_4 and deep red
(C) K_2MnO_4 and deep green (D) None of these

422. Boric acid (H_3BO_3) is [IIT-2003]
(A) tribasic and weak lewis acid
(B) monobasic and strong lewis acid
(C) monobasic and weak lewis acid
(D) diabasic and weak lewis acid

423. $(CH_3)_2SiCl_2 \xrightarrow{\text{Hydrolysis}}$ [IIT-2003]

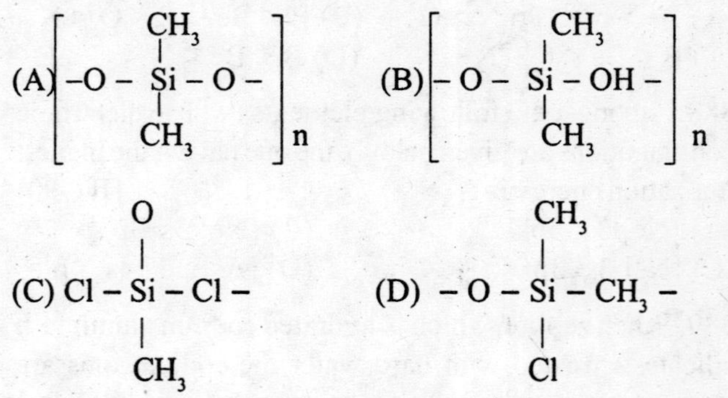

424. KO_2 (potassium super oxide) is used in oxygen cylinders in space and submarines because it. [AIEEE-2002]
(A) Absorbs CO_2 and increase O_2 content
(B) Eliminates moisture
(C) Absorbs CO_2 (D) Produces ozone.

425. Alum helps in purifying water by [AIEEE-2002]
(A) Forming helps in purifying water by
(B) Sulphate part which combines with the air and removes it.
(C) Coagulating the mud particles
(D) Making mud water soluble.

426. Most common oxidation states of Ce (cerium) are. [AIEEE-2002]
(A) +2, +3 (B) +2, +4 (C) +3, +4 (D) +3,+5

427. A metal M readily forms its sulphate MSO_4 which is water soluble. It forms its oxide MO which becomes inert on heating. It forms an insoluble hydroxide $M(OH)_2$ which is soluble in NaOH solution. Then M is. [AIEEE-2002]
(A) Mg (B) Ba (C) Ca (D) Be

428. According to the Periodic law of elements, the variation in properties of elements is related to their. [AIEEE-2003]
(A) Atomic masses (B) Nuclear masses
(C) Atomic numbers
(D) Nuclear neutron- proton number ratios.

429. Graphite is a soft lubricant extremely difficult to melt. The reason for this anomalous behaviour is that graphite. [AIEEE-2003]
(A) Is a non-crystalline substance
(B) Is an allotropic form of diamond
(C) Has molecules of variable molecular masses like polymers.
(D) Has carbon atoms arranged in large plates of rings of strongly bound carbon atoms with weak interplate bonds.

430. Which one of the following processes will produce hard water ? [AIEEE-2003]
(A) Saturation of water with $CaCO_3$
(B) Saturation of water with $MgCO_3$
(C) Saturation of water with $CaSO_4$
(D) Addition of Na_2SO_4 to water.

431. Which one of the following groupings represents a collection of isoelectronic species (At. nos.: Cs-55, Br-35) [AIEEE-2003]
(A) Na^+, Ca^{2+}, Mg^{2+} (B) N^{3-}, F^-, Na^+
(C) Be, Al^{3+}, Cl^- (D) Ca^{2+}, Cs^+, Br

432. The radius of La^{3+} (Atomic number of La = 57) is 1.06 Å. Which one of the following given value will be close to the radius of Lu^{3+}(Atomic number of Lu = 71) ? [AIEEE-2003]
(A) 1.60 Å (B) 1.40 Å (C) 1.06 Å (D) 0.85 Å

433. In curing cement plasters water is sprinkled from time to time. This helps in [AIEEE-2003]
(A) Keeping in cool
(B) Developing interlocking needlelike crystals of hydrated silicates.
(C) Hydrating sand and gravels mixed with cement.
(D) Converting sand into silicic acid.

434. The correct order of increasing basic nature for the based NH_3, CH_3NH_2 and $(CH_3)_2NH$ is [AIEEE-2003]
(A) $CH_3NH_2 < NH_3 < (CH_3)_2NH$
(B) $(CH_3)_2NH < NH_3 < CH_3NH_2$
(C) $NH_3 < CH_3NH_2 < (CH_3)_2NH$
(D) $CH_3NH_2 < (CH_3)_2NH < NH_3$

435. Butene -1 may be converted to butane by reaction with [AIEEE-2003]
(A) Zn–HCl (B) Sn–HCl (C) Zn–Hg (D) Pd/H_2

436. The solubilities of carbonates decrease down the magnesium group due to a decrease in [AIEEE-2003]
(A) Lattice energies of solids
(B) Hydration energies of cations
(C) Inter-ionic attraction
(D) Entropy of solution formation

437. The substance not likely to contain $CaCO_3$ is [AIEEE-2003]
(A) A marble (B) Calcined gypsum
(C) Sea shells (D) Dolomite

438. Which pair of atomic numbers represent elements which are both 's' block elements : [EAMCET-90]
(A) 7, 15 (B) 6, 12 (C) 9, 17 (D) 3, 12

439. A transition metal 'X' has a configuration $[Ar]3d^4$ in its +3 oxidation state. Its atomic Number is : [EAMCET-90]
(A) 25 (B) 26 (C) 22 (D) 19

440. The components present in producer gas is : [EAMCET-90]
(A) $CO + N_2$ (B) $CO + H_2$ (C) $CO_2 + N_2$ (D) $CO_2 + H_2$

441. Which indicates the correct variation in electronegativities : [EAMCET-90]
(A) F > N < O > C (B) F < N > O > C
(C) F < N < O < C (D) F > N > O < C

442. The first Ionisation energy of lithium will be. [EAMCET-90]
(A) Greater than beryllium (B) Lesser than beryllium
(C) Equal to sodium (D) Equal to fluorine

443. An element 'Y' has a ground state configuration 2, 8, 8, the type of bond that exists between the atoms 'Y' is : [EAMCET-90]
(A) Ionic (B) Covalent
(C) Metallic (D) van der Waals

444. The number of Hydrogen atoms bridging the boron atoms in a diborane is ....[EAMCET-90]

445. Among the ions Cl^-, S^{2-} and Na^+, the largest Ion is .... [EAMCET-90]

446. Among the metals Na, Mg and Al the metal with highest melting point is ....[EAMCET-90]

447. A simple oxide of carbon and that of another non-metal has the same C_P/C_V ratio, but a difference of 2 in their molecular weights. The other non-metal could be .... [EAMCET-90]

448. Among the alkali metals, the metal with the highest I.P. is .... [EAMCET-91]

449. The process of slow cooling of glass is called .... [EAMCET-91]

450. Among SiO_2 and Al_2O_3 the other major ingradient in Portland cement in .... [EAMCET-91]

451. Al^{3+} has a lower ionic radius than Mg^{2+} ion because [EAMCET-92]
(A) Mg atom has less number of neutrons than Al.
(B) Al^{3+} has a higher nuclear charge than Mg^{2+}
(C) Their electro negativities are different
(D) Al has a lower ionisation potential than Mg atom.

452. Atoms of different elements having identical mass are known as [EAMCET-92]
(A) Isotopes (B) Isobars (C) Isotones (D) Isomers

453. In the long form of periodic table, the elements having lowest ionisation potential are present in [EAMCET-92]
(A) I group (B) IV group (C) VII group (D) Zero group

454. Ionisation potential of 1s electron is .... [EAMCET-92]

455. A "magic number" nucleus contains ....

number of protons and ....number of neutrons. [EAMCET-92]

456. Electronegativity of an atom is given by the average of electron affinity and .... [EAMCET-92]

457. In diborane, boron atoms undergo ....hybridisation. [EAMCET-92]

458. The correct electronic configuration of potassium is .... [EAMCET-92]

459. The most electronegative element is .... [EAMCET-92]

460. The reaction of I_2 with $Na_2S_2O_3$ gives NaI and [EAMCET-93]
(A) SO_2 (B) $Na_2S_2O_6$ (C) $Na_2S_4O_6$ (D) $Na_2S_2O_8$

461. Which of the following electronic configurations corresponds to an inert gas ? [EAMCET-93]
(A) $1s2s^22p^5$ (B) $1s^22s^22p^6$ (C) $1s^22s^1$ (D) $1s^22s^22p^63s^1$

462. Quicklime is represented by the formula [EAMCET-93]
(A) $Ca(OH)_2$ (B) CaO (C) $CaCO_3$ (D) $CaHCO_3$

463. Temporary hardness of water is due to the presence of .... of calcium & magnesium in water. [EAMCET-93]

464. Charcoal is often used in gas masks because of its high capacity for absorbing ....molecules. [EAMCET-93]

465. Hydrogen gas may be readily prepared in the laboratory by reacting water with .... [EAMCET-93]

466. According to Aufbau principle the electron has a tendency to occupy that subshell which has ....energy. [EAMCET-93]

467. The chemical formula of Plaster of Paris is .... [EAMCET-93]

468. Which one of the following is not an allotrope of carbon [EAMCET-94]
(A) Graphite (B) Diamond
(C) Soot (D) Carborundum

469. Thermite is a mixture of iron oxide and [EAMCET-94]
(A) Sodium shavings (B) Aluminium powder
(C) Zinc powder (D) Potassium metal

470. Which one of the following statements is not correct with respect to BF_3 ? [EAMCET-94]
(A) It is an electron deficient compound
(B) It is a Lewis acid
(C) It is an ionic compound (D) It forms adducts

471. Permanent Hardness of water is due to [EAMCET-94]
(A) The chlorides of Ca and Mg
(B) The sulphates of Ca & Mg
(C) Both chlorides and sulphates of Ca & Mg
(D) The bicarbonates of Ca & Mg

472. In aqueous solution, H_2O_2 oxidises H_2S to [EAMCET-94]
(A) Sulphur (B) Sulphuric acid
(C) Caro's acid (D) Marshall's acid

473. An example of a non-transitional element is [EAMCET-94]
(A) Cobalt (B) Lead (C) Cerium (D) Silver

474. Which one of the following is the largest ion ? [EAMCET-94]
(A) Na^+ (B) Mg^{2+} (C) O^{2-} (D) F^-

475. Among the following the element that has the lowest first ionization potential is. [EAMCET-94]
(A) Nitrogen (B) Oxygen (C) Fluorine (D) Neon

476. BF_3 forms an adduct with NH_3 because [EAMCET-94]
(A) Nitrogen has high electonegativity
(B) Boron has smaller atomic radius than nitrogen
(C) Boron has an empty p-orbital and nitrogen has a lone pair of electrons
(D) Boron has electropositive character

477. The gas that is used for sterilizing water is [EAMCET-94]
(A) SO_2 (B) NO_2 (C) O_3 (D) NH_3

478. When dilute H_2SO_4 is electrolysed at room temperature between Pt electrodes the substance liberated at the anode is. [EAMCET-94]
(A) S (B) SO_3 (C) H_2 (D) O_2

479. The order of decrease in atomic radii for Be, Na and Mg is [EAMCET-95]
(A) Na > Mg > Be (B) Mg > Na > Be
(C) Be > Na > Mg (D) Be > Mg > Na

480. The name of the element with atomic number 100 was adopted in honour of [EAMCET-95]
(A) Alfred Nobel (B) Enrico Fermi
(C) Dimitri Mendeleff (D) Albert Einstein

481. Which of the following is not a peroxide ? [EAMCET-95]
(A) $NaHO_2$ (B) BaO_2 (C) Na_2O_2 (B) PbO_2

482. In the following two reactions, [EAMCET-95]
(i) $2H_2O_2 \rightarrow 2H_2O + O_2$
(ii) $H_2O_2 + 2I^- + 2H^+ \rightarrow 2H_2O$ H_2O_2 acts as
(A) An oxidizing agent (B) A reducing agent
(C) Both oxidizing and reducing agent respectively
(D) Both reducing and oxidizing agent respectively

483. The glass which contains PbO mainly is [EAMCET-95]
(A) Soda glass (B) Flint glass
(C) Pyrex glass (D) Vitreosil

484. Beryllium shows diagonal relationship with aluminium. Which of the following similarity is incorrect? [EAMCET-95]
(A) Be_2C like Al_4C_3 yields methane on hydrolysis
(B) Be like Al is rendered passive by HNO_3
(C) $Be(OH)_2$ like $Al(OH)_3$ is basic
(D) Be forms beryllates and Al forms aluminates

485. The compound formed, during the setting of the plaster of paris is [EAMCET-95]
(A) Cement (B) Gypsum (C) Anhydrite (D) Mortar

486. The following statement is not true as far as BF_3 is concerned : [EAMCET-95]
(A) Electron deficient compound (B) Lewis acid
(C) Ionic compound (D) Covalent compound

487. Ordinary glass is a combination of the following [EAMCET-95]
(A) SiO_2, Na_2SiO_3, $CaSiO_3$ (B) SiO_2, Na_2CO_3, $CaCO_3$
(C) Na_2CO_3, K_2CO_3, SiO_2 (D) SiO_2, $PbCO_3$, $CaSiO_3$

488. Water gas essentially is a mixture of[EAMCET-95]
(A) CO_2, H_2 (B) CO, H_2 (C) CO, N_2, H_2 (D) CO_2, N_2

489. Which one of the following would you expect to have highest electonegativity [EAMCET-95]
(A) Mg (at. No. 12) (B) S (At. No. 16)
(C) B (At. No. 5) (D) Te (At. No. 52)

490. The State of hybridization of Boron in BCl_3 is [EAMCET-96]
(A) sp^3 (B) sp^2 (C) sp (D) spd^3

491. The pair of elements that have similar chemical properties are [EAMCET-96]
(A) Beryllium and Boron (B) Aluminium and Magnesium
(C) Carbon and Nitrogen (D) Lithium and Magnesium

492. Hg sticks to the surface of the glass when it comes in contact with [EAMCET-96]
(A) H_2O (B) HNO_3 (C) O_3 (D) I_2

493. The temperature at which heavy water has maximum density is [EAMCET-96]
(A) 1.16°C (B) 4°C (C) 4.6°C (D) 11.6°C

494. Poisonous gas present in the exhaust fumes of an automobile is [EAMCET-96]
(A) C_2H_6 (B) CO_2 (C) CO (B) CH_4

495. Superphosphate is a mixture of [EAMCET-96]
(A) $Ca(H_2PO_4)$. H_2O + $CaCl_2$. $2H_2O$
(B) $Ca(H_2PO_4)_2$. H_2O + $2CaSO_4$. $2H_2O$
(C) $Ca_3(PO_4)_2$. H_2O + $2CaSO_4$.$2H_2O$
(D) $Ca_3(PO_4)_2$ .H_2O + $CaCl_2$.$2H_2O$

496. Sodium amalgam is useful as [EAMCET-96]
(A) Oxidising agent (B) Catalyst
(C) Reducing agent (D) Bleaching agent

497. The oxidation state of sodium in sodium amalgam is [EAMCET-96]
(A) + 1, (B) -1 (C) zero (D) +2

498. The alum used for purifying water is[EAMCET-96]
(A) Ammonium alum (B) Chrome alum
(C) Ferric alum (D) Potash alum

499. The inert form of carbon is [EAMCET-96]
(A) Diamond (B) Graphite (C) Coal (D) Charcoal

500. Which one of the following metals will not reduce H_2O ? [EAMCET-97]
(A) Ca (B) Fe (C) Cu (D) Li

501. Which of the following is reduced by Hydrogen peroxide in acid medium ? [EAMCET-97]
(A) Pot. permanganate (B) Pot. Iodide
(C) $FeSO_4$ (D) Pot. ferrocyanide

502. The electronic configuration of chromium (Z = 24) is [EAMCET-97]
(A) $[Ar]4d^4\ 4s^2$ (B) $[Ar]3d^5\ 4s^1$
(C) $[Ar]3d^4\ 3s^2$ (D) $[Ar]4d^5\ 4s^1$

503. Which one of the following compounds liberates CO_2 from aqueous $NaHCO_3$? [EAMCET-97]
(A) Anilinium Chloride (B) $CHCl_3$
(C) CCl_4 (D) CH_3Cl

504. Which one of the following elements has the highest first ionization potential ? [EAMCET-97]
(A) Boron (B) Carbon (C) Nitrogen (D) Oxygen

505. Of the following, the one with largest size is [EAMCET-97]
(A) Cl^- (B) Ar (C) K^+ (D) Ca^{2+}

506. Which of the following species has the highest ionization potential ? [EAMCET-98]
(A) Li^+ (B) Mg^+ (C) Al^+ (D) Ne

507. As per the modern periodic law the physical and chemical properties of elements are periodic functions of their. [EAMCET-98]
(A) Atomic number (B) Electronic configuration
(C) Atomic weight (D) Atomic size

508. Be^{2+} is isoelectronic with [EAMCET-98]
(A) Mg^{2+} (B) Na^+ (C) Li^+ (D) H^+

509. Hg sticks to the surface of the glass when it comes in contact with [EAMCET-98]
(A) H_2O (B) NHO_3 (C) Grease (D) O_3

510. Mortar is mixture of [EAMCET-98]
(A) Plaster of Paris + silica
(B) Slaked lime + plaster of Paris + H_2O

(C) $CaCO_3$ + silica + H_2O
(D) Slaked lime + silica + H_2O

511. CARBOGEN is [EAMCET-98]
(A) Pure form of carbon (B) $COCl_2$
(C) Mixture of CO & CO_2 (D) Mixture of O_2 & CO_2

512. Which of the following does not participate in the SOLVAY'S process, for the manufacture of Na_2CO_3 ? [EAMCET-98]
(A) NH_3 (B) NaCl soluation (C) CO_2 (D) H_2SO_4

513. Silica is soluble in [EAMCET-98]
(A) HCl (B) HNO_3 (C) H_2SO_4 (D) HF

514. The radii of F, F^-, O and O^{2-} are in the order [EAMCET(MBBS)-98]
(A) $O^{2-} > F^- > O > F$ (B) $O^{2-} > F^- > F > O$
(C) $F^- > O^{2-} > F > O$ (D) $O^{2-} > O^- > F > F$

515. Which of the following has highest first ionization potential ? [EAMCET(MBBS)-98]
(A) Boron (B) Nitrogen (C) Oxygen (D) Carbon

516. The purest form of coal is [EAMCET(MBBS)-98]
(A) Peat (B) Anthracite
(C) Lignite (D) Bituminous coal

517. NITROLIM is a mixture of [EAMCET(MBBS)-98]
(A) $Ca(CN)_2$ & graphite (B) CaC_2 & graphite
(C) $CaCN_2$ & graphite (D) $CaCN_2$ & N_2

518. When H_2O_2 reacts with chlorine the product obtained is [EAMCET(MBBS)-98]
(A) O_2 (B) H_2 (C) ClO_2 (D) HOCl

519. Which of the following species has the highest ionization potential ? [EAMCET(MBBS)-98]
(A) Li^+ (B) Mg^+ (C) Al^+ (D) Ne

520. Which of the following statements is NOT true? [EAMCET-99]
(A) Aluminium liberates hydrogen on treating with a base.
(B) Aluminium is used in the preparation of duralumin.
(C) Aluminium is extracted by the electrolysis of alumina in the presence of cryolite.
(D) Aluminium is a stong oxidising reagent.

521. Aqueous solution of Carnallite gives positive test for [EAMCET-99]
(A) Chloride ions only (B) Potassium ions only
(C) Potassium and chloride ions only
(D) Potassium, magnesium and chloride ions

522. Inorganic benzene contains [EAMCET-99]
(A) C,H,Al (B) C,B,H (C) B,N,H (D) C,N,H

523. At what temperature the density of heavy water will be maximum [EAMCET-99]
(A) 0^0C (B) 11.6^0C (C) 4^0C (D) 27^0C

524. Which of the set of raw materials are used in the manufacture of Na_2CO_3 by Solvey process ? [EAMCET-99]
(A) $Ca(OH)_2$, NH_3, CO_2 (B) $CaCl_2$, NH_3, CO_2
(C) NaOH, NH_3, CO_2 (D) NaCl, NH_3, CO_2

525. The following reaction is known to occur in the body
$CO_2 + H_2O \rightleftharpoons H_2CO_3 \rightleftharpoons H^+ + HCO_3^-$
If CO_2 escapes from the system [EAMCET-99]
(A) Hydrogen ion concentration increases
(B) Hydrogen ion concentration decreases
(C) H_2CO_3 concentration does not change
(D) The forward reaction will be promoted

526. O^{2-} and Si^{4+} are isoelectronic radius ions. If the ionic radius of O^{2-} is 1.4Å. the ionic radius of Si^{4+} will be [EAMCET-99]
(A) 1.4Å (B) 0.41Å (C) 2.8Å (D) 1.5Å

527. The electronic configuration of sodium is [EAMCET-99]
(A) $[Ne]3s^2$ (B) $[Ne]3s^1$ (C) $[Ar]4s^1$ (D) $[Ar]4s^2$

528. The dielectric constants of H_2O and D_2O are 82 and 80.5 respectively. Therefore the solubility of NaCl in D_2O is .... than in water. [EAMCET(MBBS)-99]
(A) equal to (B) Slightly lower
(C) slightly higher (D) very high

529. Metals having ns^1 as the valence electronic configuration [EAMCET(MBBS)-99]
(A) act as strong oxidizing reagents
(B) are highly electronegative
(C) are highly electropositive
(D) have a first ionization potential of more then 10 eV/atom

530. A mixture of carbon monoxide, hydrogen and hydrocarbons is known as [EAMCET(MBBS)-99]
(A) water gas (B) carburetted water gas
(C) semi water gas (D) producer gas

531. How many electrons are present in the M-shell of the atom of an element with atomic number 24 ? [EAMCET(MBBS)-99]
(A) 5 (B) 6 (C) 12 (D) 13

532. The number of periods present in the long form of periodic table is [EAMCET(MBBS)-99]
(A) 6 (B) 7 (C) 8 (D) 18

533. Which of the following is isotope of $_{32}Ge^{76}$? [EAMCET-2000]
(A) $_{33}As^{77}$ (B) $_{32}Ge^{77}$ (C) $_{34}Se^{77}$ (D) $_{35}Br^{80}$

534. Which of the following reacts with BCl_3 to form diborane ? [EAMCET-2000]
(A) NH_3 (B) $LiAlH_4$ (C) NaHg (D) $K_2Cr_2O_7$

535. What is the approximate calorific value (in Kcal m^{-3}) of water gas ? [EAMCET-2000]

(A) 500 (B) 2700 (C) 4000 (D) 1000

536. Which of the following is used in the preparation of aerated water (Soda) ? [EAMCET-2000]
(A) CO (B) CO_2 (C) SO_2 (D) HCl

537. What is the molecular formula of gypsum ? [EAMCET-2000]
(A) $CaSO_4.H_2O$ (B) $CaSO_4.2H_2O$
(C) $2CaSO_4.H_2O$ (D) $CaSO_4.3H_2O$

538. In which of the following reactions hydrogen peroxide acts as a reducting agent ? [EAMCET-2000]
(A) $PbS + 4H_2O_2 \rightarrow PbSO_4 + 4H_2O$
(B) $Na_2SO_3 + H_2O_2 \rightarrow Na_2SO_4 + H_2O$
(C) $Ag_2O(moist) + H_2O_2 \rightarrow 2Ag + H_2O + O_2$
(D) $2KI + H_2O_2 + H_2SO_4 \rightarrow K_2SO_4 + I_2 + H_2O_2$

539. Which one of the sixth group elements has the highest catenation power ? [EAMCET-2000]
(A) O (B) Se (C) S (D) Te

540. The electronic configuration of group III elements is [EAMCET-2000]
(A) ns^1np^2 (B) ns^1np^3 (C) ns^2np^2 (D) ns^2np^1

541. Which of the following has the highest first ionization potential ? [EAMCET-2000]
(A) Al (B) Si (C) K (D) P

542. What is the oxidation state of carbon in carbon dioxide ? [EAMCET(MBBS)-2000]
(A) +2 (B) +4 (C) +6 (D) +1

543. Which of the following is the correct order of ionic radii ? [EAMCET(MBBS)-2000]
(A) $Na^+ < Mg^{2+} < Al^{3+} < Si^{4+}$ (B) $Al^{3+} < Si^{4+} > Na^+ > Mg^{2+}$
(C) $Si^{4+} < Al^{3+} > Mg^{2+} > Na^+$ (D) $Na^+ > Mg^{2+} > Al^{3+} > Si^{4+}$

544. Which of the following has the highest electronegativity ? [EAMCET(MBBS)-2000]
(A) Na (B) Cl (C) K (D) B

545. What is the solution used for the preparation of hydrogen peroxide by electrolytic procedure ? [EAMCET(MBBS)-2000]
(A) 0.2 N NaOH (B) 50% H_2SO_4
(C) 20 % NaOH (D) 5% NaCl

546. In a reaction sodium is heated with $BeCl_2$. In this reaction, sodium acts as : [EAMCET(MBBS)-2000]
(A) Oxidizing agent (B) Reducing agent
(C) Hydrolysing agent (D) Precipitating agent

547. What is the molecular formula of plaster of Paris ? [EAMCET(MBBS)-2000]
(A) $2CaSO_4.H_2O$ (B) $2CaSO_4.2H_2O$
(C) $CaSO_4.2H_2O$ (D) $CaSO_4.3H_2O$

548. What is the C–C bond length (in angstroms) in Diamond ? [EAMCET(MBBS)-2000]
(A) 1.54 (B) 3.35 (C) 2.0 (D) 5.2

549. Which of the following is calcium superphosphate ? [EAMCET(MBBS)-2000]
(A) $Ca(H_2PO_4)2H_2O + 2(CaSO_2\ 2H_2O)$
(B) $Ca_3(PO_4)_2 + 2CaSO_4$
(C) $CaSO_4 + CaO$ (D) $Ca(H_2PO_4)_2.H_2O + CaSO_4$

550. In which of the following reactions, H_2O_2 acts as a reducing reagent ? [EAMCET-2001]
(A) $PbO_2(s) + H_2O_2(aq) \rightarrow PbO(s) + H_2O(l) + O_2(g)$
(B) $Na_2SO_3(aq) + H_2O_2(aq) \rightarrow Na_2SO_4(s) + H_2O(l)$
(C) $2KI(aq) + H_2O_2(aq) \rightarrow 2KOH(aq) + I_2(s)$
(D) None

551. _____ process is used for the removal of hardness of water [EAMCET-2001]
(A) Calgon (B) Baeyer (C) Serpeck (D) Hoope

552. The electronic configuration of elements A, B and C are [He]2s'' [Ne] $3s^1$ and [Ar] $4s^1$ respectively. Which one of the following order is correct for the first ionization potentials (in kJ mol^{-1}) of A, B and C ?[EAMCET-2001]
(A) A > B > C (B) C > B > A
(C) B > C > A (D) C > A > B

553. In organic reactions, sodium in liquid ammonia is used as .... [EAMCET-2001]
(A) Reducing agent (B) Hydrolyzing agent
(C) Oxidizing agent (D) Precipitating agent

554. What is the molecular formula of Borazole ? [EAMCET-2001]
(A) B_2H_6 (B) $B_6N_6H_6$ (C) $B_3N_3H_6$ (D) $B_3N_3H_3$

555. The catenation tendency of C, Si and Ge is in the order Ge < Si < C. The bond energies (In KJ. mol^{-1}) of C–C, Si–Si and Ge–Ge bonds, respectively are . [EAMCET-2001]
(A) 167, 180, 348 (B) 180, 167, 348
(C) 348, 167, 180 (D) 348, 180, 167

556. Ionic radius (in Å) of As^{3+}, Sb^{3+} and Bi^{3+} follow the order .... [EAMCET-2001]
(A) $As^{3+} > Sb^{3+} > Bi^{3+}$ (B) $Sb^{3+} > Bi^{3+} > As^{3+}$
(C) $Bi^{3+} > As^{3+} > Sb^{3+}$ (D) $Bi^{3+} > Sb^{3+} > As^{3+}$

557. Let electrongeativity ionization energy and electron affinity be represented as EN, IP and EA respectively. Which one of the following equation is correct according to Mulliken. [EAMCET-2001]
(A) EN = IP X EA (B) EN = IP/EA
(C) EN = IP+EA/2 (D) EN = IP – EA

558. The element with atomic number 12 belongs to ___ group ___ period. [EAMCET-2001]
(A) IA third (B) IIIA, third
(C) IIA, third (D) IIA, second

559. Sodium hexa-metaphosphate is known as

[EAMCET(MBBS)-2001]

(A) Calgon (B) Permutit (C) Natalite (D) Nitrolim

560. Aluminium reacts with concentrated HCl and concentrated NaOH to liberate the gases respectively. [EAMCET(MBBS)-2001]

(A) H_2 and O_2 (B) O_2 and H_2
(C) H_2 and H_2 (D) O_2 and O_2

561. Which one of the following allotropic forms of carbon is isomorphous with crystalline silicon ? [EAMCET(MBBS)-2001]

(A) Graphite (B) Coal (C) Coke (D) Diamond

562. Which one of the following represents the correct order of electro negativity ? [EAMCET-2002]

(A) P > O > N (B) N > P > O
(C) O > N > P (D) N > O > P

563. "Fluorosis" disease is caused due to the reaction of ...... with excess of fluoride in the body.[EAMCET-2002]

(A) Ca (B) Mg (C) Fe (D) K

564. The products formed when heavy water is reacted with magnesium nitride, are .... [EAMCET-2002]

(A) NH_3, Mg (B) NH_3, Mg $(OD)_2$
(C) ND_2, Mg $(OH)_2$ (D) ND_3, Mg $(OD)_2$

565. The compound formed when gypsum is dissolved in aqueous ammonium sulphate solution is.[EAMCET-2002]

(A) $CaSO_4NH_4Cl.H_2O$ (B) $CaCl_2.(NH_4)_2SO_4.H_2O$
(C) $CaSO_4(NH_4)_2SO_4.H_2O$ (D) $CaCl_2$. $NH_4Cl.2H_2O$

566. Oxalic acid reacts with concentrated H_2SO_4 to give a mixture of two gases. When this mixture is passed through caustic potash, one of the gases is absorbed. What is the product formed by the absorbed gas with caustic potash ? [EAMCET-2002]

(A) K_2SO_4 (B) K_2HCO_3 (C) K_2CO_3 (D) KOH

567. Which one of the following is not an iso-electronic pair ? [EAMCET-2002]

(A) Mg^{+2}, C^{4-} (B) N^{3-},O^{2-} (C) N^{2-}, O^{2-} (D) F^-, Al^{3+}

568. Which one of the following elements is a non-metal? [EAMCET(MED.)-2002]

(A) Boron (B) Indium
(C) Sodium (D) Magnesium

569. Aluminium reacts with hot concentrated H_2SO_4 to liberate .... [EAMCET(MED.)-2002]

(A) SO_2 (B) H_2S (C) SO_3 (D) S (Vapour)

570. What is the oxidation state of Fe in the product formed when acidified potassium ferrocyanide is treated with H_2O_2 ? [EAMCET(MED.)-2002]

(A) +2 (B) +6 (C) +1 (D) +3

571. Which one of the following is the correct order for the decomposition temperatures of $CaCO_3$ (X), $MgCO_3$ (Y) and $BaCO_3$ (Z) ? [EAMCET(MED.)-2002]

(A) Z > X > Y (B) X > Y > Z
(C) Y > Z > X (D) X > Z > Y

572. Which of the following is a correct set for SiO_2 ? [EAMCET(MED.)-2002]

(A) Linear, acidic (B) Linear, basic
(C) Tetrahedral, acidic (D) Angular, basic

573. What are the products formed when Li_2CO_3 undergoes decomposition ? [EAMCET(MED.)-2002]

(A) Li_2O_2 + CO (B) Li_2O + CO
(C) Li_2O + CO_2 (D) LiO_2 + CO

574. An element M reacts with chlorine to form a compound X. The bond angle in X is 120° What is M ? [EAMCET(MED.)-2002]

(A) Be (B) B (C) Mg (D) N

575. Elements X and Y belong to the same group. Which one of the following set of atomic numbers represents X and Y ? [EAMCET(MED.)-2002]

(A) 15, 30 (B) 20, 31 (C) 19,55 (D) 33,16

576. Sodium metal reacts with Al_2O_3 at high temperature to give a sodium compound X. X reacts with carbon dioxide in water to form Y. Y is...[EAMCET(MED.)-2002]

(A) Na_2O_2 (B) Na_2 (C) Na_2CO_3 (D) Na_2 AlO_3

577. A mixture of sodium oxide and calcium oxide are disolved in water and saturated with excess carbon dioxide gas. The resulting solution is ---- It contains ---- [EAMCET-2003]

(A) Basic, NaOH + $Ca(OH)_2$
(B) Neutral, Na_2CO_3 + $CaCO_3$
(C) Basic, Na_2CO_3 and $CaCO_3$
(D) Acidic NaOH and $CaCO_3$

578. Which one of the following is correct about stability of the given ions ? [EAMCET-2003]

(A) Pb^{2+} > Pb^{4+} (B) Pb^{4+} >Pb^{2+}
(C) Si^{2+} > $7Si^{4+}$ (D) Sn^{4+}> Sn^{2+}

579. The product obtained at anode when 50% H_2SO_4 aqueous solution is electrolysed using platinum electrodes is [EAMCET-2003]

(A) H_2SO_3 (B) $H_2S_2O_8$ (C) O_2 (D) H_2

580. What are the products formed when an aqueous solution of magnesium bicarbonate is boiled ? [EAMCET-2003]

(A) $MgCO_3.H_2O.CO_2$ (B) $Mg(HCO_3)_2.H_2O$
(C) $Mg(OH)_2.H_2O$ (D) $MgCO_3$. H_2O

581. The property of attracting electrons by the halogen atom is called [CPMT-96]

(A) Ionisation potential (B) Electron affinity
(C) Electronegativity (D) Electronic attraction

582. Which of the following is most dense ?

[CPMT-96]
(A) Fe (B) Cu (C) B (D) Pb

583. When NaOH is made, the gas released is [CPMT-96]
(A) Cl_2 (B) H_2 (C) O_2 (D) H_2O

584. Which of the following is the strongest base ? [CPMT-96]
(A) $Al(OH)_3$ (B) $Mg(OH)_2$ (C) $Ca(OH)_2$ (D) $Ba(OH)_2$

585. Commercial 10 volume of H_2O_2 is a solution with a strength of approximately - [CPMT-96]
(A) 30% (B) 3% (C) 1% (D) 10%

586. Calcium is obtained by the- [CPMT-96]
(A) Electrolysis of molten $CaCl_2$
(B) Electrolysis of solution of $CaCl_2$ in water
(C) Reduction of $CaCl_2$ with carbon
(D) Roasting of lime stone

587. $1s^2, 2s^2, 2p^6, 3s^1$ shows configuration of -[CPMT-96]
(A) Al^{+3} in ground state (B) Ne in excited state
(C) Mg^{+1} in excited state (D) None of these

588. The long form of periodic table is based on [CPMT-97]
(A) Shape of the atom (B) Mass of the atom
(C) Number of the atom (D) Electronegativity

589. Which one of the following is a strong reducing agent? [CPMT-97]
(A) HNO_3 (B) Na^+ (C) Cl (D) Cr

590. Which substance will give amide when heated with NH_3? [CPMT-97]
(A) Potassium (B) Hydrogen
(C) Ethane (D) Nitrogen

591. Which one of the following shows bond in silicon ? [CPMT-97]
(A) –Si–O–Si–O–Si– (B) Si–C–Si–O–Si
(C) Si–C–Si–C–Si (D) Si–Si–Si–Si

592. The difference of water atoms in gypsum and plaster of Paris is- [CPMT-97]
(A) 5/2 (B) 2 (C) 1/2 (D) 1 1/2

593. Which one has maximum alkalinity ? [CPMT-97]
(A) PH_3 (B) NH_3 (C) AsH_3 (D) SbH_3

594. Which one has large radius ? [CPMT-97]
(A) Na^+ (B) F (C) F^- (D) Na

595. Nessler's reagent is [CPMT-97]
(A) $KHgI_4$ (B) $K_2HgI_4 + NH_4OH$
(C) $K_2HgI_4 + KOH$ (D) $KHgI_4 + NH_4OH$

596. Metal that causes tears - [CPMT-97]
(A) Fe (B) Sn (C) Ni (D) Sb

597. $MgCl_2 . 6H_2O$ when heated gives [CPMT-97]
(A) Magnesium oxychloride (B) Magnesium dichloride
(C) Magnesium oxide (D) Magnesium chloride

598. Wool when heated with BaO at 1,100°C it gives a compound. Identify the compound. [CPMT-97]
(A) $BaZnO_2$ (B) Ba + ZnO_2 (C) $BaCdO_2$ (D) BaO_2 + Zn

599. The number of electrons and neutrons of an element is 18 and 20 respectively. Its mass number is [CPMT-97]
(A) 17 (B) 37 (C) 2 (D) 38

600. Which of the following has maximum ionization potential ? [CPMT-97]
(A) Mg (B) Si (C) Al (D) P

601. D_2O is used in [CPMT-97]
(A) Industry (B) Nuclear reactor
(C) Medicine (D) Insecticide

602. The poisonous gas that comes out with petrol burning in a car is [CPMT-97]
(A) CH_4 (B) C_2H_5 (C) CO_2 (D) CO

603. Which one has the highest catenation power ? [CPMT-97]
(A) S (B) Se (C) Po (D) Te

604. First I.P. of Mg is .... than Al. [CPMT-97]
(A) Less (B) More (C) Equal (D) None of these

605. In K^+F^-, ionic radius of F^- is more while atomic radius of K^+ is [CPMT-97]
(A) Less than F^- (B) More than F^-
(C) Equal to F^- (D) None of these

606. The characteristic property of elements in a group is [CPMT-98]
(A) Ionisation energy (B) Electronegativity
(C) No. of electrons in outermost orbit
(D) Electron affinity

607. The basis of modern periodic law is [CPMT-98]
(A) Atomic number (B) Atomic size
(C) Atomic volume (D) Atomic mass

608. The long form of periodic table consists of [CPMT-98]
(A) 8 Horizontal and 7 vertical series
(B) 7 Horizontal and 18 vertical series
(C) 7 Horizontal and 7 vertical series
(D) 8 Horizontal and 8 vertical series

609. Which indicates the correct variation in electronegativity ? [CPMT-98]
(A) $F > O > N > C$ (B) $F > N > O > C$
(C) $F > N > O > C$ (D) $F > N > O > C$

610. Na+, Mg++, Al+++ and Si++++ are isoelectronic. The order of their ionic size is [CPMT-98]
(A) $Na^+ > Mg^{++} < Al^{+++} < Si^{++++}$

(B) $Na^+ < Mg^{++} > Al^{+++} > Si^{++++}$
(C) $Na^+ > Mg^{++} > Al^{+++} > Si^{++++}$
(D) $Na^+ < Mg^{++} > Al^{+++} < Si^{++++}$

611. Which has the highest electron affinity?[CPMT-98]
(A) F (B) Cl (C) Br (D) I

612. The correct order of electron affinity of the elements of oxygen family in the periodic table [CPMT-98]
(A) O > S > Se (B) S > O > Se
(C) S > Se > O (D) Se > O > S

613. Which of the following elements will have highest electronegativity ? [CPMT-98]
(A) Li (B) Cr (C) Ac (D) F

614. Which of the following metals will adsorb hydrogen to a highest extent ? [CPMT-98]
(A) Zn (B) Pd (C) Al (D) K

615. Change of ordinary hydrogen into atomic hydrogen is a [CPMT-98]
(A) Nuclear reaction (B) Decomposition reaction
(C) Exothermic reaction (D) Endothermic reaction

616. Tritium is an isotope of [CPMT-98]
(A) Hydrogen (B) Tellurium
(C) Titanium (D) Tantalum

617. Heavy water is used as [CPMT-98]
(A) Washing water(B) Detergent
(C) Drinking water(D) A moderator

618. Which of the following is a true peroxide ? [CPMT-98]
(A) NO_2 (B) MnO_2 (C) BaO_2 (D) SO_2

619. Structure of H_2O_2 is [CPMT-98]
(A) Linear (B) Tetrahedral
(C) Square planar (D) Unsymmetrical

620. Which one of the following will be most basic ? [CPMT-98]
(A) Na_2O (B) BaO (C) As_2O_3 (D) Al_2O_3

621. Which of the following imparts violet colouration to the Bunsen-burner non-luminous flame ? [CPMT-98]
(A) NaCl (B) $BaCl_2$ (C) $CaCl_2$ (D) KCl

622. Chili Saltpetre is [CPMT-98]
(A) KNO_2 (B) $NaNO_3$ (C) Na_2SO_4 (D) K_2SO_4

623. The correct sequence of alkali metals down the group is- [CPMT-98]
(A) Na, K, Li, Rb, Cs, Fr (B) Fr, Cs, Rb, K, Na, Li
(C) Li, Na, K, Cs, Fr, Rb (D) Li, Na, K, Rb, Cs, Fr

624. Baking soda is- [CPMT-98]
(A) $Na_2CO_3 . 10H_2O$ (B) $Na_2SO_4 . 10H_2O$
(C) Na_2SO_4 (D) $NaHCO_3$

625. On heating sodium metal in a current of dry ammonia, the compound formed is- [CPMT-98]
(A) Sodium nitride (B) Sodium hydride
(C) Sodium amide (D) Sodium azide

626. Microcosmic salt is - [CPMT-98]
(A) $Na_2HPO_4 .2H_2O$ (B) $(NH_4)_2HPO_4.2H_2O$
(C) $Na(NH_4)HPO_3$ (D) $Na(NH_4)HPO_4.4H_2O$

627. The electronic configuration of $_{29}Cu$ is-[CPMT-98]
(A) $1s^22s^22p^63s^23p^63d^94s^2$
(B) $1s^22s^22p^63s^23p^63d^{10}4s^1$
(C) $1s^22s^22p^63s^23p^63d^{10}4s^24p^4$ (D) None of these

628. Which one of the following will have highest solubility in water ? [CPMT-98]
(A) $BeSO_4$ (B) $MgSO_4$ (C) $CaSO_4$ (D) $BaSO_4$

629. The mineral of magnesium is- [CPMT-98]
(A) Bauxite (B) Malachite
(C) Carnallite (D) Haematite

630. The correct order of thermal stability of $BaCO_3$ (a), $CaCO_3$ (b) and $MgCO_3$ (c) will be- [CPMT-98]
(A) a > b > c (B) a < b < c
(C) a > b < c (D) a < b > c

631. A piece of magnesium ribbon is heated to redness in the atmosphere of nitrogen and cooled with water. The evolved gas is- [CPMT-98]
(A) H_2 (B) N_2 (C) O_2 (D) NH_3

632. Magnesium burns in- [CPMT-98]
(A) N_2 (B) CO (C) NO_2 (D) None of these

633. The formula for Plaster of Paris is- [CPMT-98]
(A) $CaSO_4$ (B) $2 CaSO_4.H_2O$
(C) $CaSO_4.3/2 H_2O$ (D) $CaSO_4.H_2O$

634. Chemically quicklime is [CPMT-98]
(A) CaO (B) $Ca(OH)_2$
(C) $CaCO_3$ (D) $CaSO_4 2H_2O$

635. Mixture of $MgCl_2$ and MgO is called- [CPMT-98]
(A) Complex salt (B) Double salt
(C) Portland cement (D) Sorel's cement

636. Which one is called inorganic benzene ?[CPMT-98]
(A) B_6H_6 (B) N_6H_6 (C) C_6H_6 (D) $B_3N_3H_6$

637. Which one of the following is most abundant in the earth's crust? [CPMT-98]
(A) Al (B) B (C) In (D) Ga

638. Aluminium appears like gold, when it is mixed with- [CPMT-98]
(A) 90% Cu (B) 50% Ni (C) 90% Sn (D) 50% Co

639. Duralumin alloy consists of [CPMT-98]
(A) Al and Mg (B) Al and Ni
(C) Al, Mg and Ni (D) Al, Mg, Mn and Cu

640. The blue coloured mineral 'Lapis-lazuli,' which is used as a semi-precious stone is a mineral of the following class- [CPMT-98]

(A) Sodium-alumino silicate (B) Zinc Cobaltate
(C) Basic copper carbonate (D) Prussian blue

641. Which of the following processes does not involve a catalyst ? [CPMT-98]

(A) Thermite process (B) Ostwald's process
(C) Haber's process (D) Contact process

642. Which one of the following is not a characteristic property of carbon ? [CPMT-98]

(A) Catenation (B) Multiple bond formation
(C) Availability of d-orbitals for bonding
(D) Highest electronegative element of the group

643. Poisonous gas present in fumes of motor car is- [CPMT-98]

(A) CH_4 (B) C_2H_2 (C) CO (D) CO_2

644. Dryice is- [CPMT-98]
(A) Ice which is completely free from the water on the suface (B) Silid SO_2 (C) Solid CO_2 (D) Solid benzene

645. Water gas is a mixture of - [CPMT-98]
(A) CO and H_2 (B) CO and CO_2
(C) CO_2 and H_2O (D) CO and steam

646. Which glass has the highest percentage of lead ? [CPMT-98]

(A) Soda glass (B) Flint glass
(C) Jena glass (D) Pyrex glass

647. White lead is [CPMT-98]
(A) $PbCO_3$ (B) $PbCO_3$. PbO
(C) $2PbCO_3$.Pb (OH_2) (D) $2PbSO_4$. PbO

648. Which one of the following elements is most abundant in nature ? [CPMT-98]
(A) Ca (B) C (C) Si (D) O

649. Which will show polymorphism ? [CPMT-98]
(A) Te (B) S (C) Se (D) All of these

650. Which one of the following halogens will have greatest electronegativity ? [CPMT-98]
(A) Iodine (B) Bromine (C) Chlorine (D) Fluorine

651. Which one of the following compounds is formed by the reaction of iodine and sodium thiosulphate ? [CPMT-98]

(A) Sodium sulphide (B) Sodium sulphite
(C) Sodium sulphate (D) Sodium tetrathionate

652. Which of the following is isoelectronic as well as has same structure as that of N_2O ? [CPMT-99]
(A) N_3H (B) H_2O (C) NO_2 (D) CO_2

653. On heating $KClO_3$, we get [CPMT-99]
(A) Cl_2O (B) ClO_2 (C) ClO_3 (D) Cl_2O_7

654. Which of the order for ionization energy is correct ? [CPMT-99]
(A) Be > B > C > N > O (B) B < Be < C < O < N
(C) B < Be < C < N < O (D) B < Be < N < C < O

655. Which of the following is most electronegative [CPMT-99]

(A) Carbon (B) Silicon (C) Lead (D) Tin

656. A metal ion M^{3+} loses 3 electrons, its oxidation number will be [CPMT-99]
(A) +3 (B) +6 (C) 0 (D) - 3

657. The configuration of chromium atom in ground state is [CPMT-99]
(A) $[Ar]3d^44s^2$ (B) $[Ar]3d^54s^1$
(C) $[Ar]3d^64s^2$ (D) $[Ar]3d^14s^2$

658. The oxidation number of oxygen in H_2O_2 is - [CPMT-2000]

(A) - 1 (B) - 2 (C) - 3 (D) -1/2

659. The bond length in LiF will be - [CPMT-2000]
(A) Less than that of NaF (B) More than that of KF
(C) Equal to that of KF (D) Equal to that of NaF

660. The equivalent weight of H_2O_2 in [CPMT-2000]
$2MnO_4^{1-} + 5H_2O_2 + 6H^+ \rightarrow 2Mn^{2+} + 5O_2 + 8H_2O$ is
(A) 17 (B) 34 (C) 68 (D) 85

661. Plaster of Paris is used [CPMT-2000]
(A) In surgery and dentistry (B) As a white wash
(C) As a constituent of toothpaste
(D) For the preparation of black board chalks

662. Which one of the following atoms has no neutron in its nucleus ? [CPMT-2000]
(A) Lithium (B) Helium (C) Protium (D) Tritium

663. The first ionization potential will be maximum for [CPMT-2000]

(A) Lithium (B) Hydrogen (C) Uranium (D) Iron

664. The correct sequence of atomic radii is[CPMT-2000]
(A) Na > Mg > Al > Si (B) Si > Al > Mg > Na
(C) Al > Si > Na > Mg (D) Si > Al > Na > Mg

665. Water gas is a mixture of [CPMT-2000]
(A) CO and H_2 (B) CO_2 and H_2O
(C) CO and N_2 (D) CO, H_2 and N_2

666. Heavy water is [CPMT-2000]
(A) Water containing sulphates of Ca and Mg
(B) Water containing heavy ions
(C) Water at 4° (D) D_2O

667. The constituents of Portland cement are [CPMT-2001]

(A) $CaCO_3$ + clay + sand (B) Ca Co + aluminium silicate
(C) $CaCO_3$ + clay + SiO_2(D) CaCO3 + gypsum + clay

668. Which of the following reacts as both reducing and oxidizing agent ? [CPMT-2001]
(A) H_2SO_4 (B) H_2O_2 (C) KOH (D) $KMnO_4$

669. Which of the following is most stable ?[CPMT-2001]
(A) Sn^{++} (B) Ge^{++} (C) Si^{++} (D) Pb^{++}

670. Bauxite is an ore of [CPMT-2001]
(A) Zinc (B) Copper (C) Aluminium (D) Iron

671. In lab H_2O_2 is prepared by [CPMT-2002]
(A) cold $H_2SO_4 + BaO_2$ (B) $HCl + BaO_2$
(C) conc $H_2SO_4 + Na_2O_2$ (D) $H_2 + O_2$

672. Fusion mixture is [CPMT-2002]
(A) $Na_2CO_3 + K_2CO_3$ (B) $Na_2CO_3 + NaHCO_3$
(C) $Na_2CO_3 + NaOH$ (D) $Na_2CO_3 + K_2SO_4$

673. White lead is [CPMT-2002]
(A) Pb_3O_4 (B) PbO
(C) $2PbCO_3.Pb(OH)_2$ (D) $Pb(CH_3COO)_2.Pb(OH)_2$

674. Nessler's reagent is [CPMT-2002]
(A) Potassium in mercuric iodide
(B) $TiCl_4$ (C) Al_2O_3/Cr^2O_3 (D) anhydrous $AlCl_3$

675. Which is a noble gas ? [CPMT-2003]
(A) Hydrogen (B) Neon (C) Halogen (D) Oxygen

676. The number of neturons in tritium is [CPMT-2003]
(A) 1 (B) 2 (C) 3 (D) 0

677. Tritium is the isotope of [CPMT-2003]
(A) Hydrogen (B) Oxygen
(C) Carbon (D) Sulphur

678. Which is insoluble in water ? [CPMT-2003]
(A) H_2S (B) $HgCl_2$ (C) $Ca(NO_3)_2$ (D) CaF_2

679. Ortho and para hydrogen differ [CPMT-2003]
(A) in the number of protons (B) in the molecular weight
(C) in the nature of spin of protons
(C) in the nature of spin of electrons.

680. The most electro-positive amongst the alkaline earth metals is: [PMT-93]
(A) Berelium (B) Magnesium
(C) Calcium (D) Barium

681. Which element of the following has the largest first ionization energy: [PMT-93]
(A) Boron (B) Carbon (C) Nitrogen (D) Oxygen

682. Which of the following salts is insoluble in water at room temp. but soluble in boiling water : [PMT-93]
(A) $CaCl_2$ (B) $BaCl_2$ (C) $SrCl_2$ (D) $PbCl_2$

683. Which of the following is the smallest cation : [PMT-93]
(A) Na^+ (B) Mg^{2+} (C) Ca^{2+} (D) Al^{3+}

684. Which one of the following is most metallic: [PMT-93]
(A) P (B) As (C) Sb (D) Bi

685. An element having electronic configuration (Ar) $3d^2\ 4s^2$ belongs to : [PMT-93]
(A) s-block elements (B) p-block elements
(C) d-block elements (D) f-block elements

686. The electronic configuration of chromium is : [PMT-93]
(A) Ne $3s^2\ 3p^6\ 3d^4\ 4s^2$ (B) Ne $3s^2\ 3p^6\ 3d^5\ 4s^1$
(C) Ne $3s^2\ 3p^6\ 4s^2\ 4p^4$ (D) Ne $3s^2\ 3p^6\ 3d^1\ 4s^2\ 4p^3$

687. Na_2CO_3 can be manufactured by Solvey's Process but K_2CO_3 can not be prepared because : [PMT-93]
(A) K_2CO_3 is more soluble (B) K_2CO_3 is less soluble
(C) $KHCO_3$ is more soluble than $NaHCO_3$
(D) $KHCO_3$ is less soluble than $NaHCO_3$

688. Of the following electronegativity of beryllium is approximately equal to that of : [PMT-93]
(A) Aluminium (B) Boron
(C) Magnesium (D) Sodium

689. Element with electronic configuration $1s^2\ 2s^2\ 2p^6\ 3s^2$ belongs to which of the following : [PMT-93]
(A) Metal (B) Non-metal
(C) Metalloid (D) Inert-gas

690. On the addition of a solution containing CrO_4^{2-} ions to the solution of Ba^{2+} Sr^{2+} and Ca^{2+} ions, the precipitate obtained first will be of : [PMT-93]
(A) $CaCrO_4$ (B) $SrCrO_4$
(C) $BaCrO_4$ (D) A mixture of all the three

691. Nessler's reagent is : [PMT-93]
(A) K_2HgI_4 (B) K_2HgI_4 + KOH
(C) K_2HgI_2 + KOH (D) K_2HgI_4 + Kg

692. The reactivity of the alkali metal sodium with water, is made use of [PMT-94]
(A) in drying of alcohols (B) in drying of benzene
(C) in drying of ammonia solution
(D) as a general drying agent

693. Which of the following is isoelectronic with carbon atom [PMT-94]
(A) Na^+ (B) Al^{3+} (C) O^{2-} (D) N^+

694. The screeing effect of inner electrons of the nucleus causes [PMT-94]
(A) a decrease in the ionisation potential
(B) an increase in the ionisation potential
(C) no effect on the ionisation potential
(D) an increase in the attraction of the nucleus to the electrons

695. An atom of an element has electronic configuration 2, 8, 1. Which of the following statement is correct ?

[PMT-94]

(A) the element's valency is 7
(B) the element exists as a diatomic moleucle
(C) the element is of non-metallic nature
(D) the element forms a basic oxide

696. Which one of the following is a good conductor of elctricity ? [PMT-94]
(A) diamond (B) graphite
(D) silicon (C) amorphous carbon

697. The total number of electrons in one molecule of CO_2 is [PMT-94]
(A) 22 (B) 44 (C) 66 (D) 88

698. The element with atomic number 55 belong to which block of the periodic table ? [PMT-95]
(A) s-block (B) p-block (C) d-block (D) f-block

699. Which of the following has largest size ?[PMT-95]
(A) Al (B) Al^+ (C) Al^{2+} (D) Al^{3+}

700. With the increase in atomic weights, melting points of the alkali metals [PMT-95]
(A) increase (B) decrease
(C) remain constant (D) do not show definite trend

701. Which of the alkaline earth meatals is strongest reducing agent ? [PMT-95]
(A) Ca (B) Sr (C)Ba (D) Mg

702. Electronic configuration of Cu (29) is [PMT-95]
(A) $1s^22s^22p^63s^23p^63d^{10}4s^1$ (B) $1s^22s^22p^63s^23p^53d^{11}4s^1$
(C) $1s^22s^22p^63s^23p^53d^{10}4s^2$ (D) $1s^22s^22p^63s^13p^63d^{10}4s^1$

703. White lead is [PMT-95]
(A) $PbCO_3$ (B) $PbCO_3$. PbO
(C) $Pb(OH)_2$. $2PbCO_3$ (D) $PbSO_4$. PbO

704. Alkali metals in each period have [PMT-95]
(A) smallest size (B) lowest ionization potential
(C) highest ionization potential
(D) highest electronegativity

705. The electronic configuration in the outermost shell of the halogen aom is [PMT-96]
(A) s^2p^5 (B) s^2p^3 (C) s^2p^6 (D) s^2p^4

706. Element of atomic number 23 is placed in periodic table in : [PMT-96]
(A) s-block (B) p-block (C) d-block (D) f-block

707. Alkali metals are : [PMT-96]
(A) Li, Na, Be, Mg, Cs (B) Li, Na, K, Rb, Cs
(C) Na, K, Mg, Ca, Rb (D) K, Rb, Cs, Ba, Sr

708. The atomic number of an element is 11. Its oxide will be : [PMT-96]
(A) Acidic (B) Basic
(C) Acid and Basic both (D) Neutral

709. The commercial production of sodium carbonate is done by : [PMT-96]
(A) Lead Chamber process (B) Haber's process
(C) Solvay's process (D) Castner'sprocess

710. The formula of washing soda is : [PMT-96]
(A) Na_2CO_3. $7H_2O$ (B) Na_2CO_3. H_2O
(C) $Na_2CO_3.10H_2O$ (D) Na_2CO_3

711. The metal which reacts with water at room temperature is : [PMT-96]
(A) Copper (B) Iron (C) Magnesium (D) Sodium

712. Alkaline earth metals are : [PMT-96]
(A) Li, Be, K, Mg, Ca (B) Be, Mg, Ca, Sr, Ba
(C) Be, K, Mg, Ca, Sr (D) Be, Mg, Ca, K, Rb

713. The composition formula of gypsum is : [PMT-96]
(A) $CaSO_4$ (B) $CaSO_4$. $2H_2O$
(C) $(CaSO_4)_2$. H_2O (D) $2CaSO_4$. H_2O

714. The formula of a phosphate of a metal, M is MPO_4. The formula of its nitrate will be : [PMT-96]
(A) MNO_3 (B) $M(NO_3)_2$ (C) $M(NO_3)_3$ (D) $M_2(NO_3)_3$

715. The formula of a phosphate of a metal, M, is MNH_4PO_4.The formula of its sulphate will be :[PMT-96]
(A) MSO_4 (B) $M(SO_4)_2$ (C) M_2SO_4 (D) $M_2(SO_4)_3$

716. The atomic number of an element is 22. The highest oxidation state exhibited by it in its compounds is : [PMT-96]
(A) 1 (B)2 (C) 3 (D) 4

717. How many unpaired electrons are present in an atom of electronic configuration $1s^2\ 2s^2\ 2p^4$? [PMT-96]
(A) 1 (B) 2 (C) 3 (D) 4

718. The order of the magnitude of first ionisation potentials Be, B, N and O is [PMT-96]
(A) N > O > Be > B (B) N > Be > O > B
(C) Be > B > N > O (D) B > Be > O > N

719. The order of magnitude of ionic radii of ions Na^+, Mg^{2+}, Al^{3+} and Si^{4+} is [PMT-96]
(A) $Na^+ < Mg^{2+} < Al^{3+} < Si^{4+}$ (B) $Mg^{2+} > Na^+ > Al^{3+} > Si^{4+}$
(C) $Al^{3+} > Na^+ > Si^{4+} > Mg^{2+}$ (D) $Na^+ > Mg^{2+} > Al^{3+} > Si^{4+}$

720. The order of the magnitude of ionic radii of ions N^{3-}, O^{2-} and F^- is [PMT-96]
(A) $N^{3-} > O^{2-} > F^-$ (B) $N^{3-} < O^{2-} < F^-$
(C) $N^{3-} > O^{2-} > F^-$ (D) $N^{3-} < O^{2-} < F^-$

721. Elements of the same group in periodic table have same [PMT-96]
(A) Atomic number (B) Electronic configuration
(C) Atomic weight
(D) Number of electrons in the outermost shell

722. Which of the properties remains unchanged on descending a group in the periodic table - [PMT-97]
(A) Atomic size (B) Density

(C) Valence electrons (D) Metallic character

723. The atomic radii of alkali metals (M) lie in the order Li <Na< K < Rb but the radii of M^+ ions in aqueous solution lie in the reverse order $Li^+ > Na^+ > K^+ > Rb^+$. What is the reason for this reverse order (on going from Li -Rb) [PMT-97]
(A) Gradual increase in ionisation energy
(B) Increasing weakness of the metallic bond
(C) Increasing electropositive character
(D) Decreasing degree of hydration

724. Which of the following can be represented by the configuration [Kr] $5s^2$? [PMT-97]
(A) Ca (B) Sr (C) Ba (D) Ra

725. Point out the incorrect statement regarding Be (group-II A) - [PMT-97]
(A) It forms an ionic carbide
(B) Its carbonate decomposes on heating
(C) Its halides are covalent
(D) It is easily attacked by water

726. Beryllium differs from rest of the members of its family (Groups- II A) in many ways. The reason for this is its : [PMT-97]
(A) Small size and higer electronegativity
(B) Small size and lower electronegativity
(C) Large size and lower ionisation energy
(D) Large size and largest ionic radius

727. Which of the following tendencies remains unchanged on going down in the nitrogen family - [PMT-97]
(A) Highest oxidation state (B) Non-metallic character
(C) Stability of hydrides (D) Physical state

728. Which of the following properties increases on going down from F to I group VII-A of the pereiodic table - [PMT-97]
(A) Electronegativity (B) Volatile nature
(C) Ionic radius (D) Oxidising power

729. The number of electrons in the outer most shell of the 3d-transition elements generally remains - [PMT-97]
(A) $(n - 1) d^n$ (B) nd^n (C) ns^2 (D) $(n -1) s^2$

730. Which among the following metals is most abundant in the earth's crust [PMT-97]
(A) Na (B) Mg (C) Al (D) Fe

731. Aluminium has a great affinity for oxygen and its oxidation is an exothermic process. This fact is made use of in [PMT-97]
(A) Preparing thin foils of aluminium
(B) Making utensils
(C) Preparing duralumin alloy (D) Thermite welding

732. Point out the wrong statement. In a given period of the periodic table the s-block element has in general a lower value of - [PMT-97]
(A) Ionisation energy (B) Electronegativity
(C) Atomic redius (D) Electron affinity

733. Which of the following is the correct order of gradually decreasing basic nature of the oxides - [PMT-97]
(A) Al_2O_3, MgO, Cl_2O_7, SO_3 (B) MgO, Al_2O_3, SO_3, Cl_2O_7
(C) Cl_2O_7, SO_3, Al_2O_3, MgO (D) SO_3, Cl_2O_7, MgO, Al_2O_3

734. The metal that does not give the borax bead test is [PMT-99]
(A) Chromium (B) Nickel (C) Lead (D) Manganese

735. What is the oxidation number of chromium in potassium dichromate ? [PMT-99]
(A) +2 (B) +3 (C) +6 (D) –4

736. Which one among the following nonmetals is a liquid at 25°C ? [PMT-99]
(A) Bromine (B) Carbon (C) Phosphorus (D) Sulphur

737. Which of the following is a non-metal ? [PMT-99]
(A) Gallium (B) Indium (C) Boron (D) Aluminium

738. In the reaction $3Mg + N_2 \rightarrow Mg_3N_2$ [PMT-99]
(A) Magnesium is reduced (B) Magnesium is oxidized
(C) Nitrogen is oxidized (D) None of these

739. When sodium chloride is dissolved in water, the sodium ion is : [PMT-99]
(A) Oxidized (B) Hydrated
(C) Reduced (D) None to these

740. Which one of the following is an s-block element ? [PMT-99]
(A) Aluminium (B) Chromium
(C) Niobium (D) Potassium

741. Which one of the following elements has the highest electronegativity ? [PMT-99]
(A) Chlorine (B) Fluorine (C) Oxygen (D) Nitrogen

742. Element showing the phenomenon of allotropy is : [PMT-99]
(A) Aluminium (B) Tin (C) Lead (D) Copper

743. In the modern periodic table the place of the element with atomic number 31 is in : [PMT-99]
(A) s-block (B) d-block (C) p-block (D) f-block

744. On reduction of $KMnO_4$ by oxalic acid in acidic medium the oxidation number of Mn changes. What is the magnitude of this change ? [PMT-2000]
(A) from 7 to 2 (B) from 6 to 2
(C) from 5 to 2 (D) from 7 to 4

745. The ionic radii of N^{3-}, O^{2-}, F^- and Na^+ are in the order: [PMT-2000]
(A) $N^{3-} > O2^- > F^- > Na^+$ (B) $N^{3-} > Na^+ > O^{2-} > F^-$
(C) $Na^+ > O^{2-} > N^{3-} > F^-$ (D) $O^{2-} > F^- > Na^+ > N^{3-}$

746. The first ionisation potential of Na, Mg, Al and Si are in the order : [PMT-2000]
(A) Na < Mg > Al < Si (B) Na > Mg > Al < Si
(C) Na > Mg > Al > Si (D) Na < Mg < Al < Si

747. Which of the following represents the electronic confguration of the most clectropostive elements ? [PMT-2000]
(A) [He] $2S^1$ (B) [He] $2S^2$ (C) [Xe] $6S^1$ (D) [Xe] $2S^2$

748. Mark the oxide which is amphoteric in character : [PMT-2000]
(A) CO_2 (B) SiO_2 (C) SnO_2 (D) CaO

749. In which of the following the inert pair effect is most prominent ? [PMT-2000]
(A) C (B) Si (C) Ge (D) Pb

750. The alkali metal that reacts with nitrogen directly to form nitride is : [PMT-2000]
(A) Li (B) Na (C) K (D) Rb

751. Zinc and Mercury do not show variable valency like d-block elements because : [PMT-2000]
(A) they are soft (B) their d-shells are complete
(C) they have only two electrons in the outer most subshell
(D) their d-shells are incomplete

752. Which of the following is the smallest cation ? [PMT-2001]
(A) Na^+ (B) Mg^{2+} (C) Ca^{2+} (D) Al^{3+}

753. A solution of sodium metal in liquid ammonia is a strong reducing agent due to the presence of : [PMT-2001]
(A) Sodium atoms (B) Sodium hydride
(C) Sodium amide (D) Solvated electron

754. In India on the occasion of marriage the fireworks are used to give coloured flames. The salt of which one of the following metals is used to obtain green flame for this purpose ? [PMT-2001]
(A) Na (B) K (C) Ba (D) Ca

755. Which of the following does not exhibit the periodicity in properities of the elements ? [PMT-2001]
(A) Ionisation energy (B) n / p ratio
(C) Electronegativity (D) Atomic radius

756. Which of the following elements exhibits the most basic properities ? [PMT-2001]
(A) F (B) Cl (C) Br (D) I

757. The correct sequence of decrease in size of the following species is : [PMT-2001]
(A) $I > I^+ > I^-$ (B) $I > I^- > I^+$
(C) $I^+ > I^- > I$ (D) $I^- > I > I^+$

758. The decreasing order of the ionisation potential in the following elements is : [PMT-2001]
(A) Ne > Cl > P > S > Al > Mg
(B) Ne > Cl > P > S > Mg > Al
(C) Ne > Cl > S > P > Mg > Al
(D) Ne > Cl > S > P > Al > Mg

759. Which one of the following is an incorrect statement? [PMT-2001]
(A) The ionisation potential of nitrogen is greater than that of oxygen
(B) The electron affinity of fluorine is greather than that of chlorine
(C) The ionisation potential of beryllium is greater than that of boron
(D) The electronegativity of fluorine is greater than that of chlorine

760. The electronegativity of an element in a period of periodic table from left to right [PMT-2002]
(A) Increases (B) Remains constant
(C) Decreases (D) Changes irregularly

761. Electron affinity depends on [PMT-2002]
(A) Atomic size (B) Nuclear charge
(C) Atomic number(D) Atomic size and nuclear charge

762. Which of the following orders is correct for the first ionization potential of B, C and N ? [PMT-2002]
(A) B > C > N (B) N > C > B
(C) N > C < B (D) N < C < B

763. The ionic radii of Li^+, Na^+, K^+ are in the following order [PMT-2002]
(A) $K^+ > Na^+ > Li^+$ (B) $K^+ > Na^+ < Li^+$
(C) $K^+ < Na^+ < Li^+$ (D) $Li^+ > Na^+ < K^+$

764. Which one of the following elements is most metallic [PMT-2002]
(A) P (B) As (C) Sb (D) Bi

765. In view of their low ionization energies the alkali metals are [PMT-2002]
(A) Weak oxidizing agents (B) Strong reducing agents
(C) Strong oxidizing agents (D) Weak reducing agents

766. The electron affinity of noble gas elements is [PMT-2002]
(A) Low (B) Nearly zero (C) High (D) Very high

767. The thermal stability of alkaline earth metal carbonates $MgCO_3$, $CaCO_3$, $BaCO_3$ and $SrCO_3$ decrease as [PMT-2002]
(A) $CaCO_3 > SrCO_3 > MgCO_3 > BaCO_3$
(B) $BaCO_3 > SrCO_3 > MgCO_3 > CaCO_3$
(C) $BaCO_3 > SrCO_3 > CaCO_3 > MgCO_3$
(D) $MgCO_3 > CaCO_3 > SrCO_3 > BaCO_3$

768. The nucleus of tritium contains [PMT-2002]
(A) 1 proton + 1 neutron (B) 1 proton + 3 neutrons
(C) 1 proton + 0 neutron (D) 1 proton + 2 neutrons

769. Number of unpaired electrons in electronic configu-

ration $1s^2$, $2s^2$, sp^4 is [PMT-2002]
(A) 2 (B) 3 (C) 4 (D) 6

770. In Cl^- ion the number of electrons is : [PMT-2003]
(A) 19` (B) 17 (C) 18 (D) 40

771. The correct order of second ionization potential of C, N, O and F is [PMT-2003]
(A) O > N > F > C (B) C > N > O > F
(C) O > F > N > C (D) F > O > N > C

772. Manganese belongs to [PMT-2003]
(A) 'p' block (B) 's' block
(C) 'd' block (D) 'f' block

773. Which one is the correct order of size of the iodine species. [PMT-2003]
(A) $I > I^- > I^+$ (B) $I > I^- > I^-$
(C) $I^+ > I^- > I$ (D) $I^- > I > I^+$

774. Which element has the highest electronegativity ? [PMT-2003]
(A) F (B) H (C) Na (D) Ne

775. With respect to chlorine hydrogen will be. [PMT-2003]
(A) Electropositive (B) Electronegative
(C) Neutral (D) None of the above

776. Which of the following hydroxide is the strongest base ? [PMT-2003]
(A) $Mg(OH)_2$ (B) $Be(OH)_2$ (C) $Ca(OH)_2$ (D) $Ba(OH)_2$

777. Diagonal relationship occurs between. [PMT-2003]
(A) B and Al (B) B and C (C) B and Ge (D) B and Si

778. Which of the following has the largest size ? [PMT-2003]
(A) Al (B) Al^{2+} (C) Al^{1+} (D) Al^{3+}

779. Which of the following does not show the variable oxidation state? [PMT-2003]
(A) Cu (B) Fe (C) Zn (D) Mn

780. What change occurs in atomic volumes on moving from left to right in series of transitional elements ? [PMT-2003]
(A) Volumes increase (B) Volumes decrease
(C) Volumes remain same (D) None

781. Which of the following is the formula of Saltpetre? [PMT-2003]
(A) KNO_3 (B) NaCl (C) $NaNO_3$ (D) $NaCO_3$

782. Lithium tertrahydrido aluminate is correctly represented as : [PMT-2003]
(A) $Al_2 [LiH_4]_3$ (B) $Al [LiH_4]$
(C) $Li [AlH_4]$ (D) $Li [AlH_4]_2$

783. Which of the following does not respond to the borax bead test ? [PET-93]
(A) Cobalt salts (B) Copper salts
(C) Aluminium salts (D) Nickel salts

784. Ca has atomic no. 20 and atomic weight 40 which of the following statements is not correct about Ca atom ? [PET-93]
(A) The number of electrons is same as the number of neutrons
(B) The number of nucleons is double of number of electrons
(C) The number of protons is half of the number of neutrons
(D) The number of nucleons is double of the atomic number

785. Which one of the following is the biggest ion ? [PET-93]
(A) Al^{+3} (B) Ba^{+2} (C) Mg^{+2} (D) Na^+

786. Aluminium is diagonally related to (in periodic table) [PET-93]
(A) Li (B) C (C) B (D) Be

787. Which of the following does not illustrate the anomalous properties of lithium ? [PET-93]
(A) The melting point and boiling point of Li are comparatively high
(B) Li is much softer than the other Group 1 metals
(C) Li forms a nitride Li_3N, unlike Group I metals
(D) The ion of Li and its compounds are more heavily hydrated than those of the rest of the group

788. The right order of the solubility of sulphates of alkaline earth metals in water is [PET-93]
(A) Be > Ca > Mg > Ba > Sr
(B) Mg > Be > Ba > Ca > Sr
(C) Be > Mg > Ca > Sr > Ba
(D) Mg > Ca > Ba > Be > Sr

789. The substance used in the thermite process of reducing metal ores is [PET-93]
(A) Aluminium (B) Thorium
(C) Heated platinum (D) Carbon

790. In the periodic table, the metallic character of elements [PET-93]
(A) Decreases from left to right across a period and on descending a group
(B) Decreases from left to right across a period and increases on descending a group
(C) Increases from left to right across a period and on descending a group
(D) Increases from left to right across a period and decreases on descending a group

791. Which of the following sets of elements has the strongest tendency to form anions ? [PET-93]
(A) N, O, F (B) P, S, Cl (C) As, Se, Br (D) Sb, Te, I

792. Electron affinity is the [PET-93]
(A) Energy absorbed when an electron is added to an

isolated atom in the gaseous state
(B) Energy released when an electron is added to an isolated atom in the gaseous state
(C) Energy required to take out an election from an isolated gaseous atom
(D) Power of an atom to attract an electron to itself

793. Which of the following is most soluble in water [PET-94]
(A) $MgSO_4$ (B) $CaSO_4$ (C) $SrSO_4$ (D) $BaSO_4$

794. Which of the following statements is not correct about the electronic configuration of gaseous chromium atom [PET-94]
(A) It has 5 electrons in 3d and one electron in 4s orbitals
(B) The principal quantum numbers of its valence electrons are 3 and 4
(C) It has 6 electrons in 3d orbital
(D) Its valence electrons have quantum no. 1, 0 and 2

795. Radius of the isoelectronic species [PET-94]
(A) Increases with the increase of nuclear charge
(B) Decreases with the increase of nuclear charge
(C) Is the same for all
(D) First increases and then decreases

796. In which of the following ionic radii of chromium would be smallest [PET-94]
(A) $K_2Cr_2O_4$ (B) CrO_2 (C) $CrCl_3$ (D) CrF_3

797. Which of the following has highest first ionization energy [PET-94]
(A) Sulfur (B) Oxygen (C) Nitrogen (D) Phosphorus

798. The electronic configuration $1s^2, 2s^22p^6, 3s^23p^6, 3d^6$ corresponds to : [PET-94]
(A) Mn^{2+} (B) Fe^{2+} (C) CO (D) Ge

799. The energy evolved is highest for which of the following reactions [PET-94]
(A) $F + e^- \rightarrow F^-$ (B) $Cl + e^- \rightarrow Cl^-$
(C) $S + 2e^- \rightarrow S^{2-}$ (D) $O + 2e^- \rightarrow O^{2-}$

800. The decomposition temperatures is maximum for [PET-94]
(A) $BaCO_3$ (B) $SrCO_3$ (C) $CaCO_3$ (D) $MgCO_3$

801. On Pauling scale which of the following does not have electronegativity ≥ 3.0 [PET-94]
(A) Oxygen (B) Nitrogen (C) Chlorine (D) Bromine

802. Which of the following oxides is most basic [PET-94]
(A) Na_2O (B) Al_2O_3 (C) SO_3 (D) SO_2

803. Which of the following has density greater than water [PET-94]
(A) Li (B) Na (C) K (D) Rb

804. In the periodic table, the elements are arranged in order of [PET-95]
(A) Increasing volume (B) Increasing atomic number
(C) Increasing mass (D) Increasing density

805. The electron affinity is largest for [PET-95]
(A) F (B) Cl (C) Br (D) I

806. The atomic radii in periodic table among elements from right to left [PET-95]
(A) Decreases (B) Increases
(C) Remain constant (D) First decrease and then increases

807. Element having atomic number 17 is placed in - [PET-95]
(A) I-group (B) V-group
(C) VIII-group (D) VII-group

808. Lithium shows diagonal relationship with [PET-95]
(A) Ne (B) Mg (C) Be (D) Ce

809. In Borax bead test deep blue coloured bead is obtained due to [PET-95]
(A) Ni (B) Co (C) Cr (D) Cu

810. Electronic configuration of chromium element (Atomic Number 24) is - [PET-95]
(A) $1s^22s^22p^63s^23p^63d^6$ (B) $1s^22s^22p^63s^23p^64s^2p^4$
(C) $1s^22s^22p^63s^23p^64s^23d^4$ (D) $1s^22s^22p^63s^23p^64s^13d^5$

811. The valence shell electronic configuration of alkali metals is [PET-96]
(A) ns^2np^1 (B) ns^1
(C) $(n-1)\ p^6ns^2$ (D) $(n-1)\ d^2ns^2$

812. The ratio between radii of He^+ ion and H atom is [PET-96]
(A) 1/2 (B) 1 (C) 3/2 (D) 2

813. The maximum number of electrons accommodated in 5f orbitals are [PET-96]
(A) 5 (B) 10 (C) 14 (D) 18

814. Of the following the ion with the smallest ionic radius is [PET-96]
(A) K^+ (B) Ca^{2+} (C) Ti^{3+} (D) Ti^{4+}

815. Lithium shows diagonal relationship with [PET-96]
(A) Magnesium (B) Beryllium
(C) Aluminum (D) Boron

816. The first ionization energies of alkaline earth metals are higher than those of the alkali metals. This is because [PET-96]
(A) There is increase in the nuclear charge of the alkaline earth metals
(B) There is a decrease in the nuclear charge of the alkaline earth metals
(C) There is no change in the nuclear charge
(D) None of the above

817. Elements in which 4f orbitals are progressively filled

are called as [PET-96]
(A) Transition elements (B) Lanthanides
(C) Actinides (D) Inert gases

818. All the elements in a group in the periodic table have the same [PET-96]
(A) Atomic weight (B) Atomic number
(C) Mass number (D) Number of electrons for bonding

819. The outermost electronic configuration of the most electronegative element is [PET-96]
(A) ns^2np^3 (B) ns^2np^4 (C) ns^2np^5 (D) ns^2np^6

820. The metal which gives hydrogen on treatment with acid as well as sodium hydroxide is [PET-96]
(A) Iron (B) Zinc
(C) Copper (D) None of the above

821. The property of hydrogen which distinguishes it from other alkali metals is [PET-96]
(A) Its electropositive character
(B) Its affinity for non-metals
(C) Its reducing character (D) Its non-metallic character

822. The ground state electronic configuration of Chromium is : [PET-97]
(A) $[Ar]3d^5\ 4s^1$ (B) $[Ar]3d^4\ 4s^2$
(C) $[Ar]3d^6\ 4s^0$ (D) $[Ar]4d^5\ 4s^1$

823. Who developed the long from of periodic table ? [PET-97]
(A) Lothar Meyer (B) Niels Bohr
(C) Mendeleef (D) Moseley

824. The correct sequence of elements in decreasing order of first ionization energy is : [PET-97]
(A) Na > Mg > Al (B) Mg > Na > Al
(C) Al > Mg > Na (D) Mg > Al > Na

825. Which of the following statements is correct ? [PET-97]
(A) X^- ion is larger in size than X atom
(B) X^+ ion is larger in size than X atom
(C) X^+ ion is larger in size than X^- ion
(D) X^+ and X– ions are equal in size

826. The correct order of electron affinity of B, C, N, O is [PET-97]
(A) O > C > N > B (B) B > N > C > O
(C) O > C > B > N (D) O > B > C > N

827. Which of the following ionic radius would be maximum ? [PET-97]
(A) C^{4-} (B) N^{3-} (C) O^{2-} (D) Mg^{2+}

828. The atomic radius of elements of which of the following series would be nearly the same [PET-97]
(A) Na, K, Rb, Cs (B) Li, Be, B, C
(C) Fe, Co, Ni, Cu (D) F, Cl ,Br, I

829. Which of the following electronic configurations is of a transition element ? [PET-97]
(A) $1s^22s^22p^63s^23p^64s^2$ (B) $1s^22s^22p^23s^23p^63d^{10}4s^24p^1$
(C) $1s^22s^22p^63p^63d^{10}4s^24p^6$ (D) $1s^22s^22p^63s^23p^63d^24s^2$

830. Nessler's reagent is : [PET-97]
(A) $KHgI_4$ (B) $K_2HgI_4 + NH_4OH$
(C) $K_2HgI_4 + KOH$ (D) $KHgI_4 + NH_4OH$

831. Which metal salt gives a violet coloured bead in the borax bead test ? [PET-97]
(A) Fe^{2+} (B) Ni^{2+} (C) CO^{2+} (D) Mn^{2+}

832. Which of the following gives a precipitate with $Pb(NO_3)_2$ but not with $Ba(NO_3)_2$? [PET-97]
(A) Sodium chloride (B) Sodium acetate
(C) Sodium nitrate (D) Sodium hydrogen phosphate

833. What is the electronic configuration of Cu^{2+} (z = 29) of least position [PET-98]
(A) $[Ar]\ 4s^13d^8$ (B) $[Ar]\ 4s^23d^{10}4p^1$
(C) $[Ar]\ 4s^13d^{10}$ (D) $[Ar]\ 3d^9$

834. On going from right to left in a period in the periodic table the electronegativity of the elements : [PET-98]
(A) Increases (B) Decreases
(C) Remains unchanged
(D) Decreases first then increases

835. In the periodic table, the element with atomic number 16 will be placed in the group : [PET-98]
(A) Third (B) Fourth (C) Fifth (D) Sixth

836. Metals belonging to the same group in the periodic table are : [PET-98]
(A) Magnesium and sodium (B) Magnesium and copper
(C) Magnesium and barium
(D) Magnesium and potassium

837. In the following, the element with the highest electronegativity is : [PET-98]
(A) Carbon (B) Magnesium
(C) Oxygen (D) Sulphur

838. The electronic configuration in the outermost shell of halogens is : [PET-98]
(A) s^2p^4 (B) s^2p^5 (C) s^2p^3 (D) s^2p^6

839. The ionic radii N^{3-}, O^{2-} F^- and Na^+ follow the order: [PET-98]
(A) $N^{3-} > O^{2-} < F^- > Na^+$ (B) $N^{3-} > Na^+ > O^{2-} > F^-$
(C) $Na^+ > O^{2-} > N^{3-} > F^-$ (D) $O^{2-} > F^- > Na^+ > N^{3-}$

840. In the following, the element with the highest electropositivity is : [PET-98]
(A) Copper (B) Cesium (C) Barium (D) Chromium

841. The stability of the following alkali metal chlorides follows the order : [PET-98]
(A) LiCl > KCl > NaCl > CsCl
(B) CsCl > KCl > NaCl > LiCl

(C) NaCl > KCl > LiCl > CsCl
(D) KCl > CsCl > NaCl > LiCl

842. The electronic configuration of a metal M is $1s^2 2s^2 2p^6 3s^1$. The formula of its oxide will be : [PET-98]
(A) MO (B) M_2O (C) M_2O_3 (D) MO_2

843. Which of the following metals reacts with water at room temperature ? [PET-98]
(A) Nickel (B) Silver (C) Copper (D) Sodium

844. The compounds of alkaline earth metals have the following magnetic nature : [PET-98]
(A) Diamagnetic (B) Paramagnetic
(C) Ferromagnetic (D) Antiferromagnetic

845. In certain matters Lithium differs from other alkali metals, the main reason for this is : [PET-98]
(A) Small size of Li atom and Li^+ ion
(B) Extremely high electropositivity of Li
(C) Greater hardness of Li (D) Hydration of Li^+ ion

846. For which one of the following properties of halogens the sequence F > Cl > Br > I holds good ? [PET-98]
(A) Electronaffinity (B) Electronegativity
(C) Atomic radius (D) Boiling point

847. Which of the following metals is preserved in water? [PET-99]
(A) Phosphorus (B) Sodium (C) Zinc (D) Iron

848. Which has the smallest size ? [PET-99]
(A) Na^+ (B) Mg^{2+} (C) Al^{3+} (D) P^{5+}

849. Which of the following metals exhibits more than one oxidation state? [PET-99]
(A) Na (B) Mg (C) Fe (D) Al

850. Acidified solution of chromic acid on treatment with hydrogen peroxide yields: [PET-99]
(A) $CrO_3 + H_2O + O_2$ (B) $Cr_2O_3 + H_2O + O_2$
(C) $CrO_5 + H_2O$ (D) $H_2Cr_2O_7 + H_2O + O_2$

851. Which ion is detected by Nessler's reagent ? [PET-99]
(A) NH_4^+ (B) MnO_4^- (C) PO_4^{3-} (D) CrO_4^{2-}

852. Epsom salt is: [PET-99]
(A) $MgSO_4 .7H_2O$ (B) $CaSO_4 .H_2O$
(C) $MgSO_4 .2H_2O$ (D) $BaSO_4 .2H_2O$

853. Among the alkaline earth metals the element forming predominantly covalent compound is: [PET-99]
(A) Barium (B) Strontium (C) Calcium (D) Beryllium

854. Which of the following chemicals contains chlorine? [PET-99]
(A) Fischer salt (B) Epsom salt
(C) Fremy's salt (D) Spirit of salt

855. Molecular formula of Glauber's salt is : [PET-2000]
(A) $MgSO_4\ 7H_2O$ (B) $CuSO_4.5H_2O$
(C) $FeSO_4.7H_2O$ (D) $Na_2SO_4.10H_2O$

856. Which one of the following is correct order of the size of iodine species ? [PET-2000]
(A) $I > I^- > I^+$ (B) $I > I^+ > I^-$
(C) $I^+ > I^- > I$ (D) $I^- > I > I^+$

857. Which of the following substances consist of only one element ? [PET-2000]
(A) Marble (B) Sand (C) Diamond (D) Glass

858. Which of the following is obtained when nitrogen reacts with calcium carbide ? [PET-2000]
(A) Calcium Carbonate (B) Calcium cyanamide
(C) Calcium acetate (D) Calcium cyanate

859. The correct order of the increasing ionic character is : [PET-2000]
(A) $BeCl_2 < MgCl_2 < CaCl_2 < BaCl_2$
(B) $BeCl_2 < MgCl_2 < BaCl_2 < CaCl_2$
(C) $BeCl_2 < BaCl_2 < MgCl_2 < CaCl_2$
(D) $BeCl_2 < CaCl_2 < MgCl_2 < BeCl_2$

860. In the equation
$4M + 8CN^- + 2H_2O + O_2 \rightarrow 4\,[M(CN_2)]^- + 4OH^-$
the metal M is : [PET-2000]
(A) Copper (B) Iron (C) Gold (D) Zinc

861. The second most abundant element in the universe is : [PET-2000]
(A) Silicon (B) Nitrogen (C) Nitrogen (D) Helium

862. Which statement is not true ? [PET-2000]
(A) $Ni(CO)_4$ is diamagnetic
(B) BI_3 is a stronger Lewis acid then BF_3
(C) Graphite conducts electricity whereas diamond does not
(D) CCl_4 is hydrolysed whereas BCl_3 is inert

863. From which of the following the hydration energy of Mg^{2+} is larger ? [PET-2000]
(A) Na^+ (B) Al^{3+} (C) Be^{2+} (D) Cr^{3+}

864. Fenton reagent is : [PET-2000]
(A) $FeSO_4 + H_2O_2$ (B) Zn + HCl
(C) Sn + HCl (D) None of the above

865. Which will be the proper alternative in place of A in the following equation ? [PET-2000]
$2Fe^{3+}(aq) + Sn^{2+}(aq) \rightarrow 2\,Fe^{2+}(aq) + A$
(A) Sn^{4+} (B) Sn^{3+} (C) Sn^{2+} (D) Sn^0

866. Which of the following acts both as oxidizing and reducing agent ? [PET-2000]
(A) HNO_3 (B) $K_2Cr_2O_7$ (C) $KMnO_4$ (D) H_2O_2

867. Aluminium is diagonally related to : [PET-2001]
(A) Li (B) Be (C) C (D) B

868. Alkaline earth metals belong to the : [PET-2001]
(A) s- block in periodic table
(B) p- block in periodic table

(C) d-block in periodic table (D) f-block in periodic table

869. An element of atomic number 29 belongs to the block in periodic table : [PET-2001]
(A) s - block (B) p - block (C) d - block (D) f- block

870. Which of the following has the largest radius ? [PET-2001]
(A) Mg^{2+} (B) Na^+ (C) O^{2-} (D) F^-

871. The ionization potential is lowest for the :[PET-2001]
(A) Halogens (B) Inert gases
(C) Alkaline earth metals (D) Alkali metals

872. Which of the following elements has the maximum electron affinity ? [PET-2001]
(A) F (B) Cl (C) Br (D) I

873. The lightest metal is : [PET-2001]
(A) Li (B) Mg (C) Ca (D) Na

874. The strongest reducing agent is : [PET-2001]
(A) K (B) Al (C) Mg (D) Br

875. In the following the plaster of Paris is : [PET-2001]
(A) $CaSO_4.H_2O$ (B) $CaSO_4.2H_2O$
(C) $CaSO_4.\frac{1}{2}H_2O$ (D) $CaSO_4.1\frac{1}{2}H_2O$

876. Which of the following statements is false for alkali metals ? [PET-2001]
(A) Li is the strongest reducing agent
(B) Na is amphoteric in nature (C) Li^+ is exceptionally small
(D) All alkali metals give blue solution in liquid ammonia

877. Which of the following ions has largest heat of hydration ? [PET-2001]
(A) Ba^{2+} (B) K^+ (C) Li^+ (D) Be^{2+}

878. Which catalyst is used in Friedel-Crafts synthesis [PET-2001]
(A) Finely divided Fe (B) Platinized Pd
(C) Anhydrous $AlCl_3$ (D) Alcoholic KOH

879. The correct configuration of copper with atomic number 29 is [PET-2001]
(A) [Ar] $4s^1$ (B) [Ar] 4s2
(C) [Ar] $3d^{10}4s^1$ (D) [Ar] $3d^9 4s^2$

880. An aqueous solution carbonate is alkaline because sodium carbonate is a salt of [PET-2002]
(A) Weak acid and weak base
(B) Strong acid and weak base
(C) Weak acid and strong base
(D) Strong acid and strong base

881. Which gives green colour in fireworks? [PET-2002]
(A) Sodium (B) Potassium (C) Barium (D) Calcium

882. The chemical composition of 'Cryolite' mineral is [PET-2002]
(A) Al_2O_3 (B) $Al_2O_3.12H_2O$ (C) $KAlSi_3O_8$ (D) Na_3AlF_6

883. Which of the following has the least ionization potential [PET-2002]
(A) Lithium (Li) (B) Helium (He)
(C) Nitrogen (N) (D) Zinc (Zn)

884. Which of the following are insoluble in water [PET-2002]
(A) Na_2CO_3 (B) $CaCO_3$ (C) $ZnCO_3$ (D) $Al_2(CO_3)_3$

885. Paramagnetism is exhibited by molecules[PET-2003]
(A) Not attracted in a magnetic filed
(B) Containing only paired electrons
(C) Carrying a positive charge
(D) Containing unpaired electrons

886. Which of the following metal is present in chlorophyll [PET-2003]
(A) Chromium (B) Cobalt (C) Magnesium (D) Iron

887. The number of urrpaired electrons is maximum in (Atomic no. Ti. 22 V.23 Cr. 24 Fe.26 [PET-2003]
(A) Fe (B) Cr (C) Ti (D)V

888. The number of electron and proton in the third alkaline earth metal will be : [PET-2003]

	e	p		e	p		e	p		e	p
(A)	20	20	(B)	20	18	(C)	18	18	(D)	20	19

889. Which of the following pair of atomic number represent s-block element [PET-2003]
(A) 7, 15 (B) 6, 12 (C) 9, 17 (D) 3, 12

890. The element whose ionizatioin potential second is less than its ionization potential first [PET-2003]
(A) Be (B) Li (C) He (D) None of the above

891. Increasing order of density is [PET-2003]
(A) Li<Na<K,Rb<Cs (B) Li<k<Na<Rb<Cs
(C) K<Li<Na<Rb<Cs (D) Cs<Rb<K<Na

892. Assertion (A) : Physical absorption of molecules takes place on surface only. Reason (R) : In this process, the bonds of the absorbed molecules are broken.[PET-2003]
(A) A (B) B (C) C (D) D (E) E

893. Assertion (A) : The boiling and melting points of amides are higher than corresponding acids. Reason (R) : It is due to strong intermolecular hydrogen bonding in their molecules [PET-2003]
(A) A (B) B (C) C (D) D (E) E

894. The ligand called π acid is - [PET-2003]
(A) CO (B) NH_3 (C) $C_2O_4^{2-}$ (D) Ethylene diamine

895. The approximate composition of ordinary glass is [PET-2003]
(A) Na_2O CaO $6SiO_2$ (B) Na_2O CaO SiO_2
(C) K_2O CaO $6SiO_2$ (D) Na_2O CaO SiO_2

896. Arrange the following in order of their decreasing thermal conductivity [PET-2003]

(A) Al, Ag, Cu (B) Cu, Ag, Al (C) Ag, Cu, Al (D) Al, Cu, Ag

897. What are the products formed when an aqueous solution of magnesium bicarbonate is boiled ?[PET-2003]
(A) $MgCO_3$, H_2O, CO_2 (B) $Mg(HCO_3)_2$,H_2O
(C) $Mg(OH)_2$, H_2O (D) $MgCO_3$, H_2O

898. Correct order of polarising power is [BHU-2003]
(A) $Cs^+<K^+<Mg^{2+}<Al^{3+}$ (B) $K^+<Cs^+<Mg^{2+}<Al^{3+}$
(C) $Cs^+<K^+<Al^{3+}<Mg^{2+}$ (D) $K^+<Cs^+<Al^{3+}<Mg^{2+}$

899. Carborundum is [BHU-2003]
(A) Al_2O_3 (B) SiC (C) $AlCl_3$ (D) $CaCO_3$

900. Correct statement is [BHU-2003]
(A) K=$4s^1$, Cr=$3d^54s^1$, Cu=$3d^{10}4s^1$
(B) K=$4s^2$, Cr=$3d^54s^1$, Cu=$3d^{10}4s^1$
(C) K=$4s^2$, Cr=$3d^44s^2$, Cu=$3d^94s^2$
(D) K=$4s^1$, Cr=$3d^44s^2$, Cu=$3d^94s^2$

901. The structure of H_2O_2 is [AFMC-2003]
(A) Planar (B) Linear
(C) Non-planar (D) Three dimensional

902. Hydrogen burns in air with a [AFMC-2003]
(A) Yellow flame (B) Green flame
(C) Light bluish flame (D) None of these

903. CO is isoelectronic with [AFMC-2003]
(A) N_2 (B) O_2 (C) NH_3 (D) NO_2

904. Which of the following does not exist?[AFMC-2003]
(A)PbF_4 (B)$PbCl_4$ (C)$PbBr_4$ (D) PbI_4

905. Elements having same number of nucleons and different number of protons are known as[AFMC-2003]
(A)Isotopes (B) Isobars
(C) Isotones (D) None of these

906. Which of the following is a non-metal?[AFMC-2003]
(A) Iridium (B) Boron (C) Gallium (D) Aluminium

907. The element with the highest electron affinity is [AFMC-2003]
(A) F (B) Cl (C) Br (D) I

908. Diamond and emerald are formed of[AFMC-2003]
(A) Carbon-Carbon (B) Silica-Silica
(C) Carbon-Silica (D) Silica-Carbon

909. Proton is likely to have high [AFMC-2003]
(A) Hydration energy (B) Electron affinity
(C) Atomic size (D) Atomic mass

910. Number of elements present in fifth period is [AFMC-2003]
(A) 8 (B) 18 (C) 24 (D)32

911. Among the following, the pair in which the two species are not isostructural is [CBSE-PMT-2004]
(A) SiF_4 and SF_4 (B) IO_3^- and XeO_3
(C) BH_4^- and NH_4^+ (D) PF_6^- and SF_6

912. Which one of the following statements about the zeolites is false ? [CBSE-PMT-2004]
(A) They are used as cation exchangers
(B) They have open structure which enables them to take up small molecules
(C) Zeolites are aluminosilicates having three dimensional network
(D) Some of the SiO_4^{4-} units are replaced by AlO_4^{5-} and AlO_6^{9-} ions in zeolites

913. Ionic radii are [CBSE-PMT-2004]
(A) inversely proportional to effective nuclear charge
(B) inversely proportional to square of effective nuclear charge
(C) directly proportional to effective nuclear charge
(D) directly proportional to square of effective nuclear charge

914. H_2O is dipolar, whereas BeF_2 is not. It is because [CBSE-PMT-2004]
(A) the electronegativity of F is greater than that of O
(B) H_2O involves hydrogen bonding whereas BeF_2 is a discrete molecule
(C) H_2O is linear and BeF_2 is angular
(D) H_2O is angular and BeF_2 is linear

Periodic Table; Group 1, 2, 13 and 14 - Discussions

1. (B)
2. (B)
3. (C)
4. (A)
5. (C)
6. (C)
7. (B)
8. (D)
9. (B)
10. (A)
11. (D)
12. (D)
13. (D)
14. (D)
15. (B)
16. (D)
17. (A)
18. (B)
19. (D)
20. (C)
21. (A)
22. (B)
23. (B)
24. (D)
25. (C)
26. (C)
27. (A)
28. (C)
29. (B)
30. (D)
31. (C)
32. (B)
33. (D)
34. (A)
35. (B)
36. (C)
37. (B)
38. (B)
39. (B)
40. (D)
41. (A)
42. (A)
43. (A)
44. (A)
45. (A)
46. (A)
47. (B)
48. (B)
49. (B)
50. (A)
51. (D)
52. (D)
53. (D)
54. (B)
55. (D)
56. (B)
57. (B)
58. (B)
59. (C)
60. (B)
61. (A)
62. (A)
63. (B)
64. (D)
65. (A)
66. (A)
67. (D)
68. (D)
69. (B)
70. (A)
71. (A)
72. (A)
73. (A)
74. (A)
75. (B)
76. (B)
77. (C)
78. (A)
79. (B)
80. (A)
81. (C)
82. (A)
83. (B)
84. (B)
85. (B)
86. (D)
87. (D)
88. (C)
89. (C)
90. (D)
91. (C)
92. (B)
93. (A)
94. (A)
95. (D)
96. (D)
97. (D)
98. (C)
99. (B)
100. (C)
101. (A)
102. (C)
103. (A)
104. (A)
105. (A)
106. (C)
107. (A)
108. (D)
109. (C)
110. (B)
111. (B)
112. (C)
113. (A)
114. (C)
115. (A)
116. (C)
117. (D)
118. (A)
119. (A)
120. (A)
121. (D)
122. (B)
123. (D)
124. (A)
125. (A)
126. (D)
127. (B)
128. (A)
129. (C)
130. (C)
131. (D)
132. (C)
133. (B)
134. (B)
135. (A)
136. (A)
137. (A)
138. (A)
139. (A)
140. (A)
141. (A)
142. (B)
143. (D)
144. (A)
145. (D)
146. (B)
147. (A)
148. (A)
149. (B)
150. (D)
151. (B)
152. (D)
153. (B)
154. (A)
155. (C)
156. (B)
157. (C)
158. (C)
159. (D)
160. (A)
161. (C)
162. (A)
163. (D)
164. (D)
165. (C)
166. (B)
167. (B)
168. (D)
169. (B)
170. (A)
171. (A)
172. (C)
173. (C)
174. (B)
175. (A)
176. (A)
177. (A)
178. (C)
179. (C)
180. (A)
181. (D)
182. (A)
183. (B)
184. (A)
185. (D)
186. (A)
187. (C)
188. (D)
189. (B)
190. (C)
191. (C)
192. (C)
193. (D)
194. (D)
195. (C)
196. (B)
197. (C)
198. (D)
199. (D)
200. (B)
201. (C)
202. (D)
203. (B)
204. (A)
205. (A)
206. (B)
207. (C)
208. (A)
209. (D)
210. (C)
211. (A)
212. (D)
213. (A)
214. (A)
215. (A)
216. (D)
217. (A)
218. (D)
219. (B)
220. (C)
221. (A)
222. (B)
223. (C)
224. (D)
225. (D)
226. (D)
227. (A)
228. (D)
229. (B)
230. (D)
231. (B)
232. (B)
233. (B)
234. (D)
235. (C)
236. (D)
237. (A)
238. (D)
239. (B)
240. (C)
241. (A)
242. (A)
243. (C)
244. (A)
245. (D)
246. (B)
247. (C)
248. (D)
249. (A)
250. (A)
251. (C)
252. (A)
253. (D)
254. (C)
255. (C)
256. (A)
257. (D)
258. (C)
259. (C)
260. (A)
261. (B)
262. (B)
263. (C)
264. (A)
265. (C)
266. (B)
267. (B)
268. (C)
269. (C)
270. (B)
271. (A)
272. (D)
273. (A)
274. (B)
275. (B)
276. (C)
277. (A)
278. (B)
279. (C)
280. (D)
281. (D)
282. (D)
283. (B)
284. (C)
285. (C)
286. (A)
287. (A)
288. (C)
289. (D)
290. (C)
291. (C)
292. (B)
293. (A)
294. (A)
295. (B)
296. (C)
297. (B)
298. (A)
299. (D)
300. (C)
301. (D)
302. (C)
303. (C)
304. (A)
305. (B)
306. (D)
307. (B)
308. (A)
309. (B)
310. (A)
311. (B)
312. (A)

313.	(B)	361.	(C)	409.	*	457.	*	505.	(A)	553.	(A)
314.	(D)	362.	(C)	410.	(B)	458.	*	506.	(A)	554.	(C)
315.	(C)	363.	(A)	411.	(B)	459.	*	507.	(A)	555.	(D)
316.	(D)	364.	(C)	412.	(D)	460.	(C)	508.	(C)	556.	(D)
317.	(C)	365.	(D)	413.	(B)	461.	(B)	509.	(D)	557.	(C)
318.	(A)	366.	(A)	414.	(B)	462.	(B)	510.	(D)	558.	(C)
319.	(B)	367.	(B)	415.	(C)	463.	*	511.	(D)	559.	(A)
320.	(D)	368.	(C)	416.	(B)	464.	*	512.	(D)	560.	(A)
321.	(A)	369.	(B)	417.	(C)	465.	*	513.	(D)	561.	(D)
322.	(B)	370.	(D)	418.	(D)	466.	*	514.	(A)	562.	(C)
323.	(D)	371.	(B)	419.	(A)	467.	*	515.	(B)	563.	(A)
324.	(C)	372.	(B)	420.	(B)	468.	(D)	516.	(B)	564.	(D)
325.	(C)	373.	(C)	421.	(C)	469.	(B)	517.	(C)	565.	(C)
326.	(A)	374.	(B)	422.	(C)	470.	(C)	518.	(A)	566.	(C)
327.	(B)	375.	(A)	423.	(A)	471.	(C)	519.	(D)	567.	(C)
328.	(A)	376.	(C)	424.	(A)	472.	(A)	520.	(D)	568.	(A)
329.	(D)	377.	(B)	425.	(C)	473.	(B)	521.	(D)	569.	(A)
330.	(B)	378.	(B)	426.	(C)	474.	(C)	522.	(C)	570.	(D)
331.	(D)	379.	(A)	427.	(D)	475.	(B)	523.	(B)	571.	(A)
332.	(B)	380.	(C)	428.	(C)	476.	(C)	524.	(D)	572.	(C)
333.	(A)	381.	(B)	429.	(D)	477.	(C)	525.	(B)	573.	(C)
334.	(A)	382.	(B)	430.	(B)	478.	(D)	526.	(B)	574.	(B)
335.	(C)	383.	(B)	431.	(B)	479.	(A)	527.	(B)	575.	(C)
336.	(D)	384.	(C)	432.	(D)	480.	(B)	528.	(B)	576.	(C)
337.	(A)	385.	(C)	433.	(C)	481.	(B)	529.	(C)	577.	(C)
338.	(B)	386.	(A)	434.	(C)	482.	(A)	530.	(B)	578.	(A)
339.	(A)	387.	(B)	435.	(D)	483.	(B)	531.	(D)	579.	(B)
340.	(A)	388.	(B)	436.	(B)	484.	(B)	532.	(B)	580.	(A)
341.	(A)	389.	(B)(D)	437.	(B)	485.	(B)	533.	(B)	581.	(C)
342.	(A)	390.	(B)	438.	(D)	486.	(C)	534.	(B)	582.	(D)
343.	(B)	391.	(B)	439.	(A)	487.	(A)	535.	(B)	583.	(B)
344.	(C)	392.	(C)	440.	(A)	488.	(B)	536.	(B)	584.	(D)
345.	(C)	393.	(A)(C)	441.	(A)	489.	(B)	537.	(B)	585.	(B)
346.	(B)	394.	(B)	442.	(B)	490.	(B)	538.	(C)	586.	(A)
347.	(A)	395.	(B)(D)	443.	(D)	491.	(D)	539.	(C)	587.	(C)
348.	(C)	396.	*	444.	*	492.	(C)	540.	(D)	588.	(C)
349.	(D)	397.	*	445.	*	493.	(D)	541.	(D)	589.	(B)
350.	(C)	398.	*	446.	*	494.	(C)	542.	(B)	590.	(A)
351.	(B)	399.	*	447.	*	495.	(B)	543.	(D)	591.	(A)
352.	(D)	400.	(B)	448.	*	496.	(C)	544.	(B)	592.	(D)
353.	(B)	401.	(B)	449.	*	497.	(C)	545.	(B)	593.	(B)
354.	(C)	402.	(D)	450.	*	498.	(D)	546.	(B)	594.	(D)
355.	(B)	403.	(A)	451.	(B)	499.	(A)	547.	(A)	595.	(C)
356.	(A)	404.	(C)	452.	(B)	500.	(C)	548.	(A)	596.	(B)
357.	(C)	405.	(C)	453.	(A)	501.	(A)	549.	(A)	597.	(C)
358.	(D)	406.	(B)	454.	*	502.	(B)	550.	(A)	598.	(A)
359.	(D)	407.	(B)	455.	*	503.	(A)	551.	(A)	599.	(D)
360.	(C)	408.	*	456.	*	504.	(C)	552.	(A)	600.	(D)

396. higher effective nuclear charge
397. R_3 - Si - O - SiR_3
398. $3d^5 4s^1$
399. three centre bond
408. RbCl and ICl
409. Inert pair effect
444. 2
445. S^{2-}
446. Aluminium
447. Nitrogen
448. Lithium
449. Annealing
450. Cao
454. Greater
455. even
456. I.P.
457. sp^3
458. $1s^2 2s^2 2p^6 3s^2 3p^6 4s^1$
459. Fluorine
463. Bicarbonate
464. Poisonous
465. Alkalimetal
466. Lowest
467. $CaSO_4 . 1/2H_2O$

601. (B)
602. (D)
603. (A)
604. (B)
605. (B)
606. (C)
607. (A)
608. (B)
609. (A)
610. (C)
611. (B)
612. (C)
613. (D)
614. (B)
615. (D)
616. (A)
617. (D)
618. (C)
619. (D)
620. (A)
621. (D)
622. (B)
623. (D)
624. (D)
625. (C)
626. (D)
627. (B)
628. (A)
629. (C)
630. (A)
631. (D)
632. (A)
633. (B)
634. (A)
635. (D)
636. (D)
637. (A)
638. (A)
639. (D)
640. (A)
641. (A)
642. (C)
643. (C)
644. (C)
645. (A)
646. (B)
647. (C)
648. (D)
649. (D)
650. (D)
651. (D)
652. (D)
653. (B)
654. (B)
655. (A)
656. (B)
657. (B)
658. (A)
659. (B)
660. (A)
661. (A)
662. (C)
663. (B)
664. (A)
665. (A)
666. (D)
667. (A)
668. (B)
669. (C)
670. (C)
671. (A)
672. (A)
673. (C)
674. (A)
675. (B)
676. (B)
677. (A)
678. (D)
679. (C)
680. (D)
681. (C)
682. (D)
683. (D)
684. (D)
685. (C)
686. (B)
687. (C)
688. (A)
689. (A)
690. (C)
691. (B)
692. (B)
693. (D)
694. (B)
695. (D)
696. (B)
697. (A)
698. (A)
699. (A)
700. (B)
701. (C)
702. (A)
703. (C)
704. (B)
705. (A)
706. (C)
707. (B)
708. (B)
709. (C)
710. (C)
711. (D)
712. (B)
713. (B)
714. (C)
715. (A)
716. (D)
717. (B)
718. (A)
719. (A)
720. (A)
721. (D)
722. (C)
723. (B)
724. (B)
725. (D)
726. (A)
727. (D)
728. (C)
729. (D)
730. (C)
731. (D)
732. (A)
733. (B)
734. (C)
735. (C)
736. (A)
737. (C)
738. (B)
739. (B)
740. (D)
741. (B)
742. (B)
743. (C)
744. (A)
745. (A)
746. (A)
747. (C)
748. (C)
749. (D)
750. (A)
751. (B)
752. (D)
753. (D)
754. (C)
755. (B)
756. (D)
757. (D)
758. (B)
759. (B)
760. (A)
761. (D)
762. (B)
763. (A)
764. (D)
765. (B)
766. (B)
767. (C)
768. (D)
769. (A)
770. (C)
771. (C)
772. (C)
773. (D)
774. (A)
775. (A)
776. (D)
777. (D)
778. (A)
779. (C)
780. (D)
781. (A)
782. (C)
783. (C)
784. (C)
785. (B)
786. (D)
787. (B)
788. (C)
789. (A)
790. (B)
791. (A)
792. (B)
793. (A)
794. (C)
795. (B)
796. (A)
797. (C)
798. (B)
799. (B)
800. (A)
801. (D)
802. (A)
803. (D)
804. (B)
805. (B)
806. (B)
807. (C)
808. (B)
809. (B)
810. (D)
811. (B)
812. (A)
813. (C)
814. (D)
815. (A)
816. (A)
817. (B)
818. (D)
819. (C)
820. (B)
821. (D)
822. (D)
823. (B)
824. (D)
825. (A)
826. (A)
827. (A)
828. (C)
829. (D)
830. (C)
831. (D)
832. (D)
833. (D)
834. (B)
835. (D)
836. (C)
837. (C)
838. (B)
839. (A)
840. (B)
841. (B)
842. (B)
843. (D)
844. (A)
845. (A)
846. (B)
847. (A)
848. (D)
849. (C)
850. (B)
851. (A)
852. (A)
853. (D)
854. (D)
855. (D)
856. (D)
857. (C)
858. (B)
859. (A)
860. (C)
861. (D)
862. (D)
863. (A)
864. (A)
865. (A)
866. (D)
867. (B)
868. (A)
869. (C)
870. (C)
871. (D)
872. (B)
873. (A)
874. (A)
875. (C)
876. (B)
877. (D)
878. (C)
879. (C)
880. (C)
881. (C)
882. (D)
883. (A)
884. (B)
885. (D)
886. (C)
887. (B)
888. (A)
889. (D)
890. (D)
891. (B)
892. (C)
893. (A)
894. (A)
895. (A)
896. (B)
897. (A)
898. (A)
899. (A)
900. (A)
901. (C)
902. (C)
903. (A)
904. (D)
905. (B)
906. (B)
907. (B)
908. (A)
909. (A)
910. (B)
911. (A)
912. (D)
913. (A)
914. (D)

NITROGEN FAMILY - 15 th GROUP

N & P-non metals; As, Sb-metalloids; Bi-metallic.

1. **Atomic and Ionic Sizes:** don't increase regularly as expected because the d-electrons do not shield the nuclear charge effectively.

2. **I.P.:** Decreases

3. **Valency:** N exhibit -3 to +5 valency. Others +3 and +5 mainly.

4. **m.p. and b.p.s:** not regular as other properties. Many of them exist as polyatomic molecules e.g. N≡N; P_4 and As_4. b.p. C: 4620 (N -78); Si: 2873°C (P 533.5)

5. **Catenation:** decreases. N_2 forms compounds containing a maximum of 8 nitrogen atoms.

6. **Allotropy:** Red phosphorus has the least reactivity.

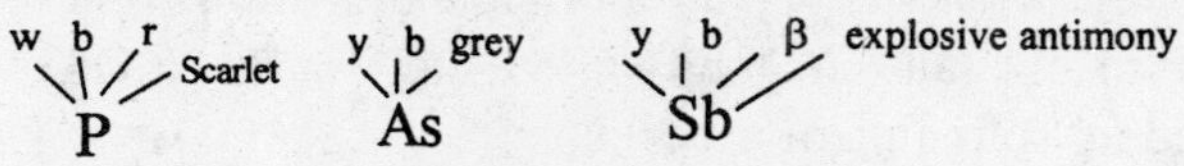

7. **Reactivity:** N with EN of 3.0 must be expected to be one of the most active elements but reacts with other elements only under extreme temp & pressure because of N-N triple bond which needs very high energy to be decomposed- 946 $kJmol^{-1}$.

8. **Atomicity:** N-diatomic; P, As, Sb all exist as discrete tetra atomic tetrahedral molecules.

9. **Hydrides :**

-Basic character : $NH_3 > PH_3 > AsH_3 > SbH_3$

-Thermal stability : $NH_3 > PH_3 > AsH_3 > SbH_3 > BiH_3$

-Bond angle : $NH_3 > PH_3 > AsH_3 > SbH_3$

Bond angle is directly proportional to E.N. of the central atom. It is also due to the repulsion between nonbonding and bond pair electrons.

-Boiling point : $NH_3 > PH_3 > AsH_3 > SbH_3$

-Reducing strength : $NH_3 < PH_3 < AsH_3 < SbH_3$

10. **Oxides :** Oxides of N and P are strongly acidic; As & Sb - amphoteric; Bi-basic.

-Common oxidation states : -3,+3,+5 generally important. The oxidation state of nitrogen is shown in the bracket :

$N_2O(+1)$, $NO(+2)$, $N_2O_3(+3)$, $N_2O_4(+4)$, $N_2O_5(+5)$, $N_2(0)$, $NH_2OH(-1)$, $N_2H_4(-2)$, $NH_3(-3)$.

In +5 state, these elements behave predominantly as non metals. The tendency to form covalent bonds increases with increasing oxidation state.

R_2O_3 type: form -ous acids in water; acidity decreases down the group. Except As, all others form RO_2 oxides also.

N_2O_3	P_2O_3	As_2O_3	Sb_2O_3	Bi_2O_3
↓ H_2O	↓	↓	↓	↓
HNO_2	H_3PO_3	H_3AsO_3	H_3SbO_3	insoluble

R_2O_5 type: forms -ic acids in water

N_2O_5	P_2O_5	As_2O_5	Sb_2O_5	Bi_2O_5
↓ H_2O	↓	↓	↓	↓
HNO_3	H_3PO_4	H_3AsO_4	H_3SbO_4	H_3BiO_4

P_2O_5 is a dehydrating agent: $CH_3CONH_2 \xrightarrow{P_2O_5} CH_3CN$

11. **Halides:** Form tri and pentahalides. Nitrogen cannot form pentahalides due to the absence of d-orbitals. NF_3 is the only stable halide of nitrogen. Except NF_3 & PF_3 all others undergo hydrolysis.

$$NCl_3 + 3H_2O \longrightarrow NH_3 + 3HOCl;$$
$$PCl_3 + 3H_2O \longrightarrow H_3PO_3 + 3HCl$$

The unstable nature of NCl_3, NBr_3, NI_3 is due to low polarity of N-X bond and a large difference in size of N and halogens.

12. **Oxyacids:** Nitrogen : HNO: hyponitrous acid; HNO_2: nitrous acid; HNO_3: nitric acid

HNO_4: peroxynitric acid

Stability : $HNO_3 > HNO_2$; acidity : $HNO_3 > HNO_2$

HNO_2 is a strong reducing agent also acts as an oxidising agent. It is also a complex forming agent.

$$2HNO_2 \longrightarrow H_2O + 2NO + O$$

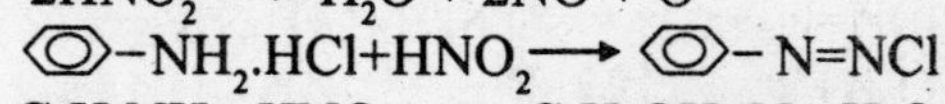

$$C_2H_5NH_2 + HNO_2 \longrightarrow C_2H_5OH + N_2 + H_2O$$
$$H_2N\text{-}NH_2 + HNO_2 \longrightarrow \underset{\text{hydrazoic acid}}{N_3H} + 2H_2O$$

HNO_2 exists as a tautomeric mixture of -

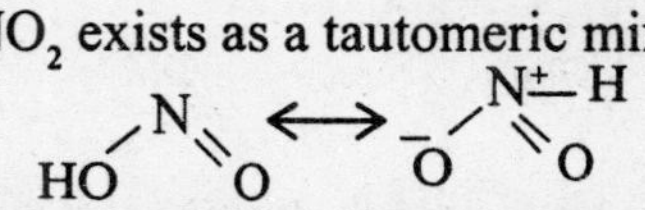

HNO_3 (i) Birkland and Eyde Process :

$N_2 + O_2 \rightleftharpoons 2NO$; $\Delta H = 180.6\ k.J.mol^{-1}$; endothermic; high temp. favourable.

$$2NO + O_2 \longrightarrow 2NO_2;$$
$$4NO_2 + 2H_2O + O_2 \longrightarrow 4HNO_3$$

(ii) **Ostwald's Process:** (Now invariably manufactured by this process).

$$4NH_3 + 5O_2 \underset{}{\overset{pt\text{-}rh}{\rightleftharpoons}} 4NO + 6H_2O + 21.6\ k.cal.$$
$$2NO + O_2 \longrightarrow 2NO_2 + 27\ k.cal.$$
$$3NO_2 + H_2O \longrightarrow 2HNO_3 + NO + 32\ k.cal.$$

Fuming HNO_3: Obtained by distilling the conc. acid with starch

Conc.HNO_3 + starch $\longrightarrow NO_2 \xrightarrow{\text{dissolved in } HNO_3}$ strong oxidising agent

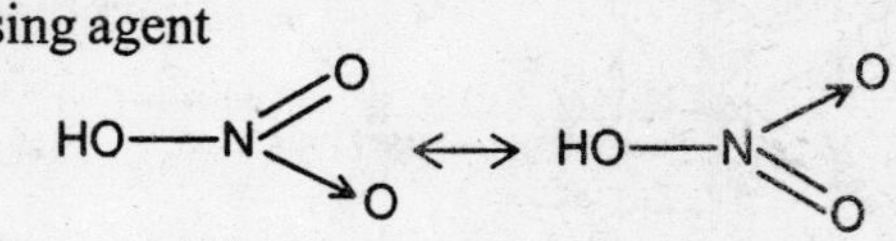

(a) Reaction of HNO_3 with less reactive metals :

Cu, Ag, Hg, Pb ⟶ conc. NO_2 ; dil. NO ; v.dil. N_2O

(b) More reactive metals :

Zn, Fe, Sn ⟶ dil. ⟶ N_2O ; cold & v.dil. ⟶ NH_4NO_3 or NH_2OH

(c) Mg & Mn with (very dil) cold and dilute acid liberate H_2

(d) Noble metals : Au, Pt, Rh, Ir dissolve in aquaregia, HCl + HNO_3 (3:1)

Aquaregia forms nascent chlorine which attacks these metals.

$$HNO_3 + 3HCl \longrightarrow NOCl + 2H_2O + 2Cl;$$

$Au + 3Cl \longrightarrow AuCl_3 \xrightarrow{HCl} HAuCl_4$

- HNO_3 + protein $\longrightarrow$ Xanthoprotein (yellow coloured)

Oxy acids of Phosphorus :

Hypophosphorus acid H_3PO_2 (mono basic)

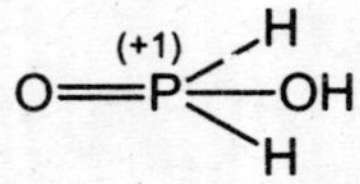

Phosphorous acid H_3PO_3 (dibasic)

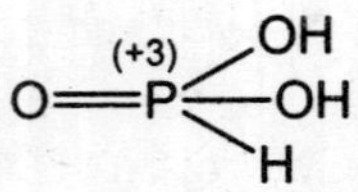

Orthophosphoric acid H_3PO_4 (Tribasic)

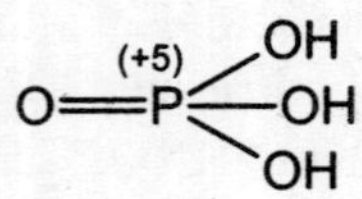

Metaphosphoric acid HPO_3 (Monobasic)

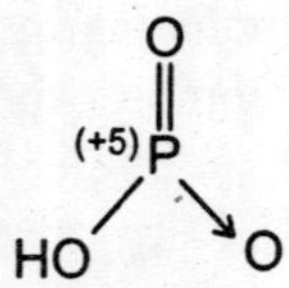

Hypophosphoric acid $H_4P_2O_6$ (Tetrabasic)

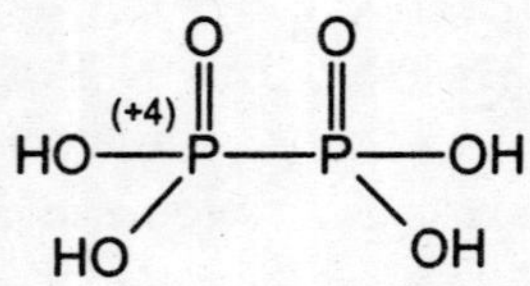

Pyrophosphoric acid $H_4P_2O_7$ (Tetrabasic)

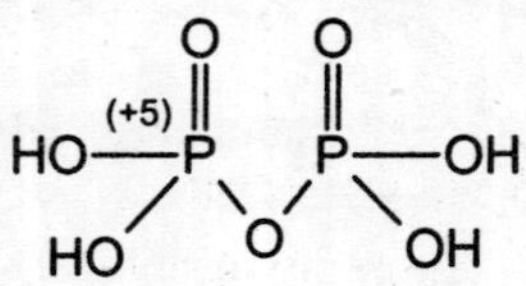

Peroxomonophosphoric acid H_3PO_5 (Tribasic)

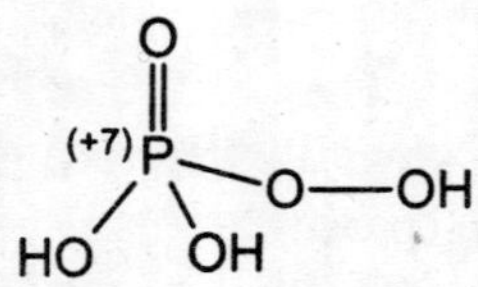

-$H_3PO_4 + 12(NH_4)_2MoO_4 + 21HNO_3 \longrightarrow (NH_4)_3PO_4 \cdot 12MoO_3 + 21\ NH_4NO_3 + 12H_2O$

ammonium phosphomolybdate (canary yellow)

13. **Nitrides:** Nitrides with transitional metals e.g. Fe,Mn,W,Mo etc. are interstitial compounds.

Oxides of Nitrogen:

N_2O, NO, N_2O_3, NO_2, N_2O_5

N_2O nitrous oxide : Laughing gas (hysterical laughter)

$NH_4NO_3 \xrightarrow{\Delta} 2H_2O + N_2O$;

$NaNH_2 + N_2O \longrightarrow NaN_3 + H_2O$ (sodium azide)

$\overset{-}{N}{=}\overset{+}{N}{=}O \longleftrightarrow N{\equiv}\overset{+}{N}{-}\overset{-}{O}$

NO: Nitric Oxide; paramagnetic

NO reacts with Fe^{+2} salts to form complex

$FeSO_4 + NO + 5H_2O \longrightarrow [Fe(NO)(H_2O)_5]SO_4$

in brown ring test, hydrated nitrosyl complex forms.

$N_2O_4 \rightleftharpoons 2NO_2$; NO_2: paramagnetic, brown colour gas

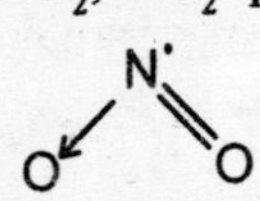

N_2O_5:anhydrideof HNO_3;Ionic solid $NO_2^+\ NO_3^-$. Its vapours contain symmetrical molecule

$O_2N{-}O{-}NO_2$;

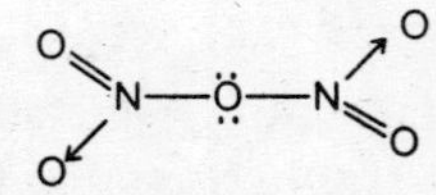

pentoxides are more acidic than trioxides.

NH_3:

(i) $Mg_3N_2 + 6H_2O \longrightarrow 3Mg(OH)_2 + 2NH_3$

(ii) $H_2NCONH_2 + H_2NCONH_2 \xrightarrow{\Delta} H_2NCONHCONH_2 + NH_3$ (biuret)

Laboratory preparation:

$2NH_4Cl + Ca(OH)_2 \longrightarrow 2NH_3 + CaCl_2 + 2H_2O$

Haber's Process:

$3H_2 + N_2 \xrightarrow[Mo]{Fe} 2NH_3$-High pressure& low temp.favourable

NH_3-pyramidal shape; NH_4^+ - Tetrahedral

$2K_2HgI_4 + NH_3 + 3KOH \longrightarrow H_2NHgOHgI + 7KI + 2H_2O$

Iodide of Millon's base

- Graham's salt (Calgon) $Na(PO_3)_6$ or $Na_2[Na_4P_6O_{18}]$ is used for softening of water
- White phosphorus shows phosphorescence and with hot NaOH, PH_3 is formed.

PH_3 is used for making Holme's Signals.

- Scheele's green $HCuAsO_3$ (Cupric arsenate)
- Tartar Emetic, potassium antimonyl tartarate

CH(OH)COOK

CH(OH)COO(SbO).$\frac{1}{2}H_2O$

It is used in the treatment of Kala-Zar.

- $Ca_3(PO_4)_2 + 2H_2SO_4 + 5H_2O \longrightarrow 2CaSO_4.2H_2O + Ca(H_2PO_4)_2.H_2O$

Calcium superphosphate (Super Phosphate of Lime)

- Paris green $(CH_3COO)_2Cu.3Cu(AsO_2)_2$ or $Cu_4(CH_3O)_2(AsO_3)_2$
- Sb_2S_3 - used in safety matches
- Red P is applied to the sides of match box.
- PCl_5, chlorinating agent in organic chemistry
- NH_4NO_3 mainly used for making explosives such as -

Amatol : NH_4NO_3+TNT

Ammonal : NH_4NO_3+Al powder

- As_2O_3 for killing rats
- N_2, CO, CN^- - isoelectronic
- persons working with phosphorus develop Phossy Jaw disease.
- P^{32} is employed in the treatment of leukemia and other blood disorders.

OXYGEN FAMILY - 16th GROUP

O, S, Se, Te - Chalcogens (ore forming)

Se and Te are semi conductors. Oxygen most abundant (46.6%) in the earth crust. Po is radioactive and short lived and is distinctly metallic.

1. **Atomic Size:** Increases
2. **Electronic Configuration :** ns^2p^4
3. **I.P.:** do not form cations.
4. **E.A.:** S > Se > O
5. **Oxidation state:** Valency of oxygen never exceeds -2. Oxygen shows positive oxidation state in O_2F_2, +1;OF_2, + 2; in peroxides H_2O_2, -1. O has no d-orbitals.

Sulphur G.S. 

$SCl_2 \longrightarrow sp^3$

$SCl_4 \longrightarrow sp^3d$ Trigonal bipyramidal

$SCl_6 \longrightarrow sp^3d^2$ Octahedral

In +4 state, the elements show both oxidising and reducing properties. But in +6 state, they are only oxidising agents.

6. **Magnetic property :** O is paramagnetic (others diamagnetic) due to the presence of unpaired electrons.

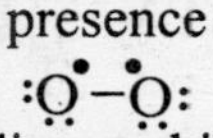

7. **Bonding :** p_π-d_π bonding prohibits the polymerisation of oxyanions. In silicates, there is little p_π-d_π bonding so that the SiO_4 units polymerise into enormous variety of structures linked by Si-O-Si bonds. Si<P<S<Cl

8. **Catenation:** O and S can catenate. Sulphur forms a puckered ring of 8 atoms. O-diatomic.

9. **Allotropy:** O,S,Se,Te exhibits. O_2, O_3; S rhombic, monclinic, plastic sulphur

$$S_{rhombic} \underset{}{\overset{95.6^\circ C}{\rightleftharpoons}} S_{monoclinic}$$

10. **Hydrides :**

Acidicty: $H_2O<H_2S<H_2Se<H_2Te$

Reducing character: $H_2S<H_2Se<H_2Te$

H_2O is not a reducing agent.

Bond angle : $H_2O>H_2S>H_2Se>H_2Te$

Boiling point : $H_2O>H_2Te>H_2Se>H_2S$

Thermal stability : $H_2O>H_2S>H_2Se>H_2Te$

11. **Oxides:** MO_2, MO_3 types are well known MO_2 type most stable.

Dioxides: SO_2 ; SeO_2 ; TeO_2

$\downarrow$ H_2SO_3 ; H_2SeO_3 ; H_2TeO_3

SO_2 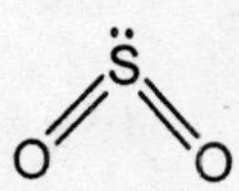 sp^2

Trioxides :

SO_3 ; SeO_3 ; TeO_3

$\downarrow$ H_2SO_4 ; H_2SeO_4 ; H_2TeO_4

SO_3 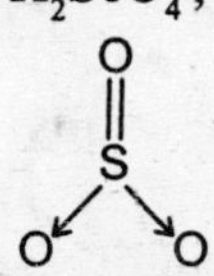 sp^2

SO_3^{2-} Pyramidal

Acidic Strength : $H_2SO_3>H_2SeO_3>H_2TeO_3$;

$H_2SO_4>H_2SeO_4>H_2TeO_4$

Oxidising Capabilities : $H_2TeO_4>H_2SeO_4>H_2SO_4$

Acidic Oxides : CO_2, B_2O_3, SiO_2, NO_2, N_2O_5, P_4O_{10}, SO_2, Cl_2O_7, I_2O_5, CrO_3

Neutral : CO, H_2O, N_2O, NO

Amphoteric : ZnO, Al_2O_3, BeO, Sb_2O_3, Cr_2O_3, PbO, As_2O_3, SnO, SnO_2 etc.

Suboxides : C_3O_2, Pb_2O, N_2O (% of oxygen less than present in normal oxides)

Superoxides : KO_2, CsO_2, (O_2^- ion)

12. **Oxyacids :**

H_2SO_3 Sulphurous acid

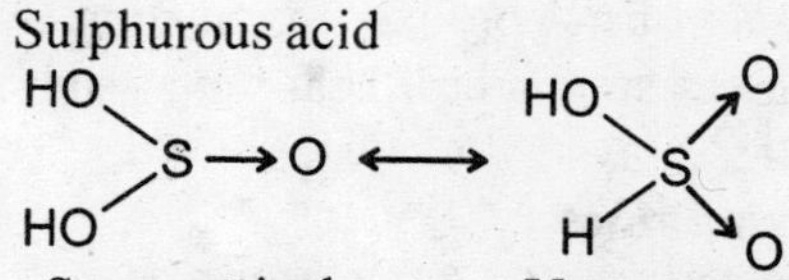

Symmetrical (more evidence in favour)   Unsymmetrical

H_2SO_4 - Sulphuric acid 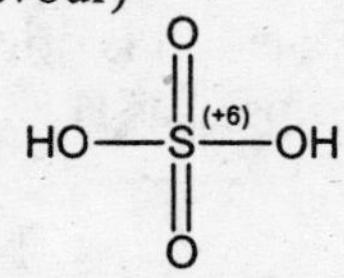

-Oil of vitriol

-$H_2SO_4 \longrightarrow H_2O + SO_2 + O$ (oxidising nature)

-Pickling agent : H_2SO_4 is used to remove basic oxide from the layers before electroplating, galvanising, soldering etc.

H_2SO_5 - Caro's acid

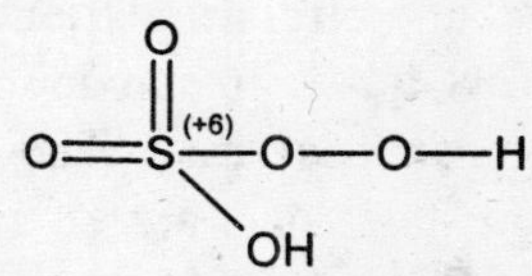

peroxymono-sulphuric acid

(Caros acid)

$H_2S_2O_7$ - Oleum ($H_2SO_4 + SO_3$); Disulphuric acid (pyrosulphuric acid)

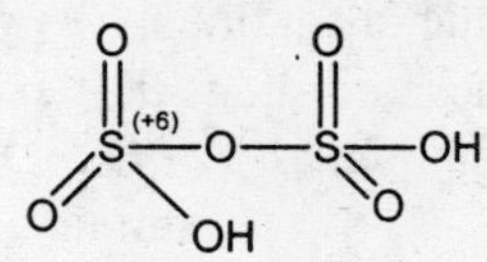

$H_2S_2O_8$ - Marshall's acid 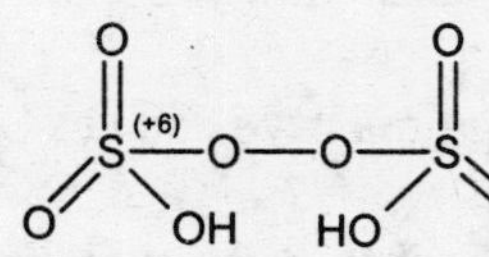

$H_2S_2O_3$ -Thiosulphuric acid 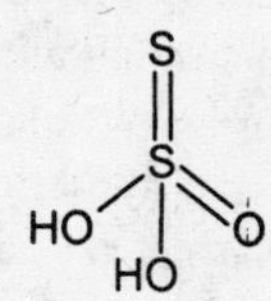

$H_2S_2O_6$ - Dithionic acid 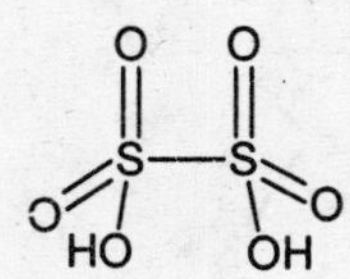

Sulphuric Acid, H_2SO_4 :

(a)Contact Process :

$$2SO_2+O_2 \overset{V_2O_5}{\rightleftharpoons} 2SO_3 + 46\ \text{k.cal.}$$

Excess of O_2; low temp favourable.

SO_3 is absorbed in H_2SO_4 producing $H_2S_2O_7$(oleum)

$H_2S_2O_7 + H_2O \longrightarrow 2H_2SO_4$

(b)Lead Chamber Process :

$SO_2 + NO_2 \rightleftharpoons SO_3 + NO + \text{Heat}$

NO_2 not only acts as a catalyst but also as oxygen carrier.

$SO_3 + H_2O \longrightarrow H_2SO_4$

It is cheaper to produce H_2SO_4 by this method.

-H_2SO_4 should be added to water for diluting

purposes.

-Conc H_2SO_4 acid can't be used for drying hydrogen (conc H_2SO_4 gives so much of heat with water that the hydrogen will ignite).

13. **H_2S and sulphides :**

In Laboratory : $FeS + \text{dil. } H_2SO_4 \longrightarrow H_2S + FeSO_4$

Pure H_2S : $Sb_2S_3 + 6HCl \longrightarrow 2SbCl_3 + 3H_2S$

H_2S is a weak acid; It is dried over $CaCl_2$ or P_2O_5 while H_2SO_4 can't be used as H_2SO_4 is reduced by H_2S.

-$(NH_4)_2S + H_2S \longrightarrow (NH_4)_2S_{x+1} + xH_2$

poly yellow ammonium sulphide

-Sulphides precipitated in acidic medium :

Hg,Ag,Pb,Cu,Bi,Cd,As,Sb & Sn.

-Sulphides precipitated in alkaline medium :

Zn,Co,Ni,Mn & Fe.

-H_2S acts as reducing agent, while SO_2 can act as a reducing agent and an oxidising agent. In H_2S, the oxidation state of S is -2. It can increase only its oxidation state and can't decrease. The maximum and minimum oxidation states of S are +6 and -2. In SO_2, oxidation state of S is +4. It can be increased as well as decreased.

14.**Ozone:** Upper layer of atmosphere contains a little larger quantity (due to exposure of air to UV light); Sea shores also (due to continuous evaporation of H_2O).

Ozone becomes poisonous if it exceeds $1:2 \times 10^6$.

Oxidation Potential : $F > O > OF_2 > O_3$

atomic

Heavier than air; liquid O_3 paramagnetic; strong bleaching agent

Oxidising property is due to $O_3 \longrightarrow O_2 + O$

nascent

-Oxygen is absorbed by alkaline pyrogallol

Tailing of Hg : In contact with O_3, Hg looses its miniscus and starts sticking on the glass surface

$2Hg + O_3 \longrightarrow Hg_2O + O_2$

-Moist iodine $I_2 + H_2O + 5O_3 \longrightarrow 2HIO_3 + 5O_2$

Iodic acid

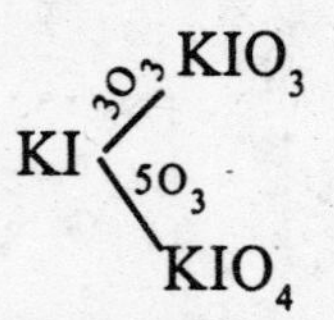

HALOGENS - 17th GROUP

Astatine is rare and artificially prepared. I_2 can be obtained from sea weeds (Laminaria species). Due to weak intermolecular forces the halogens are volatile in nature.

1.Colour : Due to absorption of visible light they are coloured. F(Py), Cl (Gy), Br(Or), I (Violet).

2. **Electronic Configuration :** ns^2np^5

3. **Electronegativity :** F highest.

4. **E.A. :** Cl > F > Br > I

5.**Oxidation State:** Normally -1 oxidation state. Due to the absence of 'd' orbitals, F can't have more than one unpaired electron. Other halogens can also exhibit +1, +3, +5 and +7 states.

6.**Oxidising Power:** F extremely powerful oxidising agent. F reacts with H_2O to give HF not HOF. F decomposes water readily even at low temp. and in dark forming a mixture of $O_2 + O_3$.

7. **Bond Dissociation Energy :** $Cl_2 > Br_2 > F_2 > I_2$

The low dissociation energy of F_2 is due to high inter electron repulsions between non bonding electrons in the 2p orbitals.

8. **Reducing Power :** $F^- < Cl^- < Br^- < I^-$. F^- doesn't show any reducing nature.

9. **Reactivity :** F > Cl > Br > I. F reacts with hydrogen in dark, Cl in sunlight, Br on heating and I only in the presence of catalyst. More reactive halogen atom can substitute less reactive one from its ionic compounds.

10. **Acids :** Acidic strength : HI>HBr>HCl>HF,(due to their dissociation constants HI & HBr are fully ionised. HCl & HF are least ionised.)

Boiling point: H_2F_2>HI>HBr>HCl (H_2F_2 in vapour state)

Heat of formation: HF > HCl > HBr > HI

Bond length: HF < HCl < HBr < HI

Solubility in water at 0^0C: HF>HI>HBr>HCl

11.**Oxyacids:** Increasing oxygen content leads to

(a) increase in thermal stability
$HClO_4 > HClO_3 > HClO_2 > HClO$

(b) increase in the strength of the acid
$HClO_4 > HClO_3 > HClO_2 > HClO$

(c) a decrease in oxidising power
$HOCl > HOCl_2 > HOCl_3 > HOCl_4$

Acidity: $HClO_3 > HBrO_3 > HIO_3$

F doesn't form oxyacids due to the absence of d-orbitals.

I_2 reacts with conc. nitric acid to give HIO_3 while other halogens do not react.

$I_2 + 10\,HNO_3(\text{conc}) \longrightarrow 2HIO_3 + 4H_2O + 10NO_2$

12. **Reaction with glass:** HF can't be stored in glass bottles.

HF is used for etching of glass.

$SiO_2 + 4HF \longrightarrow SiF_4 + 2H_2O$

$SiF_4 + 2HF \longrightarrow H_2SiF_6$ (hydrofluosilicic acid)

13.**Interhalogen Compounds:** AB-ClF, IBr; AB_3-ICl_3, ClF_3; AB_5-IF_5; AB_7-IF_7 types possible.

14.**Polyhalides:** The ability of halide ion in solution or in solid state to dissociate with molecular halogens or interhalogen compounds C S I Br Cl; $KI_3.H_2O$ etc.is possible.

15. **Action with $AgNO_3$:** AgF is not precipitated due to its abnormal solubility in H_2O.

16.**Action with $CaCl_2$:** With HF white precipitate of CaF_2 is formed. HCl, HBr, HI give no ppt.

17. **With $Pb(CH_3COO)_2$:** HF no ppt. Others form ppt.

18.**Pseudohalogens:** Each halide ion has a corresponding halogen. In the same manner some others which exist are as follows:

CN^-	SCN^-	$SeCN^-$	$SCSN_3^-$
$(CN)_2$	$(SCN)_2$	$(SeCN)_2$	$(SCSN_3)_2$
cyanogen	thiocyanogen	seleno-cyanogen	azido carbon disulphide

19. **Oxides:** OF_2, O_2F_2 (oxygen flourides)

OF_2 OCl_2 OBr_2 sp^3

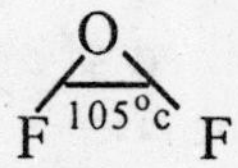

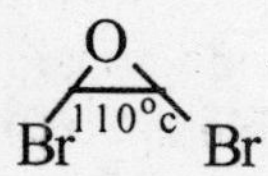

Chlorine :

1. $NaClO$ (Sod. hypochlorite) + $2HCl \longrightarrow NaCl + H_2O + Cl_2$
2. $CaOCl_2 + H_2SO_4 \longrightarrow CaSO_4 + H_2O + Cl_2$
   Cal. hypochlorite (bleaching powder)
3. $MnO_2 + 4HCl \longrightarrow MnCl_2 + 2H_2O + Cl_2$
   (Weldon's Process)
4. $4HCl + O_2 \xrightarrow[400^\circ c]{CuCl_2} 2H_2O + 2Cl_2$ (Deacon's Process)
5. Electrolytic Process : From NaCl (Sea water)
   (Down's Process)

- Bleaching property of chlorine is due to nascent O.

$H_2O + Cl_2 \longrightarrow HCl + HOCl$; $HOCl \longrightarrow HCl + O$

- **Bleaching Powder**: $Ca(OH)_2 + Cl_2 \longrightarrow CaOCl_2$
  slaked lime (Beckmann's Process)

Bleaching powder is prepared by Hasenclever method also.

Oxidising property :

$2FeSO_4 + CaOCl_2 + H_2SO_4 \longrightarrow Fe_2(SO_4)_3 + CaCl_2 + H_2O$

$Na_3AsO_3 + CaOCl_2 \longrightarrow Na_3AsO_4 + CaCl_2$

HCl, HBr, HI :

HCl : $2NaCl + H_2SO_4 \longrightarrow Na_2SO_4 + 2HCl$

HBr and HI can't be prepared by this method because HBr and HI are strong reducing agents and reduce H_2SO_4 to SO_2 and get themselves oxidised to Br_2 and I_2. Therefore they are prepared as follows:

$3KBr + H_3PO_4 \longrightarrow K_3PO_4 + 3HBr$

$3KI + H_3PO_4 \longrightarrow K_3PO_4 + 3HI$

Available Chlorine : The amount of chlorine obtained from a sample of bleaching powder by treatment with excess of dilute acids or CO_2 is called available chlorine.

-$2Na_2S_2O_3 + I_2 \longrightarrow 2NaI + Na_2S_4O_6$
hypo Sod. tetrathionate

-Freon CF_2Cl_2 is used in refrigerators and cold storage plants.

-Phosgene $COCl_2$.

-I_2 is used for treating Goitre.

-UF_6 is used to separate the isotopes of Uranium.

-ICl (Iodine monochloride) : used for the estimation of iodine number of oils and fats.

-AgBr is used in photography.

-Tear gas O_2NCCl_3

-I_2O_5 is used for the estimation of CO in the atmosphere.

-Tincture Iodine : I_2 + KI + Rectified spirit.

NOBLE GASES - 18th GROUP

The possible existence of noble gases couldn't be predicted by Mendeleev. The volume of inert gases consitute around 1% of atmosphere. Argon is the principle constituent of air among inert gases.

Ar	:	Lazy
Kr	:	Hidden one
Xe	:	Stranger

1. **Electronic Configuration** : ns^2p^6
2. **I.E.** : Decreases
3. **E.A.** : Nearly zero
4. **b.p.** : b.p. of He lowest of any known element.
5. **Atomic radii** : The radii for noble gases are only the van der Waal's radii.
   Molecular mass = Atomic mass
   Mole mass = 2 x V.P.
   Therefore, atomic mass = 2 x V.P.
6. **Forces of attraction between atoms of noble gases:** Noble gases are monoatomic Cp/Cv = 1.66. The fact that these gases can be liquified shows the existence of some forces of attraction between the atoms (Electrostatic attraction : Nuclei with the electrons of the same and the other. Electrostatic repulsion : repulsion between nuclei and between electrons respectively. The balance of these forces is van der Waal's forces).
   Higher the van der Waal's forces, higher the boiling point. The van der Waal's forces are proportional to the polarisability and inversely proportional to the ionisation potentials of the atoms. The polarisability increases down the group.
7. **Chemical behaviour :**

(a) Compounds formed under excited conditions.

(i) In a glow discharge tube $HgHe_{10}$ is formed.

(ii) Electron bombardment of Tungsten in He atmosphere forms WHe_2.

(iii) Some metals when employed as electrodes in discharge tube containing inert gases Pt_3He, PdHe, BiHe etc. are formed. They are considered as the adsorption compounds of the inert gases on the dispersed metal.

(b) Clathrate compounds : These are compounds in which the noble gases are trapped or caged within holes in the crystalline lattices of suitable organic or inroganic compound e.g. o-quinol

He and Ne don't form clathrates because of their smaller size which don't fit into the cage. Ar, Kr and Xe form clathrates: $Ar(C_6H_6O_2)_3$; $Kr(C_6H_6O_2)_4$; $Xe(C_6H_6O_2)_{3.3}$.

Clathrates are normally non-stoichiometic compounds.

The noble gas hydrates : They are in reality clathrate compounds. When H_2O is solidified in the presence of Ar, Kr or Xe, H_2O forms a cubic structure in which 46 H_2O molecules are present per unit cell. Each unit cell contains 8 holes in which the noble gas atoms are trapped e.g. $Ar.6H_2O$; $Kr.6H_2O$; $Xe.6H_2O$

No chemical bonds are present in clathrate compounds.

The molecules are being simply caged by the host lattice.

Compounds of noble gases :

Pauling (1933) : Heavier noble gases Kr, Xe and Rn are likely to form compounds with oxygen and fluorine because of large difference of electro negativities.

	Rn	Xe	O	F	
E.N	Rn (2.00)	Xe (2.25)	O(3.5)	F(4.00)	
I.P.	1037	1170	1410	1681	$kJmol^{-1}$

The first true compound of noble gases was prepared in 1962 by Neil Bartlett and Lohmann.

They first prepared a red solid by reaction between PtF_6 and O_2.

$$\underset{\text{deep red vapour}}{PtF_6} + O_2\ (1:1) \longrightarrow \underset{\text{red solid}}{O_2^+(PtF_6)^-}$$

Since the I.P. of O is close to that of Xe, Xe should also be oxidised by PtF_6.

$$\underset{\text{deep red vapour}}{PtF_6} + Xe \longrightarrow \underset{\text{yellow solid}}{Xe^+(PtF_6)^-}$$

Chemistry of Xenon :

$$Xe + F_2 \xrightarrow{1:2 (v/v)} XeF_2$$

$$Xe + F_2 \xrightarrow{1:5; 400°C; 6\ atm.} XeF_4$$

$$Xe + F_2 \xrightarrow{1:20; 300°C, 60\ atm} XeF_6$$

They are all solids. They are stored in nickel containers. XeF_2 and XeF_4 can be stored in quartz vessels also.But Xeon hexafluoride can't be as it reacts with quartz-

$$2XeF_6 + SiO_2 \longrightarrow 2XeOF_4 + SiF_4$$

- KrF_2 is the only known compound of Kr.

Reactions :

(i) As fluorinating agents

$$XeF_4 + C_6H_6 \longrightarrow Xe + C_6H_5F$$

$$XeF_4 + Pt \longrightarrow Xe + PtF_4$$

(ii) As Oxidising agent

$$XeF_4 + 4KI \longrightarrow Xe + 4KF + I_2$$

$$XeF_2 + \underset{Ce(III)}{Ce_2(SO_4)_3} \longrightarrow \underset{Ce(IV)}{Ce(SO_4)_2}$$

(iii) With H_2O

$$XeF_2 + H_2O \longrightarrow Xe + 2HF + \tfrac{1}{2}O_2$$

$$3XeF_4 + 6H_2O \longrightarrow 2Xe + XeO_3 + 12HF + 1\tfrac{1}{2}O_2$$

$$XeF_6 + 3H_2O \longrightarrow XeO_3 + 6HF$$

$$XeF_6 + H_2O(\text{partial hydrolysis}) \longrightarrow XeOF_4 + 2HF$$

Structure of Xenon compounds :

XeF_2	+2	Linear (true)	sp^3d	3	lone pairs Trigonal bipyramidal (Pseudo)
XeF_4	+4	Square planar	sp^3d^2	2	Octahedral
XeF_6	+6	distorted octahedral	sp^3d^3	1	Pentagonal bipyramidal
XeO_3	+6	pyramidal	sp^3	1	Tetrahedral
XeO_4	+8	Tetrahedral		0	

Sources of noble gases :

Helium : Sun (Jenson), atmosphere, natural gas, Clevite, Monozite, Pitch blende.

Isolation :

1.Chemico-physical method (Dewar's method) :

Adsorbed on activated coconut charcoal at different temperatures.

Coconut Charcoal at -100°C Ar, Kr, Xe get adsorbed

Coconut Charcoal at -180°C Ne adsorbed

2.By fractionation of liquid air (Rayleigh & Ramsay): Used for large scale preparation.

Uses :

He : Filling air ships and weather balloons

In diving apparatus: N_2 can't be used as it will dissolve in blood at high pressure, therefore ($He + O_2$) is used. Coisson or bends disease is common in sea divers. Used in the metallurgy of reactive metals like Ti; Welding of Mg, Al, etc.

Ne : Neon glow lamps are used for advertising purpose, in music beat indicators of deck, helipad, telephone tower etc. Orange red Neon tubes are employed in televisions and plug testers.

Ar : For filling incandescent metal filament electric bulbs.

Rn : In the treatment of cancer.

1. Which of the following is monoatomic ? [CBSE-90]
(A) Oxygen (B) Hydrogen (C) Helium (D) Fluorine

2. Which of the following is strongest acid ? [CBSE-90]
(A) HF (B) HI (C) HBr (D) HCl

3. How do we form pure N_2 gas ? [CBSE-90]
(A) $NH_3 + NaNO_2$ (B) $NH_4Cl + NaNO_2$
(C) $N_2O + Cu$ (D) $(NH_4)_2Cr_2O_7$

4. P_4O_6 is heated with H_2O to give [CBSE-90]
(A) hypophosphorous acid (B) orthophosphorous acid
(C) hypophosphoric acid (D) orthophosphoric acid

5. Which of the following is formed by sea salt ? [CBSE-90]
(A) Halogen (B) Chlorine (C) Oxygen (D) Sulphur

6. PH_4I + NaOH forms [CBSE-91]
(A) PH_3 (B) NH_3 (C) P_4O_6 (D) P_4O_{10}

7. Which of the following has more oxidising power ? [CBSE-91]
(A) Br (B) Cl (C) I (D) As

8. Oleum is [CBSE-91]
(A) Castor oil (B) Oil of vitriol
(C) Fuming H_2SO_4 (D) None of these

9. The reaction [CBSE-91]
$2SO_2 + O_2 + 2H_2O \longrightarrow 2H_2SO_4$ is an example of
(A) synthesis (B) analysis
(C) displacement (D) double decomposition

10. Chloride of metal is MCl_2. The formula of its phosphate will be [CBSE-91]
(A) M_2PO_4 (B) $M_3(PO_4)_2$ (C) $M(PO_4)_3$ (D) MPO_4

11. Basicity of orthophosphoric acid is [CBSE-91]
(A) 2 (B) 3 (C) 4 (D) 5

12. Number of electrons shared in the formation of nitrogen molecule is [CBSE-92]
(A) 6 (B) 10 (C) 2 (D) 8

13. Elements of which one of the following groups will form anions most readily ? [CBSE-92]
(A) Oxygen group (B) Nitrogen group
(C) Halogens (D) Alkali metals

14. If the valency shell electronic structure for an element is $ns^2 np^5$ This element will belong to the group of [CBSE-92]
(A) alkali metals (B) alkaline earth metals
(C) noble gases (D) halogens

15. Which one of the following would quickly absorb oxygen ? [CBSE-92]
(A) Alkaline solution of pyrogallic acid
(B) Concentrated sulphuric acid
(C) Lime water (D) Alkaline solution of copper sulphate

16. When chlorine is passed over dry slaked lime at room temperature, the main reaction product is [CBSE-92]
(A) $Ca(ClO_2)_2$ (B) $CaCl_2$ (C) $CaOCl_2$ (D) $Ca(OCl)_2$

17. In the manufacture of bromine from sea water, the mother liquor containing bromide is treated with [CBSE-92]
(A) carbon dioxide (B) chlorine
(C) iodine (D) sulphur dioxide

18. Nitrogen is a relatively inactive element because [CBSE-92]
(A) its atom has a stable electronic configuration
(B) it has a low atomic radius
(C) its electronegativity is fairly high
(D) dissociation energy of its molecule is fairly high

19. A solution containing ammonium chloride and sodium nitrite on boiling produces [CBSE-92]
(A) ammoina (B) nitric oxide (C) nitrogen (D) nitrous oxide

20. Ammonium dichromate on heating gives [CBSE-93]
(A) N_2 (B) N_2O (C) NO (D) NO_2

21. A solution of potassium bromide is treated with each of the following. Which one would liberate bromide ? [CBSE-93]
(A) Hydrogen iodide (B) Sulphur dioxide
(C) Chlorine (D) Iodine

22. A solution containing NH_4Cl and sodium nitrite on boiling produces [CBSE-93]
(A) Nitrogen peroxide (B) Ammonium
(C) Nitrogen (D) Nitrous oxide

23. Nitrogen can be purified from the impurities of oxides of nitrogen and ammonia by passing through.[CBSE-93]
(A) a solution of $K_2Cr_2O_7$, acidified with H_2SO_4
(B) conc. HCl (C) a solution of KOH
(D) alkaline solution of pyrogallol

24. Which of the following fertilizer has the highest Nitrogen percentage ? [CBSE-93]
(A) Ammonium sulphate (B) Calcium cynamide
(C) Urea (D) Ammonium nitrate

25. Which of the following sets have strongest tendency to form anions ? [CBSE-93]
(A) Ga, In, Tl (B) Na, Mg , AI
(C) N, O, F (D) V, Cr, Mn

26. Some Cl_2 is liberated when we heat [CBSE-93]
(A) $K_2Cr_2O_7$ and MnO_2 (B) $K_2Cr_2O_7$ + HCl + MnO_2
(C) $KMnO_4$+ NaCl (D) $Pb_2(NO_3)_4 + MnO_2$

27. Strongest acid from the following is [CBSE-94]
(A) HBr (B) H_2S (C) HF (D) H_3P

28. Which of the following will displace Br_2 from an acqueous solution containing bromide ions ? [CBSE-94]

(A) Cl_2 (B) I_3^- (C) I_2 (D) Cl^-

29. About 20km above the earth there is an ozone layer, each one of the following statements about ozone and ozone layer is true ? [CBSE-95]
(A) Ozone layer is beneficial to us because ozone cuts out the ultraviolet radiation of the sun.
(B) The conversion of ozone to oxygen is an endothermic reaction
(C) Ozone has a triatomic linear molecule.
(D) Ozone layer is harmful to us because it cuts out a radiation useful for photosynthesis.

30. Which of the following represents calcium chlorite ? [CBSE-96]
(A) $CaClO_2$ (B) $Ca(ClO_4)_2$
(C) $Ca(ClO_3)_2$ (D) $Ca(ClO_2)_2$

31. One of the following has greatest electron affinity. Identify it [CBSE-96]
(A) F (B) Cl (C) I (D) Br

32. Among the following oxides, the lowest acidic is [CBSE-96]
(A) P_4O_6 (B) P_4O_{10} (C) As_4O_6 (D) As_4O_{10}

33. Reducton of thiosulphate with iodine gives [CBSE-96]
(A) sulphate ion (B) sulphite ion.
(C) tetrathionate ion (D) sulphide ion.

34. The basic character of hydrides of the 15th group elements decreases in the order : [CBSE-96]
(A) $SbH_3>PH_3>AsH_3>NH_3$ (B) $NH_3>SbH_3>PH_3>AsH_3$
(C) $NH_3>PH_3>AsH_3>SbH_3$ (D) $SbH_3>AsH_3>PH_3>NH_3$

35. Among the following which one is paramagnetic? (At.no.: Be=4, Ne=10, As=33, Cl=17) [CBSE-98]
(A) Be^- (B) Ne^{+2} (C) Cl^- (D) As^+

36. A one litre flask is full of bromine vapour. The intensity of brown colour of vapour will not decrease appreciably on adding to the flask some. [CBSE-98]
(A) pieces of marble (B) carbon tetrachloride
(C) animal charcoal powder (D) carbon disulphide

37. Repeated use of which of the following fertilizer would increase the acidity of soil ? [CBSE-98]
(A) Urea (B) Potassium nitrate
(C) Ammonium sulphate (D) Superphosphate of lime

38. Which of the following oxides is the most acidic? [CBSE-99]
(A) N_2O_5 (B) P_2O_3 (C) As_2O_3 (D) Sb_2O_5

39. Which of the following phosphorus is the most reactive? [CBSE-99]
(A) Red phosphorus (B) White phosphorus
(C) Scarlet phosphorus (D) Violet phosphorus

40. Which of the following is used in the preparation of chlorine ? [CBSE-99]
(A) Only MnO_2 (B) Only $KMnO_4$
(C) Both MnO_2 and $KMnO_4$
(D) Either MnO_2 and $KMnO_4$

41. The decomposition of organic compounds in the presence of oxygen and without the development of odoriferous substances is called. [CBSE-99]
(A) decay (B) N_2-fixation
(C) nitrification (D) denitrification

42. Which of the following at ordinary conditions of temperature and pressure exist as linear polymer molecule due to extensive intermolecular force ? [CBSE-2000]
(A) H_2O (B) HCl (C) HF (D) NH_3

43. Which is planar molecule ? [CBSE-2000]
(A) XeO_4 (B) XeF_4 (C) $XeOF_4$ (D) XeO_2F_2

44. Which one of the following pairs has the isostructural species ? [CBSE-2001]
(A) XeF_2, IF_2 (B) NH_3+ BF_3
(C) CO_3^{-2} , SO_3^{-2} (D) PCl_5, ICl_5

45. Nitrogen forms N_2, but as soon as phosphorus forms P_2 it is immediaetly converted into P_4. The reason is - [CBSE-2001]
(A) Triple bond present between phosphorus atoms
(B) $p\pi-p\pi$ bonding is weak
(C) $p\pi-p\pi$ bonding is strong
(D) Multiple bonds are formed easily

46. Which reaction is not feasible [CBSE-2002]
(A) $2KI + Br_2 \rightarrow 2\ KBr + I_2$
(B) $2\ KBr + I_2 \rightarrow 2\ KI + Br_2$
(C) $2\ KBr + Cl_2 \rightarrow 2\ KCl + Br_2$
(D) $2H_2O + 2F_2 \rightarrow 4\ HF + O_2$

47. Which of the following statements is true[CBSE-2002]
(A) silicon exhibits 4 coordination number in its compound
(B) bond energy of F_2 is less than Cl_2
(C) Mn(III) oxidation state is more stable than Mn(II)in aqueous state
(D) elements of 15th gp shows only +3 and +5 oxidation states.

48. Zn gives H_2 gas with H_2SO_4 & HCl but not with HNO_3 because [CBSE-2002]
(A) Zn act as oxidising agent when reacted with HNO_3
(B) HNO_3 is weaker acid than H_2SO_4 & HCl
(C) In electrochemical series Zn is above hydrogen
(D) NO_3^- is reduced in preference to hydronium ion

49. Which of the following statements is not true ? [CBSE-2003]
(A) HF is a stronger acid than HCl
(B) Among halide ions, iodide is the most powerful reducing agent

(C) Fluorine is the only halogen that does not show a variable oxidation state
(D) HOCl is a stronger acid than HOBr.

50. The oxidation state of sulphur in the anions SO_3^{2-}, $S_2O_4^{2-}$ and $S_2O_6^{2-}$ follow the order [CBSE-2003]
(A) $S_2O_6^{2-} < S_2O_4^{2-} < SO_3^{2-}$ (B) $S_2O_4^{2-} < SO_3^{2-} < S_2O_6^{2-}$
(C) $S_3^{2-} < S_2O_4^{2-} < S_2O_6^{2-}$ (D) $S_2O_4^{2-} < S_2O_6^{2-} < SO_3^{2-}$

51. A mole of N_2H_4 loses ten moles of electron to form a new compound Y. Assuming that all the nitrogen appears in the new compound, what is the oxidation state of nitrogen in Y ? (There is no change in the oxidation number of hydrogen [AIIMS-93]
(A) –1 (B) –3 (C) +3 (D) +5

52. The statement which is not correct for nitrogen is ? [AIIMS-93]
(A) Shows greater stability as a free element
(B) Has a very small atomic radius
(C) Forms covalent compounds like NH_3, NF_3 etc.
(D) Is unable to show greater covalency than 3

53. Which of the following statements is wrong ? [AIIMS-93]
(A) Red phosphorus has polymerized structure
(B) White phosphorus has a higher internal energy than red phosphorus
(C) Red phosphorus is more active than white phosphorus
(D) White phosphorus is metastable at all temperatures upto the melting point of red phosphorus

54. Ammonia reacts with excess of chlorine to form [AIIMS-93]
(A) NH_4Cl (B) NCl_2 and HCl
(C) N_2 and NH_4Cl (D) NCl_3 and HCl

55. HBr can be prepared by treating [AIIMS-93]
(A) KBr with sulphuric acid (B) KBr with nitric acid
(C) KBr with phosphoric acid
(D) Bromine with sulphuric acid

56. With reference to chlorine which one of the following statements is false ? [AIIMS-93]
(A) It has the outer electronic configuration of ns^2np^5
(B) It has the highest electronegativity
(C) It has the highest electron affinity in the group
(D) It is a better oxidizing agent than bromine

57. Ozone is [AIIMS-93]
(A) A compound of oxygen (B) An allotrope of oxygen
(C) An isotope of oxygen (D) An isomer of oxygen

58. It is possible to obtain oxygen from air by fractional distillation because oxygen [AIIMS-93]
(A) Has higher boiling point than nitrogen
(B) Has lower density than nitrogen
(C) Is in a different group of periodic table than nitrogen
(D) Is more reactive than nitrogen

59. Assertion : Halogens do not occur in free state. Reason : Halogens are highly reactive. [AIIMS-94]
(A) A (B) B (C) C (D) D

60. Assertion : Reaction of conc. H_2SO_4 on NaBr and NaI does not give HBr and HI. Reason : Both HBr and HI are oxidized by conc. H_2SO_4 to Br_2 and I_2. [AIIMS-94]
(A) A (B) B (C) C (D) D

61. Which of the following gas is produced in the reaction $KO_2 + CO_2$? [AIIMS-94]
(A) H_2 (B) N_2 (C) O_2 (D) CO

62. The oxidation number of phosphorus and basicity of acid in pyrophosphoric acid respectively are [AIIMS-95]
(A) + one and four (B) + three and one
(C) + four and three (D) + five and four

63. Assertion : Sulphur dioxide and chloride are both bleaching agents. Reason : Both are reducing agents. [AIIMS-95]
(A) A (B) B (C) C (D) D

64. The oxidation number of oxygen atom in O_2^{2-} ion is [AIIMS-95]
(A) –1 (B) –2 (C) –3 (D) –5

65. Which of the following oxyacids does not exist [AIIMS-95]
(A) $HBiO_3$ (B) H_3BiO_4 (C) H_3SbO_3 (D) H_3AsO_4

66. The element, with atomic number 118, will be [AIIMS-95]
(A) Alkali (B) Noble gas
(C) Transition element (D) Alkaline earth metal

67. Which of the following is used in photography ? [AIIMS-95]
(A) AgCl (B) AgBr (C) Ag_2S (D) $Ag_2C_2O_4$

68. Bismuth chloride ($BiCl_3$), on hydrolysis, yields a white precipitate of [AIIMS-96]
(A) Bismuth oxyacid (B) Bismuth oxychloride
(C) Bismuth hydoxide (D) None of these

69. Chlorine oxidizes sodium thiosulphate to form [AIIMS-96]
(A) Na_2O (B) Na_2CO_3 (C) Na_2SO_3 (D) Na_2SO_4

70. Which of the following metal does not liberate H_2 from dilute H_2SO_4 ? [AIIMS-96]
(A) Al (B) Mg (C) Zn (D) Cu

71. Assertion : Amongst the halogens, fluorine can oxidize the elements to the highest oxidation state. Reason : Due to small size of fluoride ions, it is difficult to oxidize fluoride ions to fluorine. Hence reverse reaction takes place more easily. [AIIMS-96]
(A) A (B) B (C) C (D) D

72. In the reaction : $2P_2O_5 + HNO_3 \rightarrow P_4O_5 + x$, the

term x is [AIIMS-96]

(A) H_2O only (B) PNO_3 (C) N_2O_4 (D) N_2O_5

73. The aqueous solution of which of the following salts will have the lowest pH ? [AIIMS-96]

(A) NaClO (B) $NaClO_2$ (C) $NaClO_3$ (D) $NaClO_4$

74. The configuration $1s^2 2s^2 2p^5 3s^1$ shows [AIIMS-97]

(A) Ground state of fluorine atom

(B) Excited state of fluorine atom

(C) Excited state of Neon atom

(D) Excited state of ion O_2^-

75. Assertion : H_3PO_3 is dibasic compound. Reason : Two hydrogen atoms are directly attached to the P. [AIIMS-97]

(A) A (B) B (C) C (D) D

76. Assertion : NO_3^- is planar while NH_3 is pyramidal. Reason : N in NO_3^- is sp^2 and in NH_3 is sp^3 hybridisation. [AIIMS-97]

(A) A (B) B (C) C (D) D

77. Assertion : Liquid ammonia is used for refrigeration. Reason : It vapourises quickly. [AIIMS-97]

(A) A (B) B (C) C (D) D

78. Assertion : N_2 and NO^+ both are diamagnetic substance. Reason : NO^+ is iso-electronic with N_2 [AIIMS-97]

(A) A (B) B (C) C (D) D

79. Assertion : The bond angle of PBr_3 is greater than PH_3 but the bond angle of NBr_3 is less than NH_3. Reason : Size of Br is less than hydrogen. [AIIMS-97]

(A) A (B) B (C) C (D) D

80. Assertion : Di-nitrogen is chemically unreactive at ordinary temperature and is very stable. Reason : The bond dissociation energy is 946 kJ mol^{-1} [AIIMS-98]

(A) A (B) B (C) C (D) D

81. Assertion : Both chlorine and sulphur dioxide act as bleaching agents. Reason : Chlorine bleaches by vigorous oxidation, while sulphur dioxide bleaches by mild reduction. [AIIMS-98]

(A) A (B) B (C) C (D) D

82. Which of the following is the strongest acid ? [AIIMS-99]

(A) HClO (B) $HClO_2$ (C) $HClO_3$ (D) $HClO_4$

83. Which of the following is the correct statement for PH_3 ? [AIIMS-99]

(A) It is less basic than NH_3

(B) It is less poisonous than NH_3

(C) Electro-negativity of PH_3 is greater than NH_3

(D) It does not show reducing properties

84. Which oxide of nitrogen is obtained on heating ammonium nitrate at 250°C ? [AIIMS-99]

(A) Nitric oxide (B) Nitrous oxide

(C) Nitrogen dioxide (D) Dinitrogen tetraoxide

85. Which of the following exhibits the weakest intermolecular forces ? [AIIMS-99]

(A) He (B) HCl (C) NH_3 (D) H_2O

86. Assertion : Reaction of conc.H_2SO_4 on NaBr and NaI does not give HBr and HI. Reason : HBr and HI are oxidized by conc.H_2SO_4 to Br_2 and I_2. [AIIMS-99]

(A) A (B) B (C) C (D) D

87. Assertion : Tailing of Hg occurs on passing O_3 through it. Reason : Due to oxidation of Hg. [AIIMS-99]

(A) A (B) B (C) C (D) D

88. Assertion : Xenon forms fluorides. Reason : Because 5 d-orbitals are available for valence shell expansion. [AIIMS-99]

(A) A (B) B (C) C (D) D

89. Which is not hydrolysed : [AIIMS-2000]

(A) PCl_3 (B) NCl_3 (C) $AsCl_3$ (D) $SnCl_3$

90. Which of the following is true about ClO_2 ? [AIIMS-2000]

(A) It is paramagnetic (B) It dimerises in liquid phase

(C) It is an angular molecule (D) It is very reactive

91. Which of the following is paramagnetic ? [AIIMS-2000]

(A) NO_2 (B) N_2O (C) N_2O_3 (D) N_2O_5

92. Oxidation state of sulphur in S_8, H_2S and S_2F_2 is [AIIMS-2000]

(A) 0, −2, −1 (B) 1, + 2, −1

(C) 1, −2, +1 (D) 0, −2, +1

93. The deep colour produced when iodine dissolve in a potassium iodide solution is due to the presence of [AIIMS-2000]

(A) I^+ (B) I^- (C) I_2^- (D) I_3^-

94. Which of the following is not paramagnetic ? [AIIMS-2000]

(A) ClO_2 (B) ClO_3 (C) ClF_3 (D) BrO_2

95. Which of the following is the strongest acid ? [AIIMS-2000]

(A) HOCl (B) $HClO_2$ (C) $HClO_3$ (D) $HClO_4$

96. Size of the iodine species increases in the order [AIIMS-2000]

(A) $I^+ < I < I^-$ (B) $I^+ < I^- < I$

(C) $I^- < I^+ < I^-$ (D) $I < I^- < I^+$

97. Polarizability of halide ions increases in the order [AIIMS-2000]

(A) F^-, I^-, Br^-, Cl^- (B) Cl^-, Br^-, I^-, F^-

(C) I^-, Br^-, Cl^-, F^- (D) F^-, Cl^-, Br^-, I^-

98. KHX_2 here X is halogen of following acid [AIIMS-2001]
(A) HF (B) HCl (C) HBr (D) HI

99. In which pπ-dπ overlapping occurs : [AIIMS-2001]
(A) CO_3^{-2} (B) PO_4^{-3} (C) O_2 (D) N_2

100. Assertion : At room temperature oxygen exists as a diatomic gas, sulphur exists as solid. Reason : The catenated –O–O–O– changes are less stable as compared to O = O molecule. [AIIMS-2001]
(A) A (B) B (C) C (D) D

101. The correct order of solubility in water for He, Ne, Ar, Kr, Xe is [AIIMS-2002]
(A) He > Ne > Ar > Kr > Xe (B) Ne > Ar > Kr > He > Xe
(C) Xe > Kr > Ar > Ne > He (D) Ar > Ne > He > Kr > Xe

102. Which of the following compounds is tribasic acid ? [AIIMS-2002]
(A) H_3PO_2 (B) H_3PO_3 (C) H_3PO_4 (D) $H_4P_2O_7$

103. Assertion (A) : Di negative anion of oxygen (O^{2-}) is quite common but di negative anion of sulphur (S^{2-}) is less common. Reason (R) : Covalency of oxygen is two. [AIIMS-2002]
(A) A (B) B (C) C (D) D (E) E

104. Assertion (A) : The fluorine has lower reactivity. Reason (R) : F-F bond has low bond dissociation energy. [AIIMS-2002]
(A) A (B) B (C) C (D) D (E) E

105. Assertion (A) : The halogens absorb visible light. Reason (R) : All halogens are coloured. [AIIMS-2002]
(A) A (B) B (C) C (D) D (E) E

106. The true statement for the acids of phosphorus H_3PO_2, H_3PO_3 and H_3PO_4 is - [AIIMS-2003]
(A) The order of their acidity is $H_3PO_4 > H_3PO_3 > H_3PO_2$
(B) All of them are reducing in nature
(C) All of them are tribasic acids
(D) The geometry of phosphorus is tetrahedral in the three

107. The mixture of concentrated HCl and HNO_3 made in 3:1 ratio contains - [AIIMS-2003]
(A) ClO_2 (B) NOCl (C) NCl_3 (D) N_2O_4

108. (A) NF_3 is a weaker ligand than $N(CH_3)_3$.
(R) NF_3 ionizes to give F ions in aqueous solution. [AIIMS-2003]
(A) A (B) B (C) C (D) D

109. (A) PbI_4 is stable compound . [AIIMS-2003]
(R) Iodide stabilizes higher oxidation state.
(A) A (B) B (C) C (D) D

110. (A) Hemoglobin is an oxygen carrier.[AIIMS-2003]
(R) Oxygen binds as O^{2-} to Fe of hemoglobin.
(A) A (B) B (C) C (D) D

111. (A) Photochemical smog is produced by nitrogen oxides.
(R) Vehicular pollution is a major source of nitrogen oxides. [AIIMS-2003]
(A) A (B) B (C) C (D) D

112. The strongest Bronsted base among the following is [JIPMER-90]
(A) ClO^- (B) ClO_3^- (C) ClO_2^- (D) ClO_4^-

113. Iron is rendered passive by treatment with conc. : [JIPMER-90]
(A) H_2SO_4 (B) H_3PO_4 (C) HCl (D) HNO_3

114. In which of the following compounds does the ratio of anion size to cation size have the lowest value : [JIPMER-90]
(A) CsI (B) LiI (C) CsF (D) LiF

115. Compound having lowest boiling point is : [JIPMER-90]
(A) HF (B) HCl (C) HBr (D) HI

116. Yellow phosphorus is kept in : [JIPMER-90]
(A) Water (B) Kerosene (C) Ether (D) Alcohol

117. The molecule having one unpaired electron is : [JIPMER-90]
(A) NO (B) Co (C) CN^- (D) O_2

118. The ion isoelectronic with CO : [JIPMER-90]
(A) CN^- (B) O_2 (C) N_2^{2+} (D) O_2^{2-}

119. Which of the following does not represent a salt derived from H_3PO_4 : [JIPMER-90]
(A) NaH_2PO_3 (B) NaH_2PO_4
(C) Na_2HPO_4 (D) Na_3PO_4

120. Which of the following substances react with both dilute and conc.H_2SO_4 to give the same product : [JIPMER-90]
(A) Cu (II) Oxide (B) $NaNO_3$ (C) Sucrose (D) Zinc

121. Which of the following hydrides has the lowest boiling point : [JIPMER-90]
(A) NH_3 (B) PH_3 (C) SbH_3 (D) AsH_3

122. Which is not oxidised by MnO_2 : [JIPMER-90]
(A) F^- (B) Cl^- (C) Br^- (D) I^-

123. Which has maximum percentage of chlorine : [JIPMER-91]
(A) Pyrene (B) PVC
(C) Gammexane (D) Ethylidene chloride

124. Which is most stable to heat : [JIPMER-91]
(A) HCl (B) HOCl (C) HBr (D) HI

125. Nitrogen is chemically inert because :[JIPMER-91]
(A) Its molecular size is small
(B) Its electronic configuration is stable

(C) Its electronegativity is high
(D) Its bond dissociation energy is high

126. With hot water, P_2O_5 gives : [JIPMER-91]
(A) H_3PO_3 (B) H_3PO_4 (C) H_3PO_3 (D) HPO_3

127. Which of the following has positive and negative oxidation states : [JIPMER-91]
(A) Na (B) I (C) F (D) Ne

128. Most chemically active form of phosphorus is : [JIPMER-91]
(A) Black (B) Red (C) Yellow (D) Violet

129. The p–p–p bond angle in white phosphorus is : [JIPMER-91]
(A) 120° (B) 90° (C) 60° (D) 109°28

130. Which of the following does not dissolve in aqua regia? [JIPMER-92]
(A) Pt (B) Pd (C) Ir (D) Ag

131. Nitrogen dioxide can not be obtained by heating ? [JIPMER-92]
(A) KNO_3 (B) $Cu(NO_3)_2$ (C) $Pb(NO_3)_2$ (D) $AgNO_3$

132. Which of following acid possesses oxidizing, reducing and complex forming properties : [JIPMER-92]
(A) HNO_3 (B) H_2SO_4 (C) HCl (D) HNO_2

133. Phosphorus is manufactured from ? [JIPMER-92]
(A) Ca_3P_2 (B) $CaHPO_3$ (C) P_2O_5 (D) $Ca_3(PO_4)_2$

134. Upon electrolyzing a dilute solution of H_2SO_4 between platinum electrodes, the gas evolved at the anode is : [JIPMER-92]
(A) SO_2 (B) SO_3 (C) O_2 (D) H_2

135. Oxygen molecule shows ? [JIPMER-92]
(A) Paramagnetism (B) Diamagnetism
(C) Ferromagnetism (D) None

136. H_3PO_4 dissociates in : [JIPMER-92]
(A) 1 stages (B) 2 stages (C) 3 stages (D)4 stages

137. Which of the following is mono basic :[JIPMER-92]
(A) H_3PO_3 (B) H_3PO_2 (C) H_3PO_4 (D) $H_4P_2O_7$

138. Ammonia reacts with excess of chlorine to form : [JIPMER-92]
(A) NOCl (B) $NOCl_2$ (C) NO_2Cl_2 (D) N_2OCl

139. Which of the following is generally bleached by bleaching power : [JIPMER-92]
(A) Straw (B) Ivory (C) Roll of cotton (D) None

140. The noble gas which does not form clathrate compounds is : [JIPMER-93]
(A) Xenon (B) Krypton (C) Argon (D) Neon

141. To an aqueous solution of I^- and Br^-, I_2 and Br_2 are added. The products of the reaction would be ($E°Br_2/Br^- = 1.09$ V, $E°I_2/I^- = 0.74$ V) ? [JIPMER-93]
(A) Br_2, I^- (B) I_2, Br_2 (C) I_2, Br^- (D) I^-, Br^-

142. Sulphuric acid dehydrates formic acid to give [JIPMER-93]
(A) C (B) CO (C) CO_2 (D) CO and CO_2

143. Deacon's process is used for the large scale preparation of : [JIPMER-93]
(A) N_2 (B) P_4 (C) O_3 (D) Cl_2

144. Of the different allotropic forms of phosphorus, the one which is most reactive is : [JIPMER-93]
(A) Violet phosphorus (B) Scarlet phosphorus
(C) Red phosphorus (D) White phosphorus

145. The halogen having some metallic character is: [JIPMER-93]
(A) F (B) I (C) Cl (D) Br

146. When copper is heated with conc. H_2SO_4, it gives sulphur dioxide. In this reaction, sulphuric acid acts as : [JIPMER-93]
(A) A strong oxidizing agent (B) A strong reducing agent
(C) Dehydrating agent (D) Catalyst

147. Which of the following has lowest boiling point : [JIPMER-93]
(A) H_2O (B) H_2S (C) H_2Se (D) H_2Te

148. Colour of solid iodine is [JIPMER-94]
(A) Red (B) White (C) Violet (D) Steel grey

149. Select a compound in which oxidation number of chlorine is +5 [JIPMER-94]
(A) HClO (B) $HClO_2$ (C) $HClO_3$ (D) $HClO_4$

150. Hyponitrous acid is [JIPMER-94]
(A) HNO (B) HNO_2 (C) HNO_3 (D) HNO_4

151. Which of the following is a pseudohalide ? [JIPMER-94]
(A) CN^- (B) ICl (C) IF_5 (D) I_3^-

152. The noble gas forming maximum number of compounds is [JIPMER-94]
(A) Ne (B) Xe (C) Kr (D) Rn

153. Basicity of sulphuric acid is [JIPMER-94]
(A) 1 (B) 2 (C) 3 (D) 4

154. Noble gas used in radiotherapy is [JIPMER-94]
(A) Kr (B) Ar (C) Rn (D) Xe

155. Bromine is liberated when an aqueous KBr solution is treated with [JIPMER-94]
(A) Cl_2 (B) I_2 (C) HI (D) SO_2

156. The ratio of C_P/C_V for inert gases is [JIPMER-94]
(A) 1.33 (B) 1.66 (C) 2.13 (D) 1.99

157. The oxidation state of oxygen atom in potassium superoxide is [JIPMER-94]
(A) 0 (B) –½ (C) –1 (D) –2

158. Which of the following is called sulphuric anhydride ? [JIPMER-95]

(A) $H_2S_2O_7$ (B) SO_3 (C) $H_2S_2O_3$ (D) SO_2

159. Five valence electrons of P_{15} are labelled as

AB	X	Y	Z
3s	3p		

If the spin quantum of B and Z is +1/2. The group of electrons with three of the quantum number same are [JIPMER-95]

(A) XYZ, AZ. (B) AB,XYZ, BY
(C) (AB) (D) AB, XYZ.

160. The chief source of iodine, in which it is present as sodium iodate is [JIPMER-95]

(A) Carnallite (B) Sea weeds
(C) Caliche (D) Iodine never exists as sodium iodate

161. Which of the following does not have S–S bond [JIPMER-96]

(A) $S_2O_3^{2-}$ (B) $S_1O_5^{-2}$ (C) $S_2O_4^{2-}$ (D) $S_2O_7^{2-}$

162. Out of the following hydrogen halides, which one has the highest boiling point ? [JIPMER-96]

(A) Hl (B) HBr (C) HCl (D) HF

163. Which of the following triatomic molecules has the largest angle around the central atom : [JIPMER-96]

(A) H_2O (B) SO_2 (C) H_2S (D) H_2Se

164. An element X has five electrons in p-orbitals of valence shell and also belongs to fifth period. Its molecular formula should be : [JIPMER-96]

(A) X (B) X_2 (C) X_3 (D) Unpredictable

165. Hydrogen produced by displacement from dilute sulphuric acid by zinc is (i) collected over water and (ii) over mercury. [JIPMER-97]

(A) The amount of hydrogen collected over mercury was greater.
(B) The amount of hydrogen collected over mercury was the same as the amount of the gas collected over water.
(C) The amount of hydrogen collected over water was greater
(D)The volume of gas collected was the same.

166. Which of the following silver compounds finds maximum use in photography ? [JIPMER-98]

(A) AgCl (B) AgBr (C) AgI (D) $AgNO_3$

167. Red phosphorus is chemically much less reactive than white phosphorus because. [JIPMER-98]

(A) It has a polymeric structure
(B) It does not contain tetrahedral P_4 molecules
(C) It does not contain p-p bonds
(D) It has a linear structure

168. The carbonate which does not leave a residue on heating is [JIPMER-98]

(A) Na_2CO_3 (B) Ag_2CO_3 (C) $CuCO_3$ (D) $(NH_4)_2CO_3$

169. Which is the strongest of the following acids ? [JIPMER-98]

(A) $HClO_4$ (B) H_2SO_4 (C) HCl (D) HNO_3

170. The electronic configuration of sulphide ion is. [JIPMER-99]

(A) $1s^22s^22p^63s^23p^6$ (B) $1s^22s^22p^63s^23p^5$
(C) $1s^22s^22p^63s^23p^4$ (D) $1s^22s^22p^63s^23p^3$

171. Of the following hydrides which one has the lowest boiling point ? [JIPMER-99]

(A) HF (B) HCl (C) HBr (D) HI

172. Which of the following does not give nitrogen dioxide on heating ? [JIPMER-99]

(A) KNO_3 (B) $Pb(NO_3)_2$ (C) $Cu(NO_3)_2$ (D) $AgNO_3$

173. Hypophosphorous acid (H_3PO_2) is [JIPMER-99]

(A) Dibasic acid (B) Tribasic
(C) Monobasic acid (D) Neutral

174. Oxygen is obtained by heating [JIPMER-99]

(A) Ammonium dichromate (B) Sodium peroxide
(C) Potassium permanganate (D) Bleaching power

175. Bromine water cannot be decolouised by [JIPMER-99]

(A) $Na_2S_2O_3$ (B) NaOH (C) $H_4P_2O_5$ (D) $H_4P_2O_7$

176. Ozone is easily detected by the use of[JIPMER-99]

(A) Ag (B) Hg (C) AgCl (D) H_2O_2

177. Bromine is liberated when aqueous solution of potassium bromide is treated with. [JIPMER-99]

(A) I_2 (B) Cl_2 (C) SO_2 (D) dil. H_2SO_4

178. Which one of the following oxides of chlorine is obtained by passing dry chlorine over silver perchlorate at 190°C ? [JIPMER-99]

(A) Cl_2O (B) ClO_2 (C) ClO_3 (D) ClO_4

179. H_2S reacts with halogens, the halogens [JIPMER-2000]

(A) Form sulphur halides (B) Are oxidised
(C) Are reduced (D) None

180. Sulphuric acid reacts with PCl_5 to yield [JIPMER-2000]

(A) Sulphur monochloride (B) Phosphoric acid
(C) Sulphuryl chloride (D) Thionyl chloride

181. Which of the following is not oxidized by MnO_2 [JIPMER-2000]

(A) Cl^- (B) I^- (C) Br^- (D) F^-

182. Which of the following halogen has basic properties [JIPMER-2000]

(A) Chlorine (B) Iodine (C) Bromine (D) Fluorine

183. The compound HCl behaves as . . . in the reaction,

$HCl + HF \rightarrow H_2^+Cl + F^-$ [JIPMER-2001]
(A) Weak base (B) Weak acid
(C) Strong base (D) Strong acid

184. H_2SO_4 acts as dehydrating agent in its reaction with [JIPMER-2001]
(A) $H_2C_2O_4$ (B) $Ba(OH)_2$ (C) KOH (D) Zn

185. Bromine is liberated when an aqueous solution of KBr is treated with [JIPMER-2001]
(A) Dilute H_2SO_4 (B) I_2 (C) Cl_2 (D) SO_2

186. Which burns to form an oxide which is gas at room temperature [JIPMER-2001]
(A) S (B) P (C) Sodium (D) Hydrogen

187. The nitrate of which metal leaves metallic globule on heating strongly [JIPMER-2001]
(A) $Pb(NO_3)_2$ (B) $NaNO_3$ (C) $AgNO_3$ (D) $Cu(NO_3)_2$

188. What is the structure of NH_3 [JIPMER-2002]
(A) Tetrahedral (B) Octahedral
(C) Square planer (D) Pyramidal

189. Arrange S, P, As in order of increasing ionization energy. [JIPMER-2002]
(A) $S < P < As$ (B) $P < S < As$
(C) $As < S < P$ (D) $As < P < S$

190. Which of the following arrangements of molecules is correct on the basis of their dipole moments ? [JIPMER-2002]
(A) $BF_3 > NF_3 > NH_3$ (B) $NF_3 > BF_3 > NH_3$
(C) $NH_3 > BF_3 > NF_3$ (D) $NH_3 > NF_3 > BF_3$

191. Which of the following is isoelectronic as well as has same structure as that of N_2O ? [JIPMER-2002]
(A) N_3H (B) H_2O (C) NO_2 (D) CO_2

192. Which of the following noble gases is least polarisable? [JIPMER-2002]
(A) He (B) Xe (C) Ar (D) Ne

193. When lead nitrate is heated it gives[JIPMER-2002]
(A) NO_2 (B) NO (C) N_2O_5 (D) N_2O

194. Soild PCl_5 exits as [JIPMER-2002]
(A) PCl_5 (B) PCl_4^+ (C) PCl_6^- (D) PCl^+_4 and PCl^-_6

195. Zn on reacting with cold dil. HNO_3 gives [BHU-95]
(A) NO_2 (B) NH_4NO_3 (C) $ZnNO_3$ (D) NO

196. The sides of match box have coating of [BHU-95]
(A) Potassium chlorate and red lead
(B) Potassium chlorate and antimony sulphide
(C) Antimony sulphide and red phosphorus
(D) Antimony sulphide and red lead

197. The cynamide process is used in the formation of [BHU-95]
(A) N_2 (B) PH_3 (C) NH_3 (D) HNO_3

198. Moist ammonia gas is dried over [BHU-96]
(A) P_2O_5 (B) Soda lime
(C) Conc.H_2SO_4 (D) Anhydrous $CaCl_2$

199. Iodine dissolves readily in [BHU-96]
(A) Water (B) Alcohol
(C) Potassium iodide (D) Carbon tetrachloride

200. Oxidation number of N in HNO_3 is [BHU-97]
(A) −3.5 (B) +3.5 (C) −3, +5 (D) +5

201. If HNO_3 changes into N_2O, the oxidation number is changed by [BHU-97]
(A) +2 (B) 6 (C) 1 (D) +4

202. When $CuSO_4$ reacts with KI, the oxidation number of Cu changes by [BHU-97]
(A) 0 (B) −1 (C) 1 (D) 2

203. The size of particles in suspension, true solution and colloidal solution varies in the order [BHU-97]
(A) Suspension > colloidal > true solution
(B) Suspension > colloidal > true solution
(C) True solution > suspension > colloidal
(D) None of these

204. What is correct sequence of bond order ?[BHU-97]
(A) $O_2^+ > O_2^- > O_2$ (B) $O_2^+ > O_2 > O_2^-$
(C) $O_2^+ > O_2^{-1} > O_2^{+1}$ (D) $O_2^{-1} > O_2^{+1} > O_2$

205. Which of the following statements is correct ? [BHU-97]
(A) Only chlorine and bromine form oxyacid
(B) All halogens form oxyacid
(C) All halogens except fluorine form oxyacid
(D) Only iodine form oxyacid.

206. Copper turning, when heated with concentrated sulphuric acid, gives [BHU-99]
(A) O_2 (B) SO_2 (C) SO_3 (D) H_2S

207. Sides of safety matches contain [BHU-2000]
(A) P_4S_3 (B) $KClO_3$, KNO_3, Sulphur + Antimony
(C) Red phosphorus + Sand powder
(D) $Ca_3(PO_4)$ + Glass pieces

208. Which of the following statement is true ? [BHU-2000]
(A) All metal nitrates are soluble in H_2O
(B) All metal nitrates are insoluble in H_2O
(C) Solubility depends upon temperature
(D) Either (A) or (B)

209. Iodine test is shown by [BHU-2000]
(A) Polypeptides (B) glycogen (C) Starch (D) glucose

210. KI and $CuSO_4$ solutions when mixed give [BHU-2000]
(A) $CuI_2 + K_2SO_4$ (B) $Cu_2I_2 + K_2SO_4$
(C) $Cu_2I_2 + I_2 + K_2SO_4$ (D) $CuI_2 + I_2 + K_2SO_4$

211. Which solutions are mixed to form a buffer solution? [BHU-2000]
(A) A strong acid and its salt of strong base
(B) Strong acid and its salt of weak base
(C) Weak acid and its salt of strong base
(D) Weak acid and its salt of weak base

212. P_2O_5 is a [BHU-2000]
(A) Reducing agent (B) Oxidizing agent
(C) Buffer (D) None of these

213. The strongest reducing agent amongst the following is [BHU-2002]
(A) $P_2O_4^{4-}$ (B) H_3PO_4 (C) $H_2PO_2^-$ (D) $P_2O_7^{4-}$

214. Which of the following compound is less soluble in acidic solution than in pure water ? [BHU-2002]
(A) AgBr (B) $CaCO_3$ (C) $Ni(OH)_2$ (D) $Ca(PO_4)_2$

215. Sulphur molecule is [AFMC-91]
(A) Diatomic (B) Triatomic (C) Tetratomic(D) Octatomic

216. Euchlorine is a mixture of [AFMC-91]
(A) O_2 and Cl_2 (B) O_2 and ClO_2
(C) Cl_2 and ClO_2 (D) ClO_2 and Cl_2O

217. SO_3 is formed by heating [AFMC-92]
(A) $PbSO_4$ (B) NSO_4 (C) $FeSO_4$ (D) $CuSO_4$

218. Nitrogen dioxide is released by heating[AFMC-92]
(A) $PbNO_3$ (B) KNO_3 (C) $Fe_2(NO_3)_3$ (D) $AgNO_3$

219. XeF_4 has a shape of [AFMC-92]
(A) Spherical (B) Trigonal bipyramidal
(C) Square planar (D) Tetrahedral

220. The valency of inert gases is [AFMC-92]
(A) Zero (B) One (C) Three (D) Two

221. When Cl_2 is dissolved in KOH, the product formed is [AFMC-92]
(A) KClO (B) $KClO_2$ (C) $KClO_3$ (D) $KClO_4$

222. Which of the following phosphorus is most stable ? [AFMC-92]
(A) Red (B) White (C) Black (D) All stable

223. Which of the following is both oxidizing and reducing agent ? [AFMC-92]
(A) HNO_3 (B) HNO_2 (C) H_2SO_4 (D) HCl

224. Non-formation of meniscus in mercury is due to the formation of [AFMC-92]
(A) Mercuric oxide (B) Mercurous oxide
(C) Mercuric chloride (D) Mercurous chloride

225. The catalyst used in Haber's process is[AFMC-93]
(A) Finely divided iron (B) Finely divided nickel
(C) Finely divided platinum (D)Finely divided molybdenum

226. Pure nitrogen can be prepared from [AFMC-93]
(A) NH_4OH (B) Ca_3N_2 (C) NH_4NO_2 (D) BaO

227. The gas used in electric discharges for decoration is [AFMC-93]
(A) Helium (B) Neon (C) Radon (D) Krypton

228. Which of the following is most polarized among Noble gases ? [AFMC-94]
(A) He (B) Xe (C) Kr (D) Rn

229. The reducing agent is [AFMC-94]
(A) HF (B) HI (C) HCl (D) HBr

230. Which of the following elements always contain -1 oxidation state ? [AFMC-94]
(A) Carbon (B) Oxygen (C) Fluorine (D) Nitrogen

231. The solubility of I_2 increase in water in presence of [AFMC-95]
(A) KI (B) H_2SO_4 (C) $KMnO_4$ (D) H_2S

232. Which of the following phosphorus is most reactive? [AFMC-95]
(A) Red phosphorus (B) White phosphorus
(C) Scarlet phosphorus (D) Violet phosphorus

233. H_2S react with O_2 to form [AFMC-95]
(A) $H_2O + S$ (B) $H_2O + SO_2$
(C) $H_2O + SO_3$ (D) $H_2SO_4 + S$

234. Bromine water reacts with SO_2 to form[AFMC-95]
(A) H_2O and HBr (B) H_2SO_4 + HBr
(C) HBr and S (D) S and H_2O

235. In which NH_3 is not used ? [AFMC-95]
(A) Cold storage (B) Anaeshthetic
(C) Manufacture of rayon and plastic (D) None of these

236. In NH_3 and PH_3 common is [AFMC-95]
(A) Odour (B) Combustibility
(C) Basic - nature (D) None of these

237. Nitrous oxide is [AFMC-95]
(A) Soluble in cold water (B) Soluble in hot water
(C) Acidic in nature (D) None of these

238. XeF_4 on partial hydrolysis produces [AFMC-95]
(A) XeF_2 (B) $XeOF_2$ (C) $XeOF_4$ (D) XeO_3

239. In photography HYPO is used as [AFMC-96]
(A) Oxidizing agent (B) Reducing agent
(C) Complexing agent (D) All of the above

240. Sea-diver goes deep in sea with mixture of which of the following gas ? [AFMC-96]
(A) O_2 and He (B) O_2 and Ar
(C) O_2 and CO_2 (D) CO_2 and Ar

241. Glass is soluble in [AFMC-96]
(A) HF (B) H_2SO_4 (C) $HClO_4$ (D) Aqua regia

242. The test of O_3 can be done by [AFMC-97]
(A) Ag (B) Hg (C) Au (D) None of these

243. Chlorine is liberated when we heat [AFMC-98]

(A) $KMnO_4$ + NaCl (B) $K_2Cr_2O_7$ + MnO_2
(C) $Pb_2(NO_3)_4$ + MnO_2 (D) $K_2Cr_2O_7$ + HCl

244. The correct order of acidic strength is [AFMC-98]
(A) HF < HCl < HBr < HI (B) HCl > HBr < HF > HI
(C) HBr < HCl < HI < HF (D) HI < HBr < HCl < HF

245. The number of electrons shared by each outermost shell of N_2 is [AFMC-98]
(A) 2 (B) 3 (C) 4 (B) 5

246. Nitrogen can be obtained from air by removing [AFMC-98]
(A) Oxygen (B) Hydrogen
(C) Carbon dioxide (D) Both (A) and (C)

247. Ozone is prepared by passing silent electric discharge through oxygen. In this reaction [AFMC-98]
(A) Energy is given out (B) Energy is absorbed
(C) Oxygen is loaded with energy
(D) Oxygen is dissociated into atoms

248. In the preparation of sulphuric acid, V_2O_5 is used in the reaction. It is [AFMC-99]
(A) $SO_2 + H_2O \rightarrow H_2SO_4$ (B) $S + O_2 \rightarrow SO_2$
(C) $N_2 + 3H_2 \rightarrow 2NH_3$ (D) $2SO_2 + O_2 \rightarrow 2SO_3$

249. The acid formed, when iodine reacts with concentrated nitric acid, is [AFMC-99]
(A) Periodic acid (B) Hydroiodic acid
(C) Iodic acid (D) Nitrous acid

250. The correct order of reactivity of halogen is [AFMC-99]
(A) F < Cl < Br > I (B) F > Cl < Br < I
(C) F < Cl > Br > I (D) F < Cl < Br < I

251. Which of the following halogens is solid at room temperature ? [AFMC-99]
(A) Fluorine (B) Chlorine (C) Iodine (D) Bromine

252. In the reaction : $HNO_3 + P_4O_{10} \rightarrow 4HPO_3 + X$, the product X is [AFMC-99]
(A) NO_2 (B) N_2O_3 (C) N_2O_5 (D) H_2O

253. Which of the following set has the strongest tendency to form anions ? [AFMC-99]
(A) N, O and F (B) V, Cr and Mn
(C) Ga, In and T (D) Na, Mg and Al

254. The oxidation number of oxygen in O_2PtF_6 is [AFMC-2000]
(A) Zero (B) + 1/2 (C) + 1 (D) - 1/2

255. Which of the noble gas has highest polarisability ? [AFMC-2001]
(A) He (B) Ar (C) Kr (D) Xe

256. Which of the following have changeable valency ? [AFMC-2002]
(A) He (B) Br (C) B (D) None of these

257. Orthophosporic acid is [AFMC-2002]
(A) Monobasic (B) Dibasic
(C) Tribasic (D) Tetrabasic

258. The density of neon will be highest at [IIT-90]
(A) STP (B) 0°C, 2 atm
(C) 273°C, 1 atm. (D) 273°C, 2 atm.

259. The valence atomic orbitals on C in silver accetylide is ....hybridised. [IIT-90]

260. There is no S–S bond in [IIT-91]
(A) $S_2O_4^{2-}$ (B) $S_2O_5^{2-}$ (C) $S_2O_3^{2-}$ (D) $S_2O_7^{2-}$

261. In P_4O_{10}, the number of oxygen atoms bonded to each phosphorus atom is .... [IIT-92]

262. The Lead Chamber process involves oxidationof SO_2 by atmospheric oxygen under the influence of ....as catalysts. [IIT-92]

263. Solubility of iodine in water is greatly increased by the addition of iodide ions because of the formation of .... [IIT-94]

264. Hydrolysis of one mole of peroxodisulphuric acid produces. [IIT-96]
(A) two moles of sulphuric acid
(B) two moles of peroxo monosulphuric acid
(D) one mole of sulphuric acid, one mole of peroxomono sulphuric acid and one mole of hydrogen peroxide

265. KF combines with HF to form KHF_2. The compound contains the species. [IIT-96]
(A) K^+, F^- and H^+ (B) K^+, F^- and HF
(C) K^+ and $[HF_2]^-$ (D) $[KHF]^+$ and F^-

266. The following acids have been arranged in the order of decreasing acid strength. Identify the correct order. [IIT-96]
ClOH (I), BrOH (II), IOH (III)
(A) I > II > III (B) II > I > III
(C) III > II > I (D) I > III > II

267. Which one of the following species is not a pseudohalide ? [IIT-97]
(A) CNO (B) RCOO (C) OCN (D) NNN

268. Among PCl_3, CH_3^+, NH_2^- and NF_3, ... is least reactive towards water. [IIT-97]

269. The angle P–P–P in P_4 molecule is .... degree. [IIT-97]

270. Silver jewellery items tarnish slowly in the air due their reaction with .... [IIT-97]

271. White phosphorus (P_4) has [IIT-98]
(A) Six P-P single bonds (B) Four P-P singlebonds
(C) Four lone pairs of electrons (D) PPP angle of 60°

272. One mole of calcium phosphide on reaction with

excess water gives. [IIT-99]
(A) one mole of phosphine
(B) two moles of phosporic acid
(C) two moles of phosphine
(D) one mole of phosphorus pentoxide

273. On heating ammonium dichromate, the gas evolved is [IIT-99]
(A) oxygen (B) ammonia
(C) nitrous oxide (D) nitrogen

274. Ammonia, on reaction with hypochlorite anion, can form. [IIT-99]
(A) NO (B) NH_4Cl (C) N_2H_4 (D) HNO_2

275. The number of P–O–P bonds in cyclic metaphosphoric acid is [IIT-2000]
(A) zero (B) two (C) three (D) four

276. The correct order of acidic strength is [IIT-2000]
(A) $Cl_2O_7 > SO_2 > P_4O_{10}$ (B) $CO_2 > N_2O_5 > SO_3$
(C) $Ba_2O > MgO > Al_2O_3$ (D) $K_2O > CaO > MgO$

277. Ammonia can be dried by [IIT-2000]
(A) conc. H_2SO_4 (B) P_4O_{10}
(C) CaO (D) anhydrous $CaCl_2$

278. H_2O, H_2S, H_2Se and H_2Te, the one with the highest boiling point is [IIT-2000]
(A) H_2O because of hydrogen bonding
(B) H_2Te because of higher molecular weight
(C) H_2S because of hydrogen bonding
(D) H_2Se because of lower molecular weight

279. The reaction, $3ClO^-(aq) \rightarrow ClO_3^-(aq) + 2Cl^-(aq)$, is an example of [IIT-2001]
(A) oxidation reaction (B) reduction reaction
(C) disproportionation reaction
(D) decomposition reaction

280. The set with correct order of acidity is [IIT-2001]
(A) $HClO < HClO_2 < HClO_3 < HClO_4$
(B) $HClO_4 < HClO_3 < HClO_2 < HClO$
(C) $HClO < HClO_4 < HClO_3 < HClO_2$
(D) $HClO_4 < HClO_2 < HClO_3 < HClO$

281. A gas 'X' is passed through water to form a saturated solution. The aqueous solution on treatment with silver nitrate gives a white precipitate. The saturated aqueous solution also dissolves magnesium ribbon with evolution of a colourless gas 'Y'. Identify 'X' and 'Y' [IIT-2002]
(A) $X = CO_2$, $Y = Cl_2$ (B) $X = Cl_2$, $Y = CO_2$
(C) $X = Cl_2$, $Y = H_2$ (D) $X = H_2$, $Y = Cl_2$

282. Anhydrous ferric chloride is prepared by [IIT-2002]
(A) heating hydrated ferric chloride at a high temperature in stream of air
(B) heating metallic iron in a stream of dry chlorine gas
(C) reaction of ferric oxide with hydrochloric acid
(D) reaction of metallic iron with hydrochloric acid.

283. Polyphosphates are used as water softening agents because they [IIT-2002]
(A) form soluble complexes with anionic species
(B) precipitate anionic species
(C) form soluble complexes with cationic species
(D) precipitate cationic species

284. Specify the coordination geometry around and hybridization of N and B atoms in a 1:1 complex of BF_3 and NH_3 [IIT-2002]
(A) N: tetrahedral, sp^3; B:tetrahedral, sp^3
(B) N:pyramidal, sp^3; B:pyramidal, sp^3
(C) N: pyramidal, sp^3; B:Planar, sp^2
(D) N:pyramidal, sp^3; B: tetrahedral, sp^3

285. $(X) + \text{dil. } H_2SO_4 \rightarrow (Y)$
(irritating smelling gas)
$(Y) + K_2Cr_2O_7 + \text{dil. } H_2SO_4 \rightarrow$ Green transparent solution
X and Y are [IIT-2003]
(A) SO_3^{2-}, SO_2 (B) HCl , KNO_2
(C) KCl , SO_3 (D) None of these

286. H_3PO_4 and H_3PO_3 [IIT-2003]
(A) H_3PO_3 is reducing agent and diabasic
(B) H_3PO_3 is reducing agent and tribasic
(C) H_3PO_3 is reducing agent and monobasic
(D) None of these

287. In XeF_2, XeF_4, XeF_6 the number of lone pairs on Xe are respectively. [AIEEE-2002]
(A) 2, 3, 1 (B) 1, 2, 3 (C) 4, 1, 2 (D) 3, 2, 1

288. In case of nitrogen, NCl_3 is possible but not NCl_5 while in case of phosphorous, PCl_3 as well as PCl_5 are possible. It is due to. [AIEEE-2002]
(A) Availability of vacant d orbitals in P but not in N.
(B) Lower electronegativity of P than N
(C) Lower tendency of H-bond formation in P than N
(D) Occurrence of P in solid while N in gaseous state a room temperature.

289. Which of the following statements is true ? [AIEEE-2002]
(A) HF is less polar than HBr
(B) Absolutely pure water does not contain any ions
(C) Chemical bond formation take place when forces of attraction overcome the forces of repulsion
(D) In covalency transference of electron takes place.

290. Which one of the following is an amphoteric oxide ? [AIEEE-2003]
(A) ZnO (B) Na_2O (C) SO_2 (D) B_2O_3

291. What may be expected to happen when phosphine gas is mixed with chlorine gas ? [AIEEE-2003]
(A) The mixture only cools down
(B) PCl_3 and HCl are formed and the mixture cools down

(C) PCl_5 and HCl are formed and the mixture cools down
(D) PH_3Cl_2 is formed with warming up

292. Which one of the following nitrates will leave behind a metal on strong heating ? [AIEEE-2003]
(A) Ferric nitrate (B) Copper nitrate
(C) Manganese nitrate (D) Silver nitrate

293. For making good quality mirrors, plates of float glass are used. These are obtained by floating molten glass over a liquid metal which does not solidify before glass. The metal used can be [AIEEE-2003]
(A) Mercury (B) Tin (C) Sodium (D) Magnesium

294. An element 'R' is in group III. Which is true in respect of 'R' : [EAMCET-90]
(A) It is a gas at room temperature
(B) It has an oxidation State of +4
(C) It forms an oxide of type R_2O_3
(D) It forms a halide of the type RX_2

295. The chemical reaction between $Na_2S_2O_3$ and I_2 in aqueous medium produces. [EAMCET-90]
(A) $Na_2S + NaI + H_2S$ (B) $Na_2SO_3 + NaI + H_2O$
(C) $Na_2SO_3 + NaI + H_2O$ (D) $Na_2S_4O_6 + NaI$

296. An element 'X' belongs to the group I or II or V. Its oxide reacts with water to produce highly acidic solution. The element 'X' belongs to group .... [EAMCET-90]

297. Which of the following statements about the halogen is correct : [EAMCET-90]
(A) They are all diatomic and forms univalent ions
(B) They are all diatomic and forms divalent ions
(C) They all capable of exhibiting several oxidation states
(D) They can mutually displace each other from the solution of their compound with metals.

298. Which of the following oxides of nitrogen is the anhydride of HNO_3. [EAMCET-91]
(A) N_2O_3 (B) N_2O_4 (C) N_2O_5 (D) N_2O

299. The reagent that has action on glass is [EAMCET-91]
(A) Aquaregia (B) Hydrofluoric acid
(C) Oleum (D) Fuming HNO_3

300. Among the SO_2 and H_2SO_4 and sodium thiosulphate, sulphur has the highest oxidation state in .... [EAMCET-91]

301. Krypton nucleus has a ....radius than that of barium. [EAMCET-92]

302. The formula of plaster of paris is .... [EAMCET-92]

303. When NH_4NO_3 is heated, the gas formed is .... [EAMCET-92]

304. In which of the following does sulphur have the lowest oxidation number ? [EAMCET-93]
(A) H_2SO_4 (B) SO_2 (C) H_2SO_3 (D) H_2S

305. The oxidation number of manganese in $KMnO_4$ is [EAMCET-93]
(A) 3 (B) 5 (C) 7 (D) 9

306. The catalyst used in the manufacture of sulphuric acid by contact process is [EAMCET-93]
(A) Al_2O_3 (B) Cr_2O_3 (C) V_2O_5 (D) MnO_2

307. The reaction of magnesium bromide with acid solution produces .... [EAMCET-93]

308. The rare gas which is more abundant in the atmosphere is .... [EAMCET-93]

309. The halogen with highest ionisation potential is .... [EAMCET-93]

310. Glass is soluble in a soloution of ....acid . [EAMCET-93]

311. The deep colour produced when iodine is dissolved in a solution of potassium iodide is caused by the presence of .... ion. [EAMCET-93]

312. The chemical formula of bleaching powder is .... [EAMCET-93]

313. White phosphorous forms PCl_3 nitrogen does not form NCl_3 because [EAMCET-94]
(A) Nitrogen is more electronegative than Phosphorous
(B) Nitrogen atom is smaller than Phosphorous atom
(C) The ionization energy of nitrogen is greater than that of phosphorous
(D) Unlike phosphorous, nitrogen has no d-orbital in its valence shell.

314. Meta phosphoric acid is represented by [EAMCET-94]
(A) H_3PO_3 (B) H_2PO_3 (C) HPO_3 (D) $H_4P_2O_7$

315. Which of the following nitrogen halides is most stable ? [EAMCET-95]
(A) NI_3 (B) NBr_3 (C) NCl_3 (D) NF_3

316. The oxyacid of phosphorus in which phosphorus has the lowest oxidation state is. [EAMCET-95]
(A) Hypophosphorus acid (B) Orthophosphoric acid
(C) Pyrophosphoric acid (D) Metaphosphoric acid

317. Which of the following is correct with respect to the V group elements ? [EAMCET-95]
(A) All the elements exhibit allotropy
(B) Except nitrogen all elements exhibit allotropy
(C) All elements except nitrogen and bismuth exhibit allotropy
(D) All elements except nitrogen and arsenic exhibit allotropy

318. Catalyst used in the manufacture of HNO_3 by

Ostwald process [EAMCET-95]
(A) Platinum gauze (B) V_2O_5 (C) FeMO (D) MnO_2

319. Preparation of fluorine by direct electrolysis of liquid HF is not practical because [EAMCET-95]
(A) Very high temperature is needed
(B) Electrolytic products are different from fluorine
(C) Very low temperature is required
(D) HF is a non electrolyte

320. The oxidation state of Iodine in ICl_2^- is [EAMCET-95]
(A) +1 (B) -1 (C) +2 (D) -3

321. When 6 volumes of Oxygen undergoes complete reaction to ozone, the number of moles of ozone formed are [EAMCET-95]
(A) 6 (B) 3 (C) 4 (D) 2

322. The bond angle of H_2X (where X is a sixth group element) as one goes down the group [EAMCET-96]
(A) Increases (B) Decreases
(C) Does not change (D) Changes irregularly

323. With respect to both oxygen and ozone which one of the following statements is NOT CORRECT [EAMCET-96]
(A) Both have similar reactivity with water
(B) Oxygen is colourless and ozone is coloured
(C) Oxygen valency is 2 in both
(D) Oxygen has 2 bonds and ozone has 3 bonds

324. The element which can displace three other halogens from their compounds is [EAMCET-96]
(A) Cl (B) Br (C) I (D) F

325. Bleaching action of SO_2 is due to its [EAMCET-96]
(A) Oxidising action (B) Ability to hydrolyse
(C) Acidic nature (D) Reducing action

326. Which one of the following halogens does not exhibit positive oxidation state in its compounds. [EAMCET-97]
(A) I (B) Br (C) Cl (D) F

327. Which one of the following is NOT a base [EAMCET-97]
(A) N_2H_4 (B) NH_2OH (C) $(CH_3)_3N$ (D) HN_3

328. Which one of the following compounds in aqueous solution gives a white ppt. with perchloric acid ? [EAMCET-97]
(A) NaCl (B) KCl (C) $MgCl_2$ (D) $FeCl_3$

329. Which of the following sequence is correct with reference to the oxidation number of iodine ? [EAMCET-97]
(A) $I_2 < ICl < HI < HIO_4$ (B) $HIC_4 < ICl < I_2 < HI$
(C) $I_2 < HI < ICl < HIO_4$ (D) $HI < I_4 < ICl < I_2 < HIO_4$

330. Which of the following reacts with concentrated sulphuric acid ? [EAMCET-98]
(A) Au (B) Ag (C) Pt (D) Pb

331. The element which can displace three other halogens from their compounds is [EAMCET-98]
(A) Cl (B) F (C) Br (D) I

332. Which of the following is the strongest Lewis acid ? [EAMCET-98]
(A) BI_3 (B) BBr_3 (C) BCl_3 (D) BF_3

333. The element which can displace three other halogens from their compounds is [EAMCET(MBBS)-98]
(A) Cl (B) Br (C) I (D) F

334. Which one of the following hydrides is more acidic ? [EAMCET(MBBS)-98]
(A) H_2S (B) H_2O (C) H_2Se (D) H_2Te

335. The oxide of chlorine which is a colourless oily liquid is [EAMCET(MBBS)-98]
(A) Cl_2O (B) Cl_2O_6 (C) ClO_2 (D) Cl_2O_7

336. The industrial preparation of nitric acid by Ostwald's process involves [EAMCET-99]
(A) Reduction of NH_3 (B) Oxidation of NH_3
(C) Hydrogenation of NH_3 (D) Hydrolysis of NH_3

337. In which of the following reactions ozone acts as reducing reagent ? [EAMCET-99]
(A) $BaO_2 + O_3 \rightarrow BaO + 2O_2$
(B) $2HCl + O_3 \rightarrow Cl_2 + H_2O + O_2$
(C) $PbS + 4O_3 \rightarrow PbSO_4 + 4O_2$
(D) $2KI + O_3 + H_2O \rightarrow 2KOH + I_2 + O_2$

338. Which of the following is a weak acid in its aqueous solution. [EAMCET-99]
(A) H_2Te (B) H_2Se (C) H_2S (D) H_2Po

339. Which of the following is the anhydride of HNO_3 ? [EAMCET-99]
(A) NO (B) N_2O_3 (C) N_2O_4 (D) N_2O_5

340. The oxidation number of N in HN_3 is : [EAMCET-99]
(A) $+\frac{1}{3}$ (B) 0 (C) $-\frac{1}{3}$ (D) 1

341. Which of the following form of sulphur can be prepared by pouring liquid sulphur into cold water ? [EAMCET(MBBS)-99]
(A) monoclinic (B) rhombic (C) plastic (D) colloidal

342. Hydrolysis of SCl_4 forms [EAMCET(MBBS)-99]
(A) H_2SO_4 (B) H_2SO_3 (C) SO_2 (D) H_2SO_6

343. Which of the following is used in the preparation of 'ammonal' ? [EAMCET(MBBS)-99]
(A) Fe_2O_3 (B) Al powder
(C) $CaCN_2$ + C (graphite) (D) NaOH + CaO

344. What is formed when 3 parts of HCl is mixed with

one part of HNO_3? [EAMCET(MBBS)-99]
(A) NO (B) NCl_3 (C) NOCl (D) NH_4Cl

345. SiO_2 is soluble in [EAMCET(MBBS)-99]
(A) H_2SO_4 (B) HNO_3 (C) HF (D) HCl

346. Which of the following does not exhibit allotropy? [EAMCET(MBBS)-99]
(A) P (B) Bi (C) Sb (D) As

347. Which of the following can form compounds with Xenon? [EAMCET(MBBS)-99]
(A) I_2 (B) F_2 (C) Br_2 (D) Al

348. Which of the following is most stable? [EAMCET(MBBS)-99]
(A) HF (B) HI (C) HCl (D) HBr

349. Which of the following is the strongest oxidizing agent? [EAMCET-2000]
(A) F_2 (B) Br_2 (C) Cl_2 (D) I_2

350. What is the oxidation state of chlorine in hypochlorous acid? [EAMCET-2000]
(A) +1 (B) +3 (C) +5 (D) +7

351. What is the catalyst used in the manufacture of sulphuric acid by contact process? [EAMCET-2000]
(A) Ni Powder (B) Platinised asbestos
(C) Anhydrous Al_2O_3 (D) Al Powder

352. What is the molecular formula of white phosphorous? [EAMCET-2000]
(A) P_2 (B) P_4 (C) P_5 (D) P_{16}

353. Which of the following is an acid? [EAMCET-2000]
(A) Ca $(OH)_2$ (B) $P(OH)_3$ (C) NH_4OH (D) NaOH

354. What is the electrolyte used in the electrolytic method of preparation of fluorine? [EAMCET(MBBS)-2000]
(A) NaF + $CaCl_2$ (B) KHF_2 + LiF
(C) KF + $LiHF_2$ (D) $CaCl_2$ + KF

355. Bleaching action of chlorine occurs in the presence of: [EAMCET(MBBS)-2000]
(A) O_2 (B) Moisture (C) Dry air (D) Sun light

356. Which of the following has more acidic character? [EAMCET(MBBS)-2000]
(A) H_2O (B) H_2Te (C) H_2S (D) H_2Se

357. In which of the following compounds, oxyen exhibits an oxidation state of +2? [EAMCET(MBBS)-2000]
(A) H_2O (B) H_2O_2 (C) OF_2 (D) H_2SO_4

358. What are the products obtained when ammonia is reacted with excess chlorine? [EAMCET-2001]
(A) N_2 and NCl_3 (B) N_2 and NCl
(C) N_2 and NH_4Cl (D) NCl_3 and HCl

359. What are the products formed when chlorine is passed through aqueous hypo solution? [EAMCET-2001]
(A) Na_2SO_3 + HCl +S (B) Na_2SO_3 + SO_3 +HCl
(C) Na_2SO_4 + HCl +S (D) Na_2SO_4 + HCl + SO_2

360. Which one of the following order is correct for the bond energies of halogen molecules? [EAMCET-2001]
(A) $I_2 > Cl_2 > Br_2$ (B) $Br_2 > Cl_2 > I_2$
(C) $I_2 > Br_2 > Cl_2$ (D) $Cl_2 > Br_2 > I_2$

361. Which one of the following halogen liberates oxygen, when passed through hot concentrated KOH solution? [EAMCET-2001]
(A) I_2 (B) Cl_2 (C) Br_2 (D) F_2

362. The number of lone pair of electrons present on Xe in XeF_2 is [EAMCET-2001]
(A) 3 (B) 4 (C) 2 (D) 1

363. Which one of the following is mainly responsible for depletion of ozone layer? [EAMCET-2001]
(A) Methane (B) Carbon dioxide
(C) Water (D) Chloro flurocarbons

364. 3 moles of ethanol reacts with one mole of phosphorus tribromide to form 3 moles of bromo-ethane and one mole of X. Which of the following is X? [EAMCET-2001]
(A) H_3PO_4 (B) H_3PO_2 (C) HPO_3 (D) H_3PO_3

365. Bond dissociation energies of HF, HCl, HBr follow the order - [EAMCET-2001]
(A) HCl > HBr > HF (B) HF > HBr > HCl
(C) HF > HCl > HBr (D) HBr > HCl > HF

366. What are the oxidation numbers of nitrogen in NH_4NO_3? [EAMCET-2001]
(A) +3, –5 (B) –3, +5 (C) +3, –6 (D) +2, +2

367. What is the product formed when ozone reacts with mercury? [EAMCET(MBBS)-2001]
(a) HgO (B) Hg_2O_2 (C) Hg_2O (D) HgO_2

368. What is the Oxidation state of chlorine in hypochlorous acid? [EAMCET(MBBS)-2001]
(A) +7 (B) +5 (C) +3 (D) +1

369. In the presence of Cobalt Chloride ($CoCl_2$), bleaching powder decomposes to form [EAMCET(MBBS)-2001]
(A) $CaCO_3$ & O_3 (B) ClO_2 & CaO
(C) Cl_2O & CaO (D) $CaCl_2$ & O_2

370. What is the atomic number (Z) of the noble gas that reacts with fluorine? [EAMCET(MBBS)-2001]
(A) 54 (B) 10 (C) 18 (D) 2

371. Which one of the following regions of atmosphere contains Ozone? [EAMCET(MBBS)-2001]
(A) Troposphere (B) Thermosphere
(C) Mesosphere (D) Stratosphere

372. The electronic configuration of four different elements is given below. Identify the group IV element among these.

[EAMCET(MBBS)-2001]

(A) [He] $2s^1$ (B) [Ne] $3s^2$
(C) [Ne] $3s^23p^2$ (D) [Ne] $3s^23p^5$

373. Which one of the following elements show different oxidation states ? [EAMCET(MBBS)-2001]
(A) Sodium (B) Fluorine (C) Chlorine (D) Potassium

374. What is the gas liberated when alkaline formaldehyde solution is reacted with hydrogen peroxide in the presence of pyrogallol ? [EAMCET(MBBS)-2001]
(A) H_2 (B) O_2 (C) CH_4 (D) CO_2

375. Aqueous sodium hydroxide reacts with white phosphorous to form phosphine and : [EAMCET(MBBS)-2001]
(A) NaH_2PO_2 (B) P_2O_5 (C) Na_3PO_3 (D) P_2O_3

376. Which one of the following is used in the industrial preparation of nitric acid by Ostwald's process ? [EAMCET(MBBS)-2001]
(A) N_2 (B) NH_4NO_3 (C) N_2O_5 (D) NH_3

377. Which one of the following is an oxyacid ? [EAMCET(MBBS)-2001]
(A) $Ba(OH)_2$ (B) $Mg(OH)_2$ (C) H_3PO_3 (D) HCl

378. Which one of the following statements is correct with respect to basic character ? [EAMCET-2002]
(A) $PH_3 > P(CH_3)_3$ (B) $PH_3 = NH_3$
(C) $PH_3 > NH_3$ (D) $P(CH_3)_3 > PH_3$

379. What are the products formed when ammonia reacts with excess chlorine ? [EAMCET-2002]
(A) N_2 and CHI_3 (B) NCl_3 and HCl
(C) N_2 and NH_4Cl (D) N_2 and HCl

380. Which one of the following is a correct pair with respect to molecular formula of Xenon compound and hybridization state of Xenon in it ? [EAMCET-2002]
(A) XeF_4, sp^3 (B) XeF_2, sp
(C) XeF_2, sp^3d (D) XeF_4, sp^2

381. When bauxite powder is mixed with coke and reacted with nitrogen at 2075 K, carbon monoxide and X are formed. What is the gas formed when X is reacted with water ? [EAMCET-2002]
(A) NH_3 (B) N_2 (C) N_2O (D) O_2

382. Fluorine oxidizes HSO_4^- to[EAMCET(MED)-2002]
(A) $S_2O_3^{-2}$ (B) $S_2O_8^{2-}$ (C) $S_4O_6^{2-}$ (D) SO_2

383. One mole of A (g) is heated to 300°C in a closed one litre vessel till the following equilibrium is reached. $A(g) \rightleftharpoons B(g)$ The equilibrium constant of this reaction at 300°C is 4. What is the concentration of B(in mol. lit^{-1}) at equilibrium? [EAMCET(MED)-2002]
(A) 0.2 (B) 0.6 (C) 0.8 (D) 0.1

384. The number of unpaired electrons present in the first excited state of chlorine atom is.[EAMCET(MED)-2002]
(A) 1 (B) 3 (C) 5 (D) 2

385. Which one of the following elements does not form the compound, M_4O_{10} (M = element) ? [EAMCET(MED)-2002]
(A) P (B) Sb (C) As (D) Bi

386. When ammonium nitrate is gently heated, an oxide of nitrogen is formed. What is the oxidation state of nitrogen in this oxide ? [EAMCET(MED)-2002]
(A) +4 (B) +2 (C) +3 (D) +1

387. What are the products formed in reaction of xenon hexafluoride with silicon dioxide ?[EAMCET(MED)-2002]
(A) $XeSiO_4$ + HF (B) $XeF_2 + SiF_4$
(C) $XeOF_4 + SiF_4$ (D) $XeO_3 + SiF_2$

388. Which one of the following compounds undergoes hydrolysis during distillation to yield hydrogen peroxide ? [EAMCET(MED)-2002]
(A) $H_2S_2O_8$ (B) $H_2S_2O_6$ (C) HNO_3 (D) $H_4P_2O_7$

389. The reagent used in the detection of fluoride in water is. [EAMCET(MED)-2002]
(A) Zirconium - alizarin S (B) Phenolphthalein
(C) Bayer's reagent (D) Soda lime

390. Water is considered to be polluted, if the dissolved oxygen (DO) contant is less than ____ ppm. [EAMCET-2003]
(A) 5 (B) 10 (C) 15 (D) 100

391. Compound x is anhydride of sulphuric acid. The number of σ bonds and the number of π bonds respectively are [EAMCET-2003]
(A) 3,3 (B) 4,2 (C) 2,4 (D) 4,3

392. The oxidation state of Xe in XeO_3 and the bond angle in it respectively are [EAMCET-2003]
(A) +6, 109° (B) +8, 103° (C) +6, 103° (D) +8, 120°

393. Sulphur trioxide dissolved in heavy water to form a compound X. The hybridisation state of sulphur in X is [EAMCET-2003]
(A) sp^2 (B) sp^3 (C) sp (D) dsp^2

394. Two oxides of nitrogen NO and NO_2 react together at 253k and form a compound of nitrogen X. X reacts with water to yield another compound of nitrogen Y. The shape of the anion of Y molecule is [EAMCET-2003]
(A) Tetrahedral (B) Triangular planar
(C) Square planar (D) Pyramidal

395. Chlorine atom in the third excited state reacts with fluorine to form a compound X. The formula and the shape of X are [EAMCET-2003]
(A) ClF_5, pentagonal (B) ClF_4, Tetrahedral
(C) ClF_4, pentagonal (D) ClF_7, pentagonal bipyramidal

396. Which gas is obtained when urea is heated with

HNO_2 ? [CPMT-96]

(A) N_2 (B) H_2 (C) O_2 (D) NH_3

397. At. No of N is 7 the at. no. of IIIrd member of Nitrogen family [CPMT-96]

(A) 23 (B) 15 (C) 33 (D) 43

398. No. of unpaired electrons in inert gas is [CPMT-96]

(A) Zero (B) 8 (C) 4 (D) 18

399. Which of the following conditions is used to find atomic Cl from molecular Cl_2? [CPMT-96]

(A) High temp., high pressure
(B) Low temp., high pressure
(C) High temp., low pressure (D) Low temp, low pressure

400. What is formed when HNO_3 is heated ?[CPMT-96]

(A) N_2O_5 (B) N_2O_4 (C) NO_2 (D) NO

401. X in the given reaction is

$HNO_3 + 2P_2O_5 \rightarrow P_4O_{10} + X$ [CPMT-97]

(A) N_2O_4 (B) N_2O_5 (C) NO_2 (D) H_2O

402. White phosphorus + Caustic soda $\rightarrow$ Phosphine+ NaH_2PO_2. The above reaction is - [CPMT-97]

(A) Oxidation reaction (B) Reduction reaction
(C) Neutralisation (D) Redox reaction

403. Which one is an oxidising substance ? [CPMT-97]

(A) CH_2O_2 (B) CO (C) H_2S (D) CO_2

404. When iodine reacts with NaF, NaBr and NaCl - [CPMT-97]

(A) It gives mixture of F_2, Cl_2 and Br_2
(B) It gives chlorine
(C) It gives bromine (D) None of these

405. Identify the correct statement when following compounds are given HF, HBr, H_2Te, H_3P [CPMT-97]

(A) HF (strong acid) (B) H_2Te (strong alkali)
(C) HBr (Strong acid) (D) H_3P (strong alkali)

406. Fluorine has lower electron affinity than chlorine because of - [CPMT-97]

(A) Smaller radius of fluorine, high density
(B) Smaller radius of chlorine, high density
(C) Bigger radius of fluorine, less density
(D) Smaller radius of chlorine, high density

407. $KO_2 + CO_2 \rightarrow$? (gas) [CPMT-97]

(A) H_2 (B) N_2 (C) O_2 (D) CO

408. When lead nitrate is heated it gives [CPMT-97]

(A) NO_2 (B) NO (C) N_2O_5 (D) N_2O

409. The correct representation of sulphur is[CPMT-98]

(A) S (B) S_2 (C) S_4 (D) S_8

410. Bleaching powder is a compound having a formula - [CPMT-98]

(A) $CaCl_2$ (B) CaClO (C) $CaOCl_2$ (D) $CaOCl_3$

411. The ammonia is dried over- [CPMT-98]

(A) The slaked lime (B) Quick-lime
(C) Calcium chloride (D) Phosphorus pentachloride

412. Which one of the following is amphoteric ? [CPMT-98]

(A) P_2O_5 (B) As_2O_3 (C) N_2O_3 (D) N_2O_5

413. Which one of the following oxides of nitrogen is the anhydride of HNO_3? [CPMT-98]

(A) NO (B) N_2O_3 (C) N_2O_4 (D) N_2O_5

414. 'Aqua regia' is a mixture of - [CPMT-98]

(A) 3HCl + HNO_3 (B) H_3PO_4+ H_2SO_4
(C) $3HNO_3$ + HCl (D) HCl + CH_3COOH

415. If phosphoric acid is allowed to react with sufficient quantity of NaOH, the product obtained is- [CPMT-98]

(A) Na_3PO_4 (B) Na_2HPO_4
(C) NaH_2PO_4 (D) $NaHPO_3$

416. Mixture used on the tips of match sticks is- [CPMT-98]

(A) (S+K) (B) Sb_2S_3
(C) ($K_2Cr_2O_7$ + S + white P) (D) ($K_2Cr_2O_7$ + K + S)

417. Ozone is obtained from oxygen- [CPMT-98]

(A) By oxidation at high temp.
(B) By oxidation in the presence of a catalyst
(C) By heating at high pressure
(D) By silent electric discharge

418. Which one of the following burns to form an oxide which is gaseous at room temperature ? [CPMT-98]

(A) H_2 (B) He (C) Na (D) S

419. Which one is Caro's acid ? [CPMT-98]

(A) H_2SO_5 (B) $H_2S_2O_8$ (C) H_2SO_3 (D) H_2SO_4

420. Which one of the following shows only one oxidation state ? [CPMT-98]

(A) F (B) Cl (C) Br (D) I

421. The strongest reducing agent is- [CPMT-98]

(A) HF (B) HI (C) HCl (D) HBr

422. Which one of the following is strongest acid ? [CPMT-98]

(A) HF (B) HCl (C) HBr (D) HI

423. The reagent that has action on glass is-[CPMT-98]

(A) Aquaregia (B) HF acid
(C) Oleum (D) Fuming HNO_3

424. Which one of the following is most volatile ? [CPMT-98]

(A) HI (B) HCl (C) HBr (D) HF

425. Which one of the following is a pseudohalide ? [CPMT-98]

(A) CN^- (B) ICl (C) IF_5 (D) I_3^-

426. O_2 molecule is [CPMT-99]
(A) Paramagnetic (B) Diamagnetic
(C) Ferromagnetic (D) None of these

427. Which of the following are isoelectronic species ? [CPMT-99]
I - $\overset{+}{C}H_3$, II-NH_2^-, III NH_4^+, IV-NH_3
(A) I, II, III (B) II, III, IV (C) I, II, IV (D) II and I

428. Among Cl^-, Br^-, I^- the correct order for being oxidised to dihalogen is [CPMT-99]
(A) $I^- > Cl^- > Br^-$ (B) $Cl^- > Br^- > I^-$
(C) $I^- > Br^- > Cl^-$ (D) $Br^- > I^- > Cl^-$

429. Orthophosphoric acid is - [CPMT-99]
(A) Monobasic (B) Dibasic
(C) Tribasic (D) Tetrabasic

430. Which of the following ions does not have S-S linkage ? [CPMT-99]
(A) $S_2O_8^{2-}$ (B) $S_2O_6^{2-}$ (C) $S_2O_5^{2-}$ (D) $S_2O_3^{2-}$

431. The number of unpaired electrons in oxygen molecule is - [CPMT-2000]
(A) Zero (B) 1 (C) 2 (D) 3

432. Which is most reactive ? [CPMT-2000]
(A) Red phosphorus (B) White phosphorus
(C) Violet Phosphorus (D) Black phosphorus

433. The bond angles decrease in the order[CPMT-2000]
(A) $NH_3 > Ph_3 > AsH_3 > SbH_3$
(B) $PH_3 > AsH_3 > SbH_3 > NH_3$
(C) $SbH_3 > AsH_3 > PH_3 > NH_3$
(D) $NH_3 > SbH_3 > AsH_3 > PH_3$

434. The shape of XeF_2 molecule is [CPMT-2000]
(A) Tetrahedral (B) Pyramidal
(C) Trigonal planar (D) Linear

435. White phosphorus contains [CPMT-2000]
(A) P_2 molecules (B) P_6 molecules
(C) P_4 molecules (D) P_8 molecules

436. Hybridisation and structure of XeF_4 is[CPMT-2001]
(A) sp^3d, trigonal bipyramidal (B) sp^3 , tetrahedral
(C) sp^3d^2 , ocatahedral (D) sp^4d^2, hexagonal

437. Which of the following about fluorine is not correct? [CPMT-2001]
(A) The electron affinity of chlorine is greater than fluorine
(B) The bond energy of fluorine is less than chlorine
(C) Fluorine does not from oxyacids
(D) Fluorine cannot be prepared from electrolysis of metal fluorides

438. The maximum valency shown by nitrogen and phosphorus is [CPMT-2001]
(A) 3 and 5 (B) 4 and 3 (C) 5 and 5 (D) 5 and 3

439. The weakest acid among the following is [CPMT-2001]
(A) HI (B) HBr (C) HCl (D) HF

440. Ammonia is dried over [CPMT-2002]
(A) quick lime (B) slaked lime
(C) anhydrous $CaCl_2$ (D) none of these

441. Sea weeds are an important source of [CPMT-2002]
(A) F_2 (B) Br_2 (C) Cl_2 (D) I_2

442. Bleaching powder is obtained by the interaction of Cl_2 and [CPMT-2002]
(A) dry slaked lime (B) hot $Ca(OH)_2$
(C) cold $Ca(OH)_2$ (D) conc. $CaCl_2$

443. O_2 molocule is [CPMT-2002]
(A) diamagnetic (B) ferromagnetic
(C) paramagnetic (D) none of these

444. On hydrolysis PCl_3 gives [CPMT-2003]
(A) H_3PO_4 (B) $POCl_3$ (C) H_3PO_3 (D) PH_3

445. Phosphine is prepared by the reaction between [CPMT-2003]
(A) white phosphorus and excess of NaOH solution
(B) white phosphorus and dilute solution of NaOH
(C) red phosphorus and excess of NaOH solution
(D) red phosphorus and dil. solution of NaOH.

446. Which shows maximum valency ? [CPMT-2003]
(A) Phosphorus (B) Tin (C) Antimony (D) Bismuth

447. Sodium pyrophosphate is [CPMT-2003]
(A) $Na_2P_2O_7$ (B) $Na_4P_2O_7$ (C) $NaPO_4$ (D) Na_2PO_2

448. On boiling phosphorus with KOH solution product formed is [CPMT-2003]
(A) Potassium phosphate (B) Phosphorus pentaoxide
(C) Phosphorus hydroxide (D) Phosphine

449. On reaction with Mg very dilute Nitric acid produces [CPMT-2003]
(A) NH_3 (B) Nitrous oxide
(C) Nitric oxide (D) Hydrogen

450. Which of the following is a pseudo halogen : [PMT-93]
(A) IF_7 (B) $(CN)_2$ (C) ICl_2 (D) I_3

451. Deacon's Process is used for the manufacture of [PMT-93]
(A) Bleaching Powder (B) Sulphuric Acid
(C) Nitric Acid (D) Chlorine

452. HBr and Hg reduce sulphuric acid, HCl can reduce $KMnO_4$ and HF can reduce : [PMT-93]
(A) H_2SO_4 (B) $KMnO_4$ (C) $K_2Cr_2O_7$ (D) none of these

453. When ammonium dichromate is heated the gas formed is : [PMT-93]
(A) N_2 (B) O_2 (C) H_2 (D) NH_3

454. NO_2 gas can not be obtained by heating :[PMT-93]
(A) KNO_3 (B) $Pb(NO_3)_2$ (C) $Cu(NO_3)_2$ (D) $AgNO_3$

455. Calcium carbide reacts with nitrogen to form : [PMT-93]
(A) $CaCN_2$ and C (B) $CaCN_2$
(C) $CaCN_2$ and N_2 (D) none of these

456. The element which liberates oxygen gas from water is : [PMT-93]
(A) P (B) Na (C) F (D) I

457. Which one of the following noble gases is not found in atmosphere : [PMT-93]
(A) Rn (B) Kr (C) Ne (D Ar

458. Helium is added to the oxygen supply used by deep sea divers because : [PMT-93]
(A) it is less soluble in blood than nitrogen at high pressure
(B) it is lighter than nitrogen
(C) it is readily miscible with oxygen
(D) it is less poisonous than nitrogen

459. Of the following one having a square planar structure is [PMT-94]
(A) NH_4^+ (B) BF_4^- (C) XeF_4 (D) CCl_4

460. Which one of the following is the anhydride of nitric acid ? [PMT-94]
(A) NO (B) NO_2 (C) N_2O_3 (D) N_2O_5

461. Fluorine forms chemical compounds with [PMT-94]
(A) He (B) Ne (C) Ar (D) Xe

462. The bleaching action of the bleaching powder is due to the liberation of [PMT-94]
(A) chlorine (B) molelcular oxygen
(C) nascent oxygen (D) calcium carbonate

463. The passivity of iron in concentrated nitric acid is due to [PMT-94]
(A) ferric nitrate coating on the metal
(B) ammonium nitrate coating on the metal
(C) a thin oxide layer coating on the metal
(D) a hydride coating on the metal

464. Which element from V group gives most basic compound with hydrogen ? [PMT-95]
(A) Nitrogen (B) Bismuth (C) Arsenic (D) Phosphorus

465. Which is the weakest acid out of HF,HCl,HBr and HI ? [PMT-95]
(A) HF (B) HCl (C) HBr (D) HI

466. Which element out of He, Ar, Kr, and Xe forms least number of compounds ? [PMT-95]
(A) He (B) Ar (C) Kr (D) Xe

467. The sulphate of a metal has the formula $M_2(SO_4)_3$. The formula for its phosphate will be [PMT-95]
(A) $M(HPO_4)_2$ (B) $M_3(PO_4)_2$
(C) $M_2(PO_4)_3$ (D) MPO_4

468. When zinc reacts with very dilute nitric acid it produces [PMT-95]
(A) NO (B) NH_4NO_3 (C) NO_2 (D) H_2

469. The electronic configuration of an element is
$1s^2\ 2s^2\ 2p^6\ 3s^2\ 3p^6\ 3d^{10}\ 4s^2\ 4p^3$
Its properties would be similar to which of the following elements ? [PMT-95]
(A) Boron (B) Oxygen (C) Nitrogen (D) Chlorine

470. The formula of an oxyacid of phosphorus is H_3PO_3. It is a : [PMT-96]
(A) dibasic acid (B) monobasic acid
(C) tribasic acid (D) tetrabasic acid

471. The molecular formula of sulphur is : [PMT-96]
(A) S (B) S_2 (C) S_4 (D) S_8

472. The strongest oxidising agent amoungst F_2, Cl_2, Br_2 and I_2 is : [PMT-96]
(A) F_2 (B) I_2 (C) Cl_2 (D) Br_2

473. Point out in which of following properties oxygen differs from the rest of the members of its family (Group - VIA)- [PMT-97]
(A) High value of ionisation energies
(B) Oxidation states (2,4,6)
(C) Polymorphism (D) Formation of hydrides

474. The electron affinities of halogens are F=322, Cl=349, Br=324, I=295 KJ mol^{-1}. The higher value for Cl as compared to that of F is due to - [PMT-97]
(A) Weaker electron-electron repulsion in Cl
(B) Higher atomic radius of F
(C) Smaller electronegativity of F
(D) More vacant P-subshell in Cl

475. The last member of inert gas elements is :[PMT-99]
(A) Helium (B) Neon (C) Argon (D) Radon

476. Which of the following elements is paramagnetic ? [PMT-99]
(A) Nitrogen (B) Hydrogen (C) Oxygen (D) Chlorine

477. A deep brown gas is formed by mixing two colourless gases which are : [PMT-2000]
(A) NO_2 and O_2 (B) N_2O and NO
(C) NO and O_2 (D) NH_3 and HCl

478. Going from Fluorine, to chlorine, bromine and iodine the electronegativity : [PMT-2000]
(A) increases (B) decreases
(C) first decreases then increases (D) changes randomly

479. In the Nitrogen family the H-M-H bond angle in the hydrides (MH_3) gradually becomes closer to 90° on going from N to Sb. This shows that : [PMT-2000]
(A) the basic strength of hydrides increases.
(B) the bond energies of M-H bonds increase.

(C) almost pure orbitals are used for M-H bonding.
(D) the bond pairs of electrons come nearer to the central atom.

480. The halogen that is most easily reduced is : [PMT-2000]
(A) F_2 (B) Cl_2 (C) Br_2 (D) I_2

481. The oxidation state of nitrogen is highest in [PMT-2001]
(A) N_3H (B) NH_2OH (C) N_2H_4 (D) NH_3

482. The oxidation state of oxygen is zero in : [PMT-2001]
(A) CO (B) O_3 (C) SO_2 (D) H_2O_2

483. What causes nitrogen to be chemically inert ? [PMT-2001]
(A) Multiple bond formation in the molecule
(B) Absence of bond polarity
(C) Short internuclear distance (D) High bond energy

484. Which of the following compounds is explosive in nature ? [PMT-2001]
(A) Phosphorous trichloride (B) Nitrogen trichloride
(C) Hyponitrous acid (D) Nitrosyl chloride

485. To obtain bromine in the manufacture of bromine from sea water the mother liquor containing bromide is treated with : [PMT-2001]
(A) CO_2 (B) Cl_2 (C) I_2 (D) H_2O

486. At room temperature HCl is a gas while HF is a low boiling liquid. This is because : [PMT-2001]
(A) H-F bond is covalent (B) H-F bond is ionic
(C) H-F has metallic bond (D) H-F has hydrogen bond

487. Which of the following is the life saving mixture for an asthma patient ? [PMT-2001]
(A) Mixture of helium and oxygen
(B) Mixture of neon and oxygen
(C) Mixture of xenon and nitrogen
(D) Mixture of argon and oxygen

488. Which of the following inert gases forms a large number of compounds ? [PMT-2001]
(A) Ne (B) Ar (C) Kr (D) Xe

489. Of the following compounds the most acidic is [PMT-2002]
(A) AS_2O_3 (B) P_2O_5 (C) Sb_2O_3 (D) Bi_2O_3

490. The electronic configuration is $1s^2 2s^2 2p^6 3s^2 3p^5 4s^1$ shows : [PMT-2003]
(A) Lowers state of chlorine atom
(B) Excited state of argon
(C) Excited state of chlorine atom
(D) Lowest state of potassium

491. The ionization energy of nitrogen is more than that of oxygen because [PET-93]
(A) Nitrogen has half-filled p-orbitals
(B) Nitrogen is left to the oxygen in the same period of the periodic table
(C) Nitrogen contains less number of electrons
(D) Nitrogen is less electronegative

492. Which of the following fluorides does not exist ? [PET-93]
(A) NF_5 (B) PF_5 (C) AsF_5 (D) SbF_5

493. The highest catenation ability is shown by [PET-93]
(A) Oxygen (B) Sulphur (C) Selenium (D) Tellurium

494. Which of the following pairs is not correctly matched ? [PET-93]
(A) A halogen which is liquid at room temperature-bromine
(B) The most electronegative element-fluorine
(C) The most reactive halogen-fluorine
(D) The strongest oxidizing halogen-iodine

495. Which of the following statements is not correct for a noble gas ? [PET-93]
(A) Ar is used in electric bulb
(B) Kr is obtained during radioactive disintegration
(C) Half-life of Rn is only 3-8 days
(D) He is used in producing very low temperatures

496. Which of the following has highest boiling point [PET-94]
(A) NH_3 (B) PH_3 (C) AsH_3 (D) SbH_3

497. Which of the following dissociates to give H^+ most easily [PET-94]
(A) H_2O (B) H_2S (C) H_2Te (D) H_2Se

498. Which of the following oxidises H_2O to oxygen [PET-94]
(A) Chloride (B) Fluorine (C) Bromine (D) Iodine

499. Which of the following is not obtained by direct reaction of constituent elements [PET-94]
(A) XeF_2 (B) XeF_4 (C) XeO_3 (D) XeF_6

500. In solid argon the atoms are held together [PET-95]
(A) By ionic bonds (B) By hydrogen bonds
(C) By van der Waals forces
(D) By hydrophobic bonds

501. Oxygen is soluble in [PET-95]
(A) Alcohol (B) Acidic pyrogallol
(C) Acid (D) Alkaline pyrogallol

502. The laughing gas is [PET-95]
(A) Nitric oxide (B) Nitrous oxide
(C) Nitrogen peroxide (D) Nitrogen pentaoxide

503. The triple bond between N atoms of nitrogen molecule ($N \equiv N$) consists of [PET-96]
(A) Three σ bonds (B) Two σ-bonds and one π-bond

(C) One σ-bond and two π-bonds (D) Three π-bonds

504. Chlorine cannot displace [PET-96]
(A) Fluorine from NaF (B) Iodine form NaI
(C) Bromine from NaBr (D) None of the above

505. Iron when treated with conc. nitric acid [PET-96]
(A) Readily reacts (B) Slowly reacts
(C) Becomes passive (D) Gives ferrous nitrate

506. Which of the following noble gases does not have an octet of electrons in its outermost shell [PET-96]
(A) Neon (B) Radon (C) Argon (D) Helium

507. An element (X) forms compounds of the formula XCl_3, X_2O_5 and Ca_3X_2 but does not form XCl_5 which of the following is the element X ? [PET-97]
(A) B (B) Al (C) N (D) P

508. Which of the following halogens does not form oxyacid ? [PET-97]
(A) Fluorine (B) Chlorine (C) Bromine (D) Iodine

509. The strongest of the following acids is : [PET-97]
(A) HBr (B) HCl (C) HF (D) HI

510. Helium is used with oxygen in the apparatus of divers because it : [PET-97]
(A) is lighter than nitrogen
(B) Is not soluble in blood at high pressure
(C) Is easily available
(D) Is less reactive than nitrogen

511. Gradual addition of electronic shells in the noble gases causes a decrease in their : [PET-97]
(A) Ionization energy (B) Atomic radius
(C) Boiling point (D) Density

512. Very dilute nitric acid reacts with zinc to form zinc nitrate and : [PET-97]
(A) Ammonium nitrate (B) NO_2 (C) NO (D) N_2O

513. Unlike other halogens fluorine does not show higher oxidation states because : [PET-97]
(A) It is highly electronegative
(B) It has no d-orbitals
(C) Its atomic radius is very small
(D) The F^- ion is stable and isoelectronic with neon

514. Which of the following hydrides has the lowest boiling point ? [PET-97]
(A) H_2O (B) H_2S (C) H_2Se (D) H_2Te

515. In the Nitrogen family the H-M-H bond angle in the hydrides MH_3 gradually becomes closer to 90° on going from N to Sb. This shows that gradually : [PET-98]
(A) The basic strength of hydrides increases
(B) Almost pure p-orbitals are used for M-H bonding
(C) The bond energies of M-H bond increase
(D) The bond pairs of electrons become nearer to the central atom

516. The molecular formula of sulphur is : [PET-98]
(A) S_2 (B) S_4 (C) S_6 (D) S_8

517. All the elements of oxygen family are : [PET-98]
(A) Non-metals (B) Metalloids
(C) Radioactive (D) Polymorphic

518. Iodine is liberated when KI solution is added to an aqueous solution of : [PET-98]
(A) $ZnSO_4$ (B) $CuSO_4$ (C) $(NH_4)2SO_4$ (D) Na_2SO_4

519. Elements with outer electronic configuration ns^2np^6 are : [PET-98]
(A) Alkaline earth metals (B) Transition elements
(C) Chalcogens (D) Noble gases

520. In the Ostwald's process for the manufacture of HNO_3 the catalyst used is : [PET-99]
(A) Mo (B) Fe (C) Ni (D) Pt

521. Who among the following first prepared a stable compound of noble gas ? [PET-99]
(A) Rutherford (B) Rayleigh (C) Ramsay (D) Neil Bartlett

522. Which of the following noble gases is least polarizable? [PET-99]
(A) He (B) Ne (C) Ar (D) Xe

523. In which of the following is Oxygen collected ? [PET-99]
(A) Ethanol (B) Water (C) Mercury (D) Kerosene oil

524. Which of the following halogens is solid at room temperature ? [PET-99]
(A) Chlorine (B) Iodine (C) Bromine (D) Fluorine

525. Which of the following is manufactured from the molecular nitrogen by bacteria ? [PET-99]
(A) NO_3 (B) NO_2 (C) Amino acids (D) Ammonia

526. XeF_6 on hydrolysis gives : [PET-2000]
(A) XeO_3 (B) XeO (C) XeO_2 (D) Xe

527. The metal that is extracted from sea water is : [PET-2000]
(A) Mg (B) Ca (C) Fe (D) Ni

528. The most basic element is : [PET-2000]
(A) Fluorine (B) Iodine (C) Chlorine (D) Bromine

529. Which is the best reducing agent ? [PET-2000]
(A) F^- (B) Cl^- (C) Br^- (D) I^-

530. The oxidation number of oxygen in ozone is : [PET-2000]
(A) −3 (B) −2 (C) 0 (Zero) (D) + 3

531. In which of the following, oxygen has an oxidation state of + 2 ? [PET-2000]
(A) H_2O_2 (B) CO_2 (C) H_2O (D) OF_2

532. Aqua regia is a mixture of : [PET-2001]
(A) 3HCl + 1HNO$_3$ (B) H_3PO_4 + H_2SO_4
(C) 3HNO$_3$ + 1HCl (D) HCl + CH_3COOH

533. In the following the strongest base is : [PET-2001]
(A) NH_3 (B) PH_3 (C) AsH_3 (D) SbH_3

534. $FeSO_4$ forms brown ring with : [PET-2001]
(A) NO_3 (B) NO_2 (C) NO (D) N_2O_3

535. Oxidation state +1 for phosphorus is found in [PET-2001]
(A) H_3PO_3 (B) H_3PO_4 (C) H_3PO_2 (D) $H_4P_2O_7$

536. A glass bulb is filled with NO_2 gas and immersed in an icebath at 0°C, which becomes colourless after some time. This colourless gas will be [PET-2002]
(A) NO_2 (B) N_2O (C) N_2O_4 (D) N_2O_5

537. Liquid Ammonia is used for refrigeration because [PET-2002]
(A) It has a high dipole moment
(B) It has a high heat of vapourisation
(C) It is basic (D) It is a stable compound

538. When Ammonia is passed over heated copper oxide metallic copper is obtained. The reaction shows that ammonia is [PET-2002]
(A) A dehydrating agent (B) An oxidizing agent
(C) A reducing agent (D) A nitrating agent

539. Oleum is chemically known as [PET-2002]
(A) H_2SO_3 (B) H_2SO_5 (C) $H_2S_2O_7$ (D) $H_2S_2O_8$

540. Which of the following group shows the highest boiling point [PET-2002]
(A) H_2O (B) H_2S (C) H_2Se (D) H_2Te

541. Which of the following halogens exhibits only one oxidation state in its compounds [PET-2002]
(A) Bromine (B) Chlorine (C) Fluorine (D) Iodine

542. Bleaching action of SO_2 is due to its [PET-2002]
(A) Oxidising property (B) Acidic property
(C) Reducing property (D) Basic property

543. Sea weeds are an important source of [PET-2002]
(A) Iron (B) Chlorine (C) Iodine (D) Bromine

544. The coloured discharge tube for advertisements mainly contains [PET-2002]
(A) Xenon (Xe) (B) Argon (Ar)
(C) Helium (He) (D) Neon (Ne)

545. The correct sequence of decrease in the bond angle of the following hydrides is [PET-2002]
(A) $NH_3 > PH_3 > AsH_3 > SbH_3$
(B) $NH_3 > AsH_3 > PH_3 > SbH_3$
(C) $SbH_3 > AsH_3 > PH_3 > NH_3$
(D) $PH_3 > NH_3 >$ AsH3 $> SbH_3$

546. Which of the following in most stable hydride. [PET-2003]
(A) NH_3 (B) AsH_3 (C) PH_3 (D) BiH_3

547. For which of the following the bond angle will be minimum? [PET-2003]
(A) H_2S (B) H_2O (C) H_2Se (D) H_2Te

548. The ion formed with the reaction of nitrous acid and hydrochloride acid is [PET-2003]
(A) Nitrosonium ion (B) Nitronium ion
(C) Nitrate ion (D) Nitracidium ion

549. Which of the follcwing is strongest acid ?[PET-2003]
(A) HF (B) HI (C) HBr (D) HCl

550. Red phosphorus is chemically denser and less reactive because : [PET-2003]
(A) It is tetrahedral in structure
(B) It has polymeric structure
(C) It is red in colour (D) It has octahedral structure

551. When potassium dichromate is heated with conc. H_2SO_4 and soluble metal chloride, the orange vapours are produced. These are due to : [PET-2003]
(A) Cr_2O_3 (B) $CrOCl_2$ (C) CrO_4^{-2} (D) CrO_2Cl_2

552. $Na_2S_2O_3 + I_2 \rightarrow$ Product is [BHU-2003]
(A) $Na_2S_4O_6$ (B) NaI (C) Na_2S (D) S_2

553. Which of the following has the highest % of nitrogen? [AFMC-2003]
(A) Ammonium sulphate (B) Nitrolim
(C) Calcium ammonium nitrate (D) Urea

554. Which of the following forms of phosphorus is most reactive? [AFMC-2003]
(A) Red (B) White (C) Violet (D) Scarlet

555. SO_2 reacts with Cl in sunlight to form[AFMC-2003]
(A) Sulphur dioxide (B) Sulphuryl chloride
(C) Sulphanyl chloride (D) None of these

556. Argon gas was discovered by [AFMC-2003]
(A) Rayleigh (B) Ramsay
(C) Lockerey (D) None of these

557. Aqua regia is a mixture of HCl and HNO_3 in the ratio of [AFMC-2003]
(A) 1:3 (B) 3:1 (C) 3:2 (D) 2:3

558.Which is the best description of the behaviour of bromine in the reaction given below?[CBSE-PMT-2004]

$H_2O + Br_2 \longrightarrow HOBr + HBr$
(A) Proton acceptor only (B) Both oxidized and reduced
(C) Oxidized only (D) Reduced only

559.I^- reacts with MnO_4^- in alkaline medium. I^- change to [IIT-2004]

(A) I_2 (B) IO_4^- (C) IO_2^- (D) IO_3^-

560. The number of lone pairs in $XeOF_4$ is [IIT-2004]
(A) 0 (B) 1 (C) 2 (D) 3

561. The O–O linkage is present in [IIT-2004]
(A) $H_2S_2O_3$ (B) $H_2S_2O_8$ (C) $H_2S_4O_6$ (D) $H_2S_2O_6$

Group 15, 16, 17 and 18 - Discussions

1.	(C)	49.	(A)	97.	(D)	145.	(B)	193.	(A)	241.	(A)
2.	(B)	50.	(B)	98.	(A)	146.	(A)	194.	(D)	242.	(B)
3.	(B)	51.	(C)	99.	(B)	147.	(B)	195.	(B)	243.	(D)
4.	(B)	52.	(D)	100.	(A)	148.	(C)	196.	(C)	244.	(A)
5.	(A)	53.	(C)	101.	(C)	149.	(C)	197.	(C)	245.	(B)
6.	(A)	54.	(D)	102.	(C)	150.	(A)	198.	(B)	246.	(D)
7.	(B)	55.	(C)	103.	(B)	151.	(A)	199.	(C)	247.	(B)
8.	(C)	56.	(B)	104.	(E)	152.	(B)	200.	(D)	248.	(D)
9.	(B)	57.	(B)	105.	(A)	153.	(B)	201.	(D)	249.	(C)
10.	(B)	58.	(A)	106.	(D)	154.	(C)	202.	(A)	250.	(C)
11.	(B)	59.	(A)	107.	(B)	155.	(A)	203.	(A)	251.	(C)
12.	(A)	60.	(A)	108.	(C)	156.	(B)	204.	(B)	252.	(C)
13.	(C)	61.	(C)	109.	(D)	157.	(C)	205.	(C)	253.	(A)
14.	(C)	62.	(D)	110.	(C)	158.	(A)	206.	(B)	254.	(C)
15.	(A)	63.	(C)	111.	(B)	159.	(C)	207.	(C)	255.	(D)
16.	(C)	64.	(B)	112.	(A)	160.	(C)	208.	(A)	256.	(B)
17.	(B)	65.	(B)	113.	(D)	161.	(D)	209.	(C)	257.	(C)
18.	(D)	66.	(B)	114.	(C)	162.	(D)	210.	(C)	258.	(B)
19.	(C)	67.	(B)	115.	(B)	163.	(B)	211.	(C)	259.	*
20.	(A)	68.	(B)	116.	(A)	164.	(B)	212.	(B)	260.	(D)
21.	(C)	69.	(D)	117.	(A)	165.	(A)	213.	(C)	261.	*
22.	(A)	70.	(D)	118.	(A)	166.	(B)	214.	(A)	262.	*
23.	(C)	71.	(B)	119.	(A)	167.	(A)	215.	(D)	263.	*
24.	(C)	72.	(D)	120.	(A)	168.	(D)	216.	(C)	264.	(C)
25.	(C)	73.	(D)	121.	(B)	169.	(A)	217.	(C)	265.	(C)
26.	(B)	74.	(C)	122.	(A)	170.	(A)	218.	(D)	266.	(A)
27.	(A)	75.	(C)	123.	(A)	171.	(B)	219.	(C)	267.	(B)
28.	(A)	76.	(A)	124.	(B)	172.	(A)	220.	(A)	268.	*
29.	(A)	77.	(A)	125.	(D)	173.	(C)	221.	(A)	269.	*
30.	(C)	78.	(A)	126.	(B)	174.	(C)	222.	(C)	270.	*
31.	(B)	79.	(C)	127.	(B)	175.	(D)	223.	(B)	271.	(A)
32.	(C)	80.	(A)	128.	(C)	176.	(B)	224.	(B)	272.	(C)
33.	(C)	81.	(A)	129.	(C)	177.	(B)	225.	(A)	273.	(D)
34.	(C)	82.	(D)	130.	(C)	178.	(B)	226.	(C)	274.	(B)
35.	(B)	83.	(A)	131.	(A)	179.	(C)	227.	(B)	275.	(C)
36.	(A)	84.	(B)	132.	(D)	180.	(C)	228.	(B)	276.	(A)
37.	(C)	85.	(A)	133.	(D)	181.	(D)	229.	(B)	277.	(C)
38.	(A)	86.	(A)	134.	(C)	182.	(B)	230.	(C)	278.	(A)
39.	(B)	87.	(A)	135.	(A)	183.	(A)	231.	(A)	279.	(C)
40.	(C)	88.	(A)	136.	(B)	184.	(A)	232.	(B)	280.	(A)
41.	(C)	89.	(B)	137.	(B)	185.	(C)	233.	(A)	281.	(C)
42.	(C)	90.	(A)	138.	(A)	186.	(A)	234.	(B)	282.	(B)
43.	(B)	91.	(A)	139.	(C)	187.	(C)	235.	(B)	283.	(C)
44.	(A)	92.	(D)	140.	(D)	188.	(D)	236.	(C)	284.	(A)
45.	(B)	93.	(D)	141.	(C)	189.	(C)	237.	(A)	285.	(A)
46.	(B)	94.	(C)	142.	(A)	190.	(D)	238.	(D)	286.	(A)
47.	(B)	95.	(D)	143.	(D)	191.	(D)	239.	(C)	287.	(D)
48.	(D)	96.	(A)	144.	(D)	192.	(A)	240.	(A)	288.	(A)

259. sp
261. four
262. oxidation of nitrogen
263. I_3^-
268. NF_3
269. 60°
270. H_2S

289.	(C)	337.	(A)	385.	(D)	433.	(A)	481.	(A)	529.	(D)
290.	(A)	338.	(C)	386.	(D)	434.	(D)	482.	(B)	530.	(C)
291.	(B)	339.	(D)	387.	(C)	435.	(C)	483.	(D)	531.	(D)
292.	(D)	340.	(C)	388.	(A)	436.	(C)	484.	(B)	532.	(A)
293.	(A)	341.	(C)	389.	(A)	437.	(D)	485.	(A)	533.	(A)
294.	(C)	342.	(B)	390.	(A)	438.	(A)	486.	(D)	534.	(C)
295.	(D)	343.	(B)	391.	(A)	439.	(D)	487.	(A)	535.	(C)
296.	*	344.	(C)	392.	(C)	440.	(A)	488.	(D)	536.	(C)
297.	(A)	345.	(C)	393.	(B)	441.	(D)	489.	(B)	537.	(B)
298.	(C)	346.	(B)	394.	(B)	442.	(C)	490.	(B)	538.	(C)
299.	(B)	347.	(B)	395.	(D)	443.	(C)	491.	(A)	539.	(C)
300.	*	348.	(A)	396.	(A)	444.	(C)	492.	(A)	540.	(A)
301.	*	349.	(A)	397.	(C)	445.	(A)	493.	(B)	541.	(C)
302.	*	350.	(A)	398.	(A)	446.	(A)	494.	(D)	542.	(C)
303.	*	351.	(B)	399.	(B)	447.	(B)	495.	(D)	543.	(C)
304.	(D)	352.	(B)	400.	(C)	448.	(D)	496.	(D)	544.	(D)
305.	(C)	353.	(B)	401.	(C)	449.	(D)	497.	(C)	545.	(A)
306.	(C)	354.	(B)	402.	(D)	450.	(B)	498.	(B)	546.	(A)
307.	*	355.	(B)	403.	(C)	451.	(D)	499.	(C)	547.	(D)
308.	*	356.	(B)	404.	(D)	452.	(D)	500.	(C)	548.	(C)
309.	*	357.	(C)	405.	(C)	453.	(A)	501.	(D)	549.	(B)
310.	*	358.	(D)	406.	(A)	454.	(A)	502.	(B)	550.	(B)
311.	*	359.	(C)	407.	(C)	455.	(A)	503.	(C)	551.	(D)
312.	*	360.	(D)	408.	(A)	456.	(C)	504.	(A)	552.	(A)
313.	(D)	361.	(D)	409.	(D)	457.	(A)	505.	(C)	553.	(D)
314.	(C)	362.	(A)	410.	(C)	458.	(A)	506.	(D)	554.	(B)
315.	(D)	363.	(D)	411.	(B)	459.	(C)	507.	(C)	555.	(B)
316.	(A)	364.	(D)	412.	(B)	460.	(D)	508.	(A)	556.	(B)
317.	(C)	365.	(C)	413.	(D)	461.	(D)	509.	(D)	557.	(B)
318.	(A)	366.	(B)	414.	(A)	462.	(A)	510.	(B)	558.	(B)
319.	(B)	367.	(C)	415.	(A)	463.	(C)	511.	(A)	559.	(D)
320.	(A)	368.	(D)	416.	(C)	464.	(A)	512.	(A)	560.	(B)
321.	(C)	369.	(D)	417.	(D)	465.	(A)	513.	(B)	561.	(B)
322.	(B)	370.	(A)	418.	(D)	466.	(A)	514.	(B)		
323.	(A)	371.	(D)	419.	(A)	467.	(D)	515.	(B)		
324.	(D)	372.	(C)	420.	(A)	468.	(B)	516.	(D)		
325.	(D)	373.	(C)	421.	(B)	469.	(C)	517.	(D)		
326.	(D)	374.	(A)	422.	(D)	470.	(A)	518.	(B)		
327.	(D)	375.	(A)	423.	(B)	471.	(D)	519.	(D)		
328.	(B)	376.	(D)	424.	(B)	472.	(A)	520.	(D)		
329.	(D)	377.	(C)	425.	(A)	473.	(B)	521.	(D)		
330.	(B)	378.	(D)	426.	(A)	474.	(A)	522.	(A)		
331.	(B)	379.	(B)	427.	(B)	475.	(D)	523.	(B)		
332.	(A)	380.	(C)	428.	(C)	476.	(C)	524.	(B)		
333.	(D)	381.	(A)	429.	(C)	477.	(C)	525.	(A)		
334.	(D)	382.	(B)	430.	(A)	478.	(B)	526.	(A)		
335.	(D)	383.	(C)	431.	(C)	479.	(C)	527.	(A)		
336.	(B)	384.	(B)	432.	(B)	480.	(A)	528.	(B)		

---

296. v
300. H_2SO_4
301. Lesser
302. $CaSO_4 . 1/2H_2O$
303. N_2O
307. HBr
308. Argon
309. Flourine
310. HF
311. I_3
312. $CaOCl_2$

CHEMISTRY OF TRANSITION METALS

d - subshell partly filled

1st transition series: 3d : Sc^{21} - Zn^{30}.

2nd transition series: 4d : Y^{39} - Cd^{48}.

3rd transition series: 5d : La^{57}, Hf^{72} - Hg^{80}.

4th transition series: 6d : Ac^{89},Rf^{104},Db^{105},Sg^{106},Hs^{108}, Mt^{109}, Dt^{110} (Darmstadtium)

Transition elements are located between the s and p-block elements and their properties represent transition between the highly reactive metallic elements of s-block which also give ionic compounds and the elements of p-block, which form mainly covalent compounds.

The terminal members i.e. Cu, Ag, Au and Zn, Cd, Hg have d^{10} configuration. They form complexes which is a common feature of transition elements.

1.**Electronic configuration** : $(n-1)d^{1-10}\, ns^{1 or 2}$

* Cr^{24} [Ar] $4s^1$ $3d^5$;
* Mo^{42} [Kr] $5s^1 4d^5$; W^{74} [Xe] $4f^{14} 6s^2 5d^4$;
* Cu^{29} [Ar] $4s^1$ $3d^{10}$; * Ag^{47} [Kr] $5s^1$ $4d^{10}$;
* Au^{79} [Xe] $4f^{14}$ $6s^1$ $5d^{10}$; Ru^{44} [Kr] $5s^1$ $4d^7$;
  Rh^{45} [Kr] $5s^1$ $4d^8$; Pd^{46} [Kr] $4d^{10}$ $4s^0$;
  Pt^{78} [Xe] $4f^{14}$ $5d^{10}$ $6s^0$;

The energies of the 3d and 4s orbitals in neutral atoms are quite close so that interchange of electrons between these two orbitals is possible in some cases where it could lead to a more stable configuration. This is due to additional stability of half filled and completely filled d-orbitals.

The anomaly in other cases is probably due to factors like nuclear-electron and electron-electron forces.

2. **Metallic character** : All transition elements are metals, good conductors of heat & electricity, ductile, malleable and form alloys (the alloy formation is due to the similar size of the d-block elements, the atoms of one metal can substitute the atoms of other metal in its crystal lattice).

Sc - lightest; Os - heaviest

Greater the number of unpaired electrons greater is the number of covalent bonds and therefore greater is the strength of the bonds. Cr, Mo, W are hard metals while Zn, Cd, Hg are softer. (Mn not harder?)

3. **Atomic and ionic radii** : Generally decreases.

The radii of the elements from Cr to Cu are very close to one another. This is due to the fact that the successive addition of d-electrons screens the outer 4s-electrons from the inward pull of the nucleus.

$Mn^{2+} > Fe^{2+} > Ni^{2+} > Cu^{2+} > Zn^{2+}$; $Fe^{+3} < Fe^{2+}$

4. **m.p. & b.p.**s : normally higher due to

(i) interatomic attraction (ii) covalent bonding

5.**Ionisation energy** : The transition metals are poor reducing compared to s-block and 13th group elements due to high heats of sublimation, high I.P. and low heats of hydration of their ions.

6.**Oxidation state** : The relative stability of various oxidation states of transition elements can be correlated with their electronic configuration. d^0, d^5 and d^{10} configuration are more stable.e.g.

Sc^{+3},Mn^{+7}etc.(d^0); Cr^{3+}(d^3); Mn^{2+},Fe^{3+}(d^5); Cu^+, Zn^{+2}(d^{10}).

Sc	$3d^1\, 4s^2$	+2, +3
Ti	$3d^2\, 4s^2$	+2, +3, +4
V	$3d^3\, 4s^2$	+2, +3, +4, +5
Cr	$3d^5\, 4s^1$	+1, +2, +3, +4, +5, +6
Mn	$3d^5\, 4s^2$	+2, +3, +4, +5, +6, +7

After Mn, the increasing nuclear charge binds the electrons more strongly and stable oxidation state is determined by weakly held electrons.

Fe	$3d^6\, 4s^2$	+2, +3, +4, +5, +6
Co	$3d^7\, 4s^2$	+2, +3, +4, +5
Ni	$3d^8\, 4s^2$	+2, +3, +4
Cu	$3d^{10}\, 4s^1$	+1, +2 , +3 (e.g.$KCuO_2$)
Zn	$3d^{10}\, 4s^2$	+2

7. **Oxides** : Cr_2O_3; CrO_2; CrO_3; ZnO_2;
amphoteric acidic acidic amphoteric

Mn_2O_3 ; MnO_2 ; Mn_2O_7
basic amphoteric acidic

8. **Standard electrode (Oxidation) potential :** Except copper, all others have higher electrode potentials than that of the standard hydrogen electrode. Cu : -0.34V. Therefore all except Cu would evolve H_2 gas when reacted with acid solutions. With conc. acids Fe, Cr, Al etc. become passive due to the formation of a thin impervious layer of an inert oxide. As the oxidation potentials are high, they are expected to oxidize to their ions easily.

9. **Colour** : The transition metal ions which have completely filled d-orbitals are colourless. There are no vacant d-orbitals to permit promotion of electrons. Completely empty d-orbitals also lead to colourless nature. d-d electron excitation is responsible for the colour mainly.

Munshel Colour Wheel :

V ╲ ╱ B opposite colours are called complimentary
R ─ ─ G colours (absorbed vs observed).
O ╱ ╲ Y

In the case of $KMnO_4$ and $K_2Cr_2O_7$, the colour is due to charge transfer spectrum.

The colour of some salts can also be explained on the basis of percentage ionic character of the bond, e.g., AgCl(80%), AgBr(24%), AgI(15%), Ag_2S(4%).

If the ionic character is below 20%, they will be coloured.

Green : V^{3+}, Fe^{2+}, Ni^{2+}; Cu^{2+} blue; Violet : Cr^{3+}, Mn^{3+};
Pink : Co^{2+}, Mn^{2+}; MnO_4^- purple
Zn^{2+}, Cd^{2+}, Hg^{2+}, Sc^{3+} etc. are colourless.

10.**Magnetic property :** Presence of unpaired electrons in atoms, ions or molecules leads to paramagnetic property.

Total magnetic moment $\mu_{eff} = \sqrt{n(n+2)}$ B.M.
(Bohr Magneton)

n = number of unpaired electrons.

In the case of Fe, Co, Ni, the unpaired electron spins are exceptionally more pronounced resulting in reinforcement of paramagnetism. These are said to be Ferromagnetic. Magnetic moment is determined by Guoy balance.

-Ti^{2+}, Cr^{2+}, Mn^{2+}, Fe^{2+}, Cu^{2+} etc.: paramagnetic

-Sc^{+3}, Zn^{2+}, Cu^{+}, Ti^{4+} etc. : diamagnetic

11.**Catalytic properties** : Because of variable valency character, the transition elements act as good catalysts.

- Pt : in contact process (H_2SO_4);
- V_2O_5 : in contact process.

MnO_2 - in the decomposition of $KClO_3$ for the preparation of oxygen.

Fe or Fe_2O_3 in the manufacture of NH_3 - Haber's process.

Ni - in the hydrogenation of fats and oils.

12.**Non-Stoichiometric compounds** : Variable valencies and defects in the solid structure of the elements are responsible for this variation.

$FeO^{\cdot}$ not FeO ; $FeO_{0.84}$ to $FeO_{0.98}$

13. **Metal-metal bonding** : Heavier transition elements of 4d and 5d-block tend to form strong M-M bonds more readily than 3d.

14. **Interstitial compounds** : Many transition metals form a number of interstitial compounds in which they take up atoms of small size e.g. H, C, B and N in the vacant spaces in their lattices and also form bonds with them.

15.**Complex formation** : The cations of transition elements form complex ions due to (a) small size (b) high positive charge density and (c) the presence of vacant orbitals. e.g. Fe^{+3} forms stronger complexes than Fe^{+2}.

16. **Formation of ionic and covalent compounds** :The ionic character of the bond decreases with the increase in the oxidation state. In higher oxidation state covalent character increases. The tendency to show trivalency decreases in the order $Fe^{3+} > Co^{3+} > Ni^{3+}$. ($Fe^{+3}$ more stable due to half- filled orbitals.)

- $CuSO_4.5H_2O$: blue vitriol;
- $FeSO_4.7H_2O$: green vitriol;
- $ZnSO_4.7H_2O$: white vitriol;
- $Cr_2O_3.2H_2O$: Guigrets green;
- $FeSO_4.(NH_4)_2SO_4.6H_2O$: Mohr's salt.
- $4FeCl_3 + 3K_4[Fe(CN)_6] \longrightarrow Fe_4[Fe(CN)_6]_3 + 12KCl$
  Prussian blue
- $K_4[Fe(CN)_6] + FeSO_4 \longrightarrow K^+ + Fe^{3+}[Fe(CN)_6]^{4-}$
  Turn bulls blue
- $Na_2[Fe(NO)(CN)_5].2H_2O$: sodium nitroprusside
- $CuSO_4$+ Lime : Bordeaux mixture
- $HgCl_2$: Corrossive sublimate (Hg^{2+});
- Hg_2Cl_2: Calomel (Hg_2^{2+}) Cl-Hg-Hg-Cl and not as HgCl

$Hg_2Cl_2 + 2NH_3 \longrightarrow Hg(NH_2)Cl + Hg + NH_4Cl$
black ppt.

-Rolled gold : Cu + Al; Philosopher's wool : ZnO.

-Rinman's green : $CoZnO_2$.

-Brass : Cu+Zn; Bronze : Cu+Sn; Bell metal : Cu+Sn;

-Gun metal : Cu+Sn+Zn.

-German Silver : Cu+Zn+Ni; Type metal : Pb+Sb+Sn

-Monel metal : Cu+Ni+Fe+Mn

-$FeSO_4 \cdot 7H_2O$, $ZnSO_4 \cdot 7H_2O$, $MgSO_4 \cdot 7H_2O$ are isomorphs.

-The purity or fineness of gold is expressed in carats (pure gold 24 carats).

-$CuSO_4 + Na_2S_2O_3 \longrightarrow CuS_2O_3 + Na_2SO_4$

-$CuS_2O_3 + Na_2S_2O_3 \longrightarrow Cu_2S_2O_3 + Na_2S_4O_6$

-$3Cu_2S_2O_3 + 2Na_2S_2O_3 \longrightarrow Na_4[Cu_6(S_2O_3)_5]$
Sod. cuprothiosulphate

-Kroll process : extraction of Ti

-Van Arkel method : purification of Ti

-Jantesh process : for preparing V metal

-Beard & Crook's process : reduction of V_2O_5 to V

-Mond's process : for pure Ni extraction.

LANTHANIDES & ACTINIDES

Lanthanides are not scarce in the earth's crust as the name implies (rare earths) Ce^{58} - Lu^{71} (14).

Lanthanide minerals
- Cerium sub group 57-62
- Yttrium sub group 63-71

-Cerium group minerals :

(i) Monazite (mainly phosphates of rare earths)

(ii)Bastanesite: mixed fluorides & carbonates, $RFCO_3$

-Yttrium sub group minerals :

(i) Euxenite R[Nb,Ta] $TiO_6.xH_2O$

(ii) Gadolinite $R_2M^{II}_3Si_2O_{10}$

1.**Electronic configuration** : $4f^n\ 5d^1\ 6s^2$ or $4f^{n+1}\ 6s^2$

The 5d and 4f orbitals are so close in energy, it is very difficult to decide whether the electron first enters 5d or 4f. However from the chemical point of view, it hardly matters whether the configuration is the former or the later as 'f ' electrons are sufficiently shielded by intervening electron shell 5 or 6 and not involved for bonding.

2. **Oxidation state** : +3 is the characteristic oxidation state. High values of reduction electrode potential indicate that the lanthanones are powerful reducing agents. +4 is important in Cerium. Ce^{4+} is a powerful oxidising agent. +2 state is also possible in lanthanides.

3.**Magnetic property :** La^{3+} and Lu^{3+} are diamagnetic. In lanthanides the paramagnetic character is dependent on the electron spin and orbital motion.

4. **Basicity of ions** : Basicity decreases as ion size decreases. This slight regular change is made use of in the separation of Lanthanones.

5. **Separation of Lanthanides** : A synthetic ion exchange resin with functional group like -COOH or $-SO_3H$ is packed in columns. A buffered solution of citric acid/ ammonium citrate is then used for elution. Smaller lanthanide ions form stronger complexes with the citrate

ions.

6. **Colour** : La^{3+}, Ce^{3+}, Gd^{3+} -colourless

Unpaired electrons 0 1 7

-Neodynium & Praseodynium oxides are used for making coloured glasses for goggles.

-Misch-metal : Rare earth element (94-95%), Iron(5%) and traces of S,C,Ca and Al.

-The pyrophoric alloys containing rare earth metals are used in the preparation of ignition devices e.g. tracer bullets and flints for lighters.

Actinides : Th^{90} - Lr^{103} (14); 5f - block elements

-Pitchblende : UO_2

-Plutonium is obtained in large quantities from uranium fuel in nuclear reactors.

-Uranium is used in nuclear reactors.

-Actinides are used in gas mantle industry.

COORDINATION CHEMISTRY

-Chlorophyll (Mg complex); Haemoglobin (Fe complex)

Simple salts : undergo ionisation in water for e.g.NaCl gives tests for Na^+ and Cl^- ions.

Addition or Molecular Compounds : When two or more simple salts are present in a solution in stoichiometric (i.e. simple molecular) proportions and when allowed to evaporate, crystals obtained are of addition or molecular compounds. The molecular or addition compounds are of two types :

1. Double salts or Lattice compounds

(i) Intact only in crystal lattice

(ii)In water, they decompose to individual salts which in turn dissociate into ions.

(iii) The properties are those of the individual salts, e.g.

$FeSO_4.(NH_4)_2SO_4.6H_2O$: Mohr's salt

$K_2SO_4.Al_2(SO_4)_3.24H_2O$: Alum

$KCl.MgCl_2.6H_2O$: Carnalite

2. Complex compounds

(i) When $Fe(CN)_2$ and KCN in molecular rato 1:4 are heated $K_4[Fe(CN)_6]$ is obtained.

(ii) When dissolved in water, K^+ and $[Fe(CN)_6]^{4-}$ ions are produced.

Complex Compounds : They are of two types :

(i) Normal Complexes : Those which break or ionize in solution into their components

e.g. $Cd(CN)_4^{2-} \rightleftharpoons Cd^{++} + 4CN^-$

(ii) Penetration Complexes : They exist intact in solution also.

e.g. $[Fe(CN_6)^{4-}$; $[Co(NH_3)_6]^{3+}$ etc.

Coordination Compound : Compound containing a central atom or ion to which are attached molecules or ions whose number exceeds the number corresponding oxidation number or valence of the central atom or ion (the central atom or ion is usually a transition element).

The molecules or ions which attach to the central metal ion are called ligands. They are electron pair donors,and thus act as Lewis bases.

Ligands can be

(i)Neutral e.g. NH_3,H_2O,CO,en (ethylenediamine), NO etc.

(ii) -vely charged e.g. CN^-,Cl^-,NO_2^-,ONO^- etc.

(iii) +vely charged e.g. NO^+,NO_2^+,NH_4^+,$N_2H_5^+$ etc.

Depending on the presence of the number of coordinating atoms, the ligands are classified as :

(i) Monodentate NH_3,H_2O,Cl^-,CN^- etc.

(ii) Bidentate phen; oxalato 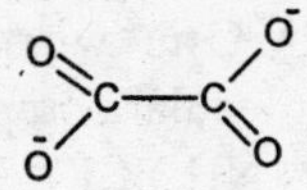

(iii) Tridentate $H_2NH_2C\text{-}CHNH_2\text{-}CH_2NH_2$

(iv) Quadridentate EDAA

(v) Hexadentate EDTA

Ambidentate Ligands : Ligands which have two or more atoms capable of attaching to the central metal ion through either of the atoms it possess.

e.g. M ⟵ CN ; M ⟵ NC
cyano isocyano

M⟵ NO_2; M⟵ ONO ; M⟵ SCN ; M ⟵ NCS
nitro nitrito thiocyano isothiocyano

Many complexes are mono nuclear i.e. they contain only one metal atom. However polynuclear complexes are also known.

Coordination Sphere : The ligands together with the central atom enclosed in the square bracket.

Coordination number : The total number of ligands surrounding and bonded to a central atom is coordination number. Coordination number 4 and 6 are most common.

e.g. $[Ag(NH_3)_2]^+$: 2 ; $[FeCl_4]^-$: 4 ; $[Fe(CN)_6]^{3-}$: 6.

Primary and Secondary Valencies : As per the Werner's coordination theory (1893), every element exhibits two types of valencies, primary and secondary valency.

Primary Valency : Primary valencies are satisfied by negative ions. It is represented by ····· lines. Ionisable and tells about oxidation state.

Secondary Valency : Satisfied either by negative ions or neutral molecules. Shown by — thick lines. Tells about the coordination number.

e.g. $[Co(NH_3)_5Cl]Cl_2$

or $[Co(NH_3)_5Cl]^{2+}$

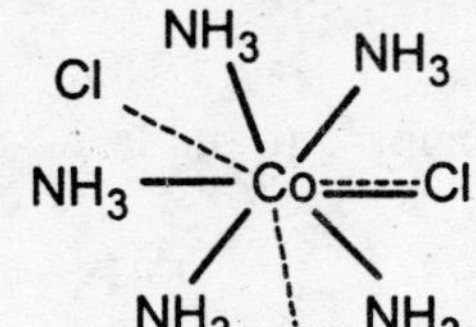

Effective Atomic Number Rule (EAN-Rule) :

Sidgwick gave the concept of EAN Rule.

The metal atom or ion will continue to accept electron pairs till it achieves the next inert gas configuration. This is known as EAN Rule. The resultant number of electrons with the metal atom or ion after gaining electrons from the ligands is effective atomic number.

e.g. $[Co(NH_3)_6]^{3+}$;

Atomic number of Co is 27

Number of electrons in Co^{3+} = 24

EAN= 24+ (2 x 6) = 36.

Using EAN Rule, write the structure of $Mn_2(CO)_{10}$.

Nomenclature of Complex Compounds :

(1) While naming the ionic compound, cation is named first and then anion.

NaCl - Sodium chloride.

$[Cu(NH_3)_4]SO_4$ - tetraamminecopper(II)sulphate.

(2) One word name is given to non-ionic and molecular complexes e.g.

$[Pt(NH_3)_2Cl_4]$ diamminetetrachloroplatinum(IV)

(3) Ligands are named before the central metallic ion.

$[Co(NH_3)_6]Br_3$ hexaamminecobalt(III)bromide.

(4) Ligands should be named in alphabetical order irrespective of the charge or complexity.

$[Pt(NH_3)_4(NO_2)Cl]CO_3$ tetraamminechloronitro platinum(IV)carbonate.

(5) When more than one simple ligand of a particular type is present then di, tri, tetra, penta, hexa is used for 2,3,4,5,6

e.g. $[Cr(H_2O)_4Cl_2]NO_3$ tetraaquodichlorochromium(III) nitrate.

(Ligands are named as e.g. NH_2^--amido; CN^-- cyano; CH_3COO^- - acetato; SO_4^{2-}sulphato; $C_2O_4^{2-}$-oxalato; CO_3^{2-} -carbonato; CO-carbonyl; NO - nitrosyl; NO^+ - nitrosylium; NH_2-NH_3^+-hydrazinium; NH_3 - ammine; OH^- - hydroxo, ethylene diamine - en; 1,10-phenanthroline-phen etc.

Complex negative ions always end in the suffix -ate e.g. $K_4[Fe(CN)_6]$ potassiumhexacyanoferrate(II).

When more than one complex ligand is present, then bis, tris, tetrakis, pentakis, hexakis etc. are used for 2,3,4,5,6 etc.

e.g.$[Ni(dmg)_2]$- bis(dimethylglyoximato)nickel(II)

(6) Ligands acting as bridge between two central metallic ions are indicated by the Greek letter μ. The letter μ is repeated before the name of each different kind of bridging group.

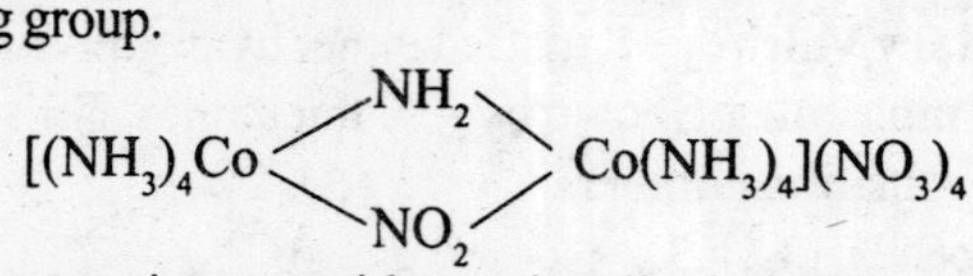

octaammine-μ-amido-μ-nitrodicobalt(III) nitrate.

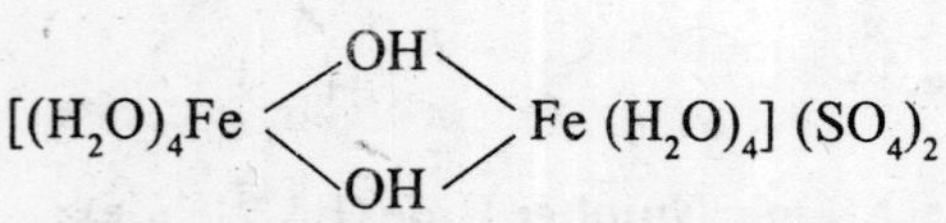

octaaquo-μ-dihydroxodiiron(III) sulphate.

Some examples :

Complexes containing complex cationic ion

(1) $[Co(NH_3)_4H_2O\ Cl\ ]Cl_2$
tetraammineaquochlorocobalt(III)chloride.

(2) $[Co(en)_2Cl_2]Br$
dichlorobis(ethylenediamine)cobalt(III)bromide.

(3) $[Fe(H_2O)_4(C_2O_4)]NO_3$
tetraaquooxalatoiron(III)nitrate.

(4) $[Cr(NH_3)_4(ONO)Cl]NO_3$
tetraamminechloronitritochromium(III)nitrate.

(5) $[Co(NH_3)_5(NCS)]Cl_2$
pentaammineisothiocyanatocobalt(III)chloride.

(6) $[\{(C_6H_5)_3P\}_3Rh]Cl$
tris(triphenylphosphine)rhodium(I)chloride.

Complexes containing complex anionic ion

(7) $K_4[Fe(CN)_6]$: potassiumhexacyanoferrate(II).

(8) $K_3[Cr(C_2O_4)_3]$: potassiumtrioxalatochromate(III).

(9) K_2HgI_4 : potassiumtetraiodomercurate(II).

(10) $K_3[Fe(CN)_5NO]$:
potassiumpentacyanonitrosylferrate(II).

Complexes containing complex cation and anionic ion:

(11) $[Pt(NH_3)_4]\ [CuCl_4]$
tetraammineplatinum(II)tetrachlorocuprate(II)

(12) $[Pt(Py)_4]\ [PtCl_4]$
tetrapyridineplatinum(II)tetrachloroplatinate(II)

(13) $[Cr(NH_3)_6\ [Co(C_2O_4)_3]$
hexaamminechromium(III)trioxalatocobaltate(III)

Complexes containing non-ionic compounds :

(14)$Fe(Co)_5$ pentacarbonyliron(0)

(15)$Ni(dmg)_2$ bis(dimethylglyoximato)nickel(II)

Bridged complex :

(16)

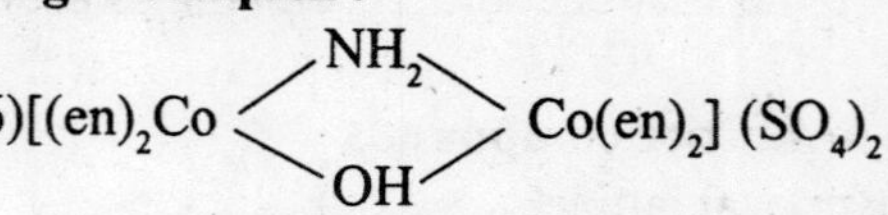

bis(ethylenediamine)cobalt(III)μ-amido-μ-hydroxobis(ethylenediamine)cobalt(III) sulphate.

Isomerism

Isomerism is of two types :

(1) Structural Isomerism

(2) Stereo isomerism : It is of two types :

(i) Geometrical isomerism

(ii) Optical isomerism

Structural Isomerism

(a) Ionisation isomerism : Identical composition but produces different ions in water e.g. $[Co(NH_3)_5SO_4]Cl$; $[Co(NH_3)_5Cl]SO_4$

(Test with $AgNO_3$ and $BaCl_2$ for Cl^- and SO_4^{2-} ions)

(b) Linkage isomerism : They are different compounds containing a ligand capable of attaching to the central metal ion through either of the atoms it possess. They are known as ambidentate ligands: (NO_2^-,CN^-,CO,SCN^-,$S_2O_3^{2-}$ etc.,)

e.g. $[Co(NH_3)_5NO_2]Cl_2$; $[Co(NH_3)_5ONO]Cl_2$

(c) Hydrate isomerism : They differ in the number of H_2O molecules inside and outside the coordination sphere. e.g.$[Cr(H_2O)_6]Cl_3$ -violet ;$[Cr(H_2O)_5Cl]Cl_2.H_2O$- green; $[Cr(H_2O)_4Cl_2]Cl.2H_2O$ -green.

Inside the coordination sphere-water of coordination; Outside the coordination sphere- water of crystallisation.

(d) Ligand isomerism : Some ligands themselves exhibit isomerism. When they are present in the complex, they lead to ligand isomerism. e.g. $[Co(pn)_2]Cl_2]^+$; $[Co(tn)_2]Cl_2]^+$

pn = 1,2 diaminopropane $\quad CH_2\text{-}CH\text{-}CH_3$ (with NH_2 on first and second carbon)

tn = 1,3 diaminopropane $H_2N\ CH_2\text{-}CH_2\text{-}CH_2NH_2$

(e) Coordination isomerism : Occurs in compounds having complex cation and complex anion of metals with same coordination number, e.g.

$[Co(NH_3)_6]^{3+}\ [Cr(CN)_6]^{3-}$ - $[Co(CN)_6]^{3-}\ [Cr(NH_3)_6]^{3+}$

(f) Polymerisation isomerism : They have same empirical composition but different molecular formulae. e.g. $[Pt(NH_3)_2Cl_2]$; $[Pt(NH_3)_4]\ [PtCl_4]$.

Geometrical isomerism : Cis and trans isomerism is not possible for complexes having coordination number 2 and 3. The tetrahedral complexes also do not show geometrical isomerism. This is due to the fact that in these arrangements, all the positions are adjacent to each other.

Coordination number 4: square planar complexes of the type MA_2B_2 type exhibit geometrical isomerism. e.g. $[Pt(NH_3)_2Cl_2]$

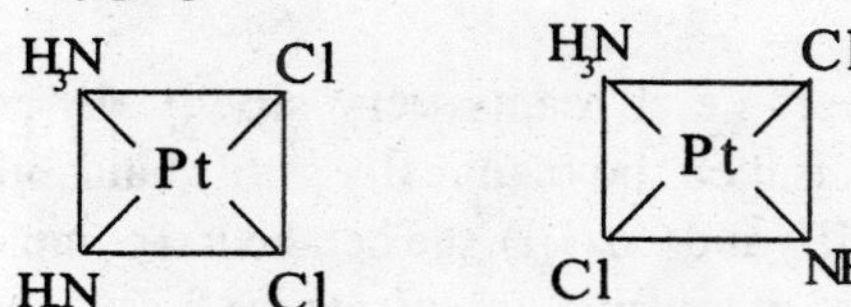

cis $\qquad$ trans

Coordination number 4, MA_2BC type

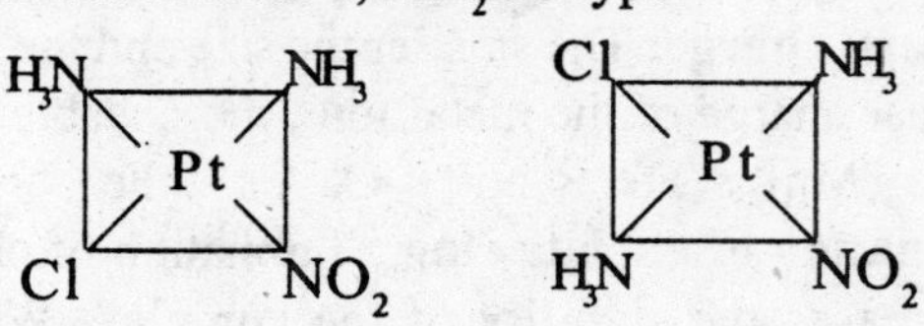

MABCD can also show geometrical isomerism. (MA_4,MB_4,MA_3B or MAB_3 type do not show geometrical isomerism)

With Coordination number 6, three types of arrangements are possible :

(a) Plane hexagonal

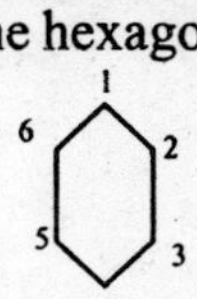

(b) Trigonal prismatic

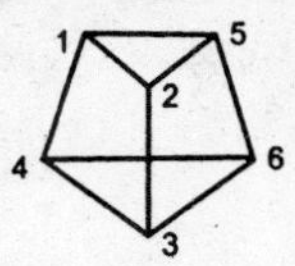

(c) Octahedral

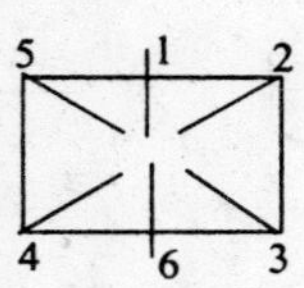

Plane hexagonal and trigonal prismatic arrangements lead to three isomers of compounds1,2; 1,3 and 1,4 respectively.

In octahedral complexes of MA_6 (or MB_6),MAB_5 or MA_5B type, no isomerism is possible. MA_4B_2 or MA_2B_4 type shows e.g. $[Co(NH_3)_4Cl_2]^+$

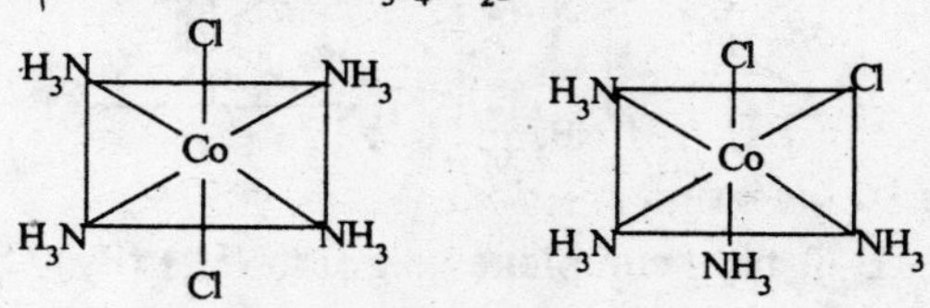

1,6 arrangement - trans $\qquad$ 1,2-arrangement - cis

MA_3B_3 also exhibits eg. $[Co(NH_3)_3Cl_3]^+$

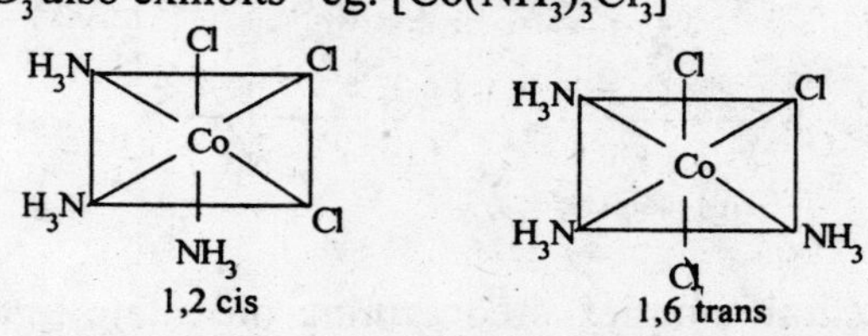

A bidentate ligand (xy) can span only two cis positions. Hence $M(xy)\ A_3B$; $M(xy)\ A_2\ B_2$and $M(xy)_2A_2$ type complexes also exhibit geometrical isomerism.

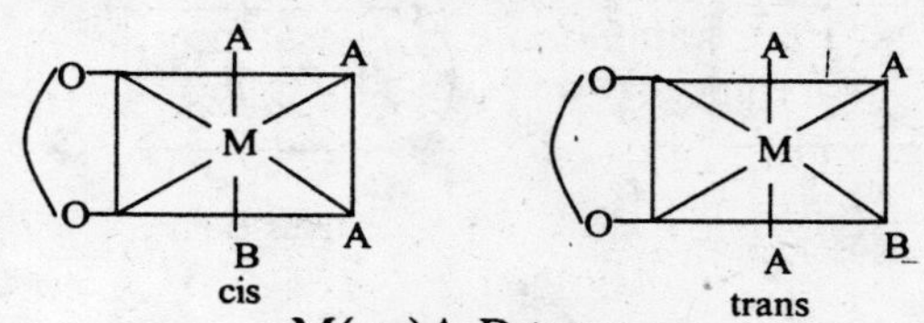

M(xy)A_3B type

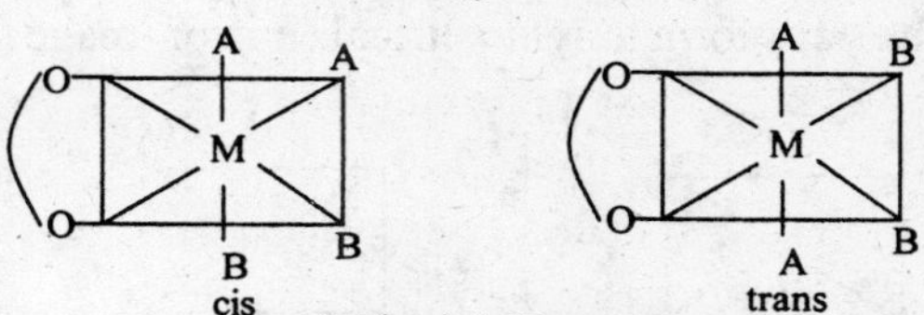

M(xy)A_2B_2 type

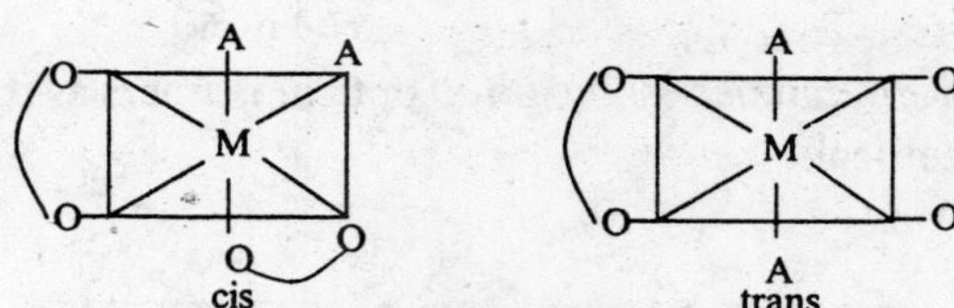

$M(xy)_2A_2$ type

"Cis and transisomers can be distinguished by dipole moment measurement". For trans it is zero.

Opitcal isomerism:

Optical isomerism among coordination number 4 complexes is encountered only in tetrahedral complexes. Square planar complexes cannot exhibit optical activity as they possess plane of symmetry i.e. here the ligands and the central metal lie in the same plane. Optical isomerism in square planar complexes is at present limited to a few examples. Optical isomers of tetrahedral complexes having four different ligands can not be isolated due to the labile nature of the complexes (labile means unstable; apt for change). Tetrahedral complexes of Be,

Cu (II), Zn (II) etc. containing unsymmetrical bidentate ligands have been resolved into optically active isomers.

Coordination number 6, MABCDEF type octahedral complexes e.g. $[Pt(NO_2)(NH_3)(Py)(Cl)(Br)(I)]$.

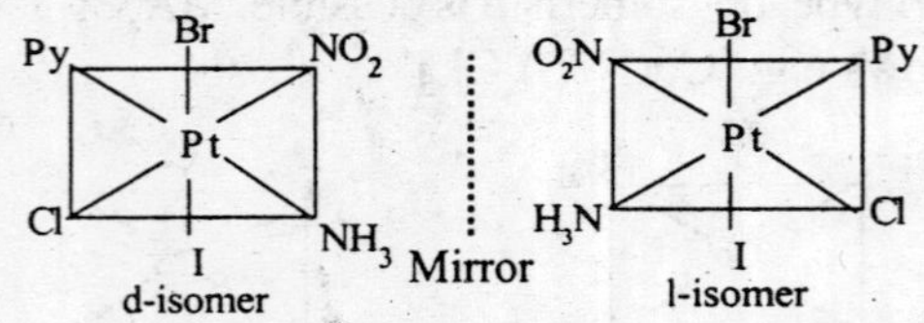

$MA_2B_2C_2$ type complexes e.g. $[Pt(NH_3)_2(Py)_2Cl_2]^{2+}$

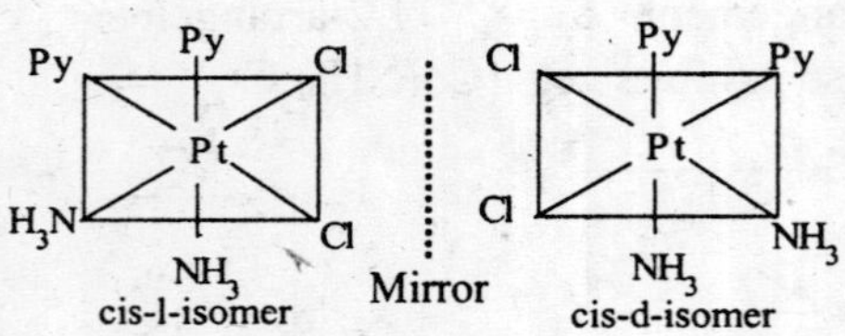

Octahedral complexes containing bidentate ligands $M(xy)_2A_2$ type e.g. $[Co(en)_2Cl_2]^+$ also exhibit,

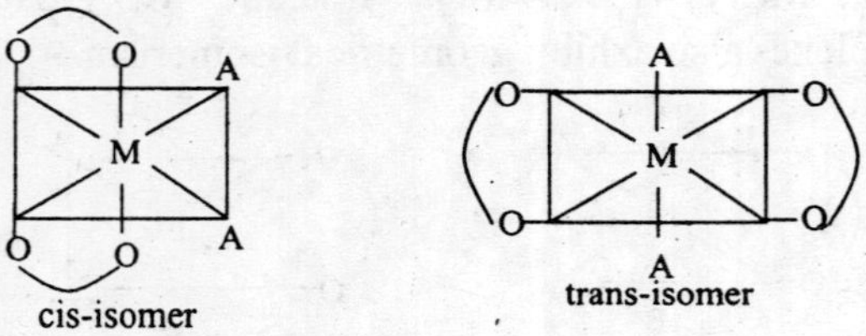

Trans form cannot be resolved into optical isomers because the trans form is symmetrical (mirror image the same)

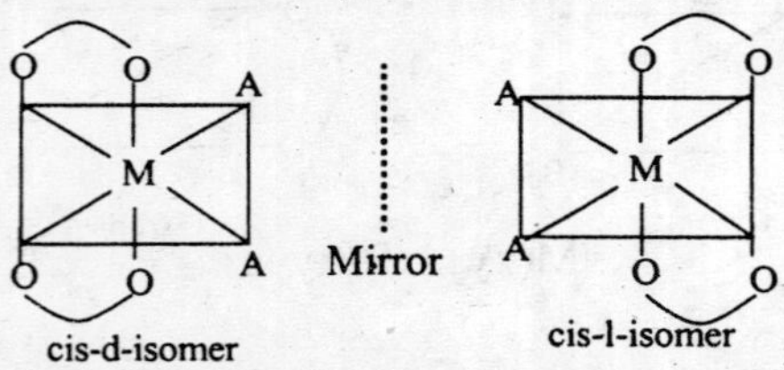

The cis form can be resolved into optical isomers as it is unsymmetrical.

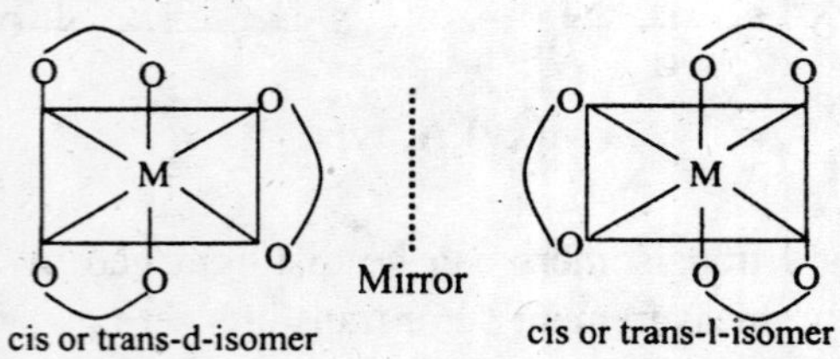

Octahedral complexes with three bidentate ligands $M(xy)_3$ can also be resolved, e.g. $[Cr(gly)_3]$.

Structure of coordination compounds:

Coordination number 2 : sp hybridized, linear or diagonal e.g. $[Ag(CN)_2]^-$; $[Ag(NH_3)_2]^+$

Coordination number 3 : sp^2 hybridized, trigonal planar e.g. $[HgI_3]^-$

Coordination number 4 :

(i) sp^3, tetrahedral e.g. $[Zn(NH_3)_4]^{2+}$; $[Ni(CO)_4]$.

(ii) dsp^2, square planar.

e.g. $[Cu(NH_3)_4]^{2+}$; $[Ni(NH_3)_4]^{2+}$; $[Ni(CN)_4]^{2-}$

Note: In the presence of strong ligands, pairing up of electrons takes place (NH_3, CO, CN etc. are strong ligands; F, H_2O etc. are weak ligands).

Coordination number 5:

(i) sp^3d trigonal bipyramidal e.g. SbF_5; IF_5

(ii) dsp^3- square pyramidal e.g. $[Cu\ Cl_5]^{3-}$; $[Fe(CO)_5]$

Coordination number 6:

(i) sp^3d^2 outer octahedral complexes e.g. $[CoF_6]^{3-}$; $[FeF_6]^{3-}$; $[Ni(NH_3)_6]^{2+}$

(ii) d^2sp^3 inner octahedral complexes e.g. $[Fe(CN)_6]^{3-}$; $[Co(NH_3)_6]^{3+}$

Stability of complexes :

(i) Thermodynamic stability : It relates to the position of equilibrium between the products and reactants. The complex is said to be thermodynamically stable if the equilibrium constant for the decomposition of the complex is very less.

e.g. $[Ni(CN)_4]^{2-} \rightleftharpoons Ni^{2+} + 4CN^-$ $\quad k = 10^{-22}$

(ii) Kinetic stability: It refers to the rate at which a substance reacts. A complex that reacts rapidly is called labile. Thermodynamic and kinetic stabilities do not necessarily correlate with each other. $[Co(NH_3)_6]^{3+}$ is thermodynamically unstable towards acid

$$[Co(NH_3)_6]^{3+} + 6H_3O^+ \rightleftharpoons [Co(H_2O)_6]^{3+} + 6NH_4^+ \quad K = 10^{25}$$

yet it is inert i.e it reacts very slowly. In contrast $[Ni\ (CN)_4]^{2-}$ is thermodynamically stable but labile.

Except Cr (III) and Co (III), the octahedral complexes of the 3-d series of transition elements are fairly labile.

Cr (III) - d^3 species; Co (III) - d^6 species.

Factors affecting stability of complexes

(a) Size and charge: The smaller the size and the larger the nuclear charge of the metal ion , more stable is the complex. $Mn^{2+} < Fe^{2+} < Ni^{2+} < Cu^{2+}$; $Fe^{3+} > Fe^{2+}$

(b) Formation of chelate ring : Formation of chelate ring increases the stability of the complex. When a bidentate, tridentate or polydentate ligand forms coordinate bonds with the same metal ion, the ring structure formed is called chelate and the ligand is chelating agent.e.g.

(i) 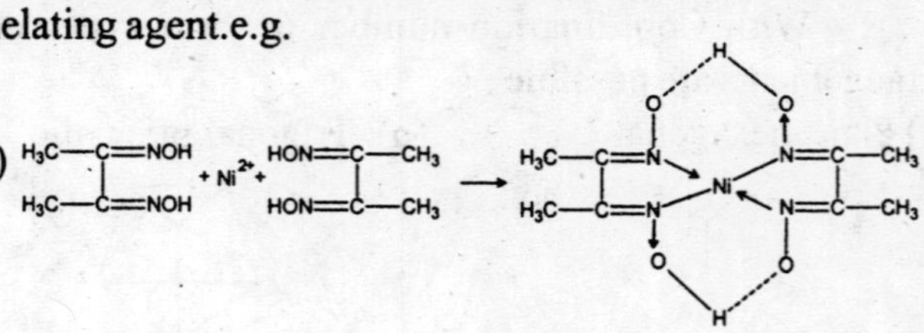

(ii) 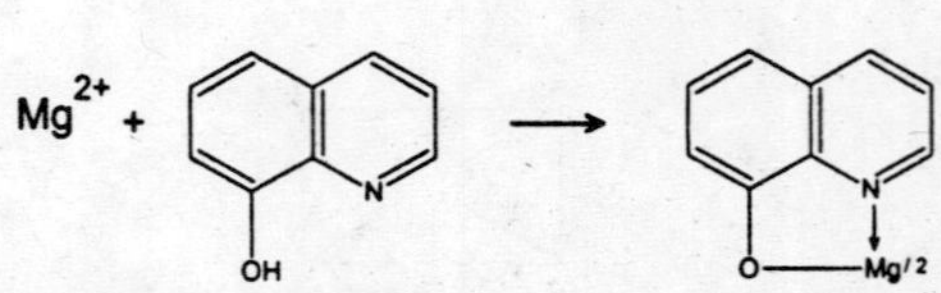

8- hydroxy quinoline

(iii) 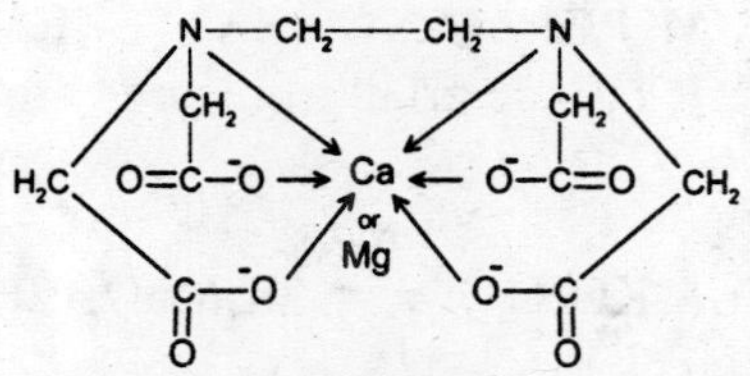

EDTA complex

Note: In chelate compounds, the cental metal ion is attched with any organic compound either by replacement of hydrogen atom or by accepting electron pairs from the coordinating group.

Uses of chelates:
(i) As ionizers.
(ii) EDTA to remove some harmful radioactive material from the body.
(c) Type of ligand and oxidation potential: If one and the same ligand forms two oxidation states, then higher oxidation state is stabler than the one with lower oxidation state.

CLASSIFICATION AND BONDING IN ORGANOMETALLICS

Organometallic compounds cantain at least one direct metal-carbon bond e.g. R Mg X, Ferrocene,$(C_2H_5)_4$ Pb etc.

Bonding: Five metal-carbon bond types are found in organometallic compounds:
(1) Ionic bonds.
(2) Sigma covalent bonds.
(3) Electron deficient bonds.
(4) Delocalized bonds in polynuclear systems.
(5) Dative bonds with d-orbital participation.

Ionic bonds: They occur only in the compounds of alkali and alkaline earth metals. The metal is found in cationic form $M^{n+}$and the organic group is carbanion.

e.g. CH_3 - CH^-_2 Na^+;  Na^+

The negative charge of the carbanion can either be localized at a particular carbon atom or can be delocalized over several carbon atoms. The organometallic molecule in this case is formed by ion pairing. Some degree of covlency is found even in these cases, particularly if no donor solvent is present to produce ion pairing.

Sigma -covalent bonds:
These are typical for all non-transition elements but are also encountered in transition metal derivatives. These are classical covalent bonds formed by pairing of two electrons of opposite spins. Thus the bonds Sn-C; Pb-C and Al-C are σ covalent bonds with polar character of the type $M^{\delta+}$ $C^{\delta-}$depending on the electronegativity differences.

The compounds with weak M-C bonds (for e.g. Cd R_2,HgR_2, PbR_2,BiR_3) rapidly polarize to deposit the metals. These are thermodynamically unstable but kinetically stable.

Some M-C bonds are rapidly hydrolyzed

$M\text{-}C+H_2O \longrightarrow MOH+R\text{-}H$

Hydrolytic stability depends upon the polarity of M-C bond.

Electron Deficient Bonds:
The elements with
(i) less than half-occupied valence shell(Be has 2 and Al has 3 electrons in a shell of 4-sp^3 orbitals).
(ii) an easily polarizable metal atom (due to low ratio of charge: atomic radius).
The possible formation of M....CH_3.....M type bridges (for e.g. with M= Be, Al) can be understood in terms of an electron pair bonding three atoms. Such bonds are weaker than the more common covalent bonds and form when the valence electrons available for bond formation are too few to fill all bonding orbitals with elecrton pairs.

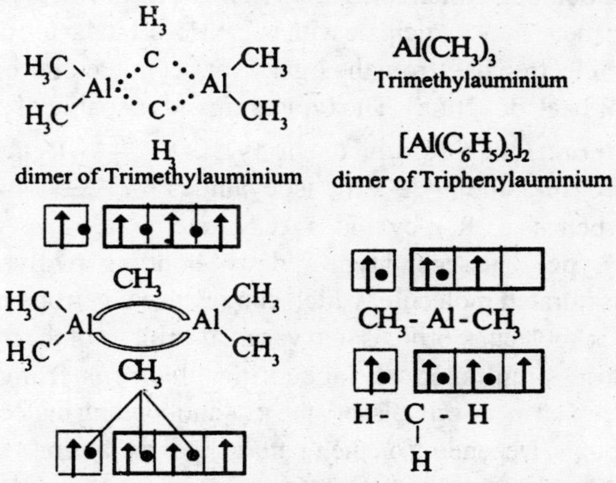

Localized, three centre bielectronic bonds are electron deficient e.g. dimethylberylium Be....CH_3 ....Be.....CH_3.

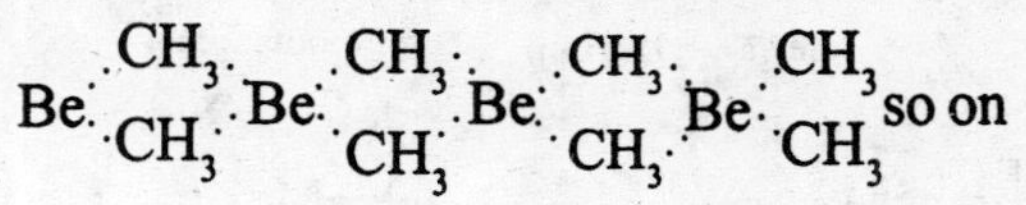

Delocalized bonds in poly nuclear systems:
The bonds between the lithium atoms in the Li_4 tetrahedra or Li_6 polyhedra in the polymeric lithium alkyls[$(LiR)_n$ n= 4,6] are delocalised metal-metal bonds. The organic groups are attached simultaneously to several lithium atoms(usually 3 on a polyhedral triangular face).

The $[Li(CH_3)]_4$ tetramer consists of an Li_4 tetrahedron with methyl groups located above the centre of the trigonal faces of the tetrahedron, simultaneously attached through polycentric bonds to 3 Li atoms.

The formation of Li_4 tetrahedra and Li_6 polyhedra can be interpreted by the sharing of Lithium orbitals collectively. The carbon sp^3-orbital electrons are collectively shared and delocalized in the cage formed by the Li-atoms.Alternatively, the bond between the +vely charged Li_n and the organic group is essentially ionic.

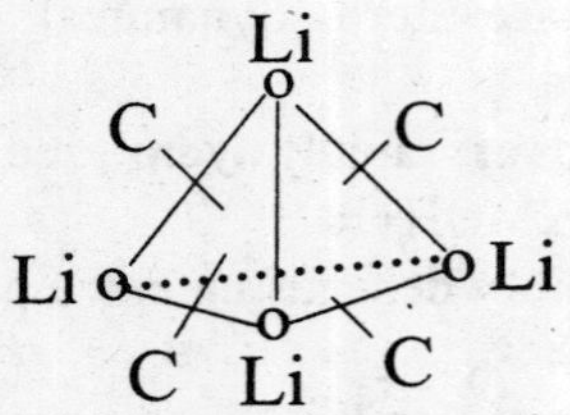

In this case isolated Li-C bonds are not found. A similar situation occurs in clusters present in metal carbonyls and carboranes. In these molecules several atoms are held together by collectivizing the valence orbitals and electrons.

Dative Bonds with d-orbital participation:

There are two types

1st type: The first type of dative bonding occurs when the carbon atom of a ligand acts as an electron pair donor and the ligand molecule can accept back some of the electron density from the metal in its vacant molecular antibonding orbital (back donation). In these interactions the metal may participate with vacant 'd' orbitals (to accept the electron pair from the ligand) or occupied d- orbitals (for back donation). The typical ligands capable of such interactions include carbonyl, CO(:C$\equiv$O);carbon monosulphide (:C$\equiv$S) ; isocyanides (:C$\equiv$N—R); carbenes (:C R_2); cyanides (:CN^-) etc.

2nd type: The second type of dative bonding involves an unsaturated molecule, which can release electrons from its π-molecular orbitals into vacant d-orbitals of the metal while simultaneously accepting electrons from the occupied metal orbitals into the π^* - antibonding molecular orbitals. Depending on their nature, ligands can contribute a variable number of electrons.

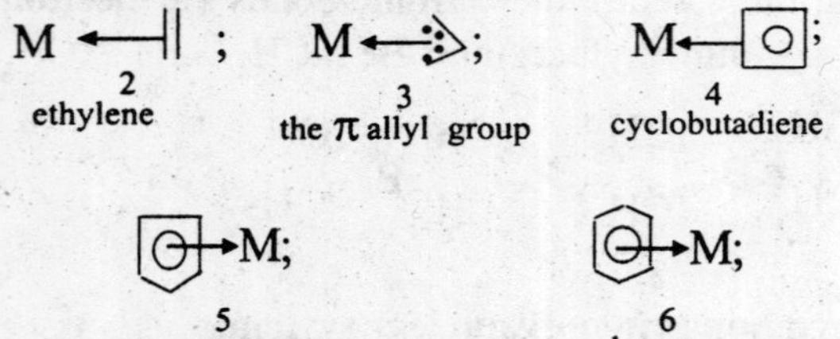

2 ethylene; 3 the π allyl group; 4 cyclobutadiene; 5 cyclopentadienyl group; 6 benzene

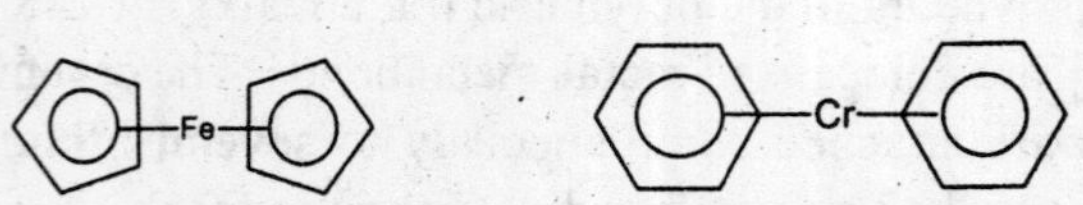

Ferrocene — Dibenzene Chromium

In order to participate in this type of bonding, each carbon atom must be sp^2- hybridized (thus posessing a electron in a P_Z- orbital) and must be part of a planar carbon atom net work (elemental planar carbon net work).

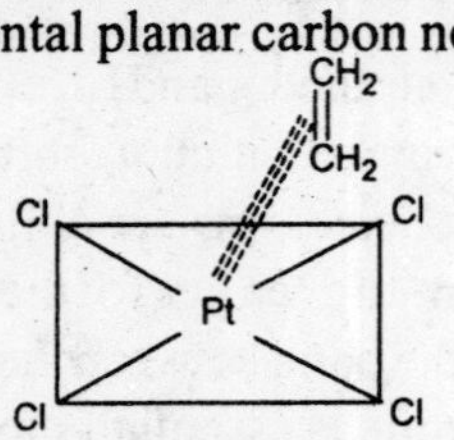

Zeisse's salt

POTASSIUM PERMANGANATE - $KMnO_4$

Preparation : It is industrially prepared from MnO_2 (pyrolousite)

1st step :

$$2MnO_2+4KOH+O_2 \xrightarrow{KNO_3} 2K_2MnO_4+2H_2O$$

2nd step :

$$3K_2MnO_4+2H_2O+4CO_2 \longrightarrow 2KMnO_4+MnO_2+4KHCO_3$$

or by oxidation by passing chlorine

$$3K_2MnO_4+Cl_2 \longrightarrow 2KMnO_4+ 2\ KCl$$

Properties : Deep purple crystals having metallic lusture. Moderately soluble in H_2O forming pink solution.

1 On heating : It decomposes

$$2KMnO_4 \xrightarrow{240^0C} K_2MnO_4+ MnO_2 + O_2$$

2 With conc. H_2SO_4 :

$$3KMnO_4+H_2SO_4 \longrightarrow Mn_2O_7 + K_2SO_4 + H_2O$$

$$2Mn_2O_7 \text{ (explosive nature)} \longrightarrow 4MnO_2 + 3O_2$$

When heated gently with conc. H_2SO_4, it gives

$$4KMnO_4+6H_2SO_4 \longrightarrow 2K_2SO_4+4MnSO_4 +6H_2O + 5O_2$$

3 Oxidation reactions : It is a strong oxidising agent. It is stronger oxidising agent than $K_2Cr_2O_7$. $K_2Cr_2O_7$ is used as a primary standard solution while $KMnO_4$ is used as a secondary as it can't be obtained in pure state (free from MnO_2). The distilled water used to prepare the solution may contain traces of organic matter, which reacts with some of the permanganate to reduce it to MnO_2.

Acid medium :

$$2KMnO_4+3H_2SO_4 \longrightarrow K_2SO_4+MnSO_4 +3H_2O + 5O$$

Neutral/slightly alkaline :

$$2KMnO_4+H_2O \longrightarrow 2KOH+MnO_2 + 3O$$

Strongly alkaline medium :

$$2KMnO_4+2KOH \longrightarrow 2K_2MnO_4+H_2O + O$$

The equivalent weight of $KMnO_4$ thus vary in different media.

Oxidation in acidic medium :

(a) Ferrous to Ferric salts :

$$2KMnO_4+8H_2SO_4+10FeSO_4 \longrightarrow K_2SO_4+2MnSO_4 + 5Fe_2(SO_4)_3+8H_2O$$

(b) Oxalic acid to CO_2 :

$$2KMnO_4+3H_2SO_4+5H_2C_2O_4 \longrightarrow K_2SO_4+2MnSO_4 + 8H_2O+10CO_2$$

(c) SO_2 to H_2SO_4 :

$$2KMnO_4+5SO_2+2H_2O \longrightarrow K_2SO_4+2MnSO_4 + 2H_2SO_4$$

(d) Halogen acids to halogens :

$$2KMnO_4+3H_2SO_4+10HX \longrightarrow K_2SO_4+2MnSO_4 + 8H_2O + 5X_2$$

(e) Sulphites to sulphates : $Na_2SO_3+ O \longrightarrow Na_2SO_4$

(f) Nitrites to nitrates : $NaNO_2+ O \longrightarrow NaNO_3$

(g) Thiosuphates to sulphates :

$$Na_2S_2O_3+ O \longrightarrow Na_2SO_4+ S$$

Oxidation in alkaline medium :

(a) KI to potassium iodate :

$$2KMnO_4+H_2O+KI \longrightarrow 2MnO_2 + 2KOH + KIO_3$$

(b) Nitrotoulene to nitrobenzoic acid :

$$O_2N\text{-}C_6H_4\text{-}CH_3+3O \longrightarrow O_2N\text{-}C_6H_4\text{-}COOH+H_2O$$

Oxidation in neutral medium :

(a) H_2S is oxidised to sulphur:

$$2KMnO_4+4H_2S \longrightarrow 2MnS+S+K_2SO_4 + 4H_2O$$

- Alkaline solution of $KMnO_4$ is known as Baeyer's reagent. It is used to detect double bonds in organic compounds.
- In quantitative analysis, it is used to detect halides and oxalates.
- In the manufacture of Saccharin and benzoic acid.
- MnO_4^- ion is tetrahedral and diamagnetic

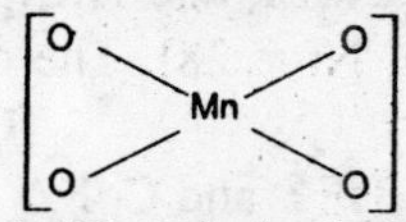

- $KMnO_4$ is Rhombohedral or Trigonal,

POTASSIUM DICHROMATE - $K_2Cr_2O_7$

It is also known as "Surkh Kahi".

Preparation :

Step I : Preparation of sodium chromate : Chrome iron stone is fused with excess of Na_2CO_3 in the presence of air in a reverboratory furnace. A small amount of lime is added to maintain the mass porous.

$$2FeO.Cr_2O_3+4Na_2CO_3+7O \longrightarrow 4Na_2CrO_4+4CO_2+Fe_2O_3$$

Step II: Conversion of chromate into dichromate

$$2Na_2CrO_4+ H_2SO_4 \longrightarrow Na_2Cr_2O_7+H_2O + Na_2SO_4$$

$Na_2Cr_2O_7$ is separated by fractional crystallisation as it is less soluble than Na_2SO_4.

Step III : Conversion of $Na_2Cr_2O_7$ to $K_2Cr_2O_7$

$$\underset{\text{hot solution}}{Na_2Cr_2O_7+2KCl} \longrightarrow K_2Cr_2O_7+ 2NaCl$$

Properties : It is an orange-red crystalline solid.

With alkalies, chromates are formed

$$K_2Cr_2O_7+2KOH \longrightarrow K_2CrO_4+ H_2O$$

Reverse reaction is

$$K_2CrO_4+H_2SO_4 \longrightarrow K_2Cr_2O_7+ K_2SO_4+H_2O$$

$$\underset{\text{orange}}{Cr_2O_7^{2-}} \rightleftharpoons \underset{\text{yellow}}{2CrO_4^{2-}} + 2H^+$$

Chromyl chloride test : When $K_2Cr_2O_7$ is heated with conc. H_2SO_4 and a soluble salt like NaCl, red vapours of chromyl chloride are obtained.

$$K_2Cr_2O_7+4NaCl+6H_2SO_4 \longrightarrow 2KHSO_4+4NaHSO_4+ \underset{\text{chromyl chloride}}{2CrO_2Cl_2}+8H_2O$$

As an Oxidising agent :

In neutral medium

$$K_2Cr_2O_7+ 4H_2O \longrightarrow 2KOH+2Cr(OH)_3 + 3O$$

In acidic medium

$$K_2Cr_2O_7+4H_2SO_4 \longrightarrow K_2SO_4+Cr_2(SO_4)_3+4H_2O+3O$$

(a) Oxidation of ferrous to ferric :

$$K_2Cr_2O_7+7H_2SO_4+6FeSO_4 \longrightarrow K_2SO_4+Cr_2(SO_4)_3+ 3Fe_2(SO_4)_3+7H_2O$$

(b) Oxidises H_2S to S :

$$K_2Cr_2O_7+4H_2SO_4+3H_2S \longrightarrow K_2SO_4+Cr_2(SO_4)_3+7H_2O+3S$$

(c) Oxidises SO_2 to H_2SO_4 :

$$K_2Cr_2O_7+3SO_2+H_2SO_4 \longrightarrow K_2SO_4+Cr_2(SO_4)_3+H_2O$$

(d) In Iodometry :

$$K_2Cr_2O_7+7H_2SO_4+6KI \longrightarrow 4K_2SO_4+Cr_2(SO_4)_3+3I_2+H_2O$$

It is used in chrome tanning; in photography; for the estimation of Fe^{2+}, I^- ions etc. and in dyeing and calicoprinting.

$Cr_2O_7^{2-}$ consists of two tetrahedra with the common oxygen atom at a common corner. Each of the chromium atom is at the centre of the tetrahedron.

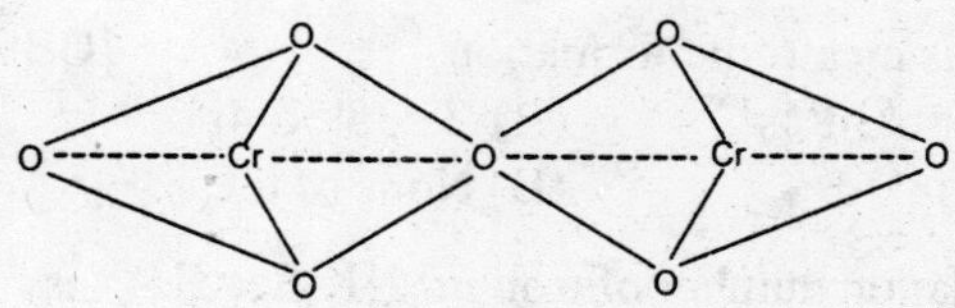

$K_2Cr_2O_7$ crystal is Triclinic

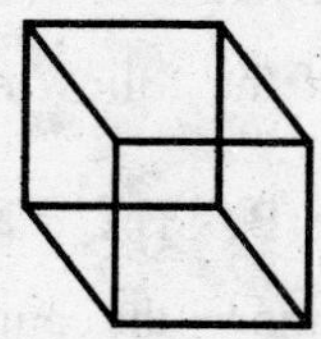

1. $K_4Fe(CN)_6$ is called [CBSE-90]
(A) potassium hexacyanoferrate (II)
(B) potassium ferricynide
(C) potassium hexacyanoferrate (III) (D) prussian blue

2. Calomel is [CBSE-90]
(A) $HgCl_2$ (B) Hg_2Cl_2
(C) $Hg+HgCl_2$ (D) Hg_2Cl_2+Hg

3. The transition elements have a general electronic configuration [CBSE-91]
(A) $ns^2\ np^6\ nd^{1-10}$ (B) $(n-1)d^{1-10}\ ns^{0-2}\ np^{0-6}$
(C) $(n-1)d^{1-10}\ ns^{0-2}$ (D) $nd^{1-10}\ ns^{0-2}$

4. Cr has electronic formation [CBSE-91]
(A) $3s^23p^63d^44s^2$ (B) $3s^23p^63d^54s^1$
(C) $3s^23p^63d^6$ (D) None of these

5. Oxidation number of iron in $[K_4Fe(CN)_6]$ is [CBSE-93]
(A) +3 (B) +2 (C) +10 (D) +1

6. Which of the following ligands is expected to be bidentate ? [CBSE-94]
(A) $C_2O_4^{2-}$ (B) $CH_3C \equiv N$ (C) Br^- (D) CH_3NH_2

7. IUPAC name for the complex compound $K_3[Fe(CN)_6]$ is [CBSE-94]
(A) potassiumhexacyanoferrate (III)
(B) potassiumferrocyanideiron (III)
(C) potassiumhexacyanoferrate (II)
(D) potassiumcyanohexaferrate (II)

8. Which one of the following transition elements shows the highest oxidation state ? [CBSE-94]
(A) Mn (B) Fe (C) V (D) Cr

9. Which one of the following has tetrahedral geometry? [CBSE-94]
(A) $[CO(NH_3)_6]^{3+}$ (B) $[Ni(CN)_4]^{2-}$
(C) $Fe(CO)_5$ (D) $[NiCl_4]^{2-}$

10. Which of the following complexes is likely to show optical activity ? [CBSE-94]
(A) Trans-$[Co(NH_3)_4Cl_2]^+$ (B) $[Cr(H_2O)_6]^{3+}$
(C) Cis-$[Co(NH_3)_2en_2]^{3+}$ (where en is ethylenediamine)
(D) Trans-$[Co(NH_3)_2\ en_2]^{3+}$ (where en is ethylenediamine)

11. Which of the following statement concerning Lanthanide elements is false. [CBSE-94]
(A) Lanthanides are separated from one another by ion exchange method.
(B) The ionic radii of trivalent lanthanides steadily increase with increase in the atomic number.
(C) All lanthanides are highly dense metals.
(D) Most characteristic oxidation state of lanthanide elements is +3.

12. In the metal carbonyls of the general formula $M\ (CO)_2$ where M = metal: x = 4, the metal is bonded to[CBSE-95]
(A) carbon and oxygen (B) Carbon
(C) Oxygen (D) $C \equiv 0$ triple bond

13. The number of geometrical isomers for $[Pt\ (NH_3)_2Cl_2]$ is [CBSE-95]
(A) Two (B) one (C) Three (D) four

14. Coordination number and oxidation state number of Cr in $K_3Cr(C_2O_4)_3$ are repsectively [CBSE-95]
(A) 4 and +2 (B) 6 and +3 (C) 3 and +3 (D) 3 and 0

15. Amongst TiF_6^{2-}, CoF_6^{3-}, Cu_2Cl_2 and $NiCl_4^{2-}$ (at nos. Ti = 22, Co = 27, Cu = 29, Ni = 28). The colourless species are [CBSE-95]
(A) CoF_6^{3-} and $NiCl_4^{2-}$ (B) TiF_6^{2-} and CoF_6^{3-}
(C) Cu_2Cl_2 and $NiCl_4^{2-}$ (D) TiF_6^{2-} and Cu_2Cl_2

16. The electronic configuration of transition elements is exhibited by [CBSE-96]
(A) $ns^2(n-1)d^{1-10}$ (B) $ns^2(n-1)d^{10}$
(C) ns^1 (D) $ns^2\ np^5$

17. When calomel reacts with NH_4OH, we get [CBSE-96]
(A) $HgNH_2Cl$ (B) NH_2-Hg-Hg-Cl
(C) Hg_2O (D) HgO

18. A transition element x has a configuration $(Ar)3d^4$ in its +3 oxidation state. Its atomic number is [CBSE-96]
(A) 25 (B) 26 (C) 22 (D) 19

19. CrO_3 dissolves in aqueous NaOH to give[CBSE-97]
(A) CrO_4^{2-} (B) $Cr(OH)_3$ (C) CrO_7^{2-} (D) $Cr(OH)_2$

20. The number of geometrical isomers of the complex $[Co(NO_2)_3(NH_3)_3]$ is [CBSE-97]
(A) 2 (B) 4 (C) 3 (D) 0

21. The formula dichlorobis(urea)copper (II) is [CBSE-97]
(A) $[Cu\{O = C(NH_2)_2\}]\ Cl_2$
(B) $[CuCl_2\ \{O = C(NH_2)_2\}_2]$
(C) $[Cu\ \{O = C(NH_2)_2\}Cl]Cl$
(D) $[CuCl_2]\ \{O = C(NH_2)_2H_2\}$

22. The electronic configuration of gadolinium (at.no.64) is [CBSE-97]
(A) $[Xe]\ 4f^8\ 5d^9\ 6s^2$, (B) $[Xe]\ 4f^7\ 5d^1\ 6s^2$,
(C) $[Xe]\ 4f^3\ 5d^5\ 6s^2$, (D) $[Xe]\ 4f^6\ 5d^2\ 6s^2$,

23. The lanthanide contraction is responsible for the fact that [CBSE-97]
(A) Zr and Y have about the same radius
(B) Zr and Nb have similar oxidation state
(C) Zr and Hf have about the same radius
(D) Zr and Zn have the same oxidation state

24. Which of the following does not represent the correct of the property indicated ? [CBSE-97]
(A) $Se^{3+} > Cr^{3+} > Fe^{3+} > Mn^{3+}$ (B) Se < Ti < Cr < Mn

(C) $Mn^{2+} < Mi^{+2} < Co^{2+} < Fe^{2+}$ (D) FeO>CaO>MiO>CuO

25. The total number of possible isomers for the complex compound $[Cu^{II}(NH_3)_4]$ is $[Pt^{II}Cl_4]$ is [CBSE-98]
(A) 4 (B) 3 (C) 6 (D) 5

26. A coordination complex compound of cobalt has the molecular formula containing five ammonia molecules, one nitro-group and two chlorine atoms for one cobalt atom. One mole of this compound produce three ions in an aqueous solution. On reacting this solution with excess of $AgNO_3$ solution, we get two moles of AgCl precipitate. The ionic formula for this complex would be.[CBSE-98]
(A) $[CO(NH_3)Cl]\ [Cl(NO_2)]$
(B) $[CO(NH_3)_5(NO_2)]Cl_2$
(C) $[CO(NH_3)_5]\ [(NO_2)_2Cl_2]$
(D) $[CO(NH_3)_4(NO_2)\ Cl]\ [(NH_3)\ Cl]$

27. Which one of the following species will impart colour to an aqueous solution ? [CBSE-98]
(A) Cu^+ (B) Zn^{2+} (C) Cr^{3+} (D) Ti^{4+}

28. IUPAC name of the $[Pt(NH_3)_3\ (Br)\ (NO_2)\ (Cl)]Cl$ is [CBSE-98]
(A) triamminenitrochlorobromoplatinum (IV) chloride
(B) Triamminebromochloronitroplatinum (IV) chloride
(C) Triamminebromonitrochloroplatinum (IV) chloride
(D) Triamminechlorobromonitrochloroplatinum (IV) chloride

29. Which one of the following elements shows maximum number of different oxidation state in its compounds ? [CBSE-98]
(A) Am (B) Gd (C) La (D) Eu

30. Which of the following has more unpaired d-electrons? [CBSE-99]
(A) Zn^+ (B) Fe^{2+} (C) N^{3+} (D) Cu^+

31. Bell metal is an alloy of [CBSE-99]
(A) Cu + Pb (B) Cu + Sn (C) Cu + Zn (D) Cu + Ni

32. In which of the following compounds transition metal has zero oxidation state ? [CBSE-99]
(A) $[Fe(CO)_5]$ (B) $NH_2.NH_2$
(C) NiS (D) CrO_5

33. Which of the following exhibits highest oxidation state? [CBSE-2000]
(A) Sc (B) Fe (C) Zn (D) Mn

34. Which of the following will give 4 isomers ? [CBSE-2000]
(A) $[CO(en)_2(H_2C)_2]Cl_3$ (B) $[CO(en)_3]Cl_2$
(C) $[CO(en)_3]Cl_3$ (D) $[CO(en)_2(NO_2)_2]Cl$

35. Which of the following will not give coloured ion? [CBSE-2000]
(A) Sc^{+3} (B) Ti^{+3} (C) V^{+3} (D) Cr^{+3}

36. Which of the following will give maximum number of isomers ? [CBSE-2001]
(A) $[\ CO(NH_3)_4Cl_2]$ (B) $Ni[(en)(NH_3)_4]^{+2}$
(C) $[Ni(C_2O_4)(en)_2]^{-2}$ (D) $[Cr(SCN)_2(NH_3)_4]^+$

37. Coordinatiion no. of Ni in $[Ni(C_2O_4)_3]^{-4}$ is [CBSE-2001]
(A) 3 (B) 6 (C) 4 (D) 2

38. Which of the following organometallic compounds is both δ and π bonded ? [CBSE-2001]
(A) $[Fe(n^5\text{-}C_5H_5)_2]$ (B) $K[PtCl_3\ (n^2\text{-}C_2H_4)]$
(C) $[Co(CO)_5NH_3]^{+2}$ (D)Fe $(CH_3)_3$

39. Which statement is incorrect ? [CBSE-2001]
(A) $Ni(CO)_4$ Tetrahedral, paramagnetic
(B) $Ni(CN)_4^{-2}$ -Sq. planar diamagnetic
(C) $Ni(CO)_4$ - Tetrahedral, diamagnetic
(D) $[Ni(Cl)_4]^{-2}$ - Tetrahedral, paramagnetic

40. General electronic configuration of lanthanides is [CBSE-2002]
(A) $(n-2)f^{1-14}\ (n-1)s^2p^6d^{0-1}\ ns^2$
(B) $(n-2)f^{10-14}\ (n-1)\ d^{0-1}\ ns^2$
(C) $(n-2)f^{0-14}(n-1)\ d^{10}\ ns^2$ (D) $(n-2)d^{0-1}(n-1)f^{1-14}\ ns^2$

41. An atom has elcctronic configuration $1s^22s^22p^63s^23p^63d^34s^2$, it is [CBSE-2002]
(A) V (B) Sc (C) Ti (D) Mn

42. Which of the following shows maximum number of oxidation states [CBSE-2002]
(A) Cr (B) Fe (C) Mn (D) V

43. Atomic number of Cr and Fe are respectively 25 and 26, which of the following is paramagnetic with the spin of electron [CBSE-2002]
(A) $[Cr(CO)_6]$ (B) $[Fe(CO)_5]$
(C) $[Fe(CN)_6]^{-4}$ (D) $[Cr(NH_3)_6]^{+3}$

44. The hypothetical complex chlorodiaquatriammine cobalt (III) chloride can be represented as [CBSE-2002]
(A) $[CoCl(NH_3)_3\ (H_2O)_2]Cl_2$ (B) $[Co(NH_3)_3(H_2O)Cl_3]$
(C) $[Co(NH_2)_3(H_2O)_2Cl]$ (D) $[Co(NH_3)_3(H_2O)_3]Cl_3$

45. In the silver plating of copper, $K[Ag(CN)_2)]$ is used instead of $AgNO_3$. The reason is [CBSE-2002]
(A) A thin layer of Ag is formed on Cu
(B) More voltage is required
(C) Ag^+ ions are completely removed from solution
(D) Less availability of Ag^+ ions, as Cu can not displace Ag from $[Ag(CN)_2]^-$ ion

46. $CuSO_4$ when reacts with KCN forms CuCN which is insoluble in water. It is soluble in excess of KCN, due to formation of the following complex [CBSE-2002]
(A) $K_2[Cu(CN)_4]$ (B) $K_3\ [Cu(CN)_4]$
(C) $CuCN_2$ (D) Cu $[KCu(CN)_4]$

47. Which one of the following octahedral complexes will not show geometrical isomerism ? (A and B are

monodentate ligands) [CBSE-2003]
(A) $[MA_5B]$ (B) $[MA_2B_4]$ (C) $[MA_3B_3]$ (D) $[MA_4B_2]$

48. The basic character of the transition metal monoxides follows the order [CBSE-2003]
(A) TiO > VO > CrO > FeO
(B) VO > CrO > TiO > FeO
(C) CrO > VO > FeO > TiO
(D) TiO > FeO > VO > CrO
(Atomic nos. Ti = 22, V = 23, Cr = 24, Fe = 26)

49. Which one of the following characteristics of the transition metals is associated with their catalytic activity? [CBSE-2003]
(A) Variable oxidation states
(B) High enthalpy of atomisation
(C) Paramagnetic behaviour (D) Colour of hydrated ions

50. The number of unpaired electrons in the complex ion $[CoF_6]^{3-}$ is [CBSE-2003]
(A) Zero (B) 2 (C) 3 (D) 4
(Atomic no. Co = 27)

51. The correct order of ionic radii of Y^{3+}, La^{3+}, Eu^{3+} and Lu^{3+} is [CBSE-2003]
(A) $La^{3+} < Eu^{3+} < Lu^{3+} < Y^{3+}$
(B) $Y^{3+} < La^{3+} < Eu^{3+} < Lu^{3+}$
(C) $Y^{3+} < Lu^{3+} < Eu^{3+} < La^{3+}$
(D) $Lu^{3+} < Eu^{3+} < La^{3+} < Y^{3+}$
(Atomic nos. Y = 39, La = 57, Eu = 63, Lu = 71)

52. Among the following, which is not the π-bonded organometallic compound? [CBSE-2003]
(A) $(CH_3)_4Sn$ (B) $K[PtCl_3(\eta^2-C_2H_4)]$
(C) $Fe(\eta^2-C_5H_5)_2$ (D) $Cr(\eta^6-C_6H_6)_2$

53. NH_3 is a complexing agent which forms stable silver diamine and copper (ii) tetramine cations in aqueous solutions of Ag (I) and Cu (II) respectively. In each of the complex ions : [AIIMS-93]
(A) NH_4^+ ions are present which are held to the metal atom by atom dipole intreraction
(B) NH_2^- ions are present which are held to the metal cations by ion-ion interactions
(C) NH_3 molecules are present which are held to the metal cations by hydrogen bonding
(D) NH_3 molecules are present which are held to metal cations by ion dipole interaction.

54. Number of water molecules in Mohr's salt is [AIIMS-94]
(A) 5 (B) 6 (C) 7 (D) 8

55. Which metal is present in brass, bronze and german-silver? [AIIMS-94]
(A) Al (B) Cu (C) Mg (D) Zn

56. Assertion : A solution of $KMnO_4$ is decolourised by SO_2. Reason : SO_2 is acidic in character. [AIIMS-94]
(A) A (B) B (C) C (D) D

57. Which of the following is called Baeyer's reagent [AIIMS-94]
(A) Saturated $KMnO_4$ solution
(B) Neutral $KMnO_4$ solution
(C) Alkaline $KMnO_4$ solution
(D) Acidic $KMnO_4$ solution

58. Which of the following molecules or ions is a bidentate ligand? [AIIMS-95]
(A) Br_2^+ (B) $CH_3-C\equiv N$ (C) $C_2O_4^{2-}$ (D) CH_3NH_2

59. Which of the following complexes has square planar structure? [AIIMS-95]
(A) $Ni(CO)_4$ (B) $[NiCl_4]^{2-}$
(C) $[Ni(Cl)_4]^{2-}$ (D) $[Zn(NH_3)_4]^{2+}$

60. Which of the following compound turns black on the addition of ammonium hydroxide? [AIIMS-95]
(A) $PbCl_2$ (B) Hg_2Cl_2 (C) $CuCl_2$ (D) AgCl

61. Which of the following compound is known as white vitriol? [AIIMS-96]
(A) $CaCO_2.2H_2O$ (B) $MgSO_4.7H_2O$
(C) $ZnSO_4.7H_2O$ (D) $Na_2SO_4.7H_2O$

62. Which of the following is organometallic compound? [AIIMS-97]
(A) $Ti(C_2H_4)_4$ (B) $Ti(OC_2H_5)_4$
(C) $Ti(OCOOH_3)_4$ (D) $Ti(OC_6H_5)_4$

63. The oxidation state of Fe in brown ring complex? [AIIMS-97]
(A) + 1 (B) + 2 (C) +3 (D) + 4

64. In the compound Lithium tetrahydroaluminate, the ligand is [AIIMS-97]
(A) H^+ (B) H (C) H^- (D) None of these

65. The following ions/atoms have five electrons in the outermost orbit, the one which has not the value of l azimuthal quantum no. equal to 2 is [AIIMS-97]
(A) Pt(O) (B) Fe(III) (C) Cr(III) (D) Mn(II)

66. Which of the following compounds is not coloured? [AIIMS-97]
(A) Na_2CuCl_4 (B) Na_2CdCl_4
(C) $K_4Fe(CN)_6$ (D) $K_3Fe(CN)_6$

67. The number of unpaired electrons in $Ni(CO)_4$ [AIIMS-97]
(A) Zero (B) One (C) Three (D) Five

68. Which one of the following shows maximum paramagnetic character? [AIIMS-98]
(A) $[Cr(H_2O)_6]^{3+}$ (B) $[Fe(CN)_6]^{4-}$
(C) $[Fe(CN)_6]^{3-}$ (D) $[Cu(H_2O)_6]^{2+}$

69. Assertion : The 19th electron in potassium atom enters into the 4s-orbital and not into the 3d-orbital. Reason :

(n+1) rule is followed for determining the orbitals of the lowest energy state. [AIIMS-98]
(A) A (B) B (C) C (D) D

70. Which of the following metal is present in brass, bronze and german-silver ? [AIIMS-99]
(A) Na (B) Mg (C) Cu (D) Al

71. $CuSO_4$ reacts with KCN solution and forms [AIIMS-99]
(A) Cu(CN) (B) $Cu(CN)_2$
(C) K_3 $[Cu(CN)_4]$ (D) K_4 $[Cu(CN)_6]$

72. Which of the following transition element shows the highest oxidation state ? [AIIMS-99]
(A) Mn (B) Fe (C) V (D) Cr

73. Which of the following is the best scientific method to test presence of water in a liquid ? [AIIMS-99]
(A) Taste (B) Smell (C) use of litmus paper
(D) Use of anhydrous copper sulphate

74. In dichromate dimer [AIIMS-2000]
(A) 4 Cr–O bonds are equal (B) 6 Cr–O bonds are equal
(C) All Cr–O bonds are equal
(D) All Cr–O bonds are unequal

75. $CuSO_4$ dissolves in NH_3 due to the formation of [AIIMS-2000]
(A) $Cu(OH)_2$ (B) $Cu[(NH_3)]SO_4$
(C) $Cu[(NH_3)_4(OH)_2$ (D) CuO

76. When an excess of KCN solution is added to a copper salt solution [AIIMS-2000]
(A) Copper is precipitated
(B) A precipitate of $Cu(CN)_2$ is formed
(C) A solution containing $[Cu(CN)_2]^-$ is formed
(D) A solution containing $[Cu(CN)_4]^{2-}$ is formed

77. Which one of the following is an example of π complex ? [AIIMS-2000]
(A) $[Cu(NH_3)_4]^{2+}$ (B) $[NiCl_4]^{2-}$
(C) $COCl_2$. $2NH_3$ (D) $Ni(CO)_4$

78. Helium is used in the balloons in the place of hydrogen because it is [AIIMS-2000]
(A) Incombustible (B) Lighter than hydrogen
(C) More abundant than hydrogen
(D) Radioactive and can be detected easily

79. $[HgI_3]^{1-}$ is [AIIMS-2000]
(A) Paramagnetic (B) Planar trigonal
(C) sp^2-hybridised (D) All

80. Lead poisoning in the body may be removed by [AIIMS-2000]
(A) Zeisse's salt (B) Ferrocene
(C) DMG (D) Ca-salt of E.D.T.A.

81. How many sigma and pi bonds are there in methane tetra-carbnitile molecule ? [AIIMS-2000]
(A) 8–σ and 8– π (B) 12– σ and 4– π
(C) 4 – σ and 12 – π (D) 12 – σ and 12 – π

82. In which dsp^2 hybridisation occur ? [AIIMS-2001]
(A) $[CuCl_4]^{2-}$ (B) $[PtCl_4]^{2-}$
(C) $[NiCl_4]^{2-}$ (D) $[CoCl_4]^{2-}$

83. Which complex compound forms max. no of isomers ? [AIIMS-2001]
(A) $[Ru(NH_3)_4Cl_2]^{2+}$ (B) $[Co(en)_2Cl_2]^{2+}$
(C) $[Cu(NH_3)_4]^{2+}$ (D) $[PtCl_4]^{2-}$

84. Which compound is not possible ? [AIIMS-2001]
(A) $[Be(OH)_6]^{4-}$ (B) $[Sn(OH)_6]^{4-}$ (C) SF_6 (D) IF_7

85. The charge on $[Ag(CN)_2]^-$ complex is [AIIMS-2001]
(A) –1 (B) +1 (C) +2 (D) +3

86. The electronic configuration of transition elements is exhibited by : [AIIMS-2001]
(A) ns^1 (B) ns^2np^5
(C) $ns^2(n-1)d^{10}$ (D) $(n-1)d^{1-10}$ ns^{0-2}

87. Assertion (A) : Cuprous ion (Cu^+) has unpaired electrons while cupric ion (Cu^{++}) does not. Reason (R) : Cuprous ion (Cu^+) is colourless whereas cupric ion (Cu^{++}) is blue in the aqueous solution. [AIIMS-2002]
(A) A (B) B (C) C (D) D (E) E

88. The complex used as an anticancer agent is - [AIIMS-2003]
(A) $[Co(NH_3)_3 Cl_3]$ (B) cis-$[Pt Cl_2(NH_3)_2]$
(C) cis-$K_2[PtCl_2Br_2]$ (D) Na_2CoCl_4

89. The colourless species is - [AIIMS-2003]
(A) VCl_3 (B) $VOSO_4$
(C) Na_3VO_4 (D) $[V(H_2O)_6]$ SO_4H_2O

90. MnO_4^{2-} (1mole) in neutral aqueous medium is disproportionate to - [AIIMS-2003]
(A) 2/3 mole of MnO_4 and 1/3 mole of MnO_2
(B) 1/3 mole of MnO_4 and 2/3 mole of MnO_2
(C) 1/3 mole of Mn_2O_7 and 1/3 mole of MnO_2
(D) 2/3 mole of Mn_2O_7 and 1/3 mole of MnO_2

91. Lanthanide for which +II and +III oxidation states are common is - [AIIMS-2003]
(A) La (B) Nd (C) Ce (D) Eu

92. What is the co-ordination number of sodium in Na_2O ? [AIIMS-2003]
(A) 6 (B) 4 (C) 8 (D) 2

93. (A) Solution of Na_2CrO_4 in water is intensely coloured. (R) Oxidation state of Cr in Na_2CrO_4 is +VI. [AIIMS-2003]
(A) (B) (C) (D)

94. Ce (Z = 58) is a member of : [JIPMER-90]
(A) s-block element (B) p- block elements

(C) d-block elements (D) f-block elements

95. Acidified $KMnO_4$ solution is decolourised by : [JIPMER-91]
(A) Bleaching powder (B) Mohr's salt
(C) Microcosmic salt (D) Laughing gas

96. How many unpaired electrons are there in Ni^{2+} : [JIPMER-91]
(A) 0 (B) 2 (C) 4 (D) 8

97. NH_3 can enter into complex formation due to : [JIPMER-92]
(A) Hydrogen bonding (B) Co-ordinate bond
(C) Covalent bond (D) Ionic bond

98. The lanthanide contraction relates to : [JIPMER-93]
(A) valence electrons (B) the densities
(C) nuclear masses (D) ionic radii

99. Which of the following is an acidic oxide ? [JIPMER-93]
(A) Mn_2O_3 (B) MnO_2 (C) Mn_2O_7 (D) MnO

100. The complex $CoCl_3.4NH_3$ ionises to give : [JIPMER-93]
(A) 1Cl^- ion (B) 2Cl^- ion (C) 3Cl^- ion (D) no Cl^- ion

101. Which of the following molecule or molecular ion is paramagnetic : [JIPMER-93]
(A) O_2^{2-} (B) N_2 (C) B_2 (D) CN^-

102. Which of the following is used in photography ? [JIPMER-94]
(A) Ag Br (B) $CuSO_4$ (C) Ag_2O (D) AgF

103. What is the oxidation state of iron in sodium nitroprusside ? [JIPMER-94]
(A) Zero (B) +1 (C) +2 (D) +3

104. A characteristic feature of transition elements is that [JIPMER-95]
(A) They show variable oxidation states
(B) They show inert pair effect
(C) They do not form complex ions
(D) They all exhibit diamagnetism

105. Which of the following ions has the highest magnetic moment ? [JIPMER-95]
(A) Sc^{3+} (B) Ti^{3+} (C) Mn^{2+} (D) Zn^{2+}

106. Which of the following complexes will give white precipitate with $BaCl_2$(aq) ? [JIPMER-95]
(A) $[Cr(NH_3)_5SO_4]Cl$ (B) $[CO(NH_2)_4SO_4]NO_2$
(C) $[Cr(NH_3)_5Cl]SO_4$ (D) Both (A) and (C)

107. The co-ordination and oxidation number of X in the compound respectively $[X(SO_4)(NH_3)_5]$ Cl will be [JIPMER-95]
(A) 10 and 3 (B) 6 and 4 (C) 2 and 6 (D) 6 and 3

108. Ammonia is bubbled through copper sulphate solution, and a deep blue solution results. The deep blue colour is due to formation of [JIPMER-95]
(A) $Cu(OH)_2$ (B) Cu^{2+} ions
(C) $[Cu(NH_3)_4]^{2+}$ (D) None of these

109. Which of the following ions in aqueous medium has orange colour? [JIPMER-96]
(A) Cr^{+3} (B) $Cr_2O_7^{2-}$ (C) MnO_4^- (D) MnO_4^{2-}

110. The most stable oxidation state of Fe is [JIPMER-96]
(A) +2 (B) + 3 (C) –2 (D) –3

111. The most abundant transition element is [JIPMER-96]
(A) Cr (B) Fe (C) W (D) Mn

112. Which of the following is used for galvanizing iron [JIPMER-96]
(A) Cr (B) Zn (C) Ni (D) Sn

113. In which of the following state of matter the average distance between the molecules lies between 10^{-5} cm to 10^{-7} cm ? [JIPMER-96]
(A) Liquid (B) Gas (C) Solid (D) none of these

114. The hybrid state of sulphur is SO_4^{-2} is : [JIPMER-96]
(A) sp^3 (B) sp^2 (C) sp (D) none

115. Which of the following compound would exhibit co-dination isomerism [JIPMER-96]
(A) $[Cr(en)_2]NO_2$ (B) $[Cr(H_2O)_6]Cl_3$
(C) $[Cr(NH_3)_6]$ $[Co(CN)_6]$ (D) None of the above

116. The blue complex formed on addition of conc. NH_4OH solution to a Cu^{2+} salt solution has the structure [JIPMER-96]
(A) $[Cu(NH_3)_2]^{2+}$ (B) $[Cu(NH_3)_4]^{2+}$
(C) $[Cu(NH_4)_4]^{2+}$ (D) $[Cu(NH_4)_2]^{2+}$

117. The oxidation number of nickel in $K_4[Ni(CN)_4]$ is [JIPMER-98]
(A) –2 (B) –1 (C) +3 (D) 0

118. The element which forms ions in the dimeric state is. [JIPMER-98]
(A) Copper (B) Nickel (C) Mercury (D) Silver

119. Which of the following transition elements shows the highest oxidation state ? [JIPMER-2001]
(A) Cr (B) V (C) Fe (D) Mn

120. Number of water molecules in Mohr's salt is [JIPMER-2001]
(A) 8 (B) 7 (C) 6 (D) 5

121. The number of unpaired electrons in ferrous ion is [JIPMER-2001]
(A) 5 (B) 4 (C) 3 (D) 2

122. Green vitriol is [JIPMER-2002]

(A) $FeSO_4.7H_2O$ (B) $FeSO_4.5H_2O$

(C) $CuSO_4.5H_2O$ (D) $CaSO_4.7H_2O$

123. An element is in M^{3+} form. Its electronic configuration is [Ar] $3d^1$ what is its atomic number [JIPMER-2002]

(A) Ti^{3+} (B) Ti^{4+} (C) Ca^{2+} (D) Sc^+

124. Which law directly explains the law of conservation of mass ? [JIPMER-2002]

(A) Dalton's law (B) Avogadro's law

(C) Beizlius' law (D) Hund's rule.

125. Which complex has square planar structure ? [JIPMER-2002]

(A) $Ni(CO)_4$ (B) $[NiCl_4]^{2-}$

(C) $[Ni(H_2O)_6]^{2+}$ (D) $[Cu(NH_3)_4]^{2+}$

126. The chemical formula for lunar caustic is [JIPMER-2002]

(A) Ag_2S (B) Ag_2SO_4 (C) $AgNO_3$ (D) AgCl

127. Which of the following represents a chelating ligand [JIPMER-2002]

(A) H_2O: (B) OH^- (C) DMG (D) Cl^-

128. The highest oxidation state shown by Mn is [JIPMER-2002]

(A) +3 (B) +2 (C) +4 (D) +7

129. Which of the following is used in cancer therapy ? [BHU-95]

(A) Po (B) Co (C) Ni (D) Rn

130. Which of the following represents the correct pattern of electron filling in chromium ? [BHU-95]

(A) [↑↓][↑↓][↑][ ][ ] [↑] (B) [↑][↑][↑][↑][ ] [↑↓]

(C) [↑][↑][↑][↑][↑] [↑] (D) [↑↓][↑][↑][↑][↑] [ ]

131. Which of the following metals acts as the most efficient catalyst ? [BHU-95]

(A) Alkali (B) Transition

(C) Alkaline earth (D) Coloured metals

132. The square planar shape is for [BHU-95]

(A) BF_4 (B) $[Cu(NH_3)_4]^{2+}$

(C) $[CrF_6]^{3-}$ (D) None of these

133. The electronic structure of Fe^{3+} is [BHU-95]

(A) $[Ar]3d^6$ (B) $[Ar]3d^5$ (C) $[Ar]2d^4$ (D) $3d^64s^2$

134. Which of the following is correct ? [BHU-95]

(A) Gun metal - Cu + Zn + Sn

(B) Duralumin - Al + Cu + Mg + Ag

(C) German silver - Cu + Zn + C (D) Solder-Pb + Al

135. Aqueous solution of Cu (II) gives an intense blue solution with ammonia, which is due to [BHU-96]

(A) $[Cu(H_2O)_6]^{2+}$ (B) $Cu(OH)_2$

(C) $[Cu(NH_3)_4]^{2+}$ (D) $Cu(NH_3)_2$

136. Gold is not used in [BHU-96]

(A) Photography (B) Jewellery (C) Medicine (D) Coins

137. Which of the following is a radical of Nessler's reagent ? [BHU-96]

(A) HgI_2^{-4} (B) HgI_4^{-2} (C) HgI_4 (D) Hg_2I_2

138. Which of the following ions is coloured ? [BHU-97]

(A) Cu^+ (B) Cu^{2+} (C) Ti^{4+} (D) V^{3+}

139. Formula of green vitrol is [BHU-97]

(A) $FeSO_4.7H_2O$ (B) $MgSO_4.7H_2O$

(C) $ZnSO_4.7H_2O$ (D) $CuSO_4.5H_2O$

140. Electronic configuration of Sc^{21} is [BHU-97]

(A) $1s^2, 2s^2p^63s^2, 3p^64s^2, 3d^1$

(B) $1s^2, 2s^2, 2p^63s^2, 3p^64s^1, 3d^2$

(C) $1s^2, 2s^2, 2p^63s^2, 3p^64s^0$, 3d3

(D) $1s^2, 2s^2p^63s^2, 3p^64s^2$, 3d2

141. What is the oxidation number of Cr in $K_2Cr_2O_7$? [BHU-98]

(A) +6 (B) –6 (C) +3 (D) –5

142. Which compound volatiles on heating ? [BHU-98]

(A) $MgCl_2$ (B) $HgCl_2$ (C) $ZnCl_2$ (D) None of these

143. One of the following is false for Hg [BHU-98]

(A) It can evolve hydrogen from H_2S

(B) It is metal (C) It has high specific heat

(D) It is less reactive than hydrogen.

144. What is the oxidation number of Co in $[Co(NH_3)_4ClNO_2]$ [BHU-99]

(A) +2 (B) + 3 (C) +4 (D) +5

145. Maximum value of paramagnetic behaviour is shown by [BHU-2000]

(A) $[Fe(CN)_6]^{3-}$ (B) $[Sc(CN_6]^{3-}$

(C) $[Cr(CN)_6]^{3-}$ (D) $[Co(CN)_6]^{3-}$

146. The unpaired electron in $Ni(CO)_4$ are [BHU-2000]

(A) Zero (B) One (C) Three (D) Four

147. Which of the following has square planar structure? [BHU-2000]

(A) $[Ni](CO_4)]$ (B) $[Ni(Cl_4)]^{-2}$

(C) $[Ni(CN)_4]^{2+}$ (D) All of these

148. The most stable oxidation state of Nickel is [BHU-2001]

(A) +1 (B) +2 (C) +3 (D) 0

149. The effective nuclear charge on $[Co(NH_3)Cl_3$ is [BHU-2001]

(A) 3 (B) 2 (D) 6 (D) 0

150. Essential condition for the formation of coordination compound is [BHU-2001]

(A) Small size (B) High charge density

(C) Both the above reasons are correct

(D) Presence of transition element

151. The oxidation number of chromium in $K_2Cr_2O_7$ is [AFMC-91]

(A) 4 (B) 6 (C) 5 (D) 3

152. The first element of rare earth metals is [AFMC-92]

(A) Cerium (B) Actinum (C) Uranium (D) Lanthanide

153. When copper nitrate is strongly heated it gets converted into [AFMC-92]

(A) Copper metal (B) Cupric oxide
(C) Cuprous oxide (D) Copper nitrite

154. The compound having blue colour is [AFMC-92]

(A) $CuSO_4$ (B) $CuSO_4.5H_2O$
(C) $PbSO_4$ (D) $HgSO_4$

155. Copper nitrate $Cu(NO_3)_2$ on heating gives [AFMC-93]

(A) Copper metal (B) Cupric oxide
(C) Cuprous oxide (D) Cuprous nitrite

156. Brass is an alloy of copper (Cu) and [AFMC-93]

(A) Al (B) Al and Ni (C) Zn (D) Sn

157. The oxide which reacts with both NaOH and HCl is [AFMC-93]

(A) CaO (B) ZnO (C) N_2O_5 (D) CO_2

158. Oxidation state of chromium in $K_2Cr_2O_7$ is [AFMC-93]

(A) 4 (B) 6 (C) 7 (D) 8

159. Hypo is used in photography because it is a [AFMC-93]

(A) Strong oxidizing agent (B) Strong reducing agent
(C) Strong complexing agent (D) None of the above

160. Complexion is shown by [AFMC-94]

(A) Ag (B) Au (C) Cu (D) All of these

161. Chromium has most stable oxidation state in [AFMC-94]

(A) +5 (B) + 3 (C) + 2 (D) + 4

162. The most common lanthanide is [AFMC-95]

(A) Lanthanum (B) Cerium
(C) Samarium (D) Plutonium

163. Which of the following is paramagnetic ? [AFMC-97]

(A) $[Ni(CO)_4]^{2+}$ (B) $CO[(NH_3)_6]^{2+}$
(C) $[Ni(CN)_4]^{2-}$ (D) $[NiCl_4]^{2-}$

164. Which of the following as square planar structure ? [AFMC-97]

(A) $[Ni(Cl)_4]^{2+}$ (B) $[Ni(CO)_4]^{2+}$
(C) $[Ni(CN)_4]^{2-}$ (D) None of these

165. Which lanthanide is most commonly used ? [AFMC-97]

(A) Lanthanum (B) Nobilium (C) Thorium (D) Cerium

166. When cold potassium permanganate ($KMnO_4$) is added to ethylene gives [AFMC-98]

(A) Glycerol (B) Ethanol
(C) Methanol (D) Ethylene glycol

167. Which of the following is Calomel ? [AFMC-98]

(A) $HgCl_2$ (B) $HgBr_2$ (C) Hg_2Cl_2 (D) $HgSO_4$

168. Which of the following metal is present in brass, bronze and German silver ? [AFMC-98]

(A) Zn (B) Mg (C) Cu (D) Al

169. The oxidation state of chromium in potassium dichromate ($K_2Cr_2O_7$) is [AFMC-99]

(A) -5 (B) +6 (C) +2 (D) -2

170. $K_4[Fe(CN)_6]$ is a/an [AFMC-2000]

(A) Complex salt (B) Double salt
(C) Alum (D) None of these

171. The variable valency is the characteristic of [AFMC-2000]

(A) Halogen (B) Transition elements
(C) Alkali metals (D) Noble gas

172. Which of the following is not correct for transition metals ? [AFMC-2001]

(A) Their melting and boiling points are high
(B) Their compounds are generally coloured
(C) They can form ionic or covalent compounds
(D) They do not exhibit variable valency

173. $K_3[Al(C_2O_4)_3]$ is called [AFMC-2001]

(A) Potassium alumino oxalate
(B) Potassium alumino (III) oxalate
(C) Potassium trioxalatoaluminate
(D) Potassium trioxalatoaluminate (III)

174. Oxidation number of iron in $[K_4Fe(CN)_6]$ is [AFMC-2002]

(A) +3 (B) +2 (C) + 10 (D) + 1

175. Which of the following will give colour to an aqueous solution ? [AFMC-2002]

(A) Cu^+ (B) Zn^{-2} (C) Cr^{+3} (D) Ti^{+4}

176. The aqueous solution of following salts will be coloured in the case of . [IIT-90]

(A) $Zn(NO_3)_2$ (B) $LiNO_3$ (C) $Ca(NO_3)_2$ (D) $CrCl_3$

177. Among $Ni(CO)_4$, $[Ni(CN)_4]^{2-}$ and $NiCl_4^{2-}$ [IIT-91]

(A) $NI(CO)_4$ and $NiCl_4^{2-}$ are diamagnetic and $[Ni(CN)_4]^{2-}$ is paramagnetic
(B) $Ni(CO)_4$ and $[Ni(CN)_4]^{2-}$ are diamagnetic and $NiCl_4^{2-}$ is paramagnetic
(C) $NiCl_4^{2-}$ and $[Ni(CN)_4]^{2-}$ are diamagnetic and $Ni(CO)_4$ is paramagnetic
(D) $Ni(CO)_4$ is diamagnetic and $NiCl_4^{2-}$ and $[Ni(CN)_4]^{2-}$

are paramagnetic.

178. Among the following which one has the highest paramagnetism ? [IIT-93]

(A) $[Cr(H_2O)_6]^{3+}$ (B) $[Fe(H_2O)_6]^{2+}$
(C) $[Cu(H_2O)_6]^{2+}$ (D) $[Zn(H_2O)_6]^{2+}$

179. Which of the following alloys contain Cu and Zn? [IIT-93]

(A) Bronze (B) Brass
(C) Gun metal (D) Type metal.

180. The type of magnetism exhibited by $(Mn(H_2O]^{2+}$ ion is .... [IIT-94]

181. The IUPAC name of $[Co(NH_3)_6]Cl_3$ is .... [IIT-94]

182. Which compound is formed when excess of KCN is added to aqueous solution of copper sulphate ? [IIT-96]

(A) $Cu(CN)_2$ (B) $K_2[Cu(CN)_4]$
(C) $K[Cu(CN)_2]$ (D) $K_3[Cu(CN)_4]$

183. Which of the following has maximum number of unpaired electrons ? [IIT-96]

(A) Mg^{2+} (B) Ti^{3+} (C) V^{3+} (D) Fe^{2+}

184. Which of the following compounds is expected to be coloured ? [IIT-97]

(A) Ag_2SO_4 (B) CuF_2 (C) MgF_2 (D) CuCl

185. Which of the following is an organometallic compound ? [IIT-97]

(A) Lithium methoxide (B) Lithium acetate
(C) Lithium dimethylamide (D) Methyl lithium

186. Which of the following statement (s) is (are) correct with reference to the ferrous and ferric ions ? [IIT-98]

(A) Fe^{3+} gives brown colour with potassium ferricyanide.
(B) Fe^{2+} gives blue precipitate with potassium ferricyanide
(C) Fe^{3+} gives red colour with potassim thiocyanate
(D) Fe^{2+} gives brown colour with ammonium thiocyanate

187. The complex ion which has no d electrons in the central metal atom atom is [IIT-2001]

(A) $[MnO_4]^-$ (B) $[Co(NH_3)_6]^{3-}$
(C) $[Fe(CN)_6]^{3-}$ (D) $[Cr(H_2O)_6]^{3-}$

188. When $KMnO_4$ acts as an oxidizing agent and ultimately forms $[MnO_4]^{-1}$, MnO_2, Mn_2O_3, Mn^{+2} then the number of electrons transferred in each case respectively is. [AIEEE-2002]

(A) 4, 3, 1, 5 (B) 1, 5, 3, 7
(C) 1, 3, 4, 5 (D) 3, 5, 7, 1

189. A square planar complex is formed by hybridization of which atomic orbitals ? [AIEEE-2002]

(A) s, p_x, p_y, d_{yz} (B) s, p_x, p_y, $d_{x^2-y^2}$
(C) s, p_x, p_y, d_{z^2} (D) s, p_y, p_z, d_{xy}

190. The type of isomerism present in nitropentamine chromium (III) chloride is. [AIEEE-2002]

(A) Optical (B) Linkage
(C) Ionization (D) Polymerisation

191. Which of the following ions has the maximum magnetic moment ? [AIEEE-2002]

(A) Mn^{2+} (B) Fe^{2+} (C) Ti^{2+} (D) Ge^{2+}

192. Arrange Ce^{3+}, La^{3+}, Pm^{3+} and Yb^{3+} in increasing order of their ionic radii. [AIEEE-2002]

(A) $Yb^{3+} < Pm^{3+} < Ce^{3+} < La^{3+}$
(B) $Ce^{3+} < Yb^{3+} < Pm^{3+} < La^{3+}$
(C) $Yb^{3+} < Pm^{3+} < La^{3+} < Ce^{3+}$
(D) $Pm^{3+} < La^{3+} < Ce^{3+} < Yb^{3+}$

193. Which is the correct order of ionic sizes ? [AIEEE-2002]

(A) Ce > Sn > Yb > Lu (B) Sn > Ce >Lu >Yb
(C) Lu > Yb > Sn > Ce (D) Sn > Yb>Ce>Lu
(At. Nos.: Ce=58, Sn=50, Yb = 70 and Lu 71)

194. The most stable ion is. [AIEEE-2002]

(A) $[Fe(OH)_3]^{3-}$ (B) $[Fe(Cl)_6]^{3-}$
(C) $[Fe(CN)_6]^{3-}$ (D) $[Fe(H_2O)_6]^{3+}$

195. The atomic numbers of vanadium (V), chromium (Cr), manganese (Mn) and iron (Fe) are respectively 23, 24, 25 and 26. Which one of these may be expected to have the highest second ionization enthalpy ? [AIEEE-2003]

(A) V (B) Cr (C) Mn (D) Fe

196. A red solid is insoluble in water. However it becomes soluble if some KI is added to water. Heating the red solid in a test tube results in liberation of some violet coloured fumes and droplets of a metal appear on the cooler parts of the test tube. The red solid is [AIEEE-2003]

(A) $(NH_4)_2Cr_2O_7$ (B) HgI_2 (C) HgO (D) Pb_3O_4

197. The number of d-electrons retained in Fe^{2+} (At. no. of Fe = 26) ion is [AIEEE-2003]

(A) 3 (B) 4 (C) 5 (D) 6

198. What would happen when a solution of potassium chromate is treated with an excess of dilute nitric acid ? [AIEEE-2003]

(A) Cr^{3+} and $Cr_2O_7^{2-}$ are formed
(B) $Cr_2O_7^{2-}$ and H_2O are formed
(C) CrO_7^{2-} is reduced to +3 state of Cr
(D) CrO_4^{3-} is oxidized to +7 state of Cr

199. In the coordination compound, $K_4[Ni(CN)_4]$, the oxidation state of nickel is [AIEEE-2003]

(A) –1 (B) 0 (C) +1 (D) +2

200. Ammonia forms the complex ion $[Cu(NH_3)_4]^{2+}$ with copper ions in alkaline solutions but not in acidic solutions. What is the reason for it ? [AIEEE-2003]

(A) In acidic solution hydration protects copper ions

(B) In acidic solution protons coordinate with ammonia molecules forming NH_4^+ ions and NH_3 molecules are not available

(C) In alkaline solutions insoluble $Cu(OH)_2$ is precipitated which is soluble in excess of any alkali

(D) Copper hydroxide is an amphtoeric substance.

201. One mole of the complex compound $Co(NH_3)_5Cl_3$, give 3 moles of ions on dissolution in water. One mole of the same complex reacts with two moles of $AgNO_3$ solution to yield two moles of AgCl (s). The structure of the complex is [AIEEE-2003]

(A) $[Co(NH_3)_5Cl]Cl_2$ (B) $[Co(NH_3)_3Cl].2NH_3$

(C) $[Co(NH_3)_4Cl_2]Cl.NH_3$ (D) $[Co(NH_3)_4Cl]Cl_2.NH_3$

202. Which one of the following $CuCl_2$, or $CaCl_2$ or $CdCl_2$ is paramagnetic in character ....[EAMCET-90]

203. The number of ions produced from one molecule of potassium ferricyanide in aqueous solution are .... [EAMCET-90]

204. Among the ions Sc^{3+}, V^{3+} and Cr^{3+} the ion that gives colourless compounds is .... [EAMCET-90]

205. Cuprous ion is colourless while cupric ion is coloured because [EAMCET-92]

(A) Both have unpaired electrons in the d-orbital

(B) Cuprous ion has a complete d-orbital and cupric ion has an incomplete d-orbital

(C) Both have half filled p and d-orbitals

(D) Cuprous ion has incomplete d-orbital and cupric ion has a completed d-orbital.

206. The oxidation number of manganese in $KMnO_4$ is .... [EAMCET-92]

207. Which of the following alloys contains Cu & Zn ? [EAMCET-93]

(A) Bronze (B) Brass (C) Gunmetal (D) Type metal

208. The 4f level is successively filled up in [EAMCET-93]

(A) Alkali metals (B) Rare gases

(C) Lanthanides (D) Actinides

209. Unpaired electrons in .... in transition metals leads to paramagnetism . [EAMCET-93]

210. Transition elements exhibit variable valency on account of .... d-orbitals.[EAMCET-93]

211. Inner transition elements exhibit different coloured compounds on account of unfilled .... [EAMCET-93]

212. A solution of $Cr(NO_3)_3$ slowly turns green when concentrated hydrochloric acid is added due to formation of .... [EAMCET-93]

213. The ion that is likely to be colourless in aqueous solution is [EAMCET-94]

(A) Ti^{+2} (B) V^{+3} (C) Mn^{+2} (D) Zn^{2+}

214. The oxidation state of Mo in $MO_2Cl_8^{4-}$ ion is [EAMCET-94]

(A) –4 (B) –2 (C) +6 (D) +2

215. The metal which is not ferromagnetic is [EAMCET-95]

(A) Iron (B) Manganese (C) Cobalt (D) Nickel

216. Blue prints are based on the formation of a transition metal cyanide complex. The metal is [EAMCET-95]

(A) Manganese (B) Cobalt (C) Chromium (D) Iron

217. Which of the following is paramagnetic ? [EAMCET-95]

(A) N_2O_4 (B) NO (C) N_2O_3 (D) N_2

218. The element that can exhibit highest number of oxidation states amongst the following is [EAMCET-96]

(A) V (B) Mn (C) Ni (D) CO

219. The number of moles of AgCl precipitated when excess $AgNO_3$ is mixed with one mole of $[Cr(NH_3)_4Cl_2]Cl$ is [EAMCET-96]

(A) 0 (B) 1 (C) 2 (D) 3

220. The effective atomic number of Cr in $[Cr(NH_3)_6]Cl_3$ is [EAMCET-98]

(A) 35 (B) 36 (C) 27 (D) 33

221. The number of moles of AgCl precipitated when excess $AgNO_3$ is added to one mole of $[Cr(NH_3)_4Cl_2]Cl$ is [EAMCET-98]

(A) Zero (B) 1.0 (C) 2.0 (D) 3.0

222. Which of the following sets of elements does not belong to transitional element set ? [EAMCET-98]

(A) Fe, CO, Ni (B) Cu, Ag, Au

(C) Ti, Zr, Hf (D) Ga, In, Tl

223. $KMnO_4$ oxidizes oxalic acid in acid medium. The number of CO_2 molecules produced as per the balanced equation is [EAMCET(MBBS)-98]

(A) 10 (B) 8 (C) 6 (D) 3

224. The oxidation number of cobalt in $K_3[Co(NO_2)_6]$ is [EAMCET(MBBS)-98]

(A) +6 (B) –3 (C) +3 (D) –6

225. Heating an ore in the absence of air below its melting point is called [EAMCET(MBBS)-98]

(A) Calcination (B) Smelting (C) Roasting (D) Leaching

226. In the electrolytic extraction of aluminum fused cryolite is added. Its role is [EAMCET(MBBS)-98]

(A) To protect the anodes (B) To increase Al production

(C) It acts as a reducing agent

(D) To make a conducting solution of Bauxite

227.In which one of the following transition metal complexes does metal NOT exist in zero oxidation state? [EAMCET(MBBS)-98]

(A) $[Co(NH_3)_6]Cl_3$ (B) $[Fe(H_2O)_6]SO_4$
(C) $[Ni(CO)_4]$ (D) $[Fe (H_2O)_6]X_3$

228. What is the oxidation state of 'Fe' in the product formed when acidified potassium ferro-cyanide ($K_4[Fe(CN)_6]$) is treated with hydrogen peroxide ? [EAMCET-99]

(A) +2 (B) +3 (C) +1 (D) +6

229. A burning strip of magnesium is introduced into a jar containing a gas. After sometime the walls of the container are coated with carbon. The gas in the container is [EAMCET-99]

(A) O_2 (B) N_2 (C) CO_2 (D) H_2O

230. Which of the following pairs of ions contain the same number of unpaired electrons ? [EAMCET-99]

(A) Ni^{2+}, CO^{2+} (B) Mn^{2+}, Fe^{3+}
(C) Mn^{2+}, Ni^{2+} (D) Ti^{3+}, CO^{2+}

231. The oxidation number of Mn in potassium permanganate is [EAMCET-99]

(A) +6 (B) +7 (C) +5 (D) +8

232. Which one of the following ions has same number of unpaired electrons as that of V^{3+} ion ? [EAMCET(MBBS)-99]

(A) Cr^{3+} (B) Mn^{2+} (C) Ni^{2+} (D) Fe^{3+}

233. What is oxidation number of Cr in potassium dichromate ? [EAMCET(MBBS)-99]

(A) +7 (B) +5 (C) +6 (D) +4

234. Which of the following ions is colourless in its aqueous solution ? [EAMCET-2000]

(A) Ti^{3+} (B) Cu^{2+} (C) Ni^{2+} (D) Zn^{2+}

235. Brass is an alloy of [EAMCET-2000]

(A) Ag and Cu (B) Sn and Zn
(C) Cu and Sn (D) Cu and Zn

236. Which of the following reactions does not involve change in the oxidation state of metal ?[EAMCET-2000]

(A) $VO^{2+} \rightarrow V_2O_3$ (B) $Na \rightarrow Na^+$
(C) $CrO_4^{2-} \rightarrow Cr_2O_7^{2-}$ (D) $Zn^{2+} \rightarrow Zn$

237. Which one of the following ions exhibits colour in aqueous solution ? [EAMCET(MBBS)-2000]

(A) Sc^{3+} (B) Ni^{2+} (C) Ti^{+4} (D) Zn^{+2}

238. What is the volume (in ml) of 0.1 M Potassium permanganate solution required to completely oxidize 100ml of 0.5 M ferrous sulphate solution in acid medium? [EAMCET(MBBS)-2000]

(A) 20 (B) 200 (C) 50 (D) 100

239. Which one of the following ions exhibits highest magnetic moment ? [EAMCET-2001]

(A) Cu^{2+} (B) Ti^{3+} (C) Ni^{2+} (D) Mn^{2+}

240. In the ground state, an element has 13 electrons in its "M-shell". The element is ..... [EAMCET-2001]

(A) Copper (B) Chromium (C) Nickel (D) Iron

241. Which one of the following is a diamagnetic ion ? [EAMCET(MBBS)-2001]

(A) Co^{2+} (B) Cu^{2+} (C) Mn^{2+} (D) Sc^{3+}

242. The coordinaton number of Fe (II) in Oxyhaemoglobin is .... [EAMCET(MBBS)-2001]

(A) 6 (B) 4 (C) 8 (D) 10

243. Which one of the following pairs of ions have the same electronic configuration ?[EAMCET(MBBS)-2001]

(A) Cr^{3+}, Fe^{3+} (B) Fe^{3+}, Mn^{2+}
(C) Fe^{3+}, CO^{3+} (D) Sc^{3+}, Cr^{3+}

244. The calculated magnetic moment (in Bohr magnetons) of Cu^{2+} ion is [EAMCET-2002]

(A) 1.73 (B) Zero (C) 2.6 (D) 3.4

245. How many 'd' electrons are present in Cr^{2+} ion ? [EAMCET(MED)-2002]

(A) 4 (B) 5 (C) 6 (D) 3

246. Ferrous ion changes to <u>X</u> ion on reaction with acidified hydrogen peroxide. The number of d-electrons present in <u>X</u> and in magnetic moment (in BM) are respectively [EAMCET-2003]

(A) 6 and 6.93 (B) 5 and 5.92
(C) 5 and 4.9 (D) 4 and 5.92

247. At. No. of an element is 26. The element shows [CPMT-96]

(A) Ferromagnetism (B) Diamagnetism
(C) Paramagnetism (D) None of these

248. Which of the following contains maximum number of unpaired electrons ? [CPMT-96]

(A) Mn^{2+} (B) Co^{2+} (C) CO^{3+} (D) Fe^{2+}

249. Zn does not show variable valency because of [CPMT-96]

(A) Complete 'd' subshell (B) Intert pair effect
(C) Have $4s^2$ subshell (D) It show variable valency

250. When silver nitrate is heated what is formed ? [CPMT-96]

(A) Ag (B) Ag_2O (C) Ag_2O_3 (D) AgO_2

251. Which of the following metals will not react with a solution of $CuSO_4$? [CPMT-96]

(A) Fe (B) Zn (C) Mg (D) Hg

252. Oxidation number of Ni in Ni $(CO)_4$ is- [CPMT-97]

(A) 2 (B) 3 (C) 1 (D) 0

253.Number of water atoms in Mohr's salt is[CPMT-97]

(A) 7 (B) 6 (C) 5 (D) 8

254. The formula of corrosive sublimate is- [CPMT-97]

(A) $HgCl_2$ (B) Hg_2Cl_2 (C) Hg_2O (D) Hg

255. Identify the transition element [CPMT-97]
(A) $1s^22s^22p^63s^23p^64s^2$
(B) $1s^22s^22p^63s^43p^63d^{10}4s^24p^1$
(C) $1s^22s^22p^63s^23p^63d^{10}4s^24p^6$
(D) $1s^22s^22p^63s^23p^63d^24s^2$

256. Number of unpaired electrons in Mn^{+2} is [CPMT-97]
(A) 3 (B) 5 (C) 4 (D) 1

257. Which metal is present in brass, bronze and German silver ? [CPMT-97]
(A) Zn (B) Mg (C) Cu (D) Al

258. Which are the constituents of gunmetal ? [CPMT-97]
(A) Cu,Sn (B) Cu, Sn, Zn, Pb
(C) Cu, Zn, Sb, Pb (D) Cu, Sn, Pb

259. The correct IUPAC name of $Fe_4[Fe(CN)_6]_3$ is [CPMT-98]
(A) Iron (III) hexacyanoferrate (II)
(B) Hexacyanoferrate (III, II)
(C) Ferroso ferric cyanide (D) Ferric ferrous hexacyanate

260. Blue complex formed on addition of conc. NH_4OH solution to a Cu (II) salt solution has following ion- [CPMT-98]
(A) $[Cu(NH_4)_4]^{2+}$ (B) $[Cu(NH_3)_2]^{2+}$
(C) $[Cu(NH_3)_4]^{2+}$ (D) $[Cu(NH_4)_2]^{2+}$

261. In solid $CuSO_4.5H_2O$, copper is coordinated to - [CPMT-98]
(A) Four water molecules (B) Five water molecules
(C) One suphate ion (D) One water molecule

262. When $AgNO_3$ (solid) is strongly heated, the products are - [CPMT-98]
(A) Ag, NO_2 and O_2 (B) Ag, N_2 and O_2
(C) $AgNO_2$ and O_2 (D) Ag, NO and O_2

263. Purest form of iron is- [CPMT-98]
(A) White cast iron (B) Grey cast iron
(C) Wrought iron (D) Steel

264. In IUPAC $K_4[Fe(CN)_6]$ is called - [CPMT-98]
(A) Potassium hexacyanoferrate (II)
(B) Potassium ferricyanide
(C) Potassium hexacyanoferrate (III) (D) Prussian blue

265. Mohr's salt is- [CPMT-98]
(A) Normal salt (B) Acid salt
(C) Basic salt (D) Double salt

266. $KMnO_4$ reacts with Ferrous Sulphate according to the equation [CPMT-99]
$MnO_4^- + 5Fe^{2+} + 8H^+ \rightarrow Mn^{2+} + 5\,Fe^{3+} + 4H_2O$
Here 10 ml of 0.1 M$KMnO_4$ is equivalent to
(A) 20ml of 0.1 M $FeSO_4$ (B) 30ml of 0.1 M $FeSO_4$
(C) 40ml of 0.1 M $FeSO_4$ (D) 50ml of 0.1 M $FeSO_4$

267. The number of unpaired electrons in Mn^{2+} is [CPMT-2000]
(A) 3 (B) 7 (C) 1 (D) 5

268. The oxidation state of Cr in $K_2Cr_2O_7$ is [CPMT-2000]
(A) -5 (B) +6 (C) - 6 (D) +5

269. Which group contains coloured ions out of
1. Cu^+ 2. Ti^{4+} 3. Co^{++} 4. Fe^{++} [CPMT-2001]
(A) 1,2,3,4 (B) 3,4 (C) 2,3 (D) 1,2

270. The equivalent weight of $MnSO_4$ is M/2 when it changes into [CPMT-2001]
(A) $KMnO_4$ (B) K_2MnO_4 (C) MnO_2 (D) MnO

271. If the ratio of co-ordination no. P to that of Q be Y : Z, then the formula of the soild is [CPMT-2001]
(A) P_yQ_x (B) P_xQ_y (C) $P_{1/y}Q_{1/z}$ (D) None of these

272. The general electronic configuration of transition element is [CPMT-2002]
(A) $(n-1)d^{1-10}\,ns^{1-2}$ (B) $np^{1-6}\,ns^{1-2}$
(C) $(n-1)d^{1-10}\,ns^2$ (D) $nd^{1-10}\,(n-1)s^2$

273. Which is not the property of transition elements ? [CPMT-2002]
(A) coloured salts (B) variable valency
(C) fixed valency (D) coordinate complex formation

274. Which noble gas is most soluble in water ? [CPMT-2002]
(A) He (B) Ar (C) Ne (D) Xe

275. In which compound chromium has +6 oxidation state? [CPMT-2003]
(A) $K_2Cr_2O_7$ (B) $CrCl_3$
(C) $Cr_2(SO_4)_3$ (D) none of these

276. Electronic configuration $1s^22s^22p^6, 3s^23p^63d^5, 4s^1$ is in [CPMT-2003]
(A) ground state (B) excited state
(C) anionic state (D) cationic state

277. The potassium ferricyanide produces on ionisation [CPMT-2003]
(A) 2 ions (B) 1 ion (C) 3 ions (D) 4 ions

278. $[Co(NH_3)_4Cl_2]NO_2$ and $[Co(NH_3)_4Cl.NO_2]Cl$ are isomers : [PMT-93]
(A) Geometrical (B) Optical
(C) Linkage (D) Ionization

279. IUPAC name of complex $K_3[Al(C_2O_4)_3]$ is : [PMT-93]
(A) Potassium aluminooxalate
(B) Potassium trioxalatoaluminate (III)
(C) Poatassium aluminium(III)oxalate

(D) Potassium trioxalatoaluminate (VI)

280. Which of the following ions has the least magnetic moment : [PMT-93]
(A) Cu^{2+} (B) Ni^{2+} (C) Co^{3+} (D) Fe^{2+}

281. Among the following outermost configurations of transition metals which shows the highest oxidation state is : [PMT-93]
(A) $3d^3\ 4s^2$ (B) $3d^5\ 4s^1$ (C) $3d^5\ 4s^2$ (D) $3d^6\ 4s^2$

282. The d-block element consists mostly of [PMT-94]
(A) monovalent metals (B) all non-metals
(C) elements which generally form stoichiometric metal oxide
(D) many metals with catalytic properties

283. The action of steam on heated iron is represented as [PMT-94]
(A) $3Fe + 4H_2O \rightarrow Fe_3O_4 + 4H_2$
(B) $2Fe + 3H_2O \rightarrow Fe_2O_3 + 3H_2$
(C) $Fe + H_2O \rightarrow FeO + H_2$
(D) $2Fe + H_2O + O_2 \rightarrow Fe_2O_3 + H_2$

284. The colour of $K_2Cr_2O_7$ changes from red orange to lemon yellow on treatment with aqueous KOH because of [PMT-94]
(A) the reduction of Cr^{VI} to Cr^{III}
(B) the formation of chromium hydroxide
(C) the conversion of dichromate to chromate
(D) the oxidation of potassium hydroxide to potassium

285. The electronic configuration of Mn^{2+} is [PMT-94]
(A) $[Ar]3d^4\ 4s^1$ (B) $[Ar]3d^5\ 4s^0$
(C) $[Ar]4d^5\ 5s^0$ (D) $[Ar]3d^3\ 4s^2$

286. The pair of co-ordination compounds $[Co(NH_3)_5Br]SO_4$ and $[Co(NH_3)_5SO_4]$ Br is an example of [PMT-94]
(A) Geometrical isomerism (B) Co-ordination isomerism
(C) Structural isomerism (D) Ionisation isomerism

287. The correct nomenclature for $Fe_4[Fe (CN_6)]_3$ is [PMT-94]
(A) Ferroso ferriccyanide
(B) Ferric-Ferroushexacyanate
(C) Iron (III)hexacyanoferrate(II)
(D) Hexacyanoferrate(III - II)

288. Which one of the following will give a white precipitate with $AgNO_3$ in aq. medium ? [PMT-94]
(A) $[Co(NH_3)_5Cl](NO_2)_2$ (B) $[Pt(NH_3)_2Cl_2]$
(C) $[Pt(en)Cl_2]$ (D) $[Pt(NH_3)_4]Cl_2$

289. Pick out from the following complex compounds, a poor electrolytic conductor solution [PMT-94]
(A) $K_2\ [PtCl_6]$ (B) $[Co(NH_3)_3(NO_2)_3]$
(C) $K_4[Fe(CN)_6]$ (D) $[Cu(NH_3)_4]SO_4$

290. Choose the correct answer for transition elements [PMT-95]
(A) Transition elements have low melting points
(B) Transition elements donot show catalytic activity
(C) Transition elements exhibit variable oxidation states
(D) Transition elements exhibit inert pair effect

291. Which of the ions having following electronic structure would have maximum magnetic moment ? [PMT-95]
(A) $1s^2\ 2s^2\ 2p^6\ 3s^2\ 3p^6\ 3d^3$ (B) $1s^2\ 2s^2\ 2p^6\ 3s^2\ 3p^6\ 3d^5$
(C) $1s^2\ 2s^2\ 2p^6\ 3s^2\ 3p^6\ 3d^7$ (D) $1s^2\ 2s^2\ 2p^6\ 3s^2\ 3p^6\ 3d^9$

292. Which of the following elements does not belong to first transition series ? [PMT-95]
(A) Fe (B) V (C) Ag (D) Cu

293. MnO_4^- on reduction in acidic medium forms : [PMT-95]
(A) MnO_2 (B) Mn^{++} (C)MnO_4^- (D) Mn

294. The number of unpaired electrons in Fe^{+++} (Z = 26) is : [PMT-95]
(A) 4 (B) 5 (C) 6 (D)3

295. Which type of isomers are [PMT-95]
$[Co (NH_3)_5\ Br]\ SO_4$ and $[Co (NH_3)_5\ SO_4]$ Br
(A) Hydrate isomers (B) Ionization isomers
(C) Ligand isomers (D) Coordination isomers

296. Correct I.U.P.A.C. nomenclature of $K_4[Fe (CN)_6]$ is: [PMT-95]
(A) Tetra potassiumferrouscyanide
(B) Potassium ferricyanide
(C) Potassium ferrocynide
(D) Potassium hexacyanoferrate (II)

297. Which of the following ions form most stable complex compound ? [PMT-95]
(A) Cu^{++} (B) Ni^{++} (C)Fe^{++} (D) Mn^{++}

298. Which of the following compounds turns black on addition of ammonium hydroxide ? [PMT-95]
(A) $PbCl_2$ (B) Hg_2Cl_2 (C) $HgCl_2$ (D) AgCl

299. On adding excess of ammonium hydroxide to a copper sulphate solution : [PMT-95]
(A) Blue precipitate of copper hydroxide is obtained
(B) Black precipitate of copper oxide is obtained
(C) A deep blue solution is obtained
(D) No change is observed

300. The unpaired electron in Ni^{++} is [PMT-95]
(A) zero (B) 2 (C) 4 (D) 8

301. To which block is related an element having electronic configuration $1s^2\ 2s^2\ 2p^6\ 3s^2\ 3p^6\ 3d^{10}\ 4s^1$ in the periodic table ? [PMT-95]
(A) s-block (B) p-block (C) d-block (D) f-block

302. The complex compound $[Co(NH_3)_3\ NO_2\ ClCN\ ]$ is named as : [PMT-96]
(A) Chlorocyanonitrotriammine cobalt (III)

(B) Nitrochlorocyanotriammine cobalt (III)
(C) Cyanonitrochlorotriammine cobalt (III)
(D) Triamminitrochlorcyano cobalt (III)

303. The oxidation number of chromium in $K_2Cr_2O_7$ is [PMT-96]
(A) -6 (B) +6 (C) +2 (D) -2

304. The 3d-elements show variable oxidation state. Maximum state shown by the element Mn - [PMT-97]
(A) +4 (B) +5 (C) +6 (D) +7

305. The tendency of 3d- metal ions to form stable complexes is due to their - [PMT-97]
(A) Variable oxidation state
(B) Strong electronegative nature
(C) High charge /size ratio and vacant d-orbitals
(D) Very low ionization energies

306. The 3d-metal ions are paramagnetic in nature because [PMT-97]
(A) They are reducing agents (B) They form coloured salts
(C) They have one or more paired s-electrons
(D) They have one or more unpaired d-electrons

307. Potassium permangnate acts as an oxidant in neutral, alkaline as well as acidic media. The final products obtained from it in the three conditions are, respectively - [PMT-97]
(A) MnO_2, MnO_2, Mn^{2+} (B) MnO_4^{2-}, Mn^{3+}, Mn^{2+}
(C) MnO_2 ,MnO_4^{2-}, Mn^{3+} (D) MnO, MnO_2 ,Mn^{2+}

308. Point out the correct systematic (IUPAC) name of $K_4[Fe(CN)_6]$ [PMT-97]
(A) Potassium ferrocyanide
(B) Potassium hexacyanoferrate (II)
(C) Tetrapotassium hexacyanoferrate (III)
(D) Tetrapotassium ferroushexacyanide

309. Dichlorodiammine platinum (II) complex has the formula - [PMT-97]
(A) $Pt.[Cl_2(NH_3)_2]$ (B) $Pt.[R.(NH_2)_2]\ Cl_2$
(C) $[Pt.\ Cl_2(NH_3)_2]$ (D) $[Pt.R.(NH_2)_2]\ Cl_2$

310. The pair of complex compounds $[Cr(H_2O)_6]\ Cl_3$ and $[Cr\ (H_2O)_5\ Cl]\ Cl_2H_2O$ are an example of [PMT-97]
(A) Linkage isomerism (B) Ionisation isomerism
(C) Coordination isomerism (D) Hydrate isomerism

311. The geometry of the complex $[Cu\ (NH_3)_4^{2+}]$ is - [PMT-97]
(A) Square planar (B) Tetrahedral
(C) Triangular bipyramid (D) Octahedral

312. Out of the following electronic configuratins the one of a transition element is [PMT-98]
(A) $1s^1 2s^2 p^6 3s^2 p^6 d^{10} 4s^2 p^6$ (B) $1s^2 2s^2 p^6 3s^2 p^6 d^3 4s^2$
(C) $1s^2 2s^2 p^6 3s^2 p^6 4s^2$ (D) $1s^2 2s^2 p^6 3s^2 p^6 d^{10} 4s^2 p^1$

313. In transition elements the orbitals partially filled by electrons are : [PMT-99]
(A) s-orbitals (B) p-orbitals (C) d-orbitals (D) f-orbitals

314. Which one of the following is known as Lunar Caustic when in fused state ? [PMT-99]
(A) Silver nitrate (B) Silver sulphate
(C) Silver chloride (D) Sodium sulphate

315. The correct electroninc configuration of Ti (Z=22) atom is : [PMT-99]
(A) $1s^2\ 2s^2\ 2p^6\ 3s^2\ 3p^6\ 4s^2\ 3d^2$
(B) $1s^2\ 2s^2\ 2p^6\ 3s^2\ 3p^6\ 4s^2\ 4p^2$
(C) $1s^2\ 2s^2\ 2p^6\ 3s^2\ 3p^6\ 3d^4$ (D) $1s^2\ 2s^2\ 2p^6\ 3s^2\ 3p^6\ 4s^1\ 3d^3$

316. Chloride of which of the following elements is paramagnetic ? [PMT-99]
(A) Silver (B) Mercury (C) Zinc (D) Cobalt

317. Which one of the following substances sublimes on heating ? [PMT-99]
(A) Magnesium chloride (B) Silver chloride
(C) Mercuric chloride (D) Sodium chloride

318. Which one of the following is not a transition metal? [PMT-99]
(A) Chromium (B) Titanium (C) Lead (D) Tungsten

319. The number of unpaired electrons in a nickel atom (ground state) are (Atomic No. of Ni 28) : [PMT-2000]
(A) 2 (B) 5 (C) 3 (D) 7

320. IUPAC name of $Na_3[Co(NO_2)_6]$ is : [PMT-2000]
(A) Sodium cobaltinitrite
(B) Sodium hexanitritocobaltate (III)
(C) Sodium hexanitritocobalt (III)
(D) Sodium hexanitritocobaltate (II)

321. A group of atoms can function as a ligand only when: [PMT-2000]
(A) it is a small molecule
(B) it has an unshared electron pair
(C) it is a negatively charged ion
(D) it is a positively charged ion

322. According to the postulates of Werner for coordinate compounds : [PMT-2000]
(A) Primary valency is ionizable
(B) Secondary valency is ionizable
(C) Primary and secondary valencies are non-ionizable
(D) only primary valency is nonionizable

323. Elements which generally exhibit multiple oxidation states and whose ions are usually coloured are : [PMT-2000]
(A) metalloids (B) non-metals
(C) transition elements (D) gases

324. The transition metals mostly are : [PMT-2000]
(A) diamagnetic (B) paramagnetic
(C) neither diamagnetic nor paramagnetic

(D) both diamagnetic and paramagnetic

325. The correct statement in respect of d-block elements is : [PMT-2000]
(A) they are all metals (B) they show variable valency
(C) they form coloured ions and complex salts
(D) all above statements are correct

326. The valency of Cr in the complex $[Cr(H_2O)_4Cl_2]^+$ is: [PMT-2000]
(A) 1 (B) 3 (C) 5 (D) 6

327. Which of the following is expected to be a paramagnetic complex ? [PMT-2000]
(A) $[Ni(H_2O)_6]^{2+}$ (B) $[Ni(CO)_4]$
(C) $[Zn(NH_3)_4]^{2+}$ (D) $[CO(NH_3)_6]$

328. The outer electronic configuration of chromium is : [PMT-2001]
(A) $4s^1\ 3d^5$ (B) $4s^2\ 3d^4$ (C) $4s^0\ 3d^6$ (D) $4s^2\ 3d^5$

329. Which of the following transition metals shows the highest oxidation state ? [PMT-2001]
(A) Sc (B) Ti (C) Mn (D) Zn

330. In potassium manganate, the oxidation state of manganese is : [PMT-2001]
(A) +5 (B) +6 (C) +7 (D) +8

331. Which of the following is an organo-metallic compound ? [PMT-2001]
(A) Lithium ethoxide (B) Ethyl lithium
(C) Lithium acetate (D) Lithium carbide

332. Which of the following compounds exhibits linkage isomerism ? [PMT-2001]
(A) $[Co(en)_3]Cl_3$ (B) $[Co(NH_3)_6][Cr(CN)_6]$
(C) $[Co(en)_2NO_2Cl]Br$
(D) $[Co(NH_3)_5Cl]Br_2$

333. The oxidation number of cobalt in $K[Co(CO)_4]$ is : [PMT-2001]
(A) +1 (B) -1 (C) +3 (D) -3

334. The I.U.P.A.C. name of $K_3[Ir(C_2O_4)_3]$ is : [PMT-2001]
(A) Potassium trioxalatoiridium(iii)
(B) Potassium trioxalatoiridate(iii)
(C) Potassium tris(oxalato)iridium(iii)
(D) Potassium tris(oxalato)iridate(iii)

335. Which of the following electronic configurations is that of a transition element [PMT-2002]
(A) $ls^22s^22p^63s^23p^63d^{10}4s^24p^6$
(B) $ls^22s^22p^63s^23p^63d^{10}4s^24p$
(C) $ls^22s^22p^63s^23p^63d^24s^2$ (D) $ls^22s^22p^63s^23p^64s^2$

336. A transition element [PMT-2002]
(A) Forms coloured compounds (B) Has variable valance
(C) Has incomplete d-orbitals
(D) Has all the above characteristics

337. The oxidation number of chromium in potassium dichromate is [PMT-2002]
(A) +6 (B) +3 (C) +5 (D) +7

338. In alkaline condition $KMnO_4$ reacts as $2KMnO_4 + 2KOH \rightarrow 2K_2MnO_4 + H_2O + O$ the equivalent weight of $KMnO_4$ would be (Atomic mass of K = 39, Mn = 55, O = 16) [PMT-2002]
(A) 158.0 (B) 79.0 (C) 52.7 (D) 31.6

339. According to Werner's theory of coordinate compounds [PMT-2002]
(A) Primary valency is ionisable
(B) Secondary valency is ionisable
(C) Primary and secondary valencies are non ionisable
(D) Only primary valency is non-ionisable

340. IUPAC name of $K_3[Al(C_2O_4)_3]$ is [PMT-2002]
(A) Potassium aluminooxalate
(B) Potassium trixalatoaluminate(III)
(C) Potassium aluminium(III)oxalate
(D) Potassium troxalatoaluminate(VI)

341. Which one of the following will not show geometrical isomerism ? [PMT-2002]
(A) $[Cr(NH_3]_4Cl_2)Cl$ (B) $[Co(en)_2Cl_2]Cl$
(C) $[Co(NH_3)_5NO_2]Cl_2$ (D) $[Pt(NH_3)_2Cl_2]$

342. According to Lewis the ligands are [PMT-2002]
(A) Acidic in nature (B) Basic in nature
(C) Neither acidic nor basic
(D) Some are acidic and others are basic

343. The most stable complex among the following is [PMT-2002]
(A) $K_3[Al(C_2O_4)_3]$ (B) $[Pt(en)_2]Cl_2$
(C) $Ag(NH_3)_2Cl$ (D) $K_2(Ni(EDTA)]$

344. Which one of the following is an example of non-typical transition elements ? [PMT-2002]
(A) Li, K, Na (B) Be, Al, Pb
(C) Zn, Cd, Hg (D) Ba, Ca, Se

345. The general electronic configuration of transitional elements is. [PMT-2003]
(A) $(n-1)d^{1-10}ns^1$ (B) $(n-1)d^{1-5}$
(C) $(n-1)d^{1-10}\ ns^{1\text{ or }2}$ (D) $(n-1)d^{1-10}\ ns^{0.1.2}$

346. Transitional elements are related to [PMT-2003]
(A) 'p' block (B) 's' block
(C) 'd' block (D) None of the above

347. Which transitional metal is used for the reduction of steam to hydrogen ? [PMT-2003]
(A) Se (B) Fe (C) Mg (D) Pt

348. Zn element belongs to. [PMT-2003]
(A) 11 group (B) 12 group
(C) 17 group (D) 18 group

349. In complex compounds ligand behaves as

[PMT-2003]
(A) Neither acceptor nor donor (B) Electron pair donor
(C) Electron pair acceptor (D) All happen

350. The oxidation of the 'Fe' atom in $[Fe(CO)_5]$ is [PMT-2003]
(A) –1 (B) +4 (C) –3 (D) None of these

351. The oxidation number of "Fe" in $K_4[Fe(CN)_6]$ is. [PMT-2003]
(A) 3 (B) 2 (C) 0 (D) 5

352. $K_3[Al(C_2O_4)_3]$ is called : [PMT-2003]
(A) Pot. trioxalatealuminate
(B) Pot. aluminooxalato (C) Pot. alumino(III)oxalate
(D) Pot. trioxalatoaluminate(III)

353. Which of the following pairs is not correctly matched? [PET-93]
(A) Effective atomic number of Pt in $[PtCl_6]^{2-}$ – 84
(B) Absorption peak for $[Cr^{III}(NH_3)_6]^{3+}$ – 21680 cm^{-1}
(C) Crystal field stabilization energy of d^2 in weak ligand field – (–) 0.8 Δ_o
(D) Example of weak ligand field for d^5 configuration – M $[Mn^{II}F_6]^{4-}$

354. Which one of the following is not true for metal carbonyls? [PET-93]
(A) The oxidation state of the metal in the carbonyls is zero
(B) The secondary carbonyls are obtained from photo-decomposition
(C) Metal carbonyls are single bonded species.
(D) $d_\pi - p_\pi$ overlap is observed in metal carbonyls

355. The electronic configuration (outermost) of Mn^{2+} ion (atomic no. of Mn 25) in its ground state is [PET-93]
(A) $3d^5\ 4s^0$ (B) $3d^4\ 4s^1$ (C) $3d^3\ 4s^2$ (D) $3d^2\ 4s^2\ 4p^2$

356. The highest magnetic moments is shown by the transition metal ion with the outer electronic configuration [PET-93]
(A) $3d^2$ (B) $3d^5$ (C) $3d^7$ (D) $3d^9$

357. Which one of the following transition metals exhibits the highest oxidation state? [PET-93]
(A) Pd (B) Os (C) Cr (D) Mn

358. Manganese achieves its highest oxidation state in its compound [PET-93]
(A) MnO_3 (B) Mn_3O_4 (C) $KMnO_4$ (D) K_2MnO_4

359. Which of the following is not true for transition metals? [PET-93]
(A) They are malleable and ductile
(B) They have high boiling and melting points
(C) They crystallize with body-centered cubic hexagonal close-packed structures only
(D) They show variable oxidation states, although not always

360. Which of the following is not true for ligand-metal complex? [PET-93]
(A) Larger the ligand the more stable is the metal-ligand bond
(B) Highly charged ligand forms stronger bond
(C) Larger the permanent dipole moment of ligand, the more stable is the bond
(D) Greater the ionization potential of central metal, the stronger is the bond

361. Which one of the isomeric-pairs shows ionization isomerism? [PET-93]
(A) $[Co(NH_3)_6][Cr(CN)_6]$ and $[Cr(NH_3)_6][Co(CN)_6]$
(B) $[Cr(H_2O)_6]Cl_3$ and $[Cr(H_2O)_5Cl]Cl_2.H_2O$
(C) $[Pt(NH_3)_2Cl_2]$ and $[(Pt(NH_3)_5Br][PtCl_4]$
(D) $[Co(NH_3)_5Br]SO_4$ and $[Co(NH)_3)_5SO_4]Br$

362. Which of the following compounds is colourless [PET-94]
(A) $Cu_2(CH_3COO)_4.H_2O$ (B) Cu_2Cl_2
(C) $CuSO_4.5H_2O$ (D) $[Cu(NH_3)_4]SO_4.4H_2O$

363. The number of isomers possible for square planar complex. $K_2[PdClBr_2(SCN)]$ is [PET-94]
(A) 2 (B) 3 (C) 4 (D) 6

364. I.U.P.A.C. name of $[Co(NH_3)_3(H_2O)Cl]Cl_2$ is [PET-94]
(A) Diaquochlorodiamminecobalt(III)chloride
(B) Triamminediaquochlorocobalt.(III)chloride
(C) Chlorodiamminediaquocobalt(II)chloride
(D) Diamminediaquochlorocobalt(II)chloride

365. Which of the following exhibits highest molar conductivity [PET-94]
(A) $[Co(NH_3)_6]Cl_2$ (B) $[Co(NH_3)_5Cl]Cl_2$
(C) $[Co(NH_3)_4]Cl_2Cl$ (D) $[Co(NH_3)_3Cl_3]$

366. In tris(ethylene diamine)cobalt(III)chloride the coordination number of cobalt is [PET-94]
(A) 3 (B) 4 (C) 6 (D) 7

367. Monazite is a source of [PET-95]
(A) Ne (B) Ra (C) Kr (D) Th

368. Which forms coloured ion [PET-95]
(A) Metalloids (B) Non-metals
(C) p-Block elements (D) Transition elements

369. Prussian blue is [PET-95]
(A) $Fe_4[Fe(CN)_6]_3$ (B) $Fe_2[Fe(CN)_6]$
(C) $Fe_3[Fe(CN)_6]$ (D) $Fe[Fe(CN)_6]_2$

370. The highest oxidation state is shown by that transition element wich has outermost configuration as [PET-95]
(A) d^3s^2 (B) d^5s^1 (C) d^5s^2 (D) d^8s^2

371. Coordination number of Cu in complex $[Cu(H_2O)_4]^{++}$ is - [PET-95]
(A) 4 (B) 3 (C) 2 (D) 1

372. Complex $[Pt(NH_3)_6]\ Cl_4$ furnishes [PET-95]
(A) 5 ions (B) 4 ions (C) 3 ions (D) 2 ions

373. The strongest ligand in the following is [PET-95]
(A) CN^- (B) Br^- (C) HO^- (D) F^-

374. The coordination number of Pt in $[Pt(NH_3)_4\ Cl_2]^{++}$ ion is [PET-95]
(A) 2 (B) 4 (C) 6 (D) 8

375. If the magnetic moment of $[Ag(CN_2)]^{-1}$ is zero, then the number of unpaired electrons will be [PET-95]
(A) 1 (B) 2 (C) 3 (D) Zero

376. $[Co(NH_3)_4Cl_2]NO_2$ and $[Co(NH_3)_4\ Cl\ NO_2]Cl$ are isomers [PET-95]
(A) Geometrical (B) Optical
(C) Linkage (D) Ionisation

377. Which of the following oxides is white but becomes yellow on heating [PET-95]
(A) AgO (B) Ag_2O (C) FeO (D) ZnO

378. Which of the following ions gives coloured solution- [PET-95]
(A) Cu^+ (B) Zn^{++} (C) Ag^+ (D) Fe^{++}

379. On heating $Mn(OH)_2$ with PbO_2 and conc. HNO_3 pink colour is obtained due to the formation of -[PET-95]
(A) $KMnO_4$ (B) $HMnO_4$ (C) $Pb(MnO_4)_2$ (D) $PbMnO_4$

380. $[Pt(NH_3)_4\ Cl_2]\ Br_2$ and $[Pt(NH_3)_4\ Br_2]\ Cl_2$ are related to each other as [PET-96]
(A) Optical isomers (B) Co-ordinate isomers
(C) Ionization isomers (D) Linkage isomers

381. Which of the following is the odd one out ?[PET-96]
(A) Potassium ferrocyanide
(B) Ferrous ammoniumsulphate
(C) Potassium ferricyanide
(D) Tetramminecopper(II)sulphate

382. Generally, a group of atoms can function as a ligand if [PET-96]
(A) They are positively charged ions
(B) They are free radicals
(C) They are either neutral molecules or negatively charged ions
(D) None of the above

383. Which of the following is not an organometallic compound ? [PET-96]
(A) Ethyl magnesium bromide
(B) Tetraethyl lead (C) Sodium ethoxide
(D) Trimethyl aluminium

384. Of the ions Zn^{2+}, Ni^{2+} and Cr^{3+} [At. Nos – Zn = 30, Ni = 28, Cr = 24] [PET-96]
(A) Only Zn^{2+} is colourless and Ni^{2+} and Cr^{3+} are coloured
(B) All three are colourless (C) All three are coloured
(D) Only Ni^{2+} is coloured and Zn^{2+} and Cr^{3+} are colouress

385. Common oxidation state of Scandium, a transition element is/are [At. No. - Sc = 21] [PET-96]
(A) + 4 (B) +1 (C) +2 & +3 (D) +4 & +1

386. Which of the following is not correct about transition metals ? [PET-96]
(A) Their melting and boiling points are high
(B) Their compounds are generally coloured
(C) They can form ionic or covalent compounds
(D) The do not exhibit variable valency

387. Acidified potassium dichromate is treated with hydrogen sulphide. In the reaction the oxidation number of chromium [PET-96]
(A) Increases from +3 to +6
(B) Decreases from +6 to +3
(C) Remains unchanged (D) Decreases from +6 to +2

388. When $KMnO_4$ reacts with acidified $FeSO_4$[PET-96]
(A) Only $FeSO_4$ is oxidized
(B) Only $KMnO_4$ is oxidized
(C) $FeSO_4$ is oxidized and $KMnO_4$ is reduced
(D) None of the above

389. The co-ordination number of a metal in co-ordination compounds is [PET-96]
(A) Same as primary valency
(B) Sum of primary and secondary valencies
(C) Same as secondary valency
(D) None of the above

390. In which of the following complexes oxidation state of metal is zero ? [PET-97]
(A) $[Pt(NH_3)_2Cl_2]$ (B) $[Cr(CO)_6]$
(C) $[Cr(NH_3)_3Cl_3]$ (D) $[Cr(en)_2Cl_2]$

391. Which of the following complexes shows ionization isomerism ? [PET-97]
(A) $[Cr(NH_3)_6Cl_3]$ (B) $[Cr(en)_2Cl_2]$
(C) $[Cr(en)_3Cl_3]$ (D) $[Co(NH_3)_5Br]SO_4$

392. The IUPAC name of $K_3\ [Fe(CN)_6]$ is [PET-97]
(A) Potassiumferrocyanide (II)
(B) Potassiumhexaferrocyanate (III)
(C) Potassiumferrohexacyanate (II)
(D) Potassiumhexacyanoferrate (III)

393. One mole of potassium dichromate completely oxidizes the following number of moles of ferrous sulphate in acidic medium : [PET-97]
(A) 1 (B) 3 (C) 5 (D) 6

394. Besides Zn and Cu German silver contains the metal: [PET-97]
(A) Sn (B) Ag (C) Ni (D) Mg

395. The oxidation state of nickel in Ni $(CO)_4$ will be : [PET-98]

(A) 1 (B) Zero (C) 2 (D) 4

396. The oxidation number of chromium in $K_2Cr_2O_7$ is : [PET-98]

(A) -6 (B) +6 (C) +2 (D) -2

397. Which of the following ligands forms a chelate ? [PET-98]

(A) Acetate (B) Oxalate (C) Cyanide (D) Ammonia

398. Which one of the following ions is colourless ? [PET-99]

(A) Cu^+ (B) Co^{2+} (C) Ni^{2+} (D) Fe^{3+}

399. The correct formula of permanganic acid is : [PET-99]

(A) $HMnO_4$ (B) $HMnO_5$ (C) H_2MnO_4 (D) H_2MnO_3

400. Which one of the following properties is not of transition elements ? [PET-99]

(A) Colour (B)Paramagnetism
(C) Fixed valency (D)None of the above

401. In which of the following tendency towards formation of coloured ions is maximum ? [PET-99]

(A) s-block elements (B) d-blcok elements
(C) p-block elements (D) f-block elements

402. The example of dsp^2 hybridisation is : [PET-99]

(A) $Fe(CN)_6^{3-}$ (B) $Ni(CN)_4^{2-}$
(C) $Zn(NH_3)_4^{2+}$ (D) FeF_6^{3-}

403. Which statement is not correct ? [PET-99]

(A) Potassium permanganate is a powerful oxidising substance
(B) Potassium permanganate is a weaker oxidising substarnce than potassium dichromate
(C) Potassium permanganate is a stronger oxidising substance than potassium dichromate
(D) Potassium dichromate oxidises a secondary alcohol into a ketone

404. What is the product obtained in the reaction of $HgCl_2$ and Hg $(CN)_2$? [PET-2000]

(A) $(CN)_2$ (B) Addition compound $HgCl_2$. Hg $(CN)_2$
(C) Hg (CN)Cl (D) Hg [Hg $(CN)_2Cl_2$]

405. Which of the following oxides reacts with HCl and NaOH ? [PET-2000]

(A) CaO (B) ZnO (C) N_2O_5 (D) CO_2

406. Which one is an example of octahedral complex ? [PET-2000]

(A) FeF_6^{3-} (B) Zn $(NH_3)_4^{2+}$
(C) Ni $(CN)_4^{2-}$ (D) Cu $(NH_3)_4^{2+}$

407. Variable valency is shown by : [PET-2001]

(A) Metallic elements (B) Normal elements
(C) Transition elements (D) Non-metal elements

408. Among the following outermost configurations of transition metals, which shows the highest oxidation state ? [PET-2001]

(A) $3d^3 4s^2$ (B) $3d^5 4s^1$ (C) $3d^5 4s^2$ (D) $3d^6 4s^2$

409. Manganese achieves its maximum oxidation state in its compounds : [PET-2001]

(A) MnO_2 (B) Mn_3O_4 (C) $KMnO_4$ (D) K_2MnO_4

410. An example of a double salt is : [PET-2001]

(A) Bleaching power (B) Hypo
(C) K_4 $[Fe(CN)]_6$ (D) Potash Alum

411. Which of the following elements does not form amalgam : [PET-2001]

(A) Fe (B) Silver (C) Zinc (D) Cobalt

412. The number of ions per mole of a complex $[CoCl_2.5NH_3]$ in aqueous solution will be : [PET-2001]

(A) Nine (B) Four (C) Three (D) Two

413. $[CO(NH_3)_4Cl_2]$ NO_2 and $[CO(NH_3)_4Cl\ NO_2]Cl$ are isomers of which type ? [PET-2001]

(A) Geometrical (B) Optical
(C) Linkage (D) Ionization

414. The oxidation state of Fe in $K_4[Fe(CN)_6]$ is [PET-2002]

(A) +2 (B) +6 (C) + 3 (D) +4

415. For which of the following transition metal ion is shown highest magnetic moment having outer electronic configuration [PET-2003]

(A) $3d^8$ (B) $3d^2$ (C) $3d^5$ (D) $3d^-$

416. Most powerful oxidizing property of manganese is shown by which of the following oxidation state: [PET-2003]

(A) Mn (+4) (B) Mn (+7)
(C) Mn (+2) (D) Mn (+3)

417. Co-ordination number of Fe in the complexes $[Fe(CN)_6]^{4-}$, $[Fe(CN)_6]^{3-}$ and $[FeCl_4]^-$ would be respectively. [PET-2003]

(A) 6,3,4 (B) 6,6,4 (C) 2,3,3 (D) 6,4,6

418. The effective atomic number of cobalt in the complex $[Co(NH_3)_6]^{3+}$ is : [PET-2003]

(A) 33 (B) 36 (C) 26 (D) 24

419. The correct name of $[Pt(NH_3)_4Cl_2]$ $[PtCl_4]$ is [PET-2003]

(A) Tetraaminedichloroplatinium(iv)tetrachloroplatinate(ii)
(B)Dichlorotetraamineplatinium(iv)tetrachloro platinate(ii)
(C)Tetrachloro platinium(ii)tetraaminedichloro platinate(iv)
(D) Tetrachloroplatinium(ii)dichlorotetraamineplatinate (iv)

420. The possible number of optical isomers in $[Co(en)_2Cl_2]^{+2}$ are : [PET-2003]

(A) 2 (B) 3 (C) 4 (D) 6

421. π-bonding is not involved in [PET-2003]

(A) Ferrocene (B) Dibenzene chromium
(C) Zeise salt (D) Grignard reagent

422. In which of the following the addition of $AgNO_3$ does not result precipitation of silver chloride AgCl : [PET-2003]
(A) $[Co(NH_3)_3Cl_3]$ (B) $[Co(NH_3)_5Cl]Cl_2$
(C) $[Co(NH_3)_4Cl_2]Cl$ (D) $[Cr(NH_3)_6]Cl_3$

423. Which of the following metals will not react with a solution of $CuSO_4$? [PET-2003]
(A) Fe (B) Zn (C) Mg (D) Hg

424. Correct I.U.P.A.C. nomenclature of $K_4[Fe(CN)_6]$ is [PET-2003]
(A) Tetrapotassiumferrouscyanide
(B) Potassiumferricyanide (C) Potassiumferrocyanide
(D) Potassiumhexacyanoferrate(II)

425. Of the four oxyacids of chlorine the strongest oxidising agent in dilute aqueous solution is :[PET-2003]
(A) $HClO_4$ (B) $HClO_3$ (C) $HClO_2$ (D) HOCl

426. The element whose valency is 7 would be [PET-2003]
(A) V (B) Mn (C) C (D) Cr

427. Which of the following statements is wrong? [BHU-2003]
(A) $Ni(CO)_4$ has oxidation number +4 for Ni
(B) $Ni(CO)_4$ has oxidation number 0 for Ni
(C) Both (A) & (B) (D) None of these

428. In $[NiCl_4]^{2-}$, the number of unpaired electrons is [BHU-2003]
(A) 2 (B) 3 (C) 4 (D) 4.5

429. Silvering of mirror is done by [AFMC-2003]
(A) Ag_2O_3 (B) Fe_2O_3 (C) Al_2O_3 (D) $AgNO_3$

430. When CuSO4 is hydrated, it becomes [AFMC-2003]
(A) Acidic (B) Basic (C) Neutral (D) Amphoteric

431. Formula of ferric sulphate is [AFMC-2003]
(A) $FeSO_4$ (B) Fe_2SO_4 (C) $Fe_2(SO_4)_3$ (D) $Fe(SO_4)_2$

432. In a regular octahedral molecule, MX_6 the number of X - M - X bonds at 180° is [CBSE-PMT-2004]
(A) three (B) two (C) six (D) four

433. Among K, Ca, Fe and Zn, the element which can form more than one binary compound with chlorine is [CBSE-PMT-2004]
(A) Fe (B) Zn (C) K (D) Ca

434. Which of the following is considered to be an anticancer species ? [CBSE-PMT-2004]
(A) 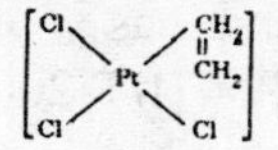 (B) 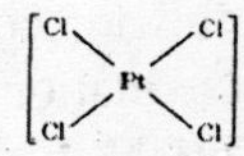
(C) 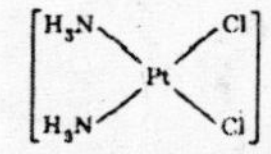 (D) 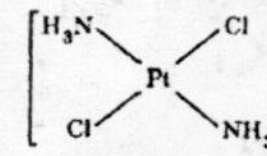

435. Which of the following coordination compounds would exhibit optical isomerism ? [CBSE-PMT-2004]
(A) pentaamminenitrocobalt(III) iodide
(B) diamminedichloroplatinum(II)
(C) trans-dicyanobis (ethylenediamine) chromium(III) chloride
(D) tris-(ethylenediamine)cobalt(III) bromide

436. Among $[Ni(CO)_4]$, $[Ni(CN)_4]^{2-}$, $[NiCl_4]^{2-}$ species, the hybridization states at the Ni atom are, respectively [CBSE-PMT-2004]
(A) sp^3, dsp^2, dsp^2 (B) sp^3, dsp^2, sp^3
(C) sp^3, sp^3, dsp^2 (D) dsp^2, sp^3, sp|
(At. no. of Ni = 28)

437. Among the following series of transition metal ions, the one where all metal ions have $3d^2$ electronic configuration is [CBSE-PMT-2004]
(A) Ti^{3+}, V^{2+}, Cr^{3+}, Mn^{4+} (B) Ti^{+}, V^{4+}, Cr^{6+}, Mn^{7+}
(C) Ti^{4+}, V^{3+}, Cr^{2+}, Mn^{3+} (D) Ti^{2+}, V^{3+}, Cr^{4+}, Mn^{5+}
(At. nos. Ti = 22; V = 23; Cr = 24; Mn = 25)

438. CN^- is a strong field ligand. This is due to the fact that [CBSE-PMT-2004]
(A) it carries negative charge (B) it is a pseudohalide
(C) it can accept electrons from metal species
(D) it forms high spin complexes with metal species

439. Considering H_2O as a weak field ligand, the number of unpaired electrons in $[Mn(H_2O)_6]^{2+}$ will be (At. no. of Mn = 25) [CBSE-PMT-2004]
(A) three (B) five (C) two (D) four

440. Which of the following does not have a metal - carbon bond ? [CBSE-PMT-2004]
(A) $Al(OC_2H_5)_3$ (B) C_2H_5MgBr
(C) $K[Pt(C_2H_4)Cl_3]$ (D) $Ni(CO)_4$

441. Lanthanoids are [CBSE-PMT-2004]
(A) 14 elements in the sixth period (atomic no. = 90 to 103) that are filling 4f sublevel
(B) 14 elements in the seventh period (atomic no. = 90 to 103) that are filling 5f sublevel
(C) 14 elements in the sixth period (atomic no. = 58 to 71) that are filling 4f sublevel
(D) 14 elements in the seventh period (atomic no. = 58 to 71) that are filling 4f sublevel

442. In an octahedral structure, the pair of d orbitals involved in d2sp3 hybridization is [CBSE-PMT-2004]
(A) $d_{x^2-y^2}$, d_{z^2} (B) d_{xz}, $d_{x^2-y^2}$
(C) d_{z^2}, d_{xz} (D) d_{xy}, d_{yz}

443. Which pair of the following compounds has elements in their highest oxidation state? [IIT-2004]

(A) $[Fe(CN)_6]^{3-}$ and $[Co(CN)_6]^{3-}$
(B) $[MnO_4]^-$ and CrO_2Cl_2
(C) $[MnO_4]^-$ and $[NiF_4]^{2-}$
(D) MnO_2 and TiO_2

444. Heating $(NH_4)_2Cr_2O_7$ yields a gas, which is also obtained by [IIT-2004]

(A) Mg_3N_2 in H_2O
(B) heating NH_4NO_2
(C) heating NH_4NO_3
(D) Na_2O_2 in H_2O

445. Which of the following complexes has tetrahedral geometry? [IIT-2004]

(A) $[Ni(CN)_4]^{2-}$
(B) $[Pd(CN)_4]^{2-}$
(C) $[NiCl_4]^{2-}$
(D) $[PdCl_4]^{2-}$

446. Spin magnetic moment in Bohr Magneton of $Hg[Co(SCN)_4]$ is [IIT-2004]

(A) $\sqrt{15}$ (B) $\sqrt{24}$ (C) $\sqrt{3}$ (D) $\sqrt{8}$

Transition metals, Lanthanides, Actinides and Coordination Chemistry

No.	Ans.	No.	Ans.	No.	Ans.	No.	Ans.	No.	Ans.	No.	Ans.
1.	(A)	50.	(D)	99.	(C)	148.	(C)	197.	(D)	246.	(B)
2.	(B)	51.	(C)	100.	(A)	149.	(A)	198.	(B)	247.	(A)
3.	(C)	52.	(A)	101.	(C)	150.	(C)	199.	(B)	248.	(A)
4.	(B)	53.	(D)	102.	(A)	151.	(B)	200.	(B)	249.	(A)
5.	(B)	54.	(B)	103.	(C)	152.	(A)	201.	(A)	250.	(A)
6.	(A)	55.	(B)	104.	(A)	153.	(B)	202.	*	251.	(D)
7.	(A)	56.	(B)	105.	(C)	154.	(B)	203.	*	252.	(D)
8.	(A)	57.	(C)	106.	(C)	155.	(B)	204.	*	253.	(B)
9.	(B)	58.	(C)	107.	(D)	156.	(C)	205.	(B)	254.	(A)
10.	(C)	59.	(D)	108.	(C)	157.	(B)	206.	*	255.	(D)
11.	(B)	60.	(B)	109.	(B)	158.	(B)	207.	(B)	256.	(B)
12.	(D)	61.	(C)	110.	(B)	159.	(C)	208.	(C)	257.	(C)
13.	(A)	62.	(A)	111.	(B)	160.	(D)	209.	*	258.	(B)
14.	(B)	63.	(A)	112.	(B)	161.	(B)	210.	*	259.	(A)
15.	(D)	64.	(C)	113.	(C)	162.	(B)	211.	*	260.	(C)
16.	(A)	65.	(A)	114.	(A)	163.	(D)	212.	*	261.	(A)
17.	(A)	66.	(B)	115.	(C)	164.	(C)	213.	(D)	262.	(A)
18.	(A)	67.	(A)	116.	(B)	165.	(D)	214.	(D)	263.	(C)
19.	(A)	68.	(C)	117.	(D)	166.	(D)	215.	(A)	264.	(A)
20.	(A)	69.	(A)	118.	(C)	167.	(C)	216.	(D)	265.	(D)
21.	(B)	70.	(C)	119.	(D)	168.	(C)	217.	(B)	266.	(D)
22.	(B)	71.	(C)	120.	(C)	169.	(B)	218.	(B)	267.	(D)
23.	(C)	72.	(A)	121.	(B)	170.	(A)	219.	(B)	268.	(B)
24.	(A)	73.	(D)	122.	(A)	171.	(B)	220.	(D)	269.	(B)
25.	(C)	74.	(C)	123.	(A)	172.	(D)	221.	(B)	270.	(A)
26.	(B)	75.	(B)	124.	(A)	173.	(D)	222.	(D)	271.	(C)
27.	(C)	76.	(C)	125.	(D)	174.	(B)	223.	(A)	272.	(A)
28.	(D)	77.	(D)	126.	(C)	175.	(C)	224.	(C)	273.	(C)
29.	(A)	78.	(A)	127.	(C)	176.	(C)	225.	(A)	274.	(D)
30.	(B)	79.	(D)	128.	(D)	177.	(C)	226.	(D)	275.	(A)
31.	(B)	80.	(D)	129.	(B)	178.	(B)	227.	(C)	276.	(A)
32.	(A)	81.	(A)	130.	(C)	179.	(B)	228.	(B)	277.	(D)
33.	(D)	82.	(A)	131.	(B)	180.	*	229.	(C)	278.	(D)
34.	(D)	83.	(B)	132.	(B)	181.	*	230.	(B)	279.	(B)
35.	(A)	84.	(A)	133.	(B)	182.	(D)	231.	(B)	280.	(A)
36.	(D)	85.	(A)	134.	(A)	183.	(D)	232.	(C)	281.	(A)
37.	(B)	86.	(D)	135.	(C)	184.	(B)	233.	(C)	282.	(C)
38.	(C)	87.	(E)	136.	(A)	185.	(D)	234.	(D)	283.	(A)
39.	(A)	88.	(B)	137.	(B)	186.	(B)(C)	235.	(D)	284.	(C)
40.	(A)	89.	(C)	138.	(B)	187.	(A)	236.	(C)	285.	(B)
41.	(A)	90.	(A)	139.	(A)	188.	(C)	237.	(B)	286.	(D)
42.	(C)	91.	(D)	140.	(A)	189.	(B)	238.	(D)	287.	(C)
43.	(D)	92.	(B)	141.	(A)	190.	(B)	239.	(D)	288.	(D)
44.	(A)	93.	(B)	142.	(B)	191.	(A)	240.	(B)	289.	(B)
45.	(D)	94.	(D)	143.	(A)	192.	(A)	241.	(D)	290.	(C)
46.	(B)	95.	(B)	144.	(A)	193.	(A)	242.	(A)	291.	(B)
47.	(A)	96.	(B)	145.	(C)	194.	(B)	243.	(R)	292.	(C)
48.	(A)	97.	(B)	146.	(A)	195.	(B)	244.	(A)	293.	(B)
49.	(A)	98.	(D)	147.	(C)	196.	(B)	245.	(A)	294.	(B)

180. paramagnetism
181. hexaamminecobalt (III) chloride
202. $CuCl_2$

203. 4
204. Sc^{3+}
206. +7

209. d - orbitals
210. Unfilled or vacant
211. f orbitals
212. $CrCl_3$

313.	(C)	364.	(B)	415.	(C)
314.	(A)	365.	(A)	416.	(B)
315.	(A)	366.	(C)	417.	(B)
316.	(D)	367.	(D)	418.	(B)
317.	(C)	368.	(D)	419.	(B)
318.	(C)	369.	(A)	420.	(A)
319.	(A)	370.	(C)	421.	(D)
320.	(B)	371.	(A)	422.	(A)
321.	(B)	372.	(A)	423.	(D)
322.	(A)	373.	(A)	424.	(D)
323.	(C)	374.	(C)	425.	(A)
324.	(B)	375.	(D)	426.	(B)
325.	(D)	376.	(D)	427.	(A)
326.	(B)	377.	(D)	428.	(A)
327.	(A)	378.	(D)	429.	(D)
328.	(A)	379.	(B)	430.	(A)
329.	(C)	380.	(C)	431.	(C)
330.	(B)	381.	(B)	432.	(A)
331.	(B)	382.	(C)	433.	(A)
332.	(C)	383.	(C)	434.	(C)
333.	(B)	384.	(A)	435.	(C)
334.	(B)	385.	(C)	436.	(B)
335.	(C)	386.	(D)	437.	(D)
336.	(D)	387.	(B)	438.	(D)
337.	(A)	388.	(C)	439.	(B)
338.	(A)	389.	(C)	440.	(A)
339.	(A)	390.	(B)	441.	(C)
340.	(B)	391.	(D)	442.	(A)
341.	(C)	392.	(D)	443.	(B)
342.	(B)	393.	(D)	444.	(B)
343.	(D)	394.	(C)	445.	(C)
344.	(C)	395.	(B)	446.	(A)
345.	(C)	396.	(B)		
346.	(C)	397.	(B)		
347.	(B)	398.	(A)		
348.	(B)	399.	(A)		
349.	(B)	400.	(C)		
350.	(D)	401.	(B)		
351.	(B)	402.	(B)		
352.	(D)	403.	(B)		
353.	(A)	404.	(A)		
354.	(C)	405.	(B)		
355.	(A)	406.	(A)		
356.	(B)	407.	(C)		
357.	(B)	408.	(C)		
358.	(C)	409.	(C)		
359.	(C)	410.	(D)		
360.	(A)	411.	(A)		
361.	(D)	412.	(D)		
362.	(B)	413.	(D)		
363.	(A)	414.	(A)		

METALLURGICAL OPERATIONS

Mineral : A solid inorganic substance occuring naturally in the earth.

Ore : Those minerals from which the metal can be profitably extracted are called ores.

Metallurgy : It is a branch of science dealing with the method of extraction of metals from the ores and preparation of alloys. Metallurgy involves : (1) concentration of ores (2) extraction of crude metal (3) refining of crude metal.

1.Concentration : Many a times, the ore is associated with foreign materials (impurities) worthless like sand and other earthy impurities called gangue. The removal of the unwanted material from the ore before actually proceeding for extraction is called concentration or dressing of the ore.

(a) Hand picking : Picking it up by hand and breaking away the adherant gangue with hammer.

(b) Washing with water:Also known as gravity separation or hydraulic washing.

(c) Magnetic separation : Impurities with magnetic property can only be separated.

(d) Froth flotation method : The crushed ore is mixed with a little oil (pine oil or fatty acids) and the whole is agitated by means of air. Here the surface of the particles is altered in such a way that the mineral particles adhere to the air bubbles and float. Sulphide ores are generally concentrated by this method. Although the sulphide ores are heavier than water and the rock powder, the oil coated ore particles stick on the walls of the bubbles of froath. NaCN and KCN are sometimes used as depressants. Potassium xanthate is used as stabilizer.

(e) Liquation : The metals with lower melting point than the impurities e.g. native copper, bismuth ore, stibmite (an ore of antimony) etc. are concentrated by this method.

(f) Leaching :In this method the powdered ore is treated with a suitable reagent that dissolves the ore but not the impurities. e.g. Bauxite contains impurities of Fe_2O_3 and silicates. Alumina dissolves in NaOH.

$$Al_2O_3 + 6NaOH \longrightarrow 2Na_3AlO_3 + 3H_2O$$

$$Na_3AlO_3 + 3H_2O \longrightarrow Al(OH)_3 + 3NaOH$$

$$2Al(OH)_3 \xrightarrow{\text{Calcination}} Al_2O_3 + 3H_2O$$

Native Ag is concentrated by leaching with dil. NaCN or KCN. The extraction is based upon formation of soluble argentocyanide ion.

$$AgCl + 2NaCN \longrightarrow Na[Ag(CN)_2] + NaCl$$

or $$Ag^+ + 2CN^- \rightleftharpoons [Ag(CN)_2]^-$$

2.Extraction of crude metal from concentrated ore : The concentrated ore must usually be converted to a compound which is suitable for reduction to the metal. If the ore consists of a carbonate, hydroxide, hydrated oxide, it is allowed to undergo calcination while roasting is employed in case of sulphide ores to convert the ore to oxide form. For thermodynamic reasons, oxide rather than a sulphide is used for reduction.

Roasting : The ore is heated strongly in excess of air without allowing the mass to melt. During this process, the volatiles are removed and the ore is changed to oxide or sulphate. Roasting is usually carried out in a reverboratory furnace in which the ore is heated by flames of a fuel. It is generally an oxidation process.

e.g. $2Cu_2S + 3O_2 \longrightarrow 2Cu_2O + 2SO_2$

or $CuS + 2O_2 \longrightarrow CuSO_4$

Calcination : The ore is strongly heated in the absence of any blast of air. The volatile impurities are removed during the process. Carbonate and oxide ores are calcinated generally when they loose moisture, CO_2 and other volatile impurities.

e.g. $CuCO_3.Cu(OH)_2 \xrightarrow{\text{Calcination}} 2CuO + CO_2 + H_2O$
malachite

$$Fe_2O_3.3H_2O \longrightarrow Fe_2O_3 + 3H_2O$$

$$Al_2O_3.2H_2O \longrightarrow Al_2O_3 + 2H_2O$$

The calcinated or roasted ore is taken for reduction to the metallic form by chemical reduction or electrolytic reduction.

(1) Chemical reduction :

(a) Smelting : The extraction of a metal from its ore by a process involving melting is termed smelting. It is often a reduction process.

(b) Using powerful reducing agents : For extracting less electropositive metals, powerful reducing agents like Na, H, Al, Mn etc. may be used

e.g. $Cr_2O_3 + 2Al \longrightarrow Al_2O_3 + 2Cr$

For Ti and Zr oxides reduction can be carried out with Na or Mg. For Mo and W metallurgy, hydrogen is used.

(c) Using carbon as reducing agent : Fe_2O_3, ZnO and PbO are reduced by carbon in reverboratory or blast furnace.

e.g. $ZnO + C \longrightarrow Zn + CO$

(2) Electrolytic reduction : In the case of highly electropositive metals, isolation by chemical reducing agent is difficult. It is mainly used for alkali, alkali earth metals, Zn and Al extraction.

(3) Refining or purification : Metals extracted by any of the reduction processes are usually contaminated with impurities.Therefore, they are subjected to refining.

(a) Liquation : Used for refining of low melting metals e.g. Sn, Pb and Bi.

(b) Distillation:Volatile metals such as Hg, Cd, Zn etc.are refined by this method.

(c) Oxidation : Oxidation methods are employed when the impurities present in the metal have more affinity for oxygen than the metal

(i) Cupellation : Pb present with Ag is removed by this method in the purification of Ag.

(ii)Bessimerization : Molten pig iron is purified by bessimerization. Impurities are oxidised by a blast of air.

(iii) Poling : The impure metal is melted and stirred with green wood poles. The hydrocarbon gases escaping from the poles reduce the oxide to metal. Blister copper contains Cu_2O as impurity. On poling the oxide is reduced to metallic copper (here it is a reduction process). In the case of refining of Sn by poling, it is an oxidation process.

Zone refining : Semi conductors e.g. Si,Ge,Ga,In etc. are purified by this method.

van Arkel method : This method is used for the purification of metals like Titanium and Zirconium.

Electrolytic refining : Metals like Cu,Ag,Pb,Cr,Ni etc are refined using appropriate aqueous solutions of their salts.

Types of furnaces used

1.Reverberatory furnace : Here, the fuel does not come in direct contact with the charge. This furnace is used for reduction as well as oxidation processes. Calcination and roasting are generally carried out in this furnace.

2. Blast furnace : Mainly employed in the metallurgy of Iron.

3. Muffle furnace : This furnace is used where high temperatures are needed and where it is desired that the fuel and the products of combustion do not come into contact with each other.

4. Heroult's furnace : Alloy steel with high m.p's are manufactured by this furnace.

5. Open-Hearth furnace : Open-hearth furnace working on the regenerative principle of heat economy is used for the manufacturing of steel.

SODIUM - Na

Sea water : NaCl; Chile Saltpetre : $NaNO_3$;
Borax : $Na_2B_4O_7.10H_2O$.

1. Down's process : $NaCl \xrightarrow{\text{Electrolysis}} Na$

$NaCl \rightleftharpoons Na^+ + Cl^-$ Anode : C

At cathode $Na^+ + e^- \longrightarrow Na$ Cathode : Fe

$Cl^- - e^- \longrightarrow \frac{1}{2}Cl_2$

Flux : Flux is a substance added to the ore either to reduce its m.p. or to combine with any infusible impurity that may be present. The flux combines with gangue to form a slag. The slag usually floats. $CaCl_2$, KCl, NaF etc. are employed as flux.

Down's process is economical due to cheaper starting material and Cl_2 as a valuable by-product.

2. Castner process : $\underset{\text{caustic soda}}{NaOH} \xrightarrow{\text{Electrolysis}} Na$

$NaOH \rightleftharpoons Na^+ + OH^-$ Cathode :Fe

At cathode $Na^+ + e^- \longrightarrow Na$ Anode : Ni

$OH^- \rightleftharpoons H^+ + O^{-2}$

$H^+ + e^- \longrightarrow \frac{1}{2}H_2$

At anode $O^{2-} - 2e \longrightarrow \frac{1}{2}O_2$

H_2 evolved protects Na formed from oxidation at anode.

Na -gives golden yellow flame
-good conductor of heat and electricity
-powerful reducing agent

Na shows two bright lines in the yellow region of the spectrum - D_1 and D_2 at 5890 Å and 5860 Å.

ALUMINIUM - Al

Bauxite : $Al_2O_3.2H_2O$; Cryolite : Na_3AlF_6;
Felspar : $K_2O.Al_2O_3.6SiO_2$
Kaolin (or China clay) : $Al_2O_3.2SiO_2.2H_2O$

1. Purification of Bauxite :

(i) When FeO as impurity (Baeyer's or Hall's process)
(ii) When silica as impurity (Serpek's process)

(i) Baeyer's process (wet process) :

Grounded ore $\xrightarrow{\text{roasted}}$ FeO is converted to Fe_2O_3

$Fe_2O_3 \xrightarrow[\text{NaOH}]{\text{Caustic Soda}} Na_3AlO_3$ (sod. aluminate)

Hall's process (dry process) :

Ore (Al_2O_3) fused with $Na_2CO_3 \longrightarrow \underset{\text{sod. meta aluminate}}{2NaAlO_2} + CO_2$

(ii) Serpek's process :

powdered bauxite + coke $\xrightarrow{\text{In the presence of } N_2\text{, Heat}}$

$SiO_2 + 2C \longrightarrow Si + 2CO$

$Al_2O_3 + 3C + N_2 \longrightarrow 2AlN + 3CO$

$AlN + 3H_2O \longrightarrow Al(OH)_3 + NH_3$

2. Electrolytic reduction of Al_2O_3 :

Al_2O_3 is infusible (m.p. 2050°C). So it is dissolved in molten cryolite (Na_3AlF_6) containing some CaF_2 and electrolysed at 900°C. Electrolysis is carried out in an iron cell lined with graphite (cathode). The anode is a series of graphite rods dipping in the melt.

$AlF_3 \rightleftharpoons Al^{3+} + 3F^-$

at cathode : $Al^{3+} + 3e^- \longrightarrow Al$

at anode : (i) $2F^- - 2e^- \longrightarrow F_2$

(ii) $2Al_2O_3 + 6F_2 \longrightarrow 4AlF_3 + 3O_2$

O_2 liberated at the anode either escapes as such or unites with the C of the anode forming CO & CO_2. So anodes which burn away have to be replaced periodically.

3. Electrolytic refining (Hoope's process) : Three fused layers : (i) top cathode layer of pure Al, (ii) middle layer of cryolite & BaF_2, (iii) bottom anode layer of an alloy of Al,Cu & Si.

-Al rapidly dissolves in alkalies with the formation of aluminates and meta aluminates.

$2Al + 2NaOH + 2H_2O \longrightarrow \underset{\text{Sodium meta aluminate}}{2NaAlO_2} + 3H_2$

$2Al + 6NaOH \longrightarrow \underset{\text{Sodium aluminate}}{2Na_3AlO_3} + 3H_2$

-$2Al + 6HCl \longrightarrow 2AlCl_3 + H_2$

-$2Al + 3CuSO_4 \longrightarrow Al_2(SO_4)_3 + 3Cu$

-Light good electrical conductor

Duralumin : Al +Mg+Cu+Mn: 95 0.5 4 0.5

Magnalium : Al+Mg 90 10

γ - alloy : Al + Cu + Mg + Ni
92.5 4 1.5 2.0

COPPER - Cu

Copper pyrites (chalcopyrites) : $CuFeS_2$;
Malachite : $CuCO_3.Cu(OH)_2$; Copper glance : Cu_2S;
Azurite : $2CuCO_3.Cu(OH)_2$

1. (a) Pyrometallurgical process :

$$Ore \xrightarrow{roasting} \begin{matrix} Cu_2O \\ FeO \end{matrix} \xrightarrow{smelting} \text{white metal (75\% Cu)}$$

$$\xrightarrow{Bessemerisation} Cu \xrightarrow[\text{or electrolytic refining}]{poling} \text{pure Cu (99.9\%)}$$

$$2CuFeS_2 + O_2 \longrightarrow Cu_2S + 2FeS + SO_2$$
$$2Cu_2S + 3O_2 \longrightarrow 2Cu_2O + 2SO_2$$

$$FeS + O_2 \longrightarrow FeO \xrightarrow{SiO_2} FeSiO_3$$
$$\left.\begin{matrix} 2Cu_2S + 5O_2 \longrightarrow 2CuSO_4 + 2CuO \\ 2Cu_2O + Cu_2S \longrightarrow 3Cu + 2SO_2 \end{matrix}\right] *$$

* Bessemerisation is a self reduction process

(b) Hydrometallurgical process :

From low grade ores

$$CuFeS_2 + O_2 + H_2O \longrightarrow CuSO_4 + FeSO_4 + \text{some } H_2SO_4$$
$$Cu + FeSO_4 \xleftarrow{\text{Fe crap}}$$

(2) Refining : (a)Poling (reduction process) :

Blister copper : 98% pure Cu

(b) Electrolytic refining : Anode : Block of impure Cu; Cathode : pure thin Cu sheets.

Electrolysis of Cu

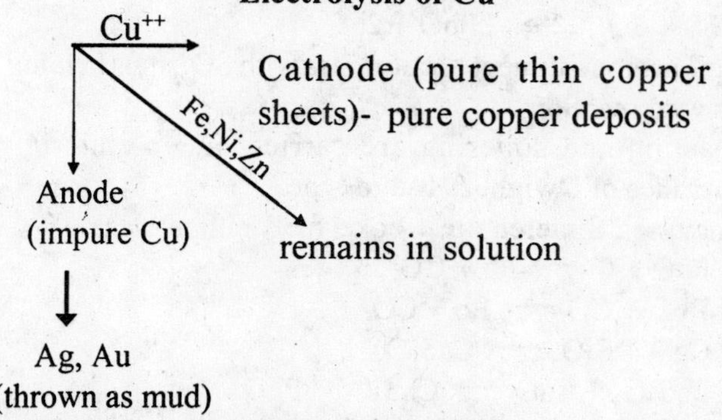

Electrolyte : Acidified $CuSO_4$ solution

- Brass : Cu + Zn ; Bronze : Cu + Sn
- Bell metal : Cu + Sn ; Gun metal : Cu + Sn + Zn
- German silver : Cu + Zn + Ni
- Monel metal : Cu + Ni + Fe + Mn
- Silver coins : Ag + Cu (92.5 + 7.5%)
- Ruby glass and enamel contain Cu_2O
- $Cu_2Cl_2.2H_2O$ cupric chloride is used as catalyst in Deacon's process
- Blue vitriol : $CuSO_4.5H_2O$
- Copper salts show deep green flame

$$2Cu + 4 \text{ dil.or conc.}HCl + O_2 \longrightarrow 2CuCl_2 + 2H_2O$$
$$2Cu + 2 \text{ dil. } H_2SO_4 + O_2 \longrightarrow 2CuSO_4 + H_2O$$
$$Cu + 2H_2SO_4(conc.) \xrightarrow{\Delta} CuSO_4 + SO_2 + 2H_2O$$
$$3Cu + 8HNO_3(dil.) \longrightarrow 3Cu(NO_3)_2 + 2NO + 4H_2O$$
$$Cu + 4HNO_3(conc.) \longrightarrow Cu(NO_3)_2 + 2NO_2 + 2H_2O$$

IRON - Fe

Haematite : Fe_2O_3; Magnetite : Fe_3O_4; Iron pyrites : FeS_2; Copper pyrites : $CuFeS_2$; Limonite : $3Fe_2O_3.3H_2O$; Siderite : $FeCO_3$

$$\underset{8\ :\ 4\ :\ 1\ (Flux)}{\text{Ore + Coke + Lime Stone}} \xrightarrow[700\text{-}900^oC]{\text{Blast furnace}} Fe + CaSiO_3$$

$$3Fe_2O_3 + CO \xrightarrow{300\text{-}500^oC} 2Fe_3O_4 + CO_2$$
$$Fe_3O_4 + CO \xrightarrow{500\text{-}700^oC} 3FeO + CO_2$$
$$FeO + CO \underset{1000^oC}{\overset{700\text{-}900^oC}{\rightleftharpoons}} Fe + CO_2$$

If $CaCO_3$ is present along with the ore

$$CaCO_3 \xrightarrow{700^oc} CaO \xrightarrow{SiO_2} CaSiO_3\,(slag)$$

The products of blast furnace are pig or cast iron

Stainless Steel : It is an alloy of Fe, Cr & Ni

1. Bessemer process : Impurities are oxidised by air

(a)Acid Bessemer process : furnace is lined with silica.

$$2Mn + O_2 \longrightarrow 2MnO; \quad Si + O_2 \longrightarrow SiO_2$$
$$MnO + SiO_2 \longrightarrow MnSiO_3(slag)$$

(b) Basic Bessmer process : The furnace is lined with lime

$$P_4 + 5O_2 \longrightarrow P_4O_{10}$$
$$6CaO + P_4O_{10} \longrightarrow \underset{\text{Thomas slag}}{2Ca_3(PO_4)_2}$$

Thomas slag is used as a fertilizer.

2. Open-Hearth process : Impurities present in the cast iron are removed by adding iron ore

Advantages over Bessemer process :

(1) superior quality of steel is obtained
(2) scrap iron and low grade pig iron can be used
(3) as no blast of air is used practically no iron is lost by slag formation

Annealing : cooling slowly

Hardening : cooling suddenly by plunging in oil or H_2O. It is also known as **quenching.**

Tempering : hardened steel is heated below redness and cooled slowly

Case-hardening : hardened steel is coated by heating with charcoal and quenching is done in oil

Nitriding : Heating steel in an atmosphere of NH_3. The surface is coated with hard iron nitride.

-With dil. HNO_3 :

$$4Fe + 10HNO_3 \longrightarrow 4Fe(NO_3)_2 + NH_4NO_3 + 3H_2O$$

-With Conc. HNO_3 : Iron becomes passive i.e. it looses its activity because of the formation of ferroso-ferric oxide thin layer, Fe_3O_4

- % of carbon in different forms of Iron :

Cast Iron : 2 - 4.5% C ; Steel : 0.15 - 1.5% C
Wrought Iron : 0.1 - 0.25% C ; Invar : 36% Ni
Allotropes : α-iron-bcc lattice; ϒ-iron-fcc; δ-iron-bcc

- Fe_3O_4 ferroso-ferric oxide :used for electrodes. It is also known as magnetic oxide of iron
- Mohr's salt : $FeSO_4.(NH_4)_2SO_4.6H_2O$
- Green vitriol : $FeSO_4.7H_2O$
- Prussian blue : $Fe_4[Fe(CN)_6]_3$

- Turnbulls blue : $Fe_3[Fe(CN)_6]_2$

SILVER - Ag

Argentite : Ag_2S; Horn silver : AgCl;
Pyrargyrite : $3Ag_2S.Sb_2S_3$ (Ruby silver)
Proustite : $3Ag_2S.As_2S_3$

1 Cyanide Process (Wet process) :

$$Ag_2S + 4NaCN \longrightarrow 2Na[Ag(CN)_2] + Na_2S$$
$$Na_2[Ag(CN)_2] \xrightarrow{Zn} Na_2[Zn(CN)_4] + Ag$$

2 From Argentiferous Lead :

(a) Parke's process :

$$\text{Agentiferous lead} \xrightarrow{Zn} \text{Ag-Zn alloy}^* \xrightarrow{\text{cupelled}} Ag \quad (\searrow Pb)$$

*Ag-Zn alloy obtained is distilled when Zn distills leaving Ag with a little amount of Pb which is removed by cupellation.

- Molten Pb and Zn are immiscible and molten Zn will float over molten Pb.
- Ag more soluble in Zn than Pb.
- Zn-Ag solidifies earlier than molten Pb.

(b) Pattison's process : Argentiferous lead is melted and allowed to cool when Pb crystallizes out. The process is repeated several times.

3 From Silver Coins (alloy of Ag + Cu) :

Coin + HNO_3 $\longrightarrow$ Remove HNO_3 by boiling $\xrightarrow{\text{add HCl}}$ AgCl

Cu remains in solution. AgCl is filtered and reduced by

(a) $4AgCl+2Na_2CO_3 \xrightarrow{Fuse} 4NaCl+4Ag+2CO_2+O_2$

(b) AgCl+H (nascent) $\xrightarrow{Zn/acid}$ Ag + HCl

(c) $2AgCl + 2NaOH \longrightarrow Ag_2O + 2NaCl + H_2O$

$$\underset{\text{Glucose}}{Ag_2O + C_6H_{12}O_6} \xrightarrow{Heat} \underset{\text{Gluconic acid}}{2Ag+C_5H_{11}O_5COOH}$$

- Colloidal silver : used in opthalmic treatment
- AgBr :used in photographic films
- $AgNO_3$: used for making glass mirrors in ammonical solutions.

ZINC - Zn

Zinc blende : ZnS; Calamine : $ZnCO_3$
Franklinite : $Zn(FeO_2)_2$;
concentrated first by gravity process & then FF.

$$\text{Ore} \xrightarrow[\text{Calcination}]{\text{Roasting or}} ZnO \xrightarrow{\text{Reduction}} Zn$$

$2ZnS + 3O_2 \longrightarrow 2ZnO + 2SO_2$
$ZnS + 2O_2 \longrightarrow ZnSO_4$
$2ZnSO_4 \longrightarrow 2ZnO + 2SO_2 + O_2$

Reduction :

(i) By Belgian process (powdered coke or coal is used)
(ii) By Silesian process (in reverboratory furnace)

Electrolysis : Cathode : Al; Anode : Pb
Electrolyte : $ZnSO_4$ solution

-With alkali : $Zn + 2NaOH \longrightarrow \underset{\text{Sod. Zincate}}{Na_2ZnO_2} + H_2$

Metals like Cu,Ag,Au etc. which are less electropositive than Zn are replaced by Zn from their salt solutions.

$$CuSO_4 + Zn \longrightarrow ZnSO_4 + Cu$$

- Philosopher's wool ::ZnO
- Rinman's green : $CoZnO_2$ (Cobalt zincate)
- White vitriol : $ZnSO_4$ $7H_2O$
- Lithopone : ZnS + $BaSO_4$

$ZnCO_3$: used for treating skin diseases.
ZnS : for making phosphorescent screens useful for X-ray & radioactive work as well as Radar & Television.

LEAD - Pb

Galena : PbS; Anglesite : $PbSO_4$; Cerussite : $PbCO_3$; Stolzite : $PbWO_4$

Purification : Due to the presence of Cu,Ag,Bi,Sb & Sr as impurities, crude lead is hard. Removal of impurities is known as softening of lead.

1 Richer ores (by self-reduction process)

$2PbO + PbS \longrightarrow 3Pb + SO_2$
$2PbS + 3O_2 \longrightarrow 2PbO + 2SO_2$
$PbS + 2O_2 \longrightarrow PbSO_4$
$PbSO_4 + PbS \longrightarrow 2Pb + 2SO_2$

2 Low grade ores (By carbon-reduction process)

Ore + Lime $\xrightarrow{\text{roasted}}$ in a sinterer in a blast of air
$2PbS + 3O_2 \underset{}{\overset{air}{\rightleftharpoons}} 2PbO + 2SO_2$

Lime acts as a diluent & combines with SO_2, thus helping forward reaction.

Roasting and sintering are carried out in sintering machines of Dwight & Lloyd type.

Smelting : Sintered ore + coke(fuel) + lime(slag)

$PbO + C \longrightarrow Pb + CO$
or $PbO + CO \longrightarrow Pb + CO_2$
$CaO + SiO_2 \longrightarrow CaSiO_3$
$PbSiO_3 + CaO \longrightarrow CaSiO_3 + PbO$

3. Refining : Bett's Process

Anode : crude lead; Cathode : pure lead sheets
Electrolyte : Lead silicofluoride $PbSiF_6$+ Hydrofluosilic acid H_2SiF_6 + a little gelatine

- Solder & Pewters : Pb + Sn

PbO : used in vulcanization of rubber
Dutch Process : For white Pb of good quality.
Sugar of Lead : $Pb(CH_3COO)_2.3H_2O$
Sindur (red lead) : Pb_3O_4

GOLD - Au

Sylvenite, a telluride of Au and Ag

From Auriferous Quartz :

1 Amalgamation process :

The ore is crushed, pulverized, washed with H_2O and allowed to flow through the screen over amalgamated cu plates. Gold particles are retained on these plates as amalgam. The amalgam from these plates is scrapped and distilled in iron retort when mercury distils over.

2 Mac Arthur-Forrest Cyanide process :

Well powdered low grade ore + NaCN(0.15%)+Blast of air $\longrightarrow$ $[Au(CN)_2]^-$
sod. aurocyanide

$[Au(CN)_2]^- + Zn \longrightarrow [Zn(CN)_4]^{2-} + Au$

3 Chlorination process :

powdered ore + moist $Cl_2 \longrightarrow AuCl_3$

$AuCl_3 + 3FeSO_4 \longrightarrow Au + Fe_2(SO_4)_3 + FeCl_3$

- Sp. gr. 19.3; the most malleable & ductile metal.
- $Au + HCl + HNO_3 \longrightarrow HAuCl_4$
  aquaregia Chlorauric acid
- Colloidal gold solution is prepared by Bredig's Arc method
- Purple of Cassius
  $2AuCl_3 + 3SnCl_2 \longrightarrow 2Au + 3SnCl_4$
- $AuCl_3$: It is used in photography for toning.

MERCURY - Hg

Cinnabar : HgS (red sulphide mineral)

$2HgS + 3O_2 \longrightarrow 2HgO + 2SO_2$

$2HgO \longrightarrow 2Hg + O_2$

$HgO + C \longrightarrow Hg + CO$

Purification : If Cu, Zn, Bi and Pb are present, they are removed by filtering through a thick canvas or chamois leather. Basic impurities are further removed as nitrates by treating Hg with dil. HNO_3. Further purification is done by vaccum distillation or fractional electrolysis.

- Sp. gr. : 13.6
- Hg reacts with HI to form complex ion HgI_4^{-2}
- Na-Hg : important reducing agent
- Mercury vapour lamp gives blue light rich in UV rays
- Ag-Hg , Au-Hg are used in dentistry
- Hg_2Cl_2-Calomel (calomel electrodes are used in potentiometric titration)
- $Hg_2Cl_2 + 2NH_3 \longrightarrow [Hg + Hg(NH_2)Cl] + NH_4Cl$
  black amino basic mercuric chloride
- $HgCl_2$ - corrosive sublimate , white of egg is antidote
- HgO + aq. $NH_3 \longrightarrow$ Yellow powder (Millons base)

TIN - Sn

Tinstone : SnO_2 Cassiterite

$SnO_2 + 2C \xrightarrow[\text{blast furnace}]{1200\ °C} Sn + 2CO$

$SnO_2 + 2C \xrightarrow[\text{anthrocite}]{} Sn + 2CO$ - Smelting

Poling :Refining is done by poling.It is an oxidation process.

Grey $\underset{}{\overset{91K}{\rightleftharpoons}}$ White $\overset{443K}{\rightleftharpoons}$ White -3 allotropic forms
cubic tetragonal rhombic

Britannia metal : Sn + Sb + Cu
93 5 2

Babbit metal : Sn + Sb + Cu
90 7 3

α-Stannic acid :

$SnCl_4 + 4NH_4OH \longrightarrow 4NH_4Cl + Sn(OH)_4$

$Sn(OH)_4 \longrightarrow SnO(OH)_2 + H_2O$

β-Stannic acid :

Sn + hot conc. $HNO_3 \longrightarrow H_2Sn_5O_{11} \cdot 4H_2O$

MAGNESIUM - Mg

Mg doesn't occur free in nature.

$MgCO_3$	-	Magnesite
$MgCO_3.CaCO_3$	-	Dolamite
$MgCl_2.KCl.6H_2O$	-	Carnalite
$MgSO_4.7H_2O$	-	Epsom salt
$MgSO_4.H_2O$	-	Kiesserite
	-	Sea Water

Mg is obtained as follows :

Sea water ($MgCl_2$or $MgSO_4$) + $Ca(OH)_2 \longrightarrow Mg(OH)_2$ ppt.

$Mg(OH)_2 + 2HCl \longrightarrow MgCl_2.6H_2O \xrightarrow[HCl\,(g)]{\Delta} MgCl_2 + 6H_2O$
anhydrous

$MgCl_2$ (35%) fused with NaCl(50%) and $CaCl_2$(15%) is electrolysed at 700° in the presence of an inert gas

Graphite - anode
Iron cell - cathode.

-Mg is present in chlorophyll and asbestos $CaMg_3(SiO_3)_4$.

1. Cinnabar is an ore of [CBSE-91]
(A) mercury (B) zinc (C) gold (D) silver

2. Which of the following elements is extracted commercially by the electrolysis of an aqueous solution of its compound ? [CBSE-93]
(A) Cl (B) Br (C) Al (D) Na

3. When $CuSO_4$ is electrolysed by platinum electrode [CBSE-93]
(A) copper is liberated at cathode, sulphur at anode
(B) copper is liberated at cathode, oxygen at anode
(C) sulphur is liberated at cathode, oxygen at anode
(D) oxygen is liberated at cathode, copper at anode

4. Suggest a purification method for obtaining highly pure silicon used as semiconductor from the following. [CBSE-94]
(A) electrochemical refining method
(B) reduction method (C) zone refining method
(D) oxidation method.

5. Sodium is made by the electrolysis of a molten mixture of about 40% NaCl and 60% $CaCl_2$ because [CBSE-95]
(A) $CaCl_2$ helps in conduction of electricity
(B) This mixture has a lower melting point than NaCl
(C) Ca^{++} can displace Na from NaCl.
(D) Ca^{++} can reduce NaCl to Na

6. The method of zone refining of metals is based on the principle of [CBSE-2003]
(A) greater solubility of the impurity in the molten state than in the solid
(B) greater mobility of the pure metal than that of the impurity
(C) higher melting point of the impurity than that of the pure metal
(D) greater noble character of the solid metal than that of the impurity

7. The important ore of aluminium is [AIIMS-95]
(A) Corundum (B) Ruby (C) Kaolin (D) Bauxite

8. Which of the following is a sulphide ore ? [AIIMS-96]
(A) Magnetite (B) Malachite
(C) Carnallite (D) Copper pyrites

9. Blister copper is [AIIMS-96]
(A) Pure copper (B) Ore of copper
(C) Alloy of copper
(D) Copper containing 1% impurity

10. Bauxite ore of aluminium is purified by [AIIMS-96]
(A) Serpek's process (B) Bayer's process
(C) Hall's process (D) All of these

11. The purification of alumina is called [AIIMS-98]
(A) Castner process (B) Bosch process
(C) Baeyer's process (D) Hoop's process

12. Which of the following is not an ore of iron ? [AIIMS-2002]
(A) Limonite (B) Magnetite
(C) Carnallite (D) None of these

13. In the extraction of copper from its sulphide ore the metal is formed by reduction of Cu_2O with - [AIIMS-2003]
(A) FeS (B) CO (C) Cu_2S (D) SO_2

14. An alloy of tin and copper is known as : [JIPMER-91]
(A) German silver (B) Brass (C) Bronze (D) Solder

15. Of the following substances the one which does not cantain oxygen is [JIPMER-95]
(A) Epsom salt (B) Bauxite (C) Cryolite (D)Dolomite

16. The iron carbonate ore is [JIPMER-97]
(A) Haematite (B) Siderite
(C) Magnesite (D) Iron pyrites

17. Cassiterite is concentrated by [JIPMER-97]
(A) Gravity separation (B) Roasting
(C) Electromagnetic separation (D) All the above

18. AgCl is soluble in [JIPMER-97]
(A) Water (B) Ether (C) Sodium cyanide (D) Ethanol

19. The alloy of copper which does not contain zinc is [JIPMER-97]
(A) Brass (B) Bronze (C) Gun metal (D) German sliver

20. An ore of potassium is [JIPMER-2001]
(A) Bauxite (B) Dolomite (C) Carnallite (D) Cryolite

21. Tin, a silvery white metal exists in [JIPMER-2001]
(A) Two allotropic forms (B) Five allotropic forms
(C) Three allotropic forms (D) Four allotropic forms

22. The substance which is added to ore in order to remove impurities during smelting is known as [JIPMER-2002]
(A) Catalyst (B) Gangue (C) Flux (D) Slag

23. The best quality of steel is obtained by [BHU-96]
(A) L-D process (B) Deacon's process
(C) Bessemer process (D) Open Hearth process

24. Froth flotation process is used for the metallurgy of [BHU-97]
(A) Oxide ores (B) Chloride ores
(C) Sulphide ores (D) Amalgams

25. H_2S is passed through an acidified solution Ag, Cu and Zn. Which forms precipitates ? [BHU-98]
(A) Ag (B) Zn (C) Cu (D) None of these

26. Cyanide process is used for extraction of [BHU-2000]
(A) Zn (B) Ni (C) Ag (D) Pt

27. During extraction of copper the walls of blast furnace are covered with [BHU-2001]
(A) CaO (B) SiO_2 (C) Graphite (D) NaCl

28. The chemical process in production of steel from haematite involves [BHU-2001]
(A) Reduction (B) Oxidation
(C) Reduction followed by oxidation
(D) Oxidation followed by reduction

29. The purest form of iron is [AFMC-91]
(A) Cast iron (B) Wrought iron
(C) Mild steel (D) Stainless steel

30. In electrorefining of copper some gold is deposited at [AFMC-95]
(A) Anode mud (B) Cathode mud
(C) Cathode (D) Electrolyte

31. Stainless steel contains iron with [AFMC-96]
(A) Cr + Ni (B) Cr +Zn (C) Sn + Pb (D) Bi + Ni

32. Which of the following impurities are present in maximum concentration in pig iron ? [AFMC-2000]
(A) Sulphur (B) Iron (C) Phosphorus (D) Sand

33. Of the following metals, the metals that can be obtained by electrolysis of the aqueous solution of their salts are : [IIT-90]
(A) Ag (B) Mg (C) Cu (D) Al

34. The major role of fluorspar (CaF_2) which is added in small quantities in the electrolytic reduction of alumina dissolved in fused cryolite (Na_3AlF_6) is : [IIT-93]
(A) as a catalyst
(B) To make fused mixture very conducting
(C) to lower the temperature of the melt
(D) to decrease the rate of oxidation of carbon at the anode.

35. Addition of high proportions of manganese makes steel useful in making rails of railroads, because manganese. [IIT-98]
(A) gives hardness to steel
(B) helps the formation of oxides of iron
(C) can remove oxygen and sulphur
(D) can show highest oxidation state of +7

36. In the commercial electrochemical process for aluminium extraction the electrolyte used is [IIT-99]
(A) $Al(OH)_3$ in NaOH solution
(B) an aqueous solution of $Al_2(SO_4)_3$
(C) a molten mixture of Al_2O_2 and Na_3AlF_6
(D) a molten mixture of AlO(OH) and $Al(OH)_3$

37. The chemical processes in the production of steel from haematite ore involve [IIT-2000]
(A) reduction (B) oxidation
(C) reduction followed by oxidation
(D) oxidation followed by reduction.

38. Electrolytic reduction of alumina to aluminium by Hall-Heroult process is carried out [IIT-2000]
(A) in the presence of NaCl
(B) in the presence of fluorite
(C) in the presence of cryolite which forms a melt with lower melting temperature.
(D) in the presence of cryolite which forms a melt with higher melting temperature.

39. The chemical composition of 'slag' formed during the smelting process in the extaction of copper is [IIT-2001]
(A) $Cu_2O + FeS$ (B) $FeSiO_3$
(C) $CuFeS_2$ (D) $Cu_2S + FeS$

40. Which of the following processes is used in the extractive metallurgy of magnesium ? [IIT-2002]
(A) fused salt electrolysis (B) self reduction
(C) aqueous solution electrolysis (D) thermite reduction

41. Impure Gold $\xrightarrow[O_2]{KCN}$ (X);
(X) $\xrightarrow{Zn}$ (Y) + other product, X and Y are [IIT-2003]
(A) $K[Au(CN)_2]$, $K+ + [Zn(CN)_4]^-$
(B) $[Au(CN)_2]^-$, $[Zn(CN)_4]^{2-}$
(C) $[Au(CN)_2]$, $[K(CN)_4]^-$ (D) None of these

42. Aluminium is extracted by the electrolysis of. [AIEEE-2002]
(A) Bauxite (B) Alumina
(C) Alumina mixed with molten cryolite
(D) Molten cryolite.

43. The metal extracted by leaching with a cyanide is [AIEEE-2002]
(A) Mg (B) Ag (C) Cu (D) Na

44. Value of gas constant R is [AIEEE-2002]
(A) 0.082 litre atm (B) 0.987 cal $mol^{-1}K^{-1}$
(C) 8.3 J mol^{-1} K^{-1} (D) 83 erg mol^{-1} K^{-1}

45. Cyanide process is used for the extraction of [AIEEE-2002]
(A) Barium (B) Aluminum (C) Boron (D) Silver

46. The transitional metal present in the alloy gun metal is .... [EAMCET-90]

47. The non metallic element present in the mineral cryolite is .... [EAMCET-91]

48. Bauxite is a mineral used for the extraction of [EAMCET-92]
(A) Ca (B) Cu (C) Al (D) Fe

49. Magnesium is mainly extracted from .... [EAMCET-92]

50. Silver is extracted from its (....) ore. [EAMCET-93]
(A) Sulphide (B) Oxide (C) Nitrate (D) Halide

51. Oxides of iron in the blast furnace get reduced due to the action of ....on haematite.[EAMCET-93]

52. The formula of Dolomite is [EAMCET-94]
(A) $MgCO_3$. $CaCl_2$ (B) $MgCO_3$. $CaCO_3$
(C) $MgCO_3$. $CaSO_4$ (D) $MgCl_2$. $CaCO_3$

53. The chief product of electrolysis of Al_2O_3 at the anode in the Hall's process is. [EAMCET-95]
(A) H_2 (B) Na (C) AlF_3 (D) CO_2

54. When sodium argentocyanide is treated with zinc dust silver precipitates because [EAMCET-96]
(A) Silver is more electropositive than zinc
(B) Zinc is more electropositive than silver
(C) Zinc forms a complex readily with cyanide
(D) Both Zn^{+2} and Ag^{+} ions have d^{10} eletronic configuration.

55. The molecular formula of Cryolite is [EAMCET-96]
(A) $3NaF.AlF_3$ (B) AlF_3 (C) $2NaF.AlF_3$ (D) $NaF.AlF_3$

56. Silver containing lead as an impurity is removed by [EAMCET-96]
(A) Poling (B) Cupellation
(C) Lavigation (D) Distillation

57. The chemical formula of diaspore, an ore of aluminium is [EAMCET-96]
(A) $Al_2O_3.3H_2O$ (B) $Al_2O_3.2H_2O$
(C) Al_2O_3. H_2O (D) Al_2O_3

58. Which of the following ores is chloride [EAMCET-97]
(A) Horn Silver (B) Zincite (C) Bauxite (D) Felspar

59. Cassiterite is concentrated by [EAMCET-98]
(A) Levigation (B) Electromagnetic separation
(C) Floatation (D) Liquation

60. Which one of the following reactions is an example of calcination process ? [EAMCET-98]
(A) $2Ag + 2HCl + (O) \rightarrow AgCl + H_2O$
(B) $2\ Zn + O_2 \rightarrow 2ZnO$
(C) $2ZnS + 3O_2 \rightarrow 2ZnO + 2SO_2$
(D) $MgCO_3 \rightarrow MgO + CO_2$

61. Silver containing lead as an impurity is removed by [EAMCET-98]
(A) Poling (B) Cupellation
(C) Levigation (D) Distillation

62. Which of the following is an oxide mineral ? [EAMCET-99]
(A) Zinc blende (B) Magnetite
(C) Dolomite (D) Carnallite

63. Tin stone is contaminated with Wolframite. Which one of the following methods can be employed to dress the tin stone ore ? [EAMCET-99]
(A) levigation (B) electromagnetic separation
(C) flotation process (D) roasting

64. Stainless steel vessels are made from steel containing [EAMCET(MBBS)-99]
(A) Ni (B) Mn (C) W (D) Cr

65. Which of the following is used to reduce pure haematite to iron ? [EAMCET(MBBS)-99]
(A) H_2 (B) CO (C) $LiAlH_4$ (D) H_2/Pd

66. "Elecktron" is an alloy of [EAMCET(MBBS)-99]
(A) Mg + Ni (B) Ni + Al (C) Mg + Al (D) Mg + Zn

67. During the extraction of zinc, ZnO is obtained from zinc blende by [EAMCET(MBBS)-99]
(A) roasting (B) cupellation
(C) poling (D) electrolytic reduction

68. Which of the following is a sulphide ore ? [EAMCET-2000]
(A) Calamine (B) Cryolite
(B) Zinc blende (D) Haematite

69. During the Serpeck's process silica is eliminated as : [EAMCET-2000]
(A) Si (solid) (B) SiO_2 (vapour)
(C) SiO_2 (soild) (D) Si (vapour)

70. In the Castner's process of extraction of sodium, cathode is [EAMCET-2000]
(A) Iron rod (B) Nickel rod
(C) Copper rod (D) Graphite rod

71. Which of the following is used in the extraction of silver ? [EAMCET(MBBS)-2000]
(A) NaCl (B) Na_2CO_3 (C) NaCN (D) Na_2SO_4

72. Lead impurity is removed from silver by ? [EAMCET(MBBS)-2000]
(A) Poling (B) Cupellation
(C) Distillation (D) Lavigation

73. What is the formula of diaspore ? [EAMCET(MBBS)-2000]
(A) $Al_2O_32H_2O$ (B) Al_2O_3 (C) $Al_2O_3.H_2O$ (D) $AlPO_4$

74. During the electrolytic reduction of alumina, the reaction at cathode is [EAMCET-2001]
(A) $2H_2O \rightarrow O_2 + 4H^+ + 4e^-$ (B) $3F^- \rightarrow 3F^- + 3e^-$
(C) $Al^{3+} + 3e^- \rightarrow Al$ (D) $2H^+ + 2e^- \rightarrow H_2$

75. Which one of the following is used to remove silicon dioxide in the Serpeck's process on purification of bauxite ? [EAMCET(MBBS)-2001]
(A) CaO (B) Na_2CO_3 (C) Coke (D) Nickel

76. What is the reaction occurring at the anode in Down's process for the extraction of sodium ? [EAMCET-2002]
(A) $4OH \rightarrow 2H_2O + O_2 + 4e^-$ (B) $Na^+ + e^- \rightarrow Na$
(C) $2\ Cl^- \rightarrow Cl + 2e^-$ (D) $NaOH \rightarrow Na^+ + OH^-$

77. Which one of the following reactions occurs at the anode in the Castner process of extracting sodium metal ? [EAMCET-2003]

(A) $H_2 \rightarrow 2H^+ + 2e^-$ (B) $2Cl^- \rightarrow Cl_2 + 2e^-$
(C) $4OH^- \rightarrow 2H_2O + O_2 + e^-$ (D) $Na^+ + e^- \rightarrow Na$

78.Which one of the following electrolytes is used in Down's process of extracting sodium metal ? [EAMCET-2003]

(A) NaCl + KCl + kF (B) NaCl
(C) NaOH + KCl + kF (D) NaCl + NaOH

79. During the electrolysis of cryolite, aluminium and fluorine are formed in__molar ratio [EAMCET-2003]

(A) 1:2 (B) 2:3 (C) 1:1 (D) 1:3

80. Alumino thermic process is used for metallurgy of [CPMT-96]

(A) Pb (B) Ag (C) Al (D) None of these

81. Which of the following pair does not distinguish by passing H_2S ? [CPMT-96]

(A) Hg, Pb (B) Cd, Pb (C) As, Cd (D) Zn, Mn

82. During smelting an additional substance is added which combines with impurities to form a fusible product which is known as - [CPMT-96]

(A) Slag (B) Mud (C) Gangue (D) Flux

83. Aluminium is most abundant in earth's crust, yet it is obtained from bauxite because - [CPMT-97]

(A) Bauxite is available in large quantity
(B) Easy extraction of aluminium from it
(C) Bauxite contains maximum aluminium
(D) Bauxite is less impure

84. Which metal can be obtained from electrolysis ? [CPMT-97]

(A) Cu (B) Mg (C) Cr (D) Ni

85. Froth flotation process is used for the concentration of [CPMT-98]

(A) Oxide ores (B) Sulphide ores
(C) Chloride ores (D) Amalgams

86. The heating of pyrites to remove sulphur is known as [CPMT-98]

(A) Bessemerisation (B) Calcination
(C) Roasting (D) Smelting

87. Which of the following are prepared by electrolytic method ? I-Mg , II-Sn, III-S, IV-F [CPMT-98]

(A) I and II (B) II and III
(C) III and IV (D) I and IV

88. In electrorefining process of a metal the impure metal is made the anode and a strip of pure metal the cathode during the electrolysis of an aqueous solution of a complex metal salt. This method cannot be used for refining of [CPMT-98]

(A) Al (B) Ag (C) Cu (D) Zn

89. Which one of the following fluxes is used to remove acidic impurities in metallurgical process ? [CPMT-98]

(A) Silica (B) Sodium chloride
(C) Lime-stone (D) Radium carbonate

90. Which process of purification is represented by the following equation ? [CPMT-98]

$$\underset{\text{(Impure)}}{Ti} + 2I_2 \xrightarrow{250^\circ C} Ti\,I_4 \xrightarrow{1400^\circ C} \underset{\text{(Pure)}}{Ti} + 2I_2$$

(A) Cupellation (B) Poling
(C) van Arkel's process (D) Zone refining

91. Sulphide ore of zinc is concentrated by [CPMT-98]

(A) Distillation (B) Electromagnetic process
(C) Froth flotation process (D) Gravity process

92. Electrometallurgical process (electrolysis of fused salt) is employed to extract [CPMT-98]

(A) Fe (B) Pb (C) Na (D) Ag

93. An important ore of iron is [CPMT-98]

(A) Pitchblende (B) Galena
(C) Malachite (D) Haematite

94. In extraction of copper we use - [CPMT-98]

(A) Cu_2S (B) Cuprite
(C) Silver argentocyanide (D) $CuFeS_2$

95. Brass is an alloy of Cu and - [CPMT-98]

(A) Pb (B) Bi (C) Zn (D) Sb

96. Bell metal is an alloy of - [CPMT-98]

(A) Zn and Cu (B) Zn and Pb
(C) Cu and Sn (D) Cu and Ni

97. The metal hydroxide which is soluble in excess of ammonium hydroxide is - [CPMT-98]

(A) $Fe(OH)_2$ (B) $Fe(OH)_3$
(C) $Al(OH)_3$ (D) $Cu(OH)_2$

98. AgCl on fusion with Na_2CO_3 gives- [CPMT-98]

(A)Ag_2CO_3 (B)Ag_2O (C)Ag (D)Silver carbide

99. The Cyanide process is used for the extraction of- [CPMT-98]

(A) Cu (B) Na (C) Ag (D) Al

100. Blister copper is- [CPMT-98]

(A) Pure copper (B) Alloy of copper
(C) Ore of copper (D) Copper containing 1%impurity

101. Which mineral is used in the isolation of radium ? [CPMT-98]

(A) Limestone (B) Pitchblende
(C) Rutile (D) Haematite

102. Cinnabar is an ore of - [CPMT-98]

(A) Copper (B) Nickel (C) Mercury (D) Tungsten

103. Calomel is - [CPMT-98]

(A) Hg_2Cl_2 and Hg (B) Hg_2Cl_2
(C) Hg and $HgCl_2$ (D) $HgCl_3$

104. 'Thomas-slag' is - [CPMT-98]
(A) $Ca_3(PO_4)_2$ (B) $MnSiO_3$ (C) $CaSiO_3$ (D) $FeSiO_3$

105. The important ore of zinc is - [CPMT-98]
(A) ZnS (B) $ZnCO_3$ (C) ZnO (D) (BaO+ZnO)

106. In the metallurgy of iron, when limestone is added to the blast furnace, the calcium ion ends up in - [CPMT-98]
(A) Slag (B) Gangue
(C) Metallic calcium (D) Calcium carbonate

107. Mond's process is used for preparing - [CPMT-98]
(A) Ni (B) NH_3 (C) Co (D) Mn

108. Turnbull's process is used for preparing- [CPMT-98]
(A) Ferric cyanide (B) Ferrous ferricyanide
(C) Ferrous cyanide (D) Ferri-ferrocyanide

109. In the extraction of Cu the reaction which takes place in Bessemer converter is [CPMT-99]
(A) $2CuFeS_2 + O_2 \rightarrow Cu_2S + FeS + SO_2$
(B) $2Cu_2S + 3O_2 \rightarrow 2Cu_2O + 2SO_2$.
(C) $2Cu_2O + Cu_2S \rightarrow 6Cu + SO_2$
(D) $2FeS + 3O_2 \rightarrow 2FeO + 2SO_2$

110. In the extraction of aluminium from bauxite cryolite is used to - [CPMT-2000]
(A) Decrease the boiling point of aluminium
(B) Purify the bauxite (C) Make the mass lighter
(D) Decrease melting point of alumina and increase conductivity

111. The final step for the extraction of copper from copper pyrites in Bessemer's converter involves the reaction [CPMT-2000]
(A) $4C_2O + FeS \rightarrow 8Cu + FeSO_4$
(B) $Cu_2S + 2Cu_2O \rightarrow 6Cu + SO_4$
(C) $2Cu_2O + 2FeS \rightarrow 4Cu + 2Fe + SO_2$
(D) $Cu_2S + 2FeO \rightarrow 2Cu + 2Fe + SO_2$

112. During the extraction of Fe, slag obtained is [CPMT-2000]
(A) FeO (B) $FeSiO_3$ (C) $MgSiO_3$ (D) $CaSiO_3$

113. Which of the following is an ore of aluminium ? [CPMT-2002]
(A) dolomite (B) azurite (C) bauxite (D) malachite

114. Blister copper is [CPMT-2002]
(A) Impure Cu (B) Cu alloy
(C) Pure Cu (D) Cu having 1% impurity

115. Refractory metals are used in construction of furnaces because [CPMT-2002]
(A) they can withstand high temperature
(B) they are chemically inert
(C) their melting point is high (D) none of these

116. Which of the following is not an ore of lead : [PMT-93]
(A) Galena (B) Anglesite (C) Calamine (D) Cerrusite

117. An example of halide ores is : [PMT-93]
(A) Galena (B) Bauxite (C) Cinnabar (D) Cryolite

118. After partial roasting the sulphide of copper is reduced by : [PMT-93]
(A) Reduction by carbon (B) Electrolysis
(C) Self-reduction (D) Cyanide Process

119. Reverberatory furnance is employed in the metallurgical process mainly for [PMT-94]
(A) reduction of oxide ores (B) smelting of sulphide ores
(C) conversion of chloride to sulphate
(D) getting magnetic materials

120. The process of ore dressing is carried out to [PMT-94]
(A) remove the siliceous materials
(B) add flux to the mineral
(C) convert the ore to oxide
(D) remove the poisonous impurities

121. In the manufacture of metallic sodium by the fused salt electrolysis (Down's process) a small amount of calcium chloride is added to [PMT-94]
(A) improve the electrical conduction
(B) increase the temperature of electrolysis
(C) bring down the melting temperature
(D) stablilize the metallic sodium

122. An important ore for iron is [PMT-94]
(A) malachite (B) pitch blende (C) galena (D) haematite

123. In the cyanide process for the extraction of silver, sodium cyanide is used to [PMT-94]
(A) convert silver into a soluble silver complex
(B) reduce silver (C) precipitate silver
(D) oxidize silver

124. The impurities associated with minerals used in metallurgy are called collectively [PMT-95]
(A) Slag (B) Flux (C) Gangue (D) Ore

125. Main process for the manufacture of sodium carbonate is [PMT-95]
(A) Castner Process (B) Solvay Process
(C) Down's Process (D) Nelson Process

126. Aluminium reacts with caustic soda to form [PMT-95]
(A) aluminium hydroxide (B) aluminium oxide
(C) sodium meta aluminate (D) sodium tetra aluminate

127. Cryolite is an ore of [PMT-96]

(A) iron (B) silver (C) zinc (D) aluminium

128. An important oxide ore of iron is [PMT-96]
(A) Haematite (B) Siderite (C) Limestone (D) Pyrites

129. The metal which is extracted by Parke's process is [PMT-96]
(A) Copper (B) Silver (C) Iron (D) Lead

130. Sodium metal is extracted by [PMT-96]
(A) Electrolysis of aqueous solution of sodium chloride
(B) Electrolysis of fused sodium chloride
(C) Heating sodium oxide with carbon
(D) Heating sodium oxide with hydrogen

131. Brass is an alloy of [PMT-96]
(A) Gold and Copper (B) Silver and Zinc
(C) Copper and Zinc (D) Copper and Aluminium

132. An ore like zinc blende is concentrated by [PMT-97]
(A) Froth flotation (B) Magnetic separation
(C) Leaching (D) Washing with water

133. Which of the following is obtained as slag in the metallurgy of copper (from copper - pyrites) ?[PMT-97]
(A) $CaSiO_3$ (B) $Ca_3(PO_4)_2$ (C) $FeSiO_3$ (D) $MnSiO_3$

134. If the impurity in a metal has a greater affinity for oxygen and is more easily oxidised than the metal, then the purification of metal may be carried out by - [PMT-97]
(A) Poling (B) Zone refining
(C) Electrolytic refining (D) Cupellation

135. Excess of NaOH solution reacts with zinc to form : [PMT-99]
(A) Zinc hydroxide (B) Zinc oxide
(C) Sodium zincate (D) Disodium zinc

136. Which one of the following is an ore of aluminium ? [PMT-99]
(A) Bauxite (B) Polomite (C) Neutral (D) Felspar

137. Permanent magnet is made from : [PMT-99]
(A) Cast iron (B) Steel
(C) Wrought iron (D) All of these

138. Heating pyrites to remove sulphur is called : [PMT-2000]
(A) Smelting (B) Calcination
(C) Liquation (D) Roasting

139. In the metallurgy of which of the following cupellation process is used ? [PMT-2000]
(A) Copper (B) Silver (C) Iron (D) Aluminium

140. Cryolite is used in the electrolytic extraction of aluminium [PMT-2000]
(A) to obtain more aluminium (B) to decompose bauxite
(C) to protect anodes (D) as a reducing agent

141. The important ore of lead is : [PMT-2000]
(A) Chalcopyrite (B) Haematite
(C) Galena (D) Bauxite

142. The iron obtained from blast furnace is : [PMT-2000]
(A) pig iron (B) wrought iron (C) soft iron (D) cast iron

143. The protection of steel by chrome plating is due to : [PMT-2001]
(A) Cathodic protection (B) Anodic protection
(C) Covering of steel surface
(D) Formation of alloy with iron

144. Which of the following metals is extracted by the electrometallurgical method ? [PMT-2001]
(A) Cu (B) Fe (C) Na (D) Ag

145. Aluminium is refined by : [PMT-2001]
(A) Serpeck's process (B) Baeyer's process
(C) Hall's process (D) Hoope's process

146. Heating of pyrite ore in air for oxidation of sulphur is called : [PMT-2001]
(A) Roasting (B) Calcination
(C) Smelting (D) Concentration

147. For which ore of the metal, froth flotation method is used for concentration [PMT-2001]
(A) Horn silver (B) Bauxite (C) Cinnabar (D) Haematite

148. The % of silver in German silver is : [PMT-2001]
(A) 100 % silver (B) 50 % silver
(C) 25 % silver (D) 0 % silver

149. At anode in the electrolysis of fused NaCl [PMT-2002]
(A) Na is oxidized (B) Cl^- is oxidized
(C) Cl is reduced (D) Na is reduced

150. At cathode the electrolysis of aqueous Na_2SO_4 gives [PMT-2002]
(A) Na (B) H_2 (C) SO_3 (D) SO_2

151. An important ore of copper is [PMT-2002]
(A) Bauxite (B) Galena (C) Siderite (D) Chalcopyrite

152. Cryolite is used in the electrolytic extraction of aluminum [PMT-2002]
(A) To obtain more aluminium (B) To reduce bauxite
(C) To decompose bauxite (D) To protect anode

153. Heating pyrite ores to remove sulphur is known as [PMT-2002]
(A) Liquation (B) Smelting (C) Calcination (D) Roasting

154. Cupellation process is used in the metallurgy of [PMT-2002]
(A) Copper (B) Silver (C) Iron (D) All ores are minerals

155. Which one of the following is a correct statement ?

[PMT-2002]

(A) A mineral cannot be an ore
(B) An ore cannot be a mineral
(C) All ores are minerals (D) all minerals are ores

156. Parke's process is used [PMT-2002]
(A) To extract silver using NaCN
(B) To extract copper using $CuFeS_2$
(C) To extract silver from argentiferrous lead
(D) To extract silver by forming amalgam

157. Calamine is a mineral which contains [PMT-2003]
(A) ZnS (B) ZnO (C) $ZnSiO_3$ (D) $ZnCO_3$

158. Which of the following methods is used for the concentration of copper pyrite? [PMT-2003]
(A) Froth flotation method (B) Gravitational method
(C) Electromagnetic method (D) All of these

159. Iron obtained from blast furnace is known as [PMT-2003]
(A) Wrought iron (B) Pig iron (C) Cast iron (D) Steel

160. The formula of Felspar is [PMT-2003]
(A) $K_2O.Al_2O_3.6SiO_2$ (B) $MgAl_2O_4$
(C) Na_3AlF_6 (D) $K_2O_3 . Al_2O_3 .6SiO_2$

161. Chili Saltpetre ore of sodium metal is. [PMT-2003]
(A) Na_2CO_3 (B) $NaNO_3$ (C) KNO_3 (D) NaCl

162. The froth flotation process is used for the concentration of [PET-93]
(A) Halide ores (B) Oxide ores
(C) Phosphate ores (D) Sulphide ores

163. In the manufacture of iron from haematite limestone is added to act as [PET-93]
(A) An oxidizing agent (B) A reducing agent
(C) Flux (D) Slag

164. The interatomic distances in H_2 and Cl_2 molecules are 74 and 198 pm respectively. The bond length of HCl is [PET-93]
(A) 272 pm (B) 136 pm (C) 124 pm (D) 248 pm

165. In the metallurgy of zinc, the zinc dust obtained from roasting and reduction of zinc sulphide contains some ZnO. How is this removed ? [PET-93]
(A) Absorbance of UV light andreemission of white light is employed
(B) Shock-cooling by contact with a shower of molten lead is done
(C) X-ray method is used (D) Smelting is employed

166. In the Down's process of obtaining sodium $CaCl_2$ is added [PET-93]
(A) To get high yield of sodium
(B) To improve the quality of sodium obtained
(C) To lower the operating temperature
(D) To act as a reducing agent

167. The reaction of $K_2Cr_2O_7$ with NaCl and conc. H_2SO_4 gives [PET-94]
(A) $CrCl_3$ (B) $CrOCl_2$ (C) $CrO_2 Cl_2$ (D) Cr_2O_3

168. Bauxite ore is concentrated by [PET-94]
(A) Froth flotation (B) Electromagnetic separation
(C) Chemical separation (D) Hydraulic separation

169. In the metallurgical extraction of zinc from ZnO the reducing agent used is [PET-94]
(A) Carbon monoxide (B) Sulphur dioxide
(C) Carbon dioxide (D) Nitric oxide

170. Cupellation is used for the purification of [PET-94]
(A) Na (B) Ag (C) Pb (D) Al

171. Chemical reduction is not suitable for converting [PET-94]
(A) Bauxite into aluminium (B) Cuprite into copper
(C) Haematite into iron (D) Zinc oxide into zinc

172. The reaction $2ZnS + 3O_2 \rightarrow 2ZnO + 2SO_2$ in the metallurgical process of zinc is called. [PET-94]
(A) Calcination (B) Cupellation
(C) Smelting (D) Roasting

173. Hoope's process is used in the refining of -[PET-95]
(A) Al (B) Zn (C) Ag (D) Cu

174. Red lead is [PET-95]
(A) Pb_3O_4 (B) PbO (C) PbO_2 (D) Pb_4O_3

175. Solder metal is [PET-95]
(A) Pb + Sb (B) Pb + Sn (C) Cu + Sn (D) Cu + Zn

176. In the manufacture of iron from haematite the function of lime stone is [PET-95]
(A) Reducing agent (B) Flux (C) Slag (D) Gangue

177. Blister copper is [PET-95]
(A) Pure copper (B) Copper ore
(C) 1% impure copper (D) Copper alloy

178. The purest form of iron is [PET-95]
(A) Cast iron (B) Pig iron (C) Wrought iron (D) Steel

179. The natural materials from which an element can be extracted economically are called [PET-96]
(A) Ores (B) Minerals
(C) Gangue (D) None of the above

180. When a metal is to be extracted from its ore, if the gangue associated with the ore is silica then [PET-96]
(A) An acidic flux is needed (B) A basic flux is needed
(C) Both acidic and basic flux are needed
(D) Neither of them is needed

181. An example of an oxide ore is [PET-96]
(A) Bauxite (B) Malachite (C) Zinc blende (D) Felspar

182. In the Hoopes process, for refining of aluminium the fused materials form three different layers and they

remain separated during electrolysis also. This is because [PET-96]
(A) The upper layer is kept attracted by the cathode and the lower layer is kept attracted by the anode
(B) There is special arrangement in the cell to keep the layers separate
(C) The 3 layers have different densities
(D) The 3 layers are maintained at different temperatures

183. In order to refine "Blister copper" it is melted in a furnance and is stirred with green logs of wood. The purpose is [PET-96]
(A) To expel the dissolved gases in blister copper
(B) To bring the impurities to surface and oxidize them
(C) To increase the carbon content of copper
(D) To reduce the metallic oxide impurities with hydrocarbon gases liberated from the wood

184. The ion that can be precipitated by HCl as well as H_2S is [PET-97]
(A) Pb^{2+} (B) Fe^{3+} (C) Zn^{2+} (D) Cu^{2+}

185. The metallurgical operation in which metal is obtained in fused state, is called : [PET-97]
(A) Froth flotation process (B) Calcination
(C) Roasting (D) Smelting

186. Which of the following metals is obtained by electrolytic reduction process ? [PET-97]
(A) Fe (B) Cu (C) Ag (D) Al

187. The metal extracted by electrolysis of its fused salt is : [PET-98]
(A) Iron (B) Lead (C) Sodium (D) Copper

188. The important oxide ore of iron is : [PET-98]
(A) Siderite (B) Haematite (C) Pyrites (D) Bauxite

189. The metal extracted by the cyanide process is : [PET-98]
(A) Silver (B) Copper (C) Iron (D) Sodium

190. Malachite is an ore of : [PET-98]
(A) Iron (B) Zinc (C) Copper (D) Mercury

191. German silver is an alloy of : [PET-98]
(A) Copper, zinc and nickel
(B) Copper, zinc and tin (C) Silver, copper and gold
(D) Iron, chromium and nickel

192. Malachite is an ore of which metal ? [PET-2000]
(A) Copper (B) Zinc (C) Iron (D) Silver

193. Which one of the following metals will not forn an amalgam ? [PET-2000]
(A) Gold (B) Silver (C) Zinc (D) Iron

194. The process of extracting the metal from its ore is called : [PET-2001]
(A) Refining (B) Concentration
(C) Leaching (D) Metallurgy

195. Copper pyrites ore is concentrated by : [PET-2001]
(A) Electromagnetic method (B) Gravity method
(C) Froth flotation method (D) All the above methods

196. The purpose of smelting an ore is to : [PET-2001]
(A) Reduce it (B) Oxidise it
(C) Obtain an alloy (D) Separate volatile impurities

197. In smelting of iron, which of the following reactions takes place in Blast furnace at 400°C - 600°C [PET-2002]
(A) $CaO + SiO_2 \rightarrow CaSiO_3$
(B) $2FeS + 3O_2 \rightarrow 2Fe + SO_2$
(C) $FeO + SiO_2 \rightarrow Fe\ SiO_3$
(D) $Fe_2O_3 + 3CO \rightarrow 2Fe + 3CO_2$

198. During extraction of silver which of the following is formed [PET-2002]
(A) $Na\ [Ag(CN)_2]$ (B) $Na_2[Ag(CN)_2]$
(C) $Na_4[Ag(CN)_2]$ (D) None of these

199. Which one of the following is correct [PET-2002]
(A) A mineral cannot be an ore
(B) An ore cannot be a mineral
(C) All minerals are ores (D) All ores are minerals

200. The common method of extraction of metals from oxide ores is [PET-2002]
(A) Reduction with carbon
(B) Reduction with hydrogen
(C) Reduction with aluminium (D)Electrolytic method

201. Which oxide of manganese is amphoteric ? [PET-2002]
(A) MnO_2 (B) Mn_2O_3 (C) Mn_2O_7 (D) MnO

202. Which of the following has maximum no. of unpaired electrons (atomic no. of Fe 26) ? [PET-2003]
(A) Fe (B) Fe (ii) (C) Fe (iii) (D) Fe (iv)

203. The number of uncoupled electrons in cobalt atom (at no. 27) is : [PET-2003]
(A) 2 (B) 3 (C) 4 (D) 8

204. Which of the follwing is not an 'ore' of magnesium? [PET-2003]
(A) Dolomite (B) Carnallite (C) Gypsum (D) Magnesite

205A process used for the concentration of ore is
(A) Froth flotation (B) Roasting [PET-2003]
(C) Electrolysis (D) None

206. Which metal is used as a reducing agent in smelting?
(A) C (B) Zn (C) Al (D) none of these [PET-2003]

207. Parke's process is used in the extraction of [PET-2003]
(A) Iron (B) Sodium (C) Silver (D) Zinc

208. Which of the followoing elements in alloyed with copper to form brass. [PET-2003]
(A) Lead (B) Bismuth (C) Zinc (D) Antimony

209. Cryolite helps in [BHU-2003]

(A) increasing the electrical conductivity
(B) decreasing the electrical conductivity
(C) lowering the melting point
(D) increasing the melting point

210. Process of heating the ore in air to remove sulphur is [AFMC-2003]

(A) Calcination (B) Roasting
(C) Smelting (D) None of these

211. In electrorefining of Copper, some gold is deposited as [AFMC-2003]

(A) Anode mud (B) Cathode
(C) Cathode mud (D) None of these

212. Copper and Tin are costituents of [AFMC-2003]

(A) Bronze (B) Brass
(C) German silver (D) None of these

213. Which of the following has the highest carbon content? [AFMC-2003]

(A) Stainless steel (B) Cast iron
(C) Wrought iron (D) None of these

214. Electrolytic refining of aluminium is done by .... process. [AFMC-2003]

(A) Serpeck's (B) Hall's
(C) Baeyer's (D) Hoop's

215.Pb and Sn metals are extracted from their ores by [IIT-2004]

(A) Self reduction and carbonreductionrespectively.
(B) Carbon reduction and self reduction respectively.
(C) Electrolytic and self reduction respectively.
(D) Self reduction and electrolytic reduction respectively.

Metallurgy

1.	(A)	51.	*	101.	(B)	151.	(D)	201.	(A)
2.	(A)	52.	(B)	102.	(C)	152.	(C)	202.	(C)
3.	(A)	53.	(D)	103.	(B)	153.	(D)	203.	(B)
4.	(C)	54.	(B)	104.	(A)	154.	(B)	204.	(C)
5.	(B)	55.	(A)	105.	(A)	155.	(C)	205.	(A)
6.	(A)	56.	(B)	106.	(A)	156.	(C)	206.	(A)
7.	(D)	57.	(C)	107.	(A)	157.	(D)	207.	(C)
8.	(D)	58.	(A)	108.	(B)	158.	(C)	208.	(C)
9.	(D)	59.	(B)	109.	(C)	159.	(B)	209.	(C)
10.	(D)	60.	(D)	110.	(D)	160.	(A)	210.	(B)
11.	(C)	61.	(B)	111.	(B)	161.	(B)	211.	(C)
12.	(C)	62.	(B)	112.	(D)	162.	(D)	212.	(A)
13.	(B)	63.	(B)	113.	(C)	163.	(C)	213.	(B)
14.	(C)	64.	(D)	114.	(D)	164.	(B)	214.	(D)
15.	(C)	65.	(B)	115.	(A)	165.	(D)	215.	(A)
16.	(B)	66.	(D)	116.	(C)	166.	(C)		
17.	(D)	67.	(A)	117.	(D)	167.	(C)		
18.	(C)	68.	(B)	118.	(C)	168.	(C)		
19.	(B)	69.	(B)	119.	(B)	169.	(A)		
20.	(C)	70.	(A)	120.	(A)	170.	(B)		
21.	(C)	71.	(C)	121.	(C)	171.	(A)		
22.	(C)	72.	(B)	122.	(D)	172.	(D)		
23.	(D)	73.	(C)	123.	(A)	173.	(A)		
24.	(C)	74.	(C)	124.	(C)	174.	(A)		
25.	(C)	75.	(C)	125.	(B)	175.	(B)		
26.	(C)	76.	(C)	126.	(C)	176.	(B)		
27.	(B)	77.	(C)	127.	(D)	177.	(C)		
28.	(A)	78.	(A)	128.	(A)	178.	(C)		
29.	(B)	79.	(B)	129.	(B)	179.	(A)		
30.	(A)	80.	(D)	130.	(B)	180.	(B)		
31.	(A)	81.	(A)	131.	(C)	181.	(A)		
32.	(C)	82.	(A)	132.	(A)	182.	(A)		
33.	(B)(D)	83.	(A)	133.	(C)	183.	(D)		
34.	(B)(C)	84.	(B)	134.	(D)	184.	(A)		
35.	(A)(C)	85.	(B)	135.	(C)	185.	(D)		
36.	(C)	86.	(C)	136.	(A)	186.	(D)		
37.	(C)	87.	(D)	137.	(B)	187.	(C)		
38.	(C)	88.	(A)	138.	(D)	188.	(B)		
39.	(B)	89.	(C)	139.	(B)	189.	(A)		
40.	(A)	90.	(C)	140.	(B)	190.	(C)		
41.	(B)	91.	(C)	141.	(C)	191.	(A)		
42.	(C)	92.	(C)	142.	(A)	192.	(A)		
43.	(B)	93.	(D)	143.	(D)	193.	(D)		
44.	(C)	94.	(D)	144.	(C)	194.	(D)		
45.	(D)	95.	(C)	145.	(D)	195.	(C)		
46.	*	96.	(C)	146.	(A)	196.	(A)		
47.	*	97.	(D)	147.	(C)	197.	(D)		
48.	(C)	98.	(C)	148.	(D)	198.	(A)		
49.	*	99.	(C)	149.	(B)	199.	(D)		
50.	(A)	100.	(D)	150.	(B)	200.	(A)		

---

46. Copper
47. Fluorine
49. Sea water
51. CO

CHEMICAL ANALYSIS

1. Detection of anions or acid radicals

(a) Carbonate (CO_3^{2-}): $Ca(OH)_2 + CO_2 \longrightarrow \underset{\text{milky}}{CaCO_3}$

When excess of CO_2 is passed it becomes colourless

$CaCO_3 + CO_2 + H_2O \longrightarrow \underset{\text{cal. bicarbonate}}{2Ca(HCO_3)_2}$

(b) Sulphite (SO_3^{-2}):

$Na_2SO_3 + H_2SO_4 \longrightarrow Na_2SO_4 + SO_2 + H_2O$

$(SO_2 + H_2O \longrightarrow H_2SO_3)$

$K_2Cr_2O_7 + 3H_2SO_3 + H_2SO_4 \longrightarrow K_2SO_4 + \underset{\text{green}}{Cr_2(SO_4)_3} + 4H_2O$

(c) Sulphide (S^{2-}):

$Na_2S + H_2SO_4 \longrightarrow Na_2SO_4 + H_2S$

$Pb(CH_3COO)_2 + H_2S \longrightarrow \underset{\text{black}}{PbS} + 2CH_3COOH$

(d) Nitrite (NO^{2-}):

$2NaNO_2 + H_2SO_4 \longrightarrow Na_2SO_4 + 2HNO_2$

$(HNO_2 \longrightarrow H_2O + NO)$

$NO + O \longrightarrow NO_2$; $FeSO_4 + NO \longrightarrow \underset{\text{brown}}{[Fe(NO)]SO_4}$

The brown colour is due to nitrosoferrous sulphate.

(e) Chloride (Cl^-):

$NaCl + H_2SO_4 \longrightarrow NaHSO_4 + HCl$

$4HCl + MnO_2 \longrightarrow MnCl_2 + 2H_2O + 2Cl_2$

$NaCl + AgNO_3 \longrightarrow AgCl + NaNO_3$

$AgCl + 2NH_4OH \longrightarrow \underset{\text{soluble complex}}{[Ag(NH_3)_2]Cl} + 2H_2O$

Chromyl chloride test :

$K_2Cr_2O_7 + 4NaCl + 6H_2SO_4 \longrightarrow \underset{\text{red vapour}}{2CrO_2Cl_2} + 2KHSO_4 + NaHSO_4 + 3H_2O$

CrO_2Cl_2 is chromyl chloride

$CrO_2Cl_2 + Pb(CH_3COO)_2 \longrightarrow \underset{\text{yellow ppt.}}{PbCrO_4}$

This test fails with chlorides of Pb, Ag, Sn and Hg. Bromide and Iodide do not respond to this test.

(f) Nitrate (NO^{3-}): Brown ring test

$NaNO_3 + H_2SO_4 \longrightarrow NaHSO_4 + HNO_3$

$Fe^{2+} + 2HNO_3 \longrightarrow Fe^{3+} + NO + HNO_2$

$FeSO_4 + NO \longrightarrow [Fe(NO)]SO_4$

(g) Sulphate (SO_4^{2-})

$Na_2SO_4 + BaCl_2 \longrightarrow 2NaCl + BaSO_4$ (white ppt.)

(h) Phosphate (PO_4^{3-}):

$Mg_3(PO_4)_2 + 6HNO_3 \longrightarrow 3Mg(NO_3)_2 + 2H_3PO_4$

$H_3PO_4 + 12(NH_4)_2MoO_4 + 21HNO_3 \longrightarrow (NH_4)_3PO_4 . 12MoO_3$ (canary yellow ppt of ammonium phosphomolybdate) $+ 12H_2O + 21NH_4NO_3$

Arsenic compounds also give this test. Therefore the test should also be performed after the second group by evaporating the filtrate.

2 Basic radicals or Cations

Basic redicals are divided into six groups on the basis of upon the solubility product.

Ist group: Ag^+, Pb^{2+}, Hg^+ (ous), precipitated as chlorides with dil. HCl.

2nd group: 2-A : Hg^{2+}, Pb^{2+}, Cd^{2+}, Bi^{3+}, Cu^{2+}

2 - B : As^{3+} & As^{5+} ; Sb^{3+} & Sb^{5+} ; Sn^{2+} & Sn^{4+}

Precipitated as sulphides in acidic medium

2 - A radicals are not soluble in yellow ammonium sulphide. Ammonium sulphide does not form complexes with stannous sulphide and therefore the precipitates are treated with yellow ammonium sulphide to get their thio-salts.

3rd group: Fe^{3+}, Cr^{3+}, Al^{3+}. They are precipitated as hydroxides (these hydroxides have lower solubility product than the corresponding hydroxides of other groups).

4th group: Ni^{2+}, Co^{2+} Mn^{2+}, Zn^{2+}. They are precipitated as sulphides in alkaline medium. The sulphides of this group have higher solubility product compared to the sulphides of the 2nd group,

5th group: Ba^{2+}, Sr^{2+} , Ca^{2+}. They are precipitated as carbonates.

6th group: Mg^{2+}. Precipitated as phosphate

TITRATIONS

Acid-Base titrations:

Typical acid-base titration uses the neutralisation reaction between oxonium ions and hydroxide ions

$H_3O^+ (aq.) + OH^- (aq.) \rightleftharpoons 2H_2O$ (l)

Phenolphthalein: It is a colourless weak acid.

$HPh \rightleftharpoons H^+ + Ph^-$. The indicator is colourless in acid and pink in alkali solution. It can be used between 8.00 - 9.8 pH range.

Methyl orange:It is a weak base: $MeOH \rightleftharpoons Me^+ + OH^-$

It remains yellow in alkaline medium while in acid solution, it gives red colour

Titration	Indicator used
SA Vs S B	Phenolphthalein, methyl red, methyl orange
W A Vs S B	Phenolphthalein,
S A Vs W B	Methyl red, methyl orange
Na_2CO_3 Vs S A	Phenolphthalein, methyl orange

Redox titrations

(i) Titration of iorn (II) with $K_2Cr_2O_7$ (VI)

$6Fe^{2+} (aq) + Cr_2O_7^{2-} (aq) + 14H^+ (aq) \rightarrow 6Fe^{3+} (aq) + 2Cr^{3+} + (aq) + 7H_2O$ (l)

(ii) Titration of oxalate ions with potassium permanganate (VII)

$5C_2O_4^{2-} (aq) + 2MnO_4^- (aq) + 8H^+ (aq) \rightarrow 10CO_2 (g) + 2Mn^{2+} (aq) + 4H_2O$ (l)

Titrations involving iodine:

(i) Iodine ions are easily oxidised

$2I^- (aq) - 2e^- \rightarrow I_2 (s)$

e.g. by iodate (V) ions

$IO_3^- (aq) + 5I^- (aq) + 6H^+ (aq) \rightarrow 3I_2 (s) + 3H_2O$ (l)

(ii) Iodine can be titrated with thiosulphate solution using starch as an indicator

$2S_2O_3^{2-} (aq) + I_2 (s) \rightarrow S_4O_6^{2-} (aq) + 2I^- (aq)$

Silver nitrate titrations

Silver ions can be titrated with chloride ions using a solution of chromate (VI) ions, CrO_4^{2-} as indicator

$Ag^+ (aq) + Cl^- (aq) \rightarrow AgCl (s)$

Solid silver chromate (Ag_2CrO_4) is red and is precipitated once the majority of chloride ions are removed.

1. 15 ml. 12 N- H_2SO_4, 30 ml. 10N- HCl, 10 ml. 2N- HNO_3 and 100 ml. 3N - NaOH were mixed and the volume was made up to 1000 ml. Find the normality of the mixture.

Sol. $\text{Normality} = \frac{(N_1V_1+N_2V_2\text{------}) - (N_3V_3+N_4V_4\text{---})}{\text{Total Volume}}$

$\frac{(15 \times 12 + 30 \times 10 + 10 \times 2) - (3 \times 100)}{1000} = \frac{200}{1000}$

$= 0.2$ or N /5

2. 1.325g., of anhydrous Na_2CO_3 was dissolved in 250 ml. of distilled water. 25 ml, of this solution required 20 ml. H_2SO_4 for complete neutralisation. Find out the normality of the acid. How much water is to be added to 450 ml. of this acid solution to make it N/12 solution?

Sol. Normality of Na_2CO_3 solution

$N = \frac{W \times 1000}{E \times V} = \frac{1.325 \times 1000}{53 \times 250} = \frac{1}{10}$

Normatity of the acid, $N_1V_1 = N_2V_2$

$= \frac{1}{10} \times \frac{25}{20} = \frac{1}{8} = 0.125$ N

Amount of water to be added, $N_2 V_2 = N_3 V_3$

$12 \times 0.125 \times 450 = V_3 = 675$ ml.

225 ml. of water is to be added to make it $\frac{N}{12}$ solution

3. A sample of Na_2CO_3, H_2O weighing o.62 g is added to 100 ml. of 0.1N- H_2SO_4 solution. Will the resulting solution be acidic, basic or neutral?

Sol. Equivalent wt. of

$Na_2CO_3.H_2O = \frac{\text{mol.wt.}}{2} = \frac{124}{2} = 62$

0.62 g. of Na_2CO_3. H_2O is equal to $\frac{0.62}{62} = \frac{1}{100}$ of eq.wt.

The amount of H_2SO_4 in 1000 ml of 0.1N H_2SO_4 is

$W = \frac{ENV}{1000} = \frac{49 \text{ X } 0.1 \text{ X } 100}{1000} = 0.49$ g.

0.49 g of $H_2SO_4 = \frac{0.49}{49} = \frac{1}{100}$ th of eq.wt.

Since the equivalent weight of Na_2CO_3. H_2O and H_2SO_4 is 1/100 th of their equivalent weights and because they are a base and an acid, the resulting solution is neutral.

4. 25C.C. of a solution of iron (II) ammonium sulphate needed excatly 20 C.C. of a solution of $K_2Cr_2O_7$ (VI) of concentration 0.05 mol dm^{-3}. What was the concentration of the iron (II) solution?

Sol: First, we work out the number of moles of dicromate (VI) ions used in the titration

1000 C.C. of soultion contains 0.05 mol.

So, 20 C.C. of solution contains, $0.05 \times \frac{20}{1000} = 0.001$ mol.

From the equation, we know that 1 mol $Cr_2O_7^{2-}$ reacts with 6 mol Fe^{2+}.

$\therefore$ the number of moles of iron (II) should be 6 times i.e. 0.006 mol

Concentration of iron (II) $= 0.006 \times \frac{1000}{25} = 0.24$ mol $=$ .24 M

5. What volume of the manganate (VII) solution of concentration 0.038 mol dm^{-3} would be needed to give the end point with 25 CC of sodium oxalate of 0.1 mol dm^{-3} concentration.

Sol: No. of moles of oxalate ions in 25 CC

$= 0.1 \times \frac{25}{1000} = 0.0025$ mol

From the eq., we know that 5 mol $C_2O_4^{2-} \equiv 2$ mol MnO_4^- or 1 mol $C_2O_4^{2-} \equiv 0.4$ mol MnO_4^-

$\therefore$ 0.0025 mol 4 0.0025 x 0.4 = 0.001 mol MnO_4^-

0.038 mol are contained in 1000 CC of solution

0.001 mol are contained in $= \frac{1000 \times 0.01}{0.038} = 26.32$

The volume of manganate (VII) solution required is 26.32 CC

6. An impure sample of iron of mass 2.55 g was dissolved in dilute H_2SO_4 and the solution made upto 250 CC. The solution contained Fe(II) ions together with the impurities. 25 CC of this solution was titrated with potassium manganate (VII) solution of 0.02 mol dm^{-3} concentration. The end point was reached at 28.50 CC titre reading. (i) How many moles of manganate (VII) solution were used in the titration? (ii) How many moles of Fe(II) ions will react with 1 mol of manganate (VII) ions? (iii) How many moles of Fe(II) must have been present in 25 CC of solution? (iv) How many moles of Fe(II) ions were there in the 250 CC stock solution? (v) what is the mass of this number of moles of Fe(II)? and (vi) What was the percentage purity of the sample of iron?

Sol: (i) No. of moles of $MnO_4^- = 0.02 \times \frac{28.50}{1000} = 5.7 \times 10^{-4}$ mol

(ii) Manganate (VII) ions take 5 electrons each Fe(II) ion can lose only one electron

$\therefore$ 1 moles of $MnO_4^- \equiv 5$ mol Fe^{2+}

(iii) There were $5 \times 5.7 \times 10^{-4} = 2.85 \times 10^{-3}$ mol

(iv) No. of mol present in 250 CC stock solution =

$2.85 \times 10^{-3} \times 10 = 2.85 \times 10^{-2}$ mol

(v) Mass $= 2.85 \times 10^{-2} \times 56 = 1.596$g

(vi) Percentage purity $= \frac{1.596}{2.55} \times 100 = 62.6\%$

7. 25 CC of a solution of potassium iodate (V), KIO_3, were placed in a titration flask. 20 CC of a solution of KI and 10 CC of dil H_2SO_4 were added. The iodine liberated required 24 CC of 0.2 mol dm^{-3} solution of sodium thio sulphate to reach the end point. What was the concentration of the original iodate solution?

Sol: From the equations, we know

1 mol $IO_3^- \equiv$ 3 mol I_2

and 1 mol $I_2 \equiv$ 2 mol $S_2O_3^{2-}$

Therefore, 3 mol $I_2 \equiv$ 6 mol $S_2O_3^{2-}$

and overall 1 mol $IO^{3-} \equiv$ 6 mol $S_2O_3^{2-}$

No. of moles of $S_2O_3^{2-}$ ions used $= 0.2 \times \frac{24}{1000} = 0.0048$mol

No. of moles of IO_3^- present $= 0.0048 \times \frac{1}{6} = 0.0008$ mol

concentration of $KIO_3 = 0.0008 \times \frac{1000}{25} = 0.032$ mol

8. 25 CC of KCl solution required 22.5 CC of 0.02 mol dm^{-3} of $AgNO_3$ to reach the end point using yellow potassium chromate (VI) as indicator. Find out the concentration of the KCl solution?

Sol: No. of moles of Ag^+ added $= 0.02 \times \frac{22.5}{1000} = 0.00045$ mol

from the equation, we know that 1 mol $Ag^+ \equiv$ 1 mol Cl^-

$\therefore$ No. of moles of ions = 0.00045

concentration of KCl solution $= 0.00045 \times \frac{1000}{25} = 0.018$ mol

1. The composition of 'Golden spangles' is [CBSE-90]
(A) $PbCrO_4$ (B) PbI_2 (C) As_2S_2 (D) $BaCrO_4$

2. When SO_2 solution is passed over which of the following it turns green ? [CBSE-93]
(A) $K_2Cr_2O_7$ (B) H_2SO_4 (C) SO_3 (D) P_2O_5

3. A bivalent metal has the equivalent weight of 12. The molecular weight to its oxide will be [AIIMS-94]
(A) 24 (B) 32 (C) 36 (D) 40

4. The weight of a metal of equivalent weight 12 which will give 0.475 g of its chloride is [AIIMS-94]
(A) 0.12 g (B) 0.16 g (C) 0.18 g (D) 0.24 g

5. The amount of zinc (at.wt.=65) necessary to produce 224 ml of H_2 by the reaction with an acid will be [AIIMS-95]
(A) 0.065 gm (B) 0.65 gm (C) 6.5 gm (D) 7.5 gm

6. Which of the following compounds is not soluble in HNO_3 ? [AIIMS-95]
(A) CuS (B) CdS (C) PbS (D) AgCl

7. Nessler's reagent is used in the test of [AIIMS-95]
(A) NH_3 (B) NH_4^+ (C) NH_4Cl (D) All of these

8. How many gm of silver will be displaced from a solution of $AgNO_3$ by 4 gm of magnesium ? [AIIMS-95]
(A) 4 gm (B) 16 gm (C) 18 gm (D) 36 gm

9. The property which can be classified as an intensive property is [AIIMS-95]
(A) Mass (B) Temperature (C) Volume (D) Heat capacity

10. The volume of carbon dioxide gas evolved at S.T.P. by heating 7.3 gm of $Mg(HCO_3)_2$ will be [AIIMS-95]
(A) 1120 ml (B) 2000 ml (C) 2240 ml (D) 2340 ml

11. The number of molecules in 4.25 g of ammonia is approximately [AIIMS-96]
(A) 0.5 x 10^{23} (B) 1.5 x 10^{23}
(C) 3.5 x 10^{23} (D) 2.5 x 10^{23}

12. The percentage of oxygen in NaOH is [AIIMS-96]
(A) 4% (B) 8% (C) 16% (D) 40%

13. Which of the following indicator is known as metal indicator ? [AIIMS-96]
(A) Phenol red (B) Methyl orange
(C) Phenolphthalein (D) Erichrome-Black-T

14. Which of the following cannot give Iodometric titration ? [AIIMS-97]
(A) Fe^{3+} (B) Cu^{2+} (C) Pb^{2+} (D) Ag^+

15. $K_2[HgI_4]$ detects the ion/group [AIIMS-97]
(A) NH_2 (B) NO (C) NH_4^+ (D) Cl^-

16. The equivalent weight of an acid is equal to [AIIMS-98]
(A) Molecular weight x acidity

(B) Molecular weight X basicity
(C) Molecular weight/basicity
(D) Molecular weight/acidity

17. The specific heat of a metal is 0.16. Its approximate atomic weight would be [AIIMS-99]
(A) 16 (B) 32 (C) 40 (D) 64

18. The compound used for gravimetric estimation of copper (II) is - [AIIMS-2003]
(A) $Cu_2(SCN)_2$ (B) Cu_2O (C) Cu_2I_2 (D) Cu_2CO_3

19. 12 g of Mg (At. wt. 24) will react completely with an acid to give : [JIPMER-90]
(A) One mole of H_2 (B) One half mole of H_2
(C) One mole of O_2 (D) None

20. How many gms of NaOH will be required to prepare 250 ml of 0.1 M solution : [JIPMER-90]
(A) 1gm (B) 4gms (C) 40gms (D) 10gms

21. Which one among the following pairs of ions can not be separated by H_2S in dilute HCl ? [JIPMER-90]
(A) Bi^{3+}, Sn^{4+} (B) Al^{3+}, Hg^{2+}
(C) Ni^{2+}, Cu^{2+} (D) Zn^{2+}, Cu^{2+}

22. 10ml of conc. H_2SO_4 (18 Molar) is diluted to 1 litre. The approximate strength of dilute acid is :[JIPMER-90]
(A) 0.18N (B) 0.36N (C) 18.0N (D) 0.09N

23. Which one of the following can be used in place of NH_4Cl for the identification of III group radicals : [JIPMER-90]
(A) NH_4NO_3 (B) $(NH_4)_2SO_4$
(C) $(NH_4)_2CO_3$ (D) NaCl

24. In alkaline condition $KMnO_4$ reacts as
$2KMnO_4 + 2KOH \rightarrow 2K_2MnO_4 + H_2O + O$
Therefore its equivalent weight will be : [JIPMER-90]
(A) 31.5 (B) 52.7 (C) 72.0 (D) 158.0

25. 30 ml of acid solution is neturalised by 15ml of 0.2 N base. The strength of acid solution is : [JIPMER-91]
(A) 0.1N (B) 0.15N (C) 0.4N (D) 0.3N

26. The volume of 0.1M H_2SO_4 that is needed to completely neutralize 30 ml of 0.2 M NaOH solution is : [JIPMER-91]
(A) 30ml (B) 15 ml (C) 40 ml (D) 60ml

27. The simplest formula of a compound containing 50% of element X (at. wt. 10) and 50% of element Y (at.wt = 20) is : [JIPMER-91]
(A) XY (B) X_2Y (C) XY_3 (D) X_2Y_3

28. The volume of 0.1M H_2SO_4 that is needed to completely neutralize 40ml of 0.2 M NaOH is : [JIPMER-91]
(A) 10ml (B) 20ml (C) 40ml (D) 80ml

29. Which of the following radicals will not be precipitated by passing H_2S in concentrated acid solution : [JIPMER-91]
(A) Cu (B) Sb (C) As (D) Cd

30. Which of the following dissolves in hot conc.NaOH solution : [JIPMER-91]
(A) Fe (B) Zn (C) Cu (D) Ag

31. 2.76 gm of silver carbonate on being strongly heated yield a residue weighing [JIPMER-92]
(A) 2.48 g (B) 2.16 g (C) 2.32 g (D) 2.64 g

32. The crystal of NaCl is : [JIPMER-92]
(A) Face centered cubi (B) Body centered cubic
(C) Hexagonal (D) Tetragonal

33. Phenolphthalein is not a good indicator for titrating [JIPMER-92]
(A) NaOH against oxalic acid (B) $FeSO_4$ against $KMnO_4$
(C) NaOH against $KMnO_4$ (D) NaOH against H_2SO_4

34. For preparing one litre of N/10 H_2SO_4 solution we need H_2SO_4 [JIPMER-92]
(A) 98 gm (B) 10 gm (C) 100 gm (D) 4.9 gm

35. What is the normality of 1 M solution of H_3PO_4 ? [JIPMER-92]
(A) 0.5 N (B) 1 N (C) 2 N (D) 3 N

36. When potassium dichromate is heated with conc. H_2SO_4 and soluble metal chloride, orange vapours are produced. These are due to : [JIPMER-93]
(A) Cr_2O_3 (B) $CrOCl_2$ (C) CrO_4^{2-} (D) CrO_2Cl_2

37. Crucible for the gravimetric estimation is made of [JIPMER-94]
(A) Silica (B) Glass (C) Iron (D) Copper

38. A bottle of dry ammonia and a bottle of dry hydrogen chloride connected through a long tube are opened simultaneously. At both ends the white ammonium chlorides formed. The first ring will be [JIPMER-95]
(A) Near the hydrogen chloride bottle
(B) At the center of the tube (C) Near the ammonia bottle
(D) Throughout the length of the tube

39. The equivalent weight of Mohr's salt is [JIPMER-97]
(A) 56 (B) 158 (C) 278 (D) 392

40. Which of the following compounds is sparingly soluble in ammonia ? [JIPMER-98]
(A) AgI (B) AgBr (C) AgCl (D) $CuCl_2$

41. The weight of sodium carbonate required to prepare 500 ml of a semi-normal solution is [JIPMER-98]
(A) 13.25g (B) 26.5 g (C) 53 g (D) 6.125g

42. 20 ml of a N solution of $KMnO_4$ just reacts with 20 ml of solution of oxalic acid. The weight of oxalic acid crystals in 1L of the solution is [JIPMER-98]
(A) 31.5g (B) 126 g (C) 63 g (D) 6.3 g

43. Cu^{2+} and Ag^{+} are both present in the same solution. To precipitate one of the two ions and leave the other in solution we should add [JIPMER-99]
(A) HNO_3(aq) (B) H_2S(aq)
(C) HCl(aq) (D) NH_4NO_3(aq)

44. On heating NaCl + $K_2Cr_2O_7$ + H_2SO_4 conc. the gas that comes out is [JIPMER-2000]
(A) O_2 (B) Cl_2 (C) $CrOCl_2$ (D) CrO_2Cl_2

45. The equivalent weight of iron in Fe_2O_3 would be [JIPMER-2001]
(A) 112 (B) 26.66 (C) 121.1 (D) 127.5

46. Vapour density of a metal chloride is 66. Its oxide contains 53% metal. The atomic weight of the metal is [JIPMER-2001]
(A) 2.706 (B) 27.06 (C) 54 (D) 21

47. Which of the following is the best scientific method to test the presence of water in a liquid [JIPMER-2001]
(A) use of anhydrous copper sulphate
(B) use of litmus paper (C) taste (D) smell

48. The ion that can be precipitated by HCl as well as H_2S is [JIPMER-2002]
(A) Pb^{2+} (B) Fe^{3+} (C) Zn^{2+} (D) Cu^{2+}

49. 100 ml of 0.2 M H_2SO_4 is added to 100 ml of 0.2 M NaOH. The resulting solution will be [BHU-96]
(A) Acidic (B) Basic (C) Neutral (D) Slightly basic

50. A sample of $Na_2CO_3.H_2O$ weighing 0.62g is added to 100ml of 0.1 N H_2SO_4 solution. What will the resulting solution be? [BHU-97]
(A) Acidic (B) Neutral (C) Basic (D) None of these

51. Specific heat of metal is 2.3 and its chloride contains 87% of chlorine. What is the molecular weight of the metal ? [BHU-98]
(A) 24 (B) 54 (C) 62 (D) 27

52. Which of the following combines with Fe(II) ions to form a brown complex ? [BHU-98]
(A) N_2O (B) NO (C) N_2O_3 (D) N_2O_5

53. In the titration of Na_2CO_3 and HCl the indicator used is [AFMC-92]
(A) Methyl orange (B) Methylene blue
(C) Litmus (D) Phenolphthalin

54. In the electrolysis of NaCl [AFMC-92]
(A) Cl is oxidised at anode
(B) Cl^- is reduced at anode (C) Cl^- is oxidised at cathode
(D) Cl^- is neither reduced nor oxidised

55. When iodine is dissolved in CCl_4 the colour that results is [AFMC-93]
(A) Brown (B) Violet (C) Colourless (D) Bluish green

56. In the titration of strong acid and weak base indicator used is [AFMC-94]
(A) Methyl orange (B) Phenolphthalein
(C) Thymolphthalein (D) None of these

57. Calcination is used in metallurgy for removal of [AFMC-95]
(A) Water and sulphide (B) Water and CO_2
(C) CO_2 and H_2S (D) H_2O and H_2S

58. Indicator used in titration of oxalic acid and NaOH is [AFMC-96]
(A) Methyl orange (B) Methyl red
(C) Phenolphthalein (D) None of these

59. In the equation
$4M+8CN^-+2H_2O+O_2 \rightarrow 4[M(CN_2)]^-+4OH^-$
identify the metal M. [AFMC-98]
(A) Copper (B) Iron (C) Gold (D) Zinc

60. The molecular formula of cryolite is [AFMC-99]
(A) $Na_2Al_2O_3$ (B) Fe_3O_4 (C) Na_3AlF_6 (D) All of these

61. How many grams of dibasic acid (Mo. wt. = 200) should be present in 100 ml of the aqueous solution to give strength N/10 ? [AFMC-99]
(A) 5g (B) 10g (C) 2g (D) 1g

62. N/10 NaOH 100 ml and N/5 HCl 50 ml are mixed then pH of solution is [AFMC-2002]
(A) 0 (B) 7 (C) Less than 7 (D) Greater than 7

63. The reagents NH_4Cl and aqueous NH_3 will precipitate [IIT-91]
(A) Ca^{2+} (B) Al^{3+} (C) Bi^{3+} (D) Mg^{2+} (E) Zn^{2+}

64. The formula of deep red liquid formed on warming dichromate with KCl in concentrated sulphuric acid is (CrO_2Cl_2) [IIT-93]

65. Which compound does not dissolve in hot dilute HNO_3? [IIT-96]
(A) HgS (B) PbS (C) CuS (D) CdS

66. In nitroprusside ion the iron and NO exist as Fe^{II} and NO^+ rather than Fe^{III} and NO. These forms can be differentiated by [IIT-98]
(A) estimating the concentration of iron
(B) measuring the concentration of CN^-
(C) measuring the solid state magnetic moment
(D) thermally decomposing the compound

67. Which of the following statement (s) is (are) correct when a mixture of NaCl and $K_2Cr_2O_7$ is gently warmed with conc.H_2SO_4? [IIT-98]
(A) a deep red vapour is evolved
(B) The vapour when passed into NaOH solution gives a yellow solution of Na_2CrO_4
(C) Chlorine gas is evolved
(D) Chromyl chloride is formed

68. Identify the correct order of solubility of Na_2S, CuS and ZnS in aqueous medium [IIT-2002]

(A) CuS > ZnS > Na_2S (B) ZnS > Na_2S > CuS

(C) Na_2S > CuS > ZnS (D) Na_2S > ZnS > CuS

69. When H_2S is passed through Hg_2S we get [AIEEE-2002]

(A) HgS (B) HgS + Hg_2S (C) Hg_2S + Hg (D) Hg_2S

70. How do we differentiate between Fe^{3+} and Cr^{3+} in group III ? [AIEEE-2002]

(A) By taking excess of NH_4OH solution

(B) By increasing NH_4^+ ion concentration

(C) By decreasing OH^- ion concentration

(D) Both (b) and (c)

71. Which one of the following statements is correct ? [AIEEE-2003]

(A) Manganese salts give a violet borax bead test in the reducing flame

(B) From a mixed precipitate of AgCl and AgI, ammonia solution dissolves only AgCl

(C) Ferric ions give a deep green precipitate on adding potassium ferrocyanide solution

(D) On boiling a solution having K^+, Ca^{2+} and HCO_3^- ions we get a precipitate of $K_2Ca(CO_3)_2$

72. Concentrated hydrochloric acid when kept in open air sometimes produces a cloud of white fumes. The explanation for it is that [AIEEE-2003]

(A) Concentrated hydrochloric acid emits strongly smelling HCl gas all the time

(B) Oxygen in air reacts with the emitted HCl gas to form a cloud of chlorine gas

(C) Strong affinity of HCl gas for moisture in air results in forming of droplets of liquid solution which appears like a cloudy smoke

(D) Due to strong affinity for water concentrated hydrochloric acid pulls moisture forms droplets of water and hence the cloud

73. Bottles containing C_6H_5I and $C_6H_5CH_2I$ lost their original labels. They were labeled A and B for testing. A and B were separately taken in test tubes and boiled with NaOH solution. The end solution in each tube was made acidic with dilute HNO_3 and then some $AgNO_3$ solution was added. Substance B gave a yellow precipitate. Which one of the following statements is true for this experiment? [AIEEE-2003]

(A) A was C_6H_5I (B) A was $C_6H_5CH_2I$

(C) B was C_6H_5I (D) Addition of HNO_3 was unnecessary

74. The weight of oxalic acid crystalline $H_2C_2O_4.2H_2O$ required to prepare 500ml of 0.2N solution is .... [EAMCET-91]

75. The weight in grams of KCl (mol. wt. = 74.5) in 100ml of a 0.1 M KCl solution is [EAMCET-96]

(A) 74.5 (B) 7.45 (C) 0.745 (D) 0.0745

76. Silver chloride dissolves in excess of NH_4OH. The cation present in this solution is [EAMCET-98]

(A) Ag^+ (B) $[Ag(NH_3)_2]^+$

(C) $[Ag(NH_3)_4]^+$ (D) $[Ag(NH_3)_6]^+$

77. How many grams of phosphoric acid is required to completely neutralise 120 grams of sodium hydroxide ? [EAMCET(MBBS)-99]

(A) 49 (B) 98 (C) 196 (D) 9.8

78. What is the volume (in litres) of 0–1 MH_2SO_4 required to completely neutralise l litre 0.5 M NaOH ? [EAMCET(MBBS)-99]

(A) 5 (B) 2.5 (C) 0.5 (D) 10

79. 250 ml of a sodium carbonate solution contains 2.65 grams of Na_2CO_3. 10ml of this solution is added x ml of water to obtain 0.001 M Na_2CO_3 solution. What is the value of x in ml ? [EAMCET-2001]

(A) 1000 (B) 990 (C) 9990 (D) 90

80. Iron sulphide is heated in air to form A, an oxide of sulphur. A is dissolved in water to give an acid. The basicity of this acid is [EAMCET-2002]

(A) 2 (B) 3 (C) 1 (D) Zero

81. In qualitative analysis NH_4Cl is added before NH_4OH [CPMT-96]

(A) To decrease $[OH]^-$ conc.

(B) To increase $[OH^-]$ conc.

(C) For making HCl (D) Statement is wrong

82. In a compound the number of atoms of Cr and O are 4.8×10^{10} and 9.6×10^{10} respectively. Its empirical formula is – [CPMT-97]

(A) Cr_2O_3 (B) CrO_2 (C) Cr_2O_4 (D) None of these

83. Identify the indicator that is used to titrate $NaCO_3$ solution with HCl [CPMT-97]

(A) Phenolphthalein (B) Dil. H_2SO_4

(C) Methyl orange (D) None of these

84. In a sample of pure compound Na = 0.0887 mole, O = 0.132 mole, C = 2.65×10^{22} atoms. The compound is [CPMT-97]

(A) Na_2CO_3 (B) $Na_3O_2C_5$

(C) $Na_{0.0877}O_{0.132}C_{2.65x10}$ (D) NaCO

85. A solution of SO_2 in water reacts with H_2S to precipitate sulphur. Here SO_2 acts as- [CPMT-98]

(A) An oxidising agent (B) A reducing agent

(C) An acid (D) A catalyst

86. In the chemical reaction-

$K_2Cr_2O_7 + XH_2SO_4 + YSO_2 \rightarrow K_2SO_4 + Cr_2(SO_4)_3 + ZH_2O$

X, Y and Z are - [CPMT-98]

(A) 1,3,1 (B) 4,1,4 (C) 3,2,3 (D) 2,1,2

87. The metal ion which is precipitated when H_2S is passed with HCl [CPMT-2001]

(A) Zn^{++} (B) Ni^{++} (C) Cd^{++} (D) Mn^{++}

88. When sodium oxalate is treated with dilute H_2SO_4 the gases produced are [CPMT-2001]

(A) CO (B) CO_2

(C) Both CO and CO_2 (D) Both CO_2 and SO_2

89. A mixture of carbonates of two bivalent metals contains 13.5% of one metal. If 2.5 g of the mixture liberates 1.33 g CO_2 the % by weight of the other metal is [CPMT-2001]

(A) 13.6 (B) 12 (C) 13.44 (D) 15

90. When $1x10^{-3}$ mole of the chloride of an element Y was completely hydrolysed it was found that the resulting solution required 20 ml of 0.1 M aqueous silver nitrate for complete precipitation of the chloride ion. Element Y could be [CPMT-2001]

(A) Aluminium (B) Phosphorous

(C) Silicon (D) Sulphur

91. Which indicator is suitable for strong acid and weak base titration? [CPMT-2002]

(A) phenolphthalein (B) methyl red

(C) methyl orange (D) litmus

92. A metal sulphide forms white precipitate with NH_4OH dissolves and again forms white ppt. with H_2S. Metal is [CPMT-2003]

(A) Ba (B) Al (C) Sr (D) Zn

93. When 100 ml N-NaOH solution and 10 ml. of 10N-sulphuric acid solution are mixed together the resulting solution will be : [PMT-93]

(A) alkaline (B) weakly acidic

(C) strongly acidic (D) neutral

94. 0.16 gm. of a dibasic acid required 25 ml. decinormal NaOH solution for complete neutralisation. The molecular weight of the acid is : [PMT-93]

(A) 32 (B) 64 (C) 128 (D) 256

95. When SO_2 is passed through aqueous H_2S sulphur is precipitated. In this reaction SO_2 acts as [PMT-94]

(A) an oxidising agent (B) a reducing agent

(C) an acid (D) a complexing agent

96. On addition of aqueous NaOH to a salt solution a white gelatinous precipitate is formed which dissolves in excess alkali. The salt solution contains [PMT-94]

(A) Chromium ions (B) Aluminium ions

(C) Barium ions (D) Iron ions

97. A mixture containing Cu^{2+} and Ni^{2+} can be separated for identification by [PMT-94]

(A) passing H_2S in acid medium

(B) passing H_2S in alkaline medium

(C) passing H_2S in neutral medium

(D) passing H_2S in dry mixture

98. The indicator used in the titration of acetic acid with sodium hydroxide for quantitative estimation is[PMT-94]

(A) Phenolphthalein (B) Methyl orange

(C) Methyl red

(D) A mixture of methyl red and methyl orange

99. What weight of Sodium Hydroxide is required to neutralise 100 ml of 0.1 N - HCl ? [PMT-94]

(A) 4.0 gm (B) 0.04 gm (C) 0.4 gm (D) 2.0 gm

100. The brown ring complex compound is formulated as $[Fe(H_2O_5)_5NO]SO_4$ the oxidation state of iron is [PMT-94]

(A) + 1 (B) - 1 (C) + 2 (D) + 3

101. On passing H_2S gas in II group some-times the solution turns milky. It indicates the presence of : [PMT-95]

(A) Acidic salt (B) Oxidising agent

(C) Thiosulphate (D) Reducing agent

102. 4.0 gram of caustic soda is dissolved in 100 cc solution. The normality of solution is : [PMT-95]

(A) 1.0 (B) 0.1 (C) 0.5 (D) 4.0

103. Chlorine atom differs from chloride ion in the number of which of the following? [PMT-95]

(A) Protons (B) Neutrons

(C) Electrons (D) Both protons and electrons

104. Which of the following compounds gives red precipitate with $AgNO_3$? [PMT-95]

(A) KI (B) K_2CrO_4 (C) NaBr (D) $NaNO_3$

105. Pb^{++}, Cu^{++} and Ni^{++} ions are present in a given acidic solution. On passing hydrogen sulphide gas through this solution the avilable precipitate will contain : [PMT-96]

(A) PbS and NiS (B) PbS and CuS

(C) CuS and ZnS (D) CuS and NiS

106. Al^{3+}, Fe^{3+}, Zn^{2+} and Ni^{2+} ions are present in an acidic solution. Excess NH_4Cl solution is added followed by addition of ammonium hydroxide solution. The avilable precipitate will contain [PMT-96]

(A) $Zn(OH)_2$ and $Ni(OH)_2$ (B) $Al(OH)_3$ and $Fe(OH)_3$

(C) $Zn(OH)_2$ and $Al(OH)_3$ (D) $Ni(OH)_2$ and $Fe(OH)_3$

107. 0.53 gm. of Na_2CO_3 has been dissolved in 100 ml of a sodium carbonate solution. The normality of the solution will be : [PMT-96]

(A) N/5 (B) N/2 (C) N/10 (D) N

108. The amount of crystalline oxalic acid required in 250 ml N/10 oxalic acid solution will be [PMT-96]

(A) 1.575 gm (B) 3.150 gm (C) 6.300 gm (D) 0.787 gm

109. An ammonium solution gives a white ppt. with

$AgNO_3$ solution. The ppt dissolves in dilute ammonia due to the formation of - [PMT-97]
(A) $Ag\ NO_3$ (B) NH_4NO_3
(C) $[Ag(NH_3)_2]\ Cl$ (D) $[Ag\ (NH_3)_2]\ Br$

110. To an acid solution of an anion a few drops of $KMnO_4$ solution are added. Which of the following if present will not decolourise the $KMnO_4$ solution - [PMT-97]
(A) NO_2^- (B) S^{2-} (C) Cl^- (D) CO_3^{2-}

111. A 100 ml solution of 0.1 - HCl was titrated with 0.2 N - NaOH solution. The titration was discontinued after adding 30 ml of NaOH solution. The remaining titration was completed by adding 0.25 N - KOH solution. The volume of KOH required for completing the titration is - [PMT-97]
(A) 16 ml (B) 32 ml (C) 35 ml (D) 70 ml

112. What weight of hydrated oxalic acid should be added for complete neutralisation of 100 ml of 0.2 N-NaOH solution - [PMT-97]
(A) 0.45 g (B) 0.90 g (C) 1.08 g (D) 1.26 g

113. How many ml of 1 M H_2SO_4 is required to neutralize 10 ml of 1 M NaOH solution : [PMT-99]
(A) 20.0 ml (B) 2.5 ml (C) 5.0 ml (D) 10.0 ml

114. Which one of the following sulphides is yellow ? [PMT-99]
(A) Zinc sulphide (B) Cadmium sulphide
(C) Nickel sulphide (D) Lead sulphide

115. How many grams of NaOH will be required to neutralize 12.2 grams of benzoic acid ? [PMT-99]
(A) 40 gms (B) 4 gms (C) 16 gms (D) 12.2 gms

116. A white solid 'A' on heating gives off a gas which truns lime water milky. The residue is yellow when hot but turns white on cooling. This solid 'A' is : [PMT-99]
(A) Zinc sulphate (B) Zinc carbonate
(C) Lead sulphate (D) Lead carbonate

117. Which of the following is not precipitated as sulphide by passing H_2S in presence of conc HCl ? [PMT-2000]
(A) copper (B) arsenic (C) cadmium (D) lead

118. Which of the following sulphate is insoluble in water? [PMT-2000]
(A) $CuSO_4$ (B) $CdSO_4$ (C) $PbSO_4$ (D) $Bi_2(SO_4)_3$

119. Phenolphthalein is most suitable indicator for the titration of : [PMT-2001]
(A) CH_3COOH and NH_4OH (B) CH_3COOH and NaOH
(C) HCl and NH_4OH (D) H_2CO_3 and NH_4OH

120. Ring test for nitrate is performed by adding freshly prepared ferrous sulphate solution and conc H_2SO_4. The brown ring in this test is due to : [PMT-2001]
(A) $FeSO_4.\ NO_2$ (B) $FeSO_4.\ NO$ (C) $FeSO_4$ (D) $Fe_2\ (SO_4)_3$

121. H_2S does not precipitate in the presence of 1 M acetic acid : [PMT-2001]
(A) Sb (iii) (B) Sn (ii) (C) Bi (iii) (D) Zn (ii)

122. An aqueous solution of an inorganic salt on treatment with HCl gives a white precipitate. This solution contains: [PMT-2001]
(A) Hg_2^{2+} (B) Hg^{2+} (C) Zn^{2+} (D) Cd^{2+}

123. Ferric ion forms a prussian blue coloured precipitate due to formation of : [PMT-2001]
(A) $K_4\ [Fe(CN)_6]$ (B) $Fe_4\ [Fe(CN)_6]_3$
(C) $Fe\ (CNS)_3$ (D) $K_3\ [Fe(CN)_6]$

124. On performing a borax-bead test with a given inorganic mixture for qualitative analysis the colour of the bead was found to be emerald green both in oxidising and reducing flame. It indicates the possibility of the presence of : [PMT-2001]
(A) Co^{2+} (B) Ni_2^{2+} (C) Cr^{3+} (D) Cu^{2+}

125. When H_2S gas is passed through the HCl containing aqueous solutions of $CuCl_2$, $HgCl_2$, $BiCl_2$ and $CoCl_2$, it does not precipitate out [PMT-2002]
(A) CuS (B) HgS (C) Bi_2S_3 (D) CoS

126. Nitrate is confirmed by ring test. The brown colour of the ring is due to the formation of [PMT-2002]
(A) Ferrous nitrate (B) Ferrous nitrite
(C) $FeSO_4.NO$ (D) $FeSO_4.NO_2$

127. Mark the correct statement [PMT-2002]
(A) I group basic radicals precipitate as chlorides
(B) IV group basic radicals precipitate as sulphides
(C) V group basic radicals precipitate as carbonates
(D) All the above statements are correct

128. What volume of water is to be added to 50 ml of 4N NaOH solution to obtain a 1 N solution ? [PMT-2002]
(A) 100 ml (B) 150 ml (C) 200 ml (D) 250 ml

129. 50 ml of 10 N H_2SO_4, 25 ml 12 N HCl and 40 ml 5 N HNO_3 were mixed together and the volume of the mixture was made upto 1000 ml by adding water. The normality of this obtained solution will be [PMT-2002]
(A) 1 N (B) 2 N (C) 3 N (D) 4 N

130. The reaction of NaCl and $K_2Cr_2O_7$ mixture on heating with conc H_2SO_4 results in [PMT-2002]
(A) CrO_2Cl_2 (B) $CrOCl_2$ (C) CrO_2Cl (D) $(CrOCl)_2$

131. Phenolphthalein is not a right indicator for which of the following titrations : [PMT-2003]
(A) Potassium hydroxide and sulphuric acid
(B) Sodium hydroxide and oxalic acid
(C) Potassium carbonate and hydrochloric acid
(D) None of these

132. In which of the following groups group reagent H_2S gas is common in the analysis of basic radicals :

[PMT-2003]

(A) II and III group (B) I and II group
(C) III and IV group (D) II and IV group

133. The brown ring test is performed for the qualitative detection of [PET-93]
(A) Bromides (B) Iodides (C) Nitrates (D) Phosphates

134. AgCl precipitate dissolves in ammonia due to the formation of [PET-93]
(A) $[Ag(NH_4)_2]OH$ (B) $[Ag(NH_4)_2]Cl$
(C) $[Ag(NH_3)_2]OH$ (D) $[Ag(NH_3)_2]Cl$

135. What is the value of carbonate hardness of water sample if 100 ml of it took 5.00 ml of 0.09 N HCl solution ? (Na_2CO_3 = 106) [PET-93]
(A) 4.50 mg – eq / litre (B) 447.00 mg – eq/litre
(C) 0.042 mg – eq/litre (D) 1.80 mg – eq/litre

136. Which of the following is soluble in yellow ammonium sulphide [PET-94]
(A) CuS (B) CdS (C) SnS (D) PbS

137. To neutralise 25 ml of 0.25 M Na_2CO_3 solution how much volume of 0.5 MHCl is required [PET-94]
(A) 12.5 ml (B) 25 ml (C) 37.5 ml (D) 50 ml

138. Sodium borate on reaction with H_2SO_4 (conc) and C_2H_5OH gives a compound 'A' which burns with green flame. The compound 'A' is [PET-94]
(A) $H_2B_4O_7$ (B) $(C_2H_5)_2B_4O_7$
(C) H_3BO_4 (D) $(C_2H_5)_3BO_3$

139. 15 ml of N/10 NaOH solution completely neutralises 12 ml of H_2SO_4 solution. The normality of H_2SO_4 solution will be - [PET-95]
(A) N/5 (B) N/10 (C) N/8 (D) N

140. In the titration of CH_3COOH with NaOH the indicator used is - [PET-95]
(A) Methyl orange (B) Methyl red
(C) Phenolphthalein (D) Litmus

141. 100 cm^3 of 0.1 NHCl solution is mixed with 100 cm^3 of 0.2 N NaOH solution the resulting solution is [PET-96]
(A) 0.1 N and the solution is basic
(B) 0.05 N and the solution is basic
(C) 0.1 N and the solution is acidic
(D) 0.05 N and the solution is acidic

142. For preparing 0.1 N solution of a compound from its impure sample of which the percentage purity is known the weight of the substance required will be [PET-96]
(A) More than the theoretical weight
(B) Less than the theoretical weight
(C) Same as theoretical weight (D) None of the above

143. A salt on treatment with dil HCl gives a pungent smelling gas and a yellow precipitate. The salt gives green flame when tested. The salt solution gives a yellow precipitate with potassium chromate. The salt is [PET-96]
(A) $NiSO_4$ (B) BaS_2O_3 (C) PbS_2O_3 (D) $CuSO_4$

144. Which of the following oxides of nitrogen reacts with ferrous sulphate to form a dark brown compound ? [PET-97]
(A) N_2O (B) NO (C) NO_2 (D) N_2O_3

145. By passing H_2S gas in acidified $KMnO_4$ solution we get [PET-97]
(A) K_2S (B) S (C) K_2SO_3 (D) MnO_2

146. Which of the following is soluble in yellow ammonium sulphide ? [PET-97]
(A) CuS (B) CdS (C) SnS (D) PbS

147. How many ml of 1(M) H_2SO_4 is required to neturalise 10ml of 1 (M) NaOH solution ? [PET-97]
(A) 2.5 (B) 5.0 (C) 10.0 (D) 20.0

148. 50ml 10N – H_2SO_4, 25ml 12 N – HCl and 40 ml 5N–HNO_3 were mixed together and the volume of the mixture was made 1000 ml by adding water. The normality of the resultant solution will be [PET-98]
(A) 1N (B) 2N (C) 3N (D) 4N

149. H_2S gas was passed through an acidic solution containing Pb^{2+}, Zn^{2+}, Cu^{2+} and Ni^{2+} ions. The precipitate will consist of : [PET-98]
(A) ZnS and PbS (B) PbS and NiS
(C) NiS and CuS (D) CuS and PbS

150. Lead sulphate is soluble : [PET-99]
(A) In conc nitric acid
(B) In conc hydrochloric acid
(C) In a solution of ammonium acetate (D) In water

151. What will be the molarity of a solution containing 5g of sodium hydroxide in 250 ml solution ? [PET-99]
(A) 0.5 (B) 1.0 (C) 2.0 (D) 0.1

152. If 100 ml of 1N-sulphuric acid were mixed with 100 ml of 1M-sodium hydroxide the solution will be: [PET-99]
(A) Acidic (B) Basic (C) Neutral (D) Slightly acidic

153. Which one of the following sulphides is only completely precipitated when the acidic solution is made dilute ? [PET-2000]
(A) HgS (B) PbS (C) CdS (D) CuS

154. Which of the following combines with Fe^{2+} ions form a brown complex ? [PET-2000]
(A) N_2O (B) NO (C) N_2O_3 (D) N_2O_5

155. 0.126 g of an acid requires 20 ml of 0.1 N NaOH for complete neutralisation. The equivalent weight of the acid is : [PET-2001]
(A) 45 (B) 53 (C) 40 (D) 63

156. The ratio of amounts of H_2S needed to precipitate all the metal ions from 100 ml of 1 M $AgNO_3$ and 100 ml of 1 M $CuSO_4$ is : [PET-2001]
(A) 1 : 2 (B) 2 : 1 (C) Zero (D) Infinity

157. Precipitate of AgCl dissolves in NH_4OH due to formation of a complex : [PET-2001]
(A) $[Ag(NH_4)_2]Cl$ (B) $[Ag(NH_4)_3]Cl$
(C) $[Ag(NH_3)_2]Cl$ (D) $[Ag(NH_3)_2OH]$

158. If NaOH is added to an aqueous solution of zinc ions a white precipitate appears and on adding excess of NaOH the precipitate dissolves. In this solution zinc exists in the [PET-2002]
(A) Cationic part (B) Anionic part
(C) Both in cationic and anionic parts
(D) There is no zinc in the solution

159. Colourless solutions of the following four salts are placed separately in four different test tubes and a strip of copper is dipped in each one of the these. Which solution will turn blue? [PET-2002]
(A) KNO_3 (B) $AgNO_3$ (C) $Zn(NO_3)_2$ (D) $ZnSO_4$

160. 20 ml of a solution of a weak monobasic acid neutralises 22.18 ml of a solution of NaOH and 20 ml of N/10 HCl neutralises 21.5 ml of the same NaOH solution. The normality of the acid is nearly [PET-2002]
(A) 10 N (B) 1 N (C) 0.10 N (D) 100 N

161. A sample of $Na_2CO_3.H_2O$ weighing 0.62g is added to 100 ml of 0.1 N H_2SO_4 solution. What will the resulting solution? [PET-2003]
(A) Acidic (B) Neutral (C) Basic (D) None of these

162. On passing 0.5 Faraday of electricity through NaCl the amount of Cl liberated at cathode is [PET-2003]
(A) 35.5 gm (B) 17.75 gm (C) 71 gm (D) 142 gm

163. When H_2S is passed through Hg_2S we get. [PET-2003]
(A) HgS (B) HgS + Hg_2S (C) Hg_2S + Hg (D) Hg_2S

164. Concentrated hydrochloric acid when kept in open air sometimes produces a cloud of white fumes. The explanation for it is that [PET-2003]
(A) Concentrated hydrochloric acid emits strongly smelling HCl gas all the time
(B) Oxygen in air reacts with the emitted HCl gas to form a cloud of chlorine gas
(C) Strong affinity of HCl gas for moisture in air results in forming of droplets of liquid solution which appears like a cloudy smoke
(D) Due to strong affinity for water concentrated hydrochloric acid pulls moisture forms droplets of water and hence the cloud

165. 250 ml of a sodium carbonate solution contains 2.65 grams of Na_2CO_3. If 10 ml of this solution is diluted to one litre what is the concentration of the resultant solution ? (molecular weight of Na_2CO_3 = 106) [PET-2003]
(A) 0.1 M (B) 0.001 M (C) 0.01 M (D) 10^{-4}M

166. The reagent used in the preparation of aspirin from salicylic acid is.... [PET-2003]
(A) $SOCl_2$/ Pyridine (B) CH_3COOH/HCl
(C) $(CH_3CO)_2O$ conc.H_2SO_4 (D) $CH_3Cl/AlCl_3$

167. 'Thermit' is a mixture of X parts of ferric oxide and Y parts of aluminium powder. X, Y respectively are : [PET-2003]
(A) 3, 1 (B) 3, 2 (C) 1,1 (D) 4.2,3

168. Bleaching powder is a compound having a formula- [PET-2003]
(A) $CaCl_2$ (B) CaClO (C) $CaOCl_2$ (D) $CaOCl_3$

169. Which of the following lead oxides is 'Sindhur' [PET-2003]
(A) PbO (B) PbO_2 (C) Pb_2O_3 (D) Pb_3O_4

170. A solution of a salt in dilute sulphuric acid produces deep blue colour with starch iodine solution. It shows the presence of which of the following [PET-2003]
(A) Br^- (B) I^- (C) NO_3^- (D) NO_2^-

171. To a 25 ml of H_2O_2 solution excess of acidified solution of KI was added. The liberated I_2 require 20ml of 0.3 M hypo solution of H_2O_2 will be : [PET-2003]
(A) 1.34ml (B) 1.52ml (C) 1.44ml (D) 2.42ml

172. Which of the folllowing has the least solubility [BHU-2003]
(A) CoS (B) Bi_2S_3 (C) PbS (D) Ag_2S

173. In NaCl, Na^+ is called [AFMC-2003]
(A) Acid radical (B) Basic radical
(C) Free radical (D) Salt

174. Precipitation of Group 4 cations takes place when H_2S is [AFMC-2003]
(A) Highly ionized (B) Less ionized
(C) Not ionized (D) None of these

175. A solid compound 'X' on heating gives CO_2 gas and a residue. The residue mixed with water forms 'Y'. On passing an excess of CO_2 through 'Y' in water, a clear solution, 'Z', is obtained. On boiling 'Z', compound 'X' is reformed. The compound 'X' is [CBSE-PMT-2004]
(A) $Ca(HCO_3)_2$ (B) $CaCO_3$ (C) Na_2CO_3 (D) K_2CO_3

176. Which of the following gives white precipitate with $MgCl_2$ on boiling? [IIT-2004]
(A) CO_3^{2-} (B) HCO_3^- (C) NO_3^- (D) SO_4^{2-}

Qualitative Analysis of Acid, Base Radicals and Volumetric Titrations

1.	(A)	51.	(D)	101.	(B)	151.	(A)
2.	(A)	52.	(B)	102.	(A)	152.	(C)
3.	(D)	53.	(A)	103.	(C)	153.	(D)
4.	(C)	54.	(A)	104.	(B)	154.	(B)
5.	(B)	55.	(B)	105.	(B)	155.	(D)
6.	(D)	56.	(A)	106.	(B)	156.	(A)
7.	(D)	57.	(B)	107.	(C)	157.	(C)
8.	(D)	58.	(C)	108.	(A)	158.	(B)
9.	(B)	59.	(C)	109.	(C)	159.	(B)
10.	(A)	60.	(C)	110.	(D)	160.	(C)
11.	(D)	61.	(D)	111.	(A)	161.	(B)
12.	(D)	62.	(B)	112.	(D)	162.	(B)
13.	(D)	63.	(B)(C)	113.	(C)	163.	(C)
14.	(A)	64.	*	114.	(B)	164.	(C)
15.	(C)	65.	(A)	115.	(B)	165.	(B)
16.	(C)	66.	(C)	116.	(B)	166.	(C)
17.	(C)	67.	(A)(B)(D)	117.	(C)	167.	(A)
18.	(A)	68.	(D)	118.	(C)	168.	(C)
19.	(B)	69.	(C)	119.	(B)	169.	(D)
20.	(D)	70.	(D)	120.	(B)	170.	(B)
21.	(A)	71.	(B)	121.	(D)	171.	(A)
22.	(B)	72.	(C)	122.	(A)	172.	(D)
23.	(A)	73.	(A)	123.	(B)	173.	(B)
24.	(D)	74.	*	124.	(C)	174.	(A)
25.	(A)	75.	(C)	125.	(D)	175.	(B)
26.	(A)	76.	(B)	126.	(C)	176.	(B)
27.	(C)	77.	(B)	127.	(D)		
28.	(C)	78.	(B)	128.	(B)		
29.	(D)	79.	(B)	129.	(A)		
30.	(B)	80.	(A)	130.	(A)		
31.	(B)	81.	(A)	131.	(D)		
32.	(B)	82.	(B)	132.	(D)		
33.	(B)	83.	(C)	133.	(C)		
34.	(D)	84.	(A)	134.	(D)		
35.	(D)	85.	(A)	135.	(A)		
36.	(D)	86.	(A)	136.	(B)		
37.	(A)	87.	(C)	137.	(B)		
38.	(A)	88.	(B)	138.	(D)		
39.	(D)	89.	(C)	139.	(C)		
40.	(A)	90.	(D)	140.	(C)		
41.	(A)	91.	(C)	141.	(B)		
42.	(C)	92.	(D)	142.	(A)		
43.	(C)	93.	(D)	143.	(C)		
44.	(D)	94.	(C)	144.	(B)		
45.	(B)	95.	(A)	145.	(B)		
46.	(B)	96.	(B)	146.	(C)		
47.	(A)	97.	(A)	147.	(B)		
48.	(A)	98.	(A)	148.	(A)		
49.	(A)	99.	(C)	149.	(D)		
50.	(B)	100.	(C)	150.	(B)		

64. CrO_2Cl_2 74. 6.3 grams

Classification and Nomenclature of Organic Compounds

Organic compounds based on structure are classified as:

Organic Compounds
- Acyclic or open Chain
  - Straight Chain $CH_3-CH_2-CH_2-CH_3$
  - Branched $CH_3-CH(CH_3)-CH_3$
- Cyclic
  - Homo Cyclic (Carbo cyclic)
    - Alicyclic 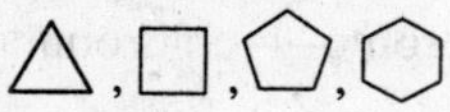
    - Aromatic
      - Benzenoid: Phenol 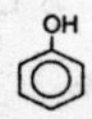, Benzoic acid 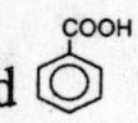
      - Non-benzenoid: Tropylium cation ,

        Tropolone 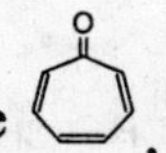, Azulene 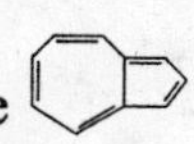
  - Hetero Cyclic : Pyridine 

Aromatic compounds follow Hückel's Rule: (4n + 2)– π electrons

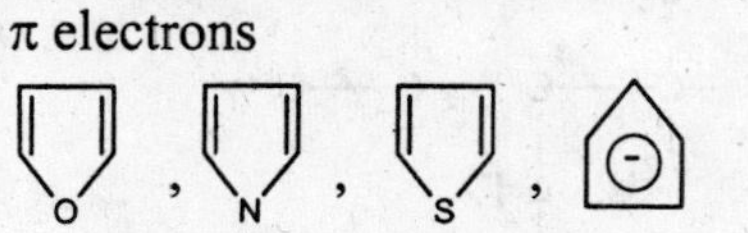

Furan, Pyrrole, Thiophene, Pentadienyl anion

In the first three cases, the lone pair(s) of electrons of hetero atom are in conjugation with the π – electrons and take part in the formation of delocalized π – orbital of six electrons.

Based on Functional groups:

1. Hydrocarbons
   - Saturated (alkanes)
   - Unsaturated
     - Alkenes
     - Alkynes
2. Haloalkanes
   - Aliphatic
     - Primary $CH_3-CH_2-CH_2X$
     - Secondary $CH_3-CHX-CH_3$
     - Tertiary $CH_3-C(CH_3)_2-X$
   - Aromatic
     - Nuclear or aryl 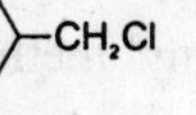
     - Side chain
3. Alkanols
   - Primary $CH_3-CH_2-CH_2-CH_2OH$
   - Secondary $CH_3-CHOH-CH_2-CH_3$
   - Tertiary $CH_3-C(CH_3)_2-OH$; $CH_2=CH-CH_2OH$ Allyl alcohol

   Aromatic alcohol ; Phenol
4. Ethers $R-\ddot{O}-R$
5. Carbonyl compounds $>C=O$
   - Alkanal $RCHO$
   - Alkanone $R-CO-R$
6. Carboxylic acids and their derivatives

   $R-C(=O)-O-H$ Carboxylic acid, $R-C(=O)-X$ Acid halide, $R-C(=O)-NH_2$ Acid amide,

   $R-C(=O)-O-R$ Ester, $R-C(=O)-O-C(=O)-R$ Acid anhydride
7. Amines
   - Primary RNH_2
   - Secondary R_2NH
   - Tertiary R_3N
   - Quaternary ammonium salt, $R_4\overset{+}{N}\,\overset{-}{O}H$ (or X^-)
8. Cyanide/ Nitrile $R-C\equiv N$

   Isocynide/ Isonitrile $R-N\rightleftarrows C$

Nomenclature

In case of saturated hydrocarbons:

1. Longest continuous carbon chain is selected e.g.

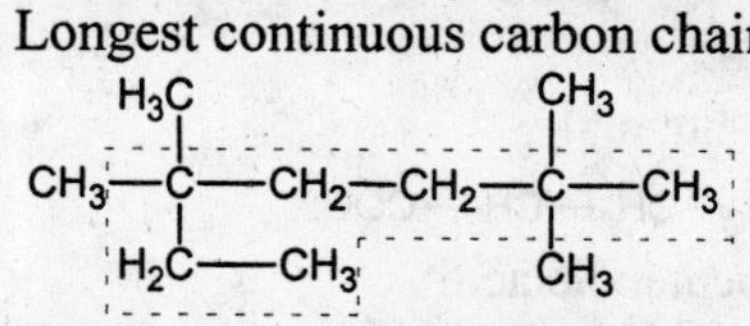

2. Numbering of the chain is done from the end of the nearest side chain

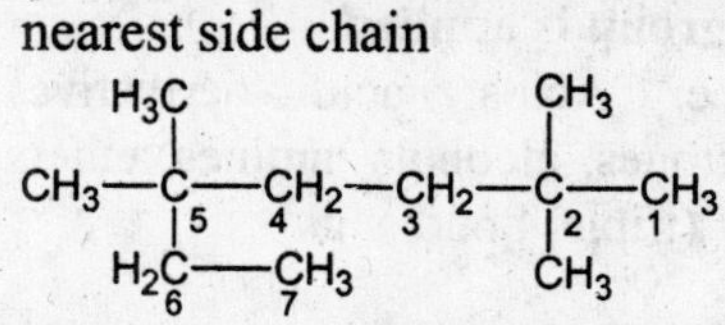

3. Side chains are mentioned in alphabetical order

1. When side chain is branched e.g.

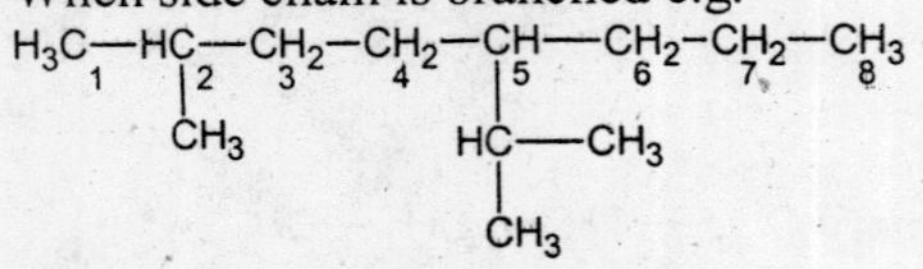

2-methyl-5(1-methyl-ethyl-octane)

When both = (double) and ≡ (triple) bonds are present together e.g.

(i) $H_3C—CH=CH—C \equiv CH$ pent–3–en–1–yne

(ii) $H_3C—C \equiv C—CH=CH_2$ pent–1–en–3–yne

(iii) $HC \equiv C—CH_2—CH=CH_2$ pent–1–en–4–yne

Note: If a molecule contains both C–C double and triple bonds, the two are treated at par in seeking the lowest number of combination. However if the sum of the numbers turns out to be the same starting from either end of the carbon chain, the lowest number is given to the C=C double bond.

When Functional groups are present

When a chain is ending with –CHO, –COOH, $–CONH_2$, –CN etc. group, it is assigned as number 1 (one). If they are not terminating groups like >C=O, $–NH_2$, –OH etc., they may not be assigned as number 1 (one). Longest carbon atom chain is selected, to include the maximum number of functional groups. Numbering of the main chain is done in such a way that the functional group is attached to the lowest numbered carbon atom. e.g.

(i) 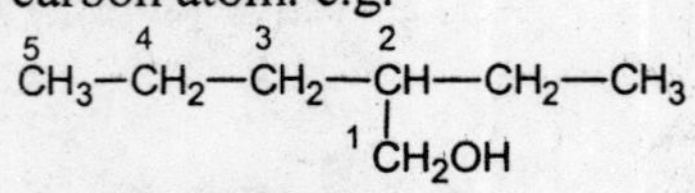

2–ethyl–pentan–1–ol

(ii) 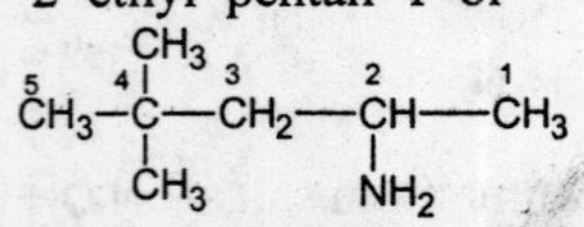

2–amino–4,4–dimethylpentane

(iii) 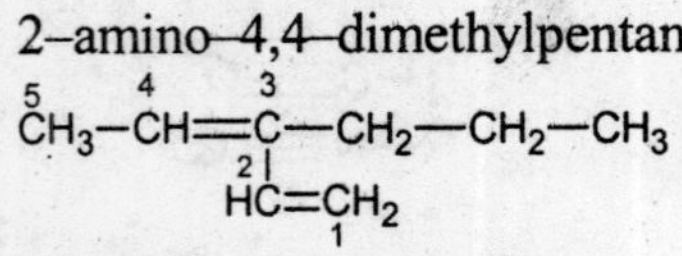

3–propyl–penta–1,3–diene

(iv) 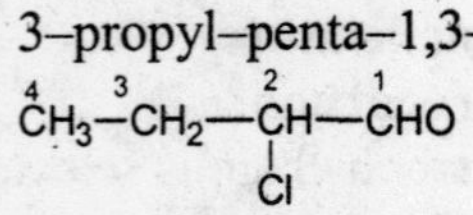

2–chloro–butanal

(v) $BrCH_2—CH_2—CH_2—CH_2—COOH$

5–bromopentanoic acid

When more than one functional group is present, the rule of principal group is applied.

Order of preference: Acids, acid derivatives, aldehydes, nitriles, ketones, alcohols, amines, ethers, = (double) and ≡ (triple) bond. e.g.

(i) 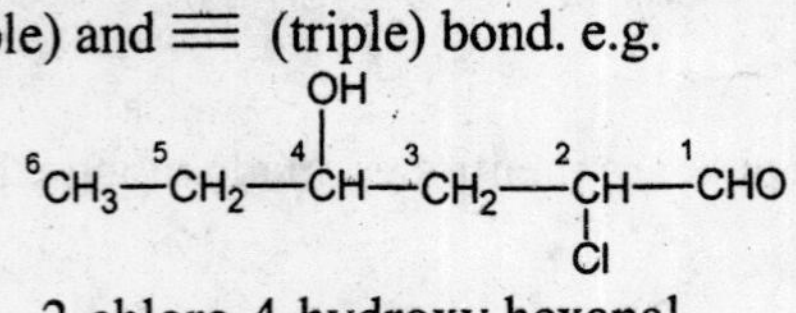

2-chloro-4-hydroxy hexanal

(ii) $H_3C—CH_2—CH=CH—COOH$

pent–2–enoic acid

(iii) 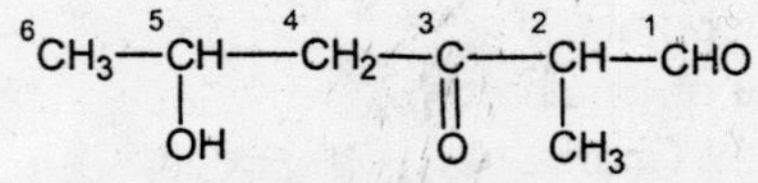

5–hydroxy–2–methyl–3–oxo–hexanal

(iv) 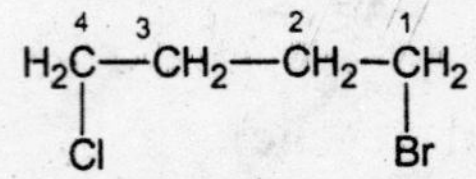

1–bromo–4–chlorobutane

(v) 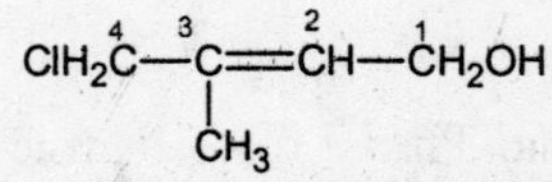

4–chloro–3–methylbut–2–en–1–ol

(vi) 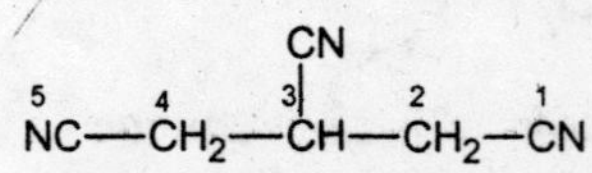

3–cyano–pentan–1,5–dinitrile

(vii) 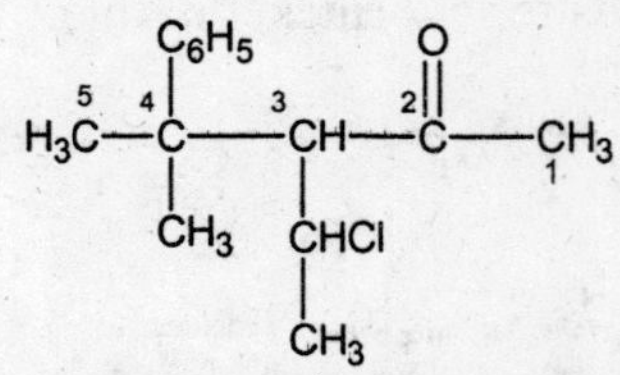

3(1–chloroethyl)4–methyl–4–phenyl–pentan–2–one

(viii) $H_3C—O—CH_2—CH_3$

methoxy ethane (In alkoxyalkanes, smaller alkyl group is taken with oxygen as alkoxy)

(ix) 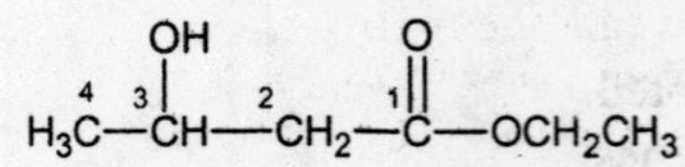

ethyl–3–hydroxy butanoate

(x) 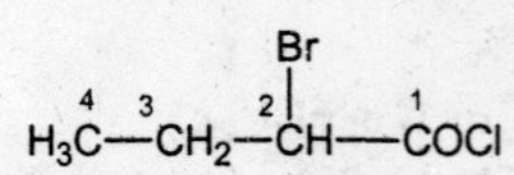

2–bromo butanoyl chloride

(xi) $HC \equiv C—CH=CH—CH=CH_2$

hex–1,3–dien–5–yne

(xii) 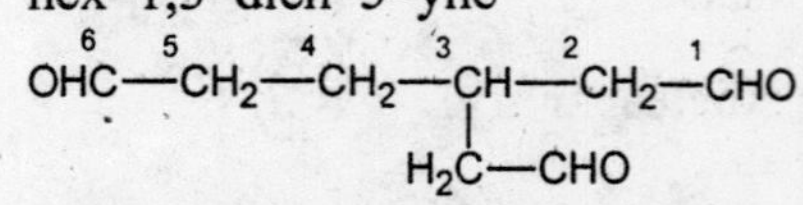

3–(formyl methyl) hexandial

(i) 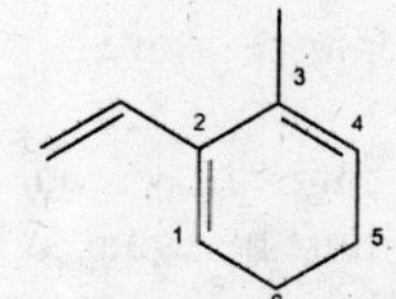

2–ethenyl–3–methyl–1,3–cyclohexadiene

(ii) 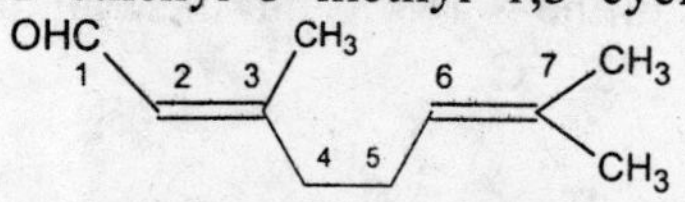

3,7–dimethyl–octa–2,6–dienal

(iii) 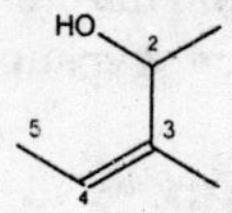

3–methylpent–3–en–2–ol

(iv) 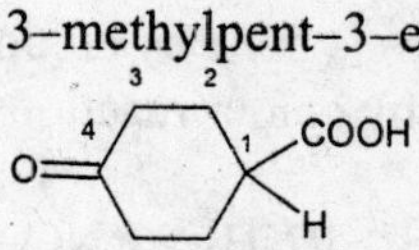

4–oxo–cyclohexane–1–carboxylic acid

(v) 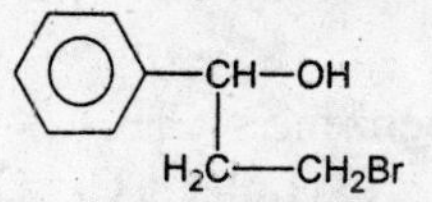

3–bromo–1–phenyl– propan–1–ol

(vi) 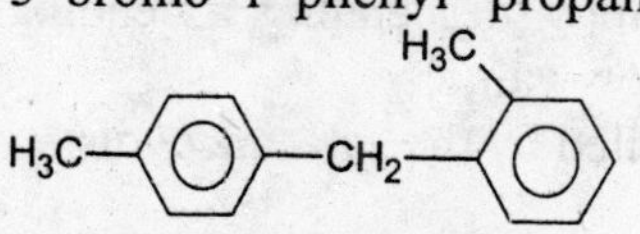

2,4′–dimethyl diphenyl methane

(vii) 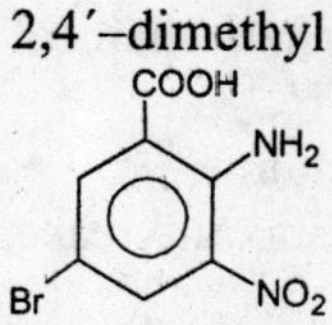

2–amino–5–bromo–3–nitrobenzoic acid

(viii) 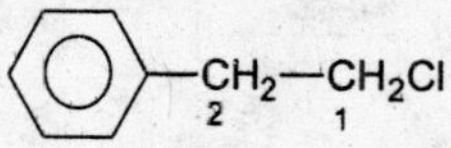

2–phenyl–1–chloro ethane

(ix) 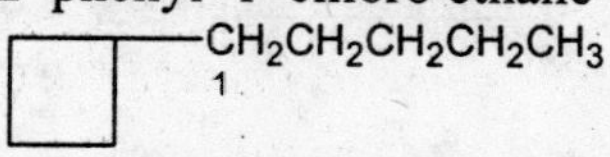

1–cyclobutyl pentane

(x) 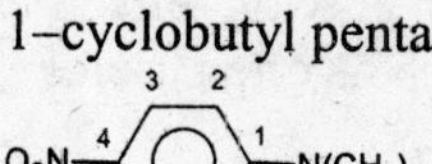

4–nitro–N–N′–dimethyl aniline

(xi) 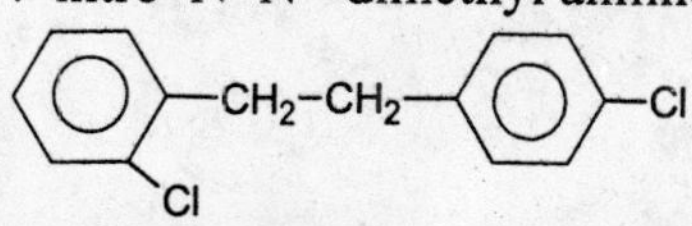

1–(2–chlorophenyl)–2–(4–chlorophenyl) ethane

(xii) 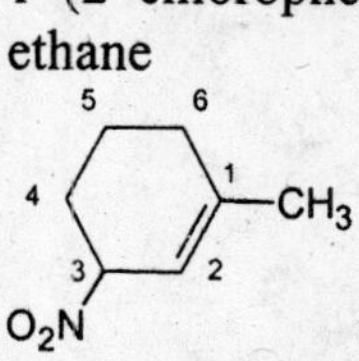

1–methyl–3–nitro cyclohexene

Nomenclature of Bicyclic Compounds: Fused or bridged rings containing compounds are called bicycloalkanes. The carbon atoms common to both the rings are known as bridge head. Each chain of carbon atoms connecting the bridge head atoms, is called bridge. For numbering, we start at one bridge head proceeding along the longest bridge to the other bridge head, then along the next longest to the first bridge head. The shortest bridge is numbered in the last. In between bicyclo and alkane, we insert in square brackets the number of carbon atoms in each bridge in the descending order.

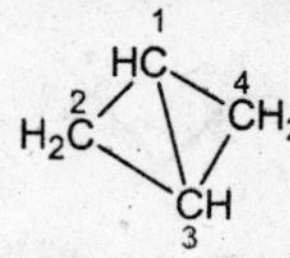

Biyclo[1,1,0]butane

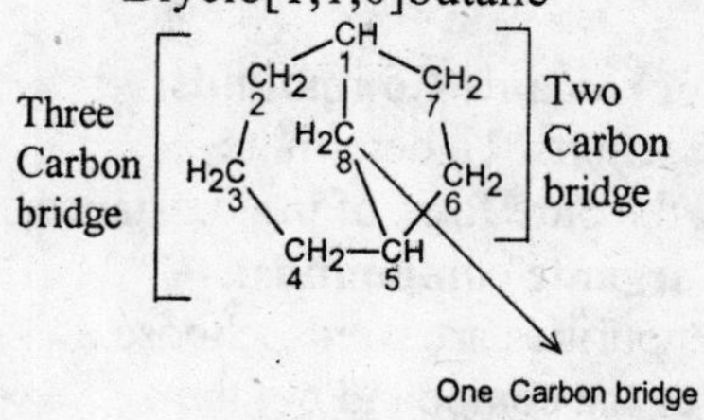

Biyclo[3,2,1]octane

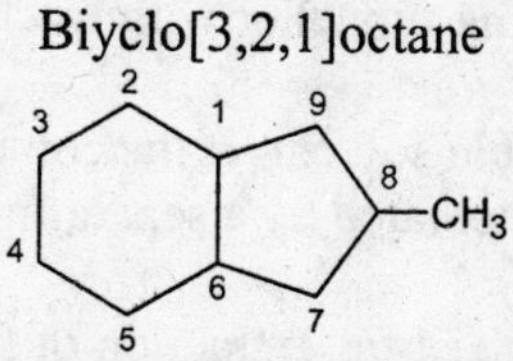

8–methyl bicyclo[4,3,0]nonane

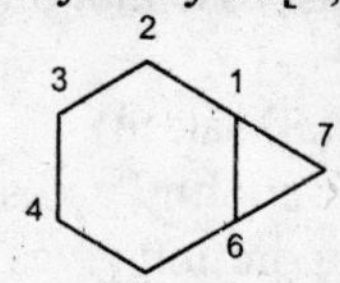

Bicyclo[4,1,0]heptane

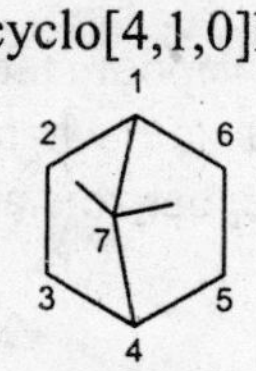

7,7–dimethyl bicyclo[2,2,1]heptane

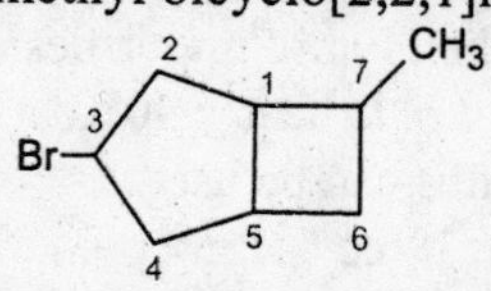

3–bromo–6–methyl bicyclo[3,2,0]heptane

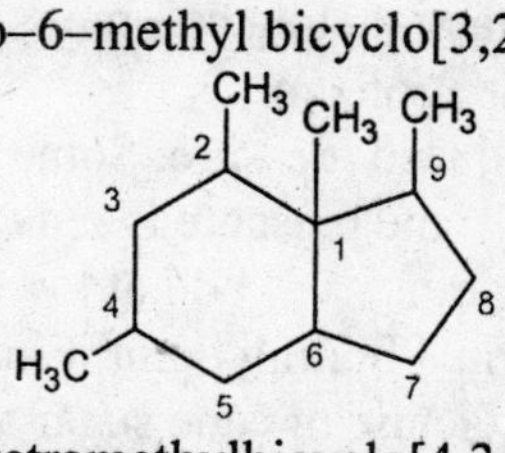

1,2,4,9–tetramethylbicyclo[4,3,0]nonane

Numbering of Spiranes

Spiranes share only one carbon atom. The numbering starts from the atom next to the spiro atom and proceeds towards the smaller ring first.

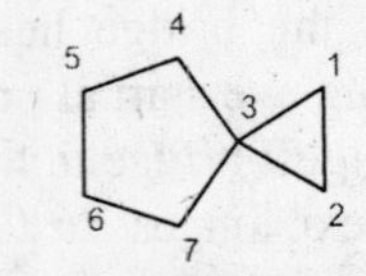

Spiro[4,2]heptane ;

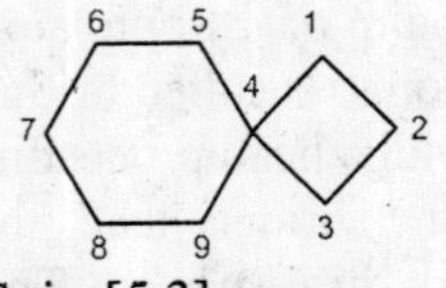

Spiro[5,3]nonane

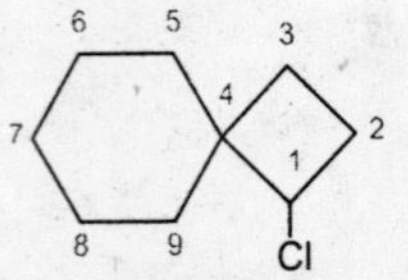

1-chlorospiro[5,3]nonane ;

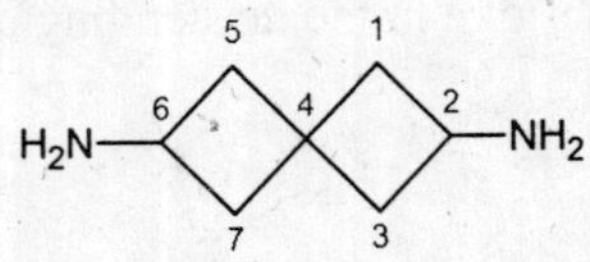

2,6-diamino[3,3]heptane

Purification of Organic Compounds

Impurities give side reactions. To determine , composition, we need the pure form of the substance.

Purification of solid organic compounds:

(i) Extraction: If the impurities are solids, choose a solvent which dissolves the compound not the impurities. Filter it by using funnel or with Buchner funnel.

(ii) Use of water immiscible solvent: If the compound is soluble in water, it is separated by a separating funnel.

(iii) Soxhlet Extraction: used for extraction of fats and oils from the seeds, alkaloids, essential oils etc. from leaves and flowers.

b). Crystallisation: (simple crystallisation)

Note: It is to be noted that quick cooling gives tiny but purer crystals whereas slow evaporation gives bigger but impure crystals.

Sublimation: Substances pass into vapour state without melting and return to solid state when vapour cools. e.g. naphthalene, anthracene, benzoic acid, camphor etc.

Purification of liquids:

(a) Distillation: Every liquid has a fixed b.p. under atmospheric pressure. When liquid impurities are present in a liquid, if the b.p. is nearly 40°C apart, a simple distillation apparatus can be used.

(b) Fractional distillation: If the boiling points of A and B are closer, then fractionating column is used e.g. benzene + toluene; petroleum.

(c) Distillation under reduced pressure: Some liquids decompose at their b.p. If the pressure is reduced during boiling, the compound will boil at lower temperature than its normal boiling point. Finds its application in the manufacture of cane sugar and glycerol.

(d) Steam distillation: This method is applicable to those substances which

(i) have non-volatile impurities
(ii) are insoluble in water
(iii) have high molecular weight
(iv) Possess a fairly high vapour pressure near about the b.p. of water

In steam distillation, at every stage, the vapour pressure P_1 and water vapour pressure P_2 must be equal to P

$P = P_1 + P_2$ or $P_1 = P - P_2$ For example the standard boiling point of aniline is 184°C but in steam distillation process it boils at 98.4°C.

Chromatography:

a. Adsorption chromatography (column chromatography)

1. Adsorbents: alumina, talc, magnesia, gypsum, silica gel, lime, kieselguhr, sugar etc.
2. The strongly adsorbed compounds move slowly.
3. Recovery of the substances is done by elution.

Characterisation is possible by colour reactions

Adsorption:

$RCOOH > RCONH_2 > ROH > RNH_2 > RNO_2 > ROCH_3 > RH$

Eluant (polarity):

Petroleum ether $< CCl_4 <$ Benzene $< CHCl_3 < C_2H_5OC_2H_5 < CH_3COOC_2H_5 <$ acetone $< C_2H_5OH < CH_3OH < HOH$

Paper chromatography (descending, ascending, circular)

R_f value (rate of flow value)

R_f= distance travelled by the substance/ distance travelled by the solvent

Decolouration: It is done with animal charcoal or activated charcoal

Chemical methods of purification:

When one of the components of mixture reacts with a reagent for e.g.

a) Diphenyl + benzoic acid $\xrightarrow{NaHCO_3}$

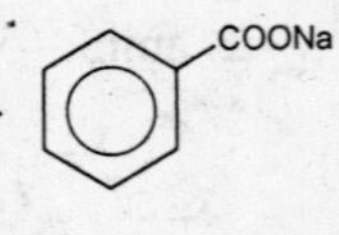

b) $CH_3OH + CH_3COCH_3 \xrightarrow{\text{anhydrous } CaCl_2} (CaCl_2.4CH_3OH)$
Crystalline Compound

Test of purity: m.p., b.p., sp. gravity, refractive index, viscosity etc. m.p. and m.m.p. is often useful.

1. Father of Chemistry is [CBSE-91]
(A) Berthold (B) Aristotle
(C) Lavoisier (D) Preistley

2. IUPAC name of this compound is [CBSE-91]

$$CH_3-\underset{CH_3}{\overset{CH_3}{\underset{|}{\overset{|}{C}}}}-CH=CH_2$$

(A) 3, 3, 3 trimethyl-1-propane
(B) 1, 1, 1 trimethyl-3-propane
(C) 3, 3 dimethyl-1-butene (D) 1, 1-dimethyl-3-butene

3. IUPAC name for CH_3CHO is [CBSE-91]
(A) acetaldehyde (B) methyl aldehyde
(C) formyl methane (D) ethanal

4. 2-methyl-2-butene will be represented as [CBSE-92]
(A) $CH_3-\underset{CH_3}{\underset{|}{CH}}-CH_2-CH_3$ (B) $CH_3-\underset{CH_3}{\underset{|}{C}}=CH-CH_3$

(C) $CH_3-CH_2-\overset{CH_3}{\overset{|}{C}}=CH_2$ (D) $CH_2=\overset{CH_3}{\overset{|}{C}}-CH=CH_2$

5. The IUPAC name of [CBSE-92]

$$CH_3-\overset{OH}{\overset{|}{CH}}-CH_2-\overset{CH_3}{\overset{|}{CH}}-CHO$$

Will be
(A) 4-Hydroxy-2-methylpentanal
(B) 4-Hydroxy-2- methylpentanal
(C) 4-Hydroxy-2-methylpentanal
(D) 2-Hydroxy-4-methylpentanal

6. Which one of the following IUPAC names is correct? [CBSE-94]
(A) 2-methyl-3-ethylpentane
(B) 2-ethyl-3-methylpentane
(C) 3-ethyl-2-methylpentane (D) 3-methyl-2-ethylpentane

7. The most suitable method of separation of a 1:1 mixture of ortho- and para-nitrophenols is [CBSE-94]
(A) sublimation (B) chromatography
(C) crystallisation (D) distillation

8. The IUPAC name for [CBSE-95]
$CH_3CH=CHCH_2\underset{NH_2}{\underset{|}{C}H}CH_2COOH$ is
(A) 5-amino-hex-2-ene-carboxylic acid
(B) 5-amino-2-heptenoic acid
(C) 3- amino-5- heptenoic acid.
(D) β-amino-δ- heptenoic acid.

9. IUPAC name of $(CH_3)_2$ $CH-CH_2-CH_2$ Br is [CBSE-96]
(A) 1-bromopentane (B) 2-methyl-4-bromobutane
(C) 1-bromo-3-methylbutane
(D) 2-methyl-3-bromopropane

10. Which of the following compounds gives benzoic acid on oxidation ? [CBSE-96]
(A) Chlorophenol (B) Chlorotoluene
(C) Chlorobenzene (D) Benzylchloride

11. Which one of the following chemical units is certainly to be found in enzyme? [CBSE-97]

(A) 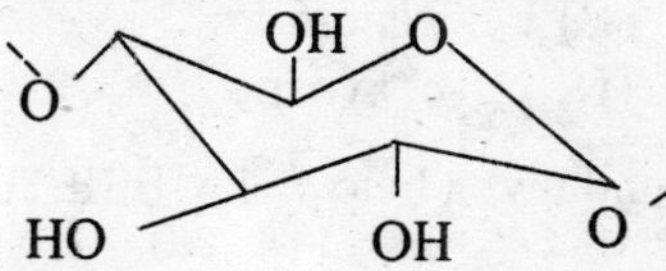 (B) 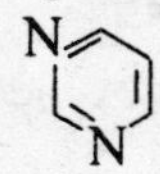

(C) 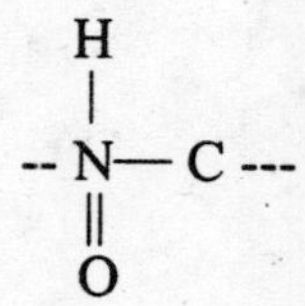 (D) 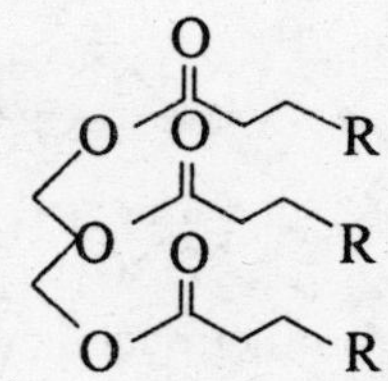

12. Which of the following techniques is most suitable for purification of cyclohexanone form a mixture containing benzoic acid, isoamylalcohol, cyclohexane and cyclohexanone ? [CBSE-97]
(A) Crystallisation (B) IR spectroscopy
(C) Sublimation (D) Evaporation

13. IUPAC name for the compounds is [CBSE-98]

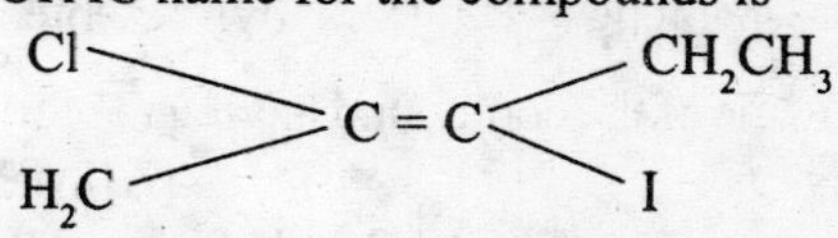

(A) trans-2-chloro-3-iodo-2-pentene
(B) cis-2-chloro-3-iodo-2-pentene
(C) trans-3-iodo-4-chloro-3-pentene
(D) cis-3-iodo-4-chloro-3-pentene

14. A compound 'A' on reduction gives 'B' which on further reaction with $CHCl_3$ and alc. KOH gives compound C which on hydrolysis gives aniline. Identify A. [CBSE-2000]
(A) Nitrobenzene (B) Methylamine
(C) Nitromethane (D) Nitrosobenzene

15. Which reagent brings about the conversion of carboxylic acids into esters. [CBSE-2000]
(A) C_2H_5OH (B) dry HCl + C_2H_5OH
(C) $LiAlH_4$ (D) $Al(OC_2H_5)_3$

16. The incorrect IUPAC name is – [CBSE-2001]
(A) $CH_3COCH(CH_3)_2$ 2-menthyl-3-butanone
(B) $(CH_3)_2CH-\underset{CH_2CH_3}{\underset{|}{CH}}-CH_3$ 2,3-dimethylpentane
(C) $CH_3-C\equiv CCH(CH_3)_2$ 4-methyl-2-pentyne

(D) CH_3-CHCl- CHBr - CH_3 2-bromo-3-chlorobutane

17. In steam distillation of toluene, the pressure of toluene vapour is - [CBSE-2001]
(A) Equal to pressure of barometer
(B) Less than pressure of barometer
(C) Equal to vapour pressure to that of toluene in simple distillation
(D) More than vapour pressure of toluene in simple distillation.

18. IUPAC name of the following is [CBSE-2002]
$CH_2 = CH - CH_2 - CH_2 - C \equiv CH$
(A) 1,5-hexenyne (B) 1-hexene-5-yne
(C) 1-hexyne-5-ene (D) 1,5-hexynene

19. Name of the compound given below is [CBSE-2003]

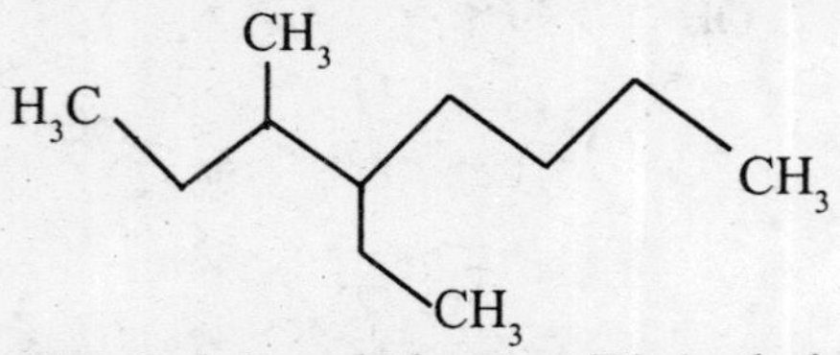

(A) 5-ethyl-6-methyloctane (B) 4-ethyl-3-methyloctane
(C) 3-methyl-4-ethyloctane (D) 2, 3-diethylheptane

20. The IUPAC name of the compound represented by the formula [AIIMS-93]
$CH_3 - CH_2 - CH_2 - CH = CH - \underset{\displaystyle CH_3}{\underset{|}{CH}} - CH_3$ is
(A) 5-isopropyl-4-pentene (B) 2-methyl-3-heptene
(C) 1-isopropyl-3-pentene (D) 6-methyl-4-heptene

21. The IUPAC name of the compound having the formula CCl_3CH_2CHO is [AIIMS-95]
(A) 1,1,1-trichloropropanal (B) 1,2,1-dichloromethanal
(C) 2,2,2-trichloropropanal
(D) 3, 3, 3, - trichloropropanal

22. Fractional distillation is used in [AIIMS-95]
(A) Crude oil (B) Coal tar
(C) Petroleum (D) All of these

23. The IUPAC name $(CH_3)_3 - C - CH = CH_2$ is [AIIMS-97]
(A) 2, 2-dimethyl-but-3-ene (B) 2, 2-dimethyl-pent-4-ene
(C) 3, 3-dimethyl-but-1-ene (D) Hex-1-ene

24. Which of the following is not aromatic ? [AIIMS-97]
(A) Benzene (B) Cyclooctatetraenyl dianions
(C) Tropylium cation (D) Cyclopentadienylanion

25. The IUPAC name of the compound [AIIMS-98]
$CH_3 - \underset{\displaystyle OH}{\underset{|}{C}} = CH - CH_2 - COOH$ is
(A) Hydroxy pentenoic acid
(B) 4-hydroxy-3-pentenoic acid
(C) 4-hydroxy-4-pentenoic acid
(D) 4-hydroxy-4-methyl-3-ene-pentanoic acid

26. Assertion : The energy of an electron is largely determined by its principal quantum number. Reason : The principal quantum number (n) is a measure of the most probable distance of finding the electrons around the nucleus. [AIIMS-98]
(A) A (B) B (C) C (D) D

27. Which ion shows aromatic character ? [AIIMS-2001]
(A) 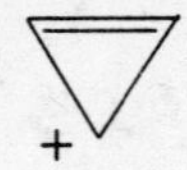 (B) 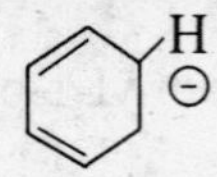
(C)  (D) 

28. IUPAC name of 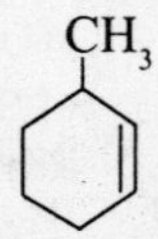 is- [AIIMS-2003]
(A) 3-methyl cyclohexene (B) 1-methyl cyclohex-2-ene
(C) 6-methyl cyclohexene (D) 1-methyl cyclohex-5-ene

29. Separation of two substances by fractional crystallisation depends upon their differences in : [JIPMER-92]
(A) Densities (B) Volalitity
(C) Solubility (D) Crystalline shape

30. The IUPAC name of the compound [JIPMER-93]
$(CH_3)_2CHCH = CHCH_2COOH$ is :
(A) 5-methyl-3-hexenoic acid
(B) 4-methlyl-2-hexenoic acid (C) 4-isopropyl-butenoic acid
(D) 5-carboxyl-2methylpentene

31. Fractional distillation is used in [JIPMER-94]
(A) Petroleum (B) Coal tar
(C) Crude oil (D) All of these

32. Aniline is usually purified by [JIPMER-95]
(A) Chromatographic technique (B) Steam distillation
(C) By addition of oxalic acid (D) Fractional crystallization

33. Which of the following methods is not used for purification of liquids ? [JIPMER-98]
(A) Chromatography (B) Steam distillation
(C) Distillation (D) Sublimation

34. The IUPAC name of the compound [JIPMER-99]
$CH_3 - \underset{\displaystyle CH_3}{\underset{|}{CH}} - CH = CH - CH_2OH$ is
(A) 4-Methyl-pent-2-en-1-ol (B) 2-Methyl-4-pentenol
(C) 4-Methyl-2-pentenol (D) 2-Methyl-4-en-1-ol

35. Distillation under reduced pressure is used to purify liquids which [JIPMER-99]
(A) are explosives (B) are highly volatile
(C) decompose below their boiling point

(D) have high boiling point.

36. A mixture of camphor and benzoic acid can be easily separated by [JIPMER-2000]
(A) Fractional crystallization (B) Chemical method
(C) Sublimation (D) Extraction with solvent.

37. A mixture of benzene and aniline can be separated by [JIPMER-2001]
(A) Hot water (B) Dil HCl (C) Dil NaOH (D) Alcohol

38. The IUPAC of name of

$$\begin{array}{ccccc} & OH & & CH_3 & \\ & | & & | & \\ CH_3- & CH & -CH_2- & CH & CHO \end{array}$$

is [JIPMER-2002]
(A) 4-Hydroxy-2-methylpentanal
(B) 4-Hydroxy-4-methylpentanal
(C) 2-Methyl-pent-4-ol-1-al (D) None of these

39. The IUPAC name of $C_2H_5CONH_2$ is [BHU-95]
(A) Methanamide (B) Benzamide
(C) Propanamide (D) Ethanamide

40. Which of the following processes is not used for the purification of solid impurities ? [BHU-96]
(A) Distillation (B) Sublimation
(C) Crystallisation (D) Steam distillation

41. IUPAC name of the compound [BHU-97]

$$\begin{array}{l} CH_3-CH=C-CH_3 \\ \qquad\qquad\;\; | \\ \qquad\quad CH_2CH_3 \end{array}$$

is
(A) 3-methyl-3-pentene (B) 2-ethyl-2-butene
(C) 3-ethyl-2-butene (D) 3-methyl-2-pentene

42. IUPAC name of [BHU-98]

$$\begin{array}{ccccc} & H & & C_4H_9 & \\ & | & & | & \\ CH_3- & C & - & C & -CH_3 \\ & | & & | & \\ & C_2H_5 & & CH_3 & \end{array}$$

is :
(A) 2-butyl-2-methyl-3-ethylbutane
(B) 2-ethyl-3,3-dimethylheptane
(C) 3,4,4-trimethylheptane (D) 3,4,4-trimethyloctane

43. The IUPAC name of the compound is [BHU-2001]

$$\begin{array}{l} CH_3-C=CH_2CH_2OH \\ \qquad\;\; | \\ \qquad CH_3 \end{array}$$

(A) 2-methyl-2-butenol (B) 2-methyl-3-butenol
(C) 3-methyl-2-butenol (D) 3-methyl-but-2-enol

44. IUPAC name for [BHU-2001]

$$\begin{array}{l} \qquad\qquad CH_3 \\ \qquad\qquad\; | \\ CH_3CH_2-CH-CH_2COCl \end{array}$$

is
(A) 3-methyl pentanoyl chloride
(B) 3-methyl butanoyl chloride (C) 1-chloro-3-ethyl butanone
(D) 1-chloro-3methyl pentanone

45. Steam distillation is used for purification of [BHU-2001]
(A) Inorganic liquids (B) Organic liquids
(C) Covalent compounds (D) Volatile liquids

46. Write the IUPAC name [BHU-2002]

$$\begin{array}{ccccc} & & & O & \\ & & & \| & \\ CH_3- & C & =CH- & C & -CH_3 \\ & | & & & \\ & CH_3 & & & \end{array}$$

(A) 4-Methylpent-3-en-2-one
(B) 2-Methylpent-3-en-2-one
(C) 3-Methylpent-2-en-1-one (D) None of these

47. Chromatography is used to separate [AFMC-93]
(A) Small samples of mixtures (B) Plant pigments
(C) Dye stuffs (D) All the above

48. Purity of organic compounds can be determined by knowing [AFMC-93]
(A) Melting point (B) Mixed melting point
(C) Solubility (D) None of the above

49. Which of the following undergoes sublimation ? [AFMC-95]
(A) $ZnCl_2$ (B) $CuCl_2$ (C) AgCl (D) NH_4Cl

50. A mixture contain benzene & chlorobenzene is separated by [AFMC-96]
(A) Sublimation (B) Separating tunnel
(C) Crystallization (D) Distillation

51. IUPAC name of the compound is [AFMC-97]

$$\begin{array}{l} CH_3-CH-CH_2-CHOH-CH_3 \\ \qquad\quad | \\ \qquad\; CH_2 \\ \qquad\quad | \\ \qquad\; CH_3 \end{array}$$

(A) 4-Methyl-3-hexanol (B) Heptanol
(C) 4-Methyl-2-hexanol (D) None of these

52. The IUPAC name of succinic acid is .... [AFMC-94]

53. Which of the following compounds has wrong IUPAC name ? [AIEEE-2002]
(A) $CH_3-CH_2-CH_2-COO-CH_2CH_3 \rightarrow$ ethyl butanoate
(B) $\begin{array}{l} CH_3-CH-CH_2-CHO \\ \qquad\quad | \\ \qquad\; CH_3 \end{array} \rightarrow$ 3- methyl -butanal
(C) $\begin{array}{l} CH_3-CH-CH-CH_3 \\ \qquad\quad |\quad\;\; | \\ \qquad\; OH\;\; CH_3 \end{array} \rightarrow$ 2-methy 1-3-butanol
(D) $\begin{array}{l} \qquad\qquad\;\; O \\ \qquad\qquad\;\; \| \\ CH_3-CH-C-CH_2-CH_3 \\ \qquad\quad | \\ \qquad\; CH_3 \end{array} \rightarrow$ 2 methyl-3pentanone

54. The IUPAC name of $CH_3COCH(CH_3)_2$ is [AIEEE-2003]
(A) Isopropylmethylketone (B) 2-methyl-3-butanone
(C) 4-methylisopropylketone (D) 3-methyl-2-butanone

55. The IUPAC name of the compound [EAMCET-90]

$$CH_3-\underset{\underset{CH_3}{|}}{\overset{\overset{H}{|}}{C}}-CH_2Br$$

is ....

56. The trivial name of 4-methylaniline is .... [EAMCET-90]

57. $$CH_3-CH_2-\underset{}{\overset{\overset{CH_3}{|}}{CH}}-CH_2-CH_2-CH_2-CH_3$$

IUPAC name is .... [EAMCET-91]

58. $$CH_3-CH_2-\overset{\overset{CH_3}{|}}{C}=CH_2$$

The IUPAC name of the compound is .... [EAMCET-92]

59. The IUPAC name for $CH_3-\underset{\underset{CH_3}{|}}{CH}-\underset{\underset{OH}{|}}{CH}-CH_3$

is .... [EAMCET-93]

60. A mixture of some solid organic compound A, B, C, D, on heating only 'C' changes from solid to vapour state. 'C' can be separated from the rest in the mixture by [EAMCET-98]
(A) Distillation (B) Sublimation
(C) Fractional distillation (D) Crystallization

61. The IUPAC name of the compound having the formula $(CH_3)_3\,C-CH=CH_2$ is [EAMCET(MBBS)-98]
(A) 3,3,3trimethyl-1-propene
(B) 3, 3dimethyl-1-butene
(C) 1,1,1trimethyl-3-propane (D) 1,1dimethyl-3-butene

62. A solid organic compound X on heating is converted directly to vapour phase which on cooling solidifies. The best method for purifying X is by : [EAMCET-99]
(A) distillation (B) distillation at reduced pressure
(C) sublimation (D) steam distillation

63. 18 C–H and 7 C–C sigma bonds are present in [EAMCET(MBBS)-99]
(A) Cyclohexane (B) 3,3dimethylpentane
(C) 2,2,3-trimethylpentane (D) n-heptane

64. The IUPAC name of $(C_2H_5)_4\,C$ is [EAMCET(MBBS)-99]
(A) tetraethyl methane (B) 2,2-dimethylpentane
(C) 3,3-diethylpentane (D) neopentane

65. Benzene is a monocyclic compound . It is aromatic because it contains [EAMCET(MBBS)-99]
(A) 6 carbons and 6 hydrogens (B) 3 pi bonds
(C) 6 p orbitals (D) $(4n+2)\,\pi$ electrons

66. A solid organic compound Y on heating is converted to vapour state directly, which on cooling solidifies. Which of the following is the correct method of purifying Y ? [EAMCET-2000]
(A) Sublimation (B) Distillation
(C) Distillation at reduced pressure (D) Steam distillation

67. The structural formula of 2-methyl-2-butene is [EAMCET-2001]
(A) $CH_3-CH(CH_3)-CH=CH_2$
(B) $CH_3-CH_2-C(CH_3)=CH_2$
(C) $CH_3-CH=CH-CH_3$
(D) $CH_3-CH=C(CH_3)-CH_3$

68. Which one of the following compounds is an isomer of 1-butanol ? [EAMCET(MBBS)-2001]
(A) 2-methyl-2-butanol (B) 2-methyl-1-butanol
(C) 3-methyl-2-butanol (D) 2-methyl-1-propanol

69. The functional group present in 3-methyl butanal is [EAMCET(MED)-2002]
(A) –CHO (B) $-\overset{\overset{O}{||}}{C}-$ (C) $-CH_2OH$ (D) –COOH

70. 2,3 Dimethyl hexane contains.......... tertiary............. secondary and............primary carbon atoms respectively. [EAMCET-2003]
(A) 2,2,4 (B) 2,4,3 (C) 4,3,2 (D) 3,2,4

71. $$CH_3-\underset{\underset{CH_3}{|}}{\overset{\overset{OH}{|}}{C}}-CH_2-\underset{\underset{CH_3}{|}}{CH}-CH_3$$

IUPAC name is [CPMT-96]
(A) 2,4-dimethylpentan-2-ol (B) 2,4dimethyl pentanol-2
(C) 2,2 dimethyl butanol-2 (D) Butanol-2

72. IUPAC name of $CH_3-CH=CH-C\equiv CH$ is [CPMT-97]
(A) Pent-2-en-4-yne (B) Pent-3-en-1-yne
(C) Pent-3-yne-1-en (D) Pent-2-yne-1-en

73. 'Activated charcoal' is employed to remove colouring matter from pure substances it does so by- [CPMT-98]
(A) Bleaching (B) Oxidation
(C) Reduction (D) Adsorption

74. IUPAC name of the following compound will be [CPMT-99]

$$CH_3-CH=\underset{\underset{CH_2-CH_2-CH_3}{|}}{C}-CH_2-CH_3$$

(A) 3-ethyl 2-hexene (B) 3-propyl 2-hexene

(C) 3-propyl-3-hexene (D) 4-ethyl-4-hexene

75. The IUPAC name of tertiary butyl chloride is- [CPMT-2000]

(A) Butyl chloride (B) Chlorobutane
(C) 2-Chlorobutane (D) 2-Chloro-2-methylbutane

76. The IUPAC name of the following is [CPMT-2001]
$CH_3C(CH_3)_2\ CH_2CH = CH_2$

(A) 2,2 dimethyl-4-pentene (B) 4,4-dimethyl-1-pentene
(C) 1,1,1-trimethyl-3-butene (D) 4,4,4-trimethy-1-butene

77. I.U.P.A.C. name of compound [CPMT-2002]

$CH_3 - CH = C(CH_2 - CH_2 - CH_3) - CH_2 - CH_3$ is

(A) 3-ethyl-2-hexene (B) 4-ethyl-2-hexene
(C) 4-ethyl-4-hexene (D) 3-propyl-2-hexene

78. The IUPAC name of the Compound [PMT-93]

$CH_3 - CH(CH_3) - CH(C_2H_5) - CH_3$ is :

(A) 2-Ethyl-3-methylbutane (B) 2,3-dimethylpentane
(C) 2-methyl-3-ethylbutane (D) 3,4-dimethylpentane

79. The I. U. P. A. C. name of [PMT-94]

$CH_3 - CH(CH_3) - CH_3$ is

(A) Butane (B) Isopropyl methane
(C) 2-methylpropane (D) Di methyl ethane

80. I.U.P.A.C. name of [PMT-95]
$CH_2 = CH - CH_2\,Cl$ is :

(A) Allyl chloride (B) 1- chloro-3-propene
(C) Vinyl chloride (D) 3-chloro-1-propene

81. The I. U. P. A. C. name of [PMT-95]

$CH_3 - CH(CH_3) - C(=O) - CH_2 - CH_2OH$ is

(A) 1-hydroxy-4-methyl-3-pentanone
(B) 2-methyl-5-hydroxy-3-pentanone
(C) 4-methyl-3-oxo-1-pentanol (D) hexan-1-ol-2-one

82. The I.U.P.A.C. name of CH_3CHO is : [PMT-96]

(A) Ethane (B) Ethanol
(C) Ethanal (D) Acetaldehyde

83. I. U. P. A. C. name of the compound [PMT-98]

$CH_3 - CH = C(CH_2 - CH_3) - CH_3$ is

(A) 2-ethyl-2-butene (B) 3-ethyl-2-butene
(C) 3-methyl-3-pentene (D) 3-methyl-2-pentene

84. The IUPAC name of crotonaldehyde is : [PMT-99]

(A) prop-2-ene-1-al (B) propenal
(C) but-2-enal (D) butenal

85. I. U. P. A. C. name of the following compound is : [PMT-2001]

$CH_3 - C(CH_3)_2 - CH = CH_2$

(A) 1,1,1trimethyl-2-propene
(B) 3,3dimethyl-1-butene
(C) 3,3,3trimethyl-1-propene
(D) 2, 2dimethyl-3-butene

86. The compound which has one isopropyl group is : [PMT-2001]

(A) 2,2,3,3 tetramethylpentane
(B) 3,3 dimethylpentane
(C) 2,2,3 trimethylpentane (D) 2-methylpentane

87. The name of $H_3C - CH(CH_3) - CH(OH) - CH_3$ according to IUPAC nomenclature system is [PMT-2002]

(A) Butanol (B) 2-Methyl-butan-3-ol
(C) 3-Methyl-butan-2-ol (D) Pentanol

88. The name of $Cl\,H_2C - C(Br) = C(Br) - CH_2Cl$ according to IUPAC nomenclature system is [PMT-2002]

(A) 2, 3-dibromo1, 4dichlorobut-2-ene
(B) 1,4-dibromo-2,3-dichlorobut-2-ene
(C) Dichlorodibromobutene (D) Dichlorodibromobutane

89. The correct IUPAC name for [PMT-2003]

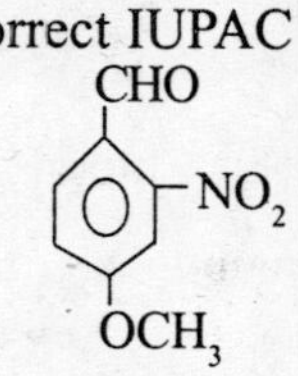

is :

(A) 4methoxy-2-nitrobenzaldehyde
(B) 4methoxy-6-nitrobenzaldehyde
(C) 4formyl-1,3-nitroanisole
(D) 2formyl-5-methoxynitrobenzene

90. The correct IUPAC name for $CH_2{=}CH{-}CH_2{-}C{\equiv}CH$ is [PMT-2003]

(A) Pent–1–en-1–yne (B) Pent–1–en-4–yne
(C) Pent–1,4–enyne (D) Pent–1,4–enyne

91. The name of compound having the structure $ClCH_2CH_2COOH$ is [PET-93]

(A) 3-chloropropanoic acid (B) 2-chloropropanoc acid
(C) 2-chloroethanoic acid (D) Chlorosuccinic acid

92. The structural formula of isobutyl chloride is [PET-93]

(A) $CH_3CH_2CH_2CH_2Cl$ (B) $(CH_3)_2CH_2CH_2Cl$
(C) $CH_3CH_2CHClCH_3$ (D) $(CH_3)_3\ CCl$

93. The correct IUPAC name for the compound [PET-94]

$$CH_3-CH_2-CH_2-CH_2-CH_2-\underset{\displaystyle CH_3}{\underset{|}{CH}}-\overset{\displaystyle CH_3}{\overset{|}{\underset{\displaystyle CH_2-CH_2-CH_3}{\underset{|}{C}}}}-CH_2-CH_3$$

(A) 3,4-dimethyl-3-n-propylnonane
(B) 6,7-dimethyl-7-ethyldecane
(C) 6,7-dimethyl-7-n-propylnonane
(D) 4,5-dimethyl-4-ethyldecane

94. The I.U.P.A.C. name for [PET-95]

$\overset{6}{C}H_3\ .\overset{5}{C}H\ .(CH_3).\overset{4}{C}H\ = \overset{3}{C}H\ - \overset{2}{C}H_2\ - \overset{1}{C}OOH$ is -

(A) 5methyl-3-hexenoic acid
(B) 5carboxyl-2-methylpent-4-ene
(C) 4isopropylbut-3-ene-1-oic acid (D) None of these

95. The IUPAC name of [PET-97]

$$CH_3-\underset{\displaystyle Cl}{\underset{|}{C}}=\underset{\displaystyle CH_3}{\underset{|}{C}}-\underset{\displaystyle C_2H_5}{\underset{|}{CH}}-CH_2-C\equiv CH \text{ is}$$

(A) 6-chloro-4-ethyl-5-methyl-hept-5-en-1- yne
(B) 6-chloro-4-ethyl-5-methyl-hept-1-yne-5- ene
(C) 2-chloro-4-ethyl-3-methyl-hept-2-en-6- yne
(D) 2-chloro-4-ethyl-3-methyl-hept-6-yne-2- ene

96. The compound  is known by which of the following names ? [PET-97]

(A) Bicyclo -[2, 2, 2] octane(B) Bicyclo - [2, 2, 1] octane
(C) Bicyclo - [1, 2, 1] octane (D) Bicyclo - [1, 1, 1] octane

97. The IUPAC name of the compound having the formula $Cl_3C.\ CH_2CHO$ is : [PET-98]

(A) 3,3,3-trichloropropanal
(B) 1,1,1-trichloropropanal
(C) 2,2,2-trichloropropanal (D) Chloral

98. The IUPAC name of the compound is: [PET-99]

$$CH_3-\underset{\displaystyle CH_3}{\underset{|}{CH}}-CH_2-CH_2-Cl$$

(A) 1-chloro-3-methylbutane (B) 2-methyl-4-chlorobutane
(C) 2-methyl-1-chlorobutane (D) 1-chlorobutane

99. The correct I.U.P.A.C. name of [PET-2001]

$$H-\overset{\displaystyle O}{\overset{||}{C}}-CHO \text{ is :}$$

(A) Formyl methanal (B) 1,2-ethanedione
(C) Formyl methanoate (D) 1,2-ethanedial

100. The IUPAC name of $CH_3OC_2H_5$ is : [PET-2003]

(A) Methylethyl ether (B)Ethylmethyl ether
(C) Methoxyethane (D) Ethoxymethane

Classification and Nomenclature

1.	(C)	46.	(A)	91.	(A)
2.	(C)	47.	(D)	92.	(B)
3.	(D)	48.	(A)	93.	(D)
4.	(B)	49.	(D)	94.	(A)
5.	(B)	50.	(D)	95.	(C)
6.	(A)	51.	(C)	96.	(A)
7.	(D)	52.	*	97.	(A)
8.	(C)	53.	(C)	98.	(A)
9.	(D)	54.	(D)	99.	(D)
10.	(D)	55.	*	100.	(C)
11.	(C)	56.	*		
12.	(B)	57.	*		
13.	(A)	58.	*		
14.	(A)	59.	*		
15.	(B)	60.	(B)		
16.	(A)	61.	(B)		
17.	(B)	62.	(C)		
18.	(B)	63.	(C)		
19.	(B)	64.	(C)		
20.	(B)	65.	(D)		
21.	(D)	66.	(A)		
22.	(D)	67.	(D)		
23.	(C)	68.	(D)		
24.	(B)	69.	(A)		
25.	(B)	70.	(A)		
26.	(A)	71.	(A)		
27.	(A)	72.	(B)		
28.	(A)	73.	(D)		
29.	(C)	74.	(A)		
30.	(A)	75.	(D)		
31.	(D)	76.	(B)		
32.	(B)	77.	(A)		
33.	(D)	78.	(B)		
34.	(A)	79.	(C)		
35.	(C)	80.	(D)		
36.	(B)	81.	(A)		
37.	(B)	82.	(C)		
38.	(A)	83.	(D)		
39.	(C)	84.	(C)		
40.	(A)	85.	(B)		
41.	(D)	86.	(D)		
42.	(D)	87.	(C)		
43.	(C)	88.	(A)		
44.	(A)	89.	(A)		
45.	(D)	90.	(B)		

52. butan - 1, 4 - dioic acid
55. 1 - Bromo - 2 - methyl propane
56. p - methyl aniline
57. 3 - methyl heptane
58. 2 - methyl - 1 - butene or 2 - methyl nut - 1 - ene
59. 3 - methyl - 2 - butanol

Empirical and Molecular formula of organic compounds

Empirical formula: Simplest ratio of the constituent elements of a molecule e.g. ($C_2H_4O_2$) & $C_6H_{12}O_6$ have CH_2O as empirical formula

Molecular formula : The actual number of the atoms of the constituent elements in a molecule e.g. CH_3COOH molecular formula $C_2H_4O_2$.

Molecular formula = empirical formula x n;

n = molecular formula weight/ empirical formula weight

Detection of elements

A. Lassaigne's test: it can be used to detect nitrogen, sulphur and halogens. Nitrogen present in diazo compounds cannot be detected by this test.

i). Na + C + N   NaCN

$FeSO_4 + 2NaOH$ $Fe(OH)_2 + Na_2SO_4$
dirty green

$Fe(OH)_2 + 6NaCN$ $Na_4[Fe(CN)_6] + 2NaOH$

$3Na_4Fe(CN)_6 + 2Fe_2(SO_4)_3$ $Fe_4[Fe(CN)_6]_3 + 6Na_2SO_4$
ferric ferrocyanide (blue)

ii). 2Na + S Na_2S

$Na_2S + Na_2[Fe(NO)(CN)_5]$ $Na_3[Fe(O{=}N{-}S{-}Na)(CN)_5]$
sodium nitroprusside   sod. thionitroprusside (violet)

when N and S are present together

Na + C + N + S   NaCNS

$3NaCNS + FeCl_3$ $Fe(CNS)_3 + 3NaCl$
ferric sulphocynanide
(blood red colour)

B. Beilstein test: It can be used for halogens. Exception: Urea, thiourea, pyridine, etc. respond to this test.

C. Phosphorous present in the organic compound is oxidised to phosphate and tested with ammonium molybdate.

$$\text{organic compound with P} \xrightarrow[\text{fuming } HNO_3]{\text{oxidised with}} H_3PO_4$$

$H_3PO_4 + 12(NH_4)_2MoO_4 + 21HNO_3$ $(NH_4)_3PO_4 \cdot 12MoO_3 + 21\,NH_4NO_3 + 12H_2O$
ammonium phosphomolybdate
(canary yellow)

Estimation

C and H	Liebig's method
N	Dumas method or kjeldahl's method
Halogens	Carius method, Stepanov's method
Sulphur	Messenger's, Carius, Pirria & Schiff, and Asboth's methods

$$\%\text{ of H} = \frac{2}{18} \times \frac{\text{wt.of } H_2O \text{ formed}}{\text{wt.of org. compound}} \times 100$$

$$\%\text{ of C} = \frac{12}{44} \times \frac{\text{wt.of } CO_2 \text{ formed}}{\text{wt.of org. compound}} \times 100$$

$$\%\text{ of } N_2 \text{ (Dumas)} = \frac{28}{22400} \times \frac{\text{vol .of } N_2 \text{ at NTP}}{\text{wt.of org. compound}} \times 100$$

$$\%\text{ of N (kjeldahls)} = \frac{1.4 \times \text{vol. of acid} \times \text{normality of acid}}{\text{wt. of organic compound}} \times 100$$

$$\%\text{ of Cl} = \frac{35.5}{143.5} \times \frac{\text{wt.of AgCl formed}}{\text{wt.of org. compound}} \times 100$$

$$\%\text{ of Br} = \frac{80}{188} \times \frac{\text{wt.of AgBr formed}}{\text{wt.of org. compound}} \times 100$$

$$\%\text{ of I} = \frac{127}{235} \times \frac{\text{wt.of AgI formed}}{\text{wt.of org. compound}} \times 100$$

$$\%\text{ of S} = \frac{32}{233} \times \frac{\text{wt.of } BaSO_4 \text{ formed}}{\text{wt.of org. compound}} \times 100$$

% of oxygen = 100 – sum of the percentages of all other elements.

some relationships for determining the molecular weights

(i). m = 2 x V.D.;

$$\text{vapour density} = \frac{\text{wt.of X C.C. of vapour at N.T.P.}}{\text{wt.of X C.C. of hydrogen at N.T.P.}}$$

(one C.C. of hydrogen at N.T.P. weighs 0.00009)

(ii). silver salt method for acids

$$\frac{\text{Equivalent wt. of silver salt}}{\text{Equivalent wt. of silver}} = \frac{\text{wt. of silver salt}}{\text{wt. of silver}}$$

$$\frac{E - 1 + 108}{108} = \frac{\text{wt. of silver salt}}{\text{wt. of silver}}$$

(iii). Volumetric method for acids

(a) 1 lit. of N– solution of a substance contains 1 g. equivalent of the substance.

(b) wt. of 22,400 C.C. vapour of any substance at NTP is equal to its molecular weight.

Some problems for practice

(1) An organic compound contains 4% sulphur. Its minimum molecule weight is (Ans: 800).

(2) The empirical formula of a compound is CH_2.One molecule of this compound has a mass of 42g. Its molecule formula is (Ans: C_3H_6)

(3) A compound (60g) on analysis gave C=24g; H=4g; 0=32g.Its empirical formula is (Ans: CH_2O)

(4) A volatile organic compound contained 10% C, 0.84% H and 89.12% Cl. In Victor Meyer'smethod 0.6g of the substance displaced 112 C.C of air at NTP. Find out the empirical and molecular formula of the compound. (Ans: $CHCl_3$)

(5) An organic monobasic acid has been analysed as follows

(a) 0.1935g, gave 0.132g of CO_2 and 0.0270g,of H_20

(b) It contains 55.04% of chlorine.

(c) 0.2709g of acid dissolved in H_20 neutralised 25 C.C of N/2 – NaOH. Calculate the empirical and molecular formula (Ans: CHClO; $C_2H_2Cl_2O_2$)

(6) 0.225g of organic nitrogenous compound on being digested with conc. H_2SO_4 followed by distillation with NaOH evolved ammonia which was absorbed in 50c.c of N/10 – H_2SO_4. After absorption the excess of acid required 10 C.C. of N/5 NaOH. Calculate the percentage

of N in the organic compound.
(Ans: 16.46)

(7) 0.76 g. of a silver salt of a dibasic acid on ignition gave 0.54 g. of silver. Calculate the molecular weight of the acid. (Ans: 90)

(8) In a Carius determination, 0.2595 g. of an organic substance gave 0.35 g. of $BaSO_4$. Calculate the % of sulphur? (Ans: 18.52)

(9) An organic compound A ($C_4H_{10}O$) is optically active, on mild oxidation it gives a compound B (C_4H_8O), but on vigorous oxidation, it gives another compound C ($C_3H_6O_2$). Write the structures of A, B and C. (Ans: A: $CH_3CH_2CHOHCH_3$;
B: $CH_3CH_2COCH_3$; C: CH_3CH_2COOH).

(10) An organic compound of vapour density 23 contains 52.2% C and 13.04% H. On reaction with conc. H_2SO_4 at 140°C, it gives a compound B, B when reacted with cold HI gives back A and another compound C which also gives A on hydrolysis. The compounds A, B, and C are: (Ans: A: C_2H_5OH ;
B: $C_2H_5OC_2H_5$; C: C_2H_5I).

(11) Two compounds A and B have the same empirical formula CH. A is a gaseous, aliphatic compound with molecular weight 26, while B is a liquid aromatic compound with molecular weight 78. Both of them burn with a sooty flame.
(i) give the names and structures. (ii) what type of reactions are generally shown by them.
(Ans: A: C_2H_2 (ethyne) and B: C_6H_6 (benzene).
(Ans: A: ethyne, C_2H_2 and B: benzene, C_6H_6).

1. Kjeldahl's method is used in the estimation of [CBSE-91]
(A) Nitrogen (B) Halogen (C) Sulphur (D) Oxygen

2. For which of the following compounds the Lassaigne's test for the detection of nitrogen will fail ? [CBSE-94]
(A) $H_2NCONHNH_2$ + HCl (B) H_2NNH_2 +2HCl
(C) H_2NCONH_2 (D) $C_6H_5NHNH_2$.2HCl

3. Haemoglobin contains- 33% of iron by weight. The molecular weight of haemoglobin is approximately 67200. The number of iron atoms { at. wt.(Fe)=56 present in one molecule of haemoglobin are [CBSE-98]
(A) 6 (B) 1 (C) 4 (D) 2

4. An organic compound containing C, H and N gave the following analysis : C = 40%, H = 13.33% and N = 46.67% its empirical formula would be [CBSE-99]
(A) CHN (B) C_2H_2N (C) CH_4N (D) C_2H_7N

5. Percentage of Se in anhydrous peroxidase enzyme is 0.5% by weight (atomic weight of Se=78.4), then minimum molecular weight of peroxidase anhydrous enzyme is - [CBSE-2001]
(A) 1.568×10^4 (B) 1.568×10^3
(C) 15.68 (D) 2.136×10^4

6. The percentage of C, H and N in an organic compound are 40%, 13.3% and 46.7% respectively then empirical formula is [CBSE-2002]
(A) $C_3H_{13}N_3$ (B) CH_2N (C) CH_4N (D) CH_6N

7. The empirical formula of an organic compound containing 75% C and 25% H is [AIIMS-93]
(A) CH_4 (B) C_2H_6 (C) C_2H_4 (D) C_3H_8

8. The molecular mass of a volatile substance may be measured by [AIIMS-94]
(A) Hoffmann's method (B) Victor Meyer's method
(C) Liebig's method (D) None of these

9. Assertion : The Dumas' method is more applicable to nitrogen containing organic compounds than the Kjeldahl's method. Reason : The Kjeldahl's method does not give satisfactory results for compounds in which nitrogen is directly linked to oxygen. [AIIMS-95]
(A) A (B) B (C) C (D) D

10. A compound with empirical formula CH_2O has a vapour density of 30. Its molecular formula is [AIIMS-98]
(A) $C_2H_2O_2$ (B) $C_2H_4O_2$ (C) $C_3H_6O_3$ (D) $C_6H_{12}O_6$

11. If a compound, on analysis was found to contain C = 18.5% H = 1.55%, Cl = 55.04% and O = 24.81%, then its empirical formula is [AIIMS-98]
(A) CHClO (B) CH_2ClO (C) C_2H_2OCl (D) $ClCH_2O$

12. The empirical formula of a compound is CH_2O. Its molecular weight is 180. The molecular formula of the compound is [AIIMS-99]
(A) $C_5H_{10}O_5$ (B) C_4HO_4 (C) $C_3H_6O_3$ (D) $C_6H_{12}O_6$

13. Beilstein test is used for the detection of [AIIMS-99]
(A) Cl (B) Na (C) N_2 (D) CO_2

14. The weight of a molecule of compound $C_{60}H_{122}$ is [AIIMS-99]
(A) 1.09×10^{-21} g (B) 1.4×10^{-21} g
(C) 5.025×10^{23} g (D) 16.023×10^{23} g

15. In sodium fusion test, the nitrogen of an organic compound is converted into. [JIPMER-99]
(A) Sodium cyanide (B) Sodium nitrate
(C) Sodamide (D) Sodium nitrite

16. Empirical formula of a hydrocarbon containing 80% carbon and 20% hydrogen is [JIPMER-2002]
(A) CH (B) CH_2 (C) CH_3 (D) CH_4

17. $CH \equiv CH \xrightarrow[\text{Pressure}]{Ni(CN)_2} X$ Here X in the reaction [JIPMER-2002]
(A) Benzene (B) Ethane
(C) Cyclo-octatetraene (D) Cyclohexane

18. 15 c.c.of a gaseous hydrocarbon required 45 c.c. of oxygen for complete combustion and 30 c.c. of CO_2 is formed. The formula of the hydrocarbon is [BHU-95]
(A) C_3H_6 (B) C_2H_6 (C) C_4H_{10} (D) C_2H_4

19. Which pair of species have same percentage of carbon? [BHU-99]
(A) CH_3COOH and C_2H_5OH
(B) CH_3COOH and $C_6H_{12}O_6$
(C) $HCOOCH_3$ and $C_{12}H_{22}O_{11}$ (D) $C_6H_{12}O_6$ and $C_{12}H_{22}O_{11}$

20. Haemoglobin contains 0.334% of iron by weight. If the molecular weight of haemoglobin is approximately 67200, then the number of iron atoms (atomic weight of Fe is 56) present in one molecule of haemoglobin are [BHU-99]
(A) 1 (B) 4 (C) 5 (D) 8

21. Aniline is purified by ? [AFMC-94]
(A) Fractional distillation (B) Steam distillation
(C) Distillation (D) None of these

22. A violet colour with sodium nitroprusside in the test of sulphur in organic compound is due to the formation of [AFMC-94]
(A) $Na_4[Fe(CN)_5NOS]$ (B) $Na_3[Fe(CN)_6]$
(C) $Na_2[Fe(CN)_5S]$ (D) None of these

23. Beilstein test is used for [AFMC-95]
(A) N_2 (B) Cl (C) Na (D) CO_2

24. If 0.2 gm of an organic compound containing carbon, hydrogen and oxygen on combustion, yields 0.147 gram of carbon dioxide and 0.12 gram of water, what will be the content of oxygen in compound ? [AFMC-98]
(A) 73.29% (B) 78.45% (C) 83.23% (D) 89.50

25. Empirical formula of a compound is CH_2O. If its molecular weight is 180, then the molecular formula of the compound is [AFMC-99]
(A) $C_6H_{12}O_6$ (B) $C_5H_{10}O_5$ (C) $C_3H_6O_3$ (D) $C_4H_8O_4$

26. The Beilstein test for organic compounds is used to detect [AFMC-99]
(A) Nitrogen (B) Sulphur (C) Carbon (D) Halogens

27. A compound has C, H, N in the following percentage C = 40%, H = 13.33%, N = 46.67%What is its empirical formula ? [AFMC-2000]
(A) CH_2N (B) C_2H_5N (C) CH_3N (D) CH_4N

28. The vapour density of a gas is 22. What is its molecular weight ? [AFMC-2000]
(A) 22 (B) 11 (C) 44 (D) 66

29. The empirical formula of an acid is CH_2O_2, the probable molecular formula of acid may be [AFMC-2000]
(A) CH_2O (B) CH_2O_2 (C) $C_2H_4O_2$ (D) $C_3H_6O_4$

30. In Kjeldahl's method, $CuSO_4$ acts as [AFMC-2001]
(A) Oxidising agent (B) Reducing agent
(C) Hydrolysing agent (D) Catalytic agent

31. Which of the following are arranged in increasing order of their bond strengths ? [AIEEE-2002]
(A) $O_2^- < O_2 < O_2^+ < O_2^{2-}$ (B) $O_2^{2-} < O_2^- < O_2 < O_2^+$
(C) $O_2^- < O_2^{2-} < O_2 < O_2^+$ (D) $O_2^+ < O_2 < O_2^- < O_2^{2-}$

32. In a compound C, H and N atoms are present in 9:1:35 by weight. Molecular weight of compound is 108. Molecular formula of compound is. [AIEEE-2002]
(A) $C_2H_6N_2$ (B) C_3H_4N (C) $C_6H_8N_2$ (D) $C_9H_{12}N_3$

33. 0.2 mole of saturated hydrocarbon 'X' on complete combustion produces 26.4 gm of CO_2. The molecular weight of 'X' is .... [EAMCET-90]

34. An organic compound having carbon and hydrogen has 80% carbon. The empirical formula of the hydrocarbon is [EAMCET-91]
(A) CH (B) CH_2 (C) CH_3 (D) CH_4

35. A compound has an empirical formula CH_2O. Its vapour density is 45. Its molecular formula is [EAMCET-95]
(A) $C_8H_6O_8$ (B) $C_3H_6O_3$ (C) $C_4H_8O_4$ (D) $C_5H_{10}O_5$

36. The molecular weight of an organic compound is 180. Its empirical formula is CH_2O. The molecular formula is [EAMCET-96]
(A) $C_6H_{12}O_6$ (B) $C_7H_{16}O_5$ (C) $C_8H_4O_5$ (D) $C_5H_8O_7$

37. The Lassaigne's solution of an organic compound with

nitrogen, sulphur and chlorine contains [EAMCET(MBBS)-99]

(A) NaCN, Na_2S, NaCl (B) KCN, K_2S, KCl
(C) AgCl, NaCN, K_2S (D) NaNC, $FeSO_4$, KCl

38. In which of the following compounds the weight percentage of C and H are 80 and 20 respectively ? [EAMCET(MBBS)-2000]

(A) CH_4 (B) C_2H_2 (C) C_2H_6 (D) C_6H_6

39. 4 grams of a hydrocarbon on complete combustion gave 12.571 grams of CO_2 and 5.143 grams of water. What is the empirical formula of the hydrocarbon ? [EAMCET-2002]

(A) CH (B) CH_2 (C) CH_3 (D) C_2H_3

40. 0.078 grams of hydrocabon occupy 22.414 ml of volume at STP. The empirical formula of the hydrocarbon is CH. The molecular formula of the hydrocarbon is [EAMCET(MED)-2002]

(A) C_2H_2 (B) C_6H_6 (C) C_8H_8 (D) C_4H_4

41. In Duma's method which of the following is used ? [CPMT-96]

(A) Nitrogen (B) Ammonium Nitrite
(C) Cynogen (D) Cyanobenzene

42. For the separation of 2 immiscible liquid which method is used ? [CPMT-96]

(A) Chromatography (B) Fractional column
(C) Fractional distillation (D) Separating funnel

43. If N and S present in an organic compound during Lassaigne test, then both change into - [CPMT-97]

(A) Na_2S and NaCN (B) NaSCN
(C) Na_2SO_3 and NaCN (D) Na_2S and NaCNO

44. The weight of an atom of a compound $C_{60}H_{22}$ is [CPMT-97]

(A) 842 gm (B) 1.4×10^{-21} gm
(C) 5.025×10^{23} gm (D) $\frac{1}{6.023 \times 10^{23}}$ gm

45. 0.765 gm of acid gives 0.55 g of CO_2 and 0.12 g of H_2O. Then find the ratio of the percentage of carbon and hydrogen ? [CPMT-2001]

(A) 19:2 (B) 18:11 (C) 20:17 (D) 1:7

46. An organic compound contains 54.6% carbon, 9.1%H and 36.56% oxygen. The empirical formula of this organic compound is [CPMT-2003]

(A) $C_2H_2O_2$ (B) C_2H_4O (C) $C_2H_4O_2$ (D) CH_2O

47. A compound has the empirical formula CH_2O. Its vapour density is 30. Its molecular formula is :[PMT-93]

(A) $C_2H_4O_2$ (B) C_2H_6O (C) $C_2H_6O_2$ (D) C_2H_4O

48. An organic compound contains C = 40%, O = 53.34% and H = 6.66%. The empirical formula of compound is [PMT-94]

(A) CH_2O (B) CHO (C) CH_4O_2 (D) C_2H_2O

49. On analysis a saturated hydrocarbon is found to contain 83.70 percent carbon and 16.30% hydrogen. The empirical formula will be: [PMT-95]
(At. wt. of C=12, At. wt. of H=1)

(A) C_3H_6 (B) C_3H_8 (C) C_3H_7 (D) C_6H_{12}

50. 64 gm of an organic compound contains 24 gm of carbon, 8 gm of hydrogen and the rest oxygen. The empirical formula of the compound is : [PMT-96]

(A) CH_2O (B) C_2H_4O (C) CH_4O (D) $C_2H_8O_2$

51. The empirical formula of a compound is CH_2. One molecule of this compound has a mass of 42 g. Its molecular formula is: [PMT-96]

(A) CH_2 (B) C_2H_2 (C) C_3H_8 (D) C_3H_6

52. The empirical formula of an organic compound is CH_2. One mole of this compound has a mass 42 gm. Its molecular formula is : [PMT-2001]

(A) CH_2 (B) C_3H_6 (C) C_2H_2 (D) C_3H_8

53. An organic compound containing carbon, hydrogen and oxygen contains 52.20% carbon and 13.04% hydrogen. Vapour density of the compound is 23. Its molecular formula will be [PMT-2002]

(A) C_2H_6O (B) C_3H_8O (C) C_4H_8O (D) $C_5H_{10}O$

54. The empirical formula of a compound is CH_2. One mole of the compound has a mass of 42 gm. Its molecular formula is [PMT-2002]

(A) CH_2 (B) C_2H_2 (C) C_3H_6 (D) C_3H_8

55. Isomers of propanoic acid are [PMT-2002]

(A) $HCOOC_2H_5$ and CH_3COOCH_3
(B) $HCOOC_2H_5$ and C_3H_7COOCH
(C) CH_3COOCH_3 and C_3H_7O (D) C_3H_7OH and CH_3COCH_3

56. What would be the molecular formula of the alkane 5.6 litres of which was found to weigh 11gm at normal temperature and pressure ? [PMT-2003]

(A) C_4H_{10} (B) C_5H_{12} (C) C_3H_8 (D) C_2H_6

57. An organic compound contains 40% carbon and 6.66% hydrogen. The empirical formula of the compound is [PET-93]

(A) CH_2 (B) CH_2O (C) CHO (D) CHO_2

58. Insulin contains 3.4% sulphur. The minimum molecular weight of insulin is [PET-93]

(A) 350 (B) 470 (C) 560 (D) 940

59. A compound has 50% carbon, 50% oxygen and approx molecular weight of 290. Its molecular formula is [PET-95]

(A) CO (B) C_4O_3 (C) $C_{12}O_9$ (D) C_3O_3

60. A hydrocarbon has C 85.72% and remaining H. The hydrocarbon is [PET-96]

(A) C_2H_4 (B) C_2H_6 (C) C_2H_2 (D) CH_4

61. Empirical formula of hydrocarbon containing 80% carbon and 20% hydrogen [PET-97]
(A) CH (B) CH_2 (C) CH_3 (D) CH_4

62. The total number of possible isomers of trimethyl benzene is [PET-97]
(A) 2 (B) 3 (C) 4 (D) 6

63. An organic compound with C = 40% and H = 6.7% will have the empirical formula : [PET-2000]
(A) CH_2 (B) CH_2O (C) $C_3H_6O_3$ (D) $C_2H_4O_2$

64. An organic compound contains 49.3% carbon, 6.84% hydrogen and its vapour density is 73. Molecular formula of the compound is : [PET-2000]
(A) $C_3H_5O_2$ (B) $C_6H_{10}O_4$ (C) $C_3H_{10}O_2$ (D) $C_4H_{10}O_2$

65. Haemoglobin contains 0.33% of iron by weight. Its molecular weight is 67200. The number of iron atoms in each molecule is : [At. wt. of Fe = 56] [PET-2000]
(A) 2 (B) 3 (C) 4 (D) 5

66. Which one has the highest percentage of nitrogen ? [PET-2001]
(A) Urea (B) Ammonium nitrate
(C) CAN (D) Calcium nitrate

67. An organic compound has been found to possess the empirical formula CH_2O and molecular weight 90. Its molecular formula is [PET-2002]
(A) $C_3H_6O_3$ (B) $C_2H_4O_2$ (C) $C_6H_{12}O_6$ (D) CH_2O

68. An organic compound has an empirical formula CH_2O. Its vapour density is 45. The molecular formula of compound is : [PET-2002]
(A) CH_2O (B) C_2H_5O (C) $C_2H_4O_2$ (D) $C_3H_6O_3$

69. The molecular formula of diphenyl methane,

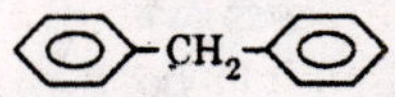

, is $C_{13}H_{12}$. How many structural isomers are possible when one of the How many structural isomers are possible when one of the hydrogens is replaced by a chlorine atom ? [CBSE-PMT-2004]
(A) 6 (B) 4 (C) 8 (D) 7

Empirical & Molecular formulae

1.	(A)	51.	(D)
2.	(B)	52.	(B)
3.	(C)	53.	(A)
4.	(C)	54.	(C)
5.	(A)	55.	(A)
6.	(C)	56.	(C)
7.	(A)	57.	(B)
8.	(B)	58.	(D)
9.	(B)	59.	(C)
10.	(B)	60.	(A)
11.	(A)	61.	(C)
12.	(D)	62.	(B)
13.	(A)	63.	(B)
14.	(B)	64.	(B)
15.	(A)	65.	(C)
16.	(C)	66.	(A)
17.	(C)	67.	(A)
18.	(D)	68.	(D)
19.	(B)	69.	(B)
20.	(B)		
21.	(B)		
22.	(C)		
23.	(B)		
24.	(A)		
25.	(A)		
26.	(D)		
27.	(D)		
28.	(C)		
29.	(B)		
30.	(D)		
31.	(B)		
32.	(C)		
33.	*		
34.	(C)		
35.	(B)		
36.	(A)		
37.	(A)		
38.	(C)		
39.	(B)		
40.	(B)		
41.	(D)		
42.	(D)		
43.	(B)		
44.	(B)		
45.	(D)		
46.	(B)		
47.	(A)		
48.	(A)		
49.	(C)		
50.	(C)		

33. 44

Theoretical Organic Chemistry
Electronic Displacements

Inductive effect:

Polarisation of electrons in σ bonds takes place. Merely, there is displacement of electrons and does not involve actual transfer of electrons. The electrons do not leave the valence shell. It is a permanent effect i.e. the effect is present even in non reacting molecules. The mode of representation of the effect is —>—.

-I effect: Electron attracting or withdrawing or pulling groups show this effect – $NO_2 > CN > F > -Cl > -Br > -I > OH > -OCH_3 > C_6H_5 > H$

+ I effect: Electron pushing or repelling groups show this effect –

$(CH_3)_3C\text{–}>(CH_3)_2CH\text{–}>CH_3\text{–}CH_2\text{–}>\text{–}CH_3>H\text{–}$

Applications:

1. Bond length: Bond length decreases with increasing inductive effect –

$H_3C\text{–}F$	$H_3C\text{–}Cl$	$H_3C\text{–}Br$	$H_3C\text{–}I$
1.38 Å	1.74 Å	1.98 Å	2.14 Å

2. Dipole moment: Due to inductive effect, the covalent bond acquires polar character and develops the property of dipole moment. The more the inductive effect, higher is the dipole moment.
3. Strength of carboxylic acids: –I effect showing group or atom increases the acidic strength whereas the +I effect showing group or atom decreases it. e.g.
   (i) $HCOOH > CH_3COOH > CH_3CH_2COOH$
   (ii) $CH_3CH_2CHClCOOH > CH_3CHClCH_2COOH > ClCH_2CH_2CH_2COOH$
4. Basic character of amines: +I effect showing group or atom increases the basicity of the amine. In general, the order of basicity found in amines is 2° > 1° > 3° > NH_3. This is probably due to steric effects or solvation effects. The order also depends upon the nature of alkyl group(s) attached.
5. Reactivity of the alkyl halides: The +I effect showing group or atom enhances the –I effect of the halide atom thus making the reactivity of alkyl halide more. e.g.

$H_3C\text{–}Cl$; 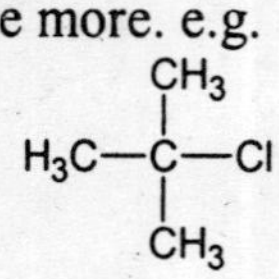 (more reacitve)

Electromeric effect: It generally involves the complete transfer of the π–electrons or lone pair of electrons. It is a temporary effect. It comes into existence in the presence of a reagent. The effect is represented by a curved arrow ↷. This effect is shown by groups like >C=C<, >C=O, —C≡N, —C≡C—, etc.

+E effect: The effect is shown by electron attracting groups. There is the transfer of electrons towards the attacking reagent.

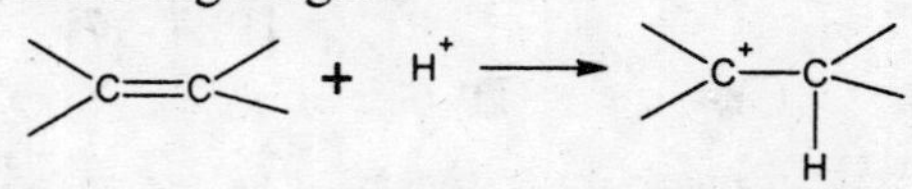

F, Cl, Br, O, –OR etc. show + E effect.

–E effect: The effect is shown by electron repelling groups. Here the electrons are repelled away from the attacking reagent.

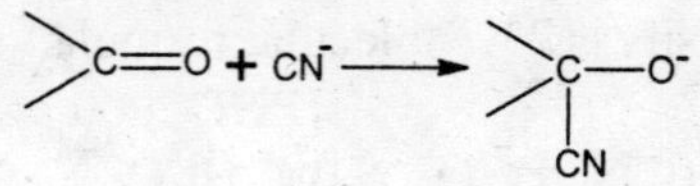

=NR, >C=O, –S, etc show –E effect.

Mesomeric effect:

Occasionally, the term mesomerism is used synonymously with resonance. Complete transfer of π or lone pair of electrons takes place. It is a permanent effect. It takes place in unsaturated and especially in conjugated system via their π – orbitals. It is better known as a conjugative effect.

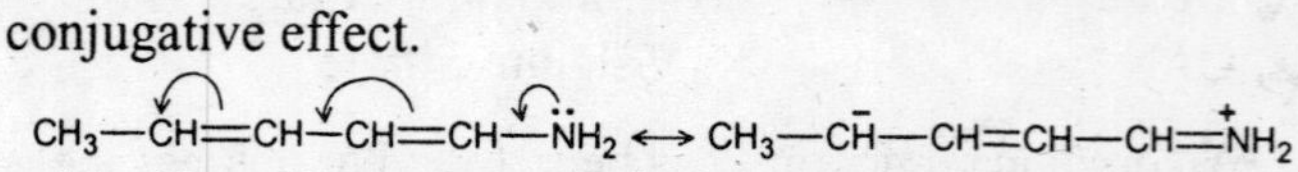

+ M effect: Here, the transfer of electron pair takes place away from the atom

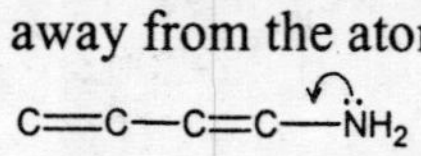

–Cl, –Br, –I, $-NH_2$, –OH, $-OCH_3$ etc. show + M effect.

– M effect: The electrons are transferred towards the atom

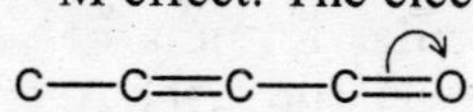

$-NO_2$, —C≡N, >C=O etc. show –M effect.

Mesomerism is applicable only in those cases where a multiple bond is in conjugation with a multiple bond or lone pair of electrons.

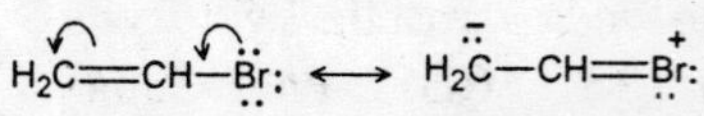

In compounds having cumulative double bonds, mesomeric effect is not possible

$H_2C{=}C{=}C{=}CH_2$

Resonance includes all types of electronic displacements

$H\text{—}H \longleftrightarrow {}^{+}H\text{—}H^{-} \longleftrightarrow H^{-}\text{—}H^{+}$; $H\text{—}Cl \longleftrightarrow {}^{+}H\ Cl^{-}$

When inductive and mesomeric effects are present together in a molecule, they may operate in the same or opposite directions. When they oppose each other, mesomeric effect overpowers.

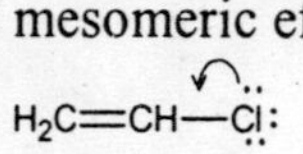

Resonance:

The properties of certain compounds cannot be explained by one structure. In benzene the C–C and C=C bond lengths on the basis of the formula should be 1.54 and 1.34 Å respectively. Actually all C–C bonds have 1.39Å length. Therefore to explain the properties, several structures are taken into account. None of the structures is a real structure. The energy difference between the resonance hybrid molecule and the most stable of the canonical structures is known as resonance energy. Greater the resonance energy, higher is the stability.

Applications:

1. Chlorine of vinyl chloride is not as reactive as the chlorine of allyl halide. Due to resonance the C–Cl bond is shortened

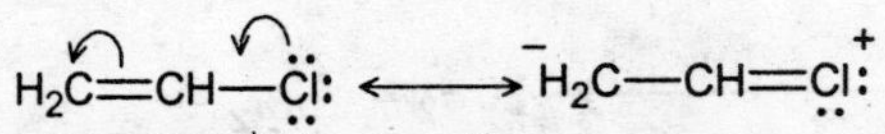

2. $CH_2=CH-{}^+CH_2$ is more stable than $CH_3-CH_2-{}^+CH_2$. This is due to the possibility of the resonance in the former.

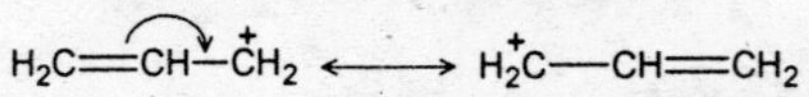

More the number of covalent bonds in a resonating structure, more is its stability.

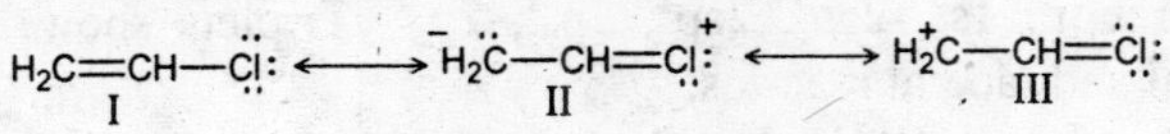

Stability: I > II > III

3. Phenol is more acidic than alcohol

Due to resonance, the electron density on the oxygen atom in phenol diminishes.

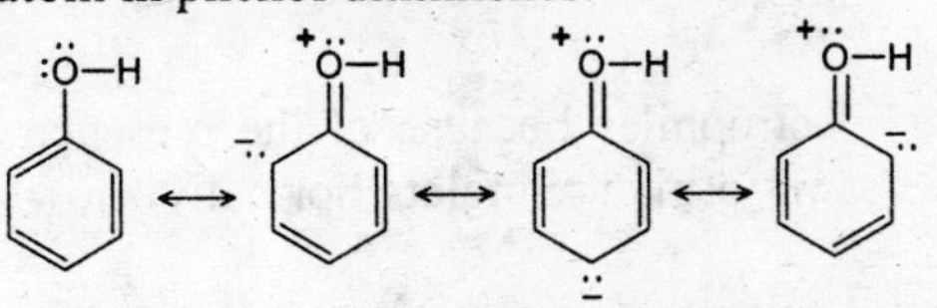

Whereas in alcohols the density of electrons increases due to +I effect of alkyl groups.

R→Ö—H

4. Aromatic amines are less basic than aliphatic amines as resonance is prominent in the former.
5. Acid amides behave as neutral while alkylamines are alkaline to litmus. It is due to resonance in the former.

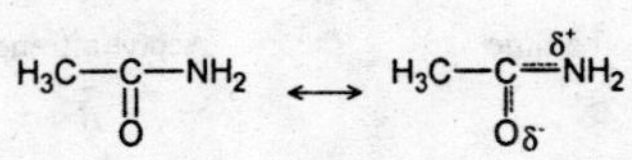

6. Carboxylic acids and their derivatives fail to form cyanohydrins, oximes, hydrazones, sodium bisulphite adducts etc. This is because of the resonance of acids and their derivatives.

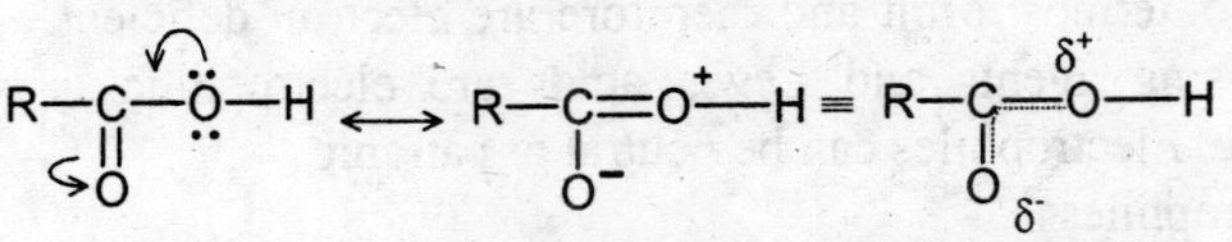

Resonance Hybrid

In compounds which exhibit resonance, bond order can be established as

Bond order = total number of bonds between two atoms/ total number of resonating structures

e.g. Bond order in benzene = 2+1/2 = 1.5 (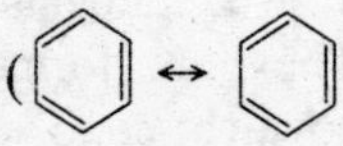 , major contributing structure)

Hyperconjugation or Baker–Nathan effect:

Conjugation of sigma electrons of a single bond with multiple bonds is observed in hyperconjugation. It depends upon the presence of hydrogen atom attached with α – carbon.

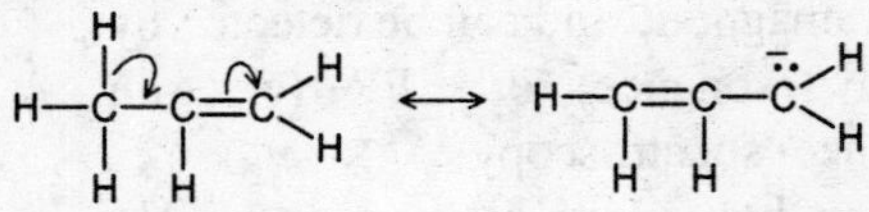

Applications:

1. Propene is more stable than ethene.
2. Reactivity of α – hydrogen atoms: more the electron attracting power of a functional group, stronger is the hyperconjugation and more reactive is the α – hydrogen

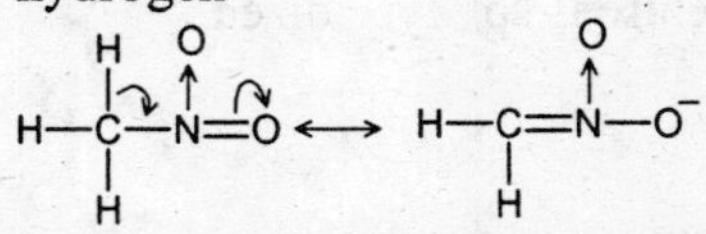

3. Shortening of C–C bond adjacent to a triple bond in acetonitrile can be represented as

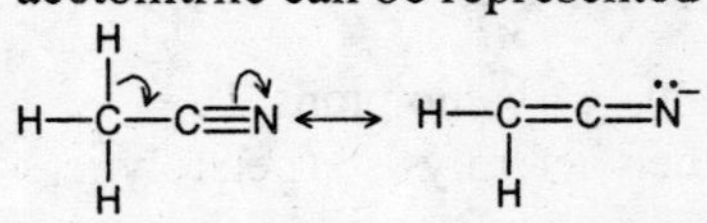

4. Stability of alkyl carbocation or alky free radicals can be explained on the basis of hyperconjugation.
5. Chlorination of toluene is much faster than the chlorination of benzene.

Steric effect:

Due to the presence of bulky groups or atoms in the space around the centre of reaction cite, the reaction is retarded completely.

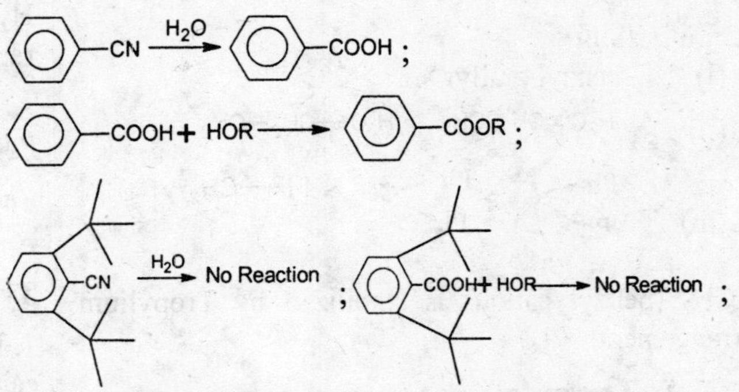

If the reacting centre is away from the bulky groups present at o – position, the reaction is not hindered.

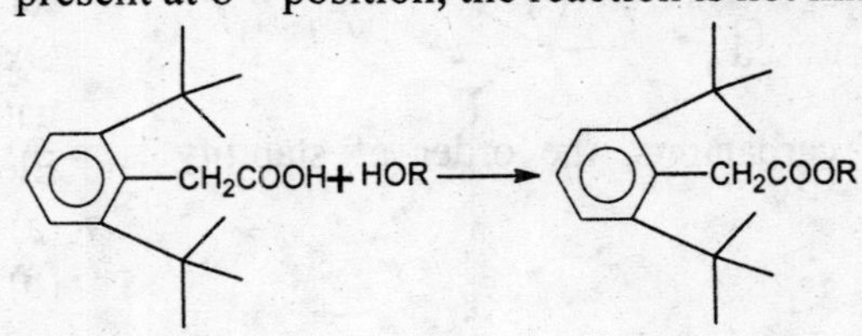

Reactive Intermediates

1. Free Radicals:

They are formed due to homolytic cleavage of a covalent bond.

$$-\overset{|}{\underset{|}{C}}-X \rightleftharpoons -\overset{|}{\underset{|}{C}}^{\bullet} + X^{\bullet} \text{ (homolytic cleavage of a } \sigma \text{ bond)}$$

$$X{=}Y \rightleftharpoons \dot{X}-Y^{\bullet} \text{ (homolytic cleavage of a } \pi - \text{bond)}$$

Free radicals are odd electron species. They are electrically neutral. Due to the presence of unpaired electrons, they are paramagnetic and can be detected by ESR (electron spin resonance) or EPR (electron paramagnetic resonance) spectroscopy.

Homolytic fission can be caused by heat and UV–light. Peroxides in gaseous state also produce free radicals.

Stability order: Benzyl > allyl > 3° > 2° > 1°

The stability can be explained on the basis of (i) +I effect of alkyl groups and (ii) hyperconjugation. The carbon with free radical centre is sp^2 hybridized

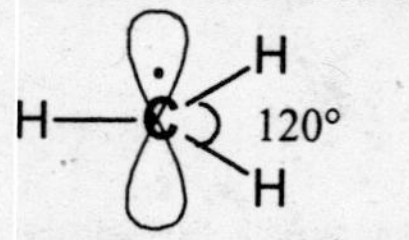

Abstraction of H–atom is one of the main reactions of free radicals, $\dot{R} + \dot{H} \rightarrow R{-}H$

2. Carbocation or carbonium / carbanion ions:

They are formed by the heterolysis (heterolytic fission) of covalent bonds.

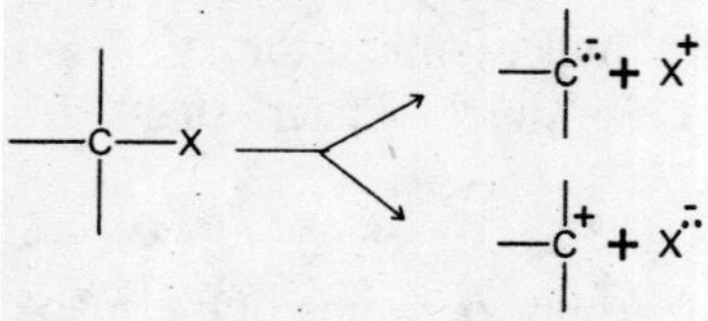

;

bipolar species

Order of stability:

(i) benzyl > allyl > 3° > 2° > 1°

(ii) $H_2C{=}CH{-}\overset{+}{C}H_2 > H_3C{-}CH_2{-}\overset{+}{C}H_2$

(iii) $Ph_3C^+ > Ph_2CH^+ > Ph{-}\overset{+}{C}H_2$

Ph^+CH_2 (benzyl cation) is stabilized by Tropylium rearrangement

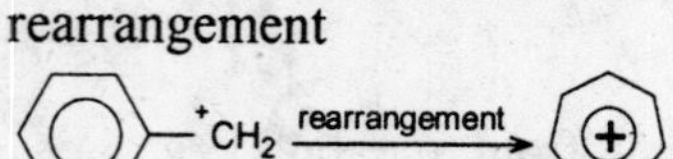

H_3C^+ is sp^2 hybridised

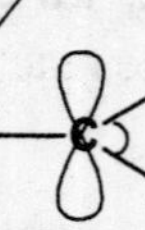

In the case of carbanions, the order of stability is

$\ddot{C}H_3 > 1° > 2° > 3°$.

–I effect showing groups or atoms increase the stability of carbanion. $:\bar{C}H_3$ is sp^3 hybridised.

3. Carbenes:

The carbon atom of carbene $-\ddot{C}-$ is neutral. Carbenes can be generated by

(i) $\underset{\text{diazomethane}}{CH_2N_2} \xrightarrow{h\nu} :CH_2 + N_2$

(ii) $Cl_3C{-}H + \bar{O}H \xrightarrow{-H_2O} Cl_3C:^- \longrightarrow \underset{\text{dichloro carbene}}{Cl_2C:} + Cl^-$

(iii) $CH_2Br_2 + Zn \xrightarrow{h\nu} :CH_2 + ZnBr_2$

The carbenes produced in the above reactions are normally triplets. Triplets have two bond pairs and two non bonding electrons.

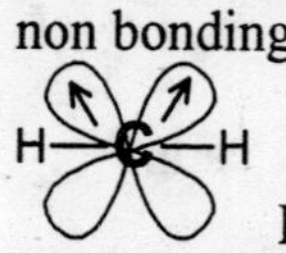

Linear, sp hybridised and has the electronic configuration, $1s^2$, $2sp^1$, $2sp^1$, $2p_y^{\ 1}$, $2p_z^{\ 1}$. Triplets show biradical and radical reactios.

In singlets, both the unshared electrons are present as a pair in the same orbital.

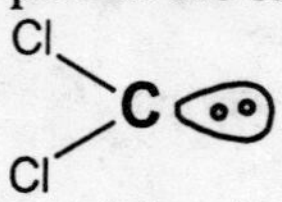

trigonal, sp^2 hybridised $1s^2$, $2(sp^2)^2$ $2(sp^2)^1$ $2(sp^2)^1$ $2p_z^{\ 0}$.

Singlets behave as electrophiles because of the presence of lone pair in the same orbital. Singlets normally show ionic reactions.

4. Nitrenes $-\ddot{N}:$

Here N atom has one σ bond and two pairs of electrons. Nitrenes are neutral. They act as electrophiles generally.

In Hofman degradation/ Bromamide reaction/ Hypobromide reaction, the formation of nitrene can be represented as

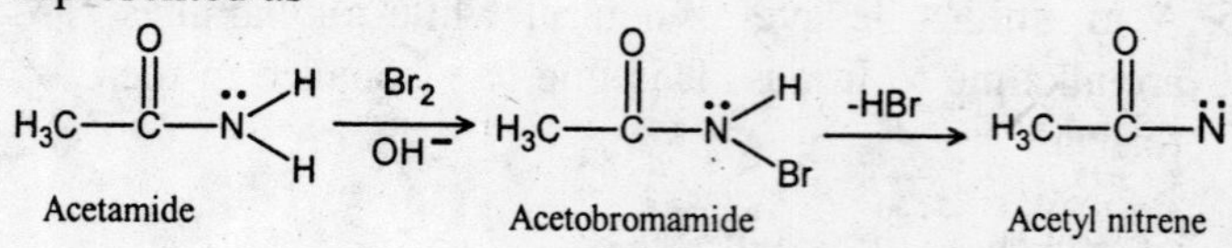

Types of Reagents

1. Free radicals: e.g. $Br^{\bullet}$, $R^{\bullet}$, $C_6H_5\dot{C}H_2$ etc. Triplet carbene can behave as a biradical.

2. Electrophiles: They have less than eight electrons in their outermost orbit and therefore are electron deficient. Oxidising agents and Lewis acids are electrophilic in nature. Electrophiles can be neutral or cationic

Electrophiles:

(a) Neutral (E) e.g. BF_3 , $:CCl_2$, SO_3 etc.

(b) Cationic (E+) e.g. R^+, RC^+O, $^+NO_2$, Cl^+ etc.

3. Nucleophiles: They are electron efficient or capable of donating electrons.

Nucleophiles

(a) Neutral (Nu) e.g. H_2O, R–OH, ROR, NH_3, amines, H_2S, RSH etc.

(b) Anionic (Nu:) e.g. carbanions, Cl^-, OH^-, $^-OCH_3$, $^-CCl_3$, R^-, H^-

Note: some species like for e.g. 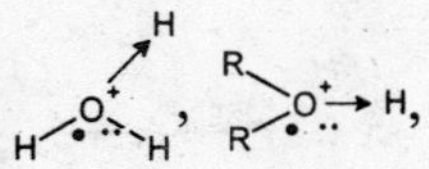

R_3N^+H, 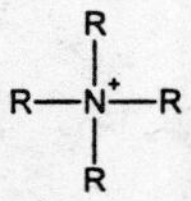 etc. can't be included in any of the above types of reagents. These species have completed octet and therefore cannot accept electron pair for the formation of a covalent bond. As they are positively charged, they cannot donate a pair of electrons for the formation of a covalent bond.

Types of Reactions

Generally the following types of reactions are encountered in organic chemistry.

1. Substitution reactions: These reactions are exhibited by saturated molecules without a heteroatom with lone pair of electrons. e.g.

$$CH_4 + Cl_2 \xrightarrow{h\nu} CH_3Cl + HCl$$

2. Addition reactions: These are shown by unsaturated compounds and saturated molecules with hetero atoms containing lone pair of electrons.

e.g.

(i) $H_3C{-}CH{=}CH_2 + HBr \longrightarrow H_3C{-}CHBr{-}CH_3$

(ii) $R{-}\ddot{O}{-}H + H^+ \longrightarrow R{-}\overset{+}{O}(\rightarrow O){-}H$

3. Elimination reactions:

$$H_3C{-}CHBr{-}CH_3 \xrightarrow{\text{Alcoholic KOH}} H_3C{-}CH{=}CH_2$$

4. Rearrangement reactions:

(i) $H_3C{-}CH(X){-}CH{=}CH_2 \longrightarrow H_3C{-}CH{=}CH{-}CH_2X$

(ii) $H_3C{-}CH_2{-}CH_2{-}CH_3 \longrightarrow H_3C{-}CH(CH_3){-}CH_3$

Mechanism of Organic Reactions

1. Free radical mechanism: it takes place generally in three steps. Let us consider a free radical addition reaction.

e.g. $R{-}CH{=}CH_2 + HBr \xrightarrow[\text{or } O_2]{\text{peroxide}} R{-}CH_2{-}CH_2Br$

Step I: Initiation $\quad R{-}O{-}O{-}R \longrightarrow 2R\dot{O}$;

$$R{-}\dot{O} + H{-}Br \longrightarrow ROH + Br^{\bullet}$$

Step II: Propagation

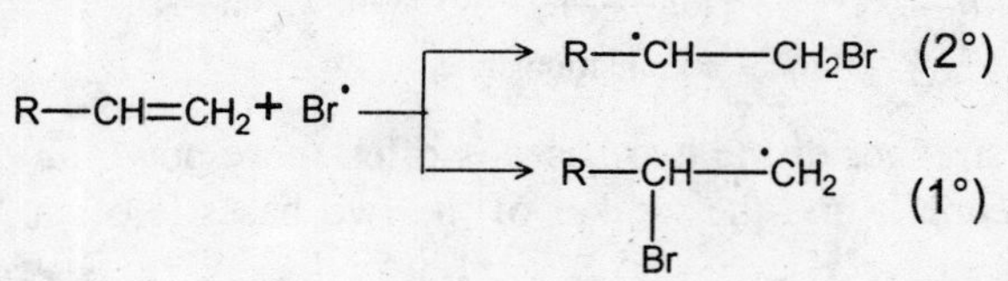

As secondary free radical is more stable, it forms predominantly and the reaction further proceeds as

$$R{-}\dot{C}H{-}CH_2Br + H{-}Br \longrightarrow R{-}CH_2{-}CH_2Br + Br^{\bullet}$$

This type of reaction is possible only with HBr. HF, HCl and HI do not show this type of addition (peroxide effect).H–F (135 kJ/mole) and H–Cl (103 kJ/mole) are highly polar due to high electronegativity of F and Cl respectively. Therefore these bonds undergo preferably heterolysis and not homolysis. H–I (71 kJ/mole) should undergo homolysis very easily in the presence of a peroxide (R–O–O–R). Contrary to this, Iodine atom formed on homolysis of HI due to its size is lethargic and moves very slowly. Therefore, the chances of its effective collisions with alkene are very low i.e. reactivity is very low.

2. Ionic mechanism: Ionic mechanism may be due to nucleophilic or electrophilic ones.

Nucleophilic addition: Addition reaction of carbonyl compounds are nucleophilic addition reactions.

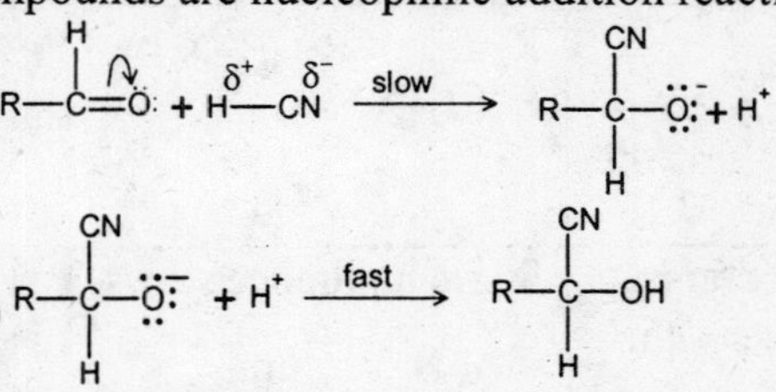

Intermediate oxide ion

The typical $>C{=}O$ group reactions of aldehydes and ketones are not shown by acids and their derivatives as a result of resonance hybrid.

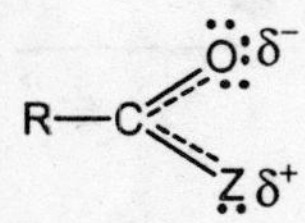

Nucleophilic substitution:

S_N1 : It takes place in two steps

(i) $R{-}X \xrightarrow{\text{slow}} R^+ + \bar{X}$

(ii) $R^+ + \dot{N}u \xrightarrow{\text{fast}} R{-}Nu$

The slow step is the rate determining step. Only one reacting species i.e. R–X is undergoing change in the concentration. Therefore it is o a first order reaction.

S_N^2: This can be represented as

$$Nu^{-}: + R—X \longrightarrow [Nu\text{------}R\text{------}X]^{-} \xrightarrow{-\ddot{X}^{-}} Nu—R$$

Transition state

The stronger of the two bases is often more powerful nucleophile and the weaker of the two bases is often the better leaving group. The differences between S_N^1 and S_N^2 can be better understood as follows:

S_N^1	S_N^2
It is a two step process	One step process
Carbocation / Carbonium ion forms	Transition state formation occurs
Concentration of the nucleophile does not effect the rate of reaction	Concentration affects the rate
Low concentration of nucleophile leads for S_N^1	High concentration of nucleophile leads for S_N^2
Weak nucleophiles such as H_2O, ROH etc. favour S_N^1	Powerful nucleophiles such as OH^-, RO^-, etc. favour S_N^2
Optically active substrate on S_N^1 forms racemic mixture	Due to Walden inversion opposite enantimorph forms
Polar conditions favour S_N^1 (due to carbocation formation) $R_3C–X > R_2CH–X > RCH_2–X$ (where 'X' is the same)	Polar conditions less favourable for S_N^2 $R_3C–X < R_2CH–X < RCH_2–X$
Greater steric hindrance on 'C' atom enhances the rate of S_N^1 reaction R–I > R–Br > R–Cl > R–F (where 'R' is same)	Steric hindrance slows the S_N^2 reaction R–I < R–Br < R–Cl < R–F
The order of reactivity in the case of halides is 3° > 2° > 1° > CH_3X	The order is 3° < 2° < 1° < CH_3X

Electrophilic addition:

It can be represented as

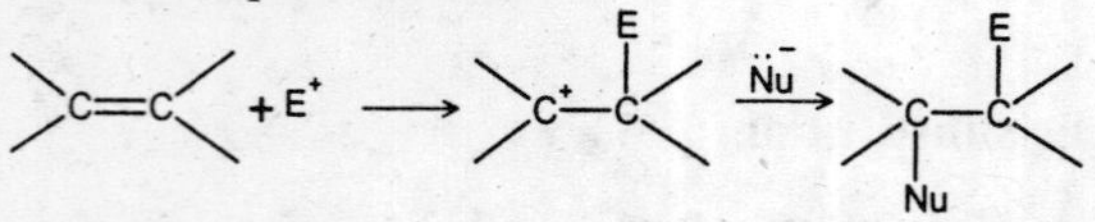

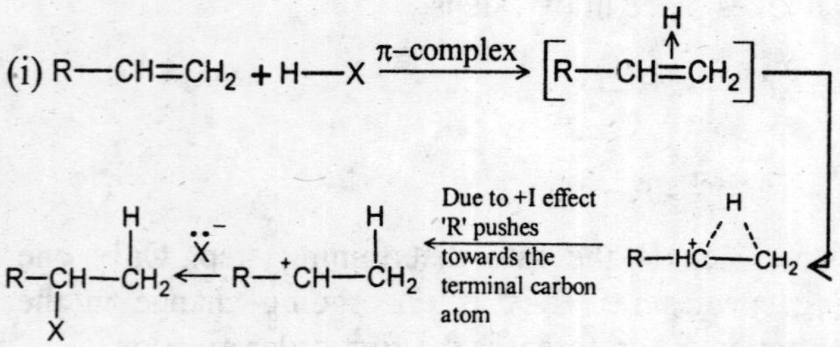

(ii) $R—CH=CH_2 + H—X \xrightarrow{X^-} R—\overset{+}{C}H—CH_3 \xrightarrow{X^-} R—CH(X)—CH_3$

Electrophilic substitution

S_E^1 it takes place in two steps

(i) $R—X \xrightarrow{slow} R^- + X^+$

(ii) $R^- + E^+ \xrightarrow{fast} R—E$

it is a first order reaction.

S_E^2 : It can be represented as

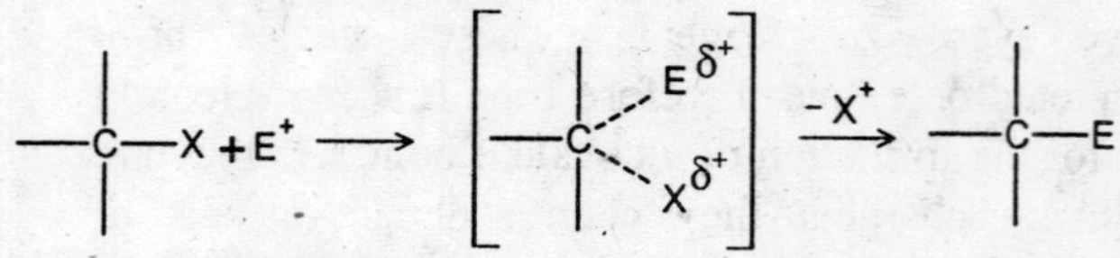

it can be exemplified as

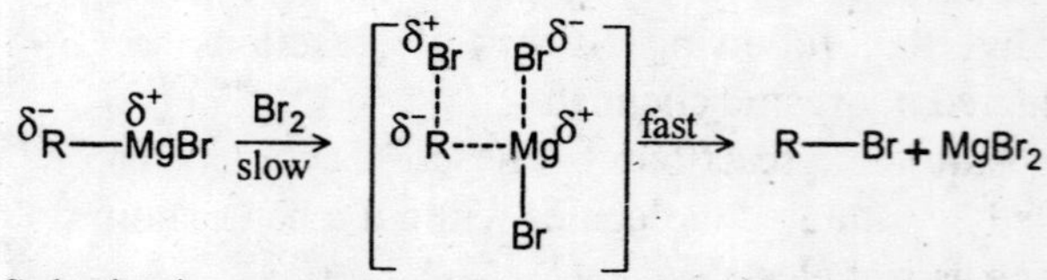

Substitution at unsaturated or aromatic carbon atoms (for e.g. nitration, halogenation, sulphonation, alkylation, acylation etc.) takes place by electrophilic substitution. Halogenation of aldehydes, ketones, acids (Hell–Volhard–Zelinsky reaction) are electrophilic substitution reactions.

Isomerism

Structural Isomerism	Stereoisomerism
They differ in the manner in which the constituent groups or atoms are linked to one another within the molecule.	Same molecular formula, same structural formula but different three dimensional arrangement.
a). Nuclear or chain isomerism	a). Conformational isomerism
b). Positional isomerism	b). Configurational isomerism
c). Functional isomerism	i). Diastereomerism
d). Metamerism	ii). Enantiomerism
e). Tautomerism	

Structural Isomerism

a). Nuclear or chain isomerism: one of the isomers is a normal chain one, while other(s) are branched chain ones. e.g.

(i) C_4H_{10} Butane — $CH_3-CH_2-CH_2-CH_3$ n- butane ; $CH_3-CH(CH_3)-CH_3$ 2-methylpropane (isobutane)

(ii) C_5H_{12} Pentane ; $CH_3-CH_2-CH_2-CH_2-CH_3$ n- pentane ; $CH_3-CH(CH_3)-CH_2-CH_3$ 2-methylbutane (isopentane) ;

$CH_3-C(CH_3)_2-CH_3$ 2, 2-dimethylpropane (neopentane)

(iii) C_6H_{14} Hexane — $CH_3-CH_2-CH_2-CH_2-CH_2-CH_3$ n- Hexane

$CH_3-CH(CH_3)-CH_2-CH_2-CH_3$ 2-methylpentane ; $CH_3-CH_2-CH(CH_3)-CH_2-CH_3$ 3-methylpentane

$CH_3-CH(CH_3)-CH(CH_3)-CH_3$ 2, 3-dimethylbutane ; $CH_3-C(CH_3)_2-CH_2-CH_3$ 2, 2-dimethylbutane

(iv) $H_2C{=}CH-CH_2-CH_3$ but - 1 - ene ; $H_2C{=}C(CH_3)-CH_3$ 2-methylpropene

(v) $CH_3-CH_2-CH_2-CH_2NH_2$ 1-aminobutane ; $CH_3-CH(CH_3)-CH_2NH_2$ 1-amino-2-methyl-propane ;

$CH_3-C(CH_3)_2-NH_2$ 2-amino-2-methyl propane

b). Positional isomerism: The position of the functional group differs in the isomers

e.g. (i) $CH_3-CHOH-CH_3$ 2–propanol;
$CH_3-CH_2-CH_2OH$ 1–propanol

(ii) $CH_2{=}CH-CH_2-CH_3$ but–1–ene;
$H_3C-CH{=}CH-CH_3$ but–2–ene

(iii) 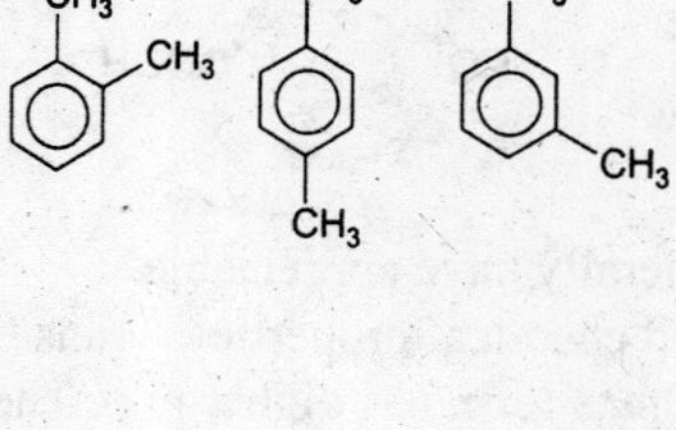 xylenes

(iv) 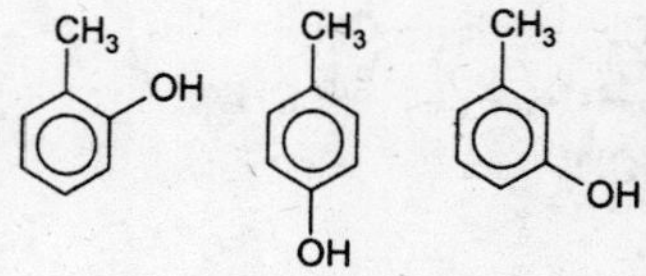 cresols

c). Functional isomerism: The isomers have different functional groups

(i) $C_nH_{2n+2}O$: For this formula alcohols and ethers are possible.

(a) C_2H_6O : H_3C-CH_2OH; $H_3C-O-CH_3$

(b) $C_4H_{10}O$: $H_3C-CH_2-CH_2-CH_2OH$;
$H_3C-CH(CH_3)-CH_2OH$; $H_3C-C(CH_3)_2-OH$;
$H_3C-CHOH-CH_2-CH_3$; $H_3C-CH_2-O-CH_2-CH_3$;
$H_3C-O-CH_2-CH_2-CH_3$; $H_3C-O-HC(CH_3)_2$

(ii) $C_nH_{2n}O$: For this formula aldehydes, ketones and unsaturated alcohols are possible.
C_3H_6O: H_3C-CH_2-CHO ; $H_3C-CO-CH_3$;
$H_2C{=}CH-CH_2OH$

(iii) $C_nH_{2n}O_2$: For this formula Acids and Esters are possible.
$C_3H_6O_2$: H_3C-CH_2-COOH; $H_3C-COO-CH_3$;
$H-COO-C_2H_5$

(iv) C_3H_9N : $H_3C-CH_2-CH_2-NH_2$ primary amine;
$H_3C-CH_2-NH-CH_3$ secondary amine;
$CH_3N(CH_3)_2$ tertiary amine.

Ring–Chain isomerism: They are treated as functional isomers e.g.

$H_3C-CH{=}CH_2$ propene ; 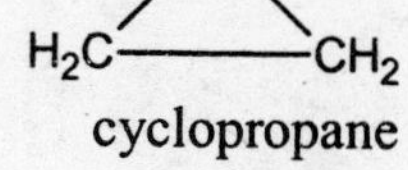 cyclopropane

d). Metamerism: Same functional group is present in all the isomers e.g.

(a) $C_4H_{10}O$: $H_3C-CH_2-O-CH_2-CH_3$;
$H_3C-O-CH_2-CH_2-CH_3$;
$H_3C-O-HC(CH_3)_2$

All are ethers. The arrangement of alkyl groups is different.

(a) $C_4H_{11}N$: $H_3C—CH_2—NH—CH_2—CH_3$;
$H_3C—NH—CH_2—CH_2—CH_2—CH_3$;

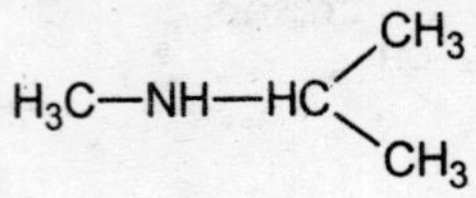

e). Tautomerism: it is also known as] desmotropism or cationotrophy. Here the oscillation of a hydrogen atom between two polyvalent atoms takes place. e.g.

(i) $H—C\equiv N \rightleftharpoons C\equiv N—H$

(ii) 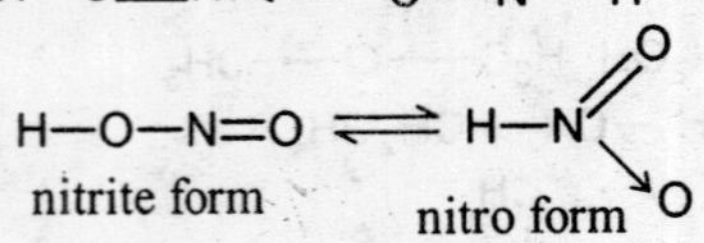

nitrite form nitro form

(iii) $H_3C-\overset{O}{\overset{||}{C}}-CH_2-\overset{O}{\overset{||}{C}}-OC_2H_5 \rightleftharpoons H_3C-\overset{OH}{\overset{|}{C}}=CH-\overset{O}{\overset{||}{C}}-OC_2H_5$

acetoacetic ester

At room temperature acetoacetic ester is a mixture of 92.5% (keto) + 7.5% (enol). Above room temperature, the keto form is more stable.

Stereoisomerism

a). Conformation: It is due to the free rotation of C–C single bond. In all conformations, the bond angle and bond length remains the same. Eclipsed and staggered forms are the two extremes. The others (many possible) are skew forms. e.g. ethane, butane etc.
Stability: Staggered > Skew > Eclipsed
The conformational isomers or rotomers cannot be separated into individual ones as the energy barrier is very less (E > Staggered by 12.6 kJ/mole).
Cyclohexane shows chair and boat conformations. The chair form is more stable and less in energy (27 kJ/mole) than the boat form. In the boat form, many hydrogens on adjacent carbon correspond to the unfavourable eclipsed conformation.

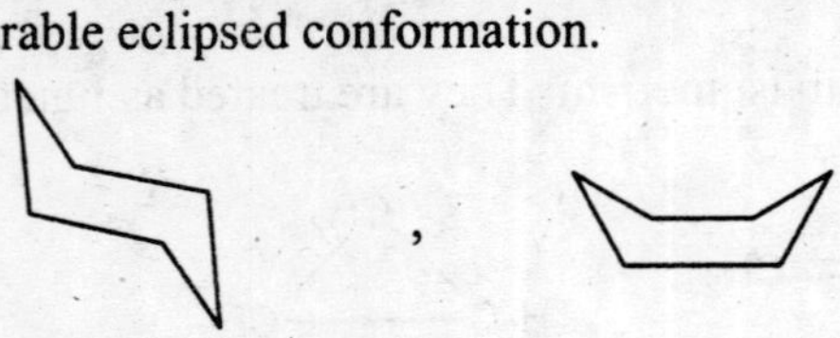

Chair Boat

b) Geometrical isomerism: It is possible if C=C double bond is present and different groups or atoms are attached to carbon atoms. They are not converted to one another due to the absence of free rotation about the double bond. If a ring system is present between carbon atoms, then also the free rotation is restricted and therefore geometrical isomerism is possible.

e.g. (i)

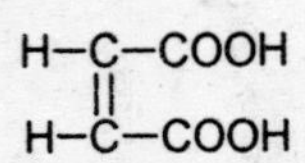

maleic acid **(cis)**

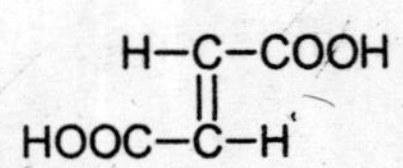

fumaric acid **(trans)**

(ii)

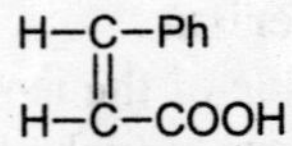

cinnamic acid **(cis)**

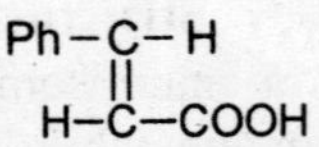

cinnamic acid **(trans)**

(iii)

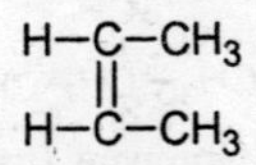

but-2-ene **(cis)**

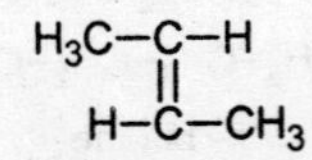

but-2-ene **(trans)**

In cis form, the identical groups or atoms are on the same side of the double bond while in trans form, they are on the opposite sides.

(iv)

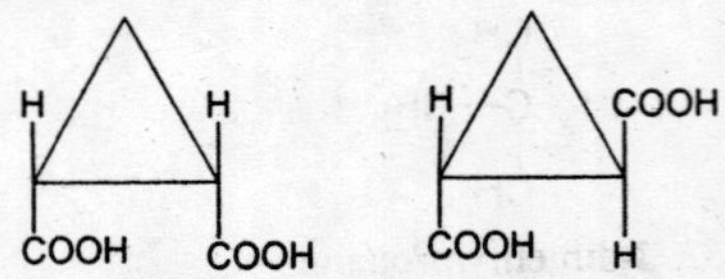

cis **trans**

cyclopropane-dicarboxylic acid

The compounds having a carbon to nitrogen double bond and nitrogen to nitrogen double bond also exhibit geometrical isomerism.

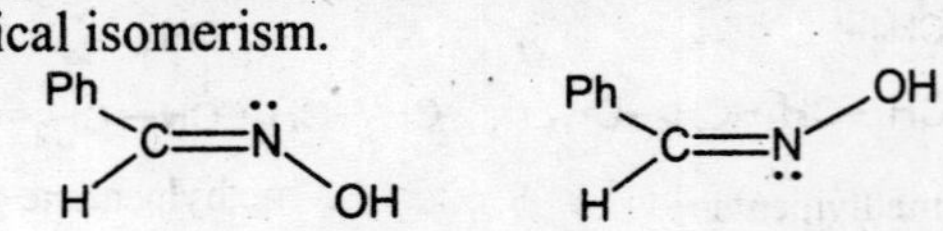

syn **anti**

benzaldoxime

In case of aldoximes, the syn-isomer is the one where hydrogen and hydroxyl groups lie on the same side of the double bond, whereas in the anti-isomer the two are on the opposite side. Azobenzene can be represented as

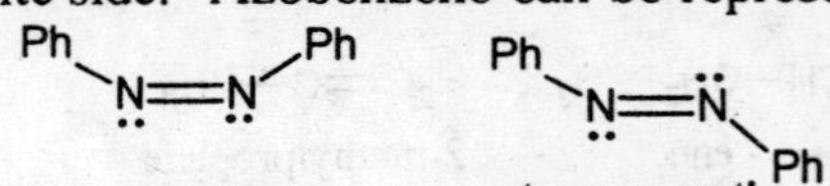

cis or **syn** **trans** or **anti**

azobenzene

If the alkene is tri or tetra substituted, it is not possible to decide whether they are cis or trans because no two groups are same. To designate such systems E – Z system has been used. In this system, the groups or atoms attached are given priority on the basis of atomic number of the atom joined to carbon atom involved in double bond formation. In case two or more atoms are same, then atoms next to them are taken into consideration (sequence rules). If the groups of the same priority are on the same side of the double bond, the alkene is designated as Z form (Z = Zusammen i.e. together) while in the reverse case, the alkene is designated as E form (E = Entgegen i.e. opposite). e.g.

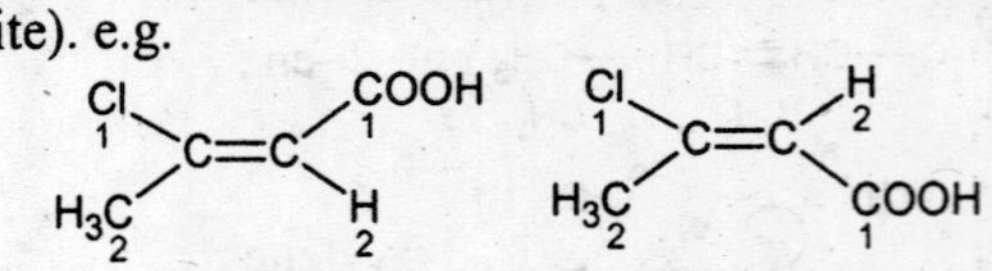

Z - form **E - form**

Geometrical isomers generally have appreciable difference in physical and chemical properties. Trans form has no dipole moment. Trans form has higher m.p. The cis form has higher refractive index, b.p., solubility, density etc. The trans form is more stable. Trans form has higher λ_{max} and ε_{max} values. Maleic acid forms anhydride as both the carboxylic groups are on the same side.

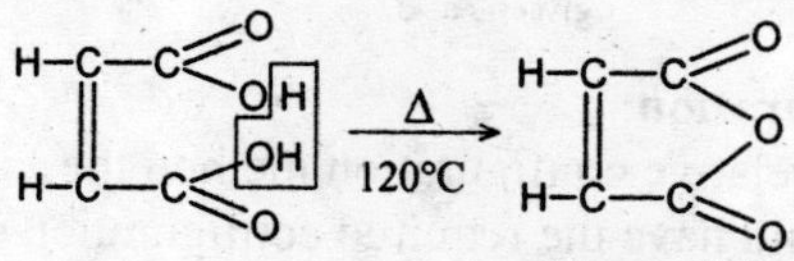

maleic acid maleic anhydride

Enantiomerism (optical isomerism)

For a molecule to show optical activity, it should be a chiral one. A chiral molecule is one which dose not possess:

(a). plane of symmetry
(b). an alternating axis of symmetry
(c). centre of symmetry

A molecule which does not possess any kind of symmetry is asymmetric. A molecule with no plane of symmetry is dissymmetric. Dissymmetric molecules may have an axis of symmetry but the main criteria is that a given molecule should not be super-imposable on its mirror image. Therefore, we may say that all asymmetric molecules are dissymmetric molecules but all dissymmetric molecules are not necessarily asymmetric. Non-dissymmetric molecules are super-imposable upon their mirror images.
The presence of an asymmetric or chiral carbon atom may lead to the chirality of the molecule but not necessarily always. A chiral carbon atom is one whose four valences are satisfied by different groups or atoms. The number of possible isomers can be derived from 2^n (n = number of asymmetric carbon atoms). A chiral molecule depending upon its configuration may rotate the plane polarised light towards right (dextro rotatory) or left (laevo rotatory). If H and OH groups are present in the molecule, one can know the type of rotation by connecting H to OH via. the group or atom whose molecular mass is less. If the direction is clockwise, it is dextro and anti-clockwise, it is laevo rotatory. e.g.

(i) 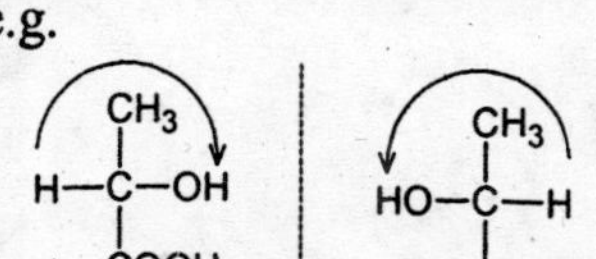

d-lactic acid or (+) lactic acid | l-lactic acid or (-) lactic acid

In the case of lactic acid, we have d, l and dl (racemic mixture) varieties. Racemic mixture is optically inactive as it is a mixture of 50% d and 50% l molecules, leading to inter-molecular or external compensation of the rotations. Dextro (d) and laevo (l) are notations of rotations.

(ii) 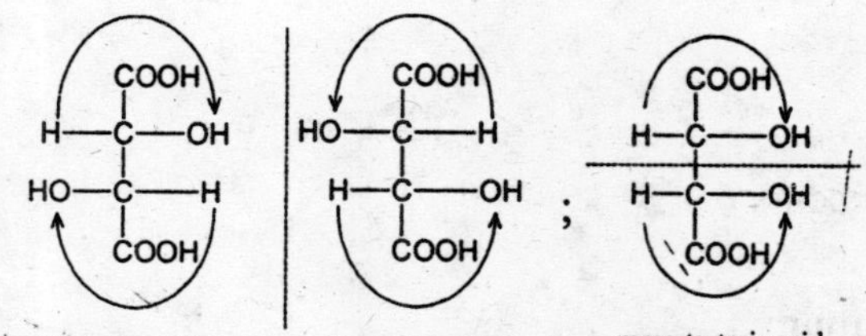

d-tartaric acid or (+) tartaric acid | l-tartaric acid or (-) tartaric acid ; meso-tartaric acid

In the case of tartaric acid we have optically active d, l and optically inactive meso and dl varieties. The meso tartaric acid is inactive due to intra molecular or internal compensation. Due to the presence of plane of symmetry meso tartaric acid is achiral, although it contains two chiral carbon atoms.

(iii) 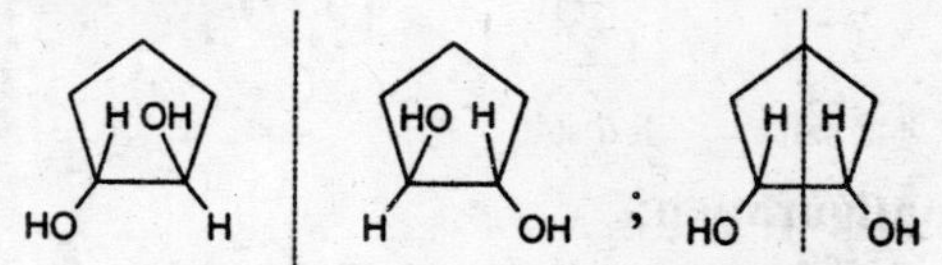

active ; inactive

1,2 –cyclopentanediol

Chiral molecules which do not possess chiral carbon atoms like allenes, spiranes, biphenyls etc. exhibit optical isomerism. e.g.

(iv) 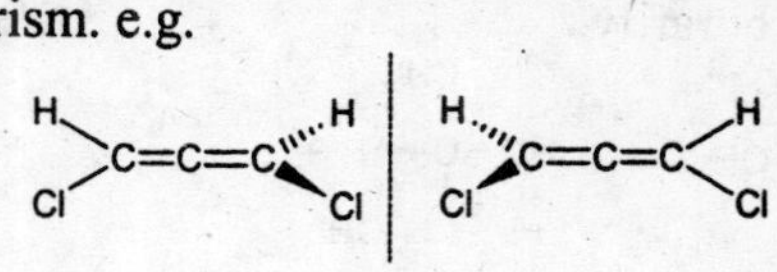

allenes

Enantiomers:
Mirror reflections of each other but not super imposable are enantiomers. e.g. d and l tartaric acids. Enanatiomers have same m.p., b.p., refractive index, solubility etc. They have the same chemical properties generally. They differ in rotation. They have different biological properties.

Diastereomers:
Stereo isomers which are not mirror reflections of each other are diastereomers. e.g. d and meso or l and meso tartaric acids. Geometrical isomers are also diastereomers.

Racemisation:
Conversion of d or l form into dl (racemic mixture). Here 50 % of the starting active enantiomorph is converted into the other enantiomorph thus giving a 50% d and 50% l mixture.

Walden inversion:
Here, the starting active enantiomorph is converted into the other active enantiomorph completely due to the inversion of configuration. The inversion involves S_N2 mechanism. e.g.

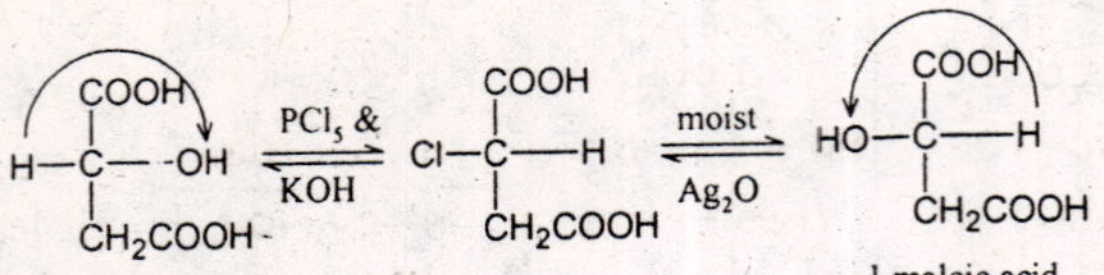

d-maleic acid   l-maleic acid

Resolution:

Separation of racemic mixture into d and l forms is resolution. One of the most important methods involves the salt formation. e.g. let us say the racemic mixture of lactic acid is treated with dextro rotatory N–Brucin (an alkaloid). The d and l forms, form d–d salt and l–d salt which differ in their properties, thus allowing their separation.

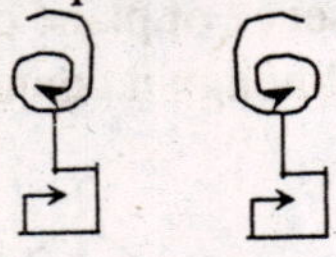

d–d salt   l–d salt

Relative configuration:

Arrangement of groups or atoms in space in a molecule is referred to as its configuration. To indicate the stereo chemical relationship than merely to indicate the actual direction of rotation, glyceraldehyde was chosen (Rosanoff) as standard and the configurations were assigned as D and L glyceraldehyde arbitrarily.

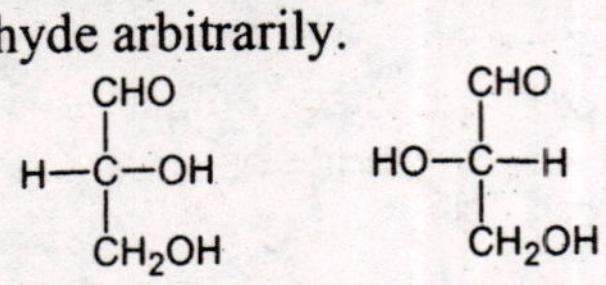

D-glyceraldehyde   L-glyceraldehyde

All those compounds which can be prepared from or converted to D-glyceraldhyde belong to D-series. Similarly, those which have configurational similarity to L-glyceraldehyde belong to L-series. The following configurations are labelled as D and L series.

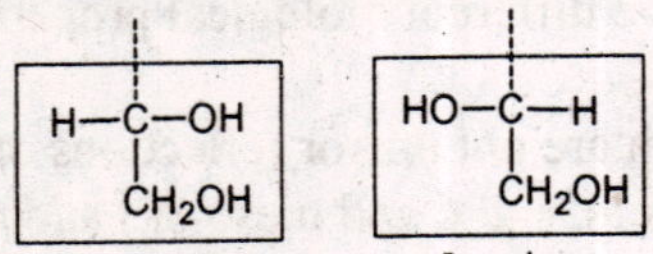

D-series   L-series

D and L are concerned only to show the configuration at chiral carbon atom. D and L and d and l do not have any interrelationship. This can be understood as follows.

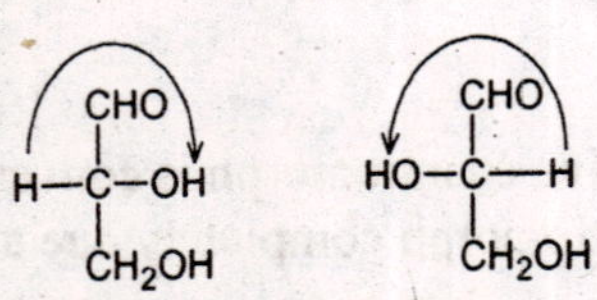

D (+) or D, d glyceraldehyde   L(+) or L, l glyceraldehyde

If D-glyceraldehyde is subjected to mild oxidation the glyceric acid formed belongs to D-series but has laevo rotation.

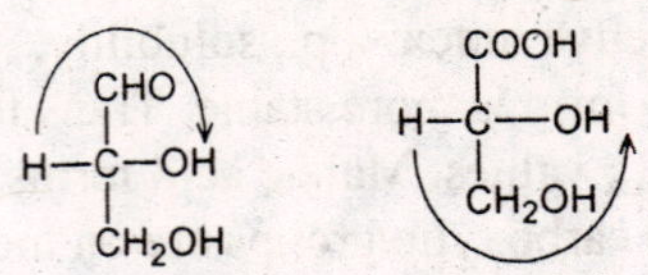

D (+) or D, d glyceraldehyde   D (-) or D, l glyceric acid

Absolute Configuration:

According to the relative configuration method the compound(s) should have the required configuration of H and OH around the chiral carbon. In the case of compounds which do not have these groups, the configuration can't be assigned. Therefore, a new convention for specifying the configuration of chiral carbon atom was given by Cahn, Ingold and Prelog. Here the order of priority of the atoms directly attached to the chiral carbon is given on the basis of the atomic numbers (sequence rules). e.g.

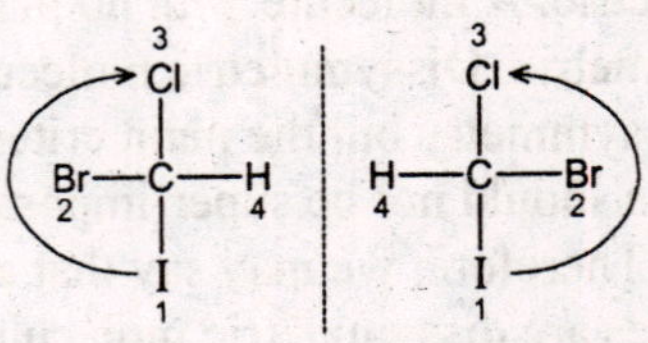

R-form   S-form

The isomer is viewed from the side opposite to the group of lowest priority and the atoms or groups are joined from top priority group to the second and then the third priority group. If it is in clockwise direction, the configuration is considered as R and in the case of anti clockwise direction, it is termed as S.

1. Among the following compounds the one which exhibits geometrical isomerism is [CBSE-90]

(A) 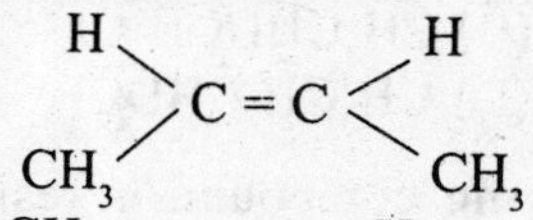

(B) 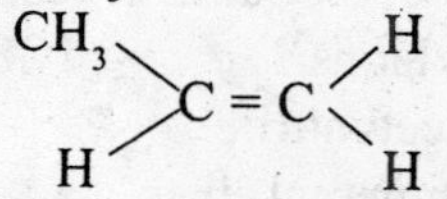

(C) 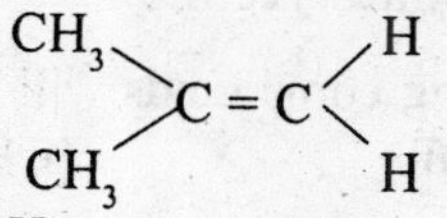

(D) 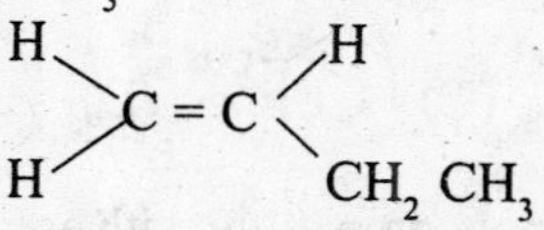

2. Which of the following compounds can exist as optical isomers ? [CBSE-90]

(A) 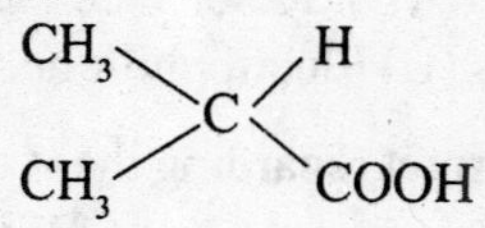 (B) 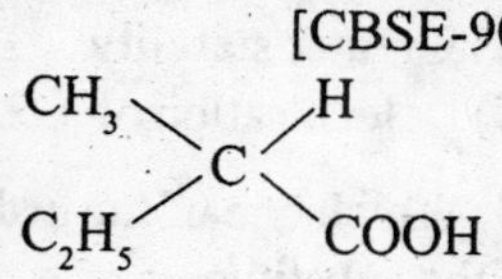

(C) 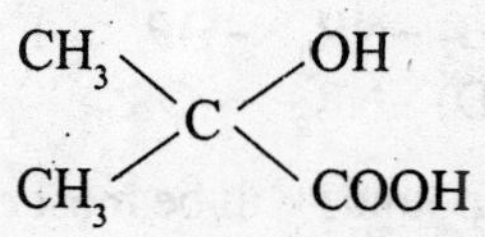 (D) 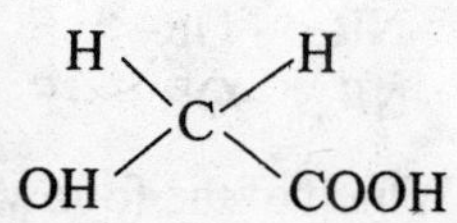

3. Ethoxy ethane and methoxy propane are [CBSE-90]
(A) geometrical isomers (B) optical isomers
(C) functional group isomers (D) metamers

4. Which one of the following is not found in alkenes ? [CBSE-90]
(A) Chain isomerism (B) Geometrical isomerism
(C) Mesomerism (D) Position isomerism

5. In the following molecule the two carbon atoms marked by asterisk * possess the following type of hybridization orbitals. [CBSE-91]

$$H_3{*}C - {*}C \equiv {*}C - CH_3$$

(A) sp^3 orbitals (B) sp^2 orbitals
(C) sp orbitals (D) s orbitals

6. Which of the following is the most stable ? [CBSE-91]
(A) $CH_3CH_2^+$ (B) $(CH_3)_2\ CH^+$
(C) $(CH_3)\ C^+$ (D) $C_6H_5CH_2^+$

7. Isomers must have the same [CBSE-91]
(A) structural formulae (B) molecular formulae
(C) chemical properties (D) physical properties

8. Geometrical isomerism will be shown by [CBSE-92]
(A) 2-butene (B) 1-butene (C) ethylene (D) propylene

9. The most reactive compound for electrophilic nitration will be [CBSE-92]
(A) toluene (B) benzoic acid
(C) nitrobenzene (D) benzene

10. Find the strongest acid from amongst the following [CBSE-92]
(A) CF_3COOH (B) CBr_3COOH
(B) $CH_3\ COOH$ (D) CCl_3COOH

11. On bromination propionic acid yields two isomeric 2-bromopropionic acids. This pair will be an example of. [CBSE-92]
(A) optical isomers (B) cis-trans isomers
(C)chain isomers (D) position isomers

12. The restricted rotation about carbon-carbon double bond in 2-butene is due to [CBSE-93]
(A) overlap of one s and one sp^2 hybridized orbitals
(B) overlap of two sp^2 hybridized orbitals
(C) overlap of one p and one sp^2 hybridized orbitals
(D) overlap of two p orbitals

13. Which is the correct symbol relating the two Kekule benzene ? [CBSE-93]
(A) $\rightleftarrows$ (B) $\rightarrow$ (C) $\rightleftharpoons$ (D) $\leftrightarrow$

14. Chiral molecules are those which are [CBSE-94]
(A) unstable molecules (B) show geometrical somerism
(C) superimposable on their mirror images
(D) not superimposable on their mirror images

15. Which of the following statements is false about resonance contributing structures ? [CBSE-94]
(A) Contributing structures contribute to the resonance hybrid in proportion of their energies
(B) Equivalent contributing structures make the resonance very important
(C) Contributing structures represent hypothetical molecules having no real existence
(D) Contributing structures are less stable than the resonance hybrid

16. Which of the following compounds reacts slower than benzene in electrophilic brom ? [CBSE-94]

(A) 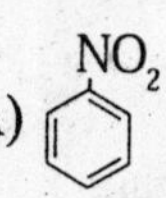 (B) 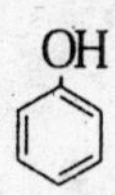 (C) 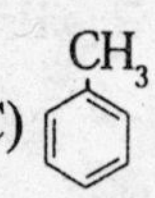 (D) 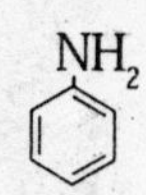

17. Which of the following orders is incorrect with respect to the property indicated ? [CBSE-94]
(A) Formicacid >Aceticacid > Propanoicacid (Acid-strength)
(B) Cyclohexanol < Phenol < Benzoic acid (Acid-strength)
(C) Benzamide<Cyclohexylamine<Aniline(Base-strength)
(D) Fluoroacetic acid>Chloroacetic acid>Bromoacetic acid (Acid-strength)

18. Meso-tartaric acid is optically inactive due to the presence of [CBSE-94]
(A) geometrical isomerism (B) two chiral carbon atoms
(C) molecular symmetry (D) external compensation.

19. In which of the compounds below is there more than one kind of hybridization (sp, sp^2, sp^3) for carbon?

[CBSE-95]

(A) $CH_3CH_2CH_2CH_3$ (B) CH_3- CH = CH -CH_3
(C) CH_2 = CH-CH= CH_2 (D) H-C ≡ C-H

20. Examine the following common chemical structures to which simple functional groups are often attached.

(i) 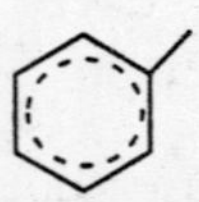 (ii) 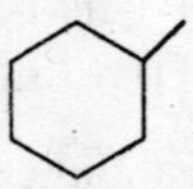 (iii) 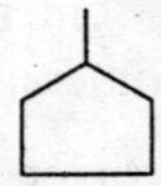

(iv) $CH_3CH_2CH_2CH_2$– (v) $H_2C = CH_2$

Which of these systems have essentially planar geometry? [CBSE-95]

(A) (i) and (v) (B) (ii) and (iii)
(C) (ii), (iii) and (iv) (D) (iv)

21. The number of isomers in C_4H_{10} will be [CBSE-96]

(A) 5 (B) 7 (C) 6 (D) 3

22. Which of the following has zero dipole moment? [CBSE-96]

(A) cis 2-butene (B) trans 2-butene
(C) 1-butene (D) 2-methyl-1-propene.

23. The reaction

$CH_2 = CH - CH_3 + HBr \rightarrow CH_3CHBr - CH_3$ is [CBSE-96]

(A) nucleophilic addition (B) electrophilic addition
(C) electrophilic substitution (D) free radical addition.

24. Electrophile in the case of chlorination of benzene in the presence of $FeCl_3$ is [CBSE-96]

(A) Cl^+ (B) Cl^- (C) Cl (D) $FeCl_3$

25. The reaction is described as [CBSE-97]

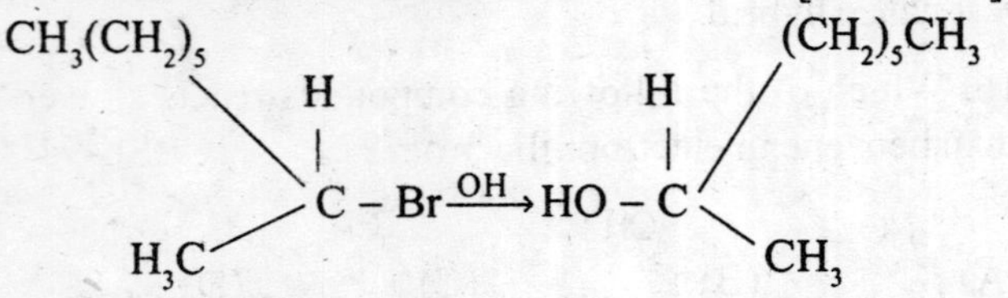

(A) S_E^2 (B) S_N^1 (C) S_N^2 (D) S_N^0

26. Tautomerism will be exhibited by [CBSE-97]

(A) $(CH_3)_3CNO$ (B) $(CH_3)_2NH$
(C) R_3CNO_2 (D) RCH_2NO_2

27. The most stable comformation of n-butane is [CBSE-97]

(A) skew boat (B) eclipsed
(C) gauche (D) staggered-anti

28. Among the following compounds (I-III) the correct reaction with electrophilic is [CBSE-97]

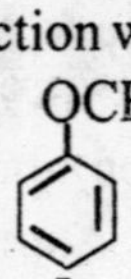 I  II 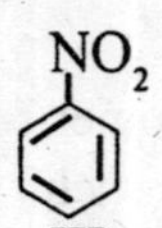 III

(A) II>III>I (B) III<I<II (C) I>II>III (D) I=II>III

29. Which one of the following compounds will be most easily attacked by an electrophile [CBSE-98]

(A) $DCH_2CH_2CH_2Cl$ (B) CH_3CHDCH_2Cl
(C) $CH_3CHClCH_2D$ (D) CH_3CH_2CHDCl

30. Which of the following compound is resistant to nucleophilic attack by hydroxyl ions? [CBSE-98]

(A) Methyl acetate (B) Acetonitrile
(C) Acetamide (D) Dimethyl ether

31. Which one of the following compounds will be most easily attacked by an electrophile [CBSE-98]

(A) 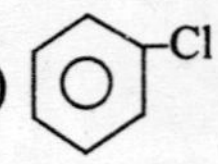 (B)  (C) 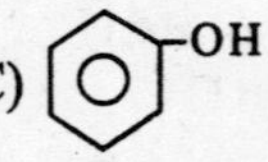 (D) 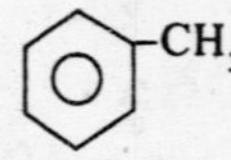

32. Which one of these is not compatible with arenes? [CBSE-98]

(A) Greater stability (B) Resonance
(C) Delocalsation of π–electrons (D) Electrophilic addition

33. Which one of the order is correct regarding the -I effect of the substituents ? [CBSE-98]

(A) $-NR_2 < -OR < -F$ (B) $-NR_2 > -OR < -F$
(C) $-NR_2 < -OR > -F$ (D) $-NR_2 > -OR > -F$

34. Which of the following compounds will be most easily attacked by an electrophile ? [CBSE-99]

(A)  (B) 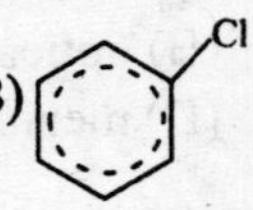 (C) 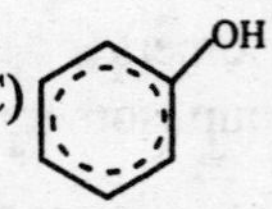 (D) 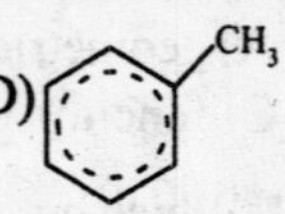

35. R & S forms of a pair of enantiomers differ from each other in. [CBSE-2000]

(A) Optical radiation of polarized light
(B) Solubility in racemic mixture
(C) Reaction with racemic mixture (D) None of these

36. 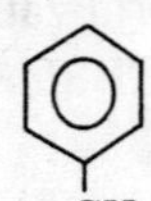 I 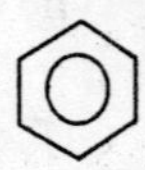 II 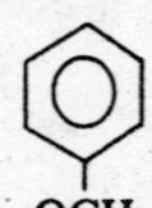 III 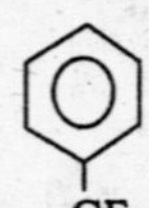 IV

Arrange in order of electrophilic substitution reactions. [CBSE-2000]

(A) I > II > III > IV (B) IV > I > II > III
(C) III > IV > II > I (D) II > IV > III > I

37. Bond angle (dihydral H–C–H) in staggered conformation of C_2H_6 is. [CBSE-2000]

(A) 120° (B) 60° (C) 0° (D) 90°

38. Polarization of acrolein may be written as. [CBSE-2000]

(A) $\overset{-\delta}{C}H_2 = CH - \overset{\delta+}{C}H = O$ (B) $\overset{-\delta}{C}H_2 = CH - \overset{\delta+}{C}H = O$
(C) $\overset{\delta-}{C}H_2 = \overset{\delta+}{C}H - CH = O$ (D) $\overset{\delta-}{C}H_2 = CH - \overset{-\delta}{C}H = O$

39. A compound of molecular formula C_7H_{16} shows optical isomerism the compound will be [CBSE-2001]

(A) 2,3-dimethyl pentane (B) 2,2-dimethyl butane

(C) 2-methyl hexane (D) none of these

40. 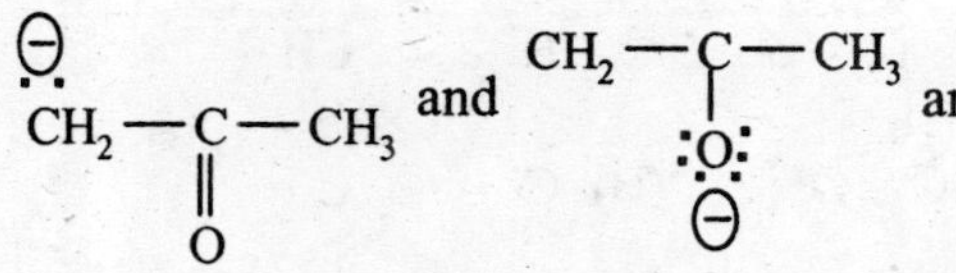 are :-

[CBSE-2002]

(A) Resonating structures (B) Tautomers
(C) Geometrical isomers (D) Optical isomers

41. Geometrical isomers differ in [CBSE-2002]
(A) Position of functional group (B) Position of atoms
(C) Spatial arrangement of atoms
(D) Length of carbon chain

42. The correct order of reactivity towards the electrophilic substitution of the compounds aniline(I), benzene(II) and nitrobenzene(III) is [CBSE-2003]
(A) I > II > III (B) III > II > I
(C) II > III > I (D) I < II > III

43. Which one of the following is a free-radical substitution reaction ? [CBSE-2003]
(A) $CH_3CHO + HCN \rightarrow CH_3CH(OH)CN$
(B) 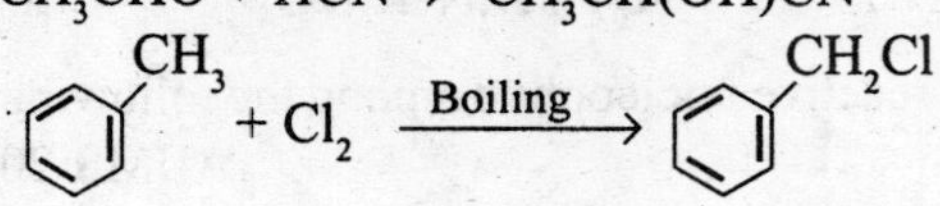
(C) 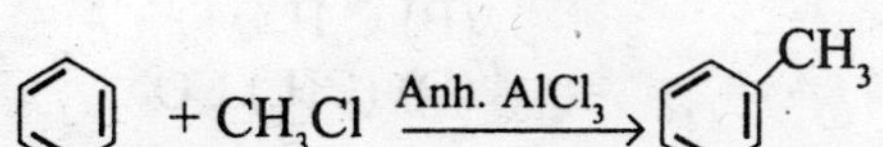
(D) 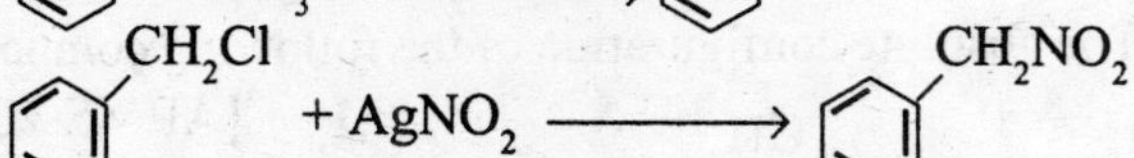

44. Which of the follwing pairs of compounds are enantiomers? [CBSE-2003]
(A) 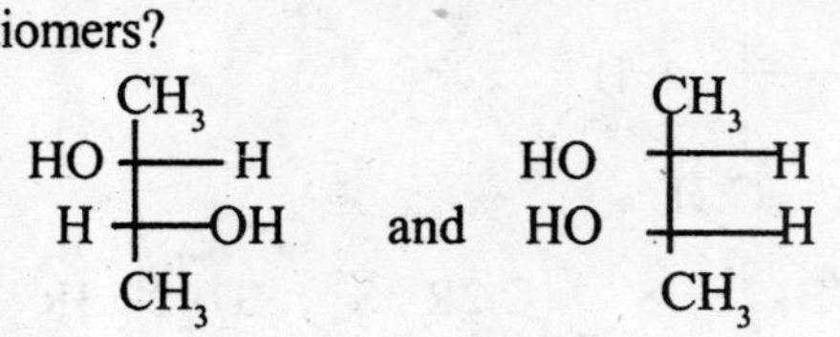
(B) 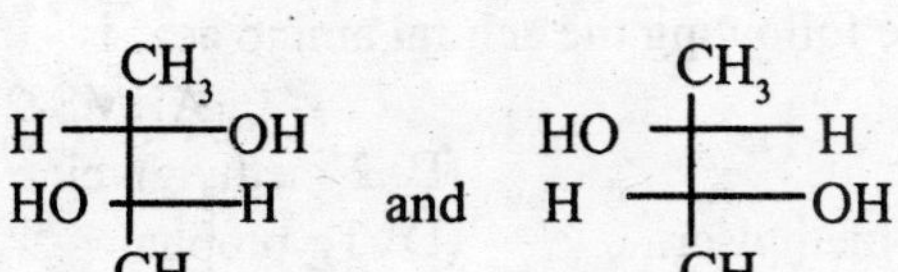
(C) 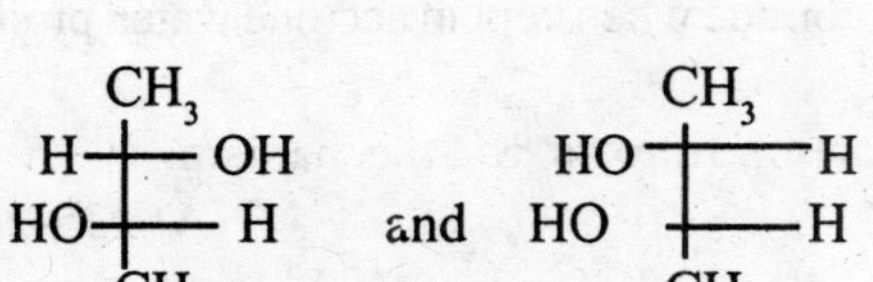
(D) 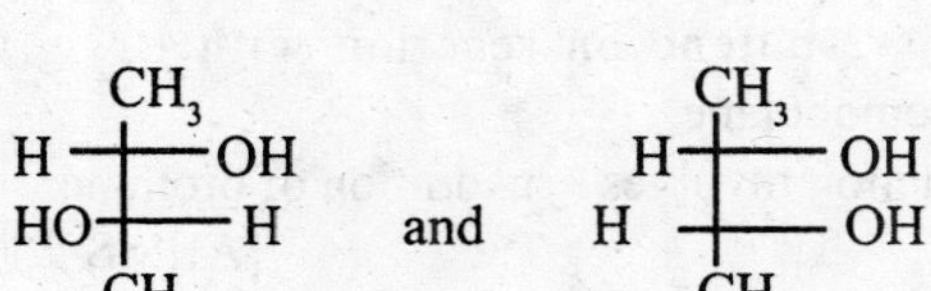

45. Which type of isomerism among the following, correctly describes the isomeric relationship between the following two isomers (i and ii) [AIIMS-93]
i. $CH_3 - CH - OHCH_2CH_2 - CH_3$
ii. $CH_3 - CH_2 - OCH_2 - CH_2 - CH_3$
(A) Chain isomerism (B) Functional isomerism
(C) Position isomerism (D) Carbon-skeleton isomerism

46. Indicate which olefin amongst the following has the E-(ie.trans) configuration. [AIIMS-93]
(A) 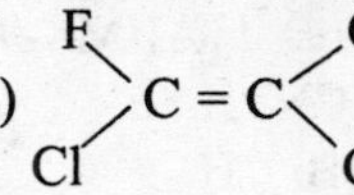 (B) 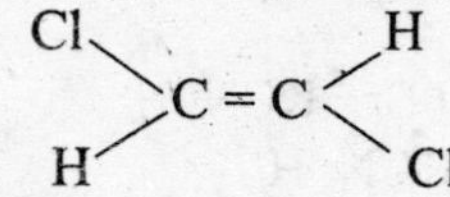
(C) 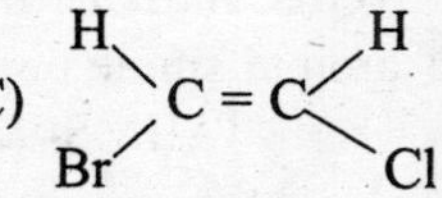 (D) 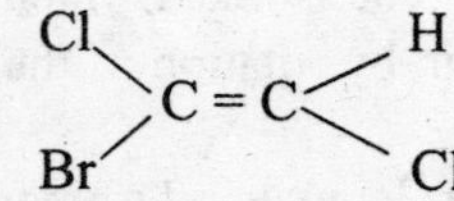

47. Which of the following compounds would be the most susceptible to electrophilic attack by nitronium ions ? [AIIMS-93]
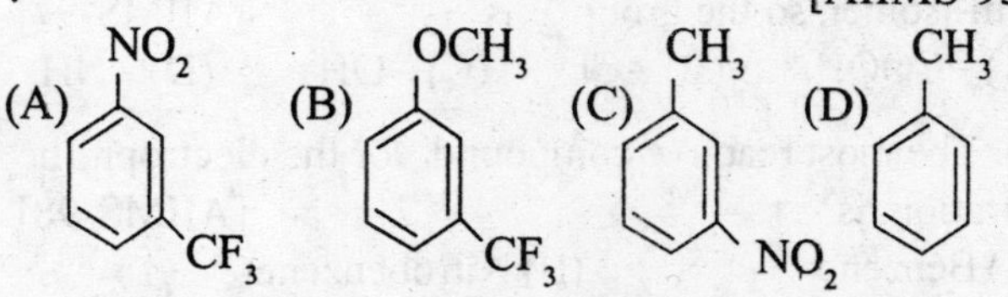

48. Heterolytic fission of an organic covalent bond give only [AIIMS-94]
(A) Free radicals (B) Cation
(C) Anion (D) both (b) and (c)

49. The type of isomerism not exhibited by alkenes is [AIIMS-94]
(A) Metamerism (B) Stereoisomerism
(C) Chain isomerism (D) Position isomerism

50. Which of the following compounds will be most easily attacked by an electrophile ? [AIIMS-94]
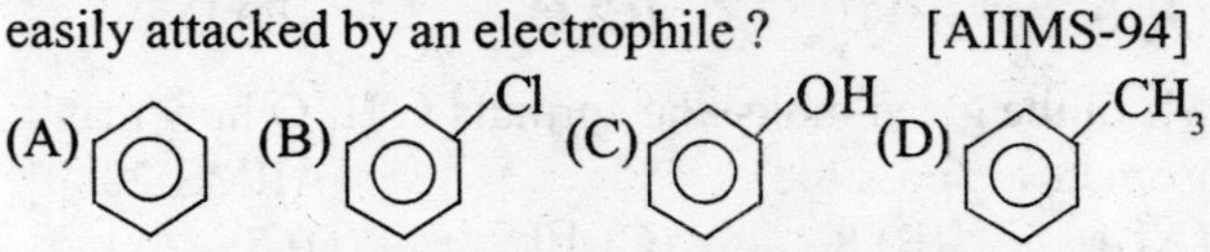

51. Which of the following carbonium ion will be most stable ? [AIIMS-96]
(A) CH_3^+ (B) $(CH_3)_3C^+$
(C) $(CH_3)_2CH^+$ (D) $CH_3CH_2^+$

52. Which of the following alkanes is optically active ? [AIIMS-96]
(A) Propane (B) 2-methylbutane
(C) 3-methylhexane (D) 2,3,4-trimethylpentane

53. The number of enantiomers of the compound $CH_3CHBrCHBrCOOH$ is [AIIMS-97]
(A) 0 (B) 1 (C) 3 (D) 4

54. The major product (70% to 80%) of the reaction between m-dinitrobenzene with NH_4HS is [AIIMS-97]
(A) 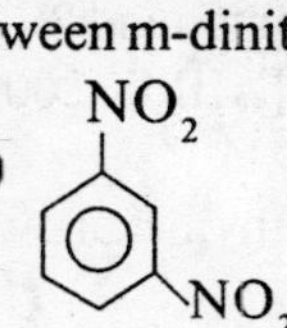 (B) 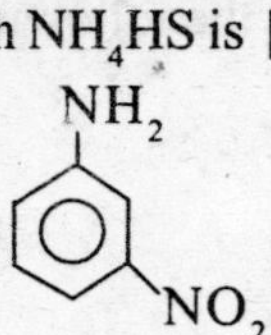

(C) 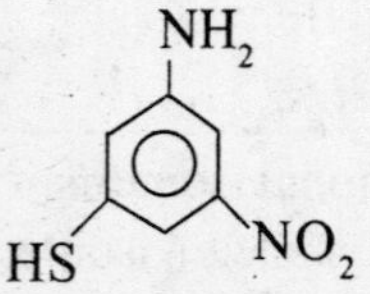 (D) 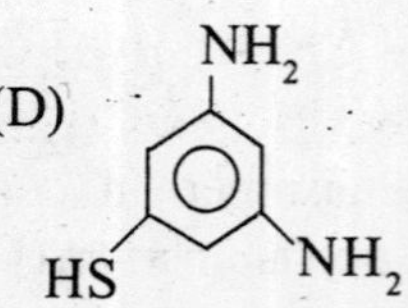

55. R–CH_2–CH_2OH can be converted into RCH_2CH_2COOH. The correct sequence of the regents is [AIIMS-97]

(A) PBr_3, KCN, H^+ (B) PBr_3, KCN, H_2
(C) KCN, H^+ (D) HCN, PBr_3, H^+

56. The isomers which can be converted into another form by rotation of the molecules around single bond are [AIIMS-97]

(A) Geometrical isomers (B) Conformers
(C) Enantiomers (D) Diastereomers

57. In a reaction of C_6H_5Y, the major product (> 60%) is m-isomer, so the group Y is [AIIMS-97]

(A) –COOH (B) –Cl (C) –OH (D) –NH_2

58. The most reactive compound, for the electrophillic nitration, is [AIIMS-98]

(A) Benzene (B) Nitrobenzene
(C) Benzoic acid (D) Toluene

59. The meta-directing group in the following is [AIIMS-99]

(A) NO_2 (B) CH_3 (C) OH (D) NH_2

60. Assertion : Metamers can also be chain insomers or position isomers. Reason : The term tautomerism was introduced by Maxwell in order to explain the chemical reactivity of a substance according to two possible structures. [AIIMS-99]

(A) A (B) B (C) C (D) D

61. In the given molecular formula $C_4H_{10}O$ how many isomers are possible. [AIIMS-2000]

(A) 4 (B) 8 (C) 10 (D) 7

62. Which of the following compound has optical isomer ? [AIIMS-2000]

(A) CH_3–CH(Cl)–CH_3 (B) 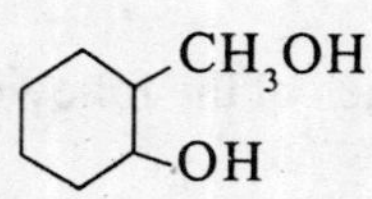

(C) CH_3–CH(OH)–CH(=O) (D) CH_3–CH_2–CH(OH)–CH_2CH_3

63. Which carbo-cation is more stable ? [AIIMS-2001]

(A) $\overset{\oplus}{C}H_3$ (B) $\overset{\oplus}{C}H_2CH_3$ (C) $(CH_3)_2\overset{\oplus}{C}H$ (D) $(CH_3)_3\overset{\oplus}{C}$

64. Which of the following is a chiral compound ? [AIIMS-2002]

(A) Hexane (B) Methane
(C) n-butane (D) 2,3,4-trimethylhexane

65. Which of the following has the highest dipole moment ? [AIIMS-2002]

(A) 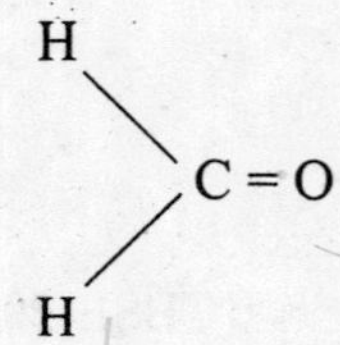 (B) 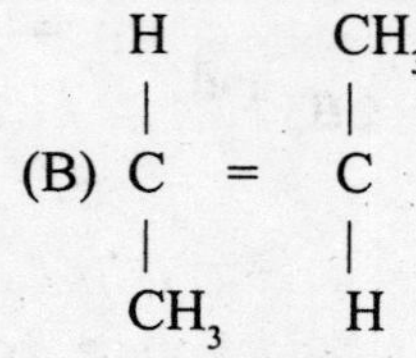

(C) 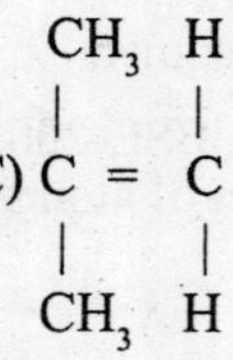 (D) 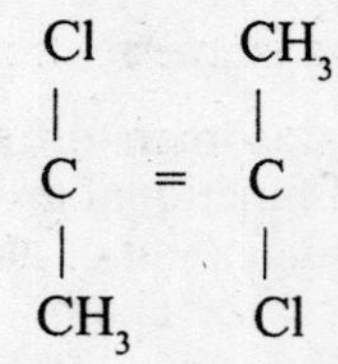

66. Among the following the strongest acid is - [AIIMS-2003]

(A) CH_3COOH (B) C_6H_5COOH
(C) m-$CH_3OC_6H_4COOH$ (D) p-$CH_3OC_6H_4COOH$

67. Among the following the weakest base is - [AIIMS-2003]

(A) $C_6H_5CH_2NH_2$ (B) $C_6H_5CH_2NHCH_3$
(C) $O_2NCH_2NH_2$ (D) CH_3NHCHO

68. The most reactive nucleophile among the following is - [AIIMS-2003]

(A) CH_3O (B) C_6H_5O
(C) $(CH_3)_2CHO$ (D) $(CH_3)_3CO$

69. The absolute configuration of the following compound is : [AIIMS-2003]

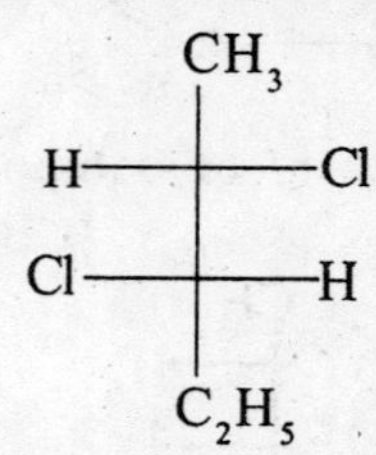

(A) 2 S,3R (B) 2S,3S (C)2R,3S (D) 2R,3R

70. Among the following the achiral amino acid is - [AIIMS-2003]

(A) 2-ethylalanine (B) 2-methylglycine
(C) 2-hydroxymethylserine (D) Tryptophan

71. (A) Benzyl bromide when kept in acetone water produces benzyl alcohol.

(R) The reaction follows $S_N{}^2$ mechanism. [AIIMS-2003]

(A) (B) (C) (D)

72. (A) Trans-2-butene on reaction with Br_2 gives meso-2,3-dibromobutane.

(R) The reaction involves synaddition of bromine. [AIIMS-2003]

(A) (B) (C) (D)

73. (A) Cis-1,3-dihydroxycyclohexane exists in boat conformation.

(R) In the chair form there will not be hydrogen bonding

beyween the two hydroxyl groups. [AIIMS-2003]

(A) (B) (C) (D)

74. The strongest electrophile is : [JIPMER-90]

(A) H_3O^+ (B) RN_2 (C) NO^+ (D)NO_2^+

75. A diastereo isomer : [JIPMER-90]

(A) Is optically inactive (B) Is optically active
(C) Is mirror image of an enantiomer
(D) Doesnot contain a chiral centre.

76. A molecule is divided into two halves which are mirror images of each other by : [JIPMER-90]

(A) A centre of symmetry (B) A plane of symmetry
(C) An axis of symmetry (D) None

77. Which of the following is not a nucleophile : [JIPMER-90]

(A) OH^- (B) NH_3 (C) BF_3 (D) HSO_2^-

78. The weakest among the following is : [JIPMER-91]

(A) CH_3COOH (B) $ClCH_2COOH$
(C) Cl_2CHOOH (D) Cl_3CCOOH

79. Which of the following exhibits optical activity [JIPMER-91]

(A) Malonic acid (B) Maleic acid
(C) Glycollic acid (D) Malic acid

80. Molecular formula C_3H_8O represents : [JIPMER-91]

(A) 2 Isomers (B) 3 Isomers
(C) 4 Isomers (D) 5 Isomers

81. Ethyl alcohol is obtained, when ethyl chloride is boiled with : [JIPMER-91]

(A) Alcoholic KOH (B) Aqueous KOH
(C) H_2O (D) H_2O_2

82. The molecules that will have dipole moments are : [JIPMER-91]

(A) 2,2-dimethylpropane (B) 2,2,3,3-tetramethylbutane
(C) 3-hexene (D) Trans-2-pentene

83. If n is the number of chiral atoms in a compound, the maximum number of stereoisomers that can exist is : [JIPMER-91]

(A) n^2 (B) 2^n (C) 2^{n+2} (D) 2^{n-1}

84. Optical isomerism is shown by : [JIPMER-91]

(A) Butan-1-ol (B) Butan-2-ol
(C) Isobutanol (D) Tertiary butylalcohol

85. The compound that is most reactive towards electrophilic nitration is : [JIPMER-91]

(A) Toluene (B) Benzene
(C) Benzoic acid (D) Nitrobenzene

86. Number of optical isomers possible for 2,3 butandial is : [JIPMER-92]

(A) 2 (B) 3 (C) 4 (D) 5

87. The chain isomers represented by formula C_6H_{14} are : [JIPMER-92]

(A) 5 (B) 4 (C) 3 (D) 6

88. Molecular formula $C_4H_{10}O$ represents isomers : [JIPMER-92]

(A) 5 (B) 7 (C) 2 (D) 4

89. Walden inversion is due to : [JIPMER-92]

(A) SN^1 mechanism (B) Formation of carbanions
(C) SN^2 displacement (D) Formation of carbonium ion

90. Which of the following is most acidic? [JIPMER-93]

(A) CH_3COOH (B) $CH_2(Cl)COOH$
(C) $ClCH_2CH_2COOH$ (D) CH_3CH_2COOH

91. The total number of isomers for the compound of the formula $C_4H_{10}O$ are : [JIPMER-93]

(A) 3 (B) 4 (C) 6 (D) 7

92. How many conformations are there for ethane ? [JIPMER-94]

(A) 2 (B) 3 (C) 5 (D) 6

93. Which of the following statements is not characteristic of free radical chain reaction ? [JIPMER-95]

(A) It is usually sensitive to changes in solvent polarity
(B) It gives major product derived from most stable free radical
(C) It proceeds in three main steps like initiation, propagation and termination
(D) It may be initiated by U.V. light.

94. Which of the following cannot be given to exemplify chiral structure ? [JIPMER-95]

(A) A screw (B) A shoe
(C) A screw driver (D) None of these

95. Which of the following is expected to be optically active ? [JIPMER-95]

(A) $CH_3CH = CHCH_3$ (B) $(CH)_4C$
(C) $C_2H_5CH(CH_3)C_3H_7$ (D) (C_2H_2) $CHCH_3$

96. A mixture of equal parts of enantiomers is called [JIPMER-96]

(A) Configuration (B) Resolution
(C) Racemic mixture (D) Meso mixture

97. The hybrid state of positive charged carbon in vinyl ($CH_2= CH^+$) cation is : [JIPMER-96]

(A) sp^3 (B) sp^3 (C) sp (D) unpredictable

98. Which of the following statements is wrong ? [JIPMER-97]

(A) In tautomerism, the tautomers exist as an equilibrium mixture
(B) In resonance, the resonating structures exist as an

equilibrium mixture

(C) The dextro and laevo forms of an optically active substance are not super-imposable

(D) A branched chain isomer has a lower boiling point than the corresponding straight chain compound

99. Which of the following is not a nucleophile ? [JIPMER-97]
(A) NH_3 (B) H_2O (C) CH_3OH (D) SO_3

100. In CH_3CH_2OH the bond that undergoes heterolytic cleavage most readily is. [JIPMER-99]
(A) C – C (B) C – O (C) C – H (D) O – H

101. The reaction of alkanes with Br_2 in the presence of ultraviolet light is [JIPMER-99]
(A) An elimination reaction (B) An addition reaction
(C) A substitution reaction (D) A rearrangement reaction

102. Carbanions initiate [JIPMER-2000]
(A) Substitution reactions (B) Addition reactions
(C) Both (D) None

103. Which of the following is not a nucleophile [JIPMER-2001]
(A) HSO^-_3 (B) BF_3 (C) Ammonia (D) OH^-

104. Which of the following molecules contains asymmetric carbon atom ? [JIPMER-2002]
(A) $CH_3CHClCOOH$ (B) $CH_3–CH_2–COOH$
(C) $CHCH_2CH_2COOH$ (D) $Cl_2CHCOOH$

105. The most stable carbonium ion among the following is [JIPMER-2002]
(A) $C_6H_5\dot{C}HC_6H_5$ (B) $C_6H_5CH_2$
(C) CH_3CH_2 (D) $C_6H_5CH_2CH_2$

106. In which of the following resonance will be possible ? [JIPMER-2002]
(A) $CH_2 = CH–CH_2–CHO$ (B) $CH_2 = CH–CH = O$
(C) $CH_3–CO–CH_3$ (D) $CH_2 = CH–CH_2–CH = CH_2$

107. Reaction of ethyne with HCN in the presence of $Ba(CN)_2$ is [JIPMER-2002]
(A) Electrophilic addition reaction
(B) Nucleophilic addition reaction
(C) Free radical addition reaction
(D) Electrophilic substitution reaction

108. How many isomeric forms of C_7H_9N contain a benzene ring ? [JIPMER-2002]
(A) 4 (B) 5 (C) 6 (D) 7

109. The order of stability of carbonium ion is [BHU-95]
(A) $CH_3^+ > 1° > 2° > 3°$ (B) $3° > 2° > 1° > CH_3^+$
(C) $3° > 1° > 2° > CH_3^+$ (D) $2° > 3° > 1° > CH_3$

110. The four bonds in methane are directed in space with an angle of [BHU-96]
(A) 120° (B) 90° (C) 109°28' (D) 180°

111. Which of the following is optically active ? [BHU-97]
(A) Ethylene glucose (B) Oxalic acid
(C) Glycerol (D) Tartaric acid

112. Cis and trans 2-butenes are [BHU-98]
(A) Conformational isomers (B) Optical isomers
(C) Position isomers (D) Geometrical isomers

113. Which is the chiral molecule ? [BHU-98]
(A) CH_3Cl (B) CH_2Cl_2 (C) $CHBr_3$ (D) CHClBrI

114. Which of the following is an electrophile ? [BHU-98]
(A) H_2O (B) NH_3 (C) $AlCl_3$ (D) $C_2H_5NH_2$

115. Aromatic compounds undergo [BHU-98]
(A) Nucleophilic addition reaction
(B) Electrophilic addition
(C) Electrophilic substitution (D) None of these

116. Total number of acyclic isomers including stereoisomers of C_4H_7Cl are [BHU-2000]
(A) 9 (B) 10 (C) 12 (D) 13

117. Which of the following alkane is optically active ? [BHU-2001]
(A) Propane (B) 2-methylbutane
(C) 3-methylhexane (D) 2,3,4-trimethylpentane

118. Meso-tartaric acid is [BHU-2001]
(A) Optically inactive due to external compensation
(B) Optically active because of asymmetric carbon atom
(C) Optically inactive isomer
(D) Optically active because of molecular symmetry

119. Formation of cyanohydrin is [BHU-2001]
(A) Nucleophilic substitution reaction
(B) Nucleophilic addition reaction
(C) Electrophilic substitution reaction
(D) Electrophilic addition reaction

120. In the nitration of benzene with a mixture of conc. HNO_3 and conc. H_2SO_4, the active species involved is [BHU-2001]
(A) NO_3 (B) NO_2 (C) NO_2 (D) NO_2^+

121. Which is the correct order of increasing basicity ? [BHU-2002]
(A) $CH_4 < NH_3 < H_2O < HF$
(B) $HF < H_2O < NH_3 < CH_4$
(C) $HF < NH_3 < N_2O < CH_4$
(D) $HF < NH_3 < H_2O = CH_4$

122. How many isomeric amines with formula C_3H_9N are possible ? [BHU-2002]
(A) 2 (B) 3 (C) 4 (D) 5

123. Conversion of CH_4 to CH_3Cl is an example of [BHU-2002]
(A) Electrophilic substitution reaction

(B) Free radical substitution reaction
(C) Nucleophilic substitution reaction
(D) Free radical addition reaction

124. Tertiary alkyl free radicals [BHU-2002]
(A) Have three hyperconjugative forms
(B) Have six hyperconjugative forms
(C) Have seven hyperconjugative forms
(D) Have nine hyperconjugative forms

125. Which of the following groups has maximum +ve inductive effect ? [BHU-2002]
(A) $(CH_3)_3C^-$ (B) $CH_3CH_2^-$
(C) $(CH_3)_2CH^-$ (D) $-CH_3$

126. Formic acid is considered as a resonance hybrid of four

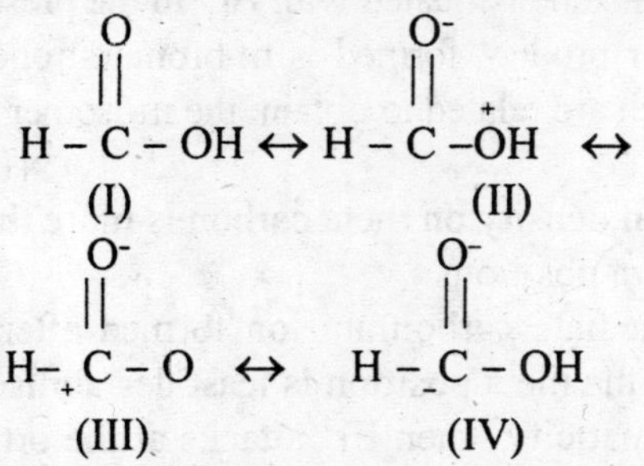

Which of the following order is correct for the stability of the four contributing structures ? [BHU-2002]
(A) I > II > III > IV (B) I > II > IV > III
(C) I > III > II > IV (D) I > IV > III > II

127. Nucleophile substitution reaction is more easier in [AFMC-91]
(A) Nitrobenzene (B) Aniline (C) Phenol (D) Benzene

128. The free radical formed is due to [AFMC-91]
(A) Heterolytic (B) Homolytic
(C) Electrophlic (D) Nucleophilic attack

129. Which of the following is optically active ? [AFMC-93]
(A) Ethanol (B) Succinic acid
(C) Butanoic acid (D) Lactic acid

130. Isomers which can be interconverted through rotation around a single bond are [AFMC-94]
(A) Conformers (B) Diastereomers
(C) Enantiomers (D) Positional isomers

131. The isomerism exhibit by alkyl cyanide and alkyl isocyanide is [AFMC-95]
(A) Functional (B) Positional
(C) Tautomerism (D) Metamerism

132. Optical activity is measured by [AFMC-95]
(A) Polarimeter (B) Refractometer
(C) Spectrograph (D) Tracer technique

133. Aromatic hydrocarbon shows mostly [AFMC-97]
(A) Addition electrophilic (B) Addition nucleophlic
(C) Substitution electrophilic
(D) Substitution nucleophilic

134. $C_3H_8 + Cl_2 \xrightarrow{\text{Light}} C_3H_7Cl + HCl$
is an example of [AFMC-97]
(A) Substitution (B) Elimination
(C) Addition (D) Rearrangement reaction

135. Which of the following shows geometric isomerism? [AFMC-97]
(A) 1-Butene (B) 2-Butene
(C) Propene (D) 1,1-dichlorobutane

136. Number of isomers of C_4H_9 is [AFMC-97]
(A) 2 (B) 3 (C) 4 (D) None of these

137. Which of the following is the strongest acid ? [AFMC-98]
(A) CF_3COOH (B) CBr_3COOH
(C) CH_3COOH (D) CCl_3COOH

138. The compounds $CH_3 - O - C_3H_7$ and $C_2H_5 - O - C_2H_5$ exhibit [AFMC-99]
(A) Optical isomerism (B) Cis-trans isomerism
(C) Metamerism (D) Chain isomerism

139. The reaction $CH_3CH{=}CH_2 + HBr \rightarrow CH_3CHBrCH_3$ [AFMC-99]
(A) Electrophilic addition reaction
(B) Nucleophilic addition reaction
(C) Free radical addition reaction
(D) Electrophilic substitution reaction

140. Which of the following will not show resonance ? [AFMC-2000]
(A) H_2O_2 (B) Allene (C) Ozone (D) Oxygen

141. Neutral divalent carbon species released as reaction intermediate in reaction of ethylene is [AFMC-2000]
(A) Carbocation (B) Carbene
(C) Carbanion (D) Free radical

142. Gamma tartaric acid is inactive due to [AFMC-2000]
(A) Internal compensation (B) External compensation
(C) Molecular dissymmetry (D) None of these

143. Compounds having same molecular formula but differ in their structure are called [AFMC-2001]
(A) Homomers (B) Mesomers
(C) Isotopes (D) Isomers

144. How many isomers of $C_5H_{11}OH$ will be primary alcohols ? [AFMC-2002]
(A) 5 (B) 4 (C) 2 (D) 3

145. Isomers which can be interconverted through rotation around a single bond are : [IIT-92]
(A) Conformers (B) Diastereomers
(C) Enantiomers (D) Positional isomers.

146. What is the decreasing order of strength of bases ?

OH^-, NH_2^-, $H—C \equiv C^-$ and $CH_3—CH_2^-$ [IIT-92]
(A) $CH_3—CH_2^- > NH_2^- > H—C{\equiv}C^- > OH^-$
(B) $H–C{\equiv}C^- > CH_3–CH_2^- > NH_2^- > OH^-$
(C) $OH^- > NH_2^- > H—C{\equiv}C^- > CH_3–CH_2^-$
(D) $NH_2^- > H–C{\equiv}C^- > OH^- > CH_3–CH_2^-$

147. In the following groups

–OAC	–OMe	$–OSO_2Me$	$–OSO_2CF_3$
I	II	III	IV

The order of leaving group ability is [IIT-97]
(A) I > II > III > IV (B) IV > III > I > II
(C) III > II > I > IV (D) III > II > IV > I

148. In the following compounds :-

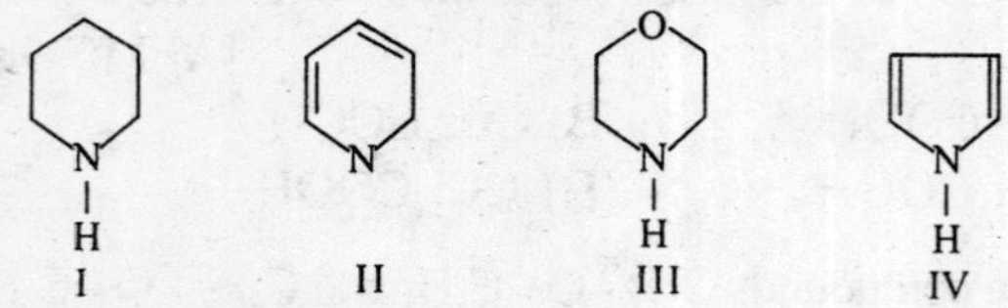

the order of basicity is [IIT-97]
(A) IV > I > III > II (B) III > I > IV > II
(C) II > I > III > IV (D) I > III > II > IV

149. The number of possible enantiomeric pairs that can be produced during monochlorination of 2-methylbutane is [IIT-97]
(A) 2 (B) 3 (C) 4 (D) 1

150. The correct statement about compounds A, B and C [IIT-97]

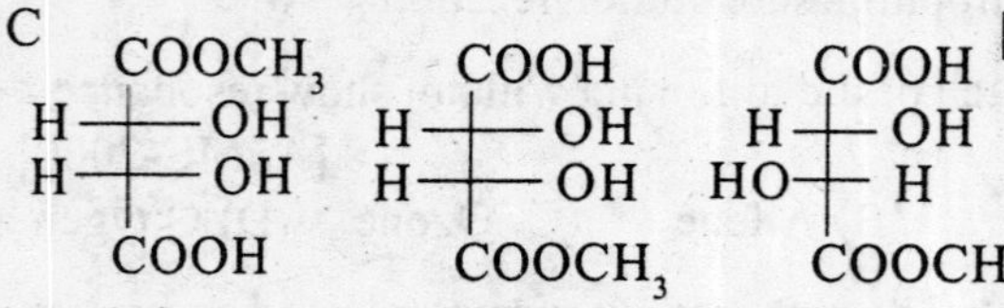

(A) (B) (C)

(A) A and B are identical
(B) A and B are diastereomers
(C) A and C are enantiomers
(D) A and B are enantiomers

151. How many optically active stereoisomers are possible for butane-2,3-diol ? [IIT-97]
(A) 1 (B) 2 (C) 3 (D) 4

152. When cyclohexane is poured on water, flots, because: [IIT-97]
(A) cyclohexane is in 'boat' form
(B) cyclohexane is in 'chair' form
(C) cyclohexane is in 'crown' from
(D) cyclohexane is less dense than water.

153. The enol form of acetone,after treatment with D_2O, gives : [IIT-99]
(A) $CH_3 – C(OD) = CH_2$ (B) $CD_3 – C(=O) – CD_3$
(C) $CH_2 = C(OH) – CH_2D$ (D) $CD_2 = C(OD) – CD_3$

154. The optically active tartaric acid is named as D – (+) –tartaric acid because it has a positive [IIT-99]
(A) optical rotation and is derived from D–glucose
(B) pH in organic solvent
(C) optical rotation and is derived from D–(+)–glyceraldehyde
(D) optical rotation only when substituted by deuterium

155. A solution of (+)–2-chloro-2-phenylethane in toluene racemises slowly in the presence of small amount of $SbCl_5$ due to the formation of [IIT-99]
(A) carbanion (B) carbene
(C) free radical (D) carbocation

156. The nitrobenzene is treated with Br_2 in the presence of $FeBr_3$, the major product formed is m-bromonitrobenzene. Statements which are related to obtain the m-isomer are : [IIT-92]
(A) The electron density on meta carbon is more than that on ortho and para positions
(B) The intermediate carbonium ion formed after initial attack of Br^+ at the meta position is least destabilised
(C)Loss of aromaticity when Br^+ attacks at the ortho and para positions and not at meta position
(D) Easier loss of H^+ to regain aromaticity from the meta position than from ortho and para positions

157. The molecules that will have dipole moment are: [IIT-92]
(A) 2,2-Dimethylpropane (B) Trans-2-pentene
(C) Cis-3-hexene (D) 2,2,3,3-Tetramethylbutane

158. Tautomerism is exihibited by. [IIT-98]

(A) 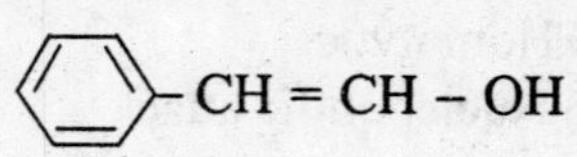 (B) 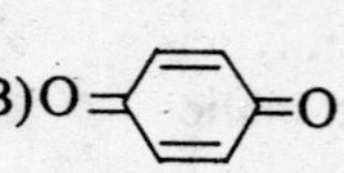

(C) 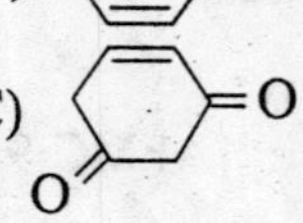 (D) 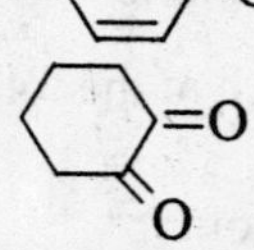

159. Which of the following compounds will show geometrical isomerism ? [IIT-98]
(A) 2-butene (B) propene
(C) 1-phenylpropene (D) 2-methyl-2-butene

160. An aromatic molecule will [IIT-99]
(A) have 4n π electrons (B) have (4n+2) π electrons
(C) be planar (D) be cyclic

161. Toluene, when treated with Br_2/Fe gives p-bromotoluene as the major product because the CH_3 group [IIT-99]
(A) is para directing (B) is meta directing
(C) activates the ring by hyperconjugation
(D)deactivates the ring

162. Which of the following compounds will exhibit geometrical isomerism ? [IIT-2000]
(A) 1-phenyl-2-butene (B) 3-phenyl-1-butene
(C) 2-phenyl-1-butene (D) 1,1-diphenyl-1-propene

163. Which of the following has the highest nucleophilicity? [IIT-2000]

(A) F^- (B) OH^- (C) CH_3^- (D) NH_2^-

164. The number of isomers for the compound with molecular formula $C_2BrClFI$ is [IIT-2001]

(A) 3 (B) 4 (C) 5 (D) 6

165.

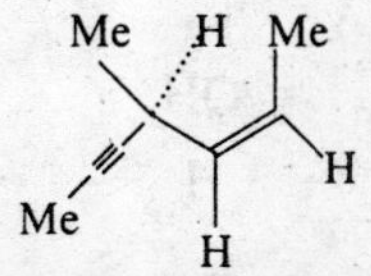

Hydrogenation of the above compound in the presence of poisoned palladium catalyst gives [IIT-2001]

(A) an optically active compound
(B) an optically inactive compound
(C) a racem ic mixture (D) a diastereomeric mixture

166. An S_N2 reaction at an asymmetric carbon of a compound always gives [IIT-2001]

(A) an enantiomer of the substrate
(B) a product with opposite optical rotation
(C) a mixture of a diastereomers (D) a single stereoisomer

167. A : Addition of bromine to trans-2-butene yields meso-2, 3-dibromobutane R : Bromine addition to an alkene is an electrophilic addition. [IIT-2001]

(A) A (B) B (C) C (D)

168. Which of the following compounds exhibits stereoisomerism ? [IIT-2002]

(A) 2-methylbutene-1 (B) 3-methylbutyne-1
(C) 3-methylbutanoic acid (D) 2-methylbutanoic acid

169. Identify the correct order of reactivity in electrophilic substitution reaction of the following compounds [IIT-2002]

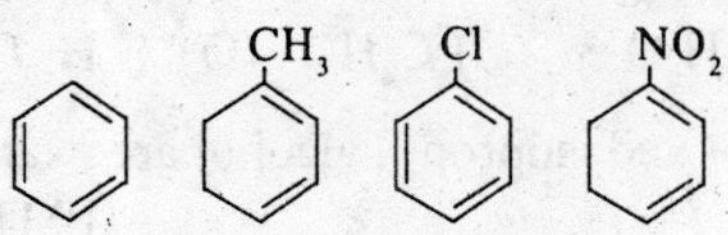

1 2 3 4

(A) 1 > 2 > 3 > 4 (B) 4 > 3 > 2 > 1
(C) 2 > 1 > 3 > 4 (D) 2 > 3 > 1 > 4

170. Reaction of enantiomerically pure acid with one chiral carbon and racemic alcohol with 1 chiral carbon gives [IIT-2003]

(A) Racemic mixture (B) Mesomeric form
(C) d-form (D) l-form

171. Arrangement of $(CH_3)_3$–C–, $(CH_3)_2$–CH–, CH_3–CH_2– when attached to benzyl or an unsaturated group in increasing order of inductive effect is [AIEEE-2002]

(A) $(CH_3)_3$ –C– < $(CH_3)_2$ –CH – < CH_3 – CH_2
(B) CH_3 – CH_2– < $(CH_3)_2$ – CH– < $(CH_3)_3$–C–
(C) $(CH_3)_2$ –CH– < $(CH_3)_3$ –C– < CH_3 –CH_2
(D) $(CH_3)_3$ – C– < CH_3 –CH_2– < $(CH_3)_2$ –CH–

172. In which of the following species is the underlined carbon having sp^3 hybridisation ? [AIEEE-2002]

(A) $CH_3\underline{C}OOH$ (B) $CH_3\underline{C}H_2OH$
(C) $CH_3\underline{C}OCH_3$ (D) $CH_2 = \underline{C}H– CH_3$

173. Racemic mixture is formed by mixing two [AIEEE-2002]

(A) Isomeric compounds (B) Chiral compounds
(C) Meso compounds (D) Optical isomers

174. A similarity between optical and geometrical isomerism is that [AIEEE-2002]

(A) Each forms equal number of isomers for a given compound
(B) If in a compound one is present then so is the other
(C) Both are included in stereoisomerisms
(D) They have no similarity.

175. Which of the following does not show geometrical isomerism ? [AIEEE-2002]

(A) 1,2 dichloro-1-pentene (B) 1,3 dichloro-2-pentene
(C) 1,1 dichloro-2-pentene (D) 1,4 dichloro-2-pentene

176. The reaction [AIEEE-2002]

$$(CH_3)_3C - Br \xrightarrow{H_2O} (CH_3)_3 - C - OH$$

(A) Elimination reaction (B) Substitution reaction
(C) Free radical reaction (D) Displacement reaction

177. Among the following four structures I to IV,

I. C_2H_5—CH(CH_3)—C_3H_7 II. CH_3 — C(=O) — CH(CH_3) — C_2H_5

III. H — $C^{\oplus}$(H)(H) IV. C_2H_5—CH(CH_3)—C_2H_5

It is true that [AIEEE-2003]

(A) All four are chiral c om pound s
(B) Only I and II are chiral compounds
(C) Only III is a chiral compound
(D) Only II and IV are chiral compounds

178. Maleic and fumaric acids are a pair of .... isomers. [EAMCET-90]

179. An isomer of ethanol is [EAMCET-93]

(A) Methanol (B) Diethylether
(C) Acetone (D) Dimethylether

180. The number of structural isomers possible for $C_3H_5Cl_3$ is [EAMCET-95]

(A) 11 (B) 8 (C) 5 (D) 6

181. The number of isomeric structures possible for a molecule having molecular formula C_5H_{12} [EAMCET-96]

(A) 2 (B) 3 (C) 4 (D) 5

182. An isomer of ethanol is [EAMCET-96]

(A) Methanol (B) Dimethylether
(C) Diethylether (D) Ethylene glycol

183. The alkane that yields two isomeric monobromo derivaties [EAMCET-97]
(A) Neo derivaties (B) Ethene
(C) Methane (D) Propane

184. n-octane and n-nonane have [EAMCET-99]
(A) same molecular formula (B) same molecular weight
(C) similar chemical properties (D) same boiling point

185. Which one of the following pairs of compounds are functional isomers ? [EAMCET-2001]
(A) $CH_3CH_2CH_2OH, CH_3CH(OH)CH_3$
(B) $CH_3CH_2CH_2CH_2OH, (CH_3)_2CHCH_2OH$
(C) $CH_3CH_2CH_2OH, CH_3CH_2CH_2Cl$
(D) $CH_3CH_2CH_2OH, CH_3-O-CH_2CH_3$

186. What is X in the following reaction
$C_2H_5Cl + \underline{X} \rightarrow C_2H_5OH + KCl$
[EAMCET(MBBS)-2001]
(A) $KHCO_3$ (B) Alcoholic KOH
(C) Aqueous KOH (D) K_2CO_3

187. Which one of the following reactions proceeds via a secondary free radical ? [EAMCET-2003]
(A) $CH_3-CH=CH_2 \xrightarrow{HBr} CH_3-\underset{Br}{\underset{|}{C}H}-CH_3$
(B) $CH_3-CH=CH_2 \xrightarrow[UV\ /\ light]{HBr} CH_3-CH_2-CH_2-Br$
(C) $C_6H_6 \xrightarrow[FeBr_3]{Br_2\ /} C_6H_5Br$
(D) $C_2H_6 \xrightarrow[light]{Br_2\ /\ UV} CH_3-CH_2Br$

188. Which of the following contains asymmetric centre ? [CPMT-96]
(A) Butene-2 (B) 2,2-dimethylpropane
(C) Hexyne-2 (D) None of these

189. Number of isomers of C_7H_9N is- [CPMT-97]
(A) 3 (B) 5 (C) 4 (D) 6

190. Which of the following is the example of S_N^2 reaction? [CPMT-99]
(A) $CH_3Br + OH \longrightarrow CH_3OH + Br$
(B) $CH_3\underset{Br}{\underset{|}{C}H}CH_3 + OH^- \longrightarrow CH_3\underset{OH}{\underset{|}{C}H}CH_3 + Br^-$
(C) $CH_3CH_2OH \xrightarrow{-H_2O} CH_2=CH_2$
(D) $CH_3-\overset{CH_3}{\overset{|}{\underset{Br}{\underset{|}{C}H}}}-CH_3 + OH^- \longrightarrow CH_3-\overset{CH_3}{\overset{|}{\underset{Br}{\underset{|}{C}H}}}-CH_3 + Br$

191. C_7H_9N has how many isomeric forms that contain a benzene ring ? [CPMT-99]
(A) 4 (B) 5 (C) 6 (D) 7

192. Which is the optically active molecule ? [CPMT-2000]
(A) $C_6H_5-\underset{O}{\underset{\|}{C}}-OH$ (B) $CH_3-\overset{H}{\overset{|}{\underset{OH}{\underset{|}{C}}}}-C_2H_5$
(C) $C_6H_5-\overset{H}{\overset{|}{\underset{H}{\underset{|}{C}}}}-OH$ (D) $C_6H_5-\overset{H}{\overset{|}{\underset{CH_3}{\underset{|}{C}}}}-CH_3$

193. Methane reacts with chlorine in presence of ultraviolet light. The reaction is known as - [CPMT-2000]
(A) Nucleophilic reaction (B) Electrophilic reaction
(C) Free radical reaction (D) None of these

194. Which will show geometrical isomerism ? [CPMT-2001]
(A) $(CH_3)_2C = C(CH_3)_2$ (B) $(CH_3)_2C = CHCH_3$
(C) $(CH_3)_2C = CH_2$ (D) $CH_2 = CH_2$

195. Which of the following is optically active ? [CPMT-2001]
(A) 1-chloropropane (B) 2-chlorobutane
(C) 1-chlorobutane (D) 2-chloropentane

196. Which compound shows cis-trans isomerism ? [CPMT-2002]
(A) 1 butene (B) 2 propene (C) 2 butene (D) benzene

197. Among the following, the one which does not exhibit functional group isomerism is : [PMT-93]
(A) C_2H_6O (B) C_3H_8O (C) C_4H_8 (D) $C_4H_{10}O$

198. n-propyl alcohol and isopropyl alcohol are examples of [PMT-94]
(A) Position isomerism (B) Chain isomerism
(C) Tautomerism (D) Geometrical isomerism

199. Optical isomerism is shown by [PMT-94]
(A) Butanol-1 (B) Butanol-2
(C) Butene-1 (D) Butene- 2

200. The tetrahedral nature of carbon was first given by [PMT-94]
(A) Kekule (B) vant Hoff and Lebel
(C) Lewis (D) Wohler

201. Which of the following types of isomerism is shown by pentanone ? [PMT-95]
(A) Chain isomerism (B) Position isomerism
(C) Functional isomerism (D) All of these

202. Which of the following is the strongest acid ? [PMT-97]
(A) Acetic acid (B) Monochloroacetic acid

(C) Dichloroacetic acid (D) Trichloroacetic acid

203. How many isomeric butanes are there ? [PMT-97]
(A) 2 (B) 3 (C) 4 (D) 5

204. Which of the following will show geometrical isomerism : [PMT-97]
(A) 1-butene (B) 2-butene
(C) 2-methyl propene (D) Propene

205. Which one of the following is most stable- [PMT-97]
(A) $CH_3\,{}^{+}CH_2$ (B) $(CH_3)_2\,{}^{+}CH$
(C) $(CH_3)_3\,C^{+}$ (D) $(CH_3)_2\,CH-CH_2-{}^{+}CH_2$

206. Which of the following compounds exhibits optical isomerism ? [PMT-99]
(A) CH_3CH_2COOH (B) $CH_3CHOHCOOH$
(C) $CH_3CH_2CH_2OH$ (D) $CH_3CHOHCH_3$

207. The number of possible isomers for compound $C_2H_3Cl_2Br$ is : [PMT-99]
(A) 2 (B) 3 (C) 4 (D) 5

208. In the mixture of conc. H_2SO_4 and HNO_3 the nitrating species is : [PMT-2000]
(A) N_2O_4 (B) NO_2 (C) NO_2^{+} (D) NO_2^{-}

209. Example of sp^2 hybridization is [PMT-2000]
(A) CH_3^{+} (B) CH_3^{-} (C) $C_2H_5^{+}$ (D) $C_2H_5^{-}$

210. Which one of the following will exhibit optical isomerism ? [PMT-2000]
(A) CH_3CH_2COOH (B) $CH_3CHOHCOOH$
(C) $CH_3CHOHCH_3$ (D) $CH_3CH_2CH_2OH$

211. Which of the following compounds shows tautomerism ? [PMT-2001]
(A) HCHO (B) $CH_3\,CHO$
(C) $CH_3\,COCH_3$ (D) HCOOH

212. The number of tertiary carbon atoms in the compound $(CH_3)_2\,CHCH_2CH\,(CH_3)_2$ is : [PMT-2001]
(A) 2 (B) 3 (C) 1 (D) 4

213. Carbon - carbon distance in benzene is nearly : [PMT-2001]
(A) 1.54 Å (B) 1.40 Å (C) 1.33 Å (D) 1.25 Å

214. The functional isomer of ethyl alcohol is [PMT-2002]
(A) CH_3OCH_3 (B) CH_3COCH_3
(C) CH_3COOH (D) CH_3CH_2CHO

215. Compounds $CH_3CH_2OCH_2CH_3$ and $CH_3OCH_2CH_2CH_3$ are [PMT-2002]
(A) Tautomers (B) Metamers
(C) Functional isomers (D) Optical isomers

216. In the formation of methane molecule, carbon makes use of [PMT-2002]
(A) sp hydridised orbitals (B) sp^2 hybridised orbitals
(C) sp^3 hybridised orbitals (D) unhybridised orbitals

217. Metamers of ethyl propionate are : [PMT-2003]
(A) C_4H_9COOH and $C_3H_3\,COOCH_3$
(B) C_4H_9COOH and $HCOOC_4H_9$
(C) CH_3COOCH_3 and $C_3H_4COOC_3H_7$
(D) $CH_3COOC_3H_7$ and $C_3H_7COOCH_3$

218. Which of the following compounds does not show geometrical isomerism : [PMT-2003]

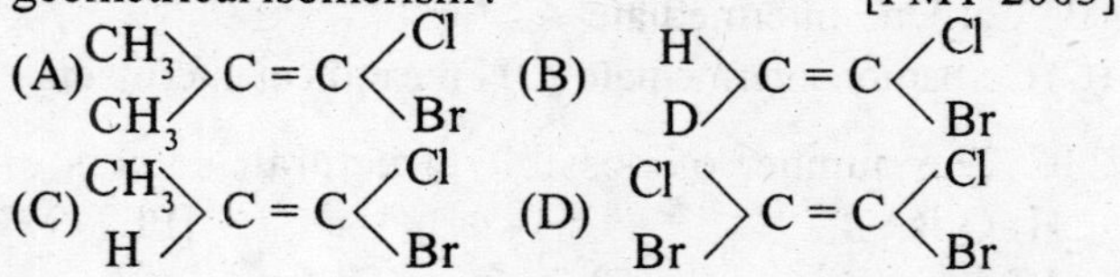

219. The most acidic compound among the following is [PET-93]
(A) CH_3CH_2OH (B) C_6H_5OH
(C) CH_3COOH (D) $CH_3CH_2CH_2OH$

220. An isomer of ethanol is [PET-93]
(A) Diethyl ether (B) Dimethyl ether
(C) Ethanal (D) Methanol

221. Halogenation of alkanes is an example of [PET-93]
(A) Electrophilic substitution
(B) Nucleophilic substitution
(C) Free-radical substitution (D) Oxidation

222. Diethylether and methyl-n-propyl ether are [PET-94]
(A) Position isomers (B) Functional isomers
(C) Metamers (D) Chain isomers

223. The number of optical isomers of an organic compound having n asymmetric carbon atoms will be [PET-94]
(A) 2^{n+1} (B) n^2 (C) 2^n (D) 2^{n-1}

224. An isomer of ethanol is - [PET-95]
(A) Methanol (B) Dimethyl ether
(C) Diethyl ether (D) Ethylene glycol

225. How many isomers alkane C_6H_{14} has [PET-95]
(A) 4 (B) 5 (C) 6 (D) 7

226. Which of the following will have geometrical isomers ? [PET-96]
(A) 2-Methylpropene (B) 2-Butene
(C) 1-Butene (C) Propene

227. Which statement is true for cyclohexane ? [PET-96]
(A) It has two possible isomers
(B) It has three conformations
(C) Boat conformation is most stable
(D) Chair and boat conformations differ in energy by 44KJ/mol

228. The order of stability of carbonium ions is [PET-96]

(A) Methyl > ethyl > iso-propyl > tert-butyl
(B) Tert-butyl > iso-propyl > ethyl > methyl
(C) Iso-propyl > tert-butyl > ethyl > methyl
(D) Tert-butyl > ethyl > isop-ropyl > methyl

229. Reimer-Tiemann reaction involves a : [PET-97]
(A) Carbonium ion intermediate
(B) Carbene intermediate
(C) Carbanion intermediate (D) Free radical intermediate

230. The number of possible structural isomers of $C_4H_{10}O$ is : [PET-97]
(A) 5 (B) 6 (C) 7 (D) 8

231. Which of the following compounds will exhibit geometrical isomerism ? [PET-97]
(A) 2-Butene (B) 2-Butyne
(C) 2-Butanol (D) Butanal

232. Point out incorrect statement about resonance : [PET-97]
(A) Resonance structures should have equal energy
(B) In resonance structures, the constituent atoms should be in the same position
(C) In resonance structures, there should not be the same number of electron pairs
(D) Resonance structures should differ only in the location of electrons around the constituent atoms

233. Amongst following compounds which is most reactive towards nitration ? [PET-98]
(A) Benzene (B) Nitrobenzene
(C) Toluene (D) Chlorobenzene

234. Heterolytic fission of an organic covalent bond gives only : [PET-98]
(A) Free radicals (B) Both cation and anion
(C) Only cation (D) Only anion

235. The compound most reactive in reaction towards electrophilic reagent is : [PET-98]
(A) Toluene (B) Benzene
(C) Nitrobenzene (D) Acetobenzene

236. Optical active isomers but not mirror images are called : [PET-99]
(A) Enantiomers (B) Mesomers
(C) Tautomers (D) Diastereoisomers

237. An organic compound will show optical isomerism if : [PET-99]
(A) Four groups attached to C atom are different
(B) Three groups attached to C atom are different
(C) Two groups attached to C atom are different
(D) All the groups attached to C atom are same

238. The compound $C_4H_{10}O$ can show : [PET-2000]
(A) Metamerism (B) Functional isomerism
(C) Position isomerism (D) All types

239. How many carbon atoms in the molecule $HCOO-(CHOH)_2-COOH$ are asymmetric ? [PET-2001]
(A) 1 (B) 2 (C) 3 (D) None of these

240. Which one among the following is the strongest acid? [PET-2001]
(A) Acetic acid (B) Propionic acid
(C) Mono-chloroacetic acid (D) Butyric acid

241. The shape of methane molecule is : [PET-2001]
(A) Tetrahedral (B) Triangular
(C) Planar (D) Octahedral

242. Which one of the following is not isomeric with diethyl ether? [PET-2001]
(A) n-propylmethyl ether (B) 1-butanol
(C) 2 methyl-2proranol (D) Butanone

243. What is the maximum number of open chain structures possible for C_4H_8 ? [PET-2001]
(A) 2 (B) 3 (C) 4 (D) 1

244. $\begin{matrix} H_3C \\ H \end{matrix}\!\!>C=C<\!\!\begin{matrix} CH_3 \\ H \end{matrix}$ and $\begin{matrix} H_3C \\ H \end{matrix}\!\!>C=C<\!\!\begin{matrix} H \\ CH_3 \end{matrix}$
exhibit which isomerism [PET-2002]
(A) Position isomerism (B) Geometrical isomerism
(C) Optical isomerism (D) Functional isomerism

245. Which of the following undergoes substitution reactions ? [CBSE-90]
(A) Alkanes (B) Akenes (C) Alkynes (D) Any of the above

246. The compound which will not react with sodium is [CBSE-94]
(A) $CH_3-CHOH-CH_3$ (B) CH_3-O-CH_3
(C) CH_3-COOH (D) C_2H_5OH

247. Which of the following hydride is capable of showing CONFORMATIONS ? [JIPMER-2002]
(A) NH_2-NH_2 (B) B_2H_6 (C) CH_4 (D) None of these

248. The order of reactivities of the following alkyl halides for a S_N2 reaction is [CPMT-99]
(A) RF > RCI > RBr >RI (B) RF > RBr > RCI >RI
(C) RCI > RBr > RF > RI (D) RI > RBr > RCI >RF

249. Which of the following is most acidic ? [BHU-2003]
(A) HCOOH (B) $C_6H_5CH_2COOH$
(C) $ClCH_2COOH$ (D) CH_3COOH

250. Diamond exists as [BHU-2003]
(A) Tetrahedral (B) Octahedral (C) Planar (D) Hectahedral

251. Most stable carbanion is [BHU-2003]
(A) 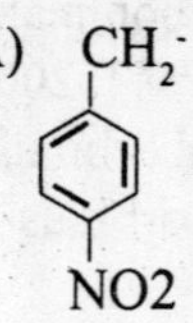

(B) 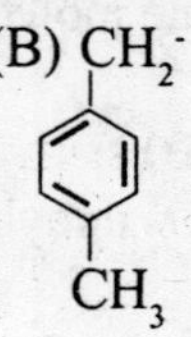

(C) CH_3^- (D) CH_4^-

252. Which of the following is the strongest nucleophile? [BHU-2003]

(A) ROH (B) CH_3O^- (C) RNH_2 (D) $C_6H_5O^-$

253. $(CH_3)_2C{=}CHCH_3 \xrightarrow{\text{catalyst/H}_2}$

Number of optical isomers possible are [BHU-2003]

(A) 0 (B) 4 (C) 2 (D) 3

254. To study the relative arrangement of atoms in a molecule, we study [AFMC-2003]

(A) Molecular formula (B) Structural formula
(C) Emperical formula (D) None of these

255. Which of the following is least reactive in a nucleophilic substitution reaction? [CBSE-PMT-2004]

(A) $(CH_3)_3C - Cl$ (B) $CH_2 = CHCl$
(C) CH_3CH_2Cl (D) $CH_2 = CHCH_2Cl$

256. One of the conformations of n–butane is drawn in the given figure.

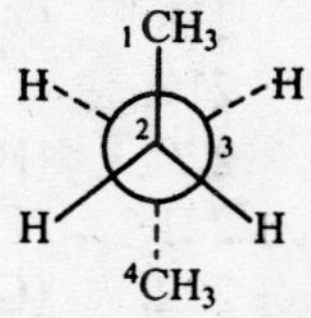

Anticlockwise rotation of C_2 around C_2–C_3 bond by 120° will lead to [IIT-2004]

(A) gauche (B) staggered
(C) partially eclipsed (D) fully eclipsed

257. The number of chiral compounds obtained on monochlorination of 2–methyl butane is [IIT-2004]

(A) 2 (B) 4 (C) 6 (D) 8

Theoretical Organic Chemistry

1.	(A)	51.	(B)	101.	(C)	151.	(B)	201.	(D)	251.	(A)
2.	(B)	52.	(C)	102.	(C)	152.	(D)	202.	(D)	252.	(D)
3.	(D)	53.	(B)	103.	(B)	153.	(A)	203.	(A)	253.	(A)
4.	(C)	54.	(B)	104.	(A)	154.	(C)	204.	(B)	254.	(B)
5.	(C)	55.	(A)	105.	(A)	155.	(D)	205.	(C)	255.	(B)
6.	(C)	56.	(B)	106.	(B)	156.	(A) (D)	206.	(B)	256.	(A)
7.	(B)	57.	(A)	107.	(B)	157.	(B) (C)	207.	(B)	257.	(B)
8.	(B)	58.	(D)	108.	(B)	158.	(A) (C) (D)	208.	(B)		
9.	(A)	59.	(A)	109.	(B)	159.	(A) (C)	209.	(A)		
10.	(A)	60.	(C)	110.	(C)	160.	(B) (C) (D)	210.	(B)		
11.	(A)	61.	(D)	111.	(D)	161.	(A) (C)	211.	(C)		
12.	(D)	62.	(C)	112.	(D)	162.	(A)	212.	(A)		
13.	(D)	63.	(D)	113.	(D)	163.	(C)	213.	(B)		
14.	(D)	64.	(D)	114.	(C)	164.	(D)	214.	(A)		
15.	(B)	65.	(A)	115.	(C)	165.	(B)	215.	(B)		
16.	(A)	66.	(C)	116.	(A)	166.	(B)	216.	(C)		
17.	(C)	67.	(C)	117.	(C)	167.	(C)	217.	(D)		
18.	(C)	68.	(D)	118.	(C)	168.	(D)	218.	(A)		
19.	(B)	69.	(B)	119.	(B)	169.	(C)	219.	(C)		
20.	(A)	70.	(C)	120.	(D)	170.	(A)	220.	(B)		
21.	(D)	71.	(A)	121.	(B)	171.	(B)	221.	(C)		
22.	(B)	72.	(C)	122.	(C)	172.	(B)	222.	(C)		
23.	(B)	73.	(A)	123.	(B)	173.	(D)	223.	(C)		
24.	(A)	74.	(D)	124.	(D)	174.	(C)	224.	(B)		
25.	(C)	75.	(B)	125.	(A)	175.	(C)	225.	(B)		
26.	(D)	76.	(B)	126.	(A)	176.	(B)	226.	(B)		
27.	(D)	77.	(C)	127.	(B)	177.	(B)	227.	(D)		
28.	(C)	78.	(A)	128.	(B)	178.	*	228.	(B)		
29.	(A)	79.	(D)	129.	(D)	179.	(D)	229.	(B)		
30.	(D)	80.	(B)	130.	(A)	180.	(C)	230.	(C)		
31.	(C)	81.	(B)	131.	(A)	181.	(B)	231.	(A)		
32.	(D)	82.	(C)	132.	(A)	182.	(B)	232.	(C)		
33.	(A)	83.	(B)	133.	(C)	183.	(D)	233.	(C)		
34.	(C)	84.	(B)	134.	(A)	184.	(C)	234.	(B)		
35.	(A)	85.	(A)	135.	(B)	185.	(D)	235.	(A)		
36.	(A)	86.	(B)	136.	(A)	186.	(C)	236.	(D)		
37.	(B)	87.	(A)	137.	(A)	187.	(B)	237.	(A)		
38.	(D)	88.	(B)	138.	(C)	188.	(D)	238.	(D)		
39.	(A)	89.	(C)	139.	(A)	189.	(B)	239.	(B)		
40.	(A)	90.	(B)	140.	(B)	190.	(A)	240.	(C)		
41.	(C)	91.	(D)	141.	(B)	191.	(B)	241.	(A)		
42.	(A)	92.	(B)	142.	(A)	192.	(B)	242.	(D)		
43.	(B)	93.	(A)	143.	(D)	193.	(C)	243.	(B)		
44.	(B)	94.	(C)	144.	(B)	194.	(B)	244.	(B)		
45.	(B)	95.	(C)	145.	(A)	195.	(B)	245.	(A)		
46.	(B)	96.	(C)	146.	(A)	196.	(C)	246.	(B)		
47.	(B)	97.	(C)	147.	(B)	197.	(C)	247.	(A)		
48.	(D)	98.	(B)	148.	(D)	198.	(A)	248.	(D)		
49.	(A)	99.	(D)	149.	(A)	199.	(B)	249.	(C)		
50.	(C)	100.	(D)	150.	(A)	200.	(B)	250.	(A)		

178. geometrical

Hydrocarbons

Hydrocarbons
- Aliphatic
  - Saturated
  - Unsaturated
    - Alkenes
    - Alkynes
- Alicyclic
- Aromatic

Alkanes

C_nH_{2n+2} ; $C_{20}H_{42}$–Eisocane ; $C_{30}H_{62}$ – Triacontane

1. Würtz reaction:

$$R{-}X + 2Na + X{-}R \xrightarrow{\text{ether}} R{-}R + 2NaX$$

Methane can't be prepared by this method. Hydrocarbons with odd number of carbon atoms can't be prepared in pure state. The reaction fails with tertiary alkyl halides.

Reactive intermediates: alkyl free radicals (as presented by A) or organometallic compounds (as represented by B)

A) (i) $H_3C{-}CH_2{-}X + \dot{Na} \longrightarrow NaX + H_3C{-}\dot{C}H_2$

(ii) $H_3C{-}\dot{C}H_2 + H_2\dot{C}{-}CH_3 \rightarrow H_3C{-}CH_2{-}CH_2{-}CH_3$

B) (i) $H_3C{-}I + 2Na \rightarrow H_3\bar{C}\cdots \overset{+}{N}a + NaI$

(ii) $H_3C{-}I + H_3\bar{C}\cdots \overset{+}{N}a \rightarrow H_3C{-}CH_3 + NaI$

disproportionation of ethyl free radical:

$$H_5\dot{C}_2 + \dot{C}_2H_5 \rightarrow C_2H_6 + C_2H_4$$

If Zn is used instead of Na, it is called Frankland reaction.

Core–House reaction

$$R'{-}X + R_2CuLi \rightarrow R'{-}R + Cu{-}R + Li{-}X$$

2. Kolbe's electrolytic reaction: Electrolysis of K or Na salt of fatty acid

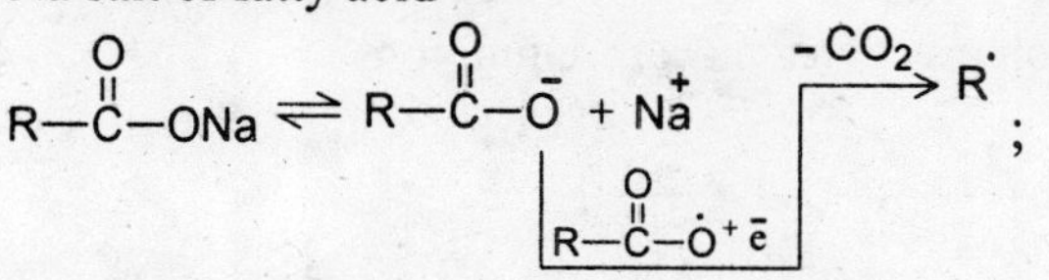

$$\dot{R} + \dot{R} \rightarrow R{-}R$$

Methane can't be prepared by this method. Ions and free radicals are formed, as intermediates.

3. Reduction of carbonyl compounds:

(i) Clemmensen's reduction *for compounds sensitive to bases*

$$>C{=}O \xrightarrow[\text{Conc. HCl}]{\text{Zn - Hg /}} >CH_2 + H_2O$$

(ii) Wolff-Kishner reduction *for compounds sensitive to acids*

$$>C{=}O \xrightarrow{H_2N\text{-}NH_2} >C{=}N{-}NH_2 \xrightarrow[\sim 150°C]{RONa} >CH_2 + N_2$$

(iii) With HI in presence of red Phosphorous

$$>C{=}O + HI \xrightarrow[\sim 150°C]{\text{Red P}} >CH_2 + H_2O + I_2$$

Red phosphorous is employed to react with I_2 to form PI_3 so that it doesn't react with the alkane formed which is otherwise possible ($3I_2 + 2P \quad 2PI_3$).

4. Sabatier and Senderens reaction:

$$\text{Alkene or Alkyne} \xrightarrow{\text{Ni; Pt; Pd / }H_2} \text{alkanes}$$

5. Grignard reagent reaction: Alkyl magnesium halides with compounds having active hydrogen produce alkane

$$RMgX + HOH \rightarrow R{-}H + Mg(OH)X$$

($R{-}OH$; RSH; RNH_2; R_2NH etc. can also be employed)

6. Decarboxylation:

$$RCOONa + \underset{\text{Soda Lime}}{NaOH / CaO} \longrightarrow RH + Na_2CO_3$$

(the alkane will have one carbon atom less than the acid salt employed)

Properties:

C_1 to C_4 gases ; C_5 to C_{17} liquids

Straight chain alkanes have higher boiling points than their isomeric branched ones (due to more space available for van der Waal's forces to act).

Alkanes with odd no. of carbon atoms have lower m.p. than the comparable one with even no. of carbon atoms.

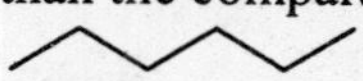 (6) ;  (7)

Alkanes do not show ionic reactions generally as they are non-polar in nature.

1.Substitution:It is one of the main reactions of alkanes as they are saturated molecules (only C and H atoms are present. No hetero atom with lone pairs is present).

Ease of substitution: 3° C–H > 2° C–H > 1° C–H substitution takes place by free radical mechanism. With I_2, the reaction is reversible

$$CH_4 + I_2 \rightleftharpoons CH_3 + HI$$

2.Aromatisation:

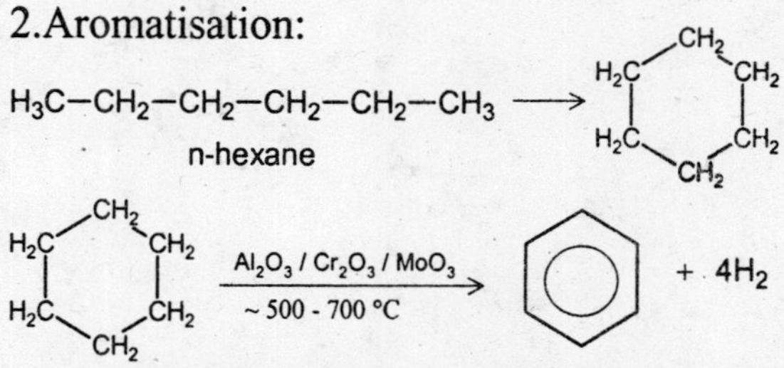

Dehydrogenation produces aromatics.

Since in the process, hydrogen is mixed with charge vapours (n-hexane) to maintain the activity of the catalyst, it is also called hydroforming or catalytic reforming.

n-heptane ⟶ toulene ;

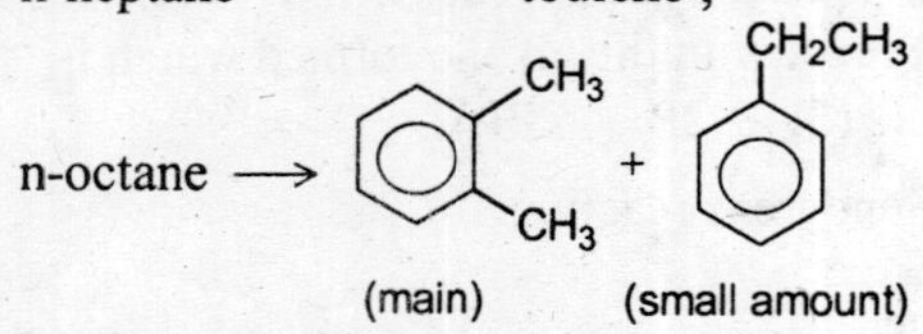

3.Isomerisation:

$$H_3C-CH_2-CH_2-CH_3 \xrightarrow[Al_2(SO_4)_3\,/\,H_2SO_4]{AlCl_3 + HCl\,/\,AlBr_3 + HBr\,/} H_3C-\underset{}{CH}(CH_3)-CH_3$$

Straight chain alkanes are converted to their branched isomers and this process is used to increase the efficiency of petroleum fractions.

4.Reed reaction:This reaction is very useful to prepare detergents industrially.

$$R-H + SO_2 + Cl_2 \xrightarrow{h\nu} R-SO_2Cl + HCl$$

$$R-SO_2Cl + NaOH \xrightarrow{-NaCl} RSO_3H \xrightarrow{NaOH} \underset{\text{alkane sulphonates}}{RSO_3^- \, Na^+} + H_2O$$

(where R ~ C_9 to C_{15})

- Carbon black for printing purposes is prepared from methane.
- Natural gas (methane is the chief constituent; C_2H_6 ; C_3H_8 are present in small amounts).
- LPG (butane chief constituent); Gobar gas contains methane mainly
- Marsh gas is mainly CH_4.
- Paraffin wax (C_{25} upward hydrocarbons); Bee wax (an ester) mainly is mericyl palmitate ($C_{15}H_{31}COOC_{30}H_{61}$).

Cycloalkanes:

They are also known as polymethylenes. C_nH_{2n}
They are isomeric with alkenes.
Preparation:

1.By Freund's reaction /Intramolecular Würtz reaction: α or ω dihalides give this reaction

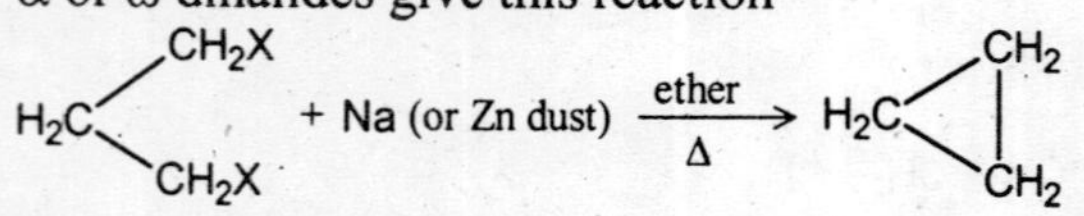

2.By addition of carbenes to alkenes

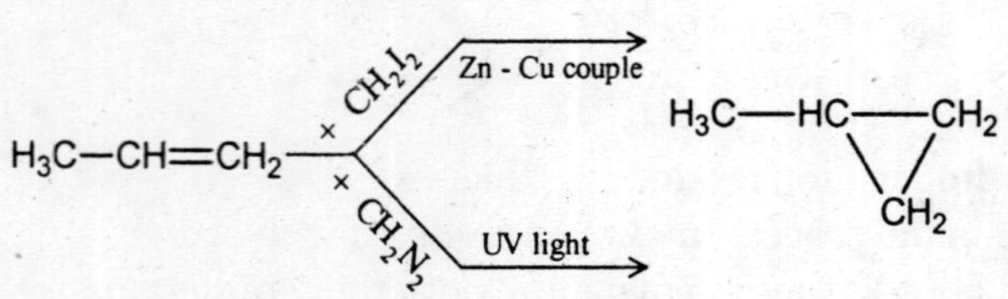

Reactions:

Cycloalkanes give both substitution and addition reactions.

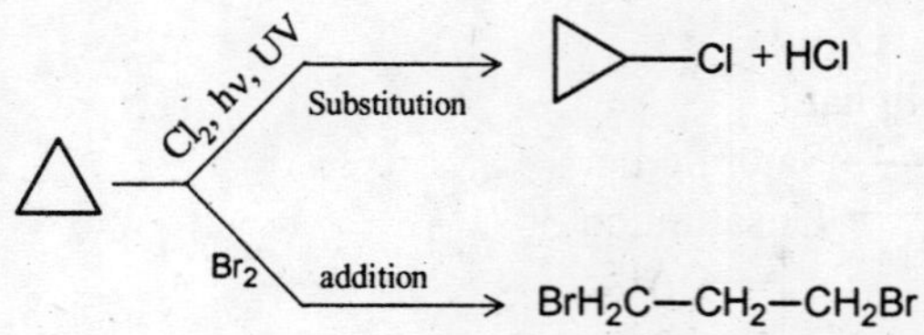

Oxidation:

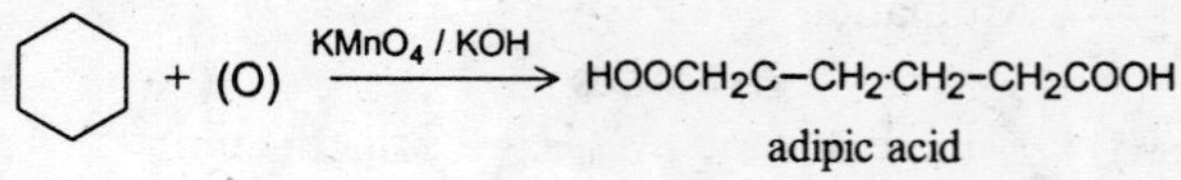

Ring Strain

Bayer's strain theory: Strain in cycloalkanes is explained in terms of tetrahedral angle. More deviation from this angle causes more strain.

S = ½ (109° 28')– bond angle in a particular cycloalkane

This theory fails to explain the more stability and less reactivity of cyclohexane compared to cyclopentane. Stability of larger ring systems also can't be explained. According to Sachse and Mohr, in cyclohexane or in larger ring cycloalkanes, the carbon atoms in ring have non-planar arrangement and therefore assume puckered structures in which the tetrahedral angle is retained.

Alkenes, C_nH_{2n}

1. Dehydration of alcohols:

$$H_3C-CH_2-CH_2-CH_2OH \xrightarrow{\sim 75\%\ H_2SO_4} H_3C-CH_2-CH=CH_2$$

$$H_3C-CH_2-CHOH-CH_3 \xrightarrow{\sim 60\%\ H_2SO_4} H_3C-CH=CH-CH_3$$

$$H_3C-C(CH_3)_2-OH \xrightarrow{\sim 20\%\ H_2SO_4} H_3C-C(CH_3)=CH_2$$

Dehydration proceeds via carbonium ions

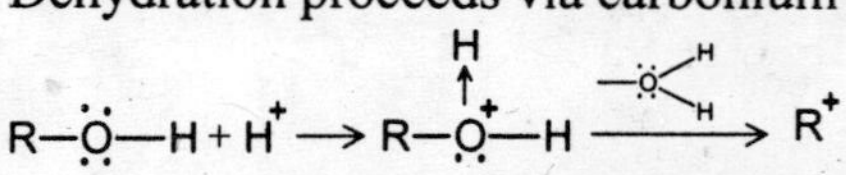

∴ The rate or ease of dehydration is 3°>2°>1°

H_3PO_4, Al_2O_3, P_2O_5 etc. also act as dehydration agents. Dehydration is not possible if there is no β–hydrogen e.g.

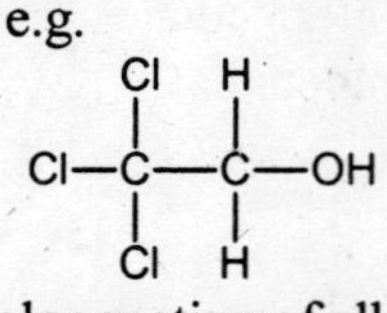

2. Dehydrohalogenation of alkyl halides :

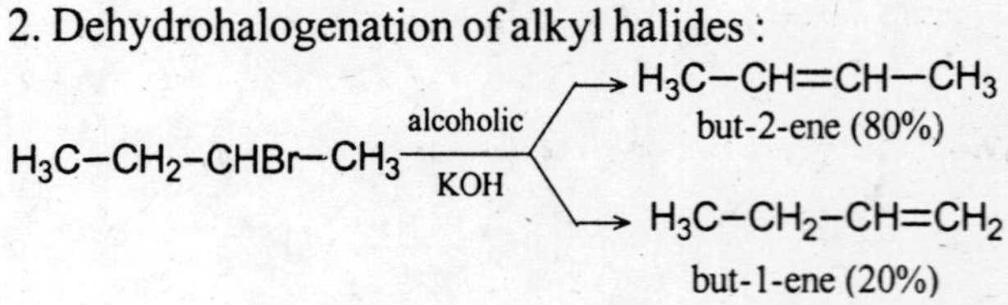

Saytzeff's rule: When two alkenes may be formed by dehydrohalogenation of alkyl halide, the one which is more substituted predominates.

Ease of formation of alkenes/ Stability of alkenes:
$R_2C{=}CR_2 > R_2C{=}CHR > R_2C{=}CH_2$,
$RHC{=}CHR > RHC{=}CH_2 > H_2C{=}CH_2$

Rate of hydrogenation increases in the order (increase in bulk or number of R groups responsible).
In tertiary halides, elimination follows E_1 mechanism. e.g.

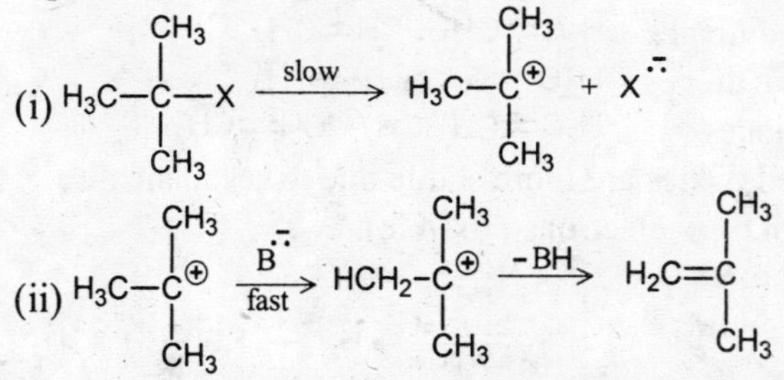

In primary, the elimination follows E_2 mechanism.
Rate of dehydrohalogenation: 3°>2°>1° ;
RI>RBr>RCl

3. Kolbe's Method: Sodium or potassium salt of a dibasic acid on electrolysis gives alkene.

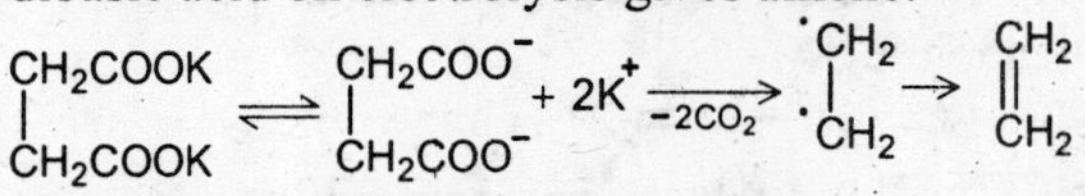

4. Reduction of alkyne: Controlled reduction of alkynes produce alkenes.

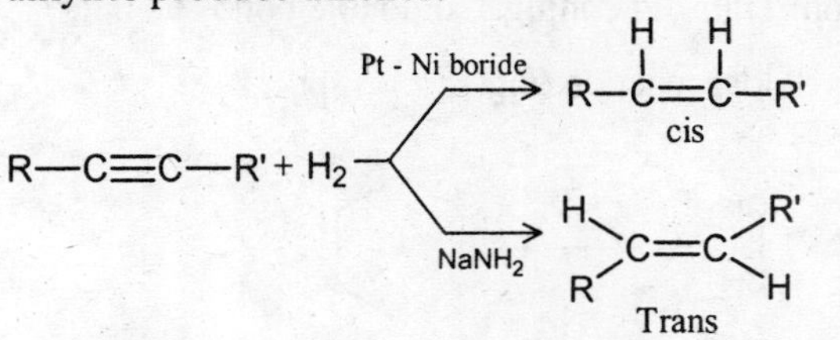

Properties: C_2 to C_4 gases; C_5 to C_{17} liquids. The dipole moment of trans is zero. Stability and m.p. of trans is more than cis.

Reactions:

1. Addition of halogens on alkene is trans addition.

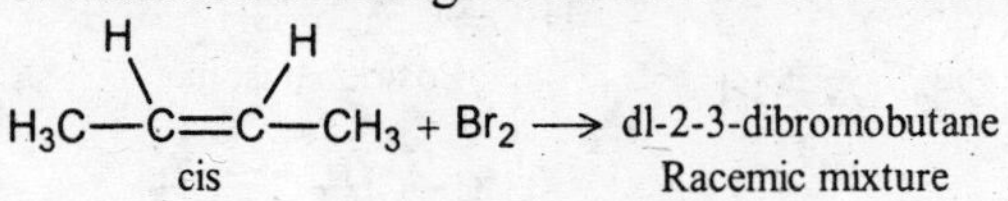

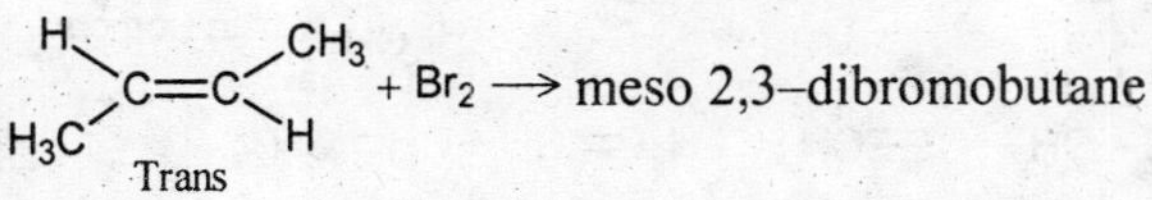

2. Addition of halogen acids: Markownikoff's addition: The negative part of the reagent will add to that carbon atom which is attached to least number of hydrogen atoms. Due to polarisation effect

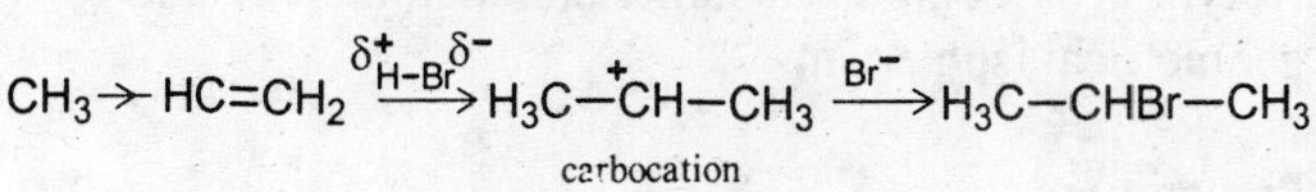

Addition of H^+ first takes place. Carbonium ion or carbocation intermediate.

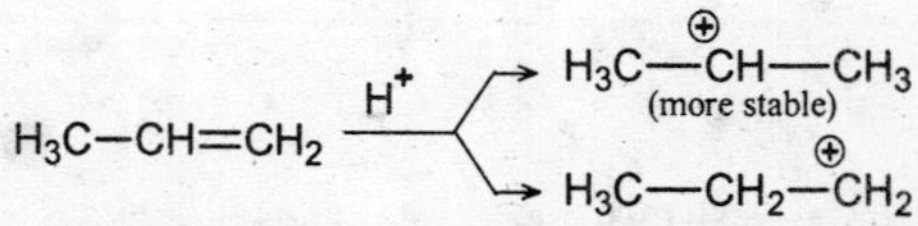

Addition in case of symmetric alkenes e.g. $H_2C{=}CH_2$, $H_3C{-}CH{=}CH{-}CH_3$ will produce same products.

Peroxide or Kharasch effect: In the presence of air, peroxide etc., HBr addition takes place via free radical mechanism. (HI>HBr>HCl)

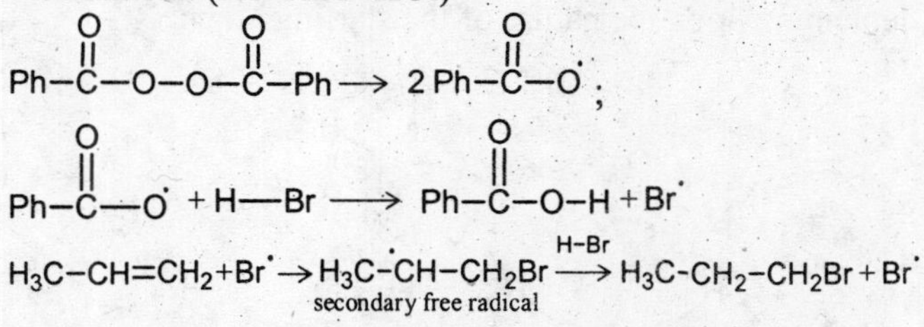

Kharasch effect does not apply to HF, HCl and HI additions. H–F and H–Cl bonds are too strong to break into free radicals. H–I breaks but due to its bigger size, I is not reactive to add on a double bond, but delivers
$\dot{I} + \dot{I} \rightarrow I_2$

3. Addition of hypohalous acid:

$HOCl \rightarrow \overset{-}{OH} + \overset{+}{Cl}$ (OH^- more electronegative than Cl)

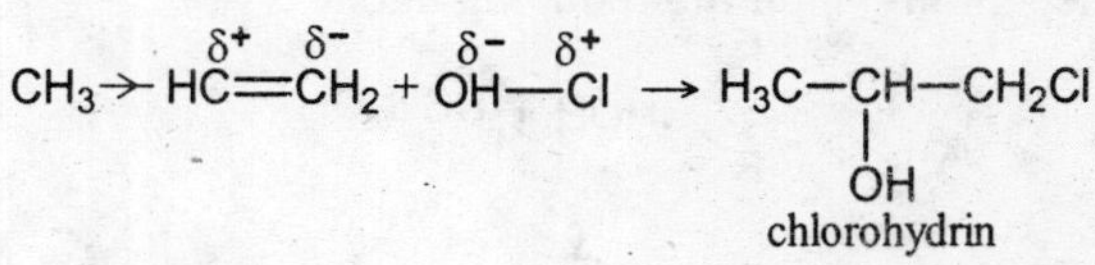

4. Epoxidation:

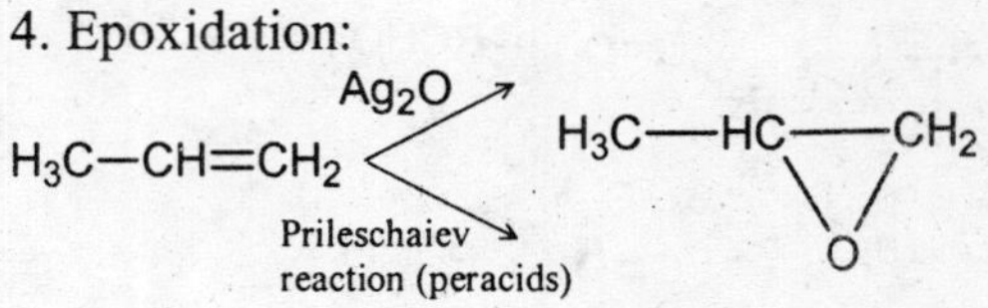

Orbital structure of ethene

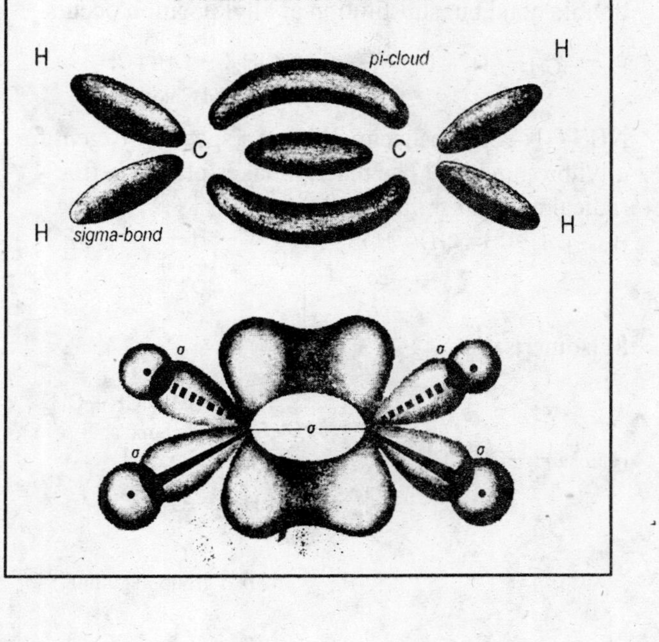

5. Oxidation:

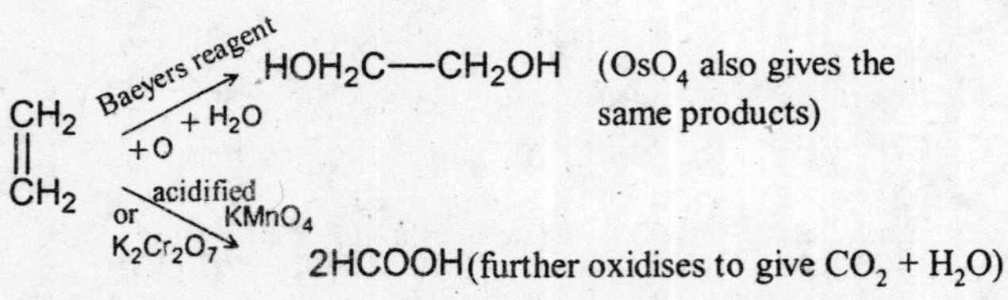

6. Ozonolysis: A variety of products will form depending upon the nature of the starting alkene e.g.

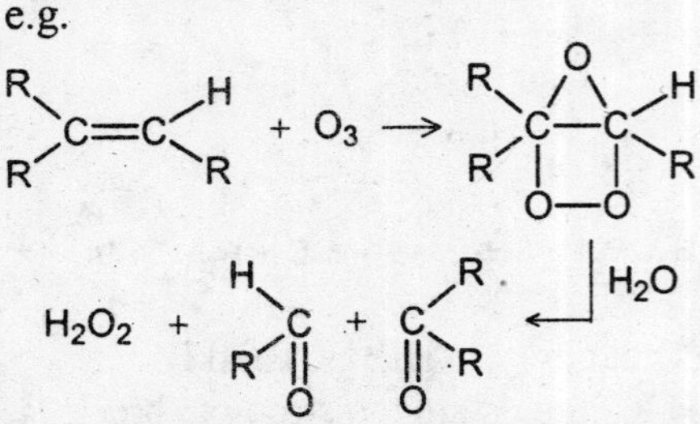

H_2O_2 formed here oxidises the aldehyde formed to the corresponding acid. It is prevented by adding Zn which decomposes H_2O_2. As the oxidation of ketones is not easy, they if formed will not be effected by H_2O_2.
Reduction of the ozonide formed delivers a variety of products depending upon the reducing agent.

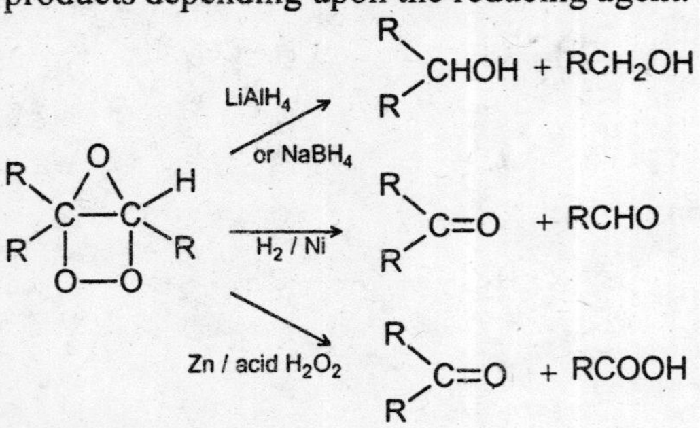

7. Substitution reaction:
At high temperature for e.g. Cl_2 will not add to the double bond but substitution at allyl position occurs.

$$H_3C-CH{=}CH_2 + Cl_2 \xrightarrow[>500°C]{\text{high temp.}} ClH_2C-CH{=}CH_2$$

Allyl Chloride

NBS– (N– bromosuccinimide) is a specific reagent for allyl bromination. The reaction takes place via free radical mechanism.

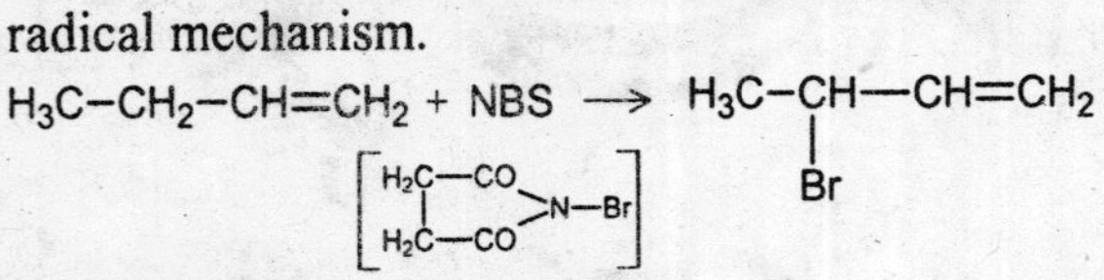

8. Isomerisation:

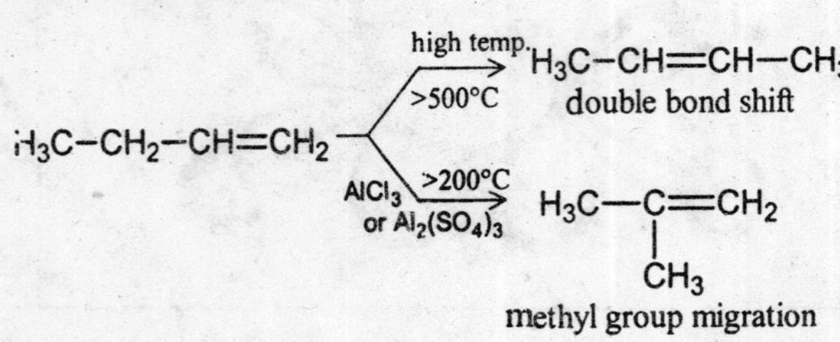

9. Mustard gas formation:

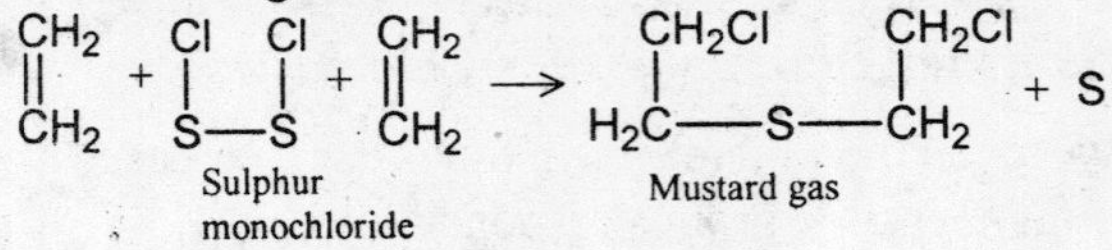

It was used as a war gas.

Dienes

Conjugated dienes $H_2C{=}CH-CH{=}CH_2$
Cumulative dienes $H_2C{=}C{=}CH-CH_3$
Isolated dienes $H_2C{=}CH-CH_2-CH{=}CH_2$
Conjugated dienes are more stable due to resonance (delocation of π–electrons possible).
(i) Addition:

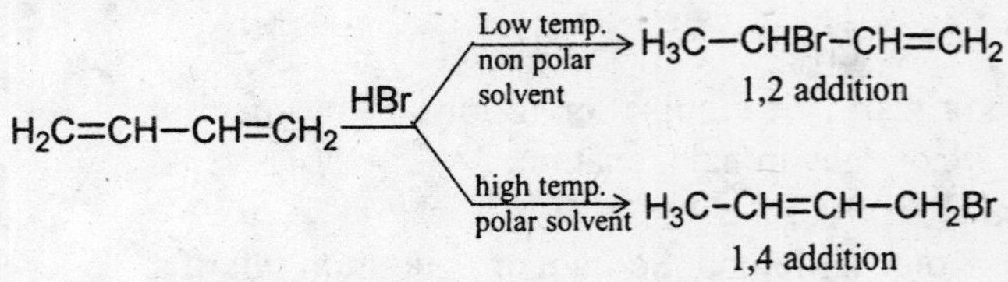

(ii) Diels' Alder reaction: (4+2) cyclo addition
Diene + Dienophile
Diene can be $H_2C{=}CH-CH{=}CH_2$, enyne $H_2C{=}CH-C{\equiv}CH$ or diyne $HC{\equiv}C-C{\equiv}CH$.
The two double bonds must be present in cisoid conformation. Transoid conjugated double bonds do not show Diel's–Alder reaction (e.g. 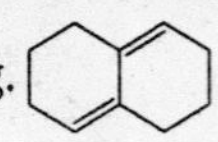).

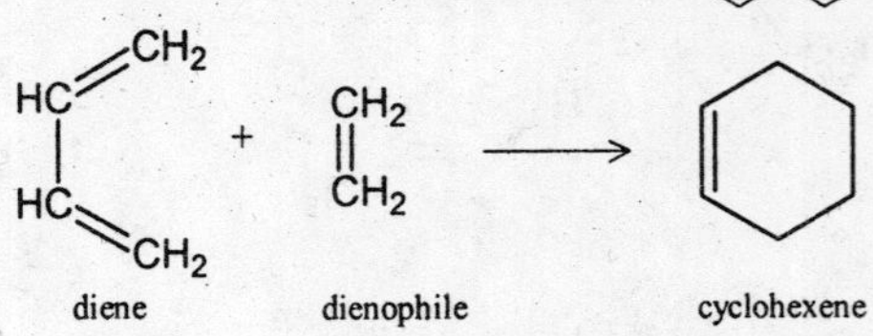

Alkynes, C_nH_{2n-2}

Depending upon the structure and the no. of carbon atoms present, acetylenes may exhibit.

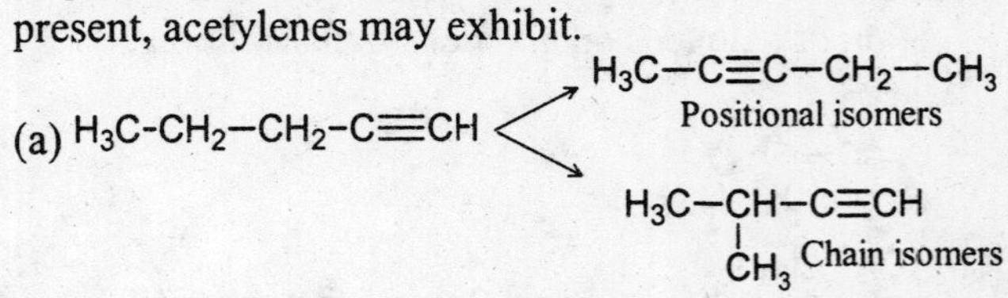

(b) $H_3C-CH_2-C{\equiv}CH$; $H_2C{=}CH-CH{=}CH_2$;

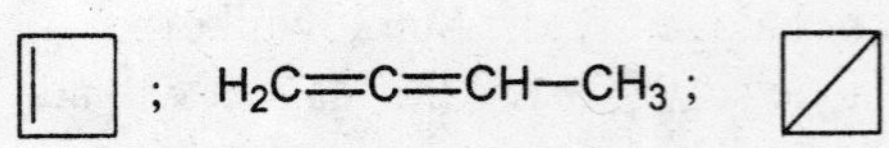

Functional isomers

Alkynes can't exhibit metamerism, tautomerism and geometrical isomerism.

Preparation:
1. Dehydrohalogenation of gem/vicinal dihalide

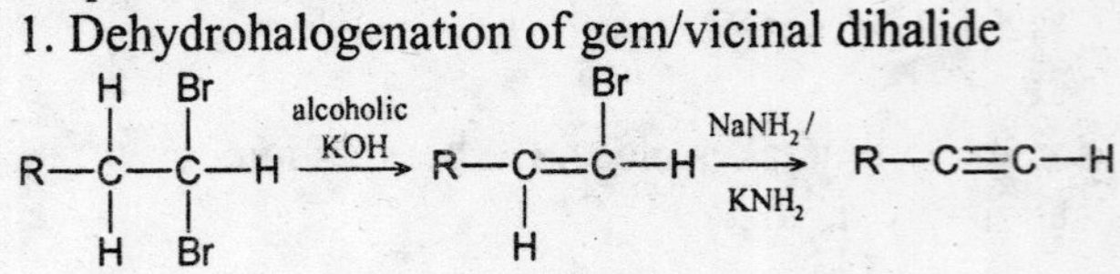

Stronger base than alcoholic KOH is required in the second step as the dehydrohalogenation is difficult in the case of vinyl halide derivative due to resonance stabilisation.

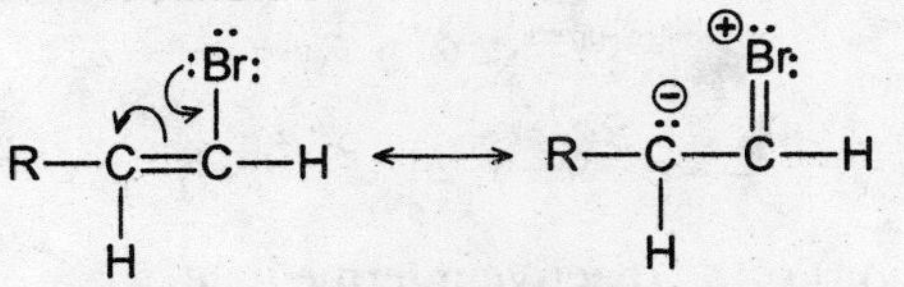

2. Dehalogenation of haloforms:

$CHI_3 + 6Ag + I_3CH \longrightarrow HC{\equiv}CH + 6AgI$

Reactivity: $CHI_3 > CHBr_3 > CHCl_3 > CHF_3$

3. Preparation of higher terminal alkyne:

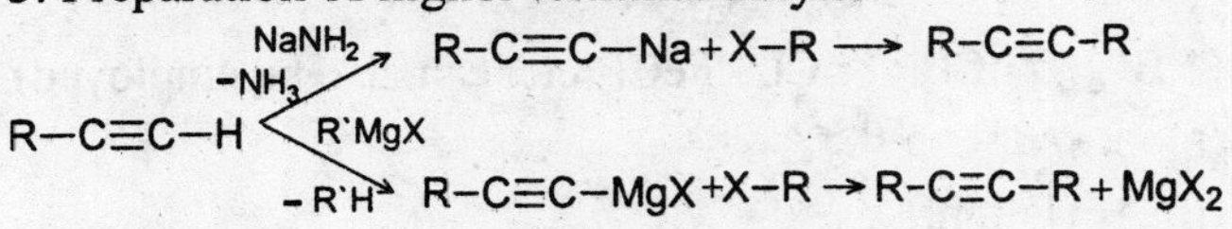

4. From CaC_2:

$CaC_2 + 2H_2O \longrightarrow HC{\equiv}CH + Ca(OH)_2$

Properties:

Acetylenic hydrogen is more acidic (terminal acetylenes are acidic whereas isomeric 2–alkynes are not). Terminal acelylenes have lower boiling points than the isomeric–2–alkynes.

Order of acidity: $HC{\equiv}CH > H_2C{=}CH_2 > H_3C{-}CH_3$

This is due to the more s–character of sp hybridised carbon atom in acetylenes. More the s–character, greatrer is the electronegativity and the acidity.

Ethyne is less reactive than ethene towards electrophilic reagents.

Ethyne is more reactive than ethene towards reagents that are themselves electron rich.

$-C{\equiv}C-$ Bond attracts nucleophiles more strongly than a $-C{=}C$ bond. Electrophilic addition of bromine to $C{=}C$ bond is much faster than to $C{\equiv}C$ bond.

This descrepency can be explained due to the more electronegative character of carbon atom of acetylene. The π–electrons are attracted by the carbon atom and therefore the electrophilic addition will be slower compared to the addition of ethene.

1. Reduction:

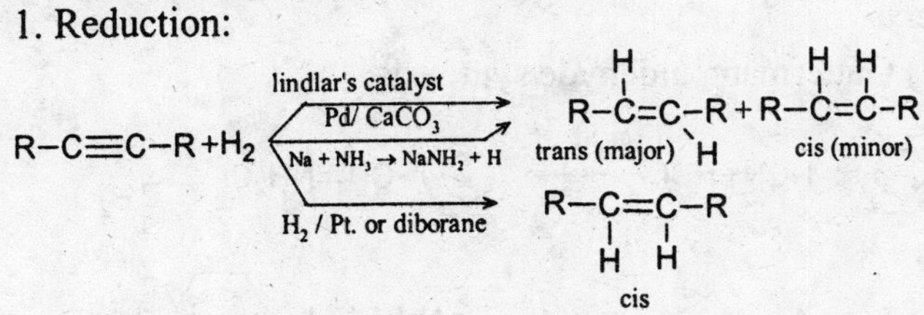

2. With halogen acids:

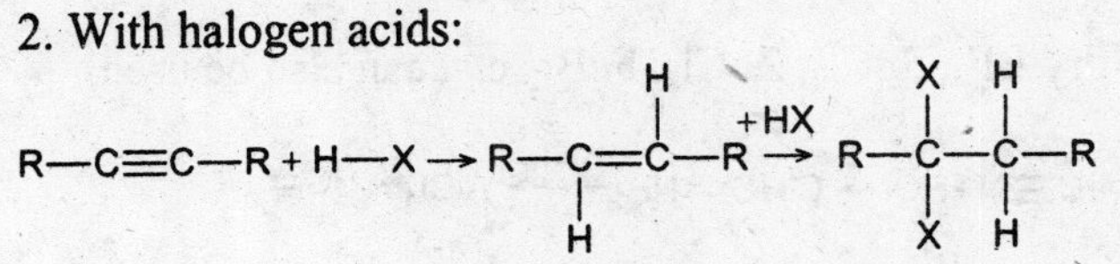

3. Addition of water /Kucherov reaction:

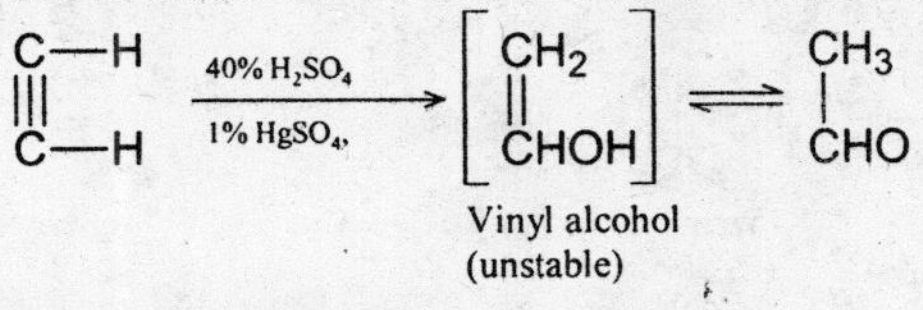

Other acetylenes give ketones

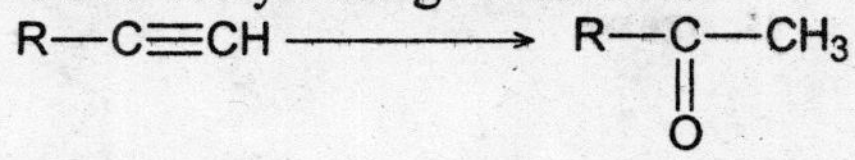

4. Ozonolysis:

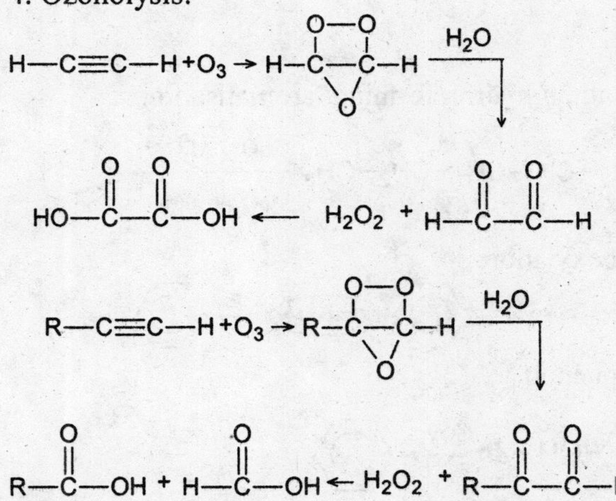

5. Oxidation:

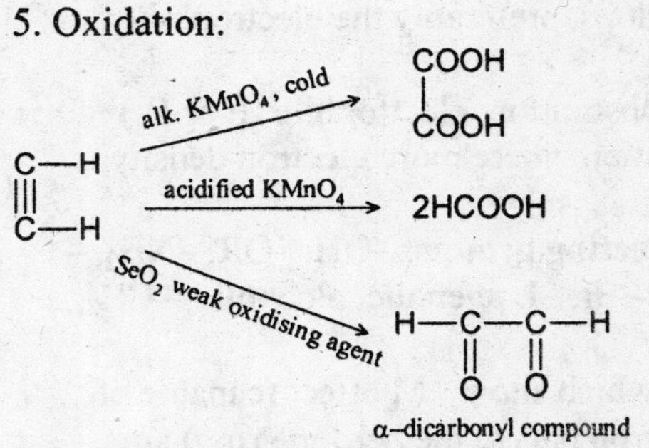

6. Westron and Westrosol:

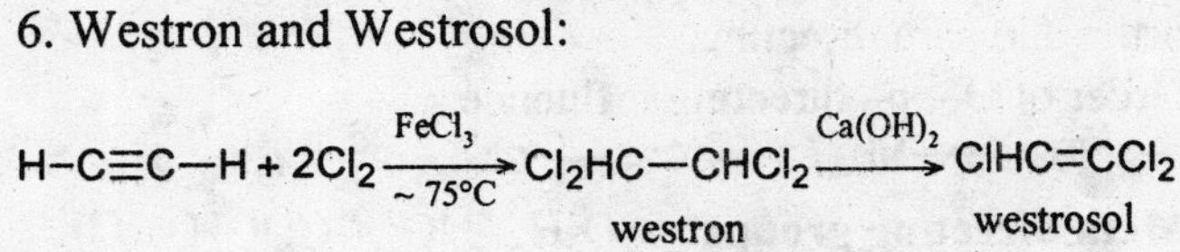

(Westron is a toxic solvent for cellulose acetate; Westrosol is a non-toxic solvent for dry cleaning)

7. Lewisite: It was used as a war gas.

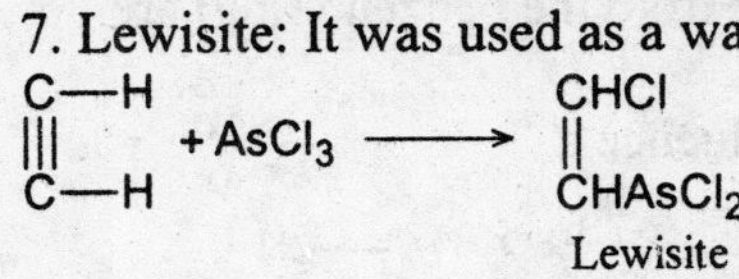

8. Reaction with Cu_2Cl_2 and $AgNO_3$: These reactions are helpful to differentiate alkynes from alkenes in one step. Only terminal acetylenes give this reaction.

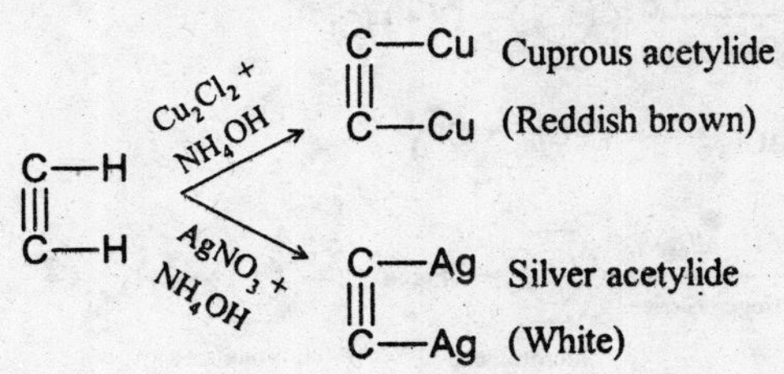

9. Cyclisation:

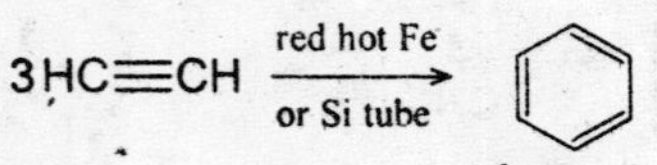

benzene

Benzene

Preparation:

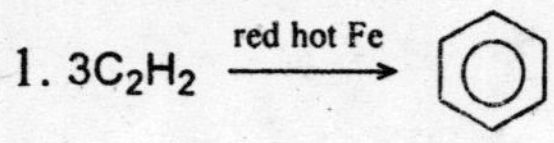

2. Cyclisation/ hydroreforming/ aromatisation

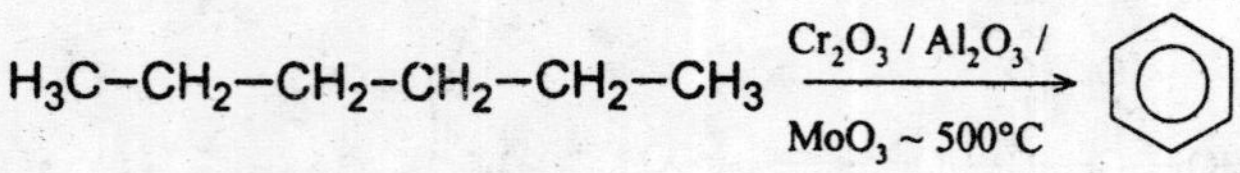

3. Decarboxylation:

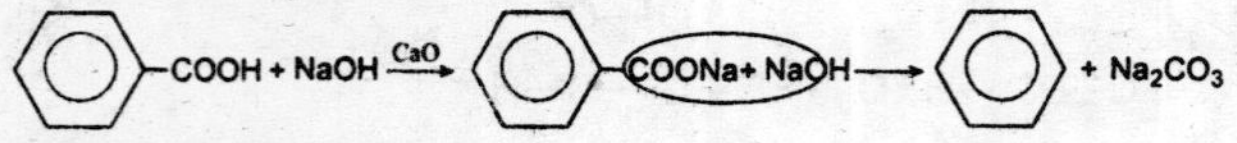

4. From phenol:

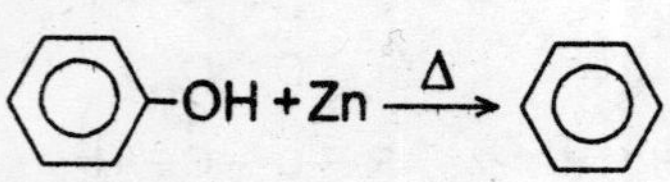

Reactions:

Substitution reactions, preferably the electrophilic are common.

In electrophilic substitution, electrophile (E or E^+) attacks at the position where more electron density is present.

Ortho and p–directing groups: –OH, –OR, –NH_2, –NHR, –NR_2, –Cl, –Br, –I, phenolic, alcoholic, –CH_3, –C_2H_5, –C_6H_5 etc.

Groups or atoms which show +M effect (capable of providing an electron pair to the benzene ring) are ortho and para directing.

Order of O–, p– directing influence:

$-\ddot{N}R_2 > -\ddot{N}H_2 > -\ddot{O}R > -\ddot{C}l:$

Meta directing groups: –SO_3H, –NO_2, –COOH, –CHO, –CN etc.

Groups or atoms which show –M effect (capable of withdrawing electron pair from the benzene ring) are meta directing.

Order of m– directing influence :

$-\overset{O}{\overset{\|}{C}}-O-R > -\overset{O}{\overset{\|}{C}}-R > -\overset{O}{\overset{\|}{C}}-H > -SO_3H > -CN > -NO_2$

1. Halogenation:

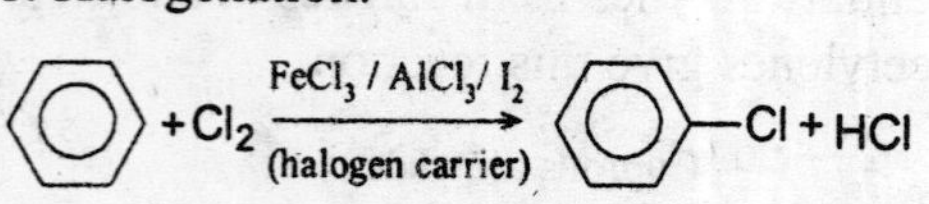

$(Cl_2 + FeCl_3 \longrightarrow FeCl_4^- + Cl^+)$

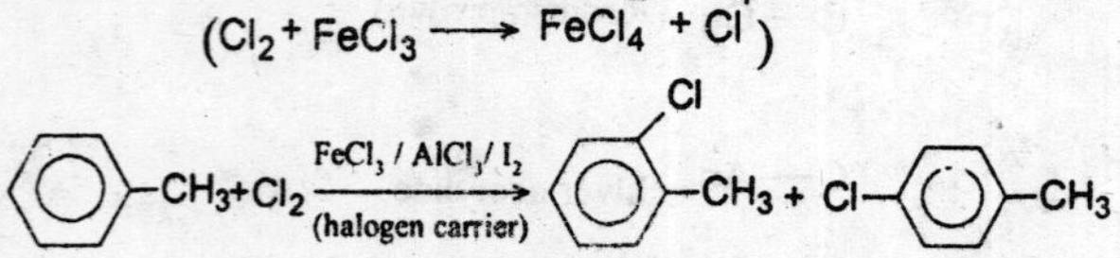

o - chlorotoluene p - chlorotoluene

2. Nitration:

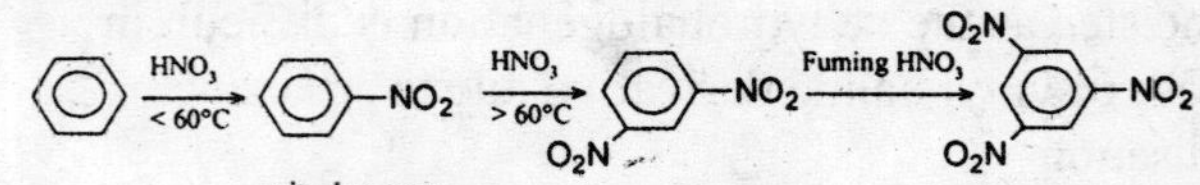

nitrobenzene (oil of Mirbane) m - dinitrobenzene trinitrobenzene

$(2H^+ HSO_4^- + OH-NO_2 \longrightarrow O=\overset{\oplus}{N}=O + 2HSO_4^- + H_3O^+)$

NO_2^+ (nitronium ion) is the reactive intermediate.

3. Fridel–Craft reaction:

(i) Alkylation:

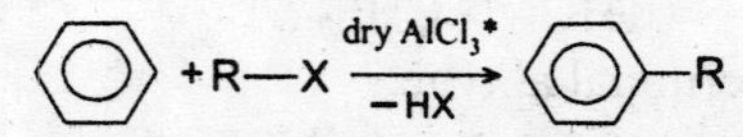

(* BF_3 / $ZnCl_2$ / $SnCl_4$ / $FeCl_3$ etc. can also be employed).

$(R-X + AlCl_3 \longrightarrow R^+ + AlCl_3X^-)$

(ii) Acylation:

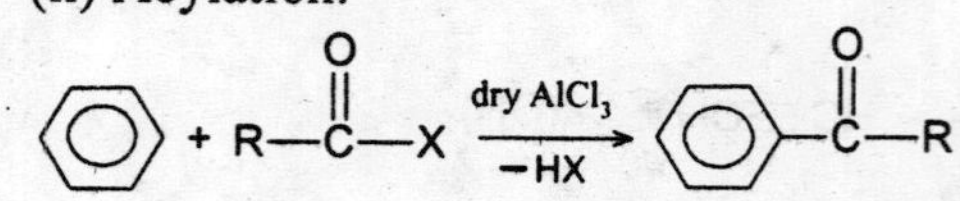

$(R-\overset{O}{\overset{\|}{C}}-X + AlCl_3 \longrightarrow R-\overset{O}{\overset{\|}{C}}{}^{\oplus} + AlCl_3X^-)$

4. Sulphonation:

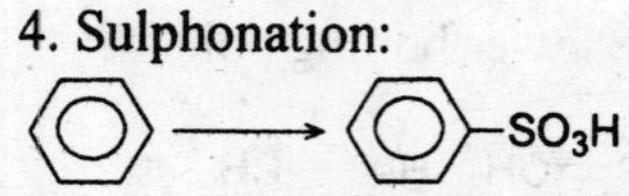

$(2H_2SO_4 \rightleftharpoons H_3O^+ + 2HSO_4^- + SO_3)$

SO_3 can be represented as

$\overset{+}{S}(=O)_2 \rightarrow O^-$ or $^+SO_2O^-$

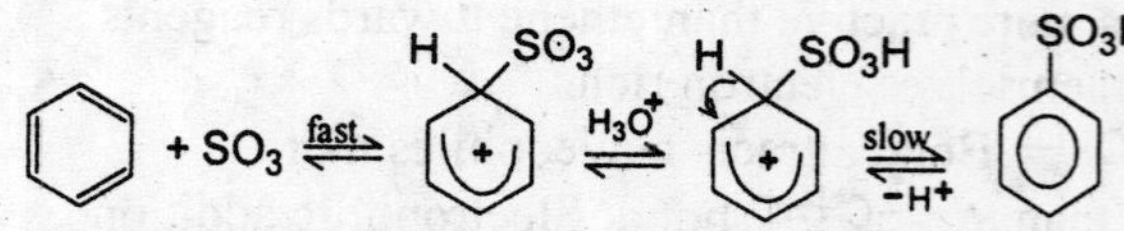

The reaction is reversible. Proton elimination step is the rate determining step.

5. (i) Gattermann Koch reaction:

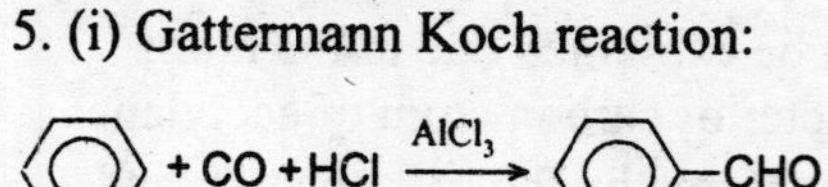

$CO + HCl \longrightarrow HCOCl \xrightarrow{+AlCl_3} AlCl_4^- + H\overset{+}{C}O$

formylchloride

(ii) Gattermann aldehyde synthesis:

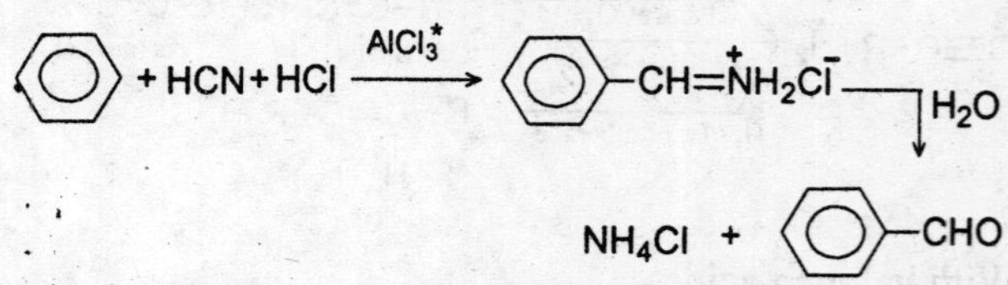

(* dry BF_3, $SnCl_2$, $ZnCl_2$, $FeBr_3$ etc. can also be used)

$H-C\equiv N + HCl \rightarrow ClHC=NH \xrightarrow{AlCl_3} AlCl_4^{\ominus} + H\overset{\oplus}{C}=NH$

Addition reactions:

1. Halogenation:

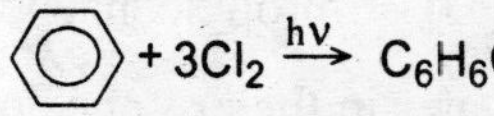

(BHC; 666)

γ–isomer of the 8 stereoisomers possible is a powerful pesticide. In γ–isomer the chlorine atoms are axial at 1, 2 and 3 positions. (Gammexene or Lindane) It is an addition reaction.

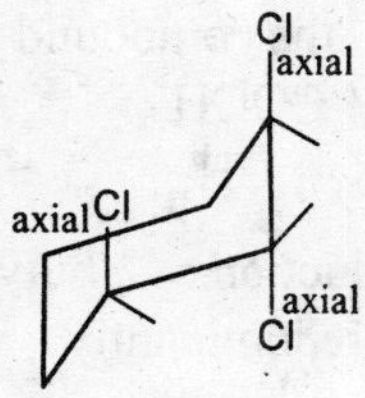

2. Hydrogenation:

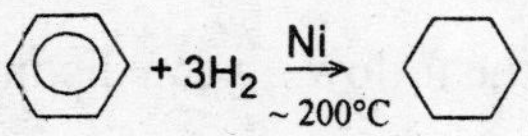

3. Brich reduction:

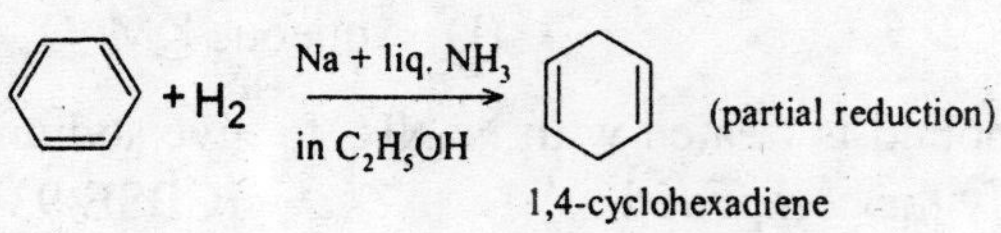

4. Oxidation:

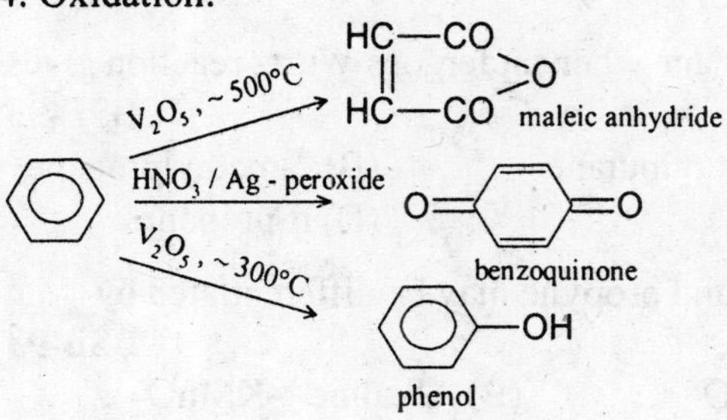

5. Ozonolysis:

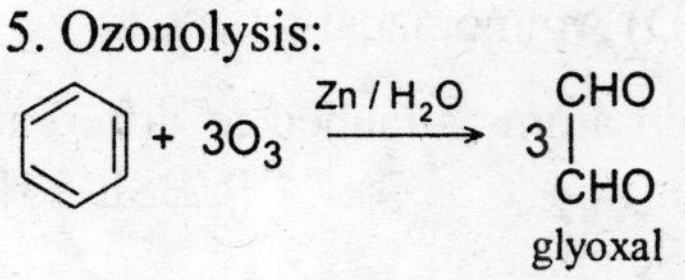

Petroleum

It is also called as lmneral oil. Cycloparaffins (cycloalkanes) are known as I nıphthenes in petroleum technology.

Petroleum is subjected to fractional distillation to separate several fractions.

Fractions: Gasoline ($C_7 - C_9$); kerosine oil ($C_{10} - C_{16}$); Diesel ($C_{16} - C_{20}$); Lubricating oil ($C_{17} - C_{20}$); Paraffin wax (C_{24} upwards); Pitch & asphalt ($C_{30} - C_{40}$).

The yield of gasoline or petrol is increased by cracking in the refineries.

Cracking: The process of decomposition of less volatile higher hydrocarbons into more volatile lower hydrocarbons with the application of heat and catalysts is known as cracking (generally the fractions like heavy oil or residual oil or higher hydrocarbons with $>C_{12}$ are subjected).

Octane number: Quality of gasoline or petrol is expressed in terms of octane numbers. The percentage by volume of iso-octane present in a mixture of n-heptane (octane number is 0 (Zero) and iso-octane (octane number is 100) is known as octane number.

Order of knocking:

Straight chain paraffin > branched chain paraffins > olefins > naphthenes > aromatics

Anti-knocking fluid:

$(C_2H_5)_4Pb$ (60%) + $CH_2Br–CH_2Br$ (26%) + $ClCH_2–CH_2Cl$ (9%) and a red dye (2%). Ethylene bromide is added to prevent deposition of lead in the cylinder.

Regular gasoline (commonly used in motorcars, scooters etc.) has octane no. 74.

The octane number of fuels can be improved by processes like alkylation, hydroforming, isomerisation, reforming etc.

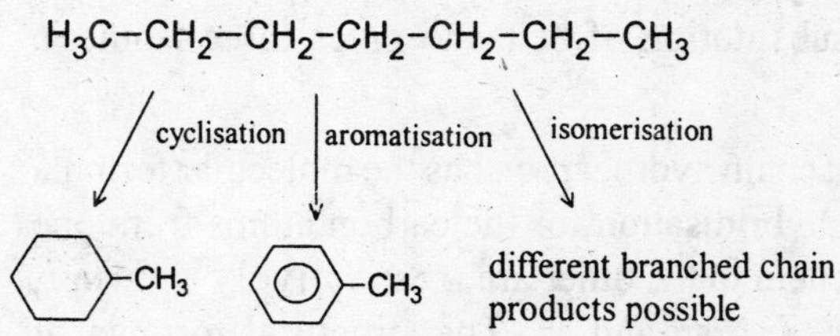

If reforming is done in presence of hydrogen, it is termed as hydroforming.

Aviation gasoline has octane number 100.

Cetane number: Quality of diesel is expressed in terms of cetane number. It is the percentage of cetane [n-hexadecane (cetane) has cetane number 100] in a mixture of cetane and α–methylnaphthalene (cetane number 0 zero). Straight chain hydrocarbons constitute a better fuel than the branched chain hydrocarbons.

Flash point: Minimum temperature at which an oil gives (produces) sufficient vapours to form an explosive mixture with air. The flash point of kerosine oil in India is 44°C while in UK it is 22.8°C.

Sweet gasoline: It is refined gasoline containing sweetening agent sodium plumbite (Na_2PbO_2). It is also known as Doctor's sweetening process.

1. Action of water on carbide forms [CBSE-91]
(A) ethane (B) ethene (C) ethyne (D) ethanol

2. $C_2H_2 \xrightarrow[H_2SO_4]{HgSO_4\ 1\%} X \xrightarrow{O} Y$
Y is known as [CBSE-91]
(A) acid (B) aldehyde (C) ketone (D) ethanol

3. Benzene reacts with CH_3COCl in the presence of $AlCl_3$ to give [CBSE-91]
(A) C_6H_5Cl (B) $C_6H_5CH_3$
(C) $C_6H_5CO_2Cl$ (D) $C_6H_5COCH_3$

4. Select the true statement about benzene from amongst the following [CBSE-92]
(A) Because of unsaturation benzene easily undergoes addition reactions
(B) There are two types of C-C bonds in benzene molecule
(C) There is cyclic delocalisation of pi eletrons in benzene
(D) Monosubstitution of benzene gives three isomeric products

5. A straight chain hydrocarbon has the molecular formula C_8H_{10}. The hybridisation for the carbon atoms from one end of the chain to the other are respectively sp^3, sp^2, sp^2, sp^3, sp^2, sp^2, sp and sp. The structural formula of hydrocarbon would be [CBSE-92]
(A) $CH_3–C≡C–CH_2–CH = CH–CH = CH_2$
(B) $CH_3–CH_2–CH = CH–CH = CH–C≡CH$
(C) $CH_3–CH = CH–CH_2–C ≡ C–CH = CH_2$
(D) $CH_3–CH = CH–CH_2–CH = CH–C≡CH$

6. On passing acetylene through a dilute solution of sulphuric acid containing mercuric sulphate, the major product obtained will be [CBSE-92]
(A) ethylene (B) ethylene glycol
(C) acetaldehyde (D) acetic acid

7. To prepare a pure sample of n-hexane using sodium metal as one reactant, the other reactant or reactants will be [CBSE-92]
(A) ethylchloride and n-butylchloride
(B) methylbromide and n-pentylbromide
(C) n-propylbromide
(D) ethylbromide and n-butylbromide

8. Baeyer's reagent is used in the laboratory for [CBSE-92]
(A) detection of double bonds (B) detection of glucose
(C) reduction (D) oxidation

9. Which one of these will react with sodium metal ? [CBSE-92]
(A) Ether (B) Ethyne (C) Ethene (D) Ethane

10. When propene is heated at 400°C in presence of chlorine it gives the following compound [CBSE-93]
(A) Polyvinylchloride (B) No reaction
(C) 1,2-Dichloropropene (D) Allylchloride

11. Benzene reacts with n-propylchloride in the presence of anhydrous $AlCl_3$ to give [CBSE-93]
(A) 3-propyl-1-chlorobenzene (B) n-propylbenzene
(C) no reaction (D) iso-propylbenzene

12. When acetylene reacts with water in the presence of $HgSO_4$, the compound actually isolated is [CBSE-93]
(A) CH_3COCH_3 (B) $CH_3–CHO$
(C) $CH_2 = CH_2$ (D) $CH_2 = CH(OH)$

13. Reduction of 2-butyne with sodium in liquid ammonia gives predominantly [CBSE-93]
(A) Cis-2-butene (B) No reaction
(C) Trans-2-butene (D) n-butene

14. Which of the following can distinguish propyne from propene ? [CBSE-93]
(A) Ammoniacal $AgNO_3$ (B) Dil. H_2SO_4
(C) $Br_2–H_2O$ (D) Aqueous $KMnO_4$

15. A compound is treated with $NaNH_2$ to give sodium salt. Identify the compound [CBSE-93]
(A) C_2H_2 (B) C_6H_6 (C) C_2H_6 (D) C_2H_4

16. Bromoethane when undergoes Wurtz reaction gives [CBSE-93]
(A) 2,3-dimethylbutane (B) 2-methylpropane
(C) n-butane (D) n-propane

17. Propene and propyne may be differentiated by [CBSE-93]
(A) $Br_2—H_2O$ (B) Alkaline—$KMnO_4$
(C) Acidic—$K_2Cr_2O_7$ (D) Ammoniacal—Cu_2Cl_2

18. Reactivity of hydrogen atoms attached to different atoms in alkanes has order [CBSE-93]
(A) Tertiary > Primary > Secondary
(B) Primary > Secondary > Tertiary
(C) Both (A) and (B) (D) Tertiary > Secondary > Primary

19. Which of the following does not observe the anti-Markownikoff's addition of HBr ? [CBSE-94]
(A) but-2-ene (B) propene (C) pent-2-ene (D) butene

20. The reactive species in nitration of benzene is [CBSE-94]
(A) NO_2^+ (B) HNO_3 (C) NO_3^- (D) NO_2^-

21. Alkene $RCH = CH_2$ reacts readily with B_2H_6 and the product on oxidation with alkaline hydrogen peroxides produces [CBSE-95]
(A) $R–CH_2–CHO$ (B) RCH_2CH_2OH
(C) $R–C(=O)–CH_3$ (D) $R–CH(OH)–CH_2(OH)$

22. Which of the following is used as an anti-knocking material ? [CBSE-96]
(A) T.E.L. (B) C_2H_5OH (C) Glyoxal (D) Freon

23. The bond length between sp^3 hybridised carbon and other carbon atom is minimum in [CBSE-96]
(A) propane (B) butane (C) propene (D) propyne

24. In the commercial gasoline, the type of hydrocarbons which are more desirable is [CBSE-97]
(A) branched hydrocarbon
(B) straight-chain hydrocarbon
(C) linear unsaturated hydrocarbon (D) toluene

25. The cylindrical shape of an alkyne is due to [CBSE-97]
(A) three sigma C-C bonds (B) three π C-C bonds
(C) two sigma C-C and one π C-C bonds
(D) one sigma C-C and two π C-C bonds

26. Which of the following reaction is expected to readily give a hydrocarbon product in good yields ? [CBSE-97]

(A) RCOOK $\xrightarrow[\text{Oxidation}]{\text{Electrolytic}}$

(B) RCOO Ag $\xrightarrow{I_2}$

(C) $CH_3CH_3 \xrightarrow[hv]{Cl_2}$

(D) $(CH_3)_2CHCl \xrightarrow{C_2H_5OH}$

27. When acetylene is passed through dil. H_2SO_4 in the presence of $HgSO_4$, the compound formed is [CBSE-99]
(A) ether (B) ketone
(C) acetic acid (D) acetaldehyde

28. Which is the order of stability for I 1-butene II cis-2-butene III trans-2-butene [CBSE-2000]
(A) I < II < III (B) I > II > III
(C) III > I > II (D) I > II < III

29. During Friedel-Crafts reaction which of the following reacts in presence of $AlCl_3$? [CBSE-2000]

(A) 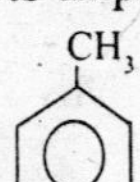 + HCl (B)  + CH_2Cl_2

(C)  + $CHCl_3$ (D) 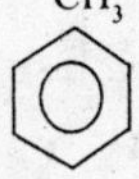 + CH_3Cl

30. Which alkene on ozonolysis gives CH_3CH_2CHO and CH_3COCH_3 ? [CBSE-2001]
(A) $CH_3CH_2CH = C(CH_3)_2$
(B) $CH_3CH_2CH = CHCH_2CH_3$
(C) $CH_3CH_2CH = CHCH_3$ (D) $CH_3 - C(CH_3) = CHCH_3$

31. Which of the following is incorrect ? [CBSE-2001]
(A) $FeCl_3$ is used in detection of phenolic group
(B) Fehling solution is used in detection of glucose
(C) Tollen's reagent is used in detection of unsaturation
(D) $NaHSO_3$ is used in detection of carbonyl group

32. In preparation of an alkene from an alcohol using Al_2O_3, which is effective factor ? [CBSE-2001]
(A) Porosity of Al_2O_3 (B) Temperature
(C) Concentration (D) Surface area of Al_2O_3

33. The anti-Markownikoff addition of HBr is not observed in [AIIMS-93]
(A) Propylene (B) 1-butene (C) 2-butene (D) 2-pentene

34. The function of anhydrous aluminium chloride in Friedel Craft's reaction is to [AIIMS-93]
(A) Absorb water (B) Absorb hydrochloric acid
(C) Produce electrophile (D) Produce nucleophile

35. Which organic structure corresponds to the product expected when excess of benzene react with CH_2Cl_2 in the presence of $AlCl_3$? [AIIMS-93]

(A) 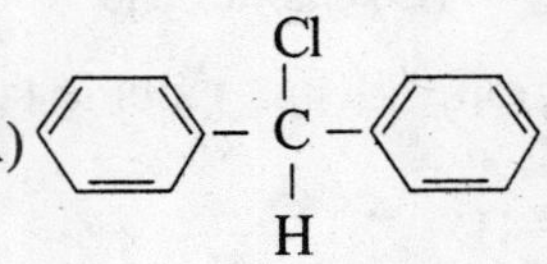 (B) 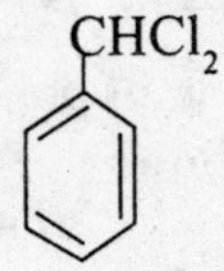

(C) 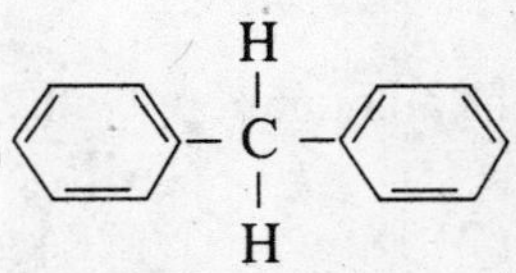 (D) 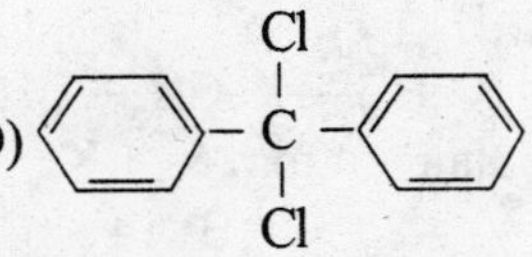

36. The formation of 2-butene as a major product by dehydrohalogenation of 2-chlorobutane is in accordance with [AIIMS-94]
(A) Huckel's rule (B) Saytzeff's rule
(C) Blanc's rule (D) Markiownikoff's rule

37. When propyne is treated with aqueous H_2SO_4, in the presence of $HgSO_4$, the product formed is [AIIMS-94]
(A) Ether (B) Propanal (C) Acetone (D) Aldehyde

38. In Wolf-Kishner reduction, the carbonyl group of aldehydes and ketones is converted into [AIIMS-94]
(A) $-CH_2-$ (B) $-CHOH-$ (C) $-CH_2OH$ (D) $-CH_3$

39. Assertion : Alkanes having more than three carbon atoms exhibit chain isomerism Reason : All carbon atoms in alkanes are sp-hybridized [AIIMS-94]
(A) A (B) B (C) C (D) D

40. Assertion : Cyclobutane is less stable than cyclopentane. Reason : Presence of 'bent bonds' causes "loss of orbital overlap ". [AIIMS-95]
(A) A (B) B (C) C (D) D

41. The complete combustion of CH_4 gives [AIIMS-95]
(A) $CO + H_2$ (B) $CO + N_2$
(C) $CO_2 + H_2O$ (D) $CO + N_2O$

42. In the presence of mercuric ion and conc. sulphuric acid, the reaction of acetylene with water produces [AIIMS-95]
(A) $CH_3 - CHO$ (B) $CH_3 - COOH$
(C) $CH_3 - CO - CH_3$ (D) $CH_3 - CH_2 - OH$

43. Ozonolysis of acetylene gives [AIIMS-96]
(A) CH_3CHO (B) CHO-CHO
(C) HCHO (D) None of these

44. Hydrolysis of beryllium carbide produces

[AIIMS-96]

(A) Methane (B) Ethene
(C) Acetylene (D) None of these

45. 1-butyne reacts with cold alkaline $KMnO_4$ to produce [AIIMS-97]

(A) $CH_3CH_2CH_2COOH$ (B) CH_5CH_2COOH
(C) $CH_3CH_2COOH + CO_2$
(D) $CH_3 - CH_2COOH + HCOOH$

46. Hexyne reacts with Na/H_2 to produce [AIIMS-97]

(A) Hexene (B) Hexylamine
(C) Trans-hexyne (D) Cis-hexyne

47. Formation of polythene from calcium carbide takes places as follows

$CaC_2 + H_2O \rightarrow Ca(OH)_2 + C_2H_2$
$C_2H_2 + H_2 \rightarrow C_2H_4$
$n\,(C_2H_4) \rightarrow (-CH_2 - CH_2-)_n$ The amount of polyethylene obtained from 64.1 Kg CaC_2 is [AIIMS-97]

(A) 7 kg (B) 14 kg (C) 21 Kg (D) 28 kg

48. If 1,3-dibromopropane reacts with Zn, NaI, the product obtained is [AIIMS-97]

(A) Propene (B) Propane
(C) Cyclopropane (D) Hexane

49. Which of the following is the most stable alkene ? [AIIMS-98]

(A) $R_2 - C = CH_2$ (B) $RCH = CHR$
(C) $RCH = CH_2$ (D) $CH_2 = CH_2$

50. Baeyer's reagent is used in the laboratory for [AIIMS-98]

(A) Oxidation process (B) Reduction process
(C) Detection of glucose (D) Detection of double bond

51. Propane reacts with HBr to form [AIIMS-99]

(A) Ethane (B) Hexane
(C) Propane (D) Bromopropane

52. If benzene reacts with Cl_2 in presence of ultraviolet light then which of the following is formed ? [AIIMS-99]

(A) C_6Cl_6 (B) CCl_4 (C) CH_3Cl (D) $C_6H_6Cl_6$

53. Which has the smallest heat of hydrogenation per mole?

(A) but-1-ene (B) cis-2-butene
(C) trans-2-butene (D) 1,3-butadiene

54. $C_6H_6 \xrightarrow[H_2SO_4]{HNO_3} X \xrightarrow[FeCl_3]{Cl_2} Y$ In the above sequence Y can be [AIIMS-99]

(A) 1-nitrochlorobenzene (B) 3-nitrochlorobenzene
(C) 4-nitrochlorobenzene (D) None of these

55. Identify X in the reaction

$CH_2 = CH_2 + H_2SO_4 \rightarrow \text{intermediate} \xrightarrow[\text{boil}]{H_2O} X$

[AIIMS-2000]

(A) CH_3OH (B) CH_3CH_2OH
(C) CH_3COCH_3 (D) CH_3OCH_3

56. A gas X decolourises bromine water and also Baeyer's reagent, it burns in air with sooty flame. When it reacts with HCl, it forms finally ethyledene chloride. X is [AIIMS-2000]

(A) C_2H_6 (B) C_3H_4 (C) C_2H_4 (D) C_2H_2

57. Ammoniacal solution of cuprous chloride gives red precipitate with [AIIMS-2000]

(A) $H-C \equiv C-CH_3$ (B) $H-\underset{H}{\underset{|}{C}} = \underset{H}{\underset{|}{C}}-H$

(C) $(C_2H_5)_2-C = CH_2$ (D) $CH_3-C \equiv C-C_2H_5$

58. Which alkene on ozonolysis gives following compounds? (i) methanal (ii) ethanal (iii) 2-oxo propanal [AIIMS-2001]

(A) 3-methyl-1,2-butadiene
(B) 3-methyl-1,3-pentadiene
(C) 3-methyl-1,3-hexadiene
(D) 2-methyl-1,3-hexadiene

59. In reaction of benzene with conc.$(HNO_3+H_2SO_4)$ which is velocity determining step ? [AIIMS-2001]

(A) $\overset{\oplus}{N}O_2$ formation (B) Carbocation formation
(C) Removal of $\overset{\oplus}{H}$ (D) All of these

60. In the presence of reductant Na+liq.NH_3 from alkyne, alkene formed is : [AIIMS-2001]

(A) Trans (B) Cis (C) Both of these (D) none of these

61. Which of the following is the most stable ? [AIIMS-2002]

(A) 1-butene (B) 2-butene
(C) 1-pentene (D) 2-pentene

62. Action of acetylene on dilute H_2SO_4 gives [AIIMS-2002]

(A) Acetic acid (B) Acetone
(C) Acetaldehyde (D) Acetoacetic ester

63. The size of C–C bond in benzene is [AIIMS-2002]

(A) 1.22 Å (B) 1.39 Å (C) 1.54 Å (D) 1.56 Å

64. Thermite is mixture of iron oxide and [AIIMS-2002]

(A) Zinc powder (B) Sodium shavings
(C) Potassium metal (D) Aluminium powder

65. The ortho/para directing group among the following is - [AIIMS-2003]

(A) COOH (B) CN (C) $COCH_3$ (D) $NHCOCH_3$

66. The treatment of benzene with isobutene in the presence of sulphuric acid gives- [AIIMS-2003]

(A) Isobutylbenzene (B) Tert-butylbenzene
(C) n-butylbenzene (D) No reaction

67. Which of the following compounds possesses the C-H bond with the lowest bond dissociation energy ? [AIIMS-2003]

(A) Toluene (B) Benzene
(C) n-pentane (D) 2,2-dimethyl propane

68. (A) Alkyl benzene is not prepared by Friedel-Crafts alkylation of benzene.
(R) Alkylhalides are less reactive than acylhalides. [AIIMS-2003]
(A) (B) (C) (D)

69. In benzene, the overlapping between carbon-carbon is of the type : [JIPMER-90]
(A) sp^3–sp^3 (B) sp^2–sp^2 (C) sp–sp (D) p–p

70. Which gives acetone on ozonolysis : [JIPMER-90]
(A) 2-Butene (B) Propene
(C) 1-Butene (D) 2-Methyl-1-propene

71. Which of the following is not a petroleum product : [JIPMER-90]
(A) Paraffin was (B) Kerosene oil
(C) Gasoline (D) Toluene

72. The nitration of nitrobenzne will give : [JIPMER-90]
(A) T.N.T. (B) 1,3 dinitrobenzene
(C) Picric acid (D) 1,4-dinitrobenzene

73. Ethyl alcohol is heated with conc. H_2SO_4. The product is : [JIPMER-91]
(A) $H_3C–CO–OC_2H_5$ (B) C_2H_6 (C) C_2H_4 (D) C_2H_2

74. Candles contain mixture of : [JIPMER-92]
(A) Bee wax and paraffin wax
(B) Bee wax and stearic acid
(C) Paraffin wax and stearic acid (D) Higher fatty acid

75. Acidic hydrogen is present in : [JIPMER-92]
(A) Ethylene (B) Ethyne (C) Benzene (D) Ethane

76. Anti-Markownikoff's addition of HBr is not observed in : [JIPMER-92]
(A) Propene (B) Butene (C) But-2-ene (D) Pent-2-ene

77. Which of the following is expected to have least boiling point : [JIPMER-92]
(A) n-pentane (B) n-hexane
(C) iso-pentane (D) neo-pentane

78. Nitration of benzene is : [JIPMER-92]
(A) Nucleophilic substitution (B) Electrophilic addition
(C) Electrophilic substitution (D) Nucleophlic addition

79. Which of the following will react with Br_2 most readily? [JIPMER-93]
(A) $CH_2 = CH_2$ (B) $CH_3CH = CHCH_3$
(C) $CH_3CH = CH_2$ (D) $(CH_3)_2C = C(CH_3)_2$

80. The weakest acid among the following is : [JIPMER-93]
(A) Phenol (B) Picric acid
(C) Acetylene (D) Acetic acid

81. Ozonolysis of an alkene A followed by decomposition with water and a reducing agent gave a mixture of two isomers of the formula C_3H_6O. The structure of alkene is : [JIPMER-93]
(A) $CH_3CH = CHCH_3$
(B) $(CH_3)_2C = CHCH_2CH(CH_3)_2$
(C) $CH_3CH_2CH = C(CH_3)_2$ (D) $(CH_3)_2)C = C(CH_3)_2$

82. When ethyl iodide is treated with sodium in the presence of ether, it forms : [JIPMER-93]
(A) Ethylene (B) Acetylene (C) Butylene (D) Butane

83. Which one of following is NOT found in alkenes ? [JIPMER-96]
(A) Chain isomerism (B) Geometrical isomerism
(C) Tautomerism (D) Position isomerism

84. Which of the following alcohols cannot be dehydrated to give alkene [JIPMER-96]
(A) Tert-butyl alcohol (B) Neopentyl alcohol
(C) Isopropyl alcohol (D) All of these can be dehydrated

85. In alkenes, the negative part of addendum adds on to the carbon atom joined to the least number of hydrogen atoms. This statement is called [JIPMER-96]
(A) Peroxide effect (B) Baeyer's strain theory
(C) Thiele's theory (D) Markownikoff's Rule

86. $CH_3CH = C = CH_2$ has [JIPMER-97]
(A) Only sp hybridised carbon atoms
(B) Both sp and sp^2 hybridised atoms
(C) Only sp^2 hybridised carbon atoms
(D) sp, sp^2 and sp^3 hybridised atoms

87. Conversion of n-hexane to benzene in the presence of chromic oxide on Al_2O_3 support is called [JIPMER-97]
(A) Hydrogenation (B)Dehydrogenation
(C) Isomerisation (D) Aromatisation

88. When 2-ethylanthraquinol dissolved in a mixture of benzene and cyclohexanol is oxidised, the product is [JIPMER-98]
(A) Ethanol (B) Hydrogen peroxide
(C) Anthracene (D) None of these

89. When ethyne is passed through dilute H_2SO_4 containing $HgSO_4$, it forms [JIPMER-98]
(A) Ethanol (B) Ethanoic acid (C) Ethanal (D) Ethene

90. Which of the following compounds will react with sodium ? [JIPMER-99]
(A) C_2H_6 (B) C_2H_4 (C) CH_3Cl (D) C_2H_2

91. Propene can be converted to 1-propanol. Indicate which set of reagents among the following is ideal for effecting the above conversion. [JIPMER-99]
(A) Osmium tetraoxide (B) C_2H_6 and alkaline H_2O_2
(C) Alkaline $KMnO_4$ (D) Ozone and Zn dust

92. Which of the following reagents would form 2,2-dibromo-propane on interaction ? [JIPMER-99]

(A) $CH_3CH = CHBr + HBr$ (B) $CH = CH + 2HBr$
(C) $CH_3CH = CH_2 + HBr$ (D) $CH_3C \equiv CH + HBr$

93. Upon hydration with $HgSO_4$ and dilute H_2SO_4, propyne gives mainly [JIPMER-99]
(A) Propanal (B) Propane
(C) Propyl alcohol (D) Propanone

94. The relationship between acetylene and benzene is comparable to the relationship between propyne and [JIPMER-99]
(A) Dimethyl benzene (B) Neoprene
(C) Mesitylene (D) Propyl benzene

95. In the case of homologous series of alkanes, which one of the following statements is incorrect [JIPMER-2000]
(A) The members of the series are isomers of each other
(B) The members of the series have similar chemical properties
(C) The members of the series have the general formula C_nH_{2n+2}, where n is an integer
(D) The difference between any two successive members of the series correcponds to 14 unit of relative atomic mass

96. Butyne-1 and butyne-2 can be distinguished by [JIPMER-2000]
(A) Ammoiniacal cuprous chloride
(B) Alkaline $KMnO_4$ (C) Benedict reagent (D) Br_2 water

97. The distinguishing test for triple bond contaning acidic hydrogen is [JIPMER-2000]
(A) $Ag(NH_3)_2^+$ (B) Br_2 in CCl_4
(C) Alkaline $KMnO_4$ (D) $Ag(NH_3)_2^+$

98. 1,3-butadiene has [JIPMER-2000]
(A) sp and sp^2 hybridised C-atoms
(B) sp, sp^2 and sp^3 hybridised C-atoms
(C) Only sp^2 hydridised C-atoms
(D) Only sp hybridised C-atoms

99. Conjugated double bond is present in [JIPMER-2001]
(A) 1,3-butadiene (B) Butylene
(C) Isobutylene (D) Propylene

100. A hydrocarbon X adds on one mole of hydrogen to give another hydrocarbon and decolourises bromine water. X reacts with $KMnO_4$ in presence of acid to give two moles of the same carboxylic acid. The structure of X is [JIPMER-2001]
(A) $CH_2 = CH–CH_2CH_2CH_3$
(B) $CH_3CH_2CH_2–CH = CHCH_3$
(C) $CH_3CH_2\,CH = CHCH_2CH_3$
(D) $CH_3CH = CHCH_2CH_3$

101. Gammaxene is the name of [JIPMER-2001]
(A) benzene hexachloride (B) chlorobenzene
(C) benzoyl chloride (D) DDT

102. Most reactive hydrocabon is [JIPMER-2002]
(A) Ethene (B) Ethyne (C) Ethane (D) Methane

103. $CH \equiv CH \xrightarrow{H_2O/Hg^{2+}} X \xrightarrow[H_2SO_4]{LiAlH_4} Y \xrightarrow{P_4/Br_2}$ Z.Here Z is [JIPMER-2002]
(A) Ethylene bromide (B) Ethanol
(C) Ethyl bromide (D) Ethylidene bromide

104. Tetraethyl lead is a [BHU-95]
(A) Solvent (B) Petroleum additive
(C) Oxidising agent (D) Fire extinguisher .

105. The complete combustion of CH_4 gives [BHU-95]
(A) $CO_2 + H_2O$ (B) $CO_2 + H_2$
(C) $COCl_2$ (D) $CO + H_2O$

106. Ethene can be produced from ethanol by [BHU-96]
(A) Dehydrohalogenation (B) Dehydrogenation
(C) Dehydration with conc. H_2SO_4 at 170°C
(D) Reduction with hydrogen iodide

107. Iodoform (CHI_3) reacts with Ag powder to produce [BHU-96]
(A) Ethene (B) Acetylene (C) Methane (D) Butane

108. Benzene shows mainly [BHU-96]
(A) Electrophilic substitution (B) Electrophilic addition
(C) Nucleophilic substitution (D)Nucleophilic addition

109. When ethyl iodide and propyl iodide react with Na in the presence of ether, they form [BHU-97]
(A) One alkane (B) Two alkanes
(C) Four alkanes (D) Three alkanes

110. Propene reacts with HBr in presence of peroxides, the product is [BHU-98]
(A) 2-bromopropane (B) 1-bromopropane
(C) 3-bromopropane (D) None of these

111. When acetylene is passed through dil H_2SO_4 in the presence of $HgSO_4$ the compound formed is [BHU-98]
(A) C_2H_5OH (B) Acetone
(C) Carbide of Hg (D) Acetadehyde

112. Alkyl halides react with metallic sodium in the presence of anhydrous ether [BHU-98]
(A) Wurtz's reaction (B) Aldol's reaction
(C) Perkin's reaction (D) None of these

113. To prepare a pure sample of n-hexane using sodium metal as one reactant, the other reactant or reactants will be [BHU-99]
(A) n-propyl bromide
(B) Ethyl bromide and n-butyl bromide
(C) Ethyl chloride and n-butyl bromide
(D) Methyl bromide and n-pentyl chloride

114. The decreasing order of boiling points is [BHU-99]
(A) n-pentane > iso-pentane > neo-pentane
(B) iso-pentane > n-pentane > neo-pentane
(C) neo-pentane > iso-pentane > n-pentane

(D) n-pentane > neo-pentane > iso-pentane

115. Heating a mixture of sodium benzoate and soda lime gives [BHU-99]
(A) Benzene (B) Aniline
(C) Toluene (D) None of these

116. Tetrabromoethane on heating with Zn gives [BHU-2000]
(A) Ethyl Bromide (B) Ethane (C) Ethene (D) Ethyne

117. Clemmensen reduction of a ketone is carried out in presence of [BHU-2000]
(A) H_2 with Pd as catalyst (B) Glycol with KOH
(C) $LiAlH_4$ in ether (D) Zn-Hg with HCl

118. The reaction [BHU-2001]
$C_6H_6 + CH_3Cl \xrightarrow{AlCl_3\ (anhyd.)} C_6H_5CH_3 + HCl$
(A) Friedel Craft's reaction (B) Wurtz reaction
(C) Sandmeyer's reaction (D) Wurtz-Fittig reaction

119. 1,3-butadiene reacts with ethene to form [BHU-2001]
(A) Benzene (B) Cyclohexane
(C) Cyclohexene (D) 2,3-dimethylbutane

120. The more electronegative part of the addendum goes to less protonated carbon atom. This is the statement of [BHU-2001]
(A) Saytzeff's Rule
(B) Markownikoff's Rule
(C) Anti-Markownikoff's Rule (D) None of these

121. The function of anhydrous $AlCl_3$ in Friedelcraft's reaction is to [BHU-2001]
(A) Absorb H_2O (B) Absorb HCl
(C) To produce attacking electrophile
(D) To produce nucleophile

122. Hybridization of carbon atom in benzene is [BHU-2001]
(A) sp^3 (B) sp (C) sp^2 (D) dsp^2

123.
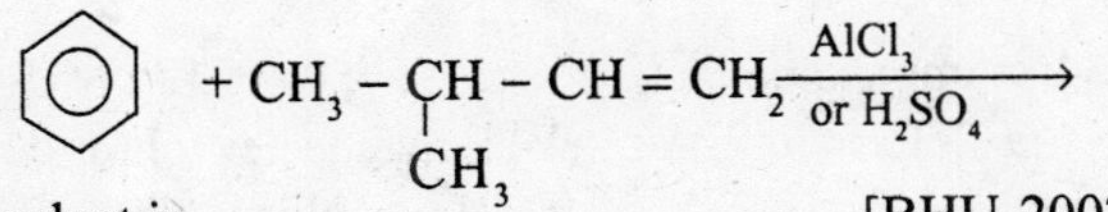
the product is [BHU-2002]

(A)
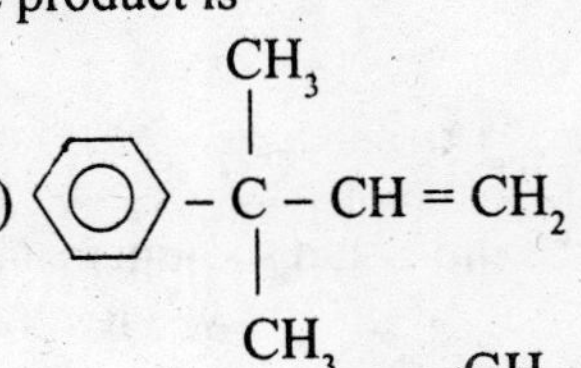
(B)
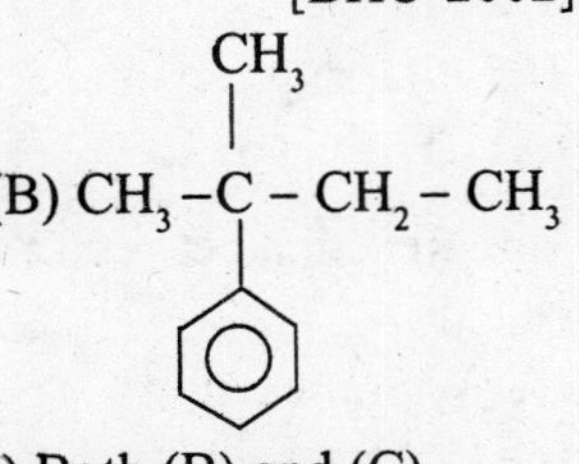
(C)
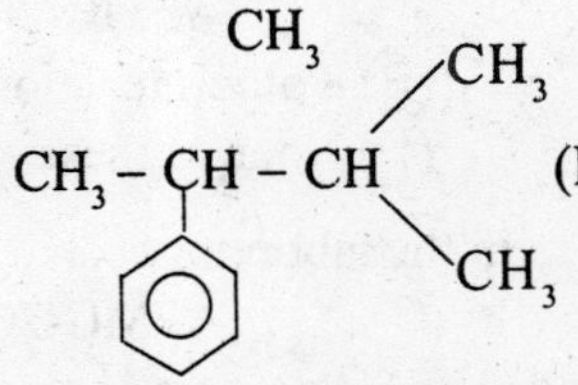
(D) Both (B) and (C)

124. Addition of HCN to Ethyne in presence of $Ba(CN)_2$ as catalyst gives [AFMC-91]
(A) 1,1-dicyanoethane (B) Ethylcyanide
(C) Vinylcyanide (D) Di vinylcyanide

125. Benzene vapour mixed with air when passed over V_2O_5 catalyst at 775 K gives [AFMC-91]
(A) Glyoxal (B) Oxalic acid
(C) Maleic anhydride (D) Fumaric acid

126. C-C bond length is minimum in [AFMC-91]
(A) Ethane (B) Ethylene (C) Benzene (D) Ethyne

127. When propyne is treated with dil. H_2SO_4 in presence of $HgSO_4$, the major product is [AFMC-91]
(A) Propanal (B) Propanone
(C) Acetone (D) Propyl H_2SO_4

128. Kerosene oil is a mixture of [AFMC-92]
(A) Aromatic compounds (B) Plant oils
(C) Gasoline (D) Hydrocarbons

129. Which of the following is a hydrocarbon ? [AFMC-92]
(A) Urea (B) Benzene
(C) Ammonium cyanate (D) Phenol

130. Electrolysis of sodium salt of acetic acid liberates at the anode [AFMC-93]
(A) Hydrogen (B) Ethane
(C) Ethylene (D) Acetylene

131. The hydrogen in which one of the following can be replaced by a metal [AFMC-93]
(A) Ethylene (B) Ethyne
(C) Dimethyl acetylene (D) Benzene

132. Gammexane is [AFMC-93]
(A) Chlorobenzene (B) Benzene hexachloride
(C) Bromobenzne (D) Benzyl chloride

133. Kerosene is a mixture of [AFMC-93]
(A) Alkenes (B) Alkanes (C) Alkynes (D) Arenes

134. Petroleum is mixture of [AFMC-94]
(A) Alkanes (B) Alkenes
(C) Alkynes (D) Alkylhalides

135. The first fractional product of petroleum is [AFMC-94]
(A) Petroleum ether (B) Kerosene
(C) Diesel (D) None of these

136. When acetylene is reacted with hydrochloric acid in the presence of mercuric chloride, the product obtained is [AFMC-94]
(A) Methyl chloride (B) Acetaldehyde
(C) Chloride (D) Formaldehyde

137. Fischer - Tropsch process is used in the manufacture of [AFMC-94]
(A) Synthetic petrol (B) Ethanol
(C) Benzene (D) Ethanoic acid

138. The formula of freon is [AFMC-95]

(A) CCl_2F_2 (B) $CFCl_3$ (C) $CClF_3$ (D) None of these

139. Structure of benzene is [AFMC-95]
(A) Planar (B) Linear
(C) Trigonal bipyramidal (D) Tetrahedral

140. A war gas Lewisite is formed by reaction of arsenic chloride with [AFMC-96]
(A) CH_4 (B) C_6H_6 (C) C_2H_2 (D) C_2H_4

141. Kerosene oil has compounds having [AFMC-96]
(A) $C_5 - C_8$ (B) $C_8 - C_{12}$ (C) $C_{12} - C_{16}$ (D) $C_{16} - C_{18}$

142. Which of the following is used as a refrigerant ? [AFMC-96]
(A) CO_2 (B) $CHCl_3$ (C) SiC (D) CF_2Cl_2

143. Higher percentage of carbon is found in [AFMC-96]
(A) Anthracene (B) Bituminous (C) Peat (D) All equal

144. Lead tetraethyl is used as [AFMC-96]
(A) Fire extinguisher (B) Pain killer
(C) Anti-knock agent (D) Mosquito repellent

145. Ethane and Methane are prepared in one step by [AFMC-96]
(A) CH_3Br (B) C_2H_5Br (C) C_2H_5OH (D) C_2H_5CHO

146. Which of the following structures is planar ? [AFMC-96]
(A) CH_4 (B) C_2H_4 (C) C_2H_2 (D) C_2H_6

147. Constituent of light oil is [AFMC-96]
(A) Benzene (B) Phenol (C) Anthracene (D) Aniline

148. Vanaspati ghee is manufactured by [AFMC-96]
(A) Hydrogenation of oil (B) Oxidation of oil
(C) Reduction of oil (D) None of these

149. Natural gas is [AFMC-96]
(A) C_6H_6 (B) CH_4 (C) C_3H_8 (D) C_2H_2

150. Iodoethane reacts with sodium in presence of ether. What is the product ? [AFMC-97]
(A) Pentane (B) Propane (C) Butene (D) Butane

151. In Friedel-Craft's reaction, besides $AlCl_3$ what are the reactants ? [AFMC-97]
(A) $C_6H_6 + CH_3Cl$ (B) $C_6H_6 + CH_4$
(C) $C_6H_6 + NH_2$ (D) None of these

152. Marsh gas contains [AFMC-97]
(A) C_2H_2 (B) CH_4 (C) H_2S (D) CO

153. A compound is treated with $NaNH_2$ to give sodium salt. Identify the compound : [AFMC-98]
(A) C_2H_2 (B) C_6H_6 (C) C_2H_6 (D) C_2H_4

154. Both methane and ethane can be prepared in one step by the reaction of [AFMC-98]
(A) C_2H_4 (B) CH_3OH (C) CH_3Br (D) CH_3CH_2OH

155. In which of the following hydrocarbons, hydrogen is most acidic ? [AFMC-99]
(A) $CH \equiv CH$ (B) $CH_3 - CH_3$
(C) $CH_2 = CH_2$ (D) None of these

156. Which of the following has the highest boiling point? [AFMC-99]
(A) Pentane (B) Ethane (C) n-octane (D) Butane

157. Heating a mixture of sodium benzoate and soda lime gives [AFMC-99]
(A) Methane (B) Sodium benzoate
(C) Benzene (D) Calcium benzoate

158. When acetylene reacts with arsenic trichloride in the presence of anhydrous aluminum chloride, it produces [AFMC-99]
(A) β-chlorovinyl dichloroarsine (B) Lewisite
(C) Nitrobenzene

159. Baeyer's reagent is used in the laboratory for [AFMC-99]
(A) Detection of double bond (B) Reduction process
(C)Oxidation process (D) Detection of glucose

160. $C_6H_6Cl_6 \xrightarrow[\text{KOH (alc.)}]{\text{Heat or}}$? [AFMC-2000]
(A) Benzene (B) Phenol
(C) $C_6H_5(OH)_6$ (D) Chlorobenzene

161. The reaction of benzene with ethyl chloride, in the presence of aluminium chloride, is called [AFMC-2000]
(A) Wurtz reaction (B) Carbylamine reaction
(C) Perkin's reaction (D) Friedel-Crafts reaction

162. Solutions having same osmotic concentration are called [AFMC-2001]
(A) Hypotonic (B) Hypertonic
(C) Isotonic (D) Normal

163. Antiknocking character of gasoline can be improved by addition of [AFMC-2001]
(A) Lead sulphate (B) Tetraethyl lead
(C) Lead bromide (D) Diethyl lead

164. Gammexane is [AFMC-2001]
(A) D.D.T. (B) Benzene hexachloride
(C) Chloral (D) Hexachloroethane

165. Marsh gas contains [AFMC-2001]
(A) CH_4 (B) CO_2 (C) C_2H_6 (D) N_2

166. Reaction between alkyl halides and sodium metal is called [AFMC-2001]
(A) Wurtz reaction (B) Kolbe's reaction
(C) Clemmenson's reaction (D) None of these

167. Baeyer's reagent is used in the laboratory for [AFMC-2002]
(A) Detection of double bonds (B) Detection of glucose
(C) Reduction (D) Oxidation

168. Propene and propyne may be differentiated by

[AFMC-2002]

(A) Br_2 - H_2O (B) Alkaline - $KMnO_4$
(C) Acidic - $K_2Cr_2O_7$ (D) Ammoniacal - Cu_2Cl_2

169. Propyne on polymerization gives [AFMC-2002]
(A) Mesitylene (B) Benzene
(C) Ethyl benzene (D) Propyl benzene

170. $KMnO_4$ will oxidise acetylene to [AFMC-2002]
(A) Ethylene glycol (B) Ethyl alcohol
(C) Oxalic acid (D) Acetic acid

171. The compounds used as refrigerant are [IIT-90]
(A) NH_3 (B) CCl_4 (C) CF_4 (D) CF_2Cl_2 (E) CH_2F_2

172. The valence atomic orbitals on C in silver acetylide is ....hybridised. [IIT-90]

173. Kolbe electrolysis of potassium succinate gives CO_2 and .... [IIT-93]

174. Which of the following has the smallest heat of hydrogenation per mole ? [IIT-93]
(A) 1–butene (B) trans–2–butene
(B) cis–2–butene (D) 1, 3–butadiene

175. Nitrobenzene can be prepared from benzene by using a mixture of conc. HNO_3 and conc. H_2SO_4. In the nitrating mixture, HNO_3 acts as a [IIT-97]
(A) base (B) acid
(C) reducing agent (D) catalyst

176. Among the following statements on the nitration of aromatic compounds, the false one is [IIT-97]
(A) the rate of nitration of benzene is almost the same as that of hexadeuterobenzene
(B) the rate of nitration of toluene is greater than that of benzene
(C) the rate of nitration of benzene is greater than that of hexadeuterobenzene
(D) nitration is an electrophilic substitution reaction.

177. 1, 3-Butadiene with bromine in molar ratio generates predominantly .... [IIT-97]

178. The intermediate during the addition of HCl to propene in the presence of peroxide is [IIT-97]
(A) $CH_3CH^+CH_2Cl$ (B) $CH_3CH^+CH_3$
(C) $CH_3CH_2{}^+CH_2$ (D) $CH_3CH_2\dot{C}H_2$

179. The reaction of $CH_3CH = CH$ 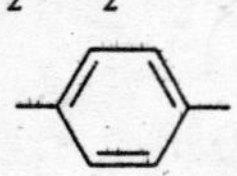 OH with HBr gives [IIT-98]
(A) $CH_3CHBrCH_2$ 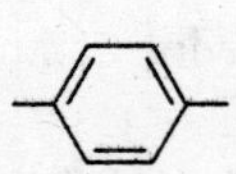 OH
(B) CH_3CH_2CHBr 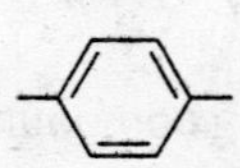 OH
(C) $CH_3CHBrCH_2$ 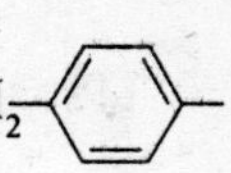 OH
(D) CH_3CH_2CHBr 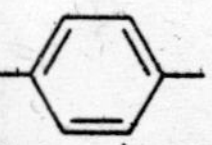 OH

180. The product(s) obtained via oxymercuration ($HgSO_4$+H_2SO_4) of 1-butyne would be [IIT-99]
(A) $CH_3 - CH_2 - \overset{\overset{O}{\|}}{C} - CH_3$
(B) $CH_3 - CH_2 - CH - CHO$
(C) $CH_3 - CH_2 - CHO + HCHO$
(D) $CH_3 - CH_2 - COOH + HCOOH$

181. The appropriate reagent for the following transformation

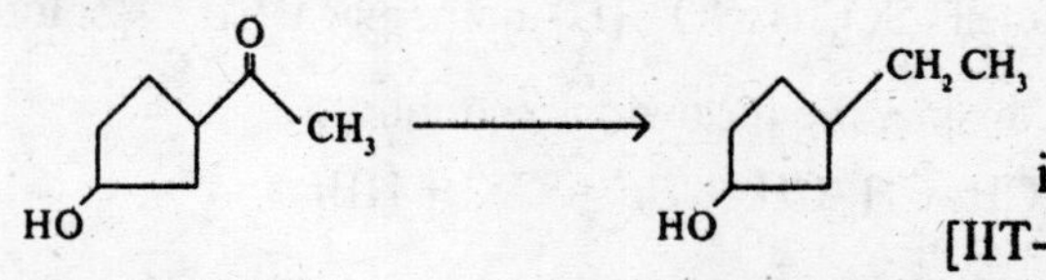

is [IIT-2000]
(A) Zn (Hg), HCl (B) NH_2NH_2,OH^-
(C) H_2 / Ni (D) $NaBH_4$

182. Propyne and propene can be distinguished by [IIT-2000]
(A) conc. H_2SO_4 (B) Br_2 in CCl_4
(C) dil. $KMnO_4$ (D) $AgNO_3$ in ammonia

183. Which one of the following alkenes will react fastest with H_2 under catalytic hydrogenation condition ? [IIT-2000]

(A) 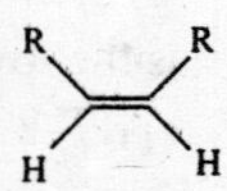

(B) 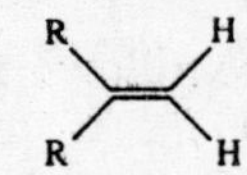

(C) 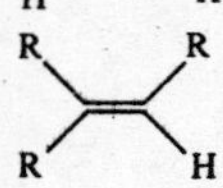

(D) 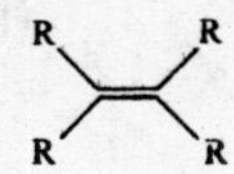

184. A : 1-butene on reaction with HBr in the presence of peroxide produces 1-bromobutane [IIT-2000]
R: It involves the formation of a primary radical
(A) A (B) B (C) C (D) D

185. In the presence of peroxide, hydrogen chloride and hydrogen iodide do not give anti-Markownikoff addition to alkenes because [IIT-2001]
(A) both are highly ionic
(B) one is oxidising and the other is reducing
(C) one of the steps is endothermic in both the cases
(D) all the steps are exothermic in both the cases

186. The reaction of propene with HOCl proceeds via the addition of [IIT-2001]
(A) H^- in the first step (B) Cl^+ in the first step
(C) OH^- in the first step (D) Cl^- and OH^- in a single step

187. A : Dimethylsulphide is commonly used for the Reduction of an ozonide of an alkene to get the carbonyl compounds. R : It reduces the ozonide giving water soluble dimethyl sulphoxide and excess of it evaporates.[IIT-2001]
(A) A (B) B (C) C (D) D

188. The nodal plane in the π-bond of ethene is located in [IIT-2002]

(A) the molecular plane
(B) a plane parallel to the molecular plane
(C) a plane perpendicular to the molecular plane which bisects the carbon-carbon σ-bond at right angle
(D) a plane perpendicular to the molecular plane which contains the carbon-carbon σ-bond

189. Identify a reagent from the following list which can easily distinguish between 1-butyne and 2-butyne [IIT-2002]

(A) bromine,CCl (B) Lindlar catalyst
(C) dilute H_2SO_4, $HgSO_4$ (D) ammoniacal Cu_2Cl_2 solution

190. Consider the following reaction

$H_3C-CH(D)-CH(CH_3)-CH_3 + Br^\bullet \rightarrow$ 'X' + HBr

Identify the structure of the a major product 'X' [IIT-2002]

(A) $H_3C-CH(D)-CH(CH)-\dot{C}H_2$ (B) $H_3C-CH(D)-\dot{C}(CH_3)-CH_3$

(C) $H_3C-\dot{C}(D)-CH(CH_3)-CH_3$ (D) $H_3C-CH-\dot{C}H(CH_3)-CH_3$

191. Which of the following hydrocarbons has the lowest dipole moment ? [IIT-2002]

(A) 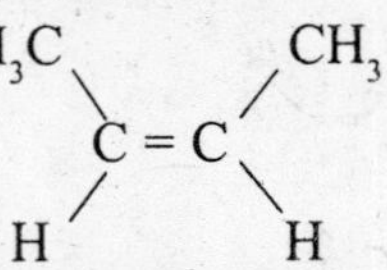 (B) $CH_3C \equiv CCH_3$

(C) $CH_3CH_2C \equiv CH$ (D) $CH_2 = CH-C \equiv CH$

192. $C_6H_6 - C \equiv C - CH_3 \xrightarrow[H_2SO_4]{HgSO_4}$ A;A is [IIT-2003]

(A) 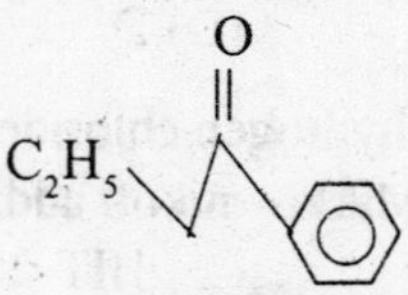 (B) $C_6H_5CH_2CHO$

(C) $C_6H_5CHO+CH_3CHO$ (D) $C_6H_5 - CO - CH_3$

193. 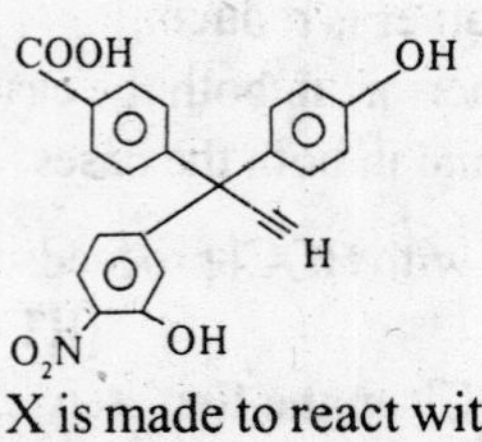

When X is made to react with 2 eq. of $NaNH_2$ the product formed will be [IIT-2003]

(A) 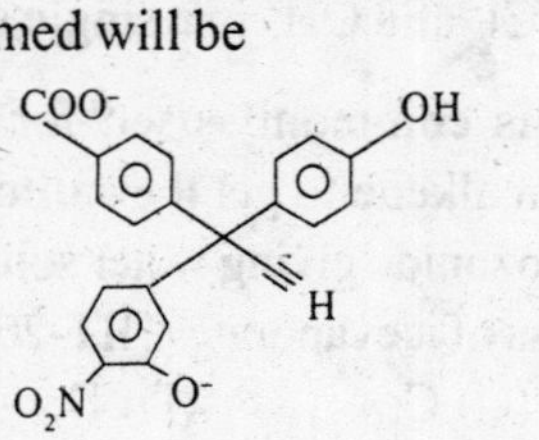 (B) 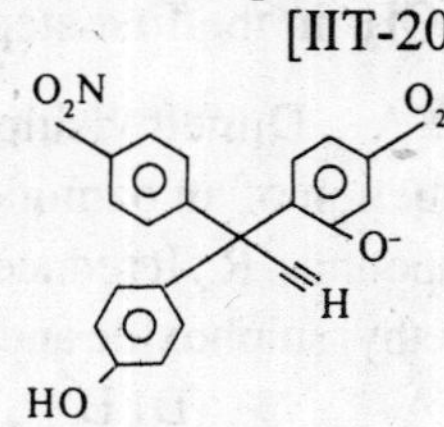

(C) 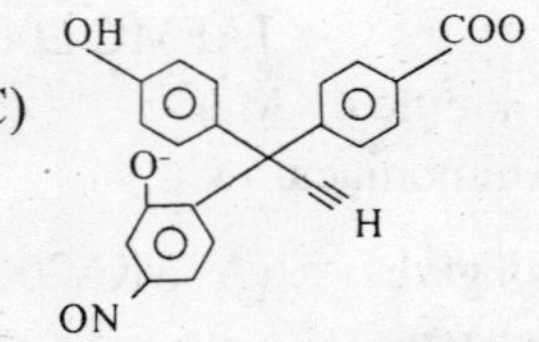 (D) None of these.

194. Dilute $KMnO_4$ oxidises acetylene to .... [EAMCET-92]

195. Ethyl bromide on boiling with alcoholic soluton of sodium hydroxide forms .... [EAMCET-92]

196. Benzene molecule contains [EAMCET-94]

(A) 3pi and 9 sigma bonds (B) 6pi and 12 sigma bonds
(C) 6pi and 6 sigma bonds (D) 3pi and 12 sigma bonds

197. Among the following the hydrocarbon which decolourises bromine water but does not give precipitate with ammoniacal silver nitrate is [EAMCET-94]

(A) CH_4 (B) C_2H_4 (C) C_2H_2 (D) C_6H_6

198. The method of converting high boiling hydrocarbons into low boiling hydrocarbons is known as [EAMCET-94]

(A) Pyrolysis (B) Isomerization
(C) Cracking (D) Inversion

199. On heating with bleaching powder ethyl alcohol is converted into [EAMCET-94]

(A) CH_3COOH (B) CH_3CHO
(C) CH_3COOH (D) $CHCl_3$

200. Which one of the following compounds is acidic [EAMCET-94]

(A) C_3H_8 (B) C_2H_6 (C) C_2H_4 (D) C_2H_2

201. The compound which gives n-hexane on heating with sodium in dry ethanol [EAMCET-95]

(A) Ethyl bromide (B) n-propyl bromide
(C) Methyl bromide (D) n-Butyl bromide

202. In the following reaction, the product "R" is

$CaC_2 \xrightarrow{H_2O} P \xrightarrow[Tube]{Hot\ Iron} Q \xrightarrow[AlCl_3]{CH_3Cl} R$ [EAMCET-95]

(A) Benzene (B) Toluene
(C) Ethyl benzene (D) n-propyl benzene

203. Action of Zinc dust on tetrabromoethane gives [EAMCET-96]

(A) CH_3OH (B) $CH \equiv CH$
(C) $CH = CH_2$ with CH_2OH (D) $CH_3 - CH_3$

204. Which one of the following compounds is most acidic [EAMCET-96]

(A) CH_3Cl (B) CH_3OH (C) $H_2C = CH_2$ (D) $HC \equiv CH$

205. Which of the following compounds does not form an ozonide ? [EAMCET-97]

(A) Ethene (B) Propyne (C) Propene (D) Propane

206. The number of pi-bonds in the product formed by passing acetylene through dil. H_2SO_4 containing mercuric sulphate is : [EAMCET-97]
(A) Zero (B) One (C) Two (D) Three

207. A gas decolorizes Br_2 in CCl_4 and forms a precipitate with ammoniacal silver nitrate. The gas is[EAMCET-98]
(A) C_2H_2 (B) C_2H_4 (C) C_2H_6 (D) CH_4

208. The homologue of ethyne is [EAMCET-98]
(A) C_2H_2 (B) C_2H_6 (C) C_3H_8 (D) C_3H_4

209. In Wurtz reaction the reagent used is[EAMCET-98]
(A) Na (B) Na/liq. NH_3
(C) Na/dry ether (D) Na/dry ethanol

210. The homologue of ethyne is [EAMCET-98(MBBS)]
(A) C_2H_4 (B) C_3H_6 (C) C_3H_3 (D) C_3H_4

211. In Wurtz reaction the reagent used is [EAMCET-98(MBBS)]
(A) Na (B) Na/liq. NH_3
(C) Na/dry ether (D) Na/ dry alcohol

212. Benzene is obtained by fractional distillation of [EAMCET-98(MBBS)]
(A) Light oil (B) Middle oil
(C) Anthracene oil (D) Heavy oil

213. The gaseous substance which decolorizes bromine but does not give a precipitate with ammoniacal silver nitrate is [EAMCET-98(MBBS)]
(A) C_3H_6 (B) C_2H_4 (C) C_2H_2 (D) CH_4

214. The hybridization involved in the six carbon atoms of benzene is [EAMCET-98(MBBS)]
(A) Three sp^3 & three sp^2 (B) Three sp^3 & three sp
(C) All six sp (D) All six sp^2

215. The number of C–C sigma bonds present in 1-butyne is [EAMCET-99]
(A) 2 (B) 3 (C) 4 (D) 5

216. Ethyne is treated with 30% H_2SO_4 in the presence of 1% mercuric sulphate at 70°C. What is the product formed ? [EAMCET-99]
(A) ethene (B) ethanol
(C) ethanal (D) ethylene glycol

217. Which of the following alkane CANNOT be prepared by Wurtz reaction ? [EAMCET-99]
(A) C_2H_6 (B) C_3H_8 (C) CH_4 (D) C_4H_{10}

218. The test used for identifying carbon-carbon double bond in an alkene is [EAMCET-99(MBBS)]
(A) Baeyer's (B) Carbylamine
(C) Lassaigne (D) Silver mirror

219. Which of the following compounds liberates hydrogen gas on reacting with sodium metal ? [EAMCET-99(MBBS)]
(A) C_6H_{14} (B) C_2H_2 (C) C_2H_4 (D) C_2H_6

220. Ethylene is reacted with Br_2/CCl_4 to form X. When X is reacted with alcoholic KOH Y is formed, X and Y are respectively [EAMCET-2000]
(A) $BrCH_2–CH_2Br$ and C_2H_2 (B) C_2H_5Br and C_2H_4
(C) C_2H_5Br and C_6H_6 (D) $C_2H_3Br_3$ and C_2H_4

221. Which of the following reacts with ammoniacal cuprous chloride ? [EAMCET-2000]
(A) CH_4 (B) C_2H_2 (C) C_2H_6 (D) C_6H_6

222. Which of the following hydrocarbons has least carbon-carbon bond length ? [EAMCET-2000]
(A) C_2H_4 (B) C_2H_4 (C) C_6H_6 (D) C_2H_2

223. Methyl benzene can be prepared by reacting benzene with bromo methane in the presence of : [EAMCET(MBBS)-2000]
(A) Anhydrous $AlCl_3$ (B) Br_2/CCl_4
(C) Ni/H_2m 70°C (D) Dil. H_2SO_4

224. Benzene is used in the preparation of : [EAMCET(MBBS)-2000]
(A) Polythene (B) Gammexane
(C) Chloroform (D) Vinegar

225. Consider the following reactions

$$C_2H_2 \xrightarrow[\text{Iron tube, 500°C}]{\text{Red hot}} A$$

$$A \xrightarrow[\text{70°C}]{\text{Conc. } HNO_3 \text{ / Conc. } H_2SO_4} B$$

$$B \xrightarrow{LiAlH_4} C_6H_5N = NC_6H_5$$, A and B are [EAMCET-2001]

(A) $A = C_2H_4; B = C_6H_6$ (B) $A = C_2H_6; B = C_6H_5NH_2$
(C) $A = C_2H_4; B = C_6H_5NH_2$
(D) $A = C_6H_6; B = C_6H_5NO_2$

226. Which one of the following is used in the preparation of styrene ? [EAMCET-2001]
(A) CH_3CHO (B) P_2O_5 (C) CH_4 (D) C_6H_6

227. ____test is used for detecting unsaturation in hydrocarbons. [EAMCET(MBBS)-2001]
(A) Silver mirror (B) Lassaigne's
(C) Carbylamine (D) Baeyer's

228. In the following reaction

$$C_2H_2 \xrightarrow[HgSO_4/H_2SO_4,\ 60°C]{H_2O} \underline{X} \rightleftharpoons CH_3CHO$$ what is $\underline{X}$? [EAMCET(MBBS)-2001]
(A) CH_3CH_2OH (B) $CH_3 – O – CH_3$
(C) CH_3CH_2CHO (D) $H_2C = CHOH$

229. In the reaction, $C_6H_6 + R – Cl \xrightarrow{\underline{A}} C_6H_5 – R + HCl$, $\underline{A}$ is ----- [EAMCET(MBBS)-2001]
(A) P_2O_5 (B) PCl_5 (C) NaOH (D) $FeCl_3$

230. Acetylene has .... bonds. [EAMCET(MBBS)-2001]
(A) 2σ, 2π (B) 2σ, 3π (C) 3σ, 2π (D) 3σ, 3π

231. What is the minimum quantity (in grams) of methyl iodide required for preparing one mole of ethane by Wurtz reaction ? (Atomic weight of iodine = 127) [EAMCET-2002]

(A) 142 (B) 568 (C) 326 (D) 284

232. In the following reaction, A and B respectively are [EAMCET-2002]

$A \xrightarrow{HBr} C_2H_5Br \xrightarrow{B} A$

(A) C_2H_4 and alcoholic KOH/Δ
(B) C_2H_5Cl and aq. KOH/Δ
(C) C_2H_5OH and aq. KOH/Δ (D) C_2H_2 and PBr_3

233. The reagent used for converting acetylene to oxalic acid is : [EAMCET-2002]

(A) $HgSO_4$/aqueous H_2SO_4 (B) $HgSO_4/CH_3COOH$
(C) $KMnO_4$/KOH, 25°C (D) Cr_2O_3/H_2SO_4

234. In the following reaction, X and Y are...

$2C_2H_6 + 7O_2 \xrightarrow{\text{Combustion}} X + Y + \text{heat}$ [EAMCET(Med.)-2002]

(A) $2C_2H_5OH$, $6H_2O$ (B) 4HCHO, $5H_2O$
(C) 4 CO, $10H_2O$ (D) CO_2, $6H_2O$

235. In the following reaction, A and B respectively are

$C_2H_6 \xrightarrow{A} C_2H_5Cl \xrightarrow{Zn/HCl} B$ [EAMCET(med.)-2002]

(A) Cl_2/uv light and C_2H_6 (B) PCl_3 and C_2H_4
(C) HCl and C_2H_6 (D) Cl_2 and C_2H_2

236. In the following reaction X and Y are respectively

$C_2H_2 \xrightarrow{X} C_6H_6 \xrightarrow{Y} C_6H_5SO_3H$ [EAMCET(med.)-2002]

(A) iron tube/heating, Na_2SO_4 (B) Zn, conc. H_2SO_4
(C) red hot iron tube, fuming H_2SO_4
(D) H_2/Pd. $BaSO_4$, dil. H_2SO_4

237. When acetylene gas is passed through ....solution, a white precipitate is formed. [EAMCET(med.)-2002]

(A) Aqueous silver nitrate (B) Ammoniacal cuprous chloride (C) Ammoniacal silver nitrate
(D) Aqueous potassium permanganate

238. The reacting ion in the nitration of benzene is [EAMCET(med.)-2002]

(A) NO_2^- (B) NO_2^+ (C) NO_3^- (D) O_2^-

239. An organic compound x gives a red precipitate on heating with Fehling's solution. Which one of the following reactions yields x as a minor product ? [EAMCET-2003]

(A) $HCHO \xrightarrow[2.\ H_2O]{1.\ CH_3MgI}$
(B) $C_2H_5Br + AgOH \xrightarrow{\Delta}$
(C) $2C_2H_5Br + Ag_2O \xrightarrow{\Delta}$
(D) $C_2H_2 + H_2O \xrightarrow[1\%\ HgSO_4,\ 60°C]{40\%\ H_2SO_4}$

240. When acetylene is passed through red hot iron tube, compound X is formed. Which one of the following reactions will yield X as the major product ? [EAMCET-2003]

(A) $C_6H_5OH + Zn \xrightarrow{\text{distiliation}}$
(B) $C_6H_5SO_3H + NaHCO_3 \longrightarrow$
(C) $C_6H_{12} + 3H_2 \xrightarrow{Ni}$
(D) $C_6H_5Cl + H_2O \xrightarrow{\Delta}$

241. Separation of Pentene-1 and Pentyne-1 is done by [CPMT-96]

(A) $Ag(NH_3)^+$ (B) Conc. H_2SO_4
(C) HCl (D) Br_2

242. $CH_2 = CH - (CH_2)_5\ COOH \xrightarrow[HBr]{Peroxide} ?$ [CPMT-96]

(A) $CH_2 = CH - (CH_3)_5\ COOH$ (with Br on the CH)
(B) $BrCH_2 - (CH_2)_6COOH$
(C) $CH_2 = CH - (CH_2)_5\ CH_2OH$ (D) C_6H_5COOH

243. Kerosene is used as fuel because it is [CPMT-96]

(A) Less volatile (B) More volatile
(C) Cheap (D) Abundantly available

244. When sodium propanoate is heated with soda lime which of the following is formed ? [CPMT-96]

(A) CH_4 (B) CH_3CH_3 (C) Butane (D) Propane

245. If HBr is added over $CH(CH_3) = CH - CH_3$, what is formed ? [CPMT-96]

(A) $CH_2(Cl) - CH(CH_3) - CH_3$ (B) $CH_3 - C(Cl)(CH_3) - CH_3$
(C) $CH_2 = C(CH_2Cl) - CH_3$ (D) None of these

246. Which of the following gases is used in metal welding? [CPMT-96]

(A) Acetylene (B) Ethane (C) Butane (D) Iso-Butene

247. Daily use candles contain (Paraffin wax) - [CPMT-96]

(A) Higher carbon saturated hydrocarbons
(B) Lower saturated hydrocarbons
(C) Higher unsaturated hydrocarbons
(D) Lower unsaturated hydrocarbons

248. Propane is obtained from propene by - [CPMT-97]

(A) Catalytic hydrogenation (B) Wurtz reaction
(C) Dehydrogenation (D) Frankland reaction

249. Constitution of petrol is- [CPMT-97]

(A) C_3–C_5 (B) C_{12}–C_{18} (C) C_5–C_{12} (D) C_8–C_{10}

250. Number of (pi) and (sigma) bonds in benzene is [CPMT-97]
(A) $3\pi + 12\sigma$ (B) $12\pi + 12\sigma$ (C) $6\pi + 12\sigma$ (D) $6\pi + 6\sigma$

251. $C_6H_6 + Cl_2 \xrightarrow{\text{uv light}} ?$ [CPMT-97]
(A) CCl_3CHO (B) $C_6H_6Cl_6$
(C) $C_6H_{12}Cl_6$ (D) $C_6H_9Cl_2$

252. Benzene + Ozone → Y. Y is [CPMT-97]
(A) Benzene monoozide (B) Benzene diozide
(C) Benzene triozide (D) All of these

253. $CH_3–CH = CH_2 + HI \rightarrow X$. X is [CPMT-97]
(A) $CH_3–CH_2–CH_2I$ (B) $CH_3–CHI–CH_3$
(C) CH_3CH_2CHI (D) None of these

254. In which of the following the bond length between carbon and carbon atom is equal ? [CPMT-97]
(A) 2–Butene (B) Benzene
(C) 1–Butene (D) 1–Propyne

255. Heat and light are both produced when coal gas is heated. This is due to- [CPMT-98]
(A) CO and H_2 (B) CH_4 and H_2
(C) Hydrocarbons (D) CO_2 and CH_4

256. Bond length of Ethane (I), Ethene (II), Acetylene (III) and Benzene (IV) follows the order [CPMT-99]
(A) I > II > III > IV (B) I > II > IV > III
(C) I > IV > II> III (D) III > IV > II > I

257. $KMnO_4$ will oxidise acetylene to [CPMT-99]
(A) Ethylene glycol (B) Ethyl alcohol
(C) Oxalic acid (D) Acetic acid

258. Which of the following methods is most appropriate for the manufacture of methane? [CPMT-2000]
(A) Wurtz reaction (B) By reduction of CH_2Cl_2
(C) Liquefaction of natural gas
(D) None of these

259. $CaC_2 + H_2O \rightarrow A \xrightarrow[HgSO_4]{H_2SO_4} B$ [CPMT-2000]
Identify A and B in the above reaction
(A) C_2H_2 and CH_3COOH (B) C_2H_4 and CH_3COOH
(C) CH_4 and HCOOH (D) C_2H_2 and CH_3CHO

260. Which one is used as an anti-knock in petrol fuel ? [CPMT-2000]
(A) Basic lead carbonate (B) Tetraethyl lead
(C) Lead tetra acetate (D) Basic lead sulphate

261. Two organic compounds 'A' and 'B' both containing carbon and hydrogen only, on quantitative analysis gave same percentage of C and H. 'A' can decolourise bromine water whereas 'B' cannot. 'A' and 'B' respectively are [CPMT-2000]
(A) C_2H_2 and C_6H_6 (B) C_2H_4 and C_6H_6
(C) C_2H_6 and C_6H_6 (D) C_2H_2 and C_2H_6

262. Which will yield 2-chlorobutene and 3-chlorobutene on chlorination ? [CPMT-2001]
(A) $CH_2 = C = CH – CH_3$ (B) $HC = C(CH_3)–CH = CH_2$
(C) $CH_2 = CH – CH = CH_2$ (D) $CH \equiv C–CH = CH_2$

263. Which of the following carbon atom is most electronegative ? [CPMT-2001]

$$\overset{III}{CH_3} – \overset{II}{CH_2} – \overset{I}{C} \equiv CH$$

(A) I (B) II
(C) III (D) All are equally electronegative

264. When benzene is oxidised in the presence of V_2O_5, the product is an anhydride of [CPMT-2001]
(A) Saturated monocarboxylic acid
(B) Saturated dicarboxylic acid
(C) Unsaturated monocarboxylic acid
(D) Unsaturated dicarboxylic acid

265. Function of $AlCl_3$ in Friedel Craft's reaction is to [CPMT-2002]
(A) produce nucleophile (B) make new bond
(C) produce electrophile (D) break bond

266. How many σ and π bonds are present in
C - C ≡ C - C = C [CPMT-2002]
(A) 10σ , 3π (B) 4σ , 2π
(C) 7σ , 3π (D) 6σ , 7π

267. The hybridisation in methane, ethene and ethyne respectively is [CPMT-2003]
(A) sp^3, sp^2 and sp (B) sp^3, sp, sp^2
(C) sp^2, sp^3 and sp (D) sp, sp^2, sp^3

268. Most stable alkene is [CPMT-2003]
(A) $CH_2 = CH_2$ (B) RCH = CHR
(C) $R_2C = CR_2$ (D) $RCH = CH_2R$

269. The product of the following reaction is :
$C_6H_6 + Cl_2 \xrightarrow{\text{sun light}} ?$ [PMT-93]
(A) C_6H_5Cl (B) O - C_6H_4 Cl_2
(C) $C_6H_6Cl_6$ (D) P - C_6H_4 Cl_2

270. The compound formed on the hydrolysis of calcium carbide is : [PMT-93]
(A) Acetylene (B) Ethane
(C) Ethylene (D) Ethylene glycol

271. Which one of the following is an explosive ? [PMT-93]
(A) Nitrobenzene (B) Nitrophenol
(C) Nitromethane (D) Trinitrobenzene

272. When treated with ammoniacal cuprous chloride, which one among the following forms copper derivative ? [PMT-93]
(A) C_2H_6 (B) C_2H_4 (C) C_2H_2 (D) C_6H_6

273. Out of the following fractions of petroleum the one having the lowest boiling point is : [PMT-93]
(A) Kerosene oil (B) Diesel oil
(B) Gasoline (D) Heavy oil

274. Which one among the following is assigned an octane number of Zero ? [PMT-93]
(A) Iso-octane (B) n-Heptane
(C) Iso-heptane (D) 2-Methyl octane

275. The hybrid orbitals used by carbon in benzene belong to : [PMT-93]
(A) sp^3 (B) sp^2 (C) sp (D) dsp^3

276. Calcium carbide reacts with water to produce [PMT-94]
(A) Methane (B) Ethane (C) Ethylene (D) Acetylene

277. Which of the following is produced when acetylene is passed in dilute H_2SO_4 in the presence of $HgSO_4$ [PMT-94]
(A) C_2H_5OH (B) CH_3COCH_3
(C) CH_3CHO (D) Mercury carbide

278. Baeyer's reagent is [PMT-94]
(A) Alkaline $KMnO_4$ (B) Ammoniacal $AgNO_3$
(C) Ammoniacal $CuSO_4$ (D) Acidic $CaSO_4$

279. Which of the following is added to fuel as an anti knock agent [PMT-94]
(A) Lead bromide (B) Zinc bromide
(C) Lead tetra ethyl and ethyl bromide
(D) Lead oxide

280. Aromatic properties of benzene are proved by [PMT-94]
(A) Aromatic Sextet Theroy (B) Resonance Theroy
(C) Molecular Orbital Theory (D) All of the above

281. The reaction between an alkyl halide and sodium metal in dry ether is called [PMT-94]
(A) Clemmensen Reduction (B) Kolbe's Reaction
(C) Wurtz's Reaction (D) Cannizaro's Reaction

282. Marsh gas primarily contains [PMT-94]
(A) C_2H_2 (B) CH_4 (C) H_2S (D) CO

283. General formula of alkenes is [PMT-94]
(A) C_nH_{2n+2} (B) C_nH_{2n} (C) C_nH_{2n-2} (D) $C_{2n+2}H_n$

284. $-C \equiv C-$ is present in [PMT-94]
(A) Ethene (B) Butylene (C) Ethyne (D) Glycerine

285. The number of sigma (σ) bonds in 1-butene is : [PMT-95]
(A) 8 (B) 10 (C) 11 (D) 12

286. A sample of gasoline contains 81% iso-octane and 19% n-heptane. It's octane number will be [PMT-95]
(A) 19 (B) 81 (C) 100 (D) 62

287. Natural Petroleum contains : [PMT-95]
(A) Saturated hydrocarbons
(B) Cyclic saturated hydrocarbons
(C) Compound of sulphur (D) All of these

288. The reaction of benzene with methyl chloride in presence of anhydrous aluminium chloride is called as : [PMT-95]
(A) Gattermann Koch reaction
(B) Friedel Craft's reaction
(C) Riemer-Tiemann reaction (D) None of these

289. The preparation of ethane by electrolysis of aqueous solution of potassium acetate is called as [PMT-95]
(A) Wurtz reaction (B) Sabatier-Senderen's reaction
(C) Kolbe's synthesis (D) Grignard reaction

290. Formation of 2-butene as major product by dehydration of 2-butanol is according to [PMT-95]
(A) Markownikoff's rule (B) Saytzeff's rule
(C) Peroxide effect (D) Anti Markownikoff's rule

291. 1,2-Butadiene contains [PMT-96]
(A) only sp hybridised carbon atoms
(B) only sp^2 hybridised carbon atoms
(C) sp and sp^2 hybridised carbon atoms
(D) sp, sp^2 and sp^3 hybridised carbon atoms

292. Benzene is prepared in laboratory from which one of the following compounds ? [PMT-96]
(A) $C_6H_5N_2Cl$ (B) C_6H_5OH
(C) C_6H_5COONa (D) $C_6H_5SO_3H$

293. A gas decolourises alkaline $KMnO_4$ but does not give precipitate with ammoniacal silver nitrate, the gas is [PMT-96]
(A) Methane (B) Ethane
(C) Ethylene (D) Acetylene

294. When ethyl alcohol is heated with conc. H_2SO_4 the product obtained is [PMT-96]
(A) $H_3C-\overset{\overset{\displaystyle O}{\|}}{C}-OC_2H_5$ (B) C_2H_6
(C) C_2H_4 (D) C_2H_2

295. The reaction

$$CH_2=CH_2+H_2 \xrightarrow[250^\circ-300^\circ C]{Ni} CH_3-CH_3$$

is called [PMT-96]
(A) Wurtz reaction (B) Kolbe's reaction
(C) Sabatier Senderen's reaction
(D) Carbylamine reaction

296. In the reaction sequence [PMT-96]

$$CaC_2 \xrightarrow{H_2O} A \xrightarrow[Hg^{++}]{dil\ H_2SO_4} B \xrightarrow[Ni]{H_2} C$$

the product C is,
(A) CH_3OH (B) CH_3CHO (C) C_2H_5OH (D) C_2H_4

297. Which one of the following substances is used an an

anti-knock compound ? [PMT-96]
(A) Tetra ethyl lead (B) Lead tetrachloride
(C) Lead acetate (D) Ethyl acetate

298. In benzene, the total number of σ bonds is - [PMT-97]
(A) 3 (B) 6 (C) 9 (D) 12

299. Which of the following is a cycloalkane [PMT-97]
(A) C_6H_{14} (B) C_6H_{12} (C) C_6H_{10} (D) C_6H_8

300. The number of sp^3 hybridized carbon atoms in cyclohexene are - [PMT-97]
(A) 2 (B) 3 (C) 4 (D) 6

301. Ethene gives with acidic $KMnO_4$ solution - [PMT-97]
(A) Ethylene glycol (B) Ethylene oxide
(C) Formaldehyde (D) Acetaldehyde

302. Ammoniacal silver nitrate solution reacts will C_2H_2 from - [PMT-97]
(A) Silver mirror (B) Silver oxide
(C) Silver formate (D) Silver acetylide

303. The reaction [PMT-97]
$C_{10}H_{22} \xrightarrow{900K} C_4H_8 + C_6H_{14}$ is called
(A) Pyrolysis (B) Dissociation
(C) Cracking (D) Fractionation

304. Acetylene reacts with H_2SO_4 (dilute) and Hg^{2+} to give the stable product [PMT-97]
(A) Vinyl alcohol (B) Acetaldehyde
(C) Formaldehyde (D) Acetic acid

305. Isopropyl alcohol is obtained by reacting which of the following alkenes with conc. H_2SO_4 and H_2O ? [PMT-99]
(A) Ethylene (B) Propylene
(C) 2-Methyl propene (D) Isoprene

306. The compound C_3H_4 has a triple bond, which is indicated by its reaction with : [PMT-99]
(A) bromine water (B) Baeyer's reagent
(C) Fehling solution (D) ammoniacal silver nitrate

307. Anti-Markownikoff addition of HBr is not observed for : [PMT-99]
(A) propene (B) but-1-ene
(C) but-2-ene (D) pent-2-ene

308. Gasoline is obtained from crude petroleum oil by its : [PMT-99]
(A) fractional distillation (B) vacum distillation
(C) steam distillation (D) pyrolysis

309. Which of the following does not give alkane ? [PMT-99]
(A) Reaction of CH_3I with Na in ether
(B) Reaction of sodium acetate with soda-lime
(C) Electrolysis of concentrated sodium acetate solution
(D) Reaction of ethylchloride with alcoholic KOH

310. LPG is a mixture of : [PMT-99]
(A) $C_6H_{12} + C_6H_6$ (B) $C_4H_{10} + C_3H_8$
(C) $C_2H_4 + C_2H_2$ (D) $C_2H_4 + CH_4$

311. The number of possible alkynes with molecular formula C_5H_8 is : [PMT-2000]
(A) 2 (B) 3 (C) 4 (D) 5

312. Toluene has : [PMT-2000]
(A) 6σ and 3π bond (B) 9σ and 3π bond
(C) 9σ and 6π bond (D) 7σ and 3π bond

313. Which alkene on ozonolysis yields acetone ? [PMT-2000]
(A) $CH_2 = CH - CH_2 - CH_3$
(B) $CH_3 - CH = CH_2$ (C) $CH_3 - CH = CH - CH_3$
(D) $(CH_3)_2 - C = C - (CH_3)_2$

314. In chlorination of benzene the reactive species is : [PMT-2000]
(A) Cl^+ (B) Cl^- (C) Cl_2 (D) Cl_2

315. The number of moles of proton which can be easily given by butyne-1 (1 mole) is : [PMT-2000]
(A) 1 (B) 2 (C) 3 (D) 6

316. The addition of HBr is easiest with : [PMT-2000]
(A) $CH_2 = CHCl$ (B) $ClCH = CHCl$
(C) $CH_3 - CH = CH_2$ (D) $(CH_3)_2 C = CH_2$

317. An organic compound has a triple bond and not double bond. It can be tested by : [PMT-2000]
(A) bromine water (B) Baeyer's reagent
(C) Fehling solution (D) Ammoniacal silver nitrate

318. Which of the following has highest octane number ? [PMT-2000]
(A) n-Hexane (B) n-Heptane
(C) n-Pentane (D) 2, 2, 4-Trimethyl pentane

319. Which of the following is an antiknocking compound? [PMT-2000]
(A) $Zn(C_2H_5)_2$ (B) $Hg(CH_3)_2$
(C) $Pb(C_2H_5)_4$ (D) $Sn(C_2H_5)_4$

320. For which one of the following the octane number is zero ? [PMT-2001]
(A) n-heptane (B) n-hexane
(C) Cyclohexane (D) Isooctane

321. Benzene reacts with chlorine to form benzene hexachloride in the presence of : [PMT-2001]
(A) Sunlight (B) Zn (C) $AlCl_3$ (D) Ni

322. The insecticide gammexane is a formation based on: [PMT-2001]
(A) DDT (B) Benzene hexachloride
(C) Hexachloroethane (D) Chloral

323. A gas decolourises alkaline $KMnO_4$ but does not give precipitate with ammoniacal $AgNO_3$. The gas is : [PMT-2001]
(A) C_2H_6 (B) C_2H_2 (C) C_2H_4 (D) CH_4

324. Liquid hydrocarbon is converted into a mixture of gaseous hydrocarbons by : [PMT-2001]
(A) Cracking (B) Hydrolysis (C) Oxidation
(D) Distillation under reduced pressure

325. General formula of alkanes is [PMT-2002]
(A) C_nH_{2n+1} (B) C_nH_{2n+2} (C) C_nH_{2n-1} (D) C_nH_{2n}

326. Ethane is formed by the reaction of methyl iodide and sodium metal in dry ether solution. This reaction is known as [PMT-2002]
(A) Clemmensen reduction (B) Kolbe's reaction
(C) Wurtz reaction (D) Cannizzaro's reaction

327. Which one of the following compounds cannot be prepared by Wurtz reaction ? [PMT-2002]
(A) CH_4 (B) C_2H_6 (C) C_3H_8 (D) C_4H_{10}

328. The reaction
$CH_3.CH = CH_2 \xrightarrow[H^+]{CO + H_2O} CH_3 - \underset{COOH}{\underset{|}{C}H} - CH_3$
is known as [PMT-2002]
(A) Wurtz reaction (B) Koch reaction
(C) Clemmensen reduction (D) Kolbe's reaction

329. In the reaction
$CH_3CH = CH_2 + H_2O + [O] \xrightarrow[Acid]{KMNO_4}$
$CH_3 - \underset{OH}{\underset{|}{C}H} - \underset{OH}{\underset{|}{C}H_2} \xrightarrow{[O]} X + HCOOH$, X is [PMT-2002]
(A) CH_3CH_2COOH (B) CH_3COOH
(C) CH_3CH_2CHO (D) CH_3CH_2OH

330. Successive alkanes differ by [PMT-2002]
(A) $>CH_2$ (B) $\geq CH$ (C) $-CH_3$ (D) C_2H_4

331. The process in which higher hydrocarbons are broken down into lower hydrocarbons by controlled pyrolysis, is called [PMT-2002]
(A) Hydrolysis (B) Cracking
(C) Oxidation (D) Reduction

332. A sample of petrol has an octane number of 40, it means that its knocking power matches with the following mixture. [PMT-2003]
(A) 40% petrol + 60% iso-octane
(B) 40% n-heptane + 60%iso-octane
(C) 60% n-heptane + 40% iso-octane
(D) 60% petrol + 40% iso-octane

333. Number of σ and π bonds in butenyne would be respectively [PMT-2003]
(A) 4σ and 5π (B) 4σ and 3π
(C) 7σ and 5π (D) 7σ and 3π

334. The chief constituent of Marsh gas is : [PMT-2003]
(A) Butane (B) Propane (C) Methane (D) Isobutene

335. Lewisite, the poisonous gas is obtained by reacting the following : [PMT-2003]
(A) $CH \equiv CH$ and $AsCl_3$ (B) $CH_2 = CH_2$ and $AsCl_3$
(C) $CH_2 = CH_2$ and NOCl (D) $CH \equiv CH$ and NOCl

336. Major product obtained by passing the vapours of benzene over V_2O_5 at 450°C would be : [PMT-2003]
(A) CHCOOH ‖ CHCOOH
(B) 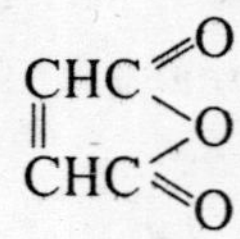
(C) 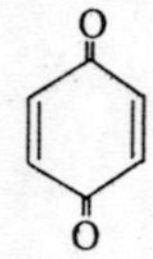
(D) 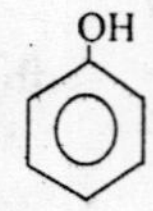

337. Which of the following compounds on ozonolysis followed by hydolysis will provide CH_3COCH_3 and CH_3CHO only ? [PMT-2003]
(A) $CH_3 - CH = C(CH_3)_2$ (B) $CH_3 - CH = CH - CH_3$
(C) $CH_3 - \underset{CH_3}{\underset{|}{C}} = CH - CH_2 - CH = CH - CH_3$
(D) $CH_3 - \underset{CH_3}{\underset{|}{C}} = \underset{CH_3}{\underset{|}{C}} - CH_3$

338. Best reagent to distinguish between propene and propyne would be : [PMT-2003]
(A) Solution of bromine in CCl_4 (B) Bromine water
(C) Alkaline solution of $KMnO_4$
(D) Ammoniacal solution of $AgNO_3$

339. Bromination of propene by N-bromosuccinimidy yields [PMT-2003]
(A) $CH_3 - \underset{Br}{\underset{|}{C}} = CH_2$ (B) $BrCH_2 - CH = CH_2$
(C) $CH_3 - \underset{Br}{\underset{|}{C}H} - CH_2Br$ (D) $BrCH_2 - CH = CH_2Br$

340. Nitro group in nitrobenzene is [PET-93]
(A) Ortho-directing (B) Meta-directing
(C) Para-directing (D) Ortho-and para-directing

341. The shape of ethylene molecule is [PET-93]
(A) Tetrahedral (B) Pyramidal (C) Planar (D) Linear

342. Halogenation of alkanes is an example of [PET-93]
(A) Electrophilic substitution
(B) Nucleophilic substitution
(C) Free-radical substitution (D) Oxidation

343. Carbon black, which is used in making printer's ink, is obtained by decomposition of [PET-93]

(A) Acetylene (B) Benzene
(C) Carbon tetrachloride (D) Methane

344. The addition of tetraethyl lead to petrol [PET-93]
(A) Lowers its octane number
(B) Raises its octane number
(C) May raise or lower the octane number
(D) Has no effect on octane number

345. The order of increasing reactivity towards HCl of the following compounds
1. $CH_2 = CH_2$ 2. $(CH_3)_2\ CH = CH_2$
3. $CH_3CH = CHCH_3$ will be [PET-94]
(A) $1 < 2 < 3$ (B) $1 < 3 < 2$
(C) $3 < 2 < 1$ (D) $2 < 1 < 3$

346. In the reaction

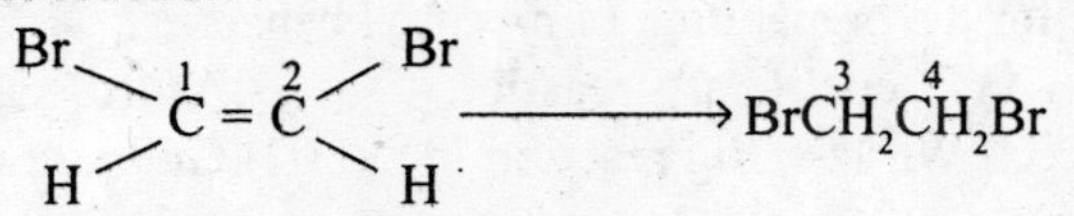

The hybridization states of carbon atoms 1, 2, 3 and 4 are [PET-94]
(A) 1 and 2 sp^2; 3 and 4 sp^3 (B) 1 and 2 sp^2; 3 and 4 sp
(C) 1, 2, 3 and 4 sp (D) 1, 2 sp^3; 3, 4 sp^2

347. The reagent which is used to distinguish between propene and propyne is [PET-94]
(A) Bromine (B) Alkaline $KMnO_4$
(C) Ammoniacal $AgNO_3$(D) Ozone

348. Which of the following reactions would be the best for the formation of 2-bromobutane ? [PET-94]

(1) $CH_3CH = CH\ CH_2CH_3 \xrightarrow{HBr}$

(2) $CH_3CH_2CH_2 = CH_2 \xrightarrow{HBr}$

(3) $CH_3CHCH = CHCH_3 \xrightarrow{Br_2}$

(4) $CH_3CH_2CH = CH_2 \xrightarrow[Peroxide]{HBr}$

(A) 1 (B) 2 (C) 3 (D) 4

349. In order to complete the reaction 1-pentyne to [PET-94]
(a) 4-octyne (b) cis-4-octene, a and b will be

(a) (b)
1. $NaNH_2$: CH_3CH_2Br : H_2(one mole) Pd or Ni
2. $NaNH_2$: $CH_3CH_2CH_2Br$: H_2 : (two moles) Pd or Ni
3. $NaNH_2$: $CH_3CH_2CH_2Br$: H_2 (one mole) Pd or Ni
4. $NaNH_2$: $CH_3CH_2CH_2Br$: $BH_3.H_2O_2.OH^-$
(A) 1 (B) 2 (C) 3 (D) 4

350. A sample of petrol contains 22% n-heptane and 78% iso-octane. Its octane no. is [PET-94]
(A) 22 (B) 78 (C) 56 (D) 44

351. In the reaction
$R–CH_2–CCl_2–R \xrightarrow{Reagent} R–C≡C–R$
the reagent is - [PET-95]
(A) Na amalgam (B) HCl and H_2O
(C) Alcoholic KOH (D) Aqueous KOH

352. The hybridization of carbon atom in C - C single bond of $HC \equiv C - CH = CH_2$ is [PET-95]
(A) sp^3 - sp^3 (B) sp^2 - sp^3
(C) sp - sp^2 (D) sp^3 - sp

353. Name of the reaction
$C_{10}H_{22} \xrightarrow{900\ K} C_4H_8 + C_6H_{14}$ is - [PET-95]
(A) Alkylation (B) Cracking
(C) Dehydrogenation (D) Fractionation

354. Ethylene reacts with alkaline $KMnO_4$ to form - [PET-95]
(A) Oxalic acid (B) HCHO
(C) Ethyl alcohol (D) Ethylene glycol

355. Which of the following compound is used as a refrigerant - [PET-95]
(A) Acetone (B) CCl_4 (C) CF_4 (D) CCl_2F_2

356. The insecticide Gammexane is - [PET-95]
(A) D. D. T. (B) Benzene hexachloride
(C) Parathion (D) Chloral

357. Which of the following statements is false for iso-pentane? [PET-96]
(A) It has three CH_3 groups (B) It has one CH_2 group
(C) It has one CH group
(D) It has a carbon which is not bonded to hydrogen

358. Which of the following has highest octane number ? [PET-96]
(A) n-Hexane (B) n-Heptane (C) iso-Octane (D) n-Heptane and iso-Octane mixed in ratio 50 : 50

359. Which of the following is obtained at lowest temperature by fractional distillation of petroleum ? [PET-96]
(A) Ligroin (B) Gasoline
(C) Kerosene (D) Diesel oil

360. Which of the following gases does not give a precipitate with ammoniacal solution of silver nitrate but decolourizes $KMnO_4$ (neutral of slightly alkaline) ? [PET-96]
(A) C_2H_2 (B) C_2H_4 (C) C_2H_6 (D) CH_4

361. In the nitration of benzene with conc. H_2SO_4 + HNO_3, the attack on ring is made by [PET-96]
(A) NO_2^+ (B) NO_2 (C) NO_2^- (D) NO_3^-

362. Acetylene reacts with HCl and $HgCl_2$ to give [PET-96]
(A) Vinyl chloride (B) Dichloro ethane
(C) Ethyl chloride (D) Ethylidene chloride

363. In the reaction

$$CH_3-C\equiv C-CH_3 \xrightarrow[(ii)\ H_2O/Zn]{(i)\ X} CH_3-\underset{\|}{\overset{}{C}}(=O)-C(=O)-CH_3$$

X is [PET-97]

(A) HNO_3 (B) O_2 (C) O_3 (D) $KMnO_4$

364. Chloroprene is used in making : [PET-97]

(A) Synthetic rubber (B) Plastic
(C) Petrol (D) All of the above

365. Gammexane is : [PET-97]

(A) D.D.T. (B) Benzene hexachloride
(C) Chloral (D) Hexachloroethane

366. Ethyl bromide reacts with lead-sodium alloy to form : [PET-97]

(A) Tetraethyl lead (B) Tetraethyl Bromide
(C) (A) and (B) both (D) None of the above

367. Which of the following substances is used as an anti-knock compound ? [PET-97]

(A) Tetraethyl lead (B) Lead tetrachloride
(C) Lead acetate (D) Ethyl acetate

368. The product of the following reaction :

$$CH_3-C(CH_3)_2-CH=CH_2 \xrightarrow[(ii)\ NaBH_4]{(i)\ Hg(OAC)_2,\ H_2O}$$

is : [PET-97]

(A) $CH_3-C(CH_3)_2-CH(OH)-CH_3$

(B) $CH_3-C(CH_3)_2-CH_2-CH_2OH$

(C) $CH_3-C(OH)(CH_3)-CH(CH_3)-CH_3$

(D) $HOCH_2-C(CH_3)_2-CH_2-CH_3$

369. The shape of methane molecule is : [PET-97]

(A) Tetrahedral (B) Octahedral
(C) Linear (D) Triangular

370. Which one is false in the following statements ? [PET-97]

(A) Each carbon in ethylene is in sp^2 hybridization
(B) Each carbon in acetylene is in sp^3 hybridization
(C) Each carbon in benzene is in sp^3 hybridization
(D) Each carbon in ethane is in sp^3 hybridization

371. An unknown compound A has a molecular formula C_4H_6. When A is treated with an excess of Br_2, a new substance B with formula $C_4H_6Br_4$ is formed. A forms a white precipitate with ammoniacal silver nitrate solution. A may be : [PET-98]

(A) Butyne-1 (B) Butyne-2
(C) Butene-1 (D) Butene-2

372. Which one of the following reactions is most suitable for the preparation of n-propyl benzene ? [PET-98]

(A) Friedel-Crafts reaction (B) Wurtz reaction
(C) Wurtz-Fittig reaction (D) Grignard reaction

373. Which one of the following reactions is a method for the conversion of a ketone into a hydrocarbon ? [PET-98]

(A) Aldol condensation (B) Reimer-Tiemann reaction
(C) Cannizzaro reaction (D) Wolff - Kishner reduction

374. The arrangement of hydrogen atoms around carbon in the methane molecule is : [PET-99]

(A) Triangular (B) Square planar
(C) Tetrahedral (D) Hexagonal

375. An alkyne has the general formula ? [PET-99]

(A) C_nH_{2n} (B) C_nH_{2n+2} (C) C_nH_{2n-2} (D) C_nH_{2n+1}

376. In its reaction with silver nitrate, acetylene shows : [PET-99]

(A) Oxidising property (B) Reducing property
(C) Basic Property (D) Acidic Property

377. A gas decolourises alkaline KMnO4 solution but does not give precipitate with $AgNO_3$, it is [PET-99]

(A) CH_4 (B) C_2H_4 (C) C_2H_2 (D) C_2H_6

378. Heating a mixture of sodium benzoate and soda lime gives : [PET-99]

(A) Benzene (B) Methane
(C) Sodium benzoate (D) Calcium benzoate

379. Octane number is zero for : [PET-99]

(A) n-heptane (B) iso-octane
(C) n-hexane (D) iso-heptane

380. For preparing an alkane, a concentrated aqueous solution of sodium or potassium salt of a saturated carboxylic acid is subjected to : [PET-99]

(A) Hydrolysis (B) Oxidation
(C) Hydrogenation (D) Electrolysis

381. Benzene reacts with chlorine to form benzene hexachloride in presence of : [PET-99]

(A) Nickel (B) $AlCl_3$

(C) Bright Sunlight (D) Zinc

382. Which of the following alkyl halides is used as a methylating agent ? [PET-99]
(A) C_2H_5Cl (B) C_2H_5Br (C) C_2H_5I (D) CH_3I

383. Gammexane is : [PET-99]
(A) D.D.T. (B) Benzene hexachloride (B.H.C.)
(C) Chloral (D) Hexachlorobenzene

384. Which of the following is most stable? [PET-2000]
(A) Methane (B) Ethane (C) Propane (D) Butane

385. The compound which gives only acetaldehyde on ozonolysis is : [PET-2000]
(A) Butene-1 (B) Butene-2
(C) Ethylene (D) Propylene

386. Which of the following is used to distinguish ethylene and acetylene ? [PET-2000]
(A) Alkaline $KMnO_4$ (B) Bromine water
(C) Ammoniacal cuprous chloride (D) Conc. H_2SO_4

387. Carbons in the compound 2-butene-4-yne are [PET-2000]
(A) sp hybridized (B) sp^2 hybridized
(C) sp and sp^2 hybridized (D) sp, sp^2 and sp^3 hybridized

388. In nitration of Benzene with conc. HNO_3 and conc. H_2SO_4 the species which takes active part is : [PET-2000]
(A) Nitrite ion (B) Nitrate ion
(C) Nitronium ion (D) Nitrogen peroxide

389. The petrol having octane number 80 has : [PET-2000]
(A) 20% normal heptane + 80% iso-octane
(B) 80% normal heptane + 20% iso-octane
(C) 20% normal heptane +80% normal octane
(D) 80% normal heptane +20% normal octane

390. The reaction between an alkyl halide and sodium metal in dry ether is called : [PET-2001]
(A) Clemmensen reduction (B) Kolbe's reaction
(C) Wurtz reaction (D) Cannizaro's reaction

391. Which of the following is an antiknock compound ? [PET-2001]
(A) Tetraethyl lead (B) Lead tetrachloride
(C) Lead acetate (D) Ethyl acetate

392. The "middle oil" fraction of coal tar distillation contains: [PET-2001]
(A) Benzene (B) Anthracene
(C) Naphthalene (D) Xylene

393. Ethylene reacts with bromine to form: [PET-2001]
(A) $Br - CH_2 - CH_3$ (B) $CH_3 - CBr_3$
(C) $Br - CH_2 - CH_2Br$ (D) $CHBr_3$

394. Mustard gas is obtained by [PET-2002]
(A) The action of dilute acids on mustard seeds
(B) Treating ethylene with mustard oil
(C) Treating sulphur chloride with ethylene
(D) None of these

395. Wurtz reaction involves the interaction of alkyl halides in dry ether with [PET-2002]
(A) Sodium (B) Zinc (C) Copper (D) Platinum

396. Which process converts olefins into paraffins [PET-2002]
(A) Halogenation (B) Dehydration
(C) Hydrogenation (D) Hydrolysis

397. The H — C — H bond angle in CH_4 is [PET-2002]
(A) 109°28 (B) 107°28 (C) 90° (D) 180°

398. Which one has the lowest octane number [PET-2002]
(A) Iso-octane (B) n-heptane
(C) n-hexane (D) n-hexadecane

399. Which gas is used for cutting and welding the metals [PET-2002]
(A) Acetylene (B) Methane
(C) Carbon dioxide (D) Neon

400. Lewisite is [PET-2002]
(A) $Cl.CH = CH\,As\begin{smallmatrix}Cl\\Cl\end{smallmatrix}$ (B) $CH_2 = CH.As\begin{smallmatrix}Cl\\Cl\end{smallmatrix}$
(C) $CH_2 = CH.As\begin{smallmatrix}Cl\\Cl\\Cl\end{smallmatrix}$ (D) $AsCl_3$

401. How is 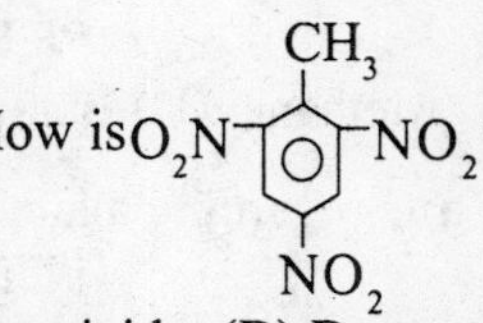 widely used [PET-2002]
(A) Insecticide (B) Drug (C) Explosive (D) Dye

402. $C_6H_6 + Cl_2 \rightarrow C_6Cl_6H_6$ This reaction takes place when [PET-2002]
(A) Chlorine is bubbled through benzene
(B) the mixture is kept in dark
(C) Chlorine (aq) is shaken with benzene
(D) the mixture is exposed to sunlight

403. An organic compound has been found to possess the empirical formula CH_2O and molecular weight 90. Its molecular formula is [PET-2002]
(A) $C_3H_6O_3$ (B) $C_2H_4O_2$ (C) $C_6H_{12}O_6$ (D) CH_2O

404. The compound 'A' when treated with HNO_3 (in presence of H_2SO_4) gives compound 'B', which is then reduced with Sn and HCl to aniline ? The compound 'A' is [PET-2002]
(A) Toluene (B) Benzene (C) Ethane (D) Acetamide

405. The compound 'A' treated with methyl alcohol and drops of H_2SO_4 and heating with water gave smell of wintergreen. The compound 'A' is [PET-2002]
(A) Succinic acid (B) Salicylic acid
(C) Tartaric acid (D) Oxalic acid

406. Which of the following acids adds to propene in the presence of peroxide to give anti-Markownikoff's product? [PET-2003]
(A) HCl (B) HF (C) HBr (D) HI

407. The method for the preparation of synthetic petroleum is: [PET-2003]
(A) Cannizzaro reaction (B) Skraup synthesis
(C) Fisher-Tropsch method (D) Curtius reaction

408. Which of the following compounds will be optically active [PET-2003]
(A) Chloroacetic acid (B) Meso-tartaric acid
(C) Lactic acid (D) Oxalic acid

409. For which category of the compounds, the general formula C_nH_{2n-2} is ? [PET-2003]
(A) Alkenes (B) Alkanes
(C) Alkynes (D) None of the above

410. Propyne on reaction with water in the presence of $HgSO_4$ and H_2SO_4 gives : [PET-2003]
(A) Acetic acid (B) Acetone
(C) Acetaldehyde (D) Methyl acetate

411. Which hydrocarbon will be most stable ? [PET-2003]
(A) Methane (B) Ethane (C) Propane (D) Butane

412. Which cannot be synthesized by the Wurtz method ? [PET-2003]
(A) Ethane (B) Propane (C) Methane (D) Butane

413. Which statement is not true regarding alkanes ? [PET-2003]
(A) Many alkanes are soluble in water
(B) At room temprature some alkanes are liquids some solids and some gases
(C) All alkanes have a lower density than water
(D) All alkanes burn

414. When hydrochloric gas is treated with propene in presence of benzoyl peroxide, It gives [CBSE-93]
(A) 2-chloro propane (B) allyl chloride
(C) No reaction (D) n-propyl chloride

415. C–C bond length in benzene is : [JIPMER-92]
(A) 1,54 Å (B) 1.39 Å (C) 1.34 Å (D) 1.20 Å

416. The maximum number of isomers for an alkene with the molecular formula C_4H_8 is : [JIPMER-92]
(A) 2 (B) 3 (C) 4 (D) 5

417. When benzene reacts with an alkyl chloride (RCl) in presence of anhydrous $AlCl_3$ [JIPMER-2000]
(A) $C_6H_5AlCl_3$ is produced (B) C_6H_5Cl is produced
(C) Nothing happens (D) C_6H_5R is produced

418. The reaction $C_6H_6 + CH_3Cl \xrightarrow{AlCl_3} C_6H_5CH_3 + HCl$ is called [PET-95]
(A) Wurtz Fittig reaction (B) Kolbe's reaction
(C) Friedel-Crafts reaction (D) None of these

419. Which of the following species participate in sulphonation of benzene ring ? [CPMT-99]
(A) H_2SO_4 (B) HSO_3^- (C) SO_3 (D) SO_2^-

420. The compound X on reaction with $HgSO_4 + H_2SO_4$ gives Y which on oxidation gives acetic acid. X is : [PMT-99]
(A) C_2H_2 (B) C_2H_4 (C) C_3H_4 (D) C_4H_6

421. Propene and propyne are distinguished by using [BHU-2003]
(A) $K_2Cr_2O_7$ (B) H_2SO_4 (C) $AgNO_3$ (D) NH_4OH

422. Isobutene +HBr $\xrightarrow{peroxide}$ X. X is [BHU-2003]
(A) iso-butyl bromide (B) tertiary butyl bromide
(C) tertiary butyl alcohol (D) iso-butyl alcohol

423. $CH_3CH_2C{\equiv}CH \xrightarrow{R}$ Butanone. R is [BHU-2003]
(A) $KMnO_4$ (B) Hg^{2+}/H^+ (C) $K_2Cr_2O_7$ (D) H_2SO_4

424. Gobar gas contains [AFMC-2003]
(A) Butane (B) Ethane
(C) Methane (D) Propane

425. Using anhydrous $AlCl_3$ as catalyst, which one of the following reactions produces ethylbenzene (PhEt) ? [CBSE-PMT-2004]
(A) $H_3C - CH_2OH + C_6H_6$ (B) $CH_3 - CH = CH_2 + C_6H_6$
(C) $H_2C = CH_2 + C_6H_6$ (D) $H_3C - CH_3 + C_6H_6$

426. Reaction of HBr with propene in the presence of peroxide gives [CBSE-PMT-2004]
(A) isopropyl bromide (B) 3-bromo propane
(C) allyl bromide (D) n-propyl bromide

427. 2-hexyne will give trans-2-hexene on heating with [IIT-2004]
(A) H_2, Pd/$BaSO_4$ (B) H_2, Pt
(C) $NaBH_4$ (D) Li, NH_3 in ethanol

Hydrocarbons

No.	Ans.	No.	Ans.	No.	Ans.	No.	Ans.	No.	Ans.	No.	Ans.
1.	(C)	51.	(D)	101.	(A)	151.	(A)	201.	(B)	251.	(B)
2.	(A)	52.	(D)	102.	(A)	152.	(B)	202.	(B)	252.	(C)
3.	(D)	53.	(D)	103.	(C)	153.	(A)	203.	(B)	253.	(B)
4.	(C)	54.	(B)	104.	(B)	154.	(C)	204.	(D)	254.	(B)
5.	(D)	55.	(B)	105.	(A)	155.	(A)	205.	(D)	255.	(C)
6.	(C)	56.	(D)	106.	(C)	156.	(C)	206.	(B)	256.	(C)
7.	(C)	57.	(A)	107.	(B)	157.	(C)	207.	(A)	257.	(C)
8.	(A)	58.	(B)	108.	(A)	158.	(D)	208.	(D)	258.	(C)
9.	(B)	59.	(B)	109.	(D)	159.	(A)	209.	(C)	259.	(D)
10.	(D)	60.	(A)	110.	(B)	160.	(B)	210.	(D)	260.	(B)
11.	(B)	61.	(D)	111.	(D)	161.	(D)	211.	(C)	261.	(A)
12.	(B)	62.	(C)	112.	(A)	162.	(C)	212.	(A)	262.	(B)
13.	(B)	63.	(B)	113.	(A)	163.	(B)	213.	(B)	263.	(A)
14.	(A)	64.	(D)	114.	(A)	164.	(B)	214.	(D)	264.	(D)
15.	(A)	65.	(D)	115.	(A)	165.	(A)	215.	(B)	265.	(C)
16.	(C)	66.	(B)	116.	(D)	166.	(A)	216.	(C)	266.	(A)
17.	(D)	67.	(A)	117.	(C)	167.	(A)	217.	(C)	267.	(A)
18.	(D)	68.	(D)	118.	(A)	168.	(D)	218.	(A)	268.	(C)
19.	(A)	69.	(B)	119.	(C)	169.	(A)	219.	(B)	269.	(C)
20.	(A)	70.	(D)	120.	(B)	170.	(C)	220.	(A)	270.	(A)
21.	(B)	71.	(D)	121.	(C)	171.	(A) (D)	221.	(B)	271.	(D)
22.	(A)	72.	(B)	122.	(C)	172.	*	222.	(D)	272.	(C)
23.	(D)	73.	(C)	123.	(B)	173.	*	223.	(A)	273.	(B)
24.	(A)	74.	(C)	124.	(C)	174.	(B)	224.	(B)	274.	(B)
25.	(D)	75.	(B)	125.	(C)	175.	(A)	225.	(D)	275.	(A)
26.	(A)	76.	(C)	126.	(C)	176.	(C)	226.	(D)	276.	(D)
27.	(D)	77.	(D)	127.	(B)	177.	*	227.	(D)	277.	(C)
28.	(A)	78.	(C)	128.	(D)	178.	(B)	228.	(D)	278.	(A)
29.	(D)	79.	(D)	129.	(B)	179.	(B)	229.	(D)	279.	(C)
30.	(A)	80.	(C)	130.	(B)	180.	(A)	230.	(C)	280.	(D)
31.	(C)	81.	(C)	131.	(B)	181.	(B)	231.	(D)	281.	(C)
32.	(B)	82.	(D)	132.	(B)	182.	(D)	232.	(A)	282.	(B)
33.	(C)	83.	(C)	133.	(B)	183.	(A)	233.	(C)	283.	(B)
34.	(C)	84.	(B)	134.	(A)	184.	(C)	234.	(D)	284.	(C)
35.	(B)	85.	(D)	135.	(A)	185.	(B)	235.	(A)	285.	(C)
36.	(B)	86.	(D)	136.	(B)	186.	(B)	236.	(C)	286.	(B)
37.	(C)	87.	(D)	137.	(A)	187.	(A)	237.	(C)	287.	(D)
38.	(A)	88.	(A)	138.	(A)	188.	(D)	238.	(B)	288.	(B)
39.	(C)	89.	(C)	139.	(A)	189.	(D)	239.	(D)	289.	(C)
40.	(A)	90.	(D)	140.	(C)	190.	(B)	240.	(A)	290.	(B)
41.	(C)	91.	(B)	141.	(C)	191.	(B)	241.	(A)	291.	(D)
42.	(A)	92.	(D)	142.	(D)	192.	(A)	242.	(B)	292.	(C)
43.	(B)	93.	(D)	143.	(A)	193.	(A)	243.	(B)	293.	(C)
44.	(A)	94.	(C)	144.	(C)	194.	*	244.	(B)	294.	(C)
45.	(A)	95.	(A)	145.	(A)	195.	*	245.	(D)	295.	(C)
46.	(A)	96.	(A)	146.	(B)	196.	(D)	246.	(A)	296.	(C)
47.	(D)	97.	(A)	147.	(A)	197.	(B)	247.	(C)	297.	(A)
48.	(C)	98.	(C)	148.	(A)	198.	(A)	248.	(A)	298.	(B)
49.	(A)	99.	(A)	149.	(B)	199.	(D)	249.	(C)	299.	(B)
50.	(D)	100.	(C)	150.	(D)	200.	(D)	250.	(A)	300.	(C)

172. sp
173. Ethylene
177. 1, 4 - dibromobut - 2 - ene
194. oxalic acid
195. C_2H_4, ethylene

301. (C)
302. (D)
303. (C)
304. (B)
305. (B)
306. (D)
307. (C)
308. (A)
309. (D)
310. (C)
311. (B)
312. (B)
313. (D)
314. (A)
315. (A)
316. (D)
317. (D)
318. (D)
319. (C)
320. (A)
321. (A)
322. (B)
323. (C)
324. (A)
325. (B)
326. (C)
327. (A)
328. (B)
329. (B)
330. (A)
331. (B)
332. (C)
333. (D)
334. (C)
335. (A)
336. (B)
337. (A)
338. (D)
339. (B)
340. (B)
341. (C)
342. (C)
343. (D)
344. (D)
345. (C)
346. (A)
347. (C)
348. (A)
349. (C)
350. (B)
351. (C)
352. (C)
353. (B)
354. (D)
355. (D)
356. (B)
357. (D)
358. (C)
359. (B)
360. (B)
361. (A)
362. (A)
363. (C)
364. (A)
365. (B)
366. (A)
367. (A)
368. (A)
369. (A)
370. (B)
371. (A)
372. (C)
373. (D)
374. (C)
375. (C)
376. (D)
377. (B)
378. (A)
379. (A)
380. (D)
381. (C)
382. (D)
383. (B)
384. (D)
385. (B)
386. (C)
387. (D)
388. (C)
389. (A)
390. (D)
391. (A)
392. (C)
393. (C)
394. (C)
395. (A)
396. (C)
397. (A)
398. (B)
399. (A)
400. (A)
401. (C)
402. (D)
403. (A)
404. (C)
405. (B)
406. (C)
407. (C)
408. (C)
409. (C)
410. (B)
411. (A)
412. (C)
413. (A)
414. (D)
415. (B)
416. (B)
417. (D)
418. (C)
419. (C)
420. (A)
421. (C)
422. (A)
423. (B)
424. (C)
425. (A)
426. (D)
427. (D)

Halogen Derivatives

All halogen derivatives are active and do not occur in nature. They are synthesized in the laboratory only. They show chain isomerism

$$H_3C-CH_2-CH_2-CH_2-Cl \; : \; H_3C-CH(CH_3)-CH_2Cl$$

and position isomerism

$$H_3C-CH_2-CH_2Cl \quad : \quad H_3C-CHCl-CH_3$$

Halogen atoms present can be primary, secondary and tertiary in nature.

Preparation:

1. From alkanes: This reaction takes place in the presence of diffused sunlight or in the presence of halides of certain metals e.g. $CuCl_2$ or $FeBr_3$ etc. which act as halogen carriers. Free radical mechanism.

2. Addition of halogen acids to alkenes:

$$H_3C-CH=CH_2 + HCl \longrightarrow H_3C-CHCl-CH_3$$

(carbonium ion intermediate).

In the presence of peroxides

$$H_3C-CH=CH_2 + HBr \xrightarrow{\text{peroxide}} H_3C-CH_2-CH_2Br$$

(Kharasch effect or anti-Markownikoff's addition; free radical intermediates).

3. Hunsdiecker's method: Silver salt of fatty acids on bromination give halo derivatives.

$$RCOOAg + Br_2 \longrightarrow R-Br + CO_2 + AgBr$$

Mechanism: $RCOOAg \xrightarrow{Br-Br} RCOO^{\cdot} + Br^{\cdot}$

$RCOO^{\cdot} \longrightarrow R^{\cdot} + CO_2$; $R^{\cdot} + Br^{\cdot} \longrightarrow RBr$

Free radical intermediates.

4. From alcohols (Darzen's reaction)

Halo alkanes are obtained in pure form by this method. It is nucleophilic substitution reaction.

(i) $ROH + SOCl_2 \longrightarrow RCl + SO_2 + HCl$

(ii) $ROH + HX \text{ (or } PX_3) \longrightarrow RX + H_2O$

(iii) $C_2H_5OH \xrightarrow{P + I_2} C_2H_5I$

Reactivity: T > S > P; HI > HBr > HCl

5. Rydon's method: mixture of alcohol and triphenyl phosphate is treated with halogen

$$ROH + X_2 + (C_6H_5O)_3P \longrightarrow R-X + (C_6H_5O)_2POX$$

Properties: On exposure to light, alkyl iodides liberate iodine. Because of this, the solution of iodides develop brown colour.

$$2RI \xrightarrow{\text{light}} R-R + I_2$$

Reactions:

1. Würtz reaction: $RX + 2Na + XR \longrightarrow R-R + 2NaX$

2. Williamson's synthesis: it follows S_N2 mechanism

$$C_2H_5Br + NaOR \xrightarrow{\Delta} C_2H_5OR + NaBr$$

Mechanism

$$RONa \rightleftharpoons RO^- + Na^+$$

$$RO^- + R-X \xrightarrow{\text{slow}} \left[R-\overset{\delta-}{O}\cdots R\cdots\overset{\delta-}{X}\right] \xrightarrow{\text{fast}} ROR + X^- \; ; \; Na^+ + X^- \longrightarrow NaX$$

With tertiary halides, the reaction fails.

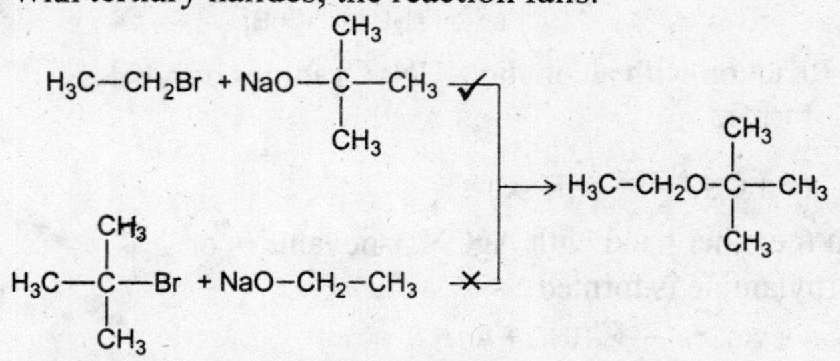

3. Preparation of Grignard reagent:

$$R-X + Mg \xrightarrow{\text{ether}} RMgX$$

4. Friedel-Craft synthesis: It is an S_E reaction. Alkylation involving primary or secondary halides is S_E2, while tertiary halides is S_E1.

S_E2: $R-Cl + AlCl_3 \rightleftharpoons R-\overset{+}{Cl}-\overset{-}{AlCl_3}$

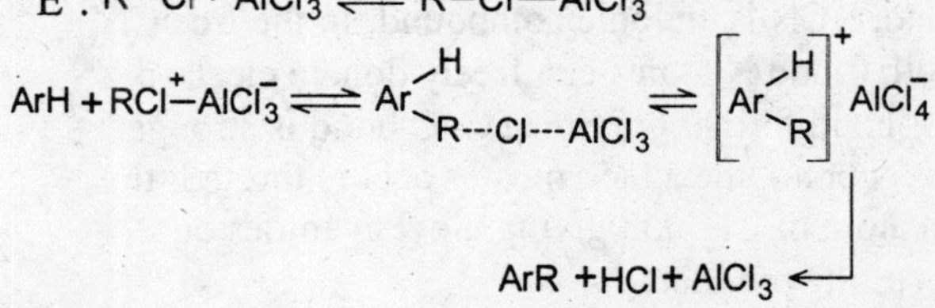

S_E1: The first step is the formation of electrophilic carbonium ion R_3C^+ which then attacks the benzene ring resulting in the formation of alkyl benzene.

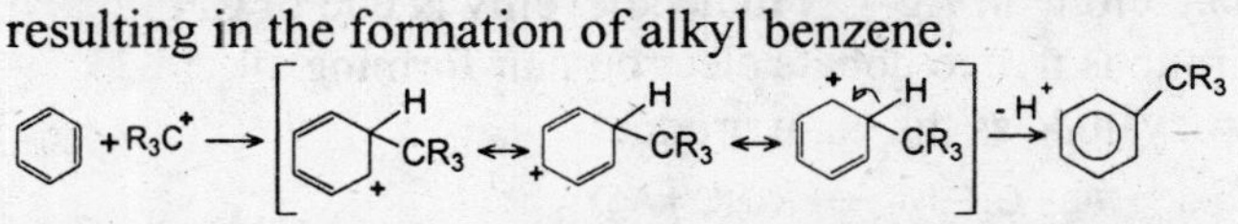

Presence of o– and p– directing group facilitate Friedel–Craft reaction while m– directing group inhibits the reaction.

5. Reaction with KOH:

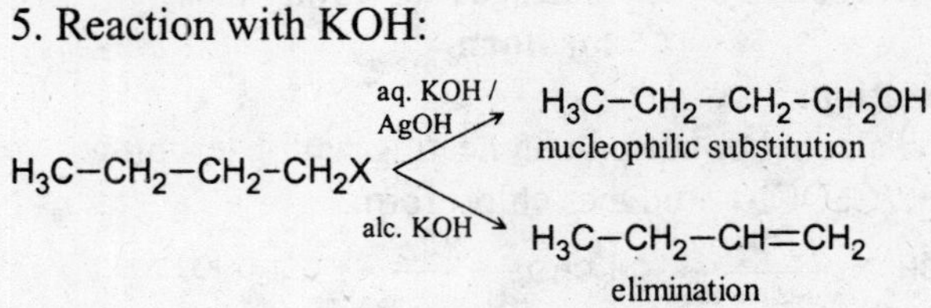

Tertiary alkyl halides undergo dehydrohalogenation reaction most readily, preferably by E_1 mechanism while primary and secondary usually by E_2 mechanism.

E_1 mechanism:

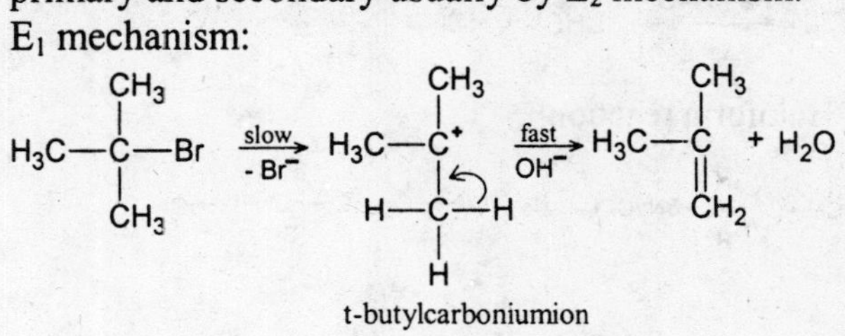

t-butylcarboniumion

Elimination in the case of t–butyl halide is of first order reaction as it is independent of OH^- concentration.

E_2 mechanism: $KOH \rightleftharpoons K^+ + OH^-$

$$OH^- + HCH_2{-}CH_2Br \xrightarrow{slow} \left[\overset{\delta-}{O}H{\cdots}H{\cdots}CH_2{=\!=}CH_2{\cdots}\overset{\delta-}{Br}\right] \text{ (transition state)} \xrightarrow{fast} H_2O + H_2C{=}CH_2 + Br^-$$

6. Reaction with alcoholic KCN: Cyanides or nitriles are formed.

$$R{-}X + KCN \xrightarrow{alcohol} R{-}CN$$

On the other hand with AgCN, isocyanides or carbylamine is formed.

$$R{-}X + AgCN \longrightarrow RNC + Ag$$

The difference in the reaction products can be explained as follows:

CN^- is an ambident nucleophile. It is a resonance hybrid of

$$:\bar{C}{\equiv}N: \longleftrightarrow :C{=}\overset{+}{\ddot{N}}: \equiv \overset{\delta-}{C}{=}\overset{\delta-}{N} \text{ (hybrid structure)}$$

It can attack the C^+ centre of R–X molecule through C or N end. KCN is an ionic compound. In the free CN^- ion, both C and N atoms can freely donate electron pair for bond formation. Since C–C bond is stronger than C–N bond, the attack mostly occurs through the carbon atom of CN^- ion giving alkyl cyanides or nitriles as the products.

$$R{-}X + CN^- \xrightarrow{KCN} R{-}C{\equiv}N$$

AgCN is predominantly a covalent compound. Therefore, in Ag–CN molecule, only N end of CN group is free to donate electron pair forming alkyl isocyanide as the main product.

$$R{-}X + Ag{-}C{\equiv}N: \longrightarrow RNC + AgX$$

- $ClH_2C{-}CH_2Cl$: used as a solvent; as paint remover and as fumigant.
- CH_3Br: used as a poisonous gas for killing rats.

Chloroform

Preparation:

1. Ethyl alcohol or acetone on heating with bleaching powder ($CaOCl_2$) produces chloroform.

$$H_3C{-}CH_2OH \xrightarrow[\text{as oxidising agent}]{Cl_2} CH_3CHO \xrightarrow[\text{as chlorinating agent}]{3Cl_2} CCl_3CHO + 3HCl$$

$$2CCl_3CHO + Ca(OH)_2 \longrightarrow 2CHCl_3 + (HCOO)_2Ca$$

2. Commercial method:

$$CCl_4 + 2[H] \xrightarrow[\Delta]{Fe/H_2O} CHCl_3 + HCl$$

3. By Haloform reaction:

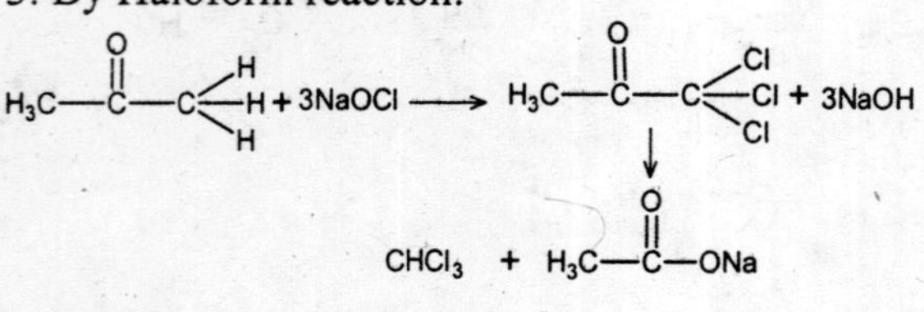

Properties:

It is stored in brown or blue coloured bottles. It is filled up to the brim of the bottle. This is to prevent the oxidation in the presence of light and air to a poisonous phosgene gas.

$$2CHCl_3 + O_2 \xrightarrow{\text{air, sunlight}} 2COCl_2 + HCl$$

$$4CHCl_3 + 3O_2 \xrightarrow[\text{sunlight}]{\text{excess air,}} 4COCl_2 + Cl_2 + 2H_2O$$

1% C_2H_5OH is added to prevent oxidation or to convert carbonyl chloride to ethylcarbonate.

$$COCl_2 + 2C_2H_5OH \longrightarrow (C_2H_5)_2CO_3 + 2HCl$$

1. Carbylamine reaction: it is regarded as a test for primary amines. Aniline also gives this test.

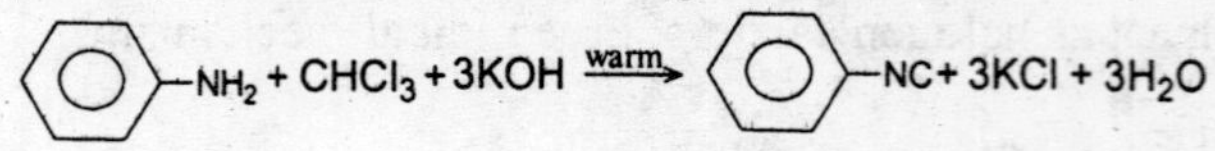

$$RNH_2 + CHCl_3 + 3KOH \xrightarrow{\Delta} RNC + 3KCl + 3H_2O$$

Mechanism:

$$H\ddot{O}^- + H{-}CCl_3 \xrightarrow{-H_2O} {}^-\!:CCl_3 \xrightarrow{-Cl^-} :CCl_2 \text{ (dichloro carbene)}$$

$$R{-}\ddot{N}H_2 + :CCl_2 \longrightarrow R{-}\overset{+}{N}H_2{-}\overset{-}{C}Cl_2 \xrightarrow{-2HCl} R{-}N{\equiv}C$$

2. Chloropicrin:

$$H{-}CCl_3 + HNO_3 \longrightarrow \underset{\text{chloropicrin}}{O_2NCCl_3} + H_2O$$

it was used as an insecticide and also as a war gas.

3. With Ag powder:

$$H{-}CCl_3 + 6Ag + Cl_3C{-}H \xrightarrow{\Delta} HC{\equiv}CH + 6AgCl$$

4. Hydrolysis: with hot aqueous KOH or NaOH

$$H{-}CCl_3 + 3NaOH \longrightarrow [H{-}C(OH)_3] \xrightarrow{-H_2O} \underset{\text{formic acid}}{H{-}COOH} \xrightarrow{NaOH} \underset{\text{sod. formate}}{H{-}COONa}$$

5. Chloretone:

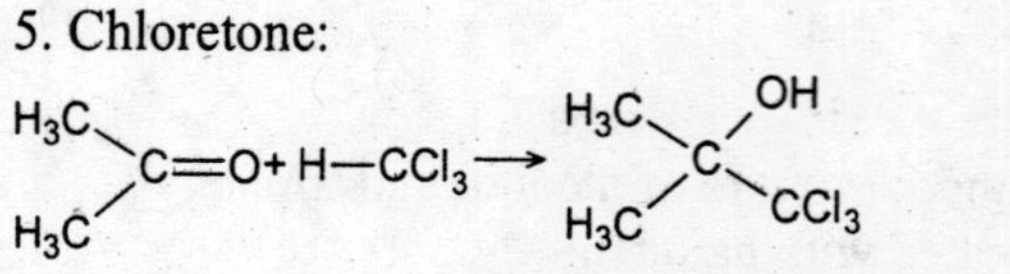

1,1,1-trichloro, 2-methyl-2-propanol

it is a hypnotic.

6. Reimer–Tiemann reaction:

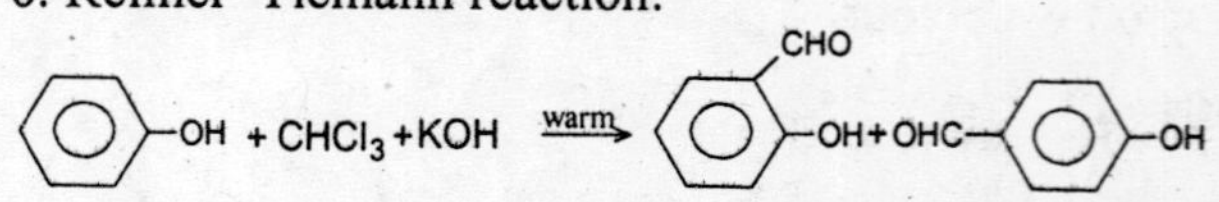

mechanism $KOH \rightleftharpoons K^+ + OH^-$; $CHCl_3 + OH^- \longrightarrow {}^-CCl_3 + H_2O$;

$${}^-CCl_3 \xrightarrow{-Cl^-} :CCl_2$$

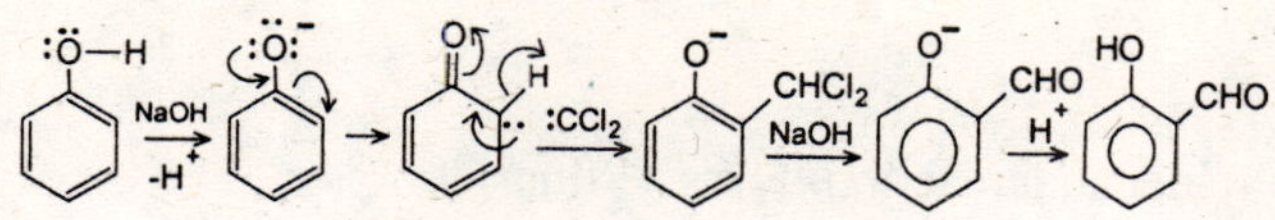

7. Reduction:

$$CHCl_3 \xrightarrow[\text{presence of } C_2H_5OH,\ 2[H]]{Zn/HCl} CH_2Cl_2 + HCl$$

$$CHCl_3 \xrightarrow[6[H]]{Zn/HCl/H_2O} CH_4 + 3HCl$$

Uses:
It is used as anaesthetic with 30% ether. It has antiseptic properties also. Purity of $CHCl_3$ (free from phosgene) is tested by the addition of $AgNO_3$.

Bromoform: It is used as medicine for whooping cough, solvent for oils, fats, waxes etc.

Iodoform: CHI_3
Laboratory method:
$C_2H_5OH + 4I_2 + 6NaOH \quad CHI_3 + HCOONa + 5NaI + 5H_2O$
Industrial method:
$CH_3COCH_3 + 3I_2 + 2Na_2CO_3 \quad CHI_3 + 3NaI + CH_3COONa + 5H_2O$
(KI on electrolysis liberates iodine which takes part in this reaction). Iodoform test is given by compounds containing CH_3CO– group or by those compounds in which this group is produced very easily. e.g. CH_3CHO, $CH_3CHOHCOOH$, CH_3COCH_3, CH_3CH_2OH etc.
Iodoform is used to distinguish

CH_3OH, C_2H_5OH ; 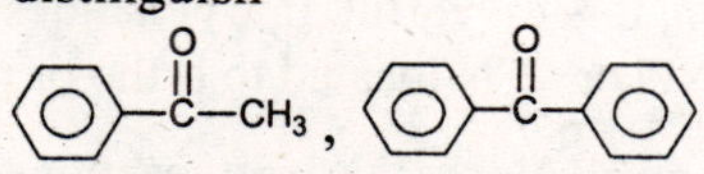

acetophenone, benzophenone

Uses: Antiseptic, to detect CH_3CO– group.

Carbon tetrachloride: CCl_4

Oxidation: $CCl_4 + H_2O \xrightarrow{500°C} \underset{\text{phosgene}}{COCl_2} + 2HCl$

Reduction: with moist iron fillings

$CCl_4 + 2[H] \xrightarrow{Fe/H_2O} \underset{\text{chloroform}}{CHCl_3} + HCl$

Reaction with phenol

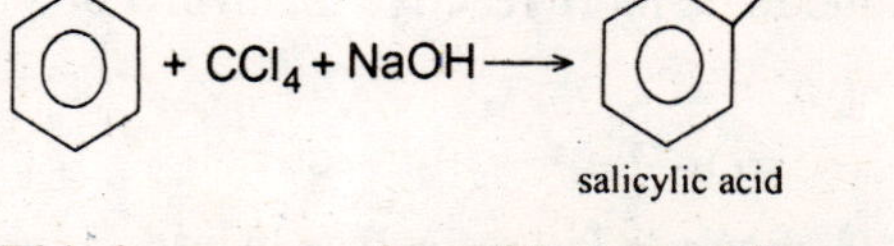

salicylic acid

This is an extension of Reimer–Tiemann reaction.
- CCl_4 is used as fire extinguisher under the name Pyrene.

- Allyl iodide undergoes nucleophilic substitution reactions by S_N1 mechanism which is due to the resonance stabilisation of the resulting allyl cation.

$$H_2C{=}CH{-}CH_2I \xrightarrow{-I^-} H_2C{=}CH{-}CH_2^+ \longleftrightarrow H_2C^+{-}CH{=}CH_2$$

- Freon: $3CCl_4 + 2SbF_3 \longrightarrow \underset{\text{freon 12}}{CCl_2F_2} + 2SbCl_3$

Used as refrigerant and propellant in aerosol sprays.
Freon 114: $ClCF_2CF_2Cl$
Teflon: Polytetrafluoroethene is unaffected by aqua regia and is used as non-stick coating in cookware.

Dichloroethers:

$H_3C{-}CH_2{-}O{-}CH_2{-}CH_3 \longrightarrow H_3C{-}CHCl{-}O{-}CHCl{-}CH_3$
α, α' dichloroethylether as lachrymator (tear gas) used in I world war

$H_3C{-}CH_2{-}O{-}CH_2{-}CH_3 \longrightarrow ClH_2C{-}CH_2{-}O{-}CH_2{-}CH_2Cl$ chlorex
β, β' dichloroethylether used to remove aromatic impurities from lubricant oils

Dihalides:

$H_3C{-}CHX_2$ (gem dihalide) $\qquad ClH_2C{-}CH_2Cl$ (vicinal dihalide)

Gem dihalides: They are prepared by the action of PCl_5 or PBr_5 on aldehydes and ketones e.g.

$$H_3C{-}CHO + PCl_5 \longrightarrow H_3C{-}CHCl_2 + POCl_3$$

Vicinal dihalides:
(i) addition of halogens to alkenes
$H_3C{-}CH{=}CH_2 + Br_2 \longrightarrow H_3C{-}CHBr{-}CH_2Br$
(ii) addition of PCl_5 on glycols

$$\begin{matrix} CH_2OH \\ | \\ CH_2OH \end{matrix} + 2PCl_5 \longrightarrow \begin{matrix} CH_2Cl \\ | \\ CH_2Cl \end{matrix} + 2POCl_3 + 2HCl$$

Gem dihalides are less active than alkyl halides: In the presence of one halogen atom with a strong –I effect, the other halogen atom can't so easily be replaced.
The reactivity of vicinal dihalides is of the same order as that of alky halides.

Why do the halogens behave differently ?

The usual pattern of events is for groups that direct ortho and para to activate the ring, and groups that direct meta to deactivate the ring. However, the halogens are exceptions. They direct ortho and para but deactivate the ring. The electron withdrawing nature of the halogens means that they withdraw electron density from the ring. This is shown for example, by the orientation of the dipole moment of chlorobenzene. As a result an incoming electrophile will find the bezene ring in chlorobenzene less attractive than in benzene itself.

Reactions:

(i) 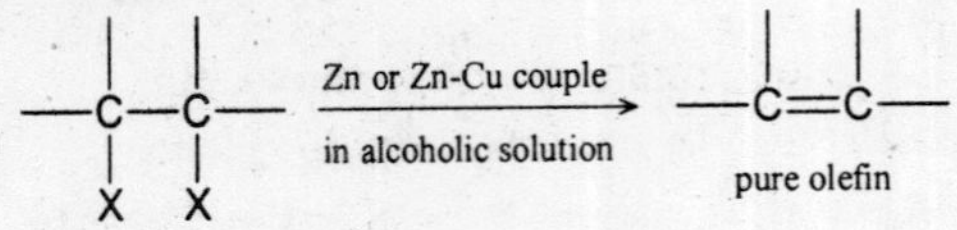

(ii) 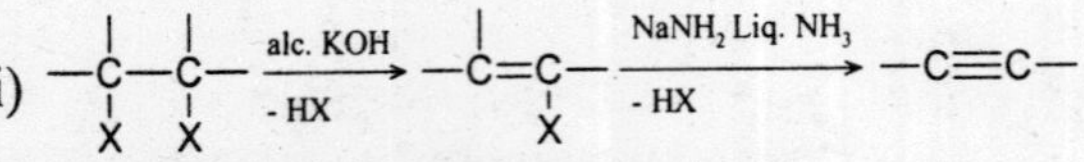

Aromatic Halogen Compounds:

Aryl halides: Halogen substituent is attached directly to an aromatic ring.

1. Addition compounds:

(i) BHC: 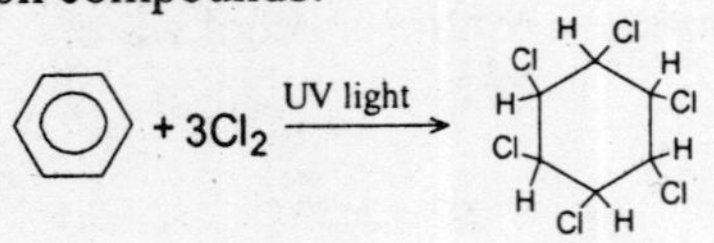

$C_6H_6Cl_6$

8. stereoisomers possible, γ-isomer (Cl atoms at position 1,2,3 are a,a,a axial) powerful insecticide (pesticide); known as Gammexene or Lindane

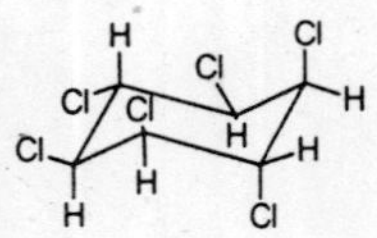

2. Side chain substitution compounds

e.g. 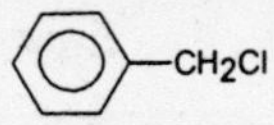

formed by free radical mechanism

(i) 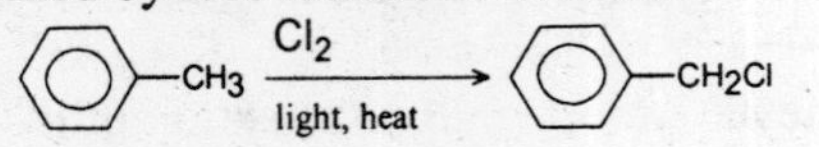

(ii) 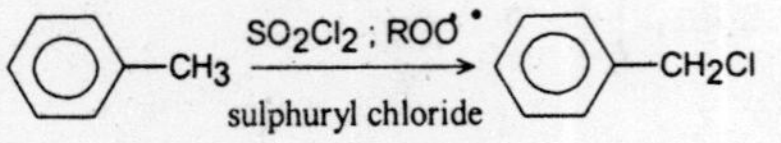

(iii) NBS: specific for allylic and benzylic position

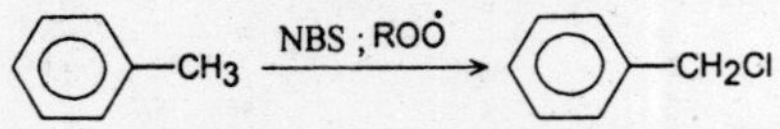

Side chain substitution compounds give reactions similar to those of alkyl halides.

3. Nuclear substitution compounds:

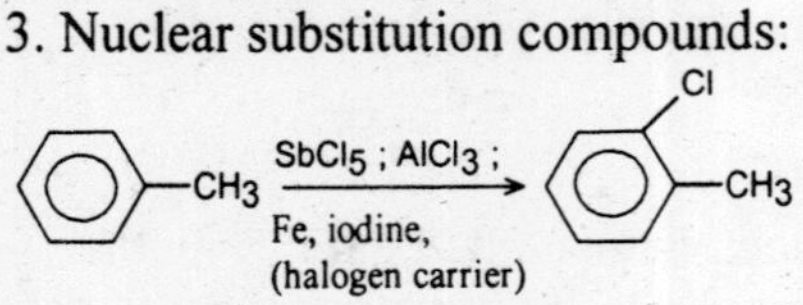

They are formed through ionic mechanism.

Preparation:

(1) Sandmeyer's reaction (Laboratory method)

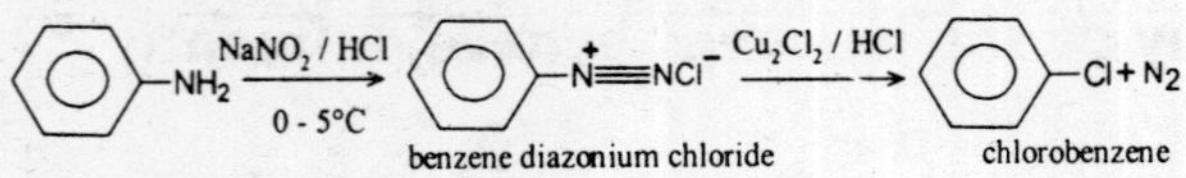

Mechanism:

(i) $C_6H_5N_2Cl + CuCl \longrightarrow C_6H_5N_2[CuCl_2]^- \longrightarrow C_6\dot{H}_5 + N_2 + CuCl_2$

(ii) $C_6\dot{H}_5 + CuCl_2 \longrightarrow C_6H_5Cl + CuCl$

(2) Gattermann reaction: Using metallic cooper or freshly precipitated copper powder (Zn dust and copper sulphate solution) as catalyst

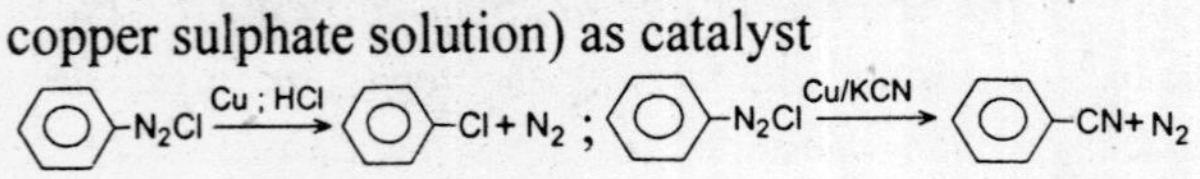

(3) Schiemann reaction: For aryl fluorides.

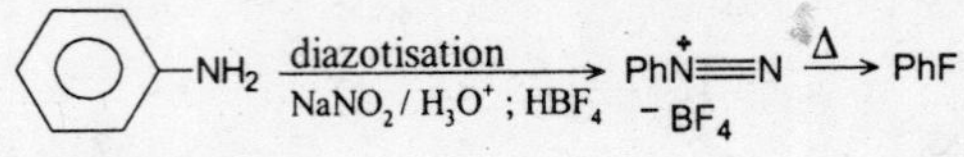

Reactions:

1. Gragnard reagent:

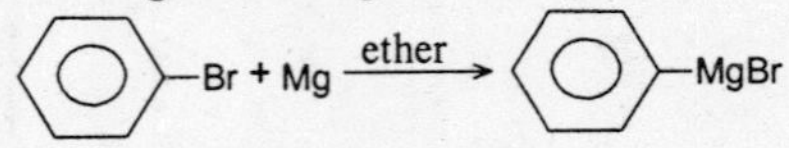

2. Würtz–fittig reaction:

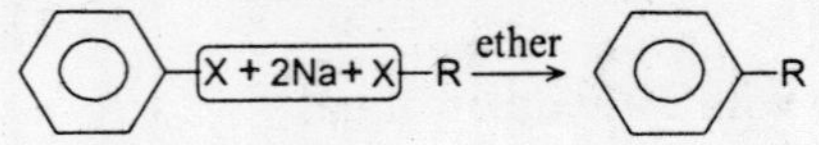

Fittig reaction:

3. Oxidation:

(i) 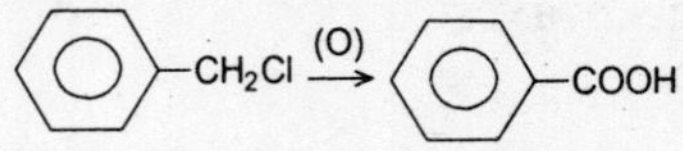

(ii) 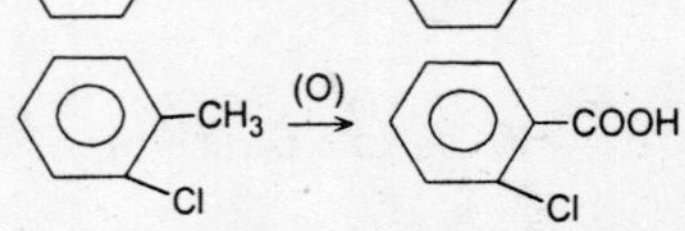

Cl attached to the side chain lois lost during oxidation.

4. Ullmann synthesis: $2C_6H_5I + 2Cu \longrightarrow$ 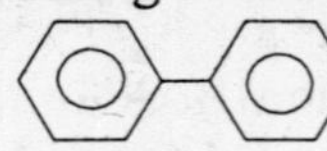

Aryl fluorides do not react in Ullmann reaction

Note: Nuclear derivatives do not give Friedel–Craft's reaction. OH^-, CN^-, NH_2^-, ^-OR can't easily replace halogen atoms in nuclear derivatives. When present in side chain, it is easy. But, if aryl halides have groups like $-NO_2$, $-NO$ or $-CN$ located to ortho or para to halogen, they can give S_N^1 or S_N^2 type reactions, characteristic of alkyl halides are not shown by aryl halides.

Aryl halides are less reactive than alkyl halides due to

(a) Delocalisation of electrons by resonance

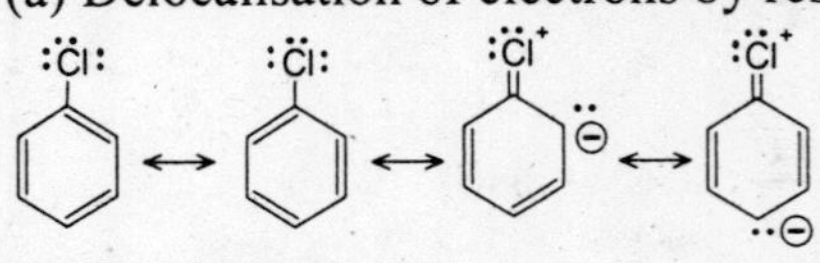

(b) Bond length

$H_3C—Cl$ 1.77 to 1.80 Å

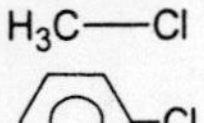 1.69 Å

(c) bond energies differ due to difference in hybridisation

$H_3C^*—X$, $*sp^3$; 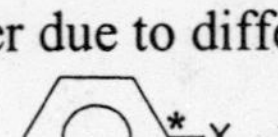 , $*sp^2$

- Allyl halides are found to be more reactive than vinyl or alkyl halides.

$H_2C{=}CH{-}CH_2{-}Br \xrightarrow{-Br^-} [H_2C{=}CH{-}\overset{+}{C}H_2 \longleftrightarrow H_2\overset{+}{C}{-}CH{=}CH_2]$

C cbonium ion resonance hybrid

The more stable the carbocation, the faster it is formed.

Resonance is responsible for the shortening of C–Cl bond in vinyl chloride. Therefore, chlorine of vinyl chloride is not as reactive as Cl in allyl halides.

$$H_2C{=}CH{-}\ddot{Cl}: \longleftrightarrow H_2\bar{C}{-}CH{=}\overset{+}{Cl}:$$

- Nitration of benzene and chlorobenzene:
Nitration of chlorobenzene is slower than the nitration of benzene.

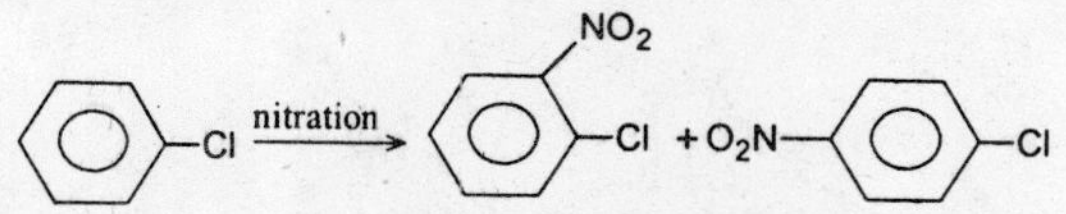

(o and p– derivatives)

Through its –I effect, halogen withdraws electrons and deactivates the entire molecule towards electrophilic attack. Through its resonance effect, halogen releases electrons and tends to activate but only towards attack at certain positions.

- DDT: Dichlorodiphenyltrichloroethane

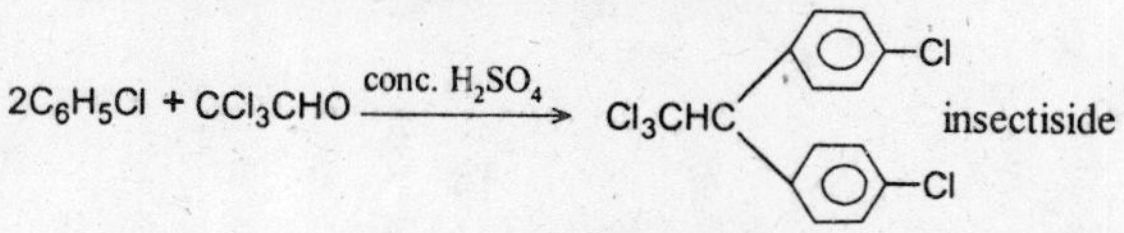

4,4'-dichloro-diphenyl-1,1,1-trichloroethane

Reactions of aryl halides

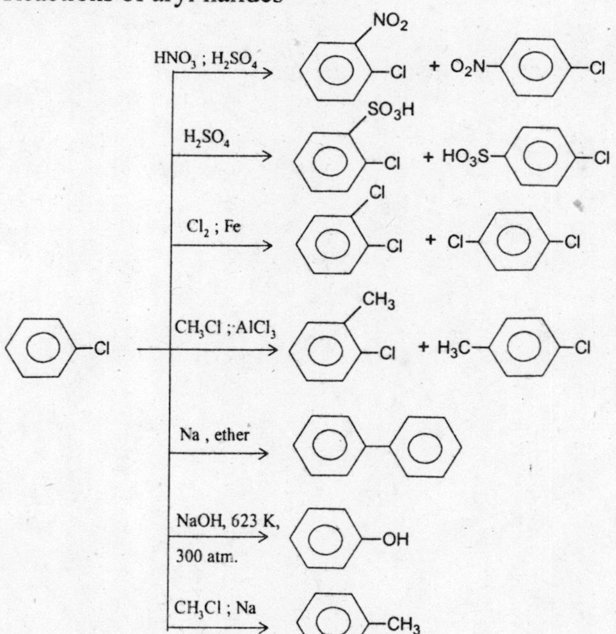

- Chlorination of toulene

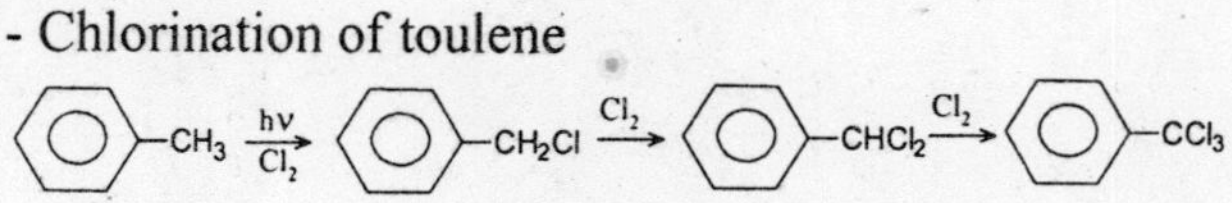

Organometallic Compounds

The organometallic compounds of Mg are known as Grignard reagents.

$$R{-}X + Mg \xrightarrow{ether} RMgX$$

In ether the structures is

$$R{-}\ddot{O}{-}R \downarrow R{-}Mg{-}X \uparrow R{-}\ddot{O}{-}R$$

In RMgX the carbon–magnesium bond is covalent but highly polar. The magnesium–halogen bond is essentially ionic.

$$R{-}Mg{-}{-}{-}X \longrightarrow \bar{R} + \overset{+}{Mg}{-}{-}{-}X$$

1. Double decomposition reactions:

(a) with compounds containing active hydrogen: Compounds having active hydrogen e.g. H_2O, ROH, –SH, NH_3, RNH_2, R_2NH etc. react with RMgX. For every active H– atom, one paraffin molecule is produced. This principle is used for the estimation of –OH, $-NH_2$, –SH etc. groups in Zeriwitinoff's method. e.g.

$$R{-}Mg{-}{-}{-}X + ROH \longrightarrow RH + Mg\begin{smallmatrix}X\\OR\end{smallmatrix};$$

$$RMgX + R_2NH \longrightarrow RH + Mg\begin{smallmatrix}X\\NR_2\end{smallmatrix}$$

(b) With alky halides: $RMgX + R{-}X \longrightarrow R{-}R + MgX_2$

(c) With acyl halides: $RMgX + RCOX \longrightarrow RCOR + MgX_2$

(d) With allyl bromide:

$$H_2C{=}CH{-}CH_2{-}Br + RMgX \longrightarrow H_2C{=}CH{-}CH_2{-}R$$

(e) With cyanogen chloride: $RMgX + ClCN \longrightarrow RCN + MgXCl$

(f) With chloroamine $RMgX + ClNH_2 \longrightarrow RNH_2 + MgXCl$

2. Nucleophilic addition reactions

(a) With oxygen: $RMgX \xrightarrow{(O)} ROMgX \xrightarrow{H_3O^+} ROH + Mg\begin{smallmatrix}X\\OH\end{smallmatrix}$

p. alcohol

with

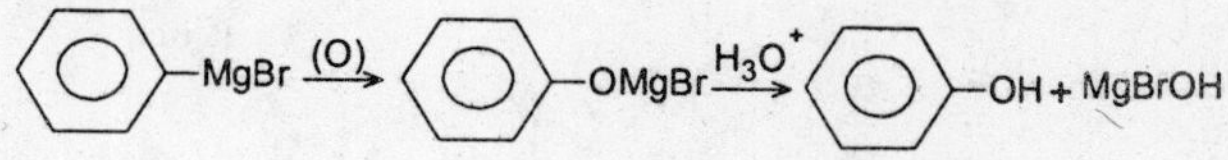

phenol

(b) With oxirane:

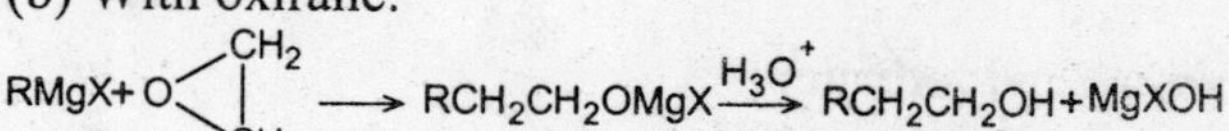

p. alcohol

(c) With Formaldehyde:

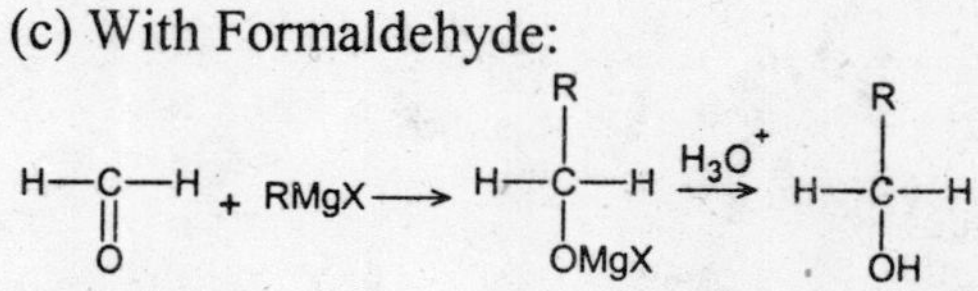

formaldehyde p. alcohol

(d) With aldehydes other than formaldehyde:

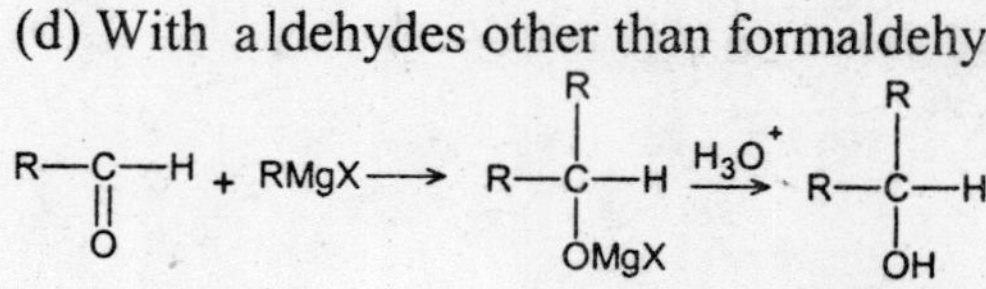

s. alcohol

(e) With ketones:

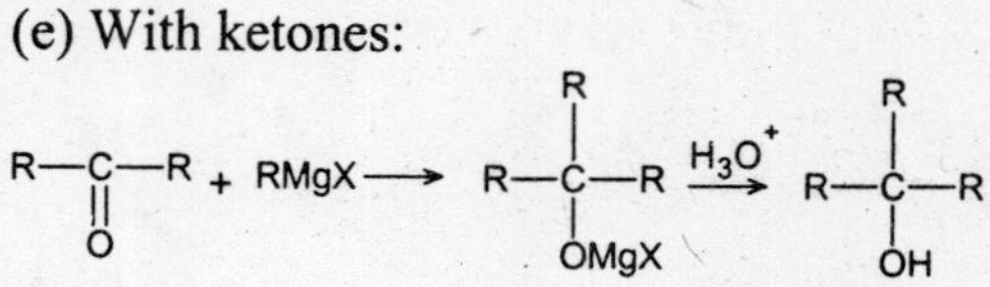

t. alcohol

(f) With carbondioxide:

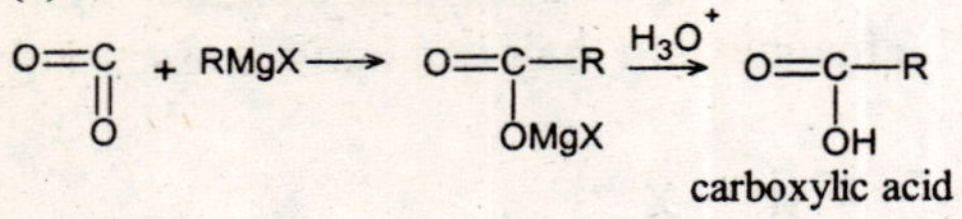

(g) With carbondisulphide:

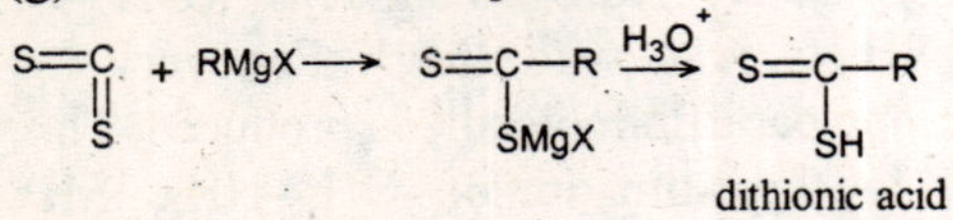

(h) With alkyl cyanide

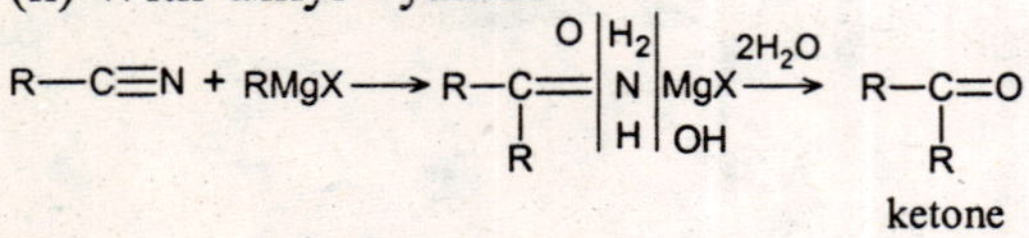

(i) With hydrogen cyanide

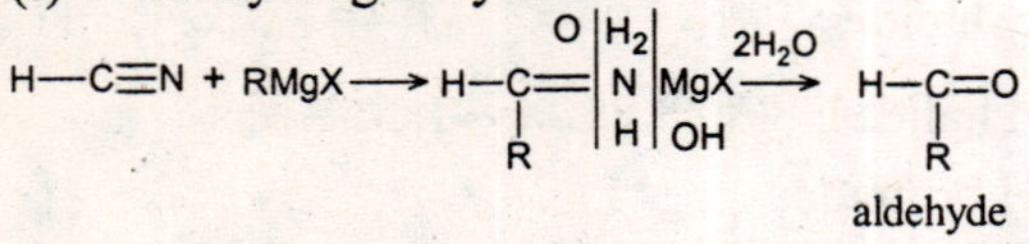

3. Double decomposition and addition reactions

(a) With amides:

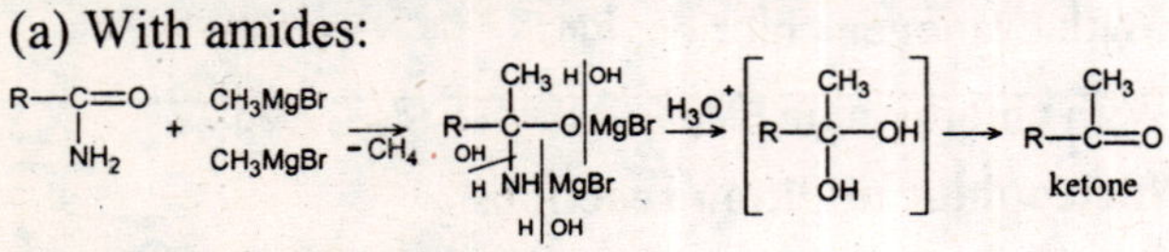

(b) With acid chlorides:

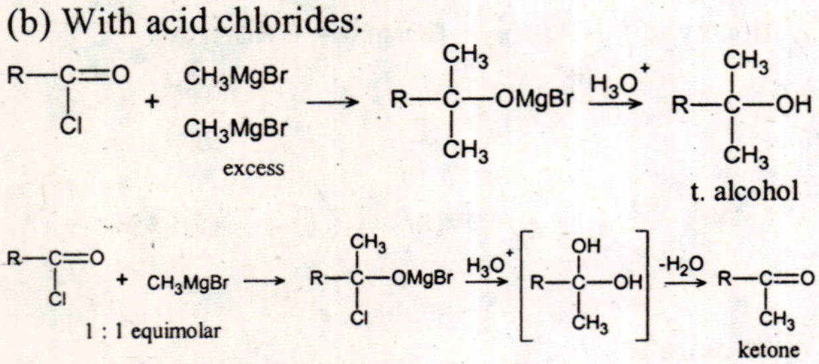

(c) With Ethyl formate:

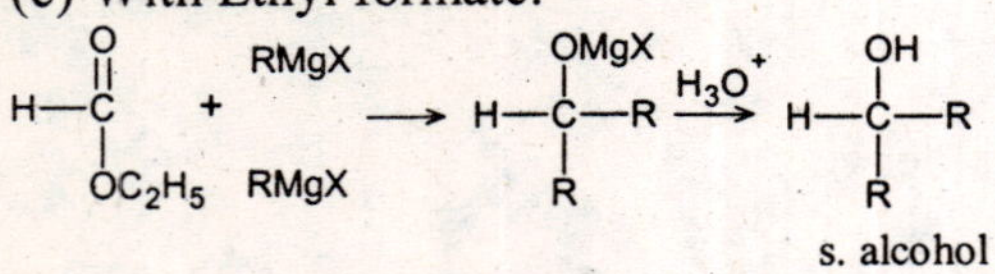

(d) With ethylacetate:

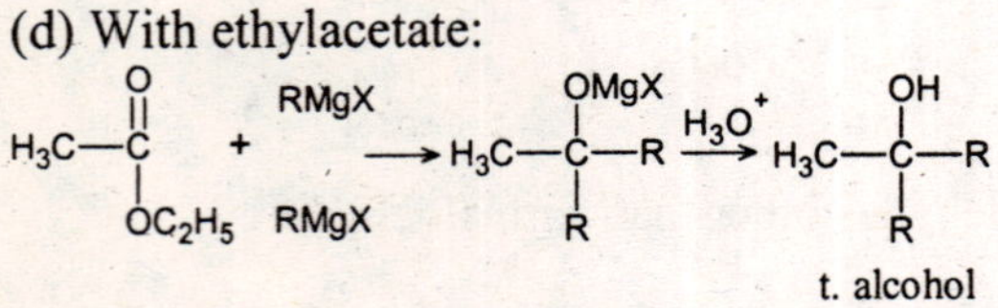

1. Identify X in the reaction [CBSE-90]

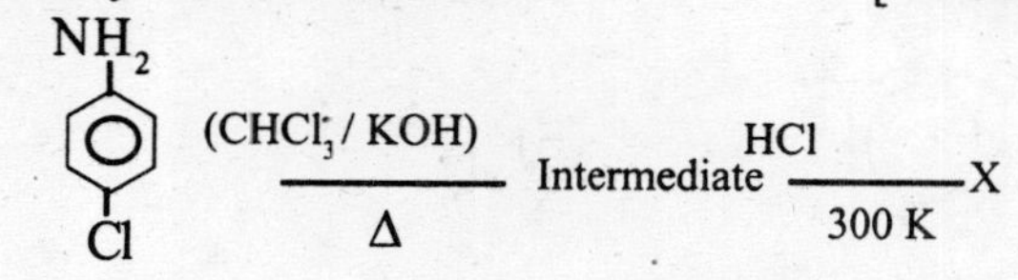

(A) 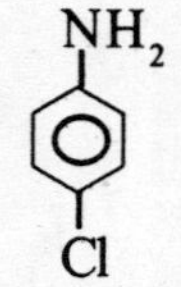
(B) 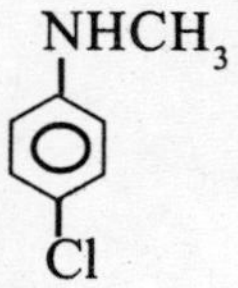
(C) 
(D) 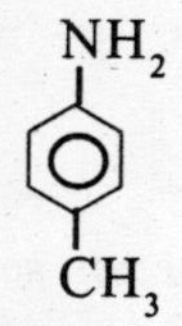

2. The reaction equation [CBSE-90]

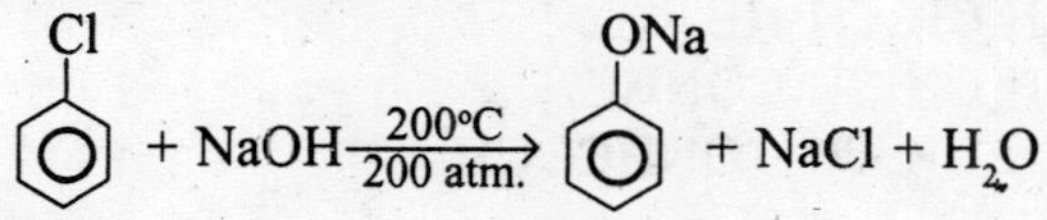

illustrates
(A) Down's process (B) Kolbe's reaction
(C) Carbylamine test (D) Halofom reaction

3. $CHCl_3$ on oxidation gives [CBSE-90]
(A) phosgene (B) formic acid
(C) chloropicrin (D) CCl_4

4. C_2H_5Br + NaOH -----→ C_2H_5OH + NaBr This reaction is known as ..... substitution [CBSE-91]
(A) electrophilic (B) nucleophilic
(C) Both (A) and (B) (D) None of these

5. Grignard reagent reacts with sulphur and forms [CBSE-91]
(A) hypo (B) thioalcohol
(C) methyl alcohol (D) formic acid

6. Isocyanide test is used for the detection of [CBSE-91]
(A) primary alcohols (B) primary amines
(C) secondary amines (D) secondary alcohols

7. Which is of the following is obtained when chloral is boiled with NaOH ? [CBSE-91]
(A) CH_3Cl (B) $CHCl_3$ (C) CCl_4 (D) All of the above

8. For the carbylamine reaction we need hot alcoholic KOH and [CBSE-92]
(A) primary amine and chloroform
(B) chloroform and silver powder
(C) a primary amine and an alkyl halide
(D) a monoalkylamine and trichloromethane

9. The reaction of toluene with chlorine in presence of ferric chloride gives mainly [CBSE-92]
(A) m-chlorotoluene (B) benzyl chloride
(C) o-and p-chlorotoluene (D) benzoyl chloride

10. Industrial preparation of chloroform employs acetone and [CBSE-93]
(A) phosgene (B) calcium hypochloride
(C) chlorine gas (D) sodium chloride

11. Chlorobenzene reacts with Mg in dry ether to give a compound (A) which further reacts with ethanol to yield [CBSE-93]
(A) phenol (B) benzene
(C) ethyl benzene (D) phenyl ether

12. A compound 'A' has a molecular formula C_2Cl_3OH. It reduces Fehling's solution and on oxidation gives a monocarboxylic acid 'B'.'A' is obtained by the action of chlorine on ethyl alcohol, 'a' is [CBSE-94]
(A) methyl chloride (B) monochloroacetic acid
(C) chloral (D) chloroform

13.The compound obtained by heating a mixture of a primary amine and chlorofrom with ethanolic potassium hydroxide (KOH) is [CBSE-97]
(A) an alkyl isocyanide (B) an alkyl halide (C) an amide
(D) an amide and nitro compound

14. The alkylhalide is converted into an alcohol by [CBSE-97]
(A) addition (B) substitution
(C) dehydrohalogenation (D) elimination

15. Replacement of Cl of chlorobenzene to give phenol requires drastic conditions but chlorine of 2,4-dinitrochloro-benzene is readily replaced since [CBSE-97]
(A) NO_2 make the ring electron rich at ortho and para
(B) NO_2 withdraws e^- from meta position.
(C) NO_2 denotes e^- at meta position
(D) NO_2 withdraws e^- from ortho/para positions

16. Reaction of $CH_2 - CH_2$ (bridged by O) with RMgX leads to formation of [CBSE-98]
(A) RCHOHR (B) $RCHOHCH_3$
(C) $R_2>CHCH_2OH$ (D) RCH_2CH_2OH

17. Iodoform test is not given by [CBSE-98]
(A) 2-Pentanone (B) 3-Pentanone
(C) Ethanal (D) Ethanol

18. 2–Bromopentene is heated with potassium ethoxide in ethanol. The major product obtained is [CBSE-98]
(A) 2-ethoxypentane (B) Pentene-1
(C) Cis-pentene-2 (D) Trans-pentene-2

19. Ethyl chloride is converted into diethyl ether by [CBSE-99]
(A) Wurtz synthesis (B) Grignard reaction
(C) Perkin's reaction (D) Williamson's synthesis

20. In Friedal-Craft's alkylation, besides $AlCl_3$ the other

reactants are [CBSE-99]
(A) $C_6H_6+NH_3$ (B) $C_6H_6+CH_4$
(C) $C_6H_6+CH_3Cl$ (D) $C_6H_6+CH_3COCl$

21. Phenyl isocyanides are prepared from which the following reactions? [CBSE-99]
(A) Rosenmund's reaction (B) Carbylamine reaction
(C) Reimer-Tieman reaction (D) Wurtz reaction

22. Gammexane is [CBSE-99]
(A) chlorobenzene (B) benzyl chloride
(C) bromobenzene (D) benzene hexachloride

23. $CH_3-CH_2-\underset{\displaystyle Cl}{\underset{|}{CH}}-CH_3$ obtained by chlorination of n-butane will be - [CBSE-2001]
(A) Meso form (B) Racemic mixture
(C) d-form (D) l-form

24. An organic compound $A(C_4H_9Cl)$ on reacting with Na/diethyl ether gives a hydrocarbon which on monochlorination gives only one chloro derivative. The A is- [CBSE-2001]
(A) t-Butyl chloride (B) Sec-butyl chloride
(C) Iso-butyl chloride (D) n-Butyl chloride

25. 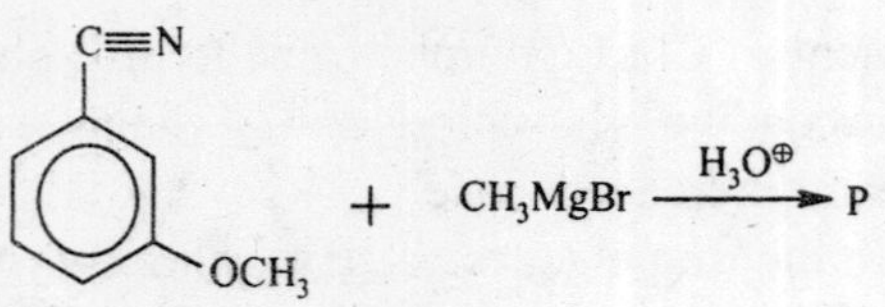

Product 'P' in the above reaction is [CBSE 2002]

(A) 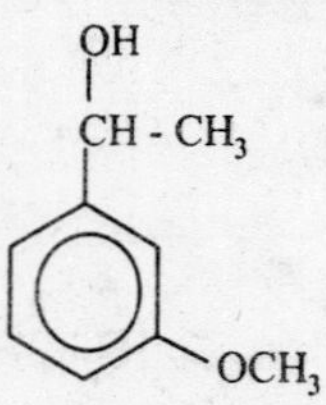
(B) 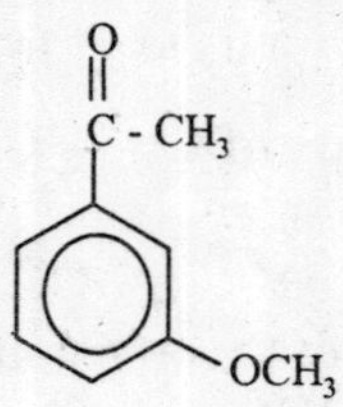
(C) 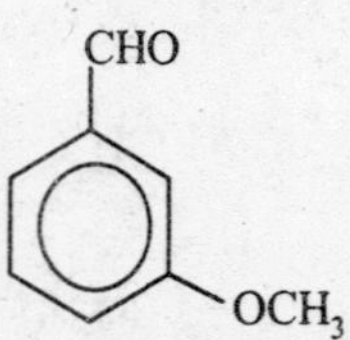
(D) 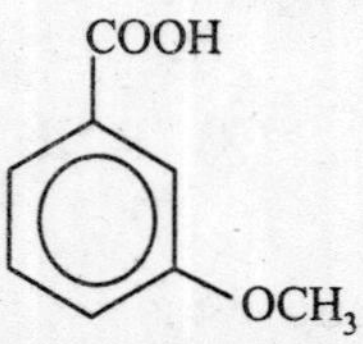

26. Reactivity order of halides for dehydrohalogenation is [CBSE 2002]
(A) R-F > R - Cl > R -Br > R -I
(B) R-I > R-Br > R-Cl > R-F
(C) R-I > R-Cl > R-Br > R-F
(D) R-F > R-I > R-Br > R-Cl

27. 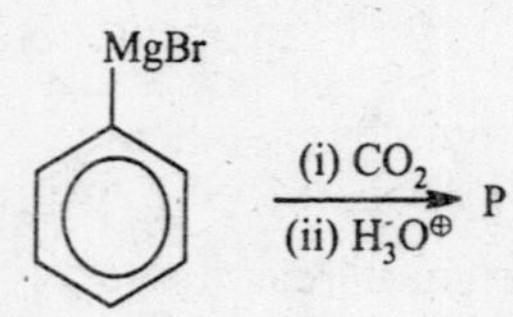 , In the above reaction product 'P' is [CBSE 2002]

(A) 
(B) 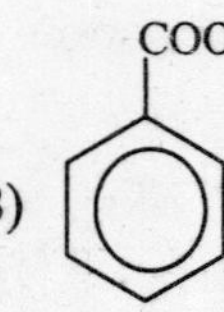
(C) 
(D) $C_6H_5-\overset{\displaystyle O}{\overset{||}{C}}-C_6H_5$

28. $CH_3CH_2Cl \xrightarrow{NaCN} X \xrightarrow{Ni/H_2} Y \xrightarrow{\text{Acetic Anhydride}} Z$

Z in the above reaction sequence is [CBSE 2002]
(A) $CH_3CH_2CH_2NHCOCH_3$
(B) $CH_3CH_2CH_2NH_2$
(C) $CH_3CH_2CH_2CONHCH_3$
(D) $CH_3CH_2CH_2CONHCOCH_3$

29. When $CH_3CH_2CHCl_2$ is treated with $NaNH_2$, the product formed is [CBSE 2002]
(A) $CH_3-CH=CH_2$ (B) $CH_3-C\equiv CH$
(C) $CH_3CH_2CH\langle^{NH_2}_{NH_2}$ (D) $CH_3CH_2CH\langle^{Cl}_{NH_2}$

30. Which of the following compounds can give haloform test ? [AIIMS-93]
(A) $C_2H_5COC_2H_5$ (B) C_2H_5CHO
(C) C_3H_7CHO (D) $CH_3COC_2H_5$

31. The compound which can be used as an antifreeze in automobile radiators is [AIIMS-93]
(A) Glyceraldehyde (B) Oxalic acid
(C) Glycol (D) Glycerol

32. The product obtained by heating propanone and the paste of beaching powder is [AIIMS-93]
(A) Trichloromethane (B) Propyl chloride
(C) Ethyl chloride (D) Chloral

33. When alkyl halides are heated, with dry Ag_2O, they give [AIIMS-94]
(A) Ester (B) Ketone (C) Diethyl ether (D) Benzene

34. When chloroform is exposed to air and sunlight, it gives [AIIMS-94]
(A) Phosgene (B) Lewisite
(C) Mustard gas (D) Carbon tetrachloride

35. Chloroform by reacting with conc. HNO_3 produces [AIIMS-95]
(A) Water gas (B) Laughing gas
(C) Tear gas (D) Producer gas

36. What happens when chloroform is left open in air in the presence of sunlight ? [AIIMS-96]
(A) Explosion takes place

(B) Phosgene, a poisonous gas is formed
(C) Polymerization takes place (D) No reaction takes place

37. Which of the following is urotropine ? [AIIMS-96]
(A) Hexamethyl-diamine (B) Hexamethylene triamine
(C) Hexamethylene tetramine (D) None of these

38. Which of the following on reaction with H_2S does not produce metallic sulphide ? [AIIMS-97]
(A) $CdCl_2$ (B) $ZnCl_2$ (C) $COCl_2$ (D) $CuCl_2$

39. The reaction of primary amine with chloroform and ethanolic KOH, is called [AIIMS-98]
(A) Carbylamine reaction (B) Kolbe's reaction
(C) Reimer-Tiemann reaction (D) None of these

40. Which of the following compounds is 2-bromotoluene? [AIIMS-98]

(A) 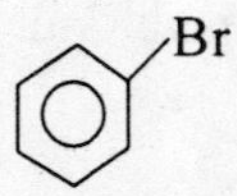

(B) 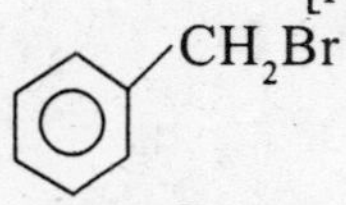

(C) 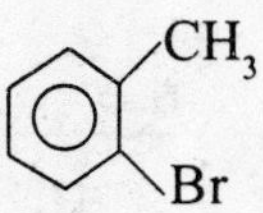

(D) 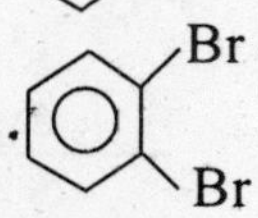

41. The reaction :
benzene + methyl halide $\xrightarrow{\text{Anhy. } AlCl_3}$ Toluene, is known as [AIIMS-98]
(A) Kolbe reaction (B) Wurtz reaction
(C) Perkin's reaction (D) Friedal Craft reaction

42. Identify X in the reaction [AIIMS-2000]
$HCHO + CH_3MgI \xrightarrow{\text{dry ether}} \text{Intermediate} \xrightarrow{H_2O / H^+}$
(A) CH_3OH (B) CH_3CH_2OH
(C) CH_3COCH_3 (D) CH_3OCH_3

43. $CH_3 - COCl + H_2 \xrightarrow{[Pd]} X$. This product X may be [AIIMS-2000]
(A) $CH_3 - CHO$ (B) $CH_3 - CH_2 - OH$
(C) $CH_3 - C(OH)(H) - Cl$ (D) Both (a) and (b)

44. At higher temperature, iodoform reaction is given by - [AIIMS-2003]
(A) $CH_3CO_2CH_3$ (B) $CH_3CO_2C_2H_5$
(C) $C_6H_5CO_2CH_3$ (D) $CH_3CO_2C_6H_5$

45. (A) Hydroxyketones are not directly used in Grignard reaction.
(R) Grignard reagents react with hydroxyl group. [AIIMS-2003]
(A) (B) (C) (D)

46. The reagent with which both acetaldehyde and acetone react easily is : [JIPMER-91]
(A) Fehling's solution (B) Grignard reagent
(C) Schiff's reagents (D) Tollen's reagent

47. Iodoform on warming with silver powder gives ? [JIPMER-92]
(A) Ethane (B) Methane
(C) Acetylene (D) Ethyl alcohol

48. The carbon-magnesium linkage in Grignard reagent is: [JIPMER-92]
(A) A metallic bond (B) Co-ordinate bond
(C) Weakly polar bond (D) Highly polar bond

49. When chloroform is exposed to air and sunlight, it gives: [JIPMER-92]
(A) CCl_4 (B) $COCl_2$ (C) CH_3Cl (D) Mustard gas

50. Chloroform on nitration gives : [JIPMER-92]
(A) Tear gas (B) Water gas
(C) Producer gas (D) Laughing gas

51. The commercially known compound gammexane is : [JIPMER-93]
(A) Freon (B) Chloral
(C) 2-phenyl ethanol (D) Benzene hexachloride

52. In the presence of air or light, chloroform changes to: [JIPMER-93]
(A) Phosgene (B) Acetylene
(C) Chloropicrin (D) Chloretone

53. Which of the following compounds is known as freon? [JIPMER-94]
(A) CCl_2F_2 (B) $CHCl_3$ (C) CH_2F_2 (D) CF_4

54. An organic halide is shaken with aqueous NaOH followed by the addition of dil. HNO_3 and silver nitrate solution gives white ppt. The substance can be [JIPMER-95]
(A) C_6H_5Cl (B) $C_6H_4(CH_3)Br$
(C) $C_6H_5CH_2Cl$ (D) None of these

55. A paste of bleaching powder on heating with ethanol gives [JIPMER-96]
(A) Chloroform (B) Ethylene
(C) Carbon tetrachloride (D) None of these

56. Chloroform when treated with acetone gives [JIPMER-99]
(A) Chloroprene (C) Chloretone
(C) Carbylamine (D) Acetylene

57. Carbon tetrachloride upon treatment with Fe and dilute HCl gives [JIPMER-99]
(A) Methane (B) Methylenechloride
(C) Methylchloride (D) Chloroform

58. Which chloro-derivative of benzene among the following would undergo hydrolysis, most readily with aqueous sodium hydroxide to furnish the corresponding hydroxy derivative ? [JIPMER-99]

(A) C_6H_6Cl (B) 

(C) 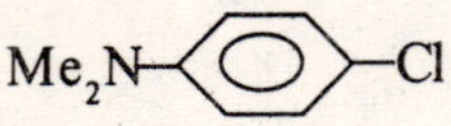 (D) 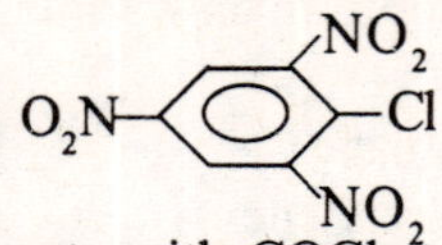

59. Which of the following reacts with $COCl_2$ to give phenyl isocyanate [JIPMER-2000]
(A) Chlorobenzene (B)Nitrobenzene
(C) Aminophenol (D)Aniline

60. Chlorobenzene is prepared commercially by [JIPMER-2000]
(A) Raschig process (B) Wurtz Fittig reaction
(C) Friedel-Craft's reaction (D) Grignard's reaction

61. Vinyl chloride reacts with HCl to form [JIPMER-2000]
(A) 1,1-dichloro ethane (B) 1,2-dichloro ethane
(C) Tetrachloro ethylene
(D) Mixture of 1,2 and 1,1-dichloro ethane.

62. HBr reacts fastest with [JIPMER-2000]
(A) 2-methylpropan-1-o1 (B) Propan 2-o1
(C) Propan-1-o1 (D) 2-methylpropan-2-o1

63. When acetone reacts with bleaching powder, the compound formed is [JIPMER-2001]
(A) CH_3COCH_3 (B) CH_3CHO
(C) $CHCl_3$ (D) CCl_4

64. Chloroform reacts with NaOH to give [JIPMER-2002]
(A) CH_3 COONa (B) Sodium Oxalate
(C) CH_3 OH (D) HCOONa

65. War gas is formed from [BHU-95]
(A) PH_3 (B) C_2H_2
(C) Zinc phosphate (D) Chloropicrin

66. Grignard's reagent, on reaction with acetone forms [BHU-95]
(A) Tertiary alcohol (B) Secondary alcohol
(C) Acetic acid (D) Acetaldehyde

67. Iodoform test is not given by [BHU-95]
(A) Acetone (B) Ethyl alcohol
(C) Acetic acid (D) None of these

68. Which of the following has highest chlorine content? [BHU-98]
(A) Pyrene (B) DDT (C) Chloral (D) Gammexane

69. DDT is prepared by reacting chlorobenzene with [BHU-98]
(A) CCl_4 (B) CCl_3-CHO (C) $CHCl_3$ (D) Ethane

70. Carbon percentage is minimum in [BHU-98]
(A) Pyrene (B) Gammexane
(C) Ethylene glycol (D) PVC

71. Chloroform on treatment with concentrated HNO_3, gives [BHU-99]
(A) CCl_3NO_2 (B) $CHCl_2HNO_3$
(C) $CHCl_2NO_2$ (D) $CHCl_2NO_3$

72. Which of the following is formed by the oxidation of chloroform? [BHU-2000]
(A) CH_3CO_2 (B) $COCl_2$ (C) CO_2 (D) O_2

73. Carboxylic acid is formed by the hydrolysis of an addition compound formed by reaction of Grignard's reagent with [BHU-2001]
(A) Methyl alcohol (B) Methanal
(C) Carbon dioxide (D) Methane

74. When chloroform is exposed to air and sunlight it gives [BHU-2001]
(A) Phosgene (B) Lewisite
(C) Mustard gas (D) Carbon tetrachloride

75. In the reaction

$$CH_3-\underset{\underset{}{}}{\overset{\overset{\displaystyle Cl}{|}}{CH}}-CH_3 \xrightarrow[\text{ether}]{\text{Mg}} A \xrightarrow{D_2O} B$$

identify B [BHU-2002]
(A) $CH_3-CH_2-CH_2-D$
(B) $CH_3-CH_2-CH_2-Mg(OD)$
(C) $CH_3-\underset{\underset{\displaystyle D}{|}}{CH}-CH_3$ (D) none reaction from A to B.

76. Which of the following does not give iodoform test? [BHU-2002]
(A) Acetophenone (B) Benzophenone
(C) $CH_3-\underset{\underset{\displaystyle CH_3}{|}}{CHOH}$ (D) $CH_3-\underset{\underset{\displaystyle CH_3}{|}}{CH}-CH_3-CH_2-CH_3$

77. Which of the following is used as refrigerant? [AFMC-92]
(A) CCl_4 (B) Acetylene
(C) Benzene (D) Ethyl alcohol

78. Gammexane is [AFMC-92]
(A) Benzene (B) Benzene hexachloride
(C) Nitrobenzene (D) Benzaldehyde

79. D.D.T. is formed by [AFMC-92]
(A) Chloral and bromobenzene
(B) Bromo and chlorobenzene (C) Chloral and benzene
(D) Chloral and chlorobenzene

80. Which of the following is used in fire extinguishers? [AFMC-93]
(A) CH_4 (B) $CHCl_3$ (C) CH_2Cl_2 (D) CCl_4

81. What is formed when in presence of alkali, chloroform is treated with acetone? [AFMC-94]
(A) Chloretone (B) Chloral
(C) Carbon tetrachloride (D) None of these

82. Gammexane is the name of [AFMC-94]
(A) DDT (B) Benzene hexachloride
(C) Chlorobenzne (D) None of these

83. When ethyl chloride and alcoholic KOH are heated, the compound obtained is [AFMC-94]
(A) C_2H_4 (B) C_2H_5OH (C) $C_2H_5OC_2H_5$ (D) C_2H_6

84. Cl_2 reacts with CS_2 in presence of I_2 to form [AFMC-95]
(A) $CHCl_3$ (B) CCl_4 (C) C_2H_5Cl (D) C_2H_6

85. CCl_4 is used as fire extinguisher because [AFMC-96]
(A) High melting point (B) Due to covalent bond
(C) Low boiling point (D) Gives incombustible vapour

86. Chloroform on warming with Ag powder gives [AFMC-96]
(A) C_6H_6 (B) C_2H_4 (C) C_2H_2 (D) None of these

87. 1-Chlorobutane on reaction with alcohlic potash (KOH), gives [AFMC-98]
(A) 1-butene (B) 1-butanol
(C) 2-butene (D) 1- butanol

88. Chloroform on treatment with concentrated HNO_3 gives [AFMC-98]
(A) $CHCl_2NO_2$ (B) $CHCl_2HNO_3$
(C) CCl_3NO_2 (D) $CHCl_2NO_3$

89. When primary alcohol is oxidised with chlorine, it gives [AFMC-99]
(A) CCl_3CHO (B) C_3H_7CHO
(C) HCHO (D) CH_3CHO

90. Chloroform on treatment with concentrated HNO_3, gives [AFMC-99]
(A) $CHCl_2NO_2$ (B) CCl_3NO_2
(C) $CHCl_2NO_3$ (D) $CHCl_2HNO_3$

91. The active constituent of bleaching powder is [AFMC-2000]
(A) $Ca(OCl)_2$ (B) $Ca(OCl)Cl$
(C) $Ca(ClO_2)_2$ (D) $Ca(ClO_2)Cl$

92. Chloroform when reacts with nitric acid gives [AFMC-2000]
(A) $C_2H_5C(NO)_5SH$ (B) CCl_3CHOH
(C) CCl_3 NO_2 (D) CCl_3NO_3

93. When aniline in hydrochloric acid is treated with ice cold solution of sodium nitrite, it gives [AFMC-2000]
(A) Chlorobenzene (B) Chloroaniline
(C) Nitroaniline (D) Benzene diazonium chloride

94. For the carbylamine reaction we need hot alcoholic KOH and [AFMC-2002]
(A) Any amine and chloroform
(B) Chloroform and silver powder
(C) A primary amine and an alkyl halide
(D) A monoalkylamine and trichloromethane

95. Grignard's reagent is prepared by the reaction between [AFMC-2002]
(A) Magnesium and alkyl halide
(B) Zinc and alkyl halide (C) Magnesium and alkane
(D) Magnesium and aromatic hydrocarbon

96. Reaction of $CH_2 - CH_2$ (bridged by O) with R-MgX leads to the formation of [AFMC-2002]
(A) R - CH - OH - R (B) R - $CHOHCH_3$
(C) R_2 > C - H - CH_2 - OH (D) R - CH_2 - CH_2 - OH

97. Which of the following will not give iodofom test ? [AFMC-2002]
(A) Ethanol (B) Ethanal
(C) Isopropyl alcohol (D) Benzyl alcohol

98. Aryl halides are less reactive towards nucleophilic substitution reaction as compared to alkyl halides due to [IIT-90]
(A) the formation of less stable carbonium ion
(B) resonance stabilization
(C) longer carbon-halogen bond
(D) the inductive effect
(E) sp^2- hybridised carbon attached to the halogen.

99. 1-Chlorobutane on reaction with alcoholic potash gives [IIT-91]
(A) 1-butene (B) 1-butanol
(C) 2-butene (D) 2-butanol

100. Butanenitrile may be prepared by [IIT-92]
(A) Propyl alcohol with KCN
(B) Butyl alcohol with KCN (C) Butyl chloride with KCN
(D) Propyl chloride with KCN

101. Arrange the following compounds in order of increasing dipole moment . Toluene (I), m-dichloro benzene (II), o-dichloro benzene (III), p-dichloro benzene(IV) [IIT-96]
(A) I < IV < II < III (B) IV < I < II < III
(C) IV < I < III < II (D) IV < II < I < III

102. During debromination of meso-dibromobutane, the major compound formed is [IIT-97]
(A) n-butane (B) 1-butene
(C) cis-2-butene (D) trans-2-butene

103. In the reaction of p-chlorotoluene with KNH_2 in liq. NH_3, the major product is : [IIT-97]
(A)o-toluidine (B) m-toluidine
(C)p-toluidine (D) p-chloroaniline.

104. $(CH_3)_3$ CMgCl on reaction with D_2O produces :

[IIT-97]

(A) $(CH_3)_3CD$ (B) $(CH_3)_3$ OD
(C) $(CD_3)_3CD$ (D) $(CD_3)_3OD$

105. Which of the following will react with water ? [IIT-98]

(A) $CHCl_3$ (B) Cl_3 CCHO
(C) CCl_4 (D) $ClCH_2$ CH_2 Cl

106. The products of reaction of alcoholic silver nitrite with ethyl bromide are: [IIT-98]

(A) Ethane (B)ethene (C) nitroethane
(D) Ethyl alcohol (E) ethyl nitrite

107. Which of the following will react with water ? [IIT-98]

(A) $CHCl_3$ (B) Cl_3CCHO (C) CCl_4 (D) $ClCH_2CH_2Cl$

108. Benzyl chloride ($C_6H_5CH_2Cl$) can be prepared from toluene by chlorination with [IIT-98]

(A) SO_2Cl_2 (B) $SOCl_2$ (C) Cl_2 (D) NaOCl

109. A new carbon-carbon bond formation is possible in [IIT-98]

(A) Cannizaro reaction (B) Friedel Craft's alkylation
(C) Clemmensen reduction (D) Reimer-Tiemann reaction

110. p–Chloroaniline and a hydrochloride can be distinguished by [IIT-98]

(A) Sandmeyer's reaction (B) $NaHCO_3$
(C) $AgNO_3$ (D) Carbylamine test

111. Benzoyl chloride is prepared from benzoic acid by [IIT-2000]

(A) Cl_2, hν (B) SO_2Cl (C) $SOCl_2$ (D) Cl_2, H_2O

112. Identify the set of reagents / reaction conditions 'X' and 'Y' in the following set of transformations [IIT-2002]

$CH_3-CH_2-CH_2Br \xrightarrow{X}$ Product $\xrightarrow{Y} CH_3-\underset{\underset{Br}{|}}{CH}-CH_3$

(A) X = dilute aqueous NaOH, 20°C; Y = HBr/acetic acid, 20°C
(B) X = concentrated alcoholic NaOH, 80°C; Y = HBr/ acetic acid, 20°C
(C) X = dilute aqueous NaOH, 20°C; Y = $Br_2/CHCl_3$, 0°C
(D) X = concentrated alcoholic NaOH, 80°C; Y = $CHCl_3$, 0°C

113. CH_3MgBr + Ethyl ester → which can be formed as product (excess) [IIT-2003]

(A) 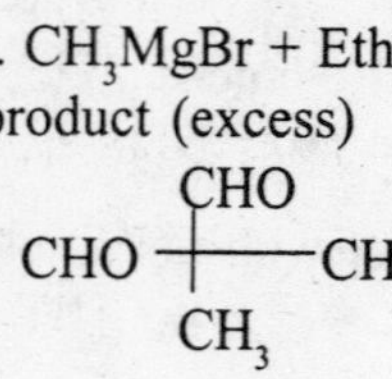 (B) 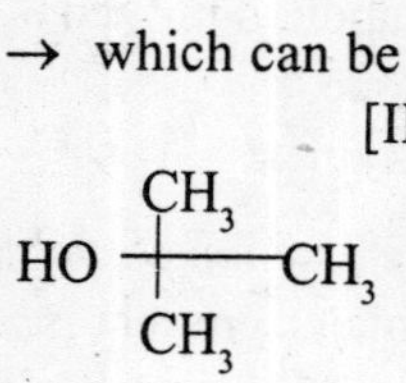

(C) 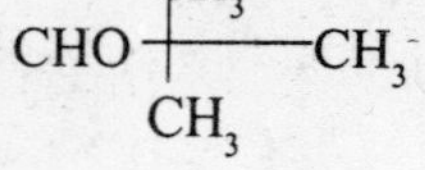 (D) 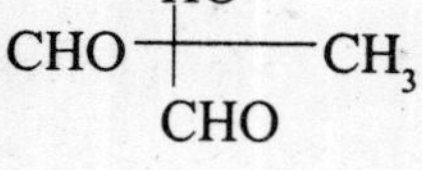

114. CH_3–Mg – Br is an organo metallic compound due to [AIEEE-2002]

(A) Mg – Br bond (B) C – Mg bond
(C) C – Br bond (D) C – H bond

115. The reaction of chloroform with alcoholic KOH and p-toluidine forms [AIEEE-2003]

(A) 

(B) 

(C) 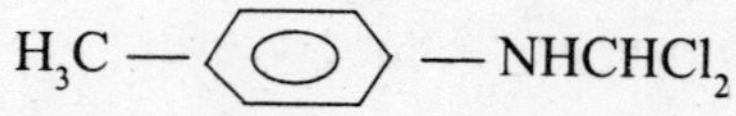

(D) 

116. Chloroform reacts with HNO_3 to give [EAMCET-91]

(A) Methane (B) Chloropicrin
(C) Nitro Methane (D) Nitrogen

117. Phosgene is formed when ....is exposed to air and sunlight. [EAMCET-91]

118. When ethyl alcohol reacts with bromine in the presence of red phosphorus the compound formed is [EAMCET-92]

(A) C_2H_6 (B) PBr_3 (C) CH_3Br (D) C_2H_5Br

119. Chloroform slowly reacts with oxygen of the air in presence of light to give .... [EAMCET-92]

120. Which of the following poisonous gases is formed when chloroform is exposed to light and moist air ? [EAMCET-93]

(A) Mustard gas (B) Phosgene
(C) Chlorine (D) Carbon monoxide

121. Which of the following compounds does not give a positive iodoform test ? [EAMCET-95]

(A) Ethanol (B) Acetone
(C) Acetaldehyde (D) Methanol

122. Which one of the following reagents and conditions convert benzene to chlorobenzene ? [EAMCET-96]

(A) Cl_2, sunlight, heat (B) HCl, heat
(C) HCl, sunlight, heat (D) Cl_2, $AlCl_3$, heat

123. Alkyl halide reacting with metallic sodium in dry ether solution is called [EAMCET-96]

(A) Friedel-Craft's reaction (B) Sandmeyer reaction
(C) Wurtz reaction (D) Gabriel's reaction

124. Aniline when heated with $NaNO_2$ and HCl at 0-5 °C the product formed is [EAMCET-96]

(A) Chloro aniline (B) Benzene diazonium chloride
(C) Chloro benzene (D) Dichloro benzene

125. The hybridization state of carbon atoms in the product formed by the reaction of ethyl chloride with aqueous KOH is [EAMCET-97]

(A) sp (B) sp^2 (C) sp^3 (D) sp^3d

126. Reaction of benzene with alkyl halide in the presence of anhydrous $AlCl_3$ is called [EAMCET-98]
(A) Friedel - Craft's reaction (B) Wurtz reaction
(C) Williamson's synthesis (D) Baeyer's reaction

127. Which one of the following reactions does not yield an alkyl halide ? [EAMCET-98]
(A) Diethyl ether + Cl_2 (B) Diethyl ether + HI
(C) Diethyl ether + PCl_5 (D) Diethyl ether $\xrightarrow{\text{Reduction}}$ SO_2Cl_2

128. Hydrolysis of trichloromethane with aqueous KOH gives [EAMCET-98]
(A) Methanol (B) Chloral
(C) Acetylene (D) Pot. formate

129. Which of the following does not participate in the carbylamine reaction ? [EAMCET-98(MBBS)]
(A) Aniline (B) Chloroform (C) Ethanol (D) KOH

130. Hydrolysis of chloroform with aqueous potassium hydroxide gives finally [EAMCET-99]
(A) $COCl_2$ (B) CH_3NC (C) HCOOK (D) $CH_2(Cl)OH$

131. Chloroethane is reacted with alcoholic potassium hydroxide. The product formed is [EAMCET-99]
(A) C_2H_6O (B) C_2H_6 (C) C_2H_4 (D) C_2H_4O

132. Aniline reacts with chloroform in the presence of alkali to give a product with very bad smell. This test is called [EAMCET-99(MBBS)]
(A) Isocyanide test (B) Silver mirror test
(C) Baeyer's test (D) Lucas test

133. Which of the following reacts with ethanol to form chloroform ? [EAMCET-2000]
(A) $SOCl_2$ (B) PCl_5 (C) HCl (D) $CaOCl_2/H_2O$

134. Chloroethane reacts with magnesium in dry ether to form X. When X is hydrolysed, a carbon compound Y and Z are formed. Which of the following is Y ? [EAMCET-2000]
(A) C_2H_4 (B) C_2H_2 (C) C_2H_6 (D) C_6H_6

135. Which of the following reacts with water to give ethane? [EAMCET-2000]
(A) CH_4 (B) C_2H_5MgBr
(C) C_2H_5OH (D) $C_2H_5OC_2H_5$

136. Aniline on heating with X in the presence of KOH gives a product with very bad smell. Which of the following is X ? [EAMCET(MBBS)-2000]
(A) CH_3Cl (B) CCl_4 (C) $CHCl_3$ (D) C_2H_5Cl

137. Chloral hydrate is dissolved in NaOH solution and distilled. What are the compounds obtained ? [EAMCET(MBBS)-2000]
(A) CH_2Cl2, NaCl (B) CH_3Cl, CH_3COONa
(C) $CHCl_3$, HCOONa, H_2O (D) C_2H_5Cl, CH_3COONa

138. Which one of the following is a secondary alkyl halide ? [EAMCET(MBBS)-2001]
(A) $(CH_3)_3CCl$ (B) C_2H_5Cl
(C) $(CH_3)_2CHCl$ (D) $(CH_3)_2CHCH_2Cl$

139. In chloroethane, the carbon bearing halogen is bonded to ... hydrogen (s). It is called... alkyl halide. [EAMCET-2002]
(A) Two, primary (B) Three, primary
(C) Two, secondary (D) One, tertiary

140. The reaction conditions used for converting 1, 2-dibromo ethane to ethylene are : [EAMCET-2002]
(A) Zn, alcohol, Δ (B) KOH, alcohol, Δ
(C) KOH, water, Δ (D) Na, alcohol, Δ

141. The reactants and reaction conditions used in the preparation of ethane are [EAMCET(med.)-2002]
(A) $2CH_3I$, Na/C_2H_5OH
(B) Electrolysis of $C_2H_5COO^-K^+$
(C) C_2H_4, H_2 27°C (D) C_2H_5I, H_2, Zn – Cu, C_2H_5OH

142. What is the density (in g. lit^{-1}) of CO_2 at 400 K and exerting a pressure of 0.0821 atm ?(R = 0.0821 lit. atm. $mol.^{-1}K^{-1}$) [EAMCET(med.)-2002]
(A) 0.01 (B) 0.11 (C) 2.5 (D) 44

143. Which one of the following compounds converts methyl magnesium iodide to methane in one step ? [EAMCET(med.)-2002]
(A) $C_2H_5OC_2H_5$ (B) C_2H_5Cl (C) C_2H_4 (D) C_2H_5OH

144. Consider the following reactions

$$X + HCl \xrightarrow[\text{addition}]{\text{Anhydrous } AlCl_3} C_2H_5Cl \xleftarrow[\text{substitution}]{\text{Anhydrous } ZnCl_2/HCl} Y$$

Y can be converted to X on heating with........... at .... temperature. [EAMCET-2003]
(A) Al_2O_3, 350°C (B) Cu, 300°C
(C) $Ca(OH)_2$, $CaOCl_2$, 60°C (D) $NaOH/I_2$, 60°C

145. Which of the following shows only one brominated compound ? [CPMT-96]
(A) Butene-2 (B) 2,2-dimethyl propane
(C) Butyne-1 (D) Butanol-3

146. When CH_3COCH_3 reacts with Cl_2 which of the following is formed ? [CPMT-96]
(A) $CHCl_3$ (B) CCl_4 (C) CCl_2H_2 (D) CH_3Cl

147. Which of the following is an example of Friedel-Craft's reaction ? [CPMT-96]
(A) $C_6H_6 + Cl_2 \xrightarrow{AlCl_3} C_6H_5Cl + HCl$
(B) $C_6H_5{+}Cl{+}C_6H_5COCl \xrightarrow{AlCl_3} C_6H_5COC_6H_5{+}Cl_2$
(C) $CH_3COOAg \xrightarrow{CCl_4} CH_3Br + CO_2 + AgBr$
(D) $C_2H_5OH \xrightarrow{H_2SO_4} C_2H_5OC_2H_5$

148. When primary alcohol is oxidized with chlorine, it gives [CPMT-97]

(A) CH_3CHO (B) CCl_3CHO
(C) CH_3COCH_3 (D) Cl_3CHO

149. Which of the following will not give iodoform test ? [CPMT-99]
(A) Ethanol (B) Ethanal
(C) Isopropyl alcohol (D) Benzyl alcohol

150. Which is the carbylamine reaction ? [CPMT-2000]
(A) $C_2H_5NH_2 + 3KOH + CHCl_3 \rightarrow C_2H_5NC + 3KCl + 3H_2O$
(B) $C_6H_6 + CH_3Cl \xrightarrow[AlCl_3]{Anhydrous} C_6H_5CH_3 + HCl$
(C) $2HCHO + NaOH \rightarrow CH_3OH + HCOONa$
(D) $2CH_3Cl + 2Na \rightarrow CH_3—CH_3 + 2NaCl$

151. Which of the following will give iodoform test [CPMT-2001]
(A) $CH_3CH_2CH_2OH$ (B) $CH_3CH_2COCH_2CH_3$
(C) $CH_3-CH(CH_3)-CO-CH_2-CH_3$
(D) $CH_3-CH_2-CH(OH)-CH_2-CH_3$

152. Which is used in the preparation of azo-dyes ? [CPMT-2001]
(A) Nitrobenzene (B) Aniline
(C) Benzaldehyde (D) Chlorobenzene

153. Which of the following species participate in sulphonation of benzene ? [CPMT-2002]
(A) SO_2 (B) SO_3H^+ (C) SO_3 (D) SO_3H^-

154. $AgNO_3$ does not give precipitate with $CHCl_3$ because [CPMT-2002]
(A) $CHCl_3$ does not ionise in water
(B) $CHCl_3$ does not react with $AgNO_3$
(C) $CHCl_3$ is chemically inert (D) none of these

155. Iodoform test is not given by [CPMT-2003]
(A) 2-pentanone (B) 3-pentanone
(C) Ethanol (C) none of these

156. When chloroform reacts with ethylamine in presence of alcoholic KOH, the compound formed is [PMT-93]
(A) Ethyl cyanide (B) Ethyl isocyanide
(C) Formic acid (D) Phosgene

157. Heating together of sodium ethoxide and ethyl iodide will give [PMT-94]
(A) Ether (B) Ethyl Alcohol
(C) Acetaldehyde (D) Acetic Acid

158. The reaction between a primary amine, chloroform and alcoholic KOH is called [PMT-94]
(A) Wurtz Reaction (B) Frankland Reaction
(C) Cannizaro's Reaction (D) Carbylamine Reaction

159. Phosgene is common name of [PMT-94]
(A) Carbon dioxide and phosgene
(B) Phosphoryl chloride
(C) Carbonyl chloride (D) Carbon tetrachloride

160. C_2H_5OH can be differentiated from CH_3OH by [PMT-94]
(A) Reaction with HCl (B) Reaction with NH_3
(C) By iodoform test (D) By solubility in water

161. Benzene hexachloride is used as [PMT-94]
(A) Dye (B) Antimalarial drug
(C) Antibiotic (D) Insecticide

162. Carbon-halogen bond is strongest among the following : [PMT-95]
(A) CH_3Cl (B) CH_3Br (C) CH_3F (D) CH_3I

163. Which of the following reacts with Phenol to give Salicylaldehyde after hydrolysis ? [PMT-95]
(A) Dichloro methane (B) Trichloro methane
(C) Methyl chloride (D) None of these

164. The C-Cl bond in chlorobenzene as compared with C-Cl bond in methyl chloride is [PMT-95]
(A) longer and weaker (B) shorter and weaker
(C) shorter and stronger (D) longer and stronger

165. The molecular formula of bleaching powder is [PMT-96]
(A) $CaCl_2$ (B) Ca_2OCl_2
(C) $CaOCl_2$ (D) $CaCl_2.CaOCl$

166. Dehydrohalogenation of an alkyl halide is [PMT-96]
(A) an addition reaction (B) a substitution reaction
(C) an elimination reaction (D) an oxidation reaction

167. Reaction of an alkyl halide with sodium alkoxide yields ? [PMT-96]
(A) an alcohol (B) a ketone
(C) an ether (D) a carboxylic acid

168. Which one of the following does NOT give iodoform test ? [PMT-96]
(A) Methyl alcohol (B) Ethyl alcohol
(C) Acetaldehyde (D) Acetone

169. The products of the reaction between $C_6H_5NH_2$, $CHCl_3$ and KOH are [PMT-96]
(A) $C_6H_5NC + KCl$ (B) $C_6H_5OH + NH_4Cl + H_2O$
(C) $C_6H_5Cl + NH_4Cl + KCl$ (D) $C_6H_5CN + KCl$

170. The chemical name of the insecticide Gammexane is [PMT-96]
(A) D. D. T. (B) Benzene hexachloride
(C) Hexachloro benzene (D) Chloral

171. Aniline reacts with chloroform in the presence of KOH to give - [PMT-97]
(A) Phenol (B) Chlorobenzene
(C) Phenyl cyanide (D) Phenyl isocyanide

172. Which of the following does not give yellow precipitate with NaOH + KI - [PMT-97]
(A) Acetone (B) Acetaldehyde
(C) Benzaldehyde (D) Acetophenone

173. Phenol reacts with $CHCl_3$ and NaOH (at 340 K) to give - [PMT-97]
(A) o-chlorophenol (B) Salicylaldehyde
(C) Benzaldehyde (D) Chlorobenzene

174. The molecular formula of DDT has - [PMT-97]
(A) 5 chlorine atoms (B) 4 chlorine atoms
(C) 3 chlorine atoms (D) 2 chlorine atoms

175. For a given alkyl group the densites of the halides follow the order : [PMT-97]
(A) RI < RBr < RCl (B) RI < RCl < RBr
(C) RBr < RI < RCl (D) RCl < RBr < RI

176. n-propyl chloride reacts with alcoholic KOH to give [PMT-97]
(A) Propylene (B) Propyl alcohol
(C) Propyne (D) 2-methyl propene

177. Acetaldehyde reacts with Cl_2 (in excess) to give [PMT-97]
(A) Chloral (B) Chloroform
(C) Acetic acid (D) Trichloroacetic acid

178. Which of the following gives a positive iodoform test? [PMT-99]
(A) CH_3OH (B) $(CH_3)_2CHOH$
(C) $(CH_3)_3CHO$ (D) C_3H_7OH

179. A compound X on reaction with chloroform and NaOH gives a compound with a very unpleasant odour. X is : [PMT-99]
(A) $C_6H_5CONH_2$ (B) $C_6H_5NH_2$
(C) $C_6H_5CH_2NHCH_3$ (D) $C_6H_5NHCH_3$

180. The reaction of CH_3MgI with acetone and hydrolysis of the resulting product gives : [PMT-99]
(A) $CH_3CH_2CH_2OH$ (B) $CH_3CH_2CH_2CH_2OH$
(C) $(CH_3)_2CHOH$ (D) $(CH_3)_3COH$

181. Iodoform on heating with KOH gives :[PMT-2000]
(A) CH_3CHO (B) CH_3COOK
(C) HCOOK (D) HCHO

182. When chloroform reacts with aniline and alcoholic KOH it gives : [PMT-2001]
(A) Phenyl cyanide (B) Phenyl isocyanide
(C) Phenyl cyanate (D) Phenyl isocyanate

183. Reduction of acetyl chloride with H_2 in presence of Pd gives : [PMT-2001]
(A) $CH_3\ COCH_3$ (B) $C_2\ H_5\ OH$
(C) $CH_3\ COOH$ (D) $CH_3\ CHO$

184. When methyl bromide is heated with Zn it gives : [PMT-2001]
(A) CH_4 (B) $C_2\ H_6$ (C) $C_2\ H_4$ (D) $CH_3\ OH$

185. A reaction of 1-bromo-3-chlorocyclobutane with Na in dioxane gives

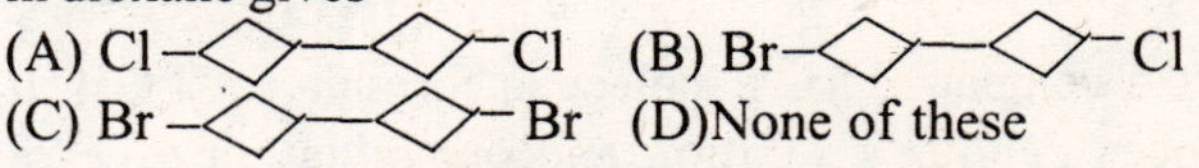

(D)None of these

186. Acetone and chloroform react in presence of a strong base to yield : [PMT-2003]
(A) $(CH_3)_2C(OH)(COOH)$ (B) $CH_3 - CO - CCl_3$
(C) $CCl_3 - \underset{\underset{O}{\|}}{C} - CCl_3$ (D) $(CH_3)_2C(CCl_3)(OH)$

187. When chloroform is exposed to light and air it gives [PET-93]
(A) Chlorine gas (B) Carbon tetrachloride
(C) Methyl chloride (D) Phosgene gas

188. Which one of the following compounds will give a positive haloform test and a positive Fehling test [PET-94]
(A) Acetone (B) Acetic acid
(C) Ethanol (D) Acetadehyde

189. Both methane and ethane may be obtained in one step reaction from [PET-95]
(A) C_2H_4 (B) CH_3I (C) CH_3OH (D) C_2H_5OH

190. $C_6H_6 + CH_3Cl \xrightarrow[\text{AlCl}_3]{\text{Anhy-drous}} C_6H_5CH_3 + CHl$ is an example of - [PET-95]
(A) Friedel Craft's reaction (B) Kolbe's synthesis
(C) Wurtz reaction (D) Grignard's reaction

191. Iodoform is formed on warming I_2 and NaOH with [PET-95]
(A) C_2H_5OH (B) CH_3OH (C) HCOOH (D) C_6H_6

192. Which of the following does not give yellow precipitate with I_2 and NaOH ? [PET-96]
(A) C_2H_5OH (B) CH_3CHO
(C) CH_3COCH_3 (D) HCHO

193. Identify Z in the following series : [PET-97]

$C_2H_5I \xrightarrow{KOH} X \xrightarrow{Br_2} Y \xrightarrow{KCN} Z$

(A) CH_3CH_2CN (B) $CN-CH_2-CH_2CN$
(C) $Br-CH_2-CH_2CN$ (D) $Br-CH = CHCN$

194. If chloroform is left open in air in the presence of sunrays it gives : [PET-97]
(A) Carbon tetrachloride (B) Carbonyl chloride
(C) Mustard gas (D) Lewisite

195. Which of the following compounds gives a ketone with Grignard's reagent ? [PET-97]
(A) Formaldehyde (B) Ethyl alcohol
(C) Methyl cyanide (D) Methyl iodide

196. Which one of the following compounds gives a posi-

tive iodoform test ? [PET-97]

(A) Pentanal (B) 1-Phenyl ethanol
(C) 2-Phenyl ethanol (D) 3- Pentanol

197. A compound has a molecular formula C_2Cl_3OH. It reduces Fehling's solution and on oxidation gives a monocarboxyclic acid (B). A is obtained by action of chlorine on ethyl alcohol. A is : [PET-97]

(A) Chloral (B) $CHCl_3$
(C) CH_3Cl (D) Chloroacetic acid

198. 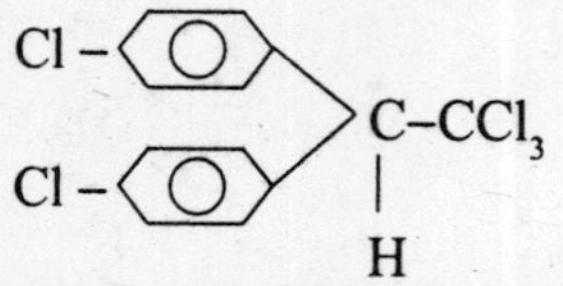

The above structural formula refers to : [PET-97]

(A) BHC (B) DNA (C) DDT (D) RNA

199. Ethyl bromide on treatment with alcoholic KOH gives : [PET-99]

(A) Ethylene (B) Ethanol (C) Acetic acid (D) Ethane

200. $AgNO_3$ does not give precipitate with chloroform because : [PET-99]

(A) $CHCl_3$ is insoluble in water
(B) $CHCl_3$ does not ionize in water
(C) $CHCl_3$ is an organic compound
(D) $AgNO_3$ is insoluble in $CHCl_3$

201. Which of the following alkyl halides is used as a methylating agent ? [PET-99]

(A) C_2H_5Cl (B) C_2H_5Br (C) C_2H_5I (D) CH_3I

202. Ethyl acetate reacts with CH_3MgBr to form : [PET-99]

(A) Secondary alcohol (B) Tertiary alcohol
(C) Primary alcohol and acid (D) Acid

203. When chloroform is left open in light and air, it forms : [PET-2000]

(A) Methylene chloride (B) Formic Acid
(C) Carbon tetrachloride (D) Phosgene

204. Distinction is made between ethyl alcohol and methyl alcohol by : [PET-2000]

(A) Chloroform test (B) Victor Meyer test
(C) Rate of esterification (D) Iodoform test

205. Chloroform on reduction with Zinc and HCl forms : [PET-2000]

(A) Formic acid (B) Chloropicrine
(C) Methylene chloride (D) Phosgene

206. From which of the following tertiary butyl alcohol is obtained by the action of methyl magnesium iodide ? [PET-2000]

(A) HCHO (B) CH_3CHO
(C) CH_3COCH_3 (D) CO_2

207. Which compound is not inflammable ? [PET-2001]

(A) CCl_4 (B) C_2H_5OH (C) CH_4 (D) C_6H_6

208. Alkyl halides react with Mg in dry ether to form : [PET-2001]

(A) Magnesium halide (B) Grignard's reagent
(C) Alkene (D) Alkyne

209. Why is chloroform put into dark coloured bottles [PET-2002]

(A) To prevent evaporation (B) To prevent from moisture
(C) To prevent it from oxidation to form phosgene
(D) To prevent its reaction with glass

210. The conversion of CH_3NH_2 into CH_3NC with alc. KOH and $CHCl_3$ is termed as [PET-2002]

(A) Friedel-Craft's reaction (B) Carbylamine reaction
(C) Wurtz reaction (D) Kolbe's reaction

211. D.D.T. stands for [PET-2002]

(A) Dichloro diphenyl trichloroethane
(B) Dihydro dichloro trichloroethane
(C) Diphenyl dimethyl trichloroethane
(D) Diamino diphenyl toluene

212. The diazonium salts are the reaction products in presence of excess of mineral acid with nitrous acid and [PET-2002]

(A) Primary aliphatic amine (B) Secondary aliphatic amine
(C) Primary aromatic amine (D) Tertiary aliphatic amine

213. Which compound requires chloral in its synthesis? [PET-2003]

(A) Mischler's ketone (B) D.D.T.
(C) Aldrin (D) Gammexane

214. The hydrolysis of alkyl substituted chlorosilanes gives .... [EAMCET-2003]

215. Vinyl chloride on reaction with dimethyl copper gives .... [EAMCET-2003]

216. The chlorobenzene is generally obtained from a corresponding diazonium salt by reacting with [AIEEE-2002]

(A) Cu_2Cl_2 (B) $CuSO_4$ (C) Cu (D) $Cu(NH_3)_4$

217. Which of the following substances doesnot give iodoform test? [BHU-2003]

(A) CH_3COCH_3 (B) $CH_3CHOHCH_3$
(C) CH_3OH (D) C_2H_5OH

218. Which of the following is responsible for depletion of the ozone layer in the upper strata of the atmosphere? [CBSE-PMT-2004]

(A)Polyhalogens (B) Ferrocene (C) Fullerenes (D) Freons

219. Chloropicrin is obtained by the reaction of [CBSE-PMT-2004]

(A) steam on carbon tetrachloride
(B) nitric acid on chlorobenzene
(C) chlorine on picric acid (D) nitric acid on chloroform

220. 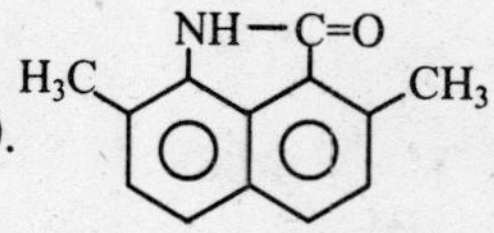 is subjected to

monobromination with Br_2/Fe. The major product is

[IIT-2004]

(A) 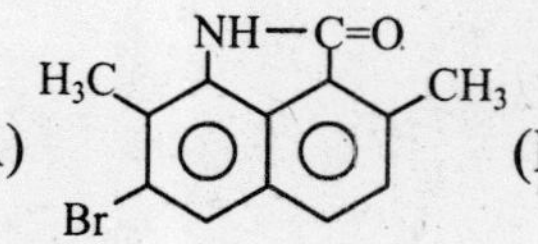 (B) 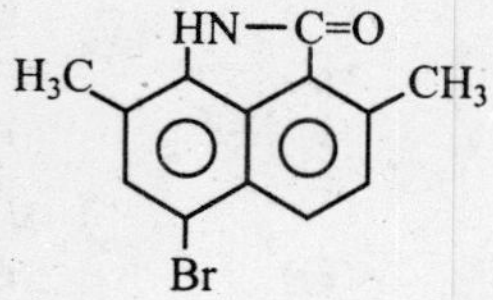

(C) 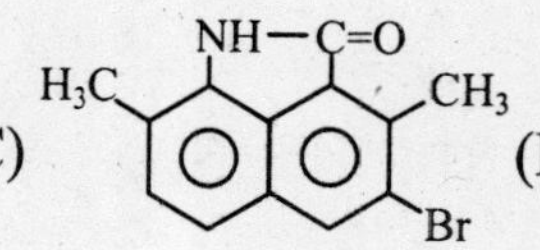 (D) 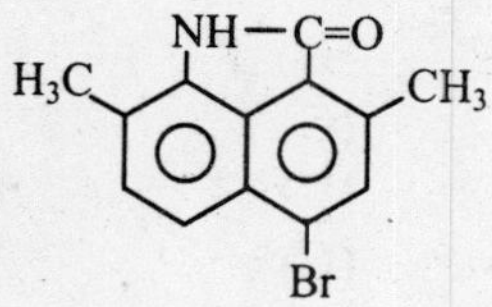

221. The order of reactivity of the following compounds with PhMgBr is

(I) PhCOPh (II) CH_3CHO (III) CH_3COCH_3

[IIT-2004]

(A) (I) > (II) > (III)
(B) (III) > (I) > (II)
(C) (II) > (I) > (III)
(D) (II) > (III) > (I)

Halogen derivatives

1.	(C)	51.	(D)	101.	(B)	151.	(C)	201.	(D)
2.	(D)	52.	(A)	102.	(D)	152.	(D)	202.	(B)
3.	(A)	53.	(A)	103.	(B)	153.	(C)	203.	(D)
4.	(B)	54.	(C)	104.	(A)	154.	(A)	204.	(D)
5.	(B)	55.	(A)	105.	(B)	155.	(B)	205.	(C)
6.	(B)	56.	(B)	106.	(C) (E)	156.	(B)	206.	(C)
7.	(B)	57.	(C)	107.	(B) (D)	157.	(A)	207.	(A)
8.	(A)	58.	(C)	108.	(B) (C)	158.	(D)	208.	(B)
9.	(C)	59.	(D)	109.	(B) (D)	159.	(C)	209.	(C)
10.	(B)	60.	(A)	110.	(C) (D)	160.	(C)	210.	(B)
11.	(B)	61.	(A)	111.	(C)	161.	(D)	211.	(A)
12.	(C)	62.	(D)	112.	(B)	162.	(C)	212.	(C)
13.	(A)	63.	(C)	113.	(B)	163.	(B)	213.	(B)
14.	(B)	64.	(D)	114.	(B)	164.	(C)	214.	*
15.	(D)	65.	(D)	115.	(D)	165.	(C)	215.	*
16.	(D)	66.	(A)	116.	(B)	166.	(C)	216.	(A)
17.	(B)	67.	(C)	117.	*	167.	(C)	217.	(C)
18.	(A)	68.	(A)	118.	(D)	168.	(A)	218.	(B)
19.	(D)	69.	(B)	119.	*	169.	(A)	219.	(D)
20.	(C)	70.	(A)	120.	(B)	170.	(B)	220.	(B)
21.	(B)	71.	(A)	121.	(D)	171.	(D)	221.	(D)
22.	(D)	72.	(B)	122.	(D)	172.	(C)		
23.	(B)	73.	(C)	123.	(C)	173.	(B)		
24.	(A)	74.	(A)	124.	(B)	174.	(A)		
25.	(B)	75.	(C)	125.	(C)	175.	(D)		
26.	(B)	76.	(B)	126.	(A)	176.	(A)		
27.	(B)	77.	(A)	127.	(A)	177.	(A)		
28.	(A)	78.	(B)	128.	(D)	178.	(B)		
29.	(B)	79.	(D)	129.	(C)	179.	(B)		
30.	(D)	80.	(D)	130.	(C)	180.	(D)		
31.	(C)	81.	(A)	131.	(C)	181.	(C)		
32.	(A)	82.	(B)	132.	(A)	182.	(B)		
33.	(C)	83.	(A)	133.	(D)	183.	(B)		
34.	(A)	84.	(B)	134.	(C)	184.	(B)		
35.	(C)	85.	(D)	135.	(B)	185.	(A)		
36.	(B)	86.	(C)	136.	(C)	186.	(D)		
37.	(C)	87.	(A)	137.	(C)	187.	(D)		
38.	(C)	88.	(C)	138.	(C)	188.	(D)		
39.	(A)	89.	(A)	139.	(A)	189.	(B)		
40.	(C)	90.	(B)	140.	(A)	190.	(A)		
41.	(D)	91.	(B)	141.	(D)	191.	(A)		
42.	(B)	92.	(C)	142.	(B)	192.	(D)		
43.	(B)	93.	(D)	143.	(D)	193.	(B)		
44.	(B)	94.	(D)	144.	(A)	194.	(B)		
45.	(A)	95.	(A)	145.	(D)	195.	(C)		
46.	(B)	96.	(D)	146.	(A)	196.	(B)		
47.	(C)	97.	(D)	147.	(B)	197.	(A)		
48.	(D)	98.	(B) (E)	148.	(B)	198.	(C)		
49.	(B)	99.	(A)	149.	(D)	199.	(A)		
50.	(A)	100.	(D)	150.	(A)	200.	(B)		

117. Chloroform
119. phosgene
214. silicones
215. propene

Alkanols

Compounds in which the –OH group is attached to a carbon which is not linked to elements other than carbon and hydrogen are called alcohols.

Monohydric alcohols:

$H_3C{-}CH_2{-}CH_2{-}CH_2OH$ (1°); $H_3C{-}CH_2{-}CHOH{-}CH_3$ (2°);

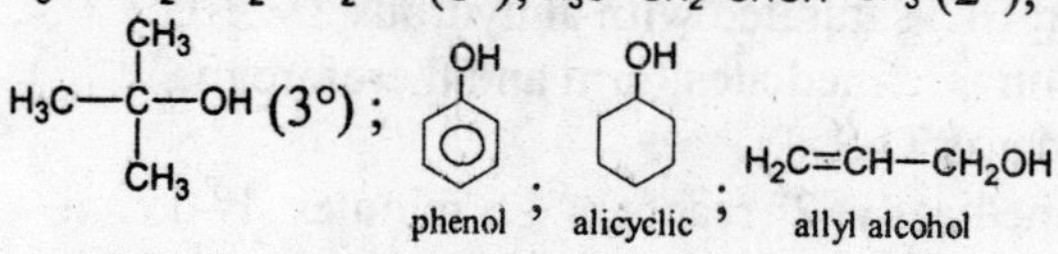

phenol; alicyclic; allyl alcohol

Alcohols generally exhibit:

Chain isomerism: $H_3C{-}CH_2{-}CH_2{-}CH_2OH$, $H_3C{-}CH(CH_3){-}CH_2OH$

Positional isomerism: $H_3C{-}CH_2{-}CH_2OH$, $H_3C{-}CHOH{-}CH_3$

Functional isomerism: $H_3C{-}CH_2OH$, $H_3C{-}O{-}CH_3$

Thioacohol: R–SH (mercaptans)

Preparation:

1. From primary amines: With nitrous acid HNO_2 ($NaNO_2$ / HCl).

$R{-}NH_2 + HONO \longrightarrow ROH + N_2 + H_2O$

Methylamine does not follow this route. The products formed depending upon the concentration of methylamine are

(i) $H_3C{-}NH_2 + HON{=}O \longrightarrow H_3C{-}O{-}N{=}O + NH_3$
methyl nitrite

(ii) $2\,H_3C{-}NH_2 + 2HNO_2 \longrightarrow H_3C{-}O{-}CH_3 + 3H_2O + N_2$
dimethyl ether

Secondary and tertiary alcohols can't be obtained by this reaction. e.g.

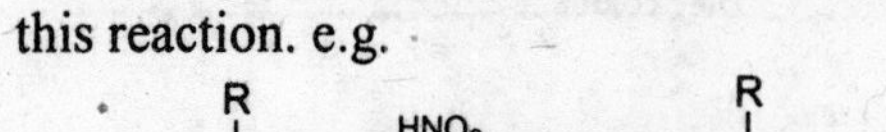

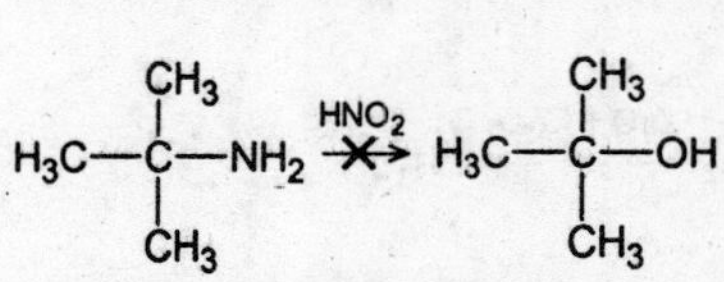

2. From Grignard reactions

Primary alcohols: Oxygen; epoxide or oxirane; or formaldehyde with RMgX.

Secondary alcohols: aldehydes other than formaldehyde; ethyl formate with RMgX.

Tertiary alcohols: Ketones; ethylacetate; acid chloride with RMgX.

3. Reduction:

(i) Bouveault–Blanc reduction:

$$\text{aldehydes; ketones; esters} \xrightarrow[\text{Na \& } C_2H_5OH]{\text{excess of}} \text{alcohol}$$

($C_2H_5OH + Na \longrightarrow C_2H_5ONa + H$ nascent hydrogen produced is responsible for reduction).

Acid derivatives also can form alcohols through this route.

(ii) With $LiAlH_4$: $LiAlH_4$ is a good reducing agent for the reduction of unsaturated ketones, aldehydes etc. as the double bond (i.e. the unsaturation) is not affected.

(iii) With $NaBH_4$: When the reduction has to be carried out in aqueous medium, it is employed.

4. Fermentation of carbohydrates:

$$C_6H_{12}O_6 + \text{Yeast} \xrightarrow[30°C]{\text{Zymase}} 2C_2H_5OH + 2CO_2$$

(as this reaction employs grains as the source, C_2H_5OH is also known as grain alcohol)

5. Oxymercuration – demercuration:

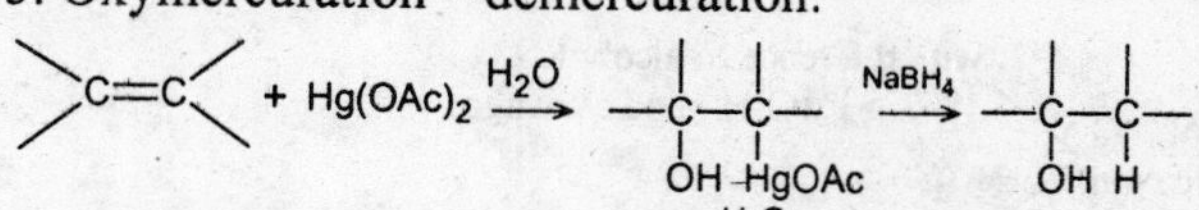

6. $H_2C{=}CH_2 + H_2SO_4 \longrightarrow H_3C{-}CH_2OSO_3H \xrightarrow[\text{boil}]{H_2O} H_3C{-}CH_2OH + H_2SO_4$

7. Hydroboration: Alkenes with diborane give an addition compound which on oxidation yield alcohols.

$$6\,H_2C{=}CH_2 + B_2H_6 \longrightarrow 2(CH_3CH_2)_3B \xrightarrow{3H_2O_2} 3H_3C{-}CH_2OH + H_3BO_3$$

Here, addition of water takes place by antiMarkownikoff's rule.

Properties and reactions:

(i) Alcohols are associated molecules due to hydrogen bonding

(ii) They have higher boiling points (among isomeric alcohols P>S>T).

(iii) The solubility in water decreases with the increase of alkyl chain. Generally the order of solubility is P>S>T. Among n–butanol, sec–butanol, tert.–butanol, tert.–butanol is miscible with H_2O in all properties (more soluble in H_2O).

In general, the –OH group of alcohols is a very reactive one. The reactions of this group can be due to C–O bond cleavage or O–H bond cleavage

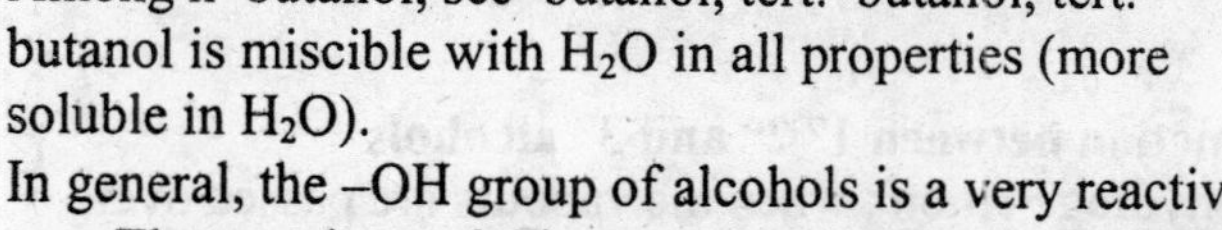

C–O bond is weaker in the case of tert.–alcohols as the cleavage is assisted by the increase in electron density at oxygen. O–H bond is weaker in primary alcohols as electron density increases between O–H bond and hydrogen tends to separate as proton.

In O–H cleavage, the order of reactivity is 1° > 2° > 3°.

In C–O cleavage, the order of reactivity is 1° < 2° < 3°.

Acidic nature: CH_3OH > 1° > 2° > 3°

1. With halogen acids:

$$ROH + HX \xrightarrow{ZnCl_2} RX + H_2O;$$

In allyl alcohol, it is always by anti-markownikoff's rule.

$$H_2C{=}CH{-}CH_2OH + HX \longrightarrow XCH_2{-}CH_2{-}CH_2OH$$

Tertiary alcohols S_N^1; Primary and secondary S_N^2

S_N^1: $R_3C{-}OH + H{-}X \longrightarrow R_3C{-}\overset{+}{O}H_2 \xrightarrow[\text{slow}]{-H_2O} R_3\overset{+}{C} \xrightarrow[\text{fast}]{+X^-} R_3C{-}X$

S_N^2 : $R{-}OH + H{-}X \rightarrow R{-}\overset{+}{O}H_2 + X^-$

$X^- + R{-}\overset{+}{O}H_2 \xrightarrow{\text{slow}} \left[\overset{\delta-}{X}{\cdots}R{\cdots}\overset{\delta-}{O}H_2\right] \xrightarrow{\text{fast}} X{-}R + H_2O$

Order of reactivity of halogen acids: HI > HBr > HCl.
Order of reactivity of alcohols: 1° < 2° < 3°

2. With organic acids: $RCOOH + HOR \xrightarrow{H_2SO_4} RCOOR + H_2O$

It is a reversible reaction. Therefore, H_2SO_4 a dehydrating agent is used.

Rates of esterification
- with reference to alcohols: 1°>2°>3° (O-H bond cleavage)
- acids: $HCOOH > CH_3COOH > (CH_3)_2CHCOOH > (CH_3)_3CCOOH$

3. Reaction with metals:
$ROH + M \longrightarrow ROM$ (M=Na, K, Mg, Al etc.)
Reactivity of alcohols: CH_3OH > 1° > 2° > 3°

4. Alcohols as acids:
$H_2O > ROH > NH_3 > RH$ (bigger the R group in ROH, lesser the acidity)
As bases: $R^- > NH_2^- > OR^- > OH^-$

5. Dehydration: In the presence of acid catalysts e.g. P_2O_5, conc. H_2SO_4, H_3PO_4 etc. or when the vapours of alcohol are passed over heated alumina, alkenes are produced. Ease of dehydration: T>S>P (formation of carbonium ion intermediate)

$R{-}\ddot{O}{-}H + H^+ \rightarrow R{-}\overset{+}{O}(H){-}H \xrightarrow{-H_2O} R^+ + H_2O$

Distinction between 1° 2° and 3° alcohols

1. Dehydrogenation : when the vapours are passed over freshly reduced copper at 300°C
P. alcohols → aldehyde; S. alcohols → Ketones
T. alcohols → alkenes

2. Oxidation:

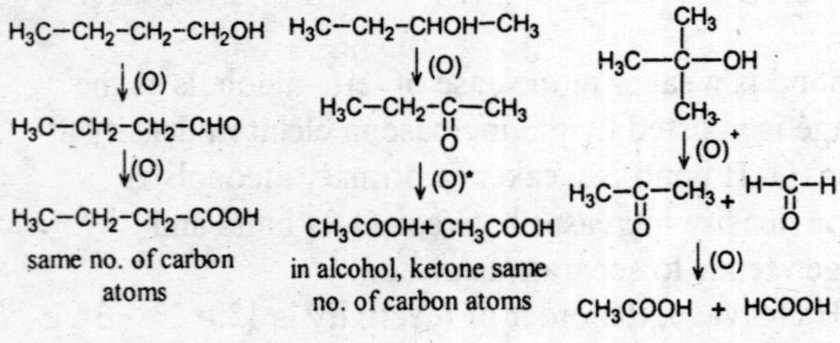

*Splitting of the ketonic chain takes place according to Popoff's rule. During the oxidation of a ketone, the carbonyl group always remains attached to the small alkyl group, while the other gets removed and oxidised e.g.

(i) $H_3C{-}CH_2{-}\underset{O}{\underset{\|}{C}}{-}CH_3 \longrightarrow CH_3COOH + CH_3COOH$

(ii) $H_3C{-}CH_2{-}CH_2{-}CO{-}CH_3 \longrightarrow CH_3CH_2COOH + CH_3COOH$

+Tertiary alcohols have no α hydrogen atoms and therefore no oxidation takes place. The reaction shown takes place only under drastic conditions.

3. Lucas test: The Lucas reagent is conc. HCl / dry $ZnCl_2$ (concentrated HCl saturated with anhydrous $ZnCl_2$). Carbonium ion intermediates form and therefore the reactivity order is T>S>P.
3° reacts immediately; 2° reacts in ~ 5 minutes; 1° do not react at room temperature.

4. Reaction with halogens: 1° and 2° undergo oxidation and halogenation. 3° do not undergo oxidation.

1° $H_3C{-}CH_2OH \xrightarrow[\text{as oxidising agent}]{Cl_2} CH_3CHO \xrightarrow[\text{as chlorinating agent}]{3Cl_2} CCl_3CHO + 3HCl$

2° $H_3C{-}CHOH{-}CH_3 \xrightarrow[\text{as oxidising agent}]{Cl_2} CH_3COCH_3 \xrightarrow[\text{as chlorinating agent}]{Cl_2} CCl_3COCH_3$

3° $(H_3C)_3C{-}OH \xrightarrow{Cl_2}$ No Reaction

5. Victor Meyer's test: (Red, Blue and White test)

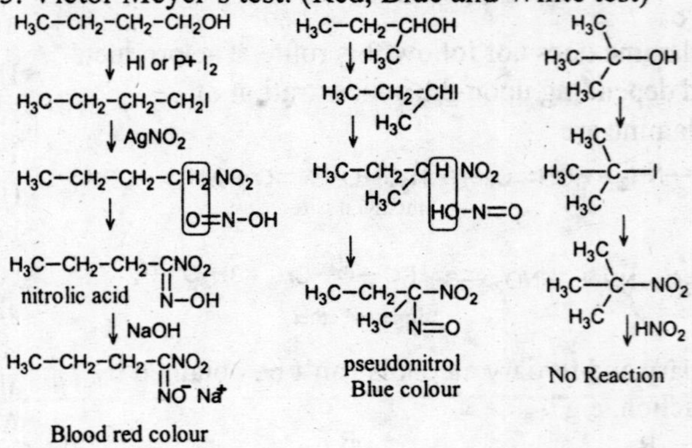

Miscellaneous

CH_3OH

$C + H_2O \xrightarrow{1300°C} CO + H_2$; $CO + 2H_2 \xrightarrow{ZnO + Cr_2O_3} CH_3OH$

C_2H_5OH :

$C_2H_4 + H_2SO_4 \xrightarrow[\text{pressure}]{70 - 80°C} C_2H_5HSO_4 \xrightarrow{C_2H_4} (C_2H_5)_2SO_4$

$C_2H_5HSO_4 + (C_2H_5)_2SO_4 + 3H_2O \xrightarrow{\Delta} 3C_2H_5OH + 2H_2SO_4$

$C_2H_5OH + H_2SO_4 \xrightarrow{\sim 100°C} C_2H_5HSO_4$
- $\xrightarrow[\text{140°C}]{C_2H_5OH \text{ excess}} C_2H_5OC_2H_5$
- $\xrightarrow[\text{excess, 160°c}]{\text{conc. } H_2SO_4} H_2C{=}CH_2$

Absolute alcohol : Moisture present is removed with metallic Mg. Rectified spirit :

Rectified spirit: $C_2H_5OH + H_2O$ (94.5 + 5.5%).

$C_2H_5OH + H_2O + \underset{\text{quick lime}}{CaO} \xrightarrow{24 \text{ hrs.}}$ distill over CaO $\rightarrow$ dried with anhydrus $CuSO_4$ $\xrightarrow{\text{distill}}$ absolute alcohol

-Azeotropic mixture : To distil it benzene is added.
-Power alcohol: petrol + absolute alcohol (4:1)
-Methylated spirit or denatured alcohol [for denaturing the light rubber distillate caoutchoucine)is added]
-Damages liver, causing cirrhosis of liver.

Dihydric alcohol: Glycol

Preparation

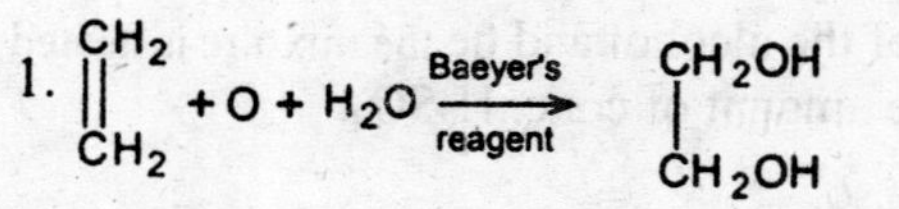

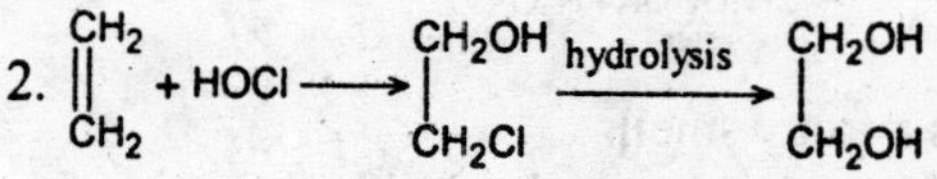

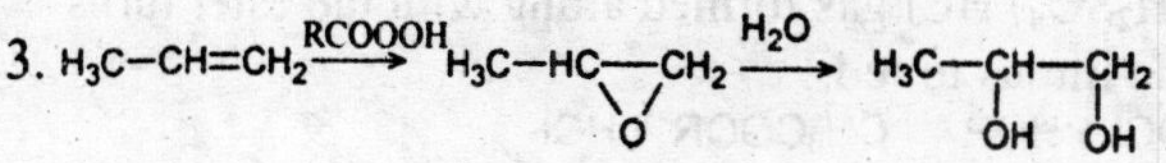

Reactions:

1. With HI or PI_3:

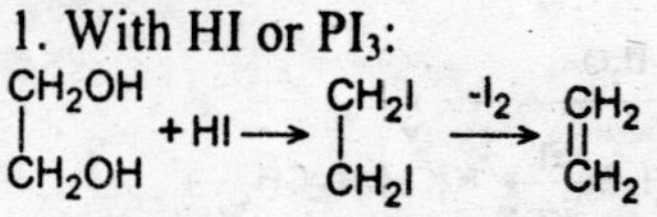

2. HIO_4 oxidation: (-CH_2OH changes to HCHO; -CHOH changes to HCOOH).

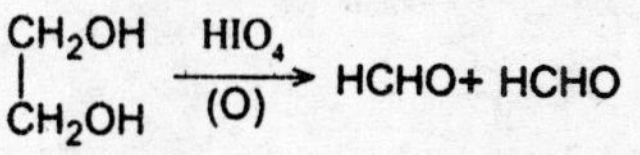

Acidic potassium dichromate converts it to formic acid.

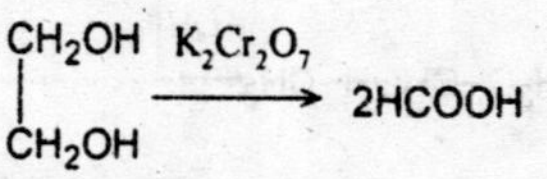

3. Dehydration:

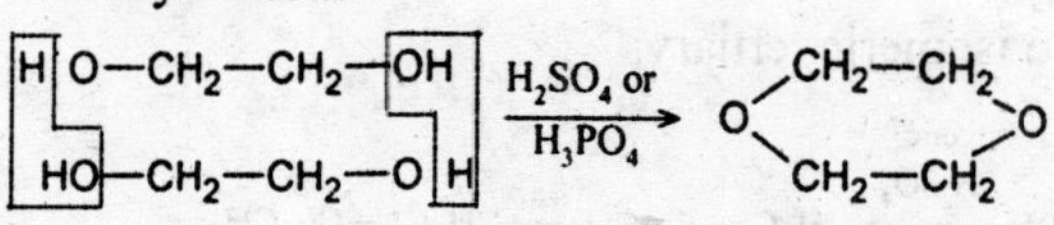

dioxan

4. Condensation with aldehydes and ketones:

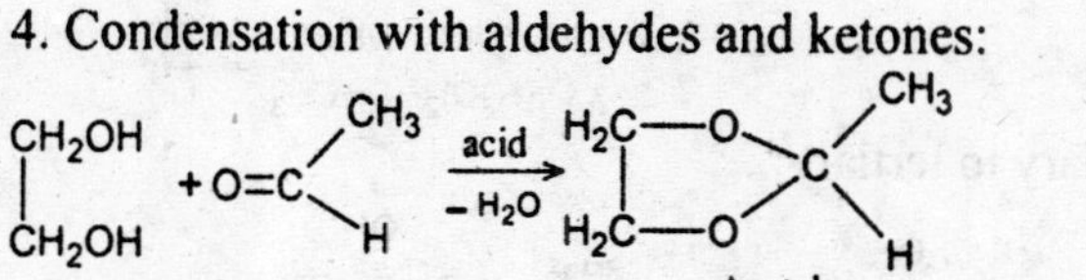

Acetal ;

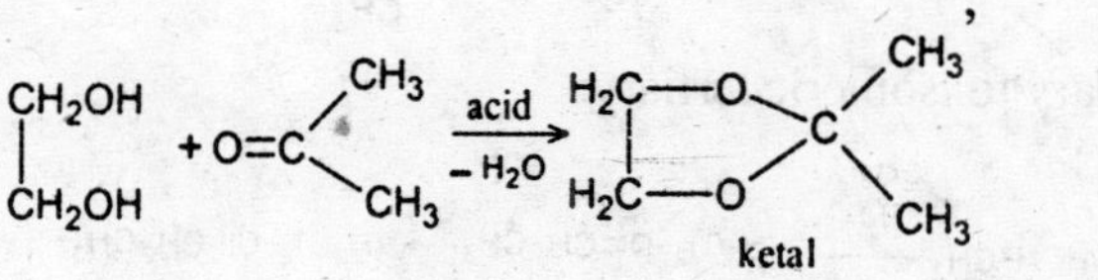

ketal

5. Polyester:

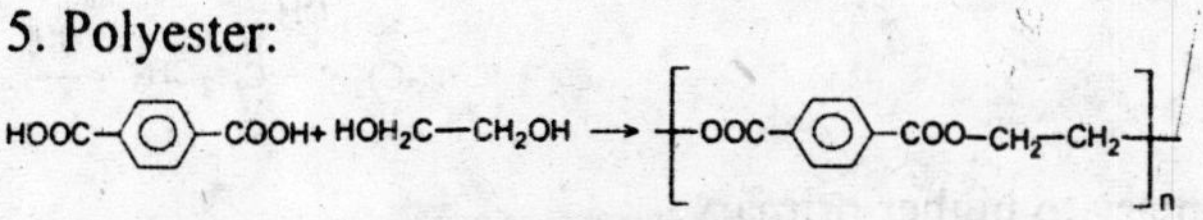

- Glycol is used as an antifreeze under the name prestone.

Trihydric alcohol, Glycerol : $\overset{1°}{HOH_2C}-\overset{2°}{CHOH}-\overset{1°}{CH_2OH}$

Preparation:

1. $C_6H_{12}O_6 \xrightarrow[\text{yeast}]{Na_2SO_3}$ glycerol + CH_3CHO + CO_2
glucose

2. In soap industry:

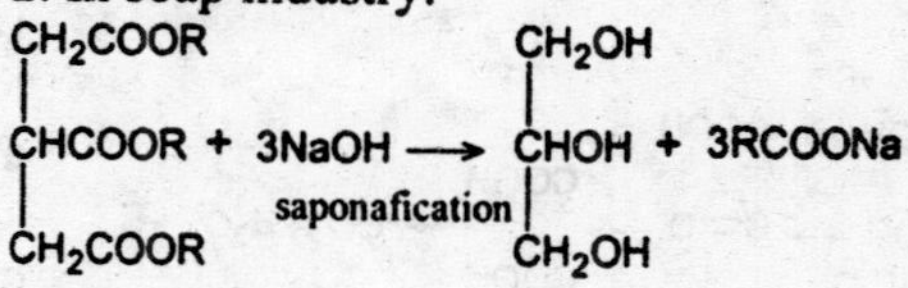

Reactions:

With Na or K only 1°, with acids 1° and 2° and with PCl_5 1° and 2° alcoholic groups react.

1. With HI or PI_3:

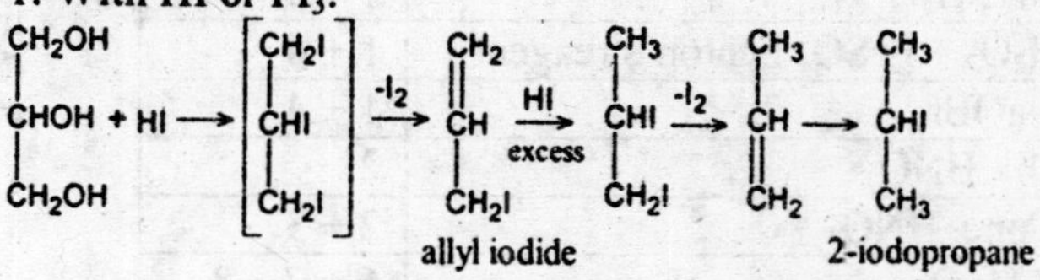

2. **With oxalic acid:** Oxalic acid reacts with glycerol at different temperatures to give different products.

2

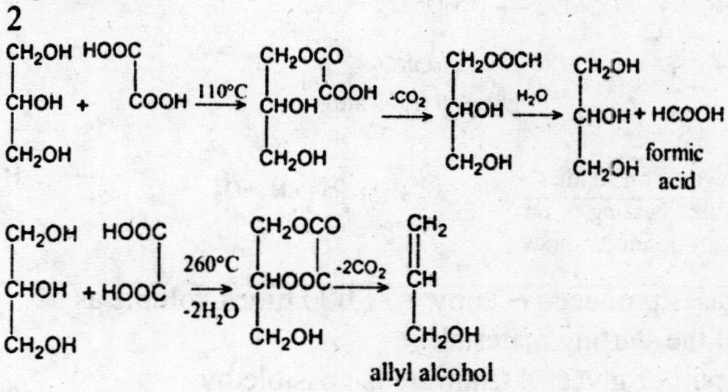

3. Dehydration: on heating alone or with dehydrating agents.

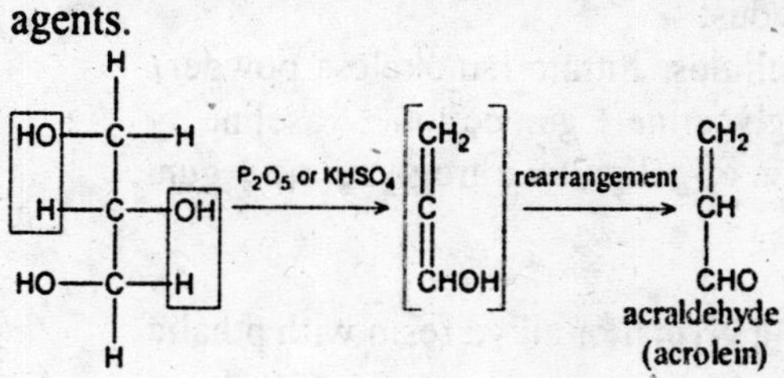

4. Oxidation with HIO_4:

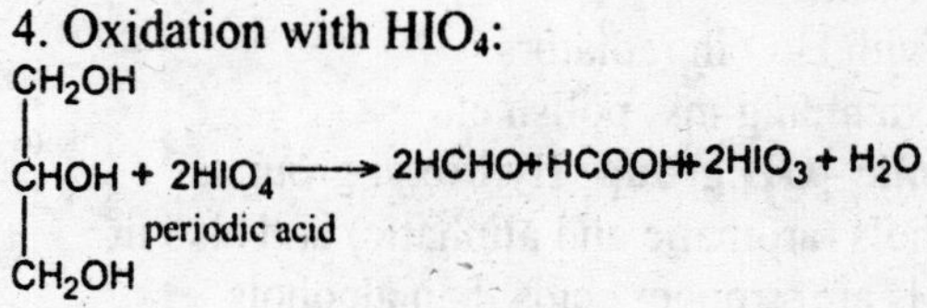

5. Oxidation

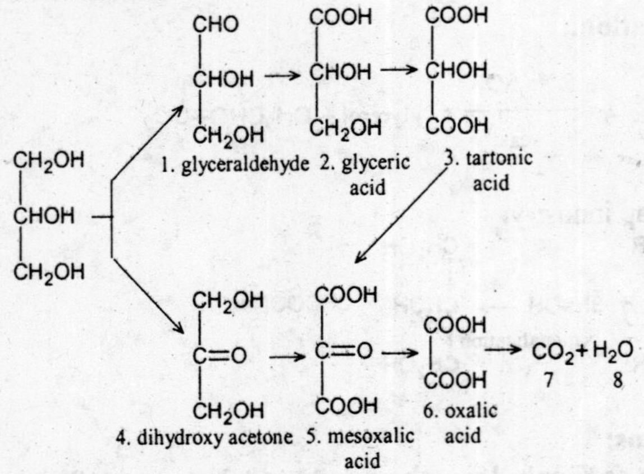

Oxidising agent	Products
1. Br_2 / H_2O	1 + 4
2. H_2O_2 + $FeSO_4$ Fenton's reagent	1 + 4
3. NaOBr	1 + 4
4. Dil. HNO_3	2
5. Conc. HNO_3	2 + 3
6. $KMnO_4$	6 + 7 + 8

6. Nitration:

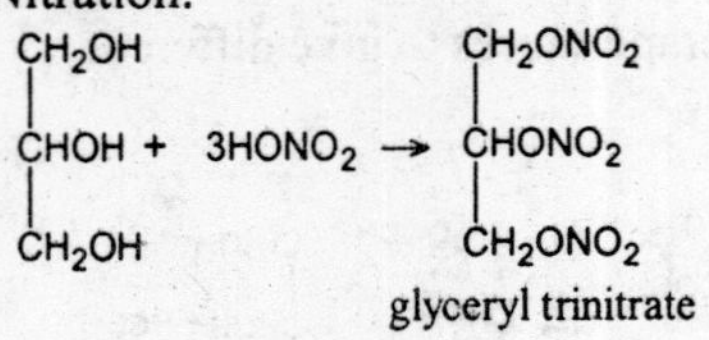

$$\text{glyceryl trinitrate} \xrightarrow[\text{heating or on sudden shock}]{\text{on sudden}} 12CO_2 + 10H_2O + 6N_2 + O_2$$

The gases produced occupy ~ 11,000 times volume as that of the starting material.

Transport of glyceryl trinitrate is possible by adsorbing it in Kieselguhr.

- Dynamite: (Alfred Nobel 1867) nitroglycerine + NH_4NO_3 + saw dust
- Gun cotton: cellulose nitrate: (smokeless powder).
- Cordite: nitroglycerine + gun cotton + vaseline
- Blasting gelatin or gelignite: nitroglycerine + gun cotton

Uses:

(a) For preparing glyptal or alkyd resin with pthalic anhydride

(b) Lubricant in watches

(c) Antifreeze with H_2O in radiators

(d) for making stamping ink, polish etc.

Test for alcoholic –OH group: Hydroxyl group is present in alcohols (aromatic and aliphatic) and also in phenols. Phenols are stronger acids than alcohols

(i) With metallic sodium: on adding a small piece of sodium metal to a little amount of dry alcohol, hydrogen gas escapes with effervescence.

$C_2H_5OH + Na \quad C_2H_5ONa + ½ H_2$

(ii) With ceric ammonium nitrate: When a few drops of ceric ammonium nitrate are added to the aqueous solution of the alcohol, red colour appears

$[(NH_4)_2Ce(NO_3)_6] + 2ROH \quad [Ce(NO_3)_4(ROH)_2] + 2NH_4NO_3$

ceric ammonium alkoxy nitrate

In this test phenols, aliphatic and aromatic amines also give coloured solution and or a precipitate.

(iii) Ester formation: A few drops of acetic acid are added to a little amount of the alcohol and he the mixture is heated after adding a little amount of conc. H_2SO_4.

$$CH_3COOH + HOR \xrightarrow[\Delta]{\text{conc. } H_2SO_4} \underset{\text{ester}}{CH_3COOR} + H_2O$$

Ester formed has a fruity smell.

If this test is performed with acetyl chloride (in place of conc. H_2SO_4) HCl gas formed along with the ester turns the blue litmus to red.

$CH_3COCl + HOR \quad CH_3COOR + HCl$

Conversions:

1. Primary to secondary:

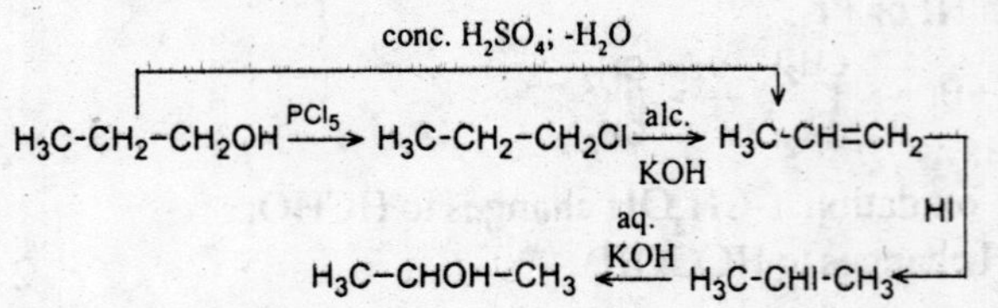

2. Primary to tertiary:

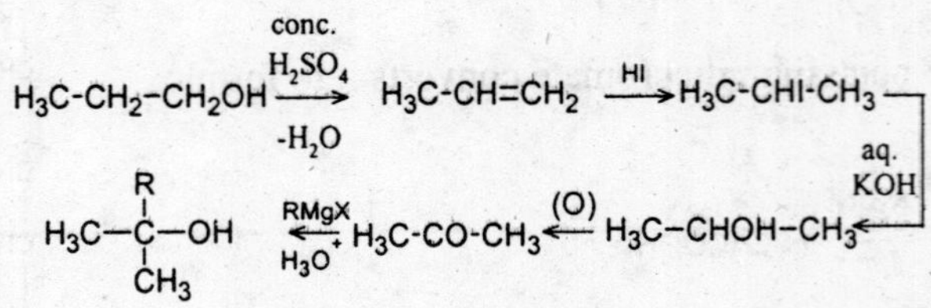

3. Primary to isomeric tertiary:

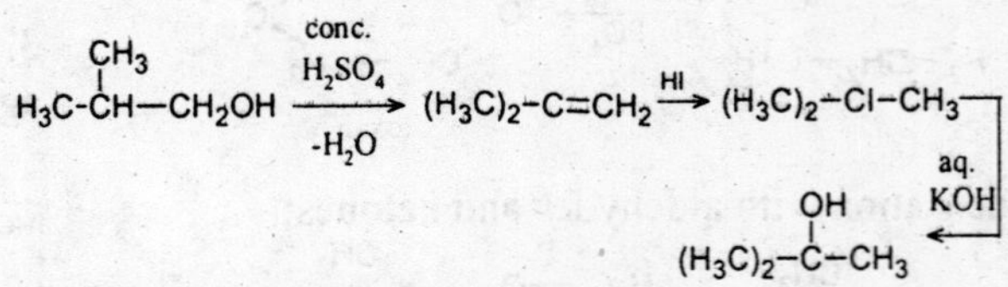

4. Secondary to tertiary:

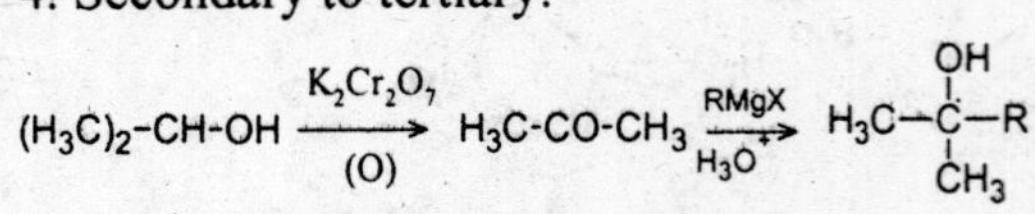

5. Secondary to isomeric tertiary:

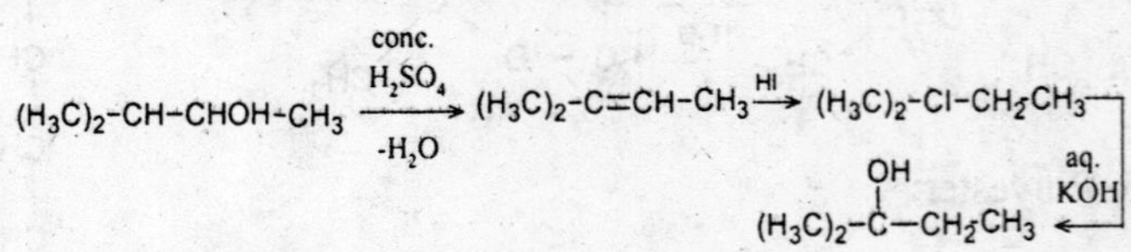

6. Primary to higher primary:

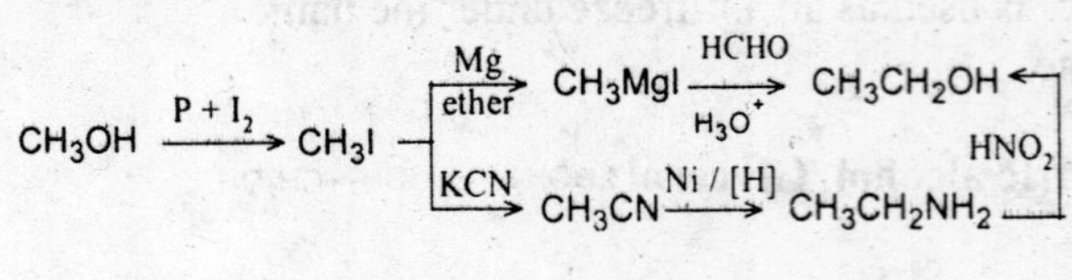

7. Primary to higher secondary:

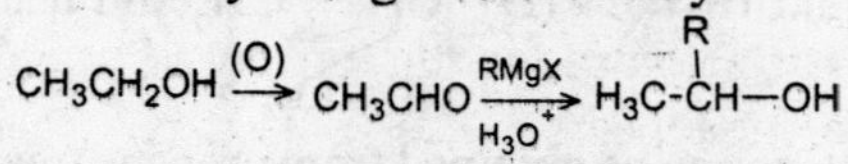

8. Secondary to higher tertiary:

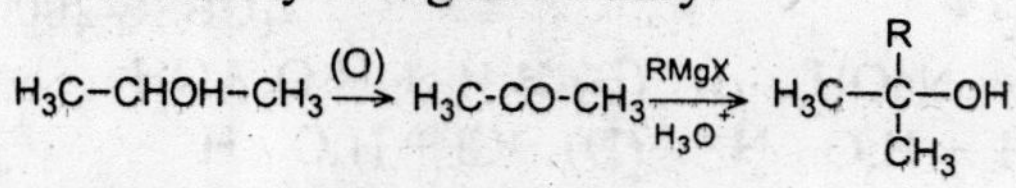

9. Primary to lower primary:

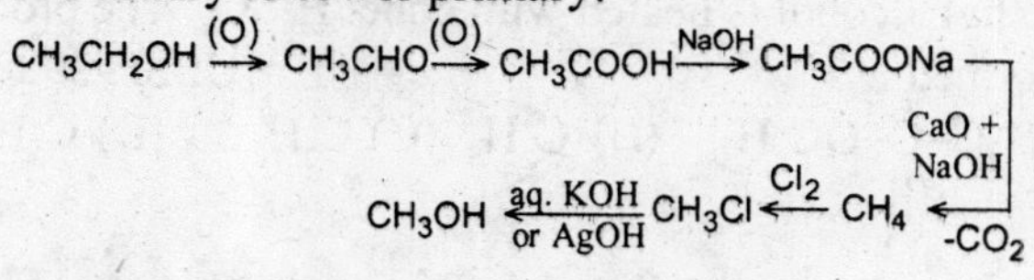

10. Higher secondary to lower primary

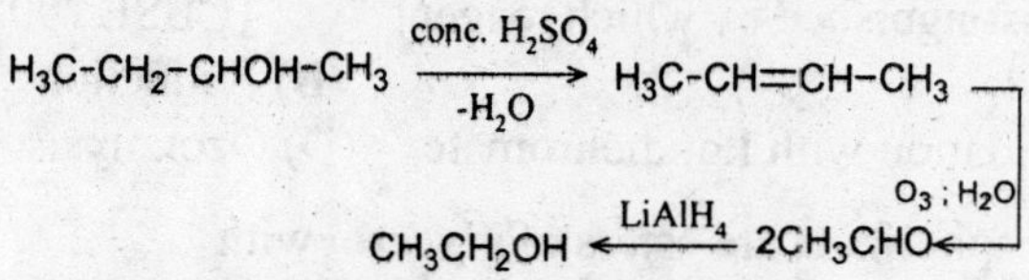

11. Tertiary to primary and secondary

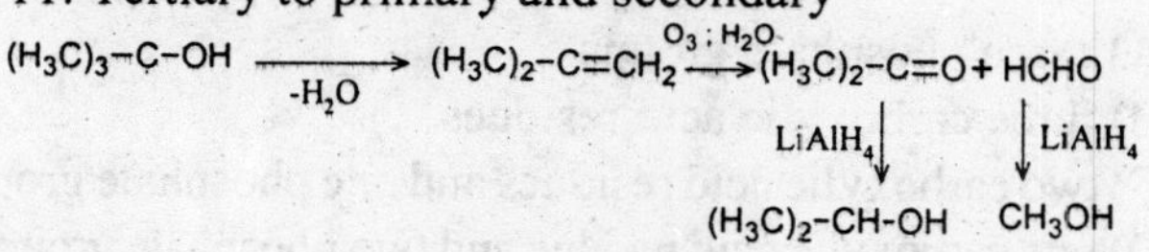

1. Lucas test is for [CBSE-90]
(A) alcohols (B) amines
(C) diethylether (D) glacial acetic acid

2. Glycerol reacts with $P_4 + I_2$ to form [CBSE-91]
(A) aldehyde (B) allyl Iodide
(C) allyl alcohol (D) acetylene

3. Methanol is industrially prepared by [CBSE-92]
(A) oxidation of CH_4 by steam at 900°C
(B) reduction of HCHO using $LiAlH_4$
(C) reacting HCHO with a solution of NaOH
(D) reduction of CO using H_2 and $ZnO\text{-}Cr_2O_3$

4. HBr will react fastest with [CBSE-92]
(A) 2-methylpropan-1-ol (B) 2-methylpropan-2-ol
(C) propan-2-ol (D) propan-1-ol

5. What is formed when primary alcohol undergoes catalytical dehydrogenation? [CBSE-93]
(A) Aldehyde (B) Ketone (C) Alkene (D) Acid

6. Which one of the following on oxidation gives a ketone? [CBSE-93]
(A) Primary alcohol (B) Secondary alcohol
(C) Tertiary alcohol (D) All of these

7. Grignard's reagent is prepared by the reaction between [CBSE-94]
(A) magnesium and alkyl halide
(B) zinc and alkyl halide (C) magnesium and alkane.
(D) magnesium and aromatic hydrocarbon.

8. On heating glycerol with conc. H_2SO_4, a compound is obtained which has bad odour. The compound is [CBSE-94]
(A) allyl alcohol (B) formic acid
(C) acrolein (D) glycerol sulphate

9. When 3,3-dimethyl-2-butanol is heated with H_2SO_4, the major product obtained is [CBSE-95]
(A) cis and trans isomers of product
(B) 2,3-dimethylbut-1-ene (C) 3,3-dimethylbut-1-ene
(D) 2,3 -dimethylbut-1-ene

10. 1-phenylethanol can be prepared by reaction of benzaldehyde with [CBSE-97]
(A) methyl bromide (B) ethyl iodide and magnesium
(C) methyl bromide and aluminium bromide
(D) methyl iodide and magnesium

11. In a reaction [CBSE-97]

$$CH_2 = CH_2 \xrightarrow[\text{acid}]{\text{Hypochlorous}} M \xrightarrow{R} \begin{array}{l} CH_2OH \\ | \\ CH_2OH \end{array}$$

M = molecules
R = reagent

M and R are
(A) CH_3CH_2Cl and NaOH
(B) CH_2ClCH_2OH and aq. $NaHCO_3$
(C) CH_3CH_2OH and HCl (D) $\underset{\diagdown O \diagup}{CH_2 - CH_2}$ and heat.

12. 1-hydroxypropane can be obtained from propene treating with [CBSE-2000]
(A) B_2H_6 + NaOH (B) $B_2H_6 + H_2O_2 / OH^-$
(C) $LiAlH_4 + H_2O / H^+$ (D) $HCl + H_2O / H^+$

13. Ethyl alcohol is heated with conc. H_2SO_4. The product formed is [CBSE-2000]
(A) $H_3C-\overset{\overset{O}{||}}{C}-OC_2H_5$ (B) C_2H_6 (C) C_2H_4 (D) C_2H_2

14. n–propyl alcohol and isopropyl alcohol can be chemically distinguished by which reagent [CBSE 2002]
(A) PCl_5 (B) Reduction
(C) Oxidation with Pot.dichromate (D) Ozonolysis

15. Phospholipids are esters of glycerol with [CBSE-2003]
(A) three phosphate groups
(B) three carboxylic acid residues
(C) two carboxylic acid residues and one phosphate group
(D) one carboxylic acid residue and two phosphate groups.

16. Which one of the following is the systematic name for allyl alcohol? [AIIMS-93]
(A) Allylol (B) Prop-2-en-1-ol
(C) Rectified spirit (D) Isopropyl alcohol

17. The compound B formed in the following sequence of reactions $CH_3CH_2CH_2 - OH \xrightarrow{PCl_5} A \xrightarrow[KOH]{Alcohol} B$ will be [AIIMS-93]
(A) Propane (B) Propene (C) Propyne (D) Propanol

18. Glycerol reacts with excess HI to give [AIIMS-93]
(A) 1,2,3-triiodopropane (B) Propene
(C) 2-iodopropane (D) Allyl iodide

19. Ethyl alcohol exhibits acidic character on reacting with [AIIMS-94]
(A) Acetic acid (B) Sodium metal
(C) Hydrogen iodide (D) All of these

20. Which one of the following can differentiate between C_2H_5OH and CH_3OH? [AIIMS-94]
(A) HCl (B) NH_3 (C) H_2O (D) I_2 + KOH

21. In the reaction sequence;
$CaC_2 \xrightarrow{H_2O} A \xrightarrow[HgSO_4]{\text{dil } H_2SO_4} B \xrightarrow{H_2/Ni} C$,
the compound 'C' is [AIIMS-96]
(A) HCHO (B) CH_3OH (C) CH_3CHO (D) C_2H_5OH

22. Glycerol is not used in [AIIMS-96]
(A) Cosmetics (B) Matches (C) Explosives (D) Soaps

23. In glycerine, [AIIMS-97]
(A) One primary - OH group is present.

(B) One secondary - OH group is present.
(C) Two secondary - OH group is present.
(D) None of these

24. Lucas test is used for the determination of [AIIMS-2002]
(A) Alcohols (B) Phenols
(C) Alkyl halides (D) Aldehydes

25. Propan-1-ol can be prepared from propene by- [AIIMS-2003]
(A) H_2O/H_2SO_4
(B) $Hg(OAc)_2/H_2O$ following by $NaBH_4$
(C) B_2H_6 following by H_2O_2 (D) CH_3CO_2H/H_2SO_4

26. The compound with the highest boiling point is : [JIPMER-91]
(A) CH_4 (B) CH_3OH (C) CH_3Br (D) CH_3I

27. The Compound which reacts faster with Lucas reagent is : [JIPMER-91]
(A) Butan-1-ol (B) Butan-2-ol
(C) 2-Methyl propan-1-ol (D) 2-Methyl propan-2-ol

28. Absolute alcohol contains : [JIPMER-91]
(A) 40% water (B) 20% water
(C) 5% water (D) No water

29. Which of the following compounds will show negative test with phenyl hydrazine : [JIPMER-92]
(A) Acetone (B) Ethyl alcohol
(C) Benzophenone (D) Acetaldehyde

30. Which of the following compounds is oxidized to methyl ethyl ketone ? [JIPMER-93]
(A) 1-Butanaol (B) 2-Butanol
(C) 2-Propanol (D) T-Butylalcohol

31. Glycerol on warming with excess of hydrogen iodide gives : [JIPMER-93]
(A) Propylene (B) Allyl iodide
(C) Isopropyl iodide (D) Nitro glycerine

32. Which of the following explains the viscous nature of glycerol ? [JIPMER-95]
(A) van der Waal forces (B) Covalent bonds
(C) Hydrogen bonds (D) Ionic forces

33. Lucas reagent is a mixture of [JIPMER-96]
(A) conc. HCl and anhydrous $ZnCl_2$
(B) conc. HCl and hydrated $ZnCl_2$
(C) conc. HNO_3 and anhydrous $ZnCl_2$
(D) conc. HNO_2 and anhydrous $ZnCl_2$

34. Acid hydrolysis of oil gives [JIPMER-96]
(A) Glycerol + detergent (B) Glycerol + soap
(C) Glycerol + fatty acid (D) Glycerol + aldehydes

35. Isopropyl alcohol on mild oxidation gives : [JIPMER-96]
(A) Acetone (B) Propanoic acid
(C) Propene (D) Propane

36. When glycerol is heated with excess of oxalic acid at 110°C, it forms [JIPMER-97]
(A) Acetic acid (B) Acetaldehyde
(C) Formic acid (D) Allyl alcohol

37. The chemical name for grain alcohol is [JIPMER-97]
(A) Methanol (B) Ethanol
(C) Amyl alcohol (D) Butyl alcohol

38. Dynamite is a mixture of nitroglycerine, [JIPMER-97]
(A) Saw dust and NH_4NO_3 (B) And cellulose nitrate
(C) And cellulose acetate
(D) Cellulose nitrate and vaseline

39. When ethylene is passed into cold dilute alkaline $KMnO_4$, the product is [JIPMER-97]
(A) Glycol (B) Methanol
(C) Acetylene (D) Two molecules of CO_2

40. The most favourable conditions for the alcoholic fermentation of sugar are [JIPMER-97]
(A) High concentration of sugar solution, low temperature and plenty of air.
(B) Low concentration of sugar solution, high temperature and absence of air.
(C) Low concentration of sugar solution, low temperature and plenty of air supply.
(D) High concentration of sugar solution, low temperature and absence of air.

41. Chlorine reacts with ethanol to give [JIPMER-98]
(A) Ethyl chloride (B) Chloroform
(C) Acetaldehyde (D) Chloral

42. The compound which reacts fastest with Lucas reagent is [JIPMER-98]
(A) Butan-1-ol (B) Butan-2-ol
(C) 2-methylpropan-1-ol (D) 2-methylpropan-2-ol

43. An organic compound P reacts with sodium to evolve hydrogen and with oxidising agents to yield a ketone. The formula of P could be [JIPMER-99]
(A) $CH_3{-}CH_2OH$ (B) $CH_3{-}CH(OH){-}CH_3$
(C) $C(CH_3)_3CHO$ (D) $OH{-}CH_2{-}CH_2{-}OH$

44. A compound X of formula C_3H_8O yields a compound C_3H_6O on oxidation. To which of the following class of compounds could X belong ? [JIPMER-99]
(A) Aldehyde (B) Secondary alcohol
(C) Alkene (D) Tertiary alcohol

45. A compound x when boiled with aqueous Na_2CO_3 gives glycol. The compound x is [JIPMER-99]
(A) ethylene (B) ethylene oxide
(C) ethylene bromide (D) ethyl hydrogen sulphate

46. Isopropyl alcohol on passing over heated copper at

300°C gives [JIPMER-2000]
(A) Acetone (B) Acetaldehyde
(C) Propylene (D) None

47. In the following sequence of reactions,

$$CH_3CH_2CH_2Br \xrightarrow{KOH\ (alc.)} (A) \xrightarrow{HBr} (B) \xrightarrow{KOH\ (aq.)} (C)$$

The product (C) is [JIPMER-2001]
(A) Propan–2–ol (B) Propan–1–ol
(C) Propyne (D) Propene

48. When ethylene glycol reacts with periodic acid compound formed is called [JIPMER-2002]
(A) Formaldehyde (B) Acetaldehyde
(C) Formic acid (D) Acetic acid

49. Which of the following is contained in alcoholic beverages ? [BHU-96]
(A) 2-propanol (B) Methanol
(C) Glycerol (D) Ethanol

50. With oxalic acid, glycerol at 530 K gives [BHU-96]
(A) Allyl alcohol (B) Formic acid
(C) Glyceraldehyde (D) Glycerol mono-oxalate

51. The relative rate of diffusion of a gas (molecular weight = 128) as compared to oxygen is [BHU-96]
(A) 2 times (B) 0.25 times
(C) 0.125 times (D) 0.5 times

52. Ethylalcohol gives ethyl chloride with the help of [BHU-97]
(A) $SOCl_2$ (B) KCl (C) Cl_2 (D) NaCl

53. Isopropyl alcohol on oxidation gives [BHU-97]
(A) Acetone (B) Acetaldehyde
(C) Ether (D) Ethylene

54. Which of the following statements is correct ? [BHU-2001]
(A) Solubility of alcohol in water increases with increase in molecular weight
(B) Lower alcohols are soluble in water and possess characteristic odour
(C) m.p. and b.p. of alcohol increases with increase in molecular weight and so is their solubility in water
(D) Higher alcohols are insoluble in water and possess characteristic odour

55. The compound which does not contain OH group is [AFMC-91]
(A) Anisol (B) Alcohol
(C) Methanoic acid (D) Methanol

56. Which is used as an antifreeze ? [AFMC-92]
(A) Glycol (B) Ethyl alcohol (C) Water (D) Methanol

57. Which of the following is not characteristic of alcohols? [AFMC-92]
(A) Lower alcohols are stronger and have bitter taste
(B) Higher alcohols are stronger and have bitter taste
(C) The boiling points of alcohols increase with increasing molecular mass
(D) The lower alcohols are soluble in water

58. Blasting of TNT is done by mixing it with [AFMC-93]
(A) NH_4Cl (B) NH_4NO_3
(C) NH_4NO_2 (D) $(NH_4)_2SO_4$

59. Which of the these statements is not true for alcohols? [AFMC-93]
(A) Lower alcohols have fiery taste and strong smell
(B) As molecular mass increases, boiling point increases
(C) Lower alcohols are water insoluble and their solubility increases with molecular weight
(D) Lower alcohols are water soluble and their solubility decreases with molecular weight.

60. Beverages are made up of [AFMC-94]
(A) Ethyl alcohol (B) Acetic acid
(C) Formic acid (D) None of these

61. Glycerol reacts with $KHSO_4$ to form [AFMC-95]
(A) Allyl alcohol (B) Acrolein
(C) Aldehyde (D) Acetone

62. Dynamite contains [AFMC-96]
(A) Nitroglycerine mixed with saw dust
(B) Nitroglycerine mixed with H_2SO_4
(C) Both of these (D) None of these

63. Which statement is not correct about alcohol ? [AFMC-97]
(A) Alcohol is lighter than water
(B) Alcohol evaporates quickly
(C) Alcohol of less no. of carbon atom is less soluble in water than alcohol of high no of carbon atoms
(D) All are correct

64. Rectified spirit has ethanol [AFMC-2000]
(A) 5% (B) 95% (C) 100% (D) 80%

65. Which one is most viscous ? [AFMC-2000]
(A) CH_3OH (B) C_2H_5OH
(C) HOH_2C-CH_2OH (D) None of these

66. Alcohol that is used as a beverage is [AFMC-2001]
(A) Propanol (B) Butanol
(C) Ethanol (D) Methanol

67. Which of the following is not true in case of reaction with heated copper at 300° C ? [AFMC-2002]
(A) Phenol → Benzyl alcohol
(B) Primary alcohol → Aldehyde
(C) Secondary alcohol → Ketone
(D) Tertiary alcohol → Olefin

68. The alcohol obtained by the hydrolysis of oils and fats is [AFMC-2002]

(A) Pentanol (B) Propanol
(C) Glycerol (D) Glycol

69. The order of reactivity of the following alcohols. [IIT-97]

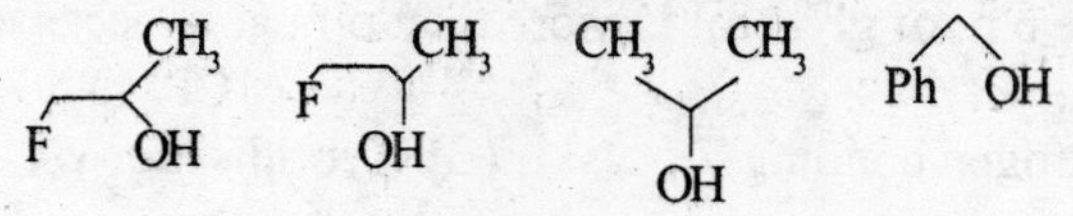

I II III IV

(A) I > II > III > IV (B) I > III > II > IV
(C) IV > III > II > I (D) IV > III > I > II

70. Among the following compounds, the strongest acid is. [IIT-98]
(A) HC≡CH (B) C_6H_6 (C) C_2H_6 (D) CH_3OH

71. Glycerine contains one .... hydroxy group. [IIT-97]

72. Which one of the following will most readily be dehydrated in acidic condition ? [IIT-2000]

(A) 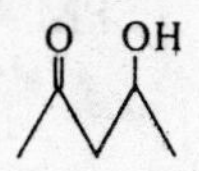 (B) 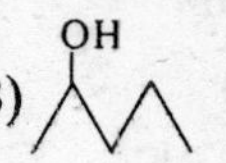 (C) 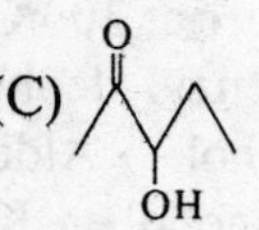 (D) 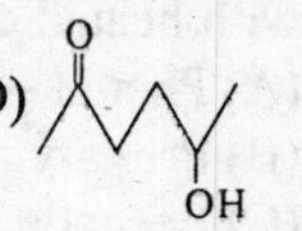

73. The compound that will react most readily with NaOH to form methanol is [IIT-2001]
(A) $(CH_3)_4N^+I^-$ (B) CH_3OCH_3
(C) $(CH_3)_3S^+I^-$ (D) $(CH_3)_3CCl$

74. 1-Propanol and 2-Propanol can be best distinguished by [IIT-2001]
(A) oxidation with alkaline $KMnO_4$ followed by reaction with Fehling's solution.
(B) oxidation with acidic dichromate followed by reaction with Fehling's solution
(C) oxidation by heating with copper followed by reaction with Fehling's solution
(D) oxidation with concentrated H_2SO_4 followed with Fehling's solution.

75. 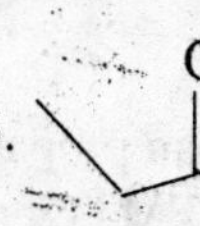 $\xrightarrow[H_2O]{H}$ (X) mixture $\xrightarrow{Br_2}$ $C_4H_8Br_2$
(5 compounds) Number of compounds in X will be [IIT-2003]
(A) 4 (B) 3 (C) 2 (D) 5

76. During dehydration of alcohols to alkenes by heating with conc. H_2SO_4 the initiation step is [AIEEE-2003]
(A) Protonation of alcohol molecule
(B) Formation of carbocation
(C) Elimination of water (D) Formation of an ester

77. Ethyl bromide on treatment with KOH(Aq) gives rise to .... [EAMCET-93]

78. The alcohol manufactured from water gas is [EAMCET-94]
(A) Ethanol (B) Butanol (C) Methanol (D) Isobutanol

79. The compound formed when 2-butene is treated with acidified $KMnO_4$ is [EAMCET-95]
(A) Acetaldehyde (B) Acetic acid
(C) CH_2OH–CH_2OH (D) CH_3–CH($C(=O)CH_3$)–CH_3

80. Compound 'A' reacts with PCl_5 to give 'B' which on treatment with KCN followed by hydrolysis gave propionic acid as the product. What is 'A' ? [EAMCET-96]
(A) Propane (B) Ethane
(C) Ethylchoride (D) Ethyl alcohol

81. In which one of the following conversions phosphorus pentachloride is used as the reagent ? [EAMCET-97]
(A) $H_2C = CH_2 \rightarrow CH_3\ CH_2\ Cl$
(B) $H_3C - O\ CH_3 \rightarrow CH_3\ Cl$
(C) $CH_3\ CH_2\ OH \rightarrow CH_3CH_2\ Cl$
(D) $CH = CH \rightarrow CH_2\ CHCl$

82. The alcohol that produces turbidity immediately with $ZnCl_2$/conc. HCl at room temperature : [EAMCET-97]
(A) 1-Hydroxybutane (B) 2-Hydroxybutane
(C) 2-Hydroxy - 2- methyl propane
(D) 1-Hydroxy - 2 - methyl propane

83. Compound 'A' reacts with PCl_5 to give 'B' which on treatment with KCN followed by hydrolysis gave propionic acid. What are A & B respectively ? [EAMCET-98]
(A) C_3H_8 & C_3H_7Cl (B) C_3H_8 & C_2H_5Cl
(C) C_2H_5Cl & C_2H_5Cl (D) C_2H_5OH & C_2H_5Cl

84. What is the product obtained when chlorine reacts with ethyl alcohol in the presence of NaOH ? [EAMCET-98(MBBS)]
(A) $CHCl_3$ (B) CCl_3CHO (C) C_2H_5Cl (D) CH_3Cl

85. The two enzymes present in yeast that are responsible for the formation of ethyl alcohol from molasses in the fermentation process are [EAMCET-99(MBBS)]
(A) invertase, zymase (B) invertase, diastase
(C) zymase, diastase (D) invertase, maltase

86. Which of the following is Lucas reagent ? [EAMCET-2000]
(A) Ammoniacal silver nitrate (B) Br_2/CCl_4
(C) $ZnCl_2$/Conc.HCl (D) Cold alkaline $KMnO_4$

87. Which of the following is a tertiary alcohol ? [EAMCET-2000]
(A) $CH_3 - CH(CH_3) - CH_2 - OH$
(B) $CH_3 - CH_2 - CH_2 - CH_2 - OH$
(C) $CH_3 - CH_2 - CH(CH_3) - CHO$
(D) $CH_3 - C(CH_3)_2 - OH$

88. Which of the following reaction conditions are used for the conversion of ethanol to ethylene ? [EAMCET(MBBS)-2000]
(A) Conc. H_2SO_4/70°C (B) Dil. H_2SO_4/140°C
(C) Dil. H_2SO_4/100°C (D) Conc.H_2SO_4/170°C

89. Which of the following is used in the manufacture of ethanol ? [EAMCET(MBBS)-2000]
(A) C_2H_6 (B) C_2H_2 (C) CH_4 (D) C_2H_4

90. What are the reaction conditions used for the formation of ethylene glycol from ethylene ? [EAMCET(MBBS)-2000]
(A) Bromine water (B) Cold alkaline $KMnO_4$
(C) SO_4/60°C (D) O_3/H_2O

91. Absolute alcohol (100% alcohol) is prepared by distilling rectified spirit over .... [EAMCET-2001]
(A) Na (B) $CaCl_2$ (C) Mg (D) $Mg(OC_2H_5)_2$

92. What is Y in the following reactions ?
$C_2H_5I + NaOC_2H_5 \rightarrow X + NaI$
$X + 2HI \rightarrow 2Y + H_2$ [EAMCET-2001]
(A) C_2H_6 (B) C_2H_5I (C) C_2H_4 (D) $C_2H_5OC_2H_5$

93. Dry distillation of a mixture of calcium acetate and calcium formate forms ...... [EAMCET-2001]
(A) Methanol (B) Ethanal (C) Ethanol (D) Acetone

94. Ethanol is heated to 170°C in the presence of concentrated H_2SO_4. What is the product formed ? [EAMCET(MBBS)-2001]
(A) CH_3COOH (B) C_2H_6 (C) $C_2H_5OC_2H_5$ (D) C_2H_4

95. Which one of the following is a tertiary alcohol ? [EAMCET(MBBS)-2001]
(A) $CH_3CH_2CH_2OH$ (B) $CH_3CH(OH)CH_3$
(C) $(CH_3)_2CHCH_2OH$ (D) $(CH_3)_3COH$

96. Which one of the following is a secondary alcohol ? [EAMCET-2002]
(A) 2–Methyl–2–Propanol (B) 1–Propanol
(C) 1–Butanol (D) 2–Pentanol

97. Which one of the following gases is liberated when ethyl alcohol is heated with methyl magnesium iodide ? [EAMCET-2003]
(A) Methane (B) Ethane
(C) Carbon dioxide (D) Propane

98. In $CH_3CH_2CH_2OH$ which bond is most easily broken ? [CPMT-96]
(A) O—H (B) C—O (C) C—C (D) C—H

99. Which of the following is not obtained when C_2H_5OH is heated with conc. H_2SO_4 ? [CPMT-96]
(A) $CH_2 = CH_2$ (B) $C_2H_5OC_2H_5$
(C) $C_2H_5SO_4H$ (D) $CH_3CH_2CH_2CH_3$

100. Which of the following reaction shows industrial method of CH_3OH ? [CPMT-96]
(A) CO + H_2 redn. (B) $CH_3Cl + H_2O$
(C) $CH_3NH_2 + HNO_2$ (D) CH_3Br + aqKOH

101. The b.p. of glycerol is more than propanol because of- [CPMT-97]
(A) Hydrogen bonding (B) Hybridisation
(C) Resonance (D) All of these

102. R – OH + HX → R – X + H_2O In the above reaction, the reactivity of different alcohol is [CPMT-97]
(A) Tertiary>Secondary>Primary
(B) Tertiary<Secondary<Primary
(C) Tertiary<Secondary>Primary
(D) Secondary<Primary<Tertiary

103. Lucas reagent is- [CPMT-98]
(A) $ZnCl_2$ + conc. HCl (B) $MnO_2 + H_2O$
(C) H_2SO_4 + HCl (D) Na + H_2O

104. Which of the following is not true in case of reaction with heated copper at 300°C ? [CPMT-99]
(A) Phenol → Benzyl alcohol
(B) Primary alcohol → Aldehyde
(C) Secondary alcohol → Ketone
(D) Tertiary alcohol → Olefin

105. The correct order of boiling point for primary (1°), secondary (2°) and tertiary (3°) alcohols is [CPMT-99]
(A) 1° > 2° > 3° (B) 3° > 2° > 1°
(C) 2° > 1° > 3° (D) 2° > 3° > 1°

106. Which of the following is the most suitable method for removing the traces of water from ethanol ? [CPMT-99]
(A) Heating with Na metal
(B) Passing dry HCl through it
(C) Distilling it (D) Reacting with Mg

107. Which of the following will dissolve in $NaHCO_3$? [CPMT-2001]
(A) CH_3CHO (B) C_2H_5OH
(C) CH_3COOH (D) CH_3COCH_3

108. The boiling point of glycerol is more than propanol because of [CPMT-2002]
(A) hybridisation (B) arrangement of molecules
(C) H-bonding (D) all of these

109. Dehydration of glycerol gives [PMT-94]
(A) Propane (B) Propene (C) Acrolein (D) Benzene

110. Which of the following is most soluble in water ? [PMT-95]
(A) Normal butyl alcohol (B) Iso butyl alcohol
(C) Tertiary butyl alcohol (D) Secondary butyl alcohol

111. Ethyl alcohol exhibits acidic chracter on reacting with [PMT-95]
(A) Acetic acid (B) Sodium metal
(C) Hydrogen iodide (D) Acidic potassium dichromate

112. Lucas reagent is [PMT-96]
(A) conc. HCl + anhydrous $ZnCl_2$
(B) dil HCl + hydrated $Zncl_2$
(C) conc. HNO_3 + anhydrous $ZnCl_2$
(D) conc. HNO_3 + anhydous $MgCl_2$

113. The percentage of ethyl alcohol present in rectified spirit is [PMT-96]
(A) 75.0 (B) 85.5 (C) 95.6 (D) 100.0

114. The enzyme, which converts glucose and fructose into ethyl alcohol is [PMT-96]
(A) diastase (B) invertase (C) zymase (D) maltase

115. Lucas reagent is [PMT-97]
(A) Conc. HCl + anhydrous $ZnCl_2$
(B) Dil. HCl + hydrous $ZnCl_2$
(C) Conc. HNO_3 + anhydrous $ZnCl_2$
(D) Conc. HNO_3 + anhydrous $MgCl_2$

116. Which of the following reacts first with Lucas reagent? [PMT-99]
(A) C_3H_7OH (B) $(CH_3)_2CHOH$
(C) $(CH_3)_3COH$ (D) C_6H_5OH

117. The reaction of ethylene glycol with PI_3 gives : [PMT-2000]
(A) I CH_2-CH_2 I (B) $CH_2 = CH_2$
(C) $CH_2 = CH$ I (D) I CH = CH I

118. The reaction of Lucas reagent is fast with : [PMT-2000]
(A) $(CH_3)_3$ COH (B) $(CH_3)_2$ CHOH
(C) CH_3 $(CH_2)_2$ OH (D) CH_3CH_2OH

119. Primary, secondary and tertiary alcohols may be distinguished by : [PMT-2001]
(A) Fehling solution test (B) Hoffmann test
(C) Beilstein test (D) Victor- Meyer test

120. Dehydrogenation of $CH_3-\underset{|}{\underset{OH}{CH}}-CH_3$ gives [PMT-2002]
(A) Acetone (B) Acetaldehyde
(C) Acetic acid (D) Acetylene

121. In the sequence of the following reactions

$$CH_3OH \xrightarrow{HI} CH_3I \xrightarrow{KCN}$$
$$CH_3CN \xrightarrow{Reduction} X \xrightarrow{HNO_2} Y$$

X and Y are respectively [PMT-2002]
(A) $CH_3CH_2NH_2$ and CH_3CH_2OH
(B) $CH_3CH_2NH_2$ and CH_3COOH
(C) CH_3CH_2OH and CH_3CHO
(D) CH_3OCH_3 and CH_3CHO

122. In the sequence of the following reactions

$$CH_3CH_2OH \xrightarrow[\text{Chromic Acid}]{[O]} X \xrightarrow[\text{Chromic Acid}]{[O]} CH_3COOH$$

X is [PMT-2002]
(A) CH_3COCH_3 (B) CH_3CHO
(C) CH_3OCH_3 (D) CH_3CH_2COOH

123. Ethyl alcohol when heated at 165-170°C in presence of excess of conc. H_2SO_4 gives. [PMT-2003]
(A) $C_2H_5HSO_4$ (B) $CH_2 = CH_2$
(C) $C_2H_5OC_2H_5$ (D) $(C_2H_5)_2SO_4$

124. In the reaction

$$C_2H_5MgBr + H_2C\underset{O}{\diagdown\!\diagup}CH_2 \xrightarrow{H_2O} A \text{ is}$$ [PET-94]

(A) $C_2H_5CH_2$ CHO (B) $C_2H_5CH_2CH_2OH$
(C) $C_2H_5CH_2OH$ (D) C_2H_5CHO

125. Lucas test is used to distinguish between [PET-94]
(A) 1°, 2° and 3° alcohols (B) 1°, 2° and 3° amines
(C) Aldehydes and ketones (D) Alkenes and alkynes

126. Lucas reagent is - [PET-95]
(A) Conc. HCl + Anhydrous $ZnCl_2$
(B) Dil. HCl + $ZnCl_2$
(C) H_2SO_4 + $ZnCl_2$ (D) Conc. HCl + Zn

127. Acetal is obtained by reacting in the presence of dry HCl an alcohol with [PET-96]
(A) Aldehyde (B) Ketone (C) Ether (D) Carboxylic acid

128. The final product obtained by distilling ethyl alcohol with excess of chlorine and $Ca(OH)_2$ is [PET-96]
(A) CH_3CHO (B) CCl_3CHO (C) $CHCl_3$ (D) $(CH_3)_2O$

129. The order of reactivity of halogen acids for reaction with C_2H_5OH is [PET-96]
(A) HCl > HBr > HI (B) HI > HBr > HCl
(C) HBr > HI > HCl (D) HBr > HCl > HI

130. The reagent used for the dehydration of an alcohol is : [PET-98]
(A) Phosphorus pentachloride (B) Calcium chloride
(C) Aluminium oxide (D) Sodium chloride

131. The alcohol which does not give a stable compound on dehydration is : [PET-98]
(A) Ethyl alcohol (B) Methyl alcohol
(C) n-Propyl alcohol (D) n-Butyl alcohol

132. Which of the following is a trihydric alcohol ? [PET-2001]
(A) Ethylene glycol (B) Trimethylene glycol
(C) Pinacol (D) Glycerol

133. H-bonding is most extensive in : [PET-2001]
(A) Ethanol (B) Diethyl ether
(C) Ethyl Chloride (D) Triethyl amine

134. Denatured spirit is mainly used as [PET-2002]
(A) Drug in medicine (B) Good fuel
(C) Solvent in preparing varnishes
(D) Material in preparing beverages

135. Which one has the highest boiling point [PET-2002]

(A) Acetone (B) Ethyl alcohol
(C) Diethyl ether (D) Chloroform

136. How many isomers of $C_5H_{11}OH$ will be primary alcohols ? [CBSE-92]
(A) 5 (B) 4 (C) 2 (D) 3

137. Which of the following tests is used to identify primary, secondary and tertiary alcohols? [AFMC-2003]
(A) Perkin's test (B) Hoffman's test
(C) Victor Meyer's test (D) All of these

138. Which one of the following can be oxidised to the corresponding carbonyl compound ?[CBSE-PMT-2004]
(A) 2-hydroxy-propane (B) Ortho-nitro-phenol
(C) Phenol (D) 2-methyl-2-hydroxy-propane

139. The OH group of an alcohol or the - COOH group of a carboxylic acid can be replaced by - Cl using [CBSE-PMT-2004]
(A) phosphorus pentachloride (B) hypochlorous acid
(C) chlorine (D) hydrochloric acid

140. Which of the following will not form a yellow precipitate on heating with an alkaline solution of iodine? [CBSE-PMT-2004]
(A) $CH_3CH(OH)CH_3$ (B) $CH_3CH_2CH(OH)CH_3$
(C) CH_3OH (D) CH_3CH_2OH

141. Which of the following product is formed on acid catalysed hydration of 2-phenyl propene? [IIT-2004]
(A) 1-phenyl-2-propanol (B) 1-phenyl-1-propanol
(C) 1-phenyl-3-propanol (D) 2-phenyl-2-propanol

Alkanols

1.	(A)	51.	(D)	101.	(A)
2.	(B)	52.	(A)	102.	(C)
3.	(D)	53.	(A)	103.	(A)
4.	(B)	54.	(B)	104.	(A)
5.	(A)	55.	(A)	105.	(A)
6.	(B)	56.	(A)	106.	(D)
7.	(A)	57.	(B)	107.	(B)
8.	(C)	58.	(B)	108.	(C)
9.	(C)	59.	(C)	109.	(C)
10.	(B)	60.	(A)	110.	(C)
11.	(B)	61.	(B)	111.	(B)
12.	(B)	62.	(A)	112.	(A)
13.	(C)	63.	(C)	113.	(C)
14.	(C)	64.	(B)	114.	(C)
15.	(C)	65.	(C)	115.	(A)
16.	(B)	66.	(C)	116.	(C)
17.	(B)	67.	(A)	117.	(B)
18.	(C)	68.	(C)	118.	(A)
19.	(B)	69.	(C)	119.	(D)
20.	(D)	70.	(D)	120.	(A)
21.	(D)	71.	*	121.	(A)
22.	(B)	72.	(A)	122.	(B)
23.	(B)	73.	(B)	123.	(B)
24.	(A)	74.	(C)	124.	(B)
25.	(C)	75.	(B)	125.	(A)
26.	(B)	76.	(B)	126.	(A)
27.	(D)	77.	*	127.	(A)
28.	(D)	78.	(C)	128.	(C)
29.	(B)	79.	(C)	129.	(B)
30.	(B)	80.	(D)	130.	(C)
31.	(C)	81.	(C)	131.	(B)
32.	(C)	82.	(C)	132.	(D)
33.	(A)	83.	(D)	133.	(A)
34.	(C)	84.	(A)	134.	(C)
35.	(A)	85.	(B)	135.	(B)
36.	(C)	86.	(C)	136.	(B)
37.	(B)	87.	(D)	137.	(C)
38.	(A)	88.	(D)	138.	(A)
39.	(A)	89.	(D)	139.	(A)
40.	(B)	90.	(B)	140.	(C)
41.	(D)	91.	(D)	141.	(D)
42.	(D)	92.	(B)		
43.	(B)	93.	(B)		
44.	(B)	94.	(D)		
45.	(C)	95.	(D)		
46.	(A)	96.	(D)		
47.	(A)	97.	(A)		
48.	(A)	98.	(A)		
49.	(D)	99.	(D)		
50.	(A)	100.	(A)		

71. secondary 77. Ethyl alcohol 139. Ethylene

Phenols

If OH group is directly attached to the benzene ring these are called Phenols.

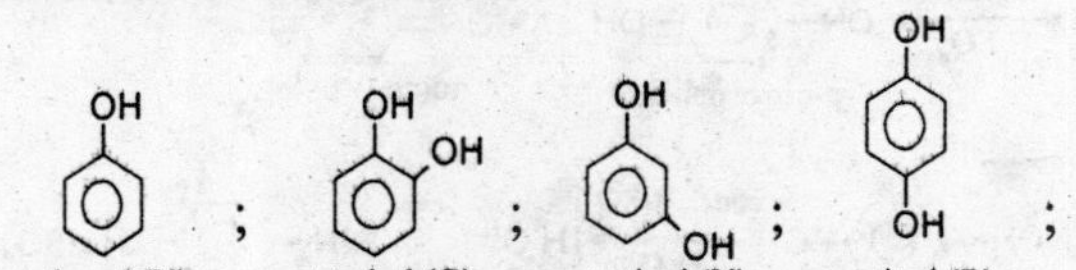

phenol (V) ; catechol (G) ; resorcinol (V) ; quinol (R) ;

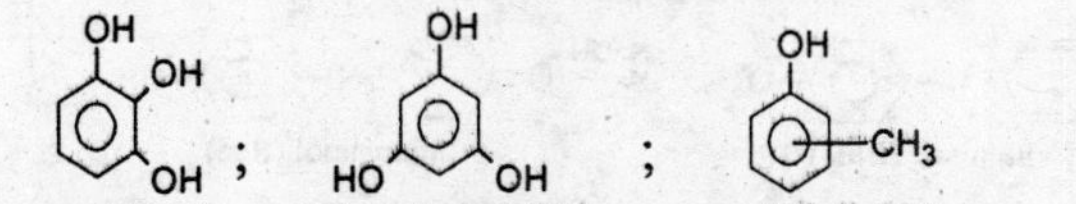

pyrogallol (R) ; phloroglucinol (DV) ; cresol (G, B, BV)

Test for phenolic –OH

(i) with $FeCl_3$: Phenols form coloured complexes with $FeCl_3$ chloride. Aqueous or alcoholic solution of the phenolic compound turns to blue, green, red or violet colours.

$6PhOH + FeCl_3 \quad [Fe(OPh)_6]^{3-} + 3H^{+} + 3HCl$

(ii) Liebermann's test:

Phenol + $NaNO_2$+ Conc. H_2SO_4 Deep green or blue colour (on dilution with H_2O) red colour (on making alkaline with NaOH) original green or blue colour retained

Preparation:

(1) 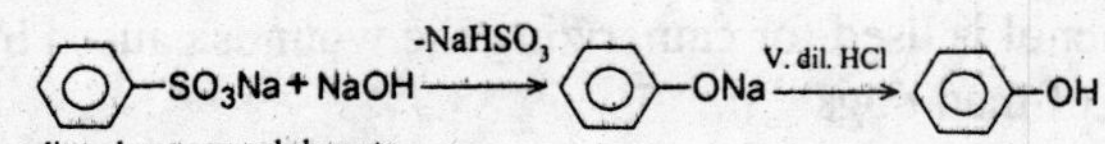

sodium benzene sulphonate

(2) 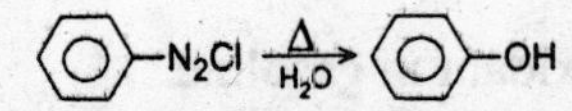

benzene diazonium chloride

(3) 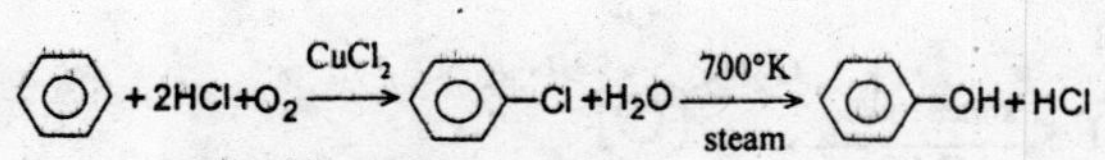

Rasching process

(4) 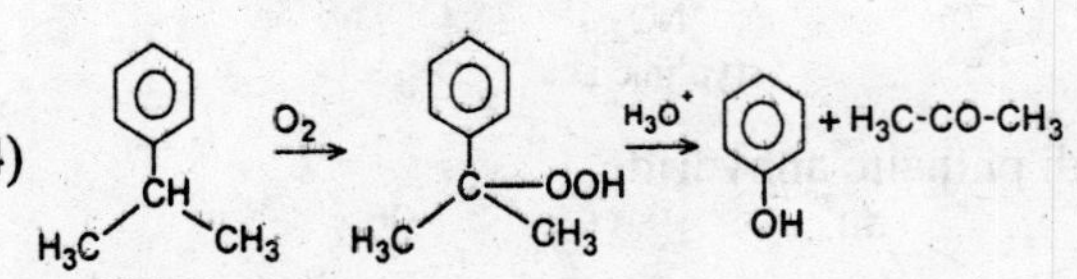

cumene

(5) 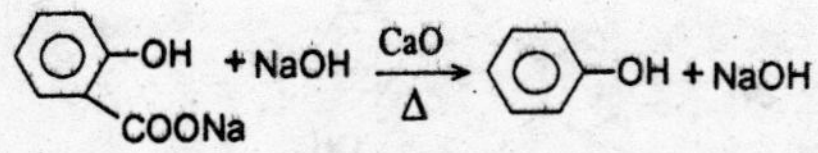

Acidity of phenols: Carboxylic acids > Phenols > Alcohols > Acetylenes

(i) 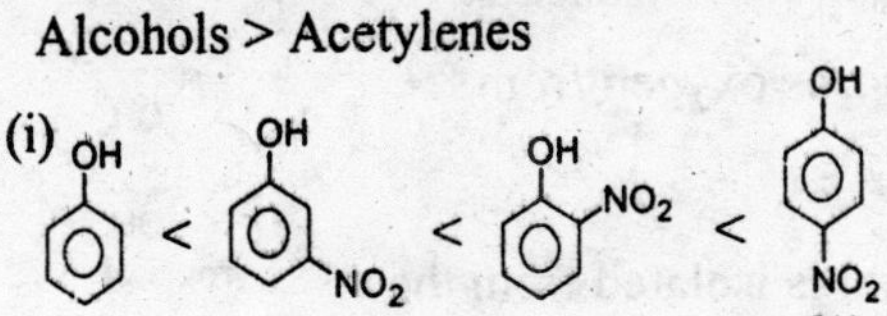

(ii) 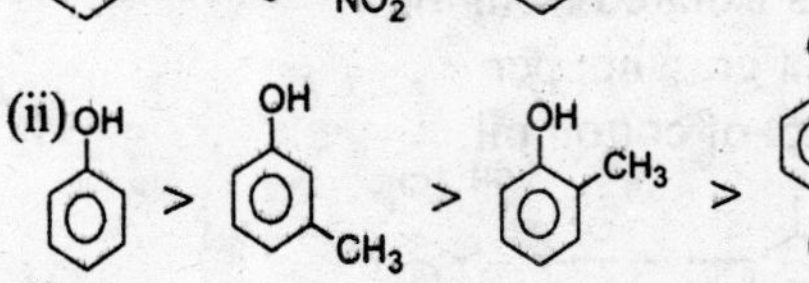

(iii) 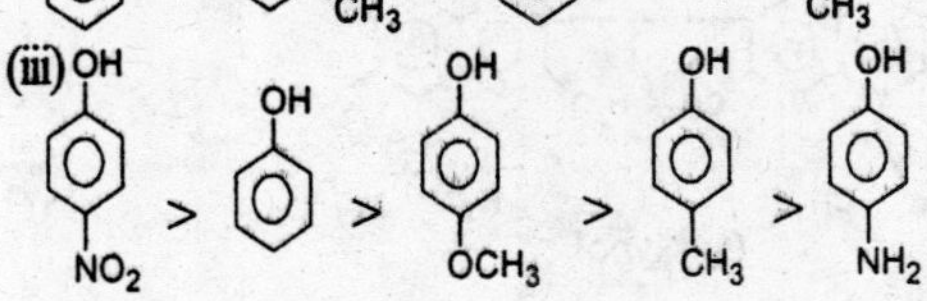

Phenols can be separated from carboxylic acids by means of their insolubility in biocarbonate solutions.

Phenols are more acidic than alcohols due to

(i) Resonance: the oxygen of O–H acquires +ve charge which helps in the release of H^+.

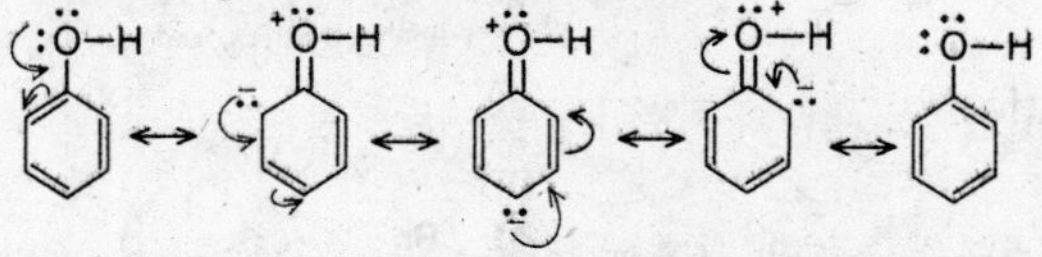

(ii) Phenol on dissociation gives phenoxide which is stabilised to a greater extent.

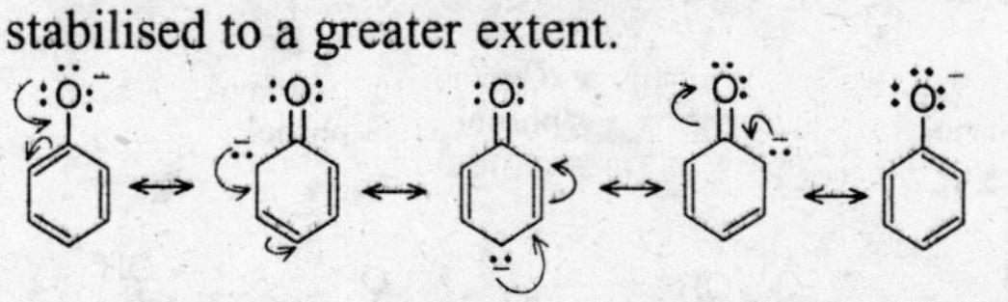

- Carbolic acid {phenol containing 5% H_2O (liquid at room temperature)}.

Solubility and boiling points:

Phenols are soluble in water in appreciable amounts because of intermolecular hydrogen bonding.

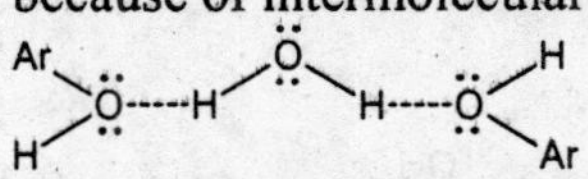

Phenols boil at higher temperature than the comparable corresponding hydrocarbons. p and m – nitrophenol have high boiling points due to intermolecular hydrogen bonding. o–nitrophenol has lower b.p. due to intramolecular hydrogen bonding.

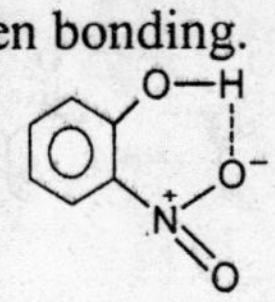

Reactions:

1. Reimer–Tiemann reaction:

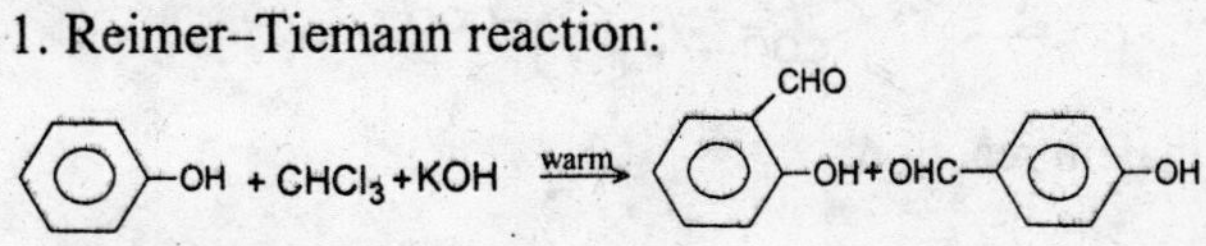

2. Friedel-craft reaction:

(i) alkylation

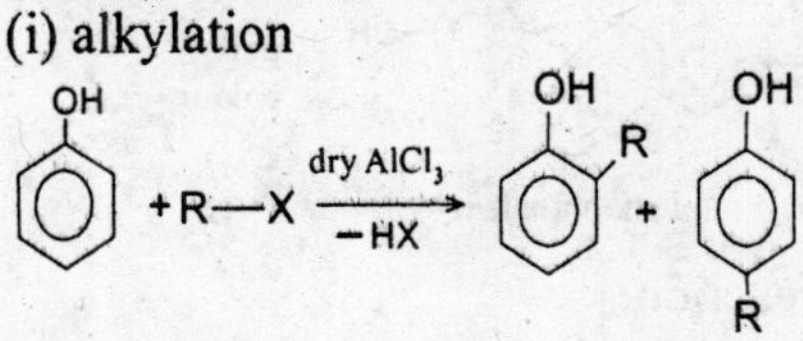

(ii) acylation

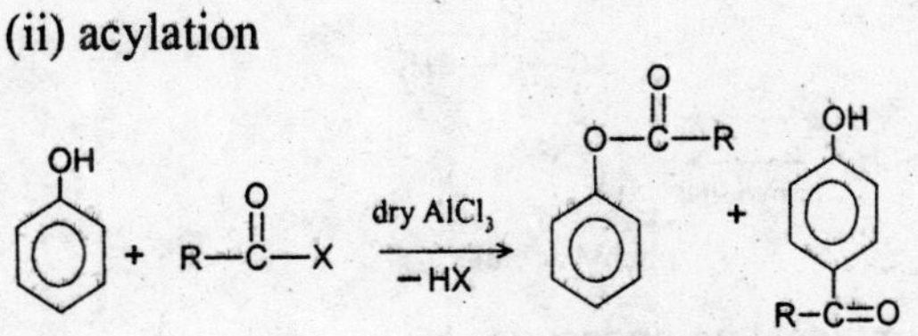

O-acylation C-acylation

3. Elbs persulphate oxidation:

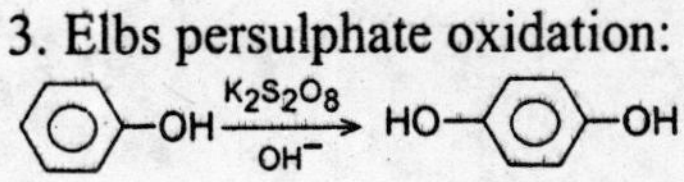

quinol

4. On exposure to air and light phenol turns pink due to slow oxidation to produce phenoquinone.

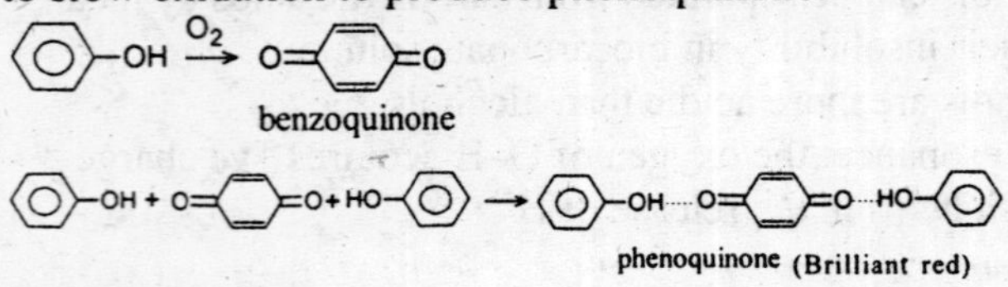

5. Halogenation:

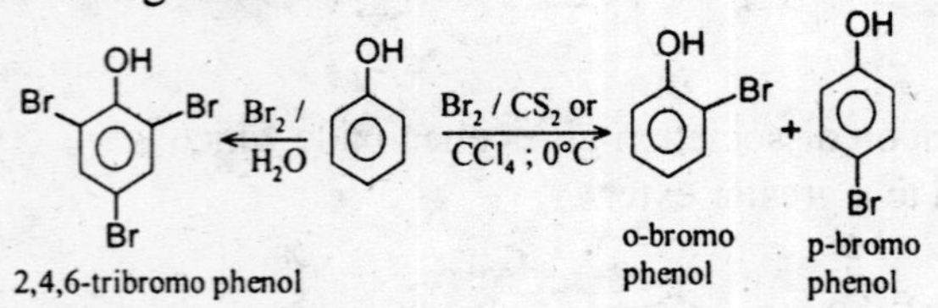

6. Nitration:

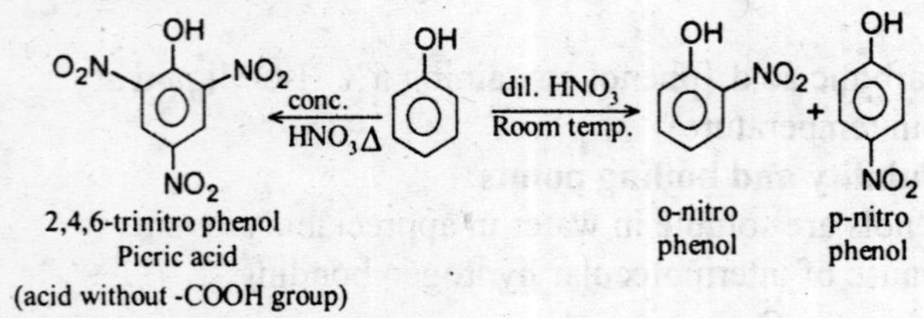

7. Kolbe – Schmidt reaction:

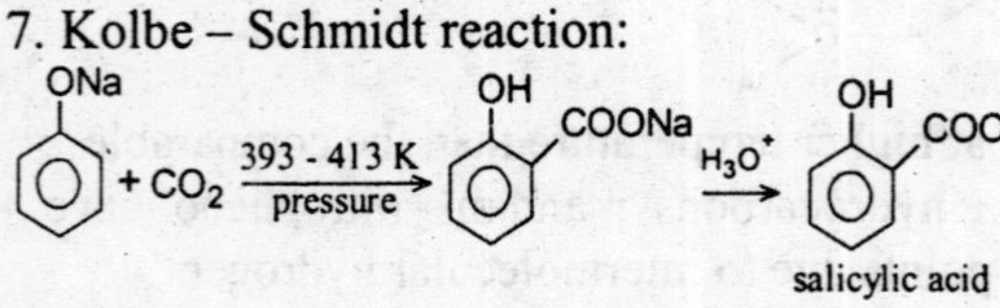

Mechanism:

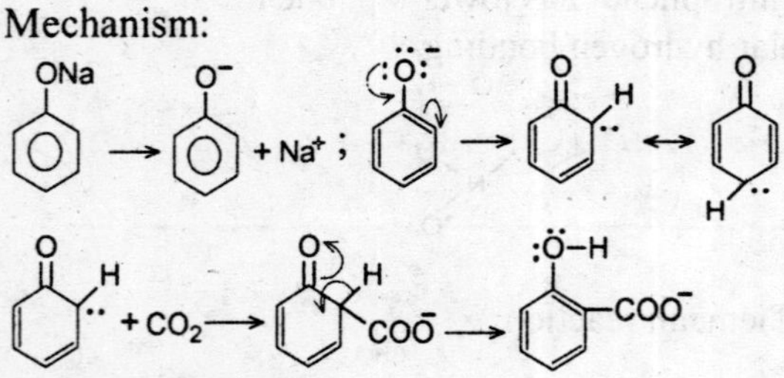

8. Phthalein reaction:

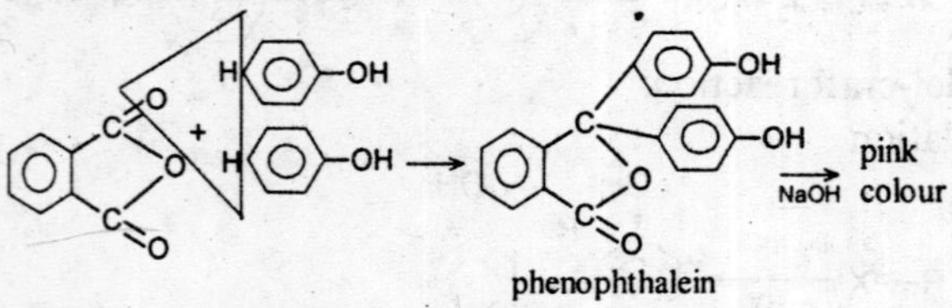

9. Schotten-Baumann reaction:

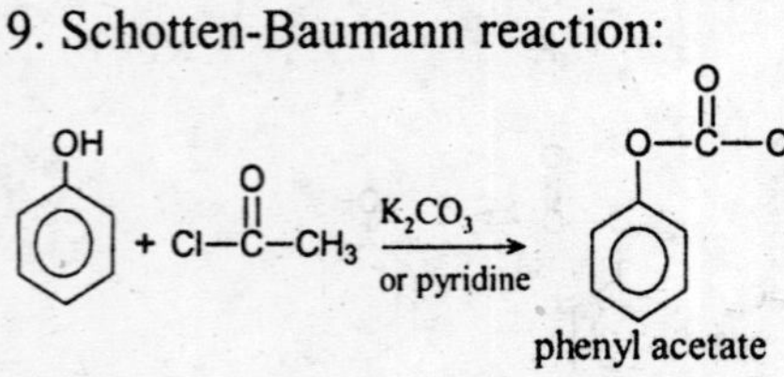

The ester, phenylacetate undergoes Fries rearrangement

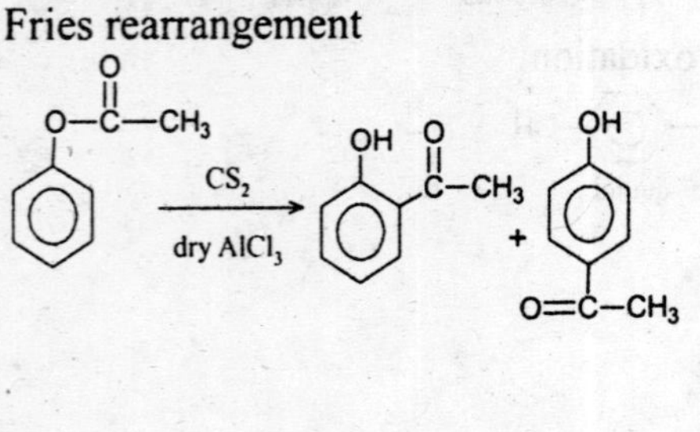

10. Liebermann's nitroso-reaction:

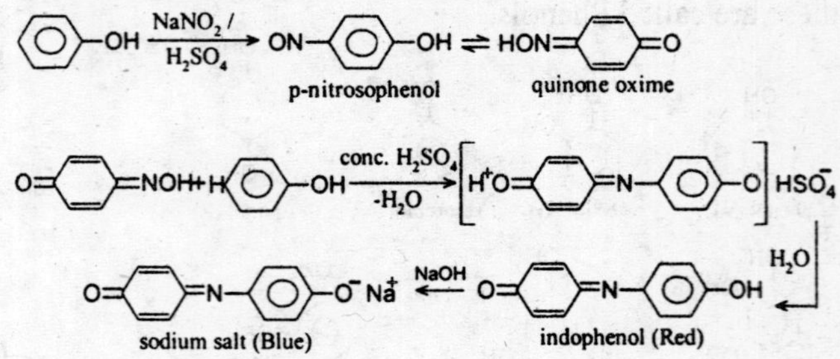

1. Reaction with HCHO

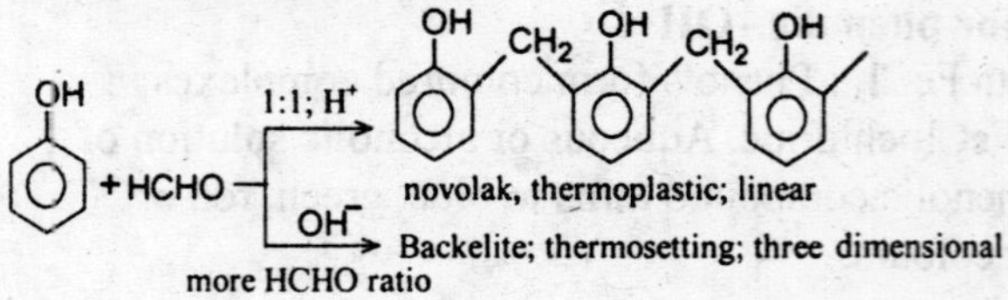

formation of methylol phenol 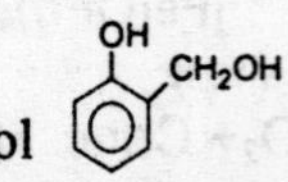 in the first step is known as Lederer–Manasse reaction.
phenol-formaldehyde resins are mainly employed for demineralisation and softening of water.

- Phenol is used for cauterizing the wounds caused by the bite of mad dogs.
- Lysol is used as disinfectant (50% soapy emulsion of cresol).
- Catechol is used as antioxidant in gasoline.

Resorcinol: it gives

(i) on nitration

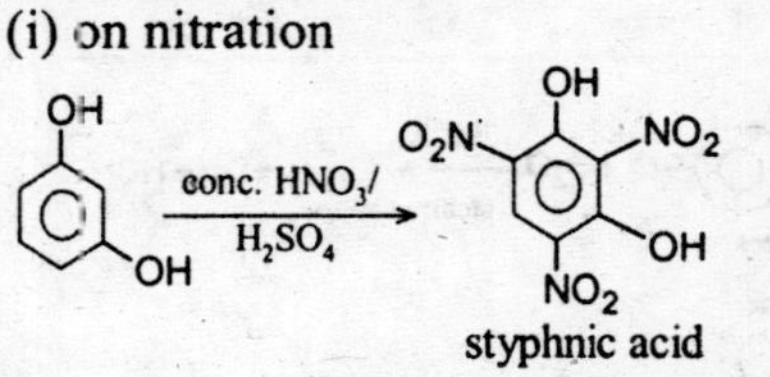

(ii) with phthalic anhydride

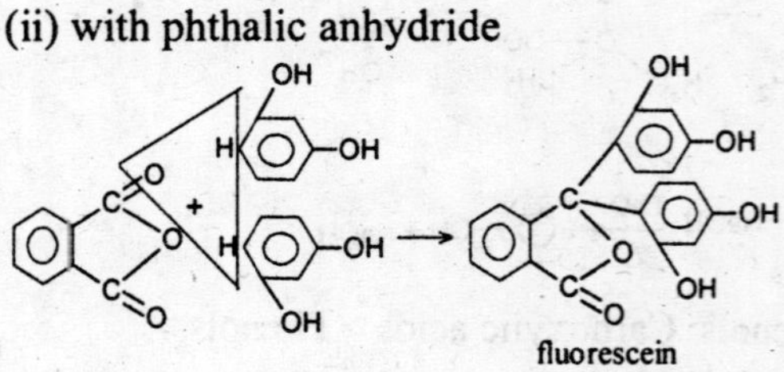

- pyrogallol: it absorbs oxygen from gaseous mixture.

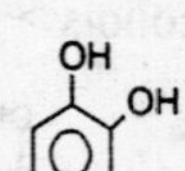

pyrogallol

- 2,5–dichlorophenol is isolated from the defensive secretion of a grasshopper
- Gossypol: By-product of cotton oil.

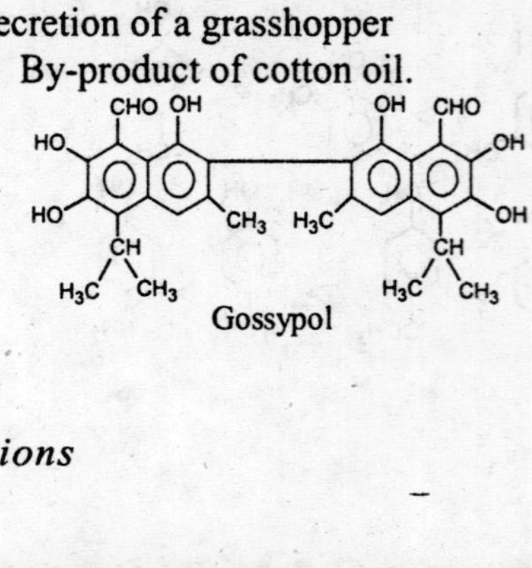

1. Identify Y in [CBSE-90]

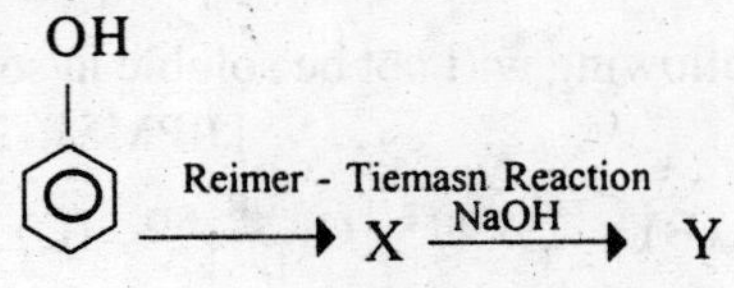

(A) 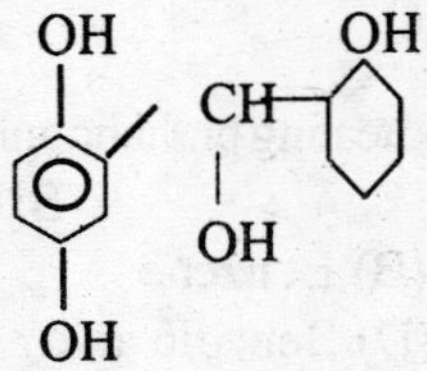

(B) 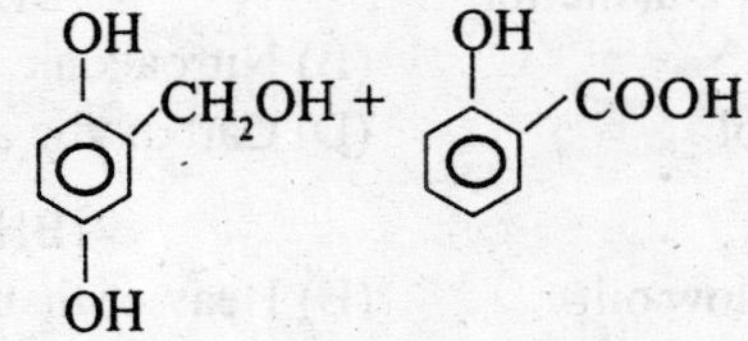

(C) 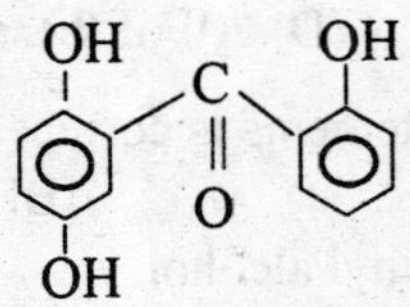

(D) 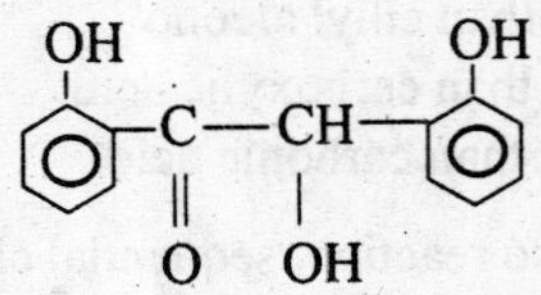

2. Sodium phenoxide reacts with CO_2 to yield a product which on hydrolysis yields a new compounds. The new compound is [CBSE-91]

(A) Phthalic acid (B) Salicylic acid
(C) Salicylaldehyde (D) Benzoic acid

3. 1, 4, 6-nitrophenol is [CBSE-91]

(A) Nucleophilic (B) Electrophilic
(C) Both (A) & (B) (D) None of these

4. Kolbe-Schmidt's reaction is used for [CBSE-91]

(A) salicyclic acid (B) salicylaldehyde
(C) phenol (D) hydrocarbon

5. When phenol is treated with excess of bromine water, the product formed would be [CBSE-92]

(A) a mixture of o- and p- bromophenol
(B) 2, 4, 6- tribromophenol
(C) m-bromophenol (D) 2, 4 dibromophenol

6. A is a higher phenol and B is an aromatic carboxylic acid Separation of a mixture of A and B can be carried out easily by [CBSE-92]

(A) sodium hydroxide (B) sodium cabonate
(C) lime (D) sodium bicarbonate

7. Bakelite is a product of the reaction between [CBSE-92]

(A) Formaldehyde and NaOH (B) Aniline and urea
(C) Phenol and methanal (D) Phenol and chloroform

8. Which one of the following compounds will give a characteristic deep colour with ferric chloride solution ? [CBSE-92]

(A) Acetone (B) Ethanol (C) Phenol (D) Acetic acid

9. The increasing order of acidity among phenol, p-methylphenol, m-nitrophenol and p-nitrophenol is [CBSE-95]

(A) m-nitrophenol, p-nitrophenol, phenol, p-methylphenol
(B) p-methyl phenol, m-nitrophenol, phenol, p-nitrophenol
(C) p-methyl phenol, phenol, m-nitrophenol, p-nitrophenol
(D) phenol, p-methylphenol, p-nitrophenol, m-nitrophenol

10. The most suitable method of the separation of a mixture of ortho and para-nitrophenol mixed in the ratio of 1 : 1 is [CBSE-99]

(A) distillation (B) crystallisation
(C) vapourisation (D) colour spectrum

11. Phenols are more acidic than aliphatic alcohols because [CBSE-2000]

(A) Phenoxide ion is stabilized by resonance
(B) Phenols are more soluble in polar solvents
(C) Phenoxide ion do not have resonance
(D) Alcohols do not loose H-atom at all

12. In the following compounds

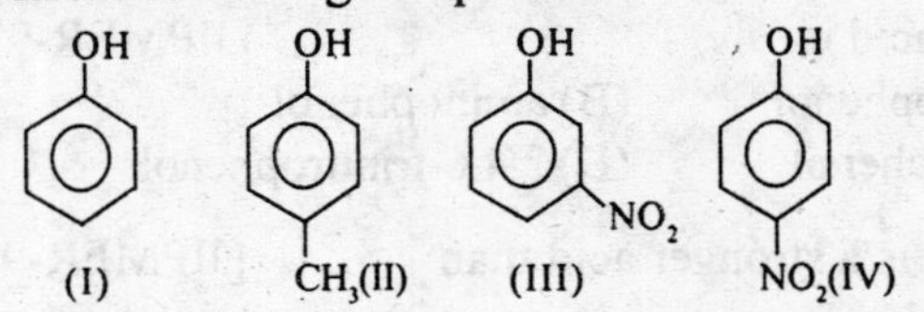

the order of acidity is [CBSE-2000]

(A) III > IV > I > II (B) I > IV > III > II
(C) II > I > III > IV (D) IV > III > I > II

13. The correct order of acidity of following is [CBSE-2001]

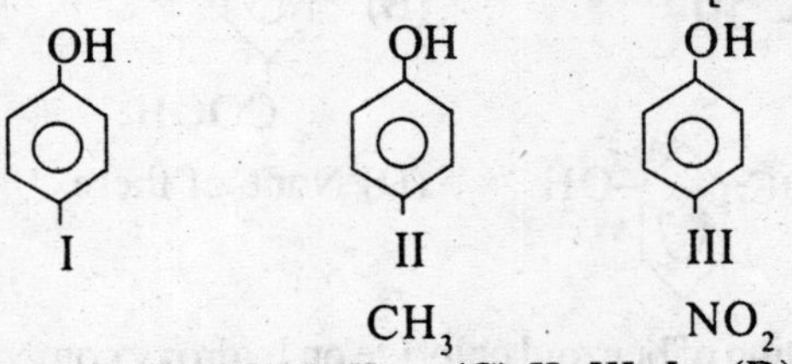

(A) I>II>II (B) III>I>II . (C) II>III>I (D) I>III>II

14. When phenol is treated with $CHCl_3$ and NaOH the product formed is [CBSE-2002]

(A) Benzaldehyde (B) Salicylaldehyde
(C) Salicylic acid (D) Benzoic acid

15. Which one of the following shows that phenol is a weak acid ? [AIIMS-93]

(A) $C_6H_5OH + 3H_2 \rightarrow C_6H_{11}OH$
(B) $C_6H_5OH +$ dil.$HNO_3 \rightarrow$ mono nitrophenol
(C) $C_6H_5OH + NH_3 \xrightarrow{ZnCl_2}$
(D) $C_6H_5OH +$ aq. $NaOH \rightarrow C_5H_5ONa + H_2O$

16. Assertion (A) : Phenol is a weaker acid than ethanol.
Reason (R) : Groups with +M effect and – I effect decrease acidity at p-position. [AIIMS-2002]
(A) A (B) B (C) C (D) D (E) E

17. Intermolecular hydrogen bonding is strongest in [AIIMS-2003]
(A) Methylamine (B) Phenol
(C) Formaldehyde (D) Methanol

18. Phenol may be distinguished from benzoic acid by reacting with : [JIPMER-90]
(A) NaOH (B) $NaHCO_3$ (C) Na_2CO_3 (D) RMgX

19. A mixture of ethanol and phenol can be separated by ? [JIPMER-92]
(A) Ion exchange (B) Sublimation
(C) Filtration (D) Fractional distillation

20. Salicylic acid upon heating with soda lime forms : [JIPMER-92]
(A) Benzoic acid (B) Benzene
(C) Benzyl alcohol (D) Phenol

21. Which of the following is most acidic ? [JIPMER-92]
(A) o-chlorophenol (B) p-nitrophenol
(C) o-aminophenol (D) chlorophenol

22. Picric acid is [JIPMER-97]
(A) m-nitrophenol (B) aminophenol
(C) p-nitrophenol (D) 2,4,6-trinitrophenol

23. Phenol is a stronger acid than [JIPMER-97]
(A) Ethanol (B) Acetic acid
(C) Chloroacetic acid (D) 2-nitrophenol

24. Sodium phenoxide when treated with CO_2 under pressure gives [JIPMER-98]
(A) 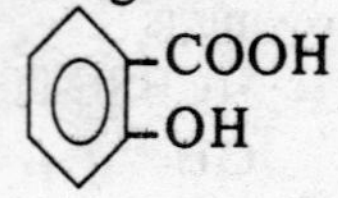
(B) 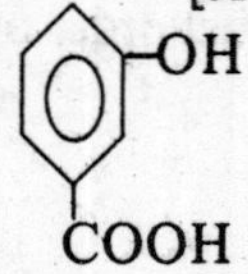
(C) 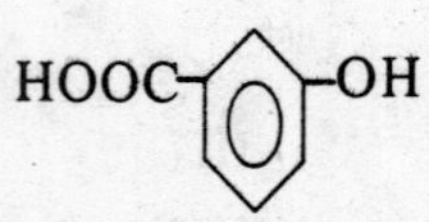
(D) None of these

25. The action of benzoyl chloride on hydroxy compounds in alkaline medium is called [JIPMER-98]
(A) Schotten-Baumann reaction
(B) Friedel-Crafts reaction
(C) Perkin's reaction (D) Kolbe's reaction

26. The reaction in which phenol condenses with chloroform and alkali is known as [JIPMER-98]
(A) Houben-Hoecsch reaction
(B) Reimer-Tiemann reaction
(C) Etard reaction (D) Cannizzaro's reaction

27. Cresol when distilled with ------- is converted into toluene. [JIPMER-99]
(A) Al_2O_3 (B) Zinc dust
(C) $LiAlH_4$ (D) Acetic anhydride

28. Which of the following will not be soluble in sodium carbonate solution ? [JIPMER-2002]
(A) 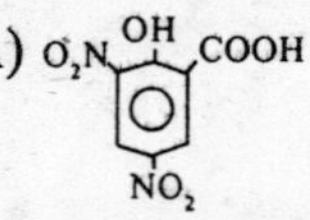
(B) 
(C) 
(D) 

29. Phenolpthalein is obtained by heating phthalic anhydride with concentrated H_2SO_4 and [BHU-96]
(A) Phenol (B) Benzene
(C) Benzyl alcohol (D) Benzoic acid

30. Picric acid is a name for [BHU-96]
(A) Phenol (B) Nitroalkane
(C) Trinitrophenol (D) Carboxylic acid

31. Cresol is [BHU-96]
(A) Greenish yellow oil (B) Heavier than water
(C) Made up of mainly cresols (D) All of these

32. Which of the following statements is correct ? [BHU-97]
(A) Phenol is less acidic than ethyl alcohol
(B) Phenol is more acidic than ethyl alcohol
(C) Phenol is more acidic than carboxylic acid
(D) Phenol is more acidic than carbonic acid

33. In Liebermann's nitroso reaction, sequential changes in the colour of phenol occurs as [BHU-99]
(A) Red → green → white (B) Green → yellow → white
(C) Red → deep blue → green
(D) Brown or red → green → red → deep blue

34. Phenol is prepared by [BHU-2001]
(A) Cumene process (B) From $C_6H_5N_2Cl$
(C) From $C_6H_5SO_3Na$ (D) All of these

35. When sodium derivative of phenol reacts with methyl chloride it form [BHU-2001]
(A) Anisole (B) Chlorobenzene
(C) Benzoic acid (D) Benoyl chloride

36. With excess bromine, phenol reacts to form [BHU-2001]
(A) 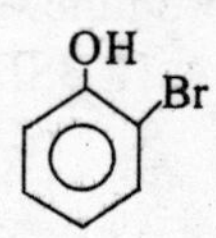
(B) 
(C) 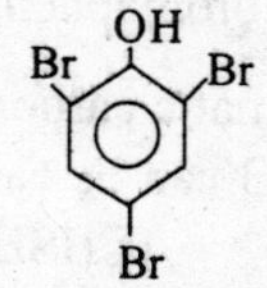
(D) mixture of A and B

37. The reaction of aromatic acyl chloride and phenol in the presence of a base NaOH or pyridine is called [AFMC-98]
(A) Kolbe's reaction (B) Perkin's reaction
(C) Sandmayer's reaction (D) Schotten Baumann reaction

38. In the Liebermann's nitroso reaction sequential changes

in the colour of phenol occurs as [AFMC-98]
(A) Brown or red → green → red→ deep blue
(B) red → deep blue → green
(C) red → green → white (D) white → red → green

39. Increasing order of acid strength is [AFMC-2002]
(A) p-Nitrophenol, p-Methoxyphenol, p-Methylphenol
(B) p-Methylphenol, p-Methoxyphenol, p-Nitrophenol
(C) p-Nitrophenol, p-Methylphenol, p-Methoxyphenol
(D) p-Methoxyphenol, p-Methylphenol, p-Nitrophenol

40. Benzene diazonium chloride on reaction with phenol in weakly basic medium gives [IIT-98]
(A) diphenyl ether (B) p-hydroxyazobenzene
(C) chlorobenzene (D) benzene

41. The most unlikely representation of resonance structures of p-nitrophenoxide ion is [IIT-99]

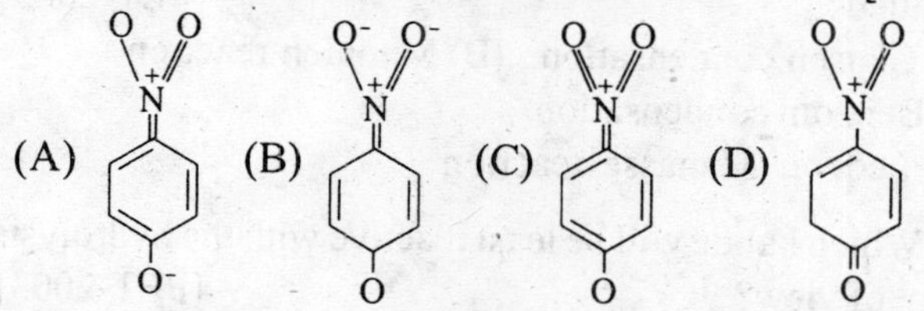

42. A : Phenol is more reactive than benzene towards electrophilic substitution reaction. R : In the case of phenol, the intermediate carbocation is more resonance stabilized.
(A) A (B) B (C) C (D) D

43. 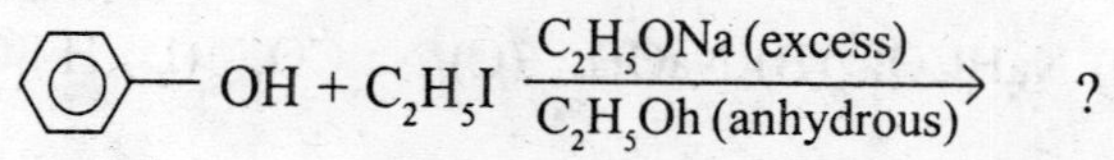
[IIT-2003]
(A) 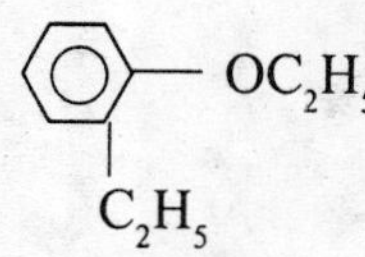 (B) 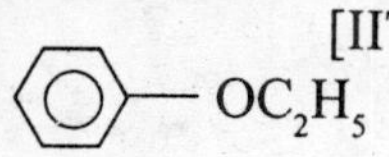
(C)  (D) C_2H_5OH

44. In which of the following reaction Phenol or Sodium phenoxide is not formed ? [CPMT-96]
(A) $C_6H_5N_2Cl$ + Alc. KOH (B) $C_6H_5OCl + NaOH$
(C) $C_6H_5N_2Cl$ + Aq NaOH (D) $C_6H_5NNCl \xrightarrow[\Delta]{H_2O}$

45. The best method to separate the mixture of ortho-and paranitrophenol (1:1) is - [CPMT-97]
(A) Distillation (B) Crystallization
(C) Vaporisation (D) Colour spectrum

46. Which acid will not form hydrocarbon ? [CPMT-97]
(A) Cinnamic acid (B) Isophthalic
(C) Salicylic acid (D) Picric acid

47. Phenol + Zn (powder) .... ? [CPMT-97]
(A) Benzene (B) Benzaldehyde
(C) Anisole (D) Phenyl acetate

48. Which of the following is most acidic? [CPMT-99]
(A) Phenol (B) Benzyl alcohol
(C) m-chlorophenol (D) Cyclohexanol

49. Cresol has [CPMT-2003]
(A) alcoholic – OH (B) Phenolic - OH
(C) –COOH (D) –CHO

50. Carbolic acid is : [PMT-93]
(A) Phenol (B) Methanol
(C) Formic acid (D) Acetic acid

51. Addition of Bromine water to an aqueous solution of phenol gives : [PMT-93]
(A) o- Bromophenol (B) p-Bromophenol
(C) m-Bromophenol (D) 2, 4, 6, - tribromophenol

52. Phenol is obtained by heating aqueous solution of [PMT-95]
(A) aniline (B) benzene diazonium chloride
(C) benzoic acid (D) none of these

53. For Phenol, which of the following statements is correct ? [PMT-95]
(A) It is insoluble in water
(B) It has lower melting point compared to aromatic hydrocarbons of comparable molecular weight
(C) It has higher boiling point than toluene
(D) It does not show acidic property

54. Which of the following does not contain - COOH group - [PMT-97]
(A) Citric acid (B) Lactic acid
(C) Malonic acid (D) Picric acid

55. The distillation of phenol with zinc dust gives : [PMT-99]
(A) C_6H_6 (B) $C_6H_5–C_6H_5$ (C) C_6H_{12} (D) $C_6H_5OC_6H_5$

56. The reaction of phenol with conc. HNO_3 gives : [PMT-99]
(A) picric acid (B) benzoic acid
(C) salicylic acid (D) phthalic acid

57. Which of the following reacts with $NaNO_2$ + HCl to give alcohol/phenol : [PMT-2000]
(A) $C_6H_5CH_2NHCH_3$ (B) $(CH_3)_2NH$
(C) CH_3NH_2 (D) $C_6H_5NH_2$

58. When phenol is distilled with Zn dust it gives : [PMT-2001]
(A) Benzaldehyde (B) Benzoic acid
(C) Toluene (D) Benzene

59. The reaction

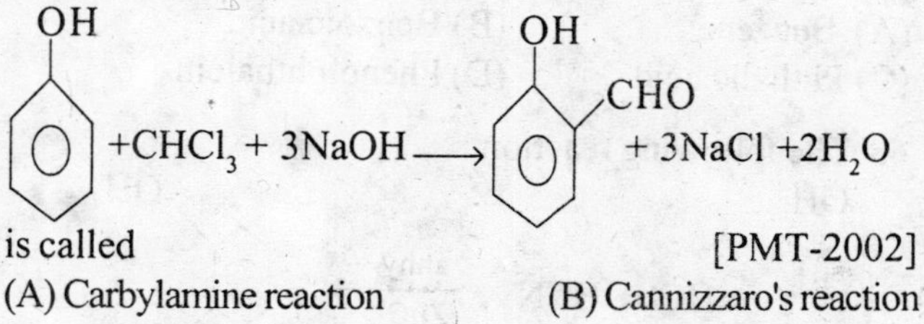

is called [PMT-2002]
(A) Carbylamine reaction (B) Cannizzaro's reaction

(C) Reimer Tiemann reaction (D) Wurtz reaction

60. The main product of deoxygenation of the carboxylated phenol obtained by Reimer Tiemann method would be : [PMT-2003]

(A) 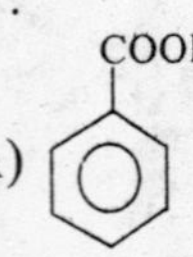 (B) 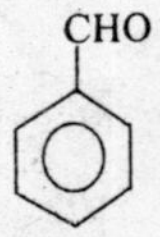 (C) 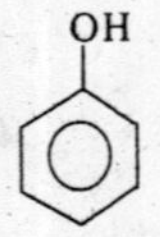 (D) 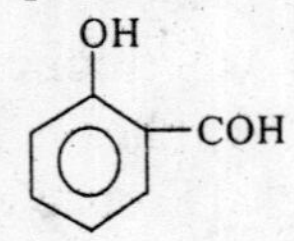

61. On Liebermann nitroso test phenol gives : [PMT-2003]
(A) Red colour (B) Blue colour
(C) Yellow colour (D) Red-brown colour

62. Phenol on reaction with bromine water will give : [PMT-2003]

(A)  (B) 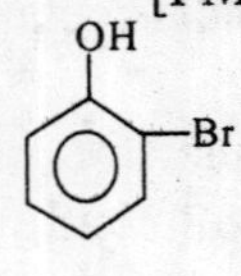

(C) 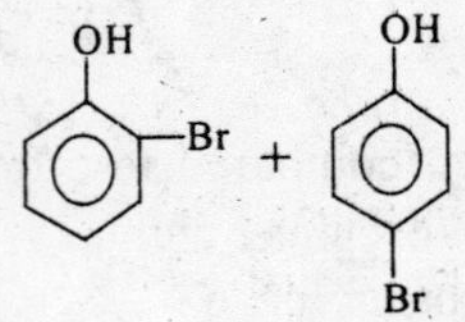 (D) 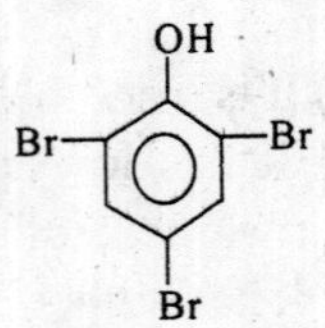

63. When phenol is treated with excess bromine water, it gives [PET-93]
(A) m-Bromophenol (B) O-and-p- Bromophenol
(C) Bromophenol (D) 2,4,6-Tribromophenol

64. An aromatic compound is insoluble in water, it dissolves in aq. NaOH but does not dissolve in aq. $NaHCO_3$. It gives deep red colouration with $FeCl_3$. The compound is [PET-94]
(A) An alcohol (B) A phenol (C) An aldehyde (D) A ketone

65. The reaction

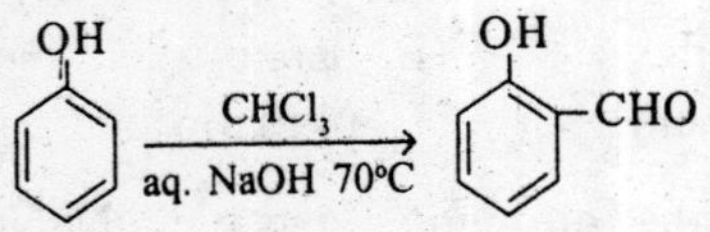

is called [PET-94]
(A) Wittig reaction (B) Williamson's Synthesis
(C) Wurtz-Fittig reaction (D) Reimer-Tiemann Reaction

66. Sodium phenoxide reacts with CO_2 at 400 K and 4-7 atm pressure to give [PET-96]
(A) Sodium salicylate (B) Salicylaldehyde
(C) Catechol (D) Benzoic acid

67. When phenol is heated with phthalic anhydride in the presence of concentrated H_2SO_4, we get : [PET-97]
(A) Benzene (B) Benzoic acid
(C) Phthalic acid (D) Phenolphthalein

68. The following reaction :

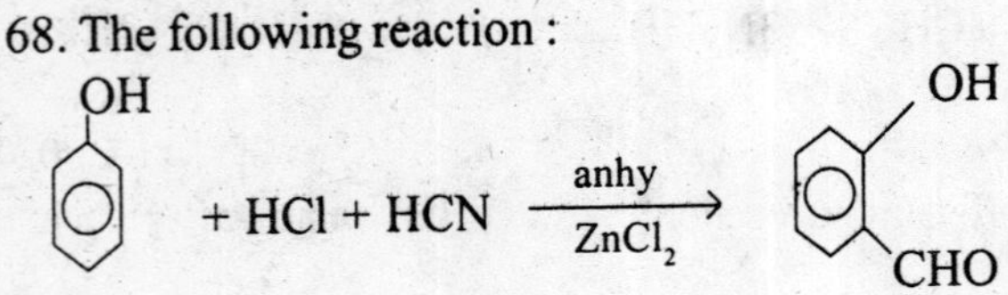

is known as : [PET-97]
(A) Perkin reaction (B) Gattermann reaction
(C) Kolbe reaction (D) Gattermann-Koch reaction

69. Carbolic acid is : [PET-98]
(A) Phenol (B) Phenyl benzoate
(C) Phenyl acetate (D) Salol

70. Carbolic acid is :- [PET-2003]
(A) Phenol (B) Ethanol
(C) Methanol (D) Benzyl alcohol

71. Reaction

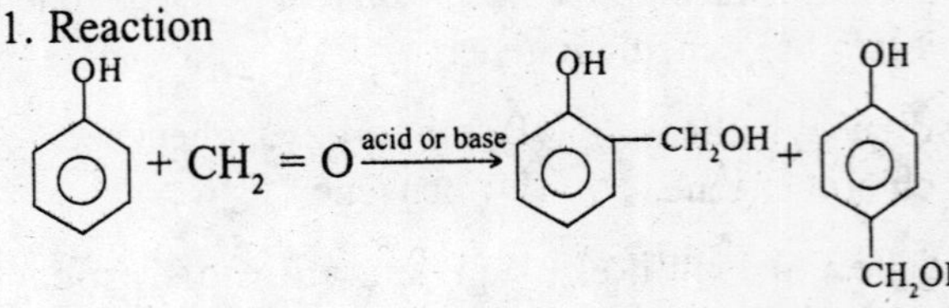

is called : [PET-2003]
(A) Claisen condensation (B) Mannich reaction
(C) Benzoin condensation
(D) Lederer Manasse reaction

72. Which halide will be least reactive with the hydrolysis point of view? [PET-2003]
(A) Methyl chloride (B) Vinyl chloride
(C) Allyl chloride (D) t- Butyl chloride

73. Phenol and benzoic acid are distinguished by [BHU-2003]
(A) $NaHCO_3$ (B) NaOH (C) Na_2CO_3 (D) H_2SO_4

Phenols

1.	(B)	51.	(D)
2.	(B)	52.	(B)
3.	(A)	53.	(C)
4.	(A)	54.	(D)
5.	(B)	55.	(A)
6.	(D)	56.	(A)
7.	(C)	57.	(C)
8.	(C)	58.	(D)
9.	(C)	59.	(C)
10.	(A)	60.	(B)
11.	(A)	61.	(A)
12.	(D)	62.	(D)
13.	(B)	63.	(D)
14.	(B)	64.	(B)
15.	(D)	65.	(D)
16.	(D)	66.	(A)
17.	(B)	67.	(D)
18.	(B)	68.	(B)
19.	(D)	69.	(A)
20.	(D)	70.	(A)
21.	(B)	71.	(D)
22.	(D)	72.	(B)
23.	(A)	73.	(A)
24.	(A)		
25.	(A)		
26.	(B)		
27.	(B)		
28.	(C)		
29.	(A)		
30.	(C)		
31.	(A)		
32.	(B)		
33.	(D)		
34.	(D)		
35.	(A)		
36.	(C)		
37.	(D)		
38.	(A)		
39.	(B)		
40.	(B)		
41.	(C)		
42.	(A)		
43.	(B)		
44.	(B)		
45.	(A)		
46.	(D)		
47.	(A)		
48.	(C)		
49.	(B)		
50.	(A)		

Ethers

Ethers can be regarded as anhydrides of alcohols. They are lewis bases (weak as alcohols).

The angle is 110°C due to lone pair-lone pair repulsion of electrons on oxygen atom. 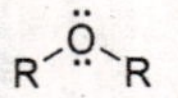 Ethers show metamerism and functional isomerism.

Preparation:

1. Williamson's synthesis: S_N^2

$$R-\boxed{X + Na}OR \xrightarrow{\text{ether}} ROR$$

Primary and secondary halides yield ethers while tertiary halides form predominantly alkenes.

2. Williamson's continuous etherification:

$$C_2H_5OH + H_2SO_4 \xrightarrow{\sim 100°C} C_2H_5HSO_4 \begin{cases} \xrightarrow[140°C]{C_2H_5OH \text{ excess}} C_2H_5OC_2H_5 \\ \xrightarrow[\text{excess, } 160°C]{\text{conc. } H_2SO_4} H_2C{=}CH_2 \end{cases}$$

$C_2H_5OC_2H_5$ is also called as sulphuric ether as H_2SO_4 is employed in its preparation.

3. Dehydration of alcohols:

Simple or symmetrical ethers can be obtained by heating excess of alcohol with conc. H_2SO_4 or H_3PO_4. Primary alcohols are mainly employed in this method as secondary and tertiary alcohols form alkenes predominantly.

$$S_N1\text{: (i) } R-\ddot{O}-H + H^+ \xrightleftharpoons{H_2SO_4} R-\overset{+}{O}(H)-H \xrightarrow{\text{slow}} R^+ + H_2O$$

$$\text{(ii) } R^+ + R-\ddot{O}-H \xrightarrow{\text{fast}} R-\overset{+}{O}(R)-H \xrightarrow{-H^+} R-\ddot{O}-R$$

$$S_N2\text{: } R-\ddot{O}-H + H^+ \xrightleftharpoons{H_2SO_4} R-\overset{+}{O}(H)-H$$

$$R-\ddot{O}(H): + R-\overset{+}{O}(H)-H \xrightarrow{\text{slow}} [R-\overset{H}{O}\cdots R\cdots\overset{H}{O}-H] \xrightarrow[-H_2O]{\text{fast}} R-\overset{+}{O}(H)-R + \overset{H}{O}-H \xrightarrow{-H^+} R-O-R$$

4. Methyl ethers:

$$\underset{}{ROH} + \underset{\text{diazomethane}}{CH_2N_2} \xrightarrow{BF_3} R-O-CH_3 + N_2$$

5. Drahowzal method:

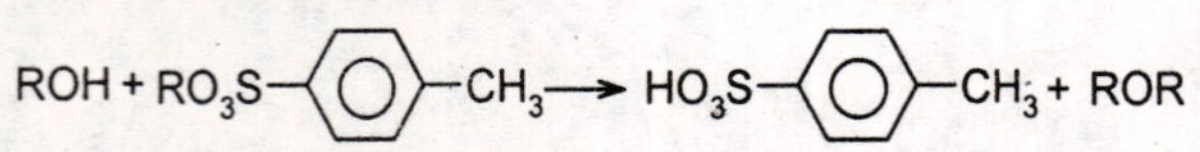

Properties: Due to the absence of any functional group or unsaturation, ethers are less reactive. Solubility of lower ethers in H_2O is mainly due to hydrogen bonding.

Reactions:

1. With HI:

(i) $CH_3OCH_3 + HI\ (\text{Cold}) \rightarrow CH_3OH + CH_3I$

(ii) $CH_3 \,\vdots\, OCH_2CH_3 + HI \xrightarrow[\text{cold}]{} CH_3I + CH_3CH_2OH$

(iii) $CH_3 \,\vdots\, O \,\vdots\, CH_3 + 2HI \xrightarrow{\Delta} 2CH_3I + H_2O$

(iv) $CH_3 \,\vdots\, O \,\vdots\, CH_2CH_3 + 2HI \xrightarrow{\Delta} CH_3I + H_2O + CH_3CH_2I$

In the case of mixed ethers, the halogen atom attaches to a smaller and less complex alkyl group.

Mechanism :

$$R-O-R + H^+ \longrightarrow R-\overset{+}{O}(H)-R$$

$$S_N1\text{ : (i) } R-\overset{+}{O}(H)-R \xrightarrow{\text{slow}} R^+ + ROH$$

$$\text{(ii) } R^+ + X^- \xrightarrow{\text{fast}} R-X$$

$$S_N2\text{ : } X^- + R-\overset{+}{O}(H)-R \xrightarrow{\text{slow}} [\overset{\delta^-}{X}\cdots R\cdots\overset{\delta^+}{O}(H)-R] \rightarrow R-X + ROH$$

Order or reactivity: HI > HBr > HCl

This reaction is the basis of Zeisel's method (Estimation of alkoxy group)

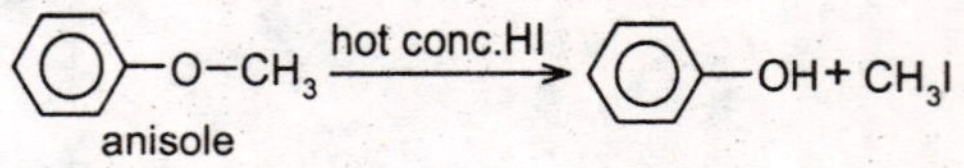

2. Oxidation:

$$CH_3CH_2OCH_2CH_3 \begin{cases} \xrightarrow[\text{sunlight}]{\text{moist air or } O_2} (C_2H_5)_2O \rightarrow O \text{ Diethyl peroxide} \\ \xrightarrow[K_2Cr_2O_7\ /\ H^+]{} CH_3CHO \xrightarrow[K_2Cr_2O_7\ /\ H^+]{} CH_3COOH \end{cases}$$

3. With acid chloride:

$$C_2H_5 \,\vdots\, OC_2H_5 + RCOCl \xrightarrow{\text{anhyd. } AlCl_3} C_2H_5Cl + RCOOC_2H_5$$

4. With acid anhydride:

$$CH_3CH_2O \,\vdots\, CH_2CH_3 + (RCO)_2O \xrightarrow{ZnCl_2} 2CH_3CH_2COOR$$

5. Hydrolysis:

$$C_2H_5OC_2H_5 + H_2O \xrightarrow{\Delta} 2C_2H_5OH$$

6. Dehydration:

$$C_2H_5OC_2H_5 \xrightarrow[380°C]{Al_2O_3} 2H_2C{=}CH_2 + H_2O$$

7. With H_2SO_4:

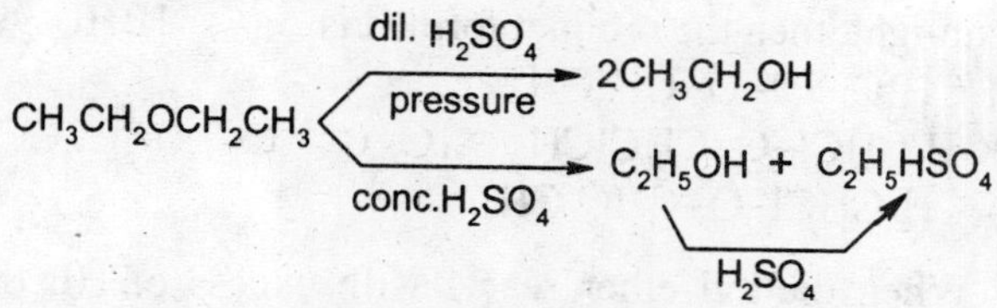

- Mixed with $CHCl_3$ and C_2H_5OH, ether is used as general anaesthetic.
- As a refrigerant with solid CO_2 (gives -110°C)
- Alcohol and ether mixture can be used as a substitute of petrol under the trade name Natalite.

Crown Ethers

Crown ethers are cyclic ethers containing several- 4,5,6 or more - oxygen atoms. Crown ethers have attracted varied applications in the field of anlytical chemistry e.g.

1. Extraction- separation.
2. Organic chemistry.
3. Ion chromatography.
4. Ion selective electrode.
5. Membrane transport.
6. Biological application etc.

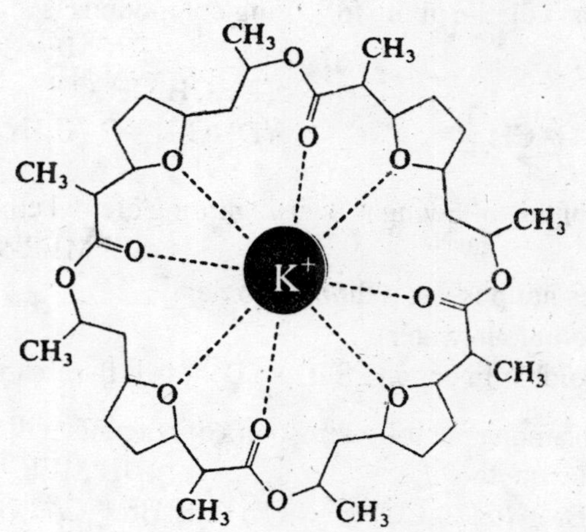

1. Reaction of t-butyl bromide with sodium methoxide produces [CBSE-94]
(A) isobutane (B) isobutylene
(C) t-buty methyl ether (D) sodium t-butoxide

2. Assertion : Ethers behave as bases in the presence of mineral acids. Reason : Due to the presence of lone pair of electrons on the oxygen. [AIIMS-94]
(A) A (B) B (C) C (D) D

3. What is the correct bond angle in dimethyl ether ? [AIIMS-95]
(A) 109° (B) 110° (C) 120° (D) 180°

4. Absolute ether is diethyl ether, which [AIIMS-96]
(A) Is mixed with absolute alcohol
(B) Contains trace of peroxide
(C) Is obtained from Williamson's synthesis
(D) Is totally free moisture and peroxides

5. Assertion (A) : Ethers behave as bases in the presence of mineral acids. Reason (R) : It is due to, the presence of lone pair of electrons on the oxygen. [AIIMS-2002]
(A) A (B) B (C) C (D) D (E) E

6. The percentage of ethyl alcohol in rectified spirit is [JIPMER-94]
(A) 100 (B) 59.6 (C) 95.6 (D) 50

7. Which of the following is used as an anaesthetic ? [JIPMER-94]
(A) CH_3OCH_5 (B) $CH_3C_2H_5$
(C) $C_2H_5OC_2H_5$ (D) $C_3H_7OC_2H_5$

8. The most volatile of the following compounds is [JIPMER-97]
(A) H_2O (B) CH_3CH_2OH
(C) $CH_3–O–CH_3$ (D) $C_2H_5–O–C_2H_5$

9. Which of the following is false regarding diethyl ether? [JIPMER-98]
(A) It does not possess a dipole moment
(B) It is soluble in water
(C) It is soluble in conc.H_2SO_4 (D) It is inflammable

10. Which among the following pairs of reagents will not give rise to an ether ? [JIPMER-98]
(A) $C_2H_5Br + (CH_3)_3CONa$ (B) $C_2H_5Br + C_2H_5ONa$
(C) $C_6H_5Br + C_3H_7ONa$ (D) $C_2H_5Br + C_6H_5ONa$

11. Diethyl ether when treated with Cl_2 in the dark gives mostly [JIPMER-99]
(A) C_2H_5Cl (B) $CH_3CHClOCH_2CH_3$
(C) $CH_2Cl.CH_2O.CH_2CH_2Cl$
(D) $CH_3CHClOCHClCH_3$

12. Diethyl ether on heating with conc. HI gives two moles of [JIPMER-2001]
(A) Methyl iodide (B) Ethyl iodide
(C) Iodoform (D) Ethanol

13. When diethyl ether is treated with Cl_2, in the presence of sunlight, then the product formed is [BHU-99]
(A) $CH_3CHCl–O–CH_2CH_3$
(B)$CH_3CHCl–O–CHClCH_3$ (C) $CCl_3CCl_2OCCl_2CCl_3$
(D) $CH_3CHCl–O–CHClCH_2$

14. When diethyl ether reacts with excess of HI, the product is [AFMC-93]
(A) Ethanol (B) Water (C) Ethyl iodide (D) Hydrogen

15. Which of the following is simple ether ? [AFMC-97]
(A) $C_2H_5OCH_3$ (B) CH_3OCH_3
(C) Phenyl-O-CH_3 (D) All are simple ethers

16. An organic compound A reacts with sodium metal and forms B. On heating with conc. H_2SO_4, A gives diethyl ether. Then A and B are [AFMC-98]
(A) C_2H_5OH and C_2H_4ONa (B) C_3H_7OH and CH_3ONa
(C) CH_3OH and CH_3ONa (D) C_4H_5OH and C_4H_2ONa

17. Aliphatic ethers are purified by shaking with a solution of ferrous salt to remove .... which are formed on prolonged standing in contact with air.[IIT-92]

18. The ether 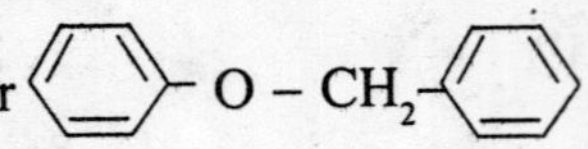 when treated with HI produces [IIT-99]
(A) 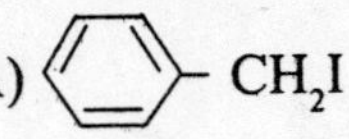 (B) 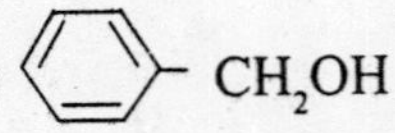
(C) 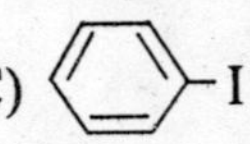 (D) 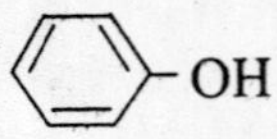

19. The general formula $C_nH_{2n}O_2$ could be for open chain [AIEEE-2003]
(A) Diketones (B) Carboxylic acids
(C) Diols (D) Dialdehydes

20. An ether is more volatile than an alcohol having the same molecular formula. This is due to [AIEEE-2003]
(A) Dipolar character of ethers.
(B) Alcohols having resonance structures.
(C) Inter-molecular hydrogen bonding in ethers.
(D) Inter-molecular hydrogen bonding in alcohols.

21. One mole of Diethyl ether on heating with con. HI. gives two moles of : [EAMCET-90]
(A) C_2H_5OH (B) CHI_3 (C) CH_3I (D) C_2H_5I

22. Old bottles of ether are dangerous because they frequently contain high concentration of .... [EAMCET-93]

23. When an alkyl halide reacts with an alkoxide the product is ? [EAMCET-96]
(A) Hydrocarbon (B) Unsaturated hydrocarbon
(C) Ether (D) Alcohol

24. Alkyl halide is not obtained from one of the following reactions. Which one is that ? [EAMCET-98(MBBS)]

(A) Diethyl ether $\xrightarrow{\text{Reduction}}$ x $\xrightarrow{SOCl_2}$
(B) Diethyl ether + HI →
(C) Diethyl ether + PCl_5 → (D) Diethyl ether + Cl_2 →

25. The solvent used in the preparation of Grignard's reagent is [EAMCET-99]
(A) dry ether (B) dry acetone
(C) dry alcohol (D) dry chloroform

26. Which of the following reaction conditions are used for preparing diethyl ether from ethyl alcohol ? [EAMCET-99(MBBS)]
(A) HI/200°C (B) Conc. H_2SO_4/140°C
(C) NaOH/70°C (D) $NaOC_2H_5$/27°C

27. Chloroethane reacts with Y to form NaCl and Z. One mole of Z reacts with two moles of HI to form water and iodoethane. Which of the following is Y ? [EAMCET-2000]
(A) CH_3COOH (B) CH_3CHO
(C) $C_2H_5OC_2H_5$ (D) C_2H_5ONa

28. Ethyl chloride reacts with X to form NaCl and Y. One mole of Y reacts with two moles of HI to form water and ethyl iodide. Which of the following is X ? [EAMCET(MBBS)-2000]
(A) C_2H_5ONa (B) $C_3H_5OC_2H_5$
(C) CH_2COOH (D) CH_3CHO

29. Chloroethane reacts with <u>X</u> to form diethyl ether. What is <u>X</u> ? [EAMCET-2001]
(A) NaOH (B) H_2SO_4 (C) C_2H_5ONa (D) $Na_2S_2O_3$

30. Williamson's synthesis is used for preparing [EAMCET(MBBS)-2001]
(A) Alcohols (B) Ketones (C) Ethers (D) Aldehydes

31. The products formed when diethyl ether is reacted with cold HI are : [EAMCET-2002]
(A) $C_2H_5I + C_2H_5OH$ (B) $2C_2H_5I + H_2O$
(C) $2C_2H_5OH$
(D) $C_2H_5 - O - O - C_2H_5 + H_2O$

32. In Williamson synthesis, which one of the following reacts with sodium ethoxide to yield a symmetrical ether ? [EAMCET(med.)-2002]
(A) Ethyl iodide (B) Methyl iodide
(C) Ethanol (D) Acetaldehyde

33. When sulphuric ether is reacted in excess of HI, what is formed ? [CPMT-96]
(A) C_2H_5OI (B) $C_2H_5I + H_2O$
(C) C_2H_5I (D) $C_2H_5OC_2H_4I$

34. When ether is reacted with O_2 it forms explosive, [CPMT-96]
(A) Peroxide (B) Acid (C) Ketone (D) TNT

35. When alkylhalide is heated with dry Ag_2O gives [CPMT-97]
(A) Ester (B) Ether
(C) Ketone (D) AgCl + Hydrocarbon

36. Which one among the following is Williamson's synthesis ? [PMT-93]

(A) $(CH_3)_2C=O \xrightarrow[\text{conc. HCl}]{\text{Zn - Hg}} CH_3\text{-}CH_2\text{-}CH_3$

(B) $CH_3 - CHO \xrightarrow[-H_2O]{\text{dil. NaOH}} CH_3 - CH = CH - CHO$

(C) $C_2H_5I + C_2H_5ONa \rightarrow C_2H_5 - O - C_2H_5 + NaI$

(D) $HCHO \xrightarrow{NaOH} HCOONa + CH_3OH$

37. In Williamson's synthesis, ethoxy ethane is prepared by [PMT-95]
(A) passing ethanol over heated alumina
(B) sodium ethoxide with ethyl bromide
(C) ethyl alcohol with sulphuric acid
(D) ethyl iodide and dry silver oxide

38. The reaction
$C_2H_5ONa + BrC_2H_5 \rightarrow C_2H_5 - O - C_2H_5 + NaBr$ is called : [PMT-2001]
(A) Frankland reaction (B) Wurtz reaction
(C) Williamson's synthesis (D) Cannizaro reaction

39. Diethyl ether on heating with conc. HI gives two moles of : [PMT-2001]
(A) Ethyl iodide (B) Ethanol
(C) Iodoform (D) Methyl iodide

40. In the following reaction
$C_2H_5OC_2H_5 + 4H \xrightarrow{\text{Red P + HI}} 2X + H_2O$ X is [PMT-2002]
(A) Ethane (B) Ethylene (C) Butane (D) Propane

41. In the reaction $Ar - OH + Rx \xrightarrow[\text{alkali}]{} A$; A is [PET-94]
(A) An aldehyde (B) An aryl chloride
(C) An ether (D) A ketone

42. The reaction of C_2H_5OH with H_2SO_4 does not give [PET-96]
(A) Ethylene (B) Diethyl ether
(C) Acetylene (D) Ethyl hydrogen sulphate

43. Methyl ter-butyl ether on heating with HI of one molar concentration gives : [PET-97]
(A) $CH_3I + (CH_3)_3COH$ (B) $CH_3OH + (CH_3)_3CI$
(C) $CH_3I + (CH_3)_3CI$ (D) None of the above

44. The product formed on heating ethyl alcohol with sulphuric acid at 140°C is [AFMC-2003]
(A) Diethyl ether (B) Ethene
(C) Ethyl sulphate (D) Ethanoyl sulphate

Ethers

1.	(C)
2.	(A)
3.	(B)
4.	(B)
5.	(A)
6.	(C)
7.	(C)
8.	(C)
9.	(B)
10.	(C)
11.	(D)
12.	(B)
13.	(C)
14.	(C)
15.	(B)
16.	(A)
17.	*
18.	(A) (D)
19.	(B)
20.	(D)
21.	(D)
22.	*
23.	(C)
24.	(D)
25.	(A)
26.	(B)
27.	(D)
28.	(A)
29.	(C)
30.	(C)
31.	(A)
32.	(A)
33.	(B)
34.	(A)
35.	(B)
36.	(C)
37.	(B)
38.	(C)
39.	(A)
40.	(A)
41.	(C)
42.	(C)
43.	(A)
44.	(B)

17. (peroxides)
22. diethyl peroxide

Aldehydes and Ketones

Preparation:

1. Friedel-craft acylation: Ketones only

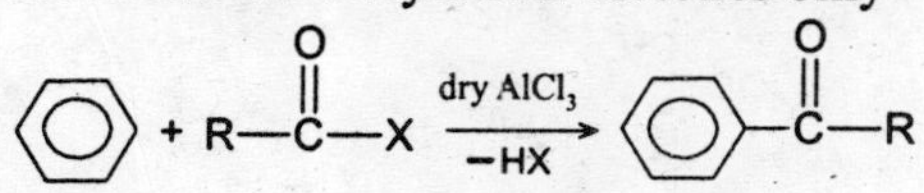

$(R-\overset{O}{\overset{\|}{C}}-X + AlCl_3 \longrightarrow R-\overset{O}{\overset{\|}{C}}{}^{\oplus} + AlCl_3X^-)$

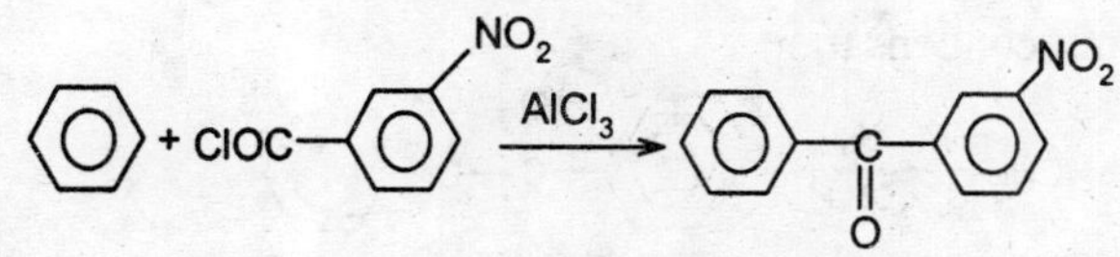

NO_2 group facilitates the formation of the reactive intermediate carbocation.

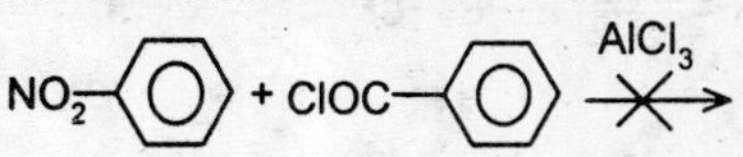 No reaction is possible since NO_2 group deactivates the ring towards nucleophilic substitution.

2. Reimer-Tiemann reaction

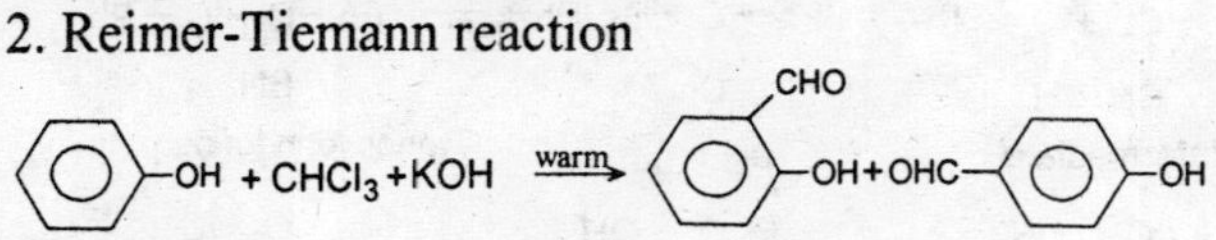

(:CCl_2 intermediate)

3. Rosenmund's reaction:

$$RCOCl + 2H \xrightarrow{Pd / BaSO_4} RCHO + HCl$$

This catalyst system is to stop further reduction of RCHO to RCH_2OH. Use of $LiAlH_4$ or $NaBH_4$ leads to the formation of alcohols.

4. Etard reaction

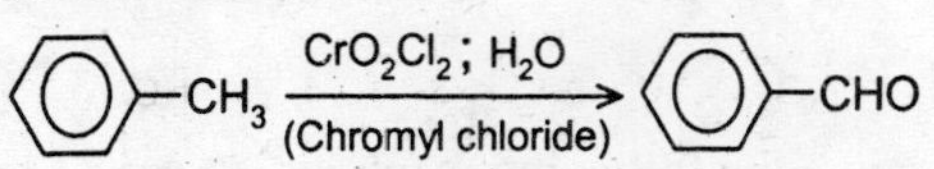

5. Grignard reaction

(i) 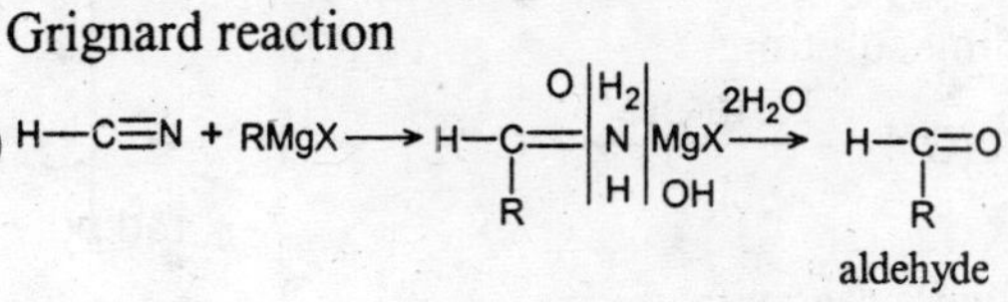

(ii) 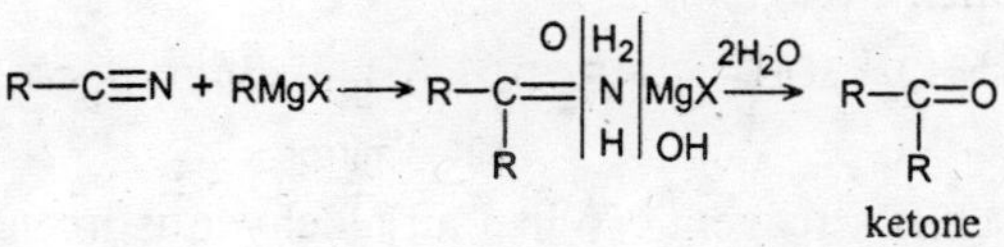

(iii) 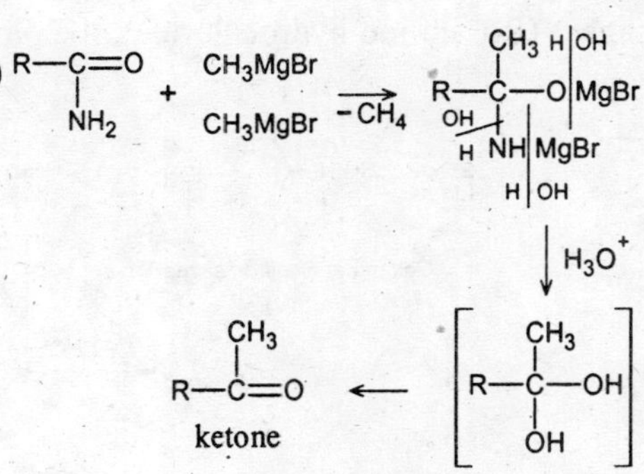

6. Stephen's reaction:

$$R-C\equiv N \xrightarrow[2[H]]{SnCl_2\ /HCl} RCH=NH.HCl \xrightarrow{H_2O} RCHO$$

7. Sommelet reaction:

8. Ozonolysis: Alkenes and alkynes depending upon their structure yield aldehydes and ketones.

9. Dehydrogenation:

(i) $CH_3CH_2CH_2OH \xrightarrow[Cu]{300^oC;} CH_3CH_2CHO$

(ii) $CH_3CHOHCH_3 \xrightarrow[Cu]{300^oC;} CH_3COCH_3$

(10) Formation of HCHO, CH_3CHO and CH_3COCH_3

(i) $CH_3OH + 1/2\ O_2 \xrightarrow{Ag} HCHO + H_2$

(ii) $(HCOO)_2Ca \xrightarrow{\Delta} 2HCHO + CaCO_3$

(iii) $CO_2 + H_2O \xrightarrow{h\nu\ Chlorophyll} HCHO$

(iv) $(RCOO)_2\ Ca \xrightarrow{\Delta} RCOR + CaCO_3$

(v) $(RCOO)_2\ Ca + (COO)_2\ Ca \xrightarrow{\Delta} RCHO + CaCO_3$

(11) Gattermann-Koch synthesis:

$$ArH + CO + HCl \xrightarrow{AlCl_3} ArCHO + HCl$$

Properties:

$>C^*=O$ The carbon atom is sp^2 hybridized. Aldehydes are easily oxidized than ketones.

Reactivity: H-CHO>CH_3CHO>CH_3COCH_3. Nucleophilic addition is the typical reaction of aldehydes and ketones.

Reactions:

1. Reduction:

(i) Clemmensen: For compounds sensitive to bases.

$$R-\overset{H}{\overset{|}{C}}=O \xrightarrow{Zn\text{-}Hg/HCl} RCH_3 + H_2O$$

(ii) Wolff-Kishner: For compounds sensitive to acids

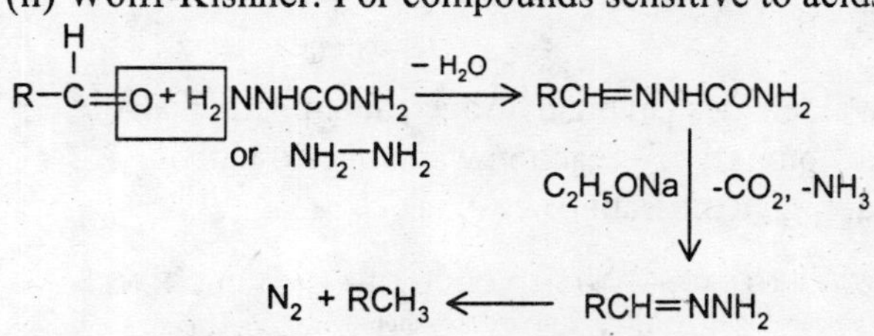

(iii) Meerwin Pondorf Verley reduction :

$$\begin{matrix}R\\R\end{matrix}\!\!>CO + [(CH_3)_2CHO)]_3Al \longrightarrow \begin{matrix}R\\R\end{matrix}\!\!>CHOH$$

(iv) Reduction with $Ni/H_2 \rightarrow$ alcohols

(v) Reduction with $LiAlH_4 \rightarrow$ alcohols

(vi) Reduction in neutral or alkaline medium:
Ketones give pinacols

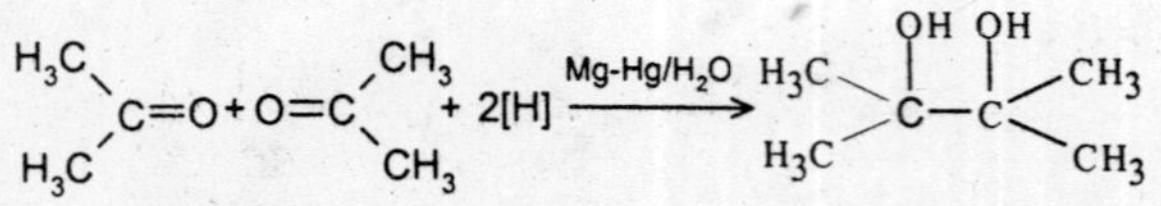

2. Tischenko reaction: aldehydes only give this reaction.

$$CH_3CHO \xrightarrow{Al(OC_2H_5)_3} CH_3COOC_2H_5$$

3. Aldol condensation: Aldehydes and ketones having α-hydrogen atom undergo this condensation. HCHO, H_6H_5CHO, $(CH_3)_3C.CHO$ do not give this reaction.

$$OH^- + H-CH_2-CHO \xrightarrow{-H_2O} :\bar{C}H_2CHO;$$

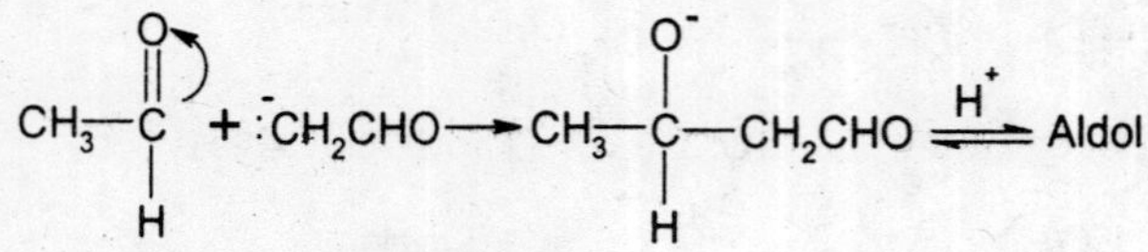

Carbanion intermediate

Reverse Aldol condensation:

$$CH_2=CH-CHO + H_2O \xrightarrow{NaOH} HCHO + CH_3CHO$$

4. Cannizzaro's reaction: Aldehydes having no α-hydrogen respond to this reaction

$$2\,HCHO + NaOH \longrightarrow HCOONa + CH_3OH$$

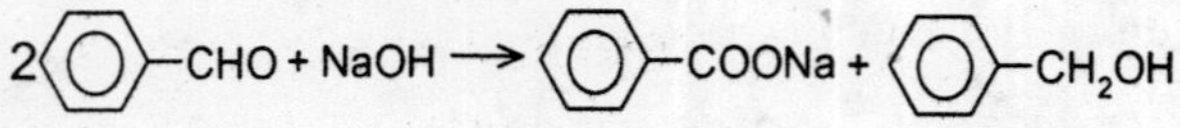

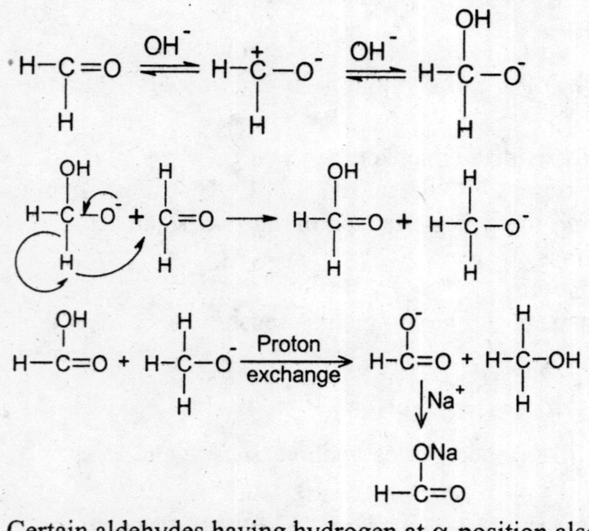

Certain aldehydes having hydrogen at α-position also undergo connizzaro's reaction when heated with NaOH at high temperature.

$$2(CH_3)_2CHCHO + NaOH \xrightarrow{200°C} (CH_3)_2CHCH_2OH + (CH_3)_2CHCOONa$$

2-methylpropanal — 2-methylpropanol — sodium salt of 2-methylpropanoic acid

Crossed cannizzaro's reaction:

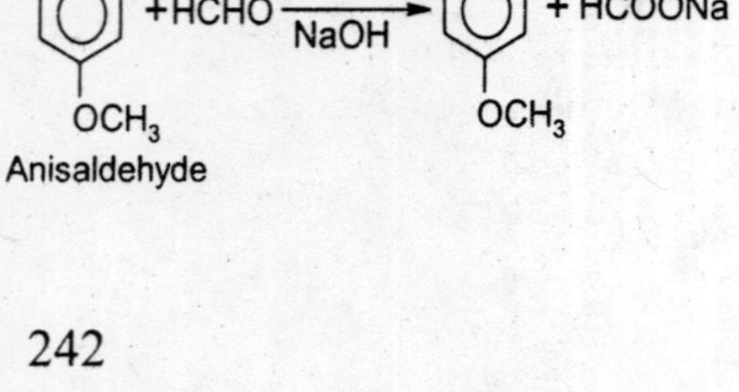

Anisaldehyde

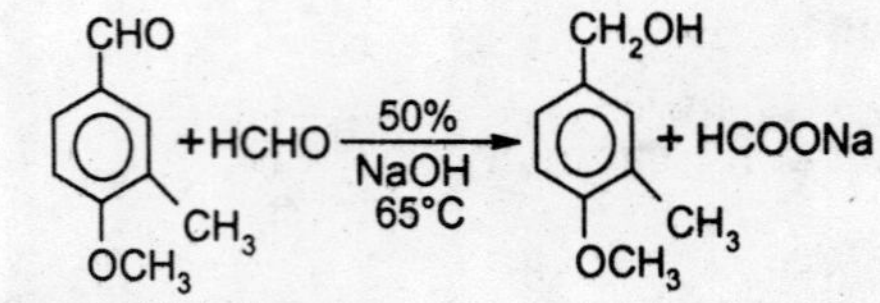

veratraldehyde

Intramolecular Cannizzaro's reaction

$$CHO.CHO \xrightarrow{NaOH} CH_2OH.COONa$$

Sod. glycollate

5. Benzoin condensation:

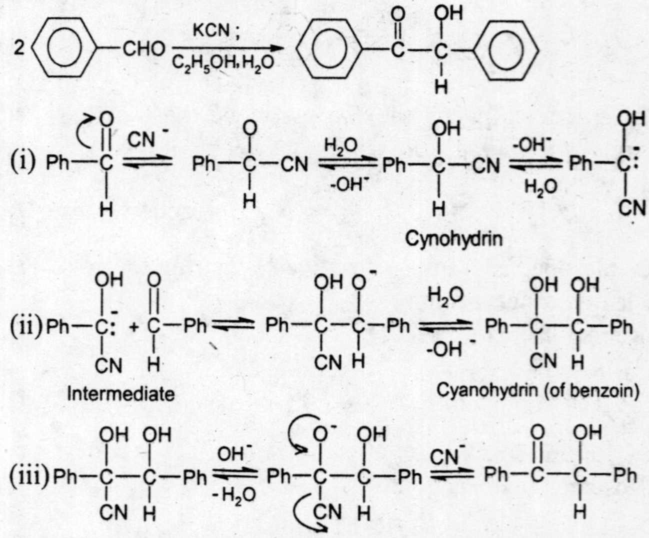

6. With 2, 4-dintrophenyl hydrazine: This test indicates the presence of >C=O. It is shown by both aldehydes & ketones

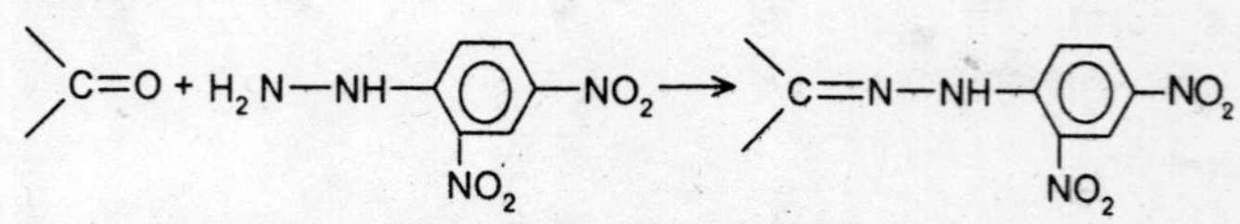

7. Aldehydes only give the following reactions

(i) With Fehling solution:

$$RCHO + 2Cu(OH)_2 \xrightarrow{Tartarate} RCOOH + 2H_2O + Cu_2O$$

red ppt

(ii) With Tollen's reagent:

$$RCHO + 2Ag(NH_3)_2^+ + 3OH^- \longrightarrow RCOO^- + 2Ag\downarrow + 4NH_3 + 2H_2O$$

silver mirror

(iii) With Schiff's reagent : When an aldehyde is treated with schiff's reagent (Rosaniline hydrochloride), the pink colour of the dye is restored.

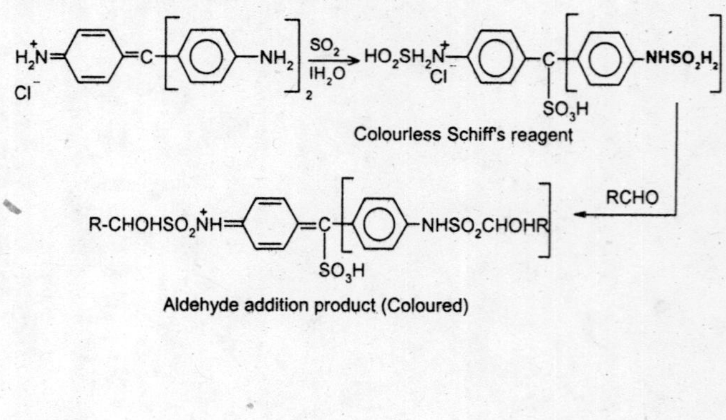

(iv) With Benedict's reagent ($CuSO_4$+sod. citrate+Na_2CO_3): Aldehydes give Cu_2O red ppt.

8. With NH_3:

(i) 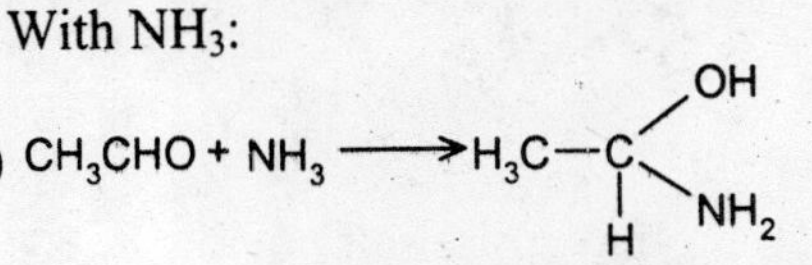

Acetaldehyde ammonia

(ii) 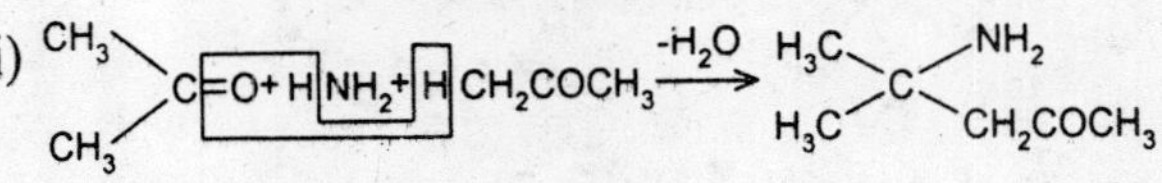

(iii) 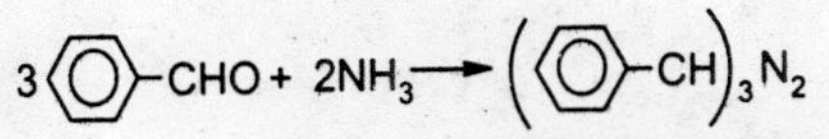

hydrobenzamide

(iv) $6HCHO + 4NH_3 \longrightarrow (CH_2)_6N_4$

hexamethylene tetramine

Urotropine: used as urinary antiseptic medicine

9. Reformatsky reaction:

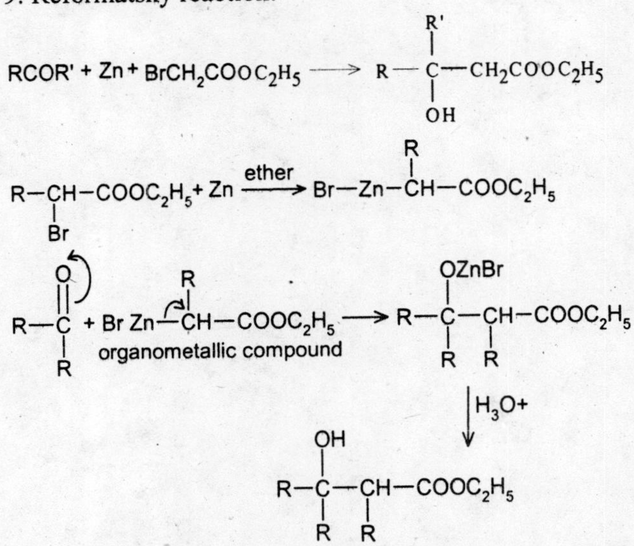

β-hydroxy ester

10. Wittig reaction: Conversion of an aldehyde or ketone into alkene

$Ph_2C = O + Ph_3P = CH_2 \rightarrow Ph_2C = CH_2$

Methylene triphenyl phosphorane

Nucleophilic addition at the carbonyl centre

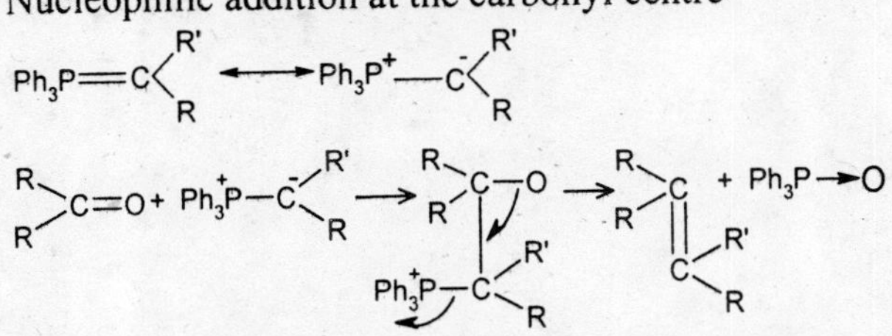

11 Perkin reaction: it is shown by aromatic aldehydes.

cinnamic acid

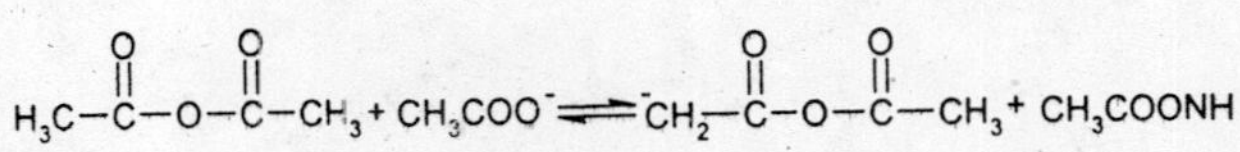

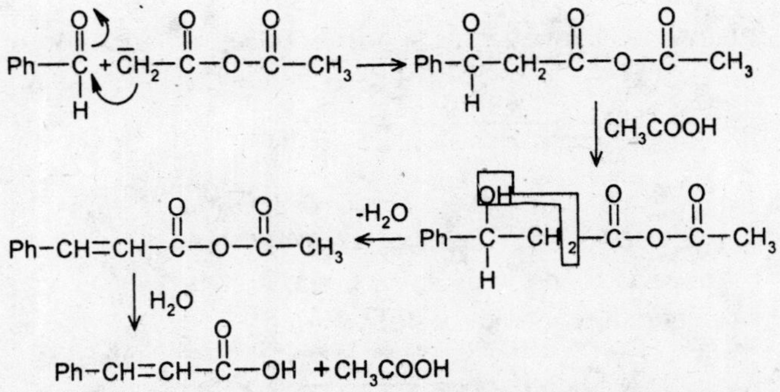

12. Knoevenagel reaction: Aldehyde or ketone with active methylene group containing compound gives α, β-unsaturated acid.

$$RCHO(\text{or ketone}) + \underset{\text{malonic ester}}{CH_2(COOC_2H_5)_2} \xrightarrow{\text{Pyridine}} RCH{=}CHCOOH$$

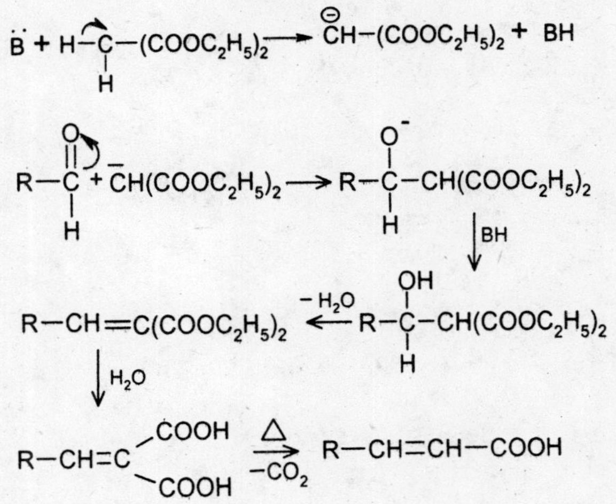

13. General reactions:

(a) Nucleophilic addition:

(i) HCN (ii) $NaHSO_3$ (iii) with grignard reagent

(b) Replacement of carbonylic oxygen:

(i) NH_2OH (ii) $NH_2\text{-}NH_2$ (iii) $C_6H_5NHNH_2$

(iv) $H_2NNHCONH_2$ (Semicarbazide)

(v) with PCl_5 (gem dihalide)

(c) Halogenation: Substitution reaction.

(d) Reductive amination:

$$CH_3CHO + NH_3 \xrightarrow{Ni/H2} CH_3CH_2NH_2$$

Acetone also gives this reaction.

(e) When formaldehyde is allowed to stand at room temperature or its 60% aqueous solution is treated with 2% H_2SO_4, trioxane is formed.

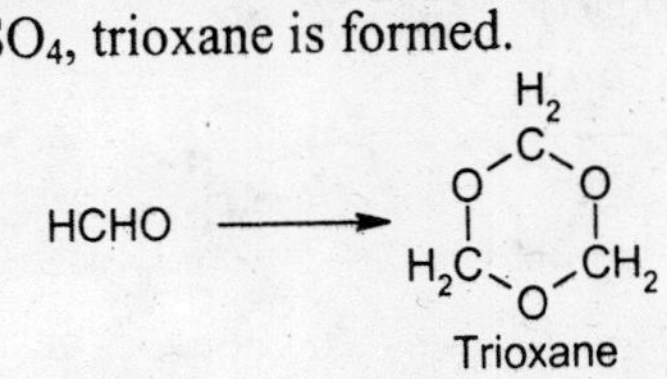

Trioxane

(f) 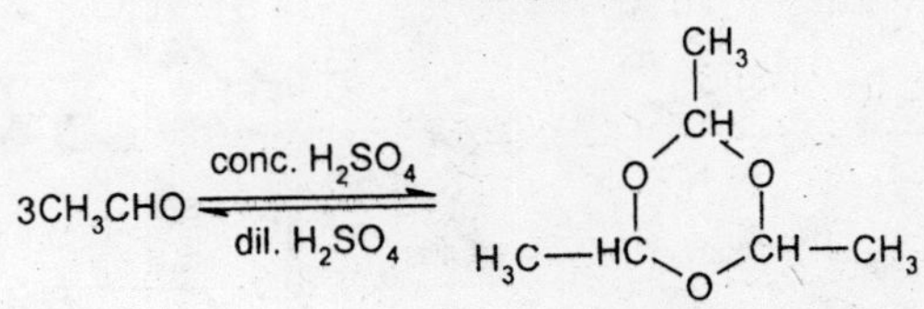

paraldehyde

It is used as hypnotic and soporific (sleep producing)

(g) 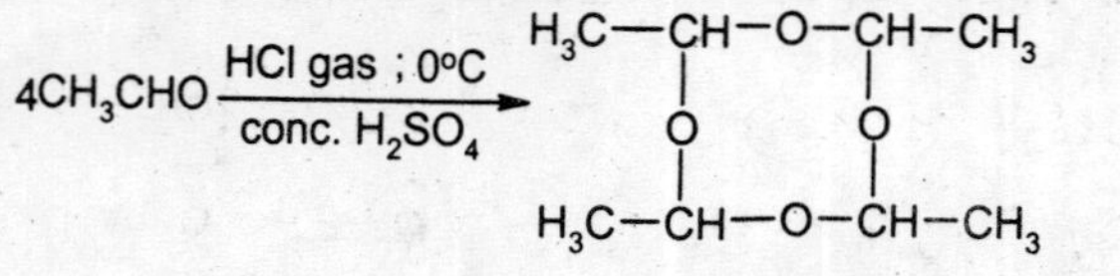

metaldehyde

It is used as solid fuel in spirit lamp.

(h) Condensation reactions of acetone:

(i) 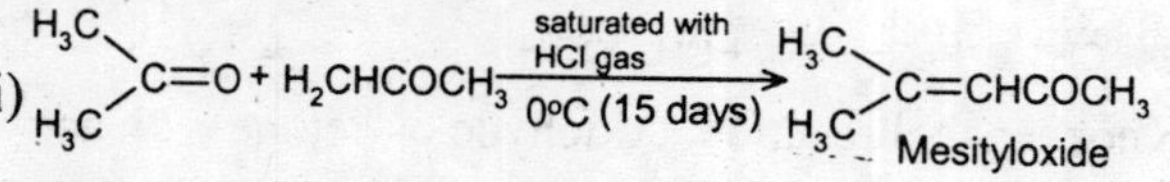

Mesityloxide

(ii) 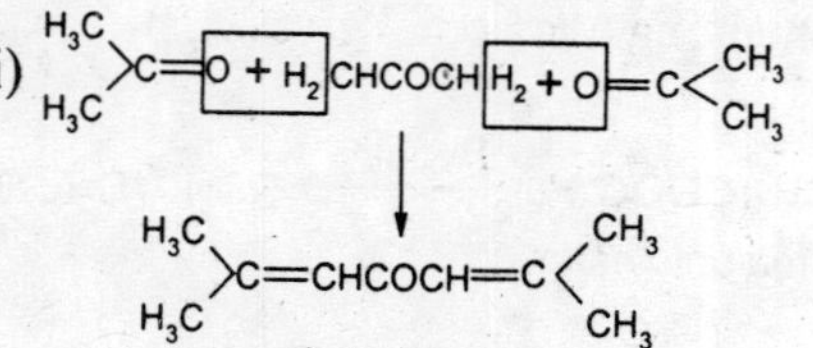

phorone

(iii) When distilled with conc. H_2SO_4 (1:2 w/w)

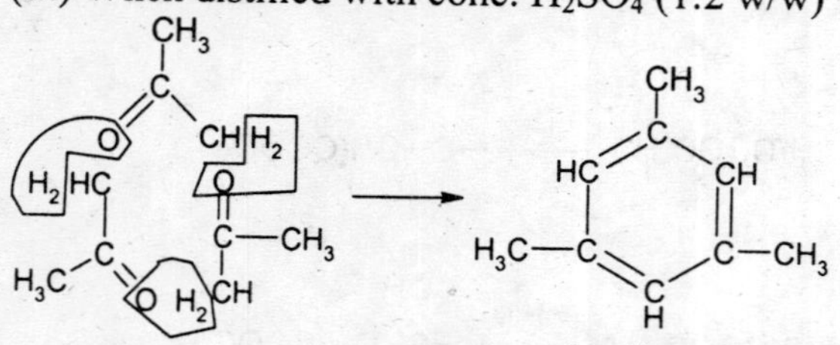

Mesitylene

14. Differences between HCHO and CH_3CHO can be made out with

a. PCl_5 b. Cl_2 c. SeO_2 d. I_2 + NaOH
e. NH_3 f. Cannizzaro g. Urea h. Phenol

Miscellaneous :

a. Formalin: 40% HCHO + 4% methanol + 56% H_2O. It is used to preserve biological specimen.

b. Galalith (milk stone): Formaldehyde+ casein

c. Formamint: Formaldehyde mixed with lactic acid-used as throat lozenges

d. Uses of acetone (i) For storing ethyne (acetylene) (ii) For the preparation of Ironone (artificial scent) (iii) For the preparation of Plexiglas (unbreakable glass) (iv) Nail polish remover.

Conversions: Formaldehyde to acetaldehyde

(i) $HCHO \xrightarrow{Ni/H_2} CH_3OH \xrightarrow{PCl_5} CH_3Cl \xrightarrow{KCN} CH_3CN$

$\xrightarrow{C_2H_5OH/Na} CH_3CH_2NH_2 \xrightarrow{NaNO_2/HCl}$

$CH_3CH_2OH \xrightarrow{K_2Cr_2O_7/H_2SO_4} CH_3CHO$

(ii) $HCHO + CH_3MgX \longrightarrow CH_3CH_2OH \xrightarrow[Cu]{300°C,} CH_3CHO$

Acetaldehyde to formaldehyde

$CH_3CHO \xrightarrow{K_2Cr_2O_7/H_2SO_4} CH_3COOH \xrightarrow{NaOH} CH_3COONa$

$\xrightarrow[(Sodalime)]{NaOH+CaO} CH_4 \xrightarrow{hv;\ Cl_2} CH_3Cl \xrightarrow{AgOH} CH_3OH \xrightarrow[300°C]{Cu} HCHO$

1. The reagents which can be used to distinguish acetophenone from benzophenone is/are [CBSE-90]
(A) 2; 4-dinitrophenylhydrazine
(B) aqueous solution of $NaHSO_3$
(C) Benedict's reagent (D) $I_2$and Na_2CO_3

2. $CH_3COCl \xrightarrow[BaSO_4]{Pd/H_2} X$. Here X is [CBSE-91]
(A) acetaldehyde (B) propionaldehyde
(C) acetone (D) acetic anhydride

3. If formaldehyde and KOH are heated together, we get [CBSE-91]
(A) methane (B) acetylene
(C) methanol (D) ethylacetate

4. Which one of the following on oxidation will not give a carboxylic acid with the same number of carbon atoms ? [CBSE-92]
(A) CH_3COCH_3 (B) CCl_3CH_2CHO
(C) $CH_3CH_2CH_2OH$ (D) CH_3CH_2CHO

5. CH_3COCH_3 can be obtained by [CBSE-92]
(A) heating acetaldehyde with methanol
(B) oxidation of propyl alcohol
(C) oxidation of isopropyl alcohol
(D) reduction of propionic acid

6. $[CH_3]_2$ C = $CHCOCH_3$ can be oxidised to $[CH_3]_2$ C = CHCOOH by [CBSE-93]
(A) chromic acid (B) NaOI (C) Cu at 300°C (D) $KMnO_4$

7. Which one of the following compounds gives positive test with Tollen's reagent ? [CBSE-94]
(A) Acetone (B) Acetic acid
(C) A and B (D) Acetaldehyde

8. Which of the following compounds will undergo self aldol condensation in the presence of cold, dilute alkali ? [CBSE-94]
(A) CH_2 = CH–CHO (B) CH ≡ C–CHO
(C) 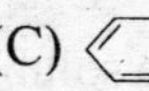–CHO (D) CH_3CH_2CHO

9. In the reaction
$CH_3CHO + HCN \rightarrow CH_3CH(OH)CN$
a chiral centre is produced. This product would be [CBSE-95]
(A) laevorotatory (B) meso compound
(C) dextrorotatory (D) racemic mixture

10. Oxidation of toluene with CrO_3 in the presence of $(CH_3CO)_2O$ gives a product 'A' which on treatment with aqueous NaOH produces [CBSE-95]
(A) C_5H_5CHO (B) $(C_6H_5CO)_2O$
(C) C_6H_5COONa (D) 2,4 diacetyl toluene

11. The general order of reactivity of carbonyl compounds for nucleophilic addition reactions is [CBSE-95]
(A) $H_2C=O>RCHO>ArCHO>R_2C=O>Ar_2C=O$
(B) $ArCHO>Ar_2C=O>RCHO>R_2C=O>H_2C=O$
(C) $Ar_2C=O>R_2C=O>ArCHO>RCHO>H_2C=O$
(D) $H_2C=O>R_2C=O>Ar_2C=O>RCHO>ArCHO$

12. Predict which one of the following molecules is most likely to organize and form a bilayer membrane structure. [CBSE-95]

I. 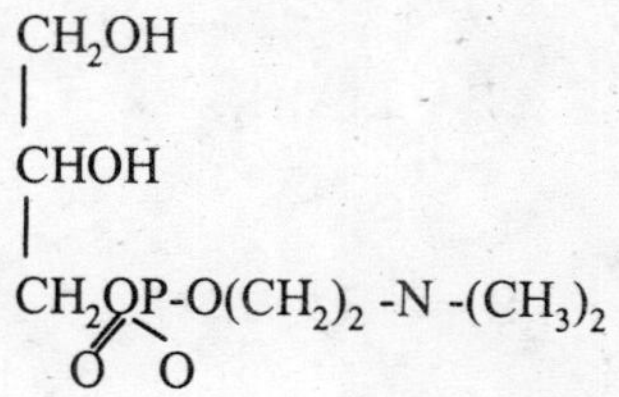

II. 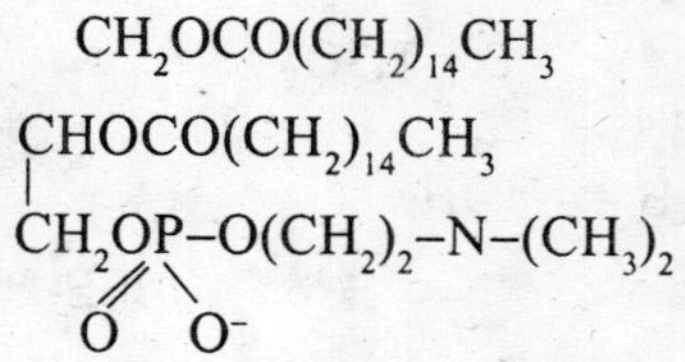

III. 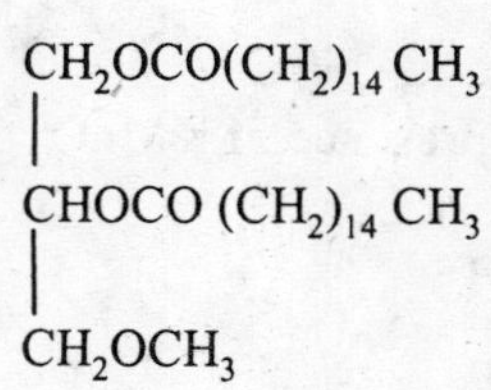 IV. 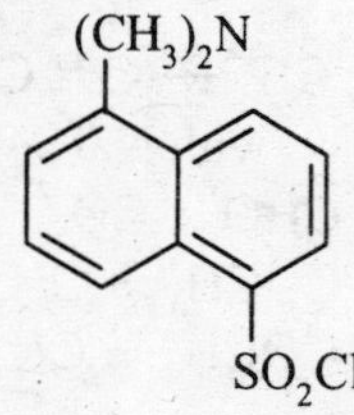

(A) II (B) I (C) III (D) IV

13. The oxidation of toluene to benzaldehyde by chromyl chloride is called [CBSE-96]
(A) Cannizzaro's reaction (B) Wurtz reaction
(C) Etard reaction (D) Riemer-Tiemann reaction

14. Aldol condensation will not take in [CBSE-96]
(A) HCHO (B) CH_3CH_2CHO
(C) CH_3CHO (D) CH_3COCH_3

15. Ketones $[R—\overset{\overset{O}{\|}}{C}—R_1]$, [CBSE-97]
Where $R = R_1$ = alkyl group can be obtained in one step by
(A) hydrolysis of esters (B) oxidation of primary alcohol
(C) oxidation of tertiary alcohol
(D) reaction of acid halide with alcohols

16. In the reaction
$CH_3CN + 2H \xrightarrow[Ether]{HCl} X \xrightarrow{Boiling\ H_2O} Y$; the term Y is [CBSE-99]
(A) Acetone (B) Ethylamine
(C) Acetaldehyde (D) Dimethylamine

17. Aldol condensation will not take place in [CBSE-99]
(A) HCHO (B) CH_3CHO
(C) CH_3COCH_3 (D) CH_3CH_2CHO

18. A and B in the following reactions are : [CBSE-2003]

$R-\underset{\underset{O}{\|}}{C}-R' \xrightarrow{HCN/KCN} A \xrightarrow{B} R-\underset{R'}{C}(OH)-CH_2NH_2$

(A) A = RR' C(CN)(OH), B = $LiAlH_4$

(B) A = RR' C(OH)(COOH), B = NH_2

(C) A = RR' C(CN)(OH), B = H3O

(D) A = RR'CH_2CN, B = NaOH

19. The compound [CBSE-2003]

$CH_3-\underset{\underset{CH_3}{|}}{C}=CH-CH_3$

on reaction with $NaIO_4$ in the presence of $KMnO_4$ gives

(A) $CH_3CHO + CO_2$ (B) CH_3COCH_3

(C) $CH_3COCH_3 + CH_3COOH$

(D) $CH_3COCH_3 + CH_3CHO$

20. When m-chlorobenzaldehyde is treated with 50% KOH solutiion, the product (s) obtained is (are) [CBSE-2003]

(A) 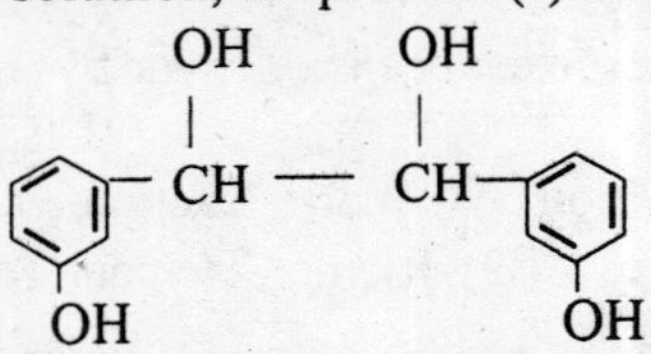

(B) 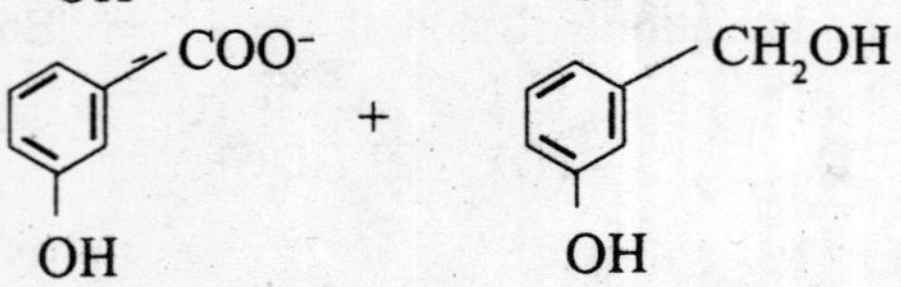

(C) 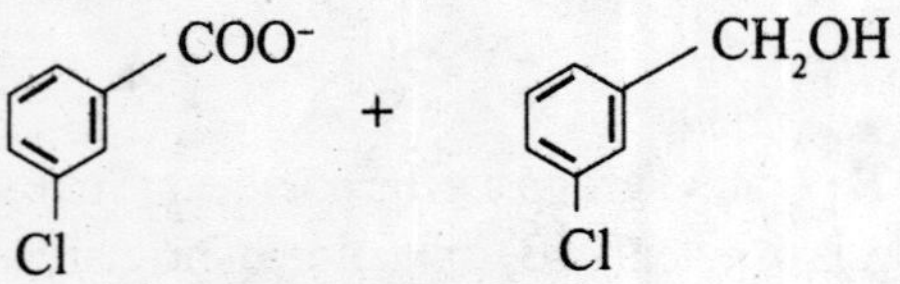

(D) 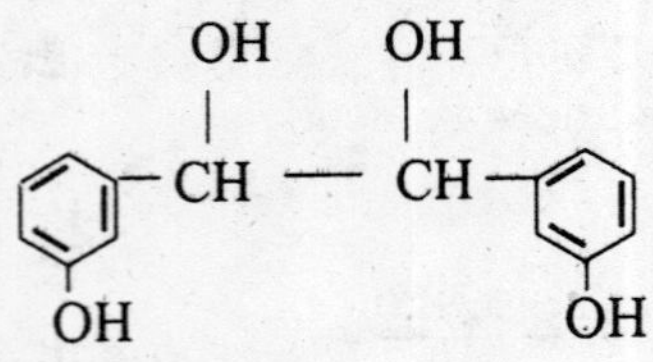

21. In this reaction

$CH_3CHO + HCN \longrightarrow CH_3CH(OH)CN \xrightarrow{H.\ OH} CH_3CH(OH)COOH$

an asymmetric centre is generated. The acid obtained would be [CBSE-2003]

(A) 20% D + 80% L-isomer (B) D - isomer

(C) L-isomer (D) 50% D + 50% L-isomer

22. Which of the following will not undergo aldol condensation ? [AIIMS-94]

(A) Acetone (B) Acetaldehyde

(C) Propionaldehyde (D) Formaldehyde

23. Which of the following compounds gives Cannizzaro's reaction ? [AIIMS-94]

(A) HCHO (B) CH_3CHO

(C) CH_3CH_2CHO (D) $(CH_3)_2CHCHO$

24. When ethanal is treated with Fehling's solution, it gives a precipitate of [AIIMS-94]

(A) Cu (B) CuO (C) Cu_2O (D) Cu_3O

25. Which of the following is formed, when benzaldehyde reacts with alcoholic KCN ? [AIIMS-96]

(A) Benzyl alcohol (B) Ethyl benzoate

(C) Benzoin (D) Benzoic acid

26. Reduction of Benzoyl chloride with Pd and $BaSO_4$ produces [AIIMS-97]

(A) Benzoyl cyanide (B) Benzaldehyde

(C) Benzoic acid (D) None of these

27. Acetaldehyde cannot show [AIIMS-97]

(A) Iodoform (B) Lucas test

(C) Benedict's test (D) Tollen's test

28. Which of the following reactions cannot produce RCOAr ? [AIIMS-97]

(A) ArCOCl + H – Ar $\xrightarrow{AlCl_3}$ (B) COCl + RMgX →

(C) R + CrO_3 → (D) None of these

29. Benzaldehyde can be prepared by the hydrolysis of [AIIMS-99]

(A) Benzyl chloride (B) Benzal chloride

(C) Benzotrichloride (D) Benzo nitrite

30. Which of the following compound will not give Cannizzaro's reaction ? [AIIMS-99]

(A) CH_3CHO (B) C_2H_5OH

(C) CH_3CH_2Cl (D) C_2H_5COOH

31. Assertion : Lower aldehydes and ketones are soluble in water but the solubility decreases as the molecular mass increases. Reason : Distinction between aldehydes and ketones can be made by Tollen's test. [AIIMS-99]

(A) A (B) B (C) C (D) D

32. Which of the following is paraldehyde ? [AIIMS-2000]

(A) 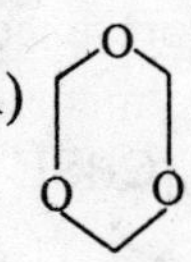

(B) 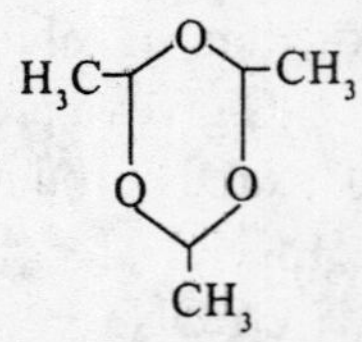

(C) A 40% HCHO solution (D) $CH_2O–CH_2O–CH_2O$

33. Which of the following will not give Cannizzaro's reaction ? [AIIMS-2001]

(A) HCHO (B) CCl_3CHO
(C) C_6H_5CHO (D) CH_3CHO

34. CH_3COCH_3 can be converted to $CH_3CH_2CH_3$ by the action of [AIIMS-2002]
(A) HIO_3 (B) HI (C) HNO_3 (D) H_3PO_3

35. In the reaction :
$C_6H_5CHO + C_6H_5NH_2 \rightarrow C_6H_5N = HCC_6H_5 + H_2O$
the compound $C_6H_5N = CHC_6H_5$ is known as [AIIMS-2002]
(A) Aldol (B) Schiff's reagent
(C) Schiff's base (D) Benedict reagent

36. The compound most suitable for the preparation of cyanohydrin is [AIIMS-2002]
(A) C_2H_5COOH (B) $C_6H_5COC_2H_5$
(C) $C_6H_5NH_2$ (D) $C_2H_5 - C_2H_5$

37. The intermediate formed in aldol condensation is [AIIMS-2002]
(A) Aldol (B) Alcohol
(C) Carbanion (D) α-hydrogen ester

38. When calcium acetate and calcium formate are heated they form : [JIPMER-91]
(A) Acetone (B) Formic acid
(C) Acetic acid (D) Acetaldehyde

39. The hydrolysis of CH_3CHCl_2 gives ? [JIPMER-91]
(A) CH_3CHO (B) CH_3COOH
(C) $CHCl_3$ (D) C_2H_5OH

40. The compound that gives acetaldehyde when heated with aqueous KOH is : [JIPMER-92]
(A) Chloroform (B) Ethylene chloride
(C) Carbon tetrachloride (D) Ethylidine chloride

41. Aniline undergoes condensation to form Schiff base on reacting with : [JIPMER-93]
(A) Acetyl chloride (B) Ammonia
(C) Acetone (D) Benzaldehyde

42. The IUPAC name of $Cl_3C–CH_2–CHO$ is [JIPMER-94]
(A) Chloral (B) Chloromethanal
(C) 3,3,3-trichloropropanal (D) 2,2,2-trichloropropanal

43. Which one of the following is used as a hypnotic ? [JIPMER-94]
(A) Formaldehyde (B) Acetaldehyde
(C) Metaldehyde (D) Paraldehyde

44. Which of the following has a smell of bitter almonds ? [JIPMER-94]
(A) Phenol (B) Benzoic acid
(C) Aniline (D) Benzaldehyde

45. The reaction is called [JIPMER-95]

$$CH_3COCl \xrightarrow[Pd/BaSO_4]{2H} CH_3CHO + HCl$$

(A) Cannizzaro's reaction
(B) Reimer and Tiemann's reaction
(C) Rosenmund's reaction (D) Reformatsky's reaction

46. The oxidation product of 2-propanol, with hot conc. HNO_3 is [JIPMER-95]
(A) Propanal (B) Ethanoic acid
(C) Propanone (D) None of these

47. When formaldehyde is heated with ammonia, the compound formed is [JIPMER-96]
(A) methylamine (B) amino formaldehyde
(C) hexamethylenetetramine (D) formalin

48. Identify the product B in the reaction :

$$CH_3CHO \xrightarrow{CH_3MgI} A \xrightarrow{Hydrolysis} B$$

[JIPMER-96]
(A) $(CH_3)_2CHOH$ (B) $(CH_3)_3COH$
(C) CH_3OH (D) CH_3CH_2OH

49. Which of the following substances possesses an odour of bitter almonds ? [JIPMER-97]
(A) Aniline (B) Acetone
(C) Benzaldehyde (D) Benzene sulphonic acid

50. Which of the following methods cannot be used for the preparation of a ketone ? [JIPMER-97]
(A) Oxidation of secondary alcohols
(B) Heating of volatile acid with MnO_2 at 100°C
(C) Hydrolysis of a cyanide
(D) Dehydrogenation of secondary alcohols

51. Acetic and formic acids do not differ in their action on [JIPMER-97]
(A) Ammoniacal $AgNO_3$ (B) $NaHCO_3$
(C) Phosphorus pentachloride (D) conc. H_2SO_4

52. Acetaldehyde on treatment with HCN forms a cyanohydrin. This on hydrolysis gives [JIPMER-97]
(A) α–hydroxy propionic acid (B) Acetic acid
(C) Acetone (D) Ethanol

53. The carbonyl compound undergoes the following reactions [JIPMER-97]
(i) With saturated $NaHSO_3$, it gives a white precipitate.
(ii) It does not give pink colour with Schiff's reagent
(iii) On reduction it gives a secondary alcohol
(iv) With iodine and sodium carbonate, it gives iodoform.
The compound is.
(A) $C_2H_5COC_2H_5$ (B) $C_2H_5COC_3H_7$
(C) $C_2H_5COCH_3$ (D) $C_3H_7COC_3H_7$

54. Benzyl alcohol is prepared form benzaldehyde by [JIPMER-97]
(A) Friedel-Crafts reaction (B) Cannizzaro's reaction
(C) Kolbe's reaction (D) Reimer-Tiemann reaction

55. Under the influence of the dilute base two molecules of an aldehyde may condense to form [JIPMER-98]

(A) An acid (B) A β-hydroxy aldehyde
(C) A β-hydroxy ketone (D) An alcohol

56. The structural formula of benzylphenyl ketone is [JIPMER-98]
(A) $C_6H_5COC_5H_5$ (B) $C_6H_5CO.COC_6H_5$
(C) $C_6H_5CH_2COC_6H_5$ (D) $C_6H_5CH_2COCH_2C_6H_5$

57. Benzaldehyde undergoes benzoin condensation in the presence of [JIPMER-98]
(A) Potassium cyanide (B) Conc.H_2SO_4
(C) Sodium metal (D) Anhydrous $ZnCl_2$

58. In the presence of HCl two molecules of acetone condense to form [JIPMER-98]
(A) Mesityloxide (B) Phorone
(C) 4-methyl-2-pentanone (D) 4-methyl-2- pentanoic acid

59. When HCHO is treated with C_6H_5CHO in presence of NaOH, the products are [JIPMER-2000]
(A) HCOONa and $C_6H_5CH_2OH$
(B) CH_3OH and C_6H_5COONa
(C) CH_3OH and HCOONa
(D) $C_6H_5CH_2OH$ and C_6H_5COONa

60. Aldehydes are isomeric with [JIPMER-2001]
(A) Ketones (B) Alcohols (C) Ethers (D) None

61. Acetaldehyde undergoes self condensation in presence of aluminium ethoxide to give ethyl acetate. This reaction is called [JIPMER-2001]
(A) Aldol condensation (B) Cannizzaro's reaction
(C) Tischenko's reaction (D) Perkin's reaction

62. The oxidation of toluene to bezaldehyde, by chromyl chloride, is called [JIPMER-2001]
(A) Reimer-Tiemann reaction (B)Etard reaction
(C) Cannizzaro's reaction (D)Wurtz reaction

63. Which of the following gives aldol condensation ? [BHU-95]
(A) Acetaldehyde (B) Formaldehyde
(C) Benzaldehyde (D) Benzophenone

64. The catalyst $SnCl_2$ / HCl is used in [BHU-95]
(A) Stephen's reduction reaction
(B) Cannizzaro's reduction reaction
(C) Clemmensen reduction reaction
(D) Rosenmund's reduction reaction

65. The reaction
$C_6H_5CHO + (CH_3CO)_2O \xrightarrow{CH_3COONa} C_6H_5CH = CH{-}COOH + CH_3COOH$ is known is [BHU-96]
(A) Perkin's reaction (B) Claisen condensation
(C) Benzoin condensation (D) Cannizzaro's reaction

66. In which reaction aromatic aldehyde is treated with acid anhydride in presence of corresponding salt of the acid to give unsaturated aromatic acid ? [BHU-98]
(A) Friedel-Crafts reaction (B) Perkin's reaction
(C) Wurtz's reaction (D) None of these

67. Formaldehyde when reacted with methyl magnesium bromide gives [BHU-98]
(A) C_2H_5OH (B) CH_3COOH
(C) CH_3CHO (D) HCHO

68. Aldehydes are oxidized to acids by [BHU-2001]
(A) $K_2Cr_2O_7/H^+$ (B) Tollen's reagent
(C) Fehling's solution (D) All of these

69. The most appropriate reagent for [BHU-2002]

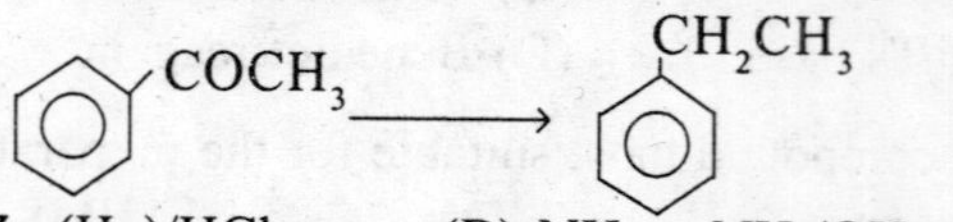

is
(A) Zn (Hg)/HCl (B) $NH_2 - NH_2/OH^-$
(C) H_2/Ni (D) $NaBH_4$

70. The catalyst used in Etard's reaction is [AFMC-91]
(A) Chromyl chloride (B) Chromic oxide
(C) CCl_4 (D) CCl_3

71. Which of the following undergoes Cannizzaro's reaction? [AFMC-91]
(A) Propanal (B) Benzaldehyde
(C) Butanal (D) None of these

72. Benzaldehyde can be prepared by the dry distillation of [AFMC-93]
(A) Calcium benzoate and formic acid
(B) Calcium benzoate and benzoic acid
(C) Calcium benzoate and acetic acid
(D) Calcium benzoate and calcium formate

73. Formalin is [AFMC-93]
(A) 40% acetaldehyde (B) 7% formic acid
(C) 40% formaldehyde (D) 7% acetic acid

74. What is formed when formaldehyde reacts with KOH? [AFMC-94]
(A) Methanol (B) Methane (C) Ethyl formate (D) Acetylene

75. Formaldehyde when treated with KOH gives [AFMC-94]
(A) CH_3OH (B) C_2H_4 (C) CH_3CHO (D) C_2H_5

76. Cyanohydrin of which of following forms lactic acid ? [AFMC-95]
(A) HCHO (B) CH_3 CO CH_3
(C) CH_3CHO (D) CH_3CH_2CHO

77. Paraldehyde is formed by polymerization of [AFMC-96]
(A) CH_3CHO (B) HCHO
(C) CH_3OH (D) CH_3CH_2OH

78. Which is mixed ketone ? [AFMC-97]
(A) Pentanone (B) Benzophenone
(C) Benzenone (D) All of these

79. Benzaldehyde reacts with ammonia to form

[AFMC-98]

(A) Aniline (B) Benzamide
(C) Hydro benzamide (D) Phenyl cyanide

80. If acetylchloride is reduced in the presence of $BaSO_4$ and Pd, then [AFMC-98]
(A) CH_3CHO is formed (B) CH_3CH_2OH is formed
(C) CH_3COOH is formed (D) CH_3COCH_3 is formed

81.The oxidation of toluene to benzaldehyde by chromyl chloride is called [AFMC-98]
(A) Wurtz reaction (B) Etard reaction
(C) Fittig reaction (D) Rosenmund reaction

82. Calcium formate, on dry heating, yields [AFMC-99]
(A) Acetone (B) Formaldehyde
(C) Acetic acid (D) Acetaldehyde

83. The oxidation of toluene to benzaldehyde by chromyl chloride is called [AFMC-99]
(A) Rosenmund reaction (B) Wurtz reaction
(C) Etard reaction (D) Fittig reaction

84. Oxidation of acetaldehyde with selenium dioxide produces [AFMC-99]
(A) Ethanoic acid (B) Methanoic acid
(C) Glyoxal (D) Oxalic acid

85. Acetaldehyde and acetone can be identified by [AFMC-2000]
(A) Schiff's test (B) Molisch test
(C) Iodoform test (D) Bromoform test

86. Condensation of two molecules of acetaldehyde is [AFMC-2000]
(A) Perkin's reaction (B) Cannizzaro's reaction
(C) Coupling (D) Aldol condensation

87. Toluene react with chromyl chloride forms [AFMC-2001]
(A) Chlorotoluene (B) Benzaldehyde
(C) Benzyl chloride (D) Benzoic acid

88. The formation of cyanohydrin from a ketone is an example of [IIT-90]
(A) Electrophilic addition (B) Nucleophilic addition
(C) Nucleophilic substitution (D) Electrophilic substitution.

89. Fehling solution A consists of an aqueous solution of copper sulphate while Fehling's solution B consists of an alkaline soultion of .... [IIT-90]

90. m-Chlorobenzaldehyde on reaction with conc. KOH at room temperature gives [IIT-91]
(A) Potassium m-chlorobenzoate and m-hydroxybenzal- dehyde
(B) m-hydroxybenzaldehyde and m-chlorobenzyl alcohol
(C) m-chlorobenzyl alcohol and m-hydroxybenzyl alcohol
(D) potassium m-chlorobenzoate and m-hydroxybenzyl alcohol

91. Hydrogenation of benzoyl chloride in the presence of Pd on $BaSO_4$ gives : [IIT-92]
(A) Benzyl alcohol (B) Benzaldehyde
(C) Benzoic acid (D) Phenol

92. The structure of intermediate product formed by the oxidation of toluene with CrO_3 and acetic anhydride whose hydrolysis gives benzaldehyde is .... [IIT-92]

93. In the following compounds

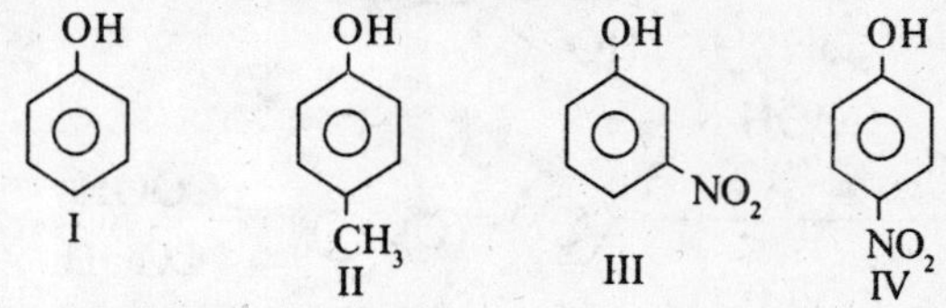

The order of acidity is [IIT-96]
(A) III > IV > I > II (B) I> IV> III> II
(C) II > I > III > IV (D) IV >III> I> II

94. In the Cannizzaro reaction given below
$Ph{-}CHO \xrightarrow{OH^-} PH{-}CH_2OH + PhCO_2^-$, the slowest step is [IIT-96]
(A) the attack of –OH at the carbonyl group
(B) the transfer of hydride to the carbonyl group
(C) the abstraction of proton from the carboxylic acid
(D) the deprotonation of $Ph{-}CH_2OH$

95. Among the following compounds, which will react with acetone to give a product containing $>C = N -$? [IIT-98]
(A) $C_6H_5\ NH_2$ (B) $(CH_3)_3\ N$
(C) $C_6\ H_5\ NHC_6H_5$ (D) $C_6H_5NHNH_2$

96. Which of the following undergo aldol condensation ? [IIT-98]
(A) acetaldehyde (B) propanadehyde
(C) benzaldehyde (D) trideutero acetadehyde

97. A mixture of benzaldehyde and formaldehyde on heating with aqueous NaOH solution [IIT-2001]
(A) benzyl alcohol and sodium formate
(B) sodium benzoate and methyl alcohol
(C) sodium benzoate and sodium formate
(D) benzyl alcohol and methyl alcohol

98. Compound 'A' (molecular formula C_3H_8O) is treated with acidified potassium dichromate to form a product 'B' (molecular formula C_3H_6O). 'B' forms a shining silver mirror on warming with ammoniacal silver nitrate. 'B' when treated with an aqeous solution of $H_2NCONHNH_2$.HCl and sodium acetate gives a product 'C'. Identify the structure of 'C' [IIT-2002]
(A) $CH_3CH_2CH = NNHCONH_2$
(B) $CH_3{-}\underset{\displaystyle CH_3}{\underset{|}{C}} = NNHCONH_2$ (C) $CH_3{-}\underset{\displaystyle CH_3}{\underset{|}{C}} = NCONHNH_2$
(D) $CH_3CH_2CH = NCONHNH_2$

99. 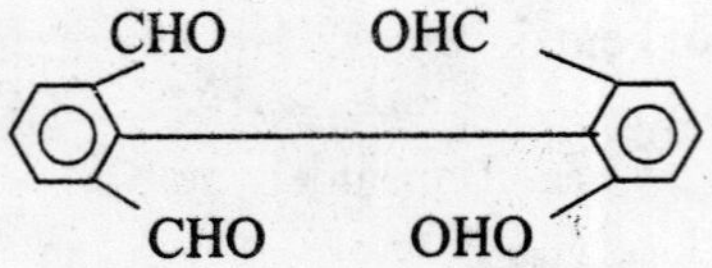

$\xrightarrow[\text{(2) H+ }2H_2O]{\text{(1) NaOH (excess) }100^\circ C}$

any one of the products formed is [IIT-2003]

(A) 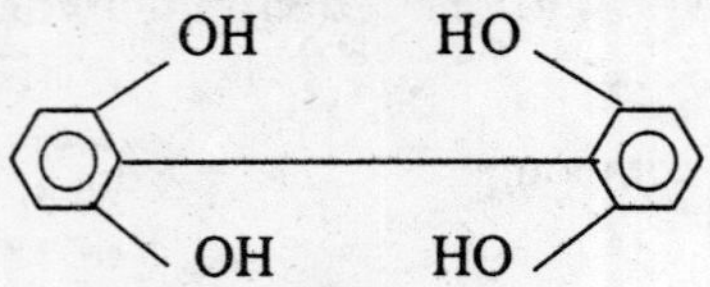

(B) 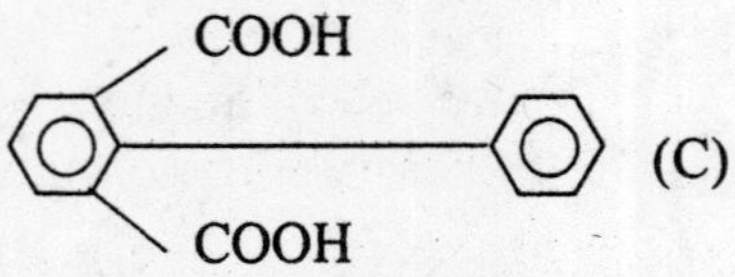 (C) 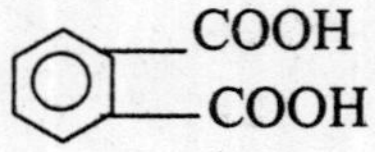

(D) 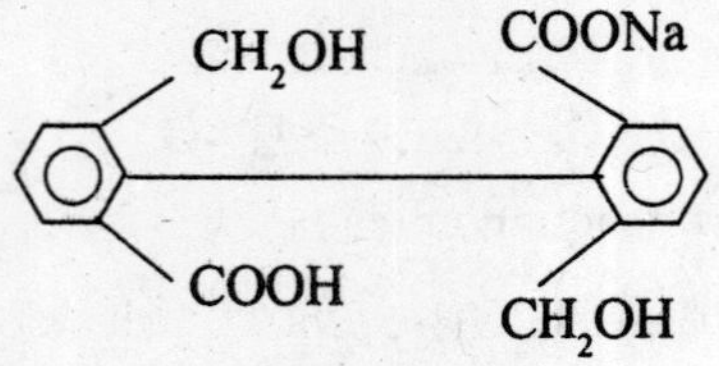

100. 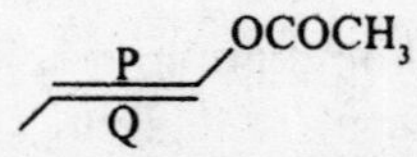 $\xrightarrow[\text{Hydrolysis}]{\text{acidic}}$

Products formed by P and Q can be differentiated by [IIT-2003]

(A) NaOH (B) Cannizaro's reaction
(C) Hydrolysis (D) Fehing's solution.

101. On vigorous oxidation by permanganate solution, $(CH_3)_2C = CH - CH_2 - CHO$ gives [AIEEE-2002,]

(A) $CH_3 - C(OH)(CH_3) - CH(OH) - CH_2CH_3$

(B) $(CH_3)_2 > COOH + CH_3CH_2COOH$

(C) $(CH_3)_2 > CH - OH + CH_3CH_2CH_2OH$

(D) $(CH_3)_2 > C = O + CH_3CH_2CHO$

102. An organic compound containing H_2 and O_2 but a single carbon atom responds positively to Tollen's reagent. The formula of the compound is .... [EAMCET-90]

103. The class of the compounds that get reduced to primary alcohols and also respond to Fehling's test is known as .... [EAMCET-90]

104. Reduction of acetaldehyde in presence of $LiAlH_4$ gives .... [EAMCET-91]

105. A few drops of an organic compound (liq.) are added to Schiff's reagent. A pink colour is produced. The organic compound is .... [EAMCET-91]

106. Ethyl alcohol when passed over copper at 300°C is dehydrogenated to give .... [EAMCET-92]

107. Ammoniacal solution of silver nitrate is called as .... [EAMCET-92]

108. The reaction of acetaldehyde by .... lead to the formation of ethanol . [EAMCET-93]

109. Acetone on reaction with chlorine gives normally [EAMCET-96]

(A) Monochloro acetone (B) Dichloro acetone
(C) Trichloro acetone (D) Hexachloro acetone

110. Acetaldehyde when treated with dilute NaOH gives [EAMCET-98]

(A) CH_3CH_2OH (B) CH_3COOH
(C) $CH_3 - CH(OH) - CH_2 - CHO$ (D) $CH_3 - CH_3$

111. C_2H_5CHO & CH_3COCH_3 can be distinguished from one another by testing with [EAMCET-98(MBBS)]

(A) Phenylhydrazine (B) 2,4 dinitrophenylhydrazine
(C) Fehling's solution (D) Sodium bisulphate

112. Which of the following compounds undergo Aldol condensation ? [EAMCET-99]

(A) CH_3CHO (B) $(CH_3)_3CHO$
(C) CCl_3CHO (D) HCHO

113. Which one of the following is acetone oxime ? [EAMCET-99(MBBS)]

(A) 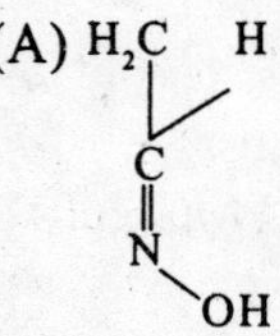 (B) 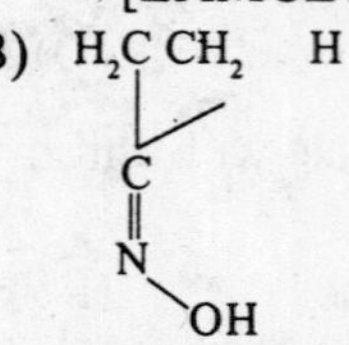

(C) 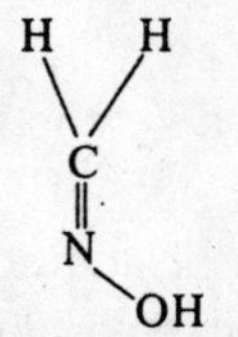 (D) 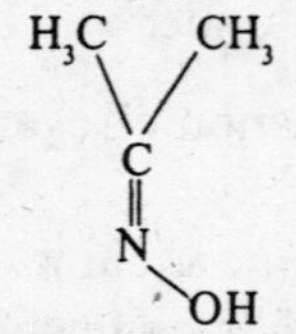

114. What is X in the following reaction ?

$2CH_3CHO \xrightarrow{\text{dil NaOH}} X$ [EAMCET-2000]

(A) $CH_3 - CH(OH) - CHO$ (B) $CH_3 - C(=O) - CH_2 - CHO$

(C) $CH_3 - CH(OH) - CH_2 - CHO$

(D) $CH_3 - CH_2 - CH(OH) - CHO$

115. Which of the following converts acetone to acetone oxime ? [EAMCET-2000]

(A) $H_2N - NH_2$ (B) 2, 4 – DNP

(C) $C_6H_5NHNH_2$ (D) NH_2OH

116. What is the compound formed when acetaldehyde is reacted with $LiAlH_4$? [EAMCET(MBBS)-2000]
(A) C_2H_5OH (B) CH_3COOH
(C) $C_2H_5OC_2H_5$ (D) CH_3COCH_3

117. Which of the following can be detected by silver mirror test ? [EAMCET(MBBS)-2000]
(A) CH_3COCH_3 (B) CH_3COOH
(C) C_2H_6 (D) CH_3CHO

118. When acetaldehyde is reacted with $LiAlH_4$ what is the product formed ? [EAMCET-2001]
(A) CH_3COOH (B) CH_3CH_2OH
(C) CH_3OH (D) HCOOH

119. Ethanal is reacted with acidified $K_2Cr_2O_7$. What is the product formed ? [EAMCET(MBBS)-2001]
(A) C_2H_5OH (B) CH_3COOH
(C) C_2H_6 (D) CCl_3CHO

120. Which one of the following compounds reacts with saturated solution of sodium bisulphite to give colourless crystalline product ? [EAMCET(MBBS)-2001]
(A) C_2H_5OH (B) C_2H_6 (C) CH_3CHO (D) $CHCl_3$

121. What is X in the following reaction ?

$2CH_3COCH_3 \xrightarrow{Ba(OH)_2} X$ [EAMCET-2002]

(A) $CH_3-C(OH)(CH_3)-CH_2-C(=O)-CH_3$

(B) $CH_3-C(OH)(CH_3)-CH_2-CH_2-CHO$

(C) $CH_3CH(CH_3)\ CH_2COCH_3$

(D) $CH_3-CH(CH_3)-CH(OH)-C(=O)-CH_3$

122. What is the gas liberated when alkaline formaldehyde solution is treated with H_2O_2 ? [EAMCET-2002]
(A) CO_2 (B) O_2 (C) CH_4 (D) H_2

123. The chemicals used for preparing acetophenone are [EAMCET-2002]
(A) C_6H_6 (B) CH_3COCH_3
(C) CH_3COCl (D) Anhydrous $AlCl_3$
(A) A, B, C (B) B, C, D, (C) A, C, D (D) A, B, D

124. Acetaldehyde is reduced with hydrogen in the presence of nickel. What is the compound formed ? [EAMCET(med.)-2002]
(A) Ethanol (B) n-propanol
(C) Methanol (D) Isopropanol

125. Acetaldehyde forms a white crystalline precipitate on mixing with a .... solution of ..... [EAMCET-2003]
(A) Acidic, Zn-Hg (B) Alcoholic, Na_2CO_3
(C) Saturated aqueous, $NaHSO_3$
(D) Aqueous, NaCl

126. When calcium acetate is heated, what product is formed ? [CPMT-96]
(A) CH_3COCH_3 (B) CH_3COOH
(C) CH_3CHO (D) HCOOH

127. What is the name of reaction when Benzaldehyde changes into Benzyl alcohol ? [CPMT-96]
(A) Friedel-Crafts reaction (B) Kolbe's reaction
(C) Wurtz reaction (D) Cannizzaro's reaction

128. Oxidation number of C in HCHO is- [CPMT-97]
(A) -2 (B) +2 (C) 0 (D) +4

129. Aldehyde and ketone can be distinguished by [CPMT-97]
(A) Fehling's solution (B) H_2SO_4 solution
(C) Alkaline $KMnO_4$ (D) NH_3

130. Benzaldehyde + NaOH →. [CPMT-97]
(A) Benzyl alcohol (B) Benzoic alcohol
(C) Hydro benzamide (D) Cinnamic acid

131. Formaldehyde + NH_3 → Y. Y is [CPMT-97]
(A) Hexamethylene tetramine (B) Formide
(C) Para-formaldehyde (D) Methanol

132. Glucose gives silver mirror test with Tollen's reagent. It shows the presence of [CPMT-2000]
(A) Acidic group (B) Alcoholic group
(C) Ketonic group (D) Aldehyde group

133. Which one gives aldol condensation reaction ? [CPMT-2001]
(A) C_6H_5-OH (B) $C_6H_5-C(=O)-C_6H_5$
(C) $CH_3CH_2-C(=O)-CH_3$ (D) $(CH_3)-C(=O)-C-C_6H_5$

134. Acetone, formaladehyde and acetaldehyde can be distinguished by [CPMT-2002]
(A) NH_2NH_2 (B) NH_3
(C) NH_2OH (D) phenyl hydrazine

135. On heating calcium acetate produces [CPMT-2003]
(A) Acetaldehyde (B) Ether
(C) Ester (D) Acetone

136. Acetophenone is prepared from [CPMT-2003]
(A) Rosenmund's reaction (B) Sandmayer's reaction
(C) Wurtz reaction (D) Friedel-Crafts reaction

137. When acetaldehyde reacts with alcohol then produce

[CPMT-2003]

(A) Acetal (B) Ketal (C) Acetone (D) None of these

138. Reduction of aldehydes and ketones to hydro-carbon takes place in the presence of [CPMT-2003]
(A) Zinc amalgam and HCl acid (B) $Pd/BaSO_4$
(C) anhydrous $AlCl_3$ (D) Ni/Pt.

139. Aldehydes and ketones decolourises [CPMT-2003]
(A) Bromine water (B) Lime stone
(C) Dil. H_2SO_4 (D) none of these

140. On oxidation, aldehyde produces [CPMT-2003]
(A) an alcohol (B) a ketone
(C) an acid (D) none of these

141. When acetone is heated with hydroxylamine, the compound formed is : [PMT-93]
(A) Cyanohydrin (B) Oxime
(C) Semicarbazone (D) Hydrazone

142. The product of the reaction between ammonia and formaldehyde is : [PMT-93]
(A) Urotropine (B) Formamide
(C) Paraformaldehyde (D) Methanol

143. When acetaldehyde reacts with PCl_5, the resulting compound is : [PMT-93]
(A) Ethyl chloride (B) Ethylene chloride
(C) Ethylidene chloride (D) Trichloro acetaldehyde

144. Reduction of an aldehyde gives [PMT-94]
(A) Primary alcohol (B) Mono-carboxylic acid
(C) Secondary alcohol (D) Tertiary alcohol

145. In aldehydes and ketones, carbon of carbonyl group is [PMT-95]
(A) sp^3 hybridised (B) sp^2 hydridised
(C) sp hybridised (D) unhybridised

146. When formaldehyde is heated with ammonia, the compound formed is [PMT-96]
(A) formaldehyde ammonia (B) methylamine
(C) hexamethylene tetramine (D) formalin

147. Which one of the following reagents can distinguish between an aldehyde and a ketone ? [PMT-96]
(A) Fehling's solution (B) H_2SO_4 solution
(C) $NaHSO_4$ (D) NH_3

148. Dry distillation of the mixture of calcium formate and calcium acetate gives [PMT-96]
(A) Acetone (B) Acetaldehyde
(C) Formaldehyde (D) Fomic acid

149. Calcium acetate on heating gives - [PMT-97]
(A) Acetaldehyde (B) Acetone
(C) Acetic anhydride (D) Aldol

150. The alkaline $CuSO_4$ containing sodium potassium tatrate does not react with - [PMT-97]
(A) CH_3CHO (B) C_2H_5CHO
(C) $C_6H_5CH_2CHO$ (D) C_6H_5CHO

151. Which of the following gives an alcohol and salt of carboxylic acid when reacted with conc. NaOH? [PMT-99]
(A) CH_3CHO (B) C_6H_5CHO
(C) CH_3COCH_3 (D) $C_6H_5COCH_3$

152. Which of the following on reactions with NH_3 gives urinary antiseptic compound ? [PMT-99]
(A) HCHO (B) CH_3CHO
(C) C_6H_5CHO (D) $C_6H_5CH_2CHO$

153. Which of the following does not give formaldehyde on heating or distillation ? [PMT-2000]
(A) Formalin (B) Trioxan
(C) Para-aldehyde (D) Para-formaldehyde

154. A cyanohydrin of a compound X on hydrolysis gives lactic acid. Compound X is : [PMT-2000]
(A) HCHO (B) CH_3CHO
(C) $(CH_3)_2CO$ (D) $C_6H_5CH_2CHO$

155. The IUPAC name of acraldehyde is : [PMT-2000]
(A) prop-2-ene-1-al (B) propenyle aldehyde
(C) but-2-ene-1-al (D) propenol

156. Acetone on heating with conc. H_2SO_4 mainly gives : [PMT-2000]
(A) Mesitylene (B) Mesityl Oxide
(C) Toluene (D) Xylene

157. Schiff reagent gives pink colour with : [PMT-2000]
(A) Acetaldehyde (B) Acetone
(C) Acetic acid (D) Methyl acetate

158. The reaction of benzaldehyde with alkali gives : [PMT-2000]
(A) Phenol + Sodium benzoate
(B) Benzene + Benzyl alcohol
(C) Benzyl alcohol + Sodium benzoate
(D) Phenol + Benzene

159. Schiff's reagent is : [PMT-2001]
(A) P-rosaniline hydrochloride solution decolourised with sulphurous scid
(B) Magenta solution decolourised with Cl_2
(C) Ammoniacal silver nitate solution
(D) Alkaline $KMnO_4$ solution

160. Tollen's reagent is [PMT-2002]
(A) Alkaline mercuric chloride solution
(B) Ammoniacal silver nitrate solution
(C) Ammonium citrate solution
(D) Alkaline potassium permanganate solution

161. The reaction

$$2CH_3CHO \xrightarrow{Al(OC_2H_5)_3} CH_3COOC_2H_5$$ is known as

[PMT-2002]
(A) Tishenko reaction (B) Cannizzaro's reaction
(C) Knoevenagel reaction (D) Wurtz's reaction

162. Formaldehyde reacts with ammonia to form Urotropine which has the formula : [PMT-2003]
(A) $(CH_2)_6N_4$ (B) $(CH_2)_4N_3$ (C) $(CH_2)N_6$ (D) $(CH_2)_3N_4$

163. The reaction of an aldehyde with hydroxylamine gives a product which is called [PET-93]
(A) Aminohydroxide (B) Hydrazone
(C) Semicarbazone (D) Oxime

164. The end product of the following reaction sequence is

$$CaO + C \xrightarrow{heat} A \xrightarrow{H_2O} B \xrightarrow[HgSO_4]{dill\ H_2SO_4} C$$

[PET-93]
(A) Acetaldehyde (B) Ethanol
(C) Ethylene (D) Ethyl hydrogen sulphate

165. Which one of the following reactions would produce secondary alcohol [PET-94]

(A) $C_6H_5-\overset{O}{\overset{\|}{C}}-CH_3 \xrightarrow[2.\ H^+]{1.\ CH_3MgBr}$

(B) $C_6H_5-\overset{O}{\overset{\|}{C}}-CH_3 \xrightarrow[2.\ H^+]{1.\ LiAlH_4}$

(C) $CH_3-CHO \xrightarrow[2.\ H^+]{1.\ LiAlH_4}$

(D) $C_6H_5-\overset{O}{\overset{\|}{C}}-CH_3 \xrightarrow[2.\ Br_2]{1.\ ^-OH}$

166. In the presence of a dilute base C_6H_5CHO and CH_3CHO react together to give a product. The product is [PET-94]
(A) $C_6H_5CH_3$ (B) $C_6H_5CH_2CH_2OH$
(C) $C_6H_5CH_2OH$ (D) $C_6H_5CH = CHCHO$

167. Which of the following reagents distinguishes between aldehyde and ketone [PET-95]
(A) Fehling's solution (B) H_2SO_4 (C) $NaHSO_3$ (D) NH_3

168. Calcium acetate on dry distillation gives [PET-95]
(A) Formaldehyde (B) Acetaldehyde
(C) Acetone (D) Acetic anhydride

169. Tollen's reagent is - [PET-95]
(A) Ammoniacal cuprous chloride
(B) Ammoniacal cuprous oxide
(C) Ammoniacal silver bromide
(D) Ammoniacal silver nitrate

170. The reaction of HCOOH with conc. H_2SO_4 gives - [PET-95]
(A) CO_2 (B) CO (C) Oxalic acid (D) Acetic acid

171. Para aldehyde is trimer of [PET-96]
(A) Formaldehyde (B) Acetaldehyde
(C) Benzaldehyde (D) Propionaldehyde

172. Which of the following on reaction with conc. NaOH gives an alcohol ? [PET-96]
(A) Methanal (B) Ethanal (C) Propanal (D) Butanal

173. Reaction of acetaldehyde with HCN followed by hydrolysis gives a compound which shows [PET-97]
(A) Optical isomerism (B) Geometrical isomerism
(C) Metamerism (D) Tautomerism

174. When ethanal is treated with Fehling's solution. It give a precipitate of : [PET-98]
(A) Cu (B) CuO (C) Cu_2O (D) $Cu_2O + Cu_2O_3$

175. The reaction in which sodium cyanide is used: [PET-98]
(A) Perkin reaction (B) Reimer-Tiemann reaction
(C) Benzoin condensation (D) Rosenmund reaction

176. If formaldehyde and KOH are heated then we get : [PET-99]
(A) Acetylene (B) Methane
(C) Methyl alcohol (D) Ethyl formate

177. Which of the following reagents reacts differently with HCHO, CH_3CHO and CH_3COCH_3 : [PET-99]
(A) HCN (B) NH_2NH_2 (C) NH_2OH (D) NH_3

178. Oxidation of toluene with chromyl chloride gives : [PET-2000]
(A) Benzyl alcohol (B) Benzaldehyde
(C) Benzoic acid (D) Acetophenone

179. Reduction of an aldehyde gives : [PET-2001]
(A) Mono-carboxylic acid (B) Primary alcohol
(C) Secondary alcohol (D) Tertiary alcohol

180. Fehling's solution for the detection of aldehyde group consists of $CuSO_4$ solution and [PET-2002]
(A) Alkaline KOH solution (B) Sodium potassium tartrate
(C) Alcoholic KOH solution
(D) Ammonium hydroxide solution

181. Benzaldehyde on reaction with acetophenone in the presence of sodium hydroxide solution gives [PET-2003]
(A) C_2H_5 - CH = CH - CO - C_6H_5
(B) $C_2H_5COCH_2C_2H_5$ (C) $C_6H_5CH = CHC_2H_5$
(D) $C_6H_5CH(OH)COC_2H_5$

182. Reaction [PET-2003]

$$\begin{matrix} R \\ R \end{matrix}\!\!>C = O \xrightarrow{H_2NNH_2} \begin{matrix} R \\ R \end{matrix}\!\!>C<\!\!\begin{matrix} H \\ H \end{matrix} + N_2$$

is called
(A) Tischenko reaction (B) Sandmeyer reaction
(C) Reformat sky reaction (D) Wolff-Kishner reaction

183. NaOH/H⁺ reacts with [BHU-2003]
(A) $C_6H_5OCH_3$ (B) CH_3OH
(C) C_2H_5OH (D) CH_3COCH_3

184. Benzophenone doesnot react with [BHU-2003]
(A) RNH_2 (B) Na_2CO_3 (C) NaOH (D) SO_3

Alkehydes & Ketones

1.	(D)	46.	(C)	91.	(B)	136.	(D)	181.	(A)
2.	(A)	47.	(C)	92.	*	137.	(A)	182.	(D)
3.	(C)	48.	(A)	93.	(D)	138.	(A)	183.	(D)
4.	(A)	49.	(C)	94.	(B)	139.	(D)	184.	(C)
5.	(C)	50.	(C)	95.	(D)	140.	(C)		
6.	(A)	51.	(B)	96.	(A)(B)(D)	141.	(B)		
7.	(D)	52.	(A)	97.	(A)	142.	(A)		
8.	(D)	53.	(C)	98.	(D)	143.	(C)		
9.	(D)	54.	(B)	99.	(D)	144.	(A)		
10.	(A)	55.	(B)	100.	(D)	145.	(B)		
11.	(A)	56.	(C)	101.	(B)	146.	(C)		
12.	(B)	57.	(A)	102.	*	147.	(A)		
13.	(C)	58.	(A)	103.	*	148.	(B)		
14.	(A)	59.	(A)	104.	*	149.	(B)		
15.	(C)	60.	(A)	105.	*	150.	(D)		
16.	(C)	61.	(C)	106.	*	151.	(B)		
17.	(A)	62.	(B)	107.	*	152.	(A)		
18.	(A)	63.	(A)	108.	*	153.	(C)		
19.	(C)	64.	(A)	109.	(C)	154.	(B)		
20.	(C)	65.	(A)	110.	(C)	155.	(A)		
21.	(D)	66.	(B)	111.	(C)	156.	(A)		
22.	(D)	67.	(A)	112.	(A)	157.	(A)		
23.	(A)	68.	(D)	113.	(D)	158.	(C)		
24.	(C)	69.	(B)	114.	(C)	159.	(A)		
25.	(C)	70.	(A)	115.	(D)	160.	(B)		
26.	(B)	71.	(B)	116.	(A)	161.	(A)		
27.	(B)	72.	(D)	117.	(D)	162.	(A)		
28.	(C)	73.	(C)	118.	(B)	163.	(D)		
29.	(B)	74.	(A)	119.	(B)	164.	(A)		
30.	(A)	75.	(A)	120.	(C)	165.	(B)		
31.	(B)	76.	(C)	121.	(A)	166.	(D)		
32.	(B)	77.	(A)	122.	(D)	167.	(A)		
33.	(D)	78.	(B)	123.	(C)	168.	(C)		
34.	(B)	79.	(C)	124.	(A)	169.	(D)		
35.	(C)	80.	(A)	125.	(C)	170.	(B)		
36.	(B)	81.	(B)	126.	(A)	171.	(B)		
37.	(C)	82.	(B)	127.	(D)	172.	(A)		
38.	(D)	83.	(C)	128.	(C)	173.	(A)		
39.	(A)	84.	(C)	129.	(A)	174.	(C)		
40.	(D)	85.	(A)	130.	(A)	175.	(C)		
41.	(D)	86.	(D)	131.	(A)	176.	(C)		
42.	(C)	87.	(B)	132.	(D)	177.	(D)		
43.	(D)	88.	(B)	133.	(C)	178.	(B)		
44.	(D)	89.	*	134.	(C)	179.	(B)		
45.	(C)	90.	(D)	135.	(D)	180.	(B)		

89. Rochelle salt
92. benzylidene acetate
102. HCOOH
103. Aldehydes
104. Ethanol
105. Aldehyde
106. Acetaldehyde
107. Tollen's reagent
108. $LiAlH_4$

Carboxylic acids and their derivatives

Acidic character is due to the presence of replaceable hydrogen in carboxylic acids. Ant contains formic acid; vinegar – acetic acid, milk – lactic acid, butter – butanoic acid, goat fat – caproic, caprylic and capric acids.

Carboxylic acids show chain, position, functional, geometrical and optical isomerism depending upon their structures.

Rancidification: Unpleasant smell which fats and oils develop on long exposure to moist air is due to rancidity. By partial saponification fatty acids are set free which have a strong smell. e.g. butter.

Reichert – Meiss value (R/M value): It represents the steam volatile fatty acids present in the oil or the fat.

Preparation:

1. From acid amides:

$$RCONH_2 + HNO_2 \longrightarrow RCOOH + N_2 + H_2O$$

2. Hydrolysis of trihalogen derivatives:

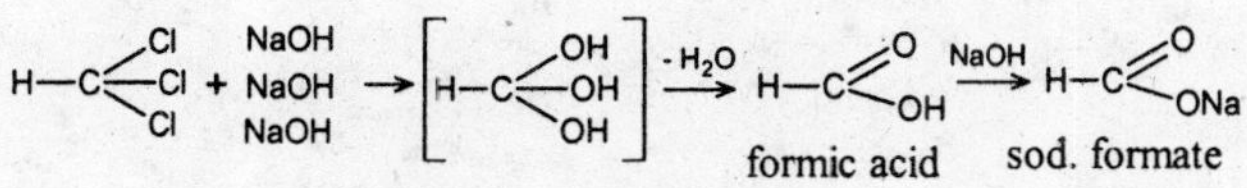

3. From Grignard reagent:

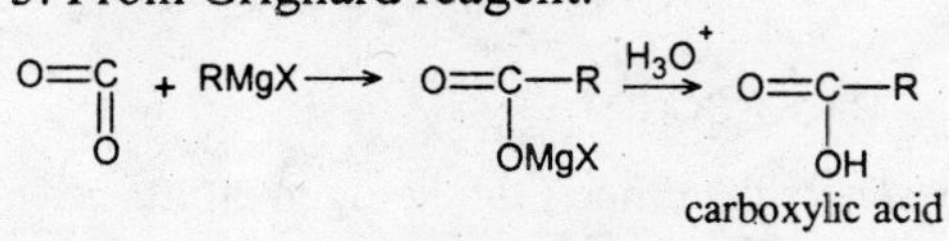

4. Hydrolysis of alkyl cyanides:

$$R-C\equiv N \xrightarrow{H_3O^+} R-\overset{OH}{\overset{|}{C}}=NH \rightleftharpoons R-\overset{O}{\overset{\|}{C}}-NH_2$$

$$R-\overset{O}{\overset{\|}{C}}-NH_2 + H_2O + HCl \longrightarrow R-\overset{O}{\overset{\|}{C}}-OH + NH_4Cl$$

5. From gem dicarboxalic acids:

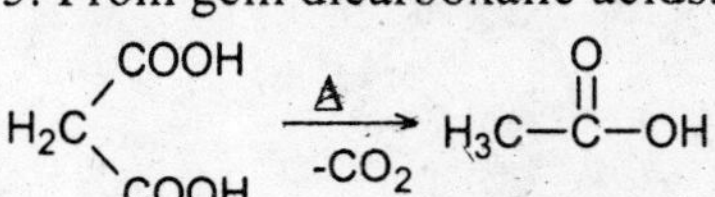

malonic acid acetic acid

6. Hydrolysis of acid derivatives:

Reactivity towards hydrolysis:

RCOX	> $(RCO)_2O$	> RCOOR	> $RCONH_2$
acid halide	anhydride	ester	amide

7. By oxidation methods:

(i) From 1°, 2° and 3° alcohols

(ii) 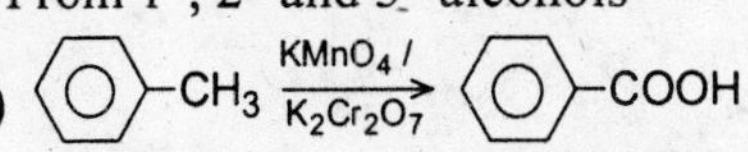

toluene benzoic acid

(iii) 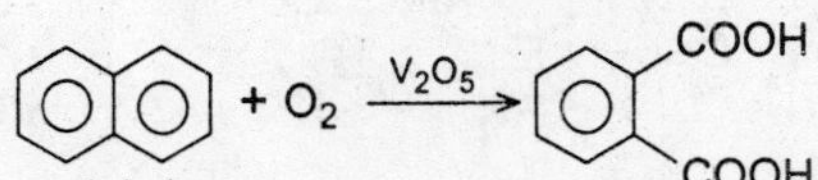

naphthalene pthalic acid

(iv) Alkenes and alkynes on oxidation with $K_2Cr_2O_7$ / $KMnO_4$ give acids.

(v) On ozonolysis alkenes and alkynes give acids.

Properties: Due to hydrogen bonding

(i) lower members are less volatile than expected.

(ii) molecular weights of lower members are more than expected.

(iii) boiling points of lower members are higher than expected.

(iv) lower members are more soluble in H_2O than expected.

(v) carboxylic acids exist in dimeric state.

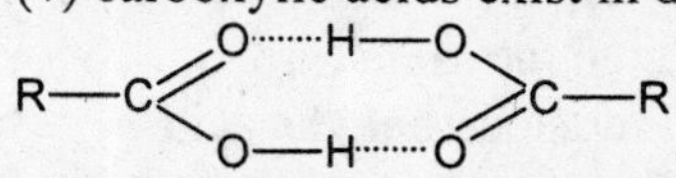

Acidic character decreases with increasing molecular weight.

The stabilisation of the carboxylate ion also contribute to the acidic nature of carboxylic acids.

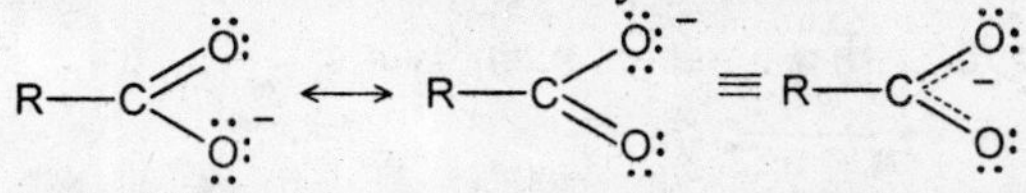

HCOOH is the strongest acid among the fatty acids.

Formic acid exhibits the properties of both aldehydes and acids.

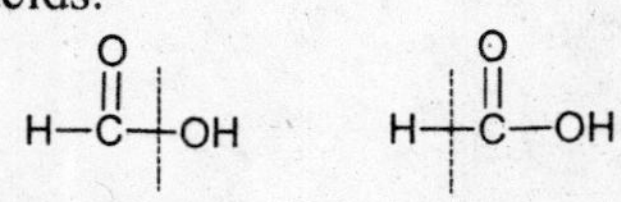

aldehyde group acid group

Order of acidity : RCOOH > HOH > ROH > $HC\equiv CH$

More the pK_a value, less the acidity.

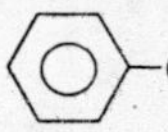 $-COOH(4.2) > H_2C=CH-COOH(4.3) > CH_3COOH(4.8)$

Reactions:

1. Hell-Volhard Zelinsky reaction:

Here α bromination of carboxylic acids takes place. It is a substitution reaction in alkyl or phenyl group. Acids without α hydrogen do not show this reaction.

$$R_2-\underset{H}{\underset{|}{C}}-COOH + Br_2 \xrightarrow{PCl_3} R_2-\underset{Br}{\underset{|}{C}}-COOH + HBr$$

2. Schmidt reaction: Primary amines are formed

$$R-COOH + \underset{\text{hydrazoic acid}}{HN_3} \xrightarrow{\text{conc. } H_2SO_4} RCON_3 \xrightarrow{H_2O} R-NH_2 + CO_2 + N_2$$

3. Kolbe's reaction: It leads to the formation of alkanes

4. Reduction:

$$R-COOH \xrightarrow{LiAlH_4\ /\ NaBH_4} RCH_2OH$$

$$R-COOH \xrightarrow{H_2\ /\ Ni} R-CH_3$$

Formic acid

Having an aldehyde group in its structure, it behaves as a reducing agent.

$$H-COOH \xrightarrow{\text{Fehling's reagent}} Cu_2O \text{ (redish brown)}$$

$$H-COOH \xrightarrow{\text{Tollens'seagent}} Ag$$

$$H-COOH \xrightarrow{HgCl_2} Hg_2Cl_2 \xrightarrow{H-COOH} Hg$$

$KMnO_4$ solution is decolourised by formic acid due to reduction.

$$2KMnO_4 + 3H_2SO_4 + 5HCOOH \longrightarrow K_2SO_4 + 2MnSO_4 + 8H_2O + 5CO_2$$

Oxalic acid:
Sucrose on oxidation in the presence of V_2O_5 gives oxalic acid.

$$\underset{\text{sucrose}}{C_{12}H_{22}O_{11}} + 18\,(O) \xrightarrow[HNO_3]{V_2O_5} 6\begin{matrix}COOH\\|\\COOH\end{matrix} + 5H_2O$$

Potassium tetraoxalate can be used for removing ink stains.

Balnc's rule: on heating

(i) Two carboxylic acid groups on the same carbon atom leads to the formation of mono carboxylic acid.

e.g.

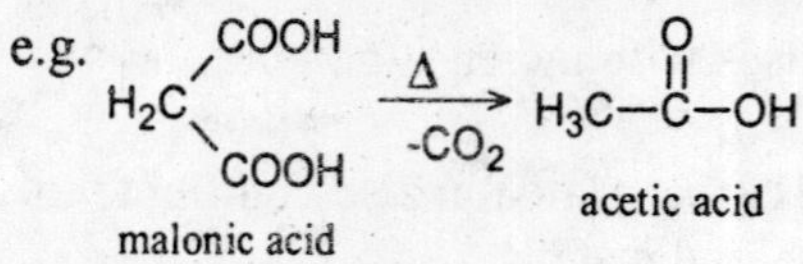

(ii) I,4 and 1,5 dicarboxylic acids form cyclic anhydrides. e.g.

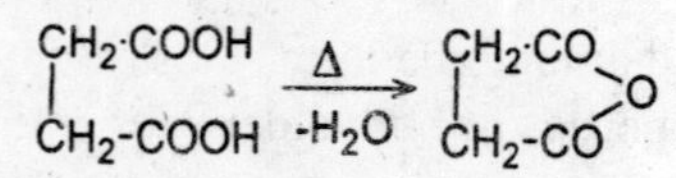

;

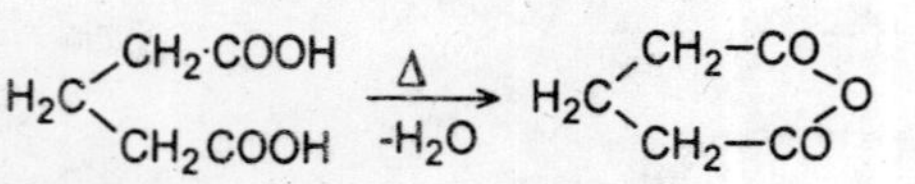

(iii) 1,6 and 1,7– dicarboxylic acids on heating with acetic anhydride form cyclic ketons.

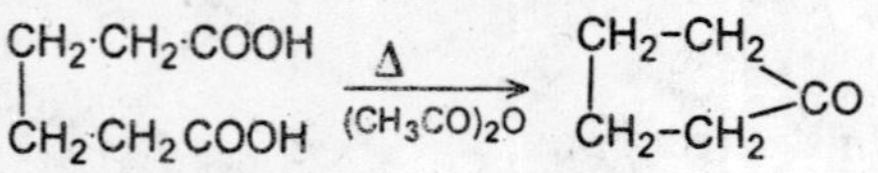

;

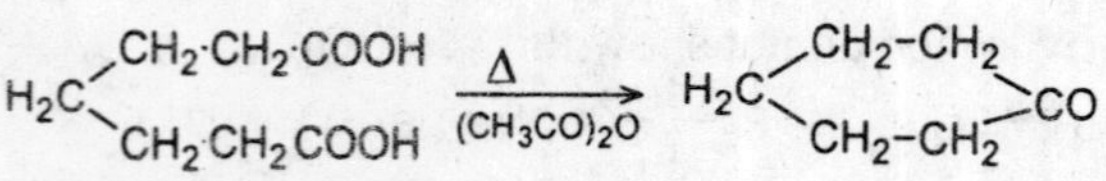

Test for COOH group

(i) $NaHCO_3$ test : When to a small amount of the carboxylic acid compound a few drops of a saturated aqueous Na_2CO_3 solution are added, effervescence due to the liberation of CO_2 gas occur.

$$RCOOH + NaHCO_3 \longrightarrow RCOONa + H_2O + CO_2\uparrow$$

(ii) Esterification with alcohol: Carboxylic acids form esters with alcohol which give fruity smell.

Carboxylic acid derivatives

All acid derivatives give acids on hydrolysis.
Rate of hydrolysis (in neutral conditions)

RCOX	> $(RCO)_2O$	> RCOOR	> $RCONH_2$
acid halide	anhydride	ester	amide

The electrophilic nature of carbonyl group also decreases in the same order. >C=O group is responsible for more reactivity of the acid derivatives compared to their alkyl derivatives without >C=O group. Here too as in aldehydes and ketones, the carbonyl group performs two functions.

(i) Provides site for nucleophilic attack.

(ii) increases the acidity of the hydrogen atoms attached to the α carbon atom.

Acid Halides

Preparation:
(i). From acids:

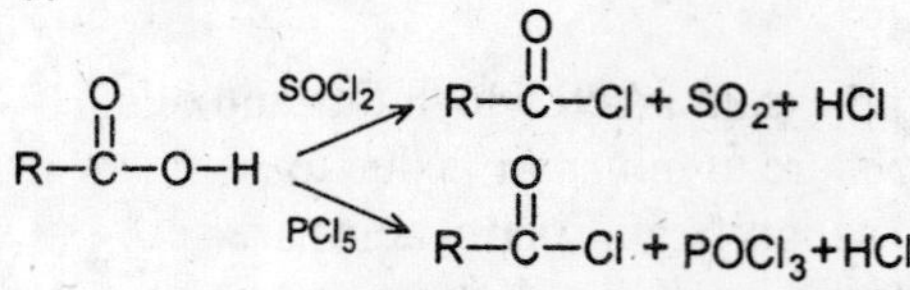

From lower carboxylic esters and anhydrides also, one can obtain the acid derivatives.

Reactions

1. Rosenmund's reaction:

$$RCOCl + 2H \xrightarrow{Pd\,/\,BaSO_4} RCHO + HCl$$

2. Fridel Carft's acylation

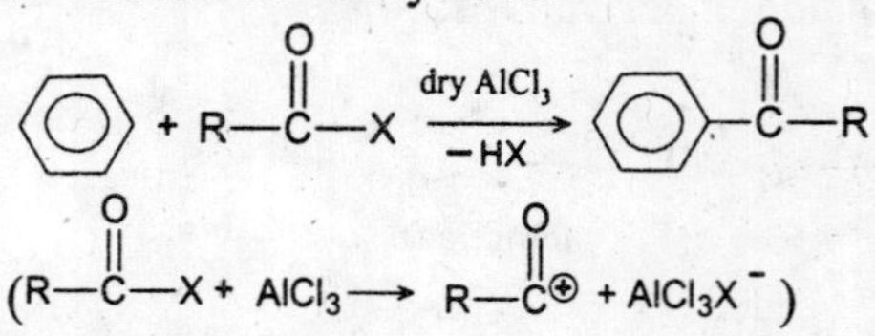

3. With ethers:

$$R{-}O{-}R + RCOX \longrightarrow RX + RCOOR$$

In the case of unsymmetrical ethers, the halogen atom attaches with the simple alkyl group.

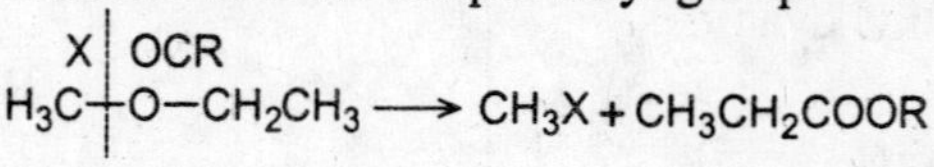

4. Esterification:

$$RCOX + HOR \longrightarrow RCOOR + HX$$

5. With salicylic acid:

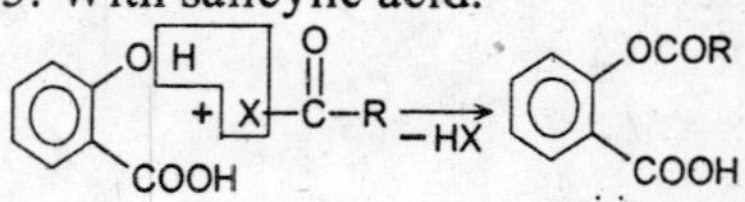

6. With Grignard reagent:

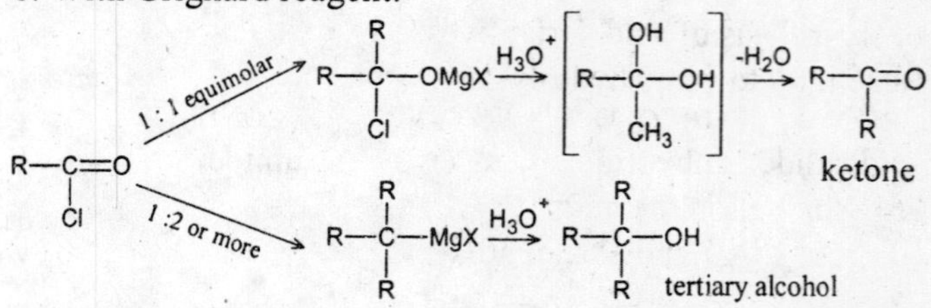

7. Curtius reaction:

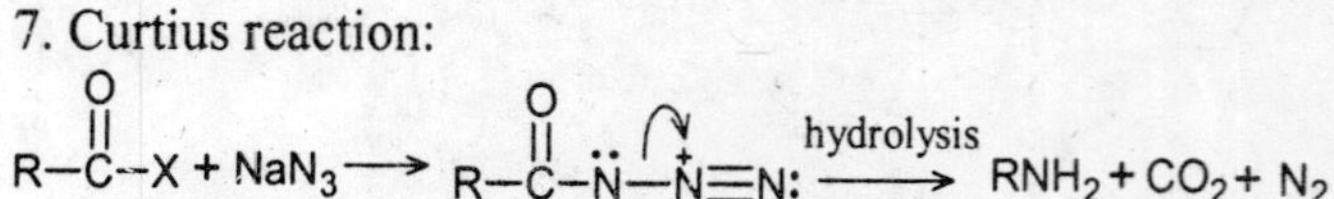

- Acetyl chloride is used to determine the number of hydroxyl, $-NH_2$, >NH, –SH etc. groups. These groups contain reactive hydrogen.
- Aromatic acid chlorides are less reactive than aliphatic acid chlorides due to resonance in the former.

Esters

Preparation

1. Tischenko reaction:

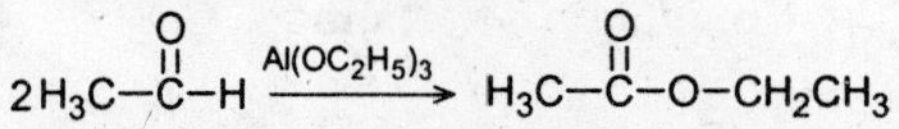

2. Claisen condensation:

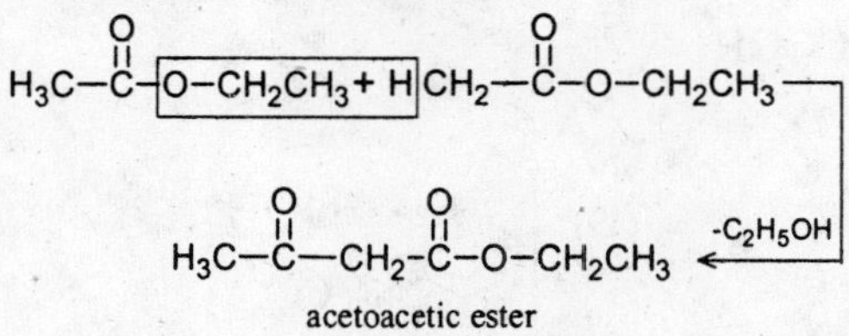

carbanion is the reactive intermediate in this reaction.

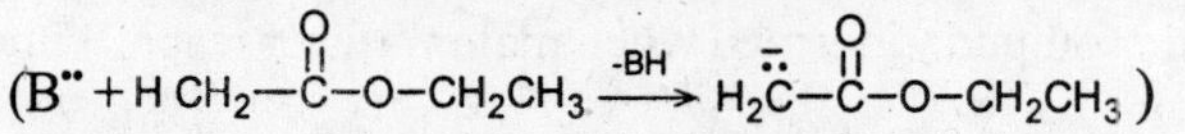

3. Baeyer – Villeger reaction: Ketones on oxidation with peracids give esters.

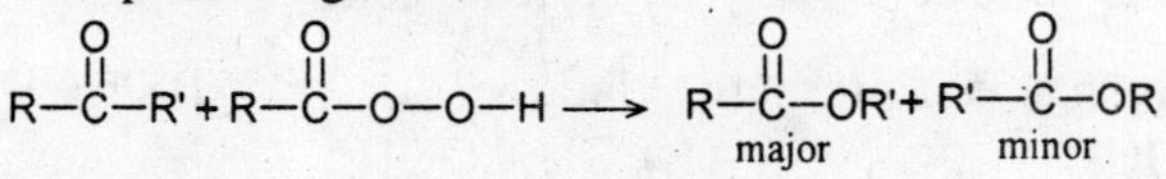

4. By Intramolecular esterification: A hydroxy acid is both alcohol and acid. In such cases, where a 5 or 6 membered ring can be formed, intramolecular esterification takes place and a lactone is formed. e.g.

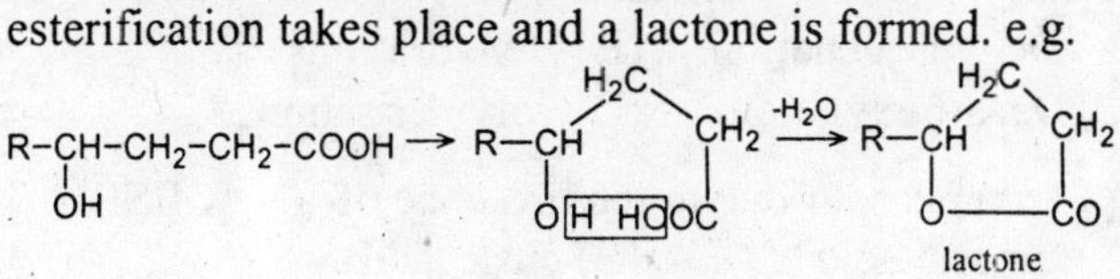

Reactions:

1. Saponification: alkaline hydrolysis of esters is known as saponification. It is a second order reaction.

$$RCOOR' + NaOH \longrightarrow RCOONa + R'OH$$

2. Bouveault and Balnc reduction:

$$RCOOR \xrightarrow{C_2H_5OH\,/\,Na} RCH_2OH$$

3. Trans esterification or ester interchange:

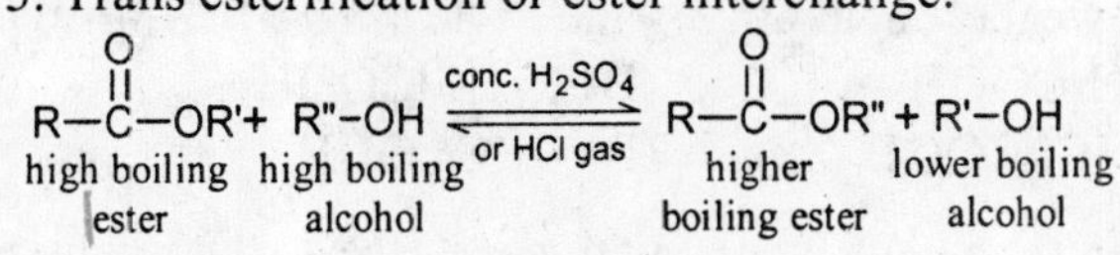

- Esters present in fruits/ flowers

Apple – Isoamyl valerate; Apricot – Amylbutyrate; Banana – Isoamyl acetate; Pineapple – Methyl butyrate; Jasmine – Benzyl butyrate; Orange – Octyl acetate; Winter green – Methyl salicylate.

- Esters with active methylene group:

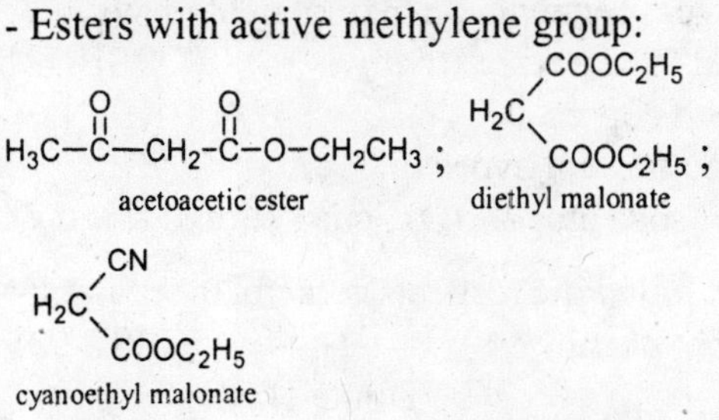

Acid Amides

Preparation:

1. Reaction of acid with urea:

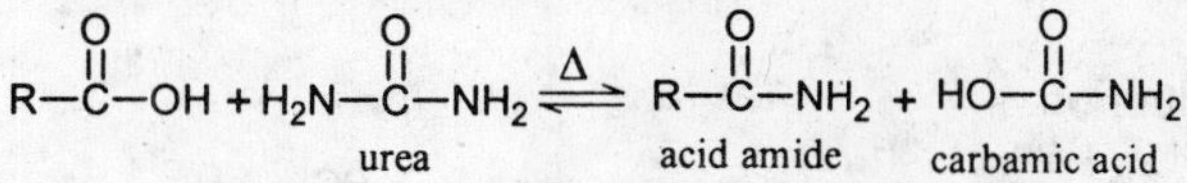

2. Reaction of acid with NH_3:

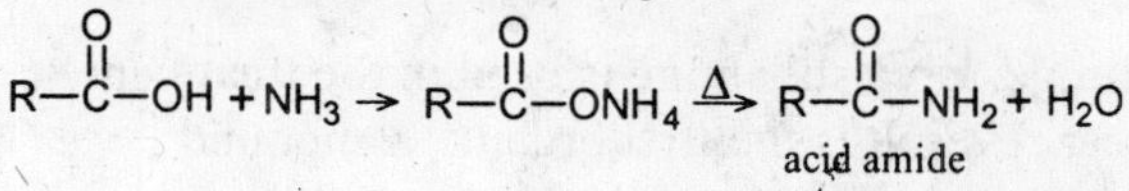

Acid derivatives also give acid amide on treatment with NH_3.

Properties:

Amides are less basic than amines due to resonance.

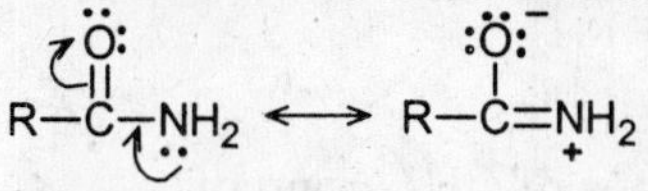

Acetamide shows keto-enol isomerism (Tautomerism).

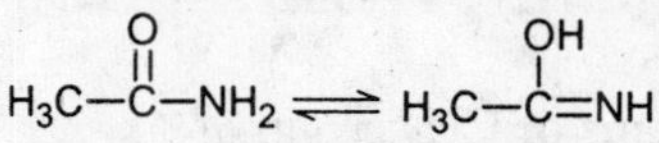

Due to intermolecular hydrogen bonding, acetamide is more soluble in water.

Reactions:

1. Nucleophilic substitution: aldehydes and ketones do not show this reaction as H^- and R^- (leaving groups respectively) are strong bases.

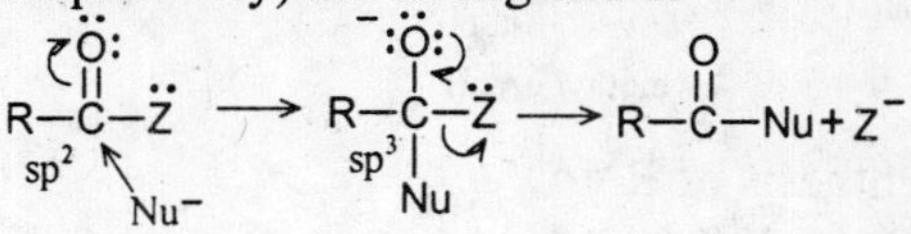

2. With nitrous acid:

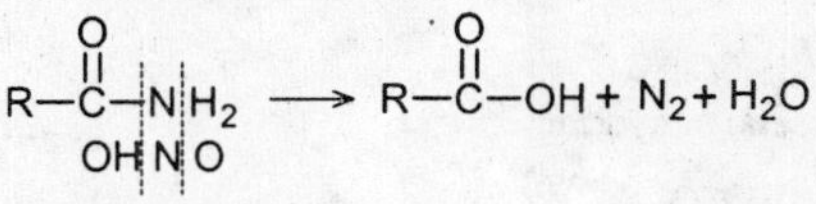

3. Dehydration:

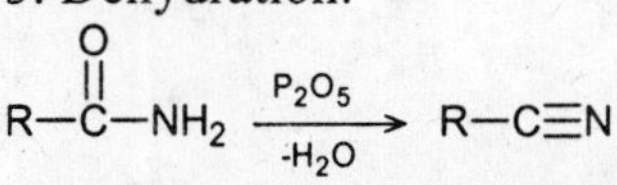

4. Hoffmann degradation/ bromamide reaction:

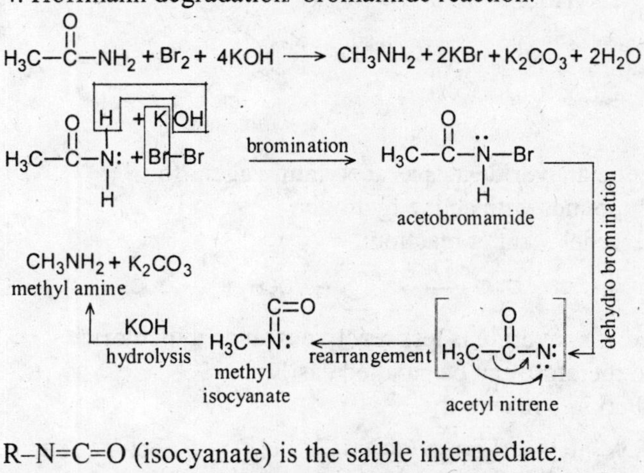

R–N=C=O (isocyanate) is the satble intermediate.

N– alkyl or N–aryl amides do not show this degradation.

5. With Grignard reagent

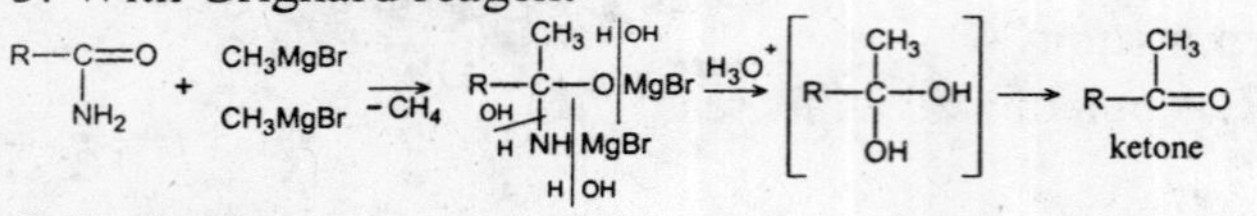

6. Reduction:

$$RCONH_2 \xrightarrow[C_2H_5OH\ /\ Na]{LiAlH_4\ /\ NaBH_4,} RCH_2OH + NH_3 \longrightarrow RCH_2NH_2$$

Urea

It is a diamide. Urea stibamine is used in the treatment of kala-azar fever. It is the first organic compound synthesised in the laboratory (Wöhler in 1828).

Reaction with malonic ester:

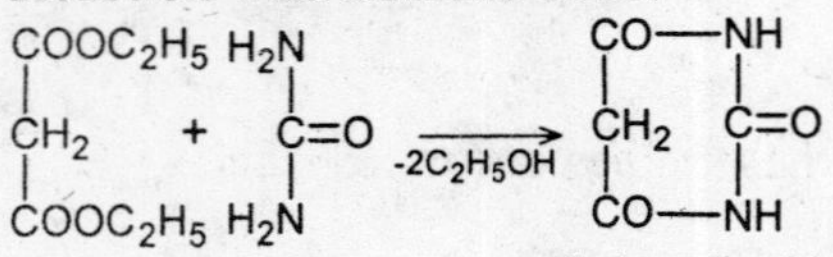

barbutyric acid

If the two active hydrogen atoms of CH_2 group of malonic ester are replaced by $(C_2H_5)_2$, Veronal is obtained.

If the two active hydrogen atoms of CH_2 group of malonic ester are replaced by C_2H_5 & C_6H_5, Luminal is obtained.

Reaction with acetoacetic ester.

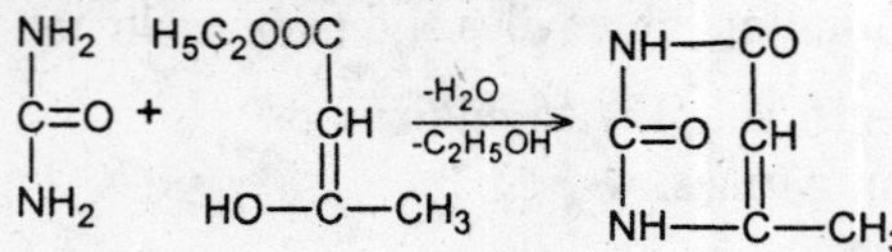

4 - methyl uracil

It is used for the synthesis of RNA.

Acid Anhydrides

Preparation:

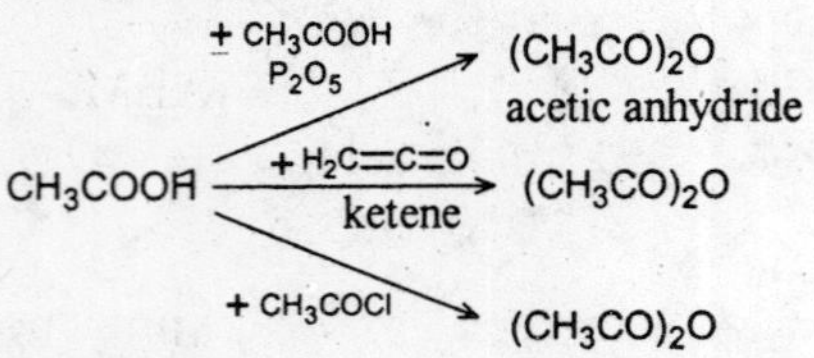

Reactions:

1. Acetylation:

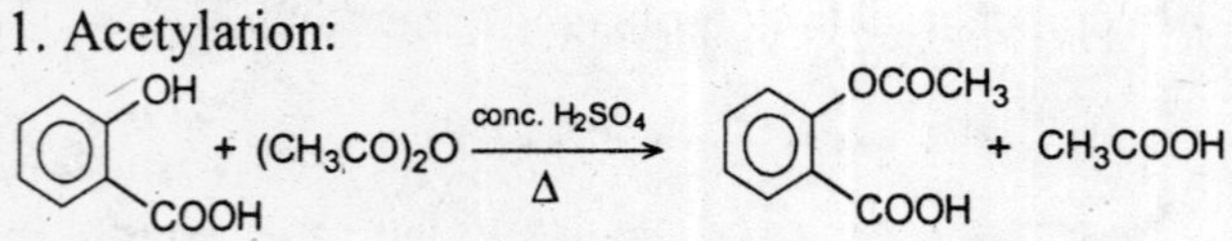

acetyl salicylic acid

Acetic anhydride acts as acetylating agent for compounds with active hydrogen.

2. Friedel–Craft's reaction:

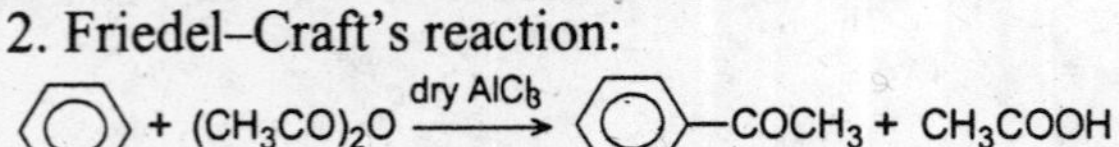

Acetic anhydride is less reactive than acetyl chloride and therefore can be handled easily.

3. Perkin's reaction:

cinnamic acid

Aromatic aldehydes only undergo this reaction.

1.Identify B in [CBSE-90]

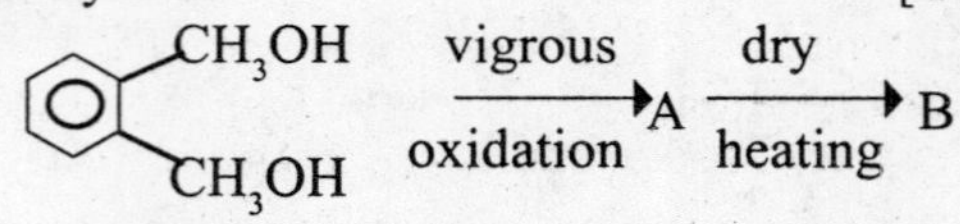

(A) 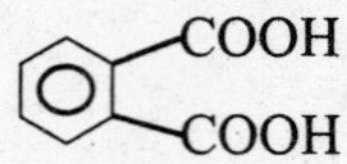 (B) 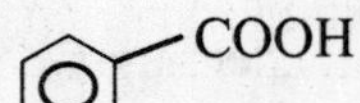

(C) 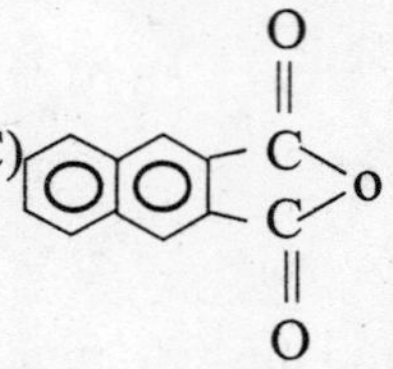 (D) 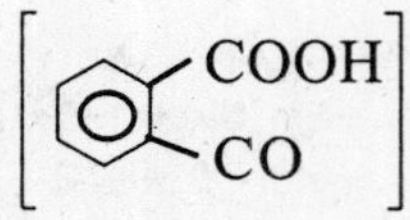

2. The compound formed when malonic ester reacts with urea is [CBSE-90]

(A) cinnamic acid (B) butyric acid
(C) barbituric acid (D) crotonic acid

3. Which acid is present is vinegar ? [CBSE-90]

(A) HCl (B) CH_3COOH
(C) Tartaric acid (D) Citric acid

4. By which of the following processes can acetic acid be converted into methane ? [CBSE-90]

(A) Dehydrogenation (B) Dehydration
(C) Decarboxylation (D) Polymerisation

5. Lemon is sour due to the presence of [CBSE-90]

(A) tartaric acid (B) citric acid
(C) sulphuric acid (D) oxalic acid

6. The general formula $(R.CO)_2O$ represents[CBSE-90]

(A) an ether (B) ketone
(C) an ester (D) an acidic anhydride

7. The first synthetic organic compound is [CBSE-90]

(A) glucose (B) methane (C) benzene (D) urea

8. The weakest acid among the following is [CBSE-91]

(A) CH_3COOH (B) $CH_2ClCOOH$
(C) $CHCl_2COOH$ (D) Cl_3CCOOH

9. When sucrose is heated with conc.HNO_3, the product is [CBSE-91]

(A) Sucrose Nitrate (B) Formic acid
(C) Oxalic acid (D) Citric acid

10. Benzoic acid gives benzene on being heated with X and phenol gives benzene on being heated with Y. Therefore X and Y are respectively [CBSE-92]

(A) Sodalime and copper
(B) Zinc dust and sodium hydroxide
(C) Zinc dust and soda lime (D) Soda lime and zinc dust

11. Which of the following reagents is useful in separating benzoic acid from phenol ? [CBSE-93]

(A) Dilute HCl (B) Dilute H_2SO_4
(C) 5% aqueous NaOH (D) 5% aqueous $NaHCO_3$

12. Among acetic acid, phenol and n-hexanol, which of

the compounds will react with $NaHCO_3$ solution to give sodium salt and carbon dioxide ? [CBSE-93]
(A) Acetic acid (B) n-hexanol
(C) Acetic acid and phenol (D) Phenol

13. Sodium formate on heating yields [CBSE-93]
(A) oxalic acid and H_2 (B) sodium oxalate and H_2
(C) CO_2 and NaOH (D) sodium oxalate

14. An acyl halide is formed by reacting PCl_5 with an [CBSE-94]
(A) acid (B) alcohol (C) ester (D) amide

15. Which one of the following esters cannot give Claisen self-condensation ? [CBSE-98]
(A) $CH_3-CH_2-CH_2-CH_2COOC_2H_5$
(B) $C_6H_5COOC_2H_5$
(C) $C_6H_{11}CH_2COOC_2H_5$ (D) $C_6H_5CH_2COOX_2H$

16. Aspirin is an acetylation product of [CBSE-98]
(A) p-dihydroxybenzene (B) o-hydroxy benzoic acid
(C) o-dihydroxybenzene (D) m-hydroxy benzoic acid

17. An ester (A) with molecular formula $C_9H_{10}O_2$ was treated with excess of CH_3 MgBr and the complex so formed was treated with H_2SO_4 to give an olefin (B). Ozonolysis of (B) gave a ketone with molecular formula C_8H_8O which shows +ve iodoform test. The structure of (A) is [CBSE-98]
(A) C_6H_5 $COOC_2H_5$ (B) $H_3COCH_2COC_6H_5$
(C) P–$H_3CO-C-C_6H_4-COCH_3$ (D) $C_6H_5COOC_6H_5$

18. Which one of the following compounds will react with $NaHCO_3$ solution to give sodium salt and carbon dioxide ? [CBSE-99]
(A) Acetic acid (B) n-hexanol
(C) Phenol (D) Both (B) and (C)

19. Amides may be converted into amines by a reaction named after [CBSE-99]
(A) Perkin (B) Claisen (C) Hoffmann (D) Kekule

20. What will happen if $LiAlH_4$ is added to an ester ? [CBSE-2000]
(A) Two units of alcohol are obtained.
(B) One unit of alcohol and one unit of acid is obtained.
(C) Two units of acids is obtained. (D) None of these

21. Intermediates formed during reaction of $RCONH_2$ with Br_2 and KOH are - [CBSE-2001]
(A) RCONHBr and RCNO (B) RNHCOBr and RNCO
(C) RNHBr and RCONHBr (D) $RCONBr_2$

22. In the following reaction product 'P' is

$$R-\underset{\underset{O}{\|}}{C}-Cl \xrightarrow[Pd\text{-}BaSO_4]{H_2} P$$ [CBSE -2002]

(A) RCH_2OH (B) RCOOH (C) RCHO (D) RCH_3

23. In a set of the given reactions, acetic acid yielded a product C. [CBSE-2003]

$$CH_3COOH + PCl_5 \longrightarrow A \xrightarrow[Anh.\ AlCl]{C_6H_6} B \xrightarrow[ether]{C_2H_5MgBr} C$$

product C would be
(A) $CH_3-\overset{\overset{C_2H_5}{|}}{C}(OH)C_6H_5$ (B) $CH_3CH(OH)C_2H_5$
(C) $CH_3COC_6H_5$ (D) $CH_3-CH(OH)C_6H_5$

24. Which one of the following orders of acid strength is correct ? [CBSE-2003]
(A) RCOOH > HC ≡ CH > HOH > ROH
(B) RCOOH > ROH > HOH > HC ≡ CH
(C) RCOOH > HOH > ROH > HC ≡ CH
(D) RCOOH > HOH > HC ≡ CH > ROH

25. When salicylic acid is distilled with sodalime, we get [AIIMS-93]
(A) Toluene (B) Carbolic acid
(C) Benzene (D) Benzoic acid

26. Which acid does not have a carboxylic group ? [AIIMS-93]
(A) Benzoic acid (B) Salicylic acid
(C) Picric acid (D) Methanoic acid

27. Which of the following is the strongest acid ? [AIIMS-93]
(A) C_2H_5COOH (B) CH_3COOH
(C) $CH_2BrCOOH$ (D) $CHBr_2COOH$

28. The product of oxidation of benzene with air in presence of V_2O_5 at 1-2 atm and at 400° C-450°C is [AIIMS-93]
(A) Benzoic acid (B) Adipic acid
(C) Succinic acid (D) Maleic anhydride

29. The most appropriate method to convert propanoic acid into 3-pentanone is [AIIMS-93]
(A) Heating the dry calcium salt of the propanoic acid out of contact with air
(B) Heating the sodium salt of the acid with ethylchloride
(C) Electrolysis of fused sodium salt of propanoic acid
(D) Heating strongly the sodium salt of propanoic acid

30. Which of the following is called as ethanoic acid ? [AIIMS-94]
(A) HCOOH (B) CH_3COOH
(C) CH_3CH_2COOH (D) $CH_3CH_2CH_2COOH$

31. Which one is the strongest of the following acids ? [AIIMS-94]
(A) HCOOH (B) CH_3CCOH
(C) CCl_3COOH (D) $CH_2ClCOOH$

32. Aspirin is an acetyation product of [AIIMS-94]
(A) o-hydroxybenzioc acid (B) o-dihydroxybenzene

(C) m-hydroxybenzoic acid (D) p-dihydroxybenzene

33. Assertion : Formic acid reduces mercuric chloride to mercurous chloride on heating, while acetic acid does not. Reason : Formic acid is a stronger acid than acetic acid. [AIIMS-95]

(A) A (B) B (C) C (D) D

34. Assertion : Aniline hydrogen sulphate, on heating, forms a mixture of ortho and para amino benzene sulphonic acids. Reason : The sulphonic acid group is electron withdrawing. [AIIMS-96]

(A) A (B) B (C) C (D) D

35. Assertion : Lactic acid shows geometrical isomerism. Reason : It has – C = C – bond [AIIMS-97]

(A) A (B) B (C) C (D) D

36. An acyl halide is formed when PCl_5 reacts with an [AIIMS-98]

(A) Acid (B) Alcohol (C) Amide (D) Ester

37. The product obtained when acetic acid is treated with phosphorus trichloride is [AIIMS-98]

(A) $CH_3-\overset{O}{\overset{||}{C}}-O-PCl_2$ (B) $CH_3-\overset{O}{\overset{||}{C}}-OCl$

(C) $CH_3-\overset{O}{\overset{||}{C}}-Cl$ (D) $CH_3-\overset{O}{\overset{||}{C}}-CH_2-OH$

38. Toluene can be oxidized to benzoic acid by [AIIMS-99]

(A) $KMnO_4$ (B) $K_2Cr_2O_7$

(C) H_2SO_4 (D) Both (a) and (b)

39. Vingear is [AIIMS-99]

(A) HCHO (B) HCOOH (C) CH_3CH (D) CH_3COOH

40. In the reaction : Phenol $\xrightarrow{NaOH}$ A $\xrightarrow{H^+}$ B

B is identified as [AIIMS-99]

(A) Benzene (B) Benzaldehyde

(C) Benzoic acid (D) Salicylic acid

41. CO_2 is liberated on adding sodium carbonate to a carboxylic acid. The carbon of CO_2 comes from [AIIMS-2000]

(A) Carboxylic group (B) Carbonate

(C) Alkyl group (D) Methyl group

42. Which is strongest acid among the following [AIIMS-2001]

(A) $CH_2\begin{matrix} \diagup COOH \\ \diagdown COOH \end{matrix}$ (B) $HO_2C-(CH_2)_2-CO_2H$

(C) $\begin{matrix} H-C-COOH \\ || \\ H-C-COOH \end{matrix}$ (D) $\begin{matrix} H-C-COOH \\ || \\ HOOC-C-H \end{matrix}$

43. Phthalic anhydride on treatment with ammonia forms : [AIIMS-2001]

(A) Phthalic acid (B) Benzoic acid

(C) Phthalimide (D) Benzaldehyde

44. $CH_3CH_2CO_2H \xrightarrow{P \text{ and } Br_2} [A] \xrightarrow{(i)\text{ alc KOH (ii) } H^+} [B]$

[B] is [AIIMS-2001]

(A) $CH_3CH_2CO_2H$ (B) CH_3CH_2CHO

(C) $CH_2=CH-CHO$ (D) $CH_2=CH-CO_2H$

45. A precipitate of calcium oxalate will not dissolve in ? [JIPMER-91]

(A) HCl (B) HNO_3 (C) Aqua regia (D) Acetic acid

46. Propionamide when heated with a mixture of bromine and caustic alkali would give : [JIPMER-93]

(A) Propylamine (B) Propanol

(C) Propanal (D) Ethylamine

47. Identify Z in the following reaction [JIPMER-93]

$$CH_3CH=CH_2 \xrightarrow[\text{Peroxide}]{HBr} X \xrightarrow[\text{Ether}]{Mg} Y \xrightarrow[(ii)\ H_2O]{(i)\ CO_2} Z$$

(A) $(CH_3)_2CHCOOH$ (B) $CH_3CH_2COCH_3$

(C) $CH_3CH_2CH_2COOH$ (D) $CH_3CH_2COOCH_3$

48. Acetic acid is reduced with $LiAlH_4$ to [JIPMER-93]

(A) CH_3CH_2OH (B) CH_3CHO

(C) CH_3CH_3 (D) CH_3OH

49. In the reaction :

$$CH_3CH_2CH_2.COOH \xrightarrow{NaOH} X \xrightarrow[\text{Heat}]{\text{Soda lime}} Y$$

is : [JIPMER-93]

(A) C_4H_{10} (B) $CH_3CH_2CH_2CH_2OH$

(C) CH_3CH_2CHO (D) C_3H_8

50. Amide may be converted into amine by reaction named after [JIPMER-94]

(A) Claisen (B) Kolbe (C) Perkin (D) Hoffmann

51. Colour obtained in Biuret test for urea is [JIPMER-94]

(A) red (B) violet (C) green (D) blue

52. Propionic acid is subjected to reduction with hydroiodic acid in the presence of a little of P. The product formed is [JIPMER-95]

(A) Butane (B) Ethane (C) Propane (D) None of these

53. Which of the following structures of carboxylic acid account for the acidic nature ? [JIPMER-95]

(A) $R-\overset{+}{C}\begin{matrix} \diagup O^- \\ \diagdown OH \end{matrix}$ (B) $R-\overset{+}{C}\begin{matrix} \diagup OH \\ \diagdown OH \end{matrix}$

(C) $R-C\begin{matrix} \diagup\!\!\diagup O \\ \diagdown OH \end{matrix}$ (D) None of these.

54. Which of the following acids will be able to give silver mirror test [JIPMER-96]

(A) Acetic acid (B) Butyric acid

(C) Formic acid (D) Carboxylic acids do not give this test

55. The compound with the smell of peppermint is

[JIPMER-97]

(A) Ethyl acetate (B) Ethyl benzoate
(C) Methyl acetate (D) Methyl salicylate

56. Which of the following is not charred when warmed with con.H_2SO_4 ? [JIPMER-98]
(A) Glucose (B) Sucrose
(C) Tartaric acid (D) Oxalic acid

57. Sodium acetate reacts with acetyl chloride to form [JIPMER-99]
(A) Acetone (B) Acetic anhydride
(C) Sodium formate (D) Acetic acid

58. Which is the correct structure of the organic product formed when acetic acid is treated with thionylchloride ? [JIPMER-99]

(A) $CH_3-\overset{O}{\overset{||}{C}}-O-\underset{O}{\overset{O}{S}}-Cl$ (B) $CH_3-\overset{O}{\overset{||}{C}}-Cl$

(C) $CH_3-\overset{O}{\overset{||}{C}}-OCl$ (D) $ClCH_2-\overset{O}{\overset{||}{C}}-O-SOCl$

59. An aldehyde can be obtained from an acid chloride by [JIPMER-99]
(A) Rosenmund's reduction
(B) Wolf-Kischner reduction (C) Clemmensen's reduction
(D) Meerwin-Pondorff-Verley reduction

60. The organic compounds A and B react with sodium metal and release H_2 gas. A and B react with each other to give ethyl acetate. The A and B are [JIPMER-2000]
(A) CH_3COOH and $HCOOH$
(B) CH_3COOH and CH_3OH (C) $HCOOH$ and C_2H_5OH
(D) CH_3COOH and C_2H_5OH

61. $CH_3CH_2\ COOH \xrightarrow[Fe]{Cl_2} X \xrightarrow[KOH]{alc.} Y.$
The compound Y is [JIPMER-2000]
(A) $CH_3CHClCOOH$ (B) $CH_2 = CHCOOH$
(C) CH_3CH_2CN (D) CH_3CH_2OH

62. When alcohol reacts with any acid, it produces [JIPMER-2001]
(A) Secondary alcohol (B) Alkane
(C) Acid with high carbon (D) Ester and water

63. Fatty acids are [JIPMER-2002]
(A) Unsaturated dicarboxylic acid.
(B) Long chain alkanonic acid.
(C) Aromatic carboxylic acid.
(D) Aromatic dicarboxylic acid.

64. $CH_3C_6H_5 \xrightarrow[H_2SO_4]{K_2Cr_2O_7} Y$ Here Y is [JIPMER-2002]
(A) Benzaldehyde (B) Toluene
(C) Benzoic acid (D) Ethyl benzene

65. What is obtained when benzoyl chloride reacts with aniline in the presence of sodium hydroxide ? [BHU-96]
(A) Acetanilide (B) Benzanilide
(C) Benzoic acid (D) Azobenzene

66. Ester and acetamide are distinguished by [BHU-96]
(A) Hydrolysis with strong acids or alkali
(B) Derivatives of fatty acids
(C) Grignard reagent (D) Both 'A' and 'B'

67. Palmitic acid is [BHU-97]
(A) $C_{16}H_{31}COOH$ (B) $C_{17}H_{35}COOH$
(C) $C_{15}H_{31}COOH$ (D) $C_{17}H_{31}COOH$

68. Chloroform reacts with NaOH to give [BHU-97]
(A) Sodium oxalate (B) CH_3COONa
(C) CH_3OH (D) $HCOONa$

69. Which of the following on hydrolysis forms acetic acid ? [BHU-97]
(A) CH_3CN (B) CH_3OH (C) C_2H_5OH (D) $C_2H_5NH_2$

70. CH_3COOH is reacted with $CH \equiv CH$ in presence of Hg^{++}, the product is [BHU-98]

(A) $\underset{HC-(OOC-CH_3)_2}{\overset{CH_3}{|}}$ (B) $\underset{H_2C-(OOC-CH_3)}{\overset{CH_3}{|}}$

(C) $\underset{CH_2(OOCCH_3)}{\overset{CH_2(OOCCH_3)}{|}}$ (D) None of these

71. The waxes are long chain compounds of fatty acids, which belong to the class of [BHU-99]
(A) Esters (B) Ethers (C) Alcohols (D) Acetic acid

72. $C_6H_5OH \xrightarrow[NaOH(aq)]{ClCOCH_3(aq)} C_6H_5C-\overset{O}{\overset{||}{C}}-CH_3$ is an example of [BHU-2000]
(A) Reimer Tiemen reaction (B) Benzoylation
(C) Schotten-Baumann reaction (D) Acetylation

73. A fruity smell is produced by the reaction of C_2H_5OH with [BHU-2000]
(A) PCl_5 (B) $CH_3-\overset{O}{\overset{||}{C}}-CH_3$
(C) $CH_3-\overset{O}{\overset{||}{C}}-OH$ (D) None of these

74. Name of Aspirin is [BHU-2000]
(A) Salicylaldehyde (B) Salicyclic acid
(C) Both (A) and (B) (D) 2 Acetoxy benzoic acid

75. Esters of amino acids are [AFMC-91]
(A) Neutra (B) Basic (C) Acidic (D) Amphoteric

76. All fats are [AFMC-91]
(A) Hydrocarbons (B) Carbohydrates
(C) Aldehydes (D) Esters

77. The acid present in tomatoes is [AFMC-91]
(A) Lactic acid (B) Oxalic acid
(C) Citric acid (D) Tartaric acid

78. Diethyl ester is not associated with which one of these isomers [AFMC-93]

(A) Butanoic acid (B) Methyl propanoate
(C) Stereoisomerism (D) Tautomerism

79. Reaction of NaOH with fats is called as [AFMC-93]
(A) Hydrolysis (B) Saponification
(C) Esterification (D) None of these

80. Formaldehyde and formic acid can be distinguished using [AFMC-93]
(A) Tollen's reagent (B) Fehling's solution
(C) Ferric chloride (D) Sodium carbonate

81. Aspirin is [AFMC-93]
(A) Phenyl salicylic acid (B) Acetyl salicylic acid
(C) Methyl salicylate (D) Salicyladehyde

82. Sour taste of lemon is due to [AFMC-93]
(A) Ascorbic acid (B) Lactic acid
(C) Citric acid (D) Tartaric acid

83. What is formed when oxalic acid is dehydrated by conc. H_2SO_4 ? [AFMC-94]
(A) C + CO_2 (B) CO (C) CO_2 (D) CO + CO_2

84. Vinegar contains how much percentage of acetic acid? [AFMC-94]
(A) 10-12% (B) 4-5% (C) 90-100% (D) 90-98%

85. Ethanamide is [AFMC-94]
(A) Feebly acidic (B) Basic
(C) Neutral (D) None of these

86. Soaps can be classified under [AFMC-94]
(A) Carbohydrates (B) Ether
(C) Organic compounds (D) Salts of fatty acid

87. Pyruvic acid is obtained by [AFMC-95]
(A) Oxidation of acetaldehyde cyanohydrin
(B) Oxidation of acetone cyanohydrin
(C) Oxidation of formaldehyde cyanohydrin
(D) None of these

88. Asprin is acetyl salicylic acid it [AFMC-96]
(A) does not ppt. in stomach but ppt. in intestine
(B) ppt. in stomach but not in intestine
(C) both of these (D) none of these

89. Oxalic acid, when heated with concentrated H_2SO_4 gives [AFMC-98]
(A) H_2O_2 and CO_2 (B) CO and CO_2
(C) H_2O_2 and CO_2 (D) CO_2 and H_2S

90. Ester and acetamide are distinguished by [AFMC-2000]
(A) Derivatives of fatty acids
(B) Hydrolysed with strong acids or alkali
(C) Grignard reagent (D) Both (A) and (B)

91. Which of the following does not form phenol or phenoxide? [AFMC-2000]
(A) C_6H_5Cl (B) C_6H_5COOH
(C) $C_6H_5N_2Cl$ (D) $C_6H_5SO_3Na$

92. Which of the following acid is present in citrus fruits ? [AFMC-2002]
(A) Citric acid (B) Tartaric acid
(C) Malic acid (D) All of these

93. Aspirin is an acetylation product of [AFMC-2002]
(A) p-dihydroxybenzene (B) o-hydroxybenzenoic acid
(C) o-dihydroxybenzene (D) o-hydroxybenzenoic acid

94. Which of the following carbon atom is likely to possess tetrahedral geometry ? [AFMC-2002]

$$H_2\overset{1}{C} = \overset{2}{C}H - \overset{3}{C}H_2 - \overset{4}{C}OOH$$

(A) 1 (B) 2 (C) 3 (D) 4

95. Chlorination of toluene in the presence of light and heat followed by treatment with aqeous NaOH gives [IIT-90]
(A) o-Cresol (B) p-Cresol
(C) 2, 4 Dihydroxytoluene (D) Benzoic acid.

96. Reaction of $R - \overset{O}{\overset{\|}{C}} - NH_2$ with a mixture of Br_2 and KOH gives $R - NH_2$ as the main product, the intermediate involved in this reaction are [IIT-92]
(A) $R - \overset{O}{\overset{\|}{C}} - NHBr$ (B) RNHBr
(C) $R - N = C = O$ (D) $R - \overset{O}{\overset{\|}{C}} - NBr_2$

97. The high melting point and insolubility in organic solvent of sulphanilic acid are due to its .... structure. [IIT-94]

98. When propionic acid is treated with aqueous sodium bicarbonate, CO_2 is liberated. The 'C' of CO_2 comes from [IIT-99]
(A) methyl group (B) carboxylic acid group
(C) methylene group (D) bicarbonate

99. Which of the following has the most acidic hydrogen? [IIT-2000]
(A) 3-hexanone (B) 2, 4-hexanedione
(C) 2, 5-hexanedione (D) 2, 3-hexanedione

100. Which of the following acids has the smallest dissociation constant ? [IIT-2002]
(A) $CH_3CHFCOOH$ (B) FCH_2CH_2COOH
(C) $BrCH_2CH_2COOH$ (D) $CH_3CHBrCOOH$

101. Identify the correct order of boiling points of the following compounds [IIT-2002]
$CH_3CH_2CH_2CH_2OH$ (1) $CH_3CH_2CH_2CHO$ (2) $CH_3CH_2CH_2COOH$ (3)
(A) 1 > 2 > 3 (B) 3 > 1 > 2
(C) 1 > 3 > 2 (D) 3 > 2 > 1

102. $CH_3CH_2COOH \xrightarrow[\text{red P}]{Cl_2} A \xrightarrow{\text{alc. KOH}} B.$

What is B? [AIEEE-2002]
(A) CH_3CH_2COCl (B) CH_3CH_2CHO
(C) $CH_2 = CHCOOH$ (D) $ClCH_2CH_2COOH$

103. When $CH_2 = CH - COOH$ is reduced with $LiAlH_4$, the compound obtained will be [AIEEE-2003]
(A) CH_3-CH_2-COOH (B) $CH_2 = CH - CH_2OH$
(C) $CH_3 - CH_2 - CH_2OH$ (D) $CH_3 - CH_2 - CHO$

104. In the anion $HCOO^-$ the two carbon-oxygen bonds are found to be of equal length. What is the reason for it? [AIEEE-2003]
(A) Electronic orbitals of carbon atom are hybridized
(B) The C = O bond is weaker than the C–O bond
(C) The anion $HCOO^-$ has two resonating structures
(D) The anion is obtained by removal of a proton from the acid molecule.

105. The formula of the main product obtained when acetic acid reacts with PCl_5 is .... [EAMCET-90]

106. The reagent used for converting ethanoic acid to ethanol [EAMCET-96]
(A) $LiAlH_4$ (B) BH_3 (C) PCl_5 (D) $K_2Cr_2O_7$

107. Hydrolysis of an ester gives a carboxylic acid which on Kolbe's electrolysis yields ethane. The ester is : [EAMCET-97]
(A) Methyl ethanoate (B) Methyl methanoate
(C) Ethyl methanoate (D) Methyl propanoate

108. Acetic acid reacts with PCl_5 to form [EAMCET-98]
(A) $CH_2ClCOOH$ (B) $CHCl_2COOH$
(C) CH_3COCl (D) CH_3COOCl

109. The reagents used for converting ethanoic acid to ethanol is [EAMCET-98]
(A) $LiAlH_4$ (B) BH_3 (C) PCl_3 (D) $K_2Cr_2O_7/H^+$

110. C_2H_5 CHO and $(CH_3)_2$ CO can be distinguished by testing with [EAMCET-98]
(A) Phenylhydrazine (B) Hydroxylamine
(C) Fehling's solution (D) Sodium bisulphite

111. Hydrolysis of trichloromethane with aqueous KOH gives [EAMCET-98(MBBS)]
(A) Methanol (B) Chloral
(C) Acetylene (D) Potassium formate

112. Oxalic acid is heated with concentrated H_2SO_4. When the resultant gases are passed over red hot carbon, X is obtained. X is. [EAMCET-99]
(A) CO (B) CH_4 (C) CO_2 (D) C_2H_4

113. Acetic acid gives acetic anhydride when treated with [EAMCET-99]
(A) PCl_5 (B) P_2O_5 (C) Cl_2/P (D) Na

114. An organic compound reacts (i) with metallic sodium to liberate hydrogen and (ii) with Na_2CO_3 solution to liberate CO_2. The compound is [EAMCET-99(MBBS)]
(A) an alcohol (B) a carboxylic acid
(C) an ether (D) an ester

115. Which of the following converts acetic acid to acetyl chloride ? [EAMCET-2000]
(A) NaCl (B) HCl (C) Cl_2/P (D) PCl_3

116. Which of the following compounds is formed when ethanol reacts with acetic acid in the presence of concentrated H_2SO_4 ? [EAMCET(MBBS)-2000]
(A) $CH_3COOC_2H_5$ (B) $C_2H_5OC_2H_5$
(C) CH_3OCH_3 (D) CH_3CH_2CHO

117. Solid sodium hydroxide is reacted with carbon monoxide under pressure. What is the product obtained ? [EAMCET(MBBS)-2000]
(A) CO_2 (B) CH_3ONa (C) HCOONa (D) HCHO

118. Acetic acid is reacted with metallic sodium to form hydrogen and X. When X is heated with soda lime, Y and sodium carbonate are formed. Y is [EAMCET-2001]
(A) C_2H_6 (B) CH_4 (C) CH_3COONa (D) CH_3CONH_2

119. What is the reagent used in the preparation of chloro-acetic acid from acetic acid ? [EAMCET(MBBS)-2001]
(A) PCl_5 (B) Cl_2/redP (C) PCl_3 (D) $SOCl_2$

120. In the following reaction X and Y are respectively.
$$CH_3COOH + NH_3 \longrightarrow X \xrightarrow{\Delta} Y + H_2O$$
[EAMCET-2002]
(A) CH_3CONH_2, CH_4 (B) CH_3COONH_4,CH_3CONH_2
(C) CH_3CONH_2,CH_3COOH (D) $CH_3NH_2$$CH_3CONH_2$

121. What is the catalyst used in the conversion of acetaldehyde to acetic acid ? [EAMCET-2002]
(A) Manganese acetate (B) $LiAlH_4$
(C) H_2/Ni (D) Na/NH_3

122. In the following reaction X, and Y respectively are
$$C_2H_5OH \xrightarrow{KmnO_4/H^+} X \xrightarrow[H_2SO_4\ \Delta]{Y} CH_3COOC_2H_5$$
[EAMCET-2002]
(A) CH_3OH, C_2H_5OH (B) CH_3CHO,CH_3OH
(C) $CH_2 = CH_2$, CH_3COOH (D) CH_3COOH, C_2H_5OH

123. In the following reaction, X and Y respectively are [EAMCET(med.)-2002]
$$X \xrightarrow{aq.\ NaOH} CH_3COOH \xrightarrow{Y} (CH_3CO)_2O \text{ Solution}$$
(A) CH_3CHO, PCl_5 (B) CH_3CN, P_2O_5
(C) CH_3CH_2OH, NaOAc (D) CH_3COCH_3; H_2SO_4

124. In the reaction $C_2H_5COC_2H_5 + CO \xrightarrow[150°C]{BF_3,\ 50\ atm} X$ what is X ? [EAMCET-2003]
(A) Diethyl Carbonate (B) Ethyl Carbonate
(C) Diethyl Peroxide (D) Ethyl Propionate

125. In the reaction sequence

$C_2H_5Cl + KCN \xrightarrow{C_2H_5OH} X \xrightarrow[\Delta]{H_3O^+}$ Y. What is the molecular formula of Y ? [EAMCET-2003]
(A) $C_3H_6O_2$ (B) C_3H_5N (C) $C_2H_4O_2$ (D) C_2H_6O

126. Which one of the following compounds give aspirin on reacting with acetic anhydride in the presence of conc. H_2SO_4 ? [EAMCET-2003]

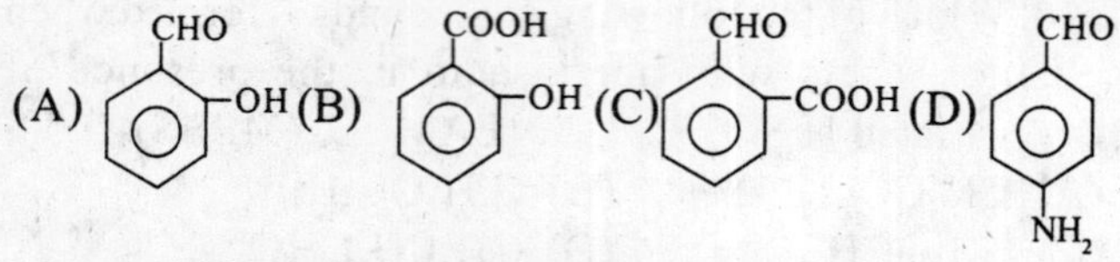

127. Acid hydrolysis of X yields two different organic compounds. Which one of the following is X [EAMCET-2003]
(A) CH_3COOH (B) CH_3CONH_2
(C) $CH_3COOC_2H_5$ (D) $(CH_3CO)_2O$

128. HCOOH show all test of aldehyde because [CPMT-96]
(A) It has one aldehyde group
(B) It is member of aldehyde
(C) All acids show test of aldehyde
(D) Does not show any test

129. To prevent rancidification of food material, which of the following is added ? [CPMT-96]
(A) Reducing agent (B) Anti-oxidant
(C) Oxidising agent (D) None of these

130. In esterification OH^- ion for making H_2O comes from [CPMT-96]
(A) Acid (B) Alcohol (C) Ketone (D) Carbohydrate

131. $CH_3CONH_2 \xrightarrow{Na + ROH} Z + H_2O$ what is Z? [CPMT-96]
(A) CH_3NH_2 (B) CH_3CH_2NC
(C) $CH_3CH_2CH_3$ (D) NH_2CONH_2

132. If glycerol is heated with oxalic acid at temperature 100-110°C, it gives an acid. The acid is- [CPMT-97]
(A) Acetic acid (B) Formic acid
(C) Propionic acid (D) Benzoic acid

133. $CO + NaOH \longrightarrow$? [CPMT-97]
(A) HCOONa (B) $C_2H_2O_4$
(C) HCOOH (D) CH_3COOH

134. Oelic, Stearic and Palmitic acids are [CPMT-97]
(A) Nucleic acid (B) Amino acid
(C) Fatty acids (D) None of these

135. $CH_3CONH_2 \xrightarrow{NaOBr}$ Z. Z is [CPMT-97]
(A) CH_3Br (B) CH_4 (C) CH_3OBr (D) CH_3NH_2

136. Which one is called an ethanoic acid ? [CPMT-97]
(A) HCOOH (B) CH_3COOH
(C) CH_3CH_2COOH (D) $CH_3CH_2CH_2COOH$

137. Which of the following compounds is obtained by heating ammonium cyanate ? [CPMT-2000]
(A) Alkane (B) Urea
(C) Ethyl amine (D) Ammonium thiocyanate

138. On sulphonation of C_6H_5Cl - [CPMT-2000]
(A) Benzene sulphonic acid is formed
(B) Metachlorobenzene sulphonic acid is formed
(C) Orthochlorobenzene sulphonic acid is formed
(D) Ortho and para chlorobenzene sulphonic acids are formed

139. $(CH_3)_2C = O \xrightarrow[(HCl)]{NaCN} A \xrightarrow[\Delta]{H_3O^+} B$

In the above sequence of reactions A and B are [CPMT-2000]

(A) $CH_3-C(OH)(CH_3)-CN$ and $CH_3-C(OH)(CH_3)-COOH$

(B) $CH_3-C(OH)(CH_3)-CN$ and $CH_3-CH(CH_3)-COOH$

(C) $CH_3-C(OH)(CH_3)-Cl$ and $CH_3-C(OH)(CH_3)-OH$

(D) $CH_3-C(OH)(CH_3)-Cl$ and $CH_3-C(CH_3)=O$

140. When acetamide is treated with Br_2 in the presence of a base, the product obtained is [CPMT-2000]
(A) CH_3OH (B) CH_3NH_2
(C) $CH_3CH_2NH_2$ (D) CH_3NC

141. The correct order of acidic strength is [CPMT-2000]
(A) $CH_3CF_2COOH > CH_3CCl_2COOH > CH_3CBr_2COOH$
(B) $CH_3CF_2COOH > CH_3CBr_2COOH > CH_3CCl_2COOH$
(C) $CH_3CBr_2COOH > CH_3CCl_2COOH > CH_3CF_2COOH$
(D) $CH_3CCl_2COOH > CH_3CBr_2COOH > CH_3CF_2COOH$

142. Which of the following is the weakest acid ? [CPMT-2001]
(A) 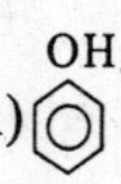 (B) CH_3COOH (C) HCOOH (D) 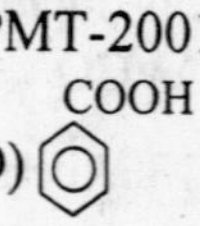

143. Picric acid and benzoic acid can be distinguished by [CPMT-2002]
(A) aq. $NaHCO_3$ (B) aq. $FeCl_3$

(C) aq. Na_2CO_3 (D) aq. NaOH

144. Which is known as oil of winter green ? [CPMT-2002]
(A) phenyl salicylate (B) ethyl salicylate
(C) methyl salicylate (D) salicylic acid

145. The general formula for saturated fatty acid is [CPMT-2003]
(A) $C_nH_{2n-2}O_2$ (B) $C_nH_{2n-1}O_2$
(C) $C_nH_{2n+1}O_2$ (D) $C_nH_{2n}O_2$

146. When acetamide reacts with KOH + Br_2, thus product formed is [CPMT-2003]
(A) CH_3NH_2 (B) CH_3COOH
(C) CH_3COOCH_3 (D) $(CH_3CO)_2O$

147. Hydrolysis of acetonitrile in acidic medium produces [CPMT-2003]
(A) CH_3CH_2OH (B) CH_3COOH
(C) CH_3NC (D) CH_3COOCH_3

148. Hoffmann's hypobromite reaction affords a method of: [PMT-93]
(A) Preparing a tertiary amine
(B) Preparing a mixture of amines
(C) Stepping down a series (D) Stepping up a series

149. Which one among the following represents an amide? [PMT-93]

(A) 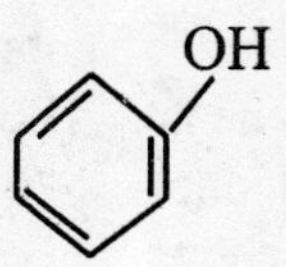 (B) 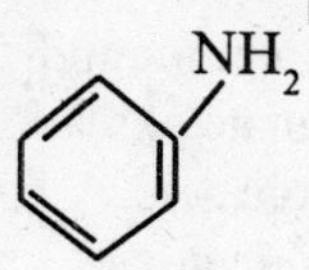

(C) 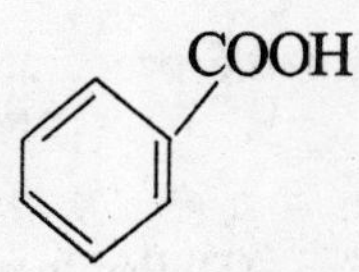 (D) 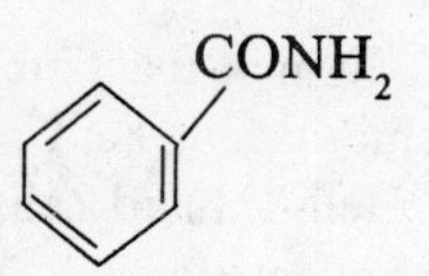

150. The reaction between an alcohol and an acid is called [PMT-94]
(A) Esterification (B) Saponification
(C) Hydrolysis (D) Hydrogenation

151. The acid present in vinegar is [PMT-94]
(A) CH_3COOH (B) H_2SO_4 (C) HCl (D) HNO_3

152. Which of the following is most acidic? [PMT-95]
(A) Formic acid (B) Chloroacetic acid
(C) Propionic acid (D) None

153. Which one of the following will react with acetamide to give methyl amine ? [PMT-96]
(A) PCl_5 (B) NaOH + Br_2 (C) Sodalime (D) H_2SO_4

154. The general $(RCO)_2O$ represents [PMT-96]
(A) a ketone (B) an ether
(C) an acid anhydride (D) an ester

155. Prolonged heating of ammonium cyanate gives [PMT-96]
(A) nitrogen (B) carbon - dioxide
(C) biuret (D) ammonium carbonate

156. Nitration of benzoic acid gives - [PMT-97]
(A) 3-nitrobenzoic acid (B) 2-nitrobenzoic acid
(C) 2,3-dinitrobenzoic acid (D) 2,4-dinitrobenzoic acid

157. Which is the strongest acid (pKa value is given) - [PMT-97]
(A) HCOOH (3.77) (B) C_6H_5COOH (4.22)
(C) CH_3COOH (4.71) (D) CH_3CH_2COOH (4.88)

158. In the presence of iodine catalyst chlorine reacts with acetic acid to form - [PMT-97]
(A) $CH_3-\overset{\overset{\displaystyle O}{\|}}{C}-Cl$ (B) $Cl-CH_2-\overset{\overset{\displaystyle O}{\|}}{C}-OH$
(C) $CH_3-\overset{\overset{\displaystyle Cl}{|}}{\underset{\underset{\displaystyle Cl}{|}}{C}}-OH$ (D) $CH_3-\overset{\overset{\displaystyle O}{\|}}{C}-CH_3$

159. Hydrolysis of an ester gives acid A and alcohol B. A reduces Fehling's solution and oxidation of B gives A. the ester is : [PMT-99]
(A) methyl formate (B) ethyl formate
(C) methyl acetate (D) ethyl acetate

160. Amongst the following acids the strongest acid is [PMT-2002]
(A) $ClCH_2COOH$ (B) Cl_3CCOOH
(C) CH_3COCH_3 (D) CH_3OCH_3

161. Hydrolysis of acetyl chloride gives [PMT-2002]
(A) C_2H_5OH (B) CH_3COOH
(C) CH_3COCH_3 (D) CH_3OCH_3

162. $2CH_3-\underset{\underset{\displaystyle O}{\|}}{C}-OC_2H_5 \xrightarrow{C_2H_5ONa}$

$CH_3-\underset{\underset{\displaystyle O}{\|}}{C}-CH_2-\underset{\underset{\displaystyle O}{\|}}{C}-OC_2H_5 + C_2H_5OH$

the above reaction is : [PMT-2003]
(A) Claisen Schmidt condensation (B) Perkin reaction
(C) Claisen condensation (D) Dieckmann reaction

163. $CH_3\underset{\underset{\displaystyle O}{\|}}{C}-OCH_3 + C_2H_5OH \xrightarrow{H^+ \text{ or } OH^-}$

$CH_3\underset{\underset{\displaystyle O}{\|}}{C}-O-C_2H_5 + CH_3OH$ [PMT-2003]

(A) Claisen Schmidt reaction (B) Schmidt reaction
(C) Esterification (D) Trans-esterification

164. Which of the following chemical transformations is

the Arndt Eistert reaction used ? [PMT-2003]
(A) $R - CH_2OH \rightarrow R - OH$
(B) $R - CH_2 - COOH \rightarrow R - COOH$
(C) $R - OH \rightarrow$ Ⓡ$- CH_2 - OH$
(D) $R - COOH \rightarrow R - CH_2 - COOH$

165. Acetamide is treated separately with the following reagents. Which one of these would give methyl amine ? [PET-93]
(A) PCl_5 (B) $NaOH + Br_2$
(C) Conc. H_2SO_4 (D) Sodalime

166. Alkaline hydrolysis of an ester is called [PET-93]
(A) Neutralisation (B) Esterification
(C) Polymerisation (D) Saponification

167. A nauseating smell in the carbylamine test for primary amines is due to the formation of [PET-93]
(A) Isocyanide (B) Chloroform
(C) Cyanide (D) D.D.T.

168. In a compound CH_3 C OCl, which type of orbitals have been used by the circled carbon in bond formation [PET-94]
(A) Sp^3 (B) Sp^2 (C) Sp (D) p

169. Oxalic acid on being heated upto 90° with conc H_2SO_4 forms [PET-94]
(A) $HCOOH + CO_2$ (B) $CO_2 + H_2O$
(C) $CO_2 + CO + H_2O$ (D) $HCOOH + CO$

170. Acetamide on being heated with P_2O_5 gives [PET-94]
(A) Methyl cyanide (B) Methyl amine
(C) Acetone (D) Acetic acid

171. Which one of the following is Friedel Crafts reaction [PET-94]
(A) $C_6H_6 + FeCl_3 + Cl_2 \rightarrow C_6H_5Cl$
(B) $C_6H_5CHO + CH_3CHO + KOH \rightarrow C_6H_5CH = CH\text{-}CHO$
(C) $C_6H_6 + CH_3COCl + AlCl_3 \rightarrow C_6H_5 - \overset{O}{\overset{||}{C}} - CH_3$
(D) $C_6H_5OH + CHCl_3 + KOH \rightarrow$ 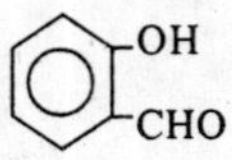

172. Identify B in the reaction [PET-95]
Acetamide $\xrightarrow[\Delta]{P_2O_5}$ A $\xrightarrow[Sn / HCl]{4H}$ B
(A) CH_3NH_2 (B) $CH_3CH_2NH_2$
(C) CH_3CN (D) CH_3COONH_4

173. Fat is a/an [PET-95]
(A) Acid (B) Alcohol (C) Ester (D) Hydrocarbon

174. The product of the reaction [PET-95]
CH_3CH_2OH + Cu (Reduced) $\xrightarrow{300°C}$? is -
(A) C_3H_6 (B) CH_3COCH_3
(C) CH_3CHO (D) CH_3COOH

175. Which of the following is aspirin [PET-95]
(A) Methyl salicylate (B) Ethyl salicylate
(C) Acetyl salicylic acid (D) Ortho-hydoxy benzoic acid

176. When acetamide is reacted with Br_2 and caustic soda, we get - [PET-95]
(A) Acetic acid (B) Bromoacetic acid
(C) Methyl amine (D) Ethane

177. Lactic acid on heating with conc. H_2SO_4 gives [PET-96]
(A) Acetic acid (B) Propionic acid
(C) Acrylic acid (D) Formic acid

178. Which of the following is most acidic? [PET-96]
(A) Acetic acid (B) Chloroacetic acid
(C) Dichloroacetic acid (D) Trichloroacetic acid

179. The product D of the reaction
$CH_3Cl \xrightarrow{KCN} (A) \xrightarrow{H_2O} (B) \xrightarrow{NH_3} (C) \xrightarrow{\Delta} (D)$ is [PET-97]
(A) $CH_3CH_2NH_2$ (B) CH_3CN
(C) $HCONH_2$ (D) CH_3CONH_2

180. What happens when 2-hydroxy benzoic acid is distilled with zinc dust ? It gives : [PET-98]
(A) Phenol (B) Benzoic acid
(C) Benzaldehyde (D) A polymeric compound

181. The compound formed as a result of potassium permanganate oxidation of ethylbenzene is : [PET-98]
(A) Benzoic acid (B) Benzyl alcohol
(C) Benzophenone (D) Acetophenone

182. Which reaction is suitable for the preparation of α-chloroacetic acid ? [PET-98]
(A) Hell-Volhard Zelinsky reaction
(B) Nef reaction (C) Stephen's reaction (D) Perkin reaction

183. Which compound is known as oil of winter green ? [PET-98]
(A) Phenyl benzoate (B) Phenyl salicylate
(C) Phenyl acetate (D) Methyl salicylate

184. Identify the product C in the series [PET-99]
$CH_3CN \xrightarrow{Na/\ C_2H_5OH} A \xrightarrow{HNO_2} B \xrightarrow{\text{Tollen's reagent}} C$
(A) CH_3COOH (B) CH_3CH_2NHOH
(C) CH_3CONH_2 (D) CH_3CHO

185. Lower carboxylic acids are soluble in water due to : [PET-99]
(A) Low molecular weight (B) Hydrogen bonding
(C) Dissociation into ions (D) Easy hydrolysis

186. An organic compound on reaction with saturated aqueous solution of $NaHCO_3$ gives effervescence, the compound would be ? [PET-2000]
(A) Alkane (B) Alkene (C) Acetic Acid (D) Ethyl Alcohol

187. In the reaction :

$$CH_3CH_2COOH \xrightarrow{Cl_2 / red\ P} A \xrightarrow{Alc.\ KOH} B$$

The compound B is : [PET-2000]

(A) CH_3CH_2OH (B) CH_3CH_2COCl

(C) $CH_2 = CHCOOH$ (D) $CH_3{-}CHClCOOH$

188. Which compound is aspirin ? [PET-2001]

(A) Acetyl salicylic acid (B) Salicylic acid

(C) Acetamide (D) Salicyl amide

189. Hydrolysis of acetamide gives: [PET-2001]

(A) Acetic acid (B) Acetadehyde

(C) Methylamine (D) Formic acid

190.CH3COOH is formed by hydrolysis of [BHU-2003]

(A) CH_3CN (B) CH_3CH_2CN

(C) CH_3NH_2 (D) $CH_3CH_2NH_2$

Carboxylic acids and their derivatives

1.	(C)	51.	(B)	101.	(B)	151.	(A)
2.	(C)	52.	(C)	102.	(C)	152.	(B)
3.	(B)	53.	(A)	103.	(B)	153.	(B)
4.	(C)	54.	(C)	104.	(C)	154.	(C)
5.	(B)	55.	(B)	105.	*	155.	(C)
6.	(D)	56.	(D)	106.	(A)	156.	(A)
7.	(D)	57.	(B)	107.	(A)	157.	(A)
8.	(A)	58.	(B)	108.	(C)	158.	(B)
9.	(C)	59.	(A)	109.	(A)	159.	(A)
10.	(D)	60.	(D)	110.	(C)	160.	(B)
11.	(D)	61.	(B)	111.	(D)	161.	(B)
12.	(A)	62.	(D)	112.	(A)	162.	(C)
13.	(B)	63.	(B)	113.	(B)	163.	(D)
14.	(A)	64.	(C)	114.	(B)	164.	(D)
15.	(B)	65.	(B)	115.	(D)	165.	(B)
16.	(B)	66.	(A)	116.	(A)	166.	(D)
17.	(A)	67.	(C)	117.	(C)	167.	(A)
18.	(A)	68.	(D)	118.	(B)	168.	(B)
19.	(C)	69.	(A)	119.	(B)	169.	(C)
20.	(A)	70.	(A)	120.	(B)	170.	(A)
21.	(A)	71.	(A)	121.	(A)	171.	(C)
22.	(C)	72.	(D)	122.	(D)	172.	(B)
23.	(A)	73.	(C)	123.	(B)	173.	(C)
24.	(C)	74.	(D)	124.	(D)	174.	(C)
25.	(B)	75.	(B)	125.	(A)	175.	(C)
26.	(C)	76.	(D)	126.	(B)	176.	(C)
27.	(D)	77.	(C)	127.	(C)	177.	(D)
28.	(D)	78.	(C)	128.	(C)	178.	(D)
29.	(A)	79.	(B)	129.	(B)	179.	(D)
30.	(B)	80.	(D)	130.	(A)	180.	(B)
31.	(C)	81.	(B)	131.	(A)	181.	(A)
32.	(A)	82.	(C)	132.	(B)	182.	(A)
33.	(B)	83.	(D)	133.	(B)	183.	(D)
34.	(D)	84.	(A)	134.	(C)	184.	(A)
35.	(D)	85.	(B)	135.	(D)	185.	(B)
36.	(A)	86.	(D)	136.	(B)	186.	(C)
37.	(C)	87.	(A)	137.	(B)	187.	(C)
38.	(D)	88.	(A)	138.	(D)	188.	(A)
39.	(D)	89.	(B)	139.	(A)	189.	(A)
40.	(D)	90.	(D)	140.	(B)	190.	(A)
41.	(B)	91.	(B)	141.	(A)		
42.	(C)	92.	(A)	142.	(A)		
43.	(C)	93.	(B)	143.	(B)		
44.	(D)	94.	(C)	144.	(C)		
45.	(D)	95.	(D)	145.	(D)		
46.	(D)	96.	(A)(C)	146.	(A)		
47.	(C)	97.	*	147.	(B)		
48.	(A)	98.	(D)	148.	(C)		
49.	(D)	99.	(B)	149.	(D)		
50.	(D)	100.	(C)	150.	(A)		

97. zwitter ionic
105. CH_3COCl

Nitrogen Compounds

I. Amines

Amines show chain, position, functional, metamerism and optical isomerism

$$H_3C-CH_2-CH_2-CH_2NH_2 \quad — \quad H_3C-CH(CH_3)-CH_2NH_2$$

chain isomers

$$H_3C-CH_2-CH_2NH_2 \begin{cases} H_3C-CH(NH_2)-CH_3 & \text{position isomers} \\ H_3C-CH_2NH-CH_3;\ H_3C-N(CH_3)-CH_3 & \text{functional isomers} \end{cases}$$

$$H_3C-CH_2NH-CH_2-CH_3;\ H_3C-NHCH_2-CH_2-CH_3;\ H_3C-NHCH(CH_3)_2$$

metamers

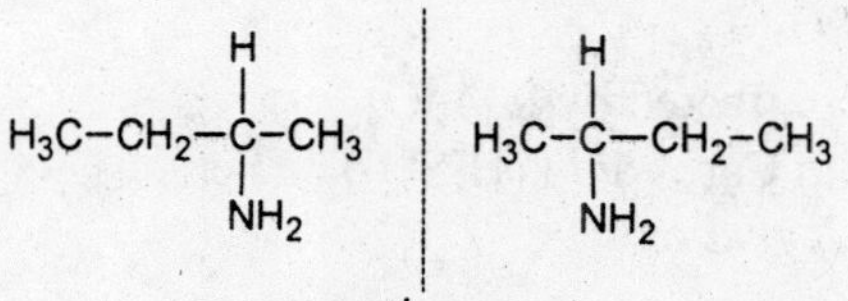

enantiomers

Preparation:

1. Hoffmann's method:

$$R-X + NH_3 \rightarrow R-NH_2 + R_2-NH + R_3-N + R_4\overset{+}{N}X^-$$

depending upon the ratio of the reactants a mixture of primary, secondary, tertiary and quaternary ammonium salt is formed.

2. Gabriel's pthalimide synthesis: only aliphatic primary amines can be prepared.

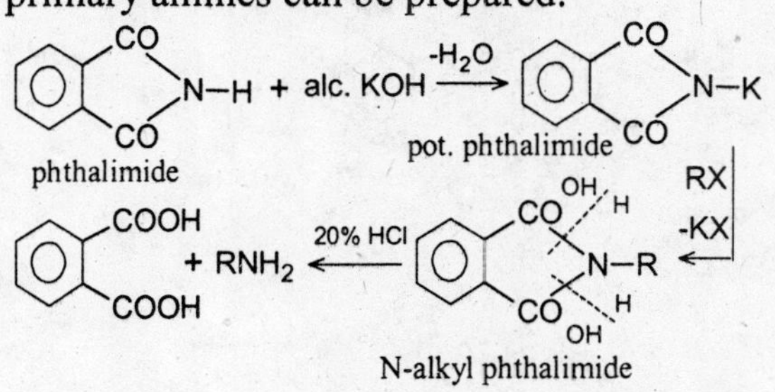

3. Curtius reaction:

$$\underset{\text{acid azide}}{R-CON_3} \xrightarrow[-N_2]{CHCl_3} \underset{\text{alkyl isocyanate}}{R-N=C=O} \xrightarrow{NaOH} R-NH_2 + Na_2CO_3 + H_2O$$

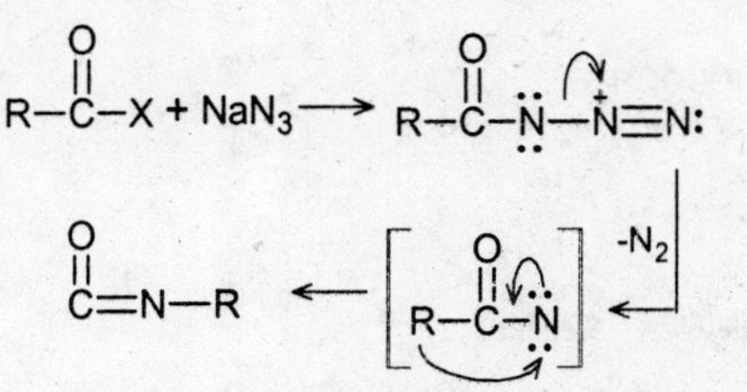

4. Schmidt reaction:

$$RCOOH + HN_3 \xrightarrow{\text{conc. } H_2SO_4} R-CON_3 \xrightarrow{H_2O} RNH_2 + CO_2 + N_2$$

5. Mendius reaction:

$$H_3C-C\equiv N + 4[H] \xrightarrow{Na/C_2H_5OH} CH_3CH_2NH_2 \text{ primary}$$

$$H_3C-N\equiv C + 4[H] \xrightarrow{Na/C_2H_5OH} CH_3NHCH_3 \text{ secondary}$$

6. Ritter reaction: Tertiary alcohols are converted into primary amines

$$H_3C-C(CH_3)_2-OH + HCN + H_2SO_4 \rightarrow H_3C-C(CH_3)_2-NH_2$$

7. Reductive amination:

$$RCHO \xrightarrow[+NH_3]{H_2/Ni} RCH_2NH_2;\quad R_2C=O \xrightarrow[+NH_3]{H_2/Ni} R_2CHNH_2$$

8. Hoffmann degradation or bromamide reaction:

$$RCONH_2 + Br_2 + KOH \rightarrow RNH_2 + KBr + H_2O + K_2CO_3$$

Separation of p, s and t – amines:

1. Hinsberg's method: The reagent employed is benzene sulphonyl chloride

$$1°:\ C_6H_5-SO_2Cl + H_2NR \xrightarrow{-HCl} C_6H_5-SO_2NHR \xrightarrow{KOH} C_6H_5-SO_2NKR$$

$$2°:\ C_6H_5-SO_2Cl + HNR_2 \xrightarrow{-HCl} C_6H_5-SO_2NR_2 \not\rightarrow \text{no reaction}$$

with 3° amines, no reaction is possible as active H atom is absent.

2. Hoffmann's method: The reagent employed is ethyl oxalate

$$\text{with } 1°\quad \begin{matrix} COOC_2H_5 \\ | \\ COOC_2H_5 \end{matrix} + \begin{matrix} H_2NR \\ \\ H_2NR \end{matrix} \rightarrow \underset{\text{dialkyl oxamide}}{\begin{matrix} CONHR \\ | \\ CONHR \end{matrix}} + 2C_2H_5OH$$

$$\text{with } 2°\quad \begin{matrix} COOC_2H_5 \\ | \\ COOC_2H_5 \end{matrix} + HNR_2 \rightarrow \underset{\text{dialkyl oxamic ester}}{\begin{matrix} CONR_2 \\ | \\ COOC_2H_5 \end{matrix}} + C_2H_5OH$$

with 3° amines no reaction is possible

Properties:

Except in tertiary amines, hydrogen bonding is possible in amines. The order of boiling point among the isomeric amines is 1°>2°>3° (due to hydrogen bonding).

Basicity: Aliphatic amines >ammonia >aromatic amines.

Because of the resonance, aromatic amines are weaker. The order of basic strength in most of the amines is 2°>1°>3°. This may be due to steric factors and salvation effects.

Reactions:

1. With HNO_2 :

(i)Primary amines generally form alkanols

$$R-NH_2 + HO-N=O \rightarrow ROH + N_2 + H_2O$$

Methyl amine behaves differently. Methyl alcohol forms in very minor quantities as it is further converted to di methyl ether.

$$CH_3NH_2 + 2HNO_2 \xrightarrow[-H_2O]{-N_2} H_3C\text{-}O\text{-}N{=}O$$
methyl nitrite

$$2CH_3NH_2 + 2HNO_2 \xrightarrow[-3H_2O]{-N_2} H_3C\text{-}O\text{-}CH_3$$
dimethyl ether (major)

Quantitative evolution of N_2 forms the basis of van Slyke method of estimating primary $-NH_2$ groups in amino acids, proteins etc.

(ii) Liebermann's nitroso reaction is shown by secondary amines

$$(CH_3)_2N\boxed{H + HO}NO \xrightarrow{-H_2O} (CH_3)_2N{-}N{=}O$$
dimethyl nitrosoamine

(iii) Tertiary amines give amine salt, which on decomposition give nitroso amine

$$R_3\text{-}N + HNO_2 \rightarrow R_3N.HNO_2 \xrightarrow{\Delta} R_2NNO + ROH$$

2. Carbyl amine reaction: Shown by primary amines only

$$R\text{-}NH_2 + CHCl_3 + 3KOH \xrightarrow{\text{warm}} R\text{-}NC + 3KCl + 3H_2O$$

3. Mustard oil reaction:

(i) $$2R\text{-}NH_2 + S{=}C{=}S \rightarrow S{=}C\langle^{NHR}_{SH.NH_2R} \xrightarrow{HgCl_2} R{-}N{=}C{=}S + HgS + 2HCl$$
N-alkyl thiocarbamic acid salt; alkyl isothiocynate (mustard oil)

(ii) $$2R_2{-}NH + S{=}C{=}S \rightarrow S{=}C\langle^{NR_2}_{SH.NHR_2} \xrightarrow{HgCl_2} \text{No reaction}$$
dialkyl dithiocarbamic acid salt

Tertiary amines do not show reaction with CS_2. In fact this test can also be used to distinguish 1°, 2° and 3° amines.

4. Hoffmann elimination:

Degradation of quaternary ammonium salts

$$R_4\overset{+}{N}\overset{-}{O}H \xrightarrow{\Delta} \text{Tertiary amine} + H_2O + \text{alkene}$$

$$H_3C{-}CH_2{-}CH_2{-}CH_2\overset{+}{N}(CH_3)_3\overset{-}{O}H \xrightarrow{\Delta} H_3C{-}CH_2{-}CH{=}CH_2 + N(CH_3)_3 + H_2O$$

5. Acylation: Acylation of primary and secondary amines is possible. Tertiary amines don't take part in acylation as there is no active hydrogen atom in them.

$$H_3C{-}CH_2NH_2 \xrightarrow[(CH_3CO)_2O]{CH_3COCl \text{ or}} H_3C{-}CH_2NH{-}\overset{O}{\overset{\|}{C}}{-}CH_3$$
N-ethyl acetamide

$$(CH_3)_2NH \xrightarrow[(CH_3CO)_2O]{CH_3COCl \text{ or}} (H_3C)_2N{-}\overset{O}{\overset{\|}{C}}{-}CH_3$$
N, N' -dimethyl acetamide

6. Oxidation:

(i) With acidic $KMnO_4$: Primary amines depending on the structure of the amine may give aldehyde or ketone

$$RCH_2NH_2 + (O) \rightarrow R{-}CH{=}NH \xrightarrow{H_2O} RCHO + NH_3$$
aldimine

$$R_2CHNH_2 + (O) \rightarrow R_2C{=}NH \xrightarrow{H_2O} RCOR + NH_3$$
ketimine

secondary amines react to give tetra alkyl hydrazine.

$$R_2NH + (O) + HNR_2 \rightarrow R_2N{-}NR_2$$

tertiary amines are not affected by $KMnO_4$

(ii) tertiary amines with Caros acid (H_2SO_5) or Fentons reagent (H_2O_2+$FeSO_4$) give oxide

$R_3N + (O) \quad R_3N \;\; O$

7. Schiff base or Anil formation: Both aliphatic and aromatic amines react with aldehydes to form schiff's bases which on reduction give secondary amines.

$$R{-}\underset{H}{C}{=}O + H_2NR \xrightarrow{-H_2O} R{-}\underset{H}{C}{=}NR \xrightarrow{\text{reduction}} RCH_2NHR$$
schiff base; secondary amine

Aniline

1. Halogenation:

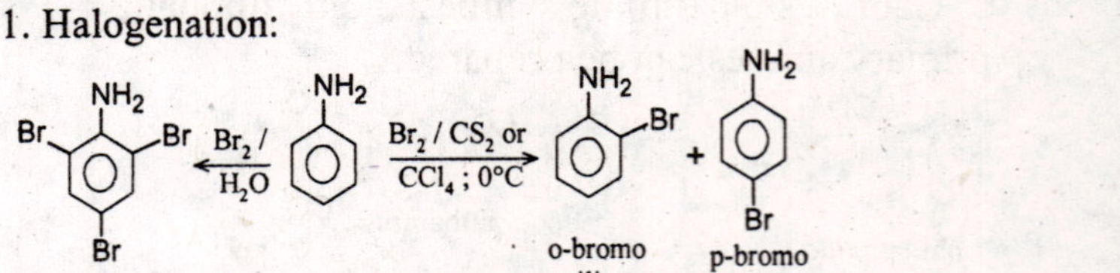

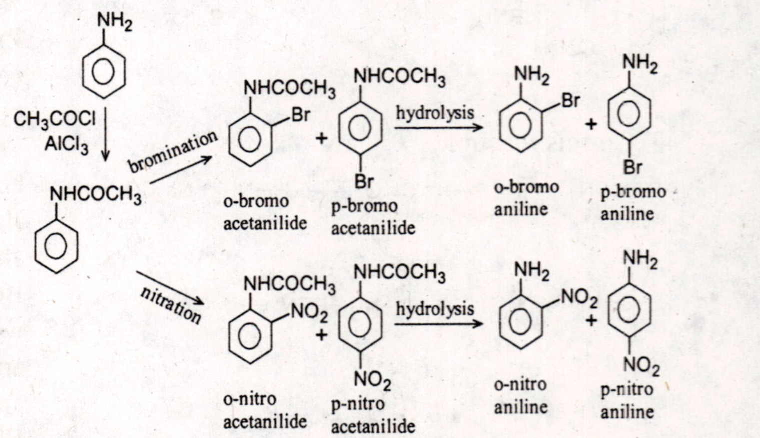

In halogenation and nitration, the amine group is protected by converting it to acetanilide as they have the oxidising property (which may decompose the benzene ring, which is a an activated one).

2 Coupling reaction: Benzene diazonium chloride couples with phenols and amines to give azo compounds. These are used as dyes or dye intermediates.

(i) 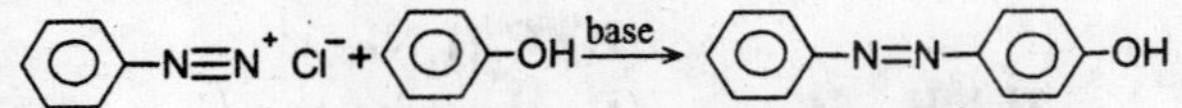

p-hydroxy azobenzene

(ii) 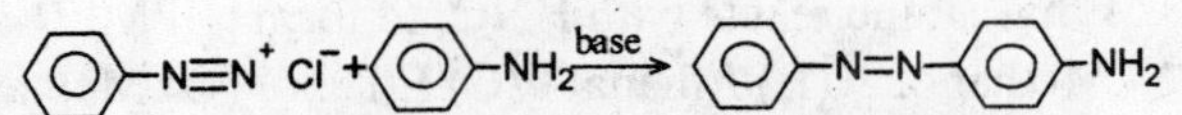

p-amino azobenzene

(iii) 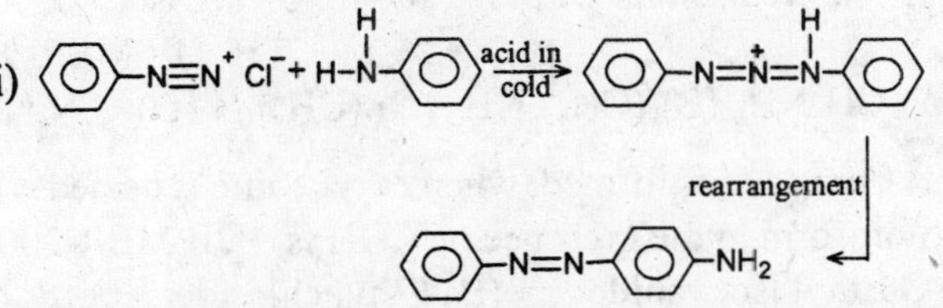

(iv) Dye test: Nitro benzene is converted into aniline by reduction with $SnCl_2$ and conc. HCl and then the dye test is performed.

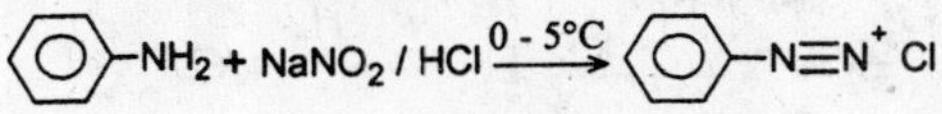

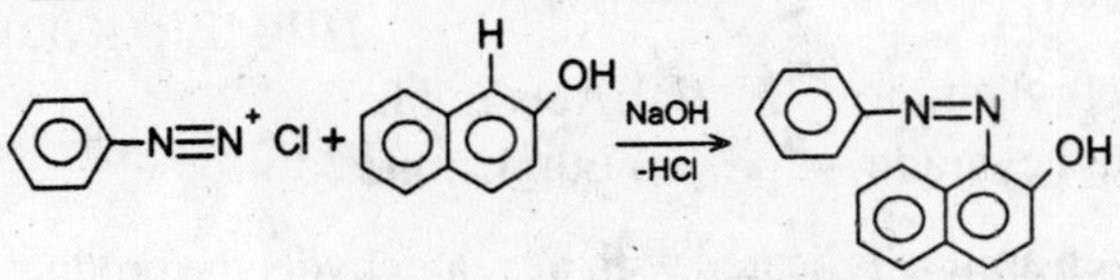

α- phenyl diazo-β- naphthol

orange red azo dye

3. Reaction with H_2SO_4 : Aniline on sulphonation gives sulphanilic acid as major product while orthanilic and metanilic acids are formed in smaller quantites.

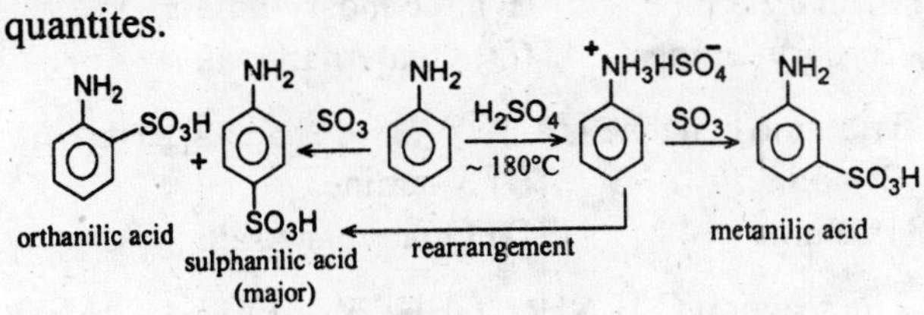

4. Oxidation:

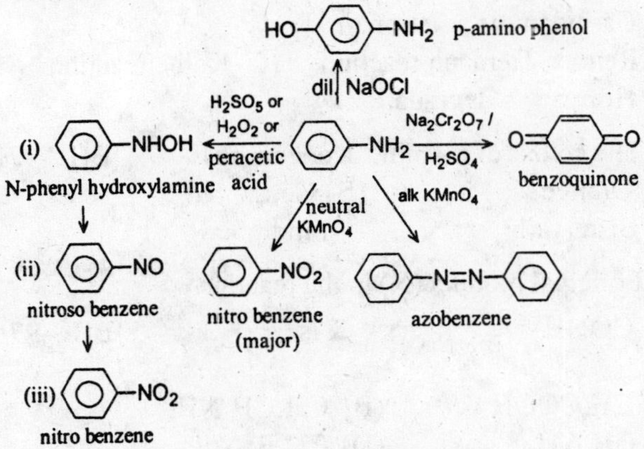

- Atoxyl: It is used in the treatment of sleeping sickness.

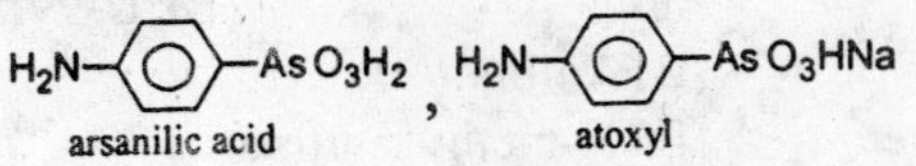

1. Identify Z in [CBSE-90]

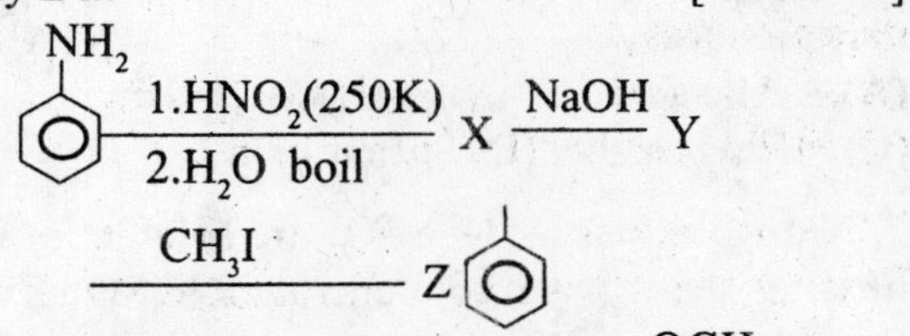

(A) 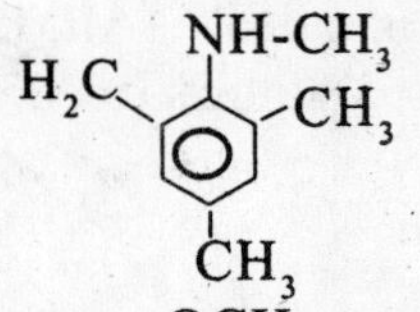 (B) 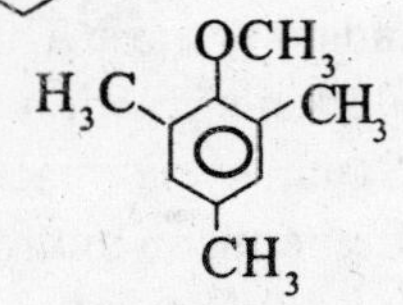

(C) 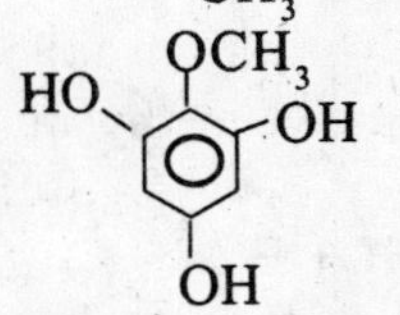 (D) 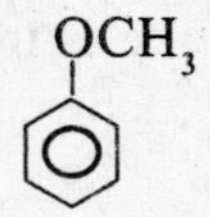

2. An aromatic amine (A) was treated with alcoholic potash and another compound (Y) when foul smelling gas was formed with formula C_6H_5NC. Y was formed by reacting a compound (Z) with Cl_2 in the presence of slaked lime. The compound (Z) is [CBSE-90]

(A) $C_6H_5NH_2$ (B) CH_3CH_2OH
(C) CH_3OCH_3 (D) $CHCl_3$

3. Identify X in the series : [CBSE-90]

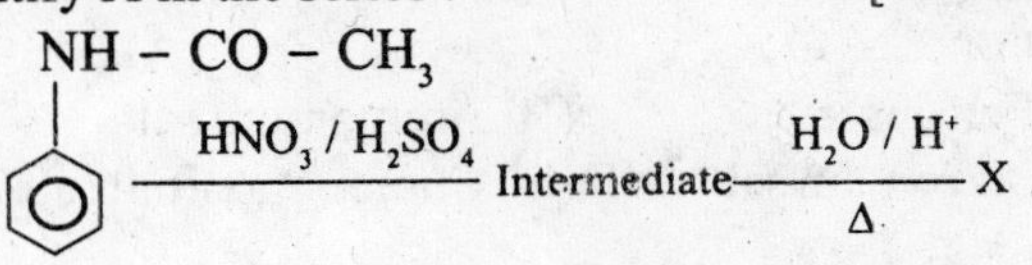

(A) 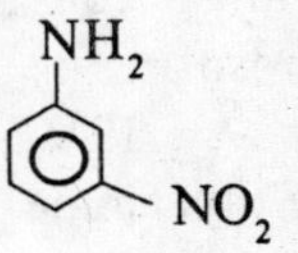 (B) 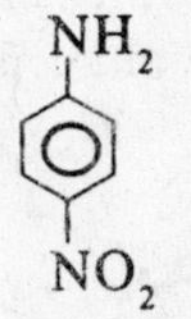

(C) 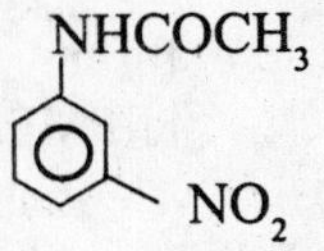 (D) 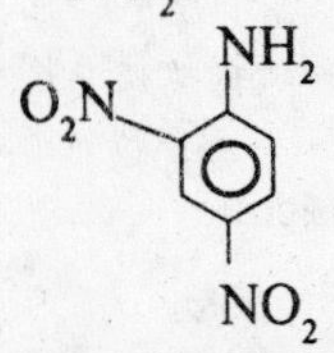

4. Aniline is reacted with bromine water and the resulting product is treated with an aqueous solution of sodium nitrite in presence of dilute HCl. The compound so formed is converted into a tetrafluoroborate which is subsequently heated dry. The final product is [CBSE-98]

(A) p-bromofluorobenzene (B) p-bromoaniline
(C) 2,4,6-tribromofluorobenzene
(D) 1,3,5-tribromobenzene

5. The final product C, obtained in this reaction [CBSE-2003]

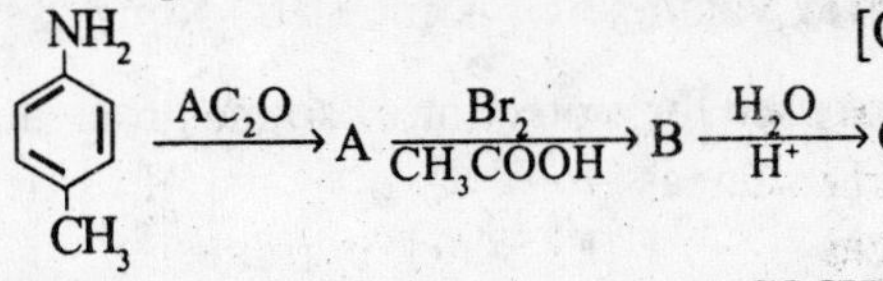 would be

(A) 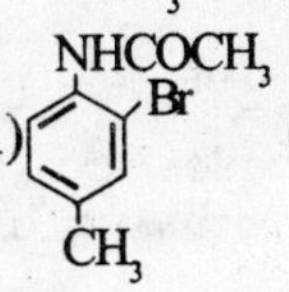 (B) 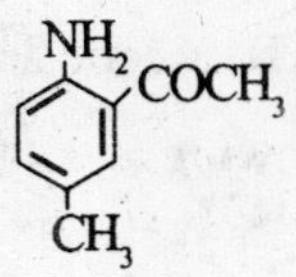 (C) 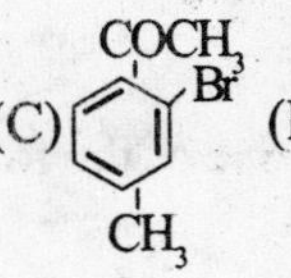 (D) 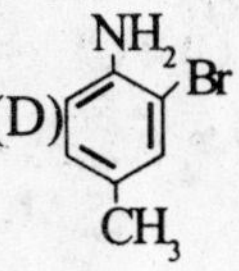

6. In the presence of an acid, hydrolysis of methyl cyanide produces [AIIMS-95]
(A) Acetic acid (B) Methylamine
(C) Methyl alcohol (D) Formic acid

7. Nitrosoamines ($R_2N-N=O$) are soluble in water. On heating them with concentrated H_2SO_4, they give secondary amines. This reaction is called [AIIMS-98]
(A) Perkin's reaction (B) Fittig's reaction
(C) Sandmeyer's reaction
(D) Liebermann niotroso reaction

8. Acetamide is treated separately with the following reagent. Which one of these would give methylamine ? [AIIMS-98]
(A) KOH (B) NaOH + Br_2
(C) NaOH (D) None of these

9. The boiling point of amines and corresponding alcohols and acids follow the order [AIIMS-2000]
(A) $RCH_2NH_2 < RCH_2OH < RCOOH$
(B) $RCH_2OH < RCOOH < RCH_2NH_2$
(C) $RCOOH < RCH_2NH_2 < RCH_2OH$
(D) $RCOOH < RCH_2OH < RCH_2NH_2$

10. Which of the following is involved in Sandmayer's reaction ? [AIIMS-2002]
(A) Ferrous salt (B) Ammonium salt
(C) Diazonium salt (D) Cuprammonium salt

11. A compound which on reaction with aqueous nitrous acid at low temperature, produces nitroso amine: [JIPMER-90]
(A) Methyl amine (B) Ethyl amine
(C) Diethyl amine (D) Triethyl amine

12. Ethylamine reacts with nitrous acid to form : [JIPMER-90]
(A) Methyl alcohol (B) Ethyl alcohol
(C) Ethane (D) Ethyl nitrate

13. $(CH_3)_3-C-NH_2$ is ? [JIPMER-90]
(A) Primary amine (B) Secondary amine
(C) Tertiary amine (D) Ammonium salt

14. Pure aniline is : [JIPMER-90]
(A) Brown coloured liquid (B) Brown coloured solid
(C) Colourless liquid (D) Colourless solid

15. An alkyl halide reacts with ammonia to give [JIPMER-94]
(A) Amide (B) Cyanide (C) Amine (D) None of these

16. The reagent used for protection of amino group during the nitration of aniline is [JIPMER-95]
(A) Acetic acid (B) $SOCl_2$/pyridine
(C) PCl_5 (D) Acetic anhydride

17. Which one of the following is not an explosive ? [JIPMER-96]
(A) nitroglycerin (B) o-aminotoluene
(C) dynamite (D) trinitrotoluene

18. Which of the following is weakest base [JIPMER-96]
(A) NH_3 (B) $C_2H_5NH_2$
(C) $C_6H_5NH_2$ (D) unpredictable.

19. Ethyl amine reacts with HNO_2 to form [JIPMER-96]
(A) Methanol (B) Ethanal (C) Ethanol (D) Methanal

20. The weakest base among the following is [JIPMER-98]
(A) CH_3NH_2 (B) $(CH_3)_2NH$ (C) $(CH_3)_3N$ (D) $C_6N_5NH_2$

21. Compounds formed, when ethylamine is heated with chloroform in the presence of KOH is [JIPMER-2000]
(A) Ethyl isocyanide (B) Ethyl chloride
(C) Ethanol (D) Diethyl ether

22. What is the end product in the following sequence of reactions, $C_2H_5NH_2 \xrightarrow{HNO_2} A \xrightarrow{PCl_5} B \xrightarrow{NH_3} C$ [JIPMER-2000]
(A) Methyl amine (B) Acetamide
(C) Ethyl cyanide (D) Ethylamine

23. When aniline is heated with benzaldehyde, the product is [JIPMER-2001]
(A) Unsaturated acid (B) Azoxy benzene
(C) Schiff's base (D) Benzoin

24. Carbylamine reaction is given by [BHU-96]
(A) Primary-amine (B) Secondary-amine
(C) Tertiary-amine (D) Quaternary salt

25. Triamino benzene is a [BHU-96]
(A) 2°-amine (B) 3°-amine
(C) 1°-amine (D) Quaternary salt

26. The reaction $C_6H_5NH_2 + CHCl_3 + 3KOH \rightarrow C_6H_5NC + 3KCl + 3H_2O$ is known is [BHU-96]
(A) Carbylamine reaction
(B) Reimer-Tiemann reaction (C) Kolbe reaction
(D) Hofmann's degradation

27. The colour of p-amino azobenzene is [BHU-97]
(A) Orange (B) Congo red
(C) Bismark brown (D) Indigo.

28. Principal product (X) in the reaction $CH_3CONH_2 \xrightarrow[Br_2+NaOH]{} X$ is [BHU-97]
(A) CH_3COOH (B) $CH_3CH_2NH_2$
(C) CH_3NH_2 (D) CH_3Br

29. Which of the following compounds is the strongest base ? [BHU-99]
(A) Ammonia (B) Aniline
(C) Methylamine (D) N-methyl aniline

30. Ethyl amine reacts with chloroform and potassium hydroxide, to form [BHU-99]
(A) Amine (B) Alcohol
(C) Aldehyde (D) Ethyl isocyanide

31. Ehylamine can be prepared by the action of bromine and caustic potash on [BHU-99]
(A) Acetamide (B) Formamide
(C) Propionamide (D) Methyl cyanide

32. C_3H_9N cannot represent [BHU-2000]
(A) 1° amines (B) 2° amines
(C) 3° amines (D) Quaternary salt

33. On reaction with HNO_2 , $C_2H_5NH_2$ produces [BHU-2000]
(A) C_2H_5OH (B) C_2H_5–NO_2 (C) CH_3–CHO (D) CH_3COOH

34. The reaction of aniline with acetyl chloride in presence of NaOH gives [BHU-2000]
(A) Acetamide (B) Aniline hydrochloride
(B) p- chloroaniline (D) None of these

35. In hydrolysis of aniline the reagent used is [AFMC-95]
(A) Dil. HCl (B) Acetal chloride
(C) CH_3OH (D) None of these

36. Increasing order of basicity of CH_3NH_2, $(CH_3)_2$ NH, $C_6H_5NH_2$, $(CH_3)_3N$ is [AFMC-97]
(A) $(CH_3)_3N < (CH_3)_2NH < CH_3NH_2 < C_6H_5NH_2$
(B) Reverse of (a)
(C) $C_6H_5NH_2 < (CH_3)N_3 < CH_3NH_2 < (CH_3)_2NH$
(D) Reverse of (c)

37. Nitroso amines (R_2N-N=O) are soluble in water. On heating them with conc. H_2SO_4, they give secondary amines. This reaction is called [AFMC-98]
(A) Perkin's reaction (B) Fries' reaction
(C) Liebermann nitroso reaction (D) Etard's reaction

38. Which of the following is the most basic in nature ? [AFMC-2001]
(A) NH_3 (B) CH_3NH_2
(C) $(CH_3)_2$ NH (D) C_6H_5N $(CH_3)_2$

39. In amines, the hybridization state of N is [AFMC-2002]
(A) sp (B) sp^2 (C) sp^3 (D) sp^3d

40. Amongst the following, the most basic compound is : [IIT-90]
(A) Benzylamine (B) p-Cresol
(C) Acetanilide (D) p-Nitroaniline

41. Examine the following two structures for the anilinium ion and choose the correct statement from the ones given below :

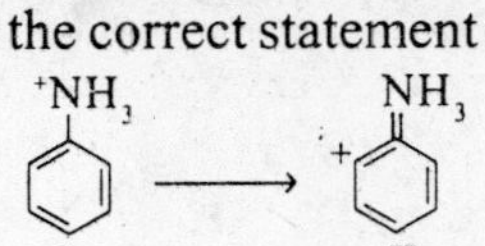

I II

(A) II is not an acceptable canonical structure because carbonium ions are less stable than ammonium ions
(B) It is not acceptable canonical structure because it is not aromatic.
(C) II is not acceptable canonical structure because nitrogen has valence electrons
(D) II is an acceptable canonical structure

42. Among the following, the strongest base is [IIT-2000]
(A) $C_6H_5NH_2$ (B) p–NO_2–$C_6H_4NH_2$
(C) m–NO_2–$C_6H_4NH_2$ (D) $C_6H_5CH_2NH_2$

43. The correct order of basicities of the following compounds is [IIT-2001]

CH_3–C(=NH)–NH_2 (1) CH_3–CH_2–NH_2 (2) $(CH_3)_2NH$ (3) CH_3–C(=O)–NH_2 (4)

(A) 2 > 1 > 3 > 4 (B) 1 > 3 > 2 > 4
(C) 3 > 1 > 2 > 4 (D) 1 > 2 > 3 > 4

44. A : In strongly acidic solutions, aniline becomes more reactive towards electrophilic reagents [IIT-2001]
R : The amino group being completely protonated in strongly acidic solution, the lone pair of electrons on the nitrogen is no longer available for resonance.
(A) A (B) B (C) C (D) D

45. When primary amine reacts with chloroform in ethanoic KOH then the product is [AIEEE-2002]
(A) An isocyanide (B) An aldehyde
(C) A cyanide (D) An alcohol

46. Which one of the following indicates carbylamine reaction : [EAMCET-90]
(A) C_6H_5(CH_3)>NH (B) CH_3NH_2
(C) $(CH_3)_2NH$ (D) $(CH_3)_3N$

47. Aniline is treated with $NaNO_2$ and HCl at 0°C, the product formed is : [EAMCET-91]
(A) Nitroaniline (B) Chloroaniline
(C) Benzenediazonium chloride salt
(D) Aminophenol

48. When aniline is treated with chloroform in alcoholic KOH product is [EAMCET-92]
(A) Benzenediazonium chloride (B) Benzanilide
(C) Schiff base (D) Phenyl isocyanide

49. Which one of the following is a tertiary amine [EAMCET-94]
(A) t-butyl amine (B) methyl ethyl amine
(C) dimethyl secondary butyl amine
(D) n-methyl amine

50. Aniline on treatment with bromine water gives a white

precipitate of [EAMCET-94]
(A) Monobromo aniline (B) Dibromo aniline
(C) Symmetrical tribromo aniline (D) Tetrabromo aniline

51. Oxidation of aniline with potassium dichromate and sulphuric acid gives [EAMCET-96]
(A) o-nitro aniline (B) N-nitroso aniline
(C) p-benzoquinone (D) o-benzoquinone

52. By which of the following reactions can one get n-methyl aniline from aniline ? [EAMCET-98]
(A) Alkylation (B) Acetylation
(C) Benzoylation (D) Bromination

53. Which of the following is a secondary amine ? [EAMCET-2000]
(A) $C_6H_5-NH_2$ (B) CH_3-NH_2
(C) $C_6H_5-\underset{\underset{CH_3}{|}}{N}-CH_3$ (D) $C_6H_5-\underset{\underset{H}{|}}{N}-CH_3$

54. Which one of the following is the molecular formula of a tertiary amine ? [EAMCET-2002]
(A) C_2H_7N (B) C_3H_9N (C) CH_5N (D) CH_3N

55. When primary amine is heated with CS_2 in presence of excess mercuric chloride, it gives iso-thiocyanate. This reaction is called [CPMT-97]
(A) Hoffman-bromide reaction
(B) Hoffman mustard oil reaction
(C) Carbylamine reaction (D) Perkin's reaction

56. Which one is less alkaline ? [CPMT-97]
(A) NO_2-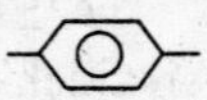$-NH_2$ (B) H_3CO-$-NH_2$
(C) C_6H_5-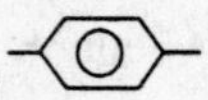$-NH_2$ (D) All of these

57. In the following reaction, X is [CPMT-99]
$X \xrightarrow{\text{Bromination}} Y \xrightarrow{NaNO_2+HCl} Z \xrightarrow[C_2H_5OH]{\text{Boiling}} \text{Tribromo Benzene}$
(A) Benzoic acid (B) Salicylic acid
(C) Phenol (D) Aniline

58. Which of the following reaction will not give primary amine ? [CPMT-99]
(A) $CH_3CONH_2 \xrightarrow{KOH,\ Br_2}$ (B) $CH_3CN \xrightarrow{LiAlH_4}$
(C) $CH_3NC \xrightarrow{LiAlH_4}$ (D) $CH_3CONH_2 \xrightarrow{LiAlH_4}$

59. In amines, the hybridisation state of N is [CPMT-99]
(A) sp (B) sp^2 (C) sp^3 (D) sp^2d

60. In test for primary amines, the amine is treated with $CHCl_3$ and KOH and a bad smelling compound is formed. If the primary amine used is ethylamine, identify the bad smelling compound formed [CPMT-2000]
(A) CH_3CN (B) CH_3CNO
(C) CH_3CH_2NC (D) CH_3NCO

61. Fusion of Na with aniline gives mainly [CPMT-2002]
(A) $Na_2NC_6H_5$ (B) $NH_2NH_2 + C_6H_6$
(C) $NaNH_2C_6H_5$ (D) $N_2 + C_6H_5Na$

62. $CH_3NO_2 \xrightarrow[HCl]{SnCl_2} CH_3X$, the 'X' contains [CPMT-2003]
(A) $-NH_2$ (B) -COOH (C) -CHO (D) $(CH_3CO)_2O$

63. Mark the correct statement [PMT-94]
(A) Methyl amine is acidic
(B) Methyl amine is less basic than NH_3
(C) Methyl amine is a stronger base than NH_3
(D) Methyl amine forms salts with bases

64. Hydrolysis of acetamide gives [PMT-94]
(A) Acetic acid (B) Acetaldehyde
(C) Methylamine (D) Formic acid

65. Which statement is not correct ? [PMT-95]
(A) Amines form hydrogen bond
(B) Ethyl amine has higher boiling point than propane
(C) Methyl amine is more basic than ammonia
(D) Dimethyl amine is less basic than methyl amine

66. Which of the following gives primary amine on reduction ? [PMT-95]
(A) $CH_3-CH_2-\overset{\overset{O}{||}}{N}\rightarrow O$
(B) $CH_3-CH_2-O-N=O$
(C) $CH_3\ CH_2\ NO_2$ (D) None

67. Primary amine and secondary amine are distinguished by [PMT-96]
(A) Br_2 / KOH (B) $HClO_4$ (C) HNO_2 (D) NH_3

68. When aniline reacts with $NaNO_2$ and dil. HCl at 0° - 5°C, the product formed is. [PMT-96]
(A) Nitroaniline (B) Benzenediazonium chloride
(C) Benzene (D) Tri-nitroaniline

69. The reaction of HNO_2 with 'A' gives quaternary ammonium salt. 'A, is - [PMT-97]
(A) Methyl amine (B) Dimethyl amine
(C) Trimethyl amine (D) Aniline

70. Identify X in the reaction : [PMT-97]
$$C_6H_5\ NH_2 \xrightarrow[280\ K]{NaNO_2/HCl} Y \xrightarrow[300\ K]{H_2O/H^+} X$$
(A) C_6H_5Cl (B) C_6H_5OH
(C) $C_6H_5N_2Cl$ (D) C_6H_5NHOH

71. The fusion of sodium with aniline gives mainly : [PMT-99]
(A) NaCN (B) NaN_3 (C) NaSCN (D) $NaNO_2$

72. HNO_2 forms an alcohol on reaction with :[PMT-99]
(A) $C_6H_5NHCH_3$ (B) $C_6H_5NH_2$
(C) $(CH_3)_2NH$ (D) CH_3NH_2

73. Which of the following is most basic ? [PMT-99]
(A) $C_6H_5NH_2$ (B) $(CH_3)_2NH$
(C) $(CH_3)_3N$ (D) NH_3

74. Which of the following **cannot** be acetylated ? [PMT-99]

(A) $C_2H_5NH_2$ (B) $C_6H_5NH_2$
(C) $(CH_3)_2NH$ (D) $(CH_3)_3N$

75. Hinsberg reagent is : [PMT-2000]

(A) $C_6H_5SO_3H$ (B) C_6H_5NO
(C) $C_6H_5SO_2Cl$ (D) $C_6H_5N_2Cl$

76. The compound which on reaction with aqueous HNO_2 at low temperature produces an oily nitrosoamine is : [PMT-2001]

(A) Methyl amine (B) Ethyl amine
(C) Diethyl amine (D) Aniline

77. Amongst the following, the most basic compound is : [PMT-2001]

(A) Benzyl amine (B) Aniline
(C) Acetanilide (D) P-nitroaniline

78. In the reaction $CH_3NH_2 + 3CHCl_3 + 3\ KOH \longrightarrow X + 3KCl + 3H_2O$ X is [PMT-2002]

(A) CH_3NC (B) CH_3CH_2NC
(C) CH_3COOH (D) CH_3CH_2COOH

79. Hinsberg's reagent is : [PMT-2003]

(A) 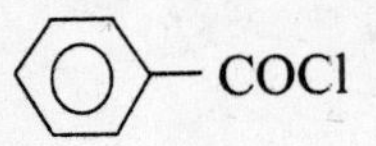 (B) 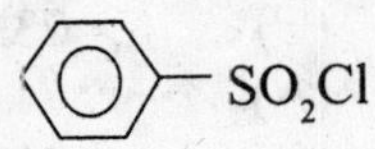

(C) 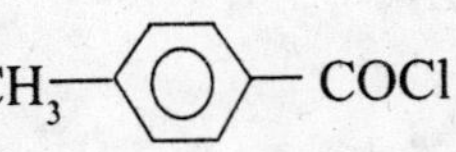 (D) 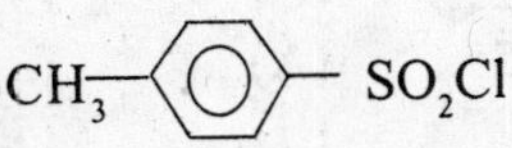

80. Ethylamine reacts with nitrous acid to give off [PET-93]

(A) Ammonia (B) Nitrous oxide
(C) Ethane (D) Nitrogen

81. In the reaction

$$ArNH_2 \xrightarrow[\text{Cold}]{NaNO_2 + HCl} A \xrightarrow{CuBr} B$$

the compound B is [PET-94]

(A) Ar - Br (B) Ar - NO_2 (C) Ar - Cl (D) Ar - Ar

82. Which of the following is converted into an alcohol on treatment with HNO_2 ? [PET-96]

(A) Methyl amine (B) Aniline
(C) Diethyl amine (D) Triethyl amine

83. Which of the following gives RNC, when reacted with $CHCl_3$ and KOH ? [PET-96]

(A) RNH_2 (B) R_2NH (C) R_3N (D) $R_4N^+Cl^-$

84. Which of the following compound gives dye test ? [PET-98]

(A) Aniline (B) Methylamine
(C) Diphenylamine (D) Ethylamine

85. Reaction of nitrous acid with aliphatic primary amine in the cold gives : [PET-98]

(A) A diazonium salt (B) An alcohol
(C) a nitrile (D) a dye

86. The compound on reaction with cold nitrous acid giving oily nitrosamine is : [PET-98]

(A) Methylamine (B) Dimethylamine
(C) Trimethylamine (D) Triethylamine

87. The amine which does not react with acetyl chloride is: [PET-99]

(A) CH_3NH_2 (B) $(CH_3)_2NH$
(C) $(CH_3)_3N$ (D) None of these

88. Ethyl amine on heating with CS_2 in presence of Hg_2Cl_2 forms : [PET-2000]

(A) C_2H_5NCS (B) $(C_2H_5)_2S$
(C) $(C_2H_5)_2\ CS$ (D) $C_2H_5\ (CS)_2$

89. Which of the following is most basic ? [PET-2000]

(A) $(CH_3)_3\ N$ (B) $(CH_3)_2\ NH$
(C) CH_3NH_2 (D) $(C_2H_5)_3\ N$

90. The number of primary amines of formula $C_4H_{11}N$ is : [PET-2000]

(A) 1 (B) 2 (C) 3 (D) 4

91. Action of HNO_2 on ethylamine gives [PET-2001]

(A) Ethane (B) Ethanol
(C) Ammonia (D) Nitromethane

92. Leakage of which gas was responsible for the Bhopal tragedy in 1984: [PET-2001]

(A) $CH_3 - N = C = O$ (B) $CH_3 - C - N = S$
(C) $CHCl_3$ (D) C_6H_5COCl

93. The action of nitrous acid on an aliphatic primary amine gives

(A) alkyl nitrite (B) secondary amine
(C) alcohol (D) nitro alkane

94. Acetic acid reacts with ethanol in the presence of H_2SO_4 to form X and water, which of the following is X ?

(A) $CH_3CH_2COOC_2H_5$ (B) CH_3COCH_3
(C) $CH_3COC_2H_5$ (D) $CH_3COOC_2H_5$

95. In the reaction $CH_3NH_2 + X + KOH \rightarrow CH_3NC$ (highly offensive odour). X is

(A) CH_2Cl_2 (B) $CHCl_3$ (C) CH_3Cl (D) CCl_4

96. The enzyme which converts glucose into ethanol is

(A) Invertase (B) Zymase (C) Diastase (D) Maltase

97. $C_2H_5Cl + KCN \rightarrow X \xrightarrow{\text{Hydrolysis}}$

X and Y are -

(A) C_2H_6 and C_2H_5CN (B) C_2H_5CN and C_2H_6
(C) C_2H_5CN and $C_2H_5CH_2NH_2$
(D) C_2H_5CN and C_2H_5COOH

98. Which test is performed to identify the secondary amine ? [PET-2003]

(A) Lucas test (B) Ethylamine
(C) Molisch test (D) Liebermann nitroso test

99. Aniline when diazotized in cold and then treated with

dimethyl aniline gives a coloured product. Its structure would be [CBSE-PMT-2004]

(A) 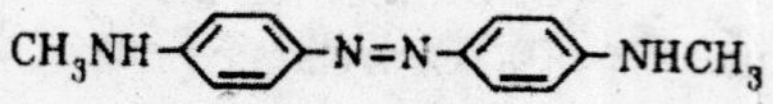

(B) 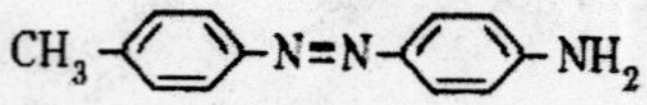

(C) 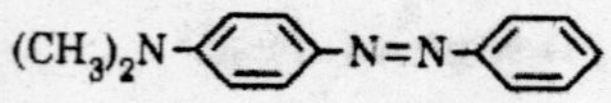

(D) 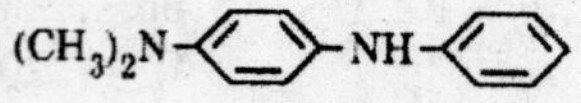

Amines

1.	(D)	51.	(C)
2.	(B)	52.	(A)
3.	(B)	53.	(D)
4.	(C)	54.	(B)
5.	(D)	55.	(B)
6.	(A)	56.	(A)
7.	(D)	57.	(D)
8.	(B)	58.	(C)
9.	(A)	59.	(C)
10.	(C)	60.	(C)
11.	(C)	61.	(A)
12.	(B)	62.	(A)
13.	(A)	63.	(C)
14.	(C)	64.	(A)
15.	(C)	65.	(D)
16.	(D)	66.	(A)
17.	(B)	67.	(C)
18.	(C)	68.	(B)
19.	(C)	69.	(C)
20.	(D)	70.	(B)
21.	(A)	71.	(A)
22.	(D)	72.	(D)
23.	(C)	73.	(B)
24.	(A)	74.	(D)
25.	(C)	75.	(C)
26.	(A)	76.	(C)
27.	(B)	77.	(A)
28.	(C)	78.	(A)
29.	(C)	79.	(B)
30.	(D)	80.	(D)
31.	(C)	81.	(A)
32.	(D)	82.	(A)
33.	(A)	83.	(A)
34.	(A)	84.	(A)
35.	(A)	85.	(B)
36.	(C)	86.	(B)
37.	(C)	87.	(C)
38.	(C)	88.	(A)
39.	(C)	89.	(B)
40.	(D)	90.	(D)
41.	(C)	91.	(B)
42.	(D)	92.	(A)
43.	(B)	93.	(C)
44.	(D)	94.	(D)
45.	(A)	95.	(B)
46.	(B)	96.	(B)
47.	(C)	97.	(D)
48.	(D)	98.	(D)
49.	(C)	99.	(C)
50.	(C)		

II. Cyanides or Nitriles / Isocyanides or Isonitriles

$C\equiv N$ bond (215 k cal / mole) is stronger than $C\equiv C$ bond (193 k cal / mole)

Preparation

1. From haloalkanes:

$$R-X \xrightarrow{\text{alc. KCN}} R-C\equiv N$$
$$R-X \xrightarrow{\text{AgCN}} R-N\equiv C$$

2. From acid amides:

$$R-\overset{O}{\overset{\|}{C}}-NH_2 \xrightarrow[-H_2O]{P_2O_5} R-C\equiv N$$

3. Carbyl amine reaction:

$$RNH_2 + CHCl_3 + 3KOH \xrightarrow{\Delta} RNC + 3KCl + 3H_2O$$

4. From Grignard reagent:

$$RMgX + \underset{\text{cyanogen chloride}}{ClCN} \longrightarrow RCN + MgXCl$$

Reactions:

1. Stephen's reaction:

$$R-C\equiv N \xrightarrow[H_2O]{SnCl_2 / HCl} RCHO + NH_4Cl$$

[R–CH=NH.HCl (aldimine hydrochloride) formed as intermediate undergoes hydrolysis].

2. Mendius reaction:

$$R-C\equiv N + 4[H] \xrightarrow{Na/C_2H_5OH} RCH_2NH_2 \quad \text{primary}$$
$$R-N\equiv C + 4[H] \xrightarrow{Na/C_2H_5OH} RNHCH_3 \quad \text{secondary}$$

3. With Grignard reagent:

$$R-C\equiv N + RMgX \longrightarrow R-\underset{R}{\underset{|}{C}}=N\,MgX \xrightarrow{2H_2O} \underset{\text{ketone}}{R-\underset{R}{\underset{|}{C}}=O}$$

Boiling point : cynaides have higher boiling points than their isomeric isocyanides.

III. Alkyl nitrites and nitro alkanes

$$\underset{\text{alkyl nitrite}}{R-\ddot{O}-\ddot{N}=\ddot{O}}\ ; \quad \underset{\text{nitro alkane}}{R-\overset{+}{N}(=\ddot{O})-\ddot{O}:^-}\ ; \quad \underset{\text{nitrite form}}{H-O-N=O} \rightleftharpoons \underset{\text{nitro form}}{H-N(=O)\rightarrow O}$$

Preparation:

1. $C_2H_5OH + HNO_2 \longrightarrow H_5C_2-O-N=O + H_2O$

$(NaNO_2 + HCl \rightarrow NaCl + HNO_2)$

In fact alkyl nitrites are esters of nitrous acid

2. (i) $R-X + KNO_2 \longrightarrow R-O-N=O + KX$

(ii) $R-X + \text{alc. } AgNO_2 \longrightarrow \underset{\text{(major)}}{\underset{\text{nitro alkane}}{RNO_2}} + \underset{\text{(minor)}}{\underset{\text{alkyl nitrite}}{R-O-N=O}}$

It is a nucleophilic substitution reaction.

Note: Nitrite ion $N(=\ddot{O})-\ddot{O}:^-$ can attack the haloalkane either through nitrogen or oxygen producing nitro alkanes or alkyl nitrites respectively. As $NaNO_2$ and KNO_2 are ionic compounds, the nitrite ion formed from them $\bar{O}-N=O$ can attack either through oxygen or nitrogen atoms giving mainly alkyl nitrites and nitroalkanes as minor products.

$$R-X + Na^+\bar{O}-N=O \longrightarrow R-O-N=O$$

Here, the attack takes place mainly through oxygen as C–O bond is stronger than C–N bond.

As $AgNO_2$ is a covalent compound, the attack can be only through nitrogen atom, consequently delivering RNO_2 (nitro alkanes)

$$R-X + N(=O)-O-Ag \longrightarrow R-\overset{+}{N}(=O)-O^- \text{ or } R-N(=O)\rightarrow O$$

- Alkyl nitrites and nitroalkanes if formed together can be separated on the basis of fractional distillation.

Reactions:

1. Reduction:

(i) $$\underset{\text{N- alkyl hydroxylamine}}{H_2O + RNHOH} \xleftarrow[\text{neutral medium}]{4[H],\ Zn + NH_4Cl} RNO_2 \xrightarrow[\text{acidic medium}]{+6[H],\ Sn + HCl} \underset{\text{primary amine}}{RNH_2 + 2H_2O}$$

Alkyl nitrites being esters undergo hydrolysis to give alcohols.

$$R-O-N=O + H_2O \xrightarrow{Sn + HCl} ROH + H-O-N=O$$

(ii) In general the reduction of nitrobenzene takes place depending upon the medium (acidic / basic/ neutral) in which it is carried out and the catalyst employed.

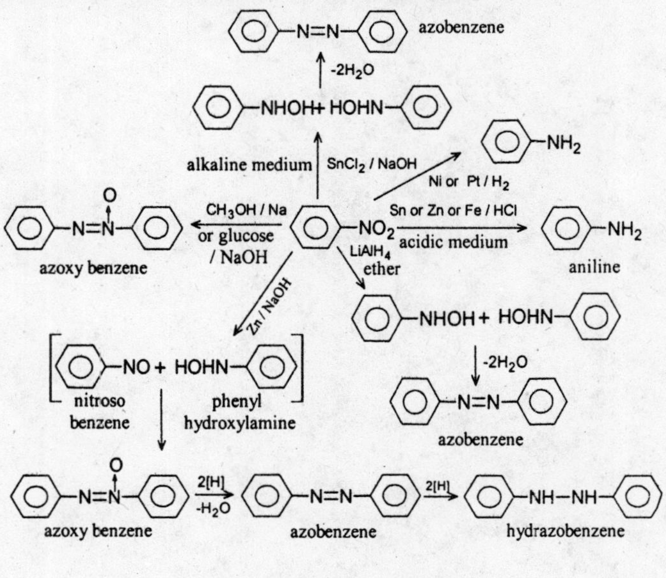

2. Mannich reaction or amino methylation

$$R-\underset{NO_2}{\underset{|}{\overset{R}{\overset{|}{C}}}}-H + HCHO + NH_4Cl \xrightarrow[-H_2O]{\Delta} R-\underset{NO_2}{\underset{|}{\overset{R}{\overset{|}{C}}}}-CH_2NH_2.HCl$$

depending upon the nature of nitroalkane employed in the reaction secondary or tertiary nitro amines are produced.

3. Oxidation: Aromatic side chain substituted nitro compounds can be distinguished from nuclear substituted nitro compounds with acidified $KMnO_4$.

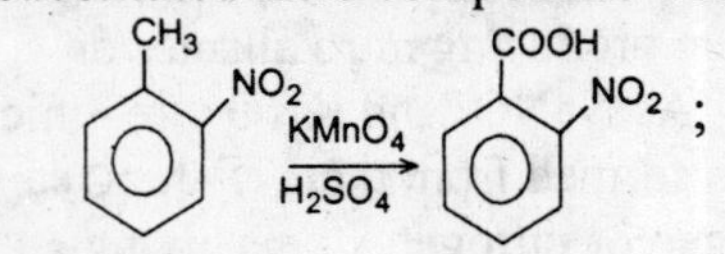

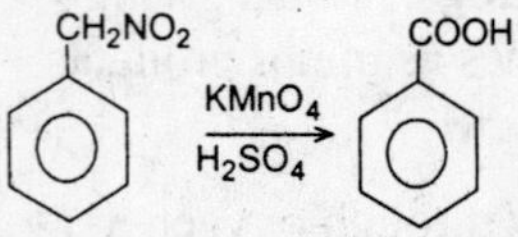

During oxidation, the side chain nitro group is lost while the nuclear substituted one remains intact.

- As nitro group is a electron pulling group, benzene ring is deactivated towards electrophilic substitution and therefore, it takes place at meta position (as o- and p- positions are deactivated). Nucleophilic substitution takes place at o- and p- positions. Nitro group increases the acidity of –OH phenolic, –COOH groups and/or groups like halogens.

- Spirit of nitre: It is 4% alcoholic solution of ethyl nitrite ($C_2H_5O–N=O$). It is used as a diuretic.

1. p-Nitrotoluene on further nitration gives [CBSE-96]

(A) 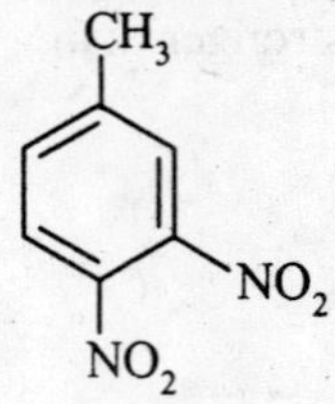

(B) 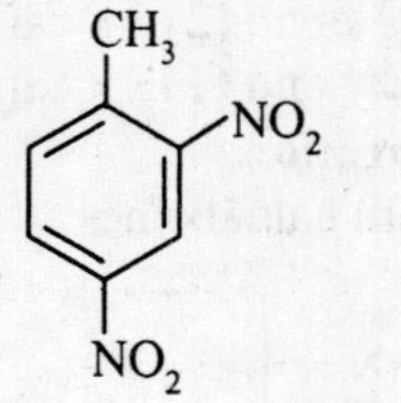

(C) 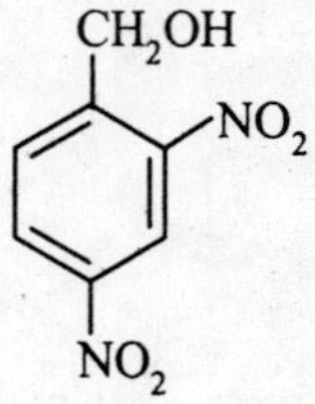

(D) 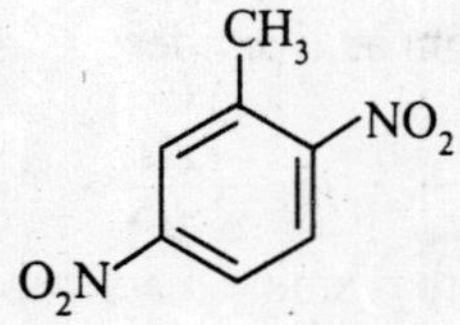

2. Nitrobenzene gives N-phenylhydroxylamine by - [AIIMS-2003]

(A) Sn/HCl (B) H_2/Pd-c (C) Zn/NaOH (D) Zn/NH_4Cl

3. Monocarboxylic acid can be obtained by hydrolysing [JIPMER-99]

(A) Aldehydes (B) Alkylcyanides (Alkane nitriles)
(C) Amides in presence of alkali (D) Ethers

4. Identify the product Z in the series,

$$CH_3CN \xrightarrow{Na / C_2H_5OH} X \xrightarrow{HNO_3} Y \xrightarrow{[O]} Z$$

[JIPMER-2001]

(A) CH_3CH_2 NHOH (B) CH_3COOH
(C) $CH_3CH_2CONH_2$ (D) CH_3CHO

5. What is formed, when nitrobenzene is reduced using zinc and alkali ? [BHU-96]

(A) Phenol (B) Aniline
(C) Nitrosobenzene (D) Hydrazobenzene

6. Nitrobenzene combines with hydrogen in the presence of platinum to yield [BHU-99]

(A) Toluene (B) Benzene
(C) Aniline (D) Azobenzene

7. Cyanide and isocyanide are isomers of type [AFMC-97]

(A) Positional (B) Functional
(C) Tautomer (D) All of these

8. Acetaldoxime reacts with P_2O_5 (Phosphorus pentoxide) to give [AFMC-97]

(A) Methyl cyanide (B) Methyl cyanate
(C) Ethyl cyanide (D) None of these

9. Which of the following reacts with chloroform and base to form phenyl isocyanide ? [AFMC-97]

(A) Aniline (B) Phenol (C) Benzene (D) Nitrobenzene

10. Identify Z in the reaction : $I \rightarrow X \rightarrow Y \rightarrow Z$ [AFMC-98]

(A) CH_3CH_2CN (B) $CNCH_2CH_2CN$
(C) $BrCH_2CH_2CN$ (D) $BrCH = CHCN$

11. 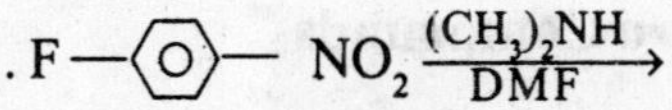

$$A \xrightarrow[\text{(ii) } H_2 \text{ Catalytic Reduction}]{\text{(i) } NaNO_2, \text{ HCl } 0-5°C} B; \text{ B is}$$

[IIT-2003]

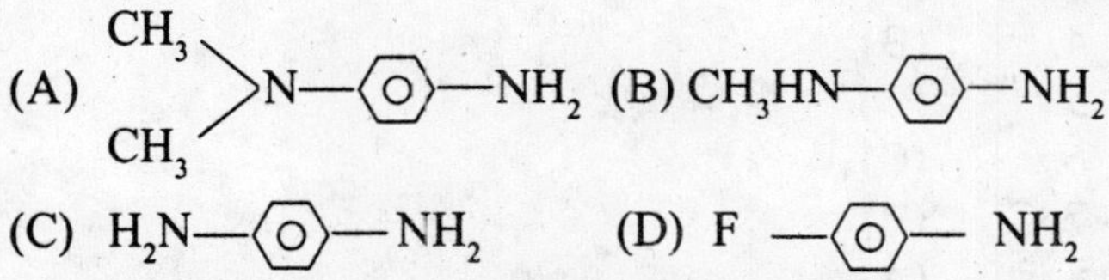

12. Ethyl isocyanide on hydrolysis in acidic medium generates [AIEEE-2003]
(A) Ethylamine salt and methanoic acid
(B) Propanoic acid and ammonium salt
(C) Ethanoic acid and ammonium salt
(D) Methylamine salt and ethanoic acid

13. $C_6H_5NO_2 \xrightarrow{Sn/HCl}$ .... [EAMCET-92]

14. The products formed when ethyl chloride is treated with silver cyanide [EAMCET-95]
(A) $C_2H_5CN + AgCl$ (B) $C_2H_5NC + AgCl$
(C) $C_2H_4 + AgCl$ (D) $C_2H_5ONO + AgCl$

15. Ethyl chloride on heating with silver cyanide forms a compound X. The functional isomer of X is [EAMCET-97]
(A) C_2H_5NC (B) C_2H_5CN
(C) $H_3C{-}NH{-}CH_3$ (D) $C_2H_5NH_2$

16. Reduction of nitrobenzene in the presence of Zn/NH_4Cl gives [EAMCET-98(MBBS)]
(A) Azobenzene (B) Aniline
(C) N-phenylhydroxylamine (D) Hydrazobenzene

17. Which of the following reagents converts nitrobenzene to N-phenyl hydroxyl amine ? [EAMCET-99]
(A) Sn/HCl (B) Zn/NaOH
(C) Zn/NH_4Cl (D) $LiAlH_4$

18. The compound formed when nitrobenzene is reduced with $LiAlH_4$ is [EAMCET-99(MBBS)]
(A) C_6HNH_2 (B) C_6H_5NHOH
(C) $C_6H_5N = NC_6H_5$ (D) $C_6H_5NH\ NHC_6H_5$

19. Nitrobenzene on reduction with Sn/HCl gives A which on reaction with acetyl chloride forms B. Which of the following is B ? [EAMCET-2000]
(A) $C_6H_5-\overset{O}{\overset{||}{C}}-CH_3$ (B) $C_6H_5-NH-\overset{O}{\overset{||}{C}}-CH_3$
(C) C_6H_5NHCl (D) $C_6H_5-O-\overset{O}{\overset{||}{C}}-CH_3$

20. Which of the following is used to convert nitrobenzene to hydrazo benzene ? [EAMCET(MBBS)-2000]
(A) Zn/NH_4Cl (B) Zn/NaOH
(C) Sn/HCl (D) $LiAlH_4$

21. Ethanol, when reacted with PCl_5 gave A, $POCl_3$ and HCl A reacts with silver nitrite to form B (major product) and AgCl A and B are respectively [EAMCET-2001]
(A) C_2H_5Cl and $C_2H_5\ OC_2H_5$
(B) C_2H_6 and $C_2H_5\ OC_2H_5$
(C) C_2H_5Cl and $C_2H_5\ NO_2$ (D) C_2H_6 and $C_2H_5\ NO_2$

22. Consider the following reaction :
$C_6H_5NO_2 \xrightarrow{Sn/HCl} X \xrightarrow{C_6H_5COCl} Y + HCl$ What is Y ? [EAMCET-2001]
(A) Acetanilide (B) Benzanilide
(C) Azo-benzene (D) Hydrazo-benzene

23. What is the product obtained when nitrobenzene is reacted with HCl in the presence of Sn ? [EAMCET(MBBS)-2001]
(A) $C_6H_5N = NC_6H_5$ (B) $C_6H_5NH_2$
(C) $C_6H_5NH - NHC_6H_5$ (D) C_6H_5NHOH

24. $C_6H_5C \equiv N$ and $C_6H_5N \equiv C$ show which type of isomerism ? [CPMT-97]
(A) Position (B) Functional
(C) Dextro isomerism (D)Positional isomerism

25. In the reduction of nitrobenzene, which of the following is the intermediate ? [CPMT-99]
(A) $C_6H_5\ N = O$ (B) $C_6H_5NH\text{-}NH\text{-}C_6H_5$
(C) $C_6H_5\text{-}N = N\text{-}C_6H_5$ (D) $C_6H_5\overset{\overset{O}{\uparrow}}{N} = N\text{-}C_6H_5$

26. When methyl cyanide is hydrolysed in presence of alkali, the product is : [PMT-93]
(A) Acetamide (B) Methane
(C) $CO_2 + H_2O$ (D) Acetic acid

27. Reduction of methyl isocyanide gives [PMT-2002]
(A) Ethylamine (B) Methylamine
(C) Dimethylamine (D) Trimethylamine

28. Which of the following is not used as an explosive ? [PET-96]
(A) Trinitrotoluene (B) Trinitrobenzene
(C) Picric acid (D) Nitrobenzene

29. In the nitration of benzene, the active nitrating agent is
(A) NO_2^- (B) NO_3^- (C) NO_2^+ (D) NO

30. Gas evolved in Bhopal gas tragedy was
(A) Methane (B) CO_2
(C) Methyl isocyanate (D) Butane

31. Heating chloroethane with ....yields a product containing an isocyanide group.
(A) AgCN (B) KCN (C) HCN (D) HNO_3

32. Identify X in the following reaction :

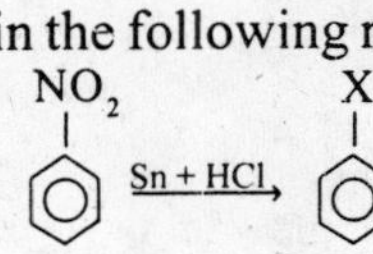

(A) -NH_2 (B) -NOH (C) -NO_3 (D) -HN_2

33. In presence of acid, hydrolysis of methyl cyanide gives:

(A) Acetic acid (B) Methylamine
(C) Methyl acid (D) Formic acid

34. Benzamide $\xrightarrow{POCl_3}$ (X). The product (X) is[IIT-2004]

(A) Aniline (B) Benzonitrile
(C) Benzylamine (D) Chlorobenzene

Nitro Compounds

1.	(B)
2.	(D)
3.	(B)
4.	(B)
5.	(D)
6.	(C)
7.	(B)
8.	(A)
9.	(A)
10.	(B)
11.	(A)
12.	(A)
13.	*
14.	(B)
15.	(B)
16.	(C)
17.	(C)
18.	(C)
19.	(B)
20.	(B)
21.	(C)
22.	(B)
23.	(B)
24.	(B)
25.	(A)
26.	(A)
27.	(C)
28.	(D)
29.	(C)
30.	(C)
31.	(A)
32.	(B)
33.	(A)
34.	(B)

13. $C_6H_5NH_2$ Aniline

Polymers

In polymers, the repeating units are held together by covalent bonds. In crystals, they are held through ionic bonds.

Classification:

1 Based on the origin:

(a) Natural polymers: They are also known as biological polymers, e.g., cellulose, starch, proteins, nucleic acids, rubber, wool etc.

(b) Synthetic polymers: phenol-formaldehyde polymer is the first synthetic one. e.g. Nylon, Acrylon, polyester (Dacron) etc.

2 Based on the chain structure:

(a) Homopolymers: They are obtained from only one type of monomer e.g. polythene; PVC, cellulose etc. Low density polythene differs from high density polythene in polymer structure. Linear polymers are well packed and have high densities. Branched chain polymers are irregularly packed and thus have low densities. High pressure polymerisation of ethene produces low density polythene (highly branched).

(b) Copolymers: Polymers are formed from two or more types of monomer units. When two monomers are polymerised together, the composition of the co-polymer will not be the same as that of monomer mixture.

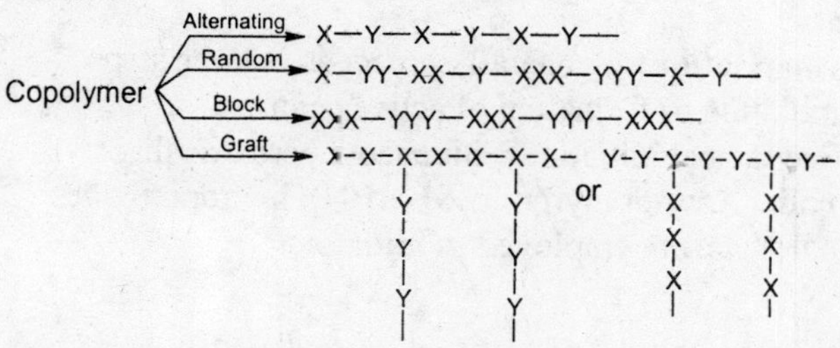

e.g. Terylene; Nylon 6, 6; Vinyon (Vinyl chloride & Venylacetate) etc.

3. Based on Polymerisation:

(a) Addition or chain polymers: The monomer unit of each polymer is given below:

(i) Polythene: CH_2=CH_2 Ethene

(ii) PVC: CH_2=CHCl Vinyl chloride

(iii) Teflon: F_2C=CF_2 (Tetrafluroethene)

(iv) Polystyrene:

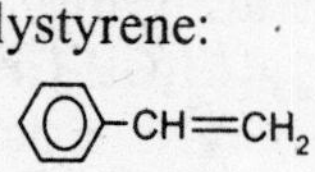

(v) PMMA :

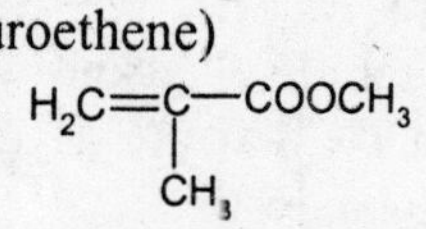

Methyl methacrylate

It is also known as (Plexiglass, perspex or Leusite). It is Used to make lenses, windowpanes etc.

(vi) Orlon or Polyacrylontrile: CH_2=CHCN

(b) Condensation polymers: Small entities like H_2O; NH_3; CH_3OH etc. are eliminated during the process. e.g. nylon, polyesters, phenol-formaldehyde, urea-formaldehyde resins etc.

(i) Nylon 6, 6:

$$nHOOC–(CH_2)_4–COOH + nH_2N(CH_2)_6NH_2 \xrightarrow{-H_2O}$$

Adipic acid   hexamethylene diamine

$$–[OC–(CH_2)_4CONH(CH_2)_6NH]–_n$$

(ii) Nylon 6 :

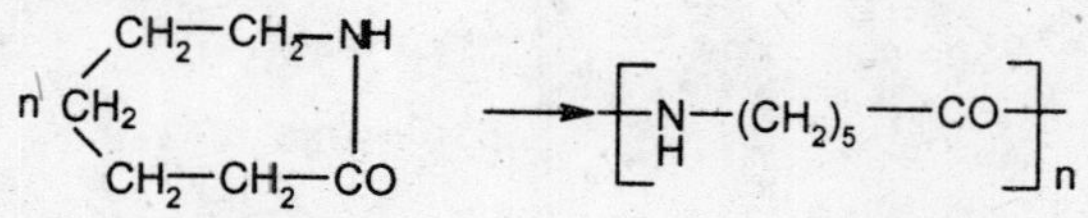

Caprolactam

Nylon is also known as a polyamide. Molecular chains are held together by hydrogen bonds.

(iii) Terylene (or Dacron, Vycron) : It is a polyester

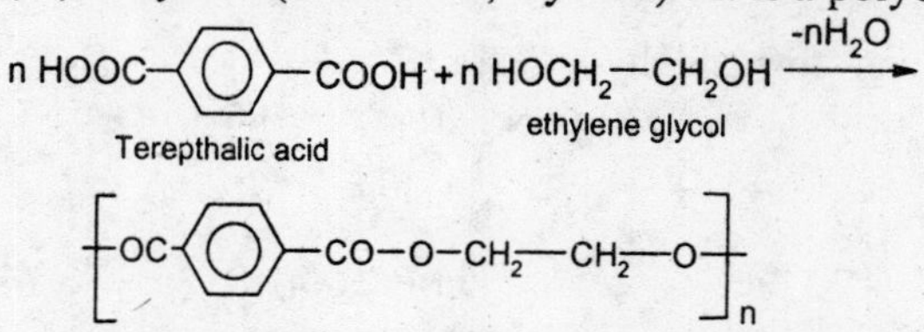

(iv) Alkyd resins or glyptals: The resins are also polyesters.

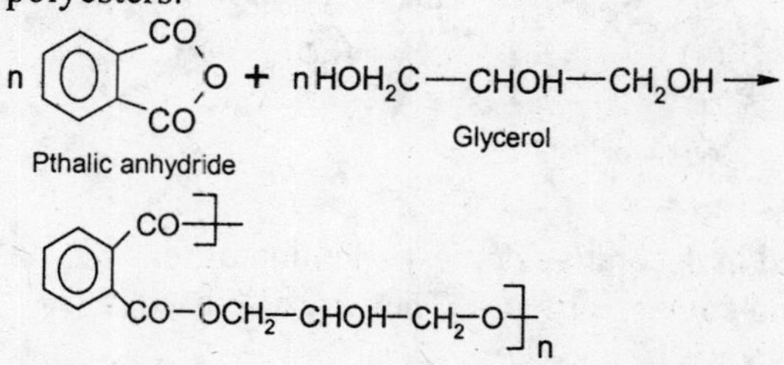

Here one can even proceed with phthalic acid and glycol. Alkyd resins are mainly employed as varnishes, paints, enamels and metal protection and decoration coatings.

4. Based on the behaviour towards heat:

(i) Thermoplastics: Polymers which soften on heating but regain their original properties on cooling. They are more or less crystalline. Linear polymers are generally thermoplastic e.g. polystyrene, PVC; polythene; Nylon, Orlon etc.

(ii) Thermosettings: polymers which are changed irreversably into hard and rigid materials on heating e.g. Araldite, Backelite; Melamine etc.

(a) Phenol-formaldehyde resin

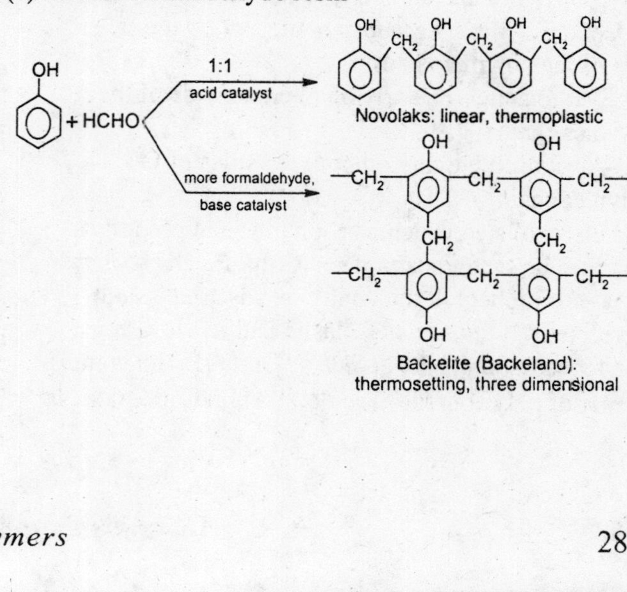

Substantial quantities of phenol-formaldehyde resins are used in the demineralisation and softening of hard water.

(b) Urea- formaldehyde resin:

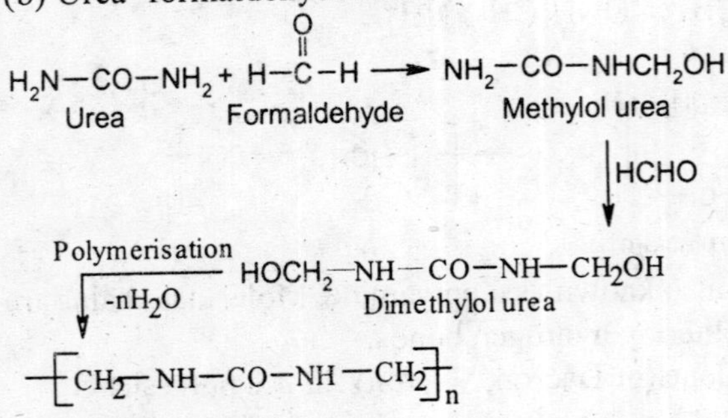

(c) Melamine- formaldehyde resin

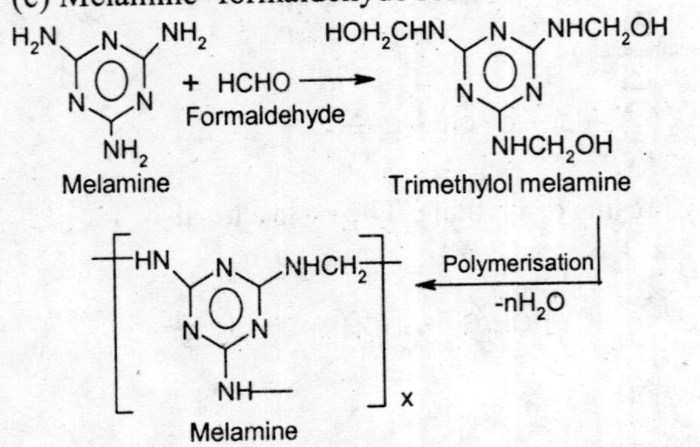

Mainly used in decorative domestic laminates on furniture and counter surfaces. They are also used to prepare crockery.

5. Based on physical parameters:

(i) Elastomers: have elasticity characteristic of rubber

(ii) Fibres: Long thin thread like with great strength e.g. cotton, wool, silk etc.(Rayon: artificial silk, prepared from cellulose by viscose process).

(iii) Plastics: Which can be extruded as sheets or pipes or moulded to form objects.

(iv) Resins or space- net work polymers: Rigid but irregular three dimensional structure.

Stereochemistry in polymers: The relative amounts of crystallinity, amorphousness and also some physical properties are either effected or dependent on the stereochemistry of the polymers. For example, in the polypropylene, the polymer can be

(i) Isotactic: Methyl groups are present on the same side of the polymer chain.

(ii) Syndiotactic: regularly on alternate side of the polymer chain.

(iii) Atactic: Randomly on both the sides of the polymer chain.

The effect of stereochemistry on physical properties may also be seen in natural and Gutta-Percha rubber. In natural rubber, all the double bonds are cis and therefore, it is amorphous, elastic and has low m.p. The repeating units are at 5.05Å. Gutta Percha is trans isomer of polyisoprene. The double bonds are trans. It is hard, highly crystalline and has high m.p. The repeating units are at 9. 15 Å.

Rubber: Rubber is in the form of colloidal solution in water (known as Latex). Rubber is coagulated by adding a little amount of acetic acid.

Vulcanisation: Raw rubber is heated with sulphur. During this process, the linear rubber molecules get cross linked through sulphur atoms. Rubber is hardened by adding carbon black (as filler for natural rubber). Ebonite is an inorganic polymer obtained by vulcanisation of rubbers. The monomer of rubber is Isoprene

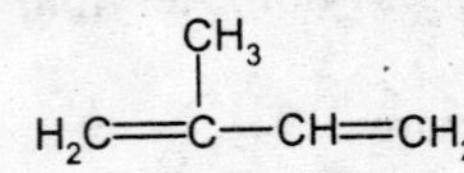

2-methyl-1,3-butadiene

Synthetic rubber:

(i) Neoprene: Monomer is 2- chloroisoprene

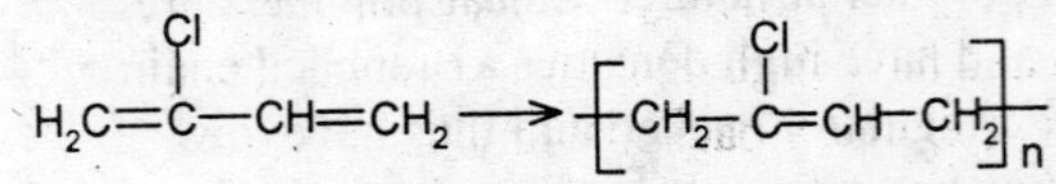

It is used for insulation of electric wires and cables.

(ii) Thiokol:

$$Cl{-}CH_2{-}CH_2{-}Cl + Na{-}S{-}S{-}Na + Cl{-}CH_2{-}CH_2Cl \xrightarrow{-2nNaCl} {-}[CH_2{-}CH_2{-}S{-}S{-}CH_2{-}CH_2{-}]_n$$

It is used for engine gaskets, tank linings for storage of oils etc and as a fuel in rocket engines with oxidising agents.

Polymerisation: It generally proceeds in three steps

(1) Initiation or formation of active centre

(a) free radical initiation : Dibenzoyl peroxide diacetyl, peroxide, Azoisobutyronitrile (AIBN), hydroperoxides (ROOH) etc. are employed as initiators.

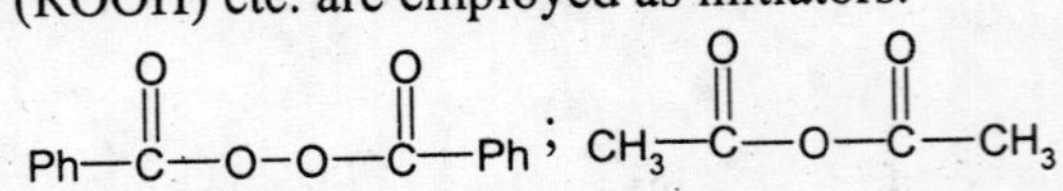

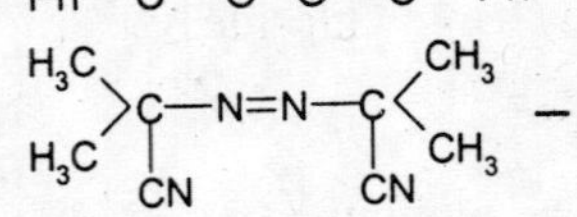 – AIBN

(b) Ionic initiation:

(i) Cationic (carbocation active centre forms). Takes place in the presence of $AlCl_3$, $SnCl_4$, BF_3, H_2SO_4, HF etc.

(ii) Anionic (Carbanion active centre forms) $Li^+NH_2^-$, $CH_3CH_2CH_2CH_2Li$, $Na^+NH_2^-$ etc. are employed.

(c) Ziegler- Natta polymerisation: Also known as coordination polymersation $TiCl_4$ and $(C_2H_5)_3$ Al is employed as Ziegler- Natta catalyst.

It gives linear polymer molecules (high density, m.p. and crystallinity more than the linear polymersation product).It permits stereochemical control.

Inhibitors: They inhibit polymerisation reaction by forming a product which is inactive (active centre no more exists). I_2, O_2,amines, phenols,quinones etc. act as inhibitors.

Telomers: Reagents employed to regulate the molcular weight of the polymer are called telomers or chain transfer reagents.

1. Cellophane used for wrapping is [CBSE-90]
(A) cellulose nitrate (B) synthetic cellulose
(C) cellulose acetate (D) regenerated cellulose

2. $CH_2 = CH_2$ is a [CBSE-91]
(A) monomer (B) polymer (C) isomer (D) equimer

3. Natural rubber is polymer of [CBSE-91]
(A) polyethylene (B) PVC (C) Isoprene (D) Insulin

4. Natural rubber is a polymer of [CBSE-92]
(A) n-pentene-1 (B) propylene
(C) butadiene (D) isoprene

5. An example of a natural biopolymer is [CBSE-94]
(A) teflon (B) rubber (C) nylon-66 (D)DNA

6. Bakelite is prepared by the reaction between [CBSE-95]
(A) urea and formaldehyde
(B) tetramethylene glycol and hexamethylenediisocyanate
(C) phenol and formaldehyde
(D) ethylene glycol and dimethyl terphthalate

7. Which one of the following is used to make 'non-stick' cookware ? [CBSE-97]
(A) PVC (B) Polystyrene
(C) Poly (ethylene terphtalate)
(D) Polyterafluoroethylene

8. Natural rubber is a polymer of [CBSE-99]
(A) butadiene (B) ethyne (C) styrene (D) isoprene

9. Terylene is a condensation polymer of ethylene glycol and [CBSE-99]
(A) benzoic acid (B) phthalic acid
(C) salicylic acid (D) terphthalic acid

10. $CF_2 - CF_2$ is a unit of [CBSE-2000]
(A) Teflon (B) Polyethene (C) Bakelite (D) Buna-S

11. Which of the following is not correctly matched ? [CBSE-2001]
(A) Neoprene $\left[-CH-\underset{Cl}{\underset{|}{C}} = CH-CH_2- \right]_n$
(B) Nylone 6,6
$\left[-NH-(CH_2)_6-NH-CO-(CH_2)_4-\overset{O}{\overset{||}{C}}- \right]_n$
(C) Terylene 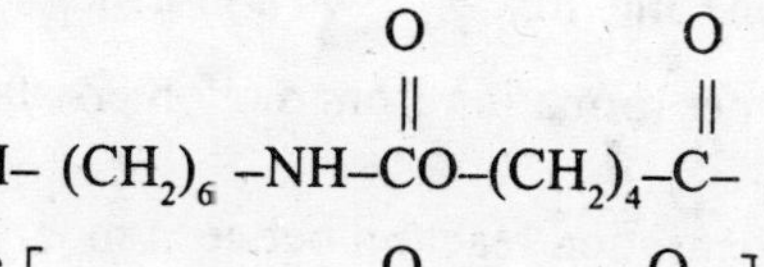
(D) PMMA $\left[-CH_2-\underset{COOH_3}{\underset{|}{\overset{CH_3}{\overset{|}{C}}}}- \right]_n$

12. Monomer of $\left[-\underset{CH_3}{\underset{|}{\overset{CH_3}{\overset{|}{C}}}}-CH_2- \right]_n$ [CBSE 2002]
(A) 2-methyl propene (B) Styrene
(C) Propylene (D) Ethene

13. Cellulose is polymer of [CBSE 2002]
(A) Glucose (B) Fructose (C) Ribose (D) Sucrose

14. Acrilan is a hard, horny and a high melting material. Which of the following represents its sturcture? [CBSE-2003]
(A) $\left[-CH_2-\underset{Cl}{\underset{|}{CH}}- \right]_n$
(B) $\left[-CH_2-\underset{CN}{\underset{|}{CH}}- \right]_n$
(C) $\left[-CH_2-\underset{COOHCH_3}{\underset{|}{\overset{CH_3}{\overset{|}{CH}}}}- \right]_n$
(D) $\left[-CH_2-\underset{}{\underset{|}{CH}}- \right]_n$

15. Which one of the following monomers gives the polymer neoprene on polymerisation ? [CBSE-2003]
(A) $CF_2 = CF_2$ (B) $CH_2 = CHCl$
(C) $CCl = CCl_2$ (D) $CH_2 = \overset{Cl}{\overset{|}{C}} - CH = CH_2$

16. Which of the following is a condensation polymer ? [AIIMS-95]
(A) Dacron (B) Neoprene (C) Teflon (D) Polystyrene

17. Which of the following statement is correct regarding the drawbacks of raw rubber ? [AIIMS-2001]
(A) It is plastic in nature (B) It has little durability
(C) It has large water-absorption capacity
(D) All of these

18. Teflon is a polymer of [AIIMS-2002]
(A) Tetrafluoroethylene (B) Tetraiodoethylene
(C) Tetrabromoethylene (D) Tetachloroethylene

19. The interparticle forces beween linear chains in Nylon-6 6 are [JIPMER-95]
(A) H-bonds (B) Covalent bonds
(C) Ionic bonds (D) Co-ordinate bonds

20. The monomer unit of polythene is [JIPMER-95]
(A) $CF_2 = CF_3$ (B) $CH_2 = CHCN$

(C) $CH_2 = CHCl$ (D) $CH_2 = CH_2$

21. Starting material for Buna-S rubber is [JIPMER-98]
(A) Butadiene and styrene (B) Styrene
(C) Butadiene and isobutylene (D) Chloroprene

22. The monomer units of a polypeptide are [JIPMER-98]
(A) Aldehydes (B) Aminoacids (C) Dienes (D) Diamines

23. In the viscose process the solvent for cellulose consists of [JIPMER-98]
(A) Ether and alcohol (B) Copper sulphate and ammonia
(C) Sodium hydroxide and carbon disulphide
(D) Acetic acid and acetic anhydride

24. The compound required for the formation of a thermosetting polymer with methanal is [JIPMER-98]
(A) Benzene (B) Phenyl amine
(C) Benzaldehyde (D) Phenol

25. Natural rubber is a polymer of which one of the following compounds ? [JIPMER-99]
(A) Ethylene (B) Acrylic acid
(C) 1,4 Butadiene (D) Isoprene

26. Which is not a macromolecule ? [BHU-98]
(A) DNA (B) Starch (C) Palmitate (D) Insulin

27. Terylene is used for making [BHU-99]
(A) Sails (B) Fabrics (C) Seat belts (D) All of these

28. Natural rubber is a polymer of [BHU-99]
(A) Ethylene (B) Isoprene
(C) Acrylic acid (D) 1,4 butadiene

29. Bakelite is prepared by the reaction between [BHU-2000]
(A) Phenol and formaldehyde
(B) Ethylene glycol and dimethyl terpthalate
(C) Urea and formaldehyde
(D) Tetramethylene glycol and hexamethylene diisocyanate

30. Orlon is a polymer of [BHU-2000]
(A) PVC (B) Bakelite (C) Acrylonitrile (D) Nylon

31. Zeigler-Natta catalyst is [BHU-2002]
(A) Mixture of titanium (III) chloride and triethyl aluminium.
(B) Mixture of zirconium (II) chloride and triethyl aluminum.
(C) Mixture of zirconium (III) chloride and dimethyl aluminium.
(D) Mixture of titanium (III) chloride and dimethyl aluminium.

32. Herculon is the polymer of [BHU-2002]
(A) Polyisobutylene (B) Polyvinylidene chloride
(C) Polymethyl methacrylate (D) Polypropylene

33. Neoprene is a polymer of [AFMC-91]
(A) Butadiene (B) Isoprene (C) Styrene (D) Chloroprene

34. The raw material to form Nylon is [AFMC-92]
(A) Adipic acid (B) Butadiene
(C) Isoprene (D) Ethylene

35. Nylon is made using [AFMC-93]
(A) Succinic acid (B) Adipic acid
(C) Benzaldehyde (D) Phenol

36. Terelyne is a condensation polymer of ethylene glycol and [AFMC-93]
(A) Benzoic acid (B) Phthalic acid
(C) Terphthalic acid (D) Salicylic acid

37. Polyamide is [AFMC-94]
(A) Nylon (B) Terylene (C) Orlon (D) Teflon

38. Natural rubber is polymer of [AFMC-96]
(A) Trans-isoprene (B) Cis-isoprene
(C) Co-cis-trans isoprene (D) None of these

39. Orlon is a polymer of [AFMC-97]
(A) Styrene (B) Tetrafluroethylene
(C) Vinyl chloride (D) Acrylonitrile

40. The polymer which has amide linkage, is [AFMC-98]
(A) Nylon-66 (B) Terylene (C) Teflon (D) Bakelite

41. Nylon-66 is made by using [AFMC-99]
(A) Phenol (B) Benzaldehyde
(C) Adipic acid (D) Succinic acid

42. Which of the following is the monomer of natural rubber? [AFMC-99]
(A) Glycol (B) Ethylene (C) Isoprene (D) Adipic acid

43. Perlon is [AFMC-2001]
(A) Rubber (B) Nylon (C) Terelene (D) Oxlon

44. Terylene is a polymer of [AFMC-2001]
(A) Adipic acid & hexamethylene diamine
(B) Terphthalic acid and ethylene glycol
(C) Phenol and formaldehyde (D) Vinyl cyanide

45. Bakelite is a product of reaction between [AFMC-2002]
(A) Formaldehyde and NaOH (B) Phenol and methanal
(C) Aniline and urea (D) Phenol and chloroform

46. Polymer formation from monomers starts by [AIEEE-2002]
(A) Condensation reaction between monomers
(B) Coordinate reaction between monomers
(C) Conversion of monomer to monomer ions by protons.
(D) Hydrolysis of monomers.

47. Nylon threads are made of [AIEEE-2003]
(A) Polyvinyl polymer (B) Polyester polymer
(C) Polyamide polymer (D) Polyethylene polymer

48. Propyne on polymerisation gives [CPMT-99]

(A) Mesitylene (B) Benzene
(C) Ethyl benzene (D) Propyl benzene

49. Propyne on polymerisation gives [CPMT-2002]
(A) mesitylene (B) benzene
(C) propylene (D) toluene

50. P-V-C is polymer of [CPMT-2003]
(A) $CH_2 == CH_2$ (B) $CH_2 == CH - Cl$
(C) $CH_2 == CH - CH_2Cl$ (D) $CH_3 - CH == CH - Cl$

51. Polymerisation in which two or more chemically different monomers take part is called : [PMT-93]
(A) Addition polymerisation (B) Co-polymerisation
(C) Chain polymerisation (D) Homo polymerisation

52. Natural rubber is basically a polymer of: [PMT-93]
(A) Neoprene (B) Isoprene
(C) Chloroprene (D) Butadiene

53. The fibre obtained by the condensation of hexamethylene diamine and adipic acid is [PMT-93]
(A) Dacron (B) Nylon 6, 6
(C) Rayon (D) Teflon

54. Which one among the following is a thermosetting plastic ? [PMT-93]
(A) PVC (B) PVA (C) Bakelite (D) Perspex

55. Starch is the polymer of [PMT-94].
(A) Glucose (B) Fructose (C) Both (D) None of them

56. Which of the following is an example of condensation polymers ? [PMT-95]
(A) Polythene (B) P. V. C. (C) Orlon (D) Terylene

57. Natural rubber is formed by polymerisaion of which of the following ? [PMT-95]
(A) Phenol (B) Isoprene (C) Chloroprene (D) Styrene

58. Which of the following is the thermosetting plastic ? [PMT-95]
(A) Nylon (B) Bakelite (C) Polythene (D) Polystyrene

59. Protein is a polymer of [PMT-96]
(A) Glucose (B) Terphthalic acid
(C) Amino acids (D) Glycol

60. Which of the following is used in the synthesis of Nylon-66 ? [PMT-99]
(A) Benzoic acid (B) Adipic acid
(C) Phthalic acid (D) Salicylic acid

61. Chloroprene is used to synthesize : [PMT-99]
(A) neoprene (B) dacron (C) nylon-6 (D) orlon

62. Natural rubber is a polymer of : [PMT-99]
(A) isoprene (B) ethylene glycol
(C) starch (D) acrylonitrile

63. Nylon-6 is made from : [PMT-2000]
(A) butadiene (B) chloroprene
(C) adipic acid (D) caprolactam

64. In polysaccharides the linkage connecting monosaccharide units is called : [PMT-2000]
(A) glycoside linkage (B) nucleoside linkage
(C) glycogen linkage (D) peptide linkage

65. Natural rubber is a polymer of : [PMT-2001]
(A) Polyethylene (B) PVC (C) Isoprene (D) Insulin

66. Which of the following is not a polymer ? [PMT-2001]
(A) ICl (B) Teflon (C) Neoprene (D) Isoprene

67. Nylon is a : [PMT-2001]
(A) Polyamide (B) Polyester
(C) Polyethylene (D) Polypropylene

68. Polythene is an addition polymer of [PMT-2002]
(A) Methane (B) Ethane (C) Ethylene (D) Acetylene

69. Monomer of the Teflon polymer is : [PMT-2003]
(A) $CF_2 = CF_2$ (B) $CF_2 = CCl_2$
(C) $CF_2 = CFCl$ (D) $CF_2 = CH_2$

70. The components of the Ziegler-Natta catalyst used for obtaining polypropylene from propylene are : [PMT-2003]
(A) $Ti\ (C_2H_5)_3 + AlCl_3$ (B) $TiCl_4 + Al\ (C_2H_5)_3$
(C) $TiCl_3 + Al\ (C_2H_5)_3$ (D) $Ti\ (C_2H_5) + AlCl_3$

71. Among the following, a natural polymer is [PET-93]
(A) Cellulose (B) P.V.C.
(C) Teflon (D) Polyethylene

72. Which one of the following can be used as monomer in a polymerization reaction ? [PET-93]
(A) CH_3CH_2Cl (B) CH_3CH_2OH (C) C_6H_6 (D) C_3H_6

73. The monomer of natural polymer Rubber is [PET-94]
(A) Isoprene (B) Ethylene
(C) Vinyl chloride (D) Acrolein

74. Which one of the following polymers will not catch fire [PET-94]
(A) $(CF_2 - CF_2)_n$ (B) $(CH_2 - CH_2)_n$
(C) $(CH(Cl) - CH(Cl))_n$ (D) $(CH_2 - CH(Cl))_n$

75. Natural rubber is a polymer derived from [PET-95]
(A) Butadiene (B) Ethylene
(C) Isoprene (D) Propylene

76. Teflon is a polymer of : [PET-98]
(A) Vinyl chloride (B) Ethylene
(C) Acetylene (D) Tetrafluoroethene

77. Methanal and phenol react in the presence of base to give : [PET-98]
(A) Bakelite (B) Polyethylene

(C) Dacron (D) Nylon - 66

78. Which of the following contains isoprene units ? [PET-98]
(A) Natural rubber (B) Nylon - 66
(C) Polyethylene (D) Dacron

79. Which of the following is not a synthetic polymer ? [PET-99]
(A) Polyethylene (B) PVC
(C) Nylon (D) Cellophane

80. What is NOT true about polymers ? [PET-99]
(A) Polymers do not carry any charge
(B) Polymers have high viscosity
(C) Polymers scatter light
(D) Polymers have low molecular weight

81. The basis of the mode of their formation the polymers can be classified : [PET-99]
(A) As addition polymers only
(B) As condensation polymers only (C) As copolymers
(D) Both as addition and condensation polymers

82. Which is used for the formation of nylon 66 ? [PET-2000]
(A) Sulphur hexafluoride (B) Adipic acid
(C) Sulphurous acid (D) Phthalic acid

83. Natural rubber is a polymer of : [PET-2001]
(A) Ethane (B) Propane (C) Styrene (D) Isoprene

84. Which polymer is used in making rayon ? [PET-2001]
(A) Polyester (B) Nylon (C) Cellulose (D) Starch

85. Which of the following is not a polymer ? [PET-2001]
(A) Silk (B) DNA (C) DDT (D) Starch

86. Which of the following is a thermoplastic polymer? [PET-2001]
(A) Bakelite (B) Melamine-formaldehyde
(C) Polythene (D) Vulcanized rubber

87. Isoprene is a valuable substance for making [PET-2002]
(A) Propene (B) Liquid fuel
(C) Synthetic rubber (D) Petrol

88. Bakelite is a product formed from [PET-2002]
(A) Reaction of formaldehyde with phenol
(B) Reaction of polyethylene with phenol
(C) Reaction of polypropylene with acid
(D) It is a natural product

89. Which of the following polymers contains nitrogen [PET-2003]
(A) PVC (B) Terylene (C) Nylon (D) Teflon

90. Which of the following compounds represent acrylonitrile ?
(A) Prop-2-ene nitrile (B) Vinyl cyanide
(C) Cyno ethene (D) All of these

91. Which one is a polymer compound ?
(A) SO_2 (B) CO_2 (C) CH_4 (D) PVC

92. Natural rubber is a
(A) Polyester (B) Polyamide
(C) Polyisoprene (D) Polysaccharide

93. The synthetic polymer 'terylene' is a
(A) Polyester (B) Polyamide (C) Polyether (D) Polythene

94. Which of the following has ester linkage? [BHU-2003]
(A) Terylene (B) PVC (C) Bakelite (D) Polythene

95. Which of the following is not a natural polymer? [AFMC-2003]
(A) Cellulose (B) PVC
(C) Protein (D) Nucleic acid

96. Nylon is a [AFMC-2003]
(A) Polyester (B) Polyamide
(C) Polysaccharide (D) None of these

97. Treatment of rubber with sulphur is [AFMC-2003]
(A) Annealing (B) Quenching
(C) Vulcanisation (D) None of these

98. Which one of the following is a chain growth polymer ? [CBSE-PMT-2004]
(A) Starch (B) Nucleic acid (C) Polystyrene (D) Protein

Polymers

1.	(C)	51.	(B)
2.	(C)	52.	(B)
3.	(C)	53.	(B)
4.	(D)	54.	(C)
5.	(D)	55.	(A)
6.	(C)	56.	(D)
7.	(D)	57.	(B)
8.	(D)	58.	(B)
9.	(D)	59.	(C)
10.	(A)	60.	(B)
11.	(B)	61.	(A)
12.	(A)	62.	(A)
13.	(A)	63.	(D)
14.	(B)	64.	(A)
15.	(D)	65.	(C)
16.	(A)	66.	(A)
17.	(D)	67.	(A)
18.	(A)	68.	(C)
19.	(A)	69.	(A)
20.	(D)	70.	(B)
21.	(A)	71.	(A)
22.	(B)	72.	(D)
23.	(C)	73.	(A)
24.	(D)	74.	(A)
25.	(D)	75.	(C)
26.	(C)	76.	(D)
27.	(D)	77.	(A)
28.	(B)	78.	(A)
29.	(A)	79.	(D)
30.	(C)	80.	(D)
31.	(A)	81.	(D)
32.	(D)	82.	(B)
33.	(D)	83.	(D)
34.	(A)	84.	(C)
35.	(B)	85.	(C)
36.	(C)	86.	(C)
37.	(A)	87.	(C)
38.	(B)	88.	(A)
39.	(D)	89.	(C)
40.	(A)	90.	(B)
41.	(C)	91.	(D)
42.	(C)	92.	(C)
43.	(B)	93.	(A)
44.	(B)	94.	(A)
45.	(C)	95.	(B)
46.	(A)	96.	(B)
47.	(C)	97.	(C)
48.	(A)	98.	(C)
49.	(A)		
50.	(B)		

Biomolecules

I. Carbohydrates: All carbohydrates respond positively to Molisch's test. Plants store energy by converting glucose into starch while animals convert glucose into glycogen (animal starch) and store. In photosynthesis, carbohydrates are prepared by the plants.

Carbohydrates are polyhydroxy aldehydes or ketones or the substances which yield them on hydrolysis. Mostly, in carbohydrates, the aldehyde or ketone groups do not occur as such but in the form of internal hemiacetal.

The carbonyl group reacts with alcohol and give hemiacetal

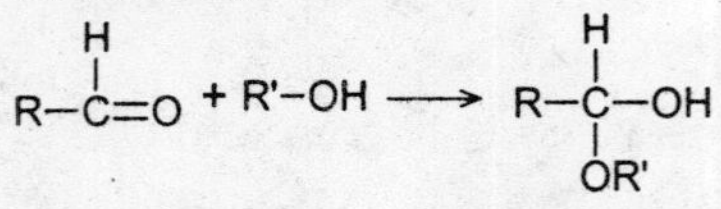

hemiacetal

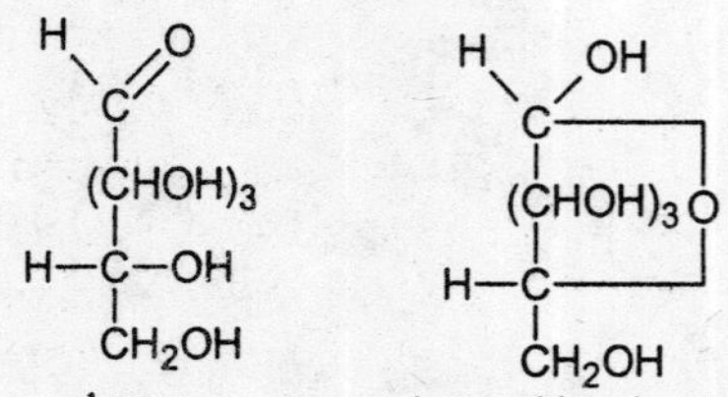

glucose internal hemiacetal

Carbohydrates with aldehyde group are called aldoses while with ketone group as ketoses.

Classification:

(a). On the basis of physical properties:

i. Sugars: Crystalline, sweet in taste, soluble in water. e.g. glucose, fructose, sucrose, lactose etc.

ii. Non sugars: Amorphous, not sweet in taste, sparingly soluble in water. e.g. Starch, cellulose etc.

(b) Based on hydrolysis products of carbohydrates:

i. Monosaccharides: They can't be hydrolysed to simpler compounds. e.g. Glucose, fructose, galactose arabinose, etc.

ii. Oligosaccharides:

- Disaccharides: on hydrolysis, they give two monosaccharide units. e.g. Sucrose, maltose, lactose, etc.

$$\underset{\text{sucrose}}{C_{12}H_{22}O_{11}} + H_2O \xrightarrow{H^+} \underset{\text{glucose}}{C_6H_{12}O_6} + \underset{\text{fructose}}{C_6H_{12}O_6}$$

- Trisachharides: on hydrolysis they give three monosaccharide units. e.g. raffinose

$$C_{18}H_{32}O_{16} + 2H_2O \xrightarrow{H^+} \text{glucose} + \text{fructose} + \text{galactose}$$

iii. Polysaccharides: on hydrolysis they give many units of monosaccharide. e.g. Starch, cellulose etc.

Starch:

The glucose units are linked by α–glycosidic linkage It is a mixture of 20% water soluble amylose and 80% water insoluble amylopectin. Starch solution is used as an indicator in iodometric and iodimetric titrations.

Amylose gives blue colour with iodine while amylopectin gives brown colour.

$$\text{Glucose} \xleftarrow{\text{Ptyalin}} \text{Starch} \xrightarrow{\text{Diastase}} \text{Maltose}$$

Cellulose $(C_6H_{10}O_5)_n + nH_2O \xrightarrow{H^+} > 1000$ glucose units

The glucose are linked by β-glycosidic linkage. In animals, cellulase enzyme converts cellulose into glucose.

Glucose: It is known as dextrose because it is optically active and dextro rotatory. It is also known as grape sugar.

Commercial preparation:

$$\underset{\text{starch}}{(C_6H_{10}O_5)_n} + nH_2O \xrightarrow{\text{dil } H_2SO_4} \underset{\text{glucose}}{n\,C_6H_{12}O_6}$$

The source of starch is potato, rice, maize etc.

Reactions:

1. Reduction:

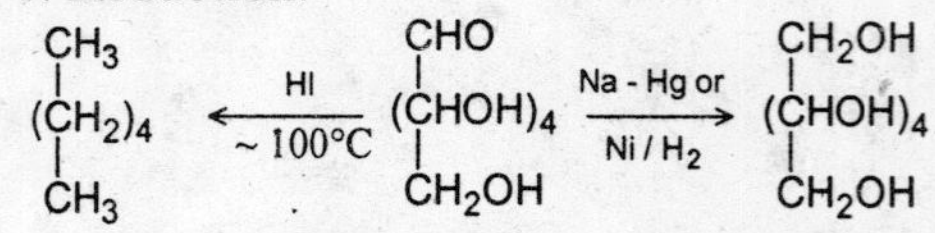

n-hexane glucose sorbitol

Fructose on reduction gives sorbitol and manintol

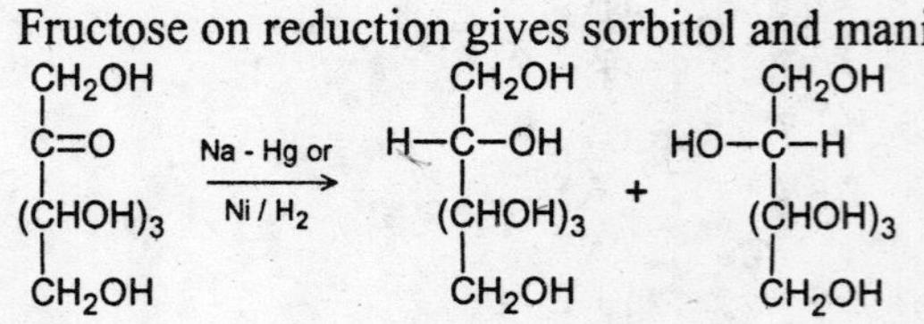

fructose sorbitol manintol

The >C=O group which is achiral on reduction gives >CHOH group, which is a chiral one, and therefore the two possibilities of configuration lead to the formation of sorbitol and manintol.

2. Oxidation:

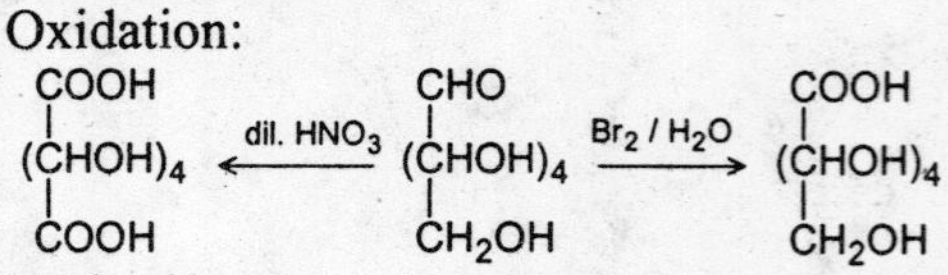

glucaric acid (saccharic acid) glucose gluconic acid

3. Osazone: Glucose, fructose and mannose give the same osazone with phenyl hydrazine

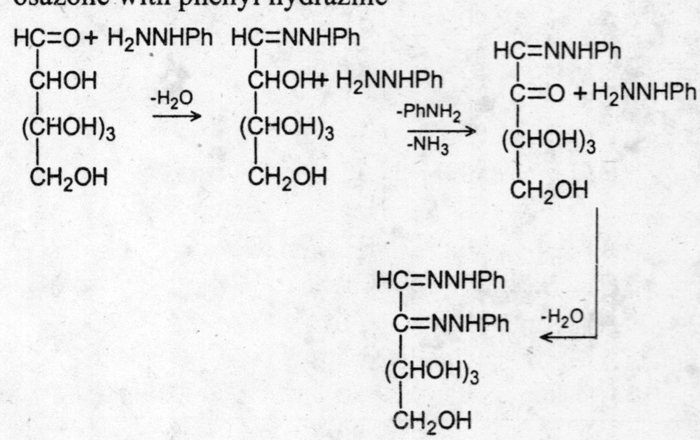

Fischer and Amadori suggested different mechanisms for the formation of osazone. In osazone formation only C_1 and C_2 positions take part in the reaction. It means that the said sugars differ in structure only at these positions.

Osazone reaction is in fact a general reaction given by α–hydroxy carbonyl compounds which may or may not be carbohydrates. Acetoin, benzoin also give this reaction.

4. Lobry de Bruyn van Ekenstein rearrangement: In the presence of dilute aqueous alkalies glucose / fructose / mannose undergoes this rearrangement and all the three are formed.

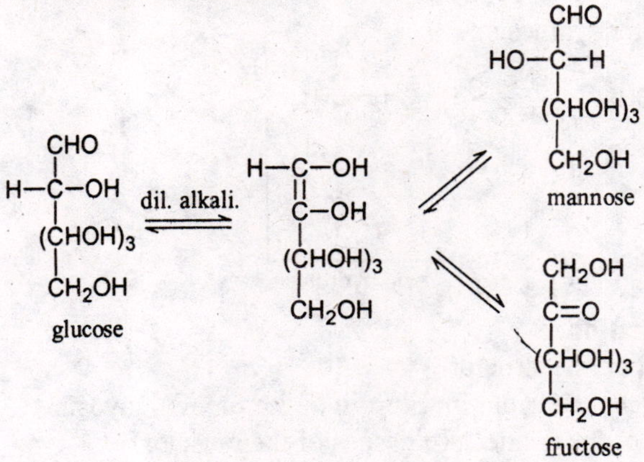

Fructose, although contains a keto group, is a reducing sugar because of this rearrangement in which glucose and mannose which contain aldehyde group are formed.

5. Fermentation:

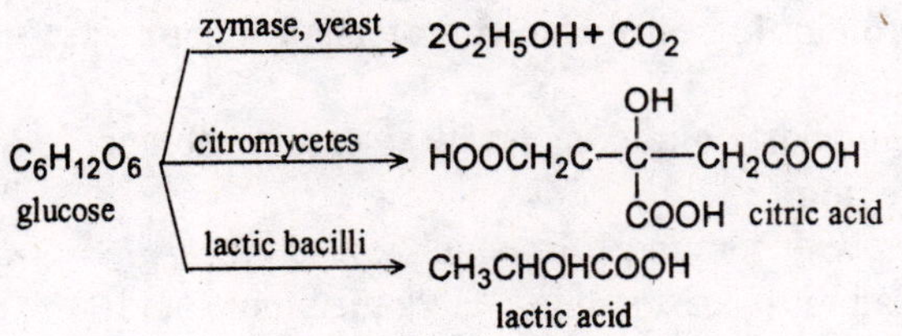

6. Reducing nature:

(i) Glucose reduces Fehling's solution to Cu_2O (reddish brown precipitate) + gluconic acid.

(ii) Glucose reduces Tollen's reagent to Ag (Silver mirror) + gluconic acid.

(iii) Glucose reacts with Benedict's solution ($CuSO_4$ + Na_2CO_3 + Sodium citrate) to give Cu_2O (reddish brown precipitate) + gluconic acid + water

(iv) Glucose reacts with Barfoed's reagent (Cupric acetate in CH_3COOH) to give Cu_2O precipitate.

Fructose also shows the reducing nature

7. Mutarotation: Glucose does not react with $NaHSO_3$ and NH_3 (reaction shown by free aldehyde group containing compounds) indicating that -CHO group is not in free state as it takes part in internal hemiacetal formation. This leads to the following two diastereomers:

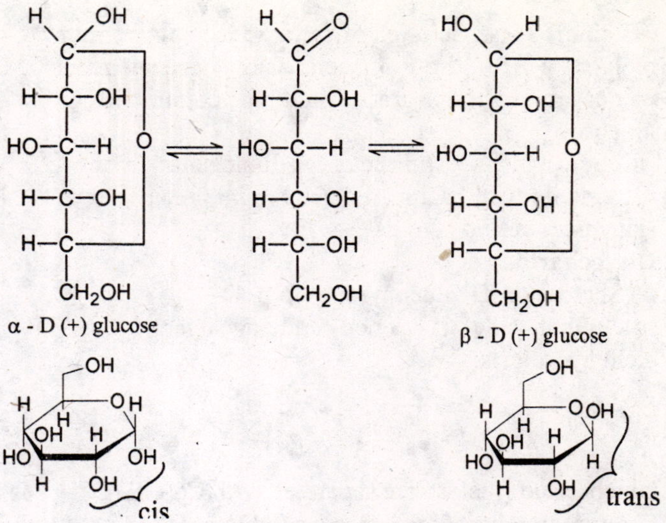

The rotation of α–D (+) glucose (freshly prepared) falls from +112° to +52.7°. The rotation of β–D (+) glucose (freshly prepared) raises from +19° to +52.7°. This phenomenon shows that the two forms show mutarotation. Equilibrium mixture contains 62% of β –form and 38% of α-form. These two forms are isomers, optically active but not mirror images. Therefore, they are diastereomers differing in configuration at C_1 position (Anomers). Glucose and fructose are functional isomers. Glucose and Mannose differ at C_2 configuration (Epimers).

8. Periodic acid oxidation:

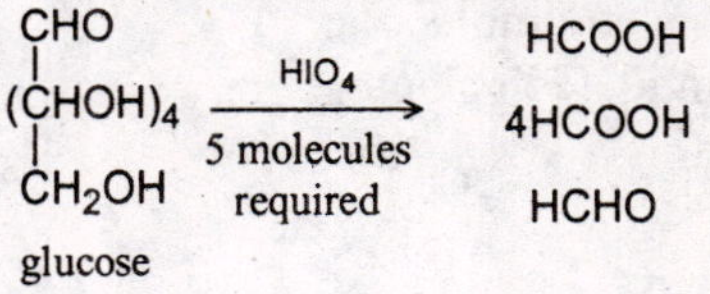

Fructose: It is also known as laevulose as it is laevorotatory. It is the sweetest of the sugars. Non sugars like Monallein, Saccharin etc. are more sweeter than sugars.

Preparation: It is commercially prepared from Inulin.

$$(C_6H_{10}O_5)_n \cdot nH_2O \xrightarrow[\Delta]{\text{dil } H_2SO_4} n\,C_6H_{12}O_6 \text{ (fructose)}$$

Reactions: Strong oxidising agents cleave the molecule at >C=O group. As fructose can be oxidised easily, it acts as a powerful reducing agent.

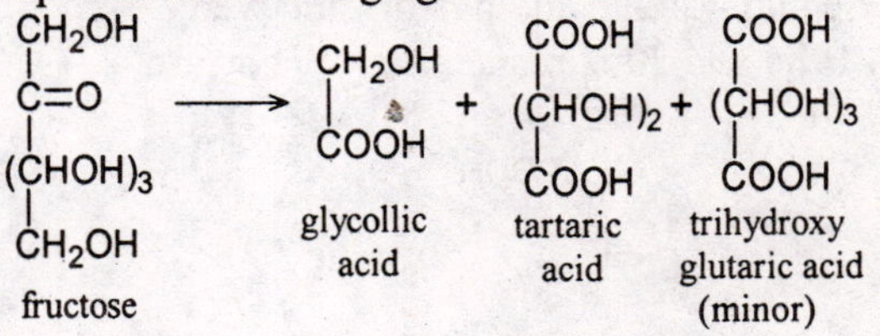

Fructose is distinguished from glucose by

(i) Furfural test: dil. solution of fructose + 1% alcoholic solution of α–naphthol + few drops of HCl $\xrightarrow{\text{boil}}$ violet colour (glucose gives no colour)

(ii) Pinoff's test: Solution of fructose with ammonium molybdate + few drops of acetic acid greenish blue colour turns to bluish green (light blue colour with glucose).

(iii) Selivanoff's test: Fructose + dilute solution of resorcinol in dil. HCl reddish brown precipitate (glucose gives no colour).

Disaccharides:

(i) Sucrose (α–D–glucopyranosyl–β–D–fructofuranoside). It is a non-reducing sugar.

Acid hydrolysis:

$$\underset{}{C_{12}H_{22}O_{11}} + H_2O \xrightarrow{H^+} \underset{\text{glucose}}{C_6H_{12}O_6} + \underset{\text{fructose}}{C_6H_{12}O_6}$$

Glucose and fructose are separated with $Ca(OH)_2$. Calcium fructosate formed is insoluble in H_2O whereas calcium glucosate is soluble.

The hydrolysis is known as inversion of cane sugar because of the formation of laevorotatory mixture from dextrorotatory sucrose.

(ii) Maltose: 4–O–(α–glucopyranosyl)D–glucopyranose. It is a reducing sugar.

$$\underset{\text{maltose}}{C_{12}H_{22}O_{11}} + H_2O \xrightarrow[\text{dil. HCl}]{\text{maltase or}} \underset{\text{D(+) glucose}}{2C_6H_{12}O_6}$$

(iii) Lactose: It is known as milk sugar. It is a reducing sugar. On hydrolysis equimolecular quantities of D(+) glucose and D(+) galactose are obtained. Human milk contains 6–8% lactose (cow milk 4%).

Proteins

Protein like

(a) Keratin serves as structural material in hair, nails etc.

(b) Haemoglobin (in blood) transport oxygen within the body.

(c) Insulin (in pancreas), a hormone regulates metabolic processes.

(d) Enzymes catalyse biological reactions.

(e) Myoglobin is oxygen storing protein of muscle.

Proteins are polymeric amides derived from monomers, α-amino acids. There are ~ 22 natural amino acids. Selenocysteine and pyrrolysine were discovered in 1986 and 2002 respectively. They are all α-amino acids.

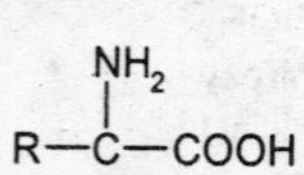

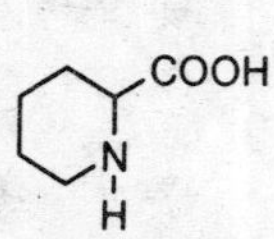

proline

Proline has secondary amino group while all others have primary amino group. They have S-configuration about α-carbon atom.

All amino acids except glycine H_2N-CH_2-COOH contain an asymmetric carbon atom.

Neurotransmitters arise from L-Tyrosine by structural modification and decarboxylation.

Essential amino acids: These are not synthesised by the body and therefore are to be supplemented in the food. They are Arginine, Histidine, Isoleucine, Leucine, Lysine, Methionine, Phenylalanine, Threonine, Tryptophane and Valine.

Preparation

1. Extension of Hell-Volhard-Zelinsky reaction:

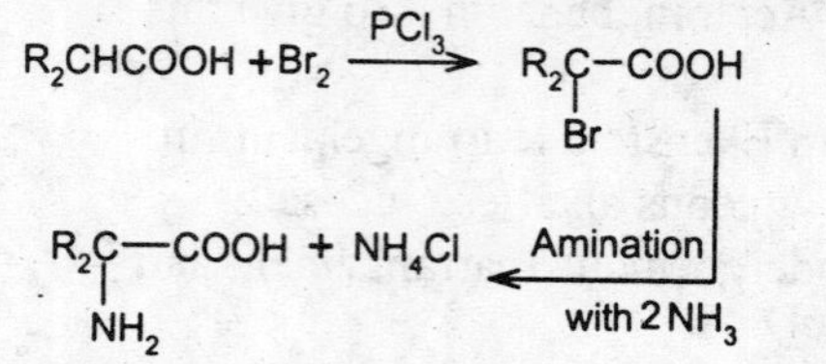

2. Strecker synthesis:

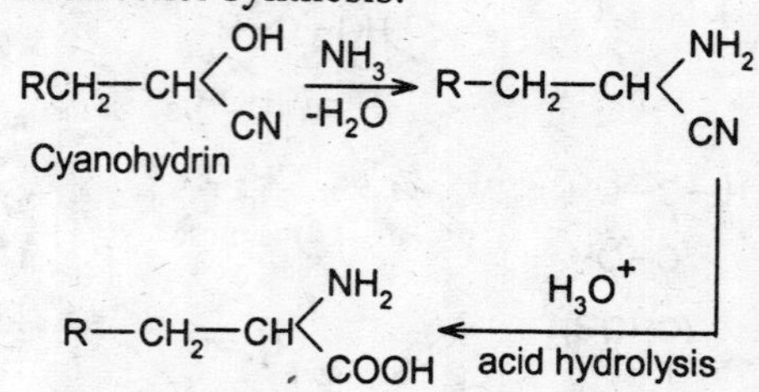

Classification:

(i) based on the structure

(a) Fibrons: e.g. Keratin; Fibroin (silkproteins), myosin (protein of the connective tissues of the muscles)

(b) Globular: e.g. Insulin; Albumin; Haemoglobin; Thyroglobulin; ACTH (Adino carticotropic hormone)

(ii) Based on the groups present

(a) Simple proteins (build up from only amino acids) e.g. Albumins; Globulins, Glutelin etc.

(b) Conjugated proteins (build up from peptide and a non peptide portion (prosthetic group) e.g. Glycoproteins; Phosphoproteins (casein); Nucleoproteins; Lipoproteins etc.

- Lecithin, a phospholipid is present in cell membrane.

Isoelectric point :

It is a point with certain pH at which ionisation is minimum and it is least soluble. Here the amino acid bears no net charge and the concentration of Zwitter ion is maximum.

Identification: Amino acids are identified by ninhydrin test. Proline gives yellow colour while all others give bluish violet colour.

Pyridine solution of protein + Ninhydrin → blusih violet

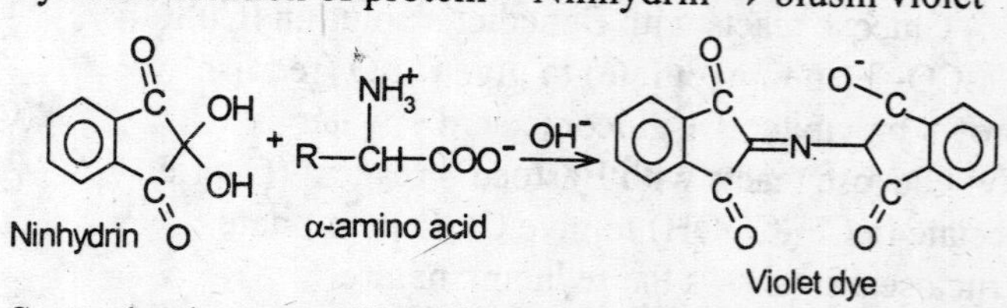

Several other tests like biuret test (given by all types of compounds having —CONH—group), Xanthoproteic test, Millons test (proteins containing phenolic-OH group), nitroprusside test (proteins containing -SH group), Hopkins- Cole test (for Tryptophan) are also used.

Structure of proteins:

For identification of N-terminal residue analysis Sanger's reagent

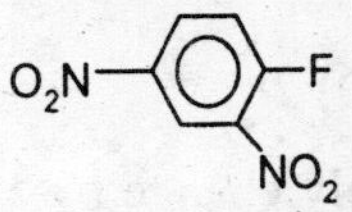

DNFB.

(2, 4-dinitrofluorobenzene) or phenyl isothio-cyanate (Ph–N=C=S) in Edman's method are used. For C–terminal residue analysis carboxy-peptidase (ex pancreas) is used.

Primary structure: Number and sequence of amino acids in the polypeptide chain is known in primary structure. (The amino acid responsible for defective haemoglobin responsible for sickle cell anaemia can be known).

Secondary structure: It deals with the arrangement of the polypeptide chain in space to form coils (α-helix and rarely β-helix), sheet like structures etc. Helical structure is mainly studied by X-ray analysis. Hydrogen bonds are present (Haemoglobin consists of α-chains (141 amino acids) and β-chains (146 amino acids).

Tertiary structure: Tells about the precise folding and bending of α-helical coil in three dimensional structure. Hydrogen, ionic, inter and intramolecular bonds are involved.

Quaternary structure: Association of various identical or different sub-units are studied. This helps in the determination of haemoglobin structure.

Enzymes

Enzymes are globular proteins produced by living cells which catalyse specific organic reactions (rarely inorganic also). They are known as biocatalysts. They are of hydrolytic and oxidative (generally conjugated proteins) nature mainly.

e.g.

(i) Starch $\xrightarrow{\text{Amylase}}$ Maltose $\xrightarrow{\text{Maltase}}$ Glucose

(ii) Sucrose $\xrightarrow{\text{Invertase}}$ Glucose + Fructose

(iii) Glucose/or Fructose $\xrightarrow{\text{Zymase}}$ C_2H_5OH

(iv) Proteins $\xrightarrow{\text{Pepsin or Trypsin}}$ Amino acids

(v) Glycerides $\xrightarrow{\text{Lipase}}$ Glycerol + Higher fatty acids
(Fixed oil or Fats)

Soyabean contains- Urease; Barley, human saliva- α-amylase; sweet potato-β-amylase.

Maltase - Hydrolyses α-glycosides

Emulsin - Hydrolyses β-glycosides

Urease - It is used for the estimation of urea.

Nucleic acids: Nucleic acids control heredity on molecular level. Nucleic acids are called polynucleotides. The nucleotide unit is composed of a base (heterocyclic nitrogenous) a sugar (Ribose or Deoxyribose) and an inorganic phosphate group. The nucleotide units are connected by covalent bonds.

The bases present in nucleic acids are purines:

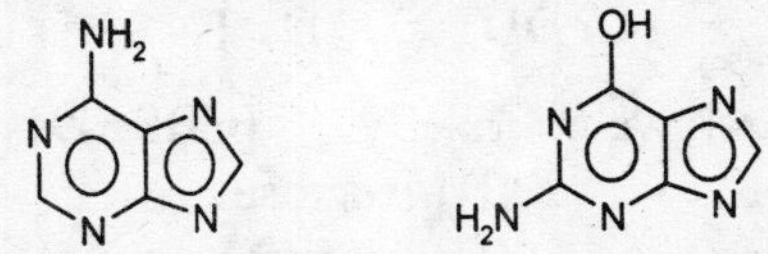

Adenine Gaunine

and pyrimidines:

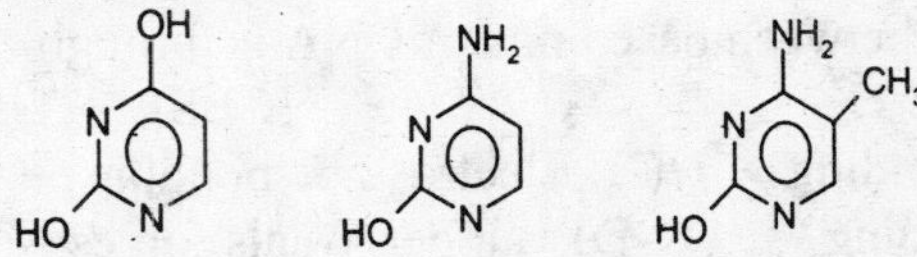

Uracil Cytosine Thymine

Sugars:

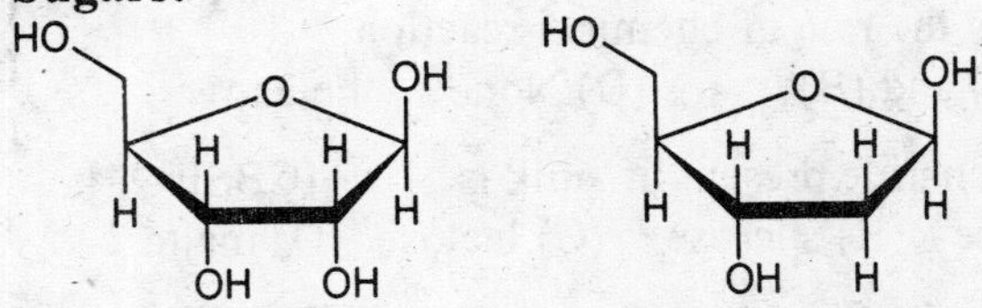

D-Ribose (β-D-ribofuranose) D-2-deoxyribose (β-2-deoxy-D-ribofuranose)

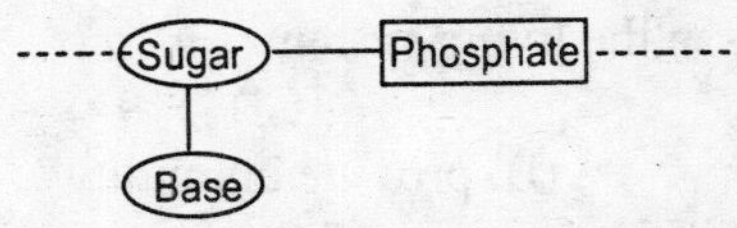

Schematic representation of a nucleotide

A nucleoside is made up of a base unit and a sugar moiety.

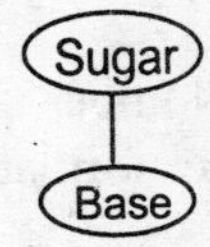

DNA:

It has unique property of replication which means the building up of another identical molecule. Due to self replication, it is responsible for heredity traits. The law of complimentarity of DNA was proposed by Watson and Crick. It is genetic material and double stranded. It controls the protein synthesis. The D-2-deoxy ribose is present as sugar unit. The bases found in DNA are Adenine, Guanine, Thymine and Cytosine.

RNA:

Ribose is present as sugar unit. Three kinds of RNA namely: Ribosomal RNA (r RNA)- 80% of RNA, transfer RNA (amino acid transport) - 15% of total RNA and messenger RNA (in ribosome & cytoplasm) are known.

- Codon is a coding unit.

1. Hereditary traits are maintained by [CBSE-90]
(A) RNA (B) DNA (C) DDT (D) TNT

2. On hydrolysis of starch we get finally [CBSE-91]
(A) glucose (B) fructose
(C) glucose and fructose (D) sucrose

3. Insulin, a hormone is chemically a [CBSE-91]
(A) fat (B) steroid (C) protein (D) carbohydrate

4. The couplings between base units of DNA is through [CBSE-92]
(A) Hydrogen bonding (B) Electrostatic bonding
(C) Covalent bonding (D) van der Waals forces

5. Enzyme takes part in reaction and [CBSE-93]
(A) decreases in rate of chemical reaction
(B) increases the rate of chemical reaction
(C) Both (A) and (B) (D) None of the above

6. The disaccharide present in milk is [CBSE-94]
(A) cellobiose (B) sucrose (C) lactose (D) maltose

7. The peptide bond is a key feature in [CBSE-94]
(A) polysaccharides (B) proteins
(C) nucleotides (D) vitamins

8. The function of enzymes in the living system is to [CBSE-94]
(A) transport oxygen (B) provide immunity
(C) catalyse biochemical reactions
(D) provide energy.

9. α-D Glucose and β-D Glucose differ from each other due to difference in one of carbon with respect to its [CBSE-95]
(A) Size of hemiacetal ring. (B) Number of OH groups.
(C) Configuration (D) Conformation.

10. Secondary structure of a protein refers to [CBSE-95]
(A) mainly denatured proteins and structures of prosthetic groups
(B) three-dimensional structure, specially the bond between amino acid residues that are distinct from each other in the polypeptide chain
(C) linear sequence of amino acid residues in the polypeptide chain
(D) regular folding patterns of continuous portions of the polypeptide chain

11. Of the following statements about enzymes which ones are true? [CBSE-95]
(i) Enzymes lack in nucleophilic groups
(ii) Enzymes are highly specific both in binding chiral substrates and in catalyzing their reactions
(iii) Enzymes catalyze chemical reactions by lowering the activation energy
(iv) Pepsin is a proteolytic enzyme
(A) (i) and (iv) (B) (i) and (iii)
(C) (ii), (iii) and (iv) (D) (i)

12. Oxidation of glucose is one of the most important reactions in a living cell. What is the number of ATP molecules generated in cells from one molecule of glucose? [CBSE-95]
(A) 38 (B) 12 (C) 18 (D) 28

13. The pH of blood does not appreciably change by a small addition of an acid or a base because blood [CBSE-95]
(A) contains serum protein which acts as buffer
(B) contains iron as a part of the molecule.
(C) can be easily coagulated. (D) it is body fluid.

14. The function of enzymes in the living system is to [CBSE-97]
(A) transport oxygen (B) provide immunity
(C) catalyse biochemical reactions
(D) provide energy

15. Reaction of O_2 and CO with Haemoglobin gives [CBSE-97]
(A) Only oxygen-haeme complex
(B) only CO-haeme complex
(C) both but oxygen-haeme is more stable
(D) both but CO-haeme complex is more stable

16. In DNA the complementary bases are [CBSE-98]
(A) uracil and adenine: cytosine and guanine
(B) adenine and thymine: guanine
(C) adenine and thymine: guanine and uracil
(D) adenine and guanine: thymine and cytosine

17. The number of molecule of ATP produced in the lipid metabolism of a molecule of palmitic acid is [CBSE-98]
(A) 86 (B) 130 (C) 56 (D)36

18. Glucose molecule reacts with X number of molecules of phenylhydrazine to yield osazone. The value of 'X' is [CBSE-98]
(A) three (B) two (C) one (D) four

19. Which of the following is the sweetest sugar? [CBSE-99]
(A) Sucrose (B) Glucose (C) Fructose (D) Maltose

20. Mg is an important component of which bimolecule occuring extensively in living world ? [CBSE-2000]
(A) Haemoglobin (B) Chlorophyll
(C) Florigen (D) ATP

21. Which of the following is correct about enzymes [CBSE-2000]
(A) It changes the equilibrium constant.
(B) It increase the activation energy.
(C) It reduces the activation energy. (D) None of these

22. Which of the following gives positive Fehling's solution test ? [CBSE-2001]
(A) Surcose (B) Glucose (C) Fats (D) Proteins

23. Which of the following is correct ? [CBSE-2001]
(A) Cycloheptane is an aromatic compound
(B) diastase is an enzyme
(C) Acetophenone is an ether (D) All the above

24. Which of the following is correct about H-bonding in nucleotide ? [CBSE-2001]
(A) A - T , G - C (B) A - G, T - C
(C) G - T, A - C (D) A - A, T - T

25. Which is correct statement ? [CBSE 2001]
(A) Starch is a polymer of α-glucose
(B) Amylose is a component of cellulose
(C) Proteins are composed of only one type of amino acid
(D) In cyclic structure of fructose there are 4C and one oxygen atom

26. Enzymes are made up of [CBSE 2002]
(A) Edible proteins
(B) Proteins with specific structure
(C) Nitrogen containing carbohydrates
(D) Carbohydrates

27. Which is not true statement [CBSE-2002]
(A) α-carbon of α-amino acid is asymmetric
(B) All proteins are found in L-form
(C) Human body can synthesize all proteins they need.
(D) At pH = 7 both amino and carboxylic groups exist in ionised form.

28. Vitamin B_{12} contains [CBSE-2003]
(A) Ca (II) (B) Fe (II) (C) Co (III) (D) Zn (II)

29. Glycolysis is [CBSE-2003]
(A) conversion of glucose to haeme.
(B) oxidation of glucose to glutamate
(C) conversion of pyruvate to citrate
(D) oxidation of glucose to pyruvate

30. Chargaff's rule states that in an organism [CBSE-2003]
(A) amounts of all bases are equal
(B) amount of adenine (A) is equal to that of thymine(T) and the amount of guanine (G) is equal to that of cytosine(C)
(C) amount of adenine (A) is equal to that of guanine (G) and the amount of thymine(T) is equal to that of cytosine (C)
(D) amount of adenine (A) is equal to that of cytosine (C) and the amount of thymine (T) is equal to that of guanine (G)

31. Skin turns yellow on contact with concentrated nitric acid because [AIIMS-93]
(A) Proteins are converted into xanthoproteins
(B) Water is removed from the skin
(C) Skin is burnt (D) Nitrocellulose is formed

32. The compound of formula $NH_2 - CH_2 - COOH$ may behave [AIIMS-93]
(A) Only as acid (B) As both acid and base
(C) Only as a base (D) As neither acid nor base

33. The functional group which is found in amino acid is [AIIMS-98]
(A) –COOH group (B) $-NH_2$ group
(C) $-CH_3$ group (D) Both (a) and (b)

34. The common molecular formula for disaccharide is [AIIMS-99]
(A) $C_{10}H_{18}O_8$ (B) $C_{10}H_{20}O_{10}$
(C) $C_{12}H_{22}O_{11}$ (D) $C_{18}H_{22}O_{11}$

35. Yeast cells derive their energy from glucose by [AIIMS-2001]
(A) Glycolysis (B) Respiration
(C) Fermentation (D) None of these

36. Which α amino acid can form cross-linked peptide chains ? [AIIMS-2001]
(A) Serine (B) Cysteine (C) Glutamine (D) Tyrosine

37. Which of the following proteins destroys the antigen when it enters in body cell ? [AIIMS-2001]
(A) Antibodies (B) Insulin
(C) Chromoprotein (D) Phosphoprotein

38. Enzymes with two binding sites are called[AIIMS-2002]
(A) Apoenzyme (B) Holoenzyme
(C) Allosteric enzyme (D) Conjugate enzyme

39. Assertion (A) : DNA molecules and RNA molecules are found in the nucleus of a cell. Reason (R) : On heating, the enzymes do not lose their specific activity. [AIIMS-2002]
(A) A (B) B (C) C (D) D (E) E

40. Subunits present in haemoglobin are- [AIIMS-2003]
(A) 2 (B) 3 (C) 4 (D)5

41. (A) Glycosides are hydrolyzed in acidic conditions.
(R) Glycosides are acetals. [AIIMS-2003]
(A) (B) (C) (D)

42. An enzyme which brings about the conversion of starch into maltose is : [JIPMER-90]
(A) Maltase (B) Zymase
(C) Invertase (D) Diastase

43. On hydrolysis of starch we get finally : [JIPMER-90]
(A) Glucose (B) Fructose
(C) Glucose and fructose (D) Sucrose

44. Which of the following base is found only in RNA and not in DNA? [JIPMER-93]
(A) Thymine (B) Guanine (C) Adenine (D) Uracil

45. The Telluric helix was given by [JIPMER-94]
(A) De Chancourtois (B) Newlands
(C) L.Meyer (D) Mendeleev

46. Which test is used for detecting the presence of sugar in urine ? [JIPMER-94]

(A) Biuret test (B) Lassaigne's test
(C) Duma's teast (D) Molisch test

47. Which of the following is a molecular disease ? [JIPMER-94]
(A) Cancer (B) Allergy
(C) Sickle cell anemia (D) German measles

48. The catalyst used in the hydrogenation of oils is [JIPMER-94]
(A) Fe (B) Ni (C) Mo (D) Pd

49. Glycine may be classified as all of the following except [JIPMER-95]
(A) An acid (B) A base
(C) A Zwitter ion (D) Optically active acid

50. The compound of formula NH_2-CH_2-COOH may behave [JIPMER-99]
(A) Only as an acid (B) As both acid and base
(C) Only as base (D) As neither acid nor base

51. The change in optical rotation with time of freshly prepared solutions of sugar is known as [JIPMER-2000]
(A) Maturation (B) Rotatory
(C) Inversion (D) Specific rotation

52. Maltose on hydrolysis gives [BHU-96]
(A) Glucose (B) Galactose + glucose
(C) Mannose + glucose (D) Mannose + fructose

53. The change in optical rotation (with time) of freshy prepared solution of sugar is known as [BHU-97]
(A) Inversion (B) Specific rotation
(C) Rotatory motion (D) Mutarotation

54. Sweetest sugar is [BHU-97]
(A) Sucrose (B) Glucose (C) Fructose (D) Maltose

55. "Reference proteins" are those obtained from [BHU-98]
(A) Meat (B) Milk (C) Egg (D) Soyabean

56. Starch is converted into maltose by [BHU-99]
(A) Maltase (B) Invertase (C) Zymase (D) Diastase

57. Which common amino acid contain sulphur ? [BHU-2002]
(A) Cysteine (B) Threonine
(C) Valine (D) Polypropylene

58. The recent discovery of cytology is that of [AFMC-91]
(A) Enzymes (B) Co-enzymes
(C) Genetic code (D) Respiration

59. Starch is converted into maltose by [AFMC-94]
(A) Maltase (B) Invertase (C) Zymase (D) Diastase

60. $\underset{\text{Sucrose}}{C_{12}O_{12}O_{11}} + H_2O \xrightarrow{\text{dil. } H_2SO_4} \underset{\text{Glucose}}{C_6H_{12}O_6} + \underset{\text{Fructose}}{C_{12}H_{12}O_6}$

In this reaction dil. H_2SO_4 is called [AFMC-97]
(A) Homogeneous catalysis (B) Homogeneous catalyst
(C) Heterogeneous catalysis (D) Heterogeneous catalyst

61. Which of the following compounds does not belong to lipids ? [AFMC-98]
(A) Fats (B) Amino acids
(C) Phospholipids (D) Carbohydrates

62. Night-blindness is caused due to the deficiency of [AFMC-98]
(A) Vitamin B_{12} (B) Vitamin A
(C) Vitamin C (D) Vitamin E

63. Galactose is converted into glucose in [AFMC-98]
(A) Mouth (B) Stomach (C) Liver (D) An intestine

64. Aspirin is obtained by reaction of salicylic acid with [AFMC-98]
(A) Acetone (B) Acetaldehyde
(C) Acetyl chloride (D) Acetic anhydride

65. The structure of DNA is [AFMC-99]
(A) Single helix (B) Double helix
(C) Triple helix (D) Linear

66. The number of essential amino acid units in man is [AFMC-2000]
(A) 8 (B) 10 (C) 18 (D) 20

67. An invert sugar is [AFMC-2000]
(A) Isorotatory (B) Dextrorotatory
(C) Laevorotatory (D) Optically inactive

68. Which of the following is not true about vitamins ? [AFMC-2001]
(A) They were named by "Funk"
(B) They are vital for life
(C) Their deficiency causes diseases
(D) They help in digestion

69. Cellulose is [AFMC-2001]
(A) Disaccharide (B) Structural Polysaccharide
(C) Structural Oligosaceharide
(D) Structural Monosaccharide

70. Bonding in DNA double helix is [AFMC-2001]
(A) Covalent (B) Ionic
(C) van der Waal's (D) Hydrogen

71. An "antibiotic" with a broad spectrum is one [AFMC-2001]
(A) That acts on a specific antigen
(B) That acts on both the antigens and antibodies
(C) That acts on different antigens
(D) That kills the antibodies

72. When conc. H_2SO_4 is brought in contact with sugar, it becomes black due to [AFMC-2001]
(A) Hydrolysis (B) Hydration
(C) Decolourisation (D) Dehydration

73. Which of the following is an aldohexose ? [AFMC-2002]

(A) Raffinose (B) Cellulose (C) Glucose (D) Sucrose

74. Amino acids are the building blocks of [AFMC-2002]

(A) Carbohydrates (B) Vitamins (C) Fats (D) Proteins

75. The base present in DNA but not in RNA is [AFMC-2002]

(A) Adenine (B) Thymine (C) Uracil (D) Guanine

76. RNA is different from DNA because RNA contains [AIEEE-2002]

(A) Ribose sugar and thymine
(B) Ribose sugar and uracil
(C) Deoxyribose sugar and thymine
(D) Deoxyribose sugar and uracil.

77. The functional group which is found is amino acid is [AIEEE-2002]

(A) –COOH group (B) $-NH_2$ group
(C) $-CH_3$ group (D) Both (a) and (b)

78. Complete hydrolysis of cellulose gives [AIEEE-2003]

(A) D-fructose (B) D-ribose
(C) D-glucose (D) L-glucose

79. The reason for double helical structure of DNA is operation of [AIEEE-2003]

(A) van der Waal's force (B) Dipole-dipole interaction
(C) Hydrogen bonding (D) Electrostatic attractions

80. The metal ion present in vitamin B_{12} is [EAMCET-2001]

(A) Co^{3+} (B) Co^{2+} (C) Fe^{2+} (D) Fe^{3+}

81. Which of the following is the sweetest sugar ? [CPMT-96]

(A) Sucrose (B) Fructose (C) Glucose (D) Lactose

82. Which of the following does not show any reducing test of aldehyde ? [CPMT-96]

(A) Sucrose (B) Fructose (C) Ketose (D) Lactose

83. Benzene is obtained from [CPMT-96]

(A) Coaltar (B) Plant (C) Animal (D) Charcoal

84. When HNO_3 is dropped into the palm and washed with water it turns into yellow. It shows the presence of [CPMT-97]

(A) NO_2 (B) N_2O (C) NO (D) N_2O_5

85. Glucose + Tollen's regent → Silver mirror, shows [CPMT-97]

(A) Presence of acidic group
(B) Presence of alkaline group
(C) Presence of ketonic group
(D) Presence of aldehyde group

86. Amino acids are the building blocks of [CPMT-99]

(A) Carbohydrates (B) Vitamins
(C) Fats (D) Proteins

87. Which of the following reagents cannot distinguish between glucose and fructose ? [CPMT-2000]

(A) Fehling's solution (B) Tollen's reagent
(C) Benedict's solution (D) All of these

88. Which carbohydrate is essential constituent of plant cell ? [CPMT-2002]

(A) starch (B) glucose (C) cellulose (D) sucrose

89. Which amino acid has aromatic ring ? [CPMT-2003]

(A) Alanine (B) Glycine (C) Tyrosine (D) Lysine

90. Haemoglobin is complex of [CPMT-2003]

(A) Fe^{3+} (B) Fe^{2+} (C) Fe^{4+} (D) Cu^{2+}

91. Enzymes are : [PMT-93]

(A) Vitamins (B) Proteins
(C) Carbohydrates (D) Lipids

92. Sucrose on hydrolysis gives : [PMT-93]

(A) Two molecules of Glucose
(B) Two molecules of Fructose
(C) One molecule each of Glucose and Fructose
(D) One molecule each of Glucose and Mannose

93. Which one of the following proteins transports oxygen in the blood stream ? [PMT-93]

(A) Myoglobin (B) Insulin
(C) Albumin (D) Haemoglobin

94. Which one of the following is the reagent used to identify glucose : [PMT-93]

(A) Neutral ferric chloride
(B) Chloroform and alcoholic KOH
(C) Ammoniacal silver nitrate (D) Sodium ethoxide

95. Glucose gives a silver mirror with Tollen's reagent. This shows the presence of [PMT-94]

(A) Aldehyde group (B) Ketonic group
(C) Alcoholic group (D) Acid group

96. RNA does not contain [PMT-94]

(A) Uracil (B) Ribose (C) Thymine (D) Phosphate

97. Hydrolysis of proteins in the presence of enzymes produces [PMT-94]

(A) Hydroxy acids (B) Dicarboxylic acids
(C) Amino acids (D) Aromatic acids

98. Which of the following is responsible for maintaining blood sugar level in human body ? [PMT-95]

(A) Riboflavin (B) Insulin (C) Fats (D) Hormone

99. Proteins are composed of [PMT-95]

(A) α – amino acids (B) carbohydrates
(C) vitamins (D) mineral salts

100. The base adenine occurs in [PMT-95]

(A) DNA only (B) RNA only
(C) DNA and RNA both (D) Protein

101. Conversion of glucose into ethyl alcohol is made by [PMT-95]
(A) Acid (B) Enzyme
(C) Hydroxylamine (D) Phenyl hydrazine

102. Enzymes are [PMT-96]
(A) Proteins (B) Mineral (C) Oils (D) Fatty acid

103. The number of atoms in the cyclic structure of D-fructose is - [PMT-97]
(A) 5 (B) 6 (C) 4 (D) 7

104. The secondary structure of protiens is due to - [PMT-97]
(A) peptide bond (B) Hydrogen bond
(C) van der Waal's forces (D) Ionic bond

105. Which of the following is the sweetest sugar - [PMT-97]
(A) Fructose (B) Glucose (C) Maltose (D) Lactose

106. Carbohydrates are stored in human body as : [PMT-99]
(A) glucose (B) glycogen (C) starch (D) fructose

107. Which of the following is **not** present in nucleic acids ? [PMT-99]
(A) Uracil (B) 2-Aminopyridine
(C) Thymine (D) Adenine

108. DNA contains the sugar : [PMT-2000]
(A) Deoxyribose (B) Ribose
(C) D-fructose (D) D-glucose

109. The enzyme pepsin hydrolyses : [PMT-2001]
(A) Fats to fatty acid (B) Glucose to ethyl alcohol
(C) Polysaccharides to monosaccharides
(D) Proteins to amino acids

110. The helical structure of proteins is stabilized by : [PMT-2001]
(A) Peptide bonds (B) Dipeptide bond
(C) Hydrogen bond (D) van der Waal's forces

111. Mutation of DNA occurs due to changes in the sequence of one of the following : [PMT-2001]
(A) Bases (B) Ribose units
(C) Phosphate (D) Sugar units

112. The optically inactive amino acid is : [PMT-2001]
(A) Lysine (B) Glycine (C) Arginine (D) Alanine

113. Proteins contain mainly [PMT-2002]
(A) C, H and O (B) C, H, O and N
(C) C, H and N (D) N, H and O

114. Proteins are hydrolysed by enzymes to give [PMT-2002]
(A) Dihydroxy acids (B) Amino acids
(C) Amines (D) Carboxylic acids

115. The most widely used and safe sugar substitute now a days is : [PMT-2003]
(A) Cyclodextrin (B) Aspartame
(C) Saccharin (D) Fructose

116. Which of the following reaction is catalysed by the enzyme maltase ? [PMT-2003]
(A) Starch→Maltose (B) Maltose→Glucose
(C) Lactose→Maltose (D) Maltose→Glucose + Fructose

117. The main structural feature of proteins is [PET-93]
(A) Ester linkage (B) Ether linkage
(C) Peptide linkage (D) Ketone group

118. On complete hydrolysis of starch we finally get [PET-93]
(A) Glucose (B) Fructose
(C) Glucose and fructose (D) Sucrose

119. Enzymes are [PET-93]
(A) Living organisms (B) Dead organisms
(C) Complex nitrogenous substances produced in living cells (D) None of these

120. The most important food reserves of animals and plants are [PET-93]
(A) Carbohydrates (B) Proteins
(C) Vitamins (D) Fats

121. For α-amino acids having the structure

$$R-\underset{\underset{NH_2}{|}}{CH}-CO_2H$$

Which of the following statements are true ? [PET-94]
(a) Water solubility is maximum at a pH when concentrations of anions and cations are equal
(b) They give ninhydrin test
(c) On reacting with nitrous acid give off N_2
(A) All (B) b and c (C) a and b (D) a

122. The linkage present in proteins and peptides is [PET-94]
(A) $-\overset{\overset{O}{\|}}{C}-O-\overset{\overset{O}{\|}}{C}-$ (B) $-\overset{\overset{O}{\|}}{C}-O-$
(C) $-NH-$ (D) $-\overset{\overset{O}{\|}}{C}-NH-$

123. Which carbohydrates has highest abundance in human blood [PET-95]
(A) d-fructose (B) d-glucose
(C) Sucrose (D) Lactose

124. Which one is not present in R.N.A.- [PET-95]
(A) Uracil (B) Thymine (C) Ribose (D) Phosphate

125. The acid showing salt-like character in aqueous so-

lution is : [PET-98]
(A) Acetic acid (B) Benzoic acid
(C) Formic acid (D) α –Amino acetic acid

126. Which substance is not present in nucleic acid [PET-98]
(A) Cytosine (B) Adenine (C) Thymine (D) Guanidine

127. The commonest disaccharide has the molecular formula : [PET-99]
(A) $C_{12}H_{22}O_{11}$ (B) $C_{10}H_{18}O_9$
(C) $C_{10}H_{20}O_{10}$ (D) $C_{18}H_{32}O_{11}$

128. Which carbohydrate is an essential constituent of plant cells ? [PET-99]
(A) Starch (B) Cellulose (C) Sucrose (D) Vitamins

129. The enzyme pepsin hydrolyses : [PET-99]
(A) Proteins to amino acids
(B) Fats to fatty acids (C) Glucose to ethyl alcohol
(D) Polysaccharides to monosaccharides

130. Which of the following gives maximum energy in metabolic processes ? [PET-99]
(A) Proteins (B) Vitamins
(C) Lipids (D) Carbohydrates

131. Enzymes in the living systems : [PET-99]
(A) Provide energy (B) Provide immunity
(C) Transport oxygen (D) Catalyse biological processes

132. The best source of vitamin - A is : [PET-2000]
(A) Oranges (B) Beans (C) Carrots (D) Wheat

133. The most important energy carrier in all the living cell is : [PET-2000]
(A) AMP (B) ATP (C) ADP (D) UDP

134. Which one gives positive silver mirror test ? [PET-2001]
(A) Sucrose (B) Glucose (C) Starch (D) Fructose

135. Which of the following statement about proteins is not true ? [PET-2001]
(A) Amino acid residues join together to make a protein molecule
(B) Proteins are polymers with formula $(C_6H_{10}O_5)n$
(C) Eggs are rich in protein
(D) Pulses are good sources of protein

136. The reaction of caustic soda with oils and fats is known as: [PET-2001]
(A) Oxidation (B) Esterification
(C) Neutralisation (D) Saponification

137. The disaccharide present in milk is : [PET-2001]
(A) Maltose (B) Lactose (C) Sucrose (D) Cellobiose

138. Iodine value is related to [PET-2002]
(A) Fats and oils (B) Alcohols
(C) Esters (D) Hydrocarbons

139. Deficiency of which vitamin causes rickets [PET-2002]
(A) Vitamin D (B) Vitamin B
(C) Vitamin A (D) Vitamin K

140. Which of the follwing statements is false in the context of enzyme ? [PET-2003]
(A) Temperature affects their functions
(B) pH affects their functions
(C) They always increase activation energy
(D) Their reactions are specific

141. The example of a protein is [PET-2003]
(A) Insulin (B) Narvone (C) Cephalin (D) Lecithin

142. Oils and fats together are called [PET-2003]
(A) Enzymes (B) Soaps (C) Polymer (D) Lipids

143. Sodium potassium salts of fatty acids are called: [PET- 2003]
(A) Terpenes (B) Proteins (C) Sugars (D) Soaps

144. Which statement is incorrect about peptide bond $\left[-\overset{\overset{\displaystyle O}{\|}}{C}-NH \right]$?
(A) C - N bond length in proteins is longer than usual bond length of C - N bond
(B) Spectroscopic analysis shows planar structure of - CO - NH - group
(C) C - N bond length in proteins is smaller than usual bond lenght of C - N bond (D) None of these

145. Assertion : A solution of sucrose in water is dextro - rotatory. But on hydrolysis in the presence of a little hydrochloric acid, it becomes laevo rotatory. Reason : Sucrose on hydrolysis gives unequal amounts of glucose and fructose. As a result of this change in sign of rotation is observed.
(A) A (B) B (C) C (D) D

146. Which enzyme breaks sucrose into glucose and fructose ?
(A) Zymase (B) Maltase (C) Diastase (D) Invertase

147. Secondary structure of proteins is due to
(A) Peptide bond (B) Hydrogen bond
(C) Covalent bond (D) Coordinate bond

148. Which of the following is not a constituent of RNA ?
(A) d-Ribose (B) Phosphate (C) Adenine (D) Pyridine

149. Chloroprene is used in making:
(A) Synthetic rubber (B) Plastic
(C) Petrol (D) All of the above

150. Which compound is not a lipid ?
(A) Lecithin (B) Lysine (C) Cerebroside (D) Cephalin

151. The enzyme which can catalyse the conversion of glucose to ethanol is :
(A) Zymase (B) Invertase (C) Maltase (D) Diastase

152. Which one of the following structures represents the peptide chain? [CBSE-PMT-2004]

(A) 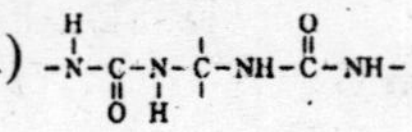 (B) 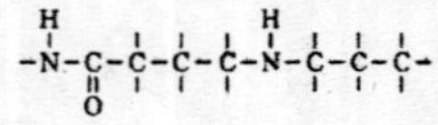

(C) 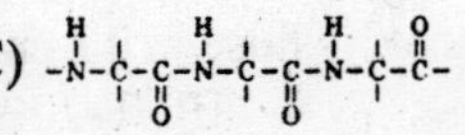 (D) 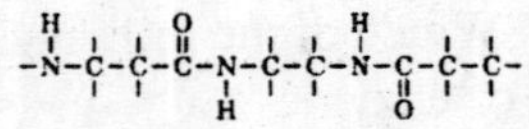

153. A sequence of how many nucleotides in messenger RNA makes a codon for an amino acid ? [CBSE-PMT-2004]

(A) Three (B) Four (C) One (D) Two

154. The hormone that helps in the conversion of glucose to glycogen is [CBSE-PMT-2004]

(A) Cortisone (B) Bile acids (C) Adrenaline (D) Insulin

155. The enzyme which hydrolyses triglycerides to fatty acids and glycerol is called [CBSE-PMT-2004]

(A) Maltase (B) Lipase (C) Zymase (D) Pepsin

156. The correct statement in respect of protein haemoglobin is that it [CBSE-PMT-2004]

(A) functions as a catalyst for biological reactions
(B) maintains blood sugar level
(C) acts as an oxygen carrier in the blood
(D) forms antibodies and offers resistance to diseases

157. Number of chiral carbons in β-D-(+)-glucose is [CBSE-PMT-2004]

(A) five (B) six (C) three (D) four

158. Acidic order of 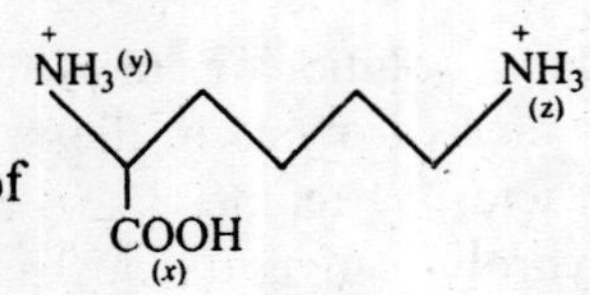 is [IIT-2004]

(A) (z) > (y) > (x) (B) (y) > (x) > (z)
(C) (x) > (y) > (z) (D) (x) > (z) > (y)

159. Which pair of compounds gives Tollen's test? [IIT-2004]

(A) Glucose & Fructose (B) Sucrose & Glucose
(C) Hexanal & Acetophenone (D) Fructose & Sucrose

Biomolecules

1.	(B)	55.	(C)	109.	(D)
2.	(A)	56.	(D)	110.	(C)
3.	(C)	57.	(A)	111.	(A)
4.	(A)	58.	(C)	112.	(B)
5.	(B)	59.	(D)	113.	(B)
6.	(C)	60.	(B)	114.	(B)
7.	(B)	61.	(D)	115.	(B)
8.	(C)	62.	(B)	116.	(B)
9.	(C)	63.	(C)	117.	(C)
10.	(D)	64.	(D)	118.	(A)
11.	(C)	65.	(B)	119.	(C)
12.	(C)	66.	(B)	120.	(A)
13.	(A)	67.	(C)	121.	(A)
14.	(C)	68.	(D)	122.	(D)
15.	(D)	69.	(B)	123.	(B)
16.	(D)	70.	(D)	124.	(B)
17.	(B)	71.	(C)	125.	(D)
18.	(A)	72.	(D)	126.	(D)
19.	(C)	73.	(C)	127.	(A)
20.	(B)	74.	(D)	128.	(B)
21.	(C)	75.	(B)	129.	(A)
22.	(B)	76.	(B)	130.	(C)
23.	(B)	77.	(D)	131.	(D)
24.	(A)	78.	(C)	132.	(C)
25.	(A)	79.	(C)	133.	(B)
26.	(B)	80.	(A)	134.	(B)
27.	(B)	81.	(B)	135.	(B)
28.	(C)	82.	(A)	136.	(D)
29.	(D)	83.	(A)	137.	(B)
30.	(B)	84.	(A)	138.	(A)
31.	(A)	85.	(D)	139.	(A)
32.	(B)	86.	(D)	140.	(C)
33.	(D)	87.	(D)	141.	(A)
34.	(C)	88.	(C)	142.	(D)
35.	(A)	89.	(C)	143.	(D)
36.	(B)	90.	(B)	144.	(A)
37.	(A)	91.	(B)	145.	(C)
38.	(C)	92.	(C)	146.	(D)
39.	(D)	93.	(D)	147.	(B)
40.	(C)	94.	(C)	148.	(D)
41.	(A)	95.	(A)	149.	(A)
42.	(D)	96.	(C)	150.	(B)
43.	(A)	97.	(C)	151.	(A)
44.	(D)	98.	(B)	152.	(C)
45.	(A)	99.	(A)	153.	(A)
46.	(D)	100.	(C)	154.	(D)
47.	(C)	101.	(B)	155.	(B)
48.	(B)	102.	(A)	156.	(C)
49.	(D)	103.	(B)	157.	(A)
50.	(B)	104.	(B)	158.	(C)
51.	(C)	105.	(A)	159.	(A)
52.	(A)	106.	(B)		
53.	(A)	107.	(B)		
54.	(C)	108.	(A)		

Chemistry in Action

1. Dyes

Chromophore: They are the colour bearing ones. e.g.

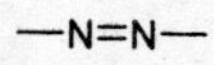
azo

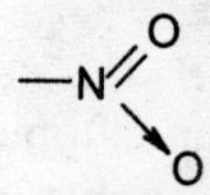
nitro

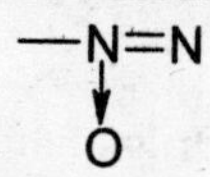
azoxy

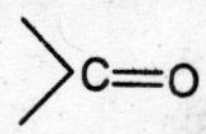
carbonyl

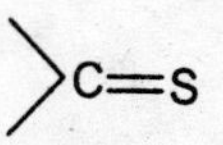
thiocarbonyl

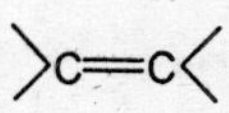

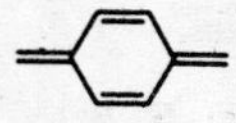
Quinoid

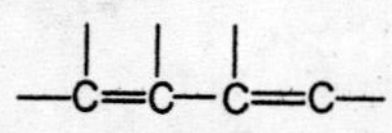
etc.

conjugated

Auxochrome: They do not produce colour themselves. When present along with chromophore, they will intensify the colour e.g.
-OH; $-SO_3H$; -COOH, $-NH_2$; -NHR; $-NR_2$ groups etc.

Classification:

(1) Based on chemical constitution:

(a) Azodyes: -N=N- group present. e.g.

(i) Methyl orange

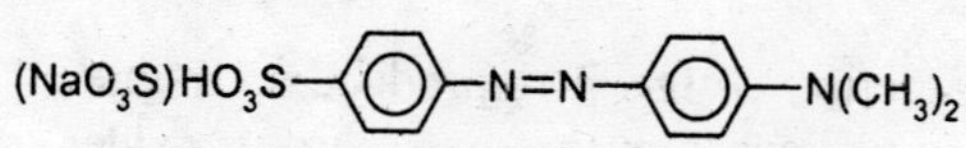

(ii) Methyl red:

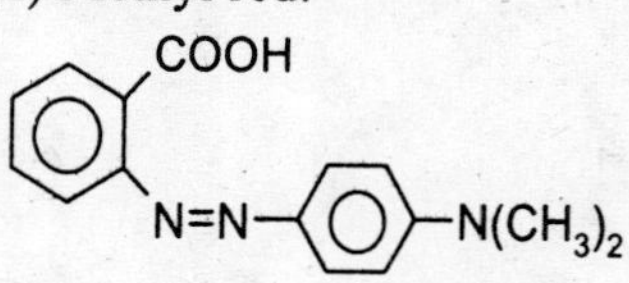

(iii) Congo red

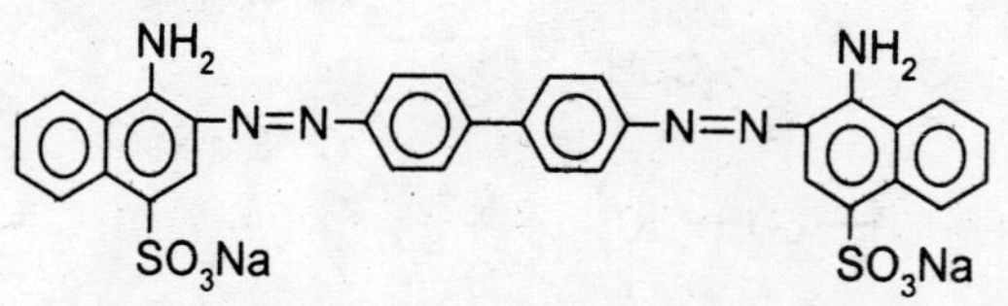

(b) Anthraquinone dyes: Anthraquinone basic moiety is present

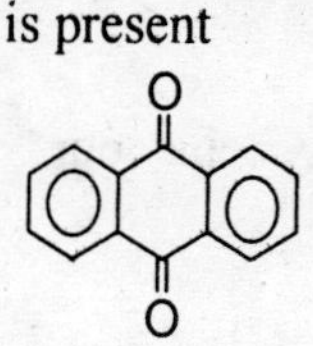
(Anthraquinone)

e.g. Alizarin

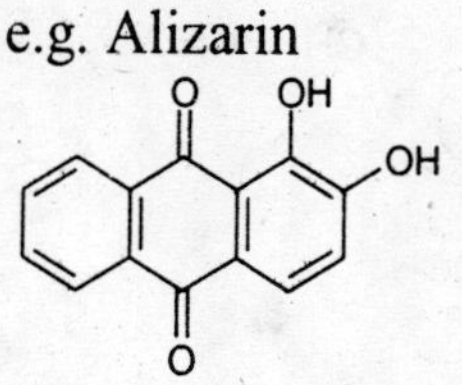

(c) Indigoid dyes: e.g. Indigo

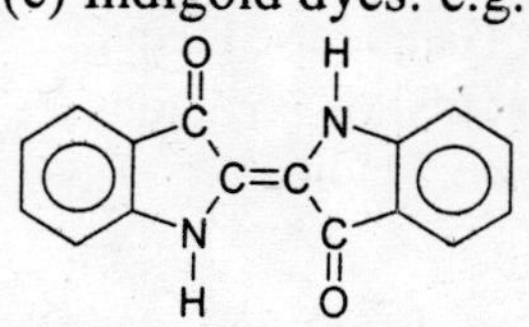

(2) Based on mode of application

(a) Acid dyes: acidic groups like phenolic -OH, -COOH, $-SO_3H$ etc. are present, e.g.

(i) Orange-I

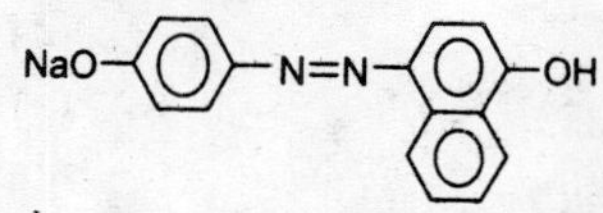

(ii) Orange-II (iii) Methyl orange (iv) Methyl red
(v) Picric acid (vi) Martius yellow

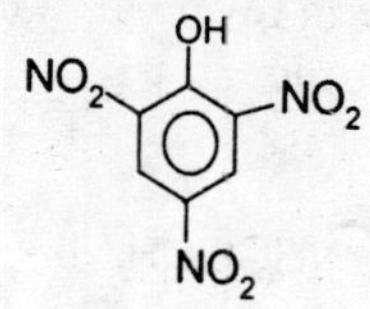

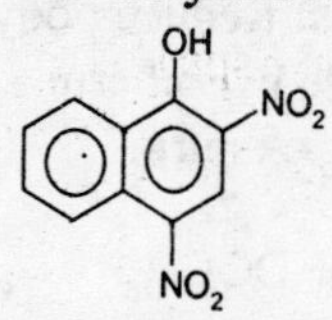
2,4-dinitro-1-naphthol

(b) Basic dyes: contain $-NH_2$ group or with substituents.

e.g. (i) Aniline yellow (ii) Malachite green

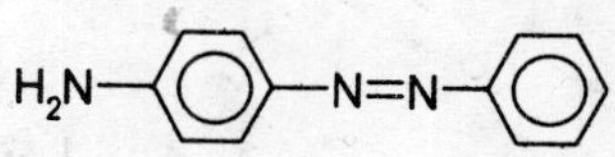

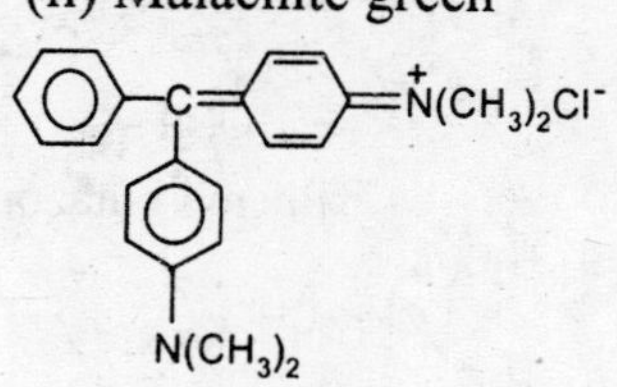

(c) Mordant dyes: These dyes have no natural affinity for the fabric and are applied to it with the help of certain addition substances known as mordants (e.g. $FeCl_3$, $Al_2(SO_4)_3$, $SnCl_4$ (metal ions Fe^{3+}, Al^{3+}, Sn^{4+}, Cr^{3+} etc. present in these act as mordants). A single dye gives different colours with different ions.

(d) Vat dyes: These dyes are insoluble in water and can't be applied directly. These dyes on reduction with $NaHSO_3$ in a vat form a soluble compound which has great affinity for cotton and other cellulose fibres, e.g. Indigo.

2. Drugs

Chemicals used for treating diseases are called drugs (or medicines). Treatment of diseases by chemical substances which do not destroy the tissues of the host while destroying the micro-organisms responsible is called chemotherapy.

(1) Analgesics: They are pain killers or pain relievers. A drug that selectively relieves pain by acting in the CNS or on peripheral pain mechanisms without significantly altering consciousness.

(a) Norcotic:

Morphine is a major alkaloid in opium.
Codeine is methylated morphine.
Heroin is diacetyl morphine.

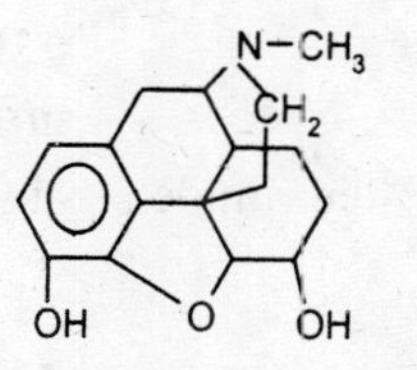
Morphine

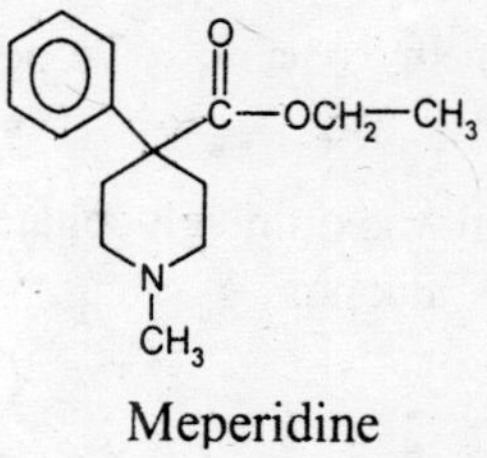
Meperidine

(b) Non Narcotic:

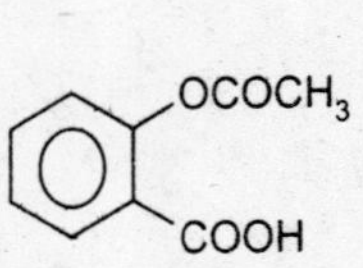

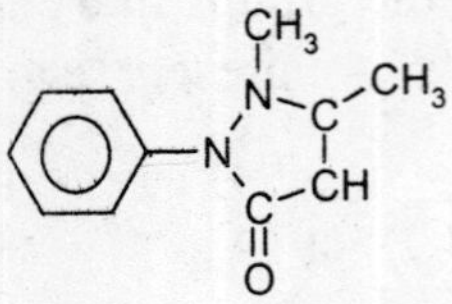

Aspirin (Acetyl salicylic acid)   Analgin

(2) Antipyretics: Reduce body temperature in fever, but do not cause hypothermia in normothermic individuals. e.g. Aspirin,

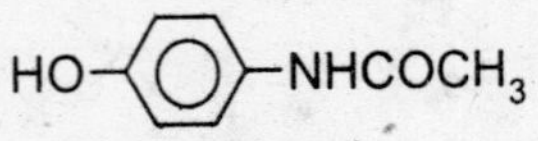 ; 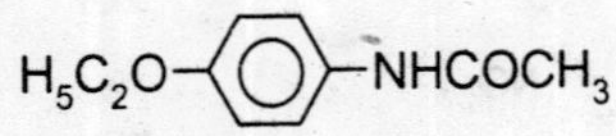

Paracetamol   Phenacetin

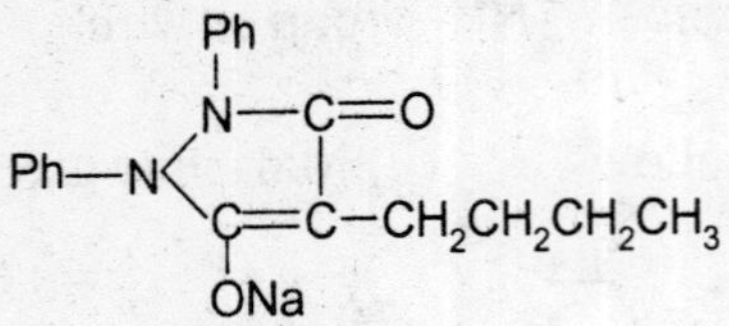

Phenyl butazone

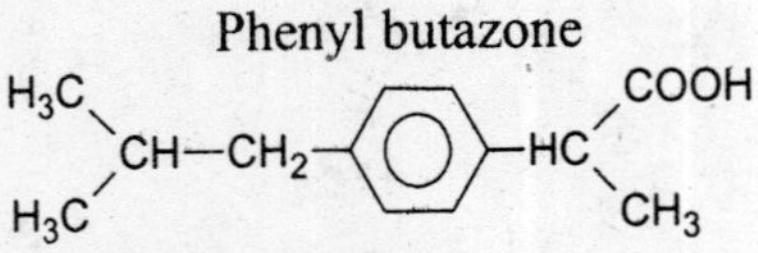

Ibuprofen

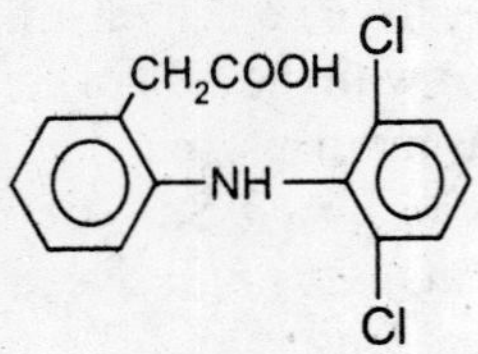

Diclofenac sodium

Most of them have anaelgesic and anti inflammatory actions also.

(3) Antiseptics and Disinfectants: The terms antiseptic and disinfectant connote an agent which inhibits or kills microbes on contact. Conventionally, agents used on living surfaces are called antiseptics while those used for inanimate objects are called disinfectants. There is considerable overlap and many agents are used either way,

Antiseptics: Dettol (4.8% solution is prepared in 9% terpinol and 13% alcohol).

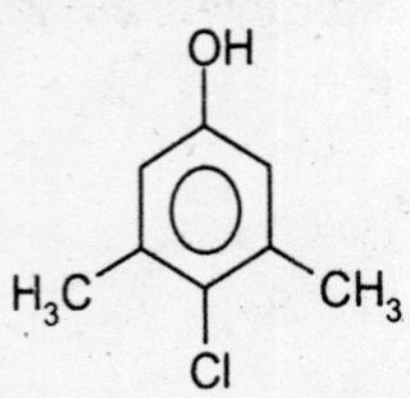 + 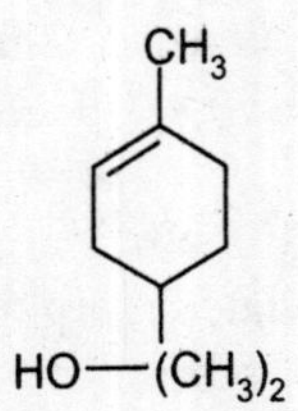

(4-chloro-3, 5-dimethyl phenol (Chloroxylenol))   Terpeneol

Savlon: 1.5% chlorhexidine gluconate + 3% cetrimide Mercurochrome; alcohol; Iodoform; Tincture of Iodine (2% I_2 in alcohol) Chlormine -T; $KMnO_4$; Boric acid etc.

Disinfectants: DDT; Phenol; O_3; H_2O_2; SO_2; $KMnO_4$ Bleaching powder etc.

(4) Anaesthetics:

(a) Local Anaesthetics: Drugs which upon topical application or local injection cause reversible loss of sensory perception, specially of pain in a restricted area of the body.

e.g.

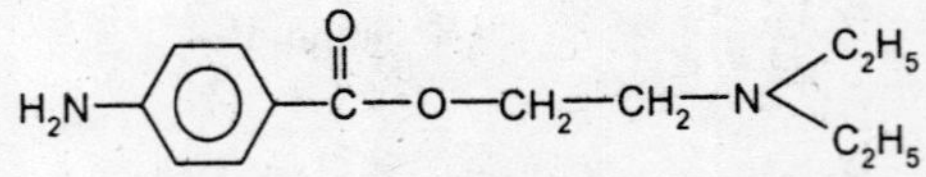

procaine (ester linkage)

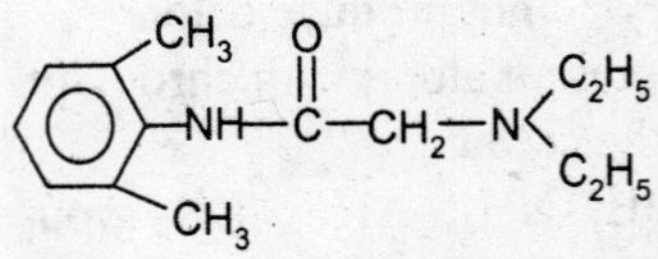

lignocaine (amide linkage)

α-eucaine; orthocaine; cocaine; tetracaine; benzocaine, xylocaine etc.

(b) General Anaesthetics: Drug which produce reversible loss of all sensation and consciousness e.g. Nitrous oxide: N_2O; Ether; Halothane F_3C-CHBrCl;

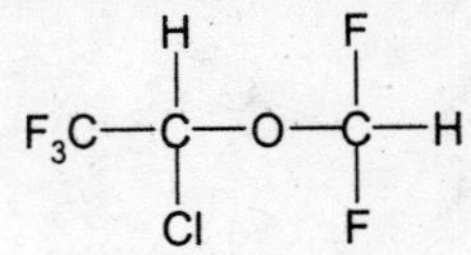

Isoflurane

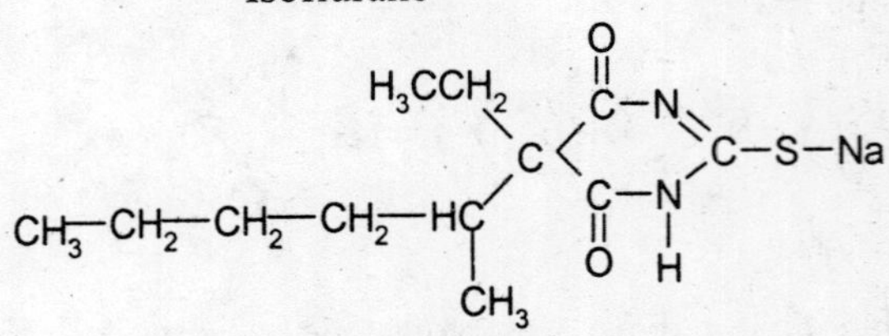

Sodium pentothal

Cyclopropane etc.

(5) Psychedelics: Drugs which alter mood, behaviour thought and manner e.g.

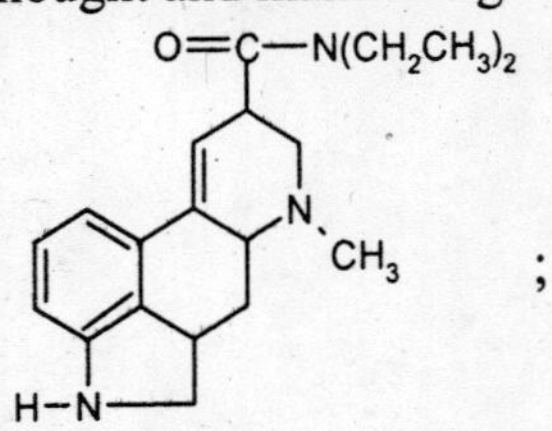 ; 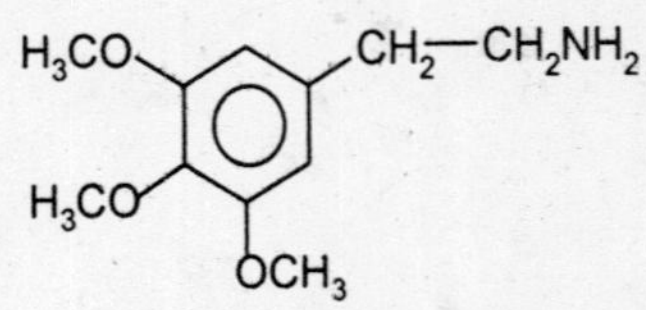

(LSD Lysergic-acid) diethylamide   Mescaline

Cannabinoids : Bhang (dried leaves) , Ganja (dried female inflorescence), Charas(resinous extract of flowering tops and flowers) are drugs of abuse [Cannabis Indica (Marijuana)]

(6) Tranquillisers: A drug which reduces mental tension and produces calmness without inducing sleep or depressing mental faculties.

(a) Sedatives: A drug that subdues excitement and calms the subjects without inducing sleep though drowsiness may be produced.

(b) Hypnotic : A drug that induces and or maintains sleep, similar to normal arousable sleep.

(c) Antidepressants : A drug which can elevate the mood in depressive illness.

(i) Barbutyrates have been popular hypnotics and sedatives

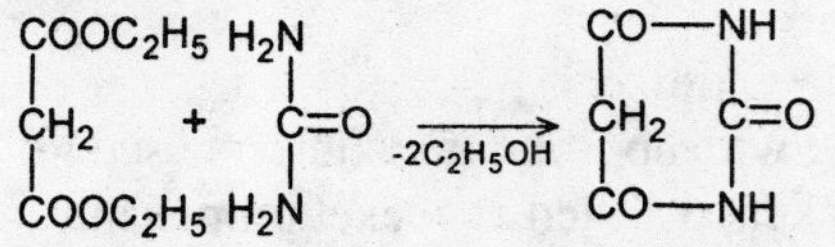

malonic ester urea barbutyric acid

If the two active hydrogen atoms of CH_2 group of malonic ester are replaced by $(C_2H_5)_2$, Veronal is obtained.

If the two active hydrogen atoms of CH_2 group of malonic ester are replaced by C_2H_5 & C_6H_5, Luminal is obtained.

(ii) Diazepam (iii) Chlorpromazine

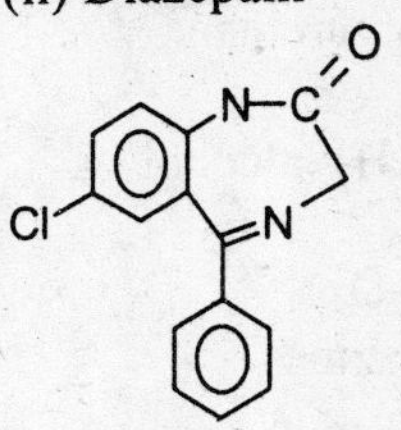

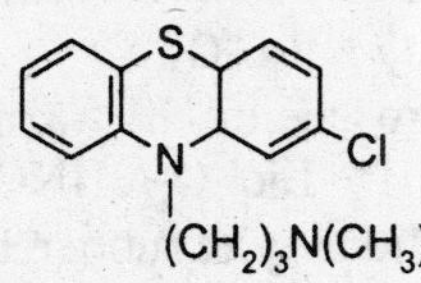

(Valium, calmpose)

(iv) Reserpine (from Rauwolfia serpentina) is used in the treatment of hypertension

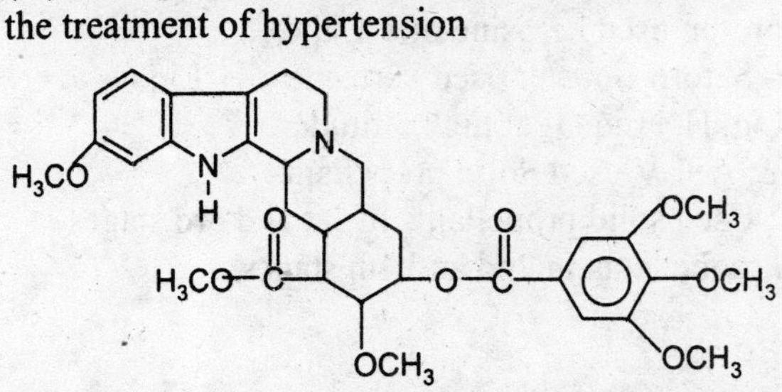

(7) Antimalarials: Malaria caused by four species of protozoal parasite plasmodium (falciparum, malariae, ovale and vivax) is endemic in most parts of India and other tropical countries

Chloroquine: It is active in blood (Single dose of 600 mg is recommended)

Primaquine is active in lever (14 days -15 mg base daily)

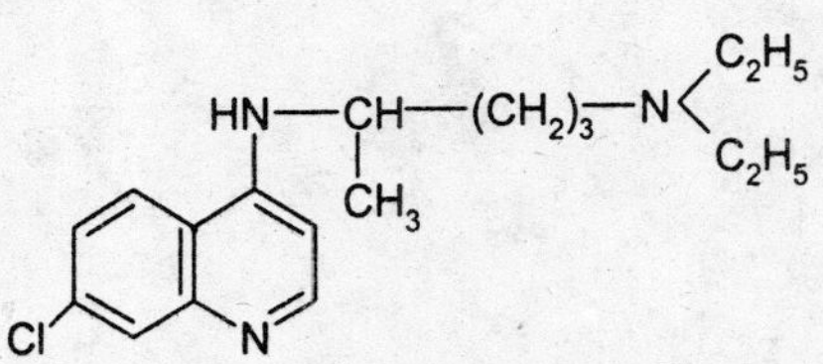

Chloroquine (4-amino quinoline moiety)

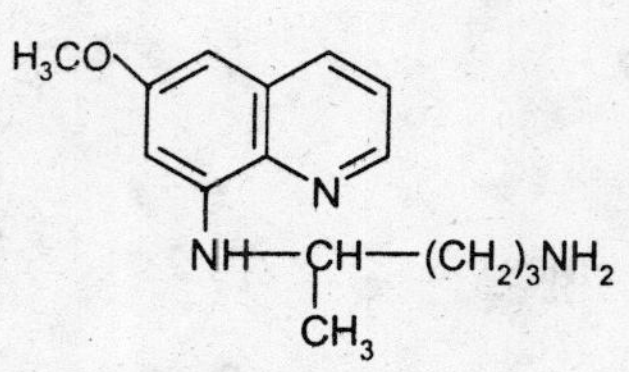

;

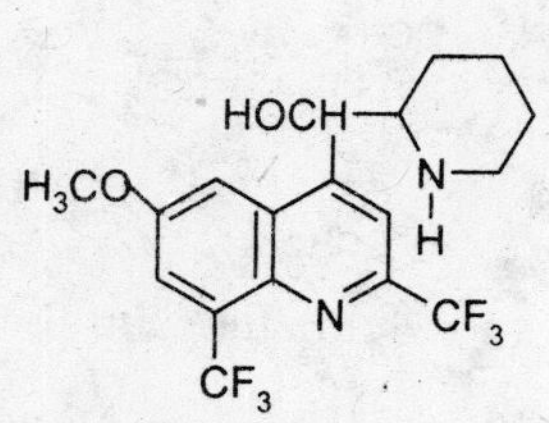

Primaquine Mefloquine

(8-amino quinoline moiety)

Mefloquine is used for chloroquine-resistant and multidrug - resistant malaria

(8) Antibiotics: Substances produced by microorganisms, which supress the growth of or kill other microorganisms at very low concentrations.

First antibotic Penicillium was discovered by Alexander Flemming (1929). Streptomycin- Waksman (1944).

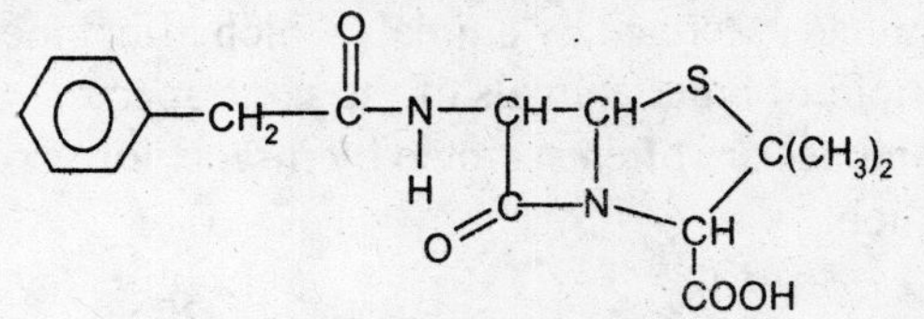

Benzyl penicillin (Penicillin-G)

If benzyl group is replaced by

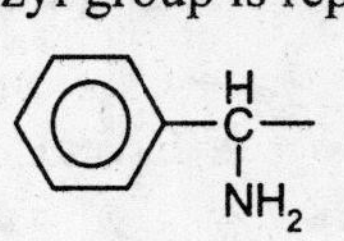

(Ampicillin)

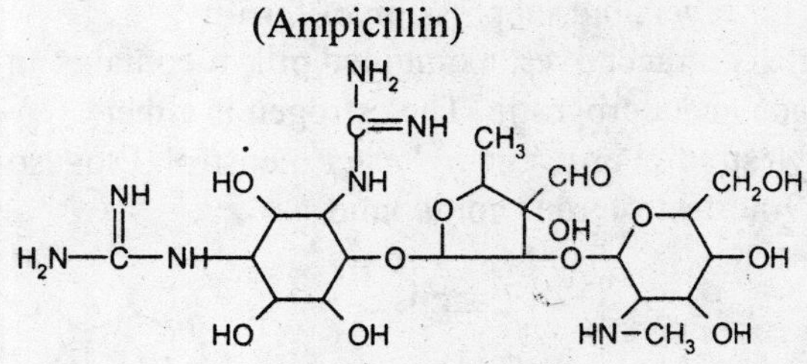

Streptomycin

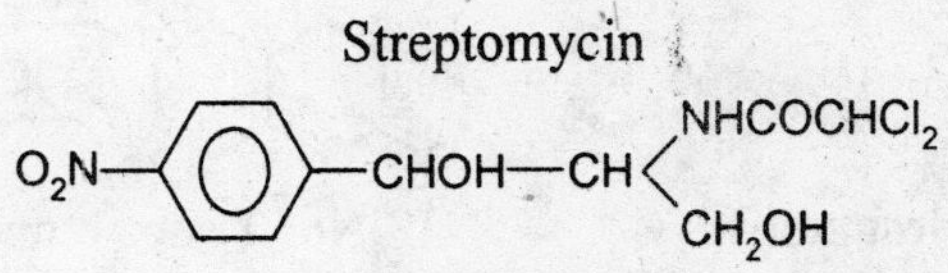

Chloramphenical

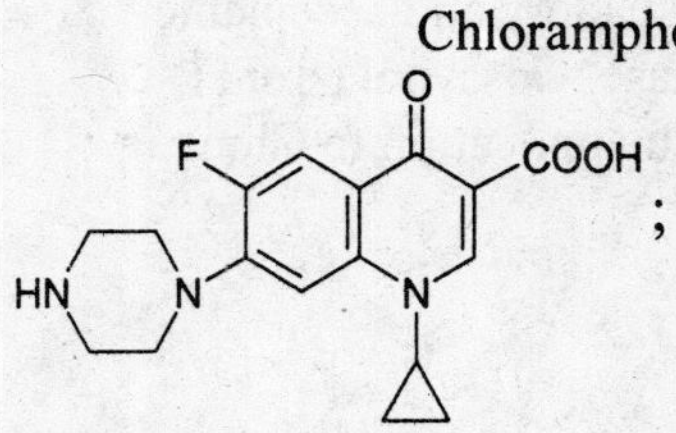

;

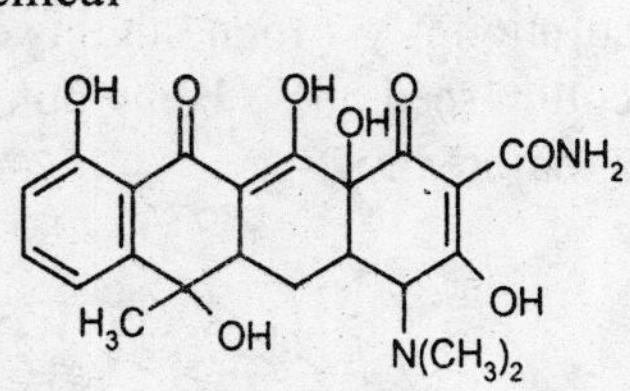

Ciprofloxacin Tetracycline

(Fluoroquinolones)

(In Norfloxacin $-C_2H_5$ is present instead of cyclopropane ring).

(9) Sulpha drugs: Sulphanilamide derivative

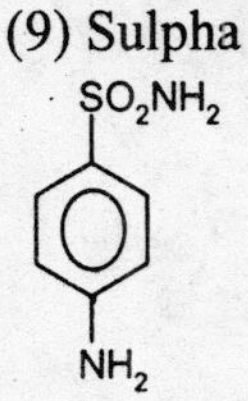

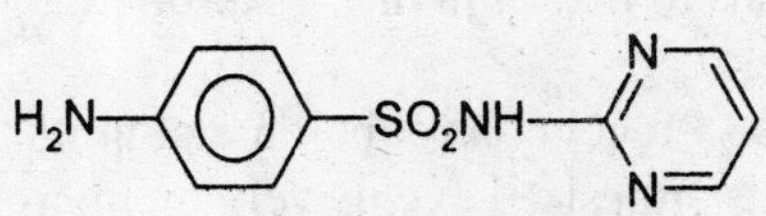

Sulphadiazine

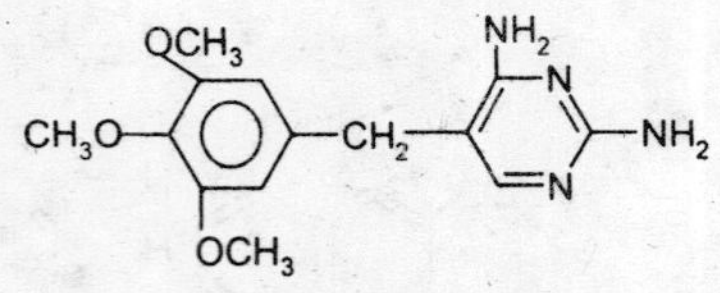

Trimethoprim (pyrimidines)

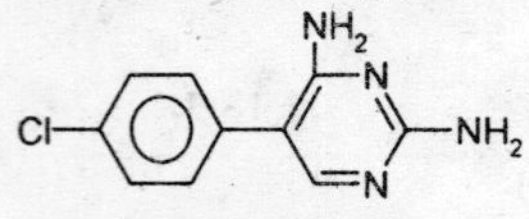

Pyrimethamine

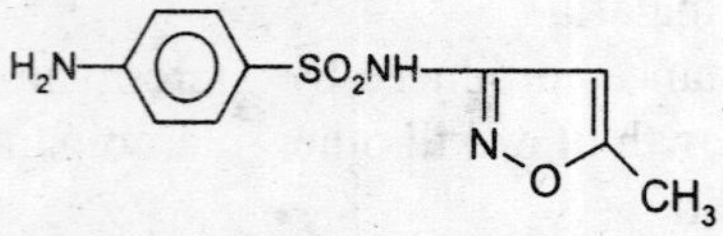

Sulphamethoxazole (Sulphonamide)

3. Pheromones:

Compounds secreted by animals which affect the behaviour of other animals of the same species. Pheromones are often secreted by insects for sexual attraction.

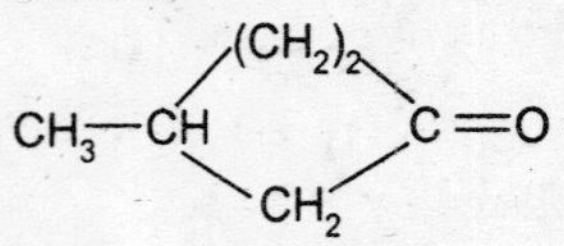

Muscone
(Tibetan musk deer)

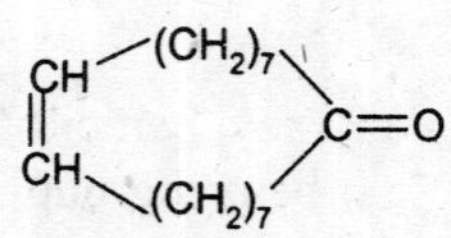

Civetone
(Civet Cat)

4. Contraceptives:

Used for reversible suppression of fertility,
Female contraceptives: Combined pill: It contains an estrogen and a progestin. The estrogen is either ethinylestradiol or its methyl ester menstrol. Progestin is a 19-nortestosterone compound.

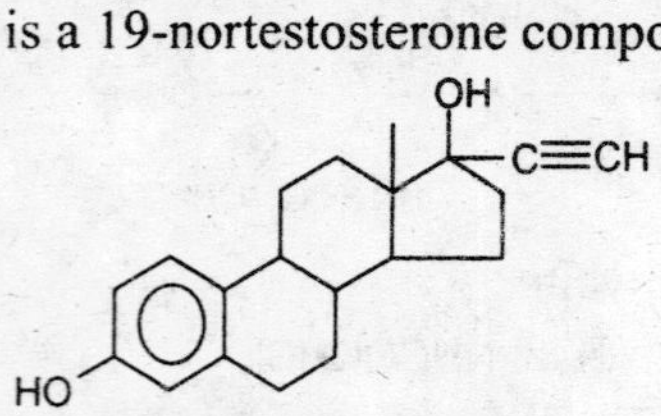

Ethinylestradiol

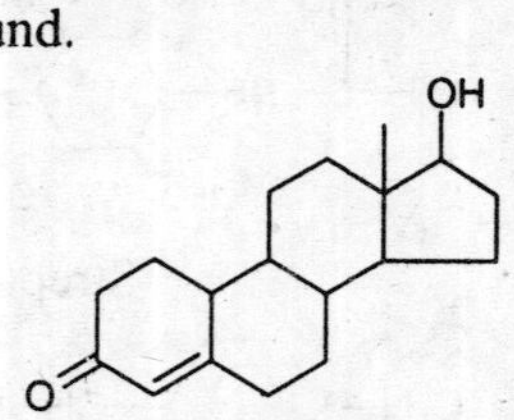

19-Nor Testosterone

Male contraceptives: No satisfactory and acceptable solution is yet found in this case. Gossypol (from cotton seed oil) is being put in application in China in a major way.

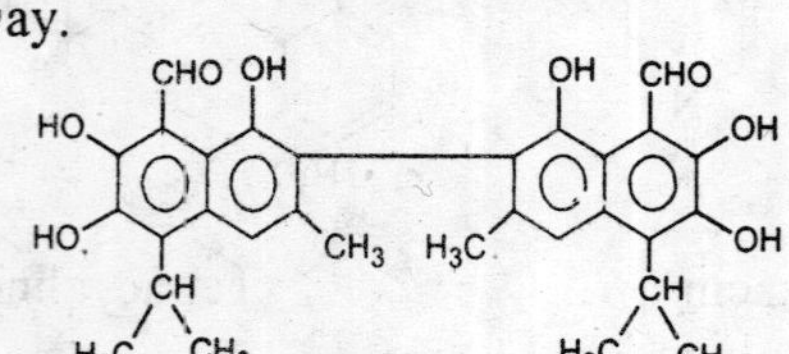

Gossypol

5. Vitamins:

(i) Vitamin A: It is derivative of β-carotene. Its deficiency leads to night blindness, xerophthalmia, keratomalacia.
(ii) Vitamin B_{12} : It contains cobalt metal in its structure. It is synthesised exclusively by micro organisms. Thus, it is absent in plants.
(iii) Vitamin C: Its deficiency leads to scurvy.
(iv) Vitamin D: It is a steroid pro-hormone. It is derived from either 7-dehydrocholesterol or ergosterol by UV radiation. Its deficiency causes rickets in children and osteomalacia in adults.
(v) Vitamin E (Tocopherol):deficiency leads to anaemia of the new born and also degeneration of reproduction power.
(vi) Vitamin K (phylloquinone) responsible for clotting of blood. Its deficiency leads to uncontrolled bleeding, hemorrhage.

6. Rocket Propellants

Propellant = Fuel +Oxidizer

Mixture of fuel along with appropriate oxidisers used in the propulsion of rockets is called as rocket propellants. Specific impulse is the measure of the energy of the propellant.

$$I_S = \sqrt{\frac{T_c}{M}}$$

I_S = specific impulse
T_c = flame temp. inside the rocket motor
M= molecular mass of the product gases.

Propellants

(a) Solid: Polybutadiene/Polyacrylic acid/ Synthetic rubbers (Thiolokol)/ cellulose nitrate, polyurethane+ $KClO_4$ / NH_4NO_3/ NH_4ClO_4.
(b) Liquid: Kerosene oil/ hydrazine/ Liq.H_2/ alcohol/ UDMH*/ MMH*+Liq. O_2/ HNO_3/ N_2O_4
(c) Hybrid: Solid acrylic rubber + Liq.N_2O_4,
*UDMH- Unsymmetrical dimethyl hydrazine
*MMH- Monomethylhydrazine

Liquid propellants are more efficient than solid propellants.
Russian- proton used kerosine oil+ Liq.O_2
American- Satorn Booster used kerosine oil+ Liq.O_2 to start and Liq. H_2+Liq.O_2 at high altitude.
(i) SLV-3& ASLV used Solid propellants.
(ii) PSLV: Used solid propellants in 1st and 3rd stages and liquid propellants in 2nd and 4th stages.

1. Aspirin is [CBSE-90]
(A) antibiotic (B) sedative
(C) antipyretic (D) psychedelic

2. Diazo coupling is useful to prepare some [CBSE-94]
(A) pesticides (B) dyes (C) proteins (D) vitamins

3. Which of the following statements is false ? [CBSE-94]
(A) Some disinfectants can be used as antiseptics in lower concentrations
(B) Sulphadiazine is a synthetic antibacterial
(C) Ampicillin is a natural antibiotic
(D) Aspirin is an analgesic and antipyretic both

4. Which of the following contains cobalt ? [AIIMS-96]
(A) Haemoglobin (B) Vitamin D
(C) Vitamin B_{12} (D) Chlorophyll

5. Methyl orange is the example of which type of dye ? [AIIMS-2002]
(A) Acid dye (B) Azo dye
(C) Mordant dye (D) Both (A) and (b)

6. An example of azo dye is : [JIPMER-93]
(A) Indigo (B) Malachite green
(C) Alizarin (D) Orange-I

7. Aspirin is : [JIPMER-93]
(A) Salicylic acid (B) Methyl salicylate
(C) Phenyl salicylate (D) Acetyl salicylic acid

8. A broad spectrum antibiotic is [JIPMER-95]
(A) Paracetamol (B) Aspirin
(C) Chloroamphenicol (D) Penicillin

9. The deficiency of Vitamin D causes [JIPMER-95]
(A) Night-blindness (B) Loss of appetite
(C) Rickets (D) Xerosis

10. Which of the following is not a water soluble vitamin ? [JIPMER-97]
(A) Thiamin (B_1) (B) Riboflavin (B_2)
(C) Ascorbic acid (C) (D) Calciferol (D)

11. Which of the following contain amino group ? [BHU-95]
(A) Vat dyes (B) Acid dyes
(C) Basic dyes (D) Mordant dyes

12. Which of the following is an antibiotic ? [BHU-96]
(A) Penicillin (B) Albumin
(C) Insulin (D) Ascorbic acid

13. Vitamin E is [BHU-96]
(A) Fat soluble (B) Water soluble
(C) Soluble in ether (D) Soluble in alcohol

14. Which one of the following vitamins is related to ergosterol ? [BHU-96]
(A) Vitamin E (B) Vitamin K
(C) Vitamin C (D) Vitamin D

15. The metal present in vitamin B_{12} is [BHU-97]
(A) Magnesium (B) Iron (C) Cobalt (D) Manganese

16. Aspirin is a/an [BHU-98]
(A) Antibiotic (B) Antipyretic
(C) Antiseptic (D) None of these

17. The chloramphenicol is an example of [BHU-99]
(A) Antibiotic (B) Antipyretic
(C) Antiseptic (D) Sulpha drug

18. 2, 4 D is a [AFMC-91]
(A) Insecticide (B) Herbicide
(C) Fungicide (D) Rodenticide

19. Which of the following is used as "Anaesthesia" [AFMC-96]
(A) CH_4 (B) CO_2 (C) N_2O (D) N_2

20. The compound 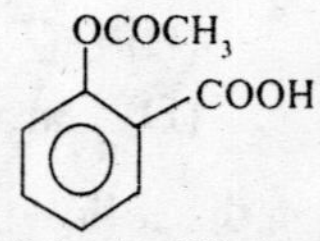

is used as. [AIEEE-2002]
(A) Antiseptic (B) Antibiotic (C) Analgesic (D) Pesticide

21. Which of the following could act as a propellant for rockets ? [AIEEE-2003]
(A) Liquid hydrogen + liquid nitrogen
(B) Liquid oxygen + liquid argon
(C) Liquid oxygen + liquid hydrogen
(D) Liquid nitrogen + liquid oxygen

22. Which one of the following can be used as an anaesthetic [EAMCET-98]
(A) N_2O (B) NO (C) NCl_3 (D) NO_2

23. Which of the following is used as rocket fuel ? [EAMCET-98(MBBS)]
(A) Nitrate $+O_3$ (B) Water gas $+O_3$
(C) Cyanogen $+C_2$ (D) Cyanogen $+O_3$

24. Which of the following is used as rocket fuel ? [EAMCET-99]
(A) F_2 (B) N_2 (C) C_2H_2 (D) CH_4

25. Night-blindness is due to [CPMT-96]
(A) Vitamin A (B) Vitamin B_5
(C) Vitamin C (D) Vitamin D

26. Metal present in blood is - [CPMT-97]
(A) Al (B) Mg (C) Cu (D) Fe

27. Scurvy is caused due to the deficiency of vitamin [CPMT-2000]
(A) B_1 (B) C (C) K (D) A

28. Cobalt is present in [CPMT-2003]
(A) Vitamin B_2 (B) Vitamin B_1
(C) Vitamin B_5 (D) Vitamin B_{12}

29. The deficiency of vitamin B_1 causes [PMT-99]
(A) beri-beri (B) scurvy (C) rickets (D) anaemia

30. Vitamin B_1 is : [PMT-2000]
(A) Riboflavin (B) Cobalamin
(C) Thiamin (D) Pyridoxine

31. The deficiency of vitamin-C causes [PMT-2000]
(A) Scurvy (B) Rickets
(C) Pyrrohea (D) Pernicious anaemia

32. Which of the following is NOT a sex hormone ? [PMT-2000]
(A) Testosterone (B) Estrone
(C) Estradiol (D) Cortisone

33. Vitamin A is present in - [PET-95]
(A) Cord liver oil (B) Carrot
(C) Milk (D) In all the above

34. Ascorbic acid is a - [PET-95]
(A) Vitamin (B) Enzyme
(C) Protein (D) Carbohydrate

35. The natural dye in the following is : [PET-2000]
(A) Alizarin (B) Methyl orange
(C) Azodyes (D) Rhodamine

36. Chloramphenicol is used in the treatment of which of the following ? [PET-2000]
(A) Typhoid (B) Pneumonia
(C) Headache and fever (D) Bronchitis

37. SLV - 3 uses propellants : [PET-2000]
(A) Solid (B) Liquid
(C) Solid-Liquid (D) Biliquid

38. Which of the following can possibly be used as analgesic without causing addiction and moon nodification ?
(A) Morphine (B) N-acetyl-para-aminophenol
(C) Drazepom (D) tetrahydrocational.

39. Azodyes are
(A) Acid dyes (B) Ingrain dyes
(C) Mordant dyes (D) Basic dyes

40. Drugs used for reducing fevers are called
(A) Anti-inflammatory (B) Analgesics
(C) Antipyretics (D) Antibiotics

Chemistry in action

1. (C)
2. (B)
3. (C)
4. (C)
5. (B)
6. (D)
7. (D)
8. (C)
9. (C)
10. (D)
11. (C)
12. (A)
13. (A)
14. (D)
15. (C)
16. (B)
17. (A)
18. (B)
19. (C)
20. (C)
21. (C)
22. (A)
23. (D)
24. (A)
25. (A)
26. (D)
27. (B)
28. (D)
29. (A)
30. (C)
31. (A)
32. (D)
33. (D)
34. (A)
35. (A)
36. (A)
37. (C)
38. (B)
39. (B)
40. (C)

ATOMIC STRUCTURE

Electrons : e^-, β^-. The basis for the idea that electricity consists of individual charges called electrons was drawn from Faraday's experiments. The cathode rays consists of negatively charged particles called electrons – J.J. Thomson (1897).

The charge of the electron=-1.602 x 10^{-19} coulomb (determined by Mullikan's oil drop method).
Mass of the electron = 1/1840 of mass of proton or 9.1 x 10^{-31} kg.
The charge/mass ratio =1.759 x 10^8 coulmbs/gram
The ratio of e/m is the same regardless of the nature of cathode or the residual gas in the discharge tube. This leads to the conclusion that electrons which are common constituents of all atoms are identical.
de Broglie first postulated the wave properties of electrons
The name "electron" was given by Stoney.

CONSTITUTION OF NUCLEUS :

1. Proton : H^1, p . Discovered by Goldstein.
charge 1.602 x 10^{-19} Coulomb/proton
mass : 1.675 x 10^{-24} gram or mass of e^- x 1840 or 1.00757 a.m.u.

The charge/mass ratio for positive ions is not constant but varies from one gas to another. The e/m ratio for positive ions is much smaller (10^5 coulombs/gram or less) than the e/m ratio of electrons (10^8 columbs/gram).

2. Neutron : n_0^1 ; Chadwick (1932)

They are electrically neutral ; mass = 1.00893 a m u on physical atomic weight scale; spin = 1/2 ;
$t_{0.5}$ = approximately 20 minutes.
Neutron decays $n_0^1 \longrightarrow H_1^1 + e^- + \nu$

OTHER UNSTABLE PARTICLES :

(1) Positron : e^+, β^+. Anderson (1932)
positive counterpart of electron.
life period ~ 1/million part of a second

(2) Antiproton : p^-. Segre & Chamberlain (1955)
mass : same as that of proton
charge : Negative charge equal in magnitude to that of proton

(3) Neutrino and anti-neutrino : neutral; mass : small mass (0.005 atomic mass unit)
The neutrino seems to be associated with and shares the energy of the electron $n_1^0 \rightarrow H_1^1 + e^- + \nu$
where as the anti-neutrino is associated with proton
$H_1^1 \rightarrow n_0^1 + e^+ + \nu^1$

(4) Meson : Yukawa (1935); positive, negative or neutral in nature. Mass : intermediate between electron and proton. They are believed to be of two types $\pi \pm$ and $\mu \pm$.
π is heavier of the two.
π has the life time 10^{-8} seconds ; $\mu = 2 \times 10^{-6}$ s
They decompose as
$\pi \pm \rightarrow \mu \pm + \nu\ (10^{-8}s)$; $\mu \pm \rightarrow e^{\pm} + 2\nu\ (2 \times 10^{-6}s)$
They were discovered from the effect of cosmic rays on matter.

Thomson's model : plum-pudding model. The electrons are distributed just as the seeds in water melon.

Rutherford model : After α-ray scattering experiment of Rutherford, the Thomson's model was found to be defective.
From Rutherford's experiment, we understand that the volume of nucleus is much smaller compared to the volume of the atom
Nucleus diameter : 10^{-15} to 10^{-14}m
Atom diameter : 10^{-10}m
Defects : The theory does not explain :
1. What is the velocity of the electron ?
2. What is the number of electrons present in each orbit ?
3. J.C. Maxwell's theory of electrodynamics : Whenever an electric charge is subjected to acceleration, it emits radiation and loses energy.
Rutherford's atomic model could not explain the emission or absorption of electromagnetic radiation and therefore was not able to account for the existance of spectral lines.

Bohr's model :
Bohr's theory is based on the postulates :
(1) An electron does not radiate energy if it stays in one orbit and therefore doesn't slow down. Such permissible orbits are known as stationary states.
(2) The motion of the electrons is so restricted that they can occupy only that orbit or energy level in which the angular momentum of the electron is an integral multiple of $h/2\pi$ i.e. $mvr = n\ h/2\pi$ (Where m = mass of electron, v= velocity of electron in orbit, r = distance from the nucleus and n= principal quantum number)
(3) When an electron moves from one orbit to another, it either radiates or absorbs energy. It emits radiation when it drops from an orbit of higher energy
E_2 to one of lower energy E_1. $\quad E_2 - E_1 = h\nu$
In the reverse process radiation or energy is absorbed.
If an electron jumps from an initial orbit 'i' to a final orbit 'f' or falls from 'f' to 'i' the different series of lines in the hydrogen spectrum obtained by varying the values of n_i and n_f are Lyman, Balmer, Paschen, Brackett, Pfund and Humphrey.

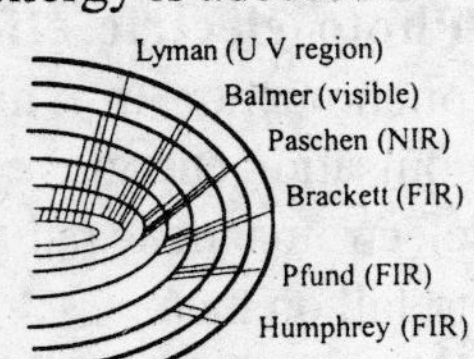

Total number of spectral lines = $\frac{n\ (n-1)}{2}$

(4) For an electron to remain in its orbit, the electrostatic attraction between the electron and nucleus which tends to pull the electron towards the nucleus must be equal to

the centrifugal force which tends to throw the electron out of its orbit

Centrifugal force = Centripetal force

$$\frac{mv^2}{r} = \frac{Ze^2}{r^2}$$

Limitations of Bohr's Theory :

(1) It cannot be extended to atoms more complex than hydrogen. In He^+, Li^{2+}, Be^{3+} etc. solitary electron is present.

(2) It is not correct to regard the electrons in atoms as discrete particle with precisely defined positions and velocities (It is not in agreement with Heisenberg's uncertainity principle).

(3) It couldnot explain the fine structure of spectral lines.

Bohr's radius for various orbits

$$r = \frac{n^2h^2}{4\pi^2Ze^2m} = n^2 \times a^0 \text{ (where } a^0 = \frac{h^2}{4\pi^2Ze^2m}\text{)}$$

$a^0 = 0.529\overset{0}{A}$ (a^0= Bohr's radius when n = 1)

or $r = \frac{\varepsilon_0 n^2h^2}{\pi Ze^2m}$ (ε_0 = permitivity of free space)

Velocity of an electron in any orbit

$$v^2 = \frac{n^2h^2}{4\pi^2m^2r^2} \quad \text{or } v = \frac{nh}{2\pi mr}$$

This relation can also be derived from the equation

$$mvr = n\frac{h}{2\pi}$$

Energy of electons :

$$E = -\frac{2\pi^2Z^2e^4m}{n^2h^2} \quad \text{or } E = -\frac{Z^2e^4m}{8\varepsilon_0^2n^2h^2}$$

Substituting the values of π, m, e and h, energy of the electron in the nth orbit is given as

$$E_n = -\frac{2.179 \times 10^{-11}}{n^2} \text{ erg/atom} = \frac{-13.6}{n^2} \text{ ev/atom}$$

$$E_n = E_1 \times \frac{z^2}{n^2}$$

$= -\frac{E_1}{n^2}$ (for hydrogen like species, where z = 1)

frequency $\nu = 3.289 \times 10^{15} \left(\frac{1}{n_1^2} - \frac{1}{n_2^2}\right)$

Wave number $\bar{\nu} = \frac{1}{\lambda} = \frac{\nu}{c} = R_H \left(\frac{1}{n_1^2} - \frac{1}{n_2^2}\right)$

$= 10.97 \times 10^6 \left(\frac{1}{n_1^2} - \frac{1}{n_2^2}\right)$

Photo electric effect : Photo electric effect is a phenomenon in which surface of metals (having low ionisation energies e.g. Na, Cs etc.) emit electrons when electromagnetic radiations of high frequency are allowed to fall on them.

Work function of the metal : When light falls on metallic surface, one photon of light is absorbed. The energy (absorbed light) interacts with the most loosely bound electron, giving it all its energy. Some of the energy is acquired by the electron to separate it from the metal and the remaining energy is the K. E. of the electron after it leaves the metal surface.

Energy of photon = Escape energy + K.E. of the ejected electron

$h\nu = p + \frac{1}{2} mv^2$ (p =escape energy; m and v are mass and velocity of the electron after it leaves the metal surface).

The escape energy which is the minimum energy required to remove an electron from the surface of a metal is called the work-function of the metal.

We know $E = h\nu$

$\therefore$ Work function = h x Threshold frequency

$$\phi = h\nu_0$$

When light of higher frequency is used, the excess energy goes as the K.E. of the ejected electron.

Energy of photon = work function + K.E. of the ejected electron

$$h\nu = \phi + K.E.$$
$$= h\nu_0 + K.E.$$
$$K.E. = h\nu - h\nu_0 = h(\nu - \nu_0)$$

Bohr-Sommerfeld theory : Bohr could not explain the existence of fine spectral lines. Sommerfeld explained it as : The movement of an electron from one orbit to another should give a single sharp line in the spectrum corresponding precisely to energy difference between the initial and final orbits. If the hydrogen spectrum is observed with a high resolution spectrometer, some of the lines show fine structure i.e. they are composed of more than one line close together. This was explained by Sommerfeld in terms of elliptical orbits which precessed in space round the nucleus. For the value of n = 1, there is a circular orbit but for n = 2 both circular and elliptical orbits are possible.

To define an elliptical orbit, a second quantum number k is needed. The shape of the ellipse is defined by ratio of the lengths of the major and minor axis.

Thus $\frac{\text{major axis}}{\text{minor axis}} = \frac{n}{k}$

k is called the azimuthal or subsidiary quantum number. This may have values 1, 2 .... Thus for n = 2, n/k may have values 2/2 (circular orbit) and 2/1 (elliptical orbit). For principal quantum number n = 3, n/k may have values 3/3 (circular), 3/2 (ellipse), 3/1 (narrow ellipse). The presence of these extra orbits which differ only slightly in their energies account for the extra lines in the spectra revealed under high resolution.

QUANTUM NUMBERS :

(1) Principal Quantum number : designated by 'n'. The energy of the electron is mainly determined by this Q number. As 'n' increases, the energy E of electron increases but becomes less electronegative. 'n' may be any integer from 1 to infinity.The maximum number of electrons 'N' in any shell or orbit can be calculated by $2n^2$.

(2) Azimuthal quantum number : Denoted by l. It tells us the shape of the electron cloud. It is equal to the number of nodal planes passing through the nucleus of the atom. It may have any value from 0 to n-1. l = 0, 1, 2, 3 are termed s, p, d, f electrons respectively. Spectral terms sharp, principal, diffuse and fundamental respectively which actually represent the four series of lines are characteristic of the atomic spectra of many elements. The azimuthal quantum number l gives a measure of the energy due to orbital angular momentum of the electron, $mvr = \frac{h}{2\pi}\sqrt{l(l+1)}$ .The sub-shell with lowest $(n+l)$ value is filled up first.

(3) Magnetic quantum number : Denoted by 'm'. Tells us the orientation of electron in the space. For every value of 'l', it possesses any integral values between -l to +l via 0. Thus for an 's' electron $l = 0$, ∴ m = 0, no orientation in the space. l = 1, m = -1, 0, +1 so on.
The total number of orbitals in a given sub shell can be obtained by the expression $m = 2l + 1$.
Positive values of m relate orbital angular momentum in the direction of the applied field whereas negative values refer to opposite direction.Magnetic quantum number explains "Zeeman" effect.
(a) Zeeman effect : When a strong magnetic field is applied to a source of a spectrum, each spectral line is split into a number of separate lines.
(b) Stark effect : Splitting of spectral lines under the influence of electric field.
(c) Compton effect : The decrease in energy or increase in wavelength of x-rays after scattering from the surface of an object is known as Compton effect.

(4) Spin quantum number : Designated by s. Spin quantum number does not follow from the wave mechanical treatment. Owing to the spin or rotation about their axes electrons possess an additional angular momentum. The spin momentum may have either + 1/2 or - 1/2 depending on the two possible directions (i.e. clockwise and anti-clockwise) of rotation of the electron.
Pauli's exclusion principle : "No two electrons in the same atom have all their four quantum numbers identical" e.g. Helium atom has two electrons. The two electrons have n=1, l =0, m=0, but one will have $s = +\frac{1}{2}$ and the other $s = -\frac{1}{2}$

Aufbau principle : The electrons enter the various orbitals in the order of increasing energy.

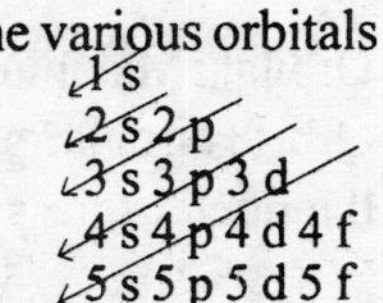

Exceptions : Cr, Cu, Mo, Ru, Rh, Ir, Pd, Ag, Au

Hund's Rule of maximum multiplicity : Electron pairing in any orbital is not possible until all the available orbitals of a given set contain one electron each.
An orbital cannot contain more than two electrons. This is based on Pauli's exclusion principle.

Orbit : Definite circular path at a definite distance from the nucleus in which electron is revolving round the nucleus.

Orbital : The maximum probability of the existence of electron.

de Broglie relation :
Dual nature of electrons : According to the models of Rutherford and Bohr, the atom has a central nucleus surrounded by electrons in definite orbits. The electron is thus regarded as a particle having mass m, charge e and momentum mvr.
Einstein has suggested (1905) that light has a dual character as wave as particle. de Broglie proposed that matter also has a dual character as wave and as particle. An expression is derived for calculating the wavelength λ, of a particle of mass 'm' moving with velocity 'v' according to which
$E = mc^2$ (c = velocity of light)
$h\nu = mc^2$ ($\because E = h\nu$)
we know that $\nu = \frac{c}{\lambda}$
$\therefore \frac{hc}{\lambda} = mc^2$
$\lambda = \frac{hc}{mc^2} = \frac{h}{mc}$
Substituting c by the velocity of electron, v, we have
$\lambda = \frac{h}{mv} = \frac{h}{p}$ (where p is the momentum of the particle)

Heisenberg's uncetainity principle : It is impossible to measure simultaneously the exact position and exact velocity of a subatomic particle. This is effective only for motion of subatomic particles of very small mass and has no importance for the motion of large bodies.
$\Delta x \times \Delta p \geq \frac{h}{4\pi}$
or $\Delta x \times m\Delta v \geq \frac{h}{4\pi}$
Δx = uncertainty in the position
Δp = uncertainty in momentum
Δv = uncertainty in velocity

Shapes of orbitals :
s-orbitals : Since an s - orbital can have only one possible orientation, it should be spherical in shape. Why ? because a sphere can be defined by a single value, i.e. its radius. The size of the orbital depends upon the value of the principal quantum number.

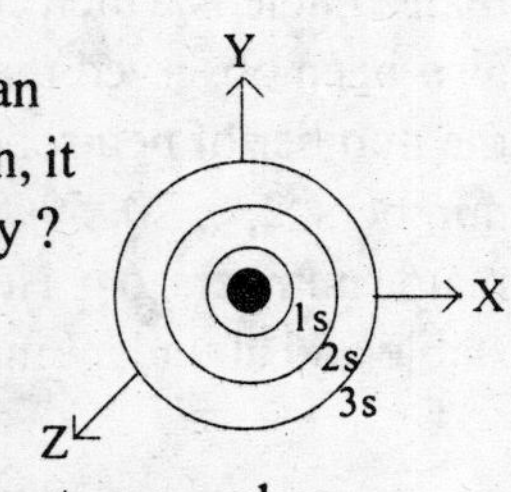

s-orbital is symmetrical along the three axes i.e.x, y & z.

p-orbitals : There are three p-orbitals. They have equal energies They are oriented along the three mutually right angled axes x, y & z. Their shape looks like a dumb-bell. The lobes of the three orbitals are oriented along the x, y, z axes.

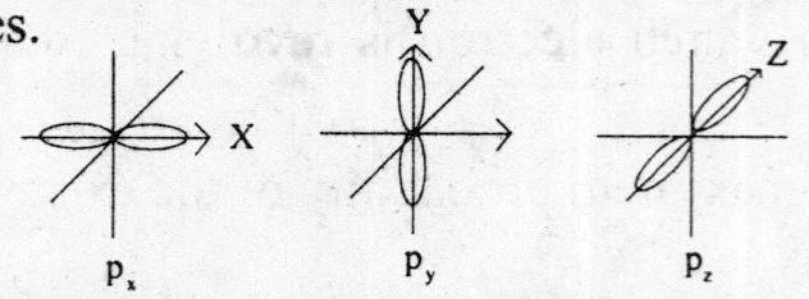

d-orbitals : There are five d-orbitals. They possess equal energy. They differ in their orientation in three dimensional space. Three of the five orbitals project in between the coordinate axes while the other two lie along the coordinate axes.

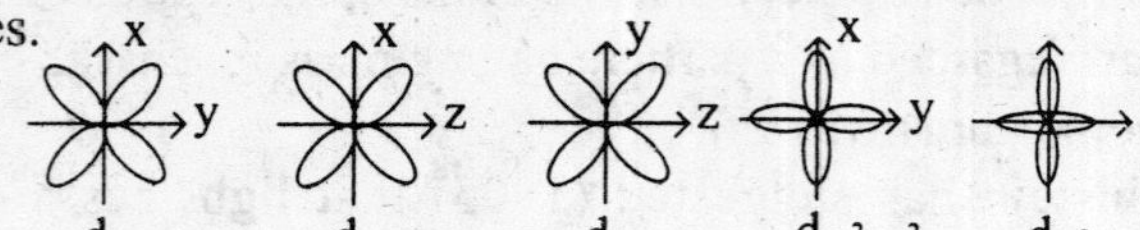

f-orbitals : There are seven f-orbitals. They start from 4n shell. m has values from -3 to +3. Therefore the number is 7. There is no unique way of representing all the seven f-orbitals. One way is to consider the seven orbitals made up of one f_{z^3}, three of f_{xz^2} f_{yx^2}, f_{zy^2} and three f_x $(x^2\text{-}3y^2)$, f_y $(y^2\text{-}3z^2)$, and f_z $(z^2\text{-}3x^2)$ orbitals.
Other way: a total of three f_{x^3} f_{y^3} and f_{z^3}; three f_x $(y^2\text{-}z^2)$, f_y $(z^2 - x^2)$ and f_z $(x^2\text{-}y^2)$ and one f_{xyz} orbitals
In the orbitals of f_x $(y^2\text{-}z^2)$, f_y $(z^2\text{-}x^2)$, f_z $(x^2\text{-}y^2)$, the lobes are directed towards their corresponding axes. In f_{xyz}, they lie between x, y and z orbitals. The orientation of f_{x^3}, f_{y^3}, f_{z^3} can be compared to that of d_{z^2}

The whole number rule and packing fraction :
The difference in atomic mass of an isotope and mass number was expressed by Aston (1927) as packing fraction

$$\text{Packing fraction} = \frac{\text{Isotopic atomic mass-mass number}}{\text{mass number}} \times 10^4$$

$$\text{packing fraction for hydrogen} = \frac{1.0078-1}{1} \times 10^4 = 78$$

$$\text{for chlorine } Cl^{35} = \frac{34.980-35}{35} \times 10^4 = -5.7$$

The magic numbers : It has been observed that atoms with even number of nucleons in their nuclei are more plentiful than those with odd numbers. This indicates that nucleus made up of even number of nucleons is more stable than a nucleus which contains odd no. of nucleons. It has also been observed that stable nuclei result when either the number of neutrons or protons is equal to one of the numbers, 2, 8, 20, 50, 82, 126. These numbers are called magic numbers. e.g. He^4_2, O^{16}_8, Ca^{40}_{20}, Pb^{208}_{82} etc. are highly stable and abundant in nature.

Things to remember

-Radius of the atom 10^{-10}m;
-Radius of the nucleus : 10^{-15} to 10^{-14}m.
-Electron : J.J. Thomson; e/m ratio constant, wave properties of electrons were first postulated by de Broglie.
-Proton : Goldstein ; e/m ratio of positive ions differ
-Neutron : Chadwick; ${}_0n^1 \rightarrow {}_1H^1 + \beta^1 + \nu$
Isotopes of the same element have identical chemial properties.
-Meson : Yukawa; binds the nucleons together.
Lyman (UV); Balmer (Visible); Paschen (NIR), Brackett (FIR), Pfund (FIR), Humphrey (FIR)

$$r = \frac{n^2h^2}{4\pi^2mze^2} \quad \text{or} \quad r = \frac{\varepsilon_0 n^2h^2}{\pi mZe^2} = n^2a^0$$

Bohr radius $(a^0) = 0.529$Å

$r = \frac{n^2}{z} \times 0.529$Å (radius for hydrogen like atoms)

$E = h\nu = h\frac{c}{\lambda} = hc\bar{\nu}$

$E_n = -\frac{z^2}{n^2} \times 13.6$

$\frac{E_1}{E_2} = \frac{n_2^2}{n_1^2}$ (in ionisation energy, electrons are removed from ground state).

ν (frequency) $= 3.289 \times 10^{15} \left(\frac{1}{n_1^2} - \frac{1}{n_2^2}\right)$

$\bar{\nu}$ (wave number) $= R_H \left(\frac{1}{n_1^2} - \frac{1}{n_2^2}\right)$; $R_H = 10.97 \times 10^6$

-Quantum numbers : Principal quantum number : n; $N=2n^2$; energy increases as 'n' increases but becomes less negative.
-Azimuthal quantum number : l; $mvr = \frac{h}{2\pi}\sqrt{l(l+1)}$; shape; $l = (n-1)$; s, p, d, f.
-Magnetic quantum number: m; orientation in space. Explains Zeeman effect.
-Spin quantum number doesn't follow from wave mechanical treatment.
-Pauli's exclusion principle : No two electrons can have all the four quantum numbers the same in the same atom.
-Aufbau principle : Electrons enter various orbitals in the order of increasing energy. It is also known as (n+l) rule
-Hund's rule of maximum multiplicity.
-de Broglie $\lambda = \frac{h}{mv} = \frac{h}{p}$
-Heisenberg's uncertainty principle : It is impossible to measure simultaneously the exact position and exact velocity of a sub atomic particle

$\Delta X \times \Delta p = \Delta X \times m \Delta v \geq \frac{h}{4\pi}$

Orbital : Maximum probability of the existence of electron.
Orbit : circular path in which electron is revolving round the nucleus. it is represented by n.

Solved examples

(1) The wave length of the line of the highest wave length of a particular series in the atomic spectrum of hydrogen is at 656.3 n m. what is the series. ($R_H = 1.097 \times 10^7 m^{-1}$)

Sol. $\frac{1}{\lambda} = R_H \left(\frac{1}{n_1^2} - \frac{1}{n_2^2} \right)$

$$\frac{1}{656.3 \times 10^{-9}} = 1.097 \times 10^7 \left(\frac{1}{2^2} - \frac{1}{x^2}\right)$$

$$\frac{1}{656.3 \times 10^{-9} \times 1.097 \times 10^7} = \left(\frac{1}{2^2} - \frac{1}{x^2}\right); x = 3$$

(2) Calculate the wave number and energy of the spectral line (emitted by an excited hydrogen atom) corresponding to n_1=3 and n_2 = 5

Sol. $\bar{v} = 1.097 \times 10^7 \left(\frac{1}{3^2} - \frac{1}{5^2}\right)$

$= 1.097 \times 10^7 \times \frac{16}{225} = 7.8 \times 10^5 m^{-2}$

$E = hv = h\frac{c}{\lambda} = hc\bar{v} = 6.626 \times 10^{-34} \times 3 \times 10^8 \times 7.8 \times 10^5$

$= 1.55 \times 10^{-19}J$

(3) The wave length of the first member of Balmer series of hydrogen is 6563 x 10^{-10}m. Calculate the wave length of its second member.

Sol. For the first member $\bar{v}_1 = R_H\left(\frac{1}{2^2} - \frac{1}{3^2}\right) = \frac{5}{36} R_H$

For the second member $\bar{v}_2 = R_H\left(\frac{1}{2^2} - \frac{1}{4^2}\right) = \frac{3}{16} R_H$

$\frac{\bar{v}_1}{\bar{v}_2} = \frac{20}{27}$ $\quad \bar{v}_2 = \frac{27}{20} \times \bar{v}_1$

$\frac{1}{\lambda_2} = \frac{27}{20} \times \frac{1}{6563 \times 10^{-10}}$

$\lambda_2 = \frac{20}{27} \times 6563 \times 10^{-10} = 4861 \times 10^{-10}$ m.

(4) What should be the kinetic energy and potential energy respectively of the electron present in the third orbit of hydrogen atom ?

Sol. We know, $E = -\frac{z^2}{n^2} \times 13.6$

Total energy of the 3rd orbit $= -\frac{1}{9} \times 13.6 = -1.5$ eV

kinetic energy = - total energy = - (-1.5) = 1.5 eV

potential energy = 2 x Total energy = 2 x -1.5 = -3.0 eV

(5) What should be the ionisation energy of He^+ (The ionisation energy of hydrogen is 13.6 ev).

Sol. we know $E_n = E_1 \times \frac{z^2}{n^2}$

$-\frac{z^2}{n^2} \times 13.6 = -\frac{2^2}{1^2} \times 13.6 = -54.4ev = 54.4ev.$

(6) What should be the emitted energy when the electron present in hydrogen atom jumps from n = 4 to n = 2?

Sol. $E_2 = -\frac{z^2}{n^2} \times 21.79 \times 10^{-19}$ joules

$= -\frac{1^2 \times 21.79 \times 10^{-19}}{2^2} = -5.4 \times 10^{-19}$ J.

$E_4 = -\frac{1^2 \times 21.79 \times 10^{-19}}{4^2} = -1.36 \times 10^{-19}$

$E_4 - E_2 = -1.36 \times 10^{-19} - (-5.4 \times 10^{-19}) = 4.04 \times 10^{-19}J$

(7) Calculate the frequency of the line in the emission spectrum of hydrogen when the atoms of the gas contain electrons in the third energy level.

Sol. When it falls from 3rd → 2nd level

$v = 3.289 \times 10^{15} \left(\frac{1}{2^2} - \frac{1}{3^2}\right) = 3.289 \times 10^{15} \times \frac{5}{36}$

$= 4.568 \times 10^{14}$ Hz

when it falls from 3rd → 1st level

$v = 3.289 \times 10^{15} \left(\frac{1}{1^2} - \frac{1}{3^2}\right) = 2.924 \times 10^{15}$Hz

(8) The ionisation energy of hydrogen atom in the ground state is 2.17 x 10^{-11} erg./atom. Calculate the wavelength of the emitted radiation in Å unit, if electron returns from 2nd to 1st orbit.

Sol. I. E. $= E_\alpha - E_1 = 0 - (2.17 \times 10^{-11}) = -2.17 \times 10^{-11}$

$\Delta E = hv = h\frac{c}{\lambda}$

$\lambda = \frac{h \times c}{\Delta E} = 1.221 \times 10^2$ Å

(9) The energy of electron in the second orbit of H atom is - 3.41 eV.Calculate the energy of second orbit of He^+ ion

Sol. Energy of an electon in n^{th} orbit of hydrogen like species is given by $E_n = -\frac{13.6z^2}{n^2}$ eV/atom

For H-atom z = 1; n=2

$E_2 = -\frac{13.6 \times 1^2}{2^2} = -3.41eV$

for He^+, z = 2; n=2;

$E_{2'} = -\frac{13.6 \times 2^2}{2^2}$

$\frac{E_{2'}}{3.41} = \frac{-13.6 \times 2^2 \times 2^2}{2^2 \times 13.6 \times 1^2}$ or $\frac{E_{2'}}{3.41} = -4$

$E_{2'} = -4 \times 3.41 = -13.6$ eV

(10) Calculate and compare the energies of two radiations one with a wave length of 800nm and the other with 400nm.

Sol. $E_1 = hv_1 = h\frac{c}{\lambda_1}$; $E_2 = hv_2 = h\frac{c}{\lambda_2}$

$\frac{E_1}{E_2} = \frac{hc}{\lambda_1} \times \frac{\lambda_2}{hc} = \frac{\lambda_2}{\lambda_1} = \frac{800 \times 10^{-9}}{400 \times 10^{-9}} = \frac{2}{1}$

$E_1 = 2E_2$

(11) The threshold wavelength for the ejection of electron from metal M is 330 nm. Calculate the photoelectric emission.

Sol. Work function $\phi = hv_0 = \frac{hc}{\lambda_0} = \frac{6.6 \times 10^{-34} \times 3 \times 10^8}{330 \times 10^{-9}}$

$= 6.0 \times 10^{-19}$

(12) The work function of gold metal is 7.7 x 10^{-19}J. Calculate the kinetic energy and velocity of the electrons ejected when sodium metal is exposed to UV radiation of

wavelength 2.0×10^5 pm.

Sol. $\phi = h\nu_o = 7.7 \times 10^{-19}$ J

$$\nu = \frac{c}{\lambda} = \frac{3 \times 10^8}{2.0 \times 10^5 \times \frac{1}{10^{12}}} = 1.5 \times 10^{15}\ s^{-1}$$

$KE = h\nu - h\nu_0$

$= 6.626 \times 10^{-34} \times 1.5 \times 10^{15} - 7.7 \times 10^{-19} = 2.239 \times 10^{-19}$J

$KE = \frac{1}{2} mv^2$

$$v^2 = \frac{2KE}{m};\ v = \sqrt{\frac{2KE}{m}} = \sqrt{\frac{2 \times 2.239 \times 10^{-19}}{9.1 \times 10^{-31}}}$$

$= \sqrt{49 \times 10^{10}} = 7 \times 10^5\ ms^{-1}$

(13) Calculate the energy of the spectral line of minimum energy in visible region obtained due to electronic transition in 1 gm atom of hydrogen.

Sol. The transition corresponds to Balmer series

$n_1 = 2;\ n_2 = 3$

$$\frac{1}{\lambda} = \bar{\nu} = 1.1 \times 10^7 \left(\frac{1}{2^2} - \frac{1}{3^2}\right) = 1.1 \times 10^7 \times \frac{5}{36}\ m^{-1}$$

$\lambda = 6.55 \times 10^{-7}$ m

$$E = h\nu = h \times \frac{c}{\lambda} = \frac{6.62 \times 10^{-34} \times 3 \times 10^8}{6.55 \times 10^{-7}} = 3.03 \times 10^{-19}\ J.$$

For 1gm atom of hydrogen

$E_{total} = 3.03 \times 10^{-19} \times 6.023 \times 10^{23} = 182.5$ K.J.

(14) An electron is moving with a kinetic energy of 4.55×10^{-25} J. calculate the de Broglie wave length for it.

Sol. Kinetic energy $\frac{1}{2} mv^2 = 4.55 \times 10^{-25}$ J

$$\frac{1}{2} mv^2 \div m = \frac{1}{2} v^2 = \frac{4.55 \times 10^{-25}}{9.1 \times 10^{-31}} = 0.5 \times 10^6$$

$v^2 = 2 \times 0.5 \times 10^6 = 1 \times 10^6$

$v = \sqrt{1 \times 10^6} = 1 \times 10^3\ ms^{-1}$

$$\lambda = \frac{h}{mv} = \frac{6.6 \times 10^{-34}}{9.1 \times 10^{-31} \times 1 \times 10^3}$$

$= 7.253 \times 10^{-7}$m or 725.3 nm

(15) Calculate the wave length of a body of mass 1 mg. moving with a velocity of 10 ms^{-1}.

$$\text{Sol. } \lambda = \frac{h}{mv} = \frac{6.6 \times 10^{-34}}{10^{-6} \times 10} = 6.6 \times 10^{-29}m.$$

(16) Calculate the uncertainity in the velocity of a cricket ball of mass 150g, if the uncertainity in its position is the order of 1Å.

$$\text{Sol. } \Delta v = \frac{h}{m\Delta x 4\pi} = \frac{6.6 \times 10^{-34}}{4 \times 3.14 \times 1 \times 10^{-10} \times 150 \times 10^{-3}}$$

$= 3.5 \times 10^{-24} ms^{-1}$

(17) An electron has a velocity of 300m s^{-1} with an uncertainity of 0.001%. what is the uncertainity in locating its position?

Sol. Velocity of electron = 300ms^{-1}

uncertainity in veocity = 0.001% $\therefore \Delta v = \frac{0.001}{100} \times 300$

$= 0.003\ ms^{-1}$

$$\Delta x = \frac{h}{m\Delta v 4\pi} = \frac{6.626 \times 10^{-34}}{4 \times 3.14 \times 9.1 \times 10^{-31} \times 3 \times 10^{-3}}$$

$= 1.932 \times 10^{-3}$m

1. In a compound with atomic number $_{29}X^{60}$; the number of electrons is [CBSE-90]

(A) 29. (B) 30. (C) 40. (D) 59.

2. Different lines in Balmer series of hydrogen spectrum lies in [CBSE-90]

(A) IR Region (B) visible region

(C) UV Region (C) Far Infrared region

3. Avagadro's number being 6×10^{23}, the number of atoms present in 6 gms of carbon is [CBSE-90]

(A) $6 \times 10^{23} \times 1$ (B) $6 \times 10^{23} \times 6$

(C) $\frac{6 \times 10^{23}}{6}$ (D) $\frac{6 \times 10^{23}}{2}$

4. No two electrons in an atom can have the same set of four quantum number. This principle was evolved by [CBSE-91]

(A) Heisenberg (B) Pauli

(C) Maxwell (D) De-Broglie

5. Energy of third orbit of Bohr's atom is [CBSE-91]

(A) –13.6 eV (B) –3.4 eV

(C) –1.5 eV (D) None of these

6. Maximum number of electrons in any orbit is [CBSE-91]

(A) n^2 (B) $2n^2$ (C) $1/2\ n^2$ (D) None of these

7. $_{20}Ca^{40}$ and $_{19}K^{40}$ are [CBSE-91]

(A) Isomers (B) Isotopes (C) Isobars (D) Isotones

8. The binding energy of an electron in the nth Bohr orbit of hydrogen atom is [CBSE-92]

(A) $13.6/n^4$ eV (B) $13.6/n^3$ eV

(C) $13.6/n^2$ eV (D) $13.6/n$ eV

9. An atom with atomic number 20 is most likely to combine chemically with the atom whose atomic number is [CBSE-93]

(A) 11 (B) 14 (C) 16 (D) 10

10. Na^+, Mg^{+2}, Al^{+3}, Si^{+4} are isoelectronic, their ionic size will follow the order of [CBSE-93]

(A) $Na^+ > Mg^{+2} < Al^{+3} < Si^{+4}$

(B) $Na^+ < Mg^{+2} > Al^{+2} > Si^{+4}$

(C) $Na^+ > Mg^{+2} > Al^{+3} > Si^{+4}$

(D) $Na^+ < Mg^{+3} < Al^{+3} < Si^{+4}$

11. One mole of CO_2 contains [CBSE-93]

(A) 6.023×10^{23} atoms of oxygen

(B) 18.1×10^{23} molecules of CO_2

(C) 3 gm atoms of CO_2

(D) 6.023×10^{23} atoms of carbon

12. If ionization potential for hydrogen atom is 13.6 eV, then ionisation potential for He^+ will be [CBSE-93]

(A) 54.4 eV (B) 6.8 eV (C) 13.6 eV (D) 24.5 eV

13. One would expect proton to have very large [CBSE-93]
(A) charge (B) ionisation potential
(C) hydration energy (D) radius

14. Ratio of the energy of a photon of wavelength 3000 Å and 6000Å is [CBSE-93]
(A) 1/2 (B) 1/3 (C) 2 (D) 1/6

15. The number of atoms is greatest in [CBSE-93]
(A) 71 gm of chlorine (B) 48 grm of magnesium
(C) 127 gm of iodine (D) 4 gm of hydrogen

16. Atomic size of an element is 7×10^{-10} mole, then size of its molecules is [CBSE-93]
(A) 7×10^{-8} (B) 7×10^{-10} (C) 7×10^{-12} (D) 7×10^{-15}

17. Bohr model postulate [CBSE-93]
(A) Angular momentum = $\frac{mn}{2\pi}$
(B) Angular momentum = $\frac{hm}{2\pi}$
(C) Angular momentum = $\frac{nh}{2\pi}$ (D) None of the above

18. When we move from left to right in second period, the atomic volume of the element. [CBSE-93]
(A) increases (B) increases with constant rate
(C) decreases (D) None of the above

19. According to Bohr's model of atom [CBSE-93]
(A) total energy of electron is quantised.
(B) angular momentum of electron is quantised
(C) Both (A) and (B) (D) None of the above

20. Be's 4th electron will have the four quantum numbers. [CBSE-93]

	n	I	m	s
(A)	2	0	0	–½
(B)	1	0	0	+½
(C)	2	1	0	+½
(D)	1	1	1	–½

21. The total number of valence electrons in 4.2 g. of N_3^- ion is [CBSE-94]
(A) 2.1 N_A (B) 4.2 N_A (C) 1.6 N_A (D) 3.2N_A
(where N_A is the Avogadro number)

22. For which one of the following sets of four quantum numbers an electron will have the highest energy ? [CBSE-94]

	n	ℓ	m	s
(A)	3	2	1	1/2
(B)	4	2	–1	1/2
(C)	4	1	0	–1/2
(D)	5	0	0	–1/2

23. How many significant figures should there be in the answer to the following problem ? [CBSE-94]

$$\frac{(29.2 - 20.2)(1.79 \times 10^5)}{1.39}$$

(A) 3 (B)1 (C) 4 (D) 2

24. The radius of hydrogen atom in the ground state is 0.53 Å. The radius of Li^{2+} ion in a similar state is [CBSE-95]
(A) 1.06 Å (B) 0.265 Å (C) 0.17 Å (D) 0.53 Å

25. The uncertainty in the position of an electron (mass = 9.1×10^{-28}g) moving with a velocity of 3.0×10^4 cms^{-1} accurate upto 0.011% will be [CBSE-95]
(A) 1.92 cm (B) 7.68 cm (C) 5.76 cm (D) 3.84 cm

26. In a Bohr's model of atom when an electron jumps from n = 1 to n = 3, how much energy will be emitted or absorbed ? [CBSE-96]
(A) 2.15×10^{-11} ergs (B) 0.1911×10^{-10} ergs
(C) 2.389×10^{-10} ergs (D) 0.239×10^{-10} ergs

27. Which of the following does not have sp^2 hybridisation? [CBSE-96]
(A) C_2H_6 (B) $BeCl_2$ (C) C_2H_4 (D) C_2H_2

28. What will be the longest wavelength line in Balmer series of spectrum ? [CBSE-96]
(A) 566 nm (B) 556nm (C) 546 nm (D) 656nm

29. Given the numbers: 161cm., 0.161cm, 0.0161cm. The number of significant figures for the three numbers is [CBSE-98]
(A) 3, 4 and 5 respectively (B) 3, 3 and 3 respectively
(C) 3, 3 and 4 respectively (D) 3, 4 and 4 respectively

30. The position of both an electron and a helium atom is known within 1.0 nm. Further the momentum of the electron is known within 5.0×10^{-26} kg. ms^{-1}. The minimum uncertainty in the measurement of momentum of helium atom is [CBSE-98]
(A) 50 kg ms^{-1} (B) 5.0×10^{-26} kg ms^{-1}
(C) 80 kg ms^{-1} (D) 80×10^{-4}

31. The number of neutrons in a parent nucleus X which give N after two successive β emissions would be [CBSE-98]
(A) 6 (B) 7 (C) 8 (D) 9

32. The Bohr orbit radius for hydrogen atom (n=1) is approximately 0.530Å. The radius for the first excited state (n = 2) orbit is (in Å) [CBSE-98]
(A) 1.06 (B) 0.13 (C) 4.77 (D) 2.12

33. Which of the following configuration is correct for iron ? [CBSE-99]
(A) $1s^2 2s^2 2p^6 3s^2 3p^6 3d^5$ (B) $1s^2 2s^2 2p^6 3s^2 3p^6 4s^2 3d^5$
(C) $1s^2 2s^2 2p^6 3s^2 3p^6 4s^2 3d^7$ (D) $1s^2 2s^2 2p^6 3s^2 3p^6 4s^2 3d^6$

34. Who modified Bohr's theory by introducing elliptical orbits for electron path ? [CBSE-99]

(A) Hund (B) Thompson

(C) Rutherford (D) Sommerfield

35. The uncertainty in momentum of an electron is 1×10^{-5} kgm/s. The uncertainty in its position will be ($h = 6.62 \times 10^{-34}$ kg-m²/s) [CBSE-99]

(A) 1.05×10^{-26} m (B) 1.05×10^{-25} m

(C) 5.27×10^{-30} m (D) 5.25×10^{-30} m

36. The de Broglie wavelength of a particle with mass 1g and velocity 100 m/s is [CBSE-99]

(A) 6.63×10^{-33} m (B) 6.63×10^{-34} m

(C) 6.63×10^{-35} m (D) 6.65×10^{-35} m

37. Half-life of active hydrogen is [CBSE-2000]

(A) Less than a second (B) 1 sec

(C) 1 min (D) 1 hour

38. The frequency emitted for which transition has higher energy in terms of light radiation ? [CBSE-2000]

(A) $6 \to 1$ (B) $2 \to 1$ (C) $1 \to 2$ (D) $3 \to 4$

39. Calculate the wavelength of light
Given Energy = 3.03×10^{-19} J, $h = 6.36 \times 10^{-34}$ J.s., $c = 3.00 \times 10^{8}$ m/s [CBSE-2000]

(A) 6.56 nm (B) 656 nm (C) 0.656 nm (D) 65.6 nm

40. The number of orbitals represented by following quantum numbers [CBSE-2001]
n = 3, I = 2, m + 2 is

(A) 1 (B) 2 (C) 3 (D)4

41. In hydrogen atom, energy of first excited state is -3.4 eV. Then find out KE of same orbit of Hydrogen atom [CBSE-2002]

(A) +3.4 eV (B) +6.8 eV (C) −13.6 eV (4) + 13.6 eV

42. The value of Planck's constant is 6.63×10^{-34} js. The velocity of light is $3.0 \times 10^{8} ms^{-1}$. Which value is closest to the wavelenght in nanometers of a quantum of light with frequency of $8 \times 10^{15} s^{-1}$? [CBSE-2003]

(A) 3×10^{7} (B) 2×10^{-25} (C) 5×10^{-18} (D) 4×10^{1}

43. In the photo-electron emission, the energy of the emitted electrons is [AIIMS-93]

(A) greater than the incident photon

(B) same as that of the incident photon

(C) smaller than the incident photon

(D) proportional to the intensity of incident proton

44. Which of the following sets (n,l,m,s) of quantum numbers is wrong ? [AIIMS-93]

(A) $2, 1, 0, +\frac{1}{2}$ (B) $3, 2, 2, +\frac{1}{2}$

(C) $2, 2, 0, +\frac{1}{2}$ (D) $3, 2, 2, -\frac{1}{2}$

45. Which of the following expression gives the de Broglie relationship ? [AIIMS-94]

(A) $\frac{h}{mv} = p$ (B) $\lambda = \frac{h}{mv}$ (C) $\lambda = \frac{h}{mp}$ (D) $\lambda m = \frac{v}{p}$

46. According to Dalton's atomic theory, the smallest particle in which matter can exist is called [AIIMS-94]

(A) An atom (B) An ion

(C) An electron (D) A molecule

47. The position of both an electron and a helium atom is known within 1.0 nm. The momentum of the electron is known within 5.0×10^{-26} kg ms^{-1}. The minimum uncertainity in the measurement of the momentum of the helium atom is [AIIMS-94]

(A) 5.0×10^{-26} kg ms^{-1} (B) 6.0×10^{-26} kg ms^{-1}

(C) 7.0×10^{-26} kg ms^{-1} (D) 8.0×10^{-26} kg ms^{-1}

48. The number of electrons and neutrons of an element is 18 and 20 respectively. Its mass number is [AIIMS-94]

(A) 17 (B) 22 (C) 37 (D) 38

49. The number of unpaired electrons in $1s^2 2s^2 2p^3$ is [AIIMS-95]

(A) 1 (B) 2 (C) 3 (D) 5

50. Rutherford's scattering experiment is related to the size of [AIIMS-95]

(A) Nucleus (B) Proton (C) Electron (D) Neutron

51. The maximum energy is possessed by an electron, when it is present [AIIMS-96]

(A) In nucleus (B) In ground energy state

(C) In first excited state

(D) At infinite distance from the nucleus

52. Avogadro's number of oxygen atoms weigh [AIIMS-96]

(A) 8g (B) 16 g (C) 32 g (D) 56 g

53. The maximum number of electrons, present in an orbit that is represented by azimuthal quantum number (l) = 3, will be [AIIMS-96]

(A) 2 (B) 6 (C) 8 (D) 14

54. Assertion : A special line will be seen for a 2px→ 2py transition. Reason : Energy is released in the form of waves of light when the electron drops form 2px to 2py orbital. [AIIMS-96]

(A) A (B) B (C) C (D) D

55. Assertion : Electrons are ejected from a certain metal when either blue or violet light strikes the metal surface. However, only violet light causes electron ejection from a second metal. Reason : The electrons in the first metal require less energy for ejection. [AIIMS-96]

(A) A (B) B (C) C (D) D

56. Which of the following species is isotonic with $_{37}Rb^{86}$? [AIIMS-96]

(A) $_{36}Kr^{87}$ (B) $_{37}Mg^{84}$ (C) $_{38}Kr^{87}$ (D) $_{19}Mg^{89}$

57. The total number of orbitals in a shell can be given in form of n the principal quantum no. is [AIIMS-97]

(A) 2n (B) n^2 (C) $2n_2$ (D) n + 1

58. Positron is [AIIMS-97]

(A) Electron with +ve charge (B) A helium nucleus
(C) A nucleus with two protons
(D) A nucleus with one neutron and one proton

59. Assertion : Hydrogen has only one electron in its orbit but it produces several spectral lines. Reason : There are many excited energy levels available. [AIIMS-97]

(A) A (B) B (C) C (D) D

60. Which of the following explains the sequence of filling the electrons in different shells ? [AIIMS-98]

(A) Hund's rule (B) Octet rule
(C) Aufbau principle (D) all of these

61. Planck's constant has the same dimesion as that of [AIIMS-98]

(A) Power (B) Work
(C) Radiant energy (D) Angular momentum

62. Which of the following is the main cause of late discovery of neutron ? [AIIMS-98]

(A) Neutron is chargeless particle
(B) Neutron is highly unstable particle
(C) Neutron is nucleus very fast (D) All of these

63. Ratio of the energy of a photon of wavelengths 3000 Å and 6000 Å is [AIIMS-98]

(A) 2 : 1 (B) 1 : 2 (C) 1 : 3 (D) 1 : 6

64. Principle quantum number of an atom is related to the [AIIMS-98]

(A) Size of the orbital (B) Spin angular momentum
(C) Orbital angular momentum
(D) Orientation of the orbital in space

65.Assertion: The energy of an electron is largely determined by its principal quantum number.
Reason: The principal quantum number (n) is a measure of the most probable distance of finding the electrons around the nucleus. [AIIMS-98]

(A) A (B) B (C) C (D) D

66. Principal, azimuthal and magnetic quantum numbers are respectively related to [AIIMS-99]

(A) Size, shape and orientation
(B) Shape, size and orientation
(C) Size, orientation and shape (D) None of these

67. If magnesium atom having atomic number 12 has an isotope with mass number 26, then the number of protons, neutrons and electrons respectively of the atom are [AIIMS-99]

(A) 12, 12, 14 (B) 12, 14, 12
(C) 14, 12, 12 (D) 12, 12, 12

68. Which of the following are $_{16}S^{32}$ and $_{15}Pb^{31}$? [AIIMS-99]

(A) Isotones (B) Isobars (C) Isotopes (D) Isomers

69. When the value of azimuthal quantum number is 3, magnetic quantum number can have values [AIIMS-2000]

(A) +3, +2, +1, 0, –1, –2, –3 (B) +2, +1, 0, –1, –2,
(C) +1, 0, –1 (D) +1, –1

70. Spectrum given due to transition of an electron form M to L shell is [AIIMS-2000]

(A) Absorption (B) Continuous
(C) Emission (D) γ - ray

71. Which of the following has no neutron in its nucleus ? [AIIMS-2000]

(A) Tritium (B) Hydrogen (C) Helium (D) Deuterium

72. Which relation is called Heisenberg's uncertainty principle ? [AIIMS-2000]

(A) $\Delta p \Delta x \geq \frac{h}{4\pi}$ (B) $\Delta u \Delta x \geq \frac{h}{4\pi m}$

(C) $\Delta E \Delta t \geq \frac{h}{4\pi}$ (D) All of these

73. Hund's rule of maximum multiplicity explains [AIIMS-2000]

(A) Filling of electrons in orbitals
(B) Filling of electrons in orbitals of atom
(C) Arrangement of electrons in orbitals
(D) Pairing of electrons in shells

74. Which is correctly matched ? [AIIMS-2001]

(A) n = 5, l = 2, m = 2, s = –1/2
(B) n = 5, l = 5, m = 2, s = +1/2
(C) n = 3, l = 2, m = 2, s = 0
(D) n = 3, l = 2, m = 3, s = +1/2

75. If energy of e^- in H-atom in some excited state is –3.4 eV, then what will be its angular momentum [AIIMS-2001]

(A) 1.8 x 10^{-30} Kgm^2s^{-1} (B) 2.1 x 10^{-34} Kgm^2s^{-1}
(C) 9.2 x 10^{-30} Kgm^2s^{-1} (D) 1.2 x 10^{-32} Kgm^2s^{-1}

76. A dye absorbs strongly in region 605-750 nm. The colour of the reflected light will be [AIIMS-2001]

(A) Blue – green (B) Blue – red
(C) Red – green (D) Green – blue

77. Assertion : Two electrons in an atom can have the same value of four quantum numbers. Reason : Two

electrons in an atom can be simultaneously in the same shell, sub-shell and orbital and have the same spin. [AIIMS-2001]

(A) A (B) B (C) C (D) D

78. Quantum numbers of an atom can be defined on the basis of [AIIMS-2002]

(A) Hund's rule (B) Aufbau's principle
(C) Pauli's exclusion principle
(D) Heisenberg's uncertainty principle

79. Which of the following has maximum energy ? [AIIMS-2002]

(A) 3s [↑↓] 3p [↑↓][↑][↑] 3d [ ][ ][ ][ ][ ]

(B) 3s [↑↓] 3p [↑][↑][↑] 3d [↑][↑][ ][ ][ ]

(C) 3s [↑↓] 3p [↑][↑][↑] 3d [ ][↑][ ][ ][ ]

(D) 3s [↑↓] 3p [↑][↑][↑] 3d [↑][ ][ ][ ][ ]

80. Spectrum of Li^{2+} is similar to that of [AIIMS-2002]

(A) H (B) He (C) Be (D) Ne

81. Azimuthal quantum number defines [AIIMS-2002]

(A) e/m ratio of electron (B) spin of electron
(C) Angular momentum of electron
(D) Magnetic momentum of electron

82. Assertion : (A) : The value of n for a line in Balmer series of hydrogen spectrum having the highest wave length is 4 and 6.Reason (R) : For Balmer series of hydrogen spectrum, the value $n_1 = 2$ and $n_2 = 3, 4, 5$ [AIIMS-2002]

(A) A (B) B (C) C (D) D (E) E

83. Assertion (A) : Absorption spectrum consists of some bright lines separated by dark spaces. Reason (R) : Emission spectrum consists of dark lines. [AIIMS-2002]

(A) A (B) B (C) C (D) D (E) E

84. The quantum number 'm' of a free gaseous atom is associated with - [AIIMS-2003]

(A) The effective volume of the orbital
(B) The shape of the orbital
(C) The spatial orientation of the orbital
(D) The energy of the orbital in the absence of a magnetic field

85. Number of sub shells in the principal quantum number 'n' are : [JIPMER-90]

(A) $2n^2$ (B) n^2 (C) 2π (D) n

86. Which of the following represent the correct set up of the four quantum numbers of 4s electron : [JIPMER-90]

(A) 4, 3, 2, +1/2 (B) 4, 2, 1, 0
(C) 4, 3, - 2, +1/2 (D) 4, 2, 1, -1/2

87. Which of the following electron transitions in a hydrogen atom will require the largest amount of energy? [JIPMER-90]

(A) From n = 1 to n = 2 (B) From n = 2 to n = 3
(C) From n = ∞ to n = 1 (D) From n = 3 to n = 5

88. The principal quantum number of an atom is related to the : [JIPMER-91]

(A) Size of the orbital (B) Spin angular momentum
(C) Orbital angular momentum
(D) Orientation of the orbital in space

89. Mass of an individual atom of sulphur is : [JIPMER-91]

(A) 32 g. (B) 5.31×10^{-23}g.
(C) $32 \times 6.023 \times 10^{23}$g (D) 32×10^{-23}g.

90. Any p-orbital can accommodate upto : [JIPMER-91]

(A) 4 electrons (B) 6 electrons
(C) Two electrons with parallel spins
(D) Two electrons with opposite spins

91. Be's 4th electron will have the four quantum numbers ? [JIPMER-92]

(A) n = 1, l = 0, m = 0, s = −½
(B) n = 1, l = 0, m = 0, s = ½
(C) n = 2, l = 0, m = 0, = −½
(D) n = 2, l = 1 m = 0, s = ½

92. Isotopes of an element have : [JIPMER-92]

(A) Different atomic numbers
(B) Same mass numbers (C) Different mass numbers
(D) Different mass and atomic numbers

93. An electron that has the quantum number n = 3 and n = 2 ? [JIPMER-92]

(A) Must have spin value +½ (B) Must have 1=1
(C) Must have 1-0, 1 or 2 (D) Must have 1=2

94. Helium atom is two times heavier than a hydrogen molecule. At 25°C the average K. E. of helium atom is : [JIPMER-92]

(A) Twice that of hydrogen (B) Same as that of hydrogen
(C) Four times that of hydrogen
(D) Half that of hydrogen

95. For a spherical shaped orbital, the value of *l* is : [JIPMER-92]

(A) 3 (B) 0 (C) 1 (D) 2

96. In a photoelectric effect, the energy of the photon striking a metallic surface is $5.6x10^{-19}$ J. The kinetic energy of ejected electrons is $12.0x10^{-20}$ J. The work function is: [JIPMER-93]
(A) $6.4x10^{-19}$J (B) $6.8x10^{-19}$J
(C) $4.4x10^{-19}$J (D) $6.4x10^{-20}$J

97. Weight of 6.02×10^{23} molecules of CO_2 is [JIPMER-94]
(A) 11g (B) 14g (C) 22g (D) 44g

98. An electronic transition from 1s orbital of an atom causes [JIPMER-95]
(A) Release of energy (B) Absorption of energy
(C) Both release and absorption of energy
(D) Unpredictable behaviour

99. The energy of a photon is [JIPMER-95]
(A) Inversely proportional to the wave number
(B) Directly proportional to its velocity
(C) Directly proportional to frequency
(D) Directly proportional to its wavelength

100. A 3d-electron specified by s = +1/2 can have a magnetic quantum number : [JIPMER-96]
(A) +3 (B) +2 (C) 4+ (D) −3

101. The K.E. of electron in He^+ is maximum in : [JIPMER-96]
(A) 2nd orbit (B) 3rd orbit
(C) 1st orbit (D) In orbit with n = ∞

102. The existence of line spectrum of hydrogen provided an evidence that : [JIPMER-96]
(A) There is absorption of energy
(B) The electrons do not lose energy while revolving in stationary states
(C) The energy of electron is quantised
(D) There is an emission of energy when hydrogen is heated

103. The continuum in an atomic spectrum is due to [JIPMER-97]
(A) Instability of the excited state
(B) Interaction of anion with an atom
(C) Wave property of an electron
(D) Ionization of the atom

104. After the emission of a ----, the daughter element is placed one group to the left of the parent element. [JIPMER-97]
(A) β-ray (B) α-ray (C) Positron (D) γ-ray

105. Energy of the atomic orbitals in a particular energy level is in the order. [JIPMER-97]
(A) s < p < d < f (B) s > p > d > f
(C) p < d < f < s (D) f > d > s > P

106. The correct set of four quantum numbers for the valence (outermost) electron of rubidium (z = 37) is [JIPMER-98]
(A) 5, 0, 0, +1/2 (B) 5, 1, 0, +1/2
(C) 5, 1, 1, +1/2 (D) 6, 0, 0, +1/2

107. Which of the following sub-atomic particles is not present in an atom ? [JIPMER-98]
(A) Neutron (B) Proton (C) Electron (D) Positron

108. p-orbitals in a given shell can accommodate upto [JIPMER-98]
(A) Four electrons (B) Two electrons with parallel spins
(C) Six electrons (D) Two electrons with opposite spins

109. Experimental evidence for the existence of the atomic nucleus comes from. [JIPMER-99]
(A) Atomic emission spectroscopy
(B) Millikan's oil-drop experiment
(C) Alpha scattering by a thin metal foil
(D) The magnetic bending of cathode rays

110. The radius of the third Bohr orbit is 4.77 Å. The radius of the second orbit will be. [JIPMER-99]
(A) 1.16 Å (B) 3.18 Å (C) 2.12 Å (D) 1.59 Å

111. An isotone of $^{30}_{14}Si$ is [JIPMER-99]
(A) $^{31}_{15}P$ (B) $^{32}_{15}P$ (C) $^{31}_{16}S$ (D) $^{34}_{17}S$

112. A subshell with ℓ = 2 can take up a maximum of ----- electrons. [JIPMER-99]
(A) 2 (B) 6 (C) 10 (D) 14

113. The de-Broglie wavelength associated with a material particle is [JIPMER-2000]
(A) Directly proportional to its energy
(B) Directly proportional to momentum
(C) Inversely proportional to its energy
(D) Inversely proportional to momentum

114. The ratio of energy of photons of 3000 Å and 6000 Å wavelength radiations is [JIPMER-2000]
(A) 4 (B) 2 (C) 1/2 (D) 1/4

115. Which property is different for neutral atoms of the two isotopes of the same element [JIPMER-2001]
(A) Number of protons (B) Atomic number
(C) Number of neutrons (D) None

116. X-rays are produced due to [JIPMER-2002]
(A) Bombarding of electrons on solids.
(B) Bombarding of α−particle on solids.
(C) Bombarding of γ-rays on solids.
(D) Bombarding of neutron on solids.

117. The uncertainty in the momentum of an electron is $1.0x10^{-5}kg.ms^{-1}$. The uncertainty in its position will be. (given, $h = 6.62 \times 10^{-34}kgm^2s^{-1}$) [JIPMER-2002]
(A) 1.05×10^{-28}m (B) 1.05×10^{-26}m

(C) 5.27×10^{-30}m (D) 5.25×10^{-28}m

118. What is the packing fraction of $^{56}_{26}Fe$? [JIPMER-2002]
(A) +14.167 (B) +73.90 (C) −14.167 (D) −73.90

119. The number of atoms in 4.25 grams of NH_3 is approximately [BHU-95]
(A) 1×10^{23} (B) 1.5×10^{23} (C) 2×10^{23} (D) 6×10^{23}

120. The azimuthal quantum number for last electron of sodium atom is [BHU-95]
(A) 1 (B) 2 (C) 3 (D) 0

121. The positron has mass equal to [BHU-95]
(A) Electron (B) Proton (C) Deutron (D) a-particle

122. Rutherford's experiment led to the discovery of [BHU-95]
(A) Nucleus (B) α-particle (C) β-particle (D) Electron

123. Ratio of masses of proton and electron is [BHU-98]
(A) Infinite (B) $1.8 \times 10^{+3}$ (C) 1.8 (D) None of these

124. An e- has magnetic quantum number as - 3, what is its principal quantum number ? [BHU-98]
(A) 1 (B) 2 (C) 3 (D) 4

125. Wavelength associated with electron motion [BHU-98]
(A) Increases with increase in speed of electron
(B) Remains same irrespective of speed of electron
(C) Decreases with speed increase of e^- (D) Is zero

126. The number of atoms in 4.25 g of NH_3 is approximately [BHU-99]
(A) 1.0×10^{22} (B) 1.0×10^{23}
(C) 2.0×10^{23} (D) 6.0×10^{23}

127. The Bohr's orbit radius for the hydrogen atom (n=1) is approximately 0.530 Å. The radius for the first excited state (n = 2) orbit is [BHU-99]
(A) 0.13Å (B) 1.06Å (C) 4.77Å (D) 2.22Å

128. Which of the following explains the sequence of filling the electrons in different shells ? [BHU-99]
(A) Octet rule (B) Hund's rule
(C) Aufbau principle (D) All of these

129. Which of the following is not possible [BHU-2000]
(A) n = 3, l = 0, m = 0 (B) n = 3, l = 1, m = −1
(C) n = 2, l = 0, m = −1 (D) n = 2, l = 1, m = 0

130. Electrons in an orbital gives rise to magnetic field [BHU-2001]
(A) In the same direction (B) In the opposite direction
(C) In either direction
(D) No magnetic field is created due to spin

131. Wavelength of spectral line emitted is inversely proportional to [BHU-2001]
(A) Quantum number (B) Energy
(C) Radius (D) Velocity

132. Which one of the following represents an impossible arrangements ? [BHU-2001]

	n	ℓ	m	n
(A)	3	2	-2	1/2
(B)	4	0	0	1/2
(C)	3	2	-3	1/2
(D)	5	3	0	1/2

133. Which of the following relates to photons having both wave as well as particle nature [BHU-2001]
(A) Interference (B) Diffraction
(C) $E = mc^2$ (D) $\lambda = c/\nu$

134. Which of the following is isotonic with $_{37}Rb^{86}$? [BHU-2001]
(A) $_{36}Kr^{84}$ (B) $_{37}Rb^{85}$ (C) $_{38}Sr^{87}$ (D) $_{39}Y^{89}$

135. For electronic transition from $n = 2 \rightarrow n = 1$. Which of the following produce shortest wavelength ? [BHU-2002]
(A) H-atom (B) D-atom (C) He^+ion (D) Li^{2+} ion

136. If uncertainty in position of an electron is zero, the uncertainty in its momentum will be [BHU-2002]
(A) Zero (B) Infinite
(C) Less than h/4π (D) greater than h/4π

137. Which of the following is correct statement for $3d_{xy}$ $3d_{yz}$ $3d_{zx}$ $3d_{x^2-y^2}$, $3d_z$? [BHU-2002]
(A) $3d_{yz}$ has maximum energy
(B) $3d_{z^2}$ has lowest energy
(C) All orbitals have same energy
(D) All orbitals have different energy

138. When the specific heat of a metallic element is 0.214 cal/gm, the atomic weight will be close to which of the following ? [AFMC-91]
(A) 6.6 (B) 12 (C) 30 (D) 68

139. Dual nature of particle is given by [BHU-2003]
(A) de Broglie equation
(B) Bohr's theory
(C) Heisenberg's uncertainity principle
(D) Thomson model

140. Compounds having same number of electrons but different mass are [AFMC-92]
(A) Isotopes (B) Isobar (C) Isotone (D) Allotropes

141. Atomic number is equal to [AFMC-94]
(A) Number of neutrons in the nucleus
(B) Number of protons in the nucleus
(C) The sum of number of neutrons and protons

(D) Atomic weight of element

142. Splitting of spectral line under the influence of magnetic field is called [AFMC-95]
(A) Stark effect (B) Zeeman's effect
(C) Photoelectric effect (D) None of these

143. The isotopes of elements was discovered by [AFMC-95]
(A) Soddy (B) Curie (C) Thomson (D) Chadwick

144. The number of electrons in the nucleus of $C1_2$ is [AFMC-95]
(A) 6 (B) 12 (C) 0 (D) 3

145. A neutral atom (Atomic No. >1) contains [AFMC-96]
(A) Proton only (B) Neutron + Proton
(C) Neutron + Electron (D) Neutron+ Proton + Electron

146. Atomic number means [AFMC-96]
(A) No. of electron (B) No. of protons
(C) No. of neutron (D) No. of nucleons

147. It is impossible to know simultaneously the position and momentum of a moving particle with absolute exactness at any instant. This is known as [AFMC-96]
(A) Aufbau principle (B) Hund's rule
(C) Heisenberg principle (D) Pauli's principle

148. Dulong and Petit's Law used specific heat for determination of atomic weight for [AFMC-96]
(A) Solid (B) Gas (C) Liquid (D) All of these

149. Positron is [AFMC-97]
(A) ${}_{-1}e^{o}$ (B) ${}_{+1}e^{o}$ (C) ${}_{1}H^{1}$ (D) None of these

150. Number of protons, neutrons and electrons in the element ${}^{231}_{89}\gamma$ is [AFMC-97]
(A) 89, 231, 89 (B) 89, 89, 242
(C) 89, 142, 89 (D) 89, 71, 89

151. If electron falls form n = 3 to n =2, then emitted energy is [AFMC-97]
(A) 10.2 eV (B) 12.09 eV (C) 1.9 eV (D) 0.64 eV

152. One mole of CO_2 contains [AFMC-98]
(A) 3 grams atoms of CO_2 (B) 18.11 x 10^{23} molecules of CO_2
(C) 6.023 x 10^{23} atoms of oxygen
(D) 6.023 x 10^{23} atoms of carbon

153. The uncertainty in the momentum of an electron is 1 x 10^{-4} kg - ms^{-1}. The uncertainty in its position will be (h = 6.62 x 10^{-34} kg-m^{-2} s^{-1}) [AFMC-98]
(A) 1.05 x 10^{-28} m (B) 1.05 x 10^{-28} m
(C) 5.27 x 10^{-30} m (D) 5.25 x 10^{-28} m

154. Aufbau principle is obeyed in which of the following electronic configuration ? [AFMC-99]
(A) $1s^2 3s^2 3p^6$ (B) $1s^2 2s^2 3s^2$ (C) $1s^2 2s^2 2p^6$ (D) $1s^2 3s^3 3p^2$

155. According to Bohr's theory of hydrogen atom, which of the following is quantised for an electron ? [AFMC-99]
(A) Angular momentum (B) Angular acceleration
(C) Acceleration (D) Velocity

156. Magnetic quantum number specifies [AFMC-99]
(A) Nuclear stability (B) Orientation of orbitals
(C) Size of orbitals (D) Shape of orbital

157. Ratio of the energy of a photon of wavelengths 3000 Å and 6000 Å is [AFMC-99]
(A) 3 : 1 (B) 2 : 1 (C) 1 : 2 (D) 1 : 3

158. The de-Broglie wavelength of a particle with mass 1 g and velocity 100 m/s is [AFMC-99]
(A) 6.63 x 10^{-35}m (B) 6.65 x 10^{-35}m
(C) 6.63 x 10^{-33} m (D) 6.63 x 10^{-34} m

159. Isotopes have same number of [AFMC-2000]
(A) Protons (B) Neutrons (C) Nucleus (D) Positrons

160. Neutron was discovered by [AFMC-2000]
(A) Chadwick (B) Rutherford
(C) Yukawa (D) Dalton

161. Which of the following laws directly explains the law of "conservation of mass" ? [AFMC-2001]
(A) Dalton's law (B) Avogadro's law
(C) Berzulis' hyporthesis (D) Hund's rule

162. Rutherford's atomic model suggests the existence of [AFMC-2001]
(A) Atom (B) Nucleus (C) α-particle (D) Mesons

163. Positron was discovered by [AFMC-2001]
(A) Rutherford (B) Fermi (C) Vulrava (D) Anderson

164. Which of the following subshells is not possible ? [AFMC-2002]
(A) 2d (B) 4f (C) 6p (D) 3s

165. Given the numbers : 161 cm, 0.161 cm, 0.0161 cm, The number of significant figures for the three numbers is [AFMC-2002]
(A) 3, 4 and 5 (B) 3, 3 and 3
(C) 3, 3 and 4 (D) 3, 4 and 4

166. Maximum electrons in a d-orbital are [AFMC-2002]
(A) 2 (B) 10 (C) 6 (D) 14

167. The mass of 1 x 10^{22} molecules of $CuSO_4.5H_2O$ is .... [IIT-91]

168. Which of the following does not characterise X rays ? [IIT-92]

(A) A radiation can ionise gases
(B) It causes ZnS to fluorescene
(C) Deflected by electric and magnetic fields
(D) Have wavelengths shorter than ultra-violet

169. Which of the following relates to photons both as wave motion and as a stream of particles ? [IIT-92]
(A) Intererence (B) $E = mc^2$ (C) Diffraction (D) $E = h\nu$

170. The light radiations with discrete quanuities of energy are called .... [IIT-93]

171. Wave functions of electrons in atoms and molecules are called .... [IIT-93]

172. Ca^{2+} has smaller ionic radius than K^+ because it has .... [IIT-93]

173. The $2p_x$, $2p_y$ and $2p_z$ orbitals of atoms have identical shapes but differ in their .... [IIT-93]

174. The orbital angular momentum of electron in 2s orbital is [IIT-96]
(A) $+\frac{1}{2}\frac{h}{2\pi}$ (B) zero (C) $\frac{h}{2\pi}$ (D) $\sqrt{2}\frac{h}{2\pi}$

175. The first use of quantum theory to explain the structure of atom was made by [IIT-97]
(A) Heisenberg (B) Bohr (C) Planck (D)Einstein

176. For a 'd'-electron, the orbital angular momentum is. [IIT-97]
(A) $\sqrt{6}h$ (B) $\sqrt{2}h$ (C) h (D) 2h

177. Which one of the following compounds has sp^2 hybridization ? [IIT-97]
(A) CO_2 (B) SO_2 (C) N_2O (D) CO

178. The number of moles of $KMnO_4$ that will be needed to react with one mole of sulphite ion in acidic solution is [IIT-97]
(A) $\frac{2}{5}$ (B) $\frac{3}{5}$ (C) $\frac{4}{5}$ (D) 1

179. The energy of an electron in the first Bohr orbit of H atom is −13.6 eV. The possible energy value (s) of the excited state (s) for electrons in Bohr orbits of hydrogen is (are). [IIT-98]
(A) −3.4 eV (B) −4.2 eV (C) −6.8 eV (D) +6.8.eV

180. Which of the following statement (is) is (are) correct ? [IIT-98]
(A) The electronic configuration of Cr is$[Ar]3d^5 4s^1$. (Atomic No. of Cr = 24)
(B) The magnetic quantum number may have negative value.
(C) In silver atom, 23 electrons have a spin of one type and 24 of the opposite type. (Atomic no. of Ag = 47)
(D) The oxidation state of nitrogen in HN_3 is −3.

181. The electrons identified by quantum numbers n and *l*, (i) l=1 (ii) n = 4,l=0 (iii) n = 3,l=2 (iv) n =3,l=1 can be placed in order of increasing energy from the lowest to highest as [IIT-99]
(A) (iv) < (ii) < (iii) < (i) (B) (ii) < (iv) < (i) < (iii)
(C) (i) < (iii) < (ii) < (iv) (D) (iii) < (i) < (iv) < (ii)

182. A signature, written in carbon pencil weighs 1 mg. what is the number of carbon atoms present in the signature? [CET-2003]
(A) 0.502×10^{20} (B) 6.02×10^{20}
(C) 5.02×10^{20} (D) 502×10^{23}

183. Ground state electronic configuration of nitrogen atom can be represented by [IIT-99]

(A) 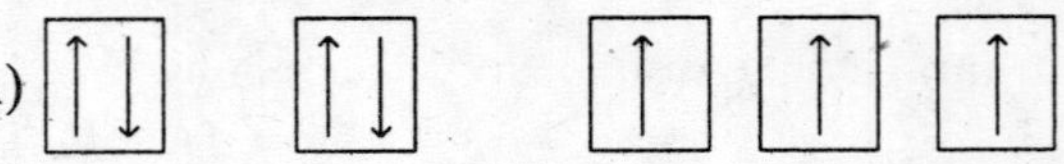

(B) 

(C) 

(D) 

184. The wavelength associated with a golf ball weighing 200 g and moving at a speed of 5 m/h is of the order [IIT-2001]
(A)10^{-10} m (B) 10^{-20} m (C) 10^{-30} m (D) 10^{-40} m

185. The quantum numbers +1/2 and −1/2 for the electron spin represent [IIT-2001]
(A) rotation of the electron in clockwise and anticlockwise direction respectively
(B) rotation of the electron in anticlockwise and clockwise direction respectively
(C) magnetic moment of the electron pointing up and down respectively
(D) two quantum mechanical spin states which have no classical analogue

186. How many moles of electron weigh one kilogram ? [IIT-2002]
(A) 6.023×10^{23} (B) $\frac{1}{9.108} \times 10^{31}$
(C) $\frac{6.023}{9.108} \times 10^{54}$ (D) $\frac{1}{9.108 \times 6.023} \times 10^8$

187. If the Nitrogen atom had electronic configuration $1s^7$. It would have energy lower than that of the normal ground state configuration $1s^2 2s^2 2p^3$, because the electrons would be closer to nucleus. Yet $1s^7$ is not observed because it violates [IIT-2002]
(A) Heisenberg uncertainty principle (B) Hund's rule
(C) Pauli exclusion principle
(D) Bohr postulate of stationary orbits.

188. In which of the following number of atoms are

maximum ? [IIT-2003]

(A) 24 gm of Mg. (B) 20 gm of Ca
(C) 1 gm of H_2 (D) 24 gm of C

189. Number of atoms in 558.5 gram Fe (at. wt. of Fe = 55.85 g mol^{-1}) is. [AIEEE-2002]

(A) Twice that in 60 g carbon (B) 6.023 x 10^{22}
(C) Half that in 8g He (D) 558.5 x 6.023 x 10^{23}

190. In a hydrogen atom, if energy of an electron in ground state is 13.6 eV, then that in the 2nd excited state is. [AIEEE-2002]

(A) 1.51 eV (B) 3.4 eV (C) 6.04 eV (D) 13.6 eV

191. Uncertainty in position of a minute particle of mass 25g in space is 10^{-5} m. What is the uncertainty in its velocity (in ms^{-1}) ? [AIEEE-2002]

(A) 2.1 x 10^{-34} (B) 0.5 x 10^{-34}
(C) 2.1 x 10^{-28} (D) 0.5 x 10^{-23}

192. In Bohr series of lines of hydrogen, spectrum, the third line from the red end corresponds to which one of the following inter-orbit jumps of the electron for Bohr orbits in an atom of hydrogen ? [AIEEE-2003]

(A) $3 \rightarrow 2$ (B) $5 \rightarrow 2$ (C) $4 \rightarrow 1$ (D) $2 \rightarrow 5$

193. The de Broglie wavelength of a tennis ball of mass 60 g moving with a velocity of 10 meters per second is approximately [AIEEE-2003]

(A) 10^{-33} meters (B) 10^{-31} meters
(C) 10^{-16} meters (D) 10^{-25} meters

Planck's constant, h = 6.63 x 10-34 Js

194. The orbital angular momentum for an electron revolving in an orbit is given by $\sqrt{l(l+1)}.\frac{h}{2\pi}$. This momentum for an s-electron will be given by [AIEEE-2003]

(A) $+\frac{1}{2}.\frac{h}{2\pi}$ (B) Zero (C) $\frac{h}{2\pi}$ (D) $\sqrt{2}.\frac{h}{2\pi}$

195. A reduction in atomic size with increase in atomic number is a characteristic of [AIEEE-2003]

(A) High atomic masses (B) d-block
(C) f-block (D) Radioactive series.

196. Which is correct with respect to 'P' orbits [EAMCET-90]

(A) They are spherical
(B) They have a strong directional character
(C) They are five fold degenerate
(D) They have no directional character.

197. The nitride ion in lithium nitride is composed of : [EAMCET-90]

(A) 7 protons + 7 electrons (B) 10 protons + 7 electrons
(C) 7 protons + 10 electrons (D) 10 protons + 10 electrons

198.The ratio between the neutrons, present in carbon and silicon with respective atomic masses of 12 and 28 is .... [EAMCET-90]

199. When 4p orbital in any atom are filled completely, the next electron goes in : [EAMCET-91]

(A) 5s (B) 3d (C) 4 d (D) 4 f

200. Which one of the following series of lines is found in the U.V. region of atomic spectrum of Hydrogen : [EAMCET-91]

(A) Balmer (B) Paschen (C) Bracket (D) Lyman

201. The energy of the electron in the Hydrogen atom is given by the expression : [EAMCET-91]

(A) $\frac{-e^2}{r^2}$ (B) $\frac{-n^2h^2}{2\pi^2z^2e^4m}$
(C) $\frac{-2p^2z^2e^4m}{n^2h^2}$ (D) $\frac{nh}{2p}$

202. One atom of $^{39}_{19}K$ contains [EAMCET-91]

(A) 19 P; 20 n and 19 e (B) 19 P; 20 n and 20 e
(C) 20 P, 19 n and 20 e (D) 20 P, 19 n and 19 e

203. When an electron in an excited hydrogen atom jumps from an energy level for which n = 5 to a lower level for which n = 2 the spectral line is observed in the .... series of the hydrogen spectrum. [EAMCET-91]

204. The max. number of electrons that can be accomodated in all the orbitals for which l = 3 is .... [EAMCET-91]

205. The magnetic quantum No. "m" for the outer most electron in the Na atom is .... [EAMCET-91]

206. Nucleides having different mass number and different atomic numbers, but the same number of neutrons are called .... [EAMCET-91]

207. The nuclei which have same number of neutrons but different number of protons are called .... [EAMCET-92]

208. Which one of the following gases contains the least number of molecules ? [EAMCET-94]

(At. wt of P = 31, Cl = 35.5, N = 14, O = 16, C = 12, H = 1)

(A) 4.0g of laughing gas (B) 3.0g of Phosphine
(C) 2.0g of marsh gas (D) 10.0g of marsh gas.

209. The constancy of e/m ratio for electrons inspite of variations of gas present in the discharge tube or of the material of the cathode shows that. [EAMCET-94]

(A) Electrons are - vely charged
(B) Electrons are universal constituents of matter
(C) Electrons are lightest of all particles
(D) Mass of the electron is 1/1838 of the mass of H-atom

210. The basic assumption of Bohrs Model of hydrogen atom is that [EAMCET-94]

(A) The energy of t he electron is quantised

(B) The angular momentum of the electron is quantised
(C) The radial distance of the electron is quantised
(D) The orbital velocity of the electron is quantised

211. Rutherford's experiment on scattering of α-particles showed for the first time that the atom has [EAMCET-95]
(A) Nucleus (B) Electron (C) Proton (D) Neutron

212. The radius of second Bohr's orbit is [EAMCET-95]
(A) 0.053nm (B) 0.106nm (C) 0.212nm (D) 0.0265nm

213. Nuclides having equal number of neutrons but differing in mass numbers and atomic numbers are called. [EAMCET-95]
(A) Isomers (B) Isotopes (C) Isotones (D) Isobars

214. The de-Broglie wavelength associated with a particle of mass 1 mg moving with a velocity of 10 ms^{-1} is [EAMCET-96]
(A) 6.63×10^{-29} m (B) 6.63×10^{-31}m
(C) 6.63×10^{-34} m (D) 6.63×10^{-22}m

215. The rule that explains the reason for Chromium to have $[Ar]3d^5 4s^1$ configuration instead of $[Ar]\ 3d^4 s^2$? [EAMCET-96]
(A) Pauli's exclusion principle (B) Aufbau Principle
(C) Hund's rule (D) Heisenberg principle

216. In the Bohr Hydrogen atom, the electronic transition emitting light of longest wave length. [EAMCET-97]
(A) n = 2 to n = 3 (B) n = 4 to n = 3
(C) n = 3 to n = 2 (D) n = 2 to n = 1

217. The de Broglie wave length of a particle with mass 1 gram and velocity 100 ms^{-1} is : [EAMCET-97]
(A) 6.63×10^{-35}m (B) 6.63×10^{-34}m
(C) 6.65×10^{-33}m (D) 6.63×10^{-32}m

218. Which one of the following set of quantum numbers is not possible for a 4p electron ? [EAMCET-98]
(A) $n = 4, \ell = 1, m = +1, m_s = +\frac{1}{2}$
(B) $n = 4, \ell = 1, m = 0, m_s = +\frac{1}{2}$
(C) $n = 4, \ell = 1, m = 2, m_s = +\frac{1}{2}$
(D) $n = 4, \ell = 1, m = -1, m_s = -\frac{1}{2}$

219. The radius of the nucleus is related to the mass number 'A' by [EAMCET-98]
(A) $R = R_0A^{1/2}$ (B) $R = R_0A$
(C) $R = R_0A^2$ (D) $R = R_0A^{1/3}$

220. The first emission line of hydrogen atomic spectrum in the Balmer series appears at (R = Rydberg constant) [EAMCET-98]
(A) $\frac{5R}{36}$ cm^{-1} (B) $\frac{3R}{4}$ cm^{-1} (C) $\frac{7R}{144}$ cm^{-1} (D) $\frac{9R}{400}$ cm^{-1}

221. The first emission line in the atomic spectrum of hydrogen in the Balmer series appears at [EAMCET-98(MBBS)]
(A) $\frac{5R}{36}$ cm^{-1} (B) $\frac{3R}{4}$ cm^{-1} (C) $\frac{7R}{144}$ cm^{-1} (D) $\frac{9R}{400}$ cm^{-1}

222. The four quantum numbers of the valence electron of potassium are [EAMCET-98(MBBS)]
(A) 4, 0, 1, (B) 4, 1, 0, (C) 4, 0, 0, (D) 4, 1, 1,

223. Which of the following has magic number of protons and neutrons ? [EAMCET-98(MBBS)]
(A) $_8O^{17}$ (B) $_3Al^{27}$ (C) $_9F^{17}$ (D) $_{20}Ca^{40}$

224. Among the first lines Lyman, Balmer, Paschen and Brackett series in hydrogen atomic spectra which has higher energy ? [EAMCET-99]
(A) Lyman (B) Balmer (C) Paschen (D) Brackett

225. Which one of the following atomic orbitals is NOT directed along the axis ? [EAMCET-99(MBBS)]
(A) P_X (B) $d_{x^2-y^2}$ (C) dxy (D) d_{z^2}

226. What are the values of n_1 and n_2 respectively for H_β line in the Lyman series of hydrogen atomic spectrum? [EAMCET-2000]
(A) 3 and 5 (B) 2 and 3 (C) 1 and 3 (D) 2 and 4

227. How many electrons are present in the M-shell of an atom of the element with atomic number Z = 24 ? [EAMCET-2000]
(A) 5 (B) 6 (C) 12 (D) 13

228. Which of the following is correct pair ? [EAMCET-2000]
(A) $BeCl_2$, Linear (B) NH_3, Linear
(C) CO_2, Tetrahedral (D) BF_3, Octahedral

229. The probability of finding an electron in an orbital is approximately ? [EAMCET(MBBS)-2000]
(A) 95% (B) 50% (C) 60% (D) 25%

230. What is the wavelength of H_β line the Balmer series of hydrogen spectrum ? (R = Rydberg constant) [EAMCET(MBBS)-2000]
(A) 36/5R (B) 5R/36 (C) 3R/16 (D) 16/3R

231. The energy of an electron present in Bohr's second orbit of hydrogen atom is ..... [EAMCET-2001]
(A) -1312 J $atom^{-1}$ (B) $-328kJmol^{-1}$
(C) -328 J mol^{-1} (D) $-164kJmol^{-1}$

232. The atomic number (Z) on an element is 25. In its ground state how many electrons are present in the "N" shell ? [EAMCET(MBBS)-2001]
(A) 13 (B) 2 (C) 15 (D) 3

233. Which one of the following is an isobar of $^{14}_{6}C$? [EAMCET(MBBS)-2001]
(A) $^{13}_{6}C$ (B) $^{12}_{6}C$ (C) $^{14}_{7}N$ (D) $^{15}_{7}N$

234. The radius of an atomic nucleus in generally expressed in units of [EAMCET(MBBS)-2001]
(A) Debye (B) Coulomb (C) Fermi (D) Tesla

235. 50 grams of calcium carbonate was completely burnt in air. What is the weight (in grams) of the residue (Atomic weights of Ca, C and O are 40, 12 and 16 respectively) [EAMCET(MBBS)-2001]
(A) 2.8 (B) 28 (C) 4.4 (D) 44

236. Which one of the following statements is not correct ? [EAMCET-2002]
(A) Rydberg's constant and wave number have same unit.
(B) Lyman series of hydrogen spectrum occurs in the ultraviolet region
(C) The angular momentum of the electron in the ground state hydrogen atom is equal to h/2p
(D) The radius of first Bohr orbit of hydrogen atom is 2.116×10^{-8}cm

237. The energy of an electromagnetic radiation is 19.875×10^{-13} ergs. What is its wave number in cm^{-1} ? ($h = 6.625 \times 10^{-27}$ erg. sec; $C = 3 \times 10^{10}$ cm sec^{-1}) [EAMCET-2002]
(A) 1,000 (B) 10^6 (C) 100 (D) 10,000

238. Which one of the following statements is correct ? [EAMCET-2002]
(A) The radius (R) of a nuclide of mass number A is given by the equation $R = R_0 (A)^{1/2}$ (R_0= constant)
(B) $^{15}_{7}N$ and $^{16}_{8}O$ are isobars
(C) The end nuclide formed in thorium (4n) series is $^{40}_{20}Bi$
(D) $^{40}_{20}Ca$ has magic number of protons and magic number of neutrons.

239. In which of the following nuclear reactions, X and Y are isotopes ? [EAMCET(med.)-2002]
(A) X (α, π) Y (B) X, (α, n) Y
(C) X (π, n) Y (D) X (H, p) Y

240. The energy of an electromagnetic radiation is 3×10^{-12} ergs. What is its wavelength in nanometers ? ($h = 6.625 \times 10^{-27}$ erg. sec; $c = 3 \times 10^{10}$ cm. sec^{-1}). [EAMCET(med.)-2002]
(A) 400 (B) 228.3 (C) 3000 (D) 662.5

241. Which one of the following statements is correct ? [EAMCET(med.)-2002]
(A) '2s' orbital is spherical with two nodal planes.
(B) The de Broglie wavelength (λ) of a particle of mass 'm' and velocity 'V' is equal to mV/n
(C) The principal quantum number (n) indicates the shape of the orbital.
(D) The electronic configuration of phosphorous is given by [Ne] 3s2, $3p_x^1$, $3p_y^1$, $3p_z^1$

242. If the electron of a hydrogen atom is present in the first orbit, the total energy of the electron is [EAMCET-2003]
(A) $\frac{-e^2}{r}$ (B) $\frac{-e^2}{r^2}$ (C) $\frac{-e^2}{2r}$ (D) $\frac{-e^2}{2r^2}$

243. If the wave length of an electromagnetic radiation is 2000 Å. What is its energy in ergs? [EAMCET-2003]
(A) 9.94×10^{-12} (B) 9.94×10^{-19}
(C) 4.97×10^{-12} (D) 4.97×10^{-19}

244. $_{20}X^{40}$ and $_{21}X^{40}$ are [CPMT-96]
(A) Isobars (B) Isotopes
(C) Isotones (D) Isosteremers

245. 1 AMU is equal to [CPMT-96]
(A) $\frac{1}{12}$ of C –12 (B) $\frac{1}{16}$ of O – 16
(C) 1 gm of H_2 (D) 1.66×10^{-23} kg

246. In ionization of Hydrogen the energy required is [CPMT-96]
(A) 13.6 eV (B) >13.6 eV (C) <13.6 eV (D) 1.5 eV

247. Electromagnetic radiation with maximum frequency is- [CPMT-96]
(A) Ultraviolet (B) Radiowave
(C) X-rays (D) Infrared

248. Which statement is not correct for n = 5, m = 3 ? [CPMT-96]
(A) $l = 4$ (B) $l = 0, 1, 2, 3$; $l = +\frac{1}{2}$
(C) $l = 3$; $m = +\frac{1}{2}$ (D) $l = 2$; $m = +\frac{1}{2}$

249. Which orbital of carbon will not emit photon after absorbing it ? [CPMT-97]
(A) 1s (B) 2s (C) 3p (D) 2p

250. van der Waal's equation is true for- [CPMT-97]
(A) Ideal gas (B) Real gas
(C) Gaseous substance (D) None of these

251. The number of electrons $[^{40}_{19}K]^{-1}$ is [CPMT-97]
(A) 19 (B) 20 (C) 18 (D) 40

252. The charge of an electron is 4.8×10^{-10} e.s.u., what is the value of charge in Li^+ ion ? [CPMT-97]
(A) 4.8×10^{-10} e.s.u. (B) 9.6×10^{-10} e.s.u.
(C) 1.44×10^{-9} e.s.u. (D) 2.4×10^{-9} e.s.u.

253. Which of the following set of quantum numbers belong to highest energy ? [CPMT-99]
(A) n = 4, l = 0, m = 0, s = + 1/2
(B) n = 3, l = 0, m = 0, s = + 1/2
(C) n = 3, l = 1, m = 1, s = + 12
(D) n = 3, l = 2, m = 1, s = + 1/2

254. Maximum electrons in a d-orbital are [CPMT-99]
(A) 2 (B) 10 (C) 6 (D) 14

255. Which quantum number will determine the shape of

the subshell ? [CPMT-99]
(A) Principal quantum number
(B) Azimuthal quantum number
(C) Magnetic quantum number (D) Spin quantum number

256. Which of the following sets of quantum numbers is not correct ? [CPMT-2000]
(A) n = 2, ℓ = 0, m = 0, s+ $\frac{12}{}$
(B) n = 4, ℓ = 3, m = 2, s= + $\frac{12}{}$
(C) n=2, ℓ = 2, m = 0, s = – $\frac{1}{2}$
(D) All of these

257. The value of 1 amu is equal to - [CPMT-2000]
(A) 1.0 gm (B) 1.66 x 10^{-24}gm
(C) 1.9924 x 10^{-23} gm (D) 12.00 gm

258. Which of the following is fundamental particle ? [CPMT-2001]
(A) Water molecule (B) Oxygen molecule
(C) Alpha particle (D) Proton

259. The presence of 3 unpaired electrons in nitrogen is explained by [CPMT-2001]
(A) Heisenberg's uncertainty (B) Afbau principle
(C) Pauli's exclusion law (D) Hund's rule

260. Which of the following quantum numbers are not possible ? [CPMT-2001]
(A) n = 2, ℓ = 1, m = – 1, s = – 1/2
(B) n = 3, ℓ = 2, m = – 3, s = + 1/2
(C) n = 2, ℓ = O, m = O, s = + 1/2
(D) n = 3, ℓ = 2, m = – 2, s = + 1/2

261. The orbital with maximum energy is [CPMT-2002]
(A) 3d (B) 5p (C) 4s (D) 6d

262. The order of density of nucleus is [CPMT-2003]
(A) 10^{-8} kg/cc (B) 10^{11} kg/cc
(C) 10^{15} kg/cc (D) 10^{-18} kg/cc

263. Which of the following is true for the value of +1/2 and -1/2 ? [CPMT-2003]
(A) Spin motion of electron in clockwise and anti-clockwise direction.
(B) Spin motion of electron in anti-clockwise and clockwise direction.
(C) Spin motion of electron in any direction.
(D) It indicates two electrons in orbitals with same spins.

264. Which of the following is correct for 2p orbitals : [PMT-93]
(A) n = 1, l = 2 (B) n = 1, l = 0
(C) n = 2, l = 1 (D) n = 2, l = 0

265. The electrons would go to lower energy levels first and then to higher energy levels according to which of the following : [PMT-93]
(A) Aufbau principle (B) Pauli's exclusion principle
(C) Hund's rule of maximum multiplicity
(D) Heisenberg's uncertainty principle

266. The shape of 'p' orbital is : [PMT-93]
(A) elliptical (B) spherical
(C) dumb-bell (D) complex geometrical

267. The atomic number of an element is always equal to [PMT-94]
(A) Atomic weight divided by 2
(B) Number of neutrons in the nucleus
(C) Weight of the nucleus
(D) Electrical charge of the nucleus

268. The possibility of finding an electron in an orbital was conceived by [PMT-94]
(A) Rutherford (B) Bohr
(C) Heisenberg (D) Schrodinger

269. For n = 3 energy level the number of possible orbitals (all kinds) are [PMT-95]
(A) 1 (B) 3 (C) 4 (D) 9

270. When the 3d orbital is complete, the new electrons will enter in [PMT-95]
(A) 4p orbital (B) 4f orbital
(C) 4s orbital (D) 4d orbital

271. The radius of an atom is of the order of [PMT-95]
(A) 10^{-10} cm (B) 10^{-13} cm (C) 10^{-15} cm (D) 10^{-8} cm

272. Which one of the following is an isotone of $^{76}_{32}Ge$? [PMT-95]
(A) $^{77}_{32}Ge$ (B) $^{78}_{33}As$ (C) $^{77}_{34}Se$ (D) $^{78}_{34}Se$

273. Which of the following describes the shape of orbital ? [PMT-96]
(A) Principal quantum number
(B) Azimuthal quantum number
(C) Magnetic quantum number (D) Spin quantum number

274. The number of orbitals in a 2p subshell is [PMT-96]
(A) 1 (B) 2 (C) 3 (D) 6

275. Which of the following expressions gives the de Broglie relationship ? [PMT-96]
(A) $h = \frac{\lambda}{mv}$ (B) $\lambda = \frac{h}{mv}$ (C) $\lambda = \frac{m}{hv}$ (D) $\lambda = \frac{v}{mh}$

276. 2p orbitals have - [PMT-97]
(A) n = 1, l = 2 (B) n = 1, l = 0
(C) n = 2, l = 1 (D) n = 2, l = 0

277. Aufbau principle is not satisfied by [PMT-97]
(A) Cr and Cl (B) Cu and Ag
(C) Cr and Mg (D) Cu and Na

278. Neutrons are found in atoms of all elements except in [PMT-97]

(A) Chlorine (B) Oxygen (C) Argon (D) Hydrogen

279. The atomic weight of an element is 39. The number of neutrons in its nucleus is one more than the number of protons. The number of protons, neutrons and electrons - respectively in its atom would be [PMT-97]

(A) 19, 20, 19 (B) 19, 19, 20
(C) 20, 19, 19 (D) 20, 19, 20

280. The constituents of an atomic nucleus are : [PMT-99]

(A) proton and electron (B) proton, neutron and electron
(C) neutron and proton (D) neutron and electron

281. The mass number of an anion, X^{3-}, is 14. If there are ten electrons in the anion, the number of neutrons in the nucleus of atom, X_2 of the element will be : [PMT-99]

(A) 10 (B) 14 (C) 7 (D) 5

282. The expression for Bohr's radius of an atom is : [PMT-99]

(A) $r = \dfrac{n^2h^2}{4\pi^2 me^4 z^2}$ (B) $r = \dfrac{n^2 h^2}{4\pi^2 m e^2 z}$

(C) $r = \dfrac{n^2 h^2}{4\pi^2 m e^2 z^2}$ (D) $r = \dfrac{n^2 h^2}{4\pi^2 m^2 e^2 z^2}$

283. The energy of an electron revolving in n^{th} Bohr's orbit of an atom is given by the expression : [PMT-99]

(A) $En = -\dfrac{2\pi^2 m^4 e^2 z^2}{n^2h^2}$ (B) $En = -\dfrac{2\pi^2 m e^2 z^2}{n^2 h^2}$

(C) $En = -\dfrac{2\pi^2 m e^4 z^2}{n^2 h^2}$ (D) $En = -\dfrac{2\pi m^2 e^2 z^4}{n^2 h^2}$

284. For Na (Z=11) atom, the four quantum numbers of last electron are : [PMT-99]

(A) $n = 3; l = 1; m = 1; s = +\frac{1}{2}$

(B) $n = 3; l = 0; m = 0; s = +\frac{1}{2}$

(C) $n = 3; l = 0; m = 1; s = +\frac{1}{2}$

(D) $n = 3; l = 1; m = 1; s = -\frac{1}{2}$

285. de-Broglie equation is : [PMT-99]

(A) $n\lambda = 2d\sin\theta$ (B) $E = h\nu$ (C) $E = mc^2$ (D) $\lambda = \dfrac{h}{mv}$

286. What is the maximum No. of electrons which can be accommodated in an atom in which the highest principal quantum No. value is 4 ? [PMT-2000]

(A) 10 (B) 18 (C) 36 (D) 54

287. In Balmer series of hydrogen atom spectrum which electronic transition causes third line ? [PMT-2000]

(A) Fifth Bohr orbit to second one
(B) Fifth Bohr orbit to first one
(C) Fourth Bohr orbit to second one
(D) Fourth Bohr orbit to first one

288. Energy of electron of hydrogen atom in second Bohr orbit is : [PMT-2000]

(A) -5.44×10^{-19} J (B) -5.44×10^{-19} KJ
(C) -5.44×10^{-19} cal (D) -5.44×10^{-19} eV

289. In 1 mole of NaCl the protons are : [PMT-2001]

(A) 6 mole (B) 11 mole (C) 17 mole (D) 28 mole

290. The radius of first Bohr's orbit for hydrogen is 0.53 Å . The radius of third Bohr's orbit would be : [PMT-2001]

(A) 0.79 Å (B) 1.59 Å (C) 3.18 Å (D) 4.77 Å

291. The third line in Balmer series corresponds to an electronic transition between which Bohr's orbits in hydrogen ? [PMT-2001]

(A) $5 \to 3$ (B) $5 \to 2$ (C) $4 \to 3$ (D) $4 \to 2$

292. The modern atomic weight scale is based on [PMT-2002]

(A) C^{12} (B) O^{16} (C) H^1 (D) C^{13}

293. Bohr model of atom is contradicted by [PMT-2002]

(A) Pauli's exclusion principle
(B) Planck's quantum theory
(C) Heisenberg's uncertainty principle
(D) All the above

294. The electronic configuration of Nitrogen has three unpaired electrons according to [PMT-2002]

(A) Hund's rule (B) Aufbau principle
(C) Pauli's exclusion principle (D) Heisenberg's principle

295. The spectrum of He is expected to be similar to that of [PMT-2002]

(A) H (B) Na (C) He^+ (D) Li^+

296. If the velocity of hydrogen molecule is 5×10^4 cm/sec the de-Broglie wave length of it will be [PMT-2003]

(A) 2A (B) 4A (C) 5A (D) 100 A

297. In form of energy, one atomic mass unit (amu) is equal to : [PMT-2003]

(A) 1000 J (B) 1.592 x 10 erg
(C) 931.1 MeV (D) 10 erg

298. If electron; hydrogen, helium and neon nuclei are all moving with the velocity of light, wavelengths associated with these particles are in the order [PET-93]

(A) Electron > hydrogen > helium > neon
(B) Electron > helium > hydrogen > neon
(C) Electron < helium < hydrogen < neon
(D) Neon < hydrogen < helium < electron

299. The four quantum numbers of the outermost elec-

tron of K (atomic no. 19) are [PET-93]

(A) $n = 2, l = 0, m = 0, s = +\frac{1}{2}$

(B) $n = 4, l = 0, m = 0, s = \frac{1}{2}$

(C) $n = 3, l = 1, m = 1, s = +\frac{1}{2}$

(D) $n = 4, l = 2, m = -1, s = +\frac{1}{2}$

300. The total number of electrons present in all the P orbitals of bromine are : [PET-94]

(A) Five (B) Eighteen (C) Seventeen (D) Thirty-five

301. The four quantum numbers for the electron in the outermost orbital of potassium are [PET-94]

(A) $n = 4, l = 1, m = 0, S = +\frac{1}{2}$

(B) $n = 4, l = 0, m = 0, S = +\frac{1}{2}$

(C) $n = 4, l = 0, m = 0, S = -\frac{1}{2}$

(D) $n = 4, l = 0, m = +1, S = -\frac{1}{2}$

302. Which of the following properties is different for neutral atoms of the isotopes of the same elements [PET-94]

(A) Mass (B) Atomic number
(C) Number of electrons (D) General chemical reactions

303. When an electron revolves in stationary orbit then [PET-94]

(A) It absorbs energy (B) It gains kinetic energy
(C) It emits radiation (D) Its energy remains constant

304. For an electron impossible combination of quantum number is - [PET-95]

(A) $n = 3, l = 2, m_l = -2, m_s = +\frac{1}{2}$

(B) $n = 3, l = 2, m_l = -3, m_s = +\frac{1}{2}$

(C) $n = 4, l = 0, m_l = 0, m_s = -\frac{1}{2}$

(D) $n = 5, l = 3, m_l = 0, m_s = \frac{1}{2}$

305. When the azimuthal quantum number has a value of $l = 0$, the shape of the orbital is - [PET-95]

(A) Rectangular (B) Spherical
(C) Dumb-bell (D) Unsymmetrical

306. The average distance of an electron in an atom from its nucleus is of the order of [PET-96]

(A) 10^{6} m (B) 10^{-6}m (C) 10^{-10} m (D) 10^{-15}m

307. The maximum probability of finding an electron in the dxy orbital is [PET-96]

(A) Along the x-axis (B) Along the y-axis
(C) Along the x-axis and y axes
(D) At an angle of 45° from the x and y axes

308. All electrons on the 4p sub shell must be characterized by the quantum number (s) [PET-96]

(A) $N = 4, m = 0, s = \pm 1/2$ (B) $l = 1$
(C) $l = 0, s = \pm 1/2$ (D) $s = \pm 1/2$

309. If the radius of first Bohr-orbit be a_o, then the radius of third orbit would be : [PET-97]

(A) $3 \times a_o$ (B) $6 \times a_o$ (C) $9 \times a_o$ (D) $1/9 \times a_o$

310. An electron has principal quantum number 3. The number of its (i) sub-shells and (ii) orbitals would be respectively : [PET-97]

(A) 3 and 5 (B) 3 and 7 (C) 3 and 9 (D) 2 and 5

311. The uncertainty principle and the concept of wave nature of matter was proposed by .... and .... respectively. [PET-97]

(A) Heisenberg, de Broglie (B) de Broglie, Heisenberg
(C) Heisenberg, Planck (D) Planck, Heisenberg

312. Which one of the following statements is incorrect ? [PET-97]

(A) Mass defect is related with binding energy
(B) 'Meson' was discovered by Yukawa
(C) The size of the nucleus is of the order of $10^{-12} - 10^{-13}$ cm
(D) Magnetic quantum number is a measure of 'orbital' angular momentum of the electron

313. The number of molecules in 16gm of methane is : [PET-98]

(A) 3.0×10^{23} (B) 6.02×10^{23}

(C) $\frac{16}{6.02} \times 10^{23}$ (D) $\frac{16}{3.0} \times 10^{23}$

314. The maximum number of electrons in an atom with $l = 2$ and $n = 3$ is : [PET-98]

(A) 2 (B) 6 (C) 12 (D) 10

315. Which of the following expressions gives the de Broglie relationship [PET-98]

(A) $\frac{h}{mv} = p$ (B) $\lambda = \frac{h}{mv}$ (C) $\lambda = \frac{h}{mp}$ (D) $\lambda m = \frac{v}{p}$

316. Heaviest particle is : [PET-99]

(A) Meson (B) Neutron (C) Proton (D) Electron

317. Which is correct statement about Proton ? [PET-99]

(A) Proton is nucleus of deuterium
(B) Proton is alpha particle
(C) Proton is ionized hydrogen molecule
(D) Proton is ionized hydrogen

318. In a given atom no two electrons can have the same values of all the quantum numbers. This is called : [PET-99]

(A) Hund's Rule (B) Aufbau Principle
(C) Uncertainty Principle (D) Pauli's Exclusion Principle

319. Magnetic quantum number specifies : [PET-99]

(A) Size of orbitals (B) Shape of orbitals
(C) Orientation of orbitals (D) Nuclear stability

320. The energy of an eleetron in nth orbit of hydrogen

atom is : [PET-99]

(A) $\frac{13.6}{n^4}$ eV (B) $\frac{13.6}{n^3}$ eV (C) $\frac{13.6}{n^2}$ eV (D) $\frac{13.6}{n}$ eV

321. If wavelength of photon is 2.2×10^{-11} m, $h = 6.6 \times 10^{-34}$ Js then momentum of photon is : [PET-99]

(A) 3×10^{-23} Kg ms^{-1} (B) 3.33×10^{22} Kg ms^{-1}
(C) 1.452×10^{-44} Kg ms^{-1} (D) 6.89×10^{43} Kg ms^{-1}

322. An electron has kinetic energy 2.8×10^{-23}J de Broglie wavelength will be nearly - ($m_e = 9.1 \times 10^{-31}$ kg) [PET-2000]

(A) 9.28×10^{-4}m (B) 9.28×10^{-7}m
(C) 9.28×10^{-8}m (D) 9.28×10^{-10}m

323. What will be de Broglie wavelength of an electron moving with a velocity of 1.20×10^5 ms^{-1}? [PET-2000]

(A) 6.068×10^{-9} (B) 3.133×10^{-37}
(C) 6.626×10^{-9} (D) 6.018×10^{-7}

324. The proton and neutron are collectively called as : [PET-2001]

(A) Deutron (B) Positron (C) Meson (D) Nucleon

325. Which of the following sets of quantum number is not possible ? [PET-2001]

(A) $n = 3 : l = +2 : m = 0 : s = +\frac{1}{2}$

(B) $n = 3 : l = +0 : m = 0 : s = -\frac{1}{2}$

(C) $n = 3 : l = +0 : m = -1 : s = +\frac{1}{2}$

(D) $n = 3 : l = +1 : m = 0 : s = -\frac{1}{2}$

326. The atomic orbitals are progressively filled in order of increasing energy. This principle is called as : [PET-2001]

(A) Hund's rule (B) Aufbau principle
(C) Exclusion principle (D) Broglie rule

327. Atomic orbitals of carbon in diamond are [PET-2002]

(A) sp hybridized (B) sp^2 hybridized
(C) sp^3 hybridized (D) Not hybridized

328. The neutron was discovered by [PET-2002]

(A) J.J. Thomson (B) G.T. Seaborg
(C) E. Rutherford (D) James Chadwick

329. The frequency of yellow light having wavelength 600 nm is [PET-2002]

(A) 5.0×10^{14} Hz (B) 2.5×10^7 Hz
(C) 5.0×10^7 Hz (D) 2.5×10^{14} Hz

330. Which one of the following pairs is not correctly matched [PET-2002]

(A) Rutherford - Proton (B) J.J. Thomsom - Electron
(C) J. Chadwick - Neutron (D) Bohr - Isotope

331. The correct order of increasing energy of atomic orbitals is [PET-2002]

(A) 5p < 4f < 6s < 5d (B) 5p < 6s < 4f < 5d
(C) 4f < 5p < 5d < 6s (D) 5p < 5d < 4f < 6d

332. The radius of first Bohr's orbit for hydrogen is 0.53 A. The size of third Bohr's orbit is [PET-2003]

(A) 4.77 A (B) 3.18 A (C) 1.59 A (D) None.

333. The frequency corresponding to transition n = 2 to n = 1 in hydrogen atom is : [PET-2003]

(A) 31.57×10^{14} Hz (B) 24.66×10^{14} Hz
(C) 17.66×10^{10} Hz (D) 40.57×10^{24} Hz

334. A 200g golf ball is moving with a speed of 5m per hour. The associated wave length is ($h = 6.625 \times 10^{-34}$J sec) [PET-2003]

(A) 10^{-40} m (B) 10^{-50} m (C) 10^{-30} m (D) 10^{-10} m

335. Orbital angular momentum for a d-electron is : [PET-2003]

(A) $\frac{12h}{2\pi}$ (B) $\frac{\sqrt{6}h}{2\pi}$ (C) $\frac{6h}{2\pi}$ (D) $\frac{\sqrt{12}h}{2\pi}$

336. General configuration of ultimate and penultimate shell is $(n-1)s^2 (n-1)p^6 (n-1)d^x ns^2$. If n = 4 & x = 5 then no. of proton in the nucleus will be : [PET-2003]

(A) > 25 (B) < 24 (C) 25 (D) None.

337. Radius of second Bohr's orbit of H atom is .... the first Bohr orbit of H atom. [BHU-2003]

(A) 4 times (B) 2 times (C)5 times (D) 3 times

338. Maximum number of orbitals in *l*-subshell is [BHU-2003]

(A) 16 (B) 11 (C) 13 (D) 14

339. a/V^2 is given for in van der Waal's equation [BHU-2003]

(A) internal pressure (B) intermolecular attraction
(C) both (A) and (B) (D) temperature correction.

340. Number of orbitals present in third shell is [AFMC-2003]

(A) 1 (B) 3 (C) 9 (D) 18

341. Bohr's theory was modified to give the concept of elliptical orbits by [AFMC-2003]

(A) Rutherford (B) Hund
(C) Sommerfeld (D)Davisson

342. Isotopes have the same [AFMC-2003]

(A) Mass number (B) Atomic number
(C) Nucleons (D) None of these

343. The frequency of radiation emitted when the electron falls from n = 4 to n = 1 in a hydrogen atom will be (Given ionization energy of H = 2.18×10^{-18} J $atom^{-1}$ and $h = 6.625 \times 10^{-34}$ Js) [CBSE-PMT-2004]

(A) 1.54×10^{15} s^{-1} (B) 1.03×10^{15} s^{-1}
(C) 3.08×10^{15} s^{-1} (D) 2.00×10^{15} s^{-1}

344. Which of the following has same Bohr's radius as hydrogen atom in n = 1 shell? [IIT-2004]

(A) n = 2, He^+ (B) n = 2 , Li^{2+}

(C) n = 3, Li^{2+} (D) n = 2, Be^{3+}

Atomic Structure

1.	(A)	51.	(D)	101.	(C)	151.	(C)	201.	(C)	251.	(B)	301.	(B)
2.	(B)	52.	(B)	102.	(C)	152.	(D)	202.	(A)	252.	(A)	302.	(A)
3.	(D)	53.	(D)	103.	(D)	153.	(C)	203.	*	253.	(D)	303.	(D)
4.	(B)	54.	(D)	104.	(C)	154.	(C)	204.	*	254.	(A)	304.	(B)
5.	(C)	55.	(A)	105.	(A)	155.	(A)	205.	*	255.	(B)	305.	(B)
6.	(B)	56.	(C)	106.	(A)	156.	(B)	206.	*	256.	(C)	306.	(C)
7.	(C)	57.	(B)	107.	(D)	157.	(B)	207.	*	257.	(C)	307.	(D)
8.	(C)	58.	(A)	108.	(C)	158.	(C)	208.	(B)	258.	(D)	308.	(B)
9.	(C)	59.	(A)	109.	(C)	159.	(A)	209.	(B)	259.	(B)	309.	(C)
10.	(A)	60.	(C)	110.	(C)	160.	(A)	210.	(B)	260.	(B)	310.	(C)
11.	(D)	61.	(D)	111.	(A)	161.	(A)	211.	(A)	261.	(D)	311.	(A)
12.	(A)	62.	(A)	112.	(C)	162.	(B)	212.	(C)	262.	(B)	312.	(D)
13.	(C)	63.	(A)	113.	(D)	163.	(D)	213.	(C)	263.	(A)	313.	(B)
14.	(C)	64.	(A)	114.	(B)	164.	(A)	214.	(A)	264.	(C)	314.	(D)
15.	(D)	65.	(A)	115.	(C)	165.	(B)	215.	(C)	265.	(A)	315.	(B)
16.	(D)	66.	(A)	116.	(A)	166.	(B)	216.	(B)	266.	(C)	316.	(B)
17.	(C)	67.	(B)	117.	(C)	167.	*	217.	(C)	267.	(D)	317.	(D)
18.	(B)	68.	(A)	118.	(C)	168.	(C)	218.	(C)	268.	(A)	318.	(D)
19.	(B)	69.	(A)	119.	(D)	169.	(D)	219.	(D)	269.	(D)	319.	(C)
20.	(A)	70.	(C)	120.	(D)	170.	*	220.	(A)	270.	(A)	320.	(C)
21.	(C)	71.	(B)	121.	(A)	171.	*	221.	(A)	271.	(D)	321.	(A)
22.	(B)	72.	(D)	122.	(A)	172.	*	222.	(C)	272.	(D)	322.	(C)
23.	(D)	73.	(C)	123.	(B)	173.	*	223.	(D)	273.	(B)	323.	(C)
24.	(C)	74.	(A)	124.	(D)	174.	(B)	224.	(A)	274.	(C)	324.	(D)
25.	(D)	75.	(B)	125.	(C)	175.	(B)	225.	(B)	275.	(B)	325.	(C)
26.	(B)	76.	(A)	126.	(D)	176.	(A)	226.	(C)	276.	(C)	326.	(B)
27.	(D)	77.	(D)	127.	(D)	177.	(B)	227.	(D)	277.	(B)	327.	(C)
28.	(D)	78.	(C)	128.	(C)	178.	(A)	228.	(A)	278.	(D)	328.	(D)
29.	(B)	79.	(B)	129.	(C)	179.	(A)	229.	(A)	279.	(A)	329.	(A)
30.	(B)	80.	(A)	130.	(B)	180.	(A)(B)(C)	230.	(D)	280.	(C)	330.	(D)
31.	(D)	81.	(C)	131.	(B)	181.	(A)	231.	(B)	281.	(C)	331.	(B)
32.	(D)	82.	(E)	132.	(C)	182.	(A)	232.	(B)	282.	(B)	332.	(A)
33.	(D)	83.	(D)	133.	(D)	183.	(A)(D)	233.	(C)	283.	(C)	333.	(B)
34.	(D)	84.	(C)	134.	(C)	184.	(C)	234.	(C)	284.	(B)	334.	(C)
35.	(C)	85.	(D)	135.	(D)	185.	(A)	235.	(B)	285.	(D)	335.	(B)
36.	(A)	86.	(D)	136.	(B)	186.	(D)	236.	(D)	286.	(C)	336.	(C)
37.	(A)	87.	(A)	137.	(C)	187.	(C)	237.	(D)	287.	(A)	337.	(A)
38.	(B)	88.	(A)	138.	(C)	188.	(D)	238.	(D)	288.	(A)	338.	(B)
39.	(B)	89.	(B)	139.	(A)	189.	(A)	239.	(D)	289.	(D)	339.	(B)
40.	(A)	90.	(D)	140.	(A)	190.	(A)	240.	(D)	290.	(D)	340.	(C)
41.	(A)	91.	(C)	141.	(B)	191.	(C)	241.	(D)	291.	(B)	341.	(C)
42.	(D)	92.	(C)	142.	(B)	192.	(B)	242.	(C)	292.	(A)	342.	(B)
43.	(C)	93.	(D)	143.	(C)	193.	(A)	243.	(A)	293.	(C)	343.	(C)
44.	(C)	94.	(B)	144.	(C)	194.	(B)	244.	(A)	294.	(A)	344.	(D)
45.	(B)	95.	(B)	145.	(D)	195.	(A)	245.	(A)	295.	(D)		
46.	(A)	96.	(C)	146.	(B)	196.	(B)	246.	(B)	296.	(B)		
47.	(A)	97.	(D)	147.	(C)	197.	(C)	247.	(C)	297.	(C)		
48.	(D)	98.	(B)	148.	(A)	198.	*	248.	(B)	298.	(A)		
49.	(C)	99.	(C)	149.	(B)	199.	(A)	249.	(A)	299.	(B)		
50.	(A)	100.	(B)	150.	(C)	200.	(D)	250.	(B)	300.	(C)		

167. 4.134
170. Photons
171. orbitals
172. higher effective nuclear charge
173. orientation in space
198. 3 : 7203. Balmer
204. 70
205. m = 0
206. Isotones
207. Isotones

CHEMICAL BOND

The attraction between atoms in a molecule is called chemical bond.

Electrovalent bond : Kossel

It involves transfer of electron (s). Electrostatic attraction between the oppositely charged ions is called the ionic bond.

Electrostatic atraction always tends to decrease the potential energy of the system. e.g. NaCl, CaF_2 etc.

The electrovalency of an element is the number of electrons transferred from or to the atom of that element in order to acquire stable s^2 p^6 configuration in the outer shell. Each ion is surrounded by electric field which is non directional i.e. it is uniformly distributed about the ion. Ionic character is directly proportional to the size of the cation.

Ions are more stable than the atoms.

Ionic compounds are soluble in water when hydration energy is greater than the lattice energy.

Decrease of energy in the formation of ionic bond :

In the formation of chemical bonds, there is a fall in potential energy e.g. formation of NaCl

Na (g) + I.E. $\rightarrow$ Na^+ (g) + e^- (498 kJ/mole)

Cl (g) + e^- $\rightarrow$ Cl^- (g) + Electron affinity (-362 kJ/mole)

A considerable amount of energy decreases due to strong electrostatic attraction between Na^+Cl^- ions.

Na^+ (g) + Cl^- (g) $\rightarrow$ Na^+Cl^- + Lattice energy (-761.7kJ/mole).

The net change in energy= 498- 362- 761.7 = -625.7 kJ/mole.

The decrease of energy due to electrostatic attraction between Na^+ and Cl^- ions i.e. the lattice energy is responsible for the formation of ionic bond.

Formation of CsCl : 376.6 - 362 - 636.1 = -621.5 kJ/mole.

Lattice energy $\propto \frac{1}{r^+ + r^-}$

Factors influencing the formation of ionic bond :

1. I.P. of the metal atoms is to be low
2. E.A. of the non metal is to be high
3. The lattice energy of the crystal of the ionic compound is to be high

The atoms of groups 13, 14, 15 and 16 don't form ionic compounds ordinarily because of high I.E. (IE_3 very high)

Properties :

1. Ionic solids do not conduct electricity in solid state.
2. They conduct electricity in molten state.

In molten state K. E. increases and thus ions overcome the forces of attraction.

3. High m. p. and b. p. . Low volatility.
4. Soluble in water but sparingly soluble in organic solvents.

Variable electrovalency

1. Outer shell configuration of ions :

Fe^{+2}	2, 8, $3s^2$, $3p^6$ $3d^6$	{ not stable configuration
Ni^{+2}	2, 8, $3s^2$, $3p^6$ $3d^8$	
Cu	2, 8, $3s^2$ $3p^6$ $3d^{10}$ $4s^1$	Cu^+ : $3d^{10}$; Cu^{2+} : $3d^9$
Au	2, 8, 18, 32, $5s^2$ $5p^6$ $5d^{10}$ $6s^1$	Au^+ : $5d^{10}$

The nuclear charge is not sufficient to keep the electrons intact in their positions. Therefore these ions will lose electrons and show variable valency.

2.Inert pair effect :Sn 2, 8, 18, $5s^2$ p^2 ;Sn^{+2} 2, 8, 18, 18, $5s^2$ only p^2 electrons take part in bonding while in C and Si the s^2p^2 electrons take part. Inert pair effect is predominant in Sn, Pb etc.

Covalent bond: Lewis

It is formed by sharing of electrons. According to spin theory of valency only the electrons with opposite spin can pair in the electric field of a nucleus.If the net result is attraction -chemical bond forms.If the net result is repulsion - no bond formation is possible.

Non polar covolent bond - e.g. Cl_2 etc.

Polar covalent bond - e.g. HCl etc.

Covalent character is inversely proportional to the size of the cation.

Variable covalency : F, O, N, having 1, 2 and 3 unpaired electrons in their outer shells show covalency of 1, 2 and 3 respectively. They do not contain d- orbitals in them. The maximum covalency is generally equal to the actual number of s and p - electrons present in the valence shell (where empty orbitals are present). Elements which don't have empty orbitals show only one covalency. Halogens except F show 1, 3, 5, 7 covalency.

Co-ordinate covalent bond or Dative bond :

This bond forms when an orbital containing lone pair of electrons overlap an empty orbital of the other atom. e.g. H_3O^+; NH_4^+. $H_3N \rightarrow AlCl_3$; $H_3N \rightarrow BF_3$ etc. Characterstics lie between electrovalent and covalent bond.Octet theory fails to explain the formation and stability of $BeCl_2$, BCl_3, $AlCl_3$, PF_5, WF_6 OsF_8, many co-ordination compounds, metal carbonyls, nitrosyls etc.

Metallic bond : Metals have low ionisation energies because valency electrons can be taken out relatively easily. Valency electrons are loosely bound to the nucleus in the metals. e.g. Li has one electron in the outer shell. In a crystal of Li, each atom is surrounded by 8 other Li atoms. Covalent bond formation is not possible. It has valency orbital 2s, $2p_x$, $2p_y$ $2p_z$ available.Hence, besides its own valency electron, the valency electrons of the neighbouring atoms can also come quite close to its nucleus.There can be complete freedom of movement of electrons in the vacant valency orbital around the nucleus of each atom of Li.

The force that binds a metal atom to a number of electrons within the sphere of influence is known as metallic bond.

Metallic bond	**Covalent bond**
not localized, non – directional, weaker, mobile	localized, directional strong, less mobile

Hydrogen bond : Here, one of the most electronegative element is sharing electrons with hydrogen. Dipole formation $H^{\delta+}$ — $F^{\delta-}$ takes place easily. In such cases electrostatic attraction between such molecules should be very strong

--- $H^{\delta+} - F^{\delta-}$--- $H^{\delta+} - F^{\delta-}$-- $H^{\delta+} - F^{\delta-}$; R–O(H)--H–O(R)---H–O(R)

HF froms 2 hydrogen bonds.

H_2O forms 4

and NH_3 forms 4

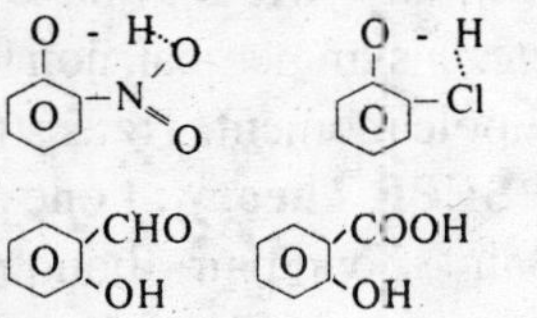

Inter & intra molecular :

HO-⟨O⟩-Cl --- HO-⟨O⟩-Cl--- HO-⟨O⟩-Cl;

Nylon, DNA, proteins also contain hydrogen bonds.

H_2O density in solid state is less. Ice contracts when heated between 0^0 to 4^0C as more holes or spaces are present in solid ice. In water H-bonds are fewer. Therefore, the molecules come closer and thus contraction occurs.

Conditions for the formation of Hydrogen bonding

1. The hetero atom is to be strongly electronegative and small in size. O, N, F are suitable hetero atoms.
(If electronegativity is not high and if the atom has large atomic size (radius), the electrostatic forces of attraction will be weak).

2. The strength of a hydrogen bond is of the order of 12 - 40 $kJmol^{-1}$. Hydrogen bonds involving F, O, N atoms have strengths of 40, 28 and 8 $kJmol^{-1}$ respectively.

Shapes of molecules (VSEPR- Theory):The theory is proposed by Sidgwick & Powell and extended by Gillespie & Nyhom. "For maximum stability of a molecule, the orbitals occupied by electrons in the valence shell of the central atom should be so arranged in space that they lie as far away from one another as possible."

In simple covalent compounds, there is one cental atom which forms covalent bonds. When the central atom forms more than one covalent bond, there will be more than one orbital occupied with electrons around the central atom. The electrons experience electrostatic repulsion.

The shape or the geometry of a covalent molecule depends upon the number of electron pairs present in the outer (valence) shell of the central atom.

If the central atom is surrounded only by orbitals containing shared pairs of electrons (bond pairs) and no orbitals containing lone pairs in the valence shell, the molecule has regular geometry, otherwise a distorted one.

Repulsion: Lone pair - Lone pair > Lone pair - Bond pair > Bond pair - Bond pair.

Hybridisation and geometry of molecules

sp	-	BeF_2 ,$HgCl_2$	Linear	180^0
sp^2	-	BF_3,	Plane triangle	120^0
sp^3	-	CH_4,	Tetrahedral	109.5^0
dsp^2	-	$[Cu(NH_3)_4]SO_4$	Square planar	90^0
sp^3d	-	PCl_5,	Trigonal bipyramidal	120^0&90^0
sp^3d^2	-	SF_6,	Octahedral	90^0
sp^3d^3	-	IF_7,	Pentagonal bipyramidal	72^0 & 90^0

Exceptions : Due to the presence of lone pairs

$SnCl_2$	-	sp^2	Plane triangle	< 120^0
H_2O	-	sp^3	Bent V shape	104.5^0
NH_3	-	sp^3	Pyramidal	107^0

The hybridisation, number of lone pairs and geometry of the molecules/ions can be deduced from

$N = 1/2\ (V + M - C + A)$; $L = N - M - D$

N = no. of orbitals in the hybridised orbital

V = no. of electrons in valence shell

M = no. of monovalent atoms

C = charge on cation A = charge on anion

L = no. of lone pairs D = no. of divalent atoms

e.g. $ClF_3 : N = \frac{1}{2}(7 + 3 - 0 + 0) = 5$; $L = 5 - 3 = 2$

∴ it is sp^3d trigonal bipyramidal. Two lone pairs are present.

VBT : Valence bond theory was given by Heitler & London (1927). It answers the questions:

1. Why a covalent bond forms ?
2. How are the electrons arranged in the space about the molecule ?
3. What is the bond energy in a covalent bond ?
4. How does the sharing electrons explain the geometry ? etc.

The calculated H-H bond energy 301.25 $kJmol^{-1}$ is less than the experimental one 435.14 $kJmol^{-1}$. ∴ resonating structures are proposed $H–H \leftrightarrow H^+–H^- \leftrightarrow H^-–H^+$

MOT : Molecular orbital theory was given by Hund and Mulliken.

VBT	**MOT**
1. Resonance plays an important role	Doesn't play any role
2. Fails to explain paramagnetic character of oxygen	offers explanation
3. Explains the difference between bonding ↑↓ and antibonding ↑↑or ↓↓ orbitals	bonding and antibonding orbtials can have paired electrons
4. Explains the existence	Fails to explain the

of noble gases existence

The order of filling molecular orbitals is : σ Is, σ* Is; σ2s, σ*2s; σ2p, $\pi_y 2p$, $\pi_z 2p$; $\pi_y *2p$, $\pi_z *2p$; σ*2p ....

Resonance energy is equal to the difference between the actual bond energy of the molecule and that of most stable of the resonating structures.

Bond order : Bond order in diatomic molecules is half of the difference between the total number of bonding electrons N_b and antibonding electrons N_a.

$$B.O. = (\frac{N_b - N_a}{2})$$

B.O. in $H_2 = 1$; $H_2^+ = 0.5$; $He_2 = 0$; $O_2 = 2$; $C_2 = 2$; $N_2 = 3$; in benzene 1.5.

Bond energy increases with increase in bond order.

-dipole moment (μ) = charge (e) x distance (d), μ = e x d

-for trans form μ = 0 ; cis form has dipole moment.

Bond energies (in kJ)

$C \equiv C > C = C > C - H > C - C$
835.1 610 413.4 347.2

$N \equiv N > O = O > H - H$
945.6 497.9 435

F - F < Cl - Cl > I - Cl ; $Cl_2 > Br_2 > F_2 > I_2$
154.8 242.7 209.2

Bond length of C - H bond : $sp < sp^2 < sp^3$

Bond energy decreases with increase in the number of lone pairs. Bond length increases with the increase in the radii of the bonded atoms. Bond angle decreases with increase in lone pairs

Types of bonds in :

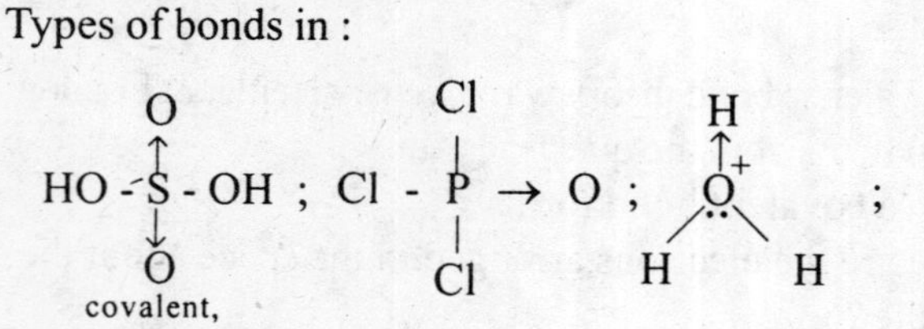

covalent, coordinate

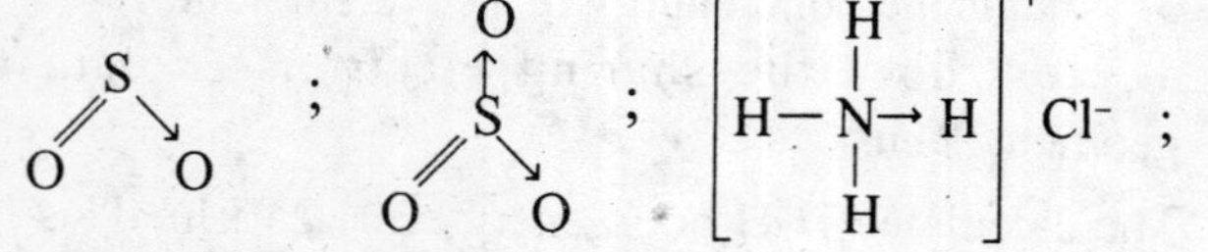

ionic,covalent, coordinate

$K^+ C^- \equiv N$; $K_4 [Fe(CN)_6]$; $Na^+ OH^-$

ionic, covalent; ionic,covalent, coordinate; ionic, covalent

Things to remember

The theory of valency was forwarded by Kossel and Lewis.

Electrovalency : Kossel. Non directional ∴ no isomerism exhibited. High bond energy due to high lattice energy: NaCl : - 625.7 ; CsCl : - 621.5 kJ/mole.

Variable electro valency : Sn^{+2}, Sn^{+4}; Pb^{+2}, Pb^{+4} (Inert pair effect)

Cu^+, Cu^{+2}; Fe^{+2}, Fe^{+3} (due to nuclear charge)

Covalency : Lewis; polar and non polar. Octet theory fails to explain the formation of $BeCl_2$, BCl_3, PCl_5, SF_6, OsF_8 etc.

Dipole moment : H_2, N_2, O_2, Cl_2, CO_2, CS_2 BeF_2, CH_4, CCl_4, BF_3 and BCl_3 (symmetrical structure the resultant μ of any two is 0) have zero μ. H_2O, SO_2, CH_3Cl, NH_3 etc. unsymmetrical, non linear polyatomic molecules have dipole moment. Trans structure has zero dipole moment.

VSEPR Theory : Lone pair - lone pair > lone pair - bond pair > bond pair - bond pair

sp : linear : CO , CO_2 , BeF_2 , C_2H_2

sp^2 : trigonal planar : SO_2, SO_3, $SnCl_2$

sp^3 : NH_3 (pyramidal) , H_2O (bent V shape) , NH_4^+ (tetrahedral)

sp^3d: PCl_5 (Trigonal bipyramidal)

sp^3d^2 : SF_6 (Octahedral)

sp^3d^3 : IF_7 pentagonal bipyramidal.

Metallic Bond : The force that binds a metal atom to a number of electrons within the sphere of influence.

Hydrogen bond : $H_2O(4)$; $NH_3(4)$; HF(2). Density of H_2O in solid state is less than water in liquid state due to hydrogen bonding.

Intramolecular (O-nitrophenol, salicylaldehyde).

Hydrogen bonding is present in Nylon, DNA, alcohol, H_2O etc.

VBT : Heitter & London ; MOT : Hund & Mulliken

Bond order : $\frac{1}{2}$ (Nb - Na)

Bond energy : decreases with the increase in number of lone pairs; decreases with increase in bond length

$C \equiv C > C = C > C - H > C - C$

- Hydration energy, $H = \frac{Z^2e^2}{2r}(1 - \frac{1}{\varepsilon})$

(ε=dielectric constant)

- Bond angle ∝ s-character of the hybridized orbitals.
- Bond angle ∝ E. N. of central atom

Types of bonds present in : NH_4Cl, NaOH, KCN; $K_4 [Fe(CN)_6]$, H_2SO_4; H_3O^+, $POCl_3$; SO_2; SO_3 etc. molecules.

1. When two atoms co-axially overlap to form a molecule then which of the following bond will be formed [CBSE-90]
(A) 1π bond (B) 2 σ bond (C) 1 σ bond (D) 2 π bond

2. The minimum bond distance is in [CBSE-90]
(A) C – C (B) C = C (C) C ≡ C (D) C – C = C – C

3. Which of the following shows hydrogen bonding ? [CBSE-90]
(A) H_2S (B) HF (C) H_2O (D) H_2S

4. Which of the following has the least bond angle ? [CBSE-90]
(A) NH_3 (B) H_2O (C) BeF_2 (D) CH_4

5. Which of the following has both ionic and cavalent bonds [CBSE-90]
(A) $CaCl_2$ (B) NH_4Cl (C) CO_2 (D) H_2O

6. In which of the following the bond angle between the two bonds is minimum ? [CBSE-90]
(A) H_2O (B) NH_3 (C) CO_2 (D) CH_4

7. In which bond, angle is the highest ? [CBSE-91]
(A) sp^3 (B) sp^2 (C) sp^3d (D) sp

8. Which type of hybridization occurs in Ethylene ? [CBSE-91]
(A) sp (B) sp^2 (C) sp^3d (D) sp^3d^2

9. dsp^2 hybridization has arrangement which is [CBSE-91]
(A) trigonal planar (B) tetrahedral
(C) square planar (D) trigonal bipyramidal.

10. Strength of hydrogen bond is intermediate between [CBSE-91]
(A) van der Waal and covalent (B) ionic and covalent
(C) ionic and metallic (D) metallic and covalent

11. Both ionic and covalent bonds are present in [CBSE-91]
(A) CH_4 (B) KCl (C) SO_2 (D) NaOH

12. The shortest carbon-carbon bond distance is found in [CBSE-91]
(A) diamond (B) ethane (C) benzene (D) acetylene

13. Which one of following is paramagnetic ? [CBSE-91]
(A) H_2O (B) NO_2 (C) SO_2 (D) CO_2

14. Hydrogen bonding is maximum in [CBSE-91]
(A) ethanol (B) diethyl ether
(C) ethyl chloride (D) triethyl amine

15. Hydrogen bond is the strongest in [CBSE-91]
(A) O – H .....S (B) S – H......O
(C) F – H.....F (D) F – H......O

16. An sp^3 hybrid orbital contains [CBSE-91]
(A) 1/4th s-character (B) 1/2th s-character
(C) 2/3rd s-character (D) 3/4th s-character

17. Which one of the following bonds will be the most polar ? [CBSE-92]
(A) N—Cl (B) O—F (C) C—F (D) N—N

18. In a molecule of carbon tetrachloride the four C—Cl bonds are directed towards the corners of a [CBSE-92]
(A) square (B) tetrahedron (C) prism (D) cube

19. Strongest hydrogen bonding is shown by [CBSE-92]
(A) Water (B) Ammonia
(C) Hydrogen fluoride (D) Hydrogen sulphide

20. A coordinate bond is formed when an atom in a molecule has [CBSE-92]
(A) electric charge on it
(B) all its valency shell electrons shared
(C) a single unshared electron
(D) one or more unshared electron pairs

21. Which one of the following species is likely to have a linear structure? [CBSE-92]
(A) SO_2 (B) CO_2 (C) CO_3^{2-} (D) SO_4^{2-}

22. At its melting point ice is lighter than water because [CBSE-92]
(A) H_2O molecules are more closely packed in solid state
(B) Ice crystals have hollow hexagonal arrangement of H_2O molecules
(C) on melting of ice the H_2O molecule shrinks in size
(D) ice forms mostly heavy water on first melting

23. The dimer of methanol will have the structure [CBSE-92]

(A)
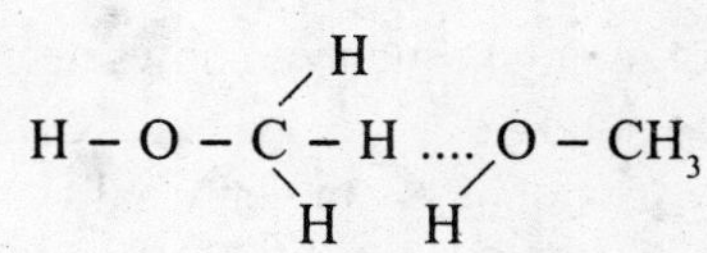

(B)
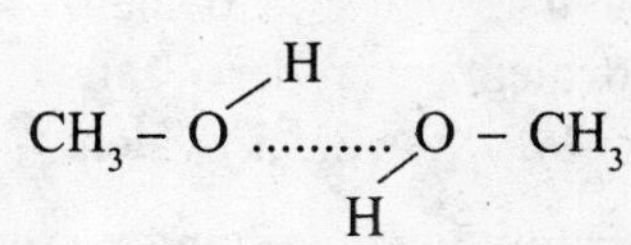

(C)
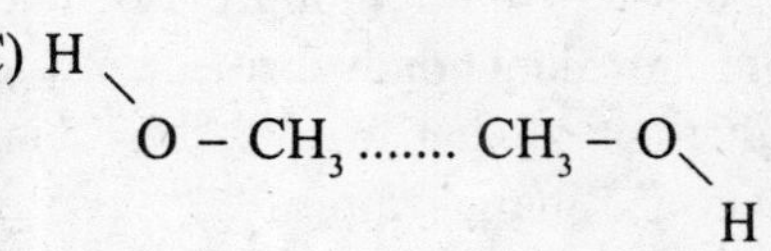

(D)
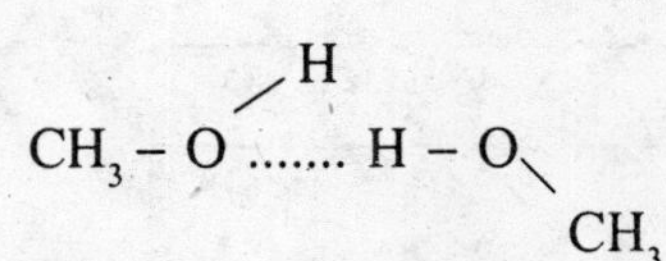

24. Ethanol and dimethyl ether are a pair of functional isomers. The boiling point of ethanol is higher than that of dimethyl ether due to the presence of [CBSE-93]
(A) CH_3 group in dimethyl ether
(B) H-bonding in ethanol

(C) H-bonding in dimethyl ether (D) CH_3 group in ethanol

25. When the hybridization state of carbon atom changes from sp^3 to sp^2 and finally to sp, the angle between the hybridized orbitals [CBSE-93]
(A) decreases gradually (B) decreases considerably
(C) is not affected (D) increases progressively

26. Which kind of valency exists in CaH_2 and C_2H_2? [CBSE-93]
(A) Electrovalency in CaH_2 and covalency in C_2H_2
(B) Electrovalency in both
(C) Covalency in CaH_2 and electrovalency in C_2H_2
(D) Covalency in both

27. Which of the following statements is not correct? [CBSE-93]
(A) Double bond is shorter than π single bond
(B) Sigma bond is weaker than π(Pi) bond
(C) Double bond is stronger than single bond
(D) Covalent bond is stronger than hydrogen bond

28. Which of the following compounds does not contain double bond or triple bond? [CBSE-93]
(A) C_2H_4 (B) H_2O (C) N_2 (D) HCN

29. Which of the following species has four lone pairs of electrons? [CBSE-93]
(A) I^- (B) O^- (C) Cl^- (D) He

30. The weakest bond is [CBSE-93]
(A) hydrogen (B) van der Waal's
(C) covalent (D) coordinate

31. Ionic bond is formed between [CBSE-93]
(A) CH_4 (B) O_3 (C) Both (A) and (B) (D) NaCl

32. Which of the following sequences shows decreasing order of strength of bonds? [CBSE-93]
(A) Electrovalent > Covalent > van der Waal's
(B) Electrovalent > van der Waal's > Covalent
(C) Covalent > Electrovalent > van der Waal's
(D) Covalent > van der Waal's > Electrovalent

33. The table shown below lists the bond dissociation energies (E_{diss}) for single covalent bonds formed between carbon atoms (C) and atoms of elements 'A', 'B', 'C' and 'D'. Which element has the smallest atoms?

Bond	E_{diss} (kJ mol^{-1})
C-A	240
C-B	328
C-C	276
C-D	485

[CBSE-94]
(A) C (B) D (C) A (D) B

34. A molecule in which sp^2 hybrid orbitals are used by the central atom in forming covalent bonds is [CBSE-94]
(A) He_2 (B) SO_2 (C) PCl_5 (D) N_2

35. Which one of the following molecules does not possess a permanent electric dipole moment? [CBSE-94]
(a) H_2S (B) SO_2 (C) CS_2 (D) SO_3

36. If a molecule MX_3 has zero dipole moment, the sigma bonding orbitals used by M (at. no.<21) are [CBSE-94]
(A) sp^2 hybrid (B) pure p-orbitals
(C) sp^3 hybrid (D) sp hybrid

37. Which of the following molecules/ions has a triangular pyramidal shape? [CBSE-94]
(A) BF_3 (B) H_3O^+ (C) NO_3 (D) CO_3^{2-}

38. Ground state electronic configuration of valence shell electrons in nitrogen molecule (N_2) is written as $(2s\sigma)^2 (2p\pi)^4 (2p\sigma)^2$. Hence the bond order in nitrogen molecule is [CBSE-95]
(A) 2 (B) 3 (C) 0 (D) 1

39. The correct order of the O-O bond length in O_2, H_2O_2 and O_3 is [CBSE-95]
(A) $O_2 > O_3 > H_2O_2$ (B) $O_3 > H_2O_2 > O_2$
(C) $H_2O_2 > O_3 > O_2$ (D) $O_2 > H_2O_2 > O_3$

40. The radius of hydrogen atom in the ground state is 0.53 Å. The radius of Li^{2+} ion in a similar state is [CBSE-95]
(A) 1.06 Å (B) 0.265 Å (C) 0.17 Å (D) 0.53 Å

41. Which of the following species is paramagnetic? [CBSE-95]
(A) CO (B) NO^- (C) O_2^{2-} (D) CN^-

42. The oxide which cannot act as a reducing agent is [CBSE-95]
(A) SO_2 (B) NO_2 (C) CO_2 (D) ClO_2

43. The structure and hybridisation of $Si(CH_3)_4$ is [CBSE-96]
(A) bent, sp (B) trigonal, sp^2
(C) octahedral, sp^3d (D) tetrahedral, sp^3

44. Which of the following bonds has the highest energy? [CBSE-96]
(A) Se-Se (B) Te-Te (C) S-S (D) O-O

45. The high density of water compared to ice is due to [CBSE-97]
(A) hydrogen bonding interactions
(B) dipole-dipole interactions
(C) dipole-induced dipole interactions
(D) induced dipole induced dipole interactions

46. Which one of the following has the highest dipole moment? [CBSE-97]
(A) NH_3 (B) PH_3 (C) SbH_3 (D) AsH_3

47. N_2 and O_2 are converted into monoanions N_2^- and O_2^- respectively,which of the following statements is wrong ? [CBSE-97]
(A) In N_2 the N-N bond weakens .
(B) In O_2 the O-O bond order increases .
(C) In NH_3 decreases .
(D) N_2^+ becomes diamagnetic .

48. The AsF_5 molecule is trigonal pyramidal. The hybrid orbitals used by the As atoms for bonding are [CBSE-97]
(A) $d_{x^2-y^2}$, d_{z^2}, s, p_x, p_y (B) dxy, s, p_x, p_y, p_z
(C) s, p_x, p_y, p_z, p_{z^2} (D) $d_{x^2-y^2}$, s, p_x, p_y, p_z

49. N_2 and O_2 are converted into monocations,N_2^+ and O_2^+ respectively. Which is wrong ? [CBSE-97]
(A) In N_2, the N-N bond weakens.
(B) In O_2, the O-O bond order increases.
(C) In O_2 paramagnetism decreases.
(D) N_2^+ becomes diamagnetic.

50. The number of antibonding electron pairs in O_2^{2-} molecular ion on the basis of molecular orbital theory is (at. no .of O = 8). [CBSE-98]
(A) 4 (B) 3 (C) 2 (D) 5

51. What is the shape of $Fe(CO)_5$ molecule ? [CBSE-2000]
(A) Tetrahedral (B) Octahedral
(C) Trigonal bipyramidal (D) Square pyramidal

52. Which of the following is electron deficient ? [CBSE-2000]
(A) NH_3 (B) BCl_3 (C) PCl_3 (D) None of these

53. Which of the following is isoelectronic? [CBSE-2002]
(A) CO_2, NO_2 (B) NO_2^-, CO_2
(C) CN^-, CO (D) SO_2, CO_2

54. Which of the following has $p_\pi-d_\pi$ bonding? [CBSE-2002]
(A) NO_3^- (B) SO_3^{2-} (C) BO_3^{3-} (D) CO_3^{2-}

55. In NO_3^- ion number of bond pair and lone pair of electrons on nitrogen atom are [CBSE-2002]
(A) 2, 2 (B) 3, 1 (C) 1, 3 (D) 4, 0

56. Which of the following statements is not correct for sigma- and pi- bonds formed between two carbon atoms ? [CBSE-2003]
(A) Sigma-bond determines the direction between carbon atoms but a pi-bond has no primary effect in this regard.
(B) Sigma-bond is stronger than a pi-bond
(C) Bond energies of sigma- and pi-bonds are of the order of 264 KJ/mol and 347 KJ/mol, respectively.
(D) Free rotation of atoms about a sigma-bond is allowed but not in case of a pi-bond

57. NH_3 and H_2O form NH_4OH by [CBSE-2003]
(A) Electrovalent bond (B) Covalent bond
(C) Coordinate bond (D) None of these

58. H_2O has a net dipole moment but BeF_2 has zero dipole moment because [AIIMS-93]
(A) Beryllium has more electronegativity than oxygen
(B) Fluorine has more electronegativity than oxygen
(C) There is lone pair of electrons in H_2O, but no lone pair in BeF_2.
(D) H_2O molecule is linear while BeF_2 molecule is bent.

59. An ethene molecule contains [AIIMS-93]
(A) 2 σ & 4 π bonds (B) 4 σ & 2 π bonds
(C) 6 σ bonds (C) 5 σ & 1 π bonds

60. Assertion : The boiling point of ethanol is much higher than that of diethyl ether. Reason : In ethanol, the molecules are associated by the formation of intermolecular hydrogen bonding, whereas in diethyl ether it is not possible. [AIIMS-94]
(A) A (B) B (C) C (D) D

61. Assertion : Ionic compounds tend to be non-volatile. Reason : The inter molecular forces in these compounds are weak. [AIIMS-94]
(A) A (B) B (C) C (D) D

62. The compound contaning coordinate bond is [AIIMS-94]
(A) O_3 (B) H_2SO_4 (C) SO_3 (D) All of these

63. Assertion : Bond order in a molecule can assume any value positive or negative ; integral or fractional, including zero. Reason : It depends on the number of electrons in the bonding and antibonding orbitals. [AIIMS-95]
(A) A (B) B (C) C (D) D

64. Which of the following molecule has regular geometry ? [AIIMS-95]
(A) PF_3 (B) SF_6 (C) H_2O (D) XeF_4

65. The boiling point of a compound is raised by [AIIMS-96]
(A) Volatility of compound
(B) Non-polarity in the molecules
(C) Intermolecular hydrogen bonding
(D) Intramolecular hydrogen bonding

66. Assertion : The atoms in a covalent molecule are said to share electrons, yet some covalent molecules are polar. Reason : In a polar covalent molecule, the shared electrons spend more time than the average near one of the atoms. [AIIMS-96]
(A) A (B) B (C) C (D) D

67. Which of the following molecules is paramagnetic ? [AIIMS-96]

(A) NO_2 (B) N_2O_5 (C) P_2O_5 (D) $H_2S_2O_7$

68. Which of the following has both polar and non-polar bonds ? [AIIMS-97]
(A) C_2H_6 (B) NH_4Cl (C) HCl (D) $AlCl_3$

69. When the hybridization state of carbon atom changes from sp^3 to sp^2 to sp, the angle between the hybridized orbitals [AIIMS-98]
(A) Decreases gradually (B) Increases progressively
(C) Decreases considerably (D) All of these

70. Ethanol and dimethyl ether form a pair of functional isomers. The boiling point of ethanol is higher than that of dimethyl ether, due to the presence of [AIIMS-98]
(A) H-bonding in ethanol
(B) H-bonding in dimethyl ether (C) CH_3-group in ethanol
(D) CH_3-group in dimethyl ether

71. Which of the following does not show hydrogen bonding ? [AIIMS-98]
(A) H_2S (B) H_2O (C) N_2O (D) Fe_3O_4

72. Assertion : Bond order is a molecular, which can assume any value positive or negative, integral or fractional including zero. Reason : Bond order depends on the number of electrons in the bonding and antibonding orbitals [AIIMS-98]
(A) A (B) B (C) C (D) D

73. Which among the following has the largest dipole moment ? [AIIMS-99]
(A) HI (B) H_2O (C) NH_3 (D) SO_3

74. Which of the following atoms would be paramagnetic ? [AIIMS-99]
(A) N (B) Be (C) Zn (D) Ca

75. Atomic orbitals of carbon dioxide are [AIIMS-99]
(A) sp-hybridised (B) sp^2-hybridised
(C) sp^3-hybridised (D) sp^3d-hybridised

76. Assertion : Bond order can assume any value number including zero. Reason : Higher the bond order, shorter is bond length and greater is bond energy. [AIIMS-99]
(A) A (B) B (C) C (D) D

77. Assertion : The dipole moment helps to predict whether a molecule is polar or non-polar. Reason : The dipole moment helps to predict geometry of molecules. [AIIMS-99]
(A) A (B) B (C) C (D) D

78. Which of the following has a tendency for covalent compound formation ? [AIIMS-2000]
(A) Na (B) Mg (C) Ca (D) Ba

79. Which of the following molecules has a tetrahedral shape ? [AIIMS-2000]
(A) Hg_2Cl_2 (B) CO_3^{-2} (C) NH_4^+ (D) $Ni(CN_4)^{2-}$

80. Fluorine molecule is formed by the overlap of [AIIMS-2000]
(A) p–p orbitals by side-to-side manner
(B) p–p orbitals by end-to-end manner
(C) s–p orbitals (D) s–s orbitals

81. Which of the following cannot exist on the basis of Molecular Orbital Theory ? [AIIMS-2000]
(A) H_2^+ (B) He_2^+ (C) He (D) C_2

82. The orbital angular momentum for a s-orbital may be [AIIMS-2000]
(A) 1 (B) zero (C) $\frac{\sqrt{2}h}{2\pi}$ (D) all of these

83. Which one is the best solvent for ionic compounds with reference to their dielectric constants (D) at 25°C [AIIMS-2000]
(A) A, (D = 78.5) (B) B, (D = 24.3)
(C) C, (D = 32.6) (D) D, (D = 20.7)

84. In which pair same structure is not possible ? [AIIMS-2001]
(A) SF_4, SiF_4 (B) $BeCl_2$, C_2H_2
(C) BF_3, SO_3 (D) H_2O, H_2S

85. CH_4,CH_3^-,CH_3^+ hybridisation respectively in them are [AIIMS-2001]
(A) sp^3, sp^3, sp^2 (B) sp^3, sp^3, sp^3
(C) sp^2, sp^2, sp^2 (D) sp^2, sp^3, sp^2

86. Correct order of bond energy : [AIIMS-2001]
(A) $C_2^{-2} < NO < O_2^+ < He_2^+$
(B) $He_2^+ < NO < O_2^+ < C_2^{-2}$
(C) $C_2^{-2} < NO < He_2^+ < O_2^+$ (D) None of these

87. Which has minimum bond length ? [AIIMS-2001]
(A) O_2^+ (B) O_2 (C) O_2^- (D) F_2

88. Assertion : Covalency of oxygen is two.
Reason : Dinegative anion of oxygen (O^{2-}) is quite common but dinegative anion of sulphur (S^{2-}) is less common. [AIIMS-2001]
(A) A (B) B (C) C (D) D

89. Which of the following molecules has highest bond energy ? [AIIMS-2002]
(A) F – F (B) C – C (C) N – N (D) O – O

90. The number of σ-and π-bonds present in pent-4-en-1-yne is [AIIMS-2002]
(A) 10,3 (B) 3,10 (C) 4, 9 (D) 9,4

91. Assertion (A) : Sigma (σ) is a strong bond, while (π) is a weak bond. Reason (R) : Atoms rotate freely about pi (π) bond. [AIIMS-2002]
(A) A (B) B (C) C (D) D (E) E

92. The ion which is not tetrahedral in shape is - [AIIMS-2003]

(A) BF_4 (B) NH_4^+ (C) $Cu(NH_3)_4^{2+}$ (D) $NiCl_4^{2-}$

93. Which of the following are arranged in the decreasing order of dipole moment ? [AIIMS-2003]
(A) CH_3Cl, CH_3Br, CH_3F (B) CH_3Cl, CH_3F, CH_3Br
(C) CH_3Br, CH_3Cl, CH_3F (D) CH_3Br, CH_3F, CH_3Cl

94. Which of the following is polar molecule? [JIPMER-90]
(A) CCl_4 (B) BCl_2 (C) CO_2 (D) HCl

95. In which of the following the central atom does not use sp^3 hybrid orbitals in its bonding : [JIPMER-90]
(A) BeF_3^- (B) OH_3^+ (C) NH_3^- (D) NF_3

96. The bond order of individual carbon bonds in benzene is : [JIPMER-90]
(A) One (B) Two
(C) Between one and two (D) One and two alternately

97. The linear structure is assumed by : [JIPMER-90]
(A) $SnCl_2$ (B) NCO^- (C) NO_2^+ (D) SO_2

98. An electrovalent compound is made up of : [JIPMER-91]
(A) Neutral atoms (B) Electrically charged molecules
(C) Electrically charged atoms or groups of atoms
(D) Neutral molecules

99. The cyanide ion, CN^- and N_2 are isoelectronic. But in contrast to CN^-, N_2 is chemically inert, because of : [JIPMER-91]
(A) Low bond energy (B) Absence of bond polarity
(C) Unsymmetrical electron distribution
(D) Presence of more number of electrons in bonding orbitals

100. The linear molecule among the following is : [JIPMER-91]
(A) CO_2 (B) NO_3 (C) SO_2 (D) ClO_2

101. Compound having zero dipole moment is : [JIPMER-91]
(A) H_2O (B) $BeCl_2$ (C) NH_3 (D) HCl

102. Which of the following have identical bond order? [JIPMER-91]
(A) C , N (B) O^{2-} , O (C) N , O^+ (D) C , N^+

103. Which molecule has pyramidal shape : [JIPMER-91]
(A) BF_3 (B) $BeCl_2$ (C) $SiCl_4$ (D) PCl_3

104. The bond angle in carbon tetrachloride is approx : [JIPMER-91]
(A) 90°C (B) 109° (C) 120° (D) 180°

105. If a molecule MX_2 has zero dipole moment, the sigma bonding orbitals used by M (Z < 21) are : [JIPMER-92]
(A) Pure p (B) sp hybrid (C) sp^2 hybrid (D) sp^3 hybrid

106. Ionic bonding is favoured by ? [JIPMER-92]
(A) Low charge on ions, large cation, small anion
(B) High charge on ion, small cation, large anion
(C) Low charge on ions, large cation, small anion
(D) High charge on ions, large cation, small anion

107. Which of the following compounds is covalent : [JIPMER-92]
(A) CaO (B) H_2 (C) Na_2S (D) KCl

108. The orbitals providing the most efficient overlap are: [JIPMER-92]
(A) sp–sp (B) s–s (C) sp^2–sp^2 (D) sp^3–sp^3

109. Which of the following contain unpaired electrons? [JIPMER-92]
(A) Co (B) N_2 (C) O_2 (D) H_2O_2

110. Bond between Cl – Cl is a [JIPMER-94]
(A) Coordinate (B) Ionic (C) Covalent (D) Metallic

111. The hybrid orbitals involved in the formation of ethane are [JIPMER-94]
(A) s (B) sp (C) sp^2 (D) sp^3

112. Which of the following species is planar ? [JIPMER-95]
(A) CO_2^{2-} (B) PCl_3 (C) NH_2 (D) None of these

113. The type of hybridization that can explain the bonding in C_2H_4 is [JIPMER-95]
(A) sp^3d (B) sp^3 (C) sp^2 (D) dsp^2

114. Which of the following molecules has shortest (C–C) single bond length ? [JIPMER-96]
(A) Propane (B) Propyne
(C) Propene (D) All will have same bond length.

115. The shape of the hydrides of Group VA elements is [JIPMER-97]
(A) Square planar (B) Pyramidal
(C) Tetrahedral (D) Trigonal planar

116. In the hydrogenation reaction given below the state of the carbon atom changes from [JIPMER-97]

$$C_2H_2 \xrightarrow{H_2/Ni} C_2H_4 \xrightarrow{H_2/Ni} C_2H_6$$

(A) sp – sp^2 – sp^3 (B) sp^3 – sp^2 – sp
(C) sp^2 – sp – sp^3 (D) sp^3 – sp – sp^2

117. With dsp^2 or sp^2d hybridisation, the geometry of the molecule will be [JIPMER-99]
(A) Trigonal planar (B) Octahedral
(C) Tetrahedral (D) Square planar

118. The non-linear molecule amongst the following is [JIPMER-99]
(A) CO_2 (B) Cl_2 (C) C_2H_2 (D) OF_2

119. Which can be described as a molecule with residual

bonding capacity [JIPMER-2000]
(A) $BeCl_2$ (B) NaCl (C) CH_4 (D) N_2

120. In which of the following the angle between the two covalent bonds is greatest [JIPMER-2001]
(A) H_2O (B) NH_3 (C) CH_4 (D) CO_2

121. Which combination is best explained by the co-ordinate covalent bond [JIPMER-2001]
(A) $H^+ + H_2O$ (B) Cl + Cl (C) Mg + O_2 (D) $H_2 + I_2$

122. The compound which forms a dative bond with ammonia [JIPMER-2001]
(A) CCl_4 (B) BCl_3 (C) $MgCl_2$ (D) NaCl

123. The structure of PF_5 molecule is [JIPMER-2001]
(A) trigonal bipyramidal (B) pentagonal bipyramidal
(C) square planar (D) tetrahedral

124. Pi bond is formed [JIPMER-2002]
(A) By the overlapping of atomic orbitals on the axis of nuclei.
(B) By transference of electrons
(C) By sidewise overlapping of half filled p-orbitals.
(D) By overlapping of s-orbitals with p-orbitals.

125. The ratio of π to σ bonds in benzene is [BHU-95]
(A) 1 : 4 (B) 1 :2 (C) 3 : 1 (D) 1 : 6

126. The bond angle of sp^2 - hybrid orbital is [BHU-95]
(A) 120° (B) 180° (C) 105° (D) 109°8"

127. In a diamond, the bond present is [BHU-96]
(A) Covalent (B) Electrovalent
(C) Hydrogen (D) Coordinate bond

128. Oxygen is paramagnetic. This is shown by [BHU-96]
(A) Valency theory (B) Molecular orbital theory
(C) Quantum theory (D) Gravitational theory

129. Which of the following statements is correct for covalent bond ? [BHU-97]
(A) Electrons are shared between two atoms
(B) It may be polar or non-polar
(C) Direction is non-polar
(D) Valency electrons are attracted.

130. Which of the following is least ionic ? [BHU-98]
(A) AgCl (B) KCl (C) $BaCl_2$ (D) $COCl_2$

131. The high density of water as compared to ice is due to [BHU-99]
(A) Hydrogen bonding interactions
(B) Dipole-dipole interactions
(C) Dipole-induced dipole interactions
(D) Induced dipole - induced dipole interactions

132. The tetrahedral structure is possessed by[BHU-99]
(A) CH_4 (B) NH_4^+ (C) BF_4^- (D) All of these

133. Ethyl alcohol is soluble in water in all proportions, because it [BHU-99]
(A) Is basic in nature (B) Is acidic in nature
(C) Dissociates in water
(D) Forms hydrogen bond with water

134. BF_3 molecule has planar structure. Correct explanation for this is [BHU-2000]
(A) BF_3 is sp^3 hybridised (B) BF^3 is sp^2 hybridised
(C) BF_3 is sp hybridised (D) BF_3 is sp^3d hybridised

135. Type of forces between the molecules of C_6H_6 are [BHU-2000]
(A) Dipole-dipole interactions (B) Dispersive forces
(C) H-bonding (D) van der Waal's forces

136. The bond length between carbon and carbon in sp^2 hybridised molecule is [BHU-2000]
(A) 1.2Å (B) 1.32Å (C) 1.54Å (D) 1.4 Å

137. Carbon atoms in benzene molecule is inclined at an angle of [BHU-2000]
(A) 120° (B) 180° (C) 109°28' (D) 60°

138. What is the ratio of π and σ bonds in benzene [BHU-2000]
(A) 1 : 4 (B) 2 : 1 (C) 1 : 1 (D) 2 : 2

139. The C–H bond distance is longest in [BHU-2001]
(A) C_2H_2 (B) C_2H_4 (C) C_2H_6 (D) C_6H_6

140. The type of hybridization of carbon atom in 1,2 butadine is [BHU-2001]
(A) sp (B) sp^3 and sp^2
(C) sp^2 and sp (D) sp, sp^2 and sp^3

141. The type of bonding present in N_2O_5 is [BHU-2001]
(A) Covalent only (B) Covalent and electrovalent
(C) Coordinate (D) Covalent and coordinate

142. During formation of a chemical bond between two atoms [BHU-2001]
(A) Only forces of attraction are operative
(B) Only forces of repulsion are operative
(C) Both of the above two forces are operative
(D) Neither of the forces are operative

143. O_2^{2+} has a bond order of [BHU-2001]
(A) 1.5 (B) 1 (C) 2.5 (D) 3

144. Hydrogen bonding is maximum in [BHU-2001]
(A) Ethanol (B) Diethyl ether
(C) Ethyl chloride (D) Triethyl amine

145. Which of the following pair posses the same structure [BHU-2001]
(A) PH_3 and BCl_3 (B) SO_2 and NH_3
(C) PCl_5 and SF_6 (D) NH_4^+ and SO_4^{2-}

146. Which one has dipole moment ? [BHU-2001]

(A) CCl_4 (B) 2-methyl propane
(C) cis-butene-2 (D) trans-butene-2

147. In which of the following reactions no change in valency occurs ? [BHU-2001]
(A) $SnCl_2 + Cl_2 \rightarrow SnCl_4$ (B) $2H_2O_2 \rightarrow 2H_2O + O_2$
(C) $Cl_2 + NaOH \rightarrow NaClO + NaCl + H_2O$
(D) $AgNO_3 + KCl \rightarrow AgCl + KNO_3$

148. The order of O–O bond length in H_2O_2, O_3 and O_2 is [BHU-2002]
(A) $O_2 > O_3 > H_2O_2$ (B) $O_3 > H_2O_2 > O_2$
(C) $H_2O_2 > O_3 > O_2$ (D) $O_2 > H_2O_2 > O_3$

149. Inorganic compounds are mostly soluble in [AFMC-91]
(A) Non-polar solvents (B) Polar (C) Acids (D) Bases

150. The number of ẟ-electrons present in naphthalene is [AFMC-91]
(A) 6 (B) 4 (C) 10 (D) 14

151. Hydrogen bonding does not play any role in boiling of [AFMC-92]
(A) Ammonia (B) H_2O (C) HI (D) C_2H_5OH

152. Iron is tougher than sodium because [AFMC-93]
(A) Iron atom is smaller
(B) Iron atoms are more close packed
(C) Metallic bonds are stronger in Iron
(D) None of the above

153. Which one of these is the weakest ? [AFMC-93]
(A) Ionic bond (B) Covalent bond
(C) Metallic (D) van der Waal's forces

154. Metallic bonds do not play a role in [AFMC-93]
(A) Brass (B) Copper (C) Germanium (D) Zinc

155. According to molecular orbital theory, which is non-existant [AFMC-93]
(A) H_2^- (B) O_2^- (C) O_2^+ (D) He_2

156. Which of the following phenomenon will occur when two atom elements of same spin will react ? [AFMC-94]
(A) Bonding will not occur
(B) Orbital overlap will not occur
(C) Both (a) and (b) (D) All of these

157. The polar compound among the following is [AFMC-94]
(A) CH_4 (B) CCl_4 (C) C_2H_6 (D) HCl

158. In formation of a molecule by an atom [AFMC-95]
(A) Attractive forces operate (B) Repulsive forces operate
(C) Both attractive & repulsive forces operate
(D) None of them

159. The structure of PF_5 molecule is [AFMC-95]
(A) Tetrahedral (B) Trigonal bipyramidal
(C) Square planer (D) Pentagonal bipyramidal

160. Which of the following is not linear ? [AFMC-95]
(A) H_2O (B) HCl (C) CO_2 (D) HF

161. Dipole moment is zero for [AFMC-96]
(A) CCl_4 (B) $CHCl_3$ (C) CH_2Cl_2 (D) CH_3Cl

162. Which cannot be formed ? [AFMC-97]
(A) He^{2+} (B) He^+ (C) He (D) He_2

163. The hybridization of carbon in diamond, graphite and acetylene is in the order of [AFMC-98]
(A) sp^3 sp sp^2 (B) sp^3 sp^2 sp
(C) sp sp^2 sp^3 (D) sp^2 sp^3 sp^2

164. Which of the following has zero dipole moment ? [AFMC-98]
(A) PH_3 (B) CH_4 (C) PH_4 (D) CH_2Cl_2

165. The compound containing coordinate bond is [AFMC-99]
(A) H_2SO_4 (B) O_3 (C) SO_3 (D) All of these

166. Which of the following is the most electrovalent ? [AFMC-99]
(A) Ni (B) Fe (C) Co (D) Cu

167. The H-O-H angle in water molecule is about [AFMC-2001]
(A) 90° (B) 180° (C) 102° (D) 105°

168. The boiling point of alcohol is higher than that of ether because of [AFMC-2001]
(A) Higher molecular weight
(B) Association of molecules
(C) The presence of hydroxyl group
(D) Hydrogen bonding between the molecules

169. Sideways overlap of p-p orbitals forms [AFMC-2001]
(A) Sigma bond (B) Pi bond
(C) Covalent bond (D) H bond

170. Shape of ClO_3^- is [AFMC-2001]
(A) Triangular pyramidal (B) Tetrahedral
(C) Triangular planar (D) Triangular bipyramidal

171. Which of the following is diamagnetic ? [AFMC-2001]
(A) O_2^+ (B) O_2 (C) O_2^- (D) O_2^{2-}

172. Strongest hydrogen bonding is shown by [AFMC-2002]
(A) Water (B) Ammonia
(C) Hydrogen fluoride (D) Hydrogen sulphide

173. A molecule in which sp^2 hybrid orbitals are used by the central atom in forming covalent bond is [AFMC-2002]

(A) He_2 (B) SO_2 (C) PCl_5 (D) N_2

174. The enolic form of acetone contains [IIT-90]
(A) 9 sigma, 1 pi bond and 2 lone pairs
(B) 8 sigma bond, 2 pi bonds and 2 lone pairs
(C) 10 sigma bonds, 1 pi bond and 1 lone pair
(D) 9 sigma bond, 2 pi bonds and 1 lone pair

175. The hybridization of carbon atoms in C–C single bond of $HC{\equiv}C{-}CH{=}CH_2$ is : [IIT-91]
(A) sp^3–sp^3 (B) sp^2–sp^3 (C) sp–sp^2 (D) sp^3–sp

176. The linear structure is assumed by [IIT-91]
(A) $SnCl_2$ (B) NCO^- (C) CS_2 (D) NO^{2+} (E) SO_2

177. The type of hybrid orbitals used by the chlorine atom in ClO_2^- is : [IIT-92]
(A) sp^3 (B) sp^2 (C) sp (D) None of these.

178. The CN^- ion and N_2 are isoelectronic. But in contrast to CN^-, N_2 is chemically inert, because of [IIT-92]
(A) low bond energy (B) absence of bond polarity
(C) unsymmetrical electron distribution
(D) presence of more number of electrons in bonding orbitals.

179. The maximum possible number of hydrogen bonds a water molecule can form is : [IIT-92]
(A) 2 (B) 4 (C) 3 (D) 1

180. Which of the following have identical bond order? [IIT-92]
(A) CN^- (B) O_2^- (C) NO^+ (D) CN^+

181. Amongst the three isomers of nitrophenol, the one that is least soluble in water is .... [IIT-92]

182. Among the following species identify the isostructural pairs : [IIT-96]
NF_3, NF_3^-, BF_3, H_3O^+, NH_3
(A) [NF_3, NO_3^-] and [BF_3, H_3O^+]
(B) [NF_3, NH_3] and [NO_3^-, BF_3]
(C) [NF_3, H_3O^+] and [NO_3^- , BF_3]
(D) [NF_3, H_3O^+] and [NH_3, BF_3]

183. The number and type of bonds between two carbon atoms in CaC_2 are ; [IIT-96]
(A) one sigma and one π (B) one sigma and two π
(C) one sigma and one and half π (D) one σ bond

184. Which of the following statements is correct for $CsBr_3$? [IIT-96]
(A) It is a covalent compound
(B) It contains Cs^{3+} and Br^- ions
(C) It contains Cs^+, and Br_3^- ions
(D) It contains Cs^+, Br^- and Br_2 molecule.

185. Which contains both polar and non-polar bonds ? [IIT-97]
(A) NH_4Cl (B) HCN (C) H_2O_2 (D) CH_4

186. The critical temperature of water is higher than that of O_2 because the H_2O molecule has [IIT-97]
(A) fewer electrons than O_2 (B) two covalent bonds
(C) V-shape (D) dipole moment

187. Among N_2O, SO_2, I_3^+ and I_3^-, the linear species are .... and .... [IIT-97]

188. The geometry and the type of hybrid orbital present about the central atom in BF_3 is [IIT-98]
(A) linear, sp (B) trigonal planar, sp^2
(D) tetrahedral, sp^3 (C) pyramidal, sp^3

189. In the compound $CH_2{=}CH-CH_2-CH_2-C{\equiv}CH$, the C_2-C_3 bond is the type, [IIT-99]
(A) sp–sp^2 (B) sp^3–sp^3 (C) sp–sp (D) sp^2–sp^3

190. The correct order of increasing C—O bond length of CO, CO_3^{2-}, CO_2 is [IIT-99]
(A) $CO_3^{2-} < CO_2 < CO$ (B) $CO_2 < CO_2 < CO$
(C) $CO < CO_3^{2-} < CO_2$ (D) $CO < CO_2 < CO_3^{2-}$

191. In the dichromate dianion [IIT-98]
(A) 4 Cr—O bonds are equivalent
(B) 6 Cr—O bonds are equivalent
(C) all Cr—O bonds are equivalent
(D) all Cr—O bonds are non-equivalent

192. The geometry of H_2S and its dipole moment are [IIT-99]
(A) angular and non-zero (B) angular and zero
(C) linear and non-zero (D) linear and zero

193. The geometry of H_2S and $Ni(PPh_3)_2Cl_2$ are [IIT-99]
(A) both square planar
(B) tetrahedral and square planar respectively
(C) both tetrahedral
(D) square planar and tetrahedral, respectively

194. In compounds of type ECl_3, where E = B, P , As or Bi, the angles Cl –E – Cl for different E are in the order [IIT-99]
(A) B > P = As = Bi (B) B > P >As > Bi
(C) B < P = As = Bi (D) B < P < As < Bi

195. The number of nodal planes in p_x orbital is [IIT-2000]
(A) one (B) two (C) three (D) zero

196. Molecular shapes of SF_4, CF_4 and XeF_4 are [IIT-2000]
(A) the same, with 2, 0 and 1 lone pairs of electrons respectively.
(B) the same,with 1, 1 and 1 lone pairs of electrons re-

spectively

(C) different, with 0, 1 and 2 lone pairs of electrons respectively

(D) different,with 1, 0 and 2 lone pairs of electrons respectively.

197. The hybridisation of atomic orbitals of nitrogen in NO_2^+, NO_3^- and NH_4^+ are [IIT-2000]

(A) sp, sp^3 and sp_2 respectively

(B) sp, sp^2 and sp^3 respectively

(C) sp^2, sp and sp^3 respectively

(D) sp^2, sp^3 and sp respectively.

198. The correct order of hybridization of the central atom in the following species. NH_3, $[PtCl_4]^{2-}$, PCl_5 and BCl_3 is [IIT-2001]

(A) dsp^2, dsp^3, sp^2 and sp^3 (B) sp^3, dsp^2, dsp^3, sp^2

(C) dsp^2, sp^2, sp^3, dsp^3 (D) dsp^2, sp^3, sp^2, dsp^3

199. The common features among the species CN^-, CO and NO^+ are [IIT-2001]

(A) bond order three and isoelectronic

(B) bond order three and weak field ligands

(C) bond order two and π-acceptors

(D) isoelectronic and weak field ligands

200. The number of S-S bonds in sulphur trioxide trimer (S_3O_9) is [IIT-2001]

(A) three (B) two (C) one (D) zero

201. Which of the following molecular species has unpaired electron(s) ? [IIT-2002]

(A) N_2 (B) F_2 (C) O_2^- (D) O_2^{2-}

202. Of the following, isoelectronic and isostructural species are NO_3^-, ClO_3^-, SO_3, CO_3^{2-} [IIT-2003]

(A) ClO_3^-, NO_3^- (B) NO_3^-, CO_3^{2-}

(C) SO^3, CO_3^- (D) ClO_3^-, CO_2^-

203. Maximum dipole moment will be of [IIT-2003]

(A) $C_2H_2Cl_2$ (B) CH_3I (C) CH_3Cl (D) $C_2H_2I_2$

204. Left to right sp^2, sp^2, sp, sp hybridisation is present in [IIT-2003]

(A) $CH_2 = CH - C \equiv N$ (B) $CH \equiv C - C \equiv N$

(C) $CH_3 - CH_2 - C \equiv N$ (D) None of these

205. In which of the following species the interatomic bond angle is 109° 28' ? [AIEEE-2002]

(A) NH_3, $(BF_4)^{-1}$ (B) $(NH_4)^+$, BF_3

(C) NH_3, BF_4 (D) $(NH_2)^{-1}$, BF_3

206. For the reaction $A + 2B \rightarrow C$, rate is given by $R = [A] [B]^2$ then the order of the reaction is [AIEEE-2002]

(A) 3 (B) 6 (C) 5 (D) 7

207. Number of sigma bonds in P_4O_{10} is [AIEEE-2002]

(A) 6 (B) 7 (C) 17 (D) 16

208. Which one of the following compounds has the smallest bond angle in its molecule ? [AIEEE-2003]

(A) SO_2 (B) H_2O (C) H_2S (D) NH_3

209. Which one of the following pairs of molecules will have permanent dipole moments for both members ? [AIEEE-2003]

(A) SiF_4 and NO_2 (B) NO_2 and CO_2

(C) NO_2 and O_3 (D) SiF_4 and CO_2

210. The pair of species having identical shapes for molecules of both species is [AIEEE-2003]

(A) CF_4, SF_4 (B) XeF_2, CO_2

(C) BF_3, PCl_3 (D) PF_5, IF_5

211. The ion that is isoelectronic with carbon monoxide is: [EAMCET-90]

(A) CN^- (B) O_2^+ (C) O_2^- (D) N_2^+

212. The type of Hybridisation of orbitals employed in the formation of SF_6 molecule is ....[EAMCET-90]

213. The bonds in $K_4[Fe(CN)_6]$ are ; [EAMCET-91]

(A) Ionic (B) Covalent (C) Ionic & covalent

(D) Ionic, covalent and coordinate covalent bond.

214. The number of σ-bonds in the ethylene molecule is .... [EAMCET-91]

215. The normal b.p. of CS_2, H_2O and CCl_4 are 41.3°C, 100°C and 77°C respectively. The liquid in which the intermolecular forces are the weakest is .... [EAMCET-91]

216. Water has a higher b.p. than corresponding hydrides H_2S, H_2Se and H_2Te. This is because of .... in water. [EAMCET-91]

217. d^2sp^3 hybridisation represents .... [EAMCET-92]

218. In which of the following compounds is hydrogen bonding strongest in the liquid phase ? [EAMCET-93]

(A) HF (B) CH_4 (C) HI (D) PH_3

219. Which of the following contains a coordinate covalent bond ? [EAMCET-93]

(A) Water (B) Ammonia

(C) Ammonium ion (D) Ethylene

220. The hybridization involved in acetylene is [EAMCET-93]

(A) sp (B) sp^2 (C) sp^3 (D) dsp^2

221. Valence bond theory of L. Pauling and J.C. Slater accounts for .... characteristics of covalent bond. [EAMCET-93]

222. When hydrogen bond is formed between two molecules it is called .... hydrogen bond. [EAMCET-93]

223. The melting point of diamond is very high, since the type of bonding in it is ....[EAMCET-93]

224. Which of the following molecules has the highest dipole moment ? [EAMCET-95]
(A) CO_2 (B) NF_3 (C) BCl_3 (D) CCl_4

225. The effect of repulsion between the two lone pairs of electrons present on oxygen in water molecule is [EAMCET-96]
(A) No change in H-O-H bond angle
(B) Increase in H-O-H bond angle
(C) Decrease in H-O-H bond angle
(D) All atoms will be in one plane

226. Which one of the following statements is true for ammonium ion ? [EAMCET-97]
(A) All bonds are ionic
(B) All bonds are coordinate covalent
(C) H atoms are situated at the corners of a square
(D) H atoms are situated at the corners of tertrahedron.

227. Among the following compounds which have more than one type of hybridization for carbon atom ? [EAMCET-98]
(a) $CH_3\ CH_2\ CH_2\ CH_3$ (b) $CH_3\ CH = CH\ CH_3$
(c) $CH_2 = CH–C \equiv CH$ (d) $H–C \equiv C–H$
(A) b & c (B) b (C) c & d (D) c

228. The state of hybridization of 'B' in BCl_3 is [EAMCET-98(MBBS)]
(A) sp^3 (B) sp^2 (C) sp (D) sp^3d^2

229. According to VSEPR theory the shape of the water molecule is [EAMCET-98(MBBS)]
(A) Linear (B) Planar triangle
(C) Tetrahedral (D) Octahedral

230. According to Fajans' rule covalent bond is favoured by [EAMCET-98(MBBS)]
(A) Small cation & large anion
(B) Small cation & small anion
(C) Large cation & large anion
(D) Large cation & small anion

231. The molecule having zero dipole moment is [EAMCET-99]
(A) $CHCl_2$ (B) H_2O (C) CCl_4 (D) HCl

232. Which of the following covalent molecules is an exception to octet rule ? [EAMCET-99]
(A) $BeCl_2$ (B) CO_2 (C) H_2O (D) CH_4

233. In which of the following compounds the angle between two covalent bonds is the highest ? [EAMCET-99(MBBS)]
(A) CH_4 (B) NH_3 (C) H_2O (D) H_2O_2

234. The number of Sigma bonds in ethane formed by the overlap of sp^3 and s orbitals is [EAMCET-99(MBBS)]
(A) 5 (B) 6 (C) 7 (D) 4

235. Which of the following has least bond energy ? [EAMCET(MBBS)-2000]
(A) F_2 (B) H_2 (C) N_2 (D) O_2

236. Which of the following has the lowest boiling point ? [EAMCET(MBBS)-2000]
(A) CH_4 (B) H_2O (C) HF (D) C_2H_5OH

237. Which of the following is an electron deficient compound ? [EAMCET(MBBS)-2000]
(A) NaCl (B) NaH (C) B_2H_6 (D) KCl

238. Which of the following statements is true ? [EAMCET-2001]
(A) Hybridisation of the central atom in NH_3 and CH_4 is sp^2
(B) $BeCl_2$ has "V" shape while SO_2 is linear
(C) SF_6 is octahedral and F–S–F bond angle is 90^0
(D) CO_2 has dipole moment

239. In which of the following substances, hydrogen bonding is absent ? [EAMCET(MBBS)-2001]
(A) HF (B) H_2O (C) CCl_4 (D) Salicylaldehyde

240. The bond energies (in kJ mol^{-1}) of P–H, As–H and N–H are respectively. [EAMCET-2002]
(A) 247, 138 & 389 (B) 247, 389 and 318
(C) 318, 389 & 247 (D) 318, 247 and 389

241. Which of the following molecules contain both ionic and covalent bonds ? [EAMCET-2002]
(A) CH_2Cl_2 (B) K_2SO_4 (C) $BeCl_2$ (D) SO_2

242. Which one of the following is the correct set with reference to molecular formula, hybridization of central atom and shape of the molecule ?[EAMCET(med.)-2002]
(A) CO_2 ,sp^2, bent (B) H_2O,sp^2, bent
(C) $BeCl_2$, sp, linear (D) H_2O, sp^3, linear

243. What is the number of sigma (σ) and pi (π) bonds present in sulphuric acid molecule ? [EAMCET(med.)-2002]
(A) 6σ, 2π (B) 6σ, 0π (C) 2σ, 4π (D) 2σ, 2π

244. Which one of the following is correct set ? [EAMCET-2003]
(A) H_2O, sp^3, angular (B) H_2O, sp^2, linear
(C) NH_4^+, dsp^3, square planar (D) CH_4,dsp^3, tetrahedral

245. If the bond length and dipole moment of a diatomic molecule are 1.25 Å and 1.0 D respectively, what is the percent ionic character of the bond ? [EAMCET-2003]
(A) 10.66 (B) 12.33 (C) 16.66 (D) 17.33

246. Which of the following have least covalent P-H

bond ? [CPMT-96]
(A) PH_3 (B) $P_2H_6^-$ (C) P_2H_5 (D) PH_6^+

247. Which of the following is not linear ? [CPMT-96]
(A) $SnCl_2$ (B) HCl (C) CO_2 (D) Hg_2Cl_2

248. Electrovalent compouds are [CPMT-96]
(A) Good conductors of electricity (B) Unopolar in nature
(C) low m.p., b.p. (D) Easily available

249. Which of the following show high m.p., b.p.? [CPMT-96]
(A) Covalent compounds (B) Electrovalent compounds
(C) Coordinate compounds (D) None of these

250. Which of the following has zero dipole moment ? [CPMT-97]
(A) CH_2Cl_2 (B) CH_4 (C) PH_4 (D) PH_3

251. Which one has high boiling point ? [CPMT-97]
(A) Butan–2–ol (B) Butane (C) Pentane (D) Ethane

252. Which of the following possesses highest melting point ? [CPMT-99]
(A) Chlorobenzene (B) o-dichlorobenzene
(C) m- dichlorobenzene (D) p-dichlorobenzene

253. Which of the following carbon atoms is likely to possess geometry ? [CPMT-99]

$$H_2C = CH–CH_2–COOH$$

(A) 1 (B) 2 (C) 3 (D) 4

254. Which of the following has maximum hydrogen bonding ? [CPMT-99]
(A) Ethylamine (B) Ammonia
(C) Ethyl alcohol (D) Diethyl ether

255. Which of the bonds is most polar ? [CPMT-99]
(A) C - O (B) C - F (C) O - F (D) N - F

256. The shape of NH_3 molecule is [CPMT-2000]
(A) Tetrahedral (B) Pyramidal
(C) Trigonal planar (D) Linear

257. The two adjacent sheets are bonded together in graphite with [CPMT-2001]
(A) van der Waal's forces (B) Ionic bonds
(C) Metallic bonds (D) Covalent bonds

258. In graphite C atom is in .... state [CPMT-2002]
(A) sp^3 (B) sp (C) sp^2 (D) none of these

259. The atomic distance in H_2 and Cl_2 are 74 and 198 pm. The bond length of HCl is [CPMT-2002]
(A) 136 pm (B) 68 pm (C) 272 pm (D) 363 pm

260. Which of the following shows linear structure ? [CPMT-2003]
(A) Ethane (B) Ethene (C) Acetylene (D) CCl_4

261. Water has high boiling point because[CPMT-2003]
(A) it has higher molecular weight
(B) it has more lattice energy (C) it is weak acid
(D) it is associated with hydrogen bonding

262. Among the bonds formed by a chlorine atom with atoms of hydrogen, chlorine, sodium and carbon, the strongest bond is formed between : [PMT-93]
(A) H - Cl (B) Cl -Cl (C) Na ...Cl (D) C - Cl

263. Which one of the following molecules is trigonal bipyramidal : [PMT-93]
(A) BF_3 (B) CH_4 (C) PCl_5 (D) SF_6

264. Ammonia molecule is formed due to the following hybridized orbitals : [PMT-93]
(A) dsp^2 (B) sp^3 (C) sp^3d (D) d^2sp

265. The high boiling point of water is due to the presence of : [PMT-93]
(A) dative bond (B) covalent bond
(C) hydrogen bond (D) van der Waal's forces

266. The high boiling point of ethanol (78.2°C) compared to dimethyl ether (-23.6°C), though both having the same molecular formula C_2H_6O, is due to : [PMT-93]
(A) Hydrogen bonding (B) Ionic bonding
(C) Coordinate covalent bonding (D) Resonance

267. In which molecule are all atoms coplanar ? [PMT-94]
(A) CH_4 (B) BF_3 (C) PF_3 (D) NH_3

268. Which of the following does not have the hydrogen bond ? [PMT-94]
(A) Phenol (B) Liquid NH_3 (C) Water (D) Liquid HCl

269. Ionic bonds are usually formed by combination of elements with [PMT-94]
(A) high ionisation potential and low electron affinity
(B) low ionisation potential and high electron affinity
(C) high ionisation potential and high electron affinity
(D) low ionisation potential and low electron affinity

270. The angle between two covalent bonds is maximum in [PMT-94]
(A) CH_4 (B) H_2O (C) CO_2 (D) SO_2

271. If a molecule MX_3 has zero dipole moment the sigma bonding orbitals used by M (At. No. 21) are [PMT-94]
(A) pure p (B) sp hybrid (C) sp^2 hybrid (D) sp^3 hybrid

272. CO_2 is isostructural with [PMT-94]
(A) $SnCl_2$ (B) SO_2 (C) $HgCl_2$ (D) all above

273. The structure of chloromethane is [PMT-95]
(A) Tetrahedral (B) Trigonal (C) Linear (D) Hexagonal

274. CO_2 is isostructural with which of the following ?

[PMT-95]

(A) $HgCl_2$ (B) H_2O (C) $SnCl_2$ (D) NO_2

275. Which one of the following molecules possesses a dipole moment ? [PMT-95]

(A) Carbon tetrachloride (B) Water
(C) Methane (D) Nitrogen

276. The octet rule is not valid for which of the following molecules ? [PMT-95]

(A) CO_2 (B) H_2O (C) CO (D) O_2

277. How many σ and π bonds are there in the molecule of tetracyano ethylene ? [PMT-95]

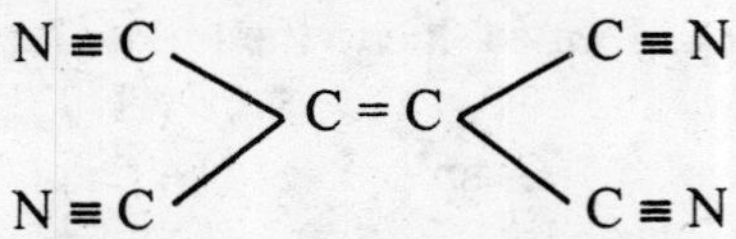

(A) nine σ , nine π (B) five σ , nine π
(C) nine σ , seven π (D) five σ , eight π

278. Which of the following molecules has a linear structure ? [PMT-95]

(A) H_2O (B) CO_2 (C) NH_3 (D) CH_4

279. Which one of the following has both electrovalency and covalency ? [PMT-95]

(A) NaCl (B) NaOH (C) H_2O (D) HCl gas

280. A covalent bond between two atoms is formed by which of the following ? [PMT-96]

(A) Electron nuclear attraction (B) Electron sharing
(C) Electron transfer (D) Electrostatic attraction

281. Hybridisation involves [PMT-96]

(A) addition of an electron pair
(B) mixing up of atomic orbitals
(C) removal of an electron pair (D) separation of orbitals

282. On the basis of Molecular Orbital theory which molecule does not exist ? [PMT-96]

(A) H_2 (B) He_2^+ (C) He_2 (D) Li_2

283. Which of the following compounds in liquid state does not have hydrogen bonding ? [PMT-96]

(A) H_2O (B) HF (C) NH_3 (D) C_6H_6

284. According to molecular orbital theory, the paramagnetism of O_2 molecule is due to presence of - [PMT-97]

(A) Unpaired electrons in the bonding σ molecular orbital
(B) Unpaired electrons in the antibonding σ molecular orbital
(C) Unpaired electron in the bonding π molecular orbital
(D) Unpaired electrons in the antibonding π molecular orbital

285. Out of the following compounds the one which forms hydrogen bond is [PMT-97]

(A) CH_4 (B) NaCl (C) H_2O (D) $CHCl_3$

286. Out of the following hybrid orbitals the one which forms the bond at angle 120° is [PMT-97]

(A) d^2sp^3 (B) sp^3 (C) sp^2 (D) sp

287. As the p character increases the bond angle in hybrid orbitals formed by s and p atomic orbitals [PMT-97]

(A) Decreases (B) Increases
(C) Doubles (D) Remains unchanged

288. As compared to covalent compounds, the electrovalent compounds generally have [PMT-97]

(A) High melting point and high boiling point
(B) Low melting point and low boiling point
(C) Low melting point and high boiling point
(D) High melting point and low boiling point

289. The number of π bonds in 3-hexyne-1-ene is : [PMT-99]

(A) 1 (B) 2 (C) 3 (D) 4

290. The hybrid orbitals used by central atoms in $BeCl_2$, BCl_3 and CCl_4 molecules are respectively : [PMT-99]

(A) sp^2, sp^3 and sp (B) sp, sp^2 and sp^3
(C) sp^3, sp and sp^2 (D) sp^2, sp and sp^3

291. Amongest ClF_3, BF_3 and NH_3 molecules the one with non-planar geometry is : [PMT-99]

(A) ClF_3 (B) NH_3 (C) BF_3 (D) None of these

292. The number of ionic, covalent and coordinate bonds in NH_4Cl are respectively [PMT-99]

(A) 1, 3 and 1 (B) 1, 3 and 2
(C) 1, 2 and 3 (D) 1, 1 and 3

293. Which one of the following compounds has zero dipole moment ? [PMT-99]

(A) Ammonia (B) Water
(C) Sulphur dioxide (D) Boron trichloride

294. The number of possible resonance structures for CO_3^{2-} is : [PMT-2000]

(A) 2 (B) 3 (C) 6 (D) 9

295. Which of the following molecules is linear ? [PMT-2000]

(A) SO_2 (B) NO_2^+ (C) NO_2^- (D) SCl_2

296. In which of the following pairs the two molecules have identical bond order ? [PMT-2000]

(A) N_2, O_2^{2+} (B) N_2, O_2^- (C) N_2^-, O_2 (D) O_2^{2+}, N_2

297. The dipole moment of HBr is 1.6×10^{-30} c.m. and interatomic spacing is 1. Å. The % ionic character of HBr is : [PMT-2000]

(A) 7 (B) 10 (C) 15 (D) 27

298. In $[Co(NH_3)_6]Cl_3$ the number of covalent bonds is : [PMT-2001]

(A) 3 (B) 6 (C) 9 (d) 18

299. Which of the following is not tetrahedral ? [PMT-2001]
(A) SCl_4 (B) SO_4^{2-} (C) $Ni(Co)_4$ (D) $NiCl_4^{2-}$

300. The bond order is not three for ? [PMT-2001]
(A) N_2^+ (B) O_2^{2+} (C) N_2 (D) NO^+

301. P in PCl_5 is hybridized as [PMT-2002]
(A) sp^3 (B) dsp^2 (C) ds^3p (D) sp^3d

302. Which of the following molecule has dipole moment zero ? [PMT-2003]
(A) Water (B) Beryllium trifluoride
(C) Sulphur dioxide (D) Ammonia

303. BCl_3 is planar molecule because in this molecule boron (B) is : [PMT-2003]
(A) sp hybridised (B) sp^2 hybridised
(C) sp^3 hybridised (D) None of these

304. The maximum number of hydrogen bonds formed by a water molecule in ice is [PET-93]
(A) 4 (B) 3 (C) 2 (D) 1

305. The number of anti-bonding electrons in O_2^{2-} is [PET-93]
(A) 4 (B) 6 (C) 8 (D) 10

306. The hybridization of carbons of C–C single bond of $HC \equiv C-CH = CH_2$ is [PET-93]
(A) $sp^3 - sp^3$ (B) $sp^2 - sp^2$ (C) $sp^3 - sp$ (D) $sp - sp^2$

307. The molecule which has the largest dipole moment is : [PET-93]
(A) CH_4 (B) $CHCl_3$ (C) CHI_3 (D) CCl_4

308. Hydrogen bonding is maximum in [PET-93]
(A) Ethanol (B) Diethyl ether
(C) Ethyl acetate (D) Triethyl amine

309. Among the following which property is commonly exhibited by a covalent compound [PET-94]
(A) High solubility in water
(B) High electrical conductance
(C) Low boiling point (D) High Melting point

310. In which of the following compounds sp^2 hybrid orbitals are used ? [PET-94]
(A) BCl_3 (B) CH_4 (C) NH_3 (D) BeH_2

311. The bond angle in water molecule is [PET-94]
(A) 120° (B) 109°, 28' (C) 90° (D) 104° 27'

312. In which of the following compounds intramolecular hydrogen bond is present [PET-94]
(A) Ethyl alcohol (B) Water
(C) Salicylaldehyde (D) Hydrogen sulphide

313. The p-p orbital overlapping is present in the following molecule [PET-94]
(A) Hydrogen (B) Hydrogen bromide
(C) Hydrogen chloride (D) Chlorine

314. Which of the following exhibits highest solubility in water [PET-94]
(A) NH_3 (B) PH_3 (C) AsH_3 (D) SbH_3

315. In which of the following pairs the difference between the covalent radii of the two metals is maximum [PET-94]
(A) K, Ca (B) Mn, Fe (C) Co, Ni (D) Cr, Mn

316. Which of the following hybridisation has maximum s-charater ? [PET-95]
(A) sp^3 (B) sp^2 (C) sp (D) None

317. The formation of a chemical bond is accompanied by [PET-95]
(A) Decrease in energy (B) Increase in energy
(C) Neither increase nor decrease in energy
(D) None of the above

318. Which of the following molecules has a linear structure [PET-95]
(A) CO_2 (B) H_2O (C) SO_2 (D) H_2O_2

319. Hydrogen bonding is formed in compounds containing hydrogen and [PET-95]
(A) Highly electro - negative atoms
(B) Highly electro - positive atoms
(C) Metal atoms with d-orbitals occupied
(D) Metalloids

320. Covalent molecules are usually held in a crystal structure by - [PET-95]
(A) Dipole-dipole attraction (B) electrostatic attraction
(C) Hydrogen bonds (D) van der Waal's attraction

321. The PCl_5 molecule is a result of the hybridization of- [PET-95]
(A) sp^2d^2 (B) sp^3d (C) spd^3 (D) sp^2d^3

322. Which is the correct electron dot structure of N_2O molecule ? [PET-96]
(A) $:N = N = \ddot{\underset{\cdot\cdot}{O}}$ (B) $:N = N^+ - \ddot{\underset{\cdot\cdot}{O}}:^-$
(C) $\ddot{\underset{\cdot\cdot}{N}} = \ddot{N} = \ddot{\underset{\cdot\cdot}{O}}$ (D) $:N = N = \underset{\cdot\cdot}{O}:$

323. The bond order of NO molecule is [PET-96]
(A) 1.5 (B) 2.0 (C) 2.5 (D) 3.0

324. The bond angle is water (H_2O) molecule is [PET-96]
(A) 109°28 (B) 180° (C) 90° (D) 105°

325. According to VSEPR theory the most probable shape of the molecule having 4 electron pairs in the outer shell of the central atom is [PET-96]
(A) Linear (B) Tetrahedral

(C) hexahedral (D) Octahedral

326. The number of a σ bonds in o-xylene is [PET-96]
(A) 6 (B) 9 (C) 12 (D) 18

327. Bond length between carbon in ethylene molecule is: [PET-97]
(A) 1.54 Å (B) 1.35 Å (C) 1.19 Å (D) 2.4 Å

328. The molecule having bond order 3, is : [PET-97]
(A) H_2 (B) N_2 (C) O^2 (D) He_2^+

329. The hydrogen bond is strongest in : [PET-97]
(A) O – H ..... S (B) S – H ..... O
(C) F – H ..... F (D) F – H ..... O

330. The geometry of ClO_3^-, according to Valence Shell Electron Pair Repulsion (VSEPR) theory will be : [PET-97]
(A) Planar Triangle (B) Pyramidal
(C) Tetrahedral (D) Square planar

331. Which of the following orbitals has a dumb-bell shape ? [PET-98]
(A) s (B) p (C) d (D) f

332. Which of the following statements regarding covalent bond is not true ? [PET-98]
(A) The electrons are shared between atoms
(B) The bond is non-directional
(C) The strength of the bond depends upon the extent of overlapping
(D) The bond formed may or may not be polar

333. Directed bonds in water form an angle of : [PET-98]
(A) 90° (B) 120° (C) 105° (D) 60°

334. Hydrogen bonding is not present in : [PET-98]
(A) Glycerine (B) Water
(C) Hydrogen (D) Hydrogen Fluoride

335. sp^3 hybridization leads to which shape of the molecule ? [PET-98]
(A) Tetrahedron (B) Octahedron
(C) Linear (D) Plane triangle

336. In a double bond connecting two atoms there is a sharing of : [PET-99]
(A) 2 electrons (B) 4 electrons
(C) 1 electrons (D) All electrons

337. The maximum number of hydrogen bonds in which a water molecule can participate is: [PET-99]
(A) 1 (B) 2 (C) 3 (D) 4

338. Which of the following is paramagnetic [PET-99]
(A) O_2 (B) CN^- (C) CO (D) NO^+

339. Which of the following will be octahedral? [PET-99]
(A) SF_6 (B) BF_4^- (C) PCl_5 (D) BO_3^{3-}

340. The bond-order in O_2^+ is : [PET-99]
(A) 2 (B) 2.5 (C) 1.5 (D) 3

341. Which one of the following should be most stable ? [PET-2000]
(A) H_2^+ (B) H^+ (C) H (D) H^-

342. Which of the following is least ionic ? [PET-2000]
(A) C_2H_5Cl (B) KCl (C) $BaCl_2$ (D) $C_6H_5N^+H_3Cl^-$

343. The number of antibonding electron pairs in O_2^{2-} molecular ion on the basis of molecular orbital theory is : [PET-2000]
(A) 4 (B) 3 (C) 2 (D) 5

344. The number of sigma and pi bonds in 1-butene 3-yne are : [PET-2000]
(A) 5 sigma and 5 pi (B) 7 sigma and 3 pi
(C) 8 sigma and 2 pi (D) 6 sigma and 4 pi

345. The bond order of individual carbon-carbon bond in benzene is : [PET-2000]
(A) One (B) Two
(C) Between one and two (D) One and two alternately

346. In which of the following molecules is the covalent bond most polar ? [PET-2000]
(A) HI (B) HBr (C) HCl (D) H_2

347. Among the following compounds which one is planar in shape ? [PET-2000]
(A) Methane (B) Acetylene (C) Benzene (D) Isobutane

348. Which molecule has a dipole moment ? [PET-2000]
(A) CCl_4 (B) C_6H_6 (C) BF_3 (D) HF

349. Which of the following species contain non directional bond ? [PET-2001]
(A) NH_3 (B) CsCl (C) NH_4 (D) BeF_2

350. According to VSEPR Theory, the most probable shape of the molecule having 4 electron pairs around the central atom is : [PET-2001]
(A) Hexahedral (B) Tetrahedral
(C) Octahedral (D) Linear

351. Two elements whose electronegativities are 1.2 and 3.0 the bond formed between them would be : [PET-2002]
(A) Ionic (B) Covalent (C) Coordinate (D) Metallic

352. In benzene, all the six C — C bonds have the same length because of [PET-2002]
(A) Tautomerism (B) sp^2 hybridisation
(C) Isomerism (D) Inductive effect

353. The value of the energy for the first excited state of hydrogen atom will be [PET-2002]
(A) –13.6 eV (B) –3.40 eV (C) –1.51 eV (D) –0.85 eV

354. The number of dative bonds in sulphuric acid molecules is [PET-2002]
(A) 0 (B) 1 (C) 2 (D) 4

355. o-nitrophenol is more volatile than p-nitrophenol due to [PET-2002]
(A) Intramolecular H-bonding
(B) Intermolecular H-bonding
(C) Resonance (D) Inductive effect

356. Which one of the following species possesses the highest bond order [PET-2002]
(A) O_2 (B) O_2^+ (C) O_2^- (D) O_2^{-2}

357. In which of the following pairs molecules have bond order three and are isoelectronic ? [PET- 2003]
(A) CN^-,CO (B) CO, O_2^+ (C) CN, O_2^+ (D) NO^+, CO

358. Which of the follwoing has dsp^2 hydridization [PET-2003]
(A) C_2H_2 (B) SCl_4 (C) NH_4^+ (D) $PtCl_4^{2-}$

359. Hydrogen bond is strongest in [PET- 2003]
(A) H - F --------H (B) H ----------- O
(C) O - H ------- O (D) H ----------- N

360. In ClF_3 the hybridized state of Cl is: [PET-2003]
(A) sp^2 (B) sp^3d^2 (C) sp^3d (D) sp^3

361. In which of the follwoing metallic bond is strongest [PET-2003]
(A) Fe (B) Cr (C) Sc (D) Mn

362. Which compound has the highest boiling point? [PET- 2003]
(A) Diethylether (B) Ethanol
(C) Methanol (D) Acetone

363. Which has a pyramidal structure ? [PET-2003]
(A) Ethanol (B) Water
(C) Acetylene (D) Trimethylamine

364. Intermolecular hydrogen - bonding is observed in [AIIMS-2000]
(A) Syn-pyridine-2-carboxaldoxime
(B) Anti-pyridine-2-carboxaldoxime
(C) Both (A) and (B) (D) Pyridine-3-carboxaldoxime

365. H_2O is liquid, H_2S is gas due to [BHU-2003]
(A) H- bonding (B) Molecular attraction
(C) H-bond and three dimensional ice structure
(D) H-bonding and molecular attraction

366. H-bonding is maximum in [BHU-2003]
(A) CH_3COCH_3 (B) C_6H_5COOH
(C) CH_3CH_2OH (D) C_6H_5OH

367. Graphite is soft while diamond is hard because [BHU-2003]
(A) Graphite is in planar form while diamond is in tetrahedral form
(B) Diamond has sp^2 hybridisation but graphite has sp^3 hybridisation
(C) Graphite is in powder form
(D) Graphite is covalent and diamond is ionic

368. The helical structure of protein is stabilized by [CBSE-PMT-2004]
(A) dipeptide bonds (B) hydrogen bonds
(C) ether bonds (D) peptide bonds

369. In BrF_3 molecule, the lone pairs occupy equatorial positions to minimize [CBSE-PMT-2004]
(A) lone pair - bond pair repulsion only
(B) bond pair - bond pair repulsion only
(C) lone pair - lone pair repulsion and lone pair - bond pair repulsion
(D) lone pair - lone pair repulsion only

370. Which of the following statements is true about O_2 and O_2^+ on the basis of M.O.T? [IIT-2004]
(A) is paramagnetic and its bond order is higher than O_2.
(B) is paramagnetic and its bond order is lower than O_2.
(C) is diamagnetic and its bond order is higher than O_2.
(D) is diamagnetic and its bond order is lower than O_2

Chemical Bonding

1.	(C)	48.	(A)	95.	(A)	142.	(C)	188.	(B)	235.	(A)	282.	(C)	329.	(C)
2.	(C)	49.	(D)	96.	(C)	143.	(D)	189.	(D)	236.	(A)	283.	(D)	330.	(B)
3.	(B)	50.	(B)	97.	(C)	144.	(A)	190.	(D)	237.	(C)	284.	(D)	331.	(B)
4.	(B)	51.	(C)	98.	(C)	145.	(D)	191.	(B)	238.	(C)	285.	(C)	332.	(B)
5.	(B)	52.	(B)	99.	(B)	146.	(C)	192.	(A)	239.	(C)	286.	(C)	333.	(C)
6.	(A)	53.	(C)	100.	(A)	147.	(D)	193.	(C)	240.	(D)	287.	(A)	334.	(C)
7.	(D)	54.	(B)	101.	(B)	148.	(C)	194.	(B)	241.	(B)	288.	(A)	335.	(A)
8.	(B)	55.	(D)	102.	(C)	149.	(B)	195.	(A)	242.	(C)	289.	(C)	336.	(B)
9.	(C)	56.	(C)	103.	(D)	150.	(C)	196.	(D)	243.	(A)	290.	(B)	337.	(D)
10.	(A)	57.	(A)	104.	(B)	151.	(C)	197.	(B)	244.	(A)	291.	(B)	338.	(A)
11.	(D)	58.	(C)	105.	(C)	152.	(C)	198.	(B)	245.	(C)	292.	(A)	339.	(A)
12.	(D)	59.	(C)	106.	(A)	153.	(D)	199.	(A)	246.	(D)	293.	(D)	340.	(B)
13.	(C)	60.	(A)	107.	(B)	154.	(C)	200.	(D)	247.	(A)	294.	(B)	341.	(D)
14.	(A)	61.	(C)	108.	(A)	155.	(D)	201.	(C)	248.	(A)	295.	(B)	342.	(A)
15.	(C)	62.	(D)	109.	(C)	156.	(D)	202.	(B)	249.	(B)	296.	(A)	343.	(A)
16.	(A)	63.	(A)	110.	(C)	157.	(D)	203.	(C)	250.	(B)	297.	(B)	344.	(B)
17.	(C)	64.	(B)	111.	(D)	158.	(C)	204.	(A)	251.	(A)	298.	(D)	345.	(C)
18.	(B)	65.	(C)	112.	(A)	159.	(B)	205.	(A)	252.	(D)	299.	(A)	346.	(C)
19.	(C)	66.	(A)	113.	(C)	160.	(A)	206.	(A)	253.	(C)	300.	(A)	347.	(C)
20.	(D)	67.	(A)	114.	(B)	161.	(A)	207.	(D)	254.	(C)	301.	(D)	348.	(D)
21.	(B)	68.	(B)	115.	(B)	162.	(D)	208.	(B)	255.	(B)	302.	(B)	349.	(B)
22.	(B)	69.	(B)	116.	(A)	163.	(B)	209.	(C)	256.	(B)	303.	(B)	350.	(B)
23.	(A)	70.	(A)	117.	(D)	164.	(B)	210.	(D)	257.	(D)	304.	(A)	351.	(A)
24.	(B)	71.	(A)	118.	(D)	165.	(D)	211.	(A)	258.	(C)	305.	(C)	352.	(B)
25.	(D)	72.	(D)	119.	(A)	166.	(D)	212.	*	259.	(A)	306.	(D)	353.	(B)
26.	(A)	73.	(B)	120.	(D)	167.	(D)	213.	(D)	260.	(C)	307.	(B)	354.	(C)
27.	(B)	74.	(A)	121.	(A)	168.	(D)	214.	*	261.	(D)	308.	(A)	355.	(A)
28.	(B)	75.	(A)	122.	(B)	169.	(B)	215.	*	262.	(C)	309.	(C)	356.	(B)
29.	(C)	76.	(B)	123.	(A)	170.	(A)	216.	*	263.	(C)	310.	(A)	357.	(A)
30.	(B)	77.	(A)	124.	(C)	171.	(D)	217.	*	264.	(B)	311.	(D)	358.	(D)
31.	(D)	78.	(B)	125.	(A)	172.	(C)	218.	(A)	265.	(C)	312.	(C)	359.	(A)
32.	(A)	79.	(C)	126.	(A)	173.	(B)	219.	(C)	266.	(A)	313.	(D)	360.	(C)
33.	(B)	80.	(B)	127.	(A)	174.	(A)	220.	(A)	267.	(B)	314.	(A)	361.	(B)
34.	(B)	81.	(C)	128.	(B)	175.	(C)	221.	*	268.	(D)	315.	(A)	362.	(C)
35.	(C)	82.	(B)	129.	(A)	176.	(B)	222.	*	269.	(B)	316.	(C)	363.	(D)
36.	(A)	83.	(A)	130.	(A)	(C)	(D)	223.	*	270.	(C)	317.	(A)	364.	(B)
37.	(A)	84.	(A)	131.	(A)	177.	(A)	224.	(B)	271.	(C)	318.	(A)	365.	(A)
38.	(B)	85.	(A)	132.	(D)	178.	(B)	225.	(C)	272.	(C)	319.	(A)	366.	(B)
39.	(C)	86.	(B)	133.	(D)	179.	(B)	226.	(D)	273.	(A)	320.	(D)	367.	(A)
40.	(C)	87.	(A)	134.	(B)	180.	(C)	227.	(A)	274.	(A)	321.	(B)	368.	(B)
41.	(B)	88.	(A)	135.	(B)	181.	*	228.	(B)	275.	(B)	322.	(B)	369.	(D)
42.	(C)	89.	(B)	136.	(B)	182.	(C)	229.	(C)	276.	(C)	323.	(C)	370.	(A)
43.	(D)	90.	(A)	137.	(A)	183.	(B)	230.	(A)	277.	(A)	324.	(D)		
44.	(C)	91.	(C)	138.	(A)	184.	(C)	231.	(C)	278.	(B)	325.	(B)		
45.	(A)	92.	(C)	139.	(C)	185.	(C)	232.	(A)	279.	(B)	326.	(D)		
46.	(A)	93.	(B)	140.	(D)	186.	(D)	233.	(A)	280.	(B)	327.	(B)		
47.	(B)	94.	(D)	141.	(D)	187.	*	234.	(B)	281.	(B)	328.	(B)		

181. O-nitrophenol
187. N_2O, I_3
212. sp^3d^2
214. 5
215. CS_2
216. Hydrogen bond
217. octahedral
221. direction
222. Intermolecular
223. covalent

RADIOACTIVITY

All heavy elements with atomic number greater than 83 i.e. from Polonium through uranium and a few of the lighter elements have natural radioactive isotopes.

Phenomenon of radioactivity was discovered by Henry Becquerel. The term radioactivity was coined by Madam Curie.

Radioactive units : Curie, I ci = 3.7×10^{10} dps
(dps = disintegration per second)
Rutherford, 1rd = 10^6 dps; Becquerel 1 BQ = 1 dps

Radioactive radiations : Radioactive substance generally gives rise to α, β and γ-rays.

Cause of Radioactivity :
If N/P ratio is 1, the nuclei are very stable. If N/P ratio is more than 1.5, they become radioactive. If N/P ratio is higher than required for stability, they have the tendency to emit β-rays i.e. transforming n_0^1 into proton. By β-emission N/P ratio decreases. If N/P ratio is less than required, it can increase by emission of an α-particle (in natural radioactive isotopes) or a position or by K-electron capture (e.g. $Au_{79}^{194} + e_{-1}^{0} \rightarrow Pt_{78}^{194}$). The later two processes generally take place in artificial radioactive isotopes.

α-rays : They have particle and wave nature. They carry +2 charge and 4 units mass. They are identical with helium nuclei. The velocity of these rays is approximately 1/10th of the velocity of light.

Penetration power : $\alpha < \beta < \gamma$ (because of the high velocity and small wavelength, γ-rays possess high penetrating power).

Ionisation power : $\alpha > \beta > \gamma$ (because of having no mass, the ionisation power of γ-ray is extremely poor).

α - rays effect a photographic plate (ZnS plate) and excite to produce luminosity. This is due to their high kinetic energy.

e.g. $U_{92}^{234} \xrightarrow{-\alpha} Th_{90}^{230} \xrightarrow{-\alpha} Ra_{88}^{226} \xrightarrow{-\alpha} Rn_{86}^{222}$

β - rays : They have particle and wave nature. They have the velocity approaching that of light (160,000 - 240,000 km/sec.). They are high speed electrons.
In radioactive reactions, β - particle emission has been described as ejection of electron from the nucleus, even though the nucleus does not contain any electrons. It is believed that electrons are produced as a result of decay of neutrons i.e. $n_0^1 \rightarrow H_1^1 + \beta + \nu$

e.g. $Th_{90}^{234} \xrightarrow{-\beta} Pa_{91}^{234} \xrightarrow{-\beta} U_{92}^{234}$

γ-rays : They are neutral, short wave electromagnetic radiations. They do not have particle nature. They have approximately the velocity of that of light. They are secondary effects of atomic disintegrations. After the emission of α and β-ray or particle, the nucleus left behind is in an excited state i.e. in a state of high energy. The excess of energy possessed by it is released in the form of a quantum γ-radiation. It means that γ-rays arise due to energy rearrangement within the nucleus after the emission of an α or a β-particle.

Secondary β-particles : The excited nucleus instead of isomeric transition, may get rid off excess of energy by ejecting an orbital electron, usually from K shell. After the ejection of conversion electron, the orbital electrons will tend to return to their normal state with emission of characteristic X-rays. These X-rays may in turn interact with orbital electrons in the atom and radiate Auger electrons which also have discrete energies. These may all be regarded as secondary β-particles.

Mass defect : The mass of an atom theoretically is to be the sum of the mass of the fundamental particles present in it. But the actual mass of an atom is generally found to be slightly less than the sum. This is due to the fact that when the nucleons combine to form a stable nucleus, a little amount of the mass is converted into energy. This energy is called the binding energy of the nucleus.

Mass defect = Sum of the mass of all fundamental particles – the actual mass of the nucleus.

Binding energy = Mass defect x 931 (1amu= 931 MeV i.e. million or mega electron volt)

The greater the value of binding energy per nucleon, more is the stability.

Packing fraction : Higher the value of packing fraction, lesser is the stability of the nucleus.

$$\text{Packing fraction} = \frac{\text{Isotopic mass} - \text{mass number}}{\text{mass number}} \times 10^4$$

Radioactive decay : group displacement law was given by Fajans, Russel and Soddy (1931).

α - decay : $Ra_{88}^{226} \xrightarrow{-\alpha} Rn_{86}^{226}$ The element formed is present two positions to the left in the periodic table. Radon formed is in 18th group while Radium is in 2nd group.

β-decay : $Pb_{82}^{210} \xrightarrow{-\beta} Bi_{83}^{210}$.
The element formed after the decay is present one position right in the periodic table. Bi formed is in 15th group while Pb is in 14th group.
Radioactive process is purely a nuclear phenomena and is independent of external conditions of temperature, pressure etc. The radioactive disintegration follows first order rate reaction.

$$\lambda = \frac{2.303}{t}\log\frac{No}{N};\quad t_{0.5} = \frac{0.693}{\lambda} \text{ or } \lambda = \frac{0693}{t_{0.5}}$$

$$t_{average} = \frac{1}{\lambda} = \frac{t_{0.5}}{0.693}$$

Amount of radioactive substance left after 'n' half life periods $N = (\frac{1}{2})^n No$

Artificial radioactivity : The process in which a stable isotope is converted into radioactive element by artificial transmutation is called artificial radioactivity or artificial transmutation.

The first nuclear transmutation was observed by Rutherford (1919)

$N_7^{14} + He_2^4 \rightarrow O_8^{17} + H_1^1$

Joliot & Irene Curie (1934) observed

$Al_{13}^{27} + He_2^4 \rightarrow P_{15}^{30} + n^{1\,0} \rightarrow Si_{14}^{30} + e^+$

Induced radioactivity by proton bombardment can be represented as $C_6^{12} + H_1^1 \rightarrow N_7^{13} \rightarrow C_6^{13} + e^+$

At present high energy nuclear projectiles are invariably obtained by man made accelerators such as Cyclotron, Synchrotron, Betatron, Bevatron etc.

Nuclear fission : The process of artificial transmutation in which heavy nucleus is broken down into lighter nuclei of nearly comparable masses with release of large amount of energy is termed nuclear fission.

$U_{92}^{235} + n_0^1 \rightarrow U_{92}^{236} \rightarrow Ba_{56}^{144} + Kr_{36}^{90} + 2n_0^1$

$$U_{92}^{235} + n_0^1 \rightarrow \begin{cases} Ba_{56}^{141} + Kr_{36}^{92} + 3n_0^1 \\ Xe_{54}^{140} + Sr_{38}^{95} + n_0^1 \quad \text{primary products} \\ Xe_{54}^{139} + Sr_{38}^{95} + 2n_0^1 \end{cases}$$

Many secondary products are also formed as shown

$Xe_{54}^{140} \xrightarrow{-\beta} Cs_{55}^{140} \xrightarrow{-\beta} Ba_{56}^{140} \xrightarrow{-\beta} La_{57}^{140} \xrightarrow{-\beta} Ce_{58}^{140}$

The chain reaction needs over critical stage for self propagation. This means that the loss of neutrons must be less than their production. The controlled chain reaction is the basis for nuclear reactors. In case of atom bomb explosions, the chain reactions are uncontrollable.

Nuclear fuels : They are mainly of two types

(a) Fissile materials : These on bombardment with slow neutrons directly produce a chain reaction leading to release of energy.

e.g. U^{235}, Pu^{239}, U^{233} etc.

U^{235} can be obtained from natural sources while the others by artificial radioactivity methods.

(b) Fertile materials : They are by themselves non fissile but can be converted into fissile materials by reactions with neutrons. e.g. U^{238}, Th^{232} etc. These are converted to Pu^{239} and U^{233} in breeder reactors.

Moderator : H_2O, graphite, D_2O (in research reactors), He (most efficient moderator) etc. are used to slow down the neutrons.

Control rods : Boron and cadmium are good neutron absorbers

$B_5^{10} + n_0^1 \rightarrow Li_3^7 + He_2^4$

Cooling system : Na & K liquid alloy, air, polyphenyls, D_2O, CO_2 etc.

Atomic reactor : The atomic reactor when used for production of electricity is known as power plant. The first nuclear reactor (Fermi) was established in USA in 1942. India established the first one in 1956 at Trombay. The first atomic bomb exploded over Hiroshima (1945) utilized U^{235}. The second one exploded over Nagasaki (1945) utilized Pu^{239}.

India tested its first atomic bomb in Pokhran on 11th May 1974 (utilized Pu^{239}). India further carried out five tests (3 tests and 2 subtonne tests) which include fission and fusion reactions for peaceful and scientific advancement purposes thus becoming a great world power on 11th May 1998.

Nuclear fission finds application in atomic bomb, nuclear reactor and power plant. The atomic reactor when used for the production of electricity is known as power plant.

Nuclear fusion : A nuclear reaction in which two lighter nuclei are fused together to form a heavier nucleus is called nuclear fusion. A fusion reaction is difficult to occur because positively charged nuclei repel each other. At very high temperatures of the order of 10^6 or 10^7 K, the nuclei may have sufficient energy to overcome the repulsive forces and fuse. It is for this reason that fusion reactions are also called thermonuclear reactions. Fusion reactions are highly exothermic.

${}_1H^2 + {}_1H^3 \rightarrow {}_2He^4 + n_0^1 + 17.8$ MeV

${}_1H^2 + H_1^2 \rightarrow He_2^4 + 24.9$ MeV

${}_1H^3 + {}_1H^3 \rightarrow He_2^4 + 2n_0^1 + 11.0$ MeV

${}_1H^1 + H_1^3 \rightarrow He_2^4 + 20.0$ MeV

${}_1H^1 + H_1^2 \xrightarrow[\text{Sun}]{\text{inside}} He_2^3 + \Upsilon$

Energy released is ~ 1000 times more than that of an atom bomb. The first explosion of hydrogen bomb was carried out in 1952.

Isotopes and their uses :

The activity of radioisotopes can be detected by means of electroscope, the electrometer, the Geiger-Muller counter or the Wilson Cloud chamber.

P^{32} - in blood cancer i.e. leukemia

As^{74} - to detect tumours

Na^{24} - blood clots ($t_{0.5}$=14.8 hrs)

I^{131} - activity of thyroid gland ($t_{0.5}$=8 days)

C^{14} - kinetics of photosynthesis ($t_{0.5}$=5577yrs.), reaction mechanism and to determine the age of fossils.

Co^{60} - testing deep seated cancer growths($t_{0.5}$=5.3 yrs)

Carbon dating : In living material the ratio of ^{14}C to C^{12}

remains relatively constant. When the tissue of an animal or plant dies, the ^{14}C decreases, because the intake and utilization of C^{14} does not occur. Therefore, the ratio of C^{14} to C^{12} would decrease, depending on the age of the tissue. The age of the dead tissue is determined as : A sample of dead tissue is burnt to carbon dioxide and the carbon dioxide is analysed for the ratio of C^{14} to C^{12}.

The age of the earth has been estimated by Uranium dating technique. A sample of uranium rock is analysed for U^{238} and Pb^{206} contents.

Radioactive series :

4n series	Thorium series	$Th^{232}_{90} \rightarrow Pb^{208}_{82}$
4n + 2	Uranium series	$U^{238}_{92} \rightarrow Pb^{206}_{82}$
4n + 3	Actinium series	$U^{235}_{92} Ac^{231}_{89} \rightarrow Pb^{207}_{82}$
4n + 1	Neptunium series	$Np^{237}_{93}(U^{235}_{92}) \rightarrow Bi^{209}_{83}$

Things to remember

Phenomenon of radioactivity - Henry Becqurel. Marie and Pierre Curie isolated Ra and Po.

Penetration power: $\alpha < \beta < \gamma$; Ionising power: $\alpha > \beta > \gamma$.

Group displacement law : Fajan - Soddy - Russell.

α - decay : $Ra^{226}_{88} \xrightarrow{-\alpha} Rn^{222}_{86}$ - Two positions to left.

β - decay : $Pb^{210}_{82} \xrightarrow{-\beta} Bi^{210}_{83}$ - one position to right.

Radioactive decay follows first order rate euqation :

$\lambda = \frac{2.303}{t} \log \frac{No}{N}$; $t_{0.5} = \frac{0.693}{\lambda}$; $t_{averge} = \frac{1}{\lambda}$; $N_t = N_o(\frac{1}{2})^n$.

Atomic bomb (Hahn & Strassman) - Nuclear fission :

$U^{235}_{92} + n^1_0 \rightarrow U^{236}_{92} \rightarrow Ba^{140}_{56} + Kr^{93}_{36} + 3n^1_0$

Hydrogen bomb - Nuclear fusion :

$H^2_1 + H^3_1 \rightarrow He^4_2 + n^1_0 + 17.8$ MeV.

Artificial radioactivity : Rutherford

Radioisotopes : C^{14} - to determine the age of fossils.

Co^{60} - to detect deep seated cancer growths.

Solved Examples

(1) How many α and β - particles will be emitted by Ra^{218}_{84} in changing to a stable isotope of lead Pb^{206}_{82}.

Sol. Let x and y be the number of α and β particles involved in bringing about the change. Then we can represent the change as

$Ra^{218}_{84} \rightarrow Pb^{206}_{82} + x^4_2 + y^0_{-1}$

comparing the mass numbers : 218 = 206 + 4x + 0y

$218 - 206 = 4x$; $x = \frac{12}{4} = 3$

comparing the atomic numbers : 84 = 82 + 2x - y

substituting the value of x : 84 = 82 + 2 x 3 - y

84 - 88 = -y

- 4 = - y ; y = 4

3 α and 4 β - particles are emitted.

(2) Calculate the packing fraction, mass defect and energy released in the formation of argon atom, Ar^{40}_{18}. Mass of the isotope of argon is 39.962384 amu.

Sol. Packing fraction = $\frac{\text{isotopic mass-mass number}}{\text{mass number}} \times 10^4$

$= \frac{39.962384 - 40}{40} \times 10^4 = -9.4025$

mass defect (ΔM) = $M^1 - M$

M^1 = 18 x 1.007825 + 22 x 1.008665 = 40.33148

M = 39.962384

ΔM = 40.33148 - 39.962384 = 0.369096

Energy released = 0.369096 x 931 = 343.628 MeV.

(3) The fission fragments in the thermal neutron fission of U^{235} are found to be Mo^{98}_{42} and Xe^{136}_{54}. What are the elementary particles released in the reaction? If the masses of U^{235}, Xe^{136}, Mo^{90} and n^1_0 are 235.044, 135.907, 97.906 and 1.0086 amu respectively, calculate the binding energy.

Sol. The balanced nuclear reaction is

$U^{235}_{92} + n^1_0 \rightarrow Mo^{98}_{42} + Xe^{136}_{54} + 4e^0_1 + 2n^1_0$

Mass defect = mass of products – mass of reactants

= (97.906 + 135.9070 + 4 x 0.005486 + 2 x 1.0086) – (235.044 + 1.0086)

= 235.854 - 236.0526 = - 0.1986 (decrease in the mass)

Energy released due to a decrease in mass = 0.1986 x 931 = 184.90 MeV.

(4) The mass defect of Cl^{35}_{17} is found to be 0.320 amu. Calculate the binding energy per nucleon.

Sol. Mass defect = 0.320

Binding energy = 0.320 x 931 = 297.92 MeV.

Binding energy per nucleon = $\frac{297.92}{35}$ = 8.512 MeV.

(5) In the fission of U^{235}_{92} with a neutron, Mo^{95}, La^{139} and two neutrons are formed. The atomic masses of U, Mo, La and n^1_0 are 235.118, 94.936, 138.950 and 1.009 amu respectively. Calculate the energy released by the fission

of 1g. of U^{235}.

Sol. We can write the balanced equation as

$U^{235}_{92} + n^1_0 \rightarrow Mo^{95} + La^{139} + 2n^1_0$

mass defect=(235.118+1.009) - (94.936+138.950+2x1.009)

= 236.127 - 235.904 = 0.223 amu.

Binding energy = 0.223 x 931 = 207.613 MeV

No. of atoms in 1g. of $U^{235} = \frac{1}{235} \times 6.623 \times 10^{23}$

Energy released by U^{235}_{92} atom = 207.613 MeV

$\therefore$ Energy released by the fission of $\frac{1}{235} \times 6.023 \times 10^{23}$ is

$= \frac{207.613 \times 6.023 \times 10^{23}}{235} = 5.321 \times 10^{23}$ MeV

(6) If half life period of an element x is 20 years, calculate the time required to reduce its activity to 1/8 of its original value.

Sol. $\frac{1}{8} = (\frac{1}{2})^x = (\frac{1}{2})^3$

Number of $t_{0.5} = 3$; time required = 3 x 20 = 60 years

(7) $t_{0.5}$ of a radioactive subtance is 10 years. What percentage of the substance will be left behind after 40 years ?

Sol. $t_{0.5}$ = 10 years or 10 years = $t_{0.5}$

$(x = \frac{\text{time passed}}{\text{half life period}})$

x = number of half lives

40 years = 4

$(\frac{1}{2})^x = (\frac{1}{2})^4 = \frac{1}{16} = 6.625\%$

(8) Half life of a radioactive isotope is 10 days. Its 125 g. is left behind. What was the original mass 40 days back ?

Sol. $t_{0.5}$ = 10 days

$x = (\frac{1}{2})^4 = \frac{x}{16} = 125$

x = 125 x 16 = 2000 g.

(9) The half life period of Po^{210}_{84} is 140 days. In how many days 1g. of this isotope is reduced to 0.25 g ?

Sol. Initial quantity of the isotope N_o = 1g.

Final quantity of isotope N = 0.25 g.

We know that $N = (\frac{1}{2})^n$ No

$\frac{1}{4} = (\frac{1}{2})^n \times 1$ or $(\frac{1}{2})^2 = (\frac{1}{2})^n$

n = 2

Time taken $t = n \times t_{0.5} = 2 \times 140 = 280$ days.

(10) A radioactive element x is spread over a room. It has $t_{0.5}$ equal to 30 days. Its activity is 50 times the permissible value. After how many days will the room be safe ?

Sol. $\lambda = \frac{0.693}{30}$

$t = \frac{2.303}{0.693} \times 30 \log \frac{50\,N}{N} = 169.38$ days

(11) If Co^{60} has a $t_{0.5}$ of 5.2 years, how much of a 10g. of a sample be radioactive after 31.2 years ?

Sol. Number of half - lives $= \frac{312}{5.2} = 6$

Sample remaining $= 10 \times (\frac{1}{2})^6 = 10 \times \frac{1}{64} = 0.156$ g.

(12) Radio-isotope P^{32}_{15} has a half life of 15 days. Calculate the time in which the radioactivity of its 1mg. quantity will fall to 10% of its initial value.

Sol. $\lambda = \frac{0.693}{t_{0.5}} = \frac{0.693}{15}$

$t = \frac{2.303 \times 15}{0.693} \log \frac{100}{10} = 49.85$ days

(13) 1g. of Mo decays by β - emission to 0.12 5g. in 200 hours. How much more time will elapse until only 0.01g of it is left ?

Sol. $\lambda = \frac{2.303}{t} \log \frac{No}{Nt} = \frac{2.303}{200} \log \frac{1}{0.125} = 0.0104\ hr^{-1}$

Now $t = \frac{2.303}{0.0104} \log \frac{0.125}{0.01} = 242.9$ hr.

(14) Calculate the ratio of U^{238} to Ra^{226} atoms in natural uranium. $t_{0.5}$ of U^{238} is 4.49×10^9 years and $t_{0.5}$ of Ra^{226} is 1622 years.

Sol. $\frac{N_1}{N_2} = \frac{(t_{0.5})_1}{(t_{0.5})_2} = \frac{4.49 \times 10^9}{1622} = 2.77 \times 10^6$

$\therefore U^{238} : Ra^{226} = 2.77 \times 10^6 : 1$

(15) A piece of wood was found to have C^{14}/C^{12} ratio 0.7 times that in a living plant. Calculate the period when the plant died. $t_{0.5}$ of C^{14} is 5760 years.

Sol. $\lambda = \frac{0.693}{t_{0.5}} = \frac{0.693}{5760}$

$t = \frac{2.303}{0.693} \times 5760 \times \log \frac{1}{0.7} = 2970$ years

(16) Calculate the half life of Na^{24} if a 2.4×10^{-5} g. sample disintegrates at the rate of 7.73×10^{12} atoms per second.

Sol. $\lambda = \frac{\text{No. of atoms disintegrating per second}}{\text{No. of atoms present}}$

$= \frac{7.73 \times 10^{12}}{2.4 \times 10^{-5} \times \frac{1}{24} \times 6.02 \times 10^{23}} = 1.28 \times 10^{-5}\ sec^{-1}$

$t_{0.5} = \frac{0.693}{1.28 \times 10^{-5}} = 5.41 \times 10^4$ sec.

(17) Uranium ore is found to contain one radium atom for every 2.78×10^6 uranium atom. Radium is known to decay with a half life of 1620 years. Find the half life of uranium.

Sol. $\lambda_1 N_1 = \lambda_2 N_2$

$\lambda_1 \times 2.78 \times 10^6 = \frac{0.693}{1620} \times 1$

$\lambda = \frac{0.693 \times 1}{1620 \times 2.78 \times 10^6}$

$t_{0.5} = \frac{0.693 \times 1620 \times 2.78 \times 10^6}{0.693} = 4.5 \times 10^9$ years

(18) Analysis of a sample of uranium ore showed that it contained 0.277 g. of Pb^{206} and 1.667 g. of U^{238}. Calculate the age of the ore assuming that all the Pb^{206} had accumulated due to decay of U^{238} . $t_{0.5}$ of U^{238} is

4.5×10^9 years.

Sol. As Pb^{206} had accumulated due to decay of U^{238}, we have to find out how much U^{238} was initially present.

If the wt. of U is 238 we obtain 206 Pb

from how much 0.277 g.Pb

$$\frac{0.277}{206} \times 238 = 0.320\text{g}.$$

Initial wt. of $U^{238} = 1.667 + 0.320 = 1.987$g.

Now we calculate $\lambda = \frac{0.693}{4.5 \times 10^9}$

$$t = \frac{2.303 \times 4.5 \times 10^9}{0.693} \log \frac{1.987}{1.667} = 1.129 \times 10^9 \text{ years}$$

(19) The amount of C^{14} isotope in a piece of wood is found to be 1/6 of its amount present in a fresh piece of wood. Calculate the age of the wood. ($t_{0.5}$ of ^{14}C is 5577 years)

Sol. $\lambda = \frac{0.693}{t_{0.5}} = \frac{0.693}{5577}$

$$t = \frac{2.303}{\lambda} \log \frac{No}{Nt} = \frac{2.303 \times 5577}{0.693} \log \frac{1}{6}$$

$$= \frac{2.303 \times 5577}{0.693} \times 0.7782$$

$= 14422.9$ years $= \sim 14423$ years

(20) Po $^{210}_{84}$ decays with α - emission to Pb $^{206}_{82}$ and has $t_{0.5}$ of 138.4 days. If 1 g. of Po^{210} is placed in a sealed tube, how much Helium will accumulate in 69.2 days?

Sol. $\lambda = \frac{0.693}{138.4} = 5.007 \times 10^{-3}$ per day

Now applying the decay constant equation

$$\lambda = \frac{2.303}{t} \log \frac{No}{Nt}, \text{ we get}$$

$$5.007 \times 10^{-3} = \frac{2.303}{69.2} \log \frac{1}{Nt}$$

$$\log \frac{1}{Nt} = \frac{5.007 \times 10^{-3} \times 69.2}{2.303} = 0.1504$$

$$\frac{1}{N} = 1.414$$

$$N = \frac{1}{1.414} = 0.707 \text{ g.}$$

Amount of Po^{210} decayed $= 1 - 0.707 = 0.293$

from the equation $Po^{210}_{84} \rightarrow Pb^{206}_{82} + He^{4}_{2}$

We can say 210 g. of Po produce He i.e. equivalent to 22,400 c.c. at STP

0.293 g. Po will produce He equivalent to

$$= \frac{22400 \times 0.293}{210} = 31.25 \text{ c.c.}$$

1. Half-life period of a substance is 1600 minutes. How much fraction of the substance will remain after 6400 minutes? [CBSE-93]
(A) 1/4 (B) 1/8 (C) 1/16 (D) 1/2

2. After the emission of one α-particle followed by one β-particle from the atom of X^{238}_{92}, the number of neutrons in the atom will be [CBSE-1995]
(A) 142 (B) 146 (C) 144 (D) 143

3. Half life for radioactive ^{14}C is 5760 years. In how many years 200 mg of ^{14}C sample will be reduced to 25 mg? [CBSE-1995]
(A) 11520 year (B) 23040 year
(C) 5760 year (D) 17280 year

4. Sulphur-35 (34.96903 a.m.u.) emits a β-particle but no γ-rays. The product is chlorine-35 (34.96885 a.m.u.).the maximum energy emitted by the β-particle is [CBSE-1999]
(A) 16.758 MeV (B) 1.6758 MeV
(C) 0.16758 MeV (D) 0.016758 MeV

5. In the following radioactive decay, $_{92}X^{232} \rightarrow {}_{89}Y^{220}$, how many α and β particles are ejected from X to form Y? [CBSE-1999]
(A) 3α and 3β (B) 5α and 3β
(C) 3α and 5β (D) 5α and 5β

6. The half-life of $_6C^{14}$ if its λ is 2.31×10^{-4}, is [CBSE-1999]
(A) 2×10^2 years (B) 3×10^2 years
(C) 3.5×10^4 years (D) 4×10^2 years

7. Arrange X-rays, γ-rays, UV rays in decreasing order of energy values for photocells. [CBSE-2000]
(A) $X < UV < \gamma$ (B) $\gamma < X < UV$
(C) $UV < X < \gamma$ (D) $\gamma > x > UV$

8. The correct relation between decay constant and half-life of a radioactive decay reaction. [CBSE-2000]
(A) $t_{1/2} = \frac{\log_e 2}{\lambda}$ (B) $t_{1/2} = \frac{2.303 \log_e 2}{\lambda}$
(C) $t_{1/2} = \lambda \log_e 2$ (D) $t_{1/2} = 2.303 \log_{10} 2\lambda$

9. Starting with 300g of a radioactive sample what amount would remain radioactive after 1 hr, if half-life of initial sample is 15 min? [CBSE-2000]
(A) No activity remains (B) No decay occurs
(C) 150 gm (D) 18.75 gm

10. $^{14}_{6}C \rightarrow {}^{14}_{5}C + {}^{0}_{+1}\beta + X$. Identify X. [CBSE-2000]
(A) Proton (B) Neutron (C) Electron (D) Neutrino

11. A human body required 0.01 activity of radioactive substance after 24 hrs. Half-life of radioactive substance is 6 hrs. Then the maximum activity of radioactive substance that can be injected is - [CBSE-2001]

(A) 0.08 (B) 0.04 (C) 0.16 (D) 0.32

12. If a ${}^{a}_{b}X$ species emits firstly a positron, then two α - and two - β particles and at last, one α- particle is also emitted and is finally converted into ${}^{c}_{d}Y$ species, So the correct relation is - [CBSE-2001]
(A) a = c + 12, d = b - 5 (B) a = c - 8, d = b -1
(C) a = c - 6, d = b (D) a = c - 4, d = b - 2

13. ${}_{92}U^{235}$, nucleus absorbs a neutron and disintegrates to ${}_{54}Xe^{139}$, ${}_{38}Sr^{94}$ and x. What will be product x :- [CBSE 2002]
(A) 3 -neutrons (B) 2-neutrons
(C) α -particle (D) β–particle

14. Atom bomb is based on the principle of [AIIMS-95]
(A) Radioactivity (B) Nuclear fusion
(C) Nuclear fission (D) Fusion and fission

15. A radioactive nuclide decays at such a rate that after 96 min only 1/8th of the original amount remains. The value of half-life of the nuclide is [AIIMS-96]
(A) 12 min (B) 24 min (C) 32 min (D) 48 min

16. Which is not emitted by radioactive substance ? [AIIMS-97]
(A) α-rays (B) β-rays (C) Positron (D) Proton

17. Half-life of radium is 1580 years. Its average life will be [AIIMS-99]
(A) 2.5×10^3 years (B) 2.275×10^3 years
(C) 8.825×10^2 years (D) 11.832×10^3 years

18. Which one is the most radioactive element ? [AIIMS-2000]
(A) Uranium (B) Thorium (C) Plutonium (D) Radium

19. The ${}_{88}Ra^{226}$ is [AIIMS-2001]
(A) n-mesons (B) μ-mesons
(C) radioactive (D) non-radioactive

20. (A) ${}_{11}Na^{22}$ emits a positron giving ${}_{12}Mg^{22}$.
(R) In β^+ emission neutron is transformed into proton. [AIIMS-2003]
(A) (B) (C) (D)

21. The radiations from naturally occurring radioactive substance, as seen after deflection by a magnetic field in one direction are : [JIPMER-90]
(A) Definitely α-rays (B) Definitely β-rays
(C) Both α and β- rays (D) Either α or β- rays

22. If uranium (mass no. 238, atomic no. 92) emits α-particle the product has mass number and atomic number as : [JIPMER-90]
(A) 236 and 92 (B) 234 and 90
(C) 238 and 90 (D) 236 and 90

23. A radioactive nucleus ${}^{218}_{84}Ra$ loses α and β particles and the daughter element has mass number 206 and atomic number. The number of α and β particles lost respectively are : [JIPMER-93]
(A) 5, 4 (B) 4, 3 (C) 5, 6 (D) 3, 4

24. $X \xrightarrow{\alpha} Y \xrightarrow{\beta} Z \xrightarrow{\beta} W$ In the above sequence of reactions, the elements which are isotopes of each other are [JIPMER-95]
(A) X and Z (B) Y and Z
(C) X and W (D) None of these

25. The basic principle involved in hydrogen bomb is [JIPMER-96]
(A) Nuclear fission (B) Nuclear fusion
(C) Nuclear disintegration (D) None of these

26. For the transformation ${}_{7}N^{14} + ? \longrightarrow {}_{6}C^{14} + {}_{1}H^{1}$. The bombarding particle is : [JIPMER-96]
(A) Proton (B) Neutron (C) Deuteron (D) Electron

27. To what radioactive family does ${}^{207}Pb$ belongs ? [JIPMER-97]
(A) ${}^{232}Th$ (B) ${}^{237}Np$ (C) ${}^{227}Ac$ (D) ${}^{238}U$

28. The unrelated member in the following group is [JIPMER-97]
(A) Positron (B) Cyclotron (C) Neutron (D) Electron

29.The radiations from a naturally occurring radio element, as seen after deflection in a magnetic field in one direction, are [JIPMER-98]
(A) Definitely α rays (B) Definitely β rays
(C) Both α and β rays (D) Either α or β rays

30. Neutrons are obtained by [JIPMER-98]
(A) Bombardment of Ra with β-particles
(B) Bombardment of Be with α-particles
(C) Radioactive disintegration of uranium
(D) None of the above

31. What kind of radioactive decay doesnot lead to the formation of a daughter nucleus that is an isobar of the parent nucleus ? [JIPMER-98]
(A) α-rays (B) β-rays (C) Positron (D)Electron capture

32. A radioactive substance has a constant activity of 2000 disintegration/minute. The material is separated into two fractions, one of which has an initial activity of 1000 disintegrations per second while the other fraction decays with $t_{1/2}$ = 24 hours. The total activity in both samples after 48 hours of separation is [JIPMER-2000]
(A) 2000 (B) 1250 (C) 1000 (D) 1500

33. Uranium decays into a stable isotope of [JIPMER-2000]
(A) Np (B) Pb (C) Rn (D) Ra

34. Radioactivity of neptunium stops when it is converted to [JIPMER-2001]

(A) Bi (B) Rn (C) Th (D) Pb

35. What happens when α particle is emitted? [JIPMER-2002]
(A) Mass no. decreases by 12 unit atomic no. decrease by 4 unit.
(B) Mass no. decreases by 4 unit atomic no. decreases by 2 units.
(C) Only mass number decreases.
(D) Only atomic number decreases.

36. A radioactive substance has half-life of 60 minutes. After 3 hours, what percentage of radioactivity will remain? [BHU-95]
(A) 175% (B) 50% (C) 25% (D) 12.5%

37. Which of the following is an n, p reaction? [BHU-95]
(A) $_5C^{13} + {}_1H^1 \rightarrow {}_6C^{14}$ (B) $_7N^{14} + {}_1H^1 \rightarrow {}_8O^{15}$
(C) $_{13}Al^{27} + {}_0n^1 \rightarrow {}_{12}Mg^{27} + {}_1H^1$
(D) $_{92}U^{235} + {}_0n^1 \rightarrow {}_{54}Xe^{140} + {}_{88}Sr^{94} + {}_{30}n^1$

38. A radioactive substance has half-life 5760 years. A sample of 100 mg containing the substance is reduced to 25 mg in [BHU-95]
(A) 11520 years (B) 2880 years
(C) 1440 years (D) 280 years

39. Half-life period of the metal is 20 days. How much fraction of metal remains after 80 days? [BHU-96]
(A) 1 (B) $\frac{1}{16}$ (C) $\frac{1}{4}$ (D) $\frac{1}{8}$

40. $_7N^{14} + {}_2He^4 \rightarrow {}_1H^1 + x$. The value of x is [BHU-96]
(A) $_8O^{16}$ (B) $_8O^{17}$ (C) $_8O^{18}$ (D) $_6O^{14}$

41. The half-life of an isotope is 10 hrs. How much will be left behind after 4 hrs in 1 gm sample? [BHU-97]
(A) 45.6×10^{23} atoms (B) 4.56×10^{23} atoms
(C) 4.56×10^{21} atoms (D) 45.6×10^{20} atoms

42. According to the nuclear reaction:
$_4Be + {}_2He^4 \rightarrow {}_6C^{12} + {}_0n^1$ the mass no of (Be) atom is [BHU-99]
(A) 4 (B) 6 (C) 7 (D) 9

43. β-particles are [BHU-2000]
(A) Charged particles emitted from nucleus
(B) Neutral particles orbiting nucleus
(C) A particle produced by splitting of e- into H+ and $_n0^1$
(D) High energy positively charged particles emitted from nucleus

44. $_{95}Am^{241}$ and $_{90}Th^{234}$ respectively belong to [BHU-2002]
(A) (4 n + 1) and 4n radioactive disintegration series
(B) (4n + 1) and (4n +3) radioactive disintegration series
(C) (4n+1) & (4n+2) radioactive disintegration series
(D) 4n and (4n +1) radioactive disintegration series

45. The reciprocal of the radioactive decay constant is called [AFMC-93]
(A) Half-life period (B) Whole life-period
(C) Mean life period (D) None of these

46. If α-particles bombard gold foil, then [AFMC-95]
(A) Most α-particles travel in straight line
(B) Some α-particles are deflected
(C) Maximum α-particles are deflected
(D) All α-particles return to original path

47. Which of the following is not a mode of dissociation of radioactive substance? [AFMC-96]
(A) Electron capture (B) α-particle emission
(C) β-emission (D) Fission

48. Which of the following is used as "Control rod"? [AFMC-96]
(A) Cadmium rod (B) Graphite rod
(C) Steel (D) All of these

49. What is the value of decay constant of a compound having half-life $t_{1/2} = 2.95$ days? [AFMC-97]
(A) $2.9 \times 10^{-5}\ s^{-1}$ (B) $2.9 \times 10^{6}\ s^{-1}$
(C) $2.9 \times 10^{-6}\ s^{-1}$ (D) $3 \times 10^{5}\ s^{-1}$

50. Which of the following atomic no. of Uranium is the most radioactive? [AFMC-97]
(A) 238 (B) 235 (C) 226 (D) 248

51. Which of the following can be used for dating a fossil? [AFMC-97]
(A) $^{14}_{6}C$ (B) $^{12}_{6}C$ (C) $^{15}_{7}N$ (D) $^{40}_{20}Ca$

52. If $_{90}Th^{228}$ disintegrates to $_{83}Bi^{212}$, then the number of α and β particles emitted are respectively [AFMC-98]
(A) 4α and 7β (B) 4α and 1β (C) 4α only (D) 7β only

53. The half-life of $_6C^{14}$, if its wavelength (λ) is 2.31×10^{-4}, is [AFMC-99]
(A) 2×10^2 year (B) 4×10^3 year
(C) 3.5×10^4 year (D) 3×10^3 year

54. Half-life of radium is 1580 years. Its average life will be [AFMC-99]
(A) 8.25×10^2 years (B) 1.8×10^3 years
(C) 2.5×10^3 years (D) 2.275×10^3 years

55. After two hours, the radioactive substance becomes 1/16th of original amount. Then half-life in minutes is [AFMC-2000]
(A) 60 minutes (B) 120 minutes
(C) 30 minutes (D) 15 minutes

56. Compounds are isobars if .... emission takes place [AFMC-2000]
(A) 1 α emission (B) 1 β emission
(C) 1 α and 1 β emission (D) 2 α and 1 β emission

57. When a slow neutron goes sufficiently close to a U^{235} nucleus, then the process is [AFMC-2000]

(A) Fusion of U^{235} (B) Fission of U^{235}

(C) Fusion of neutron (D) First (A) then (B)

58. Which of the following is used as "Moderator" in nuclear reactor ? [AFMC-2001]

(A) H_2O (B) D_2O (C) Alum (D) Any of these

59. The number of α and β particles emitted in nuclear reaction $_{90}Th^{238} \rightarrow {}_{83}Bi^{212}$ are respectively [AFMC-2001]

(A) 4, 1 (B) 3, 7 (C) 8, 1 (D) 4, 7

60. The reaction $_7N^{14} + {}_0n^1 \rightarrow {}_6C^{14} + x$ requires emission of [AFMC-2001]

(A) ^4_2He (B) ^1_1H (C) ^1_0n (D) $^0_{-1}e$

61. $^{27}_{13}Al$ is a stable isotope. $^{29}_{13}Al$ is expected to disintegrate by. [IIT-96]

(A) α-emission (B) β-emission

(C) positron emission (D) proton emission

62. Decrease in atomic number is observed during [IIT-98]

(A) alpha emission (B) beta emission

(C) positron emission (D) electron capture

63. The number of neutrons accompanying the formation of $^{139}_{54}Xe$ and $^{94}_{38}Sr$ from the absorption of a slow neutron by $^{235}_{92}U$, followed by nuclear fission is, [IIT-99]

(A) 0 (B) 2 (C) 1 (D) 3

64. Rutherford's experiment, which established the nuclear model of the atom, used a beam of [IIT-2002]

(A) β-particles which impinged on a metal foil and got absorbed.

(B) γ-rays which impinged on a metal foil and ejected electrons

(C) helium atoms, which impinged on a metal foil and got scattered

(D) helium nuclei, which impinged on a metal foil and got scattered

65. If ^{23}Na is stable then ^{24}Na will undergo [IIT-2003]

(A) β-emission (B) α-emission

(C) γ-emission (D) α, β-emission

66. If half-life of a substance is 5 yrs, then the total amount of substance left after 15 years, when initial amount is 64 grams is [AIEEE-2002]

(A) 16 Grams (B) 2 Grams (C) 32 Grams (D) 8 Grams

67. β-particle is emitted in radioactivity by [AIEEE-2002]

(A) Conversion of proton to neutron

(B) From outermost orbit

(C) Conversion of neutron to proton

(D) β-particle is not emitted.

68. The radionucleide $^{234}_{90}Th$ undergoes two successive β-decays followed by one α-decay. The atomic number and the mass number respectively of the resulting radionucleide are [AIEEE-2003]

(A) 92 and 234 (B) 94 and 230

(C) 90 and 230 (D) 92 and 230

69. The half life of a radioactive isotope is three hours. If the initial mass of the isotope were 256 g, the mass of it remaining undecayed after 18 hours would be [AIEEE-2003]

(A) 4.0g (B) 8.0g (C) 12.0g (D) 16.0g

70. A radioactive element has a half life of 20 minutes. How much time should elapse before the element is reduced to 1/8 of its original mass ? [EAMCET-90]

(A) 40 min. (B) 60 min. (C) 80 min. (D) 160 min.

71. Complete the reaction

$^{14}_7N + {}_0n^1 \longrightarrow {}_6C^{14} + {}_1H^1$ [EAMCET-90]

72. A radio isotope has a half life 10 days. If today there is 125 grams of it left what was its weight 40 days earlier? [EAMCET-91]

(A) 600 gms (B) 1000 gms (C) 1250 gms (D) 2000 gms

73. The mass number of nuclide is 125. Its nuclear radius is .... Fermi. [EAMCET-91]

74. In the reaction

$^{16}_8O + {}^2_1H \rightarrow {}^{18}_8O + {}^3_1X$, X is.............. [EAMCET-91]

75. The half life of radioactive substance is 100 sec. It's disintegration constant is .... [EAMCET-91]

76. The half-life period of a radioactive element is 140 days. After 560 days, 1 gm of the element will be reduced to [EAMCET-92]

(A) 1/2 gm (B) 1/4 gm (C) 1/8 gm (D) 1/16 gm

77. In nuclear reactions, heavy water is used as a [EAMCET-92]

(A) Fuel (B) Projectile (C) Moderator (D) Arrestor

78. A radioactive element has a decay constant of $6.93 \times 10^{-3} sec^{-1}$. Its half life is .... minutes. [EAMCET-92]

79. The energy of red stars is due to nuclear .... [EAMCET-92]

80. $_{92}U^{235} + {}_0n^1 \rightarrow 3\,{}_0n^1 + {}_{36}Kr^{90} +$...................... [EAMCET-92]

81. The cobalt isotope of mass 60 decays with a half-life of 5.3 years. How many years does it take for 7/8 of a sample of 500 g of Cobalt-60 to disintegrate ? [EAMCET-93]

(A) 10.6 (B) 4.6 (C) 15.9 (D) 9.2

82. The half-life period of a radioactive element is 120

days. After 480 days 4.0 gm of the element will be reduced to (....) gms. [EAMCET-93]
(A) 2.0 (B) 1.0 (C) 0.5 (D) 0.25

83. The emission of an alpha particle from a radioactive element causes its atomic number to decrease by .... [EAMCET-93]

84. The number, of β-particles emitted in the radioactive change $_{92}U^{238} \rightarrow {}_{82}Pb^{206} + 8\,{}_2He^4$ is [EAMCET-94]
(A) 2 (B) 4 (C) 6 (D) 8

85. The half life period of a radioactive substance is 150 days. Starting with an initial quantity of 160 mg. of the substance, the fraction of that has disintegrated in 600 days is [EAMCET-94]
(A) 1/2 (B) 1/4 (C) 1/16 (D) 15/16

86. To fill in the blanks in the following reaction $_{95}Am^{241} + X \rightarrow {}_{97}Bk^{243} + Y$, X and Y should be [EAMCET-95]
(A) X = 2 $_0n^1$ and Y = 2α^4 (B) X = α and Y = 2e
(C) X = 2e and Y = α (D) X = α and Y = $2n_0^1$

87. Which of the following is used to bombard $_{13}Al^{27}$ to give $_{15}P^{30}$ and a neutron [EAMCET-96]
(A) proton (B) α– particle
(C) β– particle (D) deuteron

88. The atomic number of a radioactive element increases by one unit in [EAMCET-97]
(A) Alpha emission (B) Beta emission
(C) Gamma emission (D) Electron capture

89. The activity of a radioactive nuclide (X^{106}) is 6.023 curie. If its disintegration constant is 3.7 x 10^4 sec^{-1}, the mass of x is [EAMCET-98(MBBS)]
(A) 10^{-23}g (B) 10^{-15}g (C) 10^{-16}g (D) 10^{-14}g

90. In Rutherford's alpha-ray scattering experiment, the alpha particles are detected using a screen coated with [EAMCET-99]
(A) Carbon black (B) Platinum black
(C) Zinc sulphide (D) Polytetrafluoroethylene

91. The activity of a radioactive element is 2 x 10^{10} disintegrations per second. After 10 minutes, the activity is reduced to 2 x 10^9 disintegrations per second. The disintegration constant of the radioactive element is (in min^{-1}) [EAMCET-99]
(A) 2 (B) 2.303 (C) 10 (D) 0.2303

92. The unit of radioactivity is : [EAMCET-99]
(A) Fermi (B) Farad (C) Curie (D) Hertz

93. In the nuclear reaction $^{27}_{13}Al + H^2_1 \xrightarrow{-\alpha} X$, is [EAMCET-99(MBBS)]
(A) $^{12}_{6}C$ (B) $^{30}_{14}Si$ (C) $^{25}_{12}Mg$ (D) $^{24}_{11}Na$

94. Radioactivity of an isotope falls to 1/8 of its original value in 48 days. Its half life period (in days) is [EAMCET-99(MBBS)]
(A) 48 (B) 16 (C) 12 (D) 8

95. The half-life of radioactive nuclide is 100 years. What is its disintegration constant (in year^{-1}) ? [EAMCET-2000]
(A) 6.93 x 10^3 (B) 6.93 (C) 6.93 x 10^{-3} (D) 0.693

96. Which of the following nuclide is unstable ? [EAMCET(MBBS)-2000]
(A) $_5B^{10}$ (B) $_4Be^{10}$ (C) $_7N^{14}$ (D) $_8O^{16}$

97. Element $_ZM^A$ emits one alpha particle followed by two beta particles. Which of the following is daughter element? [EAMCET(MBBS)-2000]
(A) z – 2M (B) z – 2M (C) zM (D) z + 2M

98. In a nucleide, one a.m.u. of mass is dissipated into energy to bind its nucleons the energy equivalent of this mass is [EAMCET-2001]
(A) 931.5 eV (B) 931.5 x 10^6 MeV
(C) 931.5 x 10^6 eV (D) 931.5 MV

99. What is X in the following reaction ?
$^{26}_{12}Mg + {}^2_1H \rightarrow {}^{27}_{12}Mg + x$ – [EAMCET-2001]
(A) γ-ray (B) ^1_0n (C) ^1_1H (D) ^2_1D

100. 10 grams of a radioactive element is disintegrated to 1 gram in 2.303 minutes. What is the half-life (in minutes) of that radioactive element ? [EAMCET-2002]
(A) $\frac{1}{0.693}$ (B) 6.93 (C) 1 (D) 0.693

101. The half-life of a radioactive nuclide is 0.693 minutes. The time (in minutes) required for the disintegration of this nuclide from 10 grams to one gram is... [EAMCET(med.)-2002]
(A) 0.693 (B) 1 (C) 6.93 (D) 2.303

102. If the mass defect of a nucleide is 3.32 x 10^{-26} g, its binding energy is ..... MeV [EAMCET-2003]
(A) 9.31 (B) 18.62 (C) 27.93 (D) 37.24

103. Half-life of 10gm of radioactive substance is 10 days. The half-life of 20 gm is [CPMT-96]
(A) 10 days (B) 20 days (C) 25 days (D) Infinite

104. Which one of the following is a nuclear reaction ? [CPMT-97]
(A) $_6C^{13} + {}_1H^1 \rightarrow {}_7H^{13} + \beta^- + v^-$
(B) $_{11}Na^{23} + {}_1H^1 \rightarrow {}_{10}Ne^{20} + {}_2He^4$
(C) $_{13}Al^{23} + {}_0n^1 \rightarrow {}_{11}Ha^{23} + e^0$
(D) $_{12}Mg^{24} + {}_2H^4 \rightarrow {}_{13}Al^{27} + {}_0n^1$

105. Half-life of a radioactive substance is one year. After 5 years what fraction of original amount will be left ? [CPMT-2000]
(A) $\frac{1}{5}$ (B) $\frac{1}{2}$ (C) $\frac{1}{8}$ (D) $\frac{1}{32}$

106. The half-life of radioactive substance is 5 years. The value of radio-constant is- [CPMT-2000]

(A) $\frac{5}{0.693}$ per year (B) $\frac{0.693}{5}$ per year
(C) 0.693 per year (D) 5 per year

107. The temperature upto 2000°K is measured by [CPMT-2001]

(A) Radiation thermometer (B) Radiation pyrometer
(C) Gas thermometer (D) Mercury thermometer

108. The binding energy of an atom is 128 MeV. If the binding energy per nucleon is 8 MeV, the number of nucleons in nucleus is [CPMT-2001]

(A) 8 (B) 12 (C) 16 (D) 24

109. A G.M. counter is used to study the radioactive process. In the absence of radioactive substance A, it counts 3 disintegration per second (dps). At the start in the presence of A, it records 23 dps; and after 10 minutes 13 dps, (1) What dPS count after 20 minutes? (2) What is half-life of A ? [CPMT-2001]

(A) 8 dps. 10 min (B) 5 dps, 10 min
(C) 5 dps, 20 min (D) 5 dps, 5 min

110. $_{90}Th^{228}$ Changes to $_{83}Bi^{212}$ by the elimination of [CPMT-2002]

(A) $4\alpha, 1\beta$ (B) $4\alpha, 2\beta$ (C) $5\alpha, 1\beta$ (D) $5\alpha, 2\beta$

111. Stable nuclides are those whose n/p ratio is : [PMT-93]

(A) $n/p = 1$ (B) $n/p = 2$ (C) $n/p > 1$ (D) $n/p < 1$

112. A radium $_{88}Ra^{224}$ isotope, on emission of an α-particle gives rise to a new element whose mass number and atomic number will be : [PMT-93]

(A) 220 and 86 (B) 225 and 87
(C) 228 and 88 (D) 224 and 86

113. The basic principle of the hydrogen bomb formation is : [PMT-93]

(A) nuclear fission (B) disintegration
(C) spallation (D) nuclear fusion

114. The radio-isotope of carbon that is used for knowing the appropriate time of death of a planet is : [PMT-93]

(A) C_6^{12} (B) C_6^{13} (C) C_6^{14} (D) C_6^{16}

115. The half life of a radioactive element is 30 minutes, one sixteenth of original quantity of element will remain unchanged after [PMT - 94]

(A) One hour (B) 16 hours (C) 4 hours (D) 2 hours

116. If the amount of radioactive substance is increased three times, the number of atoms disintegrated per unit time would [PMT - 94]

(A) be double (B) be triple
(C) remain one third (D) not change

117. On which of the following principles hydrogen bomb is based ? [PMT-95]

(A) Nuclear fission (B) Nuclear fusion
(C) Artificial radioactivity (D) Natural radioactivity

118. Half life of radioactive element is 100 years. The time in which it disintegrates to 50% of its mass, will be [PMT-95]

(A) 50 yrs. (B) 200 yrs. (C) 100 yrs. (D) 25 yrs

119. There exists on γ-rays [PMT-96]

(A) positive charge (B) negative charge
(C) no charge
(D) sometimes positive charge, some times negative charge

120. Which element is the end product of each natural radioactive series ? [PMT-96]

(A) Sn (B) Bi (C) Pb (D) C

121. For artificial transmutation of nuclei, the most effective one is [PMT-96]

(A) Proton (B) Deutron (C) Helium nuclei (D) Neutron

122. Which of the following does not take place by α-decay ? [PMT-96]

(A) $^{238}_{92}U \rightarrow ^{234}_{90}Th$ (B) $^{232}_{90}Th \rightarrow ^{228}_{88}Ra$
(C) $^{226}_{88}Ra \rightarrow ^{222}_{86}Rn$ (D) $^{213}_{83}Bi \rightarrow ^{213}_{84}Po$

123. The atomic mass of an element is 12.00719 amu. If there are 6 neutrons in the nucleus of the atom of the element, the binding energy per nucleon of the nucleus will be : [PMT-99]

(A) 7.64 MeV (B) 76.4 MeV
(C) 764 MeV (D) 0.764 MeV
($e^- = 0.00055$ amu; p = 1.00814 amu; n = 1.00893 amu)

124. In the following nuclear reactions,

$_7N^{14} + _2He^4 \rightarrow _8O^{17} + X_1$ and $_{13}Al^{27} + _1D^2 \rightarrow _{14}Si^{28} + X_2$

X_1 and X_2 are respectively [PMT-99]

(A) $_1H^1$ and $_0n^1$ (B) $_0n^1$ and $_1H^1$
(C) $_2He^4$ and $_0n^1$ (D) $_0n^1$ and $_2He^4$

125. $_{95}Am^{241}$ and $_{90}Th^{234}$ belong respectively to : [PMT-99]

(A) 4n and 4n + 1 radioactive disintegration series
(B) 4n + 1 and 4n + 2 radioactive disintegration series
(C) 4n + 1 and 4n + 3 radioactive disintegration series
(D) 4n + 1 nd 4n radioactive disintegration series

126. The half-life period of a radioactive substance is 8 years. After 16 years, the mass of the substance will reduce from starting 16.0 g to : [PMT-99]

(A) 8.0 g (B) 6.0 g (C) 4.0 g (D) 2.0 g

127. In the sequence of following nuclear reactions,

$$_{92}X^{238} \xrightarrow{-\alpha} Y \xrightarrow{-\beta} Z \xrightarrow{-\beta} L \xrightarrow{-n\alpha} _{84}M^{218}$$

the value of n will be: [PMT-99]

(A) 3 (B) 4 (C) 5 (D) 6

128. The radium and uranium atoms, in a sample of uranium mineral are in the ratio of $1 : 2.8 \times 10^6$. If half-life period of radium is 1620 years, the half-life period of uranium will be : [PMT-99]

(A) 45.3×10^9 years (B) 45.3×10^{10} years
(C) 4.53×10^9 years (D) 4.53×10^{10} years

129. The activity of carbon-14 in a piece of an ancient wood is only 12.5%. If the half-life period of carbon-14 is 5760 years, the age of the piece of wood will be : [PMT-99]

(A) 17.281×10^2 years (B) 172.81×10^2 years
(C) 1.7281×10^2 years (A) 1728.1×10^2 years
(log 2 = 0.3010)

130. The end product of (4n +1) radioactive disintegration series is : [PMT-99]

(A) $_{83}Bi^{209}$ (B) $_{84}Po^{210}$ (C) $_{82}Pb^{208}$ (D) $_{82}Pb^{207}$

131. In the nuclear reaction ^9_4Be (p, α) X, the X is : [PMT-2000]

(A) ^4_2He (B) ^6_3Li (C) ^7_3Li (D) ^8_4Be

132. Which of the following does NOT contain number of neutrons equal to that of $^{40}_{18}Ar$? [PMT-2000]

(A) $^{41}_{19}K$ (B) $^{43}_{21}Sc$ (C) $^{40}_{21}Sc$ (D) $^{42}_{20}Ca$

133. The half life of $^{90}_{38}Sr$ is 20 years. If its sample having initial activity of 8000 dis/min is taken, what would be its activity after 80 years ? [PMT-2000]

(A) 500 dis/min (B) 800 dis/min
(C) 1000 dis/min (D) 1600 dis/min

134. A wood piece is 11460 years old. What is the fraction of ^{14}C activity left in the piece ? (Half-life period of ^{14}C is 5730 years) : [PMT-2000]

(A) 0.12 (B) 0.25 (C) 0.50 (D) 0.75

135. Rutherford's α- particle scattering experiment proved that atom has [PMT-2001]

(A) Electrons (B) Neutron (C) Nucleus (D) Orbitals

136. The carbon dating is based on : [PMT-2001]

(A) $^{15}_6C$ (B) $^{14}_6C$ (C) $^{13}_6C$ (D) $^{11}_6C$

137. Which of the following is not deflected by magnetic field? [PMT-2001]

(A) Deuteron (B) Positron (C) Proton (D) Photon

138. For transmutation of the $^{232}_{90}Th$ to $^{208}_{82}Pb$ the number of α and β particles emitted are respectively : [PMT-2001]

(A) 4.6 (B) 5.5 (C) 6.4 (D) 7.3

139. Which of the following can be used to convert $^{14}_7N$ into $^{17}_8O$? [PMT-2001]

(A) Deuteron (B) Proton (C) α – particle (D) Neutron

140. Which of the following is radioactive ? [PMT-2001]

(A) Para-hydrogen (B) Protium
(C) Deuterium (D) Tritium

141. Half life of a radioactive substance which disintegrates by 75% in 60 minutes, will be [PMT-2002]

(A) 120 min (B) 30 min (C) 45 min (D) 20 min

142. The ages of most ancient formations are estimated by [PMT-2002]

(A) Potassium-argon method
(B) Carbon-14 dating method
(C) Radium-silicone method (D) Uranium-lead method

143. Hydrogen bomb is based on the principle of [PMT-2002]

(A) Nuclear fission (B) Nuclear fusion
(C) Nuclear explosion (D) None of the above

144. Penetrating power of α–particle is [PMT-2002]

(A) More than γ-rays (B) More than β-rays
(C) Less than β-rays (D) Less than γ-rays

145. A parental nucleus 'X' after consecutive expulsion of two β particles, one after the other gives $^{14}_7N$. What is the number of neutrons in X? [PMT-2003]

(A) 7 (B) 9 (C) 14 (D) 15

146. In radioactive transformation

$$R \xrightarrow{-\alpha} X \xrightarrow{-\beta} Y \xrightarrow{-\beta} Z$$

Which of the relations between the nucleus of R and Z is true? [PMT-2003]

(A) Isotopic (B) Isobaric
(C) Isoneutronic (D) Isomeric

147. Which of the following particles is used in the transformation of $^{27}_{13}Al$ into $^{31}_{15}P$: [PMT-2003]

(A) Proton (B) Neutron
(C) Alpha particle (D) Deutron

148. 87.5% dissociation of the nuclei of a radioactive substance is completed in 3 hours. What will be its half life period ? [PMT-2003]

(A) 2 Hours (B) 1 Hour
(C) 30 Minutes (D) 90 Minutes

149. $^{231}_{89}Ac$ gives $^{207}_{82}Pb$ after emission of some α and β particles. The number of such α and β particles are respectively [PET-93]

(A) 5, 6 (B) 6, 5 (C) 7, 5 (D) 5, 7

150. The decay constant of a radioactive element is 3×10^{-6} min^{-1}. Its half life is [PET-93]

(A) 2.31×10^5 min. (B) 2.31×10^6 min.
(C) 2.31×10^{-6} min. (D) 2.31×10^{-7} min.

151. The $^{14}_6C$ in upper atmosphere is generated by the

nuclear reaction [PET-93]

(A) ${}^{14}_{7}N + {}^{1}_{1}H \longrightarrow {}^{14}_{6}C + {}^{0}_{+1}e + {}^{1}_{1}H$

(B) ${}^{14}_{7}N \longrightarrow {}^{14}_{6}C + {}^{0}_{+1}e$

(C) ${}^{14}_{7}N + {}^{1}_{0}n \longrightarrow {}^{14}_{6}C + {}^{1}_{1}H$

(D) ${}^{14}_{7}N + {}^{3}_{1}H \longrightarrow {}^{14}_{6}C + {}^{4}_{2}He$

152. When a radioactive element emits a β-particle the daughter element formed will have its : [PET-94]
(A) Mass number one unit less
(B) Atomic number one unit more
(C) Atomic number one unit less
(D) Mass number one unit more

153. The half-life of radium (226) is 1620 years. The time taken to convert 10 grams of radium to 1.25 grams is [PET-94]
(A) 810 years (B) 1620 years
(C) 3240 years (D) 4860 years

154. The average life period of a radioactive element is the reciprocal of its [PET-95]
(A) Half-life period (B) Disintegration constant
(C) Number of atoms present at any time
(D) Number of neutrons

155. Uranium ultimately decays into a stable isotope of [PET-95]
(A) Radium (B) Carbon (C) Lead (D) Neptunium

156. An isotope ${}^{X}_{Y}A$ undergoes a series of m alpha and n beta disintegration to form a stable isotope ${}^{X-32}_{Y-10}B$. The value of m and n are respectively [PET-95]
(A) 6 and 8 (B) 8 and 10 (C) 5 and 8 (D) 8 and 6

157. A substance used as a moderator in a nuclear reactor is [PET-95]
(A) Cadmium (B) Uranium-235
(C) Lead (D) Heavy water

158. During a β-decay the mass of the atomic nucleus [PET-96]
(A) Decreases by one unit (B) Increases by one unit
(C) Decreases by two units (D) Remains unaffected

159. In nuclear reaction
${}^{14}_{7}N + {}^{4}_{2}He \rightarrow {}^{A}_{Z}X + {}^{1}_{1}H$ the term ${}^{A}_{Z}X$ represents [PET-96]
(A) ${}^{17}_{8}O$ (B) ${}^{16}_{8}O$ (C) ${}^{16}_{7}N$ (D) ${}^{15}_{7}N$

160. The half-life of ${}^{14}C$ is about [PET-96]
(A) 12.3 years (B) 5730 years
(C) 4.5 x 10^9 years (D) 2.52 x 10^5 years

161. Radioactive iodine is being used to diagnose the disease of [PET-96]
(A) Bones (B) Kidneys (C) Blood cancer (D) Thyroid

162. 1.0 g of a radioactive isotope was found to reduce to 125 mg after 24 hours. The half-life of the isotope is [PET-96]
(A) 8 hours (B) 24 hours (C) 6 hours (D) 4 hours

163. What is 'X' in the following nuclear reaction ?
${}_{7}N^{14} + {}_{1}H^{1} \rightarrow {}_{8}O^{15} +$ 'X' [PET-97]
(A) ${}_{0}n^{1}$ (B) ${}_{-1}e^{0}$ (C) ${}_{+1}e^{0}$ (D) γ

164. The end product of (4n + 2) radioactive disintegration series is : [PET-97]
(A) ${}^{208}_{82}Pb$ (B) ${}^{206}_{82}Pb$ (C) ${}^{207}_{82}Pb$ (D) ${}^{210}_{83}Bi$

165. The radioactive isotope which is used to estimate the age of ancient geological formations is : [PET-97]
(A) O^8 (B) Pb^{212} (C) Fe^{59} (D) C^{14}

166. The half-life period of a radioactive element is 140 days. In 560 days a sample of this element would be reduced to .... of its initial mass. [PET-97]
(A) 1/2 (B) 1/4 (C) 1/8 (D) 1/16

167. Which element is the end product of every natural radioactive series ? [PET-98]
(A) Pb (B) Sn (C) C (D) Bi

168. In radioactive decay which one of the following moves the fastest ? [PET-98]
(A) α-particle (B) β-particle (C) γ-rays (D) Positron

169. Which one of the following notations shows products incorrectly ? [PET-98]
(A) ${}^{242}_{96}Cm\ (\alpha, 2n)\ {}^{243}_{97}Bk$ (B) ${}^{10}_{5}B\ (\alpha,n)\ {}^{13}_{7}N$
(C) ${}^{14}_{7}N(n, p)\,{}^{14}_{6}C$ (D) ${}^{28}_{14}Si\ (d, n)\ {}^{29}_{15}P$

170. In terms of energy 1 a.m.u. is equal to [PET-98]
(A) 100 J (B) 931.1 MeV
(C) 931.1 kcal (D) 10^7 erg

171. An α-particle is identical with : [PET-99]
(A) Proton (B) Neutron
(C) Helium nucleus (D) Electron

172. The ratio of specific charge of a proton and an α-particle is : [PET-99]
(A) 2 : 1 (B) 1 : 2 (C) 1 : 4 (D) 1 : 1

173. The symbol X in the following equation
${}_{11}Na^{23} + {}_{1}H^{1} \rightarrow {}_{12}Mg^{23} + X$ is : [PET-99]
(A) A neutron (B) A deutron
(C) A positron (D) An α-particle

174. The half-life of ${}_{92}U^{238}$ is 4.5 x 10^9 years. After how many years, the amount of ${}_{92}U^{238}$ will be reduced to half of its present amount ? [PET-99]
(A) 9.0 x 10^9 years (B) 13.5 x 10^9 years
(C) 4.5 x 10^9 years (D) 4.5 x $10^{4.5}$ years

175. By which law, energy produced in nuclear reaction is given ? [PET-2000]
(A) Graham's law (B) Charles's law
(C) Gay Lussac's law (D) Einstein's law

176. What will be half-life period of a nucleus at the end of 42 days, $N = 0.798 N_0$? [PET-2000]
(A) 15 days (B) 10 days (C) 12.83 days (D) 20 days

177. If 2.0 g of a radioactive substance has half-life of 7 days, the half-life of 1 g sample is [PET-2000]
(A) 7 days (B) 14 days (C) 28 days (D) 35 days

178. The formation of adduct between NH_3 and BF_3 involves the formation of [PET-2001]
(A) Ionic bond (B) Covalent bond
(C) Coordinate bond (D) Hydrogen bond

179. The arrangement ABC, ABC, ABC ----- is referred as : [PET-2001]
(A) Octahedral close packing
(B) Hexagonal close packing
(C) Tetragonal close packing (D) Cubic close packing

180. If half-life of a certain radioactive nucleus is 1000 s, the disintegration constant is : [PET-2001]
(A) $6.93 \times 10^2 s^{-1}$ (B) 6.93×10^{-4} s
(C) $6.93 \times 10^{-4} s^{-1}$ (D) $6.93 \times 10^3 s$

181. Radioactive carbon dating was discovered by : [PET-2001]
(A) W.F. Libby (B) G.N. Lewis
(C) J. Willard Gibbs (D) W. Nernst

182. The nuclear reaction
$^{63}_{29}Cu + ^{4}_{2}He \rightarrow ^{37}_{17}Cl + 14\,^{1}_{1}H + 16\,^{1}_{0}n$
is referred to as [PET-2002]
(A) Spallation reaction (B) Fusion reaction
(C) Fission reaction (D) Chain reaction

183. ^{226}Ra disintegrates at such a rate that after 3160 years only one-fourth of its original amount remains. The half-life of ^{226}Ra will be [PET-2002]
(A) 790 years (B) 3160 years
(C) 1580 years (D) 6230 years

184. The proper rays for radiocarbon dating are [PET-2002]
(A) UV rays (B) IR rays (C) Cosmic ray (D) X-rays

185. Which one of the following radio isotopes is used in the treatment of blood cancer [PET-2002]
(A) P^{32} (B) Co^{60} (C) I^{131} (D) Na^{24}

186. In the transformation of $^{232}_{90}Th$ to $^{208}_{82}Pb$, the particles emitted are: [PET- 2003]
(A) $4\alpha - 3\beta$ (B) $6\alpha - 4\beta$ (C) $2\alpha - 4\beta$ (D) $4\alpha - 6\beta$

187. Which of the following can convert (transmutation) $^{14}_{7}N$ in to $^{17}_{8}O$? [PET- 2003]
(A) α- particle (B) Neutron (C) Proton (D) deutron

188. Radioactive isotope of hydrogen is : [PET- 2003]
(A) Tritium (B) Para hydrogen
(C) Deuterium (D) Ortho hydrogen

189. In a radioactive decay an emitted electron comes from [CBSE-94]
(A) the nucleus of the atom
(B) the orbit with principal quantum number 1
(C) the inner orbital of the atom
(D) the outermost orbit of the atom.

190. $U^{235} + n \rightarrow U^{235} \rightarrow$ fission product + neutron + 3.20×10^{-11} J
The energy released when 1gm of U^{235} finally undergoes fission is [CBSE-97]
(A) 12.75×10^5 kJ (B) 18.60×10^9 kJ
(C) 8.21×10^7 kJ (D) 6.55×10^6 kJ

191. Carbon-14 dating method is that the [CBSE-97]
(A) Carbon-14 fraction is the same in all objects
(B) Carbon-14 highly insoluble
(C) Ratio of carbon-14 and carbon-12 is constant
(D) All the above

192. If a $^{a}_{b}X$ species emits firstly a positron, then two α - and two - β particles and in the last one α- particle is also emitted and is finally converted into $^{c}_{d}Y$ species, So the correct relation is - [CBSE-2003]
(A) a = c + 12, d = b - 5 (B) a = c - 8, d = b -1
(C) a = c - 6, d = b (D) a = c - 4, d = b - 2

193. The radioisotope, tritium ($^{3}_{1}H$) has a half-life of 12.3 years. If the initial amount of tritium is 32 mg. How many milligrams of it would remain after 49.2 years ? [CBSE-2003]
(A) 8 mg (B) 1 mg (C) 2 mg (D) 4 mg

194. Which one of the following is not correct [PMT-97]
(A) $_{1}Li^{7} + _{1}H^{1} \rightarrow _{4}Be^{7} + _{0}n^{1}$
(B) $_{21}Se^{45} + _{0}n^{1} \rightarrow _{20}Ca^{45} + _{0}n^{1}$
(C) $_{33}As^{75} + _{2}He^{4} \rightarrow _{35}Br^{78} + _{0}n^{1}$
(D) $_{83}Bi^{209} + _{1}H^{2} \rightarrow _{84}Po^{210} + _{0}n^{1}$

195. A substance has half life period of 4 sec. After t= 20 sec, how much percentage of the substance has decayed? [BHU-2003]
(A) 96.875 (B) 3 (C) 40 (D) 50

196. In element has half life period of 6.5 hrs. Initial number of atoms is 48×10^{19}, then after 26hrs, the number of atoms will be [BHU-2003]
(A) 1.6×10^{19} (B) 3×10^{19} (C) 2.4×10^{19} (D) 5×10^{19}

197. An element has half life period of 1600 years. The mass left after 6400 years will be [AFMC-2003]

(A) 1/4 (B) 1/12 (C) 1/16 (D)1/32

198. α–particles are [AFMC-2003]

(A) Protons (B) Electrons

(C) Positrons (D) Charged helium atoms

199. Which of the following can be used for carrying out nuclear reactions? [AFMC-2003]

(A) Uranium-238 (B) Plutonium-239

(C) Neptunium-239 (D) Thorium-232

200. The radioactive isotope $_{27}Co^{60}$ which is used in the treatment of cancer can be made by (n, p)reaction. For this reaction the target nucleus is

[CBSE-PMT-2004]

(A) $_{28}Ni^{59}$ (B) $_{27}Co^{59}$ (C) $_{28}Ni^{60}$ (D) $_{27}Co^{60}$

Radioactivity

1.	(C)	49.	(C)	96.	(B)	144.	(C)	192.	(A)
2.	(D)	50.	(B)	97.	(C)	145.	(B)	193.	(C)
3.	(D)	51.	(A)	98.	(C)	146.	(A)	194.	(B)
4.	(C)	52.	(B)	99.	(C)	147.	(C)	195.	(A)
5.	(A)	53.	(D)	100.	(D)	148.	(B)	196.	(B)
6.	(B)	54.	(D)	101.	(D)	149.	(B)	197.	(B)
7.	(C)	55.	(C)	102.	(C)	150.	(A)	198.	(D)
8.	(A)	56.	(B)	103.	(A)	151.	(C)	199.	(B)
9.	(D)	57.	(B)	104.	(B)	152.	(B)	200.	(C)
10.	(D)	58.	(B)	105.	(D)	153.	(D)		
11.	(C)	59.	(A)	106.	(B)	154.	(B)		
12.	(A)	60.	(B)	107.	(A)	155.	(C)		
13.	(A)	61.	(B)	108.	(C)	156.	(D)		
14.	(C)	62.	(A)	109.	(A)	157.	(D)		
15.	(C)	(C)	(D)	110.	(A)	158.	(D)		
16.	(D)	63.	(D)	111.	(A)	159.	(A)		
17.	(B)	64.	(D)	112.	(A)	160.	(B)		
18.	(D)	65.	(A)	113.	(D)	161.	(D)		
19.	(C)	66.	(D)	114.	(C)	162.	(A)		
20.	(D)	67.	(C)	115.	(D)	163.	(D)		
21.	(D)	68.	(C)	116.	(B)	164.	(B)		
22.	(B)	69.	(A)	117.	(B)	165.	(B)		
23.	(D)	70.	(B)	118.	(C)	166.	(D)		
24.	(C)	71.	*	119.	(C)	167.	(A)		
25.	(B)	72.	(D)	120.	(C)	168.	(C)		
26.	(B)	73.	*	121.	(D)	169.	(A)		
27.	(C)	74.	*	122.	(D)	170.	(B)		
28.	(B)	75.	*	123.	(A)	171.	(C)		
29.	(D)	76.	(D)	124.	(A)	172.	(B)		
30.	(B)	77.	(C)	125.	(B)	173.	(A)		
31.	(A)	78.	*	126.	(C)	174.	(C)		
32.	(A)	79.	*	127.	(B)	175.	(D)		
33.	(B)	80.	*	128.	(C)	176.	(C)		
34.	(A)	81.	(C)	129.	(B)	177.	(A)		
35.	(B)	82.	(D)	130.	(A)	178.	(C)		
36.	(D)	83.	*	131.	(B)	179.	(D)		
37.	(C)	84.	(C)	132.	(C)	180.	(C)		
38.	(A)	85.	(D)	133.	(A)	181.	(A)		
39.	(B)	86.	(D)	134.	(B)	182.	(A)		
40.	(B)	87.	(B)	135.	(C)	183.	(C)		
41.	(B)	88.	(B)	136.	(B)	184.	(C)		
42.	(D)	89.	(B)	137.	(D)	185.	(B)		
43.	(A)	90.	(C)	138.	(C)	186.	(B)		
44.	(C)	91.	(D)	139.	(C)	187.	(A)		
45.	(C)	92.	(C)	140.	(D)	188.	(A)		
46.	(A)	93.	(C)	141.	(B)	189.	(A)		
47.	(D)	94.	(B)	142.	(D)	190.	(C)		
48.	(A)	95.	(C)	143.	(B)	191.	(C)		

---

71. $_6C^{14}$ 74. $^3_1$ T 78. 1.67 80. $_{56}Ba^{143}$

73. 7 Fm 75. 0.00693 79. fusion 83. 2 units

GASEOUS STATE

Matter may be considered anything that has mass and occupies space. All matter on the basis of physical state is divided into :

Solid, liquid and gas (Plasma state : A very hot ionised gas having approximately equal number of positive ions and of electrons, highly conducting).

Under suitable conditions of pressure and temperature, almost all substances can exist in these states.

e.g. water as ice at 0^0C & atmospheric pressure; as vapour at 100^0C & atmospheric pressure. e.g. CO_2 cooling below 31^0C & 73 atmosphere pressure as liquid; - 57°C & 5 stmospheres - solid. Under same conditions, a substance may be able to exist in all the three states simultaneously. e.g. at the freezing point of water, all the states exist together.

In fact two types of opposing molecular forces act between molecules. These are forces of attraction and thermal energy of molecules. In solids, the forces of attraction between the molecules are more than that in liquid state. Though, molecules have kinetic energy in liquids, they can't move far away because, here also forces of attraction are more than thermal energy. As a result of this, liquids have definite volume but not definite shape. In gaseous state, forces of attraction between molecules are much lesser than thermal energy i.e. the attractive forces are not sufficient to bind the molecules of a subtance together. Gases do not have definite shape and volume. Molocules move freely in this state.

A substance is considered to be a gas if under atmospheric pressure, its boiling point is below the room temperature, to be a liquid, if under atmospheric pressure, its freezing point is below the room temperature and to be a solid if under atmospheric pressure its melting point is above the room temperature.

The molecular forces of attraction in gases are very much weaker. There are, therefore, large inter-molecular voids and the gaseous molecules can have all the three types of motion viz. translatory, rotatory and vibratory motion. The molecules can also have random motion in all directions. On account of weak inter molecular forces and large voids in between the molecules, gaseous state is characterized by low density and high compressibility.

Measurable properties of gases : There are 4 important measurable properties

(1) mass (2) volume (3) pressure and (4) temperature

(1) Mass of a gas can be determined by weighing the container when it is filled with the given gas and when it is evacuated. Mass, molecular mass and moles of the gas are related as

$$\text{number of moles of gas} = \frac{\text{mass of the gas in grams}}{\text{molecular mass of the gas}}$$

One mole of gas contains Avogadro's number of molecules of the gas i.e. 6.023×10^{23}

(2) Volume of any gas occupies all the space which is available to it i.e. the space of the container. units : litre (L), cubic centimeter etc.

(3) Pressure of the gas is due to bombardment of gas molecules on the walls of the container. It is the force per unit area

$$P = \frac{\text{Force exerted by the gas on the walls of the container}}{\text{surface area of the container}}$$

$$P = \frac{F}{A}$$

The atmospheric pressure is measured by barometer and is given by $P = \rho hg$ (ρ = density of Hg; h = height of Hg column and 'g' is the acceleration due of gravity).

Pressure is expressed in atmosphere which is the pressure exerted by the mercury height of 76 cm at 0^0C and standard gravity 981 cm s^{-1}. The SI unit of pressure is Pascal, which is the pressure exerted when force of one Newton acts on $1m^2$ area.

1 atmosphere = 101.325 kilopascal (kPa)
= 101325 Nm^{-2}

1 atmosphere = 760 torr

Increase in pressure is simply due to increase in the number of gaseous molecules striking per unit area of the walls of the container in unit time.

(4) Measurement of temperature depends on expansion of substances on increasing temperature. Mercury expands and contracts uniformly with temperature, so, it is used in the thermometers to measure temperature.

Temp. in Kelvin = temp. in °C + 273.15

The volume of any gas is zero at - 273.15°C

- In solid state, vibratory mations are significant.
- Molecules possess energy of motion which is another name for kinetic energy.
- First direct evidence for molecular motion was furnished by Robert Brown, e.g. smoke (where kinetic energy of air is transferred to the carbon particles of the smoke and set them in motion)

The Gas laws : Gases behave according to certain laws. These laws relate to definite mass of gases and their temperature, pressure and volume. These laws are called gaseous laws.

Boyle's Law : At constant temp. the volume of a given mass of a gas varies inversely as the pressure.

$V \alpha \frac{1}{P}$ at constant temp. say T.

$V = K \times \frac{1}{P}$; $PV = K$

Thus at a constant temp, the product of the pressure and volume of a definite mass of gas is always a constant quantity.

Suppose at a constant temp., the initial volume of a gas is

changed from V_1 to V_2. This will be accompanied by a change in pressure as well i.e. from P_1 to P_2.

Thus $P_1V_1 = k$; $P_2V_2 = k$; $P_1V_1 = P_2V_2$

This equation is called the equation of Boyle's law.

The graph between V and $\frac{1}{P}$ which is a straight line. It verifies Boyle's law.

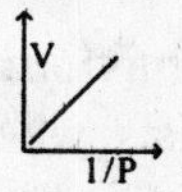

Relationship between pressure and density

Density D=Mass / Volume ; $D = M / V$

$P_1V_1=P_2V_2$

$$\frac{P_1M_1}{D_1} = \frac{P_2M_2}{D_2}$$

$$\frac{P_1}{D_1} = \frac{P_2}{D_2} \quad \because M_1 = M_2 \text{ (definite mass)}$$

or $\frac{P}{D}$ = Constant or $P \alpha D$

Thus, pressure of a definite mass of a gas is directly proportional to density.

Charles law : At constant pressure, the volume of a definite mass of a gas is directly proportional to its absolute temperature.

$V \alpha T$; $V = KT$; $\frac{V}{T} = K$

Thus at a constant pressure the ratio of volume and absolute temperature of a gas is always constant.

Suppose, the volume of gas changes from V_1 to V_2 when absolute temp. changes from T_1 to T_2

$\frac{V_1}{T_1} = K$ also $\frac{V_2}{T_2} = K$, $\therefore \frac{V_1}{T_1} = \frac{V_2}{T_2}$

The plot of volume of a definite mass of gas at pressure P_1 and P_2 against various temp. give straight lines proving the law $P_1 > P_2$.

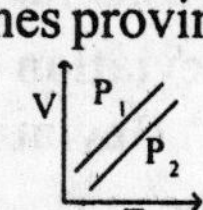

We know D = M/V or V = M/D

Substituting the value of V in Charles equation

$$\frac{M_1}{D_1 \times T_1} = \frac{M_2}{D_2 \times T_2}$$

$$\frac{1}{D_1T_1} = \frac{1}{D_2T_2} \quad \because M_1 = M_2 \text{ (definite mass)}$$

$D_1T_1 = D_2T_2$ at constant pressure

$DT = K$ or $D \alpha \frac{1}{T}$

Pressure - Temperature law :

This law was also enuntiated by Charles and involves the relationship between pressure and temperature of a definite mass of a gas at constant volume. "The pressure of a definite mass of a gas at constant volume is directly proportional to absolute temperature."

$P \alpha T$; $P = KT$;

$$\frac{P}{T} = K ; \quad \frac{P_1}{T_1} = \frac{P_2}{T_2}$$

The Gas equation : It is based on combined effect of temperature and pressure on the volume of gas.

The equation that repersents the relationship of the volume of a definite mass of gas with its pressure and temperature is called gas equation.

Boyles law $V \alpha \frac{1}{P}$

Charles law $V \alpha T$

From both the above relations, we can write $V \alpha \frac{T}{P}$

$V = K \times \frac{T}{P}$; $\frac{P_1V_1}{T_1} = \frac{P_2V_2}{T_2}$

$PV = RT$; $PV = nRT$; or $PV = \frac{W}{M} RT$

Numerical values of R

$PV = RT$; $R = \frac{PV}{T}$

There are several numerical values of R depending upon the units of pressure, temperature and volume.

(1) At NTP : P = 1atm.; T = 273K; V = 22.4lit; n = 1mole

$R = 1 \times 22.4/273 = 0.0821$ lit. - atm. mol^{-1} degree K^{-1}

(2) If P is in dynes/cm^2 ; V in ml.

$P = 76 \times 13.6 \times 981$ dynes cm^{-2} ; V = 22400 ml; T = 273K

$$R = \frac{76 \times 13.6 \times 981 \times 22400}{273} = 8.314 \times 10^7 \text{erg.degree}^{-1}\text{mol}^{-1}$$

(3) 1 joule = 10^7 ergs

R = 8.314 joules $degree^{-1}$ mol^{-1} in S.I. units

(4) 1 calorie = 4.18×10^7 ergs. in c. g. s. system

$$R = \frac{8.314 \times 10^7}{4.18 \times 10^7} = 1.90 \text{ calories degree}^{-1} \text{ mol}^{-1}$$

Graham's law of diffusion : The natural tendency of every gaseous substance is to distribute itself uniformly throughout the available space. This tendency is known as diffusion.

Another term effusion is used for the movement of a gas through a small hole when the gaseous system is subjected to pressure.

"The rates of diffusion or effusion of two different gases under similar conditions of temperature and pressure are inversely proportional to the square root of their densities or molecular weights."

$$\frac{r_1}{r_2} = \sqrt{\frac{\rho_2}{\rho_1}} = \sqrt{\frac{m_2}{m_1}}$$

$$\text{rate of diffusion } r = \frac{\text{diffused volume}}{\text{Time taken in diffusion}} = \frac{v}{t}$$

we can also write it as $\frac{r_1}{r_2} = \frac{v_1/t_1}{v_2/t_2} = \sqrt{\frac{\rho_2}{\rho_1}} = \sqrt{\frac{M_2}{M_1}}$

Grahams law of diffusion finds its application in

(1) Separation of isotopes of the same element from one another. The light isotope diffiuses out first and the heaviest one last of all.

(2) To calculate vapour densities of gases

(3) To determine molecular weights of gases

(4) Time taken in diffusion of gases

(5) Volume of diffused gases in a definite time

Dalton's Law of partial pressure

"The pressure exerted by a mixture of gases and vapours which do not react chemically with one another occupying a certain volume at a given temperature, is equal to the sum of the pressures which the constituents would exert if each had occupied separetely the volume of the mixture at that temperature."

$P = p_1 + p_2 + p_3$ etc.

Partial pressure : Whatever pressure each constituent of the gaseous mixture occupying the total volume exerts is called the partial pressure of that constituent.

Kinetic molecular theory of gases :

The main postulates of the theory are :

(1) The molecules are so small that their actual volume is a negligible fraction of the total volume or space occupied by the gas.

(2) The molecules are in a state of constant rapid motion in all directions colliding in a random manner with one another and with the walls of the container.

(3) The molecules are perfectly elastic.

(4) There are no attractive forces between molecules or molecules and the walls of the container. They move completely independent of each other.

(5) The pressure of a gas is due to the bombardment of the moving molecules on the walls of the container.

Kinetic gas equation : $PV = \frac{1}{3} mNu^2$

(m = mass; N = molecules of a gas)

(a) Most probable velocity : The most probable velocity α, is defined as velocity possessed by the maximum number of molecules of the gas.

Maxwell's distribution law equation can be treated mathematically to get $\alpha = \sqrt{\frac{2RT}{M}}$

(b) Average velocity : By using Maxwell's distribution law euqation, it can be derived mathematically

$$v = \sqrt{\frac{8RT}{\pi M}}$$

(c) Root mean square velocity : It follows directly from the kinetic gas equaltion.

$$u = \sqrt{\frac{3RT}{M}}$$

The relationship between α, v and u will be

$$\alpha : v : u = \sqrt{\frac{2RT}{M}} : \sqrt{\frac{8RT}{\pi M}} : \sqrt{\frac{3RT}{M}}$$

$$= \sqrt{2} : \sqrt{\frac{8}{M}} : \sqrt{3}$$

$$= 1.414 : 1.1595 : 1.732$$

or $\alpha : v : u = 1 : 1.128 : 1.224$

Ideal or perfect gas : Which obeys general gas equation PV = nRT and all gas laws for all values of temperature and pressure.

The molecules of an ideal gas (1) Occupy negligible volume and (2) Exert no force of attraction on each other. Experiments however confirm that no gas is ideal.

Real gas : Disobeys general gas equation PV = nRT and all other gas laws. All known gases are found to be real. Real gases (1) have a tendency to behave like an ideal gas at low pressure and high temperature. (2) show deviation from ideal gas behaviour at high pressure and low temperature.

Behaviour of real gases : A real gas is one which really exists, whether it obeys gas laws strictly or not.

Under ordinary conditions of temp. and pressure, the real gases show considerable deviation from ideal gas behaviour. A graph PV against P results as:

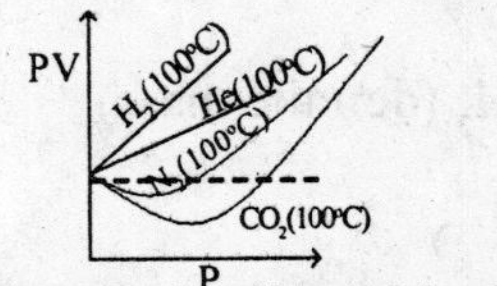

Causes of deviation from ideal gas behaviour

It is due to two incorrect assumptions of kinetic theory of gases.

(1) The total volume occupied by gas molecules is negligibly small as compared to the total volume occupied by the gas.

This assumption is valid only at low pressure & high temp. But at high pressure or low temp., when the total volume of gas decreases considerably while the volume of the molecules remains the same, the molecules being incompressible.

(2) The forces of attraction between gas molecules are negligible. At high pressure and low temp. when the total volume of a gas is small, the forces of attraction are considerable and can't be overlooked.

Deviation of real gases from ideal gas behaviour

(1) Deviation from Boyles law

The curve formed by plotting PV against P for a gas at constant temp. is called isotherm. The isotherm for ideal gas should be a straight line horizontally.

(a) At low pressures : All isotherms meet the ideal gas isotherm at zero pressure. Hence at low pressure all gases behave like ideal gases.

(b) At moderate pressure : The deviation is maximum for gases which are easily liquified e.g. CO_2 Negative deviation is shown.

(c) At high pressure : Positive deviation is shown. Gases do disobey the gas laws at all.

(d) At different temperatures : As the temp. falls, the deviation is more. Higher negative deviation is observed for N_2 and CO_2.

Alternatively, the effect of pressure and temperature on gas behaviour can also be studied in terms of compressibility factor, $Z = \frac{PV}{nRT}$

For ideal gases Z is always equal to unity. For real gases it may be less or more than 1. 50°C is called Boyles temperature. $T_B = a / bR$.

Below this, the value of Z first decreases and then increases for every gas. Above Boyles temp., Z increases sharply with rise of pressure more than unity.

(2) Deviation from Charle's Law

The coefficients volume expansion of all gases should be the same and independent of pressure. Experiments show that these coefficients approach the constant value 1/273 at low pressures but at high pressures marked deviation.

(3) Deviation from Avogadro's law

Exception : for easily liquefiable gases like NH_3, CO_2, CH_3Cl etc., the values are less. (Avogadro's law is valid only at very low pressure).

van der Waals Equation of state : J. D. van der Waal incorporates two corrections in the ideal gas equation, taking into consideration of molecular volume and the intermolecular forces

$$\left(P + \frac{an^2}{V^2}\right)(V - nb) = nRT$$

Correction due to volume of gaseous molecules

For an ideal gas PV = RT where V is the total volume occupied by one mole of the gas. But as the molecules are not compressible, the volume occupied by them remians the same irrespective of pressure. van der Waals, therefore suggested that a correction term b should be subtracted from V, the total volume, in order to get the ideal volume which is compressible. The volume which is compressible is not V but V - b. The excluded volume b is nearly four times the actual volume occupied by all the molecules contained in the gas.

Correction due to inter-molecular forces of attraction :

Consider a molecule lying somewhere in the middle of a vessel. It is attracted uniformly on all sides by the neighbouring molecules. These forces neutralize one another and there is no resultant attractive force on the molecule. However, as the molecule approaches the wall of the vessel, it experiences attractive forces from the bulk of the molecules behind it. Hence, it will strike the wall with a lower velocity and will exert a lower pressure than it would have done if there were no forces of attraction. It is therefore necessary to add a certain quantity to the pressure P of the gas in order to get the ideal pressure. The force of attraction exerted on a single molecule about to strike the wall depends upon the number of molecules per unit volume in the bulk of the gas i.e. directly upon the density of the gas. Further the number of molecules striking the wall at any given instant also depends directly upon the density of the gas.

$$\therefore \quad p \propto \rho^2$$

Density is inversely proportional to volume hence

$$p \propto \frac{1}{v^2}$$

$p = \frac{a}{v^2}$ ("a" is a constant depending upon the nature of the gas).

"a" : is higher for easily liquefiable gases e.g. SO_2, NH_3, H_2S, SO_2 etc. lower for H_2, N_2, O_2 etc.

units of a = atm. $L^2\ mol^{-2}$ SI unit : a = $Nm^4\ mol^{-2}$

b = $L\ mol^{-1}$; b = $m^3\ mol^{-1}$

Liquefiability : $SO_2 > NH_3 > H_2O > CO_2$

Ratio of molar heat capacity :

$\gamma = \frac{C_P}{C_V}$ = 1.66 monoatomic

1.40 diatomic

1.33 triatomic

C_V = heat capacity at constant volume

C_P = heat capacity at constant pressure.

$$C_V = \frac{3}{2}R + x; \quad C_P = \frac{5}{2}R$$

Things to remember

- Atmospheric pressure is measured by Barometer
- Hg expands and contracts uniformly with temperature

Boyles law : $V \alpha \frac{1}{P}$; $PV = K$; $P_1V_1 = P_2V_2$; $\frac{P}{D} = K$; $P \alpha D$

Charles law : $V \alpha T$; $\frac{V}{T} = K$; $\frac{V_1}{T_1} = \frac{V_2}{T_2}$; $D \alpha \frac{1}{T}$

Pressure Temperature law : $P \alpha T$; $\frac{P}{T} = K$; $\frac{P_1}{T_1} = \frac{P_2}{T_2}$

General gas equation, $PV = nRT$

$$\frac{P_1V_1}{T_1} = \frac{P_2V_2}{T_2}$$

Graham's law of diffusion : $\frac{r_1}{r_2} = \frac{v_1/t_1}{v_2/t_2} = \sqrt{\frac{\rho_2}{\rho_1}} = \sqrt{\frac{M_2}{M_1}}$

This principle is used in the separation of isotopes

Dalton's law of partial pressure : $P = p_1 + p_2 + p_3$

Kinetic gas equation $PV = \frac{1}{3}$ m Nu^2

Most probable velocity $\alpha = \sqrt{\frac{2RT}{M}}$

Average velocity $v = \sqrt{\frac{8RT}{\pi M}}$

Root mean square velocity $u = \sqrt{\frac{3RT}{M}}$ or $\sqrt{\frac{3PV}{M}}$ or $\sqrt{\frac{3P}{D}}$

$\alpha : v : u = \sqrt{2} : \sqrt{\frac{8}{\pi}} : \sqrt{3}$

$= 1 : 1.128 : 1.224$

Thus for a particular gas at a particular temp. $u > v > \alpha$

Boyles temp is 50°C

Van der waals equation of state : $(P + \frac{a}{v^2})(v-b) = RT$

$(P + \frac{an^2}{v^2})(v-nb) = nRT$

Solved Examples

(1) The volume of a gas with some marble measures 100ml at 760mm pressure. On raising the pressure to 1000mm, the total volume becomes 80ml. Calculate the volume occupied by the marble.

Sol. let V be the volume of the marble

Now apply $P_1V_1 = P_2V_2$

$(100 - v) \times 760 = (80-v) \times 1000$

$24v = 400$; $v = 16.6$ml.

(2) A vessel filled with cold gas can withstand a pressure of 14.9 atmosphere. The pressure gauge attached to the vessel reads 12 atmosphere at 27°C. The building in which the vessel is kept suddenly catches fire and the temperature starts rising. Calculate the temperature at which the vessel will explode.

Sol. We know at constant volume, $\frac{P_1}{T_1} = \frac{P_2}{T_2}$

$$\frac{12}{290} = \frac{14.9}{T_2}$$

$$T_2 = \frac{290 \times 14.9}{12} = 372.5k \text{ i.e. } 99.5^0C.$$

$\therefore$ The vessel will explode at any temperature beyond 99.5°C.

(3) Calculate the weight of 10 litres of nitrogen gas at 27°C and 74 cm pressure.

Sol. First we calculate the volume equivalent at NTP

we know $\frac{P_1V_1}{T_1} = \frac{P_2V_2}{T_2}$.

$$V_2 = \frac{P_1V_1T_2}{T_1P_2} = \frac{74 \times 10 \times 273}{300 \times 76} = 8.86 \text{ litres}$$

22.4 lit. of volume of nitrogen at NTP weighs 28g.

8.86 lit. of volume of nitrogen at NTP weighs

$$= \frac{8.86 \times 28}{22.4} = 11.075g.$$

(4) 2.8g. of a gas at 27°C and 180 mm pressure occupies 3.8 lit. Calculate the molecular weight of the gas

Sol. $PV = nRT$

$$= \frac{w}{m} RT$$

$$\frac{180}{760} \times 3.8 = \frac{2.8}{m} \times 0.0821 \times 300 \text{ ; } m = 76.44$$

(5) Calculate the relative rates of diffusion of $^{235}UF_6$ and $^{238}UF_6$ in the gaseous state

Sol. Molecular mass of $^{235}UF_6 = 235 + 6 \times 19 = 349$

Molecular mass of $^{238}UF_6 = 238 + 6 \times 19 = 352$

$$\frac{r_1}{r_2} = \sqrt{\frac{M_2}{M_1}} = \sqrt{\frac{352}{349}} = 1.0043$$

$r_1 : r_2 = 1.0043 : 1.0000$

(6) The rate of diffusion of a gas B with respect to another diatomic gas A, is 6 : 5. If the atomic weight of A is 36, calculate the molecular weight of B.

Sol. We know $\frac{r_B}{r_A} = \sqrt{\frac{M_A}{M_B}}$; mol. wt. of A = 2 x 36 = 72

$$\frac{6}{5} = \sqrt{\frac{72}{M_B}}$$

$$M_B = \frac{72 \times 5 \times 5}{6 \times 6} = 50$$

(7) 180ml. of a hydrocarbon diffuse in 15 minutes, whereas 120ml. of SO_2 diffuse in 20 minutes under the similar conditions. If the molecular weight of SO_2 is 64, calculate the molecular weight of the hydrocarbon.

Sol. $\frac{r_{hy}}{r_{SO_2}} = \frac{v_{hy}/t}{v_{SO_2}/t} = \sqrt{\frac{64}{M_{hy}}}$

$$\frac{180/15}{120/20} = \sqrt{\frac{64}{M_{hy}}}$$

$$\frac{12}{6} = \sqrt{\frac{64}{M_{hy}}}$$

$$M_{hy} = \frac{64 \times 6 \times 6}{12 \times 12} = 16$$

(8) At the top of a mountain, the thermometer reads 10°C and the barometer reads 700mm Hg. At the bottom of the mountain, the temp. is 27°C and the pressure is 760mm Hg. Compare the density of the air at the top with that at the bottom.

Sol. $PV = \frac{w}{m} RT$

$$P = \frac{w}{V} \times \frac{RT}{m}$$

$$= d\frac{RT}{m} \quad (\because \frac{w}{V} = d)$$

or $d = \frac{P}{T} \cdot \frac{m}{R}$

$d \; \alpha \; \frac{P}{T}$ [m/R being constant]

or $\frac{d_{top}}{d_{bottom}} = \frac{P_{top}}{T_{top}} \times \frac{T_{bottom}}{P_{bottom}} = \frac{700 \times 300}{760 \times 283} = 0.0986$

(9) A gaseous mixture of 0.336g. N_2 and 11.2g. litre O_2 is taken at NTP in a 10 lit. flask at 27°C. Calculate the partial pressure and total pressure of the gases.

Sol. Number of moles of $N_2 = \frac{\text{mass of } N_2}{\text{molecular mass}}$

$$= \frac{0.336}{28} = 0.012$$

Number of moles of $O_2 = \frac{11.2}{22.4} = 0.5$

P_{N_2} (partial pressure of N_2) $= \frac{nRT}{V}$

$$= \frac{0.012 \times 0.0821 \times 300}{10} = 0.0296 \text{ atm.}$$

$$P_{O_2} = \frac{0.5 \times 0.082 \times 300}{10} = 1.2315 \text{ atm.}$$

Total pressure = 0.0296 + 1.2315 = 1.2611 atm.

(10) A flask is filled with 250ml. of Kr at 500mm pressure. Another flask contains 500ml. He at 950mm. Both the gases are mixed at the same temperature. Calculate the combined volume and the total pressure exerted by the mixture.

Sol. Total volume = 250 + 500 = 750ml.

$$P_1V_1 = P_{kr} \times V$$

$$P_{kr} = \frac{250 \times 500}{750} = 166.6\text{mm}$$

$$P_{He} = \frac{950 \times 500}{750} = 633.3\text{mm}$$

Total pressure of the mixture $P_{kr} + P_{He} = 166.6 + 633.3$
$= 800$ mm

(11) At 28°C and 730 torr pressure, the mass of 525 cc of a gaseous compound was 0.90g. Calculate the molecular weight of the compound.

Sol. PV = nRT

$$= \frac{w}{m} RT$$

$$m = \frac{wRT}{PV} = \frac{0.90 \times 82.1 \times 301}{(730/760) \times 525} = 44.1$$

(12) Calculate the total and averge kinetic energy in joules, of the molecules in 8g. of methane at 27°C.

Sol. Average kinetic energy of one molecule $= \frac{3}{2}$ KT

$$= \frac{3}{2} \times \frac{8.314 \times 300}{6.022 \times 10^{23}} = 6.213 \times 10^{-21} \text{J.}$$

No. of moles of methane $= \frac{8}{16} = 0.5$

Total kinetic energy $= \frac{3}{2} \times 0.5 \times 8.314 \times 300 = 1870.65$ J.

(13) A gas bulb of 1 lit. capacity contains 2.0 x 10^{21} moleculas of nitrogen exerting a pressure of 7.57 x 10^3 Nm^{-2}. Calculate the root mean square speed and temperature of the gas molecules. If the ratio of most probable speed to the RMS speed is 0.82, calculate the most probable speed of these molecules at this temperature.

Sol. PV = nRT or $T = \frac{PV}{nR}$ $\quad n = \frac{2.0 \times 10^{21}}{6.023 \times 10^{23}}$ mole

$$T = \frac{10^{-3} \times 7.57 \times 10^3 \times 6.023}{2.0 \times 10^{21} \times 8.314} = 274.2\text{K}$$

$$\text{RMS speed} = \sqrt{\frac{3RT}{M}} = \frac{3 \times 8.314 \times 274.2}{28 \times 10^{-3}} = 494.22 \text{ ms}^{-1}$$

Most probable speed = RMS x 0.8
$= 494.22 \times 0.8 = 395.38 \text{ ms}^{-1}$

(14) 2 mole of NH_3 gas is kept in a closed container of 2 lit. capacity at 27°C. Calculate the pressure exerted by the gas when it behaves as (i) an ideal gas and (ii) a real gas.(a = 4.17 lit.^{2} atm. mole^2; b = 0.037 lit. mole^{-1}.)

Sol. As an ideal gas, PV = nRT

$$P = \frac{nRT}{V} = \frac{2 \times 0.0821 \times 300}{2} = 24.63 \text{ atm.}$$

As a real gas, $(P + a\frac{n^2}{V^2}) - (V - nb) = nRT$

$$P = \frac{nRT}{V-nb} - \frac{an^2}{V^2}$$

$$= \frac{2 \times 0.0821 \times 300}{2-(2 \times 0.037)} - \frac{4.17 \times 2^2}{2^2} = 21.48 \text{ atm.}$$

(15) Calculate form the van der Waal's equation the temperature at which 3 moles of SO_2 would occupy a volume of 10 litres at 15 atmospheric pressure from the given data :

a = 6.7 atm lit²/mole² ; b = 0.0564 lit/mole ;

R = 0.082 litre atm/degree/mole.

Sol. We know $(P + \frac{an^2}{V^2})(V - nb) = nRT$

$$T = (P + \frac{an^2}{V^2})(V - nb) \times 1/nR$$

$$= (15 + \frac{6.7 \times 3^2}{10^2})(10 - 3 \times 0.0564) \times \frac{1}{3 \times 0.082}$$

$$= 15.603 \times 9.8308 \times \frac{1}{0.246}$$

$$= 623.5 \text{ k}$$

$$T = 623.5 - 273 = 350.5^\circ C$$

1. The volume of CO_2 at STP obtained by heating one gram of pure $CaCO_3$ will be [CBSE-91]

(A) 1 litre (B) 22.4 litre (C) 0.224 litre (D) 11.2 litre

2. The vapour density of a gas is 11.2. The volume occupied by 11.2 gms of this gas at NTP is [CBSE-91]

(A)1 litre (B) 11.2 litre (C) 22.4 litre (D) 20 litre

3. The correct value of the gas constant R is close to [CBSE-92]

(A) 0.082 litre-atmosphere-K^{-1}-mol^{-1}

(B) 0.082 $litre^{-1}$-$atmosphere^{-1}$-K-mol

(C) 0.082 litre-atmosphere-K

(D) 0.082 litre $atmosphere^{-1}$-K-mol^{-1}

4. One litre of a gas weighs 2g at 300K and 1 atm pressure. If the pressure is made 0.75 atm., at which of the following temperatures will one litre of the same gas weigh one gram. [CBSE-92]

(A) 450 K (B) 600K (C) 800 K (D) 900K

5. The internal energy of one mole of an ideal gas is given by [CBSE-92]

(A) 3/2 RT (B) 1/2 kT (C) 1/2 RT (D) 3/2 kT

6. The density of a gas at 27°C and 1 atmosphere is d. Pressure remaining constant, at which of the following temperatures will its density become 0.75 d ? [CBSE-92]

(A) 20°C (B) 30°C (C) 400 K (D) 300K

7. Select the one correct statement. In the gas equation, PV = nRT [CBSE-92]

(A) n is the number of molecules of the gas

(B) n moles of the gas have a volume V

(C) V denotes volume of one mole of the gas

(D) P is the pressure of the gas when only one mole of gas is present

8. 3.2g of oxygen (At.wt.=16) and 0.2 g of hydrogen (At.wt.=1) are placed in a 1.12 litre flask at 0°C. The total pressure of the gas mixture will be [CBSE-92]

(A) 1 atm (B) 4 atm (C)3 atm (D) 2 atm

9. 50 ml of a gas A diffuse through a membrane in the same time as for the diffusion of 40 ml of gas B under identical pressure temperature conditions.If the molecular weight of A is 64 that of B would be [CBSE-92]

(A) 100 (B) 250 (C) 200 (D) 80

10. A closed flask contains water in all its three states, solid, liquid and vapour, at 0°C.In this situation the average kinetic energy of water molecule will be [CBSE-92]

(A) the greatest in the vapour state

(B) the same in all the three states

(C) the greatest in the solid state

(D) greater in the liquid than in the vapour state

11. An ideal gas cannot be liquefied because [CBSE-92]

(A) its critical temperature is always 0°C
(B) its molecules are relatively small in size
(C) it solidifies before becoming a liquid
(D) forces operative between its molecules are negligible.

12. The ratio amongst most most probable velocity, mean velocity and r.m.s. velocity is in order of [CBSE-93]
(A) $\sqrt{2} : \sqrt{8/\pi} : \sqrt{3}$ (B) 1 : 2 : 3
(C) $1 : \sqrt{2} : \sqrt{3}$ (D) $\sqrt{2} : \sqrt{3} : \sqrt{8/\pi}$

13. Internal energy and pressure of a gas unit volume are related as [CBSE-93]
(A) $P = \frac{2}{3} E$ (B) $P = \frac{3}{2} E$
(C) $P = \frac{1}{2} E$ (D) $P = 2 E$

14. The electronic configuration of an ideal gas is [CBSE-93]
(A) $1s^2\ 2s^2\ 2p^6$ (B) $1s^2\ 2s^2\ 2p^6\ 3s^1$
(C) $1s^2\ 2s^2\ 2p^6\ 3s^2\ 3p^5$ (D) $1s^2\ 2s^2\ 2p^1$

15. A gas "disobeys" gas law at [CBSE-93]
(A) low pressure and high temperature
(B) both low pressure and temperature
(C) high pressure and low temperature
(D) none of the above

16. 50 ml of hydrogen diffuses out through a small opening in a vessel in 20 minutes. The time needed for 40 ml of oxygen to diffuse out is [CBSE-94]
(A) 12 min (B) 64 min (C) 8 min (D) 32 min

17. The number of moles of oxygen in one litre of air containing 21% oxygen by volume, in standard conditions is [CBSE-95]
(A) 0.186 mole (B) 0.21 mole
(C) 2.10 mole (D) 0.093 mole

18. Dimensions of pressure are the same as that of [CBSE-95]
(A) Energy (B) Force
(C) Energy per unit volume (D) Force per unit volume

19. An ideal gas obeying kinetic theory of gases can be liquefied if [CBSE-95]
(A) its temperature is more than critical temperature T.
(B) it can not be liquefied at any P and T.
(C) its pressure is more than criticial pressure P.
(D) its pressure is more than P_C at a temperature less than T_C

20. At what temperature the rate of effusion of N_2 would be 1.625 times than the rate of SO_2 at 500°C? [CBSE-96]
(A) 110 K (B) 173K (C) 373K (D) 273K

21. The average kinetic energy of an ideal gas per molecule in S.I. units at 25°C will be [CBSE-96]
(A) 61.7×10^{-21} (B) 6.17×10^{-21}
(C) 6.17×10^{-20} (D) 7.16×10^{-20}

22. 0.24 gm of a volatile gas upon vapourisation gives 45 ml vapour at NTP. What will be the vapour density of the substance ? (Density of H_2 = 0.089) [CBSE-96]
(A) 95.39 (B) 5.993 (C) 95.93 (D) 59.93

23. The amount of zinc required to produce 224 ml of H_2 at STP on treatment with dilute H_2SO_4 will be [CBSE-96]
(A) 0.65 gm (B) 6.5 gm (C) 65 gm (D) 0.065 gm

24. Which of the following mixtures of gases does not obey Dalton's Law of partial pressure? [CBSE-96]
(A) O_2 amd CO_2 (B) N_2 amd O_2
(C) Cl_2 and SO_2 (D) CO_2 and He

25. One mole of an ideal gas at 300K is expanded isothermally from an initial volume of 1 litre to 10 litre. The ΔE for this process is [CBSE-98]
(R=2 cal, $mol^{-1}\ K^{-1}$)
(A) 163.7 cal (B) 1381.1 cal (C) 9 lit. atm (D) zero

26. At 25°C and 730 mm pressure, 380 ml of dry oxygen was collected. If the temperature is constant, what volume will the oxygen occupy at 760 mm pressure? [CBSE-99]
(A) 365 ml (B) 449 ml (C) 569 ml (D) 621 ml

27. Which one of the following statements is wrong for gases? [CBSE-99]
(A) Gases do not have a definite shape and volume
(B) Volume of the gas is equal to volume of container confining the gas
(C) Confined gas exerts uniform pressure on the walls of its container in all directions
(D) Mass of gas cannot be determined by weighing a container in which it is enclosed

28. The number of atoms in 4.25g of NH_3 is approximately [CBSE-99]
(A) 1×10^{23} (B) 2×10^{23} (C) 4×10^{23} (D) 6×10^{23}

29. For $BaCO_3(s) \rightarrow BaO(s) + CO_2(g)$, If 98.7g of $BaCO_3$ is decomposed at NTP the reaction on completion gives amount of CO_2 (At.wt. of Ba = 137.4)[CBSE-2000]
(A) 11.2*l* (B) 22.4*l* (C) 0.56*l* (D) 44.8*l*

30. A container contains CO and another H_2 in equal molar concentrations at the same temperature. What is the ratio of avg. molar K.E.'s of the two gases ? [CBSE-2000]
(A) Depends on volume (B) 1 : 1
(C) $KE_{CO} > KE_{n_2}$ (D) $KE_{H_2} > KE_{CO}$

31. Which has maximum molecules? [CBSE-2002]

(A) 7 gm N_2 (B) 2 gm H_2 (C) 16 gm NO_2 (D) 16 gm O_2

32. Which of the following statements is wrong ? [AIIMS-93]
(A) The gas constant per molecule is known as Boltzmann constant
(B) The product of pressure and volume of a fixed amount of gas is independent of temperature.
(C) The molecules of different gases have the same kinetic energy at a given temperature
(D) The gas equation is not valid at high pressures and low temperatures

33. At a certain temperature 3 moles of hydrogen and one mole of nitrogen are allowed to react and ammonia is formed. When 1 mole of ammonia gas is formed, the total pressure of the vessel is 15 atmospheres. The partial pressure of nitrogen in the vessel is [AIIMS-93]
(A) 7.5 atm (B) 2.5 atm (C) 9 atm (D) 6 atm

34. Which of the following gases will have the highest rate of diffusion ? [AIIMS-94]
(A) N_2 (B) O_2 (C) CO_2 (D) NH_3

35. If the density ratio of O_2 and H_2 is 16 : 1, then ratio of their Vrms will be [AIIMS-94]
(A) 1 : 4 (B) 1 : 16 (C) 4 : 1 (D) 16 :1

36. A mixture of 40 g of oxygen and 40 g of helium has a total pressure of 0.9 atm. The partial pressure of oxygen is [AIIMS-94]
(A) 0.1 atm (B) 0.2 atm (C) 0.5 atm (D) 0.9 atm

37. Two different gases enclosed in different flasks A and B at same temperature and pressure were found to contain same number of molecules. The ratio of volumes of the flasks A and B must be [AIIMS-95]
(A) 1 : 1 (B) 1 : 2 (C) 1 : 3 (D) 1 : 4

38. The temperature at which the density of O_2 at 1 atm. is the same as that of CH_4 at S.T.P. is [AIIMS-95]
(A) 100°C (B) 150°C (C) 273°C (D) 546°C

39. A certain gas diffuse four times as quickly as oxygen. The molecular weight of the gas is [AIIMS-95]
(A) 1 (B) 1.5 (C) 2 (D) 16

40. How many litres of ammonia gas at S.T.P. would be needed to prepare 100 ml of 2.5 M ammonium hydroxide solution ? [AIIMS-96]
(A) 0.056 lit (B) 0.56 lit (C) 5.6 lit (D) 11.2 lit

41. Which of the following gas molecules have equal total kinetic energy and translation kinetic energy ? [AIIMS-96]
(A) He (B) N_2 (C) O_2 (D) CH_4

42. One gram mole of a gas at N.T.P. occupies 22.4 litres. This fact was derived from [AIIMS-98]
(A) Avogadro's hypothesis (B) Dalton's atomic theory
(C) Law of gaseous volumes
(D) Law of constant proportions

43. At what pressure will a quantity of gas, which occupies 100 ml at a pressure of 720mm, occupy a volume of 84 ml ? [AIIMS-98]
(A) 736.18 mm (B) 784.15 mm
(C) 820.20 mm (D) 857.14 mm

44. In the gas equation : PV = nRT [AIIMS-98]
(A) n is the number of molecules of the gas
(B) n no. of moles of the gas have volume V
(C) V is the volume of one mole of the gas
(D) P is the pressure of one mole of the gas

45. Pressure in a mixture of 4g of O_2 and 2g of H_2 confined in a bulb of 1 litre at 0°C is [AIIMS-99]
(A) 15.210 atm (B) 25.215 atm
(C) 3.234 cm^{-1} (D) 4.616 cm^{-1}

46. Density ratio of O_2 and H_2 is 16 :1. The ratio of their r.m.s. velocities will be [AIIMS-99]
(A) 1 :4 (B) 1 : 16 (C) 4 : 1 (D) 16 : 1

47. If 300 ml of a gas at 27°C is cooled to 7°C at constant pressure, its final volume will be [AIIMS-99]
(A) 135 ml (B) 280 ml (C) 350 ml (D) 540 ml

48. A 14.0 g sample of nitrogen (N_2) occupies 8.0 litres at a pressure of 380 mm of Hg. What is temperature of N_2 under these conditions. [AIIMS-2000]
(A) 197.6 K (B) 97.6 K (C) 100.0 K (D) 297.6 K

49. For a gas, deviation from ideal behaviour is maximum at [AIIMS-2000]
(A) 0°C and 1.0 atm (B) 100° and 2.0 atm
(C) −10°C and 2.0 atm (D) −10°C and 5.0 atm

50. In a gaseous mixture at 3.0 atm pressure, 25% molecules are nitrogen, 30% are carbon dioxide and the remaining are oxygen. What is the partial pressure of oxygen ? [AIIMS-2000]
(A) 0.75atm (B) 0.90 atm (C) 1.20 atm (D) 1.35 atm

51. Assertion : The ratio by volume of gaseous reactants and products is in agreement with their molar ratio. Reason : Volume of a gas is inversely proportional to the number of moles of a gas. [AIIMS-2001]
(A) A (B) B (C) C (D) D

52. If P is pressure and ρ is density of a gas, then P and ρ are related as [AIIMS-2002]
(A) $P \propto \rho$ (B) $P \propto \rho^2$ (C) $P \propto 1/\rho$ (D) $P \propto 1/\rho^2$

53. Assertion (A) : Atoms can neither be created nor destroyed. Reason (R) : Under similar conditions of temperature and pressure, equal volumes of gases donot contain equal number of atoms. [AIIMS-2002]

(A) A (B) B (C) C (D) D (E) E

54. Which of the following is not a green house gas ? [AIIMS-2003]

(A) Carbon dioxide (B) Water vapour
(C) Methane (D) Oxygen

55. A gas has a volume of 27.3 ml at 0°C. What would be its volume at 10°C if the pressure remains unchanged : [JIPMER-90]

(A) 27.3 ml (B) 28.3 ml (C) 2.73 ml (D) 28.3 ml

56. If the pressure and temperature of 2 liters of CO_2 are doubled the volume of CO_2 would become : [JIPMER-91]

(A) 5 liters (B) 4 litres (C) 8 litres (D) 2 liters

57. When gases are heated from 20 to 40°C at constant pressure, the volumes : [JIPMER-91]
(A) Increase by the same magnitude
(B) Become double
(C) Increase in the ratio of their molecular masses
(D) Increase but to different extent.

58. Equal weights of hydrogen and helium are mixed in an empty container at 20°C. The fraction of total pressure exerted by hydrogen is [JIPMER-91]
(A) 1/2 (B) 1/3 (C) 2/3 (D) 2/3 x 273/293

59. A gas can be liquefied by pressure alone when its temperature is : [JIPMER-91]
(A) Higher than the critical temperature
(B) Lower than its critical temperature
(C) Equal to its critical temperature (D) None

60. 27 gm of Al will react completely with how many grams of oxygen : [JIPMER-92]
(A) 8g (B) 16g (C) 32g (D) 24g

61. The density of acetone vapours (CH_3COCH_3) at 27°C and 0.82 atm is : [JIPMER-93]
(A) 2.6g/l (B) 1.93g/l (C) 3.86g/l (D) 1.25g/l

62. The molar volume of methane at 819°C and 760 mm pressure is : [JIPMER-93]
(A) 22.4l (B) 44.8l (C) 89.6l (D) 5.6l

63. The kinetic energy of 1 mole of O_2 gas at O°C is [JIPMER-94]

(A) 150 Cal (B) 273.2 Cal
(C) 543.8 Cal (D) 814.3 Cal

64. At constant volume for a fixed number of moles of a gas, the pressure of a gas increses with the rise in temperature due to : [JIPMER-96]
(A) increase in molecular attraction
(B) increase in mean free path
(C) increae in average molecular speed
(D) increae in rate of collisions amongst molecules

65. The volume occupied by 1 mole of a gas at 0.5 atm and 335 K is [JIPMER-97]
(A) 501 (B) 601 (C) 551 (D) 451

66. When the same amount of Al is treated with an excess of HCl and NaOH the volumes of hydrogen evolved are in the ratio of [JIPMER-97]
(A) 1 : 1 (B) 2 : 3 (C) 1 : 2 (D) 2 : 1

67. The inversion temperature for a van der Waal's gas is given by [JIPMER-97]
(A) $Ti = \frac{2a}{Rb}$ (B) $Ti = \frac{a}{Rb}$
(C) $Ti = \frac{a}{2Rb}$ (D) Ti = 0.5 T Boyle

68. Which of the following statement on the kinetic theory of matter is incorrect ? [JIPMER-97]
(A) The velocity of the particle is proportional to the pressure of the particles.
(B) The collisions between the particles are elastic
(C) The velocity of the particle is proportional to the square root of the temperature.
(D) The collisions between the particles and the walls of the vessels are elastic.

69. A vessel is filled with a mixture of oxygen and nitrogen. At what ratio of the partial pressures will the masses of the gases be identical ? [JIPMER-98]
(A) $pO_2 = pN_2$ (B) $pO_2 = 0.5\ pN_2$
(C) $pO_2 = 0.875\ pN_2$ (D) $pO_2 = 1.14\ pN_2$

70. 1 litre of a gas is at pressure of 10^{-6} mm of mercury at 25°C. How many molecules are present in the vessel ? [JIPMER-98]
(A) 3.2×10^6 (B) 3.2×10^{10} (C) 3.2×10^{13} (D) 3.2×10^4

71. According to kinetic theory [JIPMER-98]
(A) There are intermolecular attractions
(B) There are no intermolecular attractions
(C) The velocity of the molecules decreases with each collision
(D) Molecules have considerable volume.

72. The statement, "If 0.003 moles of a gas are dissolved in 900g of water under a pressure of 1 atmosphere, 0.006 moles will be dissolved under a pressure of 2 atmospheres", illustrates [JIPMER-98]
(A) Dalton's law of partial pressure
(B) Graham's law (C) Raoult's (D) Henry's law

73. The temperature at which a real gas obeys the ideal gas laws over a wide range of pressure is called [JIPMER-99]
(A) Critical temperature (B) Inversion temperature
(C) Reduced temperature (D) Boyle's temperature

74. Joule-Thomson Expansion of an ideal gas produces. [JIPMER-99]

(A) No change in temperature (B) Heating (C) Cooling
(D) Heating above the inversion temperature

75. V versus T curves at constant pressure P_1 and P_2 for an ideal gas are shown in figure. Which is correct [JIPMER-2000]

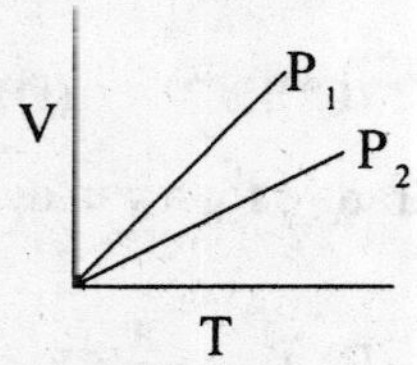

(A) $P_1<P_2$ (B) $P_1>P_2$ (C) $P_1=P_2$ (D) All

76. The volume of a gas is 100 ml at 100°C. If the pressure is held constant, at what temperature will the same have a volume of 200 ml? [JIPMER-2000]
(A) 50°C (B) 446°C (C) 473°C (D) 200°C

77. A and B are two identical vessels. A contains 15g of ethane at 298 K and 1 atm. The vessel B contains 75 g gas X_2 at the same temperature and pressure. The vapour density of X_2 is [JIPMER-2001]
(A) 300 (B) 35.5 (C) 150 (D) 75

78. Two vessels having equal volume contain molecular hydrogen at one atmosphere and helium at two atmosphere pressure respectively. If both samples are at the same temperature the mean velocity of hydrogen molecule is [JIPMER-2001]
(A) Half that of helium (B) Twice that of helium
(C) Equal to that of helium (D) $\sqrt{2}$ times that of helium

79. Rate of diffusion of gas X is twice the gas Y. If molecular weight of X is 8 what is the molecular weight of Y? [JIPMER-2002]
(A) 32 (B) 16 (C) 8 (D) None of these

80. Which of the following statements is wrong for gases? [JIPMER-2002]
(A) Gases donot have a definite shape and volume.
(B) Volume of gas is equal to volume of container confining the gas.
(C) Confined gas exerts uniform pressure on the walls of its container in all direction.
(D) Mass of gas cannot be determined by weighing a container in which it is enclosed.

81. The rate of diffusion of hydrogen gas is [BHU-96]
(A) same the He gas (B) 1.4 times to He gas
(C) two times to He gas (D) five times to he gas

82. The partial pressure of hydrogen in a flask containing 2g hydrogen 16g oxygen and 14g nitrogen is [BHU-96]
(A) Equal to total pressure (B) 1/8th of the total pressure
(C) Half of the total pressure (D) 1/3 of the total pressure

83. A closed vessel contains equal number of oxygen and hydrogen molecules at a total pressure of 740mm. If oxygen is removed from the system, the pressure will [BHU-96]
(A) Become 1/4th of 740 mm (B) Become half 740 mm
(C) Remain unchanged
(D) Become two times of 740 mm

84. The smell of agarbatti, mixed in the atmosphere, is due to [BHU-96]
(A) Osmosis (B) Diffusion
(C) Evaporation (D) Condensation

85. The number of molecules present in 1 dm^3 of hydrogen at S.T.P. is [BHU-96]
(A) 3.69×10^{22} (B) 2.69×10^{22}
(C) 1.69×10^{22} (D) 1.09×10^{22}

86. Four litres of CO_2 is kept at 27°C. What should be the volume if the temperature is lowered to 150 K at the same pressure ? [BHU-96]
(A) 0.5 litre (B) 1 litre (C) 2 litres (D) 4 litres

87. Volume of a gas at N.T.P. is 1.12×10^{-7} c.c. Calculate the number of molecules in it. [BHU-97]
(A) 30.1×10^{23} (B) 3.01×10^{20}
(C) 3.01×10^{12} (D) 3.01×10^{24}

88. A container has hydrogen and oxygen mixture in ratio of 4:1 by weight, then [BHU-98]
(A) Internal energy of the gas decreases
(B) Internal energy increases
(C) Entropy of the gases increases
(D) Entropy of the gases decreases.

89. A container contains 1 mole of a gas at 1 atm pressure and 27°C temperature and volume is 24.6 lit, if pressure is 10 atm and temperature 327°C, then new volume is approximately [BHU-98]
(A) 2 litre (B) 48 litre (C) 15 litre (D) 10 litre

90. A gas at constant temperature and pressure expands, its [BHU-98]
(A) Internal energy remains same
(B) Internal energy decreases
(C) Internal energy increases
(D) Its entropy increases and then decreases

91. 0.24 g of volatile gas, upon vaporisation, gives 45 ml vapour at STP. What will be the vapour density of the substance ? (Density of H_2 = 0.089) [BHU-99]
(A) 5.992 (B) 59.93 (C) 95.39 (D) 95.93

92. A certain mass of gas occupies a volume of 300 c.c. at 27°C and 620 mm pressure. The volume of this gas at 47°C and 640 mm pressure will be [BHU-99]
(A) 310 c.c. (B) 350 c.c. (C) 400 c.c. (D) 510 c.c.

93. The average kinetic energy of an ideal gas, per mol-

ecule in S.I. units, at 25°C will be [BHU-99]
(A) 61.7 x 10^{-21}J (B) 6.17 x 10^{-21} J
(C) 6.17 x 10^{-20} J (D) 7.16 x 10^{-20} J

94. If a gas at constant temperature and pressure expands, then its [BHU-99]
(A) Internal energy decreases
(B) Internal energy remains the same
(C) Internal energy increases
(D) Entropy increases and then decreases

95. Densities of two gases are in ratio 1:2 and their temperatures are in ratio 2:1, then ratio of their respective pressure is [BHU-2000]
(A) 1 : 1 (B) 1 : 2 (C) 2 : 1 (D) 4 : 1

96. The average speed of an ideal gas molecule at 27°C is 0.3 ms^{-1}. Calculate the average speed at 927°C. [BHU-2001]
(A) 0.4ms^{-1} (B) 0.6 ms^{-1} (C) 0.9 ms^{-1} (D) 0.2 ms^{-1}

97. Mixing of two gases by diffusion is [AFMC-91]
(A) Reversible (B) Irreversible
(C) Exothermic (D) Endothermic

98. van der Waal's forces are applied to [AFMC-92]
(A) Inert gases (B) Real gases
(C) Mixture of gases (D) Elementary gases

99. A gas of volume 100cc is kept in a vessel at pressure 104P maintained at temperature 24°C. If now the pressure is increase to 105P, keeping the temperature constant, then the volume of gas will become [AFMC-92]
(A) 10 cc (B) 100 cc (C) 10 cc (D) 1000 cc

100. The van der Waal's equation is for [AFMC-93]
(A) Ideal gases (B) Real gases
(C) Mixture of gases (D) Inert gases

101. A gas is initially at 1 atm. pressure. To compress it to 1/4th of its initial volume, pressure to be applied is [AFMC-93]
(A) 1 atm. (B) 2 atm. (C) 4 atm. (D) 1/4 atm.

102. The temperature at which a real gas behaves as an ideal gas for a wide range of pressure is [AFMC-93]
(A) Inversion temperature (B) Boyle temperature
(C) Critical temperature (D)Reduction temperature

103. The temperature at which the second virial coefficient of a real gas is zero is called [AFMC-93]
(A) Crticial temperature (B) Eutectic point
(C) Boiling point (D) Boyle's temperature

104. If the rate of diffusion of A is 5 times that of B, what will the density ratio A and B ? [AFMC-94]
(A) 1/25 (B) 1/5 (C) 25 (D) 5

105. Volume occupied by 4.4g CO_2 is [AFMC-97]
(A) 2.24 litres (B) 22.4 litres
(C) 224 litres (D) None of these

106. If the volume of 2 moles of an ideal gas at 540 K is 44.8 litre, then its pressure will be [AFMC-98]
(A) 1 atmosphere (B) 2 atmosphere
(C) 3 atmosphere (D) 4 atmosphere

107. The volume of a gas measured at 27°C and 1 atm pressure is 10 litres. What final temperature would be required to reduce the volume to 5 litres at 1 atm pressure ? [AFMC-99]
(A) 150 K (B) 327 K (C) 450 K (D) 100°C

108. PV/T = constant. A real gas will behave as ideal gas when [AFMC-2000]
(A) Both pressure and temperature are high
(B) Both pressure and temperature are low
(C) Temperature is high and pressure is low
(D) Pressure is high and temperature is low

109. The temperature below which a gas does not exist is called its [AFMC-2001]
(A) Inversion temperature (B) Critical temperature
(C) Neutral temperature (D) Curie point

110. Internal energy and pressure of a gas per unit volume are related as [AFMC-2002]
(A) P = 2/3 E (B) P = 3/2 E (C) P = 1/2 E (D) P = 2E

111. Value of universal gas constant depends upon [AFMC-2002]
(A) Temperature of the gas (B) Volume of the gas
(C) Number of moles of the gas (D) None of these

112. The rate of diffusion of methane at a given temperature is twice that of a gas X. The molecular weight of X is : [IIT-90]
(A) 64.0 (B) 32.0 (C) 4.0 (D) 8.0.

113. According to the kinetic theory of gases, for a diatomic molecule [IIT-91]
(A) the pressure exerted by the gas is proportional to the mean velocity of the molecule.
(B) the pressure exerted by the gas is proportional to the root mean square velocity of the molecule.
(C) the root mean square velocity of the molecule is inversely proportional to the temperature.
(D) the mean translational kinetic energy of the molecule is proportional to the absolute temperature.

114. At constant volume, for a fixed number of moles of a gas the pressure of the gas increases with rise of temperature due to : [IIT-92]
(A) increase in average molecular speed
(B) increased rate of collisions amongst molecules
(C) increase in molecular attraction
(D) decrease in mean free path.

115. Equal weights of ethane and hydrogen are mixed in an empty container at 25°C. The fraction of total pressure exerted by hydrogen is : [IIT-92]
(A) 1:2 (B) 1:1 (C) 1:16 (D) 15:16

116. The ratio between root mean square velocity of H_2 at 50 K and that of O_2 at 800 K is [IIT-96]
(A) 4 (B) 2 (C) 1 (D) 1/4

117. x ml of H_2 gas effuses through a hole in a container in 5 seconds. The time taken for the effusion of the same volume of the gas specified under identical conditions is [IIT-96]
(A) 10 seconds : He (B) 20 seconds : O_2
(C) 25 seconds : CO (D) 55 seconds : CO_2

118. One mole of N_2O_4(g) at 300 K is kept in a closed container under one atmosphere. It is heated to 600 K when 20% by mass of N_2O_4(g) decomposes to NO_2(g). The resultant pressure is [IIT-96]
(A) 1.2 atm (B) 2.4 atm (C) 2.0 atm (D)1.0 atm

119. The compressibility factor for an ideal gas is [IIT-97]
(A) 1.5 (B) 1.0 (C) 2.0 (D) ∞

120. The absolute temperature of an ideal gas is ....to/ than the average kinetic energy of the gas molecules. [IIT-97]

121. According to Graham's law , at a given temperature the rates of diffusion r_A/r_B of gases A and B is given by [IIT-98]
(A) $(P_A/P_B)(M_A/M_B)^{1/2}$ (B) $(M_A/M_B)(P_A/P_B)^{1/2}$
(C) $(P_A/P_B)(M_B/M_A)^{1/2}$ (D) $(M_A/M_B)(P_B/P_A)^{1/2}$
(Where P and M are pressures and molecular weights of gases A and B respectively)

122. The following statement (s) is (are) correct [IIT-99]
(A) A plot of log Kp versus I/T is linear
(B) A plot of log [X] versus time is linear for a first order reaction, $X \rightarrow P$
(C) A plot of log P versus I/T is linear at constant volume
(D) A plot of P versus I/V is linear at constant temperature

123. The rms velocity of hydrogen is √7 times the rms velocity of nitrogen. If T is the temperature of the gas, [IIT-2000]
(A) $T(H_2) = T(N_2)$ (B) $T(H_2) > T(N_2)$
(C) $T(H_2) < T(N_2)$ (D) $T(H_2) - \sqrt{7}\,T(N_2)$

124. The compressibility of a gas is less than unity at STP. Therefore [IIT-2000]
(A) $V_m > 22.4$ litres (B) $V_m < 22.4$ litres
(C) $V_m = 22.4$ litres (D) $V_m = 44.8$ litres

125. A : The heat absorbed during the isothermal expansion of an ideal gas against vacuum is zero.
R : The volume occupied by the molecules of an ideal gas is zero. [IIT-2000]
(A) A (B) B (C) C (D) D

126. A : The pressure of a fixed amount of an ideal gas is proportional to its temperature. [IIT-2000]
R : Frequency of collisions and their impact both increase in proportion to the square root of temperature
(A) A (B) B (C) C (D) D

127. The root mean square velocity of an ideal gas at constant pressure varies with density (d) as [IIT-2001]
(A) d² (B) d (C) √d (D) 1/√d

128. Which of the following volume(V)- temperature (T) plots represents the behaviour of one mole of an ideal gas at one atmospheric pressure ? [IIT-2002]

(A) 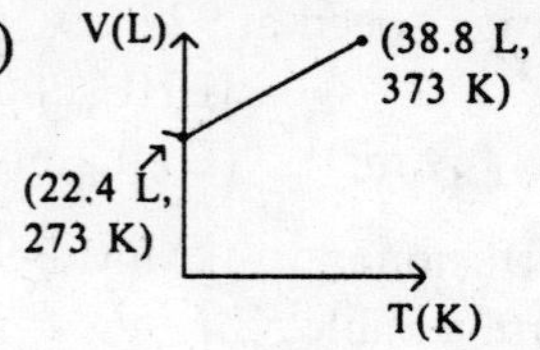

(B) 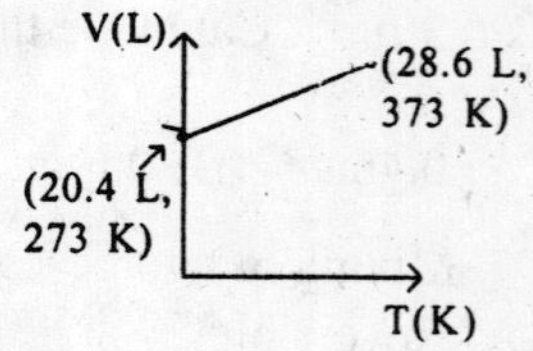

(C) 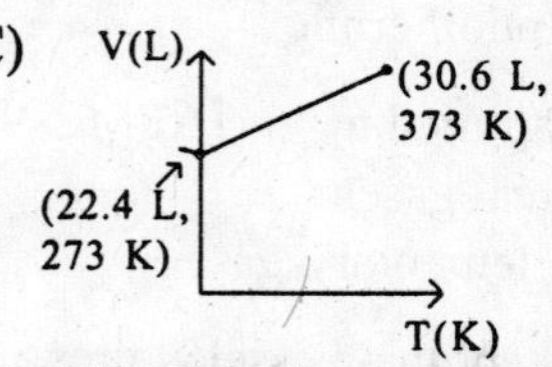

(D) 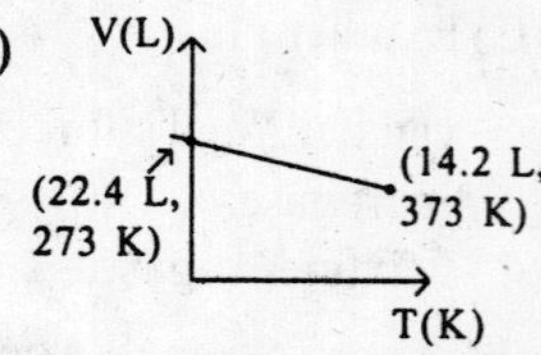

129. Which is the correct for real gases ? [IIT-2003]
(A) Shape of the gas molecule is indefinite $\frac{PV}{nRT} = 1$
(B) Shape of the gas molecule is indefinite $\frac{PV}{nRT} > 1$
(C) Gases have definite size of molecules for real gases $\frac{PV}{nRT} > 1$
(D) None of these

130. Kinetic theory of gases proves [AIEEE-2002]
(A) Only Boyle's law (B) Only Charle's law
(C) Only Avogadro's law (D) All of these

131. For an ideal gas, number of moles per liter in terms of its pressure P, gas constant R and temperature T is [AIEEE-2002]
(A) PT/R (B) PRT (C) P/RT (D) RT/P

132. According to the kinetic theory of gases, in an ideal gas, between two successive collisions a gas molecule travels. [AIEEE-2003]
(A) In a circular path (B) In a wavy path
(C) In a straight line path
(D) With an accelerated velocity

133. What volume of hydrogen gas, at 273 K and 1 atm. pressure will be consumed in obtaining 21.6 g of elemental boron (atomic mass=10.8) from the reduction of boron trichloride by hydrogen ? [AIEEE-2003]

(A) 89.6 L (B) 67.2 L (C) 44.8 L (D) 22.4 L

134. In which of the following pairs the gaseous species diffuse through a porous plug with the same rate of diffusion : [EAMCET-90]
(A) N_2, CO (B) NO, CO_2 (C) NH_3, PH_3 (D) NO, C_2H_6

135. The Root Mean square velocity at NTP of the species can not be calculated from the expression [EAMCET-90]
(A) $\sqrt{3p/d}$ (B) $\sqrt{3PV/M}$ (C) $\sqrt{3RT/M}$ (D) $\sqrt{3RT/d}$

136. The molecular weight of a gas which diffuses through a porous plug at 1/6 th of the speed of hydrogen under identical condition is .... [EAMCET-90]

137. Which of the following gases contain the same number of molecules as that of 16 grams of Oxygen : [EAMCET-91]
(A) 16 grams of O_3 (B) 32 grams of SO_2
(C) 16 grams of SO_2 (D) All

138. A gaseous mixture contains 56 grams of N_2, 44 grams CO_2 and 16 grams of CH_4. The total pressure of the mixture is 720 mm Hg. The partial pressure of CH_4 is : [EAMCET-91]
(A) 180 mm (B) 360 mm (C) 540 mm (D)720 mm

139. The R.M.S. velocity of an ideal gas at 27°C is 0.3 m/sec. Its R.M.S. velocity at 927°C is (in m/sec.) [EAMCET-91]
(A) 3.6 (B) 2.4 (C) 0.9 (D) 0.6

140. The rate of diffusion of He is .... the rate of diffusion of CH_4. [EAMCET-91]

141. If the pressure and the absolute temperature of given mass of gas are doubled, the new volume will be .... the initial volume. [EAMCET-91]

142. The K.E. of 8 grams of H_2 at 27°C is .... times more than the K.E. of 8grams of O_2 at the same temperature. [EAMCET-91]

143. A sample of gas has a volume of 0.2 lit. measured at 1 atm. pressure and 0°C. At the same pressure, but at 273°C its volume will become [EAMCET-92]
(A) 0.1 litre (B) 0.4 litre (C) 0.8 litre (D) 0.6 litre

144. For an ideal gas, the relation between the enthalpy change at constant temperature is given by [EAMCET-92]
(A) H = E + PV (B) $\Delta H = \Delta E + \Delta n\,(RT)$
(C) $\Delta H = \Delta E + P\,(\Delta V)$ (D) $\Delta H = \Delta G + T\Delta S$

145. At 27°C the ratio of root mean squre velocities of Ozone to Oxygen is [EAMCET-92]
(A) $\sqrt{3/5}$ (B) $\sqrt{4/3}$ (C) $\sqrt{2/3}$ (D) 0.25

146. A certain sample of gas has a volume of 0.2 litres measured at 1 atm. pressure and 0°C. At the same pressure but at 273°C, its volume will be [EAMCET-93]
(A) 0.4 litres (B) 0.8 litres (C) 27.8 litres (D) 55.6 litres

147. A flask contains 3g of H_2 and 24g of O_2. The partial pressure of hydrogen in the mixture is [EAMCET-94]
(A) 1/8 th the total pressure (B) 1/9th of the total pressure
(C) 1/2 of the total pressure (D) 2/3 of the total pressure

148. The molecular weight of a gas that diffuses twice as rapidly as a gas with molecular weight 64 is [EAMCET-94]
(A) 5.4 (B) 8 (C) 128 (D) 16

149. For an ideal gas a plot of $\frac{PV}{RT}$ vs T will look like [EAMCET-95]

(A)
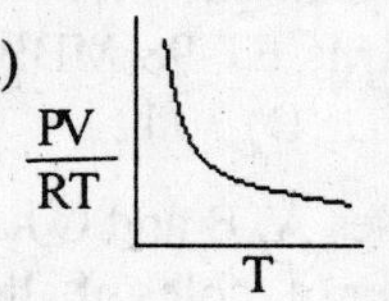

(B)
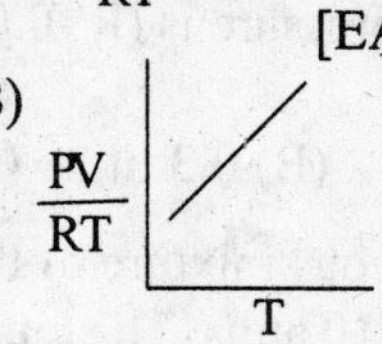

(C)
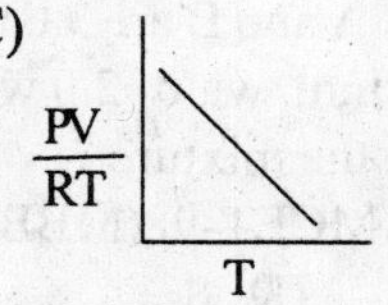

(D)
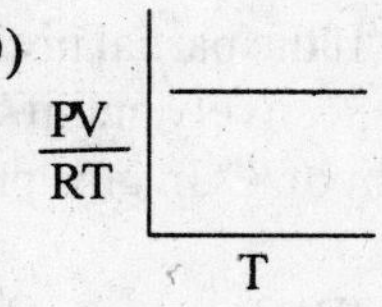

150. Which of the following is a representation of Gay-Lusac's law ? [EAMCET-95]
(A) $P_1T_1 = P_2T_2$ (B) $P_1T_2 = P_2T_1$
(C) $P_1V_1 = P_2V_2$ (D) $V_1T_2 = V_2T_1$

151. Equal masses of methane and oxygen are introduced into a vessel at 27°C. What fraction of the total pressure is due to Oxygen ? [EAMCET-95]
(A) 1/10 (B) 1/2 (C) 1/3 (D) 1/4

152. A given mass of gas obeys Boyle's law at certain temperature. Which one of the plots will not give a straight line ? [EAMCET-95]
(A) V vs 1/P (B) P vs 1/V
(C) log P vs log V (D) P vs V

153. At STP the density of a gas (molecular weight 45) in g/l is [EAMCET-96]
(A) 2 (B) 22.4 (C) 11.2 (D) 1000

154. Gases show ideal behaviour at [EAMCET-96]
(A) High pressure and High temperature
(B) Low pressure and High temperature
(C) Low pressure and Low temperature
(D) High pressure and Low temperature

155. The volume in litres of CO_2 liberated at STP when 10g of 90% pure lime stone is heated completely is [EAMCET-96]
(A) 2.016 (B) 20.16 (C) 2.24 (D) 22.4

156. The root mean square velocity of a gas is doubled when the temp. is [EAMCET-97]
(A) Reduced to half (B) Reduced to one-forth
(C) Increased four times (D) Increased two times

157. I mole of 'Ar' gas will have least density at [EAMCET-98]
(A) STP (B) 0°C & 2 atm.
(C) 273°C & 2 atm. (D) 273°C & 1 atm.

158. The volume in litres of CO_2 liberated at STP when 10 grams of 90% pure limestone is heated completely is [EAMCET-98(MBBS)]
(A) 2.016 (B) 20.16 (C) 2.24 (D) 22.4

159. The volume of 2.8g of carbon monoxide at 27°C and 0.821atm. pressure is (R=0.0821 lit. atm. $mol^{-1}k^{-1}$) [EAMCET-98(MBBS)]
(A) 1.5 lit (B) 0.3 lit (C) 3 lit (D) 30 lit

160. A gaseous mixture of three gases A, B and C has a pressure of 10 atms. The total number of moles of all the gases is 10. If the partial pressures of A and B are 3.0 and 1.0 atm. respectively and if 'C' has a mol. wt. of 2.0 what is the weight of 'C' in gms present in the mixture ? [EAMCET-98(MBBS)]
(A) 6 (B) 3 (C) 12 (D) 8

161. A mixture contains 16 grams of oxygen, 28 grams of nitrogen and 8 grams of methane. Total pressure of the mixture is 740 mm. What is the partial pressure of nitrogen in mm ? [EAMCET-99]
(A) 185 (B) 370 (C) 555 (D) 740

162. A vessel contains equal volumes of SO_2 and CH_4. Through a small hole, the gases effused into vacuum. After 200 seconds, the total volume is reduced to half. What is the ratio of SO_2 and CH_4 remaining in the vessel? [EAMCET-99(MBBS)]
(A) 1:2 (B) 2:1 (C) 1:1 (D) 1:4

163. The r.m.s. velocity of a gas at a particular temperature is found to be 12,240 cm. sec^{-1} . What is its most probable velocity (in cm.sec^{-1}) at the same temperature? [EAMCET-99(MBBS)]
(A) 10^3 (B) 10^4 (C) 11280 (D) 1128

164. At 27°C one mole of an ideal gas exerted a pressure of 0.821 atmospheres. What is its volume in litres ? (R = 0.0821 lit. atm. $mol^{-1}K^{-1}$) [EAMCET-99(MBBS)]
(A) 300 (B) 30 (C) 0.3 (D) 3

165. The total pressure of a mixture of 6.4 grams of oxygen and 5.6 grams of nitrogen present in a 2 litre vessel is 1200 mm. What is partial pressure (in mm) of nitrogen ? [EAMCET-2000]
(A) 1200 (B) 600 (C) 900 (D) 200

166. The r.m.s. velocity of an ideal gas at 300 K is 12240 cm. sec^{-1}.What is its most probable velocity (in cm.sec^{-1}) at the same temperature? [EAMCET-2000]
(A) 10000 (B) 11280 (C) 1000 (D) 12240

167. What is the volume (in lit) of carbon dioxide liberated at STP, when 2.12 grams of sodium carbonate (mol. wt. = 106) is treated with excess dilute HCl ? [EAMCET-2000]
(A) 2.28 (B) 0.448 (C) 44.8 (D) 22.4

168. The volume of 10 moles of an ideal gas is 1 litre at 27°C and 1 atm. pressure. What is the volume (in lit) of 20 moles of same gas at same temperature and pressure? [EAMCET(MBBS)-2000]
(A) 2 (B) 4 (C) 1 (D) 8

169. What is the value of gas constant R in J mol^{-1} K^{-1}? [EAMCET(MBBS)-2000]
(A) 82.1 (B) 8.314 x 10^2 (C) 8.314 (D) 0.0821

170. Heat of combustion of carbon is 96 Kcal. When some quantity of carbon is burnt in oxygen, 48 Kcal of heat is liberated. What is the volume (in lit) of oxygen at STP reacted with this carbon ? [EAMCET(MBBS)-2000]
(A) 48 (B) 22.4 (C) 11.2 (D) 1

171. What is the volume (in litres) of oxygen at STP required for complete combustion of 32 grams of CH_4? (molecular weight of CH_4 is 16). [EAMCET-2001]
(A) 44.8 (B) 89.6 (C) 22.4 (D) 179.2

172. The r.m.s. velocity of CO_2 at a temperature T (in Kelvin) is x cm.sec^{-1}. At what temperature (in Kelvin), the r.m.s. velocity of nitrous oxide would be 4x cm.sec^{-1} ? [EAMCET-2001]
(A) 16T (B) 2T (C) 4T (D) 32T

173. 'n' mole of an ideal gas at temperature T (in Kelvin) occupy "V" litres of volume, exerting a pressure of 'P' atmospheres. What is its concentration (in mol. lit^{-1}) ? (Atomic weights of C, N and O are respectively 12, 14 and 16) (R-gas constant) [EAMCET-2001]
(A) P/RT (B) PT/R (C) RT/P (D) R/PT

174. How many litres of oxygen (at STP) are required for complete combustion of 39 grams of liquid benzene (Atomic weights : C = 12; H = 1;O = 16) [EAMCET-2001]
(A) 84 (B) 22.4 (C) 42 (D) 11.2

175. What are the conditions under which the relation between volume (V) and number of moles (n) of gas plotted? (P=pressure; T= temperature) [EAMCET(MBBS)-2001]
(A) Constant P & T (B) Constant T & V
(C) Constant P & V (D) Constant n & V

176. 7.5 grams of a gas occupy 5.6 litres of volume at STP. The gas is (Atomic weights of C, N and O are 12, 14 and 16 respectively) [EAMCET(MBBS)-2001]
(A) NO (B) N_2O (C) CO (D) CO_2

177. 'X' litres of carbon monoxide is present at STP. It is completely oxidized to CO_2. The volume of CO_2 formed is 11.207 litres at S.T.P. What is the value of 'X' in litres ? [EAMCET-2002]
(A) 22.414 (B) 11.207 (C) 5.6035 (D) 44.828

178. 4 grams of an ideal gas occupies 5.6035 litres of volume at 546 K and 2 atm. pressure. What is its molecular weight? [EAMCET-2002]
(A) 4 (B) 16 (C) 32 (D) 64

179. At 27°C, a closed vessel contains a mixture of equal weights of helium (mol. wt, = 4), methane (mol. wt = 16) and sulphur dioxide (mol. wt = 64). The pressure exerted by the mixture is 210 mm. If the partial pressures of helium, methane and sulphur dioxide are P_1, P_2 and P_3 respectively, which one of the following is correct ? [EAMCET-2002]
(A) $P_3 > P_2 > P_1$ (B) $P_1 > P_2 > P_3$
(C) $P_1 > P_3 > P_2$ (D) $P_2 > P_3 > P_1$

180. Two grams of sulphur is completely burnt in oxygen to from SO_2. In this reaction what is the volume (in litres) of oxygen consumed at STP ? (At. wts of sulphur and oxygen are 32 and 16 respectively). [EAMCET(med.)-2002]
(A) $\frac{16}{22.414}$ (B) $\frac{22.414}{16}$ (C) $\frac{22.414}{32}$ (D) $\frac{32}{22.414}$

181. If the r.m.s. velocity of a gas at 100K is 10^4 cm.sec^{-1}, what is the temperature (in°C) at which the r.m.s. velocity will be 3 x 10^4 cm sec^{-1} ? [EAMCET(med.)-2002]
(A) 900 (B) 627 (C) 327 (D) 1217

182. What is the volume (in litres) of oxygen required at STP to completely convert 1.5 moles of sulphur into sulphur dioxide? [EAMCET-2003]
(A) 11.2 (B) 22.11 (C) 33.6 (D) 44.8

183. If a gas contains only three molecules that move with velocities of 100, 200, 500 ms^{-1}, what is the r.m.s. velocity of that gas in ms^{-1}. [EAMCET-2003]
(A) $100\sqrt{\frac{8}{3}}$ (B) $100\sqrt{30}$ (C) $100\sqrt{10}$ (D) $\frac{800}{3}$

184. Same mass of CH_4, H_2 is taken in a container, the partial pressure caused by H_2 is [CPMT-96]
(A) $\frac{8}{9}$ (B) $\frac{1}{9}$ (C) $\frac{1}{2}$ (D) 1

185. Gas deviates from ideal gas nature because molecules [CPMT-96]
(A) Are colourless (B) Attract each other
(C) Contain covalent bond (D) Show Brownian motion

186. NaCl and CH_3COOH kept in separate containers. If pressure is respectively P_1 and P_2, then what is the correct statement ? [CPMT-96]
(A) $P_1 > P_2$ (B) $P_1 = P_2$ (C) $P_1 < P_2$ (D) $P_1 = P_2 = 0$ atm

187. Density of methane at 25°C and 6 atm, pressure is (Given R = 0.0821 atm) - [CPMT-97]
(A) 4 gm/l (B) 8 gm/l (C) 12 gm/l (D) 16 gm/l

188. Density ratio of O_2 and H_2 is 16:1. The ratio of its V_{rms} is - [CPMT-97]
(A) 4 : 1 (B) 16 : 1 (C) 1 : 4 (D) 1 : 16

189. Value of universal gas constant 'R' depends upon [CPMT-99]
(A) Temperature of the gas (B) Volume of gas
(C) Number of moles of gas (D) None of these

190. The rate of diffusion of a gas having molecular weight just double of nitrogen gas, is 56 ml s^{-1}. The rate of diffusion of nitrogen will be - [CPMT-2000]
(A) 79.19 ml s^{-2} (B) 56 ml s^{-1}
(C) 112.0 ml s^{-1} (D) 90.00 ml s^{-1}

191. An ideal gas will have maximum density when - [CPMT-2000]
(A) P=1 atm, T=300 K (B) P=2 atm, T=150K
(C) P=0.5 atm, T=600K (D) P=1.0 atm, t=500K

192. The rate of diffusion of a gas is twice that of the other. If first gas has molecular weight 16, the molecular weight of the second is [CPMT-2001]
(A) 8 (B) 32 (C)16 (D) 64

193. A flask contains 36% nitrogen, 39% oxygen and 26% hydrogen. If the total pressure of the mixture of gases is 500 mm Hg, the partial pressure of oxygen is [CPMT-2001]
(A) 120 mm (B) 145 mm (C) 195 mm (D) 210 mm

194. Increasing order of r.m.s. velocities of H_2, O_2, N_2 and HBr is [CPMT-2002]
(A) H_2, O_2, N_2, HBr (B) HBr, O_2, N_2, H_2
(C) H_2, N_2, O_2, HBr (D) HBr, N_2, O_2, H_2

195. At what temperature will the volume of gas become 2x , if volume of this gas is x at 0°C at constant pressure ? [CPMT-2003]
(A) 0°C (B) 237°C (C) 546°C (D) 546 K

196. For and ideal gas the value of $\left(\frac{\delta E}{\delta V}\right)_T$ is [CFMT-93]
(A) Positive (B) Zero (C) Negative (D) Interchangeable

197. 2.0 litres of air formed 1.715 l ozonised air when passed through Brodiel apparatus. The volume of ozone formed is [CBSE-90]

(A) 85.0ml (B) 170.0ml (C) 225.0ml. (D) 340.0ml

198. Root mean square velocity of a gas molecule is proportional to [CBSE-90]
(A) $m^{1/2}$ (B) m^o (C) $m^{-1/2}$ (D) m

199. The vapour density of ozone would be [CBSE-90]
(A) 24 (B) 96 (C) 48 (D) 16

200. The rate of diffusion of nitrogen gas in a diffusion tube is 9 ml sec^{-1}. An unknown gas X diffuses at a rate of 6 ml sec^{-1} in the same diffusion tube. The molecular weight of X is .... gm mol^{-1}. [CBSE-97]

201. What is the numerical value of gas contant R, in J $mol^{-1}K^{-1}$? [CBSE-97]
(A) 0.0821 (B) 8.314 x 10^7 (C) 8.314 (D) 1.987

202. 'X' litrues of carbon monoxide is present at STP. It is completely oxidized to CO_2. The volume of CO_2 formed is 11.207 litres at S.T.P. What is the value of 'X' in litres ? [CBSE-2003]
(A) 22.414 (B) 11.207 (C) 5.6035 (D) 44.828

203. Which of the following is incorrect? [BHU-2003]
(A) Density of liquid = density of vapour at critical temperature
(B) $(dE/dV)_T = 0$ for an ideal gas
(C) $(dE/dV)_T \neq a/v^2$ for a real gas
(D) $(dE/dV)_T = a/2$ for a real gas.

204. Process of converting vapour into liquid is known as [AFMC-2003]
(A) Condensation (B) Vapourisation
(C) Freezing (D) Melting

205. The maximum number of molecules is present in [CBSE-PMT-2004]
(1) 15 L of H_2 gas at STP (2) 5 L of N_2 gas at STP
(3) 0.5 g of H_2 gas (4) 10 g of O_2 gas

206. An ideal monoatomic gas has molecular mass M and E is the average kinetic energy per mole, then which of the following relation holds true? [IIT-2004]

(A) $V_{rms} = \left(\frac{2E}{M}\right)^{\frac{1}{2}}$ (B) $V_{rms} = \left(\frac{M}{2E}\right)^{\frac{1}{2}}$

(C) $V_{rms} = \left(\frac{2}{3}\frac{M}{E}\right)^{\frac{1}{2}}$ (D) $V_{rms} = \left(\frac{3}{2}\frac{E}{M}\right)^{\frac{1}{2}}$

Gaseous State

1.	(C)	51.	(C)	101.	(C)	151.	(C)	201.	(C)
2.	(B)	52.	(A)	102.	(B)	152.	(A)	202.	(B)
3.	(A)	53.	(C)	103.	(D)	153.	(A)	203.	(D)
4.	(A)	54.	(D)	104.	(A)	154.	(B)	204.	(A)
5.	(A)	55.	(B)	105.	(A)	155.	(A)	205.	(A)
6.	(C)	56.	(D)	106.	(B)	156.	(C)	206.	(A)
7.	(B)	57.	(D)	107.	(A)	157.	(D)		
8.	(B)	58.	(C)	108.	(C)	158.	(A)		
9.	(A)	59.	(B)	109.	(B)	159.	(C)		
10.	(A)	60.	(D)	110.	(A)	160.	(C)		
11.	(A)	61.	(B)	111.	(D)	161.	(B)		
12.	(A)	62.	(C)	112.	(A)	162.	(B)		
13.	(A)	63.	(D)	113.	(D)	163.	(B)		
14.	(A)	64.	(C)	114.	(A)	164.	(B)		
15.	(C)	65.	(C)	115.	(D)	165.	(B)		
16.	(B)	66.	(A)	116.	(C)	166.	(A)		
17.	(D)	67.	(A)	117.	(B)	167.	(B)		
18.	(C)	68.	(A)	118.	(B)	168.	(A)		
19.	(B)	69.	(C)	119.	(B)	169.	(C)		
20.	(C)	70.	(C)	120.	*	170.	(C)		
21.	(B)	71.	(B)	121.	(C)	171.	(B)		
22.	(D)	72.	(D)	122.	(A)(B)(D)	172.	(A)		
23.	(A)	73.	(D)	123.	(C)	173.	(A)		
24.	(D)	74.	(A)	124.	(B)	174.	(A)		
25.	(B)	75.	(A)	125.	(D)	175.	(A)		
26.	(A)	76.	(C)	126.	(B)	176.	(A)		
27.	(D)	77.	(D)	127.	(D)	177.	(B)		
28.	(D)	78.	(D)	128.	(C)	178.	(B)		
29.	(A)	79.	(A)	129.	(C)	179.	(B)		
30.	(B)	80.	(D)	130.	(D)	180.	(B)		
31.	(B)	81.	(B)	131.	(C)	181.	(B)		
32.	(B)	82.	(C)	132.	(C)	182.	(C)		
33.	(B)	83.	(B)	133.	(B)	183.	(C)		
34.	(D)	84.	(B)	134.	(D)	184.	(A)		
35.	(A)	85.	(B)	135.	(D)	185.	(B)		
36.	(A)	86.	(C)	136.	*	186.	(C)		
37.	(A)	87.	(B)	137.	(B)	187.	(A)		
38.	(C)	88.	(C)	138.	(A)	188.	(C)		
39.	(C)	89.	(C)	139.	(D)	189.	(D)		
40.	(C)	90.	(A)	140.	*	190.	(A)		
41.	(A)	91.	(B)	141.	*	191.	(B)		
42.	(A)	92.	(A)	142.	*	192.	(A)		
43.	(D)	93.	(B)	143.	(B)	193.	(D)		
44.	(B)	94.	(B)	144.	(B)	194.	(B)		
45.	(B)	95.	(A)	145.	(C)	195.	(D)		
46.	(A)	96.	(B)	146.	(A)	196.	(B)		
47.	(B)	97.	(B)	147.	(D)	197.	(C)		
48.	(B)	98.	(C)	148.	(D)	198.	(C)		
49.	(D)	99.	(C)	149.	(D)	199.	(A)		
50.	(D)	100.	(B)	150.	(B)	200.	*		

120. less 140. Twice 142. Same
136. 72 141. Same as 200. 63 gm/mol.

SOLUTIONS

Modes of expressing concentration of solutions -

Molarity : Number of moles of solute present in one litre of the solution.

Molality : Number of moles of solute present in 1000 g. of the solvent.

Normality : Number of gram equivalents of the solute present in one litre of the solution.

Mole fraction : It is the ratio of number of moles of solute present in the solution to the total number of moles of the solute and solvent present in the solution.

$$\text{Mole fraction of solute} = \frac{n}{n + N}$$

Formality : It is the number of formula weights present in one litre of the solution.

Weight % : grams of solute per 100 g. of solution

$$\text{Mass percentage} = \frac{\text{Mass of solute}}{\text{Mass of solution}} \times 100$$

$$\text{Molarity} = \frac{\text{\% strength of solution x density of solution x 10}}{\text{Molecular weight}}$$

$$\text{Normality} = \frac{\text{\% strength of solution x density of solution x 10}}{\text{Equivalent weight}}$$

$$\text{\% by mass (M)} = \frac{\text{X x density x 10}}{\text{molecular weight}}$$

For determining the parameters of a solution of a substance we can use

$$W = \frac{ENV}{1000}$$ (W = wt. of the solute; E = eq.wt; N = normality; V = volume)

$$\text{Equivalent weight of an acid} = \frac{\text{Molecular weight}}{\text{Basicity}}$$

$$\text{Equivalent weight of a base} = \frac{\text{Molecular weight}}{\text{Acidity}}$$

$$\text{Equivalent weight of an oxidising agent} = \frac{\text{Molecular weight}}{\text{no. of electrons gained by one mole of the oxidising agent}}$$

$$\text{Eq.wt. of a reducing agent} = \frac{\text{Molecular weight}}{\text{no. of electrons lost or donated by the reducing agent}}$$

If normalities of two solutions of the same substance or different substances (like acid Vs base etc.) are to be compared, one can use $N_1V_1 = N_2V_2$.

To determine the normality of a solution containing a mixture of substances for e.g. acids and bases , one can ues

$$\frac{(N_1V_1 + N_2V_2) \text{ .... } -(N_3V_3 + N_4V_4)\text{.......}}{\text{Total volume}}$$

$$\text{Eq. wt. of normal salt} = \frac{\text{Molecular weight}}{\text{Total number of +ve or -ve valencies}}$$
(e.g. Na_2CO_3;$Ca_3(PO_4)_2$etc)

$$\text{Eq.wt. of an acidic salt} = \frac{\text{Molecular weight}}{\text{number of replaceable hydrogen atoms}}$$
(e.g. $NaHCO_3$;$KHSO_4$etc)

$$\text{Eq.wt. of basic salt} = \frac{\text{Molecular weight}}{\text{Number of } OH^- \text{ present}}$$
(e.g. $Fe(OH)_2Cl$;$Mg(OH)Cl$ etc)

Mole fraction and molality depend upon weights and not on volumes. Therefore they are independent of temperature.

Types of solutions: A solution is defined as a homogeneous mixture of two or more substances, the composition of which may vary within the limits. The component present in larger proportion is solvent and the component present in smaller proportion is called the solute. A solution may exist in gaseous, liquid or solid state. The solution of liquid in gas for e.g. H_2O in air, when there is mist or solid in gas for e.g. camphor in air is not possible as the constituents can not form homogeneous mixture.

Raoult's Law : The vapour pressure of a liquid diminishes when a non volatile substance is dissolved in it and the decrease is proportional to the amount of solute dissolved. Suppose, the pure liquid has a vapour pressure p and that of solution of known concentration has the vapour pressure p_s the lowering of vapour pressure is $p - p_s$. Although the value of p increases rapidly with temperature, the ratio of $p-p_s/p$ is the same for a given dilute solution. The ratio of decrease of vapour pressure to the vapour pressure of the pure liquid is known as relative lowering of vapour pressure.

The solute should be non volatile, non electrlyte and the solution should be very dilute under this law.

The relative lowering of vapour pressure is equal to (the molar fraction of the solute) the ratio of the molecules of the solute to the total number of molecules in the solution.

$$\frac{p - p_s}{p} = \frac{n}{n + N} = \frac{\frac{w}{m}}{\frac{w}{m} + \frac{W}{M}}$$

Since w/m for a dilute solution is negligible, it is neglected in the denominator

$$\frac{p - p_s}{p} = \frac{wM}{mW}$$ (w and W are the weights of solute and solvent while m and M are the molecular weights of solute and solvent respectively)

Many pairs of miscible liquids have dissimilar molecules. This results into a deviation from their ideal behaviour and do not obey Raoult's law. If the vapour pressure of the solution is lower than that of an ideal solution having the same composition, then large negative deviations are observed. In case, the liquid pair is such that the molecular interaction forces are weaker than those existing between the molecules of any of the liquids, there will be a significant positive deviation from Raoult's law i.e. the vapour pressure of the solution is higher than that of an ideal solution of the same composition.

Colligative properties : The colligative properties are the properties which depend only on the number and not the kind of solute molecules.

The various colligative properties are : (a) Relative lowering of vapour pressure (b) Osmotic pressure (c) Elevation of boiling point (ΔT_b) (d) Depression of freezing point (ΔT_f).

By dissociation, colligative property increases and by association, it decreases .

If solution of sugar, urea, glucose etc. which do not dissociate or associate in water are prepared by dissolving one mole each of the solutes in 1000 g. of water, they will show identical colligative porperties. It is because each of these solutions has the same number of molecules i.e. 6.023×10^{23}.

Molecular weight of the solute and other parameters can be calculated from the followings relationships:

By elevation of boiling point method :

$$m = \frac{1000\, k_b\, w}{\Delta T\, W}$$ (k_b = molal elevation constant or boiling point constant)

The value of k_b can be calculated from, $k_b = \frac{RT^2}{1000\, L_v}$

(L_v = heat of vapourisation)

When 1 mole of a solute that does not associate or dissociate is dissolved in 1kg. of the solvent, the boiling point elevation is called molal elevation constant.

$$\Delta T_b = k_b \times m$$

$$\Delta T_b = k_b \text{ (when molality is one)}$$

By depression of freezing point method :

$$m = \frac{1000\, k_f\, w}{\Delta T_f\, W}$$ (k_f= freezing point constant)

$$\Delta T_f = k_f . m$$

By osmotic pressure method :

Flow of molecules through semipermeable membrane takes place from a pure solvent to a solution or from a dilute to concentrate solution. The relative lowering of vapoour pressure is directly proportional to osmotic pressure i.e. $p - p_s / p \propto \pi$. Solutions which do not produce flow of solvent through the semipermeable membrane are said to be isotonic. The isotonic solutions have the same osmotic pressure at the same temperature. The isotonic solutions have the same molar concentrations :

$$\frac{w_1}{m_1 v_1} = \frac{w_2}{m_2 v_2}$$

By osmotic pressure, one can calculate the molecular weight from the following relations.

$$PV = nRT \text{ or } PV = \frac{w_B}{m_B} RT$$ (w_B= wt. of the solute m_B = molar mass)

$$P = \frac{n}{V}\ RT$$

$$P = CRT$$

The reverse of osmosis is used in desalinisation of sea water.

Abnormal molecular weights : A solution is said to be an ideal solution, if the molecules in it attract one another with equal forces irrespective of their nature. If we have an ideal solution of A in B, the forces between A and A, B and A and B and B are the same.

For an ideal solution :

(a) Total volume of the solution = vol. of A + vol. of B

i.e. $\Delta V_{mix} = 0$

(b) No heat is absorbed or evolved on mixing the components i.e. $\Delta H_{mix} = 0$

Ideal solutions : They obey Raoult's law at all concentrations and temperatures. Some examples of ideal solutions are : (a) benzene + toulene (b) CCl_4 + $SiCl_4$ (c) n-hexane + n-pentane (d) CH_2Cl_2 + CH_2Br_2 etc.

Ideal solutions can be separated by fractional distillation.

Non ideal solutions : All real solutions are non ideal.

Type I : Heat is absorbed. Temperature decreases. Positive deviation from Raoult's law is shown. Vapour pressures are greater than calculated e.g.(a) CH_3COCH_3 + CS_2 (b)Benzene+C_2H_5OH(c)$CHCl_3$+C_2H_5OH(d)$CHCl_3$+ CCl_4(e) H_2O + C_2H_5OH (f) CH_3COCH_3 + benzene etc.

Type II : Heat is evolved. Temperature increases. Negative deviation from Raoult's law is shown. Vapour pressures are less than calculated.

e.g. (a) $CHCl_3$ + CH_3COCH_3 (b) Pyridine + CH_3COOH (c) H_2O + H_2SO_4 (d) CH_3COCH_3 + C_6H_5–NH_2 (e) $CHCl_3$ + benzene etc.

When the molecular weight of certain substances are determined by one or more of the methods described, the results obtained do not agree with the normal molecular weights calculated from their accepted chemical formulae. In some cases the values obtained are too large and in other cases too small.

For substances which dissociate or associate in solution, the number of particles in the solution differs from the normal number and this causes the molecular weights to be abnormal.

$$\text{Vant Hoff factor (i)} = \frac{\text{normal molecular weight}}{\text{observed molecular weight}}$$

$$\frac{\text{normal molecular weight}}{\text{observed molecular weight}} = \frac{\text{number of solute particles after association or dissociation}}{\text{number of normal particles}}$$

Azeotropic mixture : The solution of azeotropic mixture boils at constant temperature and distil over completely at the same temperature without any change.

Azeotropic mixtures like (a) H_2O+C_2H_5OH (78.15°C) (b) pyridine + water (92.6°C) (c) C_2H_5OH +benzene (67.8°C) (d) CH_3COOH + Toulene (105.4°C) etc. show positive deviation while (a) HNO_3 + H_2O (125.5°C) (b) CH_3COOH + pyridine (139.0°C) (c) $CHCl_3$ + CH_3COCH_3 (65°C) (d) HCl + H_2O (108.6°C) show negative deviation.

Things to remember

Molarity : no. of moles of solute present in 1 lit. of the solution.

Molality : no. of moles of solute present in 1000 g. of the solvent.

Normality : no. of gram equivalents of the solute present in one litre of the solution.

Mole fraction : $\frac{n}{n+N}$

mole fraction and molality depend on weights and therefore are independent of temperature.

Raoult's Law : $\frac{p-p_s}{p} = \frac{n}{n+N} = \frac{\frac{w}{m}}{\frac{w}{m}+\frac{W}{M}} = \frac{wM}{mW}$

Colligative properties : depend only on the number of particles and not the kind. e.g. relative lowering of vapour pressure ; osmotic pressure ; elevation in b.p.; lowering in freezing point,

$m = \frac{1000\, k_b\, w}{\Delta T_b \times W}$; $k_b = \frac{RT^2}{1000 L_v}$ (L_v=Latent heat of vapourisation per gram)

$\Delta T_b = k_b \times m$

$m = \frac{1000\, k_f w}{\Delta T_f \times W}$; $\Delta T_f = k_f \times m$;

PV= nRT or $P = \frac{n}{V}$ RT or P = CRT

- Isotonic solutions : which do not produce flow of solvent through the semi permeable membrane. When two solutions have the same osmotic pressure at the same temp., they should have the same molar concentrations.

$$\frac{w_1}{m_1 v_1} = \frac{w_2}{m_2 v_2}$$

- For ideal solutions $\Delta V_{mix} = 0$ and $\Delta H_{mix} = 0$

- Vant Hoff factor $i = \frac{\text{normal molecular weight}}{\text{observed molecular weight}}$

Abnormal molecular weights : due to association or dissociation of the solute particles in solution.

$$\frac{\text{normal molecular weight}}{\text{observed molecular weight}} = \frac{\text{number of solute particles after association or dissociation}}{\text{number of normal particles}}$$

Solved Examples

1. Calculate the molality of a 20% (by weight) aqueous solution of KOH.

Sol. A 20% solution of KOH contains 20g. of KOH dissolved in 80 of H_2O.

No. of moles of KOH = $\frac{20}{56}$

$\frac{20}{56}$ moles of KOH are dissolved in 80 g. of H_2O

No. of moles of KOH in 1000g. of $H_2O = \frac{20}{56} \times \frac{1000}{80} = 4.464$

Molality = $\frac{\text{no. of moles}}{\text{no. of kg}} = \frac{4.464}{1} = 4.464$ m

2. What should be the molarity of 1 lit. solution of 93% w/w H_2SO_4, if the density is 1.84 ?

sol. Molarity = $\frac{\text{wt. in gram x density x 1000}}{\text{Molecular wt. x 100}}$

$= \frac{93 \times 1.84 \times 1000}{98 \times 100} = 17.46$

3. A solution contains 410.3 g. of H_2SO_4 per litre of solution at 20°C. If the density is 1.243 g/ml, calculate the molarity and molality of the solution.

Sol. no. of moles of $H_2SO_4 = \frac{410.3}{98} = 4.187$ M

weight of 1000 ml of H_2SO_4 solution = 1000 x 1.243
= 1243 g.

Weight of water = 1243 – 410.3 = 832.7 g.

Molality of the solution = $\frac{\text{no. of moles of } H_2SO_4}{\text{wt. of water}}$

$= \frac{4.187 \times 1000}{832.7} = 5.028$ m

4. Calculate the molality of a 1M solution of $NaNO_3$ whose density is 1.25 g cm^{-3}.

Sol. Molality = $\frac{\text{no. of moles of } NaNO_3}{\text{Mass of water in kg}}$

wt. of 1lit. of solution = 1000 x 1.25 = 1250 g.

Mass of $NaNO_3$ in the solution = 85 g. (since it is 1M solution)

Mass of water = 1250 – 85 = 1165 g.

Molality = $\frac{1}{1.165} = 0.858$

5. What volume of 95% H_2SO_4 (density = 1.85 g/cm^3) and what mass of water must be taken to prepare 100cc of 15% solution of H_2SO_4 (density = 1.10 g/cm^3).

Sol. 95% H_2SO_4 means 95 g. of H_2SO_4 is present in 100g. of solution.

Volume of the solution = $\frac{100}{1.85} = 54.05$ ($\because \frac{M}{D} = V$)

Molarity = $\frac{95}{98} \times \frac{1000}{54.05} = 17.93$

15% H_2SO_4 means 15g. of H_2SO_4 is present in 100g. of solutionvolume of this solution $\frac{100}{1.10} = 90.91$

Molarity = $\frac{15}{98} \times \frac{1000}{90.91} = 1.68$

$M_1V_1 = M_2V_2$

$V_1 = \frac{1.68 \times 100}{17.93} = 9.37$ cc.

Volume of 95% H_2SO_4 required = 9.37 c.c.
Mass of 100 c.c.of 15% H_2SO_4 to be prepared = 100 x 1.10=110g. Mass of 9.37 c.c. of 95% H_2SO_4 = 9.37 x 1.85 = 17.3 g. Mass of H_2O to be taken = 110 - 17.3 = 92.7 g.

6. Upon mixing 45ml. of 0.25 molar $Pb(NO_3)_2$ solution with 25ml. of 0.1 molar $Cr_2(SO_4)_3$ solution precipitation of $PbSO_4$ takes place. How many moles of $PbSO_4$ are formed. Calculate the molar concentrations of the species left behind in the final solution (assume that $PbSO_4$ is completely insoluble)

Sol. On balancing the reaction we get

$$3Pb(NO_3)_2 + Cr_2(SO_4)_3 \rightarrow 3PbSO_4 + 2Cr(NO_3)_3.$$

3 mole   1 mole   3 mole   2 mole

Initial concentration

$0.25 \times \frac{45}{1000}$ $0.1 \times \frac{25}{1000}$ 0   0

$= 11.25 \times 10^{-3}$moles $= 2.5 \times 10^{-3}$moles

After reaction, number of moles left unused

$=11.25 \times 10^{-3}$ $-3 \times 2.5 \times 10^{-3} = 3.75 \times 10^{-3}$.

consumed, $3\times2.5\times10^{-3}=7.5\times10^{-3}$ $2\times2.5\times10^{-3}=5.0\times10^{-3}$.

Total volume of the solution = 45 + 25 = 70ml.

molarity of $Pb(NO_3)_2$ left unused $= \frac{3.75\times10^{-3}}{70\times10^{-3}} = 0.0535$ M

molarity of $Cr(NO_3)_3$ formed $= \frac{5.0\times10^{-3}}{70\times10^{-3}} = 0.0714$ M

7. 1000 ml. of a water sample contain 1mg. of $CaCl_2$ and 1mg. of $MgCl_2$. Find the hardness of water in terms of parts of $CaCO_3$.

Sol. 1mg. of $CaCl_2 = \frac{100}{111}$ x1 of $CaCO_3$($\underset{111}{CaCl_2} \rightarrow \underset{100}{CaCO_3}$)

1mg. of $CaCl_2 = \frac{100}{95}$x1 of $CaCO_3$ ($\underset{95}{CaCl_2} \rightarrow \underset{100}{CaCO_3}$)

Total weight of $CaCO_3 = \frac{100}{111} + \frac{100}{95} = 1.9535$ mg.

Thus 1.9535 mg. of $CaCO_3$ is present in 10^6mg of water. Hardness is therefore 1.9535 ppm.

8. 10g. of a solute was dissolved in 80g. of acetone at 30°C. The vapour pressure of the solution was found to be 271 mm of mercury. Calculate the molecular weight of the solute. Vapour pressure of acetone at 30°C is 283mm. Assume that the solution is not very dilute.

Sol.

$$\frac{p - p_s}{p} = \frac{n}{n+N} = \frac{\frac{w}{m}}{\frac{w}{m}+\frac{W}{M}}$$

$$\frac{283-271}{283} = \frac{\frac{10}{m}}{\frac{10}{m}+\frac{80}{58}} \qquad m = 163.5$$

9. Air was drawn through a solution containing 38.08g. of solute in 100g. of water and through water. The loss in weight of water was 0.0551g. and total weight of water absorbed in H_2SO_4 tube was 2.2117g. Calculate the molecular weight of the dissolved substance.

Sol. $\frac{p - p_s}{p} = \frac{w}{m} \times \frac{W}{M}$

$$\frac{0.0551}{2.2117} = \frac{38.08 \times 18}{m \times 100}$$

$$m = \frac{38.08 \times 18 \times 2.2117}{0.0551 \times 100} = 267.7$$

10. Dry air was passed successively through a solution of 5g. of a solute in 80g. of water and then through pure water. The loss in weight of solution was 2.5g. and that of the pure solvent 0.04g. Calculate the molecular weight of the solute.

Sol.: $\frac{p - p_s}{p} = \frac{w}{m} \times \frac{W}{M}$

$$\frac{0.04}{2.54} = \frac{5 \times 18}{m \times 80}$$

$$m = \frac{5 \times 18 \times 2.54}{0.04 \times 80} = 71.44$$

11. The molar heat of vapourisation of water at 100°C is 40.585 kJ. At what temperature will a solution containing 5.60g. of glucose per 1000g. of water boil ? ($R=8.314 Jk^{-1}mol^{-1}$)

Sol. Molar heat of vapourisation of water at 100°C

$L_v = 40.585$ kJ = 40585 J

∴ Heat of vapourisation per gram, $L_v = \frac{40585}{18} = 2254.7$J

First we find out k_b from $k_b = \frac{RT^2}{L_v \times 1000}$

$$= \frac{8.314 \times 373 \times 373}{2254.7 \times 1000} = 0.513$$

Now we calculate ΔT from, $\Delta T = \frac{1000\, k_b\, w}{W.m}$

$$= \frac{1000 \times 0.513 \times 5.6}{1000 \times 180} = 0.016$$

Boiling point of the solution = 100 + 0.016 = 100.016°C

12. A solution of urea in H_2O has a b.p. of 100.128°C. Calculate the freezing point of the solution. k_f and k_b for water are 1.86 km^{-1} and 0.52 km^{-1} respectively.

Sol. $\Delta T_b = k_b\, m$

$$m = \frac{\Delta T_b}{k_b} = \frac{100.128° - 100°}{0.52} = 0.246 \text{ m}$$

$\Delta T_f = k_f m$

$T_f° - T_f = 1.86 \times 0.246 = 0.458°$

($T_f°$ = freezing point of the pure solvent, T_f= freezing point of the solution)

$T_f = T_f° - 0.458°$

$= 0 - 0.458 = -0.458°$

13. The latent heat of fusion of benzene is 126.6 Jg^{-1}. 0.0802g. of a substance when dissolved in 13.03g. of benzene lowered the freezing point by 0.490K. If pure benzene freezes at 278.4K, calculate the molecular weight of the substance.

Sol. First we calculate k_f from, $k_f = \frac{RT^2}{1000 \times L_f}$

$$= \frac{8.314 \times (278.4)^2}{1000 \times 126.6} = 5.088 \text{k kg. mol}^{-1}$$

Now $m = \frac{1000\, k_f\, w}{\Delta T_f . W} = \frac{1000\times5.088\times0.0802\times10^{-3}}{0.49\times13.03\times10^{-3}} = 63.92$

14. How much urea must be dissolved in 10^{-3} m^3 water to yield a solution with an osmotic pressure of 7.65×10^5 Mm^{-2} at 310 k ?

Sol. $PV = \frac{w}{m} RT$

$= \frac{PVm}{RT} = \frac{7.65 \times 10^5 \times 10^{-3} \times 60}{8.314 \times 310} = 1.781$ g.

15. A solution containg 8.6 g per litre of urea (mol. wt.60) was found to be isotonic with a 5% solution of an organic non-volatile solution. Calculate the molecular weight of the other.

Sol. The isotonic solutions have the same molar concentrations, $\frac{50}{M} = \frac{8.6}{60}$

$M = \frac{60 \times 50}{8.6} = 348.9$

16. 0.85% aqueous solution of $NaNO_3$ is apparently 90% dissociated. Calculate the osmotic pressure of the solution.

Sol. Molality $= \frac{0.85 \times 1000}{100 \times 85} = 0.1$ mol lit^{-1}.

$$NaNO_3 \rightarrow Na^+ + NO_3^-$$
$$1-x \quad\quad x \quad\quad x$$

$i = \frac{1-x+x+x}{1} = 1+x = 1+0.90 = 1.90$

$\pi = iCRT = 1.90 \times 0.1 \times 0.082 \times 300 = 4.674$ atm.

17. Acetic acid associates in benzene to form double molecules. 1.65 g. of acetic acid when dissolved in 100g. of benzene raised the b.p. by 0.36°C. Calculate the degree of association of acetic acid if the molal elevation constant of benzene is 2.57°C.

Sol. $m = \frac{1000 k_b w}{\Delta T x W} = \frac{2.57 \times 1.65 \times 1000}{100 \times 0.36} = 118$

$$\begin{array}{ccc} 1 & & 0 \\ CH_3COOH & \rightleftharpoons & (CH_3COOH)_2 \\ 1-x & & x/2 \end{array}$$

No. of molecules after association $= 1 - x + \frac{x}{2}$

$\frac{\text{normal mol. wt.}}{\text{observed mol. wt.}} = \frac{1 - x + \frac{x}{2}}{1}$

$\therefore \frac{60}{118} = \frac{1 - x + \frac{x}{2}}{1}$

$x = 0.983$

degree of association is 98.3%

18. The freezing point depression of a M/200 solution of Na_2SO_4 in H_2O was found to be 0.0265°C. Calculate the degree of dissociation of the salt at this concentration. k_f for water is 1.86°.

Sol. concentration of the solution $= \frac{M}{200} = \frac{142}{200} = 0.71$ g./l

observed molecular weight $m = \frac{1000 k_f w}{\Delta T W}$

$= \frac{1000 \times 1.86 \times 0.71}{0.0265 \times 1000}$

$= 49.84$

Let x be the dissociation of Na_2SO_4

$$\begin{array}{lccc} & Na_2SO_4 & \rightleftharpoons & 2Na^+ + SO_4^{2-} \\ \text{At eq.} & 1-x & & 2x \quad\quad x \end{array}$$

Number of molecules after dissociation $= 1-x+2x+x = 1+2x$

$\frac{\text{normal molecular weight}}{\text{observed molecular weight}} = \frac{\text{no. of particles after dissociation}}{\text{no. of normal particles}}$

$\frac{142}{49.84} = 1 + 2x$

$x = 0.925$ $\therefore$ Degree of dissociation = 92.5%

19. 0.5 percent aqueous solution of KCl was found to freeze at −0.24°C. Calculate the vant Hoff factor and degree of dissociation of the solute at this concentration (k_f for H_2O is 1.86°)

Sol. First we calculate the observed molecular weight from

$m = \frac{1000 k_f w}{\Delta T_f x W}$

$= \frac{1000 \times 1.86 \times 0.50}{100 \times 0.24} = 38.75$

Vant Hoff factor, $i = \frac{\text{normal mol. wt.}}{\text{observed mol. wt.}}$

$= \frac{74.5}{38.75} = 1.92$

Let x be the degree of dissociation of KCl

$$KCl \rightleftharpoons K^+ + Cl^-$$

concentration at equilibrium $\quad 1-x \quad\quad x \quad x$

The total no. of molecules after dissociation $= 1-x+x+x$
$= 1+x$

$\frac{\text{normal mol. wt.}}{\text{observed mol. wt.}} = \frac{\text{no. of particles after dissociation}}{\text{no. of normal particles}}$

$\frac{74.5}{38.75} = 1 + x$

$x = 0.923$ $\therefore$ Degree of dissociation = 92.3%

20. A solution containing 110 g. of $Ba(NO_3)_2$ per dm^3 of H_2O has a b.p. of 373.466 K at one atmosphere pressure. Calculate the degree of ionisation of the salt in the solution. The molal elevation constant for H_2O is 0.52 K kg mol^{-1}.

Sol. ΔT_b calculated $= \frac{1000 k_b w}{mW}$

$= \frac{1000 \times 0.52 \times 110}{261.34 \times 1000} = 0.219$

we know $\frac{\Delta T_b \text{ observed}}{\Delta T_b \text{ calculated}} = \frac{\text{no. of particles after dissociation}}{\text{no. of normal particles}}$

$$\begin{array}{ccc} Ba(NO_3)_2 & \rightleftharpoons & Ba^{+2} + 2NO_3^- \\ 1-x & & x \quad\quad 2x \end{array}$$

Let x be the degree of dissociation

$\frac{0.466}{0.219} = \frac{1+2x}{1}$

$x = 0.5635$ or 56.35%

21. The density of a solution containing 13% by mass of H_2SO_4 is 0.109 g/ml. Calculate the molarity and normality of the solution.

$$\text{Sol.: Molarity} = \frac{\text{\% strength of solution x density of solution x 10}}{\text{molecular mass}}$$

$$= \frac{13 \times 0.109 \times 10}{98} = 1.445 \text{ M}$$

Normality = molarity x n = 1.445 x 2 = 2.89 N

1. Which of the following solutions would have the highest osmotic pressure ? [CBSE-91]
(A) M/10 NaCl (B) M/10 Urea
(C) M/10 $BaCl_2$ (D) M/10 Glucose

2. Which one of the following solutions has the highest boiling point ? [CBSE-91]
(A) 0.1M glucose (B) 0.1M $BaCl_2$
(C) 0.1M NaCl (D) 0.1M Urea

3. 0.1 M solution of which one of these substances will react basic ? [CBSE-92]
(A) Sodium borate (B) Ammonium chloride
(C) Calcium nitrate (D) Sodium sulphate

4. If a 0.1 molar solution of glucose (mol.wt.=180) and 0.1 molar solution of urea(mol.wt.60) are placed on the two sides of a semipermeable membrane to equal heights, then it will be correct to say that [CBSE-92]
(A) there will be no net movement across the membrane
(B) glucose will flow across the membrane into urea solution
(C) urea will flow across the membrane into glucose solution
(D) water will flow from urea solution into glucose solution

5. How many grams of CH_3OH would have to be added to water to prepare 150 mL of solution that is 2.0 M CH_3OH ? [CBSE-94]
(A) 9.6 (B) 2.4 (C) 9.6 x 10^3 (D) 4.3 x 10^2

6. At 25°C, the highest osmotic pressure is exhibited by 0.1 M solution of [CBSE-94]
(A) $CaCl_2$ (B) KCl (C) glucose (D) Urea

7. Which one of the following salts would have the same value of the vant Hoff factor (I) as that of $K_3[Fe(CN)_6]$? [CBSE-94]
(A) $Al_2(SO_4)_3$ (B) NaCl (C) $Al(NO_3)_3$ (D) Na_2SO_4

8. Concentration units independent of temeperature would be [CBSE-95]
(A) Normality (B) Weight / volume percent.
(C) Molality (D) Molarity

9. According to Raoult's law the relative lowering of vapour pressure for a solution is equal to the [CBSE-95]
(A) mole fraction of solute (B) moles of solvent
(C) mole fraction of solvent (D) moles of solute

10. What is the molarity of H_2SO_4 solution that has a density 1.84gm/cc at 35°C and contains 98% by weight? [CBSE-96]
(A) 4.18 M (B) 8.14 M (C) 18.4 M (D) 18 M

11. Vapour pressure of CCl_4 at 25°C is 143mm Hg.0.5gm

of a non-volatile solute (mol.weight = 65) is dissolved in 100 ml CCl_4. Find the vapour pressure of the solution. (Density of CCl_4 = 1.58g/cm²) [CBSE-96]

(A) 149.93 mm (B) 94.39 mm
(C) 199.34 mm (D) 143.99 mm

12. The relationship between osmotic pressure at 273 K when 10 g glucose(P_1)10 g urea (P_2) and 10 g sucrose (P_3) are dissolved in 250 ml of water is [CBSE-96]

(A) $P_1>P_2>P_3$ (B) $P_3>P_1>P_2$ (C) $P_2>P_1>P_3$ (D) $P_2>P_3>P_1$

13. A 5% solution of cane sugar (mol.wt.=342) is isotonic with1% solution of substance X. The molecular weight of X is [CBSE-98]

(A) 171.2 (B) 68.4 (C) 34.2 (D) 136.2

14. The vapour pressure of a solvent decreased by 10 mm of mercury when a non-volatile solute was added to the solvent. The mole fraction of the solute in the solution is 0.2. What should be the mole fraction of solvent if the decrease in the vapour pressure is to be 20 mm of mercury ? [CBSE-98]

(A) 0.8 (B) 0.6 (C) 0.4 (D) 0.2

15. Without losing its concentration $ZnCl_2$ solution cannot be kept in contact with [CBSE-98]

(A) Al (B) Au (C) Pb (D) Ag

16. How many g of dibasic acid (mol. wt. 200) should be present in 100 ml of the aqueous solution to give 0.1 normality ? [CBSE-99]

(A) 1g (B) 2g (C) 10g (D) 20 g

17. If 0.15 g of a solute, dissolved in 15g of solvent, is boiled at a temperature higher by 0.216°C, than that of the pure solvent. Find out the molecular wt. of the substance.(Molal elevation constant for the solvent is 2.16°C) is [CBSE-99]

(A) 1.01 (B) 10 (C) 10.1 (D) 100

18. The vapour pressure of benzene at a certain temperature is 640 mm of Hg. A non-volatile and non-electrolyte solid, weighing 2.175g is added to 39.08 of benzene. If the vapour pressure of the solution is 600 nm of Hg, what is the molecular weight of solid substance ? [CBSE-99]

(A) 49.50 (B) 59.60 (C) 69.60 (D) 79.82

19. Which of the following solutions have lowest freezing point ? [CBSE-2000]

(A) 0.1 M $NaNO_3$ (B) 0.01 M glucose
(C) 0.01 M BCl_3 (D) 0.2 M Na_2SO_4

20. A solution contains non volatile solute of molecular mass M_2. Which of the following can be used to calculate the molecular mass of solute in terms of osmotic pressure [CBSE-2002]

(A) $M_2 = \left[\frac{m_2}{\pi}\right]VRT$ (B) $M_2 = \left[\frac{m_2}{V}\right]\frac{RT}{\pi}$

(C) $M_2 = \left[\frac{m_2}{V}\right]\pi RT$ (D) $M_2 = \left[\frac{m_2}{V}\right]\frac{\pi}{RT}$

Note: $m_2 \rightarrow$ Mass of solute
$V \rightarrow$ Volume of solution
$\pi \rightarrow$ Osmotic pressure

21. A solution containing components A and B follows Raoult's law [CBSE-2002]

(A) A-B attraction force is greater than A -A and B-B
(B) A-B attraction force is less than A-A and B-B
(C) A-B attraction force remains same as A-A and B-B
(D) Volume of solution is different from sum of volume of solute and solvent

22. Molarity of liquid HCl, if density of solution is 1.17 gm/c.c is - [CBSE-2001]

(A) 36.5 (B) 18.25 (C) 32.05 (D) 42.10

23. Pure water can be obtained from sea water by [CBSE-2001]

(A) Centrifugation (B) Plasmolysis
(C) Reverse osmosis (D) Sedimentation

24. Beans are cooked earlier in pressure cooker because [CBSE-2001]

(A) Boiling point increase with increasing pressure
(B) Boiling point decrease with increasing pressure
(C) Extra pressure of pressure cooker, softens the beans
(D) Internal energy is not lost while cooking in pressure cooker.

25. The molality of a solution having 18 gm of glucose (mol. wt. = 180) dissolved in 500 gm of water is [AIIMS-95]

(A) 0.1 M (B) 0.5 M (C) 0.2 M (D) 2.2 M

26. The molarity of pure water is [AIIMS-95]

(A) 1.16 M (B) 5.56 M (C) 18.36 M (D) 55.56 M

27. The number of moles of water present in 180 gm of water is [AIIMS-96]

(A) 5 (B) 10 (C) 18 (D) 100

28. Which of the following is the correct representation of relative lowering of vapour pressure? [AIIMS-96]

(A) $\frac{P^o - P}{P^o}$ (B) $\frac{\Delta P}{P^o} = \frac{P^o - P}{P^o}$

(C) $\frac{P^o}{\Delta P^o} = \frac{P^o - P}{P^o}$ (D) $\frac{P^o}{P^o - P}$

29. Which is not affected by temperature ? [AIIMS-97]

(A) Normality (B) Formality (C) Molarity (D) Molality

30. Assertion : 0.1 M solution of glucose has higher increment in the freezing point than 0.1 M solution of urea. Reason : K_f for both has different value. [AIIMS-97]

(A) A (B) B (C) C (D) D

31. The boiling point of water (100°C) becomes 100.52°C, if 3 grams of a non volatile solute is dissolved in 200 ml of water. The molecular weight of solute is (if K_b for water is = 0.6 K/m) [AIIMS-98]

(A) 12.2 g mol^{-1} (B) 15.4 g. mol^{-1}
(C) 17.3 g. mol^{-1} (D) 20.4 g. mol^{-1}

32. Which of the following salt has the same value of vant Hoff factor as that of $K_3[Fe(CN)_6]_2$? [AIIMS-98]

(A) $Al_2(SO_4)_3$ (B) $Al(NO_3)_3$ (C) Na_2SO_4 (D) Fe_3O_4

33. The vapour pressure of benzene at certain temperature is 640mm of Hg. A non-volatile and non-electrolyte solid weighing 2.175 g is added to 39.08 g of benzene. The vapour pressure of the solution is 600mm of Hg. What is the molecular weight of solid substance ? [AIIMS-99]

(A) 49.59 (B) 59.60 (C) 69.60 (D) 108.30

34. How many grams of sucrose should be dissolved in 500 gms of water so as to get a solution which has difference of 104°C between boiling point and freezing point (K_f = 1.86 K kg mol^{-1}, K_b = 0.52 kg mol^{-1}) [AIIMS-2000]

(A) 1.68 (B) 3.36 (C) 8.40 (D) 0.840

35. Assertion (A) : An increase in surface area increases the rate of evaporation. Reason (R) : Stronger the intermolecular attractive forces, faster is the rate of evaporation at a given temperature. [AIIMS-2002]

(A) A (B) B (C) C (D) D (E) E

36. Which one of the statements given below concerning properties of solutions, describes a colligative effect ? [AIIMS-2003]

(A) Boiling point of pure water decreases by the addition of ethanol
(B) Vapour pressure of pure water decreases by the addition of nitric acid
(C) Vapour pressure of pure benzene decreases by the addition of naphthalene
(D) Boiling point of pure benzene increases by the addition of toluene

37. Evaporation of a liquid is accompanied by : [JIPMER-90]

(A) Increase in enthalpy (B) Decrease in enthalpy
(C) No change in free energy (D) Increase in entropy

38. The number of moles of solute per kilogram of a solvent is called : [JIPMER-91]

(A) Molarity (B) Molality
(C) Normality (D) Mole fraction

39. As a result of osmosis, the volume of the solution : [JIPMER-91]

(A) Decreases (B) Increases
(C) Do not change (D) None

40. Isotonic solutions have same : [JIPMER-93]

(A) boiling point (B) freezing point
(C) concentration (D) vapour pressure

41. A 0.01 M solution of glucose in water freezes at –0.0186°C. A 0.01 M solution of NaCl in water will freeze at : [JIPMER-93]

(A) –0.0372°C (B) 0°C (C) –0.0186°C (D) 0.0186°C

42. Which of the following is not a solution ? [JIPMER-94]

(A) Air (B) Smoke (C) Gold ring (D) Salt water

43. The volume of the resulting solution obtained by mixing 25 mL of acetone and 25 mL of $CHCl_3$ will be [JIPMER-96]

(A) 50 mL (B) <50 mL (C) >50 mL (D) unpredictable

44. Out of molality (m), molarity (M), formality (F), normality (N) and mole fraction (x), the two independent of temperature are [JIPMER-96]

(A) N, m (B) m, x (C) m, N (D) F, N

45. Which of the following is observed when some solvent is added to a dilute solution of a non-volatile solute ? [JIPMER-97]

(A) Increase of vapour pressure
(B) Depression of freezing point
(C) Elevation of boiling point
(D) Increase of osmotic pressure

46. The osmotic pressures of a dilute solution of a non-volatile solute is [JIPMER-98]

(A) Directly proportional to its temperature on the centigrade scale
(B) Inversely proportional to its temperature on the Kelvin scale
(C) Directly proportional to its temperature on the Kelvin scale
(D) Inversely proportional to its temperature on the centigrade scale

47. If 15g of a solute in 100g of water makes the solution freeze at –1°C, then 30g of the same solute in 100g of water will make the solution freeze at [JIPMER-99]

(A) –2°C (B) –0.5°C (C) 0°C (D) 2°C

48. How many g are present in one mole of Ag ? [JIPMER-2000]

(A) 10.29 (B) 107.9 (C) 108.6 (D) None

49. As a result of osmosis the volume of solution [JIPMER-2000]

(A) Increases (B) Decreases

(C) Remains constant (D) Increases or decreases

50. The vant Hoff factor calculated from association data is always . . . than calculated from dissociation data [JIPMER-2000]

(A) Less (B) More (C) Same (D) More or less

51. Which salt shows maximum osmotic pressure in its 1 M solution? [JIPMER-2001]

(A) $MgCl_2$ (B) $(NH_4)_3PO_4$ (C) Na_2SO_4 (D) $AgNO_3$

52. The number of moles of hydroxide (OH^-) ion in 0.3 litre of 0.005 M solution of $Ba(OH)_2$ is [JIPMER-2001]

(A) 0.0050 (B) 0.0030 (C) 0.0015 (D) 0.0075

53. The molarity of a solution having 18g of glucose (mole wt. =180) dissolved in 500g of water is [JIPMER-2001]

(A) 2.2 M (B) 0.2 M (C) 0.5 M (D) 0.1 M

54. Azeotropic mixture of HCl and water has [JIPMER-2002]

(A) 84% HCl (B) 22.2% HCl

(C) 63% HCl (D) 20.2% HCl

55. The osmotic pressure of a 5% solution of cane sugar at 15°C is (Mol. wt of cane sugar = 342) [BHU-95]

(A) 2.35 atm (B) 3.45 atm (C) 3.57 atm (D) 4 atm

56. Molecular weight of urea is 60. A solution of urea containing 6g urea in one litre is a [BHU-96]

(A) 1 molar (B) 1.5 molar (C) 0.1 molar (D) 0.01 molar

57. Which of the following solution in water possesses the lowest vapour pressure ? [BHU-96]

(A) 0.1 M NaCl (B) 0.1 M $BaCl_2$

(C) 0.1 M KCl (D) None of these

58. 0.15 gm of a substance dissolved in 15g of solvent is boiled at a temperature higher by 0.216°C than that of the pure solvent. Find out the molecular weight of the substance. (Molal elevation constant for the solvent is 2.16°C) [BHU-97]

(A) 10 (B) 10.1 (C) 100 (D) 1.01

59. The largest number of molecules is in [BHU-97]

(A) 34 gm of water (B) 54 gm of N_2O_5

(C) 28 gm of CO_2 (D) 46 gm of CH_3OH

60. Lowering of vapour pressure is the highest for [BHU-97]

(A) 0.1M Urea (B) 0.1 M glucose

(C) 0.1 M $MgSO_4$ (D) 0.1 M $BaCl_2$

61. Increase in boiling point of a sucrose solution is 0.1 K, then what is increase in boiling point of the same concentration of NaCl solution ? [BHU-98]

(A) 0.1K (B) 0.2 K (C) 0.4 K (D) 0.58

62. 0.2 gm of fine animal charcoal is mixed with half litre of acetic acid solution and shaken for 30 minutes [BHU-98]

(A) The concentration of the solution decreases

(B) Concentration increases

(C) Concentration remains same (D) None of these

63. Units of mole fraction are [BHU-98]

(A) Moles/lit. (B) Moles/lit^2

(C) Moles lit (D) Dimensionless.

64. What will be the molarity of a solution containing 5 g of sodium hydroxide in 250 ml solution ? [BHU-99]

(A) 0.1 (B) 0.5 (C) 1 (D) 2

65. The molarity of pure water is [BHU-99]

(A) 1 (B) 18 (C) 34.3 (D) 55.3

66. Molecular weight of urea is 60. The number of moles in the solution of urea, containing 6 g urea in one litre, is [BHU-99]

(A) 0.1 (B) 0.2 (C) 0.4 (D) 0.8

67. Partial pressure of a solution component is directly proportional to its mole fraction. This statement is known as [BHU-2000]

(A) Henry's law (B) Raoult's law

(C) Distribution law (D) Ostwald's dilution law

68. Which of the following is not a colligative property ? [BHU-2000]

(A) Osmotic pressure (B) Elevation of boiling point

(C) Depression is freezing point

(D) Increase in freezing point

69. What is the mole fraction of a solute in 2.5 M aqueous solution ? [BHU-2002]

(A) 0.043 (B) 0.43 (C) 4.3 (D) 43

70. Which of the following is a colligative property ? [BHU-2002]

(A) Relative lowering in vapour pressure

(B) Elevation in boiling point

(C) Osmotic pressure (D) All of these

71. The osmotic pressure of equimolar solution of glucose, sodium chloride and barium chloride will be in the order [BHU-2002]

(A) $BaCl_2$>NaCl>Glucose (B) $BaCl_3$>Glucose>NaCl

(C) Glucose>$BaCl_2$>NaCl (D) NaCl>$BaCl_2$>Glucose

72. When acetic acid is dissolved in benzene, its molecular mass [AFMC-91]

(A) Decreases (B) Increases

(C) Either increases or decreases (D) Suffers no change

73. The Osmotic pressure in atomosphere of 10% solution of cane-sugar at 69°C is [AFMC-91]
(A) 724 (B) 824 (C) 8.21 (D) 7.21

74. The molarity of a solution of Na_2CO3 in 10.6 gm/500 ml is [AFMC-92]
(A) 0.2 M (B) 2 M (C) 20 M (D) 0.02 M

75. An example of colligative property is [AFMC-92]
(A) Boiling point (B) Osmotic pressure
(C) Vapour pressure (D) Freezing point

76. Ostwald's dilution law is applicable to [AFMC-93]
(A) Strong electrolytes only (B) Weak electrolytes only
(C) Both strong and weak electrolytes
(D) Non - electrolytes only

77. Osmotic pressure of a solution increases by [AFMC-93]
(A) Decreasing the temperature
(B) Increasing the volume
(C) Increasing the number of molecules of the solute
(D) None of the above

78. Isotonic solutions have the same [AFMC-94]
(A) Density (B) Molar concentration
(C) Normality (D) None of the above

79. Isotonic solution have same [AFMC-95]
(A) Molar concentration (B) Molality
(C) Normality (D) None of these

80. Fog is an example of [AFMC-95]
(A) Solid dispersed in gas (B) Liquid dispersed in gas
(C) Liquid dispersed in solid (D) Solid dispersed in liquid

81. Osmotic pressure can be measured by [AFMC-95]
(A) Ostwald Method (B) Berkley and Hartley Method
(C) Solvay's Method (D) Hober's Method

82. Unit of molarity is [AFMC-96]
(A) gm/lit (B) mole/lit (C) kg/lit (D) None of these

83. Osmotic pressure can be increased by [AFMC-96]
(A) Increasing temperature (B) Decreasing temperature
(C) Increasing volume (D) None of these

84. In towns, a layer of air is condensed as smoke due to pollution is called [AFMC-96]
(A) Smog (B) Mist (C) Smoke (D) Fog

85. Azeotropic mixture of HCl and H_2O has [AFMC-97]
(A) 48% HCl (B) 22.2% HCl
(C) 36% HCl (D) 20.2% HCl

86. Which has the highest weight ? [AFMC-97]
(A) $1m^3$ water (B) 10 litre of Hg
(C) a normal adult man (D) all have same weight

87. If 5.85 grams of NaCl (molecular weight = 58.5) is dissolved in 90 grams of water, the mole fraction of NaCl will be [AFMC-98]
(A) 0.01 (B) 0.1 (C) 0.0196 (D) 0.2

88. The freezing point of a solution prepared from 1.25 gm of a non-electrolyte and 20 gm of water, is 271.9 K. If molar depression constant is 1.86 K mole-1, then molar mass of the solute will be [AFMC-98]
(A) 105.7 (B) 106.7 (C) 115.3 (D) 93.9

89. Solutions with same osmotic pressure are called [AFMC-99]
(A) Hypertonic (B) Normal
(C) Isotonic (D) None of these

90. Boiling point of chloroform was raised by 0.323 K, when 0.5143 g of anthracene was dissolved in its 35 g. Molecular mass of anthracene is
(Kb for $CHCl_3$ = 3.9 k – kg mol^{-1}) [AFMC-99]
(A) 177.42 g/mol (B) 79.42 g/mol
(C) 132.32 g/mol (D) 242.32 g/mol

91. The boiling point of heavy water is [AFMC-2000]
(A) 100°C (B) 101.4°C (C) 99°C (D) 110°C

92. Water dissolved in HCl [AFMC-2001]
(A) Shows a positive deviation from Raoult's law
(B) Shows a negative deviation from Raoult's law
(C) Obeys Raoult's law (D) Is an ideal solution

93. The mass of $CaCO_3$ produced when carbon dioxide is bubbled through 500 ml of 0.5 M $Ca(OH)_2$ will be [AFMC-2001]
(A) 10 g (B) 20 g (C) 50 g (D) 25g

94. The freezing point of equimolal aqueous solution will be highest for [IIT-90]
(A) $C_9H_5NH_3Cl$ (aniline hydrochloride)
(B) $Ca(NO_3)_2$ (C) $La(NO_3)_2$ (D) $C_6H_{12}O_6$ (glucose)

95. Given that ΔT_f is the depression in freezing point of the solvent in a solution of a non-volatile solute of molaity, m the quantity Lt (ΔT_f /m) is equal to .... m→O. [IIT-94]

96. The molecular weight of benzoic acid in benzene as determined by depression in freezing point method corresponds to [IIT-96]
(A) ionization of benzoic acid
(B) dimerization of benzoic acid
(C) trimerization of benzoic acid
(D) solution of benzoic acid

97. The normality of 0.3 M phosphorus acid (H_3PO_3) is,

[IIT-99]

(A) 0.1 (B) 0.9 (C) 0.3 (D) 0.6

98. In the depression of freezing point experiment, it is found that the [IIT-99]
(A) vapour pressure of the solution is less than that of pure solvent
(B) vapour pressure of the solution is more than that of pure solvent
(C) only solute molecules solidify at the freezing point
(D) only solvent molecules solidify at the freezing point

99. At 100°C and 1 atm, if the density of liquid water is 1.0 g. cm^{-3} and that of water vapour is 0.0006 gm cm^{-3}, then the volume occupied by water in 1 litre of steam at that temperature is [IIT-2000]
(A) 6 cm^3 (B) 60 cm^3 (C) 0.6 cm^3 (D) 0.06 cm^3

100. An aqueous solution of 6.3 g oxalic acid dihydrate is made up to 250 ml. The volume of 0.1N NaOH required to completely neutralize 10 ml of this is [IIT-2001]
(A) 40 ml (B) 20 ml (C) 10 ml (D) 4 ml

101. When the temperature is increased, surface tension of water [IIT-2002]
(A) increases (B) decreases
(C) remains constant (D) shows irregular behaviour

102. When the depression in freezing point is carried out, then equilibrium exist between [HT-2003]
(A) solid solvent and solid solvent
(B) liquid solvent and solid solvent
(C) liquid solvents (D) liquid solvent and liquid solvent

103. Freezing point of an aqueous solution is (−0.186)°C. Elevation of boiling point of the same solution is K_b=0.512°C, K_f = 1.86°C, find the increase in boiling point. [AIEEE-2002]
(A) 0.186°C (B) 0.0512°C (C) 0.092°C (D) 0.2372°C.

104. With increase of temperature, which of these changes [AIEEE-2002]
(A) Molality (B) Weight fraction of solute
(C) Fraction of solute present in water
(D) Mole fraction

105. In mixture A and B components show –ve deviation as [AIEEE-2002]
(A) $\Delta V_{mix} > 0$ (B) $\Delta H_{mix} < 0$
(C) A – B interaction is weaker than A – A and B – B interaction
(D) A – B interaction is stronger than A – A and B – B interaction

106. In a 0.2 molal aqueous solution of a weak acid HX the degree of ionization is 0.3. Taking k_f for water as 1.85, the freezing point of the solution will be nearest to [AIEEE-2003]
(A) −0.480°C (B) −0.360°C (C) −0.260°C (D) +0.480°C

107. 25ml of a solution of barium hydroxide on titration with a 0.1 molar solution of hydrochloric acid gave a titre value of 35ml. The molarity of barium hydroxide solution was [AIEEE-2003]
(A) 0.07 (B) 0.14 (C) 0.28 (D) 0.35

108. A pressure cooker reduces cooking time for food because [AIEEE-2003]
(A) Heat is more evenly distributed in the cooking space.
(B) Boiling point of water involved in cooking is increased.
(C) The higher pressure inside the cooker crushes the food material.
(D) Cooking involves chemical changes helped by a rise in temperature.

109. If liquids A and B form an ideal solution [AIEEE-2003]
(A) The enthalpy of mixing is zero
(B) The entropy of mixing is zero
(C) The free energy of mixing is zero
(D) The free energy as well as the entropy of mixing are each zero

110. How many moles of KI are required to produce 0.4 moles of K_2HgI_4 : [EAMCET-90]
(A) 0.4 (B) 0.8 (C) 3.2 (D) 1.6

111. A molal solution is one that contains one mole of solute in : [EAMCET-90]
(A) 1000 gms of the solvent (B) One litre of the solution
(C) One litre of the solvent (D) 22.4 litres of solution

112. The number of moles of water present in 90.0 gms. of water is .... [EAMCET-90]

113. The relative vapour pressure is equal to the [EAMCET-91]
(A) Ratio of the number of solute molecules to the total number of molecules in solution
(B) Ratio of the number of solvent molecules to the number of solute molecules
(C) Ratio of the number of solute molecules to the no. of solvent molecules
(D) Ratio of the number of solvent mol. td the total no. of molecules in sol

114. 9.8 grams of H_2SO_4 is present in two litres of a solution. The molarity of the solution is [EAMCET-91]
(A) 0.1 m (B) 0.05 m (C) 0.2m (D) 0.01m

115. The number of moles of barium carbonate which contains 1.5 moles of O_2 atoms is .... [EAMCET-91]

116. If 20 ml of 0.4 N NaOH solution completely neutralise 40 ml of a dibasic acid, the molarity of the acid

solution is [EAMCET-92]
(A) 0.1 M (B) 0.2 M (C) 0.3 M (D) 0.4 M

117. The degree of dissociation of an acid, HA in 0.1 M solution is 0.1%, its dissociation constant is [EAMCET-92]
(A) 1×10^{-3} (B) 1×10^{-7} (C) 1×10^{-10} (D) 1×10^{-14}

118. The mole fraction of water in 20% aqueous hydrogen peroxide solution is .... [EAMCET-92]

119. 3.5 gm of Carbon monoxide at 0^oC and 760 mm pressure contains .... [EAMCET-92]

120. A solution that obeys Raoult's law is called [EAMCET-93]
(A) Normal solution (B) Molar solution
(C) Ideal solution (D) Saturated solution

121. When the same amount of Zn is treated with excess H_2SO_4 and with excess NaOH, the ratio of the volumes of H_2 evolved is [EAMCET-94]
(A) 3 : 2 (B) 1 : 2 (C) 2 :1 (D) 1 :1

122. 5.85 gms of NaCl (Mol. wt 58.5) is dissolved in water and the solution made up to 500 ml the molarity of the solution will be [EAMCET-94]
(A) 0.1 (B) 0.2 (C) 1.0 (D) 0.117

123. The mole percentage of oxygen in a mixture of 7.0 gm of nitrogen and 8.0 gm of oxygen is [EAMCET-95]
(A) 8 (B) 16 (C) 24 (D) 50

124. 10.6 g of a substance of mol. wt. 106 was dissolved in 100 ml. 10 ml. of this solution was pipetted out into a 100 ml. flask and made up to the mark with distilled water. The molarity of the resulting solution is [EAMCET-98]
(A) 1.0 M (B) 10^{-2}M (C) 10^{-3}M (D) 10^{-4}M

125. How many litres of CO_2 at STP will be formed when 100 ml of 0.1 M H_2SO_4 reacts with excess of Na_2CO_3 ? [EAMCET-98]
(A) 22.4 (B) 2.24 (C) 0.224 (D) 5.6

126. Liquid benzene burns in oxygen according to $2C_6H_6(l) + 15O_2(g) \rightarrow 12CO_2(g) + 6H_2O(g)$. How many litres of oxygen are required for complete combustion of 39gm of liquid C_6H_6 ? (At. wt. of C = 12, O = 16) [EAMCET-98]
(A) 11.2 (B) 22.4 (C) 42 (D) 84

127. The weight of $AgNO_3$ (Mol.wt.170) present in 100 ml of its 0.25 M solution is [EAMCET-98(MBBS)]
(A) 4.25 g (B) 42.5 g (C) 17 g (D) 1.7 g

128. How many moles of potassim chlorate should be decomposed completely to obtain 67.2 litres of oxygen at STP ? [EAMCET-99]
(A) 3 (B) 4 (C) 1 (D) 2

129. A non-volatile solute (A) is dissolved in a volatile solvent (B). The vapour pressure of resultant solution is Ps. The vapour pressure of pure solvent is P If X is the mole fraction, which of the following is correct ? [EAMCET-2000]
(A) $P_S = P \times X_A$ (B) $P = P_S \times X_B$
(C) $P_S = P \times X_B$ (D) $P = P_S \times X_A$

130. Which of the following has the lowest bond angle ? [EAMCET-2000]
(A) H_2O (B) CH_4 (C) H_2S (D) NH_3

131. 500 ml of a solution contains 12.6 grams of oxalic acid (mol.wt.=126) 10 ml of this solution is diluted to 100ml in a flask. What is the molarity (in mol.lot-1) of the resultant solution ? [EAMCET(MBBS)-2000]
(A) 0.2 (B) 0.1 (C) 0.02 (4) 0.01

132. The concentration of a 100 ml solution containing X grams of Na_2CO_3 (mol. wt.= 106) is Y M. The values of X and Y are respectively [EAMCET-2002]
(A) 2.12, 0.05 (B) 1.06, 0.2
(C) 1.06, 0.1 (D) 2.12, 0.1

133. One mole of fluorine is reacted with two moles of hot and concentrated KOH. The products formed are KF, H_2O and O_2. The molar ratio of KF, H_2O and O_2 respectively is : [EAMCET-2002]
(A) 1 : 1 : 2 (B) 2 : 1 : 0.5
(C) 1 : 2 : 1 (D) 2 : 1 : 2

134. Equal volumes of 0.1 M Ag NO_3 and 0.2 M NaCl solutions are mixed. The concentration of nitrate ions in the resultant mixture will be [EAMCET(med.)-2002]
(A) 0.1M (B) 0.2 M (C) 0.05 M (D) 0.5M

135. Which one of the following represents the graph between log p (on y axis) and (on x axis) p = v.p. of a liquid T = absolute temperature. [EAMCET-2003]

(A) 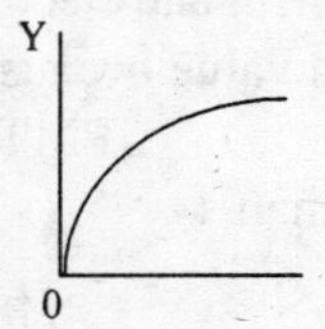
(B) 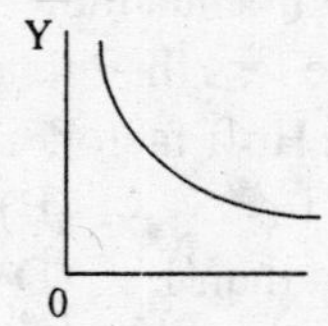
(C) 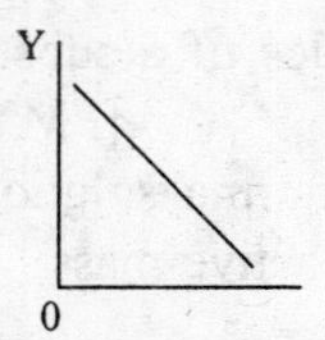
(D) 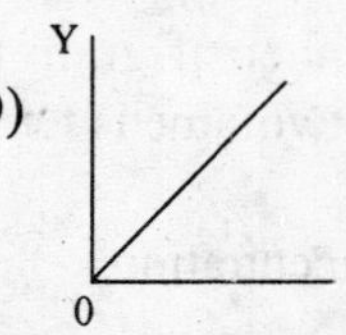

136. If one mole of substance is present in the 1 kg of solvent, which statement is correct ? [CPMT-96]
(A) Molar concentration (B) Molal concentration
(C) Normality (D) Strength gm/gm

137. 2N-HCl will have the same molar concentration as-

[CPMT-96]

(A) $0.5N - H_2SO_4$ (B) $1.0N - H_2SO_4$
(C) $2N - H_2SO_4$ (D) $4N - H_2SO_4$

138. vant Hoff factor of $Ca(NO_3)_2$ is- [CPMT-97]
(A) 1 (B) 2 (C) 3 (D) 4

139. The rise in the boiling point of a solution containing 1.8 gram of glucose in 100g of a solvent is 0.1°C. The molal elevation constant of the liquid is [CPMT-99]
(A) 0.01 k/m (B) 0.1 k/m (C) 1 k/m (D) 10 k/m

140. Number of moles of a solute per kilogram of a solvent is called [CPMT-99]
(A) Molarity (B) Molality (C) Normality (D) Formality

141. The osmotic pressure of 1 m solution at 27°C is [CPMT-99]
(A) 2.46 atm (B) 24.6 atm (C) 1.21 atm (D) 12.1 atm

142. 0.5 g of solute dissolved in 25 g of solvent. If molal elevation of boiling point is 1.82, find the molecular weight of solute if $K_b = 3.84$ [CPMT-2001]
(A) 100 (B) 200 (C) 300 (D) None of these

143. 5 moles of SO_2 and 5 moles of O_2 react to form SO_3. Only 60% SO_2 is used. Number of moles left in total [CPMT-2003]
(A) 6.5 (B) 10 (C) 8 (D) 8.5

144. A mixture of benzene and toluene form : [PMT-93]
(A) an ideal solution (B) non ideal solution
(C) suspension (D) emulsion

145. The colligative properties of a solution depends on : [PMT-93]
(A) number of solute particles present
(B) chemical nature of the solute particles present in it
(C) nature of the solvent used
(D) nature of both the solvent and solute

146. The experimental molecular weight of and electrolyte will always be less than its calculated value because the value of van't Hoff factor, 'i' is : [PMT-93]
(A) less than 1 (B) greater than 1
(C) equivalent than 1 (D) zero

147. The number of gram molercules of a substance present in unit volume is termed as : [PMT-93]
(A) activity (B) normal solution
(C) molar concentration (D) active mass

148. Colligative properties of a solution depend upon [PMT-94]
(A) nature of solute (B) nature of solvent
(C) the relative number of solute and solvent particles
(D) none of these

149. If 5.85 gms of NaCl are dissolved in 90 gms of water, the mole fraction of NaCl is [PMT-94]
(A) 0.1 (B) 0.01 (C) 0.2 (D) 0.196

150. The molality of a solution is [PMT-96]
(A) number of moles of solute per 1000 ml of the solvent
(B) number of moles of solute per 1000 gm of the solvent
(C) number of moles of solute per 1000 ml of the solution
(D) number of gram equivalents of solute per 1000 ml of the solution

151. An ideal solution is that which [PMT-96]
(A) shows positive deviation from Raoult's law
(B) shows negative deviation from Raoult's law
(C) has no connection with Raoult's law
(D) obeys Raoult's law

152. Which of the following is a colligative property ? [PMT-96]
(A) Boiling point (B) Freezing point
(C) Melting point (D) Osmotic pressure

153. Which of the following will produce the maximum depression in freezing point of its aqueous solution ? [PMT-96]
(A) 0.1 M glucose (B) 0.1 M sodium chloride
(C) 0.1 M barium chloride
(D) 0.1 M magnesium sulphate

154. One mole of solute A is dissolved in a given volume of a solvent. The association of the solute take place according to $nA \rightleftharpoons (A)_n$. The vant Hoff's factor is expressed as [PMT-97]
(A) $i = 1 - x$ (B) $i = 1 + \frac{x}{n}$
(C) $i = \frac{1 - x + \frac{x}{n}}{1}$ (D) $i = 1$

155. Freezing point of water is °C. At what temperature will one molal solution of NaCl freeze if NaCl is taken to be completely dissociated (mole depression constant of - water = 1.86) [PMT-97]
(A) - 3.72°C (B) - 1.86°C (C) 0°C (D) 3.72°C

156. The number of moles of solute in its solution is 20 and total number of moles are 80. The mole fraction of solute is [PMT-97]
(A) 2.5 (B) 0.25 (C) 1 (D) 0.75

157. What would happen to a reversible reaction at equlibrium when an inert gas is added while the pressure remains unchanged - [PMT-97]
(A) More of the product will be formed
(B) Less of the product will be formed
(C) More of the reactantswill be formed
(D) It remains unaffected

158. 200 ml of a solution contains 5.85 g dissolved sodium chloride. The concentration of the solution will be : [PMT-99]
(A) 1 molar (B) 2 molar (C) 0.5 molar (D) 0.25 molar

159. Molarity of pure water is : [PMT-99]
(A) 18 (B) 5.56 (C) 55.6 (D) 100

160. 20 ml of HCl solution requires 19.85 ml of 0.01 M NaOH solution for complete neutralization. The molarity of HCl solution is : [PMT-99]
(A) 0.0099 (B) 0.099 (C) 0.99 (D) 9.9

161. Which of the following does NOT show positive deviation from Raoults law ? [PMT-2000]
(A) Benzene - Chloroform (B) Benzene - Acetone
(C) Benzene - Ethanol (D) Benzene - Carbon tetrachloride

162. A mixture has 18 g water and 414 g ethanol. The mole fraction of water in mixture is (assume ideal behaviour of the mixture) [PMT-2000]
(A) 0.1 (B) 0.4 (C) 0.7 (D) 0.9

163. The vapour pressure of a solvent A is 0.80 atm. When a nonvolatile substance B is added to this solvent its vapour pressure drops to 0.6 atm. What is mole fraction of B in solution ? [PMT-2000]
(A) 0.25 (B) 0.50 (C) 0.75 (D) 0.90

164. A solution of 1 molal concentration of a solute will have maximum boiling point elevation when the solvent is : [PMT-2000]
(A) ethyl alcohol (B) acetone
(C) benzene (D) chloroform

165. The molarity of a solution made by mixing 50 ml of conc. H_2SO_4 (36N) with 50 ml of warter is : [PMT-2001]
(A) 36 M (B) 18 M (C) 9 M (D) 6 M

166. 171 g of cane sugar ($C_{12}H_{22}O_{11}$) is dissolved in 1 litre water. The molarity of the solution is : [PMT-2001]
(A) 2.0 M (B) 1.0 M (C) 0.5 M (D) 0.25 M

167. The vapour pressure of a pure liquid 'A' is 0.80 atm. When a non-volatile solute 'B' is dissolved in 'A' its vapour pressure become 0.60 atm. The mole fraction of 'B' in the solution is : [PMT-2001]
(A) 0.125 (B) 0.25 (C) 0.50 (D) 0.75

168. Which of the following does not show negative deviation from Raoult's law ? [PMT-2001]
(A) Acetone - chloroform (B) Acetone - Benzene
(C) Chloroform - Ether (D) Chloroform - Benzene

169. The volume of 0.1 M H_2SO_4 that is needed to completely neutralise 30 ml of 0.2 M NaOH solution is : [PMT-2001]
(a) 15 ml (B) 30 ml (C) 40 ml (D) 60 ml

170. A solution of H_2SO_4 whose 9.8 gm is dissolved in 2 litres of water has molarity [PMT-2002]
(A) 0.1 M (B) 0.05 M (C) 0.01 M (D) 0.2 M

171. Isotonic solutions have the same [PMT-2002]
(A) Density (B) Molar concentration
(C) Normality (D) Optical rotation

172. Colligative properties of a solution depend upon [PMT-2002]
(A) Nature of solvent (B) Nature of solute
(C) Relative numbers of solute and solvent in solution
(D) None of the above

173. A solute is soluble in two immiscible liquids which are present in mixture. The concentration of the solute in upper layer mixture will be [PMT-2002]
(A) Same as in the lower layer
(B) Less than in the lower layer
(C) More than in the lower layer
(D) In fixed ratio with that in the lower layer

174. 2.0 molal solution is obtained by dissolving 0.5 mol of solute in : [PMT-2003]
(A) 250 g. of solvent (B) 500 ml of solvent
(C) 250 ml of solution (D) 250 ml of solvent

175. 500 ml of solution contains 4.9 g of H_3PO_4 in water. Calculate the normality of the solution. [PMT-2003]
(A) 3.0 (B) 1.0 (C) 0.3 (D) 0.4

176. Chloroform when dissolved in ethyl alcohol a solution is obtained which shows : [PMT-2003]
(A) Ideal properties
(B) Non experimental of Henry law
(C) Negative deviation from Raoult's law
(D) Positive deviation from Raoult's law

177. Assuming the 90% dissociation of decimolar solution of each of following salt, which has the highest value of osmotic pressure. [PMT-2003]
(A) Aluminum sulphate
(B) Potassium Sulphate (C) Barium Chloride
(D) The mixture of equal volumes of (B) + (C)

178. Which of the following is not a colligative property? [PMT-2003]
(A) Lowering of vapour pressure
(B) Depression of freezing point
(C) Osmotic pressure (D) Elevation of boiling point

179. 36g of water and 828g. of ethyl alcohol form a mixture of ideal nature. What is the mole fraction of water in this mixture ? [PMT-2003]
(A) 0.9 (B) 0.7 (C) 0.4 (D) 0.1

180. For the strength "S" in gram per litre for substance

in a solution is related to its normality (N) and the eq. wt. (E) by [PMT-2003]

(A) $S = \frac{N}{E}$ (B) $S = \frac{E}{N}$ (C) $S = N \times E$ (D) None

181. The molecular weight of a tribasic acid is M. The equivalent weight of the acid will be: [PMT-2003]

(A) $\frac{M}{2}$ (B) 2M (C) $\frac{M}{3}$ (D) $\sqrt{\frac{M}{3}}$

182. The wt. of solute (W) required to prepare a standard solution of solutes of known normality in a definite volume is given by: (N- normality, E- equivalent weight of solute. V- volume of solution) [PMT-2003]

(A) $W = \frac{NEV}{1000}$ (B) $W = \frac{NE}{1000}$

(C) W = 0.01 NEV (D) $W = \frac{0.01NE}{V}$

183. Which one of the following liquid-pairs shows a positive deviation from Raoult's law ? [PET-93]

(A) Acetone-chloroform (B) Benzene-methanol
(C) Water-hydrochloric acid (D) Water-nitric acid

184. The molarity of a solution prepared by dissolving 7.1 g of Na_2SO_4 in 100 ml water is [PET-93]

(A) 2.0 M (B) 1.0 M (C) 0.5 M (D) 0.05 M

185. Acetic acid dissolved in benzene shows a molecular mass of [PET-93]

(A) 30 (B) 60 (C) 120 (D) 240

186. On dissolving 1 mole of each of the following acids in 1 litre water, the acid which does not give a solution of strength 1 N is [PET-93]

(A) Hydrochloric acid (B) Perchloric acid
(C) Nitric acid (D) Phosphoric acid

187. A molal solution is one that contains one mole of solute in : [PET-94]

(A) 1000 grams of the solvent
(B) One litre of the solution
(C) One litre of the solvent (D) 22.4 litres of the solution

188. The vapour pressure of a solution is : [PET-94]

(A) Directly proportional to the mole fraction of the solvent
(B) Inversely proportional to the mole fraction of the solvent
(C) Directly proportional to the mole fraction of the solute
(D) Inversely proportional to the mole fraction of the solute

189. Which of the following solutions will have the highest osmotic pressure ? [PET-94]

(A) 0.1 M Sucrose (B) 0.1 M Sodium chloride
(C) 0.1 M Barium chloride (D) 0.1 M Urea

190. The molarity of solution prepared by adding 7.1 grams of Na_2SO_4 (formula weight 142 amu) to water to make 100 ml volume is [PET-95]

(A) 2.0 M (B) 1.0 M (C) 0.5 M (D) 0.05 M

191. On dissolving common salt in water, the [PET-95]

(A) Boiling point of water decreases
(B) Boiling point of water increases
(C) Boiling point of water remains same
(D) Nothing can be predicted

192. The liquid pair benzene-toluene shows- [PET-95]

(A) Irregular deviation from Raoult's Law
(B) Negative deviation from Raoult's Law
(C) Positive deviation from Raoult's Law
(D) Practically no deviation from Raoult's Law

193. Which of the following is a colligative property [PET-96]

(A) Change in free energy (B) Change in pressure
(C) Heat of vaporization (D) Osmotic pressure

194. If the solution boils at a temperature T_1 and the solvent at a temperature T_2 the elevation of boiling point is given by [PET-96]

(A) $T_1 + T_2$ (B) $T_1 - T_2$ (C) $T_2 - T_1$ (D) $T_1 \div T_2$

195. The pH of 10^{-5} M aqueous solution of NaOH [PET-96]

(A) 5 (B) 7 (C) 9 (D) 11

196. Which one of the following is not an ideal solution ? [PET-97]

(A) $C_2H_5Br + C_2H_5I$ (B) $C_6H_5Cl + C_6H_5Br$
(C) $C_6H_6 + C_6H_5CH_3$ (D) $C_2H_5I + C_2H_5OH$

197. "The relative lowering of the vapour pressures is equal to the mole fraction of the solute". The law is called: [PET-97]

(A) Henry's law (B) Raoult's law
(C) Ostwald's law (D) Arrhenius's law

198. The order of osmotic pressures of equimolar solutions of $BaCl_2$, NaCl and glucose will be : [PET-97]

(A) $BaCl_2$ > NaCl > Glucose
(B) BaCl > $NaCl_2$ > Glucose
(C) Glucose > $BaCl_2$ > NaCl
(D) Glucose > NaCl > $BaCl_2$

199. Molal depression constant for water is 1.86°C. The freezing point of a 0.05 molal solution of a non-electrolyte in water is : [PET-97]

(A) −1.86°C (B) −0.93°C (C) −0.093°C (D) 0.93°C

200. Which of the following will have the highest boiling point at 1 atm pressure ? [PET-98]

(A) 0.1 M NaCl (B) 0.1 M Sucrose
(C) 0.1 M $BaCl_2$ (D) 0.1 M Glucose

201. What will be the molality of a solution having 18 g of glucose (mol. wt. = 180) dissolved in 500 g of water ? [PET-98]

(A) 1m (B) 0.5 m (C) 0.2 m (D) 2 m

202. An aqueous solution freezes at $-0.186°C$ ($k_f = 1.86°$; $k_b = 0.512°$) What is the elevation in boiling point ? [PET-98]

(A) 0.186 (B) 0.512 (C) 0.512/1.86 (D) 0.0512

203. Which one of the following is a colligative property? [PET-98]

(A) Surface Tension (B) Osmotic Pressure
(C) Viscosity (D) Refractive Index

204. A molal solution is one that contains 1 mole of a solute in: [PET-99]

(A) 1000 g of the solvent (B) One litre of the solvent
(C) One litre of the solution (D) 22.4 litre of the solution

205. Which of the following is not a colligative property? [PET-99]

(A) Osmotic pressure (B) Elevation in boiling point
(C) Vapour pressure (D) Depression in freezing point

206. Solutions with same osmotic pressure are called: [PET-99]

(A) Hypertonic (B) Hypotonic (C) Isotonic (D) Normal

207. A monoprotic acid in a 0.1 M solution ionizes to 0.001%. Its ionization constant is: [PET-99]

(A) 1.0×10^{-3} (B) 1.0×10^{-6}
(C) 1.0×10^{-8} (D) 1.0×10^{-11}

208. The molar solution of sulphuric acid is equal to: [PET-99]

(A) N solution (B) 2N solution
(C) N/2 solution (D) 3N solution

209. The solution of sugar in water contains free/hydrated: [PET-2000]

(A) Atoms (B) Molecules
(C) Ions (D) Atoms and molecules

210. Which of the following is a non-colligative property? [PET-2001]

(A) Elevation of boiling point
(B) Depression freezing point
(C) Osmotic pressure
(D) Optical activity

211. The law stating that "the relative lowering of vapour pressure is equal to the mole fraction of solute" in the solution is known as : [PET-2001]

(A) Faraday's law (B) Raoult's law
(C) vant Hoff's law (D) None of these

212. Conc. H_2SO_4 has a density of 1.98 gm/ml and is 98% H_2SO_4 by weight. Its normality is [PET-2002]

(A) 2 N (B) 19.8 N (C) 39.6 N (D) 98 N

213. The condition for an ideal solution is [PET-2002]

(A) ΔH mixing = 0 (B) ΔV mixing $\neq 0$
(C) $P_A \neq P_A^\circ \times A$ (D) ΔH mixing > 1

214. The boiling point of a solution of 0.11 gm of a substance is 15 gm of ether was found to be 0.1° C higher than that of the pure ether. The molecular weight of the substance will be [$K_b = 2.16$] [PET-2002]

(A) 148 (B) 158 (C) 168 (D) 178

215. Acetic acid dissolved in benzene shows a molecular weight of [PET-2002]

(A) 60 (B) 120 (C) 180 (D) 240

216. The vapour pressure of liquid A is 0.80 atm. On mixing a non volatile B to A, its vapour pressure becomes 0.6 atm. The mole fraction of B in the solution is : [PET- 2003]

(A) 0.125 (B) 0.25 (C) 0.75 (D) 0.50

217. Which of the follwoing mixtures does not show positive deviation from Raoult's law ? [PET- 2003]

(A) Acetone - ethanol (B) Benzene - acetone
(C) Acetone - chloroform (D) Water - ethanol

218. Equimolar solutions of $BaCl_2$, $NaCl_2$ and sucrose have osmotic pressure in the order : [PET- 2003]

(A) $BaCl_2$ > NaCl > Sucrose
(B) Sucrose > NaCl > $BaCl_2$
(C) NaCl > $BaCl_2$ > Sucrose
(D) $BaCl_2$ > Sucrose > NaCl

219. The boiling point of a solvent containing non-volatile solute is [CBSE-90]

(A) depressed (B) elevated
(C) does not change (D) none of these

220. Number of moles of the solute dissolved per litre of the solutions is known as [CBSE-90]

(A) molality (B) molarity
(C) formality (D) normality

221. 50.0ml of 0.2N HCI, 50.0 ml of 0.1 NH_2SO_4 and 100.0ml of 0.2 N HNO_3 are mixed. The normality of resulting solution is [CBSE-90]

(A) 0.10N (B) 0.15N (C) 0.175N (D) 0.20N

222. A mixture of two completely miscible non-ideal liquids which distil as such without change in its compositon at a constant temperature as if it were a pure liquid. This mixture is known as [CBSE-90]

(A) binary liquid mixture (B) azeotropic mixture
(C) eutelic mixture (D) ideal mixture

223. Which one of the following modes of expressing concentration of solutions is independent of temperature ? [CBSE-92]

(A) Normality (B) Grams per litre

(C) Molarity (D) Molality

224. Which of the following 0.10 m aqueous solution will have the lowest freezing point ? [CBSE-97]
(A) $Al_2(SO_4)_3$ (B) $C_5H_{10}O_5$ (C) KI (D) $C_{12}H_{22}O_{11}$

225. Formation of a solution from two components can be considered as [CBSE-2003]
I. Pure solvent → separated solvent molecules, ΔH_1
II. Pure solute → separated solute molecules, ΔH_2
III. Separated solvent and solute molecules → solution, ΔH_3
(A) $\Delta H_{soln} = \Delta H_3 - \Delta H_1 - \Delta H_2$
(B) $\Delta H_{soln} = \Delta H_1 - \Delta H_2 - \Delta H_3$
(C) $\Delta H_{soln} = \Delta H_1 - \Delta H_2 - \Delta H_3$
(D) $\Delta H_{soln} = \Delta H_1 - \Delta H_2 - \Delta H_3$

226. The Molarity of a solution containing 1.0 gms. NaOH in 250 ml of water is .... [EAMCET-90]

227. 6 grams of urea is dissolved in 16.2 grams of water at 25°C. What is the vapour pressure (in mm) of the resultant solution ? The vapour pressure of water at 25°C is 24 mm. [EAMCET(MBBS)-99]
(A) 2.4 (B) 21.6 (C) 23.6 (D) 18

228. In the phenomenon of osmosis, the membrane allows passage to [CBSE-2003]
(A) solute molecules only (B) solvent molecules only
(C) both solute and solvent molecules (D) none

229. T=550 K, V=49.8 l, n= 2 mol., then the pressure will be (atm.) [BHU-2003]
(A) 2 (B) 4 (C) 1.5 (D) 2.5

230. Diffusion of solvent through a semipermeable membrane is called [AFMC-2003]
(A) Osmosis (B) Diffusion
(C) Absorption (D) Plasmolysis

231. Surface water contains [AFMC-2003]
(A) Salt (B) Suspended impurities
(C) Organic compounds
(D) Salt and organic compounds

232. Camphor is often used in molecular mass determination because [CBSE-PMT-2004]
(A) it is readily available
(B) it has a very high cryoscopic constant
(C) it is volatile (D) it is solvent for organic substances

233. 0.004 M Na_2SO_4 is isotonic with 0.01 M glucose at equilibrium. What is the apparent degree of dissociation of Na_2SO_4? [IIT-2004]
(A) 25% (B) 50% (C) 75% (D) 85%

Solutions

1.	(C)	51.	(B)	101.	(B)	151.	(D)	201.	(C)
2.	(B)	52.	(B)	102.	(B)	152.	(D)	202.	(D)
3.	(A)	53.	(B)	103.	(B)	153.	(C)	203.	(B)
4.	(A)	54.	(D)	104.	(C)	154.	(C)	204.	(A)
5.	(A)	55.	(B)	105.	(B)	155.	(A)	205.	(C)
6.	(A)	56.	(C)	106.	(A)	156.	(B)	206.	(C)
7.	(C)	57.	(B)	107.	(B)	157.	(D)	207.	(D)
8.	(C)	58.	(C)	108.	(B)	158.	(C)	208.	(B)
9.	(A)	59.	(A)	109.	(A)	159.	(C)	209.	(B)
10.	(C)	60.	(D)	110.	(B)	160.	(A)	210.	(D)
11.	(A)	61.	(D)	111.	(A)	161.	(A)	211.	(B)
12.	(C)	62.	(A)	112.	*	162.	(A)	212.	(C)
13.	(B)	63.	(D)	113.	(A)	163.	(A)	213.	(A)
14.	(B)	64.	(B)	114.	(B)	164.	(D)	214.	(B)
15.	(C)	65.	(D)	115.	*	165.	(C)	215.	(B)
16.	(A)	66.	(A)	116.	(A)	166.	(C)	216.	(B)
17.	(D)	67.	(B)	117.	(B)	167.	(C)	217.	(C)
18.	(C)	68.	(D)	118.	*	168.	(B)	218.	(A)
19.	(C)	69.	(A)	119.	*	169.	(B)	219.	(B)
20.	(B)	70.	(D)	120.	(C)	170.	(B)	220.	(B)
21.	(C)	71.	(A)	121.	(D)	171.	(B)	221.	(C)
22.	(C)	72.	(B)	122.	(B)	172.	(C)	222.	(B)
23.	(C)	73.	(B)	123.	(D)	173.	(D)	223.	(D)
24.	(A)	74.	(A)	124.	(B)	174.	(A)	224.	(A)
25.	(C)	75.	(B)	125.	(C)	175.	(C)	225.	(B)
26.	(D)	76.	(B)	126.	(D)	176.	(D)	226.	*
27.	(B)	77.	(C)	127.	(A)	177.	(A)	227.	(B)
28.	(B)	78.	(B)	128.	(D)	178.	(A)	228.	(B)
29.	(D)	79.	(A)	129.	(C)	179.	(D)	229.	(C)
30.	(A)	80.	(B)	130.	(C)	180.	(C)	230.	(A)
31.	(C)	81.	(B)	131.	(C)	181.	(C)	231.	(B)
32.	(B)	82.	(B)	132.	(C)	182.	(A)	232.	(B)
33.	(C)	83.	(A)	133.	(B)	183.	(B)	233.	(C)
34.	(D)	84.	(A)	134.	(C)	184.	(C)		
35.	(C)	85.	(D)	135.	(C)	185.	(C)		
36.	(C)	86.	(A)	136.	(B)	186.	(D)		
37.	(A)	87.	(C)	137.	(C)	187.	(A)		
38.	(B)	88.	(A)	138.	(C)	188.	(A)		
39.	(B)	89.	(C)	139.	(C)	189.	(C)		
40.	(C)	90.	(A)	140.	(B)	190.	(C)		
41.	(A)	91.	(B)	141.	(B)	191.	(B)		
42.	(B)	92.	(B)	142.	(B)	192.	(D)		
43.	(B)	93.	(D)	143.	(C)	193.	(D)		
44.	(B)	94.	(D)	144.	(A)	194.	(B)		
45.	(A)	95.	*	145.	(A)	195.	(C)		
46.	(B)	96.	(B)	146.	(B)	196.	(D)		
47.	(A)	97.	(D)	147.	(C)	197.	(B)		
48.	(B)	98.	(A)(D)	148.	(C)	198.	(A)		
49.	(D)	99.	(C)	149.	(D)	199.	(C)		
50.	(A)	100.	(A)	150.	(B)	200.	(C)		

95. k_f
112. 5 moles
115. 0.5 moles
118. 55/63
119. N/8 molecules
226. 0.1 M

SOLID STATE

Molecules having random or chaotic motion and exert little or no force on one another belong to gaseous state.

Solids : They are characterized by incompressibility, rigidity and mechanical strength. This shows that the atoms, molecules or ions that make up solids are closely packed. They have well-rendered molecular arrangement.

Crystalline	Amorphous
1. have definite and regular geometry.	1. have no definite geometrical shape.
2. have sharp m. p. (abruptly changes into liquid)	2. don't have sharp m. p.
3. Anisotropic	3.isotropic

Isotropic : The properties such as electrical conductivity, thermal conductivity, mechanical strength and refractive index are the same in all directions. Amorphous substances are isotropic. Liquids and gases are isotropic.

The size of the crystal depends on the rate at which it is formed. The slower the rate, the bigger the crystal.

Interfacial angles : The size and shape of the crystals of one and the same substance may differ but the interfacial angles between any two corresponding faces of the crystals remain the same.

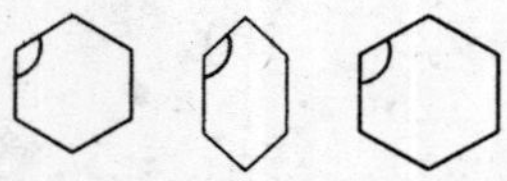

Symmetry in crystal systems :

1. Plane of symmetry : When an imaginary plane can divide a crystal into two parts, such that one is the exact mirror image of the other.

In a cubic crystal, there are (a) Rectangular plane of symmetry (b) Diagonal plane of symmetry

(a) Rectangular plane of symmetry : These are the planes situated mid way of faces. Therefore 3 are possible.

(b) Diagonal plane of symmetry : Planes touching opposite edges. There are 12 edges. Therefore 6 are possible.

2. Axis of symmetry : It is a line about which the crystal may be rotated such that, it presents the same appearance more than once during the complete revolution.

The cube presents the same appearance 4 times in one revolution. The original appearance after every 90°.

- if after 180° Diad axis
- if after 120° Triad axis
- if after 60° Hexad axis

Depending upon the number of times the crystal appears unchanged in the course of such rotation, it can have

(a) Axis of 4 fold symmetry : Imagine a line passing through the centres of two opposite faces the cube appears 4 times with the same shape. There are 6 faces. Therefore 3 such axis of 4 fold symmetry possible.

(b) Axis of 3 fold symmetry : Imagine a line passing through the opposite corners. There are 8 corners. Therfore, 4 such axis of 3 fold symmetry possible.

(c) Axis of 2 fold symmetry : Imagine a plane through the centre of the two diagonally opposite edges. There are 12 edges. Therefore, 6 such axis of 2 fold symmetry possible.

(3) Centre of symmetry : The centre of symmetry of a crystal is such a point that any line drawn through it intersects the surface of the crystal at equal distances in both the directions.

The crystal may have any number of planes or axes but it has only one centre of symmetry.

For a cube : $\overset{RPS}{3} + \overset{DPS}{6} + \overset{3FFS}{3} + \overset{4ThFS}{4} + \overset{6TFS}{6} + \overset{CS}{1} = 23$

There can be a maximum of 32 different combinations of elements of symmetry. From geometrical considerations, there can be 32 different combinations of elements of symmetry of a crystal called 32 point groups or 32 systems.

Seven crystal systems or habits

1. Cubic Space Lattices (simple, bc, fc)
2. Orthorhombic Space Lattices (simple, bc, fc, End centred.)
3. Tetragonal (simple, bc)
4. Monoclinic (simple, fc)
5. Triclinic 6. Hexagonal 7. Rhombohedral

Bravais showed that there are 14 different basic arrangements possible for arranging similar points in space. Thus the total nimber of space lattices belonging to all the seven crystal systems put together is 14.

There can be 230 different arrangements known as space groups. (230 crystal forms are possible.)

The crystals belonging to cubic system have three kinds of Bravais lattices depending upon the shape of the unit cells.

Simple cubic lattice : There are points only at the corners e.g. Po

$$8 \times \frac{1}{8} = 1$$

For a simple cubic unit cell, length of the edge of the unit cell = 2 x radius of the atom. coordination no. 6

Body centred cubic lattice : There are points at the corners as well as at the centre of each cube. $8 \times \frac{1}{8} + 1 = 2$;

For a body centred cubic unit cell, length of the edge of the unit cell = $\frac{4}{\sqrt{3}}$x radius of the atom coordination. no. 8

Face centred cubic lattice : There are points at the corners as well as at the centre of the each of the six faces of the cube.

$$8 \times \frac{1}{8} + \frac{1}{2} \times 6 = 1 + 3 = 4$$

For a face centred cubic unit cell, length of the edge of unit cell = $\sqrt{8}$ x radius of the atom. coordination no. 12.

Atoms at edge count as 1/4.

Crystal System	Bravais Lattices	Examples	Interfacial angles
1.Cubic	simple, fc, bc	NaCl, Alumn, CaF_2, ZnS	$\alpha=\beta=\gamma=90^\circ$, a=b=c
2.Hexagonal	primitive	Graphite, Ice	$\alpha=\beta=90^\circ$, $\gamma=120^\circ$, $a=b\neq c$
3.Monoclinic	primitive, Ec	$FeSO_4.7H_2O$, $Na_2SO_4.10H_2O$, $NaHCO_3$, $CaSO_4.2H_2O$	
4.Rhombic or Orthorhombic	primitive, Bc, Fc, Ec	KNO_3, $MgSO_4.7H_2O$, $BaSO_4$, $PbCO_3$, K_2SO_4	$\alpha=\beta=\gamma=90^\circ$, $a\neq b\neq c$
5.Triclinic	primitive	$K_2Cr_2O_7$, $CuSO_4.5H_2O$	
6.Rhombohedral	primitive	$KMnO_4$, Calcite	$\alpha=\beta\neq 90^\circ$, a=b=c
7.Tetragonal	primitive, Bc	$K_4[Fe(CN)_6]$, $NiSO_4$, TiO_2	

Types of Crystals :

1. Ionic crystals : Ions are the units; bad comductors. In molten state they act as good conductors of electricity. e.g. NaCl, CaF_2, Na_2SO_4, KNO_3, CsCl, ZnS etc. (NaCl - FCC; Unit cell of NaCl consists of 14 chloride ions and 13 Na ions. Each Cl^- is surrounded by six Na^+ and similarly each Na^+ is surrounded by 6 chloride ions.).

2.Covalent crystals : non volatile. e.g. graphite, diamond, SiC, BN, Quartz (SiO_2)

3. Metallic crystals : positive metal ions are dispersed in electron gas. e.g. Cu, Al, Fe, Na

4. Molecular crystals : van der Waal forces hold atoms or molecules
(a) polar molecular crystals: Electrostatic forces between dipoles + vander Waal's forces. e.g. S_2Cl_2, PCl_3, H_2O
(b) non polar molecular crystals : e.g. Ice, ammonia solid, H_2, He, CO_2 etc.

Liquid Crystals :
(i) Smectic (soap like) : e.g. Cholesteryl oleyl carbonate
(ii) Nematic (thread like) : e.g. N-p-methoxy benzylidene - p- butyl aniline
(iii) Cholesteric (thread like optically active) : For detecting tumours in the body by the method called thermography.

Polymorphism : One and the same substance can crystallise in more than one form. e.g. NaCl crystallises as cubes and also in octahedral form; Silica as Quartz, flint, opal, sand and agate.

Allotropism : If the substance showing polymorphism is the element, the process is allotropism.

Enantiotropism : allotropic forms that can undergo reversible change are said to be enantiotropic.

e.g. $S_{rh} \xrightleftharpoons{95.6^\circ C} S_{monoclinic}$

Monotropism : The allotropic forms that can not change into one another reversibly are said to be monotropic e.g. O_2 & O_3 ; Red P & Yellow P

Isomorphism (the same form) **:** When crystals of different chemical substances exhibit the same crystalline form e.g. Ag_2S - PbS; Arsenates - phosphates.

Imperfections in crystals : Ionic compounds are ordinarily non conductors in solid state at room temperature. However, when the temp is raised, there is a significant raise in conductivity e.g. conductivity of NaCl at its m. p. is 10^{10} times grater than at 298K. The increase in the conductivity with increase in temp. of ionic solids is due to the defects in the crystal structure. These defects are due to the missing of ions or out of place of ions.

Oxides and sulphides of heavy metals like PbS have enough conductivity at room temperature and such solids are classified as semiconductors.

Metals : good conductors of electricity, $1x10^4$ ohm^{-1}-cm^{-1}. Conductivity decreases with increase in temperature.

Non metals : Extremely small or undetectable

Semi conductors : Small but measurable, 10 to $10^{-5}ohm^{-1}cm^{-1}$.

Si and Ge also behave as semiconductors. Small amounts of impurities incorporated into the crystal lattice of these elements convert them from insulators to semiconductors. Extremely pure samples of Ge and Si contain virtually no free electrons as four valency electrons of each atom are tightly held in the covalent bonds formed with the neighbouring atoms. Therefore, they are non conductors. When a small amount of a foreign atom (even 1 in a million) gets introduced or doped into the crystal lattice, the situation drastically changes. e.g. an atom of As or Sb with five valency electrons can fit into the tetrahedral lattice of Ge or Si only by setting one of the five electrons relatively free. Such electrons moving under the influence of electric field give rise to n-type semiconductors (named so because negative charge flows). However, when a trivalent B atom is introduced, electron deficiency is created at the site of foreign atom. Such a defect is known as formation of positive hole in the lattice. Under the influence of an electric field, electrons from a neighbouring atom fill this hole leaving a positive hole around the atom. Such semiconductors are known as p-type semiconductors due to the moment of positive hole through the lattice in them. A semiconductor in which there is a junction between an electron rich and electron poor regions acts as a rectifier i.e. can convert A.C. to D. C. The n-p junction can be formed by traces of B and As from different sides.

Stoichiometric defects:
- **Schottky** (paired cation & anion vacancies), density changes; electrical neutrality is maintained e.g. NaCl, CsCl, KCl, KBr
- **Frenkel** (shifting of an ion into interstitial position), density remains same e.g. AgBr, ZnS, CaF_2, AgI

Non stoichiometric defects:
- F^- centres(the electrons are trapped in the anion vacancies). Excess of Li in LiCl makes it pink. Excess of K in KCl makes it violet.
- Fool's gold : Existence of Fe^{+2} and Fe^{+3} side by side results in free exchange of electrons between two such sites leading to metallic lusture. some samples of Iron pyrites shine like gold due to this. Such samples are called Fool's gold. e.g. FeS.FeO

Crystal packing : The constituent particles in the formation of crystals are either atoms, ions or molecules. These particles may be of various shapes and thus the mode of packing of these particles will change according to their shapes.

Packing fraction = v / V (v = volume occupied by atoms in unit cell, V = Total volume of the unit cell)

The two versions of hexagonal close packing are :

AB ... AB ... and ABC .... ABC .... patterns

Hexagonal close packing, hcp or AB.......AB....... AB

e.g. Mg, Zn, Mo, V, Cd etc.

Cubic close packing or ccp or ABC ABC ....

e.g. Cu, Ag, Au, Ni, Pt

b c c Li, Na, K, Rb, Cs, Ba

The number of nearest neighbour of each atom in a crystal lattice is called its coordination number.

h c p	co. no. 12
f c c	12
b c c	8
simple cubic	6

Interstitial sites in close packed lattices : In a close packing of spheres, there is always some empty space left. The empty space is called hole or void or interstitial site. Common interstitial sites in closely packed spheres are :

r^+ / r^- ratio	Coordination number	Shape of the void	example
0.155 - 0.225	3	Planar	B_2O_3
0.225 - 0.414	4 (4:4)	Tetrahedral	ZnS, ZnO
0.414 - 0.732	6 (6:6)	Octahedral	NaCl
0.732 - 1.000	8 (8:8)	bcc	CsCl
1.000		fcc	

Structure of Ionic crystals :

NaCl : Coordination number is 6. fcc arrangement (6:6 arrangement). AB type. Some NaCl type crystals are LiCl, KCl, NaI, RbI, AgF, AgCl, AgBr, etc.

CsCl : bcc structure. Eight Cs^+ at corners with a Cl^- at the centre, Some CsCl type crystals are CsBr, CsI, NH_4Cl etc. AB type ionic solids.

ZnS : c c p arrangement: It possess 4:4 coordination. Some crystals having zinc blende structure are AgI, HgS etc.

CaF_2 : Each Ca^{2+} is in contact with 8 F^- ions and each F^- is in contact with 4 Ca^{2+} ions. It has 8:4 arrangement BaF_2, SrF_2, CuF_2, PbF_2 etc have fluorite structure.

Silicates : The basic unit of all silicates is SiO_4^{4-}.

-frame work silicate : Quartz, zeolites etc.

-Pyrosilicates contain $Si_2O_7^{6-}$ anion

-Sheet silicate – Mica, Talc

Superconductivity : The phenomenon of offering no resistance to the flow of electric current is known as superconductivity. It is exhibited by most metals at 2K and 5K.

e.g. $Tl_2Ca_2Ba_2O_{10}$ (125 K); $Bi_2Ca_2Sr_2Cu_3O_{10}$ (105 K); $YBa_2Cu_3O_7$ (90 K)

1.Ferroelectric : Dipoles are polarized even when no current is passed e.g. Rochelle salt, KH_2PO_4, B_2TiO_3

2. Piezoelectric : A polar crystal which produces electricity on application of mechanical stress. e.g. Record players

3. Antiferroelectric : Inspite of having dipoles, they do not exhibit ferroelectric character (polar polyhedra in them point up & down. Therefore, net dipole moment is zero. e.g. $PbZrO_3$(lead Zirconate)

4. Pyroelectric : On heating, some of the polar crystals produce feeble electric current. e.g. Tourmalin

Law of rational Indices : A unit cell infact is the smallest repeating unit in space lattice which when repeated over and over again results in a crystal of the given substance. The unit cell infact is the smallest sample that represents the picture of the entire crystal. The crystal may be considered to consist of infinite number of unit cells.

Let OX, OY and OZ respresent the three crystallographic axes with 0 as origin. ABC is a unit plane which cuts the three axes at A B C respectively, giving unit intercepts OA = a; OB = b and OC = c. The lengths of a, b and c may or may not be equal but thier ratio OA : OB :OC = a: b: c is constant. Any other face of the crystal K L M will cut the three axes such that OL, OM and OK are multiples of a, b, c. The multiples of a, b, c are called the Weiss indices. Reciprocal of Weiss indices are called the Miller indices.

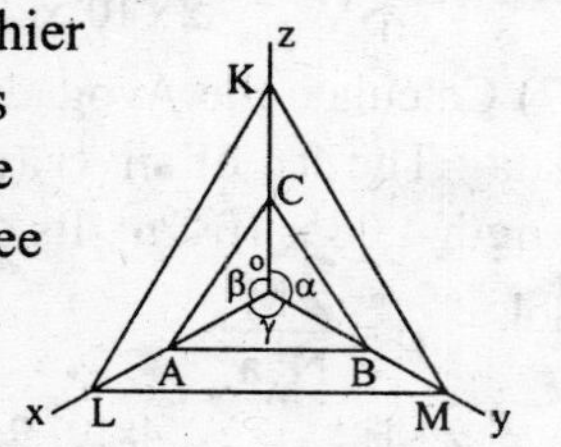

Representation : Weiss indices : e.g. 211, 110, -101 etc.

Miller indices : e.g. (211), (110), $(\bar{1}01)$ etc.

Bragg's Equation : Planes A B C and D E F be drawn perpendicular to the incident and reflected beams respectively. Drawing OL and OM perpendicular to the incident and reflected beams, it will be seen that the difference in the path lengths (say δ) of the waves reflected from the frist two planes is given by

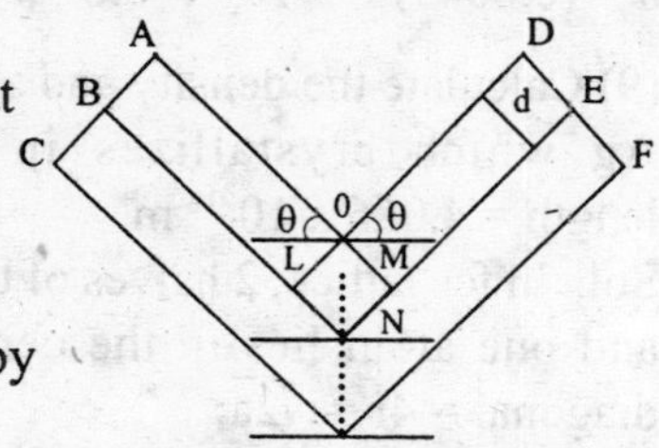

$\delta = LN + NM$

This should be equal to whole number multiple of wave length λ i.e. $n\lambda = LN + NM$

since the triangles O L N and O M N are congruent,

$LN = MN$

$\therefore n\lambda = 2LN$

or $n\lambda = 2d \sin\theta$

The distance between the parallel planes in a crystal are designated by $d_{h,k,l}$.

$$d_{h,k,l} = \frac{a}{\sqrt{h^2 + k^2 + l^2}}$$

where a = length of the cube side h, k, l are the Miller indices of the plane.

Things to remember :

Crystalline substances : definite geometry; sharp m. p.; Anisotropic e.g. NaCl, S, Sugar etc.

Amorphous : no definite geometry e.g. glass, rubber, plastics etc. Infact, they are super cooled liquids; isotropic

Elements of Symmetry : a cube has 23 elements of symmetry; there can be 32 elements of symmetry.

Crystals : There are 7 crystal habits or systems; 14 Bravais arrangements or space lattices. There can be 230 different arrangements known as space groups.

Cholesteric liquid crystals are optically active.

Polymorphism; allotropism; isomorphism; enantiotropism; monotropism.

Defects : n-type, p-type, n-p type

Stoichiometric defects
- Schottky (paired cation & anion vacancies) NaCl, CsCl, KBr
- Frenkel (shifting of an ion into interstitial position) AgBr, ZnS, CaF_2

Non stoichiometric defects
- F centres (the electrons trapped in the anion vacancies)
- Fool's gold.

Crystal packing :

hcp or AB ..... AB ..... AB ....

ccp or ABC ..... ABC ....

r^+/r^- ratio	Coordination no.	Shape	examples
0.155 - 0.225	3	Planar	B_2O_3
0.225 - 0.414	4	Tetrahedral	Zn S
0.414 - 0.732	6	Octahedral	NaCl
0.732 - 1.000	8	bcc	CsCl

Bragg's equation : $n\lambda = 2d \sin\theta$

multiples of the intercepts – weiss indices

Reciprocal of weiss indices is Miller indices

$$d_{h,k,l} = \frac{a}{\sqrt{h^2 + k^2 + l^2}}$$

a = length of the cube side
d = distance between parallel planes in a crystal
h, k, l = Miller indices of the plane

Theoretical density of the crystal can be calculated from the equation

$$\rho = \frac{nM}{N_o V} \quad \text{or} \quad \rho = \frac{zM}{N_A a^3}$$

n = no. of atoms in a unit cell
M = mol. wt.
V = vol. of the unit cell

No. of atoms in a simple cube : $8 \times \frac{1}{8} = 1$

bcc $8 \times \frac{1}{8} + 1 = 2$

fcc $8 \times \frac{1}{8} + (\frac{1}{2} \times 6) = 4$

In bcc, $r = a \times \sqrt{\frac{3}{4}}$; In fcc, $r = \sqrt{\frac{2a}{4}}$

Anisotropy

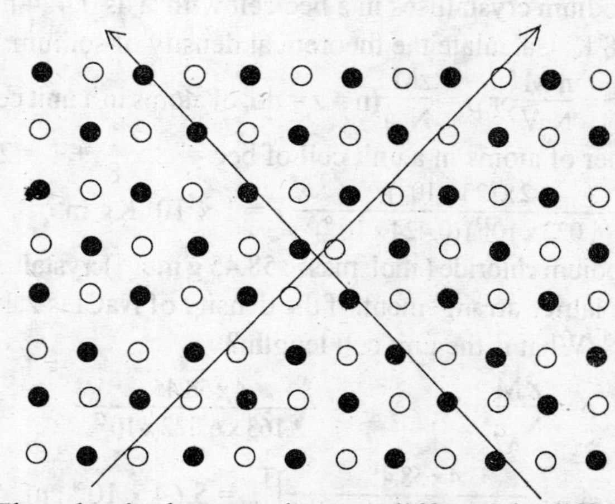

The physical properties are different in different directions in crystalline substances as they have definite geometry.

Solved Examples

(1) The three planes in a crystalline solid intersect the crystal axes at (i) 2a , b, c (ii) a, -b, c (iii) -a, b, ∝. Write the Miller indices.

Sol.

Intercepts	2, 1, 1	1, -1, 1	-1, 1, ∝
Weiss indices	2, 1, 1	1, -1, 1	-1, 1, ∝
Reciprocal of WI	1/2, 1/1, 1/1	1/1 -1/1, 1/1	-1/1,1/1,1/∝
Clear fractions	1, 2, 2	1, -1, 1	-1, 1, 0
Miller indices	(1 2 2)	$(1\,\bar{1}\,1)$	$(\bar{1}\,1\,0)$

(2) How do the spacings of three planes (100), (110) and (111) of cubic lattice vary ?

Sol. The distances between the parallel planes in a crystal are designated by $d_{h,k,l}$. It is given by

$$d_{h,k,l} = \frac{a}{\sqrt{h^2+k^2+l^2}}$$

where a = length of the cube side h, k, l are the Miller indices of the plane.

$$d_{(100)} = \frac{a}{\sqrt{1^2+0+0}} = a; \quad d_{(110)} = \frac{a}{\sqrt{1^2+1^2+0}} = \frac{a}{\sqrt{2}}$$

$$d_{(111)} = \frac{a}{\sqrt{1^2+1^2+1^2}} = \frac{a}{\sqrt{3}}$$

Thus $d_{(100)} : d_{(110)} : d_{(111)} = 1 : \frac{1}{\sqrt{2}} : \frac{1}{\sqrt{3}} = 1 : 0.707 : 0.577$

(3) Perdict the type of crystal structure of (i) NaCl and (ii) Li F. What is the coordination number of each ion in the crystal ? Ionic radii of the ions are :

Li^+ = 60 pm; Na^+= 95 pm; Cl^- = 181 pm and F^-= 136 pm

Sol. In the case of NaCl = r^+/r^- = 95/181 = 0.525

This value is less than 0.732 and greater than 0.414. As such the cation will occrpy the octahedral void with coordination no. 6. Hence, it should crystallise in ccp. In case of LiF r^+/r^- = 60/136 = 0.441. In this case also the larger F^- ions will form cubic close packing in which smaller Li^+ ions fit into the octahedral holes. Coordination number is 6.

(4) Sodium crystallises in a bcc cell with 'a' is 0.424 n m at 298 K. Calculate the theoretical density of sodium.

Sol. $\rho = \frac{n\,M}{N_o V}$ or $\rho = \frac{zM}{N_A a^3}$ (n = z = no. of atoms in a unit cell)

number of atoms in a unit cell of bcc = $8 \times \frac{1}{8} + 1 = 2$

$$\rho = \frac{2 \times (23 \times 10^{-3})}{(6.022 \times 10^{23})(0.424 \times 10^{-9})^3} = 1 \times 10^3 \text{ Kg m}^{-3}$$

(5) Sodium chloride (mol. mass : 58.45 g mol^{-1})crystallises in fcc lattice arrangement. If the density of NaCl is 2.163 g cm^{-3}. What is the unit cell length ?

Sol. $\rho = \frac{z\,M}{N_A a^3}$ $\qquad a^3 = \frac{4 \times 58.45}{2.163 \times 6.022 \times 10^{23}}$

$$a = \left(\frac{4 \times 58.45}{2.163 \times 6.022 \times 10^{23}}\right)^{\frac{1}{3}} = 5.64 \times 10^{-8} \text{ cm.}$$

(6) An element occurs in the bcc structure with a cell edge length of 228 pm. The density of the element is 7.2g cm^{-3}. How many atoms of the element does 208 g of the element contain ?

Sol. $N = \frac{M}{\rho V} = \frac{208 \times 10^{-3}}{7.2 \times 10^3 \times (228 \times 10^{-12})^3} = 24.2 \times 10^{23}$ atoms

(7) Calculate the Avogadro number from the following data : The fcc of silver (at. wt. 107.9 amu) has side of length 4.08 x 10^{-10}m. Its density is 10.53 x 10^3 kg. m^{-3}.

Sol. $\rho = \frac{z\,M}{N_A a^3}$

$$N_A = \frac{z\,M}{\rho a^3} = \frac{4 \times 107.9 \times 10^{-3}}{10.53 \times 10^3 \times (4.08 \times 10^{-10})^3} = 6.052 \times 10^{23} \text{ mol}^{-1}$$

(8) Al crystallises as fcc lattice and it has a density of 2.7x10^3 kg m^{-3}. Calculate the edge length of the unit cell.

Sol.

$$\rho = \frac{zM}{N_A a^3}\ a^3; = \frac{zM}{\rho N_A} = \frac{4 \times 27 \times 10^{-3}}{2.7 \times 10^3 \times 6.02 \times 10^{23}} = 0.6645 \times 10^{-30}$$

a = $(0.6645)^{1/3}$ x 10^{10} = 0.87 x 10^{-10} = 0.087 nm.

(9) Calculate the density and atomic radii of elementary Ag which crystallizes in a fcc with unit cell length = 4.086 x 10^{-10} m.

Sol. In fcc lattice, 2 halves of the atom lie at each corner and one atom lies in the centre of diagonal, so that diagonal = 4r = $\sqrt{2}$a;

$$r = \frac{\sqrt{2}a}{4} = \frac{\sqrt{2} \times 4.086 \times 10^{-10}}{4} = 1.445 \times 10^{-10} \text{ m.}$$

(10) NH_4Cl crystallises in a bcc lattice with a unit distance of 387 pm. Calculate (a) the distance between oppositely charged ion in the lattice and (b) the radius of the NH_4^+ ion, if the radius of the Cl^- ion is 181 pm.

Sol. (a) In bcc lattice oppositely charged ions touch each other along the cross-diagonal of the cube

So $2r_c + 2r_a = \sqrt{3}a$

$$r_c + r_a = \frac{\sqrt{3}}{2}a = \frac{\sqrt{3}}{2} \times 387 = 335.15 \text{ pm.}$$

(b) given r_a = 181 pm, r_c = 335.15 - 181.0 = 154.15 pm

(11) The unit cell cube length for LiCl(NaCl structure) is 5.14Å. Assuming anion-anion contact, calculate the ionic radius of Cl^- ion.

Sol. In a fcc lattice anions touch each other along the fcc diagonal of the cube; $4r_{Cl} = \sqrt{2}a$

$$r_{Cl} = \frac{\sqrt{2}}{4}a = \frac{\sqrt{2}}{4} \times 5.14 = 1.82 \text{ Å}$$

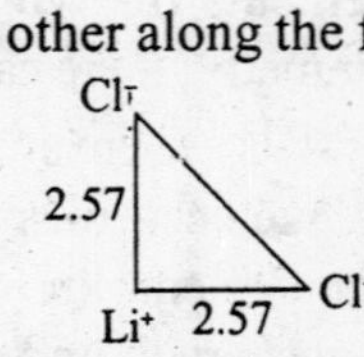

(12) An element crystallises in fcc structure having an unit cell of an edge 200 pm. Calculate its density, if 200g. of this element contains 24x10^{23} atoms.

Sol.

Length of the edge of the unit cell = 200pm = 2 x 10^{-8} cm.

Volume of the unit cell = $(2 \times 10^{-8})^3$ = 8 x 10^{-24} cm^3

$$\text{Number of unit cells} = \frac{\text{Total number of atoms}}{\text{number of atoms per unit cell}}$$

$$= \frac{24 \times 10^{23}}{4} = 6 \times 10^{23}$$

Volume of the substance = no. of unit cells x volume of a unit cell

$$= 6 \times 10^{23} \times 8 \times 10^{-24} = 4.8 \text{ cm}^3$$

$$\text{Density } \rho = \frac{\text{Mass}}{\text{Volume}} = \frac{200}{4.8} = 41.7 \text{ g / cm}^3$$

(13) x-rays of wavelength 1.54 Å strike a crystal and are observed to be deflected at an angle of 22.5°. Assuming that n=1, calculate the spacing between the planes of atoms that are responsible for the reflection.

Sol.

$$d = \frac{1.5}{2 \text{ Sin } 22.5} = \frac{1.54}{2 \times 0.383} = 2.01 \text{ Å}$$

(14) The first order reflection of a beam of x-rays of wavelength 1.54 Å from the (100) face of a crystal of the simple cubic type occurs at an angle of 11.29°. Calculate the length of the unit cell

Sol.

$$d = \frac{1.54 \times 10^{-8}}{2 \times \text{Sin } 11.29} = \frac{1.54 \times 10^{-8}}{2 \times 0.1991} = 3.68 \times 10^{-8} \text{ cm.}$$

$$d_{hkl} = \frac{a}{\sqrt{h^2 + k^2 + l^2}} = a; \quad a = 3.68 \times 10^{-8} \text{ cm.}$$

(15) The first order reflection maximum from crystals of NaCl is obtained at an angle 5.9°. If $d_{h, k, l}$ is 0.282×10^{-9} m, what is the wave length of the x-rays used.

Sol. $2 d \text{ Sin}\theta = n \lambda$

$2 \times 0.282 \times 10^{-9} \text{ Sin}\theta = 1 \times \lambda$

$\lambda = 0.058$ nm

(16) At what angle would a first order reflection be observed in the x-ray diffraction of a set of crystal planes for which d = 0.303 nm., if the x-rays used have a wavelength of 0.071 nm.

Sol.

$$\text{Sin}\theta = \frac{n\lambda}{2d} = \frac{1 \times 7.1 \times 10^{-11}}{2 \times 3.03 \times 10^{-10}} = 0.1172 \quad \therefore \theta = 6.73°$$

(17) An element occurs in bcc structure with a cell edge of 288 pm. The density of metal is 7.2 g cm^{-3}. How many atoms does 208 g of the element contain ?

Sol. Volume of unit cell = $(288 \times 10^{-10})^3 = 23.9 \times 10^{-24}$ cm^3

Volume of 208 g of the element $= \frac{208}{7.2} = 28.88$ cm^3

$$\text{No. of unit cells in } 28.88\text{cm}^3 = \frac{28.88}{23.9 \times 10^{-24}}$$

$$= 12.08 \times 10^{23} \text{ unit cells}$$

each bcc structure contains 2 atoms

∴ Total no. of atoms in 12.08×10^{23} unit cells

$$= 2 \times 12.08 \times 10^{23} = 24.16 \times 10^{23}$$

(18) Li forms bcc crystals. Calculate the atomic radius of Li, if the length of the side of a unit cel of Li is 351 pm.

Sol. In body centred cubic crystals, $r = \frac{\sqrt{3}}{4} a$

$$= \frac{\sqrt{3}}{4} \times 351 = 151.98 \text{ m}$$

(19) Find the lattice spacing in a cubic crystal of NaCl having a density of 2.16×10^3 kg.m^{-3}. Mol. wt. of NaCl is 58.

Sol. If M is the mol. wt. of NaCl and N, the Avagadro's number, since there are 2N ions or diffracting centres i.e. N ions of Na^+ and N ions of Cl^- in one g. molecule, the mass associatied with each ion is M/2 N.

As there is only one ion for each elementary cube in the system either sodium or chlorine

$$\frac{M}{2N} = \rho d^3; \; d = \left(\frac{M}{2N\rho}\right)^{1/3} = \left(\frac{58 \times 10^{-3}}{2 \times 6.023 \times 10^{23} \times 2.16 \times 10^3}\right)^{1/3}$$

$$= 2.814 \times 10^{-10} \text{ m.}$$

(20) A compound formed by elements A and B crystallizes in cubic structure where A atoms are at the corners of the cube and B atoms are at face centre. What is the formula of the compound ?

Sol. no. of A atoms in the unit cell $= 8 \times \frac{1}{8} = 1$

no.of B atoms in the unit cell $= \frac{1}{2} \times 6 = 3$

∴ The formula is AB_3.

1. The ability of a substance to assume two or more crystalline structures is called [CBSE-90]
(A) Isomerism. (B) Polymorphism.
(C) Isomorphism. (D) Amorphism

2. For orthorhombic axial distance $a \neq b \neq c$, the axial angle is [CBSE-91]
(A) $\alpha = \beta = \gamma \neq 90^o$ (B) $\alpha = \beta = \gamma = 90^o$
(C) $\alpha = \gamma = 90^o, \beta \neq 90^o$ (D) $\alpha \neq \beta \neq \gamma \neq 90^o$

3. A pure crystalline substance on being heated gradually first forms a turbid looking liquid at a constant temperture and then at a still higher constant temperature the turbidity completely disappears. This behaviour is charcteristic of substances forming [CBSE-92]
(A) isomeric crystals (B) liquid crystals
(C) isomorphous crystals (D) allotropic crystals

4. On doping Ge metal with a little amount of indium one gets [CBSE-93]
(A) rectifier (B) insulator
(C) n-type semiconductor (D) p-type semi conductor

5. For covalent solid, the units which occupy lattice point are [CBSE-93]
(A) electrons (B) atoms (C) ions (D) molecules

6. Number of atoms in one unit cell of a face centred cubic lattice is [CBSE-93]
(A) 2 (B) 6 (C) 4 (D) 1

7. The number of atoms in 100 g of an f.c.c. crystal with density d = 10.0g/cm^3 and cell-edge equal 100 pm, is equal to [CBSE-94]
(A) 3×10^{25} (B) 4×10^{25} (C) 1×10^{25} (D) 2×10^{25}

8.Ionic solids with Schottky defects contain in their structure [CBSE-94]
(A) equal number of cation and anion vacancies
(B) anion vacancies and interstitial anions
(C) cation vacancies
(D) cation vacancies vacancies and interstitial cations

9. An f.c.c. crystal contains how many atoms in each unit cell ? [CBSE-96]
(A) 4 (B) 6 (C) 3 (D) 8

10. If we mix a pentavalent impurity in a crystal lattice of germanium, what type of semiconductor formation will occur ? [CBSE-96]
(A) P-type semiconductor (B) N-type smiconductor
(C) Both (A) and (B) (D) None of these.

11. An element (atomic mass = 100g/mol) having b.c.c. structure has unit cell edge 400 pm. Then, density of the element is [CBSE-96]
(A) 10.376 g/cm^3 (B) 5.188 g/cm^3
(C) 7.289 g/cm^3 (D) 2.144 g/cm^3

12. For two ionic solids CaO and KI, indentify the wrong statement among the following. [CBSE-97]
(A) Lattice energy of CaO is much larger than that KI
(B) KI is soluble in benzene
(C) CaO has high m.p. (D) KI has high m.p.

13. The intermetallic compound LiAg crystallizes in cubic lattice in which both lithium and silver have co-ordination number of eight. The crystal class is [CBSE-97]
(A) Simple cube (B) Body centre cube
(C) Face central cube (D) None of these

14. The edge length of face centred unit cubic cell is 508 pm. If the radius of the cation is 110 pm, the radius of anion is [CBSE-98]
(A) 110 pm (B) 288 pm (C) 618 pm (D) 398 pm

15. In crystals of which of the following ionic compound would you expect maximum distance between centres of cations and anions ? [CBSE-98]
(A) CsF (B) CsI (C) LiI (D) LiF

16. Schottky defect in crystals is observed when [CBSE-98]
(A) unequal number of cations and anions are missing from the lattice
(B) equal number of cations and and anions are missing from the lattice
(C) an ion leaves its normal sites and occupies an interstitial site.
(D) density of crystal is increased

17. The second order Bragg diffraction of of X ray with $\lambda = 1.0$ Å from a set of parallel planes in a metal occurs at an angle 60°. The distance between the scattering planes in the crystal is [CBSE-98]
(A) 0.57 Å (B) 1.00 Å (C) 1.15 Å (D) 2.00 Å

18. Compounds consisting of equal number of cations and anions in a 3-d lattice separated by equal distances in space are [CBSE-2000]
(A) Molecular compounds (B) Ionic compounds
(C) Covalent (D) Amorphous

19. A compound is formed by elements A & B. This crystallizes in the cubic structure where the A atoms are at the corners of the cube and B atoms are at the body centres. The simplest formula of the compound is [CBSE-2000]
(A) AB (B) A_6B (C) A_8B_4 (D) AB_6

20. Specific volume of a cylindrical virus particle is 6.02×10^{-2} c.c/gm whose radius and lenght are 7Å and 10Å respectively. If $N_A = 6.02 \times 10^{23}$ find molecular weight of virus. [CBSE-2001]
(A) 15.4 kg/mol (B) 1.54×10^4 kg/mol
(C) 3.08×10^4 kg/mol (D) 3.08×10^3 kg/mol

21. When molten zinc is converted into solid state, it acquires h.c.p. structure. The number of nearest neighbours will be - [CBSE-2001]
(A) 6 (B) 8 (C) 12 (D) 4

22. The pyknometric density of sodium chloride crystal is 2.165×10^3 kg m^{-3} while its X-ray density is 2.178×10^3 kg m^{-3}. The fraction of unoccupied sites in sodium chloride crystal is [CBSE-2003]
(A) 5.96×10^{-3} (B) 5.96
(C) 5.96×10^{-2} (D) 5.96×10^{-1}

23. Assertion : In crystal lattice, size of the cation is larger in a tetrahedral hole than in an octahedral hole. Reason : The cations occupy more space than anions in crystals packing. [AIIMS-96]
(A) A (B) B (C) C (D) D

24. Sodium metal crystallizes as a body centred cubic lattice with cell edge 4.29 Å. What is the radius of sodium atom (in cm) ? [AIIMS-99]
(A) 1.857×10^{-8} (B) 2.371×10^{-7}
(C) 3.817×10^{-8} (D) 9.312×10^{-7}

25. A molecule contains atoms x and y so that x occurs at the corners of the cube while y at the face centre. The formula of the molecule can be [AIIMS-2000]
(A) xy_3 (B) x_3y (C) xy_2 (D) x_2y

26. Sodium chloride crystalises in the face centered cubic structure. What is the total sodium ions per unit cell in sodium chloride ? [AIIMS-2000]
(A) 2 (B) 4 (C) 6 (D) 8

27. Which relation is called Bragg's equation ? [AIIMS-2000]
(A) $mvr = \frac{nh}{2\pi}$ (B) $n\lambda = 2d \sin\theta$ (C) $\lambda = \frac{h}{mv}$ (D) none

28. In NaCl if $r_{Na} = 100$ pm then maximum size r_{cl} - will be [AIIMS-2001]
(A) 241.5 pm (B) 261.5 pm
(C) 251.55 pm (D) 271.5 pm

29. In crystalline solid which is true : [AIIMS-2001]
(A) Regular arrangement
(B) Regular arrangement with some irregularity
(C) Irregular arrangment
(D) Irregular arrangement with some regularity

30. An element (atomic mass = 100 g/mol) having b.c.c. structure has unit cell edge 400 pm. The density of element is (No. of atom in b.c.c. (Z) = 2). [AIIMS-2002]
(A) 2.144 g/cm^3 (B) 5.188 g/cm^3
(C) 7.289 g/cm^3 (D) 10.376 g/cm^3

31. An AB_2 type structure is found is [AIIMS-2002]
(A) NaCl (B) Al_2O_3 (C) CaF_2 (D) N_2O

32. Schottky defect defines imperfection in the lattice structure of a [AIIMS-2002]
(A) Solid (B) Liquid (C) Gas (D) Plasma

33. Molten sodium chloride conducts electricity due to the presence of : [JIPMER-91]
(A) Free electrons (B) Free ions
(C) Free molecules (D) Atoms of sodium and chlorine

34. If two compounds have the same crystalline structures and analogous formulae, they are called : [JIPMER-91]
(A) Isotopes (B) Allotropes
(C) Isomers (D) None of these

35. Which of the following forms molecular solid : [JIPMER-91]
(A) SO_2 (B) SiO_2 (C) SiC (D) Diamond

36. The occurrence of different chemical compounds in the same crystalline form is known as : [JIPMER-92]
(A) Isomerism (B) Isomorphism
(C) Polymorphism (D) Enantiomorphism

37. An example of a covalent crystalline solid is [JIPMER-94]
(A) NaF (B) Sillicon (C) Al (D) Ar

38. The unit cell cube length for LiCl (just like NaCl structure) is 5.14 Å. Assuming anion-anion contact, the ionic radius for chloride ion is [JIPMER-2000]
(A) 4.815 Å (B) 3.8 Å (C) 2.8 Å (D) 1.815 Å

39. The crystal structure of NaCl is [BHU-95]
(A) f.c.c. (B) b.c.c. (C) s.c.c. (D) Either 'B' or 'C'

40. Molten sodium chloride conducts electricity due to the presence of [BHU-2001]
(A) Free electrons (B) Free ions
(C) Free molecules (D) Atoms of sodium and chlorine

41. A compound formed by elements A and B, crystallize in the cubic arrangement in which A atoms are at the corner of a cube and B atoms are at the face centre. What is the formula of the compound [BHU-2002]
(A) AB_3 (B) AB_3 (C) A_2B (D) A_2B_2

42. Zeolite is [BHU-2002]
(A) Hydrated aluminium silicate
(B) Hydrated sodium silicate
(C) Hydrated sodium potassium silicate
(D) Hydrated sodium aluminium silicate

43. The force present in the crystals of Naphthalene is [AFMC-91]
(A) van der Waal's (B) Electrostatic
(C) Hydrogen bonding (D) None of these

44. Rutile is an important source for [AFMC-91]
(A) Ruthlnlum (B) Radium (C) Titanium (D) Iron

45. Glass is a/an [AFMC-93]

(A) Polymer (B) Amorphous solid
(C) Crystalline solid (D) Special type of crystal

46. Which of the following is not a semi-conductor ? [AFMC-93]
(A) Si (B) Ge (C) As (D) B

47. Example of unit cell with crystallographic dimensions $a \neq b \neq c$, $\alpha = \gamma = 90^\circ$, $\beta \neq 90^\circ$ is [AFMC-98]
(A) Calcite (B) Graphite
(C) Rhombic sulphur (D) Monoclinic sulphur

48. When two ice cubes are pressed over each other, they unite to form one cube. Which of the following forces responsible to hold them together ? [AFMC-2001]
(A) Hydrogen bond formation
(B) van der Waal's forces
(C) Covalent attraction (D) Ionic interaction

49. Number of atoms in one unit cell of a f.c.c. lattice is [AFMC-2002]
(A) 2 (B) 6 (C) 4 (D) 1

50. The edge length of face centred unit cubic cell is 508 pm. If the radius of the cation is 110 pm. The radius of anion is [AFMC-2002]
(A) 110 pm (B) 249 pm (C) 618 pm (D) 398 pm

51. Which of the following is Asbestos ? [AFMC-2002]
(A) Calcium - magnesium silicate
(B) Sodium - calcium silicate
(C) Sodium - magnesium silicate (D) None of these

52. In the sodium chloride structure, each Na^+ ion is surrounded by six Cl^- ions nearest neighbours and .... Na^+ ion nearest neighbours. [IIT-97]

53. A liquid which is permanently supercooled is frequently called a .... [IIT-97]

54. Which of the following statement (s) is (are) correct. [IIT-98]
(A) The coordination number of each type of ion in CsCl is 8.
(B) A metal that crystallizes in b.c.c. structure has a coordination number of 12.
(C) A unit cell of an ionic crystal shares some of its ions with other unit cells
(D) The length of the unit cells in NaCl is 552 pm. ($r_{Na^+} = 95$; $r_{Cl^-} = 181$ pm)

55. The coordination number of a metal crystallising in a hexagonal close-packed structure is [IIT-99]
(A) 12 (B) 4 (C) 8 (D) 6

56. In a solid 'AB' having the NaCl structure, 'A' atoms occupy the corners of the cubic unit cell. If all the face-centred atoms along one of the axes are removed, then the resultant stoichiometry of the solid is [IIT-2001]
(A) AB_2 (B) A_2B (C) A_4B_3 (D) A_3B_4

57. A : In any ionic solid [MX] with Schottky defects, the number of positive and Negative ions are same
R : Equal number of cation and anion vacancies are present. [IIT-2001]
(A) A (B) B (C) C (D) D

58. A substance A_XB_Y crystallizes in a face centered cubic (f.c.c.) lattice in which atoms 'A' occupy each corner of the cube and atoms 'B' occupy the centers of each face of the cube. Identify the correct composition of the substance A_XB_Y [IIT-2002]
(A) AB_3 (B) A_4B_3
(C) A_3B (D) Composition can not be specified.

59. Na and Mg crystallize in b.c.c. and f.c.c. type crystals respectively , then the number of atoms of Na and Mg present in the unit cell of their respective crystal is [AIEEE-2002]
(A) 4 and 2 (B) 9 and 14 (C) 14 and 9 (D) 2 and 4

60. How many unit cells are present in a cube-shaped ideal crystal of NaCl of mass 1.00 ? [Atomic masses : Na = 23, Cl = 35.5] [AIEEE-2003]
(A) 2.57×10^{21} unit cell (B) 5.14×10^{21} unit cell
(C) 1.28×10^{21} unit cell (D) 1.71×10^{21} unit cell

61. Glass is a [AIEEE-2003]
(A) Micro-cystalline solid (B) Super-cooled liquid
(C) Gel (D) Polymeric mixture

62. The co-ordination number of the ions in the CsCl crystal lattice is .... [EAMCET-91]

63. The crystalline form of sulphur that is stable at room temperature is [EAMCET-94]
(A) Engel's sulphur (B) Rhombic sulphur
(C) Monoclinic sulphur (D) Plastic sulphur

64. Quartz is a crystalline variety of [EAMCET-97]
(A) Sodium silicate (B) Silicon carbide
(C) Silicon (D) Silica

65. What is the crystal structure of cesium chloride ? [EAMCET-2000]
(A) Body centered cubic (B) Face centered cubic
(C) Tetrahedral (D) Octahedral

66. Which of the following has hexagonal close packed crystal structure ? [CPMT-98]
(A) NaCl (B) CsCl (C) RbCl (D) Mg

67. For an ionic crystal of the general formula AX and coordination number 6 the value of radius ratio will be : [PMT-93]
(A) greater than 0.73 (B) in between 0.73 and 0.41
(C) in between 0.41 and 0.22 (D) less than 0.22

68. Space lattice of CaF_2 is : [PMT-93]

(A) face centered cubic (B) body centered cubic
(C) simple cubic (D) hexagonal close packing

69. The existence of a substance in more than one solid modifications is known as : [PMT-93]
(A) Polymorphism (B) Isomorphism
(C) Allotropy (D) Enantiomorphism

70. Tetragonal crystal system has the following unit cell dimensions : [PMT-93]
(A) $a = b = c \quad \alpha = \beta = \gamma = 90^\circ$
(B) $a = b \neq c \quad \alpha = \beta = \gamma = 90^\circ$
(C) $a \neq b \neq c \quad \alpha = \beta = \gamma = 90^\circ$
(D) $a = b \neq c \quad \alpha = \beta = \gamma = 90^\circ ; \gamma = 120^\circ$

71. Crystals can be classified into .... basic crystal habits [PMT-94]
(A) 3 (B) 7 (C) 14 (D) 4

72. Hexagonal close packed arrangement of ions is decscribed as [PMT-94]
(A) ABC ABA (B) ABC ABC
(C) ABABA (D) ABBAB

73. Bragg's law is given by the equation ? [PMT-95]
(A) $n\lambda = 2\theta \sin\theta$ (B) $n\lambda = 2d \sin\theta$
(C) $2n\lambda = d \sin\theta$ (D) $n\frac{\theta}{2} = \frac{d}{2} \sin\theta$

74. How many space lattices are obtainable from the different crystal systems ? [PMT-96]
(A) 7 (B) 14 (C) 32 (D) 230

75. What type of lattice is found in potassium chloride crystal ? [PMT-96]
(A) Face centered cubic (B) Body centered cubic
(C) Simple cubic (D) Simple tetragonal

76. How many molecules are there in the unit cell of sodium chloride ? [PMT-96]
(A) 2 (B) 4 (C) 6 (D) 8

77. What type of point defect is found in crystal of sodium chloride ? [PMT-96]
(A) Frenkel defect (B) Schottky defect
(C) Linear defect (D) Impurity defect

78. Structure of NaCl crystal is [PMT-97]
(A) Tetragonal (B) Cubic
(C) Orthorhombic (D) Monoclinic

79. Bravais lattices are of [PMT-97]
(A) 8 types (B) 12 types (C) 14 types (D) 9 types

80. Point defects are present in [PMT-97]
(A) Ionic solids (B) Molecular solids
(C) Amorphous solids (D) Liquids

81. The number of unit cells in 58.5 g of NaCl is nearly : [PMT-2000]
(A) 6×10^{20} (B) 3×10^{22} (C) 1.5×10^{23} (D) 0.5×10^{24}

82. The number of octahedral sites per sphere in an f.c.c. structure is : [PMT-2000]
(A) 8 (B) 4 (C) 2 (D) 1

83. The packing factor for a body centered cube is : [PMT-2000]
(A) 0.42 (B) 0.53 (C) 0.68 (D) 0.82

84. Which one of the following has Frenkel defect ? [PMT-2000]
(A) Sodium chloride (B) Graphite
(C) Silver bromide (D) Diamond

85. In CsCl structure, the coordination number of Cs is : [PMT-2001]
(A) Equal to that of Cl, that is 6
(B) Equal to that of Cl, that is 8
(C) Not equal to that of Cl, that is 6
(D) Not equal to that of Cl, that is 8

86. The number of octahedral sites per sphere arranged in a face centred cubic structure, is : [PMT-2001]
(A) 1 (B) 2 (C) 3 (D) 4

87. 1 mole of crystalline NaCl will have how many unit cells? [PMT-2001]
(A) 1.506×10^{23} (B) 3.012×10^{23}
(C) 4.518×10^{23} (D) 6.023×10^{23}

88. Which of the following is Bragg's equation ? [PMT-2002]
(A) $n\lambda = 2\theta. \sin\theta$ (B) $n\lambda = 2d. \sin\theta$
(C) $2n\lambda = d. \sin\theta$ (D) $n.\theta/2 = d/2. \sin\theta$

89. Due to Frenkel defect the density of ionic solids [PMT-2002]
(A) Decreases (B) Increases
(C) Does not change (D) Changes

90. Body centred cube lattice has a coordination number [PMT-2002]
(A) 8 (B) 12 (C) 6 (D) 4

91. A metallic element has a cubic lattice. Each edge of the unit cell is 2 Å. The density of the metal is $2.5 gcm^{-3}$. The unit cells in 200 g of the metal are [PET-93]
(A) 1×10^{25} (B) 1×10^{24} (C) 1×10^{22} (D) 1×10^{20}

92. In the crystal of CsCl the nearest neithbours of each Cs ion are [PET-93]
(A) Six chloride ions (B) Eight chloride ions
(C) Six Cs ions (D) Eight Cs ions

93. The number of atoms in a face centred cubic unit cell of a monoatomic substance is [PET-93]
(A) 2 (B) 4 (C) 6 (D) 8

94. A match box exhibits [PET-93]
(A) Cubic geometry (B) Monoclinic geometry
(C) Orthorhombic geometry (D) Tetragonal geometry

95. An element occurring in the b.c.c. structure has 12.08×10^{23} unit cells. The total number of atoms of the element in these cells will be [PET-94]
(A) 24.16×10^{23} (B) 36.18×10^{23}
(C) 6.04×10^{23} (D) 12.08×10^{23}

96. Frenkel defect is caused due to [PET-94]
(A) An ion missing from the normal lattice site creating a vacancy
(B) An extra positive ion occupying an interstitial position in the lattice
(C) An extra negative ion occupying an interstitial position in the lattice
(D) The shift of a positive ion from its normal lattice site to an interstitial site

97. A match box exhibits [PET-95]
(A) Monoclinic geometry (B) Cubic geometry
(C) Orthorhombic geometry (D) Tetragonal geometry

98. The following is NOT a function of an impurity present in a crystal [PET-95]
(A) Establishing thermal equilibrium
(B) Having tendency to diffuse
(C) Contributing to scattering
(D) Introducing new electronic energy levels

99. In a NaCl crystal each Na^+ is surrounded by how many nearest neighbouring equally spaced Cl^- [PET-95]
(A) 2 (B) 4 (C) 6 (D) 8

100. Which is NOT a property of solids- [PET-95]
(A) Solids are always crystalline in nature
(B) Solids have high density and low compressibility
(C) The diffusion of solids is very slow
(D) Solids have definite volume

101. Due to Frenkel defect, the density of ionic solids [PET-96]
(A) Increases (B) Decreases
(C) Does not change (D) Changes

102. The coordination number of Na^+ in cubic- close packing (c.c.p.) structure of NaCl is [PET-96]
(A) 4 (B) 6 (C) 8 (D) 12

103. The correct statement in the following is : [PET-97]
(A) The ionic crystal of AgBr has Schottky defect
(B) The unit cell having crystal parameters $a = b \neq c$, $\alpha = \beta = 90^\circ$, $\gamma = 120^\circ$ is hexagonal
(C) In ionic compounds having Frenkel defect the ratio r^+/r^- is high
(D) The coordination number of Na^+ ion in NaCl is 4

104. The number of NaCl molecules in unit cell of its crystal is : [PET-97]
(A) 2 (B) 4 (C) 6 (D) 8

105. How many kinds of space lattices are possible in a crystal ? [PET-98]
(A) 23 (B) 7 (C) 230 (D) 14

106. Potassium crystallizes with a : [PET-98]
(A) Face-centred cubic lattice
(B) Body- centred cubic lattice
(C) Simple cubic lattice (D) Orthorhombic lattice

107. Diamond is an example of : [PET-98]
(A) Covalent solid (B) Electrovalent solid
(C) Solid with hydrogen bonding (D) Glass

108. How many Cl^- ions are there around Na^+ ions in NaCl crystal ? [PET-99]
(A) 3 (B) 4 (C) 6 (D) 8

109. The existence of a substance in more than one solid modifications is known as : [PET-99]
(A) Isomorphism (B) Polymorphism
(C) Amorphism (D) Allotropy

110. Which of the following substances consists of only one element ? [PET-99]
(A) Marble (B) Sand (C) Diamond (D) Glass

111. Which of the following substances can be used for drying gases ? [PET-99]
(A) Calcium carbonate (B) Sodium carbonate
(C) Sodium bicarbonate (D) Calcium oxide

112. Which of the following metals is sometimes found native in nature ? [PET-99]
(A) Aluminium (B) Copper
(C) Iron (D) Magnesium

113. Ferrous oxide has a cubic structure and each edge of the unit cell is 5.0 Å. Assuming density of the oxide as 4.0 g-cm^{-3}, the number of Fe^{2+} and O^{2-} ions present in each unit cell will be : [PET-2000]
(A) Four Fe^{2+} and four O^{2-} (B) Two Fe^{2+} and four O^{2-}
(C) Four Fe^{2+} and two O^{2-} (D) Three Fe^{2+} and three O^{2-}

114. The number of close neighbours in a body centred cubic lattice of identical sphere is : [PET-2001]
(A) 8 (B) 6 (C) 4 (D) 2

115. The number of equidistant oppositely charged ions in a sodium chloride crystal [PET-2001]
(A) 8 (B) 6 (C) 4 (D) 2

116. The interionic distance for cesium chloride crystal will be : [PET-2002]
(A) a (B) $\frac{a}{2}$ (C) $\frac{\sqrt{3}a}{2}$ (D) $\frac{2a}{\sqrt{3}}$

117. Which one of the following crystals does not exhibit Frenkel defect [PET-2002]

(A) AgBr (B) AgCl (C) KBr (D) ZnS

118. Which one of the following metal oxides is antiferromagnetic in nature [PET-2002]

(A) MnO_2 (B) TiO_2 (C) VO_2 (D) CrO_2

119. In solid CsCl each Cl is closely packed with how many Cs ? [PET- 2003]

(A) 8 (B) 6 (C) 10 (D) 2

120. Which of the following has lowest fusion temperature? [PET-2003]

(A) Naphthalene (B) NaCl (C) Diamond (D) Cr

121. Frenkel and Schottky defects are...... defects. [BHU-2003]

(A) Nucleus (B) Crystal (C) Nuclear (D) Non-crystal

122. A compound formed by elements X and Y crystallizes in a cubic structure in which the X atoms are at the corners of a cube and the Y atoms are at the facecentres. The formula of the compound is [CBSE-PMT-2004]

(A) XY_3 (B) X_3Y (C) XY (D) XY_2

Solid State

1.	(B)	51.	(A)	101.	(C)
2.	(C)	52.	*	102.	(B)
3.	(B)	53.	*	103.	(B)
4.	(D)	54.	(A)(C)	104.	(B)
5.	(D)	55.	(A)	105.	(D)
6.	(C)	56.	(D)	106.	(B)
7.	(B)	57.	(A)	107.	(A)
8.	(A)	58.	(A)	108.	(C)
9.	(A)	59.	(D)	109.	(B)
10.	(B)	60.	(C)	110.	(C)
11.	(B)	61.	(B)	111.	(D)
12.	(B)	62.	*	112.	(C)
13.	(B)	63.	(B)	113.	(A)
14.	(A)	64.	(D)	114.	(A)
15.	(A)	65.	(A)	115.	(B)
16.	(B)	66.	(D)	116.	(C)
17.	(D)	67.	(B)	117.	(C)
18.	(B)	68.	(C)	118.	(A)
19.	(A)	69.	(C)	119.	(A)
20.	(A)	70.	(B)	120.	(A)
21.	(C)	71.	(B)	121.	(B)
22.	(A)	72.	(A)	122.	(A)
23.	(D)	73.	(B)		
24.	(A)	74.	(B)		
25.	(A)	75.	(A)		
26.	(B)	76.	(B)		
27.	(B)	77.	(B)		
28.	(A)	78.	(B)		
29.	(B)	79.	(C)		
30.	(B)	80.	(A)		
31.	(C)	81.	(C)		
32.	(A)	82.	(D)		
33.	(B)	83.	(C)		
34.	(D)	84.	(C)		
35.	(D)	85.	(B)		
36.	(B)	86.	(A)		
37.	(B)	87.	(A)		
38.	(D)	88.	(B)		
39.	(A)	89.	(C)		
40.	(B)	90.	(A)		
41.	(A)	91.	(A)		
42.	(D)	92.	(B)		
43.	(A)	93.	(B)		
44.	(C)	94.	(C)		
45.	(B)	95.	(A)		
46.	(C)	96.	(D)		
47.	(D)	97.	(C)		
48.	(A)	98.	(C)		
49.	(C)	99.	(C)		
50.	(B)	100.	(A)		

52. 12   53. glass   62. 8

CHEMICAL EQUILIBRIUM

Reversible reactions :

1. $3Fe(s) + 4H_2O(g) \rightleftharpoons Fe_3O_4(s) + 4H_2(g)$

2. $N_2(g) + 3H_2(g) \rightleftharpoons 2NH_3(g)$

3. $CH_3COOH + C_2H_5OH \rightleftharpoons CH_3COOC_2H_5 + H_2O$

A reaction is said to be reversible when the composition of the reaction mixture on the approach of equilibrium at a given temperature is the same irrespective of the initial state of the system.

Irreversible Reactions :

$2KClO_3\ (s) \xrightarrow{heat} 2KCl\ (s) + 3O_2\ (g)$

$NH_4NO_2\ (g) \xrightarrow{\Delta} N_2\ (g) + 2H_2O\ (g)$

(nitrogen and water can't recombine to form NH_4NO_2 under any circumstance.)

Hence, for these reactions, law of mass action is not applicable. In irreversible reactions, generally insoluble (in the form of a precipitate) or volatile products formed serve as sink and make the reaction non directional.

Characteristics of Chemical Equilibrium :

1. Chemical equilibrium at a given temperature is characterised by constancy of concentration, pressure, density or colour.

2. Chemical equilibrium can be approached from either direction.

$$\underset{1\ mole}{H_2\ (g)} + \underset{1\ mole}{I_2\ (g)} \xrightarrow{448°C} \underset{1.56\ mole}{2HI\ (g)}$$

The equilibrium mixture contains 1.56 mole of HI, 0.22 mole of H_2 and I_2 respectively

$$H_2 + I_2 \xleftarrow{448°C} \underset{2\ mole}{2HI}$$

In the opposite direction when we take 2 mole of HI, only 0.44 mole of HI decomposes to give 0.22 mole of I_2 and 0.22 mole of H_2.

3. A catalyst can hasten the approach of equilibrium but doesn't alter the state of equilibrium i.e. the relative concentrations of the products and the reactants remain the same

$$2SO_2\ (g) + O_2\ (g) \overset{500°C}{\rightleftharpoons} 2SO_3\ (g)$$

This reaction attains equilibrium after a long time. If platinum pentoxide or V_2O_5 is used as catalyst, equilibrium is attained very rapidly.

4. Chemical equilibrium is dynamic in nature. Even after attaining the equilibrium state, forward and backward reactions are proceeding constantly but at equal rates.

Experimental proof for dynamic equlibrium :

$$H_2 + I_2 \rightleftharpoons 2HI$$

After the equilibrium is obtained, if radioactive isotope of iodine is added, after sometime the HI formed will show radioactive nature due to radioactive iodine.

5. At equilibrium the change in free energy is zero ($\Delta G=0$).

The Law of Mass Action : It is applicable to reversible and homogeneous reactions.

"The rate at which a substance reacts is proportional to its active mass and the rate of a chemical reaction is proportional to the product of the active masses of the reacting substances."

Active mass can be calculated from the relationship

$$[A] = \frac{\text{no. of gms of A/molecular wt. of A}}{\text{volume in litres}}$$

The active mass of solids is taken as unity as the molecules are very closely packed in them.

Unit of active mass is : gram molecule / litre

Consider the equilibrium $A + B \rightleftharpoons C + D$

The velocity of forward reaction $u_1 \propto [A]\ [B]$ or $u_1 = k_1\ [A][B]$

The velocity of backward reaction $u_2 \propto [C][D]$ or $u_2 = k_2\ [C][D]$

At equilibrium $u_1 = u_2$

$$\frac{k_1}{k_2} = k_C = \frac{[C]\ [D]}{[A]\ [B]}$$

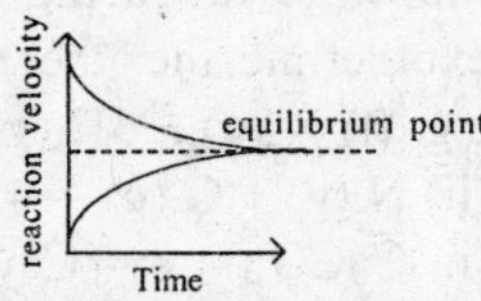

If products are made reactants and the reaction is reversed

$$k_C' = \frac{[A]\ [B]}{[C]\ [D]}\ ;\ \text{Therefore } k_C' = \frac{1}{k_C}$$

Complete form of Law of mass action :

It states that the rate of a reaction is proportional to the molar concentration of each reacting substance raised to the power of the number of its molecules taking part in the reaction as represented in the chemical equation

$$a\,A + b\,B + c\,C \rightleftharpoons l\,L + m\,M + n\,N$$

$$k_C = \frac{[L]^l\ [M]^m\ [N]^n}{[A]^a\ [B]^b\ [C]^c}$$

For gaseous reactions, it is more convenient to use partial pressure instead of concentrations

$$k_P = \frac{[pL]^l\ [pM]^m\ [pN]^n}{[pA]^a\ [pB]^b\ [pC]^c}$$

Relation between k_P and k_C :

$k_P = k_C\ (RT)^{\Delta n}$ or $k_C = k_P \times \left(\frac{1}{RT}\right)^{\Delta n}$

Δn = number of moles of product–no. of moles of reactants When $\Delta n = 0$, $k_P = k_C\ (RT)^o = k_C$

When Δn = +ve, $k_P > k_C$

When Δn = -ve, $k_P < k_C$

Standard free energy change and equilibrium constant are related by $\Delta G^o = -RT \ln k_{eq}$.

Variation of equilibrium constant with temperature as given

by Vant Hoff's equation is $\log \frac{k_2}{k_1} = \frac{\Delta H}{2.303R}\left[\frac{T_2 - T_1}{T_1 \times T_2}\right]$

Le Chatelier's principle and its applications :

The three main factors which alter the state of equilibrium are (a) concentration (b) Temperature and (c) pressure "If an equilibrium is subjected to a stress, the equilibrium shifts in such a way as to reduce the stress".

(1) Effect of change of concentration :

e.g. $\underset{\text{pale yellow}}{Fe^{3+}(aq)} + \underset{\text{colourless}}{SCN^-(aq)} \rightleftharpoons \underset{\text{dark brown}}{[Fe(SCN)]^{2-}(aq)}$

In general, in a chemical equilibrium, increasing the concentration of the rectants results in shifting the equilibrium in favour of the products while increasing the concentration of products results in shifting the equilibrium in favour of the reactants.

(2) Effect of change of temperature :

The value of equilibrium constant is higher at higher temperature in case of endothermic reactions. Equilibrium constant is lower at higher temperature in case of exothermic reactions. The chemical equilibrium involves two opposing reactions, one favouring the products and the other favouring the reactants. If one of these is exothermic, the other must be endothermic.

e.g. (i) $N_2(g) + 3H_2(g) \rightleftharpoons 2NH_3(g)$; ΔH=-92.38 kJ/mole

(ii) $N_2(g) + O_2(g) \rightleftharpoons 2NO(g)$; ΔH=1812.5 kJ/mole

(iii) $CaCO_3(s) \rightleftharpoons CaO(s) + CO_2(g)$; ΔH=177.82 kJ/mole

(3) Effect of change of pressure :

Suppose if the pressure of the system is increased, the volume of the system will decrease proportionately.

e.g. (i) $N_2(g) + 3H_2(g) \rightleftharpoons 2NH_3(g)$

(ii) $PCl_5(g) \rightleftharpoons PCl_3(g) + Cl_2(g)$

(iii) $N_2O_4(g) \rightleftharpoons 2NO_2(g)$

Suppose if the pressure is increased in the case of the equilibrium (i) more ammonia is formed (as the volume of the product NH_3 is 2 moles while the volume of reactants $N_2 + 3H_2$ is 1+3 = 4 mole)

In the case of the following equilibria, pressure has no effect as the volume of products and the reactants is the same

(a) $H_2(g) + I_2(g) \rightleftharpoons 2HI(g)$

(b) $N_2(g) + O_2(g) \rightleftharpoons 2NO(g)$

Effect of adding inert gas : There will be no effect on equilibrium state as the equilibrium constant is independent of total volume and pressure.

Effect of addition of inert gas can be understood from the following conditions :

$\Delta v = 0$; v = constant	Δn = 0,+ve or -ve	no effect
$\Delta v \neq 0$; v $\neq$ constant	$\Delta n = 0$,	no effect
$\Delta v \neq 0$; v $\neq$ constant	$\Delta n > 0$,	forward shift
$\Delta v \neq 0$; v $\neq$ constant	$\Delta n < 0$,	backward shift

Application of Le Chatelier's principle to physical equilibria :

1. Liquid $\rightleftharpoons$ Vapour

(a) Effect of temperature : As the formation of vapour proceeds with the absorption of heat, increase in the temperature shifts the equilibrium in the forward direction.

(b) Effect of pressure : Increase in pressure results in the condensation of some vapour to liquid.

2. Effect of pressure on the freezing point of a liquid:

Solid $\rightleftharpoons$ Liquid

Ice $\rightleftharpoons$ Water

In general, the volume of solid is less than liquid. Therfore, increase in pressure favours the backward reaction. In the case of Ice and water equilibrium, the volume of ice is more than water. Therefore, increase in pressure results in the melting of ice to water. In order to retain ice in equilibrium with water at higher pressure, it is necessary to lower the temperature.

3. Effect of temperature on solubility :

In most of the cases, when a solute passes into solution, heat is absorbed i.e. cooling occurs. But some salts, when they dissolve heat is evolved e.g. Ca salts of organic acids.

IONIC EQUILIBRIUM

Arrhenius concept : According to Arrhenius, an acid is a substance that dissociates to given hydrogen ions when dissolved in water,

$HCl(g) + \text{Water} \rightleftharpoons H^+(aq) + Cl^-(aq)$

hydrated ions

$CH_3COOH + \text{Water} \rightleftharpoons H^+(aq) + CH_3COO^-(aq)$

A base is one which dissociates into hydroxyl ions when disolved in water

$Ca(OH)_2 + \text{Water} \rightleftharpoons Ca^{+2}(aq) + 2OH^-(aq)$

(decomposition into ions is due to high dielectric constant of H_2O). The theory is applicable only to aqueous solutions

Concept of Lowry and Bronsted :

Acid is one which tends to donate a proton to any other substance. Base is one which tends to accept a proton from any other substance.

Acid is a proton donor and a base a proton acceptor.

This theory is applicable to both aqueous and non aqueous solutions.

Conjugated Acids and Bases :

$\underset{\text{acid}}{CH_3COOH} + \underset{\text{base}}{H_2O} \rightleftharpoons \underset{\text{acid}}{H_3O^+} + \underset{\text{base}}{CH_3COO^-}$

Such pairs of substances which can be formed from one another by the gain or loss of a proton are known as conjugated acid base pairs. Thus acetic acid is the conjugate acid of acetate ion and the acetate ion is the conjugate base of acetic acid.

(1) A substance shows its acidic character only when another substance capable of accepting a proton is present

e.g. CH_3COOH or HCl solution in benzene is not acidic.

(2) Hydrogen ion exists as hydrated ion $H^+ . H_2O$ or H_3O^+ and not as H^+ ion.

(3) Ions may also act as acids or bases

$$\underset{base}{NH_3} + \underset{acid}{H_2O} \rightleftharpoons \underset{acid}{NH_4^+} + \underset{base}{OH^-}$$

Substances that can accept proton and donate proton are called amphiprotic or amphoteric.

Strength of Acids and Bases :

$CH_3COOH > H_2CO_3 > HCN > PhOH$

1. "The stronger an acid, the weaker must be its conjugated base and vice versa".

Strength of acids and bases depends upon the degree of ionisation

Decreasing strength of acids : HCl, HNO_3, H_2SO_4, H_3PO_4, HSO_4^-, CH_3COOH, H_2CO_3, NH_4^+, HCO_3^-, HS^-, H_2O.

Decreasing strength of bases : OH^-, S^{-2}, CO_3^{2-}, NH_3, CN^-, HCO_3^-, SO_3^{2-}, CH_3COO^-, SO_4^{2-}, $H_2PO_4^-$, HSO_4^-, NO_2^-, Cl^-

2. All the strong acids in aqueous solution appear almost equally strong. Obviously, their relative strength in aqueous solutions can't be compared. This phenomenon is called Levelling Effect.

3. In liquid ammonia, HCl, HNO_3 (strong acids) and weak acids such as CH_3COOH, CH_3CH_2COOH etc. appear to be of almost equal strength because ammonia takes up protons very much

$$CH_3COOH + NH_3 \rightleftharpoons CH_3COO^- + NH_4^+$$

4. Ionisation in HF : HF has strong acidic properties and no basic i.e. no proton accepting properties.

$$\underset{base}{HNO_3} + \underset{acid}{HF} \rightleftharpoons H_2NO_3^+ + F^-$$

$$CH_3COOH + HF \rightleftharpoons CH_3COOH_2^+ + F^-$$

5. $C_6H_5NH_2$, a weak base and NaOH, a very strong base appear to be equally strong when acetic acid is used as solvent.

Lewis concept of acids and bases :

Base : Which can furnish a pair of electrons-Electron donor

Acid : Which can accept a pair of electrons -Electron acceptor

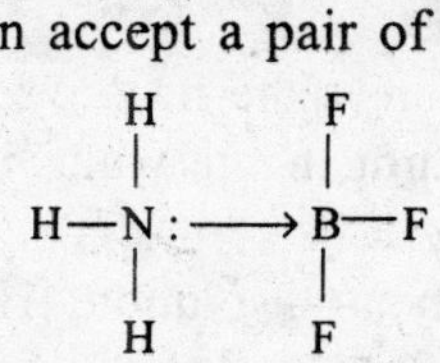

Solubility product :

$$\underset{solid}{AgCl} \rightleftharpoons AgCl\ (solution) \rightleftharpoons Ag^+ + Cl^-$$

A very small amount of AgCl will dissolve in H_2O and the rest of it remain in the solid state. Here solid AgCl is in equilibrium with Ag^+ and Cl^- ions furnished hy the dissolved AgCl.

$$\underset{solid}{AgCl} \rightleftharpoons Ag^+ + Cl^-$$

Applying the law of chemical equilibrium

$$k = \frac{[Ag^+]\,[Cl^-]}{[Ag\,Cl]}$$

Since the activity of solid is taken as unity by convention the above expression may be written as

$$k_{sp} = [Ag^+]\,[Cl^-]$$

k_{sp} is the solubility product of AgCl. It is constant at a given temperature.

Applications :

1. Predicting ionic or precipitation reactions : A substance gets precipitated when the ionic product i.e. the product of the concentration of its ions present in a solution exceeds the value of the solubility product of the substance.

2. Precipitation of an insoluble salt : AgCl is sparingly soluble so is AgI. When we add KI solution to AgCl

$$\underset{\text{sparingly soluble}}{AgCl + KI} \rightleftharpoons \underset{\text{sparingly soluble}}{KCl + AgI}$$

The compound with lower solubility product gets precipitated.

3. Precipitation of soluble salts

Precipitation of soluble salts from their saturated solutions in pure state is known as salting out.

(i) $NaCl\ (s) \rightleftharpoons \underset{\text{in solution}}{Na^+ + Cl^-}$

By passing HCl gas NaCl can be precipitated from the solution. It is due to increase of Cl^- ions.

(ii) In the precipitation of soap

$$C_{17}H_{35}COONa\ (s) \rightleftharpoons C_{17}H_{35}COO^- + Na^+$$

When some NaCl is added, the concentration of Na+ increases and thereby the soap is precipitated.

Application of solubility product in qualitative analysis :

(a) Precipitation of sulphides

H_2S is a weak acid, $H_2S \rightleftharpoons 2H^+ + S^{2-}$

The solubility product of 2nd group sulphides is lower than the solubility product of 4th group sulphides.

The Cd S has higher solubility product compared to other 2nd group sulphides. Therefore either excess of HCl acid is avoided or the solution should be diluted before passing H_2S.

(b) Precipitation of hydroxides :

The solubility product of hydroxides of Fe, Al, Cr are much lower than Mg, Zn.

Dissolution of precipitates in acid solutions :

(a) Salts of weak acids such as phosphates, carbonates, sulphides etc. are soluble in dil. HCl or H NO_3

e.g. $ZnS \rightleftharpoons \underset{\text{in solution}}{Zn^{2+} + S^{2-}}$

if dilute HCl is added H^+ combine with S^{2-} ions

$$2H^+ + S^{2-} \rightarrow H_2S \uparrow$$

As the equilibrium is disturbed, more of ZnS will pass into solution.

(b) precipitates of strong acids such as chlorides and sulphates do not dissolve in acids

e.g. $BaSO_4 \rightleftharpoons \underset{\text{in solution}}{Ba^{2+} + SO_4^{2-}}$

The H^+ ions from the HCl or HNO_3 will not be able to remove SO_4^{2-} ions since H_2SO_4 formed is itself very largely ionised and therefore, the equilibrium is not disturbed.

Common ion effect : The supression of the dissociation of a weak acid or a base on the addition of its own ions is called common ion effect.

(1) $CH_3COOH \rightleftharpoons CH_3COO^- + H^+$

On addition of CH_3COONa to this solution, dissociation of acetic acid is suppressed. CH_3COO^- ion formed from CH_3COONa pushes the equilibrium to the left.

(2) $NH_4OH \rightleftharpoons NH^{4+} + OH^-$

The dissociation of NH_4OH is supressed by the addition of NH_4Cl. Similarly addition of NaOH supresses the dessociation of NH_4OH due to OH^- ions from NaOH.

Hydrolysis of Salts :

The phenomenon of the interaction of cations and anions of the salt with the H^+ or OH^- ions furnished by water is known as salt hydrolysis.

Salts are of four types : Strong acid+Strong base ; Strong acid+Weak base ; W.A + S.B and W.A + W.B;

(1) Salts of S.A + S.B : e.g. KCl; $NaNO_3$

They do not undergo hydrolysis. Neutral in nature.(2) Salts of W.A + S.B.: e.g. CH_3COONa, KCN, Na_2CO_3, $NaHCO_3$ etc.

They undergo hydrolysis to give alkaline solutions (H^+ ions obtained from the dissociation are used up by CH_3COO^- ions while OH^- ions increase in the solution).

Hydrolysis Constant :

$CH_3COO^- + Na^+ + H_2O \rightleftharpoons CH_3COOH + Na^+ + OH^-$

$$k_h = \frac{[OH^-]\,[CH_3COOH]}{[CH_3COO^-]}$$

Relation between k_h, k_a and k_w :

It should be noted that when the equilibrium of hydrolysis reaction is established, the following equilibria have to be satisfied

$$CH_3COOH \rightleftharpoons CH_3COO^- + H^+,\ k_a = \frac{[CH_3COO^-]\,[H^+]}{[CH_3COOH]}$$

$$H_2O \rightleftharpoons H^+ + OH^-;\ k_w = [H^+]\,[OH^-]$$

$$\frac{k_w}{k_a} = k_h$$

Degree of hydrolysis : It is defined as the fraction of the total salt that has undergone hydrolysis on the attainment of equilibrium

Initial	C	large excess		O	O
	CH_3COO^-	$+ H_2O$	$\rightleftharpoons$	$CH_3COOH +$	OH^-
Eq. conc	C (1-x)			C_x	C_x

$$k_h = \frac{C_x \times C_x}{C(1-x)} = \frac{C_{x^2}}{(1-x)}$$

If x is small as compared to unity (1-x) in the equation be replaced by 1.

$$k_h = C_{x^2} \text{ or } x^2 = \frac{k_h}{C} \text{ or } x = \sqrt{\frac{k_h}{C}} = \sqrt{\frac{k_w}{k_a \times C}}$$

For a salt of a weak acid and a strong base the, p^H can be calculated as

$$p^H = 7 + \frac{1}{2}\,p\,k_a + \frac{1}{2}\log C$$

(3) Salts of strong acids and weak bases : NH_4Cl, $FeCl_3$, $AlCl_3$, $CuSO_4$ etc. The aqueous solution of these salts is acidic in nature.

$$NH_4^+ + Cl^- + \underset{\text{large excess}}{H_2O} \rightleftharpoons NH_4OH + H^+ + Cl^-$$

$$K_h = \frac{k_w}{k_b}\ ;\ \text{Degree of hydrolysis } x = \sqrt{\frac{k_h}{C}} = \sqrt{\frac{k_w}{k_b \times C}}$$

for a salt of a strong acid and a weak base, the p^H can be calculated from the relationship

$$pH = 7 - \frac{1}{2}\,p\,K_a - \frac{1}{2}\log C$$

(4) Salts of weak acids and weak bases : CH_3COONH_4, $(NH_4)_2CO_3$, NH_4HCO_3, $HCOONH_4$ etc.

$$CH_3COO^- + NH_4^+ + H_2O \rightleftharpoons CH_3COOH + NH_4OH$$

$$k_h = \frac{k_w}{k_a \times k_b}\ ;\ x = \sqrt{\frac{k_w}{k_a \times k_b}}$$

The degree of hydrolysis of salts of W.A and W.B. is unaffected by dilution as there is no concentration term in the equation.

The nature of aqueous solution of these salts can be understood as :

If $k_a = k_b$, it will be neutral; $k_a > k_b$, acidic; $k_a < k_b$, basic.

The pH of the salt of the weak acid and a weak base can be calculated from $pH = 7 + \frac{1}{2}\,pK_a - \frac{1}{2}\,pK_b$

- The degree of dissociation can also be calculated from density measurements

$$x = \frac{D - d}{d}$$

(d = observed vapour density at a particular temperature when degree of dissociation is x ; D = vapour density when there is no dissociation.)

The pH : Sorensen introduced the pH scale. Every aqueous solution whether acidic, basic or neutral contains the H^+ and OH^- ions.

The pH of a solution is the negative logarithm of the concentration of hydrogen ions which it contains

$$pH = -\log [H^+]$$

for pure water or a neutral solution, $[H^+] = 1 \times 10^{-7}$

$$pH = -\log [1 \times 10^{-7}] = 7$$

The sum of pH and pOH is equal to 14 in water or in any aqueous solution at 25°C

The conc. of H^+ and OH^- increase with increase in temp. (k_w also increases).

Buffers : A buffer is one which can resist change in its pH value on the addition of a solution of an acid or base.

Acid Buffer : mixture of a weak acid and its salt with a strong base e.g. (a) $CH_3COOH + CH_3COONa$
(b) Phthalic acid + Pot. phthalate
(c) Boric acid + Borax

Basic Buffer : mixture of a weak base and its salt with a strong acid e.g. (a) $NH_4OH + NH_4Cl$; (b) glycine + glycine hydrochloride

Acid Buffer : CH_3COOH dissociates to a very little extent while CH_3COONa dissociates almost completely. The mixture contains CH_3COOH, CH_3COO^- and Na^+ ions. If a strong acid is added H^+ ions added will be taken up by CH_3COO^- ions to form very feebly dissociated CH_3COOH.

$$CH_3COO^- + H^+ \rightleftharpoons CH_3COOH$$

If a strong base is added the OH^- ions added are neutralised by CH_3COOH present in the mixture

$$CH_3COOH + OH^- \longrightarrow CH_3COO^- + H_2O$$

Basic Buffer : weak base + its salt with a strong acid

$$NH_4OH + NH_4Cl$$

mixture contains NH_4OH as well as NH_4^+ and Cl^- ions.

$$NH_4OH + H^+ \longrightarrow NH_4^+ + H_2O$$
$$NH_4^+ + OH^- \longrightarrow NH_4OH$$

p^H of buffers:

Acid buffer : $[H^+] = k_a\left[\frac{Acid}{Salt}\right]$

$-\log[H^+] = -\log k_a + \log\left[\frac{Salt}{Acid}\right]$

$pH = p\,k_a + \log\left[\frac{Salt}{Acid}\right]$

This is known as Henderson's equation

Basic buffer : $[OH^-] = k_b\left[\frac{Base}{Salt}\right]$

$pOH = p\,k_b + \log\left[\frac{Salt}{Base}\right]$

$pH = 14 - pOH$ (since $pH + pOH = 14$)

Things to remember

Characteristics of chemical equilibrium :

(a) Constancy of concentration, pressure, density, colour etc. at equilibrium point.

(b) Catalyst can hasten the approach of equilibrium but doesn't alter the state of equilibrium.

(c) Equilibrium can be achieved from either side.

(d) It is dynamic in nature

Active mass $[A] = \frac{\text{no. of grams of A/mol. wt}}{\text{volume in litres}}$

$$k_c = k_p\left[\frac{1}{RT}\right]^{\Delta n}; \quad k_p = k_c\,(RT)^{\Delta n}$$

- variation of k with temperature is given by

$$\log\frac{k_2}{k_1} = \frac{\Delta H}{2.300R}\left[\frac{T_2 - T_1}{T_1 \times T_2}\right]$$

Le Chatelier's principle: Concentration, temperature & pressure effect the equilibrium.

Arrhenius concept: Applicable to aqueous solutions only.

Bronsted & Lowry concept: Applicable to both aqueous and non aqueous solutions. The stronger an acid, the weaker is its conjugated base and vice-versa.

Lewis concept : Substance which can accept a pair of electrons is an acid and the one which donate a pair is a base.

- Solubility product: prediction of pptation; saltingout
- Common ion effect (application to qualitative analysis)
- Salts of strong acids and strong bases do not hydrolyse.
- Heat of neutralisation of a S.A. and S.B. is always 57.32 k.J./mole.

Salts of W.A. + S.B. – basic; salts of W.B.+S.A. – acidic; Salts of W.A. + W.B. – neutral in aqueous solutions.

For W.A. + S.B: $k_h = \frac{k_w}{k_a}$; $\propto$ or $x = \sqrt{\frac{k_w}{k_a \times c}}$

For W.B. + S.A: $k_h = \frac{k_w}{k_b}$; $\propto$ or $x = \sqrt{\frac{k_w}{k_b \times c}}$

For W.B. + W.A: $k_h = \frac{k_w}{k_a \times k_b}$; $\propto$ or $x = \sqrt{\frac{k_w}{k_a \times k_b}}$

For weak electrolytes, $k = \frac{Cx^2}{1-x}$ and x or $\propto = \sqrt{\frac{k}{c}}$

For weak acids, x or $\propto = \sqrt{\frac{k_a}{c}}$ and $[H^+] = \sqrt{k_a \times c}$

For weak bases, x or $\propto = \sqrt{\frac{k_b}{c}}$ and $[H^+] = \sqrt{k_b \times c}$

$p^H = -\log[H^+]$; $p\,k_w = 14 = [H^+][OH^-]$

Buffers : $CH_3COOH + CH_3COONa$; $NH_4OH + NH_4Cl$

$[H^+] = k_a\left[\frac{Acid}{Salt}\right]$ $\quad pH = p\,k_a + \log\left[\frac{Salt}{Base}\right]$

$[OH^-] = k_b\left[\frac{Base}{Salt}\right]$ $\quad pOH = p\,k_b + \log\left[\frac{Salt}{Acid}\right]$

$pH = 14 - pOH$

The free energy change ΔG^o and Equilibrium constant, k are related by the expression $\Delta G = -2.303\,RT\log k$

ΔG^o at equilibrium is zero.

Solved Examples

(1) For the reactions $N_2 + 3H_2 \rightleftharpoons 2NH_3$

$\frac{1}{2}N_2 + \frac{3}{2}H_2 \rightleftharpoons NH_3$

write down the expression for equilibrium constants k_c and k_c'. How is k_c related to k_c'?

Sol. $N_2 + 3H_2 \rightleftharpoons 2NH_3$; for $\frac{1}{2}N_2 + \frac{3}{2}H_2 \rightleftharpoons NH_3$

$$k_c = \frac{[NH_3]^2}{[N_2][H_2]^3} \qquad k_c' = \frac{[NH_3]}{[N_2]^{1/2}[H_2]^{3/2}}$$

$$\text{squaring; } (k_c')^2 = \frac{[NH_3]^2}{[N_2][H_2]^3}$$

Thus $k_c = (k_c')^2$, $\quad \sqrt{k_c} = k_c'$

(2) At a certain temp. equilibrium constant k_c is 16 for the reaction $SO_2(g) + NO_2(g) \rightleftharpoons SO_3(g) + NO(g)$; if we take one mole each of the four gases in one lit. container, what would be the equilibrium concentration of NO and NO_2?

Sol. $SO_2 + NO_2 \rightleftharpoons SO_3 + NO$

	SO_2	NO_2	SO_3	NO
Initial conc.	1	1	1	1
Eq. conc.	1-x	1-x	1+x	1+x

$$k_c = \frac{[SO_3][NO]}{[SO_2][NO_2]} = \frac{(1+x)(1+x)}{(1-x)(1-x)} = 16$$

$$\frac{1+x}{1-x} = 4 \text{ or } 1+x = 4-4x$$

$$5x = 3;\ x = \frac{3}{5} = 0.6$$

conc. of NO_2 at equilibrium = 1-0.6 = 0.4 moles

conc. of NO at equilibrium = 1+0.6 = 1.6 moles

(3) What is the percent decomposition of H_2S if 1.0 mole of H_2S is introduced into 1.1 lit. vessel at 1000 K? k_c for the reaction $2H_2S(g) \rightleftharpoons 2H_2(g) + S_2(g)$ is 1×10^{-6}.

Sol. $2H_2S \rightleftharpoons 2H_2 + S_2$

	H_2S	H_2	S_2
Initail conc.	1	0	0
Eq. conc.	1-x	x	x / 2

$$k_c = \frac{[H_2]^2[S_2]}{[H_2S]^2} = \frac{\left(\frac{x}{1.1}\right)^2 \times \frac{x}{2.2}}{\left(\frac{1-x}{1.1}\right)^2}$$

$$1.0\times10^{-6} = \frac{x^3}{2.2\times(1-x)^2} = \frac{x^3}{2.2} \text{ (as x is very small)}$$

$x^3 = 2.2 \times 10^{-6}$

$x = 1.3 \times 10^{-2}$

% decomposition of H_2S is 1.3

(4) At 250°C and 1 atmospheric pressure, the vapour density of PCl_5 is 57.9. Calculate k_p for the reaction

$PCl_5 \rightleftharpoons PCl_3 + PCl_2$.

Sol. degree of dissociation $\alpha = \frac{D-d}{d} = \frac{104.25-57.9}{57.9} = 0.80$

At equilibrium $PCl_5 \rightleftharpoons PCl_3 + PCl_2$

PCl_5	PCl_3	PCl_2
$1-\alpha$	α	α
(1-0.80)	0.80	0.80

Total no. of moles $= (1+\alpha) = (1+0.80) = 1.80$

$$p\,PCl_5 = \frac{0.2}{1.80} \times 1 = \frac{1}{9};\ pPCl_3 = \frac{0.80}{1.80} \times 1 = \frac{4}{9};$$

$$pPCl_2 = \frac{0.80}{1.80} \times 1 = \frac{4}{9}$$

$$k_p = \frac{pPCl_3 \times pCl_2}{pPCl_5} = \frac{4/9 \times 4/9}{1/9} = \frac{16}{9} = 1.78$$

(5) One mole of nitrogen and three moles of hydrogen are mixed in a 4 lit. container. If 0.25% of nitrogen is converted to ammonia by the following reaction;

$$N_2 + 3H_2 \rightleftharpoons 2NH_3$$

calculate the equilibrium constant (k_C). What will be the value of k for the equilibrium $\frac{1}{2}N_2 + \frac{3}{2}H_2 \rightleftharpoons NH_3$

Sol. $N_2 + 3H_2 \rightleftharpoons 2NH_3$

	N_2	$3H_2$	$2NH_3$
At equilibrium	(1-x)	(3-3x)	2x (x = 0.0025)
Active masses	$\frac{(1-0.0025)}{4}$	$\frac{(3-0.0075)}{4}$	$\frac{(0.0050)}{4}$

$$k_c = \frac{[NH_3]^2}{[N_2][H_2]^3} = \frac{\left(\frac{0.0050}{4}\right)^2}{\left(\frac{0.9975}{4}\right)\left(\frac{2.9925}{4}\right)^3} = 1.49\times10^{-5}\text{ lit.}^2\text{mol}^{-2}$$

k for the reaction $\frac{1}{2}N_2 + \frac{3}{2}H_2 \rightleftharpoons NH_3$ is,

$k = \sqrt{k_c} = \sqrt{1.49 \times 10^{-5}} = 3.86 \times 10^{-3}$ lit mol^{-1},

(6) A mixture of 7.9 ml. H_2 and 33.1 ml I_2 vapour was heated at 444°C until the equilibrium was reached. If the equilibrium constant for the reaction $H_2 + I_2 \rightleftharpoons 2HI$ is 36.68, calculate the number of ml. of H_2 present when equilibrium is obtained.

Sol. Let x ml. of H_2 and I_2 be changed to HI

volume of H_2 at equilibrium = (7.9 – x) ml.

volume of I_2 at equilibrium = (33.1 – x) ml.

volume of HI formed at equilibrium = 2x

$$k = \frac{[HI]^2}{[H_2][I_2]} \qquad 36.68 = \frac{(2x)^2}{(7.9-x)(33.01-x)}$$

$4x^2 = 36.68\ (7.9\text{-}x)\ (33.1\text{-}x)$

$\frac{4}{36.68}x^2 = (7.9\text{-}x)\ (33.1\text{-}x)$

$0.109\ x^2 = 261.49 - 41x + x^2$

$x^2 - 0.109x^2 - 41x + 261.49 = 0$

x = 7.64 or 8.37. x can't be more than 7.9.

(The roots of a quadratic equation are given by the relation $x = \frac{-b\pm\sqrt{b^2-4ac}}{2a}$)

(7) When sulphur in the form of S_8 is heated at 900 K, the initial pressure of 1 atm. falls by 29% at equilibrium. This is because of conversion of some S_8 to S_2. Find the equilibrium constant.

Sol. $S_8 \rightleftharpoons 4S_2$

	S_8	$4S_2$
initial conc.	1	0
equilibrium conc.	1-x	4x
	(1- 0.29)	(4x0.29)
	= 0.71	= 1.16

$$k_p = \frac{(pS_2)^4}{(pS_8)} = \frac{(1.16)^4}{0.71} = 2.5582\text{ atm}^3$$

(8) The value of k_C for the reaction,

$N_2(g) + 3H_2(g) \rightleftharpoons 2NH_3(g)$ is 0.50 at 400°C. Find the value of k_p at 400°C when, concentrations are expressed in mol lit^{-1} and pressure in atomsphere.

Sol. $k_p = k_c (RT)^{\Delta n}$

$= 0.05 (0.082 \times 673)^{-2} = 1.64 \times 10^{-4}$

(9) 1 mole of nitrogen and 3.0 moles of PCl_5 are placed in 100 lit. vessel and heated to 227°C. The equilibrium pressure is 2.05 atmos. Assuming ideal gas behaviour, calculate the degree of dissociation for PCl_5 and k_p for the reaction $PCl_5 \rightleftharpoons PCl_3 + Cl_2$

Sol. $PCl_5 \rightleftharpoons PCl_3 + Cl_2$

	PCl_5	PCl_3	Cl_2
Initial con.	3	0	0
At equilibrium	3(1-x)	3x	3x

1 mole of nitrogen is present ; hence actual no. of moles at equlibrium = 3 (1+x) + 1

we know PV = nRT

$$n = \frac{2.05 \times 100}{0.082 \times 500} = 5$$

$3 (1+x) + 1 = 5$

$3x = 1$ or $x = 0.333$

Thus $3 (1+x) + 1 = 3 + 3 \times 0.333 + 1 = 4 + 0.999$

At equilibrium $pPCl_5 = \frac{3(1-x)}{(3x+4)} \times 2.05$; $pPCl_3 = \frac{3x}{3x+4} \times 2.05$

$$k_p = \frac{\frac{3x}{(3x-4)} \times 2.05 \times \frac{3x}{(3x+4)} \times 2.05}{\frac{3(1-x)}{(3x+4)} \times 2.05} = \frac{9x^2 \times 2.05}{3(3x+4)(1-x)}$$

$$= \frac{3x(0.333)^2 \times 2.05}{(4+0.999)(0.667)} = 0.204 \text{ atm}$$

(10) 2 gram molecules of PCl_5 were heated in a closed two litre vessel. When equilibrium is obtained the PCl_5 is 40% dissociated into PCl_3 and Cl_2. Calculate the equilibrium constant.

Sol. $PCl_5 \rightleftharpoons PCl_3 + Cl_2$

amount of PCl_5 at equilibrium $= 2 - \frac{2 \times 40}{100} = 2 - 0.8 = 1.2$ mole

amount of PCl_3 and Cl_2 at equilibrium respectively is 0.8 mole each

$[PCl_5] = \frac{1.2}{2} = 0.6$; $[PCl_3] = \frac{0.8}{2} = 0.4$; $[Cl_2] = \frac{0.8}{2} = 0.4$

$$k = \frac{[PCl_3][Cl_2]}{[PCl_5]} = \frac{[0.4][0.4]}{[0.6]} = 0.267$$

(11) 60 ml. of H_2 and 42 ml. of I_2 are heated in a closed vessel. At equilibrium the vessel contains 28 ml. HI. Calculate the degree of dissociation of HI.

Sol. $H_2 + I_2 \rightleftharpoons 2HI$

volume at equilibrium (60-x) (42-x) 2x

$2x = 28$, $x = 28/2 = 14$

$$k_c = \frac{28 \times 28}{46 \times 28} = \frac{28}{46} = 0.61$$

(12) The vapour density of N_2O_4 at a certain temp. is 30 Calculate the percentage dissociation of N_2O_4 at this temperature.

Sol. $N_2O_4 \rightleftharpoons 2NO_2$

mol. wt. of $N_2O_4 = 92$ ∴ Vapour density $D = \frac{92}{2} = 46$

$$x = \frac{D-d}{d} = \frac{46-30}{30} = \frac{16}{30} = 0.533 \text{ i.e. } 53.3\%$$

(13) A 0.01M solution of acetic acid is 5% ionised at 25°C. Calculate its dissociation constant.

Sol. $k_a = \frac{\alpha^2}{(1-\alpha)V} = \frac{0.05 \times 0.05}{(1-0.05)100} = 2.63 \times 10^{-5}$

(14) What will be the dissociation constant of 0.1N aqueous ammonia solution in terms of degree of dissociation 'α'? What will be the value if the concentration is 0.01N ?

Sol. $NH_4OH \rightleftharpoons NH_4^+ + OH^-$

$1 - \alpha \qquad \alpha \qquad \alpha$

$$k_b = \frac{\alpha^2}{(1-\alpha)10}$$

for 0.01 N solution, k_b remains same at the same temp. but degree of dissociation bemomes different.

(15) The equilibrium constant, k_p for the reaction $PCl_5 \rightleftharpoons PCl_3 + Cl_2$ at 523K is 0.82. Calculate the degree of dissociation at the same temperature and at a pressure of 5 atmosphere.

Sol. $PCl_5 \rightleftharpoons PCl_3 + Cl_2$

1-x x x x is the degree of dissociation

Total number of molecules at equilibrium = 1-x + x+x = 1+x

Total pressure is 5 atmosphere

$pPCl_5 = \frac{1-x}{1+x} \times 5$; $pPCl_3 = \frac{x \times 5}{1+x}$; $pCl_2 = \frac{x \times 5}{1+x}$

$$k_p = \frac{[pPCl_3][pCl_2]}{[pPCl_5]} \quad 0.82 = \frac{\frac{5x}{1+x} \times \frac{5x}{1+x}}{\frac{1-x}{1+x} \times 5} \quad 0.82 = \frac{5x^2}{1-x^2}$$

$0.82 - 0.82x^2 = 5x^2$; $5.82x^2 = 0.82$, $x = \sqrt{\frac{0.82}{5.82}} = 0.375 = 37.5\%$

(16) Calculate the standard equilibrium constant at 298 K. Given that the value of $\Delta H°$ and $\Delta S°$ of the reaction at 298k are 77.2 kJ mol^{-1} and 122 J K^{-1} mol^{-1}.

Sol. $\Delta G° = \Delta H° - T\Delta S° = 77200 - 298 \times 122 = 40844$ J mol^{-1}

Let the equilibrium constant $k_c°$.

We know $\Delta G° = -2.303RT \log k_c°$

$$\log k_c° = -\frac{\Delta G°}{2.303 \times 8.314 \times 298} = \frac{40844}{2.303 \times 8.314 \times 298} = -7.158$$

$k_c° = 6.95 \times 10^{-8}$

(17) A 0.0128 N solution of acetic acid has Λ=14 mhos equiv^{-1} and Λ_∞ = 391 mhos equiv^{-1} at 25°C. Calculate the dissociation constant of acetic acid.

Sol. $\alpha = \frac{\Lambda}{\Lambda_\infty} = \frac{14}{391} = 3.58 \times 10^{-2}$

$$k_a = \frac{\alpha^2}{(1-\alpha)v} = 3.58 \times 10^{-2} \times 3.58 \times 10^{-2} \times \frac{1}{0.128} = 1.64 \times 10^{-5}$$

(18) $\Delta G°$ for $\frac{1}{2}N_2(g) + \frac{3}{2}H_2(g) \rightleftharpoons NH_3(g)$ is -16.5 kJmol^{-1} Find out k_p for the reaction at 25°C. Also report k_p and $\Delta G°$ for $N_2(g) + 3H_2(g) \rightleftharpoons 2NH_3(g)$ at 25°C.

Sol. $\log k_p = \frac{-\Delta G°}{2.303RT} = \frac{-(-16.5 \times 10^3)}{2.303 \times 8.314 \times 298} = 2.8917$

$k_p = 779.41$

k_p for the reaction

$N_2(g) + 3H_2(g) \rightleftharpoons 2NH_3(g)$, $= (779.41)^2 = 6.07 \times 10^5$

$\Delta G° = -2.303 \times 8.314 \times 298 \log 6.07 \times 10^5 = -32.99$ kJmol^{-1}

(19) The equilibrium constant k_p for the reaction $A+B \rightleftharpoons C+D$ is 1×10^{-12} at 600 K and 1×10^{-7} at 800 K. Calculate the heat of reaction.

Sol. we know $\log \frac{kp_2}{kp_1} = \frac{\Delta H}{2.303R}\left[\frac{T_2 - T_1}{T_1 x T_2}\right]$

$$\log(1 \times 10^{-7}) - \log(1 \times 10^{-12}) = \frac{\Delta H}{2.303 \times 8.314}\left[\frac{800-600}{600 \times 800}\right]$$

$$-7 + 12 = \frac{\Delta H}{2.303 \times 8.314}\left[\frac{200}{480000}\right]$$

$$\Delta H = \frac{5 \times 2.303 \times 8.314 \times 480000}{200} = 229.77 \text{ kJ mol}^{-1}$$

(20) 500 ml. of a solution contains 15 g. acetic acid and 8 g. methanol. What will be their active mass ?

Sol. $[X] = \frac{\text{mass of x (g.) / molar mass of x (g mol}^{-1})}{\text{volume (l)}}$

$[CH_3COOH] = \frac{15/60}{0.5} = 0.5 \text{ mol l}^{-1}$

$[CH_3OH] = \frac{8/32}{0.5} = 0.5 \text{ mol l}^{-1}$

(21) How many moles of calcium hydroxide must be dissolved to produce 250 ml. of an aqueous solution of p^H 10.65 ? Assume complete ionisation.

Sol. $pOH = 14 - 10.65 = 3.35$

$[OH^-] = 10^{-3.35} = 10^{-4} \times 10^{0.65}$

$= 4.47 \times 10^{-4}$ M

one mole of $Ca(OH)_2$ furnishes 2 OH^- ions

Hence, concentration of $Ca(OH)_2 = \frac{4.47 \times 10^{-4}}{2}$

$= 2.235 \times 10^{-4}$ M

No. of moles in 250 ml $= \frac{2.235 \times 10^{-4}}{4} = 5.58 \times 10^{-5}$

(22) The pH of 0.1M hydrocyanic acid solution. is 5.2. What is the value of k_a for hydrocyanic acid ?

Sol. $pH = -\log[H^+]$

$\log[H^+] = -pH = -5.2$

$[H^+] = 10^{-5.2} = 10^{-6} \times 10^{-0.8} = 6.3 \times 10^{-6}$ M

'α', degree of dissociation $= \frac{[H^+]}{C} = \frac{6.3 \times 10^{-6}}{0.1} = 6.3 \times 10^{-5}$

According to Ostwald's formula for weak electrolyte,

$k_a = \alpha^2 c = 6.3 \times 10^{-5} \times 6.3 \times 10^{-5} \times 0.1 = 3.69 \times 10^{-10}$

(23) Calculate for 0.01N solution of sodium acetate (a) hydrolysis constant (b) degree of hydrolysis and (c) pH $k_a = 1.9 \times 10^{-5}$.

Sol. $k_h = \frac{k_w}{k_a} = \frac{10^{-14}}{1.9 \times 10^{-5}} = 5.26 \times 10^{-10}$

$x = \sqrt{\frac{k_h}{c}} = \sqrt{\frac{5.26 \times 10^{-10}}{0.01}} = 2.29 \times 10^{-4}$

$CH_3COO^- + H_2O \rightleftharpoons CH_3COOH + OH^-$

$[OH^-] = c \times h = 0.01 \times 2.29 \times 10^{-4} = 2.29 \times 10^{-6}$ M

$[H^+] = \frac{kw}{[OH^-]} = \frac{10^{-14}}{2.29 \times 10^{-6}} = 4.37 \times 10^{-9}$ M

$pH = -\log[H^+] = -\log(4.37 \times 10^{-9}) = 8.36$

(24) If 250 c.c. of 0.050 M $Ba(NO_3)_2$ are mixed with 250 c.c. of 0.020M NaF, will any BaF_2 precipitate ? k_{sp} of BaF_2 is 1.7×10^{-6} at 298 k.

Sol. As equal volumes of two solutions are mixed, the concentration will be halved,

$[Ba^{2+}] = \frac{0.05}{2} = 0.025$ M, $[F^-] = \frac{0.02}{2} = 0.01$ M

$BaF_2 \rightleftharpoons [Ba^{2+}] + 2F^-$

Ionic product $= 0.025 \times (0.01)^2 = 2.5 \times 10^{-6}$

Ionic product is greater than the solubility product. Hence precipitation will take place.

(25) The solubility product of magnesium hydroxide at 25°C is 1.4×10^{-11}. What is the solubility of $Mg(OH)_2$?

Sol. $Mg(OH)_2 \rightleftharpoons Mg^{2+} + 2OH^-$

Let the solubility be x moles per litre

$\therefore$ concentration of Mg^{2+} = x moles per litre

$x \times (2x)^2 = 1.4 \times 10^{-11}$

$4x^3 = 1.4 \times 10^{-11}$

$x = 1.5 \times 10^{-4}$ mole / lit.

$\therefore$ Solubility of $Mg(OH)_2 = 1.5 \times 10^{-4}$ mole / lit.

$= 58 \times 1.5 \times 10^{-4} = 0.0087$ g / lit.

(26) The solubility product of $PbBr_2$ is 8×10^{-5}. If the salt is 80% dissociaed in saturated solution, find out the solubility of the salt.

Sol. $PbBr_2 \rightleftharpoons Pb^{2+} + 2Br^-$

Let x be the number of gram moles ionised

$k_{sp} = [x] \times [2x]^2 = 4x^3 = 8 \times 10^{-5}$

$x = \sqrt[3]{\frac{8 \times 10^{-5}}{4}} = 2.714 \times 10^{-2}$

$\therefore$ we have 2.714×10^{-2} moles ionized per litre

The total number of moles per lit. $= 2.714 \times 10^{-2} \times \frac{100}{80}$

$= 3.392 \times 10^{-2}$

Molecular weight of $PbBr_2 = 207 + 160 = 367$

Solubility $= 3.392 \times 10^{-2} \times 367 = 12.45$ g / lit.

(27) Calculate the pH value of a solution obtained by mixing 50 ml of 0.2 N HCl with 50 ml. of 0.1N NaOH.

Sol.: The normality of the resulting solution

$= \frac{N_1V_1 - N_2V_2}{\text{Total Volume}}$

$= \frac{0.2 \times 50 - 0.1 \times 50}{100} = 0.05$N

As the acid is monobasic the H^+ Concentration

= 0.05 mole/lit.

pH of the solution $= -\log[H^+]$

$= -\log[0.05] = 1.30$

(28) The ionisation constant of aniline, acetic acid and water at 25°C are respectively 3.83×10^{-10}, 1.75×10^{-5} and 1.008×10^{-14}. Calculate the percentage hydrolysis of aniline acetate in decinormal solution.

Sol.

$x = \sqrt{\frac{kw}{ka \times kb}} = \sqrt{\frac{1.008 \times 10^{-14}}{1.75 \times 10^{-5} \times 3.83 \times 10^{-10}}} = 1.219$

x can't be more than 1. It means that the degree of hydrolysis in this case is very large.

$\therefore$ We have to use $\frac{x}{1-x} = 1.219$

$x = 1.219 - 1.219\,x$; $x + 1.219\,x = 1.219$

$2.219x = 1.219$

$x = \frac{1.219}{2.219} = 0.5495$ i.e. 54.95%

(29) Calculate H^+ ion concentration of a solution which has a pH of 4.5.

Sol. $pH = -\log[H^+]$

$4.5 = -\log[H^+]$

$\log[H^+] = -4.5$

$[H^+] = \text{Antilog}[-4.5] = 0.000031622 = 3.1622 \times 10^{-5}$

(30) 7.3×10^{-3} g. of HCl is dissolved in 200 ml. of water. Calculate the pH of the solution.

Sol. Molarity $= 7.3 \times 10^{-3} \times \frac{1000}{200} \times \frac{1}{36.5} = 10^{-3}$ M

$pH = -\log[H^+] = -\log 10^{-3} = 3$

(31) What will be the pH value of a solution obtained by mixing 5gm of acetic acid and 7.5 gm of sodium acetate and making the volume equal to 500 ml. Dissociation constant of acetic acid at 25°C is 1.8×10^{-5}.

Sol. We know for acid buffer $pH = pk_a + \log\frac{[Salt]}{[Acid]}$

$pk_a = -\log k_a = -\log(1.8 \times 10^{-5}) = 4.7447$

$[Salt] = \frac{1000}{500} \times \frac{7.5}{82} = 0.1829$ mole/lit.

$[Acid] = \frac{1000}{500} \times \frac{5}{60} = 0.1666$ mole/lit.

$pH = 4.7447 + \log\frac{0.1829}{0.1666} = 4.7447 + 0.0435 = 4.7882$

$pH = 4.79$

(32) A buffer solution contains 0.20 mole of NH_4OH and 0.25 mole of NH_4Cl per litre. Calculte the pH of the solution. Dissociation constant of NH_4OH at the room temperature is 1.80×10^{-5} and $k_w = 10^{-14}$.

Sol. We know $pOH = pk_b + \log\frac{[Salt]}{[Base]}$

$pk_b = -\log k_b = -\log(1.80 \times 10^{-5}) = 4.7447$

$pOH = 4.7447 + \log\frac{0.25}{0.20} = 4.8419$

$pH + pOH = 14$

$pH = 14 - 4.8419 = 9.1581$

Thus the pH of the solution is 9.16.

1. For the reaction $N_2 + 3H_2 \rightleftharpoons 2NH_3$, $\Delta H =$ [CBSE-91]

(A) $\Delta E + 2RT$ (B) $\Delta E - 2RT$ (C) $\Delta E + RT$ (D) $\Delta E - 2RT$

2. In a reversible reaction both side rates of reaction are [CBSE-91]

(A) Same (B) different

(C) One side more (D) reactant rate less

3. If $A+B \rightarrow C$ reactant takes some heat then reaction is completed at the [CBSE-91]

(A) high temperature and high pressure

(B) high temperature and low pressure

(C) low temperature and high pressure

(D) low temperature and low pressure

4. The pH of 0.001M HCl is [CBSE-91]

(A) 3 (B) 11 (C) 2 (D) 10

5. Calculate the OH^- ion concentration of NaOH of whose pH is 2 [CBSE-91]

(A) 10^{-2} (B) 10^{-9} (C) 10^{-12} (D) 10^{-7}

6. NH_4OH is a weak base because [CBSE-91]

(A) It has low vapour pressure

(B) It is only slightly ionised

(C) It is not hydroxide of metal

(D) It has low density

7. The compound whose solution has the highest pH is [CBSE-91]

(A) NH_4Cl (B) Na_2CO_3 (C) NaCl (D) $NaHCO_3$

8. Aqueous solution of ammonia consists of [CBSE-91]

(A) H^+ (B) CH^- (C) NH_4^+ (D) NH_4^+ and OH^-

9. In reversible reaction a catalyst is the substance which [CBSE-92]

(A) increase rate of the forward reaction

(B) decreases the value of enthalpy change in the reaction

(C) reduces the time required for reaching the equilibrium state in the reaction

(D) decrease the rate of the reverse reaction

10. In the reversible reaction of the type $A+B \rightleftharpoons AB$, in general [CBSE-92]

(A) neither of the reactions will be endothermic

(B) both the reactions will be exothermic

(C) the combination reaction will be endothermic and the dissociation reaction will be exothermic

(D) the combination reaction will be exothermic and the dissociation reaction will be endothermic

11. The solubility product of AgCl is 1.8×10^{-10} mol^2–litre^{-2}. On mixing the following solutions in equal volumes, in which case the precipitation of AgCl could be expected ? [CBSE-92]

(A) 10^{-10} M of Ag^+ and 10^{-10} M of Cl^-
(B) 10^{-4} M of Ag^+ and 10^{-4} M of Cl^-
(C) 10^{-5} M of Ag^+ and 10^{-5} M of Cl^-
(D) 10^{-6} M of Ag^+ and 10^{-6} M of Cl^-

12. In which one of the following solvents will AgBr have the highest solubility ? [CBSE-92]
(A) 10^{-3}M NaBr (B) 10^{-3}M NH_4 OH
(C) Pure water (D) 10^{-3}M HBr

13. Which one of the following information can be obtained on the basis of Le Chatelier principle ? [CBSE-92]
(A) Dissociation constant of a weak acid
(B) Entropy change in a reaction
(C) Equilibrium constant of a chemical reaction
(D) Shift in equilibrium position on changing value of a constraint

14. In the gas phase reaction $C_2H_4 + H_2 \rightleftharpoons C_2H_6$ the equilibrium constant can be expressed in units of [CBSE-92]
(A) $litre^{-1}$ - mol^{-1} (B) litre- mol^{-1}
(C) mol^2 - $litre^{-2}$ (D) mol - $litre^{-1}$

15. 10 ml of a strong acid solution of pH = 2.000 are mixed with 990 ml of a buffer solution of pH = 4.000. The pH of the resulting solution will be [CBSE-92]
(A) 4.002 (B) 4.000 (C) 4.200 (D) 3.800

16. At 90°C,pure water $[H_3O^+] = 10^{-6}$, what is the value of K_W at 90°C ? [CBSE-93]
(A) 10^{-12} (B) 10^{-11} (C) 10^{-8} (D) 10^{-6}

17. In which of the following the solubility of AgCl will be maximum ? [CBSE-93]
(A) 0.1 M NaBr (B) 0.1 $NaNO_3$
(C) H_2O (D) 0.1 M NaCl

18. Conjugate base of NH_4^+ ion is [CBSE-93]
(A) NH_3 (B) OH^- (C) KOH (D) NH_4OH

19. Which of the following is most soluble? [CBSE-94]
(A) $Bi_2S_3(K_{sp} = 1 \times 10^{-70})$ (B) MnS $(K_{sp} = 7 \times 10^{-16})$
(C) $CuS(K_{sp} = 8 \times 10^{-37})$ (D) Ag_2S $(K_{sp} = 6 \times 10^{-51})$

20. At 80^0 C pure distilled water has $[H_3O^+]$ concentration equal to 1×10^{-6} mol/l.The value of K_w at this temperature will be [CBSE-94]
(A) 1×10^{-8} (B) 1×10^{-12} (C) 1×10^{-14} (D) 1×10^{-6}

21. 1.1 mol of A is mixed with 2.2 mol of B and the mixture is kept in a one litre flask till the equilibrium, $A + 2B \rightleftharpoons 2C + D$ is reached. At equilibrium 0.2 mol of C is formed. The equilibrium 0.2 mol of C is formed.The equilibrium constant of the above reaction is [CBSE-94]
(A) 0.002 (B) 0.004 (C) 0.001 (D) 0.003

22. Solubility of AgCl will be minimum in [CBSE-95]
(A) 0.001M $AgNO_3$ (B) Pure water
(C) 0.01M $CaCl_2$ (D) 0.01M NaCl

23. pH of a 10M solution of HCl is [CBSE-95]
(A) less than 0 (B) 2 (C) 0 (D) 1

24. Which of the following is not a Lewis acid ? [CBSE-96]
(A) BF_3 (B) $FeCl_3$ (C) SiF_4 (D) C_2H_4

25. pH value of N/10 NaOH solution is [CBSE-96]
(A) 10 (B) 11 (C) 12 (D) 13

26. Total number of moles for the reaction
$2\,HI \rightleftharpoons H_2 + I_2$ [CBSE-96]
if α is dissocciation constant, is
(A) 2 (B) 2-α (C) 1 (D) 1-α

27. The ionic prouct of water at 250°C is 10^{-14} the ionic product at 900°C will be [CBSE-96]
(A) 1×10^{-20} (B) 1×10^{-12} (C) 1×10^{-14} (D) 1×10^{-16}

28. The equilibrium constant for the reaction $N_2+3H_2 \rightleftharpoons 2NH_3$ is K, then the equilibrium constant for the equilibrium $NH_3 \rightleftharpoons 1/2\,N_2 + 3/2H_2$ is [CBSE-96]
(A) 1/K (B) $1/K^2$ (C) $\sqrt{K}$ (D) $\sqrt{1/K}$

29. Boron compounds behave as Lewis acids because of their [CBSE-96]
(A) acidic nature. (B) covalent nature.
(C) electron deficiency (D) ionisation property.

30. In the reaction $4NH_3(g) + 5O_2(g) \rightarrow 4NO(g) + 6H_2O(l)$ when 1 mole of ammonia and 1 mole of O_2 are made to react to completion. [CBSE-98]
(A) 1.0 mole of NO will be produced.
(B) All the oxygen will be consumed.
(C) All the ammonia will be consumed.
(D) 1.0 mole of H_2O is produced.

31. If K_1 and K_2 are the respective equilibrium constant for the two reactions [CBSE-98]
$Xe_2F_6(g) + H_2O \rightleftharpoons XeOF_4(g) + 2HF(g)$
$XeO_4(g) + XeF_6(g) \rightleftharpoons XeOF_4(g) + XeO_3F_2(g)$
the equilibrium constant for the reaction
$XeO_4(g) + 2HF(g) \rightleftharpoons XeO_3F_2(g) + H_2O(g)$ will be
(A) $K_1.K_2$ (B) $K_1/(K_2)^2$ (C) K_1 / K_2 (D) K_2 / K_1

32. The concentration of $[H^+]$ and concentration of $[OH^-]$ of a 0.1 aqueous solution of 2% ionised weak acid is [ionic product of water = 1×10^{-14}] [CBSE-99]
(A) 0.02×10^{-3} M and 5×10^{-11} M
(B) 1×10^{-3} M and 3×10^{-11} M
(C) 2×10^{-3} M and 5×10^{-12} M

(D) 3×10^{-2} M and 4×10^{-18} M

33. The strongest conjugate base is [CBSE-99]
(A) NO_3^- (B) Cl^- (C) SO_4^{2-} (D) CH_3COO^-

34. The solubility of a saturated solution of calcium fluoride is 2×10^{-4} moles per litre. Its solubility product is [CBSE-99]
(A) 12×10^{-2} (B) 14×10^{-4} (C) 22×10^{-11} (D) 32×10^{-12}

35. For every diprotic acid of the type H_2X. How would you relate ionization constants Ka_1 and Ka_2 ? [CBSE-2000]
(A) $Ka_1 = Ka_2$ (B) $Ka_1 < Ka_2$
(C) $Ka_2 < Ka_1$ (D) Insufficient data

36. $ZnCO_3(s) \longrightarrow ZnO(s) + CO_2(g)$ Expression for the partial pressure constant for the above reaction would be [CBSE-2000]
(A) $K_p = \frac{[ZnO][CO_2]}{[ZnCO_3]}$ (B) $K_p = \frac{P(ZnO)P(CO_2)}{P(ZnCO_3)}$
(C) $K_p = P(ZnO)^2\ P(CO_2)$ (D) $K_p = P(CO_2)$

37. Conjugate acid of NH_2^- [CBSE-2000]
(A) NH_3 (B) NH_2^+ (C) NH_3^- (D) NH_4^+

38. In HS^-, I^-, RNH_2, NH_3 order of proton accepting tendency will be - [CBSE-2001]
(A) $I^- > NH_3 > RNH_2 > HS^-$
(B) $NH_3 > RNH_2 > HS^- > I^-$
(C) $RNH_2 > NH_3 > HS^- > I^-$
(D) $HS^- > RNH_2 > NH_3 > I^-$

39. Ionisation constant of CH_3COOH is 1.7×10^{-5} and conc. of H^+ ions is 3.4×10^{-4} then find out initial conc. of CH_3COOH molecules- [CBSE-2001]
(A) 3.4×10^{-4} (B) 3.4×10^{-4}
(C) $6.9\ 10^{-4}$ (D) 7.24×10^{-3}

40. Solubility of a M_2S salt is 3.5×10^{-16} then find out solubility product - [CBSE-2001]
(A) 1.7×10^{-6} (B) 1.7×10^{-16} (C) 1.7×10^{-18} (D) 1.7×10^{-12}

41. Reaction $BaO_2(s) \longrightarrow BaO(s) + O_2(g)$; $\Delta H = +$ ve. In equilibrium condition, pressure of O_2 depends on [CBSE-2002]
(A) Increase of mass of BaO_2
(B) Increase of mass of BaO
(C) Increase of temp. on Eq^m
(D) Increase of mass of BaO_2 and BaO both

42. Solubility of MX_2 -type electrolytes is 0.5×10^{-4} mole/lit. Then find out K_{sp} of electrolytes [CBSE-2002]
(A) 5×10^{-12} (B) 25×10^{-10}
(C) 1×10^{-13} (D) 5×10^{-13}

43. 1 M and 2.5 litre NaOH solution mixed with another 0.5 M and 3 litre NaOH solution. Then find out molarity of resultant solution [CBSE-2002]
(A) 0.80 M (B) 1.0 M (C) 0.73 M (D) 0.50 M

44. Which has highest pH [CBSE-2002]
(A) CH_3COOK (B) Na_2CO_3 (C) NH_4Cl (D) $NaNO_3$

45. Solution of 0.1 N NH_4OH and 0.1 N NH_4Cl has pH 9.25. Then find out pK_b of NH_4OH [CBSE-2002]
(A) 9.25 (B) 4.75 (C) 3.75 (4) 8.25

46. In which of the following reaction K_p is less than K_c ? [AIIMS-93]
(A) $H_2(g) + Br_2(g) \leftrightarrow 2HBr(g)$
(B) $PCl_5(g) \leftrightarrow PCl_3(g) + Cl_2(g)$
(C) $N_2(g) + 3H_2(g) \leftrightarrow 2NH_3(g)$
(D) $2SO_3(g) \leftrightarrow 2SO_2(g) + O_2(g)$

47. Equilibrium constant for the reaction $FeO(s) + CO(g) \leftrightarrow Fe(s) + CO_2(g)$ is 0.5. The equilibrium constant for the reverse reaction $Fe(s) + CO_2(g) \leftrightarrow FeO(s) + CO_2(g)$ is [AIIMS-93]
(A) −0.5 (B) 0.5 (C) 2 (D) 0.25

48. When sodium acetate is added to acetic acid, it [AIIMS-93]
(A) Increases ionization of acetic acid
(B) Decreases ionization of acetic acid
(C) Increases ionization of sodium acetate
(D) Is independent of concentration

49. Which pair of substances from the following pairs can act as a buffer solution ? [AIIMS-93]
(A) HNO_3 and KNO_3 (B) NaOH and NaCl
(C) Na_2SO_4 and H_2SO_4 (D) NH_4OH and NH_4Cl

50. What is the pH of 1 M sodium acetate ?
[K_a for acetic acid is 1.8×10^{-5} and $K_w = 1 \times 10^{-14}$ [AIIMS-93]
(A) 2.4 (B) 4.8 (C) 7 (D) 9.4

51. Given the reaction $HA_{aq} + B^-_{aq} \leftrightarrow HB_{aq} + A^-_{aq}$ Which of the following is correct, if the equilibrium constant value for the above reaction is 10^{-3} ? [AIIMS-93]
(A) HB is stronger acid than HA
(B) HA is a stronger acid than H_3O^+
(C) B^- is a stronger base than A^-
(D) K_c for the reverse reaction is 10^{-3}

52. The law of equilibrium was first given by [AIIMS-94]
(A) Goldberg (B) Waage
(C) Boyle (D) Both (A) and (B)

53. Which of the following is a Lewis-acid ? [AIIMS-94]
(A) Cl^- (B) Ag^+ (C) Br^- (D) S^-

54. In $N_2 + 3H_2 \rightleftharpoons 2NH_3$ reversible reaction, increase in pressure will favour [AIIMS-95]
(A) Forward direction (B) Backward direction
(C) Reversible reaction (D) Irreversible reaction

55. A buffer solution contains 0.1 M of acetic acid and 0.1 M of sodium acetate. What will be its pH ? (pK_a of acetic acid is 4.75) [AIIMS-95]
(A) 4.00 (B) 4.75 (C) 5.00 (D) 5.25

56. In which of the following reaction $K_p > K_c$? [AIIMS-95]
(A) $H_2 + I_2 \rightarrow 2HI$ (B) $N_2 + 3H_2 \rightarrow 2NH_3$
(C) $PCl_3 + Cl_2 \rightarrow PCl_5$ (D) $2SO_3 \rightarrow O_2 + 2SO_2$

57. If pH value of a solution is 3 and by adding water, it becomes 6, then the dilution is increased by [AIIMS-95]
(A) 10 times (B) 100 times
(C) 500 times (D) 1000 times

58. Which of the following compounds will undergo hydrolysis ? [AIIMS-96]
(A) KCl (B) NaCl (C) $NaNO_3$ (D) NH_4Cl

59. The rate, at which a substance reacts, depends upon its [AIIMS-96]
(A) Molecular mass (B) Atomic mass
(C) Equivalent mass (D) Active mass

60. The pH value of human blood is about [AIIMS-96]
(A) 2.3 to 3.4 (B) 3.5 to 4.6
(C) 4.5 to 5.6 (D) 7.3 to 7.5

61. Ostwald's dilution law is applicable to only [AIIMS-96]
(A) Solute (B) Solvent
(C) Strong electrolytes (D) Weak electrolytes

62. Which of the following is the strongest base ? [AIIMS-96]
(A) H (B) H^- (C) H_2^+ (D) HCO_3^-

63. The equilibrium constant of a reaction is 300. If the volume of a reaction flask is tripled, the equilibrium constant will be [AIIMS-96]
(A) 100 (B) 150 (C) 300 (D) 600

64. The strongest Bronsted base among the following anions is [AIIMS-96]
(A) ClO^- (B) ClO_2^- (C) ClO_3^- (D) ClO_4^-

65. The pH of 0.001 (N) acetic acid solution, which is 10% dissociated, is [AIIMS-96]
(A) 1 (B) 2 (C) 3 (D) 4

66. Which of the following is a characteristic of a reversible reaction ? [AIIMS-96]
(A) It can be influenced by a catalyst
(B) Number of moles of reactants and products are equal
(C) It never proceeds to completion
(D) It proceeds only forward direction

67. If active mass of a 6% solution of a compound is 2, its molecular weight will be [AIIMS-96]
(A) 15 (B) 22 (C) 30 (D) 60

68. In the reaction $I_2 + I^- \rightarrow I_3^-$, the Lewis base is [AIIMS-97]
(A) I_2 (B) I^- (C) I_3^- (D) None of these

69. The maximum amount of $BaSO_4$ precipitated on mixing $BaCl_2$ (0.5M) with H_2SO_4 (1M) will correspond to [AIIMS-97]
(A) 0.5 M (B) 1.0 M (C) 1.5 M (D) 2.0 M

70. Assertion : Heat of neutralization for both HNO_3 and HCl with NaOH is 53.7 kJ per mole. Reason : NaOH is a strong electrolyte/base. [AIIMS-97]
(A) A (B) B (C) C (D) D

71. An equilibrium mixture of the reaction $2H_2S\ (g) = 2H_2(g) + S_2$, had 0.5 mole H_2S, 0.10 mole H_2 and 0.4 mole S_2 in one litre vessel, the value of K_c is [AIIMS-98]
(A) 0.004 (B) 0.008 (C) 0.016 (D) 0.160

72. Which of the following information can be obtained on the basis of Le Chatelier's principle [AIIMS-98]
(A) Entropy change in a reaction
(B) Dissociation constant of a weak acid
(C) Equilibrium constant of a chemical reaction
(D) Shift in equilibrium position on changing value of a constraint

73. The ionic product of water at 25°C is 10^{-14}. Its ionic product at 90°C will be [AIIMS-98]
(A) 1×10^{-20} (B) 1×10^{-12} (C) 1×10^{-14} (D) 1×10^{-16}

74. The solubility of $BaSO_4$, in water, is 2.33×10^{-3} gram/litre. Its solubility product will be (molecular weight of $BaSO_4$ = 233) [AIIMS-98]
(A) 1×10^{-5} (B) 1×10^{-10} (C) 1×10^{-15} (D) 1×10^{-20}

75. Which of the following donot act as Lewis acid ? [AIIMS-2000]
(A) BF_3 (B) $SnCl_4$ (C) CCl_4 (D) SF_3

76. Assertion : Bond order can have positive, negative or fractional value. Reason : Bond order depends upon the bonding and antibonding orbital electrons. [AIIMS-2000]
(A) A (B) B (C) C (D) D

77. If higher is the value of K_c, then [AIIMS-2000]
(A) Greater is the thermodynamic stability of products
(B) Reaction goes to completion

(C) Lesser is the thermodynamic stability of reactants
(D) All of these

78. If the value of reaction quotient Q of a chemical reaction is greater than the value of equilibrium constant K_C, then chemical reaction will proceed [AIIMS-2000]
(A) In forward direction (B) In reverse direction
(C) In equilibrium
(D) Direction will depend upon nature of reactants and products.

79. Increasing order of base strength is SH^-, I^-, OH^-, NH_2^- is (proton affinity order) [AIIMS-2001]
(A) $SH^- < I^- < OH^- < NH_2^-$ (B) $I^- < SH^- < OH^- < NH_2^-$
(C) $NH_2^- < OH^- < SH^- < I^-$ (D) None of these

80. What is order of base strength ? [AIIMS-2001]
(A) $NH_3 > NH_2 - NH_2 > NH_2OH > N_3H$
(B) $NH_2OH > N_3H > NH_3 > NH_2 - NH_2$
(C) $NH_2 - NH_2 > NH_3 > NH_2OH > N_3H$
(D) None of these

81. Which is strongest conjugate base ? [AIIMS-2001]
(A) ClO^- (B) ClO_2^- (C) ClO_3^- (D) ClO_4^-

82. If for ammonium formate pK_a 1 = 3.78 pK_b = 4.78, then pH will be equal to [AIIMS-2001]
(A) 6.9 (B) 6.5 (C) 7.3 (D) 6.2

83. $2SO_3 (g) \rightleftharpoons 2SO_2 (g) + O_2 (g)$; If at 700K $K_p = 1.3 \times 10^{-3}$ kPa, then determine K_c [AIIMS-2001]
(A) 2.33×10^{-5} mole lit^{-1} (B) 2.33×10^{-4} mole lit^{-1}
(C) 2.33×10^{-7} mole lit^{-1} (D) 2.33×10^{-6} mole lit^{-1}

84. Assertion : Ionic product is used for any type of electrolyte whereas solubility product is applicable only to sparingly salts. Reason : Ionic product is defined at any stage of the reaction whereas solubility product is only applicable to the saturation stage. [AIIMS-2001]
(A) A (B) B (C) C (D) D

85. The solubility of CuBr is 2×10^{-4} mol/L at 25°C. The K_{sp} value of CuBr is [AIIMS-2002]
(A) 4×10^{-8} mol^2L^{-2} (B) 4×10^{-11} mol^2L^{-1}
(C) 4×10^{-4} mol^2L^{-2} (D) 4×10^{-15} mol^2L^{-2}

86. At 80°C, distilled water has (H_3O^+) concentration equal to 1×10^{-6} mole/litre. The value of K_w at this temperature will be [AIIMS-2002]
(A) 1×10^{-6} (B) 1×10^{-9} (C) 1×10^{-12} (D) 1×10^{-15}

87. The pH of solution containing 0.10 M sodium acetate and 0.03 M acetic acid is (pK_a for CH_3COOH = 4.57) [AIIMS-2002]
(A) 4.09 (B) 5.09 (C) 6.09 (D) 7.09

88. The heat of neutralization of a strong base and a strong acid is 57 kJ. The heat released when 0.5 mole of HNO_3 solution is added to 0.20 moles of NaOH solution, is [AIIMS-2002]
(A) 11.4 kJ (B) 23.5 kJ (C) 34.7 kJ (D) 58.8 kJ

89. Which one of the following is not a buffer solution ? [AIIMS-2003]
(A) 0.8M H_2S + 0.8M KHS_+
(B) 2M $C_6H_5NH_2$ + 2M $C_6H_5NH_3Br$
(C) 3M H_2CO_3 + 3M $KHCO_3$
(D) 0.05M $KClO_4$ + 0.05M $HClO_4$

90. In which of the following acid-base titration, pH is greater than 8 at equivalence point ? [AIIMS-2003]
(A) Acetic acid versus ammonia
(B) Acetic acid versus sodium hydroxide
(C) Hydrochloric acid versus ammonia
(D) Hydrochloric acid versus sodium hydroxide

91. (A) Activity of an enzyme is pH dependent.
(R) Change in pH affects the solubility of the enzyme in water. [AIIMS-2003]
(A) (B) (C) (D)

92. (A) $BaCO_3$ is more soluble in HNO_3 than in plain water.
(R) Carbonate is a weak base and reacts with the H^+ from the strong acid, causing the barium salt to dissociate. [AIIMS-2003]
(A) (B) (C) (D)

93. (A) Increasing pressure on pure water decreases its freezing point.
(R) Density of water is maximum at 273 K. [AIIMS-2003]
(A) (B) (C) (D)

94. Dissociation constant of an acid is 1×10^{-10}. The pK_a of the acid is : [JIPMER-90]
(A) 10 (B) –10 (C) 4 (D) 5

95. The value of equilibrium constant for the reaction $H_2 + I_2 \rightleftharpoons 2NI$ is 48. What would be the value of equilibrium constant for the reverse reaction ? [JIPMER-90]
(A) 0.01 (B) 0.02 (C) 0.04 (D) 0.08

96. The acid with highest pH is : [JIPMER-90]
(A) Formic acid (B) Acetic acid
(C) Monochloroacetic acid (D) Dichloroacetic acid

97. The compound that is not a Lewis acid is : [JIPMER-91]
(A) BF_3 (B) $BaCl_2$ (C) $AlCl_3$ (D) $SnCl_4$

98. The pH of 0.001 M solution of HCl is : [JIPMER-91]

(A) 1.0 (B) 2.0 (C) 3.0 (D) 10

99. The pH of a 10^{-8} M solution of HCl in water is : [JIPMER-92]
(A) 8 (B) –8
(C) Between 7 and 8 (D) Between 6 and 7

100. The pH of a solution is 4. The H^+ ion concentration of the solution in gms/litre is : [JIPMER-92]
(A) 10^{-2} (B) 10^4 (C) 10^{-4} (D) 4

101. If the solubility product of lead iodide is 0.8×10^{-9} in water, its solubility in moles per litre is : [JIPMER-92]
(A) 2×10^{-3} (B) 1.26×10^{-3}
(C) 8.94×10^{-5} (D) 4.47×10^{-5}

102. If 0.5 mole of $BaCl_2$ is mixed with 0.2 mole of Na_3PO4, the maximum number of mole of $Ba_3(PO_4)_2$ that can be formed is : [JIPMER-92]
(A) 0.7 (B) 0.5 (C) 0.3 (D) 0.1

103. The conjugate acid of NH_2^- is : [JIPMER-92]
(A) NH_3 (B) NH_4^+ (C) NH_4OH (D) N_2H_4

104. For the reaction, 2 A(a) + B(g) ↔ 3C(g) + D(g.), two moles of each of A and B were taken in a flask. The following must be always true at equilibrium : [JIPMER-93]
(A) [A] = [B] (B) [A] > [B]
(C) [B] = [D] (D) [A] < [B]

105. Solubility product of lead bromide is 3.2×10^{-5} at a certain temperature. Its solubility in moles per litre is : [JIPMER-93]
(A) 1.6×10^{-3} (B) 2.0×10^{-3} (C) 0.4×10^{-4} (D) 2.0×10^{-2}

106. The K_a of an acid is 3.2×10^{-5}. The degree of dissociation of the acid at concentration of 0.2 M is: [JIPMER-93]
(A) 6.0×10^{-2} (B) 1.6×10^{-2} (C) 4.0×10^{-4} (D) 0.04

107. A 1×10^{-6} M HCl solution is diluted to hundred times. The pH of the resulting solution is : [JIPMER-93]
(A) 8 (B) 6 (C) 7 (D) 6.96

108. The buffer solution contains equal concentrations of X^- and HX. The K_b for X^- is 10^{-10}. The pH of buffer is : [JIPMER-93]
(A) 4 (B) 7 (C) 10 (D) 14

109. The hydrogen ion concentration in pure water is 10^{-7}g ions/litre. An aqueous solution will be acidic if its hydrogen ion concentration in g ions/litre is [JIPMER-94]
(A) 10^{-6} (B) 10^{-7} (C) 10^{-8} (D) 10^{-9}

110. The compound that is not a Lewis acid is [JIPMER-94]
(A) BF (B) $AlCl_3$ (C) $BeCl_2$ (D) $SnCl_4$

111. A solution of weak acid is diluted by adding an equal volume of water. Which of the following will not change ? [JIPMER-95]
(A) pH of the solution (B) Strength of the acid
(C) The value of $[H_3O^+]$ (D) None of the these

112. Which of the following solutions can act as buffer ? [JIPMER-95]
(A) 0.1 molar aq. ammonium acetate
(B) 0.1 molar aq. $NaCl^-$
(C) 0.2 molar aq. CH_3COOH + 0.1 molar NaOH
(D) None of these

113. A solution is 5×10^{-8} M solution of HCl. Its pH will be : [JIPMER-96]
(A) 7 (B) 8–log 5 (C) 8 (D) between 6 and 7

114. The conjugate acid of CH_3NH_2 is: [JIPMER-96]
(A) $CH_3NH_3^+$ (B) CH_3NH (C) CH_3OH (D) NH_2

115. The equilibrium constant in a reversible chemical reaction at a given temperature [JIPMER-96]
(A) does not depend on the initial concentration of reactants
(B) depends on the initial concentrations of the reactants
(C) in not characteristic of the reaction
(D) depends on the concentration of one of the products at equilibrium.

116. For the basic buffer solution the pH is 9.3. The pH can be increased by [JIPMER-97]
(A) Increasing the concentration of the salt
(B) Increasing the concentration of the base
(C) Decreasing the concentration of the base
(D) Decreasing the Kb value of the base

117. The pH of an aqueous solution is 10.8. It [JIPMER-97]
(A) Turns pink with phenolphthalein
(B) Turns yellow with phenolphthalein
(C) Has no action on phenolphthalein
(D) Turns from red to yellow with phenolphthalein

118. The units for the equilibrium constant K_P for the reaction. $N_2(g) + 3H_2(g) \leftrightarrow 2NH_3(g)$ are [JIPMER-97]
(A) atm^{-1} (B) atm^{-1} (C) atm (D) atm^2

119. In the dissociation of HI (g) it is found that 20% of HI is dissociated. The equilibrium constant will be [JIPMER-97]
(A) 0.0625 (B) 16 (C) 0.0156 (D) 64

120. pK_a of a weak acid is defined as [JIPMER-98]
(A) $\log_{10} K_a$ (B) $\frac{1}{\log_{10} K_a}$
(C) $\log_{10} \frac{1}{K_a}$ (D) $-\log_{10} \frac{1}{K_a}$

121. At 3000 K the equilibrium pressures of CO_2, CO and O_2 are 0.6, 0.4 and 0.2 atmospheres respectively. K_p for the reaction. $2CO_2 \leftrightarrow 2CO + O_2$ is [JIPMER-98]
(A) 0.074 (B) 0.0533 (C) 0.133 (D) 0.177

122. pH of a solution is 5. Its hydroxyl ion concentration is [JIPMER-98]
(A) 5 (B) 10 (C) 10^{-5} (D) 10^{-9}

123. Which of the following statements regarding a chemical equilibrium is wrong ? [JIPMER-98]
(A) An equilibrium can be shifted by altering the temperature or pressure.
(B) An equilibrium is dynamic.
(C) The same state of equilibrium is reached whether one starts with the reactants or the products.
(D) The forward reaction is favoured by the addition of a catalyst.

124. In which of the following neutralisation reactions, the heat of neutralisation will be the highest ? [JIPMER-99]
(A) HCl and NaOH (B) NH_4OH and HCl
(C) NaOH and CH_3COOH (D) CH_3COOH and NH_4OH

125. When sodium acetate is added to acetic acid it [JIPMER-99]
(A) Increases ionization of acetic acid
(B) Decreases ionization of acetic acid
(C) Increases ionization of sodium acetate
(D) Is independent of concentration

126. The Lowry-Bronsted theory defines the acid as a molecule or an ion which [JIPMER-99]
(A) Contains a replaceable hydrogen
(B) Donates a proton
(C) Releases hydrogen in its reaction with metals
(D) Accepts a proton

127. How must the hydrogen ion concentration in a solution be changed for the pH of the solution to increase by 1 ? [JIPMER-99]
(A) It must be increased 10 times
(B) It must increased by 1 mole/litre
(C) It must be diminished to 1/10 th of its original value
(D) It must be diminished by 1 mole/litre

128. If α is the degree of dissociation of PCl_5 at equilibrium for the reaction $PCl_5 \leftrightarrow PCl_3 + Cl_2$ the total number of moles at equilibrium is [JIPMER-99]
(A) 1 (B) $1 - \alpha$ (C) $1 + \alpha$ (D) 3

129. For the dissociation of $N_2O_4(g) \leftrightarrow 2NO_2$, K_p is [JIPMER-99]
(A) Equal to K_C (B) Less than K_C
(C) Greater than K_C (D) None of these

130. According to Le Chatelier's principle adding heat to a solid and liquid in equilibrium with endothermic nature will cause the [JIPMER-2000]
(A) Temperature to rise (B) Temperature to fall
(C) Amount of solid to decrease
(D) Amount of liquid to decrease

131. Equilibrium concentration of HI, I_2 and H_2 is 0.7, 0.1 and 0.1 M respectively. The equilibrium constant for the reaction, $I_2 + H_2 \leftrightarrow 2HI$ is [JIPMER-2000]
(A) 36 (B) 49 (C) 0.49 (D) 0.36

132. Which may be added to one litre of water to act as a buffer [JIPMER-2000]
(A) One mole of $HC_2H_3O_2$ and 0.5 mole of NaOH
(B) One mole of NH_4Cl and one mole of HCl
(C) One mole of NH_4OH and one mole of NaOH
(D) One mole of $HC_2H_3O_2$ and one mole of HCl

133. The pH of a 0.01M solution of acetic acid having degree of dissociation 12.5% is [JIPMER-2000]
(A) 5.623 (B) 2.903 (C) 3.723 (D) 4.509

134. If the pH of a solution of an alkali metal hydroxide is 13.6, the concentration of hydroxide is [JIPMER-2000]
(A) Between 0.1M and 1M (B) More than 1M
(C) Less than 0.001M (D) Between 0.01M and 1M

135. Heat of neutralization of the acid-base reaction is 57.32 kJ for [JIPMER-2000]
(A) HNO_3 + LiOH (B) HCOOH + KOH
(C) HCl + NH_4OH (D) CH_3COOH + NaOH

136. What is the pH for a neutral solution at the normal temperature of the human body [JIPMER-2000]
(A) 7.2 (B) 14.0 (C) 6.8 (D) 6.0

137. For the equilibrium $N_2 + 3H_2 \rightleftharpoons 2NH_3$, K_c at 1000 K is 2.37×10^{-3}. If at equilibrium $[N_2] = 2M$, $[H_2] = 3M$, the concentration of NH_3 is [JIPMER-2000]
(A) 0.00358 M (B) 0.0358 M (C) 0.358 M (D) 3.38

138. A reversible reaction $H_2 + Cl_2 \rightleftharpoons 2HCl$ is carried out in one litre flask. If the same reaction is carried out in two litre flask, the equilibrium constant will be [JIPMER-2001]
(A) Decreased (B) Doubled (C) Halved (D) Same

139. For the reaction
$2NO_2(g) \rightleftharpoons 2NO(g) + O_2(g)$, $K_C = 1.8 \times 10^{-6}$ at 185°C,
At 185°C, the K_C for $NO(g) + 1/2\, O_2(g) \rightleftharpoons 2NO_2(g)$ is [JIPMER-2001]
(A) 1.95×10^{-3} (B) 1.95×10^{3}
(C) 7.5×10^{2} (D) 0.9×10^{6}

140. A solution of sodium borate has a pH of approximately [JIPMER-2001]
(A) < 7 (B) > 7 (C) =7 (D) Between 4 to 5

141. For exothermic reaction, the equilibrium constant [JIPMER-2001]

(A) Increases with increase of P
(B) Decreases with increase of P
(C) Increases with increase of temperature
(D) Decreases with increase of temperature

142. The K_C for $H_2(g) + I_2(g) \rightleftharpoons 2HI(g)$ is 64. If the volume of the container is reduced to one-half of its original volume, the value of the equilibrium constant will be [JIPMER-2001]

(A) +28 (B) 64 (C) 32 (D) 16

143. If Ksp for $HgSO_4$ is 6.4×10^{-5}, then solubility of its salt is [JIPMER-2001]

(A) 6.4×10^{-5} (B) 6.4×10^{-3}
(C) 8×10^{-6} (D) 8×10^{-3}

144. Which one of the following is acid salt ? [JIPMER-2002]

(A) Na_2S (B) Na_2SO_3 (C) $NaHSO_3$ (D) Na_2SO_4

145. The ionic product of water will increase if [JIPMER-2002]

(A) Pressure is decreased (B) H^+ are added
(C) OH^- are added (D) Temperature is increased

146. The solubility product of $BaSO_4$ is 1.5×10^{-9} at 18°C. Its solubility in water at 18°C is [BHU-95]

(A) 1.5×10^{-9} mole litre^{-1} (B) 1.5×10^{-5} mole litre^{-1}
(C) 3.9×10^{-9} mole litre^{-1} (D) 3.9×10^{-5} mole litre^{-1}

147. The pH value of a solution, whose hydronium ion concentration is 6.2×10^{-9} mole L^{-1}, is [BHU-95]

(A) 6.21 (B) 7.21 (C) 7.75 (D) 8.21

148. Which of the following is a buffer solution ? [BHU-95]

(A) $NaOH + CH_3COONa$ (B) $NaOH + Na_2SO_4$
(C) $K_2SO_4 + H_2SO_4$ (D) $NH_4OH + NH_4Cl$

149. Which of the following is a Lewis base ? [BHU-95]

(A) CH_4 (B) C_2H_5OH (C) Acetone (D) Amines

150. pH of 0.01 M NaOH solution is [BHU-97]

(A) 12 (B) 1 (C) 2 (D) 14

151. Which of the following is not Lewis acid ? [BHU-97]

(A) BF_3 (B) $AlCl_3$ (C) $FeCl_3$ (D) PH_3

152. The pH of 0.0001 M NaOH is [BHU-97]

(A) 10 (B) 4 (C) 12 (D) 11

153. The solubility of AgCl at 20°C is 1.35×10^{-5} gm/lit. The solubility product of AgCl is [BHU-97]

(A) 1.0×10^{-14} (B) 108×10^{-3}
(C) 2×10^{-16} (D) 1.035×10^{-5}

154. The solubility product of Ag_2CrO_4 is 32×10^{-12} What is the concentration of CrO_4^{--} ions in that solution ? [BHU-97]

(A) 2×10^{-4} m/s (B) 16×10^{-4}
(C) 8×10^{-4} m/s (D) 8×10^{-8} m/s

155. Which of the following is not a Bronsted acid ? [BHU-97]

(A) NH_4^+ (B) CH_3COO^- (C) H_2O (D) HSO_4

156. Equal volumes of methanoic acid and sodium hydroxide are mixed, if x is the heat of formation of water, then heat evolved on neutralisation is [BHU-98]

(A) More than x (B) Equal to x
(C) Twice x (D) Less than x

157. A reaction is $A+B \rightarrow C+D$. Initially we start with equal concentration A and B. At equilibrium we find the no. of moles of C is two times of A. What is the equilibrium constant of the reaction ? [BHU-98]

(A) 4 (B) 2 (C) 1/4 (D) 1/2

158. The neutralisation of a strong acid by a strong base liberates an amount of energy per mole of H^+ that [BHU-98]

(A) Depends upon which acid and base are involved
(B) Depends upon the temperature at which the reaction takes place
(C) Depends upon which catalyst is used
(D) Is always the same

159. 240g of urea are present in 10 litre solution the active mass of urea will be [BHU-98]

(A) 0.2 (B) 0.06 (C) 0.4 (D) 0.08

160. Degree of ionsation of a solution depends upon [BHU-98]

(A) Temperature (B) Nature of the electrolyte
(C) Nature of the solvent (D) All of these

161. Which is most acidic ? [BHU-98]

(A) Na_2O (B) MgO (C) Al_2O_3 (D) CaO

162. Heat of neutralistion of strong acid against strong base is constant and equal to [BHU-99]

(A) 13.7 kcal (B) 53.7 kcal (C) 57.3 kcal (D) 59.3 kcal

163. The conjugate acid of NH_2^- ion is [BHU-99]

(A) NH_3 (B) NH_4^+ (C) N_2H_4 (D) NH_2OH

164. For an endothermic reaction ΔS is (+) ve. The reaction is [BHU-2000]

(A) Feasible when $T\Delta S > \Delta H$
(B) Feasible when $\Delta H > T\Delta S$
(C) Feasible at all temperatures (D) Not feasible at all

165. The yield of product can be increased by [BHU-2001]

(A) Decreasing the conc. of reactant
(B) Increasing the conc. of product
(C) Carrying the reaction in a closed container vessel

(D) By escaping out the gaseous product

166. Which of the following is strongest base ? [BHU-2002]
(A) ClO^- (B) CHO_2^- (C) CHO_3^- (D) ClO_4^-

167. The pH of a 0.05 M solution of a weak monobasic acid which is 20% ionized is [BHU-2002]
(A) 2.5 (B) 1.5 (C) 5.2 (D) 2.0

168. The conjugate base of a weak acid is [AFMC-91]
(A) A strong base (B) A weak base
(C) A neutral species (D) May be a weak or strong base

169. The compound that is not Lewis acid is [AFMC-91]
(A) BF_3 (B) C_2H_5OH (C) $AlCl_3$ (D) BCl_3

170. pH of 0.002 N H_2SO_4 is [AFMC-91]
(A) 2 (B) 1 (C) 0.002 (D) 2.6990

171. Solubility of salt M_2X_3 is x gm mole/l. The solubility product of the salt will be [AFMC-91]
(A) x^5 (B) $126x^2$ (C) $96x^5$ (D) $108x^5$

172. Pure NaCl is prepared by saturating a cold staturated solution of common salt in water with HCl gas. The principle used is [AFMC-92]
(A) Le Chatelier principle (B) Displacement law
(C) The common ion effect (D)Fractional distillation

173. According to Lewis concept, an acid is [AFMC-93]
(A) Proton donor (B) Proton acceptor
(C) Electron pair donor (D) Electron pair acceptor

174. A base, as defined by Bronsted theory, is a substance which [AFMC-93]
(A) Can accept protons (B) Can lose a pair of electrons
(C) Can donate protons (D) Can gain a pair of electrons

175. Which of the following is a characteristic of a reversible reaction ? [AFMC-93]
(A) Number of moles of reactants and products are equal
(B) It can be influenced by a catalyst
(C) It never proceeds to completion
(D) None of the above

176. Which of the following have highest pH with water ? [AFMC-94]
(A) NaCl (B) Na_2CO_3 (C) $NaHCO_3$ (D) KCl

177. If the hydrogen ion concentration of an acid is decreased ten times, its pH [AFMC-94]
(A) Increases by one (B) Decreases by one
(C) Remains unchanged (D) Increases by 10

178. In which of the following reaction $K_P > K_C$? [AFMC-95]
(A) $N_2 + 3H_2 \rightleftharpoons 2NH_3$ (B) $H_2 + I_2 \rightleftharpoons 2HI$
(C) $PCl_3 + Cl_2 \rightleftharpoons PCl_5$ (D) $2SO_3 \rightleftharpoons O_2 + 2SO_2$

179. Precipitation occurs if the product of ionic concentrations is [AFMC-95]
(A) Less than solubility product
(B) More than solubility product
(C) Equal to solubility product (D) None of them

180. pH of a solution is 4, then OH^- concentration is [AFMC-96]
(A) 10×10^{-5} (B) 10^{-10} (C) 10^{-14} (D) 10^{-4}

181. In which case Kp, is less then Kc ? [AFMC-97]
(A) $PCl_5 \rightleftharpoons PCl_3 + Cl_2$ (B) $H_2 + Cl_2 \rightleftharpoons 2HCl$
(C) $2SO_2 + O_2 \rightleftharpoons 2SO_3$ (D) All of these

182. If a neutral solution has $pK_w = 13.36$ at 50° C, then pH of the solution is [AFMC-97]
(A) 6.63 (B) 7 (C) 7.63 (D) None of these

183. BF_3 is acid according to [AFMC-97]
(A) Arrhenius (B) Lewis
(C) Bronsted & Lowry (D) All of these

184. Arrange the following in order of increasing acidity H_3PO_4, H_2CO_3, HCl, HI [AFMC-97]
(A) H_3PO_4, HCl, H_2CO_3, HI
(B) H_3PO_4, H_2CO_3, HCl, HI
(C) H_2CO_3, H_3PO_4, HCl, HI (D)None of these

185. K_{sp} for $HgSO_4$ is 6.4×10^{-5}, then solubility of the salt is [AFMC-97]
(A) 8×10^{-3} (B) 8×10^{-6}
(C) 6.4×10^{-5} (D) None of these

186. The pH value of 10^{-7} M HCl is [AFMC-98]
(A) Equal to 7 (B) Less than 7
(C) More than 7 (D) Either (A) or (C)

187. pH value of a solution, whose hydronium ion concentration is 6.2×10^{-9} mol/1, is [AFMC-99]
(A) 8.21 (B) 6.21 (C) 7.75 (D) 7.21

188. A base, as defined by Bronsted theory, is a substance which can [AFMC-99]
(A) Lose a pair of electrons (B) Donate protons
(C) Gain a pair of electrons (D) Accept protons

189. In the manufacture of NH_3, the most favourable condition is [AFMC-99]
(A) Low temperature and low pressure
(B) High temperature and high pressure
(C) Low temperature and high pressure
(D) High temperature and low pressure

190. Which of the following has lowest pH value ? [AFMC-2000]
(A) 1 M HCl (B) 1 M NaOH
(C) 1 M H_2SO_4 (D) $C_3H_6O_4$

191. What happens to the yield of this reaction when temperature in increased ? $N_2 + O_2 \leftrightarrow 2NO$; 43.2 Kcal. [AFMC-2000]

(A) Increases (B) Decreases
(C) Remains same (D) None of these

192. At 25°C the pH value of a solution is 6, the pH of that solution is [AFMC-2001]
(A) Acidic (B) Neutral (C) Basic (D) Alkaline

193. A buffer solution changes its pH [AFMC-2001]
(A) Very rapidly (B) Very slowly
(C) Does not change (D) None of these

194. $FeCl_3$ solution is [AFMC-2001]
(A) Acidic (B) Basic (C) Neutral (D) Amphoteric

195. Conjugate base of NH_4^+ is [AFMC-2002]
(A) NH_3 (B) OH^- (C) KOH (D) NH_4OH

196. Which of the following is most soluble ? [AFMC-2002]
(A) Bi_2S_3 [$K_{sp} = 1 \times 10^{-70}$] (B) MnS [$K_{sp} = 7 \times 10^{-16}$]
(C) CuS [$K_{sp} = 8 \times 10^{-37}$] (D) Ag_2S [$K_{sp} = 6 \times 10^{-51}$]

197. pH of a solution can be expressed as [AFMC-2002]
(A) $-\log(H^+)$ (B) $-\log_{10}(H^+)$
(C) $-\log_e(H^+)$ (D) $\log_{10}(H^+)$

198. The reaction $2SO_2 + 2O_2 \rightleftharpoons 2SO_3$ will be favoured by [AFMC-2002]
(A) High temperature and low pressure
(B) Low temperature and high pressure
(C) low temperature and low pressure
(D) High temperature and high pressure

199. Amongst the following hydroxides, the one which has the lowest value of K_{sp} at ordinary temperature (about 25°C) is : [IIT-90]
(A) $Mg(OH)_2$ (B) $Ca(OH)_2$ (C) $Ba(OH)_2$ (D) $Be(OH)_2$

200. The reaction which proceeds in the forward direction is : [IIT-91]
(A) $Fe_2O_3 + 6HCl \leftrightarrow 2FeCl_3 + 3H_2O$
(B) $NH_3 + H_2O + NaCl \leftrightarrow NH_4Cl + NaOH$
(C) $SnCl_4 + Hg_2Cl_2 \leftrightarrow SnCl_2 + 2HgCl_2$
(D) $2CuI + I_2 + 4K^+ \leftrightarrow 2Cu^{2+} + 4KI$

201. For reaction, $PCl_5(g) \rightleftharpoons PCl_3(g) + Cl_2(g)$
The forward reaction at constant temperature is favoured by [IIT-91]
(A) introducing an inert gas at constant volume
(B) introducing chlorine gas at constant pressure
(C) introducing an inert gas at constant pressure
(D) increasing the volume of the container
(E) Introducing PCl_5 at constant volume

202. The following equilibrium is established when hydrogen chloride is dissolved in acetic acid [IIT-92]

$$HCl + CH_3COOH \rightarrow Cl^- + CH_3COOH_2^+$$

The set that characterises the conjugate acid-base pairs is:
(A) (HCl, CH_3COOH) and ($CH_3COOH_2^+$, Cl^-)
(B) (HCl, $CH_3COOH_2^+$) and CH_3COOH, Cl^-)
(C) ($CH_3COOH_2^+$, HCl) and (Cl^-, CH_3COOH)
(D) (HCl, Cl^-) and ($CH_3COOH_2^+$, CH_3COOH)

203. Which of the following solutions will have pH close to 1.0 ? [IIT-92]
(A) 100 ml of (M/10) HCl +100 ml of (M/10) NaOH.
(B) 55 ml of (M/10) HCl + 45 ml of (M/10) NaOH
(C) 10 ml of (M/10) HCl + 90 ml of (M/10) NaOH
(D) 75 ml of (M/5) HCl + 25 ml of (M/5) NaOH.

204. For a given reversible reaction, at a fixed temperature equilibrium constant K_p and K_c are related by .... [IIT-94]

205. $(CH_3OH_2)^+$ is .... acidic than .... [IIT-97]

206. For a gaseous reaction $2B \rightarrow A$, the equilibrium constant K_p is .... to/than K_c. [IIT-97]

207. In the reaction $I^- + I_2 \rightarrow I_3^-$, the lewis acid is .... [IIT-97]

208. Molar heat capacity of water in equilibrium with ice at constant pressure is [I.I.T.-97]
(A) zero (B) infinity (∞)
(C) 40.45 kJ K^{-1} mol^{-1} (D) 75.48 KJ^{-1} mol^{-1}

209. For the reaction [IIT-98]

$$CO(g) + H_2O(g) \rightleftharpoons CO_2(g) + H_2(g)$$

At a given temperature the equilibrium amount of $CO_2(g)$ can be increased by
(A) adding a suitable catalyst
(B) adding an inert gas
(C) decreasing the volume of the container
(D) increasing the amount of CO(g)

210. Which of the following statement (s) is (are) correct? [IIT-98]
(A) The pH of 2.0×10^{-8}M solution of HCl is 8.
(B) The conjugate base of $H_2PO_4^-$ is HPO_4^{2-}
(C) Autoprotolysis constant of water increases with temperature
(D) When a solution of a weak monoprotic acid is titrated against a strong base, at half neutralization point $pH = (1/2)pK_a$

211. The pH of 0.1M solution of the following salts increases in the order [IIT-99]
(A) $NaCl < NH_4Cl < NaCN < HCl$
(B) $HCl < NH_4Cl < NaCl < NaCN$
(C) $NaCN < NH_4Cl < NaCl < HCl$

(D) $HCl < NaCl < NaCN < NH_4Cl$

212. For the chemical reaction $3X(g) + Y(g) \rightleftharpoons X_3Y(g)$, the amount of X_3Y at equilibrium is affected by [IIT-99]
(A) temperature and pressure
(B) temperature only (C) pressure only
(D) temperature, pressure and catalyst

213. A buffer solution can be prepared from a mixture of [IIT-99]
(A) sodium acetate and acetic acid in water
(B) sodium acetate and hydrochloric acid in water
(C) ammonia and ammonium chloride in water
(D) ammonia and sodium hydroxide in water

214. For the reversible reaction, $N_2(g) + 3H_2(g) = 2NH_3$ at 500°C, the value of K_p is 1.44×10^{-5} when partial pressure is measured in atmospheres. The corresponding value of K_c, with concentration in mole litre^{-1}, is [IIT-2000]
(A) $1.44 \times 10^{-5} / (0.082 \times 500)^{-2}$
(B) $1.44 \times 10^{-5} / (8.314 \times 773)^{-2}$
(C) $1.44 \times 10^{-5} / (0.082 \times 773)^{2}$
(D) $1.44 \times 10^{-5} / (0.082 \times 773)^{-2}$

215. At constant temperature, the equilibrium constant (K_p) for the decomposition reaction $N_2O_4 \rightleftharpoons 2NO_2$ expressed by $K_p = (4x^2P)/(1-x^2)$, where P = pressure, x = extent of decomposition. Which one of the following statements is true ? [IIT-2001]
(A) K_p increases with increase of P
(B) K_p increases with increase of x
(C) K_p increases with decrease of x
(D) K_p remains constant with change in P and x.

216. For a sparingly soluble salt A_pB_q, the relationship of its solubility product (L_s) with its solubility (S) is [IIT-2001]
(A) $L_S = S^{p+q} . p^p . q^q$ (B) $L_S = S^{p+q} . p^q . q^p$
(C) $L_S = S^{pq} . p^p . q^q$ (D) $L_S = S^{pq} . (pq)^{p+q}$

217. Consider the following equilibrium in a closed container : $N_2O_4(g) \rightleftharpoons 2NO_2(g)$
At a fixed temperature, the volume of the reaction container is halved. For this change, which of the following statement holds true regarding the equilibrium constant (K_p) and degree of dissociation (α) ? [IIT-2002]
(A) neither K_p nor α changes (B) both K_p and α change
(C) K_p changes, but α does not change
(D) K_p does not change, but α changes.

218. A 2 litre solution (X) contains 0.02 mole of $[Co(NH_3)_5SO_4]Br$ and 0.02 mole $[Co(NH_3)_5Br]SO_4$, 1 litre of this taken [IIT-2003]
$X + AgNO_3$ (excess) $\rightarrow$ Y mole of AgBr $\downarrow$
$X + BaCl_2$ (excess) $\rightarrow$ Z mole $BaSO_4 \downarrow$
values of Y and Z are
(A) 0.01 , 0.01 (B) 0.01 , 0.02
(C) 0.02 , 0.03 (D) 0.00 , 0.01

219. If the $[S^{2-}] = 6.0 \times 10^{-2}$ mol dm^{-2} and the solubility of $[Hg^{2+}][Mn^{2+}] = [Fe^{2+}] = [Zn^{2+}] = 1 \times 1$ dm^{-2}, then which of the following will precipitate first in aqueous solution ? [Solubility products of HgS = 3×10^{-53}, MnS = 6×10^{-16}; FeS = 1×10^{-19}; ZnS = 1×10^{-22}] [IIT-2003]
(A) HgS (B) K_2S (C) MgS (D) NaS

220. 1 M NaCl and 1 M HCl are present in an aqueous solution. The solution is [AIEEE-2002]
(A) Not a buffer solution with pH < 7
(B) Not a buffer solution with pH > 7
(C) A buffer solution with pH < 7
(D) A buffer solution with pH > 7

221. Species acting as both Bronsted acid and base is [AIEEE-2002]
(A) $(HSO_4)^{-1}$ (B) Na_2CO_3 (C) NH_3 (D) OH^{-1}

222. Let the solubility of an aqueous solution of $Mg(OH)_2$ be x then its K_{sp} is [AIEEE-2002]
(A) $4x^3$ (B) $108x^5$ (C) $27x^4$ (D) 9x

223. For the reaction [AIEEE-2002]
$CO(g) + (1/2) O_2(g) \leftrightarrow CO_2(g)$, K_p / K_c is
(A) RT (B) $(RT)^{-1}$ (C) $(RT)^{-1/2}$ (D) $(RT)^{1/2}$

224. Change in volume of the system does not alter the number of moles in which of the following equilibria ? [AIEEE-2002]
(A) $N_2(g) + O_2(g) \rightleftharpoons 2NO(g)$
(B) $PCl_5(g) \rightleftharpoons PCl_3(g) + Cl_2(g)$
(C) $N_2(g) + 3H_2(g) \rightleftharpoons 2NH_3(g)$
(D) $SO_2Cl_2(g) \rightleftharpoons SO_2(g) + Cl_2(g)$

225. In which of the following reactions, increase in the volume at constant temperature does not affect the number of moles at equilibrium? [AIEEE-2002]
(A) $2NH_3 \rightarrow N_2 + 3H_2$ (B) $C(g) + (1/2)O_2(g) \rightarrow CO(g)$
(C) $H_2(g) + O_2(g) \rightarrow H_2O_2(g)$ (D) None of these

226. The solubility of $Mg(OH)_2$ is S moles/litre. The solubility product under the same condition is [AIEEE-2002]
(A) $4s^3$ (B) $3s^4$ (C) $4s^2$ (D) s^3

227. Consider the reaction equilibrium
$2SO_2(g) + O_2(g) \rightleftharpoons 2SO_3(g)$; $\Delta H° = -198$ kJ
On the basis of Le Chatelier's principle, the condition favourable for the forward reaction is [AIEEE-2003]
(A) Lowering of temperature as well as pressure.
(B) Increasing temperature as well as pressure.
(C) Lowering the temperature and increasing the

pressure.

(D) Any value of temperature and pressure.

228. For the reaction equilibrium $N_2O_4(g) \rightleftharpoons 2NO_2(g)$ the concentrations of N_2O_4 and NO_2 at equilibrium are 4.8×10^{-2} and 1.2×10^{-2} mol L^{-1} respectively. The value of K_C for the reaction is [AIEEE-2003]

(A) 3.3×10^2 mol L^{-1} (B) 3×10^{-1} mol L^{-1}

(C) 3×10^{-3} mol L^{-1} (D) 3×10^3 mol L^{-1}

229. The solubility in water of a sparingly soluble salt AB_2 is 1.0×10^{-5} mol L^{-1}. Its solubility product number will be [AIEEE-2003]

(A) 4×10^{-15} (B) 4×10^{-10} (C) 1×10^{-15} (D) 1×10^{-10}

230. Which one of the following substances has the highest proton affinity ? [AIEEE-2003]

(A) H_2O (B) H_2S (C) NH_3 (D) PH_3

231. Which one of the following statements is not true ? [AIEEE-2003]

(A) The conjugate base of $H_2PO_4^-$ is HPO_4^{2-}

(B) pH + pOH = 14 for all aqueous solutions

(C) The pH of 1×10^{-5} M HCl is 8

(D) 96,500 coulombs of electricity when passed through a $CuSO_4$ solution deposits 1 gram equivalent of copper at the cathode.

232. When rain is accompanied by a thunderstorm, the collected rain water will have a pH value [AIEEE-2003]

(A) Slightly lower than that of rain water without thunderstorm.

(B) Slightly higher than that of when the thunderstorm is not there.

(C) Uninfluenced by occurrence of thunderstorm.

(D) Which depends on the amount of dust in air.

233. Which of the following mixtures in aqueous solution acts as a buffer : [EAMCET-90]

(A) $HNO_3 + KNO_3$ (B) $H_2SO_4 + K_2SO_4$

(C) $NH_4OH + NH_4Cl$ (D) $CH_3COOH + KaCl$

234. The equilibrium concentration of x, y and yx_2 are 4, 2 and 2 moles respectively for the equilibrium $2x + y \leftrightarrow yx_2$. The value of K_C is .... [EAMCET-90]

235. In the equilibrium reaction $N_2 + 3H_2 \rightleftharpoons 2NH_3$ the sign of ΔH accompanying the reaction is .... [EAMCET-90]

236. A solution of pH 9.0 is one thousand times as basic as a solution of pH .... [EAMCET-90]

237. The pKa of a certain weak acid is 4.0 what should be the (salt) to (acid) ratio if we have to prepare a buffer with a pH = 5 using the acid and one of the salts : [EAMCET-91]

(A) 4:5 (B) 5:4 (C) 10:1 (D) 1:10

238. The conjugate base of HSO_4^- ion is .... [EAMCET-91]

239. Aqueous solution of $FeCl_3$ is acidic because of .... [EAMCET-91]

240. 0.1 M HCl solution is diluted by 100 times. The pH of the solution so formed is ....[EAMCET-91]

241. In the reaction $H_2(g) + Cl_2(g) \rightleftharpoons 2HCl(g)$ [EAMCET-92]

(A) $K_P = K_C$ (B) $K_P \neq K_C$ (C) $K_P > K_C$ (D) $K_P < K_C$

242. If in a 100 ml of an aqueous hydrochloric acid of pH 1.0, 900 ml of more distilled water is added, the pH of the resultant solution will become [EAMCET-92]

(A) 1.0 (B) 2.0 (C) 4.0 (D) 7.0

243. The pH of 10^{-2} M NaOH solution is .... [EAMCET-92]

244. The solution that is made by mixing apropriate amounts of a weak acid and its salt is called .... [EAMCET-92]

245. The pH of a 0.001 M aqueous solution of sodium hydroxide will be [EAMCET-93]

(A) 5.0 (B) 7.5 (C) 9.0 (D) 11.0

246. A 0.1 M solution of sodium bisulphate is .... in nature . [EAMCET-93]

247. The ionic product of water $[H^+] [OH^-]$ is .... [EAMCET-93]

248. The solutions which tend to keep the concentration of hydrogen ions constant even when small amounts of strong acid or strong base are added to them are known as .... solutions. [EAMCET-93]

249. The equilibrium constant (k) for the reaction $HA + B \rightleftharpoons BH^+ + A^-$ is 100, if the rate constant for the forward reaction is 10^5, the rate constant for the reverse reaction is [EAMCET-94]

(A) 10^{-3} (B) 10^{-5} (C) 10^7 (D) 10^3

250. In the equilibrium reaction $A(g)+2B(g) \rightleftharpoons C(g)+D(g)$ the equilibrium constant (K_C) is given by the expression [EAMCET-94]

(A) $K_C = \frac{[C][D]}{[A][B]}$ (B) $K_C = \frac{[C][D]^2}{[A][B]}$

(C) $K_C = \frac{[A][B]^2}{[C][D]}$ (D) $K_C = \frac{[C][D]}{[A][B]^2}$

251. The ionic product of water is 14. What is the H^+ ion concentration of a 0.1 M NaOH solution ? [EAMCET-94]

(A) 10^{-14}M (B) 14M (C) 13M (D) 10^{-13}

252. The pH of a 0.05M solution of H_2SO_4 is

[EAMCET-94]

(A) 5 (B) 1 (C) 3 (D) 0.1

253. Which of the following can act as both Bronsted acid and a Bronsted base ? [EAMCET-95]

(i) HCO_3^- (ii) NH_3 (iii) O_2^- (iv) HS^-

(A) (i) & (ii) (B) (ii) & (iii)
(C) None of them (D) All of them

254. Which of the following is not a Lewis acid ? [EAMCET-95]

(A) BF_3 (B) $AlCl_3$ (C) $BeCl_2$ (D) $SnCl_2$

255. Which one of the following is least likely to act as a Lewis base ? (At. No. of P = 15, S = 16, Cl = 17, I = 53) [EAMCET-95]

(A) PCl_3 (B) SCl_2 (C) I^- (D) I^+

256. In the equilibrium reaction $A + B \rightleftharpoons C + D$ the activation energy for the forward reaction is 25k.cal.mol^{-1} and that of the backward reaction is 15 k.cal.mol^{-1} which one of the following statements is correct ? [EAMCET-96]

(A) It is an exothermic process
(B) It is an endothermic process
(C) It is a reaction for which $\Delta H = 0$
(D) It is a sublimation process

257. The equilibrium concentration of C_2H_4 in the following gas phase reaction can be increased by
$C_2H_4(g) + H_2(g) \rightleftharpoons C_2H_6(g); \Delta H = -32.7$ Kcal. [EAMCET-96]

(A) Removal of C_2H_6 (B) Removing H_2
(C) Increase in temperature (D) Increase in pressure

258. Urine normally has a pH of 6. If a patient eliminates 1.3 litres of urine per day how many moles of H^+ ions does he urinate ? [EAMCET-96]

(A) 1.3×10^{-3} (B) 1.3×10^{-6}
(C) 1.3×10^{-7} (D) 1.3×10^{6}

259. If the pH of a solution is increased from 3 to 6 its hydrogen ion concentration will be [EAMCET-96]

(A) Reduced to half (B) Doubled
(C) Reduced by 1000 times (D) Increases by 1000 times

260. The strongest conjugate base results from : [EAMCET-97]

(A) Formic acid (B) Benzoic acid
(C) Acetic acid (D) Acetylene

261. The Value of K_p for the reaction $2H_2S(g) \leftrightarrow 2H_2(g) + S_2(g)$ is 1.2×10^{-2} at 1065°C. The value of K_c for this reaction [EAMCET-97]

(A) 1.2×10^{-2} (B) $< 1.2 \times 10^{-2}$
(C) 83 (D) $> 1.2 \times 10^{-2}$

262. The pH of the aqueous solution containing 0.49 gram. of H_2SO_4 in one litre is : [EAMCET-97]

(A) 2 (B) 1 (C) 1.7 (D) 0.3

263. The conjugate acid of NH_2^- ion is : [EAMCET-97]

(A) NH_3 (B) NH_4^+ (C) N_2H_4 (D) NH_2OH

264. Which one of the following compounds is Lewis acid [EAMCET-97]

(A) PCl_3 (B) BCl_3 (C) NCl_3 (D) $CHCl_3$

265. The pH of a solution is increased from 3 to 6. Its H^+ ion concentration will be [EAMCET-98]

(A) Reduced to half (B) Doubled
(C) Reduced by 1000 times (D) Increased by 1000 times

266. The conjugate base of NH_3 is [EAMCET-98]

(A) NH_3 (B) NH_2^- (C) N^+_4 (D) N_3^-

267. 4.5 moles each of hydrogen and iodine heated in a sealed 10 litre vessel. At equilibrium 3 moles of HI were found. The equilibrium constant for $H_2(g) + I_2(g) \rightleftharpoons 2HI(g)$ is [EAMCET-98]

(A) 1 (B) 10 (C) 5 (D) 0.33

268. Which of the following solutions cannot act as a buffer ? [EAMCET-98]

(A) $NaH_2PO_4 + H_3PO_4$ (B) $CH_3COOH + CH_3COONa$
(C) $HCl + NH_4Cl$ (D) $H_3PO_4 + Na_2\ HPO_4$

269. If the equilibrium constant for the reaction $2AB \rightleftharpoons A_2 + B_2$ is 49, what is the value of equilibrium constant for $AB \rightleftharpoons A_2 + B_2$? [EAMCET-98]

(A) 49 (B) 2401 (C) 7 (D) 0.02

270. A buffer sol. is prepared by mixing 10 ml of 1.0 M acetic acid & 20 ml of 0.5 M sodium acetate & then diluted to 100ml with distilled water. If the pk_a of CH_3COOH is 4.76, what is the pH of the buffer sol. prepared ? [EAMCET-98(MBBS)]

(A) 2.84 (B) 4.76 (C) 4.34 (D) 5.21

271. The dissociation constant of a weak monobasic acid K is 1×10^{-5}. The pH of 0.1 M of that acid would be [EAMCET-98(MBBS)]

(A) 3 (B) Between 2 to 3 (C) 1 (D) 5

272. For the reaction $C(s)+CO_2(g) \rightleftharpoons 2CO_{(g)}$, the partial pressures of CO_2 and CO are 2.0 and 4.0 atm. respectively at equilibrium. What is the value of K_p for this reaction? [EAMCET-98(MBBS)]

(A) 0.5 (B) 4.0 (C) 8.0 (D) 32

273. If K_1 & K_2 are the equilibrium constant of the equilibria (A) & (B) respectively, what is the relationship between the two constants ? [EAMCET-98(MBBS)]

(i) $SO_2(g) + 1/2\ O_2(g) \rightleftharpoons SO_3(g)\ K_1$
(ii) $2SO_3(g) \rightleftharpoons SO_2(g) + O_2,\ K_2$

(A) $K_1 = K_2$ (B) $K_1 = 1/K_2$
(C) $K_1 = K_1^2$ (D) $K_1^2 = 1/K_2$

274. Which of the following equilibria doesnot change on

increasing the pressure? [EAMCET-98]

(A) $NH_3(g) + 3Cl_2(g) \rightleftharpoons NCl_3(g) + 3HCl(g)$

(B) $N_2(g) + 3H_2(g) \rightleftharpoons 2NH_3(g)$

(C) $H_2(g) + Cl_2(g) \rightleftharpoons 2HCl(g)$ (D) None

275. Δx is given by : Δx = (hydration energy - lattice energy) For a salt, Δx is found to be 500 kcal.mol^{-1}. Then on increasing the temperature, its solubility in water [EAMCET-99]

(A) increases (B) decreases (C) remains same

(D) increases initially and then decreases

276. The pH of a 1 litre solution is 2. It is diluted with water till its pH become 4. How many litres of water is added? [EAMCET-99]

(A) 99 (B) 9 (C) 999 (D) 100

277. At 90°C, the concentration of H_3O^+ in pure water is 10^{-6} mol.lit^{-1}. What is the value of K_W at this temperature? [EAMCET-99]

(A) 10^{-4} (B) 10^{-8} (C) 10^{-14} (D) 10^{-12}

278. The conjugate acid N_3^- (azide) ion is [EAMCET-99(MBBS)]

(A) NH_3 (B) HN_3 (C) NH_2^- (D) N_2^-

279. The pH of 1 litre NaOH solution is 12. What is the amount in grams of NaOH present in this solution? [EAMCET-99(MBBS)]

(A) 40 (B) 4 (C) 0.4 (D) 20

280. The pH of an aqueous solution is 6.0. Some chlorine is bubbled through this solution. The pH of the resultant solution will be [EAMCET-99(MBBS)]

(A) 6 (B) 7 (C) less than 6 (D) more than 7

281. What is the amount of heat liberated (in k cal) when one mole of aqueous NaOH reacts with one mole of aqueous HCl solution? [EAMCET-99(MBBS)]

(A) 22.4 (B) 13.7 (C) 2.303 (D) 0.693

282. The equilibrium constant for the reaction $H_2O(g) + CO(g) \leftrightarrow H_2(g) + CO_2(g)$ is 64. If the rate constant for the forward reaction is 160, the rate constant for the backward reaction is : [EAMCET-99(MBBS)]

(A) 0.4 (B) 2.5 (C) 6.2 (D) 10.24×10^3

283. The equilibrium constant for the reaction $H_2(g) + I_2(g) \rightleftharpoons 2HI(g)$ is 64 at certain temperature. The equilibrium concentrations of H_2 and HI are 2 mol. lit^{-1} and 16 mol lit^{-1} respectively. What is the equilibrium concentration (in mol lit^{-1}) of I_2? [EAMCET-2000]

(A) 16 (B) 4 (C) 8 (D) 2

284. Which of the following is correct for acid buffer? (Salt = S; acid = A). [EAMCET-2000]

(A) $pK_a = pH + \log\frac{[S]}{[A]}$ (B) $pH = pK_a \frac{[S]}{[A]}$

(C) $pK_a = pH - \frac{A}{S}$ (D) $pH = pK_a + \log\frac{A}{S}$

285. The pH of an aqueous solution of a salt is 10. The salt is : [EAMCET-2000]

(A) NaCl (B) NH_4Cl

(C) CH_3COONa (D) $(NH_4)_2SO_4$

286. The pK_a of a weak acid is 4.8. what is the ratio of [salt]/[acid]. If a buffer of pH = 5.8 is to be prepared? [EAMCET(MBBS)-2000]

(A) 1:1 (B) 10:1 (C) 2:1 (D) 1:10

287. 4 moles of HI is taken in a 1 litre closed vessel and heated till equilibrium is reached. At equilibrium the concentration of H_2 is 1 mol. lit^{-1}. What is the equilbrium constant for? [EAMCET(MBBS)-2000]

(A) 4 (B) 0.5 (C) 2 (D) 0.25

288. The pK_a of a weak acid is 4.8. What should be the ratio of (acid)/(salt), If a buffer of pH = 5.8 is required? [EAMCET-2001]

(A) 0.1 (B) 10 (C) 1 (D) 2

289. Which of the following would produce a buffer solution when mixed in equal volumes? [EAMCET-2001]

(A) 1 M CH_3COOH & 0.5 M NaOH

(B) 1 M CH_3COOH & 0.5 M HCl

(C) 1 M NH_4OH & 0.5 M NaOH

(D) 1 M NH_4Cl & 0.5 M HCl

290. Equilibrium constant for the reaction $H_2O(g) + CO(g) \rightleftharpoons H_2(g) + CO_2(g)$ is 81. If the velocity constant of the forward reaction is 162 lit. mot^{-1} sec^{-1}, what is the velocity constant (in lit. mol^{-1} sec^{-1}) for the backward reaction? [EAMCET-2001]

(A) 131.22 (B) 2 (C) 261 (D) 243

291. The pH a 0.1 M aqueous solution of a weak acid (HA) is 3. What is its degree of dissociation? [EAMCET(MBBS)-2001]

(A) 1% (B) 10% (C) 50% (D) 25%

292. The pH range of methyl red indicator is : [EAMCET(MBBS)-2001]

(A) 4.2 to 6.3 (B) 8.3 to 10.0

(C) 8.0 to 9.6 (D) 6.8 to 8.4

293. 9.2 grams of $N_2O_4(g)$ is taken in a closed one litre vessel and heated till the following equilibrium is reached. $N_2O_4(g) \rightleftharpoons 2NO_2(g)$. At equilibrium 50% of $N_2O_4(g)$ is dissociated. What is the equilibrium constant (in mol.lit^{-1})? (Molecular weight of N_2O_4 is 92). [EAMCET(MBBS)-2001]

(A) 0.1 (B) 0.2 (C) 0.4 (D) 2

294. 75 ml of 0.2 M HCl is mixed with 25 ml of 1 M HCl. To this solution, 300 ml of distilled water is added. What

is the pH of the resultant solution ? [EAMCET-2002]
(A) 1 (B) 2 (C) 4 (D) 0.2

295. Which one of the following statements is correct ? [EAMCET-2002]
(A) Bronsted - Lowry theory could not explain the acidic nature of BCl_3.
(B) The pH of 0.01 M NaOH solution is 2.
(C) The ionic product of water at 25°C is 10^{-10} $mol^{-2}lit^{-2}$
(D) The pH of a solution can be calculated using the equation pH = log $[H^+]$

296. One mole of A(g) is heated to 200°C in a one litre closed flask, till the following equilibrium is reached. $A(g) \rightleftharpoons B(g)$ The rate of forward reaction at equilibrium is 0.02 mol. lit^{-1} min^{-1} what is the rate (in mol.lit^{-1} $min.^{-1}$) of the backward reaction at equilibrium ? [EAMCET-2002]
(A) 0.04 (B) 0.01 (C) 0.02 (D) 1

297. What is the hybridization state of the central atom in the conjugate base of NH_4^+ ion ? [EAMCET-2002]
(A) sp (B) sp^3 (C) sp^2 (D) dsp^2

298. Which one of the following is not correct ? [EAMCET(med.)-2002]
(A) $pH = \frac{1}{\log[H^+]}$ (B) $pH = \log\left[\frac{1}{[H^+]}\right]$
(C) $[H^+] = 10^{-pH}$ (D) $pH = \log[H^+]$

299. Which one of the following statements is not correct? [EAMCET(med.)-2002]
(A) Cl^- is a Lewis acid
(B) The pH of 10^{-8} M HCl solution is less than 7
(C) The ionic product of water at 25°C is 10^{-14} mol^2 lit^{-2}.
(D) Bronsted - Lowry theory could not explain the acidic character of $AlCl_3$.

300. Consider the following reaction equilibrium
$$N_2(g) + 3H_2(g) \rightleftharpoons 2NH_3(g)$$
Initially 1 mole of N_2 and 3 moles of H_2 are taken in a 2 litre flask. At equilibrium state if the no. of moles of N_2 is 0.6, what is the total of the number of moles of all gases preset in the flask ? [EAMCET-2003]
(A) 0.8 (B) 1.6 (C) 3.2 (D) 6.4

301. If the ionic product of water (k_W) is 1.96 x 10^{-14}at 35°C, what is its value at 10°C ? [EAMCET-2003]
(A) 1.96 x 10^{-14} (B) 3.92 x 10^{-14}
(C) 2.95 x 10^{-15} (D) 1.96 x 10^{-13}

302. Which one of the following equations is correct for the reaction $N_2(g) + 3H_2(g) \rightarrow 2NH_3(g)$ [EAMCET-2003]

(A) $3\frac{d[NH_3]}{dt} = 2\frac{d[H_2]}{dt}$ (B) $2\frac{d[NH_3]}{dt} = 3\frac{d[H_2]}{dt}$

(C) $2\frac{d[NH_3]}{dt} = -3\frac{d[H_2]}{dt}$ (D) $3\frac{d[NH_3]}{dt} = -2\frac{d[H_2]}{dt}$

303. If the conc. of HX and X^- are equal. If the K_b of the solution is 10^{-10} the pH of the solution is [CPMT-96]
(A) 4 (B) 10 (C) 14 (D) 8

304. What is the pH of Ba $(OH)_2$, if normality is 10 ? [CPMT-96]
(A) 4 (B) 10 (C) 7 (D) 9

305. $2SO_3 \rightleftharpoons 2SO_2 + O_2$. If K_C = 100, α = 1 half of the reaction is completed the conc. of SO_3 and SO_2 are equal, the conc. of O_2 is [CPMT-96]
(A) 0.001 M (B) $\frac{1}{2}$ of SO_2
(C) 2 time of SO_2 (D) Data incomplete

306. Solubility product of $PbCl_2$ is 1.0 x 10^{-6}. The solubility is [CPMT-96]
(A) 7.5 x 10^{-5} (B) 6.3 x 10^{-6}
(C) 6.0 x 10^{-7} (D) None of these

307. $2NO_2 \rightleftharpoons 2NO + O_2$, K= 1.6x$10^{-12}$
$2NO + O_2 \rightleftharpoons 2NO_2$, K^1 = ? [CPMT-96]
(A) $K^1 = \frac{1}{K^2}$ (B) $K^1 = \frac{1}{K}$ (C) $K^1 = \frac{1}{\sqrt{K}}$ (D) None of these

308. Which statement for equilibrium constant is true for the reaction $A+B \rightleftharpoons C$? [CPMT-97]
(A) Not changes with temperature
(B) Changes when catalyst is added
(C) Increases with temperature
(D) Changes with temperature

309. When an acid or alkali is mixed with buffer solution, then pH of buffer solution - [CPMT-97]
(A) Remains unchanged (B) Changes slightly
(C) Increases (D) Decreases

310. Which oxychloride has maximim pH ? [CPMT-97]
(A) NaClO (B) $NaClO_2$ (C) $NaClO_3$ (D) $NaClO_4$

311. The pH value of 0.1M NaOH solution is (when there is a given reaction $[H^+][OH^-] = 10^{-14}$ – [CPMT-97]
(A) 13 (B) 12 (C) 5 (D) 18

312. Which one is true ? [CPMT-97]
(A) $K_C = K_P(RT)^{\Delta n}$ (B) $K_P = K_C(RT)^{\Delta n}$
(C) $\frac{K_c}{RT} = (KP)^{\Delta n}$ (C) $\frac{K_p}{RT} = (KC)^{\Delta n}$

313. Which one is Lewis acid ? [CPMT-97]
(A) Cl^- (B) Ag^+ (C) C_2H_5OH (D) S^-

314. What is the solubility of calcium fluoride in a saturated solution if its solubility product is 3.2 x 10^{-11} ? [CPMT-97]
(A) 2.0 x 10^{-4} mole/litre (B) 12.0 x 10^{-3} mole/litre
(C) 0.2 x 10^{-4} mole/litre (D) 2 x 10^{-3} mole/litre

315. pH of HCl (10^{-12}M) is [CPMT-97]

(A) 12 (B) -12 (C) < 7 (D) 14

316. Which one is Buffer solution ? [CPMT-97]
(A) $[PO_4^-][HPO_4^-]$ (B) $[PO_4^-][H_2PO_4^-]$
(C) $[HPO_4^{--}][H_2PO_4^-]$ (D) All of these

317. pH of a solution can be expressed as [CPMT-99]
(A) $-\log_e(H^+)$ (B) $-\log_{10}(H^+)$
(C) $\log_e(H^+)$ (D) $\log_{10}(H^+)$

318. The reaction $2SO_2 + 2O_2 \rightleftharpoons 2SO_3$ will be favoured by [CPMT-99]
(A) High temperature and low pressure
(B) Low temperature and high pressure
(C) Low temperature and low pressure
(D) High temperature and high pressure

319. Which one of the following is basic buffer solution ? [CPMT-2000]
(A) $CH_3COOH + CH_3COONa$
(B) $CH_3COOH + NH_4Cl$
(C) $NH_4OH + NH_4Cl$ (D) $CH_3COONa + NH_4OH$

320. At 25°C the value of K_C for the reaction $\frac{1}{2}N_2 + O_2 \rightleftharpoons NO_2$, is 3×10^4. What will be the value of K_C for $2NO_2 \rightleftharpoons N_2 + 2O_2$? [CPMT-2000]
(A) 2×10^4 (B) $\frac{1}{4 \times 10^8}$ (C) $\sqrt{2} \times 10^2$ (D) 10^4

321. Which solution will be strongly acidic ? [CPMT-2000]
(A) When pH = 00 (B) When pOH = 4.5
(C) When pOH = 14 (D) (A) and (C) both

322. The pH of an aqueous solution of CH_3COONa will be - [CPMT-2000]
(A) 3 (B) 7 (C) 9 (D) 14

323. The ionic product of water at 500°C is 4×10^{-14}. The pH of water is [CPMT-2001]
(A) 6.3 and it is acidic (B) 7 and it is neutral
(C) 6.3 and it is almost neutral (D) 7.3 and it is basic

324. If, in the reaction $N_2 + O_2 \rightleftharpoons 2NO$, the production of NO is 40% at equilibrium. Then Kc for the reaction is [CPMT-2001]
(A) 0.2 (B) 0.25 (C) 0.26 (D) 0.44

325. The pH of 0.01 M NaOH is [CPMT-2001]
(A) 5 (B) 7 (C) 9 (D) 12

326. The pH of 0.1 M solution of the following salts increases in the order [CPMT-2001]
(A) $NaCl < NH_4Cl < NaCN < HCl$
(B) $HCl < NH_4Cl < NaCl < NaCN$
(C) $NaCN < NH_4Cl < NaCl < HCl$
(D) $HCl < NaCl < NaCN < NH_4Cl$

327. 2 mol. N_2 and 3 mol. H_2 gas are allowed to react in a 20 liters flask at 400 K and after complete conversion of H_2 into NH_3, 10 litres H_2O was added and temperature reduced to 300 K. Pressure of the gas after reaction is :
$N_2 + 3H_2 \rightarrow 2NH_3$ [CPMT-2001]
(A) 3R x 300/20 (B) 3R x 300/10
(C) R x 300/20 (D) R x 300/10

328. Which salt undergoes hydrolysis ? [CPMT-2002]
(A) NaCl (B) CH_3COONa (C) K_2SO_4 (D) KNO_3

329. Strongest Bronsted base is [CPMT-2002]
(A) ClO^- (B) ClO_3^- (C) ClO_2^- (D) ClO_4^-

330. The reaction $2SO_2 + O_2 \rightarrow 2SO_3$; ΔH = -ve is favoured by [CPMT-2002]
(A) low temperature, low pressure
(B) low temperature, high pressure
(C) high temperature, high pressure
(D) high temperature, low pressure

331. Solubility of BaF_2 in a solution of $Ba(NO_3)_2$ will be represented by [CPMT-2002]
(A) $[Ba^{2+}]$ (B) $[NO_3]^{-2}$ (C) $[F^-]$ (D) $[Ba]^2$

332. On dissolving KCN in H_2O [CPMT-2002]
(A) pH increases (B) pH remains constant
(C) pH decreases (D) none of these

333. In a chemical equilibrium the rate constant of backward reaction is 3.2×10^{-2} and the equilibrium constant is 2^{-5}. The rate constant of forward reaction is [CPMT-2003]
(A) 8×10^{-2} (B) 4×10^{-2} (C) 1×10^{-3} (D) 2×10^{-2}

334. pH of blood lies between [CPMT-2003]
(A) 7–8 (B) 6–7 (C) 8–9 (D) 13–14

335. Which is Buffer solution ? [CPMT-2003]
(A) CH_3COOH and CH_3COONa
(B) HCl and NaOH (C) NaOH and H_2SO_4
(D) Carbonic acid and Acetic acid.

336. Acetic acid is weaker acid than sulphuric acid because [CPMT-2003]
(A) it decomposes on increasing temperature
(B) it has less degree of ionisation.
(C) it has -COOH group. (D) None of these.

337. A monoprotic acid in 0.1M solution, ionises to 0.001%. Its ionisation constant is [CPMT-2003]
(A) 1×10^{-3} (B) 1×10^{-6} (C) 1×10^{-8} (D) 1×10^{-11}

338. Which of the following mixtures forms an acid buffer? [PMT-93]
(A) NaOH + HCl (B) $CH_3COOH + CH_3COONa$
(C) $NH_4OH + NH_4Cl$ (D) $H_2CO_3 + (NH_4)_2CO_3$

339. When NH_4Cl is added to NH_4OH solution, the

dissociation of ammonium hydroxide is reduced. It is due to : [PMT-93]

(A) Common ion effect (B) Hydrolysis
(C) Oxidation (D) Reduction

340. In which of the following equilibria the value of K_p is less than K_c? [PMT-93]

(A) $H_2 + I_2 \rightleftharpoons 2HI$
(B) $N_2 + 3H_2 \rightleftharpoons 2NH_3$
(C) $N_2 + O_2 \rightleftharpoons 2NO$
(D) $CO + H_2O \rightleftharpoons CO_2 + H_2$

341. According to Le Chatelier's Principle, an increase in temperature on the reaction $N_2 + O_2 \rightleftharpoons 2NO + 43200$ cal. will : [PMT-93]

(A) increase the yield of NO
(B) decrease the yield of NO
(C) not effect the yield of NO
(D) not help the reaction to proceed in forward direction

342. The pH of a 0.02 M solution of hydrochloric acid is: [PMT-93]

(A) 2.0 (B) 1.7 (C) 0.3 (D) 2.2

343. When CH_3COONa is added to aqueous CH_3COOH solution [PMT-94]

(A) pH value decreases (B) pH value increases
(C) pH value remains unchanged
(D) pH value becomes zero

344. The solubility product of AgCl is 1.44×10^{-4} at 100°C. The solubility of silver chloride in boiling water may be : [PMT-94]

(A) 0.72×10^{-4} M (B) 1.20×10^{-2} M
(C) 0.72×10^{-2} M (D) 1.20×10^{-4} M

345. The pOH of beer is 10.0. The hydrogen ion concentration will be [PMT-94]

(A) 10^{-2} (B) 10^{-10} (C) 10^{-8} (D) 10^{-4}

346. Heat of neutralisation of NH_4OH and HCl is [PMT-94]

(A) equal to 13.7 KCal. (B) more than 13.7 KCal.
(C) less than 13.7 KCal. (D) more than one is correct

347. Which of the following solutions will not hydrolyse ? [PMT-94]

(A) KNO_3 (B) KCN
(C) Pot. succinnate (D) Pot. carbonate

348. Which one of the following is a buffer solution ? [PMT-94]

(A) KOH + HCl (B) $HNO_3 + NaNO_3$
(C) $CH_3COOH + CH_3COONa$ (D) HCl + NaCl

349. The relation between Kp and Kc is correctly shown as [PMT-94]

(A) $Kc = Kp\ (RT)^{\Delta n}$ (B) $Kp = Kc\ (RT)^{-\Delta n}$
(C) $Kp = Kc\ (RT)^{\Delta n}$ (D) $Kc = Kp\ (RT)^{-\Delta n}$

350. Which of the following is insoluble in water and acid both ? [PMT-94]

(A) $CaCO_3$ (B) $Zn(NO_3)_2$ (C) $BaSO_4$ (D) $ZnCO_3$

351. Strongest Bronsted base among the following anions is [PMT-94]

(A) ClO^- (B) ClO_2^- (C) ClO_3^- (D) ClO_4^-

352. pH of the solution is 4. The hydroxide ion concentration of the solution would be [PMT-94]

(A) 10^{-4} (B) 10^{-10} (C) 10^{-2} (D) 10^{-12}

353. 120 gms of urea are present in 5 litre solution, the active mass of urea is [PMT-94]

(A) 0.2 (B) 0.06 (C) 0.4 (D) 0.08

354. For which of the following reactions Kp = Kc ? [PMT-95]

(A) $H_2 + I_2 \leftrightarrow 2HI$ (B) $2SO_2 + O_2 \leftrightarrow 2SO_3$
(C) $PCl_3 + Cl_2 \leftrightarrow PCl_5$ (D) $N_2 + 3H_2 \leftrightarrow 2NH_3$

355. Low pressure is favourable for which of the following reactions ? [PMT-95]

(A) $PCl_5 \leftrightarrow PCl_3 + Cl_2$ (B) $N_2 + 3H_2 \leftrightarrow 2NH_3$
(C) $2SO_2 + O_2 \leftrightarrow 2SO_3$ (D) $N_2 + O_2 \leftrightarrow 2NO$

356. pH of 0.001 M Na OH is [PMT-95]

(A) 10^{-3} (B) 3 (C) 10^{11} (D) 11

357. The law of mass action was enunciated by [PMT-95]

(A) Guldberg and Waage (B) Bodenstein
(C) Birthelot (D) Graham

358. If the equilibrium constant of the reaction $2HI \leftrightarrow H_2 + I_2$ is 0.25, the equilibrium contant for the reaction $H_2 + I_2$ 2HI would be [PMT-95]

(A) 1.0 (B) 2.0 (C) 3.0 (D) 4.0

359. Conjugate base of HPO_4^{2-} is [PMT-95]

(A) PO_4^{3-} (B) $H_2PO_4^-$ (C) H_3PO_4 (D) H_4PO_3

360. The solubility of $PbCl_2$ is [PMT-95]

(A) $\sqrt{Ksp}$ (B) $\sqrt[3]{Ksp}$ (C) $\sqrt[3]{\frac{Ksp}{4}}$ (D) $\sqrt{8Ksp}$

361. The ppt. of CaF_2 ($Ksp = 1.7 \times 10^{-10}$) is obtained when equal volume of which of the following are mixed ? [PMT-95]

(A) 10^{-1} M Ca^{2+} + 10^{-4} M F^- (B) 10^{-2} M Ca^{2+} + 10^{-3} M F^-
(C) 10^{-2} M Ca^{2+} + 10^{-5} M F^- (D) 10^{-3} M Ca^{2+} + 10^{-5} M F^-

362. A chemical reaction is in equilibrium when [PMT-96]

(A) the reactants are completely transformed into products
(B) the rate of forward reaction is equal to the rate of backward reaction

(C) the formation of products is minimised
(D) equal amount of reactants and products are present

363 For the following gaseous reaction, $H_2 + I_2 \rightarrow 2HI$. the equilibrium constant. [PMT-96]

(A) Kp > Kc (B) Kp < Kc (C) Kp = Kc (D) $Kp = \frac{1}{Kc}$

364. According to Le Chatelier's principle which of the following factors influence a chemical system ? [PMT-96]

(A) concentration only (B) Pressure only
(C) Temperature only
(D) Concentration, Pressure and Temperature

365. Sodium chloride is purified by passing hydrogen chloride gas in an impure solution of sodium chloride. It is based on [PMT-96]
(A) buffer action (B) common ion effect
(C) association of salt (D) hydrolysis of salt

366. A buffer solution contains 0.1 M of acetic acid and 0.1 M of sodium acetate. What will be its pH, if pKa of acetic acid is 4.75 ? [PMT-96]
(A) 4.00 (B) 4.75 (C) 5.00 (D) 5.25

367. Which of the following cannot be hydrolysed ? [PMT-96]

(A) A salt of weak acid and strong base
(B) A salt of strong acid and weak base
(C) A salt of weak acid and weak base
(D) A salt of strong acid and strong base

368. Which of the following statements is true ? [PMT-96]

(A) The conjugate base of a strong acid is a strong base
(B) The conjugate base of a weak acid is a strong base
(C) The conjugate base of a weak acid is a weak base
(D) The base and its conjugate acid react to form a neutral solution

369. For the reaction $H_2 (g) + I_2 (g) \leftrightarrow 2HI(g)$ the equilibrium constant Kp changes with [PMT-97]
(A) Total pressure (B) Catalyst
(C) Amount of H_2 and I_2 (D) Temperature

370. In the gaseous reaction $A + 2B \leftrightarrow C +$ heat, the forward reaction is favoured by - [PMT-97]
(A) Low temperature and low pressure
(B) Low pressure
(C) High pressure and low temperature
(D) High pressure and high temperature

371. The solubility of $CaCO_3$ in water is 3.05×10^{-4} moles/litre. Its solubility product will be [PMT-97]
(A) 3.05×10^{-4} (B) 9.3
(C) 6.1×10^{-4} (D) 9.3×10^{-8}

372. The degree of ionisation of a weak electrolyte [PMT-97]
(A) Decreases with dilution (B) Increases with dilution
(C) May increases or decrease with dilution
(D) Is not affected by dilution

373. pH of water is 7. When a substance Y is dissolved in water, the pH becomes 13. The substance Y is a salt of [PMT-97]

(A) Strong acid and strong base
(B) Weak acid and weak base
(C) Strong acid and weak base
(D) Weak acid and strong base

374. For the reaction $N_2 + 3H_2 \leftrightarrow 2NH_3$ + Heat - [PMT-97]
(A) $K_P = K_C$ (B) $K_P = K_C RT$
(C) $K_P = K_C (RT)^{-2}$ (D) $K_P = K_C (RT)^{-1}$

375. The common ion effect is shown by solution containing : [PMT-99]
(A) Barium chloride and Barium nitrate
(B) Sodium chloride and Hydrochloric acid
(C) Sodium cyanide and Silver cyanide
(D) Ammonium chloride and Ammonium hydroxide

376. For maximum production of SO_3 in the following chemical reaction, $2SO_2 + O_2 \leftrightarrow 2SO_3$; $\Delta H = -Q$ kcals the favourable conditions will be : [PMT-99]
(A) high temperature and high pressure
(B) low temperature and high pressure
(C) high temperature and low pressure
(D) low temperature and low pressure

377. For maximum production of NH_3 in the following chemical reaction $N_2 + 3H_2 \leftrightarrow 2NH_3 + Q$ kcals the favourable conditions will be [PMT-99]
(A) low temperature and low pressure
(B) high temperature and low pressure
(C) low temerature and high pressure
(D) high temperature and high pressure

378. Amongst the following reactions, the one which will not be affected by the change in pressure of the reactants will be : [PMT-99]
(A) $N_2 (g) + O_2 (g) \leftrightarrow 2NO (g)$
(B) $2O_3 (g) \leftrightarrow 3O_2 (g)$
(C) $2NO_2 (g) \leftrightarrow N_2O_4 (g)$
(D) $2SO_2 (g) + O_2 (g) \leftrightarrow 2SO_3 (g)$

379. Amongst the following chemical reactions the irreversible reaction is : [PMT-99]
(A) $H_2 + I_2 \leftrightarrow 2HI$
(B) $AgNO_3 + NaCl \leftrightarrow AgCl + NaNO_3$
(C) $CaCO_3 \leftrightarrow CaO + CO_2$
(D) $O_2 + 2SO_2 \leftrightarrow 2SO_3$

380. The relation between solubility (S) and solubility

product (K_{sp}) of a partially soluble binary electrolyte is : [PMT-99]

(A) $S = K_{sp}$ (B) $S = \sqrt{K_{sp}}$

(C) $S = \frac{1}{3}\sqrt{K_{sp}}$ (D) $S = \frac{K_{sp}}{2}$

381. Addition of ammoinum chloride in the solution of ammonium hydroxide results in : [PMT-99]

(A) increase of OH^- ion concentration in solution

(B) decrease of OH^- ion concentration in solution

(C) increase of NH_4^+ and OH^- ion concentration in solution

(D) increase of dissociation of NH_4OH

382. pH values of HCl and NaOH solutions each of strength will be respectively : [PMT-99]

(A) 2 and 2 (B) 2 and 12 (C) 12 and 2 (D) 2 and 10

383. 50 ml water is added to a 50 ml solution of $Ba(OH)_2$ of strength 0.01 M. The pH value of the resulting solution will be : [PMT-99]

(A) 8 (B) 10 (C) 12 (D) 6

384. Amongst the following solutions, the buffer solution is : [PMT-99]

(A) $NH_4Cl + NH_4OH$ solution

(B) NH_4Cl + NaOH solution

(C) NH_4OH + HCl solution (D) NaOH + HCl solution

385. Amongst the following the one having characteristics of Lewis acid is : [PMT-99]

(A) ClF_3 (B) BF_3 (C) NCl_3 (D) BrF_3

386. At 20°C temperature, the concentration of Ag^+ ions in a saturated solution of Ag_2CrO_4 is 1.5×10^{-4} mole/litre. The solubility product of Ag_2CrO_4 at 20°C temperature will be [PMT-99]

(A) 16.87×10^{-14} (B) 16.87×10^{-13}

(C) 16.87×10^{-12} (D) 16.87×10^{-15}

387. Aqueous solution of ferric chloride is [PMT-99]

(A) Acidic (B) Basic (C) Neutral (D) Amphoteric

388. For the following chemical reaction $2X + Y \rightleftharpoons Z$ the expression of equilibrium constant will be [PMT-99]

(A) $K_c = \frac{[X]^2 [Y]}{[Z]}$ (B) $K_c = \frac{[X] [Y]^2}{[Z]}$

(C) $K_c = \frac{[Z]}{[X]^2 [Y]}$ (D) $K_c = \frac{[Z]}{[X] [Y]^2}$

389. The pH of 10^{-8} M HCl is : [PMT-2000]

(A) 8 (B) >7 (C) < 7 (D) 7

390. How many grams of CaC_2O_4 will dissolve in distilled water to make one litre to saturated solution ? (Ksp for $CaC_2O_4 = 2.5 \times 10^{-9}$ mole^2 litre^{-2}) : [PMT-2000]

(A) 0.0064 g (B) 0.1028 g

(C) 0.1280 g (D) 0.2056

391. A1M solution of glucose reaches dissociation equilibrium according to equation given below.

6 HCHO $\rightleftharpoons$ $C_6H_{12}O_6$ What is the concentration of HCHO at equilibrium if equilibrium constant is 6×10^{22} : [PMT-2000]

(A) 1.6×10^{-8} M (B) 6.2×10^{5} M

(C) 3.2×10^{-4} M (D) 1.6×10^{-4} M

392. What is [H^+] of a solution that is 0.01 M in HCN and 0.02 M in NaCN ? (Ka for HCN = 6.2×10^{-10}) [PMT-2000]

(A) 3.1×10^{10} (B) 6.2×10^{5}

(C) 6.2×10^{-10} (D) 3.1×10^{-10}

393. Which of the following is strongest Lewis base ? [PMT-2000]

(A) CH_3^- (B) NH_2^- (C) OH^- (D) F^-

394. What is the minimum concentration of SO_4^{2-} required to precipitate $BaSO_4$ in a solution contatining 1.0×10^{-4} mol. Ba^{2+} ? Ksp for $BaSO_4 = 4 \times 10^{-11}$. [PMT-2000]

(A) 4×10^{-10} M (B) 2×10^{-7} M

(C) 4×10^{-6} M (D) 2×10^{-3} M

395. Which of the following equilibria will shift to right side on increasing the temperature ? [PMT-2000]

(A) CO (g) + H_2O (g) $\rightleftharpoons$ CO_2 (g) + H_2O (g)

(B) $2SO_2$ (g) + O_2 (g) $\rightleftharpoons$ $3SO_3$ (g)

(C) H_2O (g) $\rightleftharpoons$ H_2 (g) + 1/2 (O_2) (g)

(D) 4HCl (g) + O_2 (g) $\rightleftharpoons$ $2H_2O$ (g) + $2Cl_2$ (g)

396. When 3 mole of A and 1 mole of B are mixed in 1 litre vessel the following reaction takes place

A (g) + B (g) $\rightleftharpoons$ 2C (g). The equilibrium constant for the reaction is : [PMT-2000]

(A) 0.12 (B) 0.25 (C) 0.50 (D) 0.75

397. The equilibrium SO_2Cl_2 (g) $\rightleftharpoons$ SO_2 (g) + Cl_2 (g) is attained at 25°C in a closed container and an inert. gas, helium is introduced, which of the following statements is correct ? [PMT-2000]

(A) More chlorine is formed

(B) Concentration of SO_2 is reduced

(C) More SO_2Cl_2 is formed

(D) Concentration of SO_2Cl_2. SO_2 and Cl_2 does not change

398. Which of the following equilibriums is not shifted by increase in the pressure ? [PMT-2001]

(A) H_2 (g) + I_2 (g) $\rightleftharpoons$ 2HI (g)

(B) N_2 (g) + $3H_2$ (g) $\rightleftharpoons$ $2NH_3$ (g)

(C) 2CO (g) + O_2 (g) $\rightleftharpoons$ $2CO_2$ (g)

(D) 2C (s) + O_2 (g) $\rightleftharpoons$ 2CO (g)

399. SO_2 (g) + $\frac{1}{2}$ O_2 (g) $\rightleftharpoons$ SO_3 (g) equilibrium constant = K_1

$2SO_3\,(g) \rightleftharpoons 2SO_2\,(g) + O_2\,(g)$
equilibrium constant = K_2 . Which of the following is correct ? [PMT-2001]

(A) $K_2 = K_1^2$ (B) $K_2 = K_1^{-2}$
(C) $K_2 = K^1$ (D) $K_2 = K^{-1}$

400. The pH of water decreases with : [PMT-2001]
(A) increase in temperature (B) increase in pressure
(C) decrease in temperature (D) decrease in pressure

401. If for $H_2\,(g) + \quad S_2\,(g) \rightleftharpoons H_2S\,(g)$ and $H_2\,(g) + Br_2(g) \rightleftharpoons 2\,HBr(g)$. The equilibrium costants are K_1 and K_2 respectively. The reaction : $Br_2\,(g) + H_2S\,(g) \rightleftharpoons 2HBr\,(g) + \quad S_2\,(s)$ would have equilibrium constant : [PMT-2001]
(A) $K_1 \times K_2$ (B) K_1 / K_2 (C) K_2 / K_1 (D) K^2 / K_1

402. The aqueous solution of which one of the following is basic ? [PMT-2001]
(A) HOCl (B) $NaHSO_4$ (C) $NH_4\,NO_3$ (D) NaOCl

403. The pH of 10^{-7} M NaOH is : [PMT-2001]
(A) 7.01 (B) Between 7 and 8
(C) Between 9 and 10 (D) Greater than 10

404. Among the following the weakest base is [PMT-2002]
(A) H^- (B) CH_3^- (C) CH_3O^- (D) Cl^-

405. Acidic buffer solution is produced on mixing the following aqueous solutions of [PMT-2002]
(A) $NH_4Cl + NH_4OH$ (B) $CH_3COONa + CH_3COOH$
(C) NaCl + NaOH (D) $Na_2SO_4 + H_2SO_4$

406. A solution $MgCl_2$ in water has pH [PMT-2002]
(A) < 7 (B) > 7 (C) 7 (D) 14.2

407. The rate of reaction of which of the following is not affected by pressure? [PMT-2002]
(A) $PCl_3 + Cl_2 \rightleftharpoons PCl_5$ (B) $N_2 + 3H_2 \rightleftharpoons 2NH_3$
(C) $N_2 + O_2 \rightleftharpoons 2NO$ (D) $2SO_2 + O_2 \rightleftharpoons 2SO_3$

408. If the equilibrium constant of reaction $2AB \rightleftharpoons A_2 + B_2$ is 49, what is the value of equilibrium constant for the reaction $AB \rightleftharpoons A_2 + \quad B_2$ [PMT-2003]
(A) 0.01 (B) 2401 (C) 49 (D) 7

409. For the chemical reaction. $PCl_3(g) + Cl_2(g) \rightleftharpoons PCl_5(g)$ the value of K_c at 452°C is 26 mol^{-1}/ $litre^{-1}$. The value of K_p at this temperature is : [PMT-2003]
(A) 0.506 Atm. pressure (B) 5.06 Atm. pressure
(C) 0.0506 Atm. pressure (D) None of these

410. In gaseous reaction. $C_2H_4 + H_2 \rightleftharpoons C_2H_6$: $\Delta H = -32.7$ kcal. the concentration of C_2H_4 can be reduced by [PMT-2003]
(A) Increasing the pressure
(B) Reducing the temperature
(C) Increasing the H concentration (D) All of these

411. In the chemical reaction. $SnCl_2 + 2Cl^- \rightarrow SnCl_4 + 2e^-$ which of the following is acting as Lewis acid : [PMT-2003]
(A) $SnCl_4$ (B) Cl^- (C) $SnCl_2$ (D) None of these

412. Which one of the following is conjugated acid of water in the reaction $H_2SO_4 + H_2O \rightleftharpoons H_3O^+ + HSO_4^-$ [PMT-2003]
(A) H_2O (B) H_3O^+ (C) SO_4^{2-} (D) HSO_4^-

413. Which aqueous solutions has the pH value less than 7 ? [PMT-2003]
(A) NaOH (B) $FeCl_3$ (C) NaCN (D) KNO_3

414. 50ml of 2N CH_3COOH ($K_a = 10^{-5}$) is mixed with 10 ml of 1 N sodium acetate. pH obtained is about : [PMT-2003]
(A) 7 (B) 6 (C) 5 (D) 8

215. The K_{sp} values of CuS, Ag_2S and HgS are 10^{-31}, 10^{-42} and 10^{-54} respectively. What is the right sequence of solubility of them in water : [PMT-2003]
(A) HgS > Ag_2S > CuS (B) HgS > CuS > Ag_2S
(C) Ag_2S > CuS > HgS (D) Ag_2S > HgS > CuS

416. For NaCl, the K_{sp} = 36 $mol^2/litre^2$, the molar concentration of it will be : [PMT-2003]
(A) $\frac{1}{36}$ M (B) $\frac{1}{16}$ M (C) 36M (D) 6M

417. For the neutralization of 1 mol of H_2SO_4 with 2 mol of NaOH in dilute solution, the heat evolved is : [PMT-2003]
(A) 57.0 kJ (B) Unpredictable
(C) More then 57.0 kJ (D) Less than 57.0 kJ

418. An equilibrium mixture for the reaction $2H_2S(g) \rightleftharpoons 2H_2(g) + S_2(g)$ had 0.5 mole H_2S, 0.1 mole of H_2 and 0.4 mole of H_2 and 0.4 mole S_2 in a one litre vessel. The equilibrium constant of this reaction is given by [PET-93]
(A) 0.004 mole $litre^{-1}$ (B) 0.080 mole $litre^{-1}$
(C) 0.016 mole $litre^{-1}$ (D) 0.160 mole $litre^{-1}$

419. What conditions will favour the exothermic ammonia synthesis reaction. $N_2(g) + 3H_2(g) \rightleftharpoons 2NH_3(g)$ [PET-93]
(A) High temperature, high pressure
(B) High temperature, low pressure
(C) Low temperature, high pressure
(D) Low temperature, low pressure

420. Which of the following statements is wrong ? [PET-93]
(A) pH + pOH = 14 at room temperature

(B) The commonest primary standards for standardization of acids are Na_2CO_3 or borax

(C) $[H^+][OH^-] = 10^{-14}$ at 22°C

(D) Alkalies are generally standardized against acetic acid or borax

421.The two gaseous equilibria $SO_2(g)+\frac{1}{2}O_2(g) \rightleftharpoons SO_3(g)$ and $2SO_3(g) \rightleftharpoons SO_2(g)+O_2(g)$ have equlibrium constants K_1 and K_2 respectively at 298°K. Which of the following relationships between K_1 and K_2 is correct ? [PET-93]

(A) $K_1 = K_2$ (B) $K_2 = K_1^2$ (C) $K_2 = 1/K_1^2$ (D) $K_2 = 1/K_1$

422. Which of the following solutions will have pH close to 1.0 ? [PET-93]

(A) 100 ml of $\frac{M}{10}$ HCl + 100 ml of $\frac{M}{10}$ NaOH

(B) 55 ml of $\frac{M}{10}$ HCl + 45ml of $\frac{M}{10}$ NaOH

(C) 10 ml of $\frac{M}{10}$ HCl + 90 ml of $\frac{M}{10}$ NaOH

(D) 75 ml of $\frac{M}{5}$ HCl + 25 ml of $\frac{M}{5}$ NaOH

423. How many grams of CaC_2O_4 will dissolve in distilled water to make one litre of saturated solution? Solubility product of CaC_2O_4 is 2.5×10^{-9} mole² litre⁻² and its molecular weight is 128. [PET-93]

(A) 0.0064 g. (B) 0.0128 g. (C) 0.0032 g. (D) 0.0640 g.

424. The addition of HCl will not suppress the ionization of [PET-93]

(A) Acetic acid (B) Benzoic acid

(C) H_2S (D) Sulphuric acid

425. Which of the following mixtures in aqueous solution acts as a buffer ? [PET-94]

(A) $CH_3COOH + NaCl$ (B) $HNO_3 + KNO_3$

(C) $H_2SO_4 + K_2SO_4$ (D) $NH_4OH + NH_4Cl$

426. In which of the following gaseous equilibria K_p is less than K_c ? [PET-94]

(A) $2SO_2 + O_2 \rightleftharpoons SO_3$ (B) $2HI \rightleftharpoons H_2 + I_2$

(C) $N_2O_4 \rightleftharpoons 2NO_2$ (D) $N_2 + O_2 \rightleftharpoons 2NO$

427. The pH of a solution is 2. If its pH is to be raised to 4, then the $[H^+]$ of the original solution has to be: [PET-94]

(A) Doubled (B) Halved

(C) Increased hundred times

(D) Decreased hundred times

428. In a reversible reaction a catalyst [PET-94]

(A) Increases the rate of the forward reaction

(B) Increases the rate of the backward reaction

(C) Increases the rate of the forward and backward reactions equally

(D) None of above

429. Which of the following conditions is favourable for the production of ammonia by Haber's process ? [PET-94]

(A) High concentration of reactants

(B) Low temperature and high pressure

(C) Continuous removal of ammonia

(D) All the above

430. Which of the following is a Lewis acid? [PET-94]

(A) NH_3 (B) H_2O (C) BF_3 (D) Cl

431. The hydrogen ion concentration of a 0.006 M benzoic acid solution is ($K_a = 6 \times 10^{-5}$) [PET-94]

(A) 0.6×10^{-4} (B) 6×10^{-4}

(C) 6×10^{-5} (D) 3.6×10^{-4}

432. If the solubility products of AgCl and AgBr are 1.2×10^{-10} and 3.5×10^{-13} respectively, then the relation between the solubilities (denoted by the symbol 'S') of these salts can correctly be represented as [PET-94]

(A) S of AgBr is less than that of AgCl

(B) S of AgBr is greater then that of AgCl

(C) S of AgBr is equal to that of AgCl

(D) S of AgBr is 106 greater than that of AgCl

433. Which of the following equilibria is not affected by a pressure change ? [PET-95]

(A) $2NO_2(g) \rightleftharpoons N_2O_4(g)$

(B) $CO_2(g) + H_2O(l) \rightleftharpoons CO_2$ solution

(C) $H_2(g) + I_2(g) \rightleftharpoons 2HI(g)$

(D) $2O_3(g) \rightleftharpoons 3O_2(g)$

434. In a chemical reaction, equilibrium is established when - [PET-95]

(A) Velocity of opposing reaction is the same as that of the forward one

(B) Order of the reaction is infinite

(C) Concentrations of the reactants and products are equal

(D) Reaction ceases to generate heat

435. Consider the two gaseous equilbria involving SO_2 and the corresponding equilibrium constants at 290°K

$SO_2(g) + \frac{1}{2}O_2(g) \rightleftharpoons SO_3(g)$; K_1

$2SO_3(g) \rightleftharpoons 2SO_2(g) + O_2(g)$; K_2

The value of the equilibrium constants are related by - [PET-95]

(A) $K_2 = K_1$ (B) $K_2 = K_1^2$

(C) $K_2 = 1/K_1^2$ (D) $K_2 = 1/K_1$

436. Given that the dissociation constant for H_2O is $K_w = 1 \times 10^{-14}$ mole² litre⁻², what is the pH of a 0.001 molar KOH solution- [PET-95]

(A) 10^{-11} (B) 3 (C) 14 (D) 11

437. Which of the following can act both as Bronsted acid and a base - [PET-95]

(A) Cl^- (B) HCO_3^- (C) H_3O^+ (D) SO_4^{2-}

438. If the solubility product of $BaSO_4$ is 1.5×10^{-9} in

water, its solubility in moles per litre is- [PET-95]
(A) 1.5×10^{-9} (B) 3.9×10^{-5}
(C) 7.5×10^{-5} (D) 1.5×10^{-5}

439. For the reaction $N_2 + 3H_2 \rightleftharpoons 2NH_3$ the equilibrium constant K_p changes with [PET-96]
(A) Catalyst (B) Total pressure
(C) Temperature (D) Amount of N_2 & H_2

440. For the reaction $PCl_5(g) \rightleftharpoons PCl_3(g) + Cl_2(g)$ [PET-96]
(A) $K_p = K_c$ (B) $K_p = K_c (RT)^{-1}$
(C) $K_p = K_c (RT)$ (D) $K_p = K_c (RT)^2$

441. The conjugate base of NH_3 is [PET-96]
(A) NH_4OH (B) NH_2^- (C) NH^{2-} (D) N_2H_2

442. The strength of an acid depends on its tendency to [PET-96]
(A) Accept protons (B) Donate protons
(C) Accept electrons (D) Donate electrons

443. The solubility of AgCl in 0.2 M NaCl solution (K_{sp} for AgCl = 1.20×10^{-10}) is [PET-96]
(A) 0.2 M (B) 1.2×10^{-10} M
(C) 0.2×10^{-10} M (D) 6.0×10^{-10} M

444. The equilibrium constant of the reaction $H_2(g) + I_2(g) \rightleftharpoons 2HI(g)$ is 64. If the volume of the container is reduced to one fourth of its original volume, the value of the equilibrium constant will be [PET-96]
(A) 16 (B) 33 (C) 64 (D) 128

445. The aqueous solution having highest pH value is [PET-96]
(A) NaCl (B) CH_3COOH (C) NH_4Cl (D) Na_2CO_3

446. On the basis of Le Chatelier's principle, predict which of the following conditions would be unfavourable for the formation of SO_3. Given that : $2SO_2 + O_2 \rightleftharpoons 2SO_3$ $\Delta H = -42$ kcal [PET-97]
(A) Low temperature (B) High pressure
(C) High temperature (D) High concentration of SO_2

447. For which one of the following reactions Kp = Kc ? [PET-97]
(A) $N_2 + 3H_2 \rightleftharpoons 2NH_3$ (B) $N_2 + O_2 \rightleftharpoons 2NO$
(C) $PCl_5 \rightleftharpoons PCl_3 + Cl_2$ (D) $2SO_3 \rightleftharpoons 2SO_2 + O_2$

448. The equilibrium constant for the reversible reaction, $N_2 + 3H_2 \rightleftharpoons 2NH_3$ is K and for the reaction $\frac{1}{2}N_2 + \frac{3}{2}H_2 \rightleftharpoons NH_3$ the equilibrium constant is K', K and K' will be related as : [PET-97]
(A) K = K' (B) $K' = \sqrt{K}$
(C) $K = \sqrt{K'}$ (D) K x K' = 1

449. At 20°C the Ag^+ ion concentration in a saturated solution of Ag_2CrO_4 is 1.5×10^{-4} mole/litre at 20°C the solubility product of Ag_2CrO_4 would be : [PET-97]
(A) 3.3750×10^{-12} (B) 1.6875×10^{-10}
(C) 1.6875×10^{-12} (D) 1.6875×10^{-11}

450. For preparing a buffer solution of pH 6 by mixing sodium acetate and acetic acid, the ratio of the concentration of salt and acid should be : ($K_a = 10^{-5}$) [PET-97]
(A) 1 : 10 (B) 10 : 1 (C) 100 : 1 (D) 1 : 100

451. Of the following anions, the strongest Bronsted base is : [PET-97]
(A) ClO^- (B) ClO_2 (C) ClO_3^- (D) ClO_4^-

452. Degree of dissociation of 0.1N CH_3COOH is (dissociation const. = 1×10^{-5}) [PET-97]
(A) 10^{-5} (B) 10^{-4} (C) 10^{-3} (D) 10^{-2}

453. For the homogeneous gas reaction $4NH_3 + 5O_2 \rightleftharpoons 4NO + 6H_2O$ the equilibrium constant K_c has the dimension of : [PET-98]
(A) conc.$^{+10}$ (B) conc.$^{+1}$
(C) conc.$^{-1}$ (D) It is dimensionless

454. Formation of SO_3 takes place according to the reaction : $2SO_2 + O_2 \rightleftharpoons 2SO_3$: $\Delta H = -45.2$ kcal
Which of the following factors favours the formation of SO_3 ? [PET-98]
(A) Increase in temperature (B) Increase in pressure
(C) Removal of oxygen (D) Increase in volume

455. In the equilibrium reaction $2HI(g) \rightleftharpoons H_2 + I_2$ which of the following expressions is true ? [PET-98]
(A) $K_p = K_c$ (B) $K_c = 2K_p$
(C) $K_p > K_c$ (D) $K_c = K_p (RT)^2$

456. What will be the pH of a 10^{-8} M HCl solution ? [PET-98]
(A) 8.0 (B) 7.0 (C) 6.98 (D) 14.0

457. Which of the following can act both as Bronsted acid and a Bronsted base ? [PET-98]
(A) Na_2CO_3 (B) OH^- (C) HCO_3^- (D) NH_3

458. What will be the sum of pH and pOH in an aqueous solution ? [PET-98]
(A) 7 (B) pK_w (C) Zero (D) 1

459. Given that the dissociation constant for H_2O is $K_w = 1 \times 10^{-14}$ mol.^{2} L^{-2}, what is the pH of a 0.001 M KOH solution ? [PET-98]
(A) 11 (B) 3 (C) 14 (D) 10^{-11}

460. The pH of a 10^{-8} molar solution of HCl in water is : [PET-99]
(A) 8 (B) -8
(C) Between 7 and 8 (D) Between 6 and 7

461. Which of the following salts when dissolved in wa-

ter hydrolyse ? [PET-99]

(A) NaCl (B) NH_4Cl (C) KCl (D) Na_2SO_4

462. What is the correct representation for the solubility product of SnS_2? [PET-99]

(A) $[Sn^{2+}][S^{2-}]^2$ (B) $[Sn^{4+}][S^{2-}]^2$

(C) $[Sn^{2+}]2S^{2-}]$ (D) $[Sn^{4+}]^2S^{2-}]^2$

463. In which of the following dissociation of NH_4OH will be minimum ? [PET-2000]

(A) NaOH (B) H_2O (C) NH_4Cl (D) NaCl

464. Conjugate base of NH_4^+ is [PET-2000]

(A) NH_3 (B) H_2O (C) NH_2^- (D) H_3O^+

465. If a 4 : 1 mixture of N_2 and O_2 is taken and 1% of Argon is added to it, the vapour pressure of mixed gases will : [PET-2000]

(A) Increase (B) Decrease

(C) Remain constant (D) First increase and then decrease

466. The solution of sodium carbonate has pH : [PET-2000]

(A) Greater then 7 (B) Less than 7

(C) Equal to 7 (D) Equal to Zero

467. If the pH of a solution is 2, its normality will be : [PET-2000]

(A) 2 N (B) 0.5 N (C) 0.01 N (D) 0.02N

468. Which of the following is not used as a Lewis acid ? [PET-2000]

(A) $SnCl_4$ (B) $FeCl_3$ (C) KCl (D) BF_3

469. Ammonium hydroxide is a weak base because [PET-2000]

(A) It has low vapour pressure

(B) It is only slightly ionized

(C) It is not a hydroxide of any metal

(D) It has low density

470. The conjugate acid of NH_2^- is : [PET-2000]

(A) NH_3 (B) NH_4OH (C) NH_3 (D) N_2H_4

471. In the equilibrium : $N_2 + 3H_2 \rightleftharpoons 2\,NH_3 + 22$ kcal the formation of ammonia is favoured by : [PET-2000]

(A) Increasing the pressure

(B) Increasing the temperature

(C) Decreasing the pressure (D) Adding ammonia

472. For $N_2 + 3H_2 \rightleftharpoons 2NH_3$ + heat : [PET-2000]

(A) $K_P = K_C$ (B) $K_P = K_CRT$

(C) $K_P = K_C\,(RT)^{-2}$ (D) $K_P = K_C\,(RT)^{-1}$

473. pH of a 10^{-3} M solution of hydrochloric acid will be: [PET-2000]

(A) 1.3 (B) 2.0 (C) 3.0 (D) 4.5

474. Sodium sulphate dissolves in water with evolution of heat. Consider a saturated solution of sodium sulphate. If the temperature is raise then according to Le Chatelier principle [PET-2000]

(A) More solid will dissolve

(B) Some solid will precipitate out from the solution

(C) The solution will become supersaturated

(D) Solution concentration will remain unchanged

475. An aqueous solution of ammonia consists of : [PET-2001]

(A) H^+ (B) OH^- (C) NH_4^+ (D) NH_4^+ and OH^-

476. In a chemical reaction equilibrium is established when : [PET-2001]

(A) Opposing reaction ceases

(B) Concentration of reactants and products are equal

(C) Velocity of opposing reaction is the same as that of forward reaction

(D) Reaction ceases to generate heat

477. For the reaction $PCl_3\,(g) + Cl_2\,(g) \rightleftharpoons PCl_5\,(g)$ The value of K_C at 250° C is 26. The value of K_P at this temperature will be : [PET-2001]

(A) 0.61 (B) 0.57 (C) 0.83 (D) 0.46

478. According to Le Chatelier's principle adding heat to a solid and liquid in equilibrium will cause the [PET-2001]

(A) Amount of solid to decrease

(B) Amount of liquid to decrease

(C) Temperature to rise (D) Temperature to fall

479. The dissociation constant for water is 1×10^{-14} mol^2 litre^{-2}. What is the pH of a 0.001 M KOH solution ? [PET-2001]

(A) 10^{-11} (B) 10^{-3} (C) 3 (D) 11

480. Which buffer solution out of the following will have pH > 7 ? [PET-2001]

(A) $CH_3COOH + CH_3COONa$ (B) HCOOH + HCOOK

(C) CH_3COONH_4 (D) $NH_4OH + NH_4Cl$

481. The solubility of Ag in NaI solution is less than that in pure water because : [PET-2001]

(A) AgI forms complex with NaI

(B) Of common ion effect

(C) Solubility product of AgI is less than that of NaI

(D) The temperature of the solution decreases

482. The solubility product of a sparingly soluble salt AB at room temperature is 1.21×10^{-6}m^2 Its molar solubility is : [PET-2001]

(A) 1.21×10^{-6}M (B) 1.1×10^{-4} M

(C) 1.1×10^{-3} M (D) 1.21×10^{-3} M

483. An aqueous solution of CH_3COONa will be: [PET-2001]

(A) Acidic (B) Alkaline

(C) Neutral (D) None of these

484. The solubility of A_2X_3 is y mol dm^{-3}. Its solubility product is : [PET-2001]
(A) $6y^4$ (B) $64\ y^4$ (C) $36\ y^5$ (D) $108y^5$

485. The heat of neutralization of HCl and NaOH is : [PET-2001]
(A) Zero (B) −57.3 kJ (C) +57.3 kJ (D) None of these

486. The heat of neutralization of HCl by NaOH under certain condition is −55.9 kJ and that of HCN by NaOH is −12.1 kJ The heat of ionization of HCN is : [PET-2001]
(A) −68.0 kJ mol^{-1} (B) −43.8 kJ mol^{-1}
(C) 68.0 kJ mol^{-1} (D) 43.8 kJ mol^{-1}

487. The favourable conditions for Haber's process are [PET-2002]
(A) Low temperature and low pressure
(B) Low temperature and high pressure
(C) High temperature and low pressure
(D) High temperature and high pressure

488. The correct relationship between K_P and K_C is [PET-2002]
(A) $K_P = K_C\ [RT]$ (B) $K_P = K_C\ [\Delta n]$
(C) $K_P = K_C\ [RT]^{\Delta n}$ (D) $K_P = K_C\ [RT]^{1/\Delta n}$

489. Which of the following is not a Lewis acid [PET-2002]
(A) CO (B) $SiCl_4$ (C) SO_3 (D) Zn^{2+}

490. Which of the following is the buffer solution of strong acidic nature [PET-2002]
(A) $HCOOH + HCOO^-$ (B) $CH_3COOH + CH_3COO^-$
(C) $H_2C_2O_4 + C_2O_4^{2-}$ (D) $H_3BO_3 + BO_3^{3-}$

491. Which one of the following salts gives an acidic solution in water [PET-2002]
(A) CH_3COONa (B) NH_4Cl
(C) NaCl (D) CH_3COONH_4

492. The solubility product of $BaSO_4$ is 1.3 x 10^{-9}. The solubility of this salt in pure water will be [PET-2002]
(A) 1.69 x 10^{-9} mol $litre^{-1}$ (B) 1.69 x 10^{-18} mol $litre^{-1}$
(C) 3.6 x 10^{-5} mol $litre^{-1}$ (D) 3.6 x 10^{-9} mol $litre^{-1}$

493. A tenfold increase in pressure on the rection $N_2(g) + 3H_2(g) \rightarrow 2NH_3(g)$ at equitabrium makes K_P : [PET-2003]
(A) unchanged (B) five times
(C) two times (D) ten times

494. For equilibrium $2NO(g) \rightarrow N_2(g) + O_2(g)$ which of the following is not correct. [PET-2003]
(a) Equilibrium constant does not change with pressure at constant temperature.
(b) Equilibrium constant does not change with volume at constant temperature.
(c) At high pressure dissociation of NO is favoured.
(d) At high temperature dissociation of NO is favoured.

495. Two moles of NH_3 when introduced into a previously evacuated vessel (one liter), partially dissociate into N_2 and H_2. The equilibrium constant is ; [PET-2003]
(A) 27/64 $mol^2\ l^{-2}$ (B) 3/4 $mol^2\ l^{-2}$
(C) 27/16 $mol^2\ l^{-2}$ (D) 27/1 $mol^2\ l^{-2}$

496. The pH of 0.1 M NaOH is : [PET-2003]
(A) 11 (B) 13 (C) 10 (D) 12

497. How many grams of CaC_2O_4 (molecular weight =128) on dissolving in distilled water will give a saturated solution ? [$K_{sp}(CaC_2O_4)$ = 2.5 x 10^{-9} mol^2 $liter^{-2}$] [PET-2003]
(A) 1.2800g (B) 0.0128g (C) 0.1280g (D) 0.0064g

498. A buffer solution contains equal concentration of X− and HX the k for HX is 10^{-8}. The pH of the buffer solution is ; [PET-2003]
(A) 3 (B) 8 (C)10 (D)14

499. Which of the following makes basic solution? [PET-2003]
(A) NaOCl (B) HOCl (C) $NaHSO_4$ (D) None

500. Which of the following solutions will have pH close to 1.0 ? [PET-2003]
(A) 55ml of $\frac{M}{10}$ HCl + 45ml of $\frac{M}{10}$ NaOH
(B) 100ml of $\frac{M}{10}$ HCl +100ml of $\frac{M}{10}$ NaOH
(C) 10ml of $\frac{M}{10}$ HCl + 90ml of $\frac{M}{10}$ NaOH
(D) 75ml of $\frac{M}{10}$ HCl + 25ml of $\frac{M}{10}$ NaOH

501. which of the follwoing statements is wrong ? [PET-2003]
(A) pH+ pOH = 14 at room temperature
(B) The commonest primary standards for standardization of acids are Na_2CO_3 or borax
(C) $[H^+][OH^-] = 10^{-14}$ at 22°C
(D) Alkalies are generally standardized against acetic acid or borax.

502. To what pH must a solution containing a precipitate of $Cr(OH)_3$ be adjusted so that all of precipitate (dissolves) when Cr^{3+} = 0.1mol/L, k_{sp} = 6 x 10^{-31} [PET-2003]
(A) Upto 4.0 (B) Upto 4.1
(C) Upto 4.3 (D) Upto 4.2

503. For complete neutralization of HCl with NaOH, the heat of neutralization is. [PET-2003]
(A) +57.32 kJ mol (B) −13.70 kJ mol
(C) −57.32 kJ mol (D) +13.70 kJ mol

504. When NH_4Cl is added to a solution of NH_4OH

[CBSE-90]
(A) the dissociation of NH_4OH is increased.
(B) pH will increase (C) pH will decrease
(D) It has no effect on pH at all.

505. 28 gm of N_2 and 6gm of H_2 were kept at 400°C in 1-liltre vessel, the equilibrium mixture contained 27.54 gm of NH_3.The approximate value of K_C;for the above reaction can be in (mol^{-2} $litre^2$). [CBSE-90]
(A) 75 (B) 50 (C) 25 (D) 100

506. $H_2 + I_2 \rightleftharpoons 2HI$ [CBSE-90]
in the above equilibrium system if the concentration of the reactants at 25°C is increased, the value of K_C will
(A) Increase (B) decrease (C) remain same
(D) depend on the nature of the reactants.

507. For the reaction $H_2(g) + I_2(g) = 2HI(g)$ at 721K the value of equilibrium constant is 50. When the equilibrum concentration of both is 0.5N, value of K_P under the same conditions will be [CBSE-90]
(A) 0.02 (B) 0.2 (C) 50.0 (D) 50/RT

508. The ionisation of hydrogen atom would give rise to [CBSE-90]
(A) hydride ion (B) hydronium ion
(C) proton (D) hydroxyl ion

509. Conjugate base of HCO_3^- is [CBSE-90]
(A) H_2CO_3 (B) CO_3^{2-} (C) H (D) CO_2

510. Which of the following acts as both Bronsted acid and base ? [CBSE-90]
(A) H_2O (B) NH_3 (C) OH^- (D) H_3PO_4

511. Conjugate base for HSO_4^- is [CBSE-90]
(A) H^+ (B) H_2O (C) SO_4^{-2} (D) H_2SO_4

512. The pH of 0.001M HCl is [CBSE-91]
(A) 3 (B) 11 (C) 2 (D) 10

513. The solubility product of CuS Ag_2S and HgS are 10^{-31}, 10^{-44}, 10^{-54} respectively. The solubilities of these sulphides are in the order [CBSE-91]
(A) Ag_2S>CuS>HgS (B) AgS>HgS>CuS
(C) HgS>Ag_2S>CuS (D) CuS>Ag_2S>HgS

514. A physician wishes to prepare a buffer solution at pH = 3.85 that efficiently resist changes in pH yet contains only small concentration of the buffering agents. Which one of the following weak acid together with its sodium salt would be best to use [CBSE-97]
(A) m-chlorobenzoic acid (pKa=3.98)
(B) p-chlorocinnamic acid (pKa-4.41)
(C) 2,5-dihydroxybenzoic acid (pKa =2.97)
(D) acetoacetic acid (pka =3.58)

515. In Haber's process 30 litres of dihydrogen and 30 litres of dinitrogen were taken for reaction which yielded only 50% of the expected product. What will be the composition of gaseous mixture under the aforesaid condition in the end ? [CBSE-2003]
(A) 20 litres ammonia, 25 litres nitrogen, 15 litres hydrogen
(B) 20 litres ammonia, 20 litres nitrogen, 20 litres hydrogen
(C) 10 litres ammonia, 25 litres nitrogen, 15 litres hydrogen
(D) 20 litres ammonia, 10 litres nitrogen, 30 litres hydrogen

516. The following equilibria are given : [CBSE-2003]
$N_2 + 3H_2 \rightleftharpoons 2NH_3$ K_1
$N_2 + O_2 \rightleftharpoons 2NO$ K_2
$H_2 + \frac{1}{2} O_2 \rightleftharpoons H_2O$ K_3
The equilibrium constant of the reaction
$2NH_3 + \frac{5}{2} O_2 \rightleftharpoons 2NO + 3H_2O$
in terms of K_1, K_2 and K_3 is
(A) $K_2K_3^3/K_1$ (B) $K_1K_2K_3$ (C) K_1K_2/K_3 (D) $K_1K_3^2/K_2$

517. Which one of the following compounds is not a protonic acid ? [CBSE-2003]
(A) $SO_2(OH)_2$ (B) $B(OH)_3$
(C) $PO_2(OH)_3$ (D) $SO(OH)_2$

518. The solubility product of AgI at 25°C is 1.0 x 10^{-16} mol^2 L^{-2}. The solubility of AgI is 10^{-4} N solution of KI at 25°C is approximately (in mol L^{-1}) [CBSE-2003]
(A) 1.0 X 10^{-8} (B) 1.0 X 10^{-16}
(C) 1.0 X 10^{-12} (D) 1.0 X 10^{-10}

519. $\delta H_f(H_2O)$=X. Heat of neutralisation of CH_3COOH and NaOH will be [BHU-2003]
(A) less than X (B) less than 2X
(C) between X and 2X (D) X

520. NaOH (aq) , HCl (aq) ,NaCl (aq) , concentration of each is 10^{-3}M. The pH will be respectively- [BHU-2003]
(A) 11 , 3 , 7 (B) 10 , 2 , 6 (C) 10 , 6, 2 (D) 3 , 4 , 7

521. K_{sp} (AB)= 1× 10^{-10}. If [A+] = 1×10^{-5} , which of the following solutions of [B-] cannot be precipitated? [BHU-2003]
(A) 1×10^{-5} (B) 5 × 10^{-6} (C) 2 × 10^{-5} (D) 5 × 10^{-5}

522. pH value of water is [AFMC-2003]
(A) Pressure dependent
(B) Temperature dependent
(C) Temperature independent (D) None of these

523. The solubility product of a sparingly soluble salt AX_2 is 3.2 x 10^{-11}. Its solubility (in moles/litre) is [CBSE-PMT-2004]

(A) 5.6×10^{-6} (B) 3.1×10^{-4} (C) 2×10^{-4} (D) 4×10^{-4}

524. The rapid change of pH near the stoichiometric point of an acid-base titration is the basis of indicator detection. pH of the solution is related to ratio of the concentrations of the conjugate acid (HIn) and base (In-) forms of the indicator by the expression [CBSE-PMT-2004]

(A) $\log In^- / HIn = pK_{In} - pH$
(B) $\log HIn / In^- = pK_{In} - pH$
(C) $\log HIn / In^- = pH - pK_{In}$
(D) $\log In^- / HIn = pH - pK_{In}$

525. A weak acid HX has Ka = $1.0 ' 10^{-5}$ and it forms a salt NaX when treated with caustic soda. The degree of hydrolysis of a 0.1 M NaX solution is [IIT-2004]
(A) 0.1% (B) 0.01% (C) 0.001% (D) 1%

Chemical & Ionic Equilibria

1.	(B)	51.	(A)	101.	(B)	151.	(D)	201.	(C)(D)(E)	251.	(D)
2.	(A)	52.	(D)	102.	(D)	152.	(A)	202.	(D)	252.	(B)
3.	(A)	53.	(B)	103.	(A)	153.	(A)	203.	(D)	253.	(A)
4.	(A)	54.	(A)	104.	(B)	154.	(A)	204.	*	254.	(C)
5.	(C)	55.	(B)	105.	(D)	155.	(B)	205.	*	255.	(D)
6.	(B)	56.	(D)	106.	(D)	156.	(D)	206.	*	256.	(B)
7.	(B)	57.	(D)	107.	(D)	157.	(A)	207.	*	257.	(B)
8.	(D)	58.	(D)	108.	(A)	158.	(D)	208.	(A)	258.	(B)
9.	(C)	59.	(D)	109.	(A)	159.	(C)	209.	(D)	259.	(C)
10.	(D)	60.	(D)	110.	(C)	160.	(D)	210.	(B) (C)	260.	(D)
11.	(B)	61.	(D)	111.	(C)	161.	(D)	211.	(B)	261.	(B)
12.	(B)	62.	(C)	112.	(C)	162.	(A)	212.	(A)	262.	(B)
13.	(D)	63.	(C)	113.	(D)	163.	(A)	213.	(A)(B)(C)	263.	(A)
14.	(B)	64.	(A)	114.	(A)	164.	(A)	214.	(D)	264.	(B)
15.	(B)	65.	(D)	115.	(A)	165.	(D)	215.	(D)	265.	(C)
16.	(A)	66.	(A)	116.	(A)	166.	(A)	216.	(A)	266.	(B)
17.	(D)	67.	(C)	117.	(A)	167.	(D)	217.	(D)	267.	(A)
18.	(A)	68.	(B)	118.	(A)	168.	(A)	218.	(A)	268.	(C)
19.	(B)	69.	(A)	119.	(C)	169.	(B)	219.	(A)	269.	(C)
20.	(B)	70.	(A)	120.	(C)	170.	(D)	220.	(A)	270.	(B)
21.	(C)	71.	(C)	121.	(A)	171.	(D)	221.	(A)	271.	(B)
22.	(C)	72.	(D)	122.	(D)	172.	(C)	222.	(A)	272.	(C)
23.	(A)	73.	(B)	123.	(D)	173.	(D)	223.	(C)	273.	(D)
24.	(C)	74.	(B)	124.	(A)	174.	(A)	224.	(A)	274.	(C)
25.	(D)	75.	(C)	125.	(B)	175.	(C)	225.	(D)	275.	(B)
26.	(A)	76.	(B)	126.	(B)	176.	(B)	226.	(A)	276.	(A)
27.	(B)	77.	(D)	127.	(C)	177.	(A)	227.	(C)	277.	(D)
28.	(D)	78.	(B)	128.	(C)	178.	(D)	228.	(C)	278.	(B)
29.	(C)	79.	(B)	129.	(C)	179.	(B)	229.	(A)	279.	(C)
30.	(B)	80.	(C)	130.	(C)	180.	(B)	230.	(C)	280.	(C)
31.	(D)	81.	(A)	131.	(B)	181.	(C)	231.	(C)	281.	(B)
32.	(C)	82.	(B)	132.	(A)	182.	(A)	232.	(A)	282.	(B)
33.	(D)	83.	(C)	133.	(B)	183.	(B)	233.	(C)	283.	(D)
34.	(D)	84.	(B)	134.	(A)	184.	(C)	234.	*	284.	(B)
35.	(C)	85.	(A)	135.	(A)	185.	(A)	235.	*	285.	(C)
36.	(D)	86.	(C)	136.	(C)	186.	(B)	236.	*	286.	(B)
37.	(A)	87.	(B)	137.	(C)	187.	(A)	237.	(C)	287.	(D)
38.	(C)	88.	(A)	138.	(D)	188.	(D)	238.	*	288.	(A)
39.	(D)	89.	(D)	139.	(C)	189.	(C)	239.	*	289.	(A)
40.	(B)	90.	(B)	140.	(B)	190.	(C)	240.	*	290.	(B)
41.	(C)	91.	(C)	141.	(D)	191.	(B)	241.	(A)	291.	(A)
42.	(D)	92.	(A)	142.	(B)	192.	(A)	242.	(B)	292.	(A)
43.	(C)	93.	(C)	143.	(D)	193.	(C)	243.	*	293.	(B)
44.	(B)	94.	(A)	144.	(C)	194.	(A)	244.	*	294.	(A)
45.	(B)	95.	(B)	145.	(D)	195.	(A)	245.	(D)	295.	(A)
46.	(C)	96.	(B)	146.	(D)	196.	(B)	246.	*	296.	(C)
47.	(C)	97.	(B)	147.	(D)	197.	(B)	247.	*	297.	(B)
48.	(B)	98.	(C)	148.	(D)	198.	(B)	248.	*	298.	(A)
49.	(D)	99.	(D)	149.	(D)	199.	(D)	249.	(D)	299.	(A)
50.	(D)	100.	(C)	150.	(A)	200.	(A)	250.	(D)	300.	(C)

204. $k_p = k_c (RT)^{\Delta n}$
205. $CH_3NH_3^+$
206. less
207. I_2
234. 1/16
235. Negative
236. pH6
238. SO_4^{2-}
239. Hydrolysis
240. 3
243. 12
244. Buffer Solution
246. Neutral
247. 1.0×10^{-14} mole^2 / lit^2
248. Buffer

301. (C)
302. (B)
303. (B)
304. (B)
305. (D)
306. (D)
307. (C)
308. (D)
309. (A)
310. (A)
311. (A)
312. (B)
313. (B)
314. (A)
315. (C)
316. (C)
317. (B)
318. (B)
319. (C)
320. (B)
321. (D)
322. (C)
323. (C)
324. (D)
325. (D)
326. (B)
327. (D)
328. (B)
329. (A)
330. (B)
331. (C)
332. (A)
333. (C)
334. (A)
335. (A)
336. (B)
337. (A)
338. (B)
339. (A)
340. (B)
341. (B)
342. (B)
343. (B)
344. (B)
345. (D)
346. (C)
347. (A)
348. (C)
349. (C)
350. (C)
351. (A)
352. (B)
353. (C)
354. (A)
355. (A)
356. (D)
357. (A)
358. (D)
359. (A)
360. (C)
361. (B)
362. (B)
363. (C)
364. (D)
365. (B)
366. (B)
367. (D)
368. (B)
369. (D)
370. (C)
371. (D)
372. (B)
373. (D)
374. (C)
375. (D)
376. (B)
377. (C)
378. (A)
379. (B)
380. (B)
381. (B)
382. (B)
383. (C)
384. (A)
385. (B)
386. (C)
387. (A)
388. (C)
389. (C)
390. (A)
391. (D)
392. (D)
393. (A)
394. (C)
395. (C)
396. (A)
397. (A)
398. (A)
399. (B)
400. (A)
401. (C)
402. (D)
403. (B)
404. (D)
405. (B)
406. (C)
407. (C)
408. (D)
409. (D)
410. (D)
411. (C)
412. (B)
413. (B)
414. (B)
415. (C)
416. (D)
417. (A)
418. (C)
419. (C)
420. (D)
421. (C)
422. (D)
423. (A)
424. (D)
425. (D)
426. (A)
427. (D)
428. (C)
429. (B)
430. (C)
431. (B)
432. (A)
433. (C)
434. (A)
435. (C)
436. (D)
437. (B)
438. (B)
439. (C)
440. (C)
441. (B)
442. (B)
443. (D)
444. (C)
445. (D)
446. (C)
447. (B)
448. (B)
449. (C)
450. (B)
451. (A)
452. (D)
453. (B)
454. (B)
455. (A)
456. (C)
457. (C)
458. (B)
459. (A)
460. (D)
461. (B)
462. (B)
463. (C)
464. (A)
465. (C)
466. (A)
467. (C)
468. (C)
469. (B)
470. (A)
471. (A)
472. (C)
473. (C)
474. (B)
475. (D)
476. (C)
477. (A)
478. (A)
479. (D)
480. (D)
481. (B)
482. (C)
483. (B)
484. (D)
485. (B)
486. (B)
487. (B)
488. (C)
489. (A)
490. (A)
491. (B)
492. (C)
493. (A)
494. (C)
495. (C)
496. (B)
497. (D)
498. (B)
499. (A)
500. (C)
501. (D)
502. (D)
503. (C)
504. (C)
505. (A)
506. (C)
507. (C)
508. (C)
509. (B)
510. (A)
511. (C)
512. (A)
513. (A)
514. (C)
515. (C)
516. (A)
517. (B)
518. (C)
519. (A)
520. (A)
521. (B)
522. (C)
523. (C)
524. (C)
525. (B)

THERMODYNAMICS

Most common form of energy is heat. We have mechanical, electrical, radiant, chemical energy etc. They can be transformed from one form to the other. The study of inter relationship of the various forms of energy in a system is called thermodynamics.

Limitations :

1. The laws of thermodynamics are not applicable to small particles like individual atoms or molecules. However, the laws can be applied to macroscopic systems.

2. Thermodynamics does not give any information about the rate at which a given chemical reaction or process may proceed and also time required for the change.

Thermodynamic System : Any specified portion of the universe or matter, real or imaginary separated form the rest of the universe selected for thermodynamic treatment.

Surroundings : The rest of the universe, which may exchange matter or energy or both with the system.

Systems : There are three types of systems

1. Open System : A system which can exchange energy as well as the matter with its surroundings

e.g. Beaker containing water absorbs heat from the surroundings and evoporates. Heat and matter (water vapour) are exchanged with the surroundings.

2. Closed System : When a system can exchange only energy and not the matter with its surroundings, it is called a closed system.

e.g. a chemical reaction taking place in a closed vessel.

3.Isolated System : A system which can neither exchange matter nor energy with its surroundings.

e.g. a chemical reaction taking place in a closed vessel which is heat insulated from all the sides.

Homogeneous System : Which consists of only one phase

Macroscopic System : A macroscopic system is one that consists of large number of particles, molecules or radicals.

Macroscopic properties : e.g. pressure, temperature, volume, density, composition, viscosity, surface tension etc. They can change the original state of the system.

Extensive property : The property which depends on the quantity or amount of matter present in the system is called extensive property, e.g. mass, no. of moles, enthalpy, entropy, Gibbs free energy etc. This is an additive property.

Intensive property : The property which is independent of the amount of the substance present in the system is called intensive property e.g. pressure, temperature, density, concentration, viscosity, specific heat, surface tension, refractive index etc.

Processess and their kinds :

Whenever a system changes from one state to another, it is accompanied by change in energy.

1. Isothermal process : The temp. of the system remains constant during each step of the process (process means, the path or the operation by which a system changes from one state to another). $dT = 0$

2. Adiabatic process : No heat enters or leaves the system during any step of the process. Here, temp. rises or falls depending upon either it is exothermic or endothermic reaction $dQ = 0$

3.Isobaric process : The pressure remains constant during each step of the change in the state of the system. Here, the pressure is constant but change of volume takes place. $dP = 0$

4. Isochoric process : Volume of the system remains constant during each step of the change in the state of the system. $dV=0$

5. Cyclic process : A process during which system comes to its initial state through a number of different processes is called a cyclic process.

For a cyclic process $dE = 0$; $dH=0$

Heat of reactions : Heat of reaction is defined as the amount of heat evolved or absorbed i.e. change in enthalpy. Since the value of heat of reaction depends on pressure and temperature, the values are generally given under standard conditions (1 atmosphere and 25°C or 298K).

Change in heat during a chemical reaction : Change in heat contents or change in enthalpy of the system

ΔH= heat contents of the products - heat contents of reactants

$\Delta H = H_P - H_R$

if $H_p > H_R$ endothermic, $\Delta H = +Ve$

$H_R > H_P$ exothermic, $\Delta H = -Ve$

Thermochemical equation :

$2H_2\,(g) + O_2\,(g) \rightarrow 2H_2O\,(l) + 13663$ calories or

$2H_2\,(g) + O_2\,(g) \rightarrow 2H_2O\,(l)\,;\ \Delta H = -13663$ cal.

Heat of formation (H_f) : It is the quantity of heat evolved or absorbed when one mole of the subtance is formed from its constituent elements under given conditions of temperature and pressure.

$C\,(s) + O_2\,(g) \rightarrow CO_2\,(g);\ H_f = -94$ k cal.

Heat of Combustion : It is the amount of heat evolved when one mole of the substance is completely burnt in air or oxygen. It is always negative

$CH_4(g) + 2O_2(g) \rightarrow CO_2(g) + 2H_2O(l);\ \Delta H = -192$ k. cal.

Heat of Neutralisation : It is the amount of heat evolved when one equivalent of an acid is neutralised by one equivalent of a base in fairly dilute solution.

$HCl\,(aq) + NaOH\,(aq) \rightarrow NaCl\,(aq) + H_2O;$

$\Delta H = -13.7$ k. Cal.

It has been observed that the heat of neutralisation of a

strong acid against a strong base is always constant i.e. 13.7 k.cal. or 57 kJ/mole. It is actually the heat of formation of water $H^+ + OH^- \rightarrow H_2O$; $\Delta H = -13.7$ k. cal.

Heat of neutralisation of a weak acid or base against a strong base or acid is always less than 13.7 k. cal. Here, a part of heat evolved is used up to ionize the weak acid or base.

Heat of solution : Heat evolved or absorbed when one mole of solute is dissolved completely in excess of the solvent.

Heat of dilution : When one mole of solute is diluted to such an extent that on further dilution, no heat is evolved or absorbed.

Heat of hydration : When one mole of anhydrous or partially hydrated salt combines with the required number of moles of water to form a specific hydrate.

Heat content and enthalpy relationships:

$\Delta H = \Delta E + \Delta nRT$ or $\Delta H = \Delta E + P\,\Delta V$ ($P\Delta V$ = work done $= \Delta W$)

$\Delta E = \Delta H - \Delta nRT$

When $\Delta n = O$, $\Delta H = \Delta E$; $\Delta n > O$, $\Delta H > \Delta E$

$\Delta n < O$; $\Delta H < \Delta E$

Hess Law : It states that the amount of heat evolved or absorbed in a process, including the chemical change, is the same whether the process takes place in one or several steps.

The heat of reaction depends only on the initial reactants and final products and not at all on the intermediate products that may be formed. It is also independent of the path or the manner by which the change is brought about.

Applications of Hess Law :

1. Calculation of heats of reactions : Hess's law makes it possible to calculate heats of many reactions which can't be determined experimentally.

2. Calculation of enthalpy of formation : It is impossible to determine experimentally, the enthalpy of formation of for example benzene from its elements, carbon and hydrogen. However, it can be calculated from the heats of combustion of benzene and the enthalpies of formation of water and carbon dioxide.

- Heat of Combustion is determined by Bomb calorimeter.
- Heat of Neutralisation is measured in Dewar vacuum flask.

First Law of Thermodynamics :

The total energy of the universe is constant.

The energy can neither be created nor destroyed although it can be transformed from one form to another. This is also known as law of conservation of energy.

It is impossible to construct a perpetual motion machine i.e. a machine which can produce energy without expenditure of energy.

The modified law therfore, states that the total mass and energy of an isolated system remains unchanged.

Internal energy : Every substance is associated with a definite amount of energy which depends upon its chemical nature as well as upon its temperature, pressure and volume. This energy is known as intrinsic or internal energy.

Internal energy of a substance is made up of a number of components (a) energy of translation (b) rotational (c) vibrational (d) coulombic energy of electron and nucleus in atoms (e) interaction energy.

The internal energy of a system depends upon the state of the system and not how the state is attained.

The energy of a thermodynamic system is called Internal energy.

Energy change in relation to work and heat changes

Let E_A be the energy of a system in its state A and let E_B the energy in its state B.

Suppose a system while undergoing change from state A to state B absorbs heat Q from the surroundings and performs some work, mechanical or electrical, equal to W. The absorption of heat by the system tends to raise the energy of the system. The performance of work by the system, on the other hand, tends to lower the energy of the system.

Hence, the change of internal energy, ΔE, accompanying the process can be understood as :

1. The net energy change of a closed system is equal to heat absorbed plus the work done on the system

$\Delta E = Q + W$ or

2. The net energy change of a closed system is equal to heat absorbed minus the work done by the system.

$\Delta E = Q - W$

(i) If $\Delta E = 0$, $Q = -W$ (heat absorbed = work done by the system) or $W = -Q$ (heat given to surroundings = work done on the system)

(ii) If $W = 0$ then $\Delta E = Q$

(iii) If $Q = 0$ then $\Delta E = W$

(iv) If a gas expands against the constant external pressure, P, let the volume change be Δv then

$$w = -P\,\Delta V$$

It should be remembered that in all energy transformations, Q is given a +ve sign when heat is absorbed and a -ve sign when heat is evolved by a system. Work, W, is given a positive sign when work is done on the system and a negative sign when work is done by the system.

Work done in an isothermal expansion : The amount of work done in an isothermal reversible expansion of a gas from volume V_1 to V_2 is given by

$W = -2.0303nRT\log\frac{V_2}{V_1}$;Similarly, $W = -2.303nRT\log\frac{P_1}{P_2}$

$W_{compression} = 2.303\ nRT \log \frac{V_1}{V_2}$

Similarly $W = 2.303\ nRT \log \frac{P_2}{P_1}$

Work done in an adiabatic expansion : The work done in an abiabatic reversible expansion is given by

$$P_1V_1^{\gamma} = P_2V_2^{\gamma}$$

$\gamma = \frac{c_p}{c_v}$ c_p = heat capacity of a gas at constant pressure
c_v = heat capacity at constant volume

The difference between c_p and c_v is equal to the work done by 1 mole of gas in expansion when heated through 1°C.

work done by the gas at constant pressure = $P\Delta V$

for one mole of gas $PV = RT$

When temp. is raised by 1°C, the volume becomes $V+\Delta V$

$$P(V+\Delta V) = R\ (T+1)$$

or $$P\Delta V = R$$

Hence $C_p - C_v = P\Delta V = R$

Kirchoff's equation : It explains the variation of heat of reaction with temperature

$$\Delta C_p = C_{p'} - C_p = \frac{\Delta H_2 - \Delta H_1}{T_2 - T_1}$$

ΔC_p = difference in the heat capacities of products and reactants

$C_{p'}$ = mean molar heat capacity of the products at constant pressure

C_p = mean molar heat capacity of the reactant

ΔH_1 = heat change at temp T_1

ΔH_2 = heat change in the entire process

If the reaction takes place at constant volume then the equation is

$$\Delta C_v = C_{v'} - C_v = \frac{\Delta E_2 - \Delta E_1}{T_2 - T_1}$$

Second Law of Thermodynamics :

What is the necessity of the second law ?

First law can't explain :

(1) The heat absorbed and work done by a system in a given process puts no restriction on the source of this heat or on the direction in which it flows.

(2) Whether a reaction or a process can occur spontaneously.

(3) How one kind of energy be quantitatively converted into the other forms of energy ?

" Work can be completely converted into heat, but it is not possible to convert heat completely into work."

Carnot Cycle : The cycle determines the conversion of heat into work. It is shown by experience that heat engines (steam engine, diesel engine, petrol engine etc.) operating periodically can convert only a fraction of the absorbed heat into work.

Carnot engine works in four steps :

Step 1 : A reversible isothermal expansion from V_1 to V_2 at a temp. T_h .

Step 2 : A reversible adiabatic expansion from V_2 to V_3 with temp change from T_h to T_l.

Step 3 : A reversible isothermal compression from V_3 to V_4 at T_l.

Step 4 : A reversible adiabatic compression from V_4 to V_1 at a temperature change from T_l to T_h.

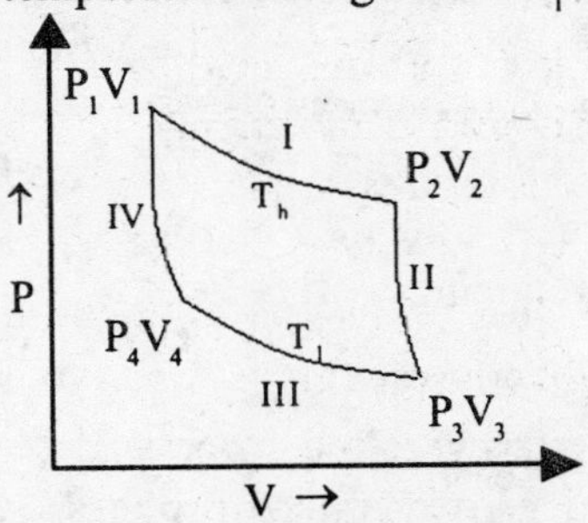

$$\frac{W}{Q_2} = \frac{Q_2 - Q_1}{Q_2} = \frac{T_2 - T_1}{T_2}$$

The total work done, W, in the cycle is equal to the total heat absorbed i.e. $W = Q_2 - Q_1$

Thus from Carnot cycle we come to conclusion that the ratio of the total work, W, obtained in the cycle to the heat Q_2 absorbed at a higher temperature is known as efficiency of Carnot engine, or heat engine

$$\text{Efficiency, } \varepsilon = \frac{W}{Q_2}$$

All thermodynamically reversible heat engines operating between two given temperatures have the same efficiency and is independent of the properties of the substances and mode of operation.

No engine can be more efficient than a reversible engine. Any engine working between two given temperatures can't be more efficient than the Carnot engine.

Entropy : There is a property of substances called entropy which measures the disorder in them. It is denoted by S. The change in entropy of system, $S_{final} - S_{initial}$ is denoted by ΔS.

$$\Delta S = \frac{Q_{rev}}{T}$$

ΔS = entropy change; Q_{rev} = heat absorbed in a reversible process

Units of entropy : Cal. degree^{-1} or Joules per degree kelvin $J K^{-1}$

Entropy change in Isothermal Expansion of an ideal gas

In an isothermal expansion of ideal gas, the temperature remains constant and therefore there will be no change of internal energy, i.e. $\Delta E = 0$ and hence the first law of thermodynamics can be treated as

$$\Delta E = Q_{rev} - W \text{ or } Q_{rev} = W$$

In such a case, the work done in the expansion of one mole of gas from volume V_1 to V_2 at constant temperature, is given by

$$W = RT \ln \frac{V_2}{V_1} \quad \therefore Q_{rev} = RT \ln \frac{V_2}{V_1} \quad (\text{since } Q_{rev} = W)$$

Hence $$\Delta S = \frac{Q_{rev}}{T} = \frac{1}{T} \times RT \ln \frac{V_2}{V_1}$$

or $\Delta S = nR \ln \frac{V_2}{V_1}$ or $\Delta S = nR \times 2.303 \log \frac{V_2}{V_1}$

Entropy change in a Reversible process :
Entropy is a state function and therefore depends only on the initial and final states of the system. The entropy change in a complete reversible cycle, when the system returns to its initial state is expressed by

$$\Delta S = \frac{Q_{rev}}{T} = 0$$

For a reversible process, the total entropy change of the system and the surroundings will be zero,

Hence $\Delta S_{System} + \Delta S_{surroundings} = \frac{Q_{rev}}{T} - \frac{Q_{rev}}{T} = 0$

Entropy change in Irreversible process :
Thermodynamically irreversible process is always accompanied by an increase in the entropy of the system and its surrounding. The total of the entropy change of the system and its surroundings during an irreversible process is therefore given by

$\Delta S_{System} + \Delta S_{surroundings} = \frac{Q_{rev}}{T} - \frac{Q_{rev}}{T}$

Since $\frac{Q_{rev}}{T} > \frac{Q_{rev}}{T}$ $\therefore \Delta S_{System} + \Delta S_{surroundings} > 0$

Entropy change for an ideal gas : The entropy change for a process involving an ideal gas thus depends on the initial and final temperatures and volumes

$$\Delta S = C_v \ln \frac{T_2}{T_1} + R \ln \frac{V_2}{V_1}$$

$$\Delta S = C_v \times 2.303 \log \frac{T_2}{T_1} + R \times 2.303 \log \frac{V_2}{V_1}$$

Entropy change at constant pressure $\Delta S_p = C_p \ln \frac{T_2}{T_1}$

Entropy change at constant volume $\Delta S_v = C_v \ln \frac{T_1}{T}$

Entropy change accompanying change of phase :
(1) From solid phase to liquid phase : Solid changes into liquid state, at its fusion point
The process requires absorption of heat (Latent heat of fusion)

$\Delta S_f = \frac{\Delta H_f}{T_f}$ (ΔH_f = molar heat of fusion)

(2) From liquid phase to vapour phase

$\Delta S_v = \frac{\Delta H_v}{T_b}$ (ΔH_v = molar heat of vapourisation)

(3) From one crystalline form to another
The change in entropy when 1 mole of a solid substance undergoes change of state from one crystalline form (say from rhombic form) to another crystalline form (say, monoclinic form) at the transition temperature T_t, is given by

$\Delta S = \frac{\Delta H_t}{T_t}$ (ΔH_t = molar heat of transition of the substance)

Free energy (G) :
Process occurs spontaneously in the direction of increasing entropy $\Delta G = \Delta H - T\Delta S$

if $\Delta G = -$ Ve - spontaneous
$= +$ Ve - non spontaneous
$= 0$ - equilibrium

ΔG is a measure of decrease in free energy (ΔG = free energy of products - free energy of reactants).

The free energy change ΔG and equilibrium constant, k are related by expression, $\Delta G = -2.303\ RT \log K$
The standard cell potential is related to the standard free energy change by the relation $\Delta G^{\circ} = -nFE^{\circ}$

Third law of Thermodynamics :
According to third law of thermodynamics, the entropy of crystalline substances at absolute zero is taken as zero.

$S = 0$ at $T = 0$ K

Clausius and Clapeyron equation : This is an important equation which deals with equilibrium between two phases of one component system and provides many informations about such equilibria.

$$\log \frac{P_2}{P_1} = \frac{\Delta H_v}{2.303 R}\left(\frac{T_2 - T_1}{T_1 \times T_2}\right)$$

Things to remember

Extensive property : The property which depends on the quantity or amount of matter present in the system e.g. energy, mass, enthalpy, entropy, Gibb's free energy, number of moles etc.
Intensive property : property, which is independent of the amount e.g. pressure, temp, density, concentration, viscosity, specific heat, surface tension, refractive index etc.
The state variables of thermodynamic system are
(a) Temperature (b) pressure (c) volume (d) number of moles (e) Internal energy (f) enthalpy (g) entropy (h) free energy etc.
- Isothermal process : Change in temp, $dT = 0$
- Adiabatic process : Change in heat, $dQ = 0$
- Isobaric process : Change in pressure, $dP = 0$
- Isochoric process : Change in volume, $dV=0$
- Cyclic process : A process during which system comes to its initial state through a number of different processes.

For a cyclic process $dE = 0$; $dH = 0$

Hess Law : The heat evolved or absorbed in a process, including the chemical change is the same whether the process takes place in one or several steps.
Heat content and enthalpy are related by

$\Delta E = \Delta H - nRT$ or $\Delta E = \Delta H - P\Delta V$
$\Delta H = \Delta E + nRT$ or $\Delta H = \Delta E + P\Delta V$
$H_P > H_R$ endothermic $\Delta H = +$Ve
$H_P < H_R$ exothermic $\Delta H = -$Ve

The heat of neutralisation of a strong acid against a strong base is always constant i.e. 13.7 k.cal. or 57 KJ/ mole.
Note : Heat of neutralisation of a weak acid or base against a strong base or acid is always less than 13.7 k. cal.. Here a part of heat is used up to ionise the weak acid or base.

I^{st} law of Thermodynamics : Also known as law of conservation of energy. $\Delta E = Q - W$

IInd law of Thermodynamics : " Work can be completely converted into heat, but it is not possible to convert heat completely into work."

- Carnot equation $\frac{W}{Q_2} = \frac{Q_2 - Q_1}{Q_2} = \frac{T_2 - T_1}{T_2}$

$\varepsilon = \frac{W}{Q_2}$ (ε = efficiency of Carnot engine)

$\Delta S = \frac{Q_{rev}}{T}$; (units of entropy JK^{-1})

Entropy change in isothermal expansion of an ideal gas is given by

$$\Delta S = nR \times 2.303 \log \frac{V_2}{V_1}$$

Entropy change in reversible processes, $\Delta S = 0$

Entropy change in irreversible process, $\Delta S > 0$

$\Delta S_f = \frac{\Delta H_f}{T_f}$ (ΔH_f = molar heat of fusion)

$\Delta S_v = \frac{\Delta H_v}{T_b}$ (H_v = molar heat of vapourisation)

$\Delta S = \frac{\Delta H_t}{T}$ (H_t = molar heat of transition of the substance)

Enthalpy and entropy are related by,

$\Delta G = \Delta H - T\Delta S$; $\Delta G = -ve$ spontaneous

$= +ve$ non spontaneous

$= 0$ is at equilibrium

The free energy change $\Delta G°$ and equilibrium constant k, are related by the expression $\Delta G° = -2.303\ RT \log k$

Cell potential and free engergy are related by

$\Delta G° = -nFE°$ cell

Third Law of Thermodynamice : "At absolute zero, the entropy of a perfect crystal is also zero." $S = 0$ at $T = 0$ K

Solved Examples

(1) The heat of combustion of CH_4 (g) at constant volume is measured in a bomb calorimeter at 298.2K and is found to be -885389 J/mole. Find the value of enthapy change.

Sol. on combustion : $CH_4(g) + 2O_2(g) \rightarrow CO_2(g) + 2H_2O(l)$

$\Delta H = \Delta E + \Delta nRT$ or $\Delta H = \Delta E + P\Delta V$

$= -885389 + (-2)(8.314) \times 298$

$= -885389\ Jmol^{-1} - 4959\ Jmol^{-1}$

$= -890348\ J\ mol^{-1}$ or - 890.348 k. J. /mole.

or $\Delta H = \Delta E$ since $\Delta V = 0$

(2) Calculate standard enthalpy change and standard internal energy change for the reaction

$OF_2(g) + H_2O(g) \rightarrow O_2(g) + 2HF(g)$

$\Delta H_f\ OF_2(g)$ 23; $H_2O(g) = -241.8$ & $HF(g) = -268.6\ kJmol^{-1}$

Sol. $\Delta H_f = \Sigma\Delta H_p$ (products) - $\Sigma\Delta H_f$ (reactants)

$= 2(-268.6) - \{23.0 - 241.8\}$

$= -537.2 + 218.8 = -318.4$

(The enthalpy of formation of O_2 is zero)

$\Delta E = \Delta H - \Delta nRT$

$= -318.4 - (1) \times 8.31 \times 298.2$

$= 320.9$ k. J. mol^{-1}

(3) Calculate the enthalpy of formation of C_2H_5OH at 25°C from the following data

(i) $C_2H_5OH(l) + 3O_2(g) \rightarrow 2CO_2(g) + 3H_2O(l)$;

$\Delta H = -1368.2$ k J

(ii) $C(s) + O_2(g) \rightarrow CO_2(g)$; $\Delta H = -393.5$ kJ

(iii) $H_2(g) + \frac{1}{2} O_2(g) \rightarrow H_2O(l)$; $\Delta H = -286.2$ kJ

Sol. We aim for the equation

$2C(s) + 3H_2(g) + \frac{1}{2} O_2(g) \rightarrow C_2H_5OH(l)$; $\Delta H = ?$

To obtain this equation, multiply equation (ii) by 2 and (iii) by 3. (iv) $2C(s) + 2O_2(g) \rightarrow 2CO_2(g)$; $\Delta H = -787.0$ kJ

(v) $3H_2(g) + 1\frac{1}{2} O_2(g) \rightarrow 3H_2O(l)$; $\Delta H = -858.5$ kJ

Add (iv) and (v) and subtract (i)

on adding $2C(s) + 3H_2(g) + 3\frac{1}{2} O_2(g) \rightarrow 2CO_2(g) + 3H_2O(l)$;

$\Delta H = -1645.5$ kJ

$C_2H_5OH(l) + 3O_2(g) \rightarrow 2CO_2(g) + 3H_2O(l)$; $\Delta H = -1368.2$kJ

$-C_2H_5OH(l) + 2C(s) + 3H_2(g) \frac{1}{2} O_2(g) \rightarrow -277.3$ kJ

finally we can write $2C(s) + 3H_2(g) + \frac{1}{2}O_2(g) \rightarrow C_2H_5OH(l)$;

$\Delta H = -277.3$ kJ

(4) Calculate the heat of neutralisation of aqueous NH_3 with aqueous HCl from the following data

(i) $NH_3(g) + aq \rightarrow NH_3(aq)$ + 8,400 cal.

(ii) $HCl(g) + aq \rightarrow HCl(aq)$ + 17,300 cal.

(iii) $NH_3(g) + HCl(g) + aq \rightarrow NH_4Cl(aq)$ + 38,200 cal.

Sol. We aim for the reaction

$NH_3(aq) + HCl(aq) \rightarrow NH_4Cl(aq) + \Delta H$

To obtain this equation, rearrange the equation (i) and (ii) and add the equation (iii) to their sum

$NH_3(aq) \rightarrow NH_3(g) + aq$ - 8,400 cal.

$HCl(aq) \rightarrow HCl(g) + aq$ - 17,300 cal.

$NH_3(aq) + HCl(aq) \rightarrow NH_3(g) + HCl(g) + aq$ - 25700 cal

add $NH_3(g) + HCl(g) + aq \rightarrow NH_4Cl(aq)$ + 38200 cal.

$NH_3(aq) + HCl(aq) \rightarrow NH_4Cl(aq) + 12500$ cal
Thus, heat of neutralisation = 12500 cal

(5) If 500 calories of heat energy are added to a system and the system does 350 calories of work, on the surroundings, what is the energy change of the system ?

Sol. $\Delta E = Q + W = 500 + (-350) = 150$ calories

(6) Calculate the work done when 1.0 mole of water at 373K vaporizes against an atmospheric pressure of 1.0 atmosphere. Assume ideal gas behaviour.

Sol. $V = \frac{nRT}{P} = \frac{1.0 \times 0.0821 \times 373}{1.0} = 31.0$ lit.

$W = -P_{ext}. \Delta V = -1 \times 31.0 = -31.0$ lit-atmo.

$= -31 \times 101.3 = -3140.3$ J

(7) Calculate the work done in the reversible compression of 10.0g. of helium gas from an initial pressure of 1.00 atmos. to a final pressure of 5.00 atmos. at a constant temp of 27°C.

Sol. $W = nRT \times 2.303 \log \frac{P_2}{P_1}$

$= \frac{10}{4} \times 1.987 \times 300 \times 2.303 \log 5 = 2400$ cal.

(8) 5 moles of oxygen are heated at constant volume from 10°C to 20°C. What will be the change in the internal energy of the gas ? The molar heat of oxygen at constant pressure, $C_p = 7.03$ cal. mol^{-} deg^{-1}and R = 8.31 J mol^{-1} deg.$^{-1}$.

Sol. We know $C_p - C_v = R$

$C_v - C_p - R = 7.03 - 1.99 = 5.04$ cal. mol^{-1} deg.$^{-1}$

Heat absorbed by 5 moles of oxygen in heating from 10°C to 20°C = $5 \times C_v \times \Delta T = 5 \times 5.04 \times 10 = 252$ cal.

since gas is heated at constant volume, no external work is done, $W = 0 \quad \therefore \Delta E = Q + W = 252 + 0 = 252$ cal.

(9) Two moles of an ideal gas initially at 27°C and one atm. are compresed reversibly to half the initial volume. Calculate Q, W, ΔE and ΔH when the process is carried on isothermally.

Sol. $W = 2.303 \, nRT \log \frac{V_2}{V_1}$

$= 2.303 \times 2 \times 1.98 \times 300 \log \frac{1}{2} = -1914$ cal.

Since the process is being carried on isothermally i.e. no change of temp., $\therefore \Delta H = 0$ and $\Delta E = 0$

According to 1st law of thermodynamics

$Q = \Delta E + P\Delta V$, or $Q = \Delta E + W$

$\therefore W = -1914 + 0 = -1914$ cal.

(10) A gas contained in a cylinder fitted with a frictionless piston expands against a constant pressure 1 atmo. from a volume of 4 lit. to 14 litres. In doing so, it absorbs 800J thermal energy from surroundings. Determine ΔE for the process.

Sol. $\Delta E = Q + W$

$W = -P \times \Delta V = -1 \times 10 = -10$ lit.- atmos.

But 0.082 lit- atmos = 1.987 cal.

$W = \frac{10 \times 1.987}{0.082} = -242.3$ cals.$= 242.3 \times 4.184 = -1013.7$J

$\Delta E = 800 - 1013.7$ J $= -213.7$ J.

(11) 0.5 g. of benzoic acid was subjected to combustion in a bomb calorimeter. The temp. of the calorimeter system including water was found to rise by 0.5°C. Calculate the heat of combustion of methane at (i) constant volume and (ii) constant pressure. The thermal capacity of calorimeter system is 17.7 kJ K^{-1}.

Sol. Heat of combustin at constant volume,

$$\Delta E = \frac{\text{Heat capacity of}}{\text{calorimeter system}} \times \text{rise in temp.} \times \frac{\text{mol. mass of a comp.}}{\text{mass of a compd.}}$$

$= 17.7 \times 0.5 \times \frac{16}{0.16} = 885$

$\Delta E = -885$ k.J.mol^{-1}

$CH_4(g) + 2O_2(g) \rightarrow CO_2(g) + 2H_2O(l)$

$\Delta H = \Delta E + \Delta nRT$

$= -885 + (-2) \times 8.314 \times 10^{-3} \times 300$

$= -885 - 4.988 = -889.988$ k J mol^{-1}.

(12) Heat supplied to a Carnot engine is 453.6 k. cal. How much useful work can be done by the engine which works between 10°C and 100°C ?

Sol.

$\frac{W}{Q_2} = \frac{T_2 - T_1}{T_2} \quad \therefore W = Q_2 \times \frac{T_2 - T_1}{T_2}$

$= 1897.86 \times \frac{373 - 283}{373}$

$= \frac{1897.86 \times 90}{373} = 457.92$ k J.

(13) The standard molar enthalpy of formation of ammonia is - 46.2 k J mol^{-1}. Given $H_{H-H} = 435$ k J mol^{-1} and $H_{N-H} = 390$ k J mol^{-1}. Calculate the bond energy in N_2(g)

Sol. $N_2(g) + 3H_2(g) \rightarrow 2NH_3(g)$

sum of bond energies of reactants = $H_{N\equiv N} + 3 \times H_{H-H}$

$= H_{N\equiv N} + 3 \times 435 = H_{N\equiv N} + 1305$

sum of bond energies of products = $2 \times 3H_{N-H}$

$= 6 \, H_{N-H} = 6 \times 390 = 2340$ kJ

ΔH = sum of bond energies of reactants - sum of bond energies of products

$-92.4 = H_{N\equiv N} + 1305 - 2340$

$H_{N\equiv N} = -92.4 - 1305 + 2340 = 942.6$ k J mol^{-1}.

(14) Calculate the temperature at which liquid water will be in equilibrium with water vapour.

$\Delta H_{vap} = 40.73$ and $\Delta S_{vap} = 0.109$ k J/ mol/ k

Sol. $\Delta G = \Delta H - T\Delta S$

At equilibrium ΔG = 0

$0 = 40.73 - T \times 0.109$

$T = \frac{40.73}{0.109} = 373.6$ k

(15) Calculate ΔG for the isothermal expansion of 2.00 moles of an ideal gas at 25°C from a pressure $P_1 = 1$ atm. to $P_2 = 0.100$ atmos.

Sol. $\Delta G = nRT \ln \frac{P_2}{P_1}$

$= 2.303 \, nRT \log \frac{P_2}{P_1}$

$= 2.303 \times 2 \times 1.987 \log \frac{0.100}{1.00} = -2720$ cals.

(16) One mole of liquid water at 100°C is placed in a thermostat maintained at 25°C. Calculate the entropy change in the process. The molar heat capacity of liquid water is 18 cal. degree^{-1}.

Sol. Entropy change of water $\Delta S_1 = 2.303\ C_p \log \frac{T_2}{T_1}$

$= 2.303 \times 18 \log \frac{298}{373}$

$= -4.042$ cal. degree^{-1}

Entropy change in the thermostat $\Delta S_2 = \frac{Q}{T_1}$

$= 18 \times 1 \times (T_2 - T_1)$

$= 18 \times 1 \times 75 = 1350$ cals.

$\Delta S_2 = \frac{1350}{298} = 4.53$ cal / deg.

Net entropy change = 4.53 - 4.042 = 0.488 cal. deg^{-1}.

(17) The enthalpy and entropy changes for the reaction C(s) diamond + $O_2 \rightarrow CO_2(g)$ at 25°C and 1 atm. are -393.4 k J and 0.006 k J/K respectively. Is the conversion at room temp. a spontaneous process ?

Sol. $\Delta G = \Delta H - T\Delta S$

$= 393.4 - (298)(0.006) = -395.2$ kJ.

Since ΔG = -ve, the reaction is spontaneous

(18) The enthalpy change for the transition of liquid water to steam is 40.8 kJ mol^{-1} at 373K. Calculate $\Delta S°$ for the process.

Sol. $\Delta S_{vap} = \frac{\Delta H_{vap}}{T} = \frac{40.8 \times 1000}{373} = 109\ J\ K^{-1}\ mol^{-1}$.

(19) Calculate the standard entropy change and the free energy change when 1 mole of water is formed from its elements at 25°C. The standard entropy of H_2, O_2 and H_2O are 31.21, 49.0 and 16.73 cal.deg^{-1}mole^{-1} respectively.

Sol. $H_2(g) + \frac{1}{2}O_2(g) \rightarrow H_2O(l)$

$\Delta S°_f = S°_{H_2O} - S°_{H_2} - \frac{1}{2}S°_{O_2}$

$= 16.73 - 31.21 - 24.50 = -38.98$

$\Delta G = \Delta H - T\Delta S = -68{,}320 - 298(-38.98) = -56{,}700$ cal mol^{-1}.

(20) ΔH and ΔS for the system $H_2O\ (l) \rightleftharpoons H_2O\ (g)$ at 1 atmos pressure are 40.63 kJ mol^{-1} and 108.8 J k^{-1} mol^{-1} respectively. Calculate the temp. at which the rates of forward and backward reactions will be the same. Predict the sign of free energy for this transformation above this temperature.

Sol. $\Delta G = \Delta H - T\Delta S$

$O = 40.63 - T \times 0.1088$

$T = \frac{40.63}{0.1088} = 373.4$ k

$\Delta G = \Delta H - T\Delta S$

$= 40.63 - 374 \times 0.1088 = 40.63 - 40.69 = -0.06$ kJ.

ΔG is negative, hence, the reaction will be spontaneous.

(21) Calculate $\Delta G°$ of the following reaction

$CO(g) + \frac{1}{2}O_2(g) \rightarrow CO_2(g)\ \Delta H° = -282.84$ kJ

given $S°_{CO_2(g)} = 213.8$ Jk^{-1}mol^{-1}, $S°_{CO(g)} = 197.9$ Jk^{-1}mol^{-1}, $S°_{O_2(g)} = 205.0$ Jk^{-1}mol^{-1}.

Sol. $\Delta S° = \Sigma S°_{products} - \Sigma S°_{reactants}$

$= 213.8 - 197.9 + \frac{1}{2} \times 205.0 = -86.6\ JK^{-1} = -0.0866\ kJK^{-1}$.

$\Delta G = \Delta H - T\Delta S$ (T=25°C = 298 K as the values of $\Delta H°$ & $\Delta S°$ are at standard conditions = -282.84 - 298 x -0.0866 $\doteq$ -282.84 - 25.8 = -308.64 kJ

(22) The standard voltage of the cell

$2H_2(g) + O_2(g) \rightleftharpoons 2H_2O(l)$ is 1.23 V at 25°C. The standard enthalpy of the reaction is -571.6 kJmol^{-1}.

(i) Calculate the $\Delta G°$, $\Delta S°$ and k_{eq} values of the reaction.

Sol. $2H^+(aq) + 2e^- \rightarrow H_2(g)$ ---- anode reaction

$2H_2O(l) \rightarrow O_2(g) + 4H^+(aq) + 4e^-$ ----- cathode reaction

for balancing purpose, we have to multiply the anode reaction by 2

(i) $\Delta G° = -nFE° = 4 \times 96485 \times 1.23 = 474706$ J

(ii) $\Delta G° = \Delta H° - T\Delta S°$

$\Delta S° = \frac{\Delta H° - \Delta G°}{T} = \frac{-571.6 \times 10^3 - (-474706)}{298} = 325.1\ Jk^{-1}$.

(iii) $\Delta G° = -2.303\ RT \log k_{eq}$

$\log k_{eq} = \frac{-\Delta G°}{2.303RT} = \frac{-(-474706)}{2.303 \times 8.314 \times 298} = 83.2$

k = Antilog of 83.2 = 1.28×10^{83}.

(23) Calculate the free energy change accompanying the compression of 1 mole of a gas at 57°C from 25 to 200 atmospheres. The fugacities of the gas at 57°C may be taken as 23 and 91 atmospheres, respectively, at pressures of 25 and 200 atmospheres.

Sol.: we know $\Delta G = nRT \times 2.303 \log \frac{P_2}{P_1}$

$= 1 \times 8.314 \times 330 \times 2.303 \log \frac{200}{25}$

$= 5702.8$ J

for more accurate value; we should use equation involving fugacities i.e. $\Delta G = nRT \times 2.303 \log \frac{f_2}{f_1}$

$= 1 \times 8.314 \times 330 \times 2.303 \log \frac{91}{23}$

$= 3774.0$ J

(24) Equilibrium constant for the reaction

$PCl_5(g) \rightleftharpoons PCl_3(g) + Cl_2(g)$ at 298 k is 1.8×10^{-7}.

Calculate the value for $\Delta G°$.

Sol. we know $\Delta G° = -2.303\ RT \log k_{eq}$

$= -2.303 \times 8.31 \times 298 \times \log(1.8 \times 10^{-7})$

$= -2.303 \times 8.31 \times 298 \times -6.9664$

$= 39.73$ k J mol^{-1}.

(25) Calculate the equilibrium constant for the reaction $2NOCl(g) \rightleftharpoons 2NO(g) + Cl_2(g)$ at 400k if $\Delta H° = 77.2$ kJmol^{-1} and $\Delta S° = 122$ J K^{-}.

Sol. $\Delta G° = \Delta H - T\Delta S$

$= 77200 - 400 \times 122 = 28400$ J mol^{-1}

we know $\Delta G = 2.303\ RT \log k_{eq}$.

$\log k_{eq} = \frac{-\Delta G}{2.303RT} = \frac{-28400}{2.303 \times 8.31 \times 400}$

$\log k_{eq} = -3.70$

k_{eq} = Antilog (-3.70) = 1.995×10^{-4}

(26) The activity of 2.5 moles of a substance changes from 0.05 to 0.35. What would be the change in its free energy at 27°C ?

Sol.: we know that $\Delta G = nRT \times 2.303 \log \frac{a_2}{a_1}$

$= 2.5 \times 8.314 \times 300 \times \log \frac{0.35}{0.05} = 12133.6$ J

1. If ΔH is the change in enthalpy and ΔE, the change in the internal energy accompanying a gaseous reaction [CBSE-90]

(A) ΔH is always greater than ΔE

(B) ΔH< ΔE only if the number of moles of the products is greater than the number of the reactants.

(C) ΔH is always less than ΔE

(D) ΔH < ΔE only if the number of moles of products is less than the number of moles of the reactants.

2. Given that for the reaction [CBSE-90]

$C_2H_4(g) + 3O_2(g) \longrightarrow 2CO_2(g) + 2H_2O$ (l) at 298°K

ΔE = –1415kJ

If R = 0.0084 kJ per degree, ΔH =

(A) –1400 kJ (B) –1420 kJ

(C) –1410 kJ (D) –1430 kJ

3. Given that [CBSE-90]

$2C(s) + 2O_2(g) \longrightarrow 2CO_2(g)$, ΔH = –787kJ

$H_2(g) + 1/2O_2(g) \longrightarrow H_2O(l)$, ΔH = – 286kJ

$C_2H_2(g) + 2½O_2(g) \longrightarrow 2CO_2(g) + H_2O(l)$, ΔH = –1310 kJ

Heat of formation of acetylene is

(A) +1802 kJ (B) –1802kJ (C) –800kJ (D) +237kJ

4. 1 mole of ice is converted into liquid state at 273K. $H_2O(s)$ and $H_2O(l)$ have entropies 38.20 and 60.01 J $mole^{-1}$ deg^{-1}. Enthalpy change in conversion will be [CBSE-90]

(A) 59.85 J/mole (B) 5958.5 J/mole

(C) 595.85 J/mole (D) 59585.0 J/mole

5. Absolute zero is defined as [CBSE-90]

(A) the temperature at which all molecular motion ceases

(B) at which liquid helium boils

(C) at which water boils at 298 K (D) All of the above

6. Heat absorbed or evolved in a reaction at constant temperature and volume is [CBSE-90]

(A) ΔH (B) ΔE (C) P ΔV (D) None of these

7. If the system undergoes contraction of volume the work done by system is [CBSE-90]

(A) positive (B) negative (C) negligible (D) zero

8. An aqueous solution whose pH = 0 is [CBSE-90]

(A) acidic (B) alkaline (C) neutral (D) amphoteric

9. Based on the following thermochemical equations

$H_2O(g) + C(s) \rightarrow CO(g) + H_2(g)$; ΔH = 131kJ

$CO(g) + 1/2\ O_2(g) \rightarrow CO_2(g)$; ΔH = –282 kJ

$H_2(g) + 1/2O_2(g) \rightarrow H_2O(g)$; ΔH = –242 kJ

$C(s) + O_2(g) \rightarrow CO_2(g)$; ΔH = x kJ

The value of x will be [CBSE-92]

(A) –393 kJ (B) –655kJ (C) 393kJ (D) 655kJ

10. Which of the following is not a state function of thermodynamic system ?

[CBSE-93]

(A) Work (B) Enthalpy
(C) Free energy (D) Internal Energy

11. Consider the following reaction occurring in an automobile: [CBSE-94]
$2\ C_8H_{18}(g) + 25O_2(g) \rightarrow 16\ CO_2(g) + 18H_2O(g)$
The sign of ΔH, ΔS and ΔG would be
(A) +, –, + (B) –, +, – (C) –, +, + (D) +, +, –

12. The latent heat of vaporisation of a liquid at 500 K and 1 atm pressure is 10.0 Kcal/mol. What will be the change in internal energy (ΔU) of 3 moles of liquid at the same temperature and pressure ? [CBSE-94]
(A) 13.0 kcal/mol (B) –13.0 kcal/mol
(C) 7.0 kcal/mol (D) –7.0 kcal/mol

13. Normal boiling point of water is 373 K (at 760mm Hg),. Vapour pressure of water at 298 K is 23 mm Hg. If the enthalpy of evaporation is 40.656kJ/mol, the boiling point of water at 23 mm Hg pressure will be [CBSE-95]
(A) 25 K (B) 75.7 K (C) 51.6 K (D) 12.5 K

14. For a reaction to occur spontaneously [CBSE-95]
(A) (Δ H-TΔS) must be negative
(B) (Δ H+TΔS) must be negative.
(C) Δ H must be negative (D) ΔS must be negative

15. The standard state Gibbs Free energy change for the isomerization reaction
cis. 2 pentene $\rightarrow$ trans-2 pentene
is -3.67kJ/mol at 400K. If more trans-2 pentene is added to the reaction vessel [CBSE-95]
(A) more cis-2 pentene is formed.
(B) equilibrium is shifted in the forward direction.
(C) equilibrium remains unaffected.
(D) additional trans-2 pentene is formed.

16. If enthalpies of formation for $C_2H_4(g)$, $CO_2(g)$ and $H_2O(l)$at 25°C and 1 atm pressure be 52 , -394 and 286 kJ mol^{-1} respectively, enthalpy of combustion of C_2H_4 (g) will be [CBSE-95]
(A) +1412 kJ mol^{-1} (B) -1412 kJ mol^{-1}
(C) +14.2 kJ mol^{-1} (D) -141.2 kJ mol^{-1}

17. If one end of a piece of a metal is heated the other end becomes hot after some time. This is due to [CBSE-95]
(A) energised electrons moving to the other part of the metal
(B) resistance of the metal
(C) mobility of atoms in the metal
(D) minor perturbation in the energy of atoms.

18. Which of the following is the correct equation ? [CBSE-96]
(A) $\Delta U = \Delta Q - W$ (B) $\Delta W = \Delta U + \Delta Q$
(C) $\Delta U = \Delta W + \Delta Q$ (D) None of these

19. Given that [CBSE-97]
$C + O_2 \rightarrow CO_2,\ \Delta H° = -x$ kJ
$2CO + O_2 \rightarrow 2CO_2,\ \Delta H° = -y$ kJ
The enthalpy of formation of carbon monoxide will be
(A) Y – 2x (B) $\frac{2x-y}{2}$ (C) $\frac{y-2x}{2}$ (D) 2x – y

20. Identify the correct statement regarding entropy [CBSE-98]
(A) At absolute zero of temp., the entropy of perfectly crystalline substance is +ve
(B) At absolute zero entropy of perfectly crystalline substance is taken to be zero.
(C) At 0°C the entropy of a perfectly crystalline substance is taken to be zero
(D) At absolute zero of temperature, the entropy of all crystalline substance is to be zero.

21. In an endothermic reaction, the value of ΔH is [CBSE-99]
(A) Zero (B) Positive (C) Negative (D) Constant

22. In the reaction : $S + 3/2\ O_2 \rightarrow SO_3 + 2x$ kcal and $SO_2 + 1/2\ O_2 \rightarrow SO_3 + y$ kcal, the heat of formation of SO_2 is [CBSE-99]
(A) (x + y) (B) (x - y) (C) (2x + y) (D) (2x - y)

23. Which of the following relations holds good to the laws of thermodynamics for the reaction? [CBSE-2000]
$C_2H_4(l) + 3O_2(g) \longrightarrow 2CO_2(g) + 2H_2O(l)$
(A) $\Delta H = \Delta E + RT$ (B) $\Delta H = \Delta E - RT$
(C) $\Delta H = \Delta E + 2RT$ (D) $\Delta H = \Delta E - 2RT$

24. For reaction to proceed spontaneously [CBSE-2000]
(A) ΔH is +ve (B) ΔE is +ve
(C) ΔG is –ve (D) G is +ve

25. For the melting of a solid at 25°C, the fusion process requires energy equivalent to 2906 Joules to be added to system. Considering the process to be reversible at fusion point, what is the entropy value of process ? [CBSE-2000]
(A) 9.76 J/K (B) 11.272 J/K
(C) 2.33 J/K (D) Insufficient data

26. Find out ΔG for the following reaction.
$2ZnS + 3O_2 \rightarrow 2ZnO + 2SO_2\ \Delta G = ?$
Data given :
$2Zn + O_2 \rightarrow 2ZnO\ \Delta G = -616$ J
$2Zn + S_2 \rightarrow 2ZnS\ \Delta G = -293$ J
$S_2 + 2O_2 \rightarrow 2SO_2\ \Delta G = -408$ J [CBSE-2000]
(A) –731 J (B) –1317 J (C) –501 J (D) +731 J

27. Change in enthalpy for reaction - $2H_2O_2(l) \rightarrow 2H_2O(l) + O_2(g)$, if heats of formation of $H_2O_2(l)$ are -188 and -288 kJ/mol respectively is [CBSE-2001]

(A) -196 kJ/mol (B) +196 kJ/mol
(C) +948 kJ/mol (D) -948 kJ/mol

28. When 1 mole of a gas is heated at constant volume, the temperature is raised from 298 to 308K. heat supplied to the gas is 500J. Then which statement is correct [CBSE-2001]

(A) $q = W = 500$ J, $\Delta U = 0$
(B) $q = \Delta U = 500$ J, $W = 0$
(C) $q = W = 500$ J, $\Delta U = 0$
(D) $\Delta U = 0$, $q = W = -500$ J

29. Enthalpy of $CH_4 + \frac{1}{2} O_2 \rightarrow CH_3OH$ is negative. [CBSE-2001]

If enthalpies of combustion of CH_4 and CH_3OH are x and y respectively then which relation is correct ?
(A) $x > y$ (B) $x < y$ (C) $x = y$ (D) $x \geq y$

30. Unit of entropy is : [CBSE-2002]
(A) $JK^{-1} mol^{-1}$ (B) $J mol^{-1}$ (C) $J^{-1}K^{-1}mol^{-1}$ (D) $JK mol^{-1}$

31. In a closed insulated container a liquid is stirred with a paddle to increase the temperature which of the following is true [CBSE-2002]
(A) $\Delta E = W \neq 0, q = 0$ (B) $\Delta E = W = q \neq 0$
(C) $\Delta E = 0, W = q \neq 0$ (D) $W = 0 , \Delta E = q \neq 0$

32. 2 moles of ideal gas at 27°C temp. is expanded reversibly from 2 lit. to 20 lit. Find entropy change (R = 2 cal/mol K) [CBSE-2002]
(A) 92.1 (B) 0 (C) 4 (D) 9.2

33. Heat of combustion $\Delta H°$ for C(s), $H_2(g)$ and $CH_4(g)$ are -94, -68 and -213 kcal/mol, then $\Delta H°$ for $C(s) + 2H_2(g) \rightarrow CH_4(g)$ is [CBSE-2002]
(A) -17 kcal (B) -111 kcal
(C) -170 kcal (D) -85 kcal

34. The molar heat capacity of water at constant pressure, C_P is 75 $JK^{-1} mol^{-1}$. When 1.0 kJ of heat is supplied to 100g of water which is free to expand, the increase in temperature of water is [CBSE-2003]
(A) 6.6 K (B) 1.2 K (C) 2.4 K (D) 4.8 K

35. What is the entropy change (in $JK^{-1} mol^{-1}$) when one mole of ice is converted into water at 0°C ? (The enthalpy change for the conversion of ice to liquid water is 6.0 KJ mol^{-1} at 0°C) [CBSE-2003]
(A) 21.98 (B) 20.13 (C) 2.013 (D) 2.198

36. The densities of graphite and diamond at 298K are 2.25 and 3.31 g cm^{-3}, respectively. If the standard free energy difference ($\Delta G°$) is equal to 1895 J mol^{-1}, the pressure at which graphite will be transformed into diamond at 298 K is [CBSE-2003]
(A) 9.92×10^5 Pa (B) 9.92×10^8 Pa
(C) 9.92×10^7 Pa (D) 9.92×10^5 Pa

37. For which one of the following equatiions is $\Delta H°_{react}$ equal to $\Delta H°_f$ for the product ? [CBSE-2003]
(A) $2 CO(g) + O_2(g) \rightarrow 2CO_2 (g)$
(B) $N_2(g) + O_3(g) \rightarrow N_2O_3(g)$
(C) $CH_4(g) + 2Cl_2(g) \rightarrow CH_2Cl_2(l) + 2HCl(g)$
(D) $Xe(g) + 2F_2(g) \rightarrow XeF_4(g)$

38. For the reaction $C_3H_8 (g) + 5O_2(g) \rightarrow 3CO_2(g) + 4H_2O(l)$ at constant temperature, $\Delta H - \Delta E$ is [CBSE-2003]
(A) – RT (B) +RT (C) – 3RT (D) + 3RT

39. Iron forms an oxide Fe_2O_3 according to the equation, $2Fe + \frac{3}{2} O_2 \rightarrow Fe_2O_3$ $\Delta H = -822$ kJ mol^{-1}
What is the ΔH, if one mole of oxygen was formed by the decomposition of Fe_2O_3 ? [AIIMS-93]
(A) +548 kJ (B) –548 kJ (C) –411 kJ (D) +411 kJ

40. The heat liberated when 1.89 g of benzoic acid is burnt in a bomb clarorimeter at 25°C increases the temperature of 18.94 kg of water by 0.632°C. If the specific heat of water at 25°C is 0.9998 cal/g-deg, the value of the heat combustion of benzoic acid is [AIIMS-94]

(A) 771.4 kcal (B) 871.2 kcal
(C) 881.1 kcal (D) 981.1 kcal

41. Which of the following statement regarding entropy is correct ? [AIIMS-94]
(A) At 0°C, the entropy of all crystalline substances is taken to be zero
(B) At 0°C, the entropy of perfect crystalline substances is taken to be zero
(C) At absolute zero temperature, the entropy of all crystalline substances is taken to be zero.
(D) At absolute zero temperature, entropy of a perfect crystalline substance is taken to be zero.

42. Assertion : Enthalpy of graphite is lower than that of diamond. Reason : Entropy of graphite is greater than that of diamond. [AIIMS-95]
(A) A (B) B (C) C (D) D

43. Assertion : The enthalpy of formation of gaseous oxygen molecules at 298 K under a pressure of one atm is zero. Reason : The entropy of formation of gaseous oxygen molecules under the same conditions is zero. [AIIMS-96]

(A) A (B) B (C) C (D) D

44. The enthalpy change of a reaction does not depend on [AIIMS-97]

(A) The state of reactants and products.
(B) Nature of reactants and products
(C) Different intermediate reaction
(D) Initial and final enthalpy change of a reaction

45. $S + \frac{3}{2}O_2 \rightarrow SO_3 + 2x$ kcal
$SO_2 + \frac{1}{2} O_2 \rightarrow SO_3 + y$ kcal.
Find out the heat of formation of SO_2 [AIIMS-97]
(A) (2x−y) (B) (2x+y) (C) (x+y) (D) 2x/y

46. Assertion : Decrease in free energy causes spontaneous reaction. Reason :Spontaneous reaction are invariably exothermic reaction. [AIIMS-97]
(A) A (B) B (C) C (D) D

47. If enthalpies of formation for $C_2H_4(g)$, $CO_2(g)$ and H_2O (l) at 25°C and 1 atm pressure are 52, −394 and −286 kJ/mol respectively, the enthalpy of combustion of $C_2H_4(g)$ will be [AIIMS-98]
(A) +14.2 kJ/mol (B) +1412 kJ/mol
(C) −141.2 kJ/mol (D) −1412 kJ/mol

48. In an endothermic reaction, the value of change in enthalpy (ΔH) is [AIIMS-98]
(A) Positive (B) Negative
(C) Zero (D) Either (B) or (C)

49. Hess' law is applicable for the determination of heat of [AIIMS-98]
(A) Reaction (B) Formation
(C) Transition (D) All of these

50. Assertion : Heat energy is completely transformed into work during the isothermal expansion of a gas.
Reason : During an isothermal process, the changes in internal energy of a gas due to decrease in pressure is nullified by the changes due to increase in volume. [AIIMS-98]
(A) A (B) B (C) C (D) D

51. Assertion : The temperature of a gas doesnot change, when it undergoes an adiabatic expansion. Reason : During an adiabatic process, heat energy is not exchanged between a system and its surroundings. [AIIMS-98]
(A) A (B) B (C) C (D) D

52. Internal energy does not include [AIIMS-99]
(A) Vibrational energy (B) Rotational energy
(C) Nuclear energy
(D) Energy arising by gravitational pull

53. Which of the following equations does not obey the law of conservation of mass ? [AIIMS-99]
(A) $H_2 + O \rightarrow 2H_2O$ (B) $2H + O_2 \rightarrow H_2O$
(C) $2H + O_2 \rightarrow 2H_2O$ (D) Both (A) and (B)

54. C (Graphite) + $O_2 \rightarrow CO_2$, $\Delta H = -a$,
$2CO + O_2 \rightarrow 2CO_2$, $\Delta H = -b$. Determine heat of formation of CO : [AIIMS-2001]
(A) $\frac{b-a}{2}$ (B) $\frac{b-2a}{2}$ (C) $\frac{a-b}{2}$ (D) $\frac{a-2b}{2}$

55. Assertion (A) : Mass and volume are extensive properties.Reason (R) : Mass/Volume is also an extensive parameter. [AIIMS-2002]
(A) A (B) B (C) C (D) D

56. Assertion (A) : Absolute value of internal energy of substances can not be determined. Reason (R) : It is impossible to determine exact values of constituent energies of the substances. [AIIMS-2002]
(A) A (B) B (C) C (D) D (E) E

57. Assertion (A) : During an adiabatic process, heat energy is not exchanged between system and its surroundings. Reason (R) : The temperature of a gas increases when it undergoes an adiabatic expansion.
[AIIMS-2002]
(A) A (B) B (C) C (D) D (E) E

58. One gram sample of NH_4NO_3 is decomposed in a bomb calorimeter. The temperature of the calorimeter increases by 6.12 K. The heat capacity of the system is 1.23 kJ/g/deg. What is the molar heat of decomposition for NH_4NO_3 ? [AIIMS-2003]
(A) −7.53 kJ/mol (B) −398.1 kJ/mol
(C) −16.1 kJ/mol (D) −602 kJ/mol

59. Which one of the following has ΔS^o greater than zero ? [AIIMS-2003]
(A) $CaO(s) + CO_2(g) \rightleftharpoons CaCO_3(s)$
(B) $NaCl(aq) \rightleftharpoons NaCl(s)$
(C) $NaNO_3(s) \rightleftharpoons Na^+(aq) + NO_3^-(aq)$
(D) $CaO(s) + CO_2(g) \rightleftharpoons CaCO_3(s)$

60. (A) The increase in internal energy (ΔE) for the vaporization of one mole of water at 1 atm and 373K is zero.
(R) For all isothermal processes $\Delta E = 0$.
[AIIMS-2003]
(A) (B) (C) (D)

61. (A) ΔH and ΔE are aimost the same for the reaction
$N_2(g) + O_2(g) \rightleftharpoons 2NO(g)$
(R) All reactants and products are gases.
[AIIMS-2003]
(A) (B) (C) (D)

62. (A) Temperature near the sea coast is moderate.
(R) Water has a high thermal conductivity.
[AIIMS-2003]
(A) (B) (C) (D)

63. An endothermic reaction is one in which :
[JIPMER-91]

(A) Heat is converted into electricity
(B) Heat is absorbed (C) Heat is given out
(D) Heat is converted into mechanical work

64. The enthalpy of solution of anhydrous $CuSO_4$ and $CuSO_4.5H_2O$ are –15.89 and 2.80 kcal mol^{-1} respectively. The enthalpy of hydration of $CuSO_4$ to $CuSO_4.5H_2O$ is : [JIPMER-93]
(A) –18.69 kcal (B) 13.09 kcal
(C) –13.09 kcal (D) 18.69 kcal

65. The enthalpy of vaporization for water is 41.2 x kJ mol^{-1}, ΔH for the reaction, $2H_2O(g) \rightarrow 2H_2O(l)$ is : [JIPMER-93]
(A) –82.4 kJ (B) –20.6 kJ (C) 82.4 kJ (D) 2.57 kJ

66. The difference between heats of reaction at constant pressure and constant volume for the reaction :
$2C_6H_6(l) + 15O_2(g) \rightarrow 12CO_2(g) + 6H_2O(l)$ at 27°C is : [JIPMER-93]
(A) –7.48 kJ (B) 3.74 kJ (C) –3.74 kJ. (D) 7.48 kJ

67. Which of the following is not correct for ideal solution ? [JIPMER-95]
(A) ΔS mixing = 0 (B) ΔV mix = 0
(C) ΔH mix = 0 (D) It obeys Raoult's law

68. If bond energies of N = N, H–H and N–H bonds are 945, 437 and 389 kJ respectively, then ΔH for the following gaseous reaction $N_2 + 3H_2 \rightarrow 2NH_3$ is [JIPMER-95]
(A) –156kJ (B) –78kJ (C) –1478kJ (D) +1478 kJ

69. Consider the reactions :
$C(s) + 2H_2(g) \rightarrow CH_4(g)$, ΔH = –x kcal
$C(g) + 4H(g) \rightarrow CH_4(g)$, ΔH = –x kcal
$CH_4(g) \rightarrow CH_3(g) + H(g)$, ΔH = –y kcal
The bond energy of C–H bond is [JIPMER-95]
(A) y kcal mol^{-1} (B) x/4 kcal mol^{-1}
(C) x_1 kcal mol^{-1} (D) x_1/4 kcalmol^{-1}

70. Which of the following quantities are equal at equilibrium under the particular set of conditions : [JIPMER-96]
(A) ΔH and ΔG (B) ΔH and ΔS
(C) ΔG and ΔS (D) none of the above

71. A chemical reaction will not occur at all if the conditions are chosen so that : [JIPMER-96]
(A) Enthalpy change and entropy change are both negative (B) Enthalpy change and entropy change are both positive (C) Enthalpy change is negative and entropy change is positive
(D) Enthalpy change is positive and entropy change is negative

72. For an endothermic reaction carried out in a closed vessel : [JIPMER-96]
(A) ΔH > 0 (B) ΔH < 0 (C) ΔH = Δ E (D) ΔH = 0

73. (i) $Na_2S + 2HCl \longrightarrow 2NaCl + H_2S$; $\Delta H = -X_1$ kJ
(ii) $H_2(g) + S(s) \longrightarrow H_2S(g)$; $\Delta H = -X_2$ kJ
(iii) $Mg(HS)_2 + 2H_2O \longrightarrow Mg(OH)_2 + 2H_2S$; $\Delta H = -X_3$ kJ
The value of ΔH_f of hydrogen sulphide is : [JIPMER-96]
(A) $-X_3$ kJ mol^{-1} (B) $-X_1$kJ mol^{-1}
(C) $-X_2$ kJ mol^{-1} (D) none of these

74. Calculate the heat of formation of $PCl_5(s)$ from the following data : [JIPMER-98]
$2P(s) + 3Cl_2(g) \rightarrow 2PCl_3(l)$; ΔH = –151.8kcal
$PCl_3(l) + Cl_2(g) \rightarrow PCl_5(s)$; ΔH = –32.8kcal
(A) –108.7 kcal (B) +108.7 kcal
(B) –184.6 kcal (D) + 184.6 kcal

75. Which of the following statements is wrong ? The enthalpy of a reaction depends upon [JIPMER-99]
(A) The physical state of the reactants and products
(B) The temperature of the reaction
(C) The number of moles of reactants
(D) The path by which the reactants are converted into products

76. Joule-Thomson expansion is [JIPMER-2000]
(A) Iosbaric (B) Isoenthalpic
(C) Isothermal (D) Isoentropic

77. In an exothermic reaction, the reactants [JIPMER-2001]
(A) Have more energy than the products
(B) Are at a higher temperature than the products
(C) Have less energy than the products
(D) Have the same energy as the products

78. If
$C_6H_6(l) + \frac{15}{2}O_2(g) \rightarrow 3H_2O(l) + 6CO_2(g)$; ΔH = –3264.6kJ
Heat evolved when 39g of C_6H_6 are burnt is [JIPMER-2001]
(A) 2448.45 kJ (B) 6528.2 kJ
(C) 1632.3 kJ (D) 816.15 kJ

79. Enthalpy of neutralization of NH_4OH and HCl is numerically. [JIPMER-2002]
(A) 57.1 kJ mol^{-1} (B) < 57.1 KJ mol^{-1}
(C) > 57.1 kJ mol^{-1} (D) Zero

80. For the reaction $C + O_2 \rightarrow CO_2$ [BHU-95]
(A) ΔH > ΔE (B) ΔH < ΔE
(C) ΔH = ΔE (D) Either 'A' or 'B'

81. For a process to be spontaneous [BHU-95]
(A) ΔG should be –ve (B) ΔG should be +ve
(C) ΔH should be –ve (D) ΔH should be +ve

82. $C + O_2 \rightarrow CO_2$; ΔH = X
$CO + \frac{1}{2}O_2 \rightarrow CO_2$; ΔH =Y, then the heat of formation of CO is [BHU-97]
(A) X–Y (B) Y–2X (C) X+Y (D) 2X–Y

83. Gibbs free energy for reaction at equilibrium is [BHU-97]

(A) Zero (B) –ve (C) +ve (D) +ve or –ve

84. The enthalpy of neutralization of HCN by NaOH is –12.13 kJ mol^{-1}. The enthalpy of ionisation of HCN will be [BHU-97]

(A) 4.519 kJ (B) 45.10 kJ (C) 451.9 kJ (D) 45.19 kJ

85. The values of heat of formation of SO_2 and SO_3 are –298.2 kJ and –98.2 kJ. The heat of reaction of this reaction will be $SO_2 + \frac{1}{2}O_2 \rightarrow SO_3$ [BHU-97]

(A) –200 kJ (B) –356.2 kJ (C) +200kJ (D) –396.2 kJ

86. Calculate the standard heat of formation of carbon disulphide (l) given that the standard heat of combustion of carbon (s), sulphur (s) and carbon disulphide (l) are –393.3, –293.72 and –1108.76 kJ mol^{-1} respectively. [BHU-97]

(A) –128.02 kJ mol^{-1} (B) +12.802 kJ mol^{-1}
(C) +128.02 kJ mol^{-1} (D) –12.802 kJ mol^{-1}

87. The enthalpy of water is 386 kJ. What is entropy of water ? [BHU-97]

(A) 0.5 kJ (B) 1.03 kJ (C) 1.5 kJ (D) 22.05 kJ

88. If 900 J/g of heat is exchanged at boiling point of water, then what is increase in entropy ? [BHU-98]

(A) 43.4 J/mole (B) 87.2 J/mole
(C) 900 J/mole (D) Zero

89. Heat exchanged in a chemical reaction at constant temperature and constant pressure is called [BHU-98]

(A) Internal energy (B) Enthalpy
(C) Entropy (D) Free energy

90. The values of heat of formation of SO_2 and SO_3 are –298.2 kJ and –98.2 kJ. The heat of formation of the reaction $SO_2 + (1/2)O_2 \rightarrow SO_3$ will be [BHU-99]

(A) –200 kJ (B) –356.2 kJ
(C) +200 kJ (D) –396.2 kJ

91. Standard state Gibbs free energy change for isomerization reaction cis-2- pentene $\rightleftharpoons$ trans-2- pentene is –3.67 kJ/mol at 25 K. If more trans-2- pentene is added to the reaction vessel, then [BHU-99]

(A) More cis-2-pentene is formed
(B) Additional trans-2-pentene is formed
(C) Equilibrium remains unaffected
(D) Equilibrium is shifted in forward direction

92. The relationship which describes the variation of vapour pressure with temperature is called [BHU-2000]

(A) Hess' Law (B) Arrhenius equation
(C) Kirchhoff's law (D) Clausius-Clapeyron equation

93. In the given reaction $K_2Cr_2O_7 + XH_2SO_4 + YSO_2 \rightarrow K_2SO_4 + Cr_2(SO_4)_3 + ZH_2O$, X, Y, Z, are [BHU-2000]

(A) 1 , 3 , 1 (B) 4 , 1 , 4 (C) 3 , 2 , 3 (D) 2 , 1 , 2

94. Given :

$C(s) + O_2(g) \rightarrow CO_2(g) + 94.2$ kcal
$H_2(g) + 1/2O_2(g) \rightarrow H_2O(l) + 68.3$ kcal
$CH_4(g) + 2O_2(g) \rightarrow CO_2(g) + 2H_2O(l) + 210.8$ kcal

The heat of formation of methane in kcal will be [BHU-2001]

(A) 45.9 (B) 47.8 (C) 20.0 (D) 47.3

95. For a reaction to occur spontaneously [BHU-2001]

(A) ΔH and ΔS must be negative with $T\Delta S > \Delta H$
(B) ΔH must be negative and ΔS + ve
(C) ΔS must be negative and ΔH + ve
(D) ΔH and ΔS both +ve, $\Delta H > T\Delta S$

96. One molecule of alkane will require how many molecules of oxygen for complete combustion ? [BHU-2002]

(A) (n + 1) x 2 (B) (3n + 1)/2
(C) (3n + 1) (D) (n + 1)/2

97. For a reversible process at equilibrium, the change in entropy many be expressed as [BHU-2002]

(A) $\Delta S = Tq_{rev}$ (B) $\Delta S = \frac{\Delta H}{T}$
(C) $\Delta S = \frac{q_{rev}}{T}$ (D) $\Delta S = T\Delta H$

98. For reaction to be non-spontaneous at all temperature, the signs of enthalpy; entropy and free-energy changes must be [BHU-2002]

(A) –, + and – (B) –, – and +
(C) +, – and + (D) +, + and –

99. Mark the incorrect statement [BHU-2002]

(A) The spontaneous process always accompanies the decrease in free energy.
(B) Entropy is a measure of disorder.
(C) Entropy is expressed in the units of JK^{-1} mol^{-1}.
(D) Entropy is not a state function.

100. Calculate the standard enthalpy change (ΔH^o) for the reaction $OF_2(g) + H_2O(g) \rightarrow O_2(g) + 2HF(g)$ given that standard enthalpies of formation ΔH_f^o of $OF_2(g)$, $H_2O(g)$ and HF(g) are +23.0, –241.8 and –268.6 kJ $mole^{-1}$ respectively [BHU-2002]

(A) –318.4 kJ (B) 318.4 kJ (C) –418.4 kJ (D) 418.4 kJ

101. Efficiency of machine A is 80% and that of B is 70%. Which will be having more frictional resistance ? [AFMC-91]

(A) B (B) A (C) Both (D) None of these

102. Heat of combustion is always [AFMC-91]

(A) Positive (B) Negative
(C) Neutral (D) All of the above

103. Hess's law of heat summation includes [AFMC-92]
(A) Initial reactants only
(B) Initial reactants and final products
(C) Final products only (D) Intermediates only

104. Which one of the following is endothermic ? [AFMC-92]
(A) $2NO + O_2 \rightarrow NO_2$
(B) $CH_3Br + NaOH \rightarrow CH_3OH + HBr$
(C) $CH_4 + 2O_2 \rightarrow CO_2 + 2H_2O$
(D) $2CH_3OH + O_2 \rightarrow 2HCHO + 2H_2O$

105. The oxidation of benzene releases 718 k.cal. If the mass of benzene formed is 39 gm, then the heat of formation of benzene is [AFMC-92]
(A) 718 kcal (B) 1359 kcal (C) 309 kcal (D) 1436 kcal

106. For an exothermic reaction to be spontaneous [AFMC-93]
(A) Temperature should be very high
(B) Temperature is zero
(C) Temperature should be low
(D) Temperature has no impact

107. At equilibrium ΔG is [AFMC-94]
(A) Positive (B) Negative
(C) Zero (D) None of the above

108. Which of the following is correct relationship [AFMC-97]
(A) $\Delta G = \Delta H + T \Delta S$ (B) $\Delta H = \Delta G - T \Delta S$
(C) $\Delta G = \Delta H - T \Delta S$ (D) $\Delta G + \Delta H = T \Delta S$

109. First law of thermodynamics is represented by the equation [AFMC-97]
(A) $\Delta E = Q + W$ (B) $\Delta E = Q - W$
(C) $W = Q + \Delta E$ (D) None of these

110. When enthalpy and entropy change for a chemical reaction are -2.5×10^3 cals and 7.4 cals deg^{-1} respectively, the reaction at 298 K would be [AFMC-98]
(A) Spontaneous (B) Reversible
(C) Irreversible (D) Non-spontaneous

111. If heat of formation of $C + O_2 \rightarrow CO_2 = -94$ kcal and $2C + O_2 \rightarrow 2CO = -110.5$ kcal, then heat of combustion of carbon is [AFMC-2000]
(A) 94 kcal (B) 110.5 kcal
(C) 16.5 kcal (D) 15 kcal

112. Which of the following reactions is endothermic ? [AFMC-2000]
(A) $N_2 + O_2 \rightarrow 2NO$
(B) $H_2 + Cl_2 \rightarrow 2HCl$
(C) $H_2SO_4 + 2NaOH \rightarrow Na_2SO_4 + 2H_2O$
(D) None of these

113. Based on the following thermochemical equations
$H_2O(g) + C(s) = CO(g) + H_2(g)$ $\Delta H = 131$ kJ/mole
$CO(g) + \frac{1}{2}O_2(g) = CO_2(g)$ $\Delta H = -282$ kJ/mole
$H_2(g) + \frac{1}{4}O_2(g) = H_2O(g)$ $\Delta H = -242$ kJ/mole
$C(s) + O_2(g) = CO_2(g)$ $\Delta H = x$ kJ/mole
the value of x will be [AFMC-2002]
(A) - 393 kJ (B) - 655 kJ (C) 393 kJ (D) 655 kJ

114. Consider the following reaction occurring in an automobile : $2C_8H_{18}(g) + 25O_2(g) \rightarrow 16CO_2(g) + 18H_2O(g)$
The sign of ΔH, ΔS and ΔG would be [AFMC-2002]
(A) +, –, + (B) –, +, – (C) –, +, + (D) +, +, –

115. The difference between heats of reaction at constant pressure and constant volume for the reaction, $2C_6H_6(l) + 15O_2(g) \rightarrow 12CO_2(g) + 6H_2O(l)$ at 25°C in kJ is [I.I.T.-91]
(A) –7.43 (B) +3.72 (C) –3.72 (D) +7.42

116. The products of combustion of an aliphatic thiol (RSH) at 298 K are: [I.I.T.-92]
(A) $CO_2(g)$, $H_2O(g)$ and $SO_2(g)$
(B) $CO_2(g)$, $H_2O(l)$ and $SO_2(g)$
(C) $CO_2(l)$, $H_2O(l)$ and $SO_2(g)$
(D) $CO_2(l)$, $H_2O(l)$ and $SO_2(g)$

117. For an endothermic reaction where ΔH represents the enthalpy of the reaction in kJ/mol, the minimum value for the energy of activation will be [I.I.T.-92]
(A) less than ΔH (B) zero
(C) more than ΔH (D) equal to ΔH

118. Identify the intensive quantities from the following: [I.I.T.-93]
(A) Enthalpy (B) Temperature
(C) Volume (D) Refractive index

119. A system is said to be .... if it can neither exchange matter nor energy with the surroundings. [IIT-93]

120. The heat content of the products is more than that of the reactants in an .... [IIT-93]

121. When Fe(s) is dissolved in aqueous hydrochloric acid in a closed vessel, the work is done ....[IIT-97]

122. Enthalpy is an ....property.[IIT-97]

123. Standard molar enthalpy of formation of CO_2 is equal to [IIT-97]
(A) zero
(B) the standard molar enthalpy of combustion of gaseous carbon
(C) the sum of standard molar enthalpies of formation of CO and O_2
(D) the standard molar enthalpy of combustion of carbon

(graphite)

124. The ΔH for $CO_2(g)$, $CO(g)$ and $H_2O(g)$ are -393.5, -110.5. and 241.8 kJ mol^{-1} respectively. The standard enthalpy change (in kJ) for the reaction $CO_2(g) + H_2(g) \rightarrow CO(g) + H_2O(g)$ is [IIT-2000]
(A) 524.1. (B) 41.2 (C) –262.5 (D) –41.2

125. In thermodynamics a process is called reversible when [IIT-2001]
(A) surroundings and system change into each other
(B) there is no boundary between system and surroundings
(C) the surroundings are always in equilibrium with the system
(D) the system change into the surroundings spontaneously.

126. Which one of the following statement is false ? [IIT-2001]
(A) Work is a state function
(B) Temperature is a state function
(C) Change in the state is completely defined when the intial and final states are specified
(D) Work appears at the boundary of the system.

127. One mole of a non-ideal gas undergoes a change of state (2.0 atm, 3.0L, 95K) → 40 atm, 5.0 L, 245K) with a change in internal energy, $\Delta U = 30.0$ L atm. The change in enthalpy (ΔH) of the process in 1atm. is [IIT-2003]
(A) 40.0 (B) 42.3 (C) 44.0
(D) not defined, because pressure is not constant.

128. Which of the following equation gives the value of heat of formation (ΔH^o) ? [IIT-2003]
(A) $H(g) + F(g) \rightarrow HF(l)$
(B) $H_2(g) + \ F_2(g) \rightarrow HF(g)$
(C) $H_2(g) + F_2(l) \rightarrow HF(g)$ (D) None of these

129. For the reactions. [AIEEE-2002]
$C + O_2 \rightarrow CO_2$; $\Delta H = -393$ J
$2\ Zn + O_2 \rightarrow 2\ ZnO$; $\Delta H = -\ 412$ J
(A) Carbon can oxidize Zn
(B) Oxidation of carbon is not feasible
(C) Oxidation of Zn is not feasible
(D) Zn can oxidize carbon

130. If an endothermic reaction is non-spontaneous at freezing point of water and becomes feasible at its boiling point, then. [AIEEE-2002]
(A) ΔH is –ve, ΔS is +ve
(B) ΔH and ΔS both are +ve
(C) ΔH and ΔS both are –ve (D) ΔH is +ve, ΔS is –ve

131. A heat engine absorbs heat Q_1 at temperature T_1 and heat Q_2 at temperature T_2. Work done by the engine is J (Q_1+Q_2) This data [AIEEE-2002]
(A) Violates 1st law of thermodynamics
(B) Violates 1st law of thermodynamics if Q_1 is –ve
(C) Violates 1st law of thermodynamics if Q_2 is –ve
(D) Does not violate 1st law of thermodynamics

132. The heat required to raise the temperature of body by 1 K is called [AIEEE-2002]
(A) Specific heat (B) Thermal capacity
(C) Water equivalent (D) None of these

133. The correct relationship between free energy change in a reaction and the corresponding equilibrium constant K_c is [AIEEE-2003]
(A) $\Delta G = RT \ln K_c$ (B) $-\Delta G = RT \ln K_c$
(C) $\Delta Go = RT \ln K_c$ (D) $-\Delta G^o = RT \ln K_c$

134. If at 298 K the bond energies of C – H, C – C, C = C and H – H bond are respectively 414, 347, 615 and 435 kJ mol^{-1}, the value of enthalpy change for the reaction $H_2C = CH_2\ (g) + H_2\ (g) \rightarrow H_3C - CH_3\ (g)$ at 298 K will be [AIEEE-2003]
(A) +250 kJ (B) –250 kJ (C) +125 kJ (D) –125 kJ

135. The enthalpy change for a reaction does not depend upon [AIEEE-2003]
(A) The physical states of reactants and products
(B) Use of different reactants for the same product
(C) The nature of intermediate reaction steps
(D) The differences in initial or final temperatures of involved substances

136. In an irreversible process taking place at constant T and P and in which only pressure- volume work is being done, the change in Gibbs free energy (δG) and change in entropy (δS), satisfy the criteria [AIEEE-2003]
(A) $(\delta S)_{V,E} < 0, (\delta G)_{T,P} < 0$ (B) $(\delta S)_{V,E} > 0, (\delta G)_{T,P} < 0$
(C) $(\delta S)_{V,E} = 0, (\delta G)_{T,P} = 0$ (D) $(\delta S)_{V,E} = 0, (\delta G)_{T,P} > 0$

137. The internal energy change when a system goes from state A to B is 40 kJ/mole. If the system goes from A to B by a reversible path and returns to state A by an irreversible path what would be the net change in internal energy ? [AIEEE-2003]
(A) 40 kJ (B) > 40 kJ (C) < 40 kJ (D) Zero

138. In which one of the following reactions does the heat change represent the heat of formation of water : [EAMCET-91]
(A) $2H_2 + O_2 \rightarrow 2H_2O$; $\Delta H = -116$ kcal
(B) $H_2 + \ O_2 \rightarrow H_2O$; $\Delta H = -\ 58$ kcal
(C) $H^+ + OH^- \rightarrow H_2O$ $\Delta H = -13.7$
(D) $C_2H_2 + 2\ \ O_2 \rightarrow 2CO_2 + H_2$; $\Delta H = -310$ K.Cal

139.Given the bond energies of N ≡ N, H–H and N – H bonds as 945, 436 and 391 kJmol^{-1} respectively, the enthalpy of the reaction $N_2\ (g) + 3H_2\ (g) \rightarrow 2NH_3\ (g)$ is [EAMCET-92]
(A) –93 kJ (B) 102 kJ (C) 90 kJ (D) 105kJ

140. At constant temperature and pressure if $\Delta G < 0$ the

precess is .... [EAMCET-92]

141. The reaction which proceeds with evolution of heat is called .... reaction. [EAMCET-93]

142. Consider the following reactions :
$C + O_2 \rightarrow CO_2$ $\Delta H = -400$ kJ mol^{-1}
$2H_2 + O_2 \rightarrow 2H_2O$ $\Delta H = -570$ kJ mol^{-1}
$CO_2 + 2H_2O \rightarrow CH_4 + 2O_2$ $\Delta H = 890$ kJ mol^{-1}
The heat of formation of methane is [EAMCET-94]
(A) –1860 kJ mol^{-1} (B) –1060 kJ mol^{-1}
(C) –720 kJ mol^{-1} (D) –80 kJ mol^{-1}

143. Given that
$CH_3CHO + O \rightarrow 2CO_2 + 2H_2O : \Delta H = -1168$ kJ mol^{-1}
$CH_3COOH + 2O_2 \rightarrow 2CO_2 + 2H_2O : \Delta H = -876$ kJmol^{-1}
ΔH for the reaction $CH_3CHO + O_2 \rightarrow CH_3COOH$ is [EAMCET-94]
(A) –292 kJ mol^{-1} (B) 378 kJ mol^{-1}
(C) 195 kJ mol^{-1} (D) –2044 kJ mol^{-1}

144. Liquid benzene burn in oxygen according to $2C_6H_6(l) + 15O_2(g) \rightarrow 12CO_2(g) + 6H_2O(g)$. How many litres of oxygen are required for complete combustion of 39 gm of liquid C_6H_6 (At.Wt. of C = 12, O = 16) [EAMCET-96]
(A) 12.2 (B) 22.4 (C) 42 (D) 84

145. The ΔH value for the reaction $H_2 + Cl_2 \rightarrow 2\,HCl$ is –4.12 kcal. If E_1 is the activation energy of the reactants and E_2 is the activation energy of the products, then for the above reaction. [EAMCET-97]
(A) $E_1 > E_2$ (B) $E_1 < E_2$ (C) $E_1 = E_2$
(D) ΔH is not related of E_1 & E_2

146. In the complete combustion of butanol C_4H_9OH (I) if ΔH is enthalphy of combustion & ΔE is the heat of combustion at constant volume, then [EAMCET-97]
(A) $\Delta H < \Delta E$ (B) $\Delta H = \Delta E$
(C) $\Delta H > \Delta E$ (D) ΔH, ΔE relation cannot to be predicted.

147. The heat of combustion of CH_4, C graphite and H_2 (g) are respectively 20 kcals –40kcals & –10 kcals. The heat of formation of CH_4 is [EAMCET-98]
(A) –40 kcals (B) +40 kcals
(C) –80 kcals (D) +80 kcals

148. In which of the following reactions, heat liberated is known as standard heat of formation of CO_2 ? [EAMCET-99]
(A) $2CO(g) + O_2(g) \rightarrow 2CO_2(g) + 135.5$ kcal
(B) $C(diamond) + O_2(g) \rightarrow CO_2(g) + 94.5$ kcal
(C) $C(graphite) + O_2(g) \rightarrow CO_2(g) + 94.05$ kcal
(D) $CH_4(g) + 2O_2(g) \rightarrow CO_2(g) + 2H_2O(l) + 212.8$ kcal

149. Calculate the heat of formation (ΔH) of CO(in kcal), from the following data : [EAMCET-2000]
(i) $C(graphite) + O_2(g) \rightarrow CO_2(g)$ $\Delta H = -94$ kcal
(ii) $CO(g) + 1/2O_2(g) \rightarrow CO_2(g)$ $\Delta H = -68$ kcal
(A) –13 (B) –26 (C) –162 (D) –81

150. In the reaction
$CH_3COOH(aq) + OH^-(aq) \rightarrow CH_3COO^-(aq) + H_2O(l)$;
$\Delta H = -55.2$ kJ.mol^{-1}.
What is the heat of dissociation (in kJ mol^{-1}) of acetic acid ? [EAMCET(MBBS)-2001]
(A) +2.1 (B) –57.3 (C) –35.1 (D) –176.1

151. Calculate the heat of combustion (in kJ) of methane from the following data : [EAMCET-2002]
(i) $C(graphite) + 2H_2(g) \rightarrow CH_4(g)$; $\Delta H = 74.8$ kJ
(ii) $C(graphite) + O_2(g) \rightarrow CO_2(g)$; $\Delta H = -393.5$ kJ
(iii) $H_2(g) + 1/2O_2(g) \rightarrow H_2O(l)$; $\Delta H = -2863.2$ kJ
(A) –891.1 (B) –816.3 (C) –965.9 (D) –1040.7

152. Which one of the following is an example of exothermic reaction ? [EAMCET(med.)-2002]
(A) $H_2(g) + Cl_2(g) \rightarrow 2HCl(g)$; $\Delta H = -184.6$ kJ
(B) $N_2(g) + O_2(g) \rightarrow 2NO(g)$; $\Delta H = +180.8$ kJ
(C) $C(graphite) + H_2O(g) \rightarrow CO(g) + H_2(g)$;$\Delta H = +131.4$ kJ
(D) $C(graphite) + 2S(s) + 91.9kJ \rightarrow CS_2(l)$

153. When 10 grams of methane is completely burnt in oxygen, the heat evolved is 560 kJ. What is the heat of combustion (in kJ/ mole) of methane [EAMCET-2003]
(A) –1120 (B) –968 (C) –896 (D) –560

154. Heat of combustion is [CPMT-96]
(A) Always -ve (B) +ve (C) Both A and B (D) 0

155. $C + \frac{1}{2}O_2 \rightarrow CO$ $\Delta H = 42$, $CO + \frac{1}{2}O_2 \rightarrow CO_2 = +24$.
The heat of formation of CO_2 is [CPMT-96]
(A) –16 kcal (B) +66 kcal
(C) + 16 kcal (D) –66 kcal

156. Which one is true ? [CPMT-97]
(A) 1 calorie > 1 joule > 1 erg
(B) 1 erg > 1 calorie > 1 joule
(C) 1 joule > 1 calorie > 1 erg
(D) 1 calorie > 1 erg > 1 joule

157. The entropy of a spontaneous reaction is always [CPMT-2001]
(A) Negative (B) Positive
(C) Zero (D) Either positive or negative

158. The energy that opposes dissolution of a solute is [CPMT-2002]
(A) hydration energy (B) lattice energy
(C) internal energy (D) bond energy

159. An isolated system is that system is which : [PMT-93]
(A) there is no exchange of energy with the surroundings
(B) there is exchange of mass and energy with the

surroundings
(C) there is no exchange of mass or energy with the surroundings
(D) there is exchange of mass with the surroundings

160. Which of the following statements is true ? The entropy of the universe : [PMT-93]
(A) increases and tends towards maximmum value.
(B) decreases and tends to be zero
(C) remains constant.
(D) decreases and increases with a periodic rate.

161. The overall heat change in a chemical reaction at constant volume is carried out : [PMT-93]
(A) in one step constant temperature and pressure
(B) in several steps at constant temperature and pressure.
(C) in one step at constant temperature only.
(D) in one or several steps at constant pressure.

162. Which of the following relation is correct ? [PMT-94]
(A) $\Delta H = \Delta G + T\Delta S$ (B) $\Delta H = -\Delta G - T\Delta S$
(C) $\Delta H = \Delta G - T\Delta S$ (D) $\Delta H = -\Delta G + T\Delta S$

163. When water is added to quick lime, the reaction is [PMT-95]
(A) exothermic (B) endothermic
(C) explosive (D) none of these

164. Heat of formation in the reaction is [PMT-95]
$H_2 + Cl_2 \rightarrow 2HCl + 44$ kcal.
(A) 44 kcal (B) 44000 kcal
(C) 22 kcal (D) 11 kcal

165. In an exothermic reaction ΔH is [PMT-95]
(A) Positive (B) Negative
(C) Zero (D) Both positive and negative

166. Which of the following is the correct equation ? [PMT-95]
(A) $\Delta G = \Delta H + T\Delta S$ (B) $\Delta G = \Delta H - T\Delta S$
(C) $\Delta G = T\Delta S - \Delta H$ (D) $\Delta G = \Delta H - S\Delta T$

167. Heat of neutralization of an acid with a base is 13.7 kcal when [PMT-95]
(A) both acid and base are weak
(B) acid is weak and base is strong
(C) acid is strong and base is weak
(D) acid and base both are strong

168. The heat change for the following reaction $C\ (s) + 2S\ (s) \rightarrow CS_2\ (1)$ is known as [PMT-96]
(A) heat of vaporization (B) heat of solution
(C) heat of fusion (D) heat of formation

169. When a solid changes into liquid entropy [PMT-96]
(A) becomes zero (B) becomes minimum
(C) increases (D) remains the same

170. Which of the following expressions is correct ? [PMT-96]
(A) $E = H + PV$ (B) $H = E + PV$
(C) $H = E - PV$ (D) $P = E + HV$

171. Which one of the following relations between free energy, enthalpy and entropy is correct [PMT-97]
(A) $\Delta H = \Delta G + T\Delta S$ (B) $\Delta H = \Delta G - T\Delta S$
(C) $\Delta H = \Delta G + \frac{\Delta S}{T}$ (D) $\Delta H = T\Delta G + \Delta S$

172. Heat of reaction at constant pressure is given by [PMT-97]
(A) $E_P - E_R$ (B) $E_R - E_P$ (C) $H_P - H_R$ (D) $H_R - H_P$

173. If enthalpy of B be greater than that of A, the reaction $A \rightarrow B$ will be [PMT-97]
(A) Exothermic (B) Endothermic
(C) Instantaneous (D) Spontaneous

174. In an isochoric process, the increase in internal energy is [PMT-97]
(A) Equal to the heat absorbed
(B) Equal to he heat evolved
(C) Equal to the work done
(D) Equal to the sum of the heat absorbed and work done

175. The positive value of ΔS indicates that [PMT-97]
(A) The system becomes less disordered
(B) The system becomes more disordered
(C) The system is in equlibrium position
(D) The system tends to reach at equilibrium position

176. Calculate standard free energy change for the reaction-
$\frac{1}{2}Cu\ (s) + \frac{1}{2}Cl_2 \rightarrow \frac{1}{2}Cu^2 + Cl^-$
taking place at (25°C) in a cell whose standard e.m.f. is 1.02 volts [PMT-97]
(A) - 98430 J (B) 98430 J (C) 96500 J (D) -49215 J

177. A chemical reaction cannot occur at all it its : [PMT-99]
(A) ΔH value is positive and ΔS value is negative
(B) ΔH value is negative and ΔS value is positive
(C) ΔH value and ΔS value are negative but $\Delta H > T\Delta S$
(D) ΔH value and ΔS value are positive but $\Delta H < T\Delta S$

178. The values of bond dissociation energies of gaseous H_2, Cl_2 and HCl are respectively 104, 58 and 103 kcals. The enthalpy of formation of gaseous HCl will be : [PMT-99]
(A) 22 kcals/mole (B) 44 kcals/mole
(C) - 22 kcals/mole (D) - 44 kcals/mole

179. For an endothermic reaction where ΔH represents

the enthalpy of reaction in kJ/mol, the minimum value for the energy of activation will be : [PMT-2000]

(A) less thanΔH (B) zero
(C) more than ΔH (D) equal to ΔH

180. Which of the following does NOT result in an increase in the entropy ? [PMT-2000]

(A) crystallization of sucrose from a solution
(B) rusting of iron (C) conversion of ice to water
(D) vaporization of camphor

181. For conversion C (graphite) $\rightarrow$C (diamond) the ΔS is : [PMT-2001]

(A) Zero (B) Positive (C) Negative (D) Unknown

182. In which of the following $\Delta E = \Delta H$? [PMT-2001]

(A) N_2O_4 (g) $\rightleftharpoons$ $2NO_2$ (g)
(B) $2SO_2$ (g) + O_2 (g) $\rightleftharpoons$ $2SO_3$ (g)
(C) H_2 (g) + I_2 (g) $\rightleftharpoons$ 2HI (g)
(D) H_2 (g) + $\frac{1}{2}O_2$ (g) $\rightleftharpoons$ H_2O (I)

183. The bond dissociation energies of H_2, Cl_2 and HCl are 104.58 and 103 kcal respectively. The enthalpy of formation of HCl is : [PMT-2001]

(A) +44 kcal (B) +22 kcal
(C) -22 kcal (D) -44 kcal

184. For melting of 3 moles of water at 0°C the $\Delta G°$ is : [PMT-2001]

(A) Zero (B) +ve (C) -ve (D) Unpredictable

185. An adiabatic expansion of an ideal gas, always has [PMT-2002]

(A) Increases in temperature (B) $\Delta H = 0$
(C) $q = 0$ (D) $w = 0$

186. Which of the following expressions represents the first law of thermodynamics ? [PMT-2002]

(A) $q = \Delta E - W$ (B) $\Delta E = q - w$
(C) $\Delta H = q + W$ (D) $\Delta E = \Delta H + P\Delta V$

187. The heat of neutralization will be highest in [PMT-2002]

(A) NH_4OH and HCl (B) NH_4OH and HCl
(C) KOH and CH_3COOH (D) KOH and HCl

188. Gibb's free energy G. enthalpy H and entropy S are interrelated as in [PMT-2002]

(A) G = H + TS (B) G = H – TS
(C) G – TS = H (D) G = S = H

189. A reaction having equal activation energies for forward and reverse reaction has [PMT-2002]

(A) $\Delta H = 0$ (B) $\Delta S = 0$
(C) Zero order (D) None of the above

190. According to Hess' law the thermal effect of a reaction depends upon : [PMT-2003]

(A) Final condition of the reacting substances
(B) Initial condition of reactions
(C) Intermediate state of the reaction
(D) Initial and final conditions of the reacting substances.

191. Which of the following equations is the correct one : [PMT-2003]

(A) G = H – TS (B) G = H – S T
(C) G = T S – H (D) G = H + T S

192. For a spontaneous reaction ΔG (free energy) is : [PMT-2003]

(A) Zero (B) Negative (C) Infinite (D) Positive

193. When ice melts into water, entropy [PET-93]

(A) Becomes zero (B) Decreases
(C) Increases (D) Remains the same

194. In exothermic reactions [PET-93]

(A) ΔE is zero (B) ΔH is negative
(C) ΔS is zero (D) ΔG is zero

195. Decrease of free energy of a reacting system indicated to a/an [PET-93]

(A) Exothermic reaction (B) Equilibrium reaction
(C) Spontaneous reaction (D) Slow reaction

196. The enthalpy change for the transition of liquid water to steam, ΔH_{vap} = 37.3 kJ mol^{-1} at 373° K. The entropy change for the process is [PET-94]

(A) 111.9 J mol^{-1} K^{-1} (B) 37.3J mol^{-1} K^{-1}
(C) 100 J mol^{-1} K^{-1} (D) 74.6 J mol^{-1} K^{-1}

197. A 500 ml solution of 2 M sodium hydroxide is added to a 500 ml solution of 2 M hydrochloric acid and the mixture is well stirred. The rise in temperature T_1 is recorded. The experiment is then repeated using 250 ml of each solution and the rise in temperature T_2 is recorded. Then which of the following statements is correct [PET-94]

(A) $T_1 = T_2$ (B) $T_1 = 2T_2$
(C) $T_1 = 4T_2$ (D) $T_2 = 2T_1$

198. For an adiabatic process [PET-95]

(A) $q = 0$ (B) $q = -w$ (C) $q = +w$ (D) $p\Delta V = 0$

where q, w, p and ΔV denote heat, work, pressure and change in volume respectively.

199. Decrease of free energy of a reacting system points to a/an - [PET-95]

(A) Exothermic reaction (B) Equilibrium reaction
(C) Spontaneous reaction (D) Isothermal reaction

200. When a solid is converted into liquid, entropy - [PET-95]

(A) Becomes zero (B) Remains the same
(C) Decreases (D) Increases

201. Carbon and carbon monoxide burn in oxygen to form carbon dioxide according to the following reactions

$C + O_2 \rightarrow CO_2$ $\Delta H = -394$ kJ mol^{-1}
$2CO + O_2 \rightarrow 2CO_2$ $\Delta H = -569$ kJ mol^{-1}
The heat of formation of 1 mol of carbon monoxide is thus [PET-95]
(A) – 219.0 kJ mol^{-1} (B) –109.5 kJ mol^{-1}
(C) – 175.0 kJ mol^{-1} (D) – 87.5 kJ mol^{-1}

202. The dependence of Gibb's free energy on pressure for an isothermal process of an ideal gas is given by [PET-96]
(A) $\Delta G_T = nRT \ln \frac{P_2}{P_1}$ (B) $\Delta G_T = nRT \ln \frac{V_2}{V_1}$
(C) $\Delta G_T = nRT \log \frac{P_1}{P_2}$ (D) $\Delta G_T = nRT \log \frac{V_1}{V_2}$

203. Which of the following expressions represents the first law of thermodynamics ? [PET-96]
(A) $q = \Delta E - w$ (B) $\Delta E = q - w$
(C) $\Delta E = q + w$ (D) $\Delta E = q + PdV$

204. The bond dissociation energies of gaseous H_2, Cl_2 and HCl are 140, 58 and 103 kcal respectively. The enthalpy of formation of HCl gas would be [PET-97]
(A) –44 kcal (B) –88 kcal
(C) –22 kcal (D) –11 kcal

205. Mark the correct statement : [PET-97]
(A) For a chemical reaction to be feasible, ΔG should be zero
(B) Entropy is a measure of order in a system
(C) For a chemical reaction to be feasible ΔG should be positive
(D) The total energy of an isolated system is constant

206. For the reaction : $CO(g) + \frac{1}{2} O_2(g) \rightarrow CO_2(g)$
Which one of the statements is correct at constant T and P ? [PET-97]
(A) $\Delta H = \Delta E$ (B) $\Delta H < \Delta E$ (C) $\Delta H > \Delta E$
(D) ΔH is independent of physical state of the reagents

207. A reaction is not feasible, if : [PET-97]
(A) ΔH is positive and ΔS is also positive
(B) ΔH is negative and ΔS is also negative
(C) ΔH is negative and ΔS is positive
(D) ΔH is positive and ΔS is negative

208. The relation between enthalpy (H), pressure (P) volume (V) and internal energy (E) is given by [PET-98]
(A) E = H + PV (B) H = E + PV
(C) H = E – PV (D) H = E + P + V

209. In an adiabatic process which of the following is true ? [PET-98]
(A) q = + w (B) q = 0 (C) $\Delta E = q$ (D) $P\Delta V = 0$

210. What is the free energy ΔG, when 1.0 mole of water at 100°C and 1 atm pressure is converted into steam at 100°C and 1 atm pressure ? [PET-98]
(A) 540 Cal (B) –9800 Cal
(C) 9800 Cal (D) 0 Cal

211. The standard Gibbs free energy change. $\Delta G°$ is related to equilibrium constant, k_p, as : [PET-98]
(A) $k_p = -RT \ln \Delta G°$ (B) $k_p = \left(\frac{e}{Rt}\right) \Delta G°$
(C) $k_p = \frac{\Delta G°}{RT}$ (D) $k_p = e^{\frac{\Delta G°}{RT}}$

212. At constant T and P which one of the following statements is correct for the reactions
$CO(g) + \frac{1}{2} O_2(g) \rightarrow CO_2(g)$ [PET-99]
(A) $\Delta H = \Delta E$ (B) $\Delta H < \Delta E$ (C) $\Delta H > \Delta E$
(D) ΔH is independent of the physical state of the reactants of that compounds

213. Evaporation of water is: [PET-99]
(A) An exothermic change (B) An endothermic change
(C) A process where no heat changes occur
(D) A process accompanied by chemical reaction

214. Heat of neutralisation of strong acid against strong base is constant and equals to : [PET-99]
(A) 13.7 kcal (B) 57 kJ
(C) 5.7×10^4 J (D) All of the above

215. The temperature coefficient of most of the reactions lies between : [PET-99]
(A) 1 and 3 (B) 2 and 3 (C) 1 and 4 (D) 2 and 4

216. The heat evolved in the combustion of benzene is given by the equation : [PET-2000]
$C_6H_6(l) + 7\frac{1}{2}O_2(g) \rightarrow 3H_2O(l) + 6CO_2(g)$
$\Delta H = 781.0$ kcal mol^{-1} Which of the following quantities of heat energy will be evolved when 39g of benzene is burnt in an open container ? [PET-2000]
(A) 122.25 kcal mol^{-1} (B) 390.5 kcal mol^{-1}
(C) 1562.0 kcal mol^{-1} (D) 586.75 kcal mol^{-1}

217. Energy required to dissociate 4g of gaseous hydrogen into free gaseous atoms is 208 kcal at 25°C. The bond energy of H–H will be : [PET-2000]
(A) 104 kcal (B) 10.4 kcal
(C) 1040 kcal (D) 52 kcal

218. Which of the following is correct for the first law of thermodynamics ? [PET-2000]
(A) $W = \Delta E - q$ (B) $\Delta E = q - W$
(C) $\Delta E = q + W$ (D) $\Delta E = q + P$

219. Who had proposed the following equation
$\Delta G = \Delta H - T\Delta S$ [PET-2000]
(A) Boltzmann (B) Faraday
(C) Gibbs-Helmholtz (D) Thomson

220. Which of the following units represents the largest

amount of energy ? [PET-2000]
(A) Calorie (B) Joule (C) Erg (D) Electron volt

221. Hess' law is an application of : [PET-2001]
(A) Ist law of thermodynamics
(B) IInd law of thermodynamics
(C) Entropy change (D) Free energy change

222. Free energy change for a reversible process is : [PET-2001]
(A) More than zero (B) Less than zero
(C) Equal to zero (D) None of these

223. Spontaneity of a chemical reaction is decided by the negative change in : [PET-2001]
(A) Internal energy (B) Enthalpy
(C) Entropy (D) Free energy

224. The favourable conditions for a spontaneous reaction are [PET-2002]
(A) $T\Delta S > \Delta H$, $\Delta H = +ve$, $\Delta S = +ve$
(B) $T\Delta S > \Delta H$, $\Delta H = +ve$, $\Delta S = -ve$
(C) $T\Delta S = \Delta H$, $\Delta H = -ve$, $\Delta S = -ve$
(D) $T\Delta S = \Delta H$, $\Delta H = +ve$, $\Delta S = +ve$

225. Heat exchanged in a chemical reaction at constant temperature and constant pressure is termed as [PET-2002]
(A) Internal energy (B) Enthalpy
(C) Entropy (D) Free energy

226. The entropy changed involved in the conversion of 1 mole of liquid water at 373K to vapour at the same temperature will be [ΔH_{vap} = 2,257 kJ/gm] [PET-2002]
(A) 0.119 kJ (B) 0.109 kJ
(C) 0.129 kJ (D) 0.120 kJ

227. Which of the following reactions is not exothermic [PET-2002]
(A) $C(s) + O_2(g) \rightarrow CO_2(g)$
(B) $C(s) + 2S(s) \rightarrow CS_2(g)$
(C) $CH_4(g) + 2O_2(g) \rightarrow CO_2(g) + 2H_2O(l)$
(D) $CO(g) + \frac{1}{2}O_2(g) \rightarrow CO_2(g)$

228. The dissociation energy of methane is 360kcal mol^{-1} and that of ethane is 620 kcal mol^{-1} the C–C bond energy (kcal mol^{-1}) is : [PET-2003]
(A) 210 (B) 130 (C) 180 (D) 80

229. For the raection C (graphite) → C (diamond) ΔS is [PET-2003]
(A) Zero (B) Positive (C) Negative (D) None

230. For which of the following $\Delta E = \Delta H$ [PET-2003]
(A) $H_2(g) + \frac{1}{2}O_2(g) \rightarrow H_2O(l)$
(B) $2SO_2(g) + O_2(g) \rightarrow 2SO_2(g)$
(C) $H_2(g) + I_2(g) \rightarrow 2HI(g)$ (D) $N_2O_4(g) \rightarrow 2NO_2(g)$

231. Which is an extensive property of the system ? [CBSE-93]
(A) Volume (B) Viscosity
(C) Temperature (D) Refractive index

232. C (diamond) + $O_2 \rightarrow CO_2$; ΔH = -395.4 kJ/mol.
C (graphite) + $O_2 \rightarrow CO_2$; ΔH = -393.5 kJ/mol.
C (graphite) → C (diamond) ; ΔH = ? [BHU-2003]
(A) + 1.9 (B) -1.9 (C) +3.8 (D) -3.8

233. $A_2(g) + B_2(g) \leftrightarrow 2AB(g)$; ΔH = +ve. [BHU-2003]
(A) it ocuurs at 1000 atm. pressure
(B) unaffected by pressure
(C) it occurs at high temperature
(D) it occurs at high temperature and high pressure

234. Which of the following is an intensive property [AFMC-2003]
(A) Temperature (B) Entropy
(C) Volume (D) Mass

235. If in acontainer mass and no heat exchange occurs, then it is [AFMC-2003]
(A) Closed system (B) Open system
(C) Imaginary system (D) Isolated system

236. Absolute temperature is the temperature at which [AFMC-2003]
(A) Mass becomes zero (B) Volume ceases to zero
(C) All molecular motion ceases (D) None of these

237. If the bond energies of H - H, Br - Br, and H - Br are 433, 192 and 364 kJ mol^{-1} respectively, the DH° for the reaction [CBSE-PMT-2004]
$H_2(g) + Br_2(g) \longrightarrow 2HBr(g)$ is
(A) -261 kJ (B) + 103 kJ (C) + 261 kJ (D) - 103 kJ

238. Standard enthalpy and standard entropy changes for the oxidation of ammonia at 298 K are -382.64 kJ mol^{-1} and -145.6 J K^{-1} mol^{-1}, respectively. Standard Gibbs energy change for the same reaction at 298 K is [CBSE-PMT-2004]
(A) - 221.1 kJ mol^{-1} (B) - 339.3 kJ mol^{-1}
(C) - 439.3 kJ mol^{-1} (D) - 523.2 kJ mol^{-1}

239. The work done during the expansion of a gas from a volume of 4 dm^3 to 6 dm^3 against a constant external pressure of 3 atm is (1 L atm = 101.32 J) [CBSE-PMT-2004]
(A) - 6 J (B) - 608 J (C) + 304 J (D) - 304 J

240. Considering entropy (S) as a thermodynamic parameter, the criterion for the spontaneity of any process is [CBSE-PMT-2004]
(A) $\Delta S_{system} + \Delta S_{surroundings} > 0$
(B) $\Delta S_{system} - \Delta S_{surroundings} > 0$

(C) $\Delta S_{system} > 0$ only (D) $\Delta S_{surroundings} > 0$ only

241. Adsorption of a gas is exothermic and a spontaneous process because [IIT-2004]

(A) enthalpy increases (B) entropy increases

(C) entropy decreases (D) free energy increases

242. 2 moles of an ideal gas expand isothermally and reversibly at 300 K from $V_1 = 1$ litre to $V_2 = 10$ litres. The change in enthalpy in kJ mol^{-1} is [IIT-2004]

(A) 11.47 (B) -11.47 (C) 0 (D) 4.8

243. Enthalpy of vaporization of a liquid is 30 kJ mol^{-1} and the entropy change during the process is 75 J mol^{-1} K^{-1}. The boiling point of the liquid is [IIT-2004]

(A) 400 K (B) 500 K (C) 750 K (D) 3

Thermodynamics

1.	(D)	51.	(D)	101.	(B)	151.	(A)	201.	(B)
2.	(B)	52.	(D)	102.	(B)	152.	(A)	202.	(A)
3.	(D)	53.	(D)	103.	(B)	153.	(C)	203.	(B)
4.	(C)	54.	(B)	104.	(D)	154.	(A)	204.	(C)
5.	(A)	55.	(B)	105.	(C)	155.	(D)	205.	(D)
6.	(B)	56.	(A)	106.	(C)	156.	(A)	206.	(B)
7.	(B)	57.	(C)	107.	(C)	157.	(A)	207.	(D)
8.	(A)	58.	(D)	108.	(C)	158.	(B)	208.	(B)
9.	(A)	59.	(C)	109.	(B)	159.	(A)	209.	(B)
10.	(A)	60.	(A)	110.	(A)	160.	(A)	210.	(D)
11.	(B)	61.	(B)	111.	(A)	161.	(D)	211.	(D)
12.	(B)	62.	(C)	112.	(A)	162.	(A)	212.	(B)
13.	(B)	63.	(B)	113.	(A)	163.	(A)	213.	(B)
14.	(A)	64.	(A)	114.	(B)	164.	(B)	214.	(D)
15.	(A)	65.	(C)	115.	(A)	165.	(A)	215.	(B)
16.	(B)	66.	(A)	116.	(B)	166.	(D)	216.	(B)
17.	(A)	67.	(A)	117.	(C)	167.	(D)	217.	(A)
18.	(A)	68.	(B)	118.	(B)(D)	168.	(D)	218.	(B)
19.	(C)	69.	(A)	119.	*	169.	(C)	219.	(C)
20.	(B)	70.	(D)	120.	*	170.	(B)	220.	(A)
21.	(B)	71.	(D)	121.	*	171.	(A)	221.	(A)
22.	(D)	72.	(A)	122.	*	172.	(C)	222.	(C)
23.	(A)	73.	(C)	123.	(D)	173.	(B)	223.	(D)
24.	(C)	74.	(A)	124.	(B)	174.	(A)	224.	(A)
25.	(A)	75.	(D)	125.	(C)	175.	(B)	225.	(B)
26.	(A)	76.	(B)	126.	(A)	176.	(A)	226.	(B)
27.	(A)	77.	(A)	127.	(C)	177.	(A)	227.	(B)
28.	(B)	78.	(C)	128.	(B)	178.	(C)	228.	(D)
29.	(B)	79.	(B)	129.	(D)	179.	(C)	229.	(C)
30.	(A)	80.	(C)	130.	(B)	180.	(A)	230.	(C)
31.	(A)	81.	(A)	131.	(A)	181.	(B)	231.	(A)
32.	(D)	82.	(A)	132.	(B)	182.	(C)	232.	(A)
33.	(A)	83.	(A)	133.	(D)	183.	(C)	233.	(B)
34.	(C)	84.	(D)	134.	(D)	184.	(A)	234.	(A)
35.	(A)	85.	(A)	135.	(C)	185.	(C)	235.	(D)
36.	(A)	86.	(B)	136.	(B)	186.	(A)	236.	(C)
37.	(D)	87.	(B)	137.	(D)	187.	(D)	237.	(D)
38.	(C)	88.	(A)	138.	(B)	188.	(B)	238.	(B)
39.	(A)	89.	(B)	139.	(A)	189.	(A)	239.	(B)
40.	(A)	90.	(C)	140.	*	190.	(D)	240.	(A)
41.	(D)	91.	(A)	141.	*	191.	(A)	241.	(C)
42.	(B)	92.	(D)	142.	(D)	192.	(B)	242.	(C)
43.	(C)	93.	(A)	143.	(A)	193.	(C)	243.	(A)
44.	(C)	94.	(C)	144.	(D)	194.	(B)		
45.	(A)	95.	(B)	145.	(B)	195.	(C)		
46.	(C)	96.	(B)	146.	(C)	196.	(C)		
47.	(D)	97.	(C)	147.	(C)	197.	(A)		
48.	(A)	98.	(A)	148.	(C)	198.	(A)		
49.	(D)	99.	(D)	149.	(B)	199.	(C)		
50.	(D)	100.	(A)	150.	(A)	200.	(D)		

119. Isolated
120. endothermic
121. O
122. Extensive
140. Spontaneous
141. Exothermic

KINETICS

For a reaction to take place there must be effective collisions. The collision frequency 'z' is the number of collisions per second per unit volume. It should be ~ 10^{25} to 10^{28}. The molecules must have the right energy.

Rate of a reaction :

A chemical reaction in fact involves breaking of bonds in reactant molecules and formation of new bonds in product molecules.

$AgNO_3 + NaCl \rightarrow AgCl\downarrow + NaNO_3$

The reactions involving ions such as precipitation of AgCl are almost instantaneous. The time required for the precipitation is 10^{-6} seconds. In the case of simple ionic reactions, no bonds are to be broken. Therefore they are very fast. The reactions involving big organic molecules proceed slowly e.g. $C_{12}H_{22}O_{11} + H_2O \rightarrow C_6H_{12}O_6 + C_6H_{12}O_6$. In these cases large number of bonds are to be broken in the reactants and a number of bonds are to be formed in the products.

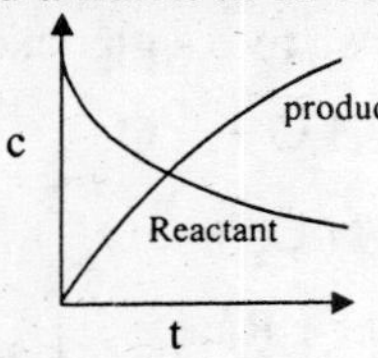

Factors affecting the rate :

1. Concentration of the reacting species : The rate of a reaction keeps on falling with time. The velocity of a reaction depends upon the concentration of the reactants. More the concentration of the reactants, greater will be the velocity at that instant. Exception : in zero order reactions the rate is independent of the concentration of the reacting species.

2. Temperature of the system : The rate of a reaction generally increases with the increase in the temperature. For every 10° Absolute, it doubles.

3. Nature of the reactants and products

(a) $5Fe^{2+}(aq) + MnO_4^-(aq) + 8H^+(aq) \rightarrow 5Fe^{3+}(aq) + Mn^{2+}(aq) + 4H_2O$

(b) $5C_2O_4^{2-}(aq) + 2MnO_4^-(aq) + 16H^+(aq) \rightarrow 10CO_2(aq) + 2Mn^{2+}(aq) + 8H_2O$

Those reactions which do not involve much bond rearrangements are generally rapid at room temperature. The rate of reaction (a) is more than (b).

4. Presence of catalyst : Catalyst normally enhances the rate of reaction by lowering the activation energy. A catalyst increases the rate of backward reaction to the same extent as the forward rection.

5. Effect of surface area of reactants : Surface area of a reactant is of importance only in the case of heterogeneous reactions. As the particle size decreases, the surface area of the same mass increases. e.g. (a) powdered sugar dissolves faster than the crystalline sugar. (b) A log of wood burns slower than the shavings of the wood.

6. Effect of radiation : Radiation generally increases the rate.

$H_2 + Cl_2 \xrightarrow{dark} 2HCl$ slow

$H_2 + Cl_2 \xrightarrow{sun\ light} 2HCl$ fast

7. Physical state : Liquid > Gases > Solids

In liquid state, the molecules not only move relatively freely but also closer to one another.

Rate	Rate constant
1. Rate keeps on changing with time i.e rate is the function of concentration of the reactants (except for zero order).	Rate constant is a constant at a particular temp. and is independent of concentration of the reactant.
2. The units of rate is most commonly moles/lit.time^{-1}.	depends on the order of the reaction.

Molecularity : It is the number of molecules of the reactants involved in the reaction.

e.g. $C_{12}H_{22}O_{11} + H_2O \rightarrow C_6H_{12}O_6 + C_6H_{12}O_6$

Molecularity is 2.

Reactions having molecularity more than 3 are rare. It is because the chances of simultaneous collisions between three or more particles is rare.

Order of a reaction : It is the number of molecules whose concentration alters appreciably as a result of chemical reaction.

e.g. (1) $CH_3COOC_2H_5 + H_2O \rightleftharpoons CH_3COOH + C_2H_5OH$ order is 1.

(2) $2N_2O_5 \longrightarrow 4NO_2 + O_2$ order is 1.

Here the order of reaction is 1, because the reaction actually takes place in two steps

step 1 : $N_2O_5 \xrightarrow{slow} N_2O_4 + O$

step 2 : $O + O \xrightarrow{fast} O_2$

The step 1 is very slow as compared to step 2. Slow step is the rate determining step.

Molecularity	Order of a reaction
1. It is a theoretical concept.	It can be determined through experiment only.
2. It is an integer, generally 1, 2 or 3.	It can have +ve, -ve, 0 and fractional values.

Pseudo - unimolecular reactions : The first order reactions whose molecularity is not one are called pseudo -unimolecular reactions.

e.g. $CH_3COOCH_3 + H_2O \rightleftharpoons CH_3COOH + CH_3OH$

Rate constant or specific reaction rate :

consider the reaction $A \longrightarrow B + C$

Here only one reactant is present, the more the concentration.

of A, the higher will be the rate of the reaction i.e. the rate of the reaction is directly proportional to the concentration of the reactant

Thus $\frac{dx}{dt} \propto C_A$ or $\frac{dx}{dt} = k.\ C_A$

The rate constant is defined as the fraction of reactant A that reacts in unit time.

If the concentration of A = 1 mole per litre i.e., $C_A = 1$

Then $\frac{dx}{dt} = k$

At unit concentration the rate constant is called specific reaction rate

Now consider $A + B \rightarrow$ product (s)

$$\frac{dx}{dt} \propto C_A \times C_B$$

$$\frac{dx}{dt} = k\ C_A \times C_B$$

$\frac{dx}{dt} = k$ (if the concentration of each reactant is unity i.e. $C_A=1$ and $C_B = 1$)

Rate constant of a reaction is defined as equal to the rate of the reaction when the concentration of each of the reactant is taken as unity.

Zero order reaction :

A reaction is said to be the zero order reaction in which the rate is independent of the concentration of any reactant in the system.

A reaction is of zero order when

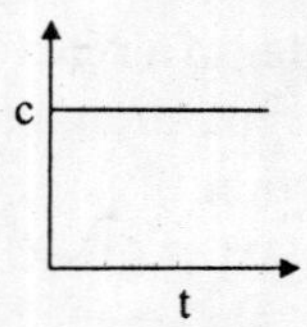

(a) Concentration of reactants does'nt change with time

(b) rate does'nt change with time

(c) The rate of a reaction becomes independent of concentration of a reactant when that reactant is taken in large excess.

$$\text{Rate} = k\ [A]^o$$
$$= k$$

Unit : mole/lit. time^{-1}. Rate law : $k = \frac{x}{t}$

Enzyme reactions are zero order reactions. Photochemical reactions also are zero order reactios.

e.g. 1. $H_2(g) + Cl_2(g) \xrightarrow{hv} 2HCl(g)$

2. $CH_3COCH_3 + Br_2 \xrightarrow{hv} CH_3COCH_2Br + HBr$

3. $2NH_3 \xrightarrow{[Mo]} N_2 + 3H_2$

Half time for zero order reaction is proportional to initial concentration.

First order :

1. unit of k : Rate = k [A]

$$\frac{(\text{mole / lit.}) \text{ time}^{-1}}{(\text{mole / lit.})} = k = \text{time}^{-1}$$

2. They all obey $k = \frac{2.303}{t} \log \frac{a}{(a-x)}$

In ester hydrolysis : $k = \frac{2.303}{t} \log \frac{T_\infty - T_o}{T_\infty - T_t}$

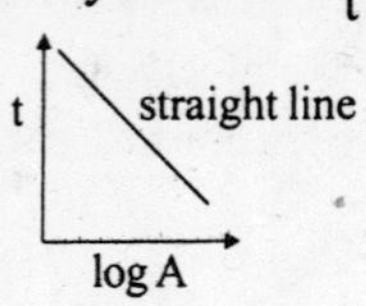

3. Time for half change for first order reaction is independent of concentration (even for any fraction)

$$t_{0.5} = \frac{0.693}{k}$$

4. Sometimes while determining the order experimentally, the initial concentration is not known definitely because the exact moment when the reaction starts is uncertain.

In such cases $k = \frac{2.303}{t_2 - t_1} \log \frac{a - x_1}{a - x_2}$ ($a - x_1$ and $a - x_2$ are concentrations at time intervals t_1 and t_2.)

$$k = \frac{2.303}{\Delta t} \log \frac{C_1}{C_2}$$

e.g. (1) Ester hydrolysis

$CH_3COOCH_3 + H_2O \overset{H^+}{\rightleftharpoons} CH_3COOH + CH_3OH$

(2) Inversion of cane sugar

$C_{12}H_{22}O_{11} + H_2O \overset{H^+}{\rightleftharpoons} C_6H_{12}O_6 + C_6H_{12}O_6$

(The solution which is dextro rotatory ultimately becomes laevo rotatory)

3. Radio active decay is of first order reaction

Radio active reactions are independent of external temperature and pressure.

(4) α - glucose to β - glucose

(5) $H_2O_2 \rightarrow H_2O + \frac{1}{2}O_2$

(6) $N_2O_5 \rightarrow N_2O_4 + \frac{1}{2}O_2$

Second order : Rate = k $[A]^2$

$$\frac{(\text{mole / lit.}) \text{ time}^{-1}}{(\text{mole / lit.})^2} = k = (\text{mole/lit})^{-1} \text{ time}^{-1} \text{ or mole}^{-1}\text{lit.sec}^{-1}$$

$$k = \frac{1}{t} \times \frac{x}{a\ (a-x)}$$

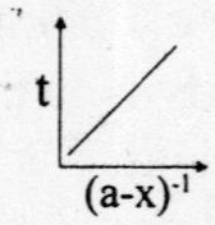

Half life of second order reaction is inversely proportional to the initial concentration.

$t_{0.5} \propto \frac{1}{a}$ (second order)

$t_{0.5} \propto \frac{1}{a^2}$ (third order)

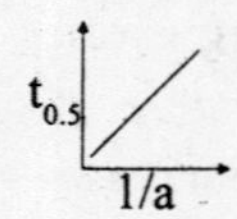

e.g. (1) Saponification of esters

$CH_3COOC_2H_5 + NaOH \rightleftharpoons CH_3COONa + C_2H_5OH$

(2) Benzoin condensation

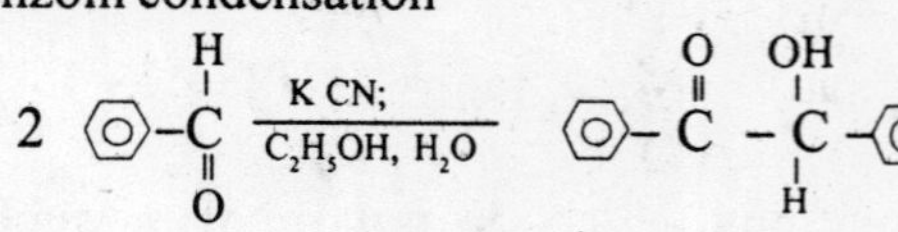

(3) $2O_3 \longrightarrow 3O_2$

(4) $H_2 + I_2 \rightleftharpoons 2HI$

(5) Interaction of alkyl halides with t-amines or pyridine

$C_2H_5I + C_6H_5N(CH_3)_2 \rightarrow C_6H_5N(CH_3)_2 + C_2H_5^+ + I^-$

The second order becomes first order if one of the reactants is present in large excess.

Rarity of higher order : Slow step is rate determing step.

(1) $2NO + 2H_2 \rightarrow N_2 + 2H_2O$ order 3 not 4

a. $2NO + H_2 \xrightarrow{slow} N_2 + H_2O_2$

b. $H_2O_2 + H_2 \xrightarrow{fast} 2H_2O$

(2) $H_2O_2 + 2HI \longrightarrow 2H_2O + I_2$

a. $H_2O_2 + I^- \xrightarrow{slow} H_2O + IO^-$ order 2 not 3

b. $H^+ + IO^- \xrightarrow{fast} HIO$

c. $HIO + H^+ + I^- \xrightarrow{fast} H_2O + I_2$

- pseudo bimolecular reactions

$HBrO_3 + 6HI \rightarrow HBr + 3H_2O + I_2$ molecularity 7

step a. $HBrO_3 + HI \rightarrow HBrO_2 + HIO$ - slow order 2

b. $HBrO_2 + 4HI \rightarrow HBr + 2H_2O + I_2$ - fast

c. $HIO + HI \rightarrow H_2O + I_2$ - fast

Determination of order of a reaction :

(1) Integration method : 0 order, $k = \frac{x}{t}$

1st order, $k = \frac{2.303}{t} \log \frac{a}{(a-x)}$; 2nd order, $k = \frac{1}{t} \times \frac{x}{a(a-x)}$

(2) Graphical method :

log (a-x) vs. t. - straight line - firts order

$\frac{1}{(a-x)}$ or $(a-x)^{-1}$ vs. t. - straight line - second order

$\frac{1}{(a-x)^2}$ or $(a-x)^{-2}$ vs. t. - straight line - third order

(3) Half life method : $t_{0.5} \propto a$ i.e. $t_{0.5} = \frac{a}{2k}$ for zero order

$t_{0.5}$ independent of 'a' : first order reaction

$t_{0.5} \propto \frac{1}{a}$: second order ; $t_{0.5} \propto \frac{1}{a^2}$: third order

$t_f \propto (\frac{1}{a^{n-1}})$ or $t_f \propto (\frac{1}{a})^{n-1}$ where 't_f' is the time required to complete a definite fraction of the reaction and 'a' is the initial concentration and n is the order of the reaction.If the initial concentrations are a_1 and a_2 with the corresponding times t_1 and t_2 in two different experiments, then

$$\frac{t_1}{t_2} = \left(\frac{a_2}{a_1}\right)^{n-1}$$

Taking the logarithm of equation

$$\log \frac{t_1}{t_2} = (n-1) \log \left(\frac{a_2}{a_1}\right)$$

$$n = 1 + \frac{\log t_1 - \log t_2}{\log a_2 - \log a_1}$$

Energy of Activation :

The minimum energy which a molecule must acquire before reaction can take place is known as energy of activation.

$k = A.\, e^{-E_A/RT}$

A = frequency factor or pre exponential factor

E_A = activation energy

R = gas constant

T = absolute temp.

Here A and E are characteristic of the reaction

$$\log k = -\frac{E_A}{2.303\,R} \times \frac{1}{T} + \log A$$

(like Y = mx + C)

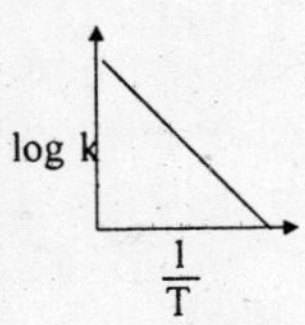

1. slope of the straight line is $-\frac{E_A}{2.303\,R}$

2. The intercept log A gives the frequency factor

Integrated form of Arrhenius equation :

$$\log \frac{k_2}{k_1} = \frac{E_A}{2.303\,R} \left(\frac{T_2 - T_1}{T_1 \times T_2}\right) \text{ or } \log \frac{k_2}{k_1} = \frac{E_A}{2.303\,R} \left(\frac{1}{T_1} - \frac{1}{T_2}\right)$$

From this equation, E_A can be determined if the specific rate constants k_1 and k_2 and two temp. T_1 & T_2 are known.Graphical relationships of E_A :

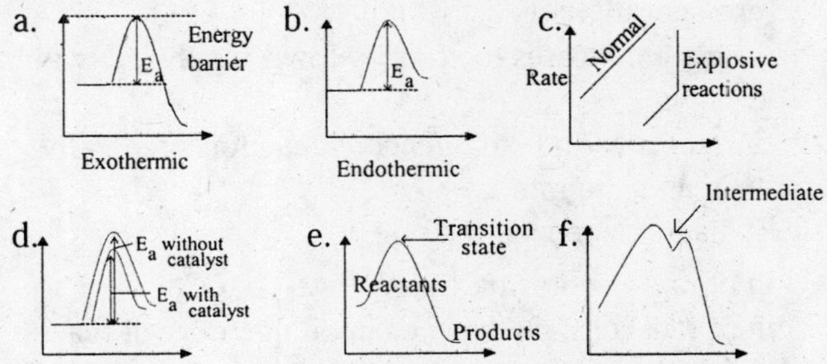

Transition state theory or absolute reaction rates or the activated complex theory was given by Eyring and Polanyi. The colliding molecules must acquire a certain energy i.e. threshold energy to make the collisions most effective. Most of the molecules have much less kinetic energy than the threshold energy. The excess energy that the reactant molecules acquire to attain the thershold energy in order to react to yield products is called activation energy. Thus Activation energy = Threshold energy - Energy posessed by the particles

$E = E_T - E_r$

Things to remember

Factors influencing the rate :

a. Concentration of reactants (exception : zero order)
b. Generally increases with increase in temperature
c. Reactions that don't involve much bond rearrangements are faster.
d. Catalyst increases the rate by lowering the activation energy.
e. Surface area in heterogeneous reactions generally increases the rate
f. Radiation increases the rate

Unit of rate = mole / lit. time^{-1}

Unit of rate constant depends upon the order of reaction

Molecularity can never be zero or fraction.

Order of a reaction can be 0, +ve, -ve or fraction.

First order reactions whose molecularity is not one are called pseudo-unimolecular reactions.

Zero order reaction : Unit: moles/lit. time^{-1}; $k = \frac{x}{t}$; $t_{0.5} \propto k$ i.e. $t_{0.5} = \frac{a}{2k}$; photochemical and enzyme reactions are generally zero order reactions.

First order : time^{-1}; $t_{0.5}$ independent of initial concentration; $k = \frac{2.303}{t} \log \frac{a}{a-x}$; ester hydrolysis, inversion of cane sugar and radioactive decay belong to first order.

Second order : (moles/lit)$^{-1}$ time^{-1} or mole^{-1} lit. sec^{-1}.; $k = \frac{1}{t} x \frac{x}{a(a-x)}$; $t_{0.5} \propto \frac{1}{a}$; Benzoin condensation, Saponification etc. are second order reactions.

Slow step in a multi step reactions is rate determinig step.

Order of a reaction can be determined from $\frac{t_1}{t_2} = (\frac{a_2}{a_1})^n$.

Arrhenius equation : $k = A.\ e^{-Ea/RT}$

$$\log k = -\frac{-E_a}{2.303R} \times \frac{1}{T} + \log A$$

$$\log\frac{k_2}{k_1} = \frac{E_a}{2.303R}\left[\frac{1}{T_1} - \frac{1}{T_2}\right]$$

Rate doubles for every 10 degrees absolute.

Solved Examples

1. For the reaction P + Q + R → products, the rate equation is : Rate = k [P] [Q] $[R]^2$.What would be the rate of reaction when the concentration of P and Q are doubled while that of R is halved ?

Sol. Rate = k [P] [Q] $[R]^2$.

Substituting the new concentrations in the equation

$$\text{Rate} = [2]\ [2]\ [\tfrac{1}{2}]^2 = 1$$

∴ It remains the same.

2. The half life of a first order chemical reaction A → B is 10 min. What percentage of A remains after 1 hour ?

Sol. $t_{0.5} = \frac{0.693}{k}$,$k = \frac{0.693}{10} = 0.0693$

using first order reaction, $k = \frac{2.303}{t} \log \frac{a}{a-x}$

$$0.0693 = \frac{2.303}{60} \log \frac{100}{C}$$

$$C = 1.565\%$$

3. For the reaction : A + B → C + D, the initial rates for different initial concentrations are found as follows :

Initial conc. [A]	[B]	Initial rate mol l^{-1} s^{-1}
1.0	1.0	2.0 x 10^{-3}
2.0	1.0	4.0 x 10^{-3}
4.0	1.0	8.0 x 10^{-3}
1.0	2.0	2.0 x 10^{-3}
1.0	4.0	2.0 x 10^{-3}

What is the order with respect to A and B ? Overall order? Write the rate law equation. Calculate the rate constant. Suggest a mechanism.

Sol. By doubling [A] and keeping [B] constant, the initial rate also doubles. Hence with respect to 'A' first order. Doubling [B] and keeping [A] constant, rate constant doesn't change, hence with respect to 'B' is zero order. Overall order 1+0 = 1 i.e. first order .

Rate law expression, r = k [A]

Rate constant,$k = \frac{r}{[A]} = \frac{4 \times 10^{-3} \text{ mol lit}^{-1} \text{ sec}^{-1}}{2.0 \text{ mol lit}^{-1}} = 2 \times 10^{-3} \text{ sec}^{-1}$

Mechanism : A → x (slow)

x + B → C + D (fast)

4. The rate constant of a reaction with respect to reactant A is 6 min^{-1}. If we start with [A] = 0.8 mol L^{-1}, when would [A] reach the value of 0.08 mol L^{-1}.

Sol. $$t = \frac{2.303}{k} \log \frac{[A]_o}{[A]_t}$$

$$= \frac{2.303}{6} \log \frac{0.8}{0.08} = \frac{2.303}{6} \log 10 = 0.38 \text{ min}$$

5. Show that in case of a first order reaction, the time required for 99.9% of the reaction to take place is 10 times that required for half of the reaction.

Sol. $$t_{50\%} = \frac{2.303}{k} \log \frac{1}{1-0.5} = \frac{2.303}{k} \log 2 \quad (a)$$

$$t_{99.9\%} = \frac{2.303}{k} \log \frac{1}{1-0.999}$$

$$= \frac{2.303}{k} \log \frac{1}{0.001} = \frac{2.303}{k} \log 10^3 \quad (b)$$

Now dividing (b) by (a), we get

$$\frac{t_{99.9\%}}{t_{50\%}} = \frac{2.303}{k} \times \frac{k}{2.303} \times \frac{\log 10^3}{\log 2} = \frac{3.000}{0.3010} = 9.967 \approx 10$$

6. For a reaction A + B + C → products, the rate of the reaction is given by rate = k $[A]^a [B]^b [C]^c$.
From the following data determine (i) the order of reaction with respect to A, B and C (ii) the overall order (iii) the rate constant

Initial conc. (mol. d m^{-3})	[A]	[B]	[C]	Initial rate (mol.$dm^{-3}s^{-1}$)
(i)	0.010	0.005	0.010	5.0×10^{-5}
(ii)	0.010	0.005	0.015	5.0×10^{-5}
(iii)	0.010	0.010	0.010	2.5×10^{-5}
(iv)	0.020	0.005	0.010	15×10^{-5}

Sol. comparing (i) and (iv) the conc. of [B] and [C] are constant. When the [A] is doubled the rate changes to 3 fold. Thus $a = \frac{3}{2}$ i.e. the order is $\frac{3}{2}$ with respect to A.

Comparing (i) and (iii) where [A] and [C] are constant. By doubling the [B], the rate changes to half. Thus, b= -1 i.e. the order is -1 with respect to B.

Comparing (i) and (ii) where [A] and [B] are constant. When [C] is changed $\frac{3}{2}$ times, the rate remains unaltered. The order with respect to C is zero.

(i) overall order = $\frac{3}{2} - 1 + 0 = \frac{1}{2}$

(ii) rate = k $[A]^{3/2} [B]^{-1} [C]^0$ = k $[A]^{3/2} [B]^{-1}$

$$k = \frac{\text{rate}}{[A]^{3/2}[B]^{-1}} = \frac{15 \times 10^{-5} \text{ mol d m}^{-3} \text{ s}^{-1}}{(0.020)^{3/2}(0.005)^{-1}}$$
$$= 2.65 \times 10^{-4} \text{ (mol. d m}^{-1})^{1/2} \text{ s}^{-1}.$$

7. Nitrogen pentoxide when allowed to decompose in an inert solvent gave the following volumes of oxygen

t(s)	0	300	600	900	∞
vol. (cc)	0	3.42	6.30	8.95	34.75

Show that it is of first order reaction

$$k = \frac{2.303}{t} \log \frac{V\infty}{V\infty - Vt}$$

Sol. $K_{300} = \frac{2.303}{300} \log \frac{34.75}{31.33} = 3.454 \times 10^{-4} \text{ s}^{-1}$

$K_{600} = \frac{2.303}{600} \log \frac{34.75}{28.45} = 3.334 \times 10^{-4} \text{ s}^{-1}$

$K_{900} = \frac{2.303}{900} \log \frac{34.75}{25.80} = 3.310 \times 10^{-4} \text{ s}^{-1}$

8. 5 ml. of methyl acetate was added to the reaction vessel containing 100 ml. of 1.89 N HCl maintained at a room temp. of 298 k, 5 ml. of reaction mixture was withdrawn and titrated with a standard alkali and the following results were obtained

Time (s)	0	900	1800	3000	∞
ml. of alkali used	11.03	11.70	12.15	12.75	14.10

from the above data, find out the order of the reaction

Sol. $k = \frac{2.303}{t} \log \frac{V\infty - Vo}{V\infty - Vt}$

$K_{900} = \frac{2.303}{900} \log \frac{14.10 - 11.03}{14.10 - 11.70} = 0.2735 \times 10^{-3} \text{ s}^{-1}$

$K_{1800} = \frac{2.303}{1800} \log \frac{14.10 - 11.03}{14.10 - 12.15} = 0.255 \times 10^{-3} \text{ s}^{-1}$

$K_{3000} = \frac{2.303}{3000} \log \frac{14.10 - 11.03}{14.10 - 12.75} = 0.2738 \times 10^{-3} \text{ s}^{-1}$

As the values of k are almost constant, the order is one.

(9) The optical rotation of sucrose in 0.5 N HCl at 308k and at various time intervals are given below. Find out the order of the reaction.

Time (min)	0	10	20	30	40	∞
Rotation (degree)	+32.4	+28.8	+25.5	+22.4	+19.6	−11.1

Sol. $k = \frac{2.303}{t} \log \frac{\theta_o - \theta_\infty}{\theta_t - \theta_\infty}$; $\theta_o - \theta_\infty = +32.4 - (-11.1) = 43.5$

$K_{10} = \frac{2.303}{10} \log \frac{43.5}{28.8-(-11.1)} = 8.636 \times 10^{-3} \text{ min}^{-1}$

$K_{20} = \frac{2.303}{20} \log \frac{43.5}{25.5-(-11.1)} = 8.638 \times 10^{-3} \text{ min}^{-1}$

$K_{30} = \frac{2.303}{30} \log \frac{43.5}{22.4-(-11.1)} = 8.706 \times 10^{-3} \text{ min}^{-1}$

$K_{40} = \frac{2.303}{40} \log \frac{43.5}{19.6-(-11.1)} = 8.658 \times 10^{-3} \text{ min}^{-1}$

Since the value of k is fairly constant, it is first order reaction.

10. The decomposition of H_2O_2 was studied by titrating it at different intervals with $KMnO_4$. Calculate the velocity constant for it from the following data assuming it to be of the first order.

Time (min)	0	10	20	50
Vol. of K MnO_4(ml)	23.05	14.90	6.15	2.6

Sol. $K_{10} = \frac{2.303}{10} \log \frac{23.05}{14.90} = 0.04362 \text{ min}^{-1}$

$K_{30} = \frac{2.303}{30} \log \frac{23.05}{6.15} = 0.04404 \text{ min}^{-1}$

$K_{50} = \frac{2.303}{50} \log \frac{23.05}{2.6} = 0.04364 \text{ min}^{-1}$

constant values of k prove that it is of first order.

11. The half change time of a second order reaction is 20 min., When the initial concentration of the single reactant is 12×10^{-2} moles. Calculate the value of k.

Sol. For second order, $t_{0.5} \propto \frac{1}{a}$

or $t_{0.5} = \frac{1}{ka}$

$$k = \frac{1}{t_{1/2} .a} = \frac{1}{20 \times 12 \times 10^{-2}} = 0.416 \text{ d m}^3 \text{ mol}^{-1} \text{ min}^{-1}.$$

12. In the saponification of ethylacetate by NaOH at 283k, y ml. of 0.043 N HCl were required to neutralize 100 ml. of the reaction mixture, t minutes after the commencement of the reaction.

t (min)	0	4.89	10.37	28.18	∝
y (ml)	61.95	50.59	42.40	29.35	14.92

find out the order of reaction.

Sol. $k = \frac{1}{t} \times \frac{x}{a(a-x)}$

t	a-x	x
0	61.95(=a)	0

$K_{4.89} = \frac{1}{61.95 \times 4.89} \times \frac{11.36}{50.59} = 7.412 \times 10^{-4} \text{ ml}^{-1} \text{ min}^{-1}$

$K_{10.37} = \frac{1}{61.95 \times 10.37} \times \frac{19.55}{42.40} = 7.177 \times 10^{-4} \text{ ml}^{-1} \text{ min}^{-1}$

$K_{28.18} = \frac{1}{61.95 \times 28.18} \times \frac{32.60}{29.35} = 6.362 \times 10^{-4} \text{ ml}^{-1} \text{ min}^{-1}$

As the k values are constant, it is second order.

13. In Arrhenius equation for a certain reaction, the value

of A and E_a are $4x10^{13}$ sec^{-1} and 98.6 kJ. mol^{-1} respectively. At what temp. the reaction will have specific rate constant 1.1 x 10^{-3} sec^{-1}?

Sol. $2.303 \log k = 2.303 \log A - \frac{E_a}{RT}$

$2.303 \log (1.1x10^{-3}) = 2.303 \log (4x10^{13}) x - \frac{98.6 x 10^3}{RT}$

$T = \frac{98.6 x 10^3}{8.314 x 2.303 x 16.56} = 310.96 K$

14. The slope of the curve found in a graph between $\frac{1}{T}$ vs. $\log_{10} k$ for the decomposition of acetaldehyde was - 9920. Calculate the energy of activation for this reaction.

Sol. we know that $\frac{E_a}{2.303R}$ = slope

E_a = slope x 2.303 x R

= -9920 x 2.303 x 8.314 = 189940 J/mole

15. A first order reaction is 50% completed in 30 minutes at 27°C and in 10 min. at 47°C. Calculate the reaction rate constant at 27°C and the energy of activation of the reaction.

Sol. At 27°C, $k_{27°C} = \frac{0.693}{30} = 0.0231 \ min^{-1}$

At 47°C, $k_{47°C} = \frac{0.693}{10} = 0.0693 \ min^{-1}$

$\log \frac{k_1}{k_2} = \frac{-Ea}{2.303R} [\frac{T_2 - T_1}{T_1 x T_2}]$

$\log \frac{0.0231}{0.0693} = \frac{-Ea}{2.303x8.314} x [\frac{320-300}{320 x 300}]$

$E_a = -\frac{19.1471 x 96000}{20} x \log 0.3333$

= - 91906 x (-0.4772) = 43857 J mol^{-1}

= 43.86 k J/ mole

16. The following values for the first order rate constant were obtained for a certain reaction :

temp. (C)	k x 10^{-5} sec^{-1}
25	3.46
35	13.50

Calculate the Arrhenius frequency factor.

Sol. $\log \frac{k_2}{k_1} = \frac{E_a}{2.303R} [\frac{T_2 - T_1}{T_1 x T_2}]$

$E_a = 2.303 x \frac{1.987 x 298 x 308}{308 - 298} \log \frac{13.5 x 10^{-5}}{3.46 x 10^{-5}}$

= 24830 calories

$\because \log k_1 = -\frac{E_A}{2.303 RT} + \log A$

$\log 3.46 x 10^{-5} = \frac{24830}{2.303 x 1.987 x 298} + \log A$

log A = 13.7491

A = 5.611 x 10^{13}

17. For the bimolecular decomposition of NO_2 into NO and O_2, the velocity constant at 600k is 83.9 and at 645k is 407. Calculate the energy of activation for 2 moles of NO_2 (given log 407 =2.6096 and log 83.9 = 1.9298)

Sol. $1.9298 - 2.6096 = \frac{E_a}{2.303 x 8.314} x (\frac{45}{645 x 600})$

$E_a = \frac{0.6798 x 2.303 x 8.314 x 645 x 600}{45} = 111.94 \ kJmol^{-1}$.

18. The time required for 10% completion of a first order reaction at 298 K is equal to that required for its 25% completion at 308 K. If the pre-exponential factor for the reaction is $3.56x10^9$ sec^{-1}, calculate its rate at 318 K and also the energy of activation.

Sol. $t = \frac{2.303}{k_{298}} \log \frac{100}{90} = \frac{2.303}{k_{308}} \log \frac{100}{75}$

$\frac{k_{308}}{k_{298}} = \frac{E_a}{2.303xR} [\frac{T_2 - T_1}{T_1 x T_2}]$

$\frac{\log 100/75}{\log 100/90} = \frac{E_a}{2.303 x 8.314} x \frac{10}{298 x 308}$

$E_a = 76.66$

Now apply $k = A e^{-E_a/RT}$

$k_{318} = 3.56 x 10^9 e^{-76.66 x 10^3/(8.314 x 318)}$

$= 9.1 x 10^{-4} sec^{-1}$.

19. Raising concentration of a reactant 4 fold increases the reaction rate 64 times. What is the kinetic order of this reaction.

Sol. $R \propto [A]^a$ --------- (i)

$64 \propto [4A]^a$ --------- (ii)

Dividing (ii) by (i)

we obtain $\frac{64R}{R} = \frac{[4A]^a}{[A]^a}$

$[4]^3 = [4]^a$

a = 3 ∴ third order reaction

20. At a certain temp. the $t_{0.5}$ of the decomposition of ammonia in contact with Tungsten were as follows:

pressure (mm of Hg)	50	100	200
$t_{0.5}$ (relative)	3.52	1.82	0.93

find the order of the reaction.

Sol. First method :

1st order $t_{0.5}$ independent of initial concentration, a

2nd order $t_{0.5} \propto \frac{1}{a}$ or t x a = constant

3rd order $t_{0.5} \propto \frac{1}{a^2}$ or t x a^2 = constant

t x a	t x a^2
3.52 x 50 = 176	3.52 x $(50)^2$ = 8800
100 x 1.82 = 182	1.82 x $(100)^2$ = 18200
200 x 0.93 = 186	0.93 x $(200)^2$ = 37200

∵ t x a is almost constant, the reaction is second order

Second method :

$\frac{t_1}{t_2} = (\frac{a_2}{a_1})^{n-1}$

$n = 1 + \frac{\log t_1 - \log t_2}{\log a_2 - \log a_1}$

I. $n = 1 + \frac{\log 3.52 - \log 1.82}{\log 100 - \log 50} = 1.95$

II. $n = 1 + \frac{\log 1.82 - \log 0.93}{\log 200 - \log 100} = 1.97$

Thus order of reaction is 2.

1. The half-life of a first order reaction is 69.35. The value of the rate constant of the reaction is [CBSE-90]

(A) 1.0 s^{-1} (B) 0.1 s^{-1} (C) 0.01 s^{-1} (D) 0.001s^{-1}

2. In which of the following cases does the reaction go farthest to completion ? [CBSE-93]

(A) K = 10 (B) K = 1 (C) K = 10^3 (D) K=10^{-2}

3. Select the rate law that corresponds to the data shown for the following reaction: [CBSE-94]

A + B → C

Exp.	$[A]_0$	$[B]_0$	Initial Rate
1.	0.012	0.035	0.10
2.	0.024	0.070	0.80
3.	0.024	0.035	0.10
4.	0.012	0.070	0.80

(A) Rate = k $[B]^3$ (B) Rate = k $[B]^4$

(C) Rate = k[A]$[B]^3$ (D) Rate = k$[A]^2[B]^2$

4. Activation energy of a chemical reaction can be determined by. [CBSE-98]

(A) evaluating rate constant at two different temperatures

(B) changing concentrations of reactants

(C) evaluating rate constant at standard temperature

(D) evaluating velocities of reaction at two different temperatures

5. For a first-order reaction, the half life period is independent of [CBSE-99]

(A) initial concentration

(B) cube root of initial concentration

(C) first power of final concentration

(D) square root of final concentration

6. For a reversible reaction if the conc. of reactants is doubled, how will the rate constant and to what extent is affected at NTP. Reaction depends only on conc. of one of the reactants [CBSE-2000]

(A) Doubles (B) Halves

(C) Remains same (D) Nothing can be inferred

7. For reactiion $2N_2O_5 \rightarrow 4NO_2 + O_2$ rate and rate constant are 1.02 x 10^{-4} and 3.4 x 10^{-5} sec^{-1} respectively . Then conc. of N_2O_5 at that time will be - [CBSE-2001]

(A) 1.732 (B) 3 (C) 1.02 x 10^{-4} (D) 3.4 x 10^5

8. When a bio-chemical reaction is carried out in laboratory (outside the human body) in absence of enzyme, the rate of reaction obtained is 10^{-6} times. The activation energy of reaction in presence of enzyme is : [CBSE-2001]

(A) $\frac{6}{RT}$ (B) P is required

(C) Different from E_a obtained in laboratory

(D) Can't say anything

9. 3A → 2B, rate of reaction $\frac{+d(B)}{dt}$ is equal to [CBSE-2002]

(A) $-\frac{3}{2}\frac{d(A)}{dt}$ (B) $-\frac{2}{3}\frac{d(A)}{dt}$

(C) $-\frac{1}{3}\frac{d(A)}{dt}$ (D) $+2\frac{d(A)}{dt}$

10. 2A → B + C It would be a zero order reaction when [CBSE-2002]

(A) The rate of reaction is proportional to square of conc. of A

(B) The rate of reaction remains same at any conc. of A

(C) The rate remains unchanged at any conc. of B and C

(D) The rate of reaction doubles if conc. of B is doubled

11. The temperature dependence of the rate constant (k) of a chemical reaction is written in terms of Arrhenius equation, $k=A.e^{-E^*/RT}$. Activation energy (E*) of the reaction can be calculated by plotting [CBSE-2003]

(A) log k vs 1/log T (B) k vs T

(C) k vs 1/ log T (D) log k vs 1/T

12. If the rate of the reaction is equal to the rate constant the order of the reaction is [CBSE-2003]

(A) 3 (B) 0 (C) 1 (D) 2

13. The activation energy for a simple chemical reaction A ⟶ B is E_a in the forward direction. The activation energy for the reverse reaction [CBSE-2003]

(A) is always double of E_a (B) is negative of E_a

(C) is always less than E_a

(D) can be less than or more than E_a

14. The reaction A → B follows first order kinetics. The time taken for 0.8 mole of A to produce 0.6 mole of B is 1 hour. What is the time taken for conversion of 0.9 mole of A to produce 0.675 mole B ? [CBSE-2003]

(A) 2 hours (B) 1 hour (C) 0.5 hour (D) 0.25 hour

15. Assertion : The molecularity of the reaction $H_2 + Br_2 = 2HBr$ is two. Reason : The order of this reaction is 3/2. [AIIMS-95]

(A) A (B) B (C) C (D) D

16. In a reaction : A → B, the rate of reaction increases two times on increasing the concentration of reactants four time, then order of reaction is [AIIMS-97]

(A) 0 (B) 2 (C) 1/2 (D) 4

17. The rate constant of a reaction is 0.69 x 10^{-2} min^{-1} and the initial concentration is 0.2 mol L^{-1}. The half-life period is [AIIMS-98]

(A) 400 sec (B) 600 sec (C) 800 sec (D) 1200 sec

18. A first order reaction which is 30% complete in 30 minutes has a half-life period of [AIIMS-98]
(A) 24.2 min (B) 58.2 min
(C) 102.2 min (D) 120.2 min

19. The rate constant of a first order reaction is 3×10^{-6} per sec. If the initial concentration is 0.10 mol, the initial rate of reaction is [AIIMS-99]
(A) 3×10^{-5} mol s^{-1} (B) 3×10^{-6} mol s^{-1}
(C) 3×10^{-8} mol s^{-1} (D) 3×10^{-7} mol s^{-1}

20. In the reaction $A + B \rightarrow C$, the potential energy as function of reaction coordinate is shown in the figure. What is the value of activation energy E_a ? [AIIMS-2000]

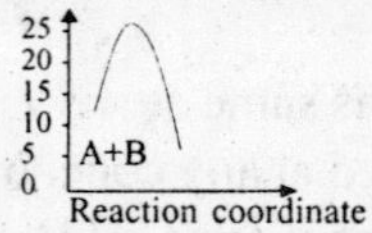

(A) 5 K cal/mole (B) 10 K cal/mole
(C) 20 K cal/mole (D) 15 K cal/mole

21. HCl or H_2SO_4 may be used for the hydrolysis of esters. The rate constant value will be. [AIIMS-2000]
(A) Same for both acids (B) Greater if HCl is used
(C) Greater if H_2SO_4 is used
(D) Will depend upon the nature of ester

22. Given $\rightarrow \frac{_{92}U^{238}}{_{82}Pb^{206}} = \frac{4}{1}$, half life of $_{92}U^{238} = 1.35 \times 10^9$ then the age of crystal will be : [AIIMS-2001]
(A) 4.2×10^8 (B) 4.2×10^9 (C) 4.2×10^{10} (D) 4.2×10^{11}

23. The potential energy diagram for a reaction R→P is given [AIIMS-2003]

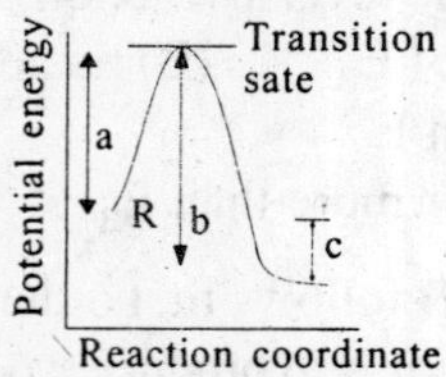

ΔH^o of the reaction corresponds to the energy -
(A) a (B) b (C) c (D) a + b

24. For the first order reaction reaction half life period is 20 min. What percentage of the reactant is left at the end of 100 minutes ? [JIPMER-93]
(A) 25% (B) 12.5% (C)10.0% (D) 3.125%

25. For a hypothetical reaction aA+bB → product, the rate law is $r = k [A]^x [B]^y$, then : [JIPMER-96]
(A) (a +b) = (x + y) (B) (a +b) < (x + y)
(C) (a +b) > (x + y)
(D) (a + b) may not be equal to (x + y)

26. The order of a reaction with respect to one of the reacting component Y is zero. It implies that : [JIPMER-96]
(A) The reaction must proceed at a constant rate.
(B) The rate of a reaction does not very with temperature.
(C) The reaction rate is independent of the conc. of Y
(D) The rate of formation of activated complex is zero.

27. For a hypothetical reaction $A \rightarrow L$, the rate expression is $r = \frac{-d[A]}{dt}$. The negative sign represents [JIPMER-96]
(A) Attractive forces between the reactants
(B) That rate is negative
(C) Decrease in concentration of reactants
(D) None is correct.

28. If the rate of a chemical reaction between gases is independent of pressure the order of the reaction is [JIPMER-97]
(A) 0 (B) 1 (C) 2 (D) 3

29. At 25°C the time for half-change for the thermal dissociation of N_2O is 5.7 hours. The specific rate constant is given by [JIPMER-97]
(A) $\frac{1}{5.7}$ $hour^{-1}$ (B) $\frac{\ln 2}{5.7}$ $hour^{-1}$
(C) $\ln \frac{2}{5.7}$ $hour^{-1}$ (D) 0.5 $hour^{-1}$

30. The alkaline hydrolysis of an ester represented by $RCOOR' + OH^- \longrightarrow RCOO^- + R'OH$ is [JIPMER-98]
(A) First order, bimolecular
(B) Second order but not bimolecular
(C) Bimolecular but zero order
(D) Bimolecular and second order

31. For the reaction A + B → Products, doubling the concentration of A the rate of the reaction is doubled, but on doubling the concentration of B rate remains unaltered. The over all order of the reaction is [JIPMER-98]
(A) 1 (B) 0 (C) 2 (D) 3

32. The activation energy for a reaction is 9.0 k cal/mol. The increase in the rate constant when its temperature is increased from 298 K to 308 K is [JIPMER-2000]
(A) 63% (B) 50% (C) 100% (D)10%

33. The rate for a first order reaction is 0.6932×10^{-2} mol l^{-1} min^{-1} and the initial concentration of the reactants is 1 mol , $T_{1/2}$ is equal to [JIPMER-2001]
(A) 6.932 min. (B) 100 min.
(C) 0.6932×10^{-3} min. (D) 0.6932×10^{-2} min.

34. For a reaction $2N_2O_5 \longrightarrow 4NO_2 + O_2$ rate and rate constant are 1.02×10^{-4} mol $l^{-1}S^{-1}$ and 3.4×10^{-5} sec^{-1}. The conc. of N_2O_5 at that time will be [JIPMER-2002]
(A) 1.732 mol l^{-1} (B) 3 mol l^{-1}
(C) 1.02×10^{-4} mol l^{-1} (D) 3.2 x 105 mol l^{-1}

35. Km value of an enzyme is the subsrate concentration of which [BHU-97]
(A) Reaction attains two times of its maximum velocity

(B) Reaction attains half of its maximum velocity
(C) Reaction attains 8 times maximum velocity
(D) Reaction attains same as maximum velocity

36. Hydroysis of sucrose is called [BHU-97]
(A) Saponification (B) Hydration
(C) Esterification (D) Inversion

37. If initial concentration is reduced to 1/4th in a zero order reaction, the time taken for half the reaction to complete [BHU-98]
(A) Remains same (B) Becomes 4 times
(C) Becomes one-fourth (D) Doubles

38. The rate constant of a first order reaction is 3 x 10^{-6} per sec. If the initial concentration is 0.10 mol, the initial rate of reaction is [BHU-99]
(A) 3 x 10^{-5} mol s^{-1} (B) 3 x 10^{-6} mol s^{-1}
(C) 3 x 10^{-7} mol s^{-1} (D) 3 x 10^{-8} mol s^{-1}

39. Hydrolysis of ester in alkaline medium is [AFMC-94]
(A) First order reaction with molecularity one
(B) Second order reaction with molecularity two
(C) First order reaction with molecularity two
(D) Second order reaction with molecularity one

40. The units of a first order rate constant are [AFMC-94]
(A) mol $dm^{-3}s^{-1}$ (B) dn^3 mol^{-1} s^{-1}
(C) s^{-1} (D) None of the above

41. If concentration of reactants is increased by 'X', then the K becomes [AFMC-97]
(A) In K/x (B) K/x (C) K + x (D) K

42. The reaction N_2O_5(in CCl_4 solution)$\rightarrow$2 NO_2 (solution) + 1/2 O_2 (g) is of first order in N_2O_5 with rate constant 6.2 x 10^{-1} s^{-1}. What is the value of rate of reaction when $[N_2O_5]$ = 1.25 mol l^{-1} ? [AFMC-98]
(A) 7.75 x 10^{-4} mole l^{-1} s^{-1} (B) 6.35 x 10^{-3} mole l^{-1} s^{-1}
(C) 5.15 x 10^{-2} mole l^{-1} s^{-1} (D) 3.85 x 10^{-1} mole l^{-1} s^{-1}

43. Rate constant depends on [AFMC-2000]
(A) Temperature (B) Time
(C) Initial concentration (D) None of these

44. For reaction A + B $\rightleftharpoons$ C + D, doubling the concentration of both the reactants increases the reaction rate by 8 times and doubling initial concentration of only B simply doubles the reaction rate. The rate law is given as [AFMC-2001]
(A) r = K [A] [B] (B) r = K $[A]^2$ [B]
(C) r = K [A] $[B]^2$ (D) r = K $[A]^{1/2}$ $[B]^{1/2}$

45. Effect of temperature on reaction rate is given by [AFMC-2001]
(A) Clausius - Clapeyron equation
(B) Arrhenius equation
(C) Gibb's-Helmholtz equation (D) Kirchhoff's equation

46. Hydrolysis of sucrose is called [AFMC-2001]
(A) Hydration (B) Saponification
(C) Esterification (D) Inversion

47. For the reaction $N_2(g)$ +$3H_2(g)$ $\rightarrow$ $2NH_3(g)$ under certain conditions of temperature and partial pressure of the reactants, the rate of formation of NH_3 is 0.001kg hr^{-1}. The rate of conversion of H_2 under the same conditions is .... kg hr^{-1}. [IIT-94]

48. The compound $YBa_2Cu_3O_7$, which shows superconductivity, has copper in oxidation state .... Assume that the rare earth element Yttrium is in its usual +3 oxidation state. [IIT-94]

49. The rate constant, the activation energy and the Arrhenius parameter of a chemical reaction at 25°C are 3.0 x $10^{-4}s^{-1}$, 104.4 kJ mol^{-1} and 6.0 x $10^{14}s^{-1}$ respectively. The value of rate constant as T$\rightarrow\infty$ is [IIT-96]
(A) 2.0 x 10^{18} s^{-1} (B) 6.0 x 10^{14} s^{-1}
(C) infinity (D) 3.6 x 10^{30} s^{-1}

50. In Arrhenius equation, k = A $e^{(-E/RT)}$, A may be termed as the rate constant at .... [IIT-97]

51. For a first order reaction [IIT-98]
(A) the degree of dissociation is equal to $(1-e^{-kt})$
(B) a plot of reciprocal concentration of the reactant vs. time gives a straight line
(C) the time taken for the completion of 75% reaction is thrice $t_{1/2}$ of the reaction
(D) the pre-exponential factor in Arrhenius equation has the dimension of time T^{-1}

52. When two reactants, A & B are mixed to give products C & D, the reaction quotient, Q, at the initial stages of the reaction [IIT-2000]
(A) is zero (B) decreases with time
(C) is independent of time (D) increases with time

53. The rate constant for the reaction, $2N_2O_5 \rightarrow 4NO_2+O_2$ is 3.0x10^{-5} sec^{-1}. If the rate is 2.40x10^{-5} mol $litre^{-1}$ sec^{-1}, then the concentration of N_2O_5 (mol $litre^{-1}$) is [IIT-2000]
(A) 1.4 (B) 1.2 (C) 0.04 (D) 0.8

54. If 'I' is the intensity of absorbed light and 'C' is the concentration of AB for the photochemical process AB + hv $\rightarrow$ AB*, the rate of formation of AB is directly proportional to [IIT-2001]
(A) C (B) I (C) I^2 (D) C.I

55. Consider the chemical reaction, $N_2(g)$ + $3H_2(g)$ $\rightarrow$ $2NH_3(g)$. The rate of this reaction can be expressed in terms of time derivative of concentration of $N_2(g)$, $H_2(g)$ or $NH_3(g)$. Identify the correct relationship amongst the

rate expressions [IIT-2002]

(A) Rate = $-d[N_2]/dt = -1/3d[H_2]/dt = 1/2d[NH_3]/dt$

(B) Rate = $-d[N_2]/dt = -3d[H_2]/dt = 2d[NH_3]/dt$

(C) Rate = $d[N_2]/dt = 1/3d[H_2]/dt = 1/2d[NH_3]/dt$

(D) Rate = $-d[N_2]/dt = -d[H_2]/dt = d[NH_3]/dt$

56. Find the rate constant (s^{-1}) of first order when the activity drops from 800 mol/dm³ to 50 mol/dm³ in time 2 sec [IIT-2003]

(A) $1.40 \times 10^{-5}\ s^{-1}$ (B) $1.30 \times 10^{-4}\ s^{-1}$

(C) $1.30 \times 10^{-8}\ s^{-1}$ (D) $1.38^{-1}\ s^{-1}$

57. Units of rate constant of first and zero order reactions in terms of molarity M unit are respectively. [AIEEE-2002]

(A) sec^{-1}, mol sec^{-1} (B) sec^{-1}, mol

(C) mol sec^{-1}, sec^{-1} (D) mol, sec^{-1}

58. The differntial rate law for the reaction $H_2 + I_2 \rightarrow 2HI$ is [AIEEE-2002]

(A) $-\frac{d[H_2]}{dt} = \frac{d[I_2]}{dt} = -\frac{d[HI]}{dt}$

(B) $\frac{d[H_2]}{dt} = \frac{d[I_2]}{dt} = \frac{1}{2}\frac{d[HI]}{dt}$

(C) $\frac{1}{2}\frac{d[H_2]}{dt} = \frac{1}{2}\frac{d[I_2]}{dt} = -\frac{d[HI]}{dt}$

(D) $-2\frac{d[H_2]}{dt} = -2\frac{d[I_2]}{dt} = \frac{d[HI]}{dt}$

59. The integrated rate equation is $Rt = \log C_o - \log C_t$ The straight line graph is obtained by plotting. [AIEEE-2002]

(A) time vs $\log C_t$ (B) $\frac{1}{time}$ vs C_t

(C) time vs C_t (D) $\frac{1}{time}$ vs $\frac{1}{C}$

60. The rate law for a reaction between the substances A and B is given by Rate = $k[A]^n[B]^m$
One doubling the concentration of A and halving the concentration of B, the ratio of the new rate to the earlier rate of the reaction will be as [AIEEE-2003]

(A) $\frac{1}{2^{(n-m)}}$ (B) (m + n) (C) (n – m) (D) $2^{(n-m)}$

61. For reaction system : $2NO\ (g) + O_2\ (g) \rightarrow 2NO_2\ (g)$ volume is suddenly reduced to half its value by increasing the pressure on it. If the reaction is of first order with respect to O_2 and second order with respect to NO, the rate of reaction will [AIEEE-2003]

(A) Diminish to one-fourth of its initial value

(B) Diminish to one-eight of its initial value

(C) Increase to eight times of its initial value

(D) Increase to four times of its initial value

62. In respect of the equation $k = Ae^{-Ea/RT}$ in chemical kinetics, which one of the following statements is correct ? [AIEEE-2003]

(A) k is equilibrium constant (B) A is adsorption factor (C) E_a is energy of activation (D) R is Rydberg's constant

63. Which order reaction obeys the expression $t_{1/2} = 1/KA$ in chemical kinetics .... [EAMCET-90]

64. A catalyst increases the rate of reaction, because it [EAMCET-92]

(A) Increases the activation energy

(B) Decreases the energy barrier for reaction

(C) Decreases the collision diameter

(D) Increases the temperature coefficient

65. The rate of reaction is generally doubled for a .... degree rise in temperature. [EAMCET-92]

66. The use of a catalyst in chemical reaction .... activation energy of the reaction. [EAMCET-93]

67. The rate of a reaction $2A + 3B + 4C \rightarrow$ products is equal to $R = [A]^2 [B]^1 [C]^0$. The overall order of the reaction is [EAMCET-95]

(A) 9 (B) 5 (C) 2 (D) 3

68. For a reaction A + B → products, the rate of the reaction was doubled when the concentration of A was doubled. When the concentration of A & B were doubled, the rate was again doubled, the order of the reaction with respect to A & B are [EAMCET-95]

(A) 1, 1 (B) 2, 0 (C) 1, 0 (D) 0, 1

69. The units of rate of a first order reaction are [EAMCET-95]

(A) sec^{-1} (B) $lit.mol^{-1}$ (C) $mole\text{-}lit^{-1}\ sec^{-1}$ (D) No units

70. The time for half life of a first order reaction is 1 hr. What is the time taken for 87.5% completion of the reaction ? [EAMCET-95]

(A) 1 hour (B) 2 hours (C) 3 hours (D) 4 hours

71. 75% of a first order reaction is complete in 32 min. 50% of the reaction would have been completed in [EAMCET-98]

(A) 24min (B) 16min (C) 18min (D) 23min

72. After how many seconds will the concentration of the reactant in a first order reaction be halved if the rate constant is $1.155 \times 10^{-3}\ sec^{-1}$? [EAMCET-98(MBBS)]

(A) 600 (B) 100 (C) 60 (D) 10

73. The half life of a first order reaction is 10 sec. What its rate constant (in sec^{-1}) [EAMCET-99]

(A) 0.0693 (B) 0.693 (C) 6.93 (D) 69.3

74. From the following data for the decomposition of N_2O_5 at 30°C, find out the rate constant (in min^{-1})
Volume of O_2 after 10 minutes of the reaction = 90 ml
Volume of O_2 after completion of the reaction = 100 ml
[EAMCET-99(MBBS)]

(A) 2.303 (B) 0.2303 (C) 0.02303 (D) 23.03

75. What is the rate of the reaction for $2A \rightarrow B$? [EAMCET(MBBS)-2000]

(A) 1/2 (–d[A]/dt) (B) –d[A]/dt
(C) –d[B]/dt (D) +d[A]/dt

76. Consider the following reaction $A \rightarrow$ products. This reaction is completed in 100 minutes. The rate constant of this reaction at t_1 = 10 min is 10^{-2} min^{-1} What is the rate constant (in min^{-1}) at t_2 = 20 minutes ?[EAMCET-2001]
(A) 2 x 10^{-2} (B) 10^{-2} (C) 5 x 10^{-3} (D) 0.1

77. The rate of the reaction for $A \rightarrow$ products is 10 mol. lit^{-1} min^{-1} at time t_1=2 minutes. What will be the rate at time t_2 = 12 minute? [EAMCET(MBBS)-2001]
(A) more than 10 (B) less than 10 (C) 10 (D) 20

78. Which of the following is correct for a first order reaction ? (k = rate constant $t_{1/2}$ = half-life) [EAMCET(med.)-2002]

(A) $t_{1/2}$ = 0.693 x k (B) $k.t_{1/2}$ = 1/0.693
(C) $k.t_{1/2}$ = 0.693 (D) 6.93 x k x $t_{1/2}$ = 1

79. The minimum energy required for molecules to enter into the reaction is called : [PMT-93]
(A) potential energy (B) kinetic energy
(C) nuclear energy (D) activation energy

80. The inversion of cane-sugar into glucose and fructose, according to the equation :

$$\underset{\text{cane sugar}}{C_{12}H_{22}O_{11}} + H_2O \rightarrow \underset{\text{glucose}}{C_6H_{12}O_6} + \underset{\text{fructose}}{C_6H_{12}O_6}$$

is an example of : [PMT-93]

(A) First order reaction (B) Second order reaction
(C) Third order reaetion (D) Zero order reaction

81. The reaction rate at a given temperature becomes slower when : [PMT-93]
(A) the free energy of activation is higher
(B) the free energy of activation is lower
(C) entropy changes
(D) initial concentration of the reactants remains constant

82. The unit for specific reaction rate of second order reaction is : [PMT-94]
(A) sec^{-1} (B) mole $litre^{-1}$ sec^{-1}
(C) $litre^2$ $mole^{-2}$ sec^{-1} (D) litre $mole^{-1}$ sec^{-1}

83. Half life period of second order reaction is [PMT-94]
(A) Proportional to initial concentration of reactants
(B) Independent of initial concentration of reactants
(C) Inversely proportional to initial concntration of reactants
(D) Inversely proportional to square of initial concentration of reactants

84. The term – $\left(\frac{dc}{dt}\right)$ in a rate equation refers to the [PMT-96]
(A) concentration of the reactant
(B) decreases in concentration of the reactant with time
(C) increase in concentration if the reactant with time
(D) velocity constant of the reaction

85. If the rate expression for a chemical reaction is given by Rate = k $[A]^m$ $[B]^n$ [PMT-96]
(A) the order of the reaction is m
(B) the order of the reaction is n
(C) the order of the reaction is m + n
(D) the order of the reaction is m - n

86. Unit of first order rate constant is [PMT-96]
(A) mole litre second (B) $mole^{-1}$ $litre^{-1}$ $second^{-1}$
(C) mole $litre^{-1}$ $second^{-1}$ (D) $second^{-1}$, $minute^{-1}$ etc.

87. Radioactive decay is a reaction of [PMT-97]
(A) Zero order (B) Second order
(C) Third order (D) First order

88. The rate at which substances react depends on their [PMT-97]

(A) Atomic weight (B) Molecular weight
(C) Equivalent weight (D) Active mass

89. When temperature is raised, the rate of the reaction increases because of [PMT-97]
(A) Lowering of activation energy
(B) Increase in the number of collisions
(C) Decrease in the number of activated molecules
(D) Decrease in the number of collisions

90. The half-life period of a first order reaction is 100 sec. The rate constant of the reaction is [PMT-97]
(A) 6.93 x 10^{-3} sec^{-1} (B) 6.93 x 10^{-4} sec^{-1}
(C) 0.693 sec^{-1} (D) 69.3 sec^{-1}

91. Amongst the following chemical reactions, the unimolecular reaction is : [PMT-99]
(A) $H_2 + Cl_2 \rightarrow 2HCl$ (B) $PCl_3 + Cl_2 \rightarrow PCl_5$
(C) $N_2O_5 \rightarrow N_2O_4 + 1/2\ O_2$ (D) $2HI \rightarrow H_2 + I_2$

92. The unit of rate constant of a second order chemical reaction is : [PMT-99]

(A) $mole^{-1}$ $litre^{-1}$ $second^{-1}$ (B) mole $litre^{-1}$ $second^{-1}$
(C) mole litre second (D) $mole^{-1}$ litre $second^{-1}$

93. The unit of rate constant of a first order chemical reaction is : [PMT-99]
(A) mole litre (B) second $^{-1}$
(C) mole $^{-1}$ second $^{-1}$ (D) mole $^{-1}$ second $^{-2}$

94. The expression for rate constant of a first order chemical reaction is : [PMT-99]

(A) $k = \frac{x}{t}$ (B) $k = \frac{2.303}{t} \log_{10}\left(\frac{a}{a - x}\right)$

(C) $k = \frac{1}{t} \cdot \frac{x}{a(a-x)}$ (D) $k = \frac{1}{2t}\left[\frac{1}{(a-x)^2} - \frac{1}{a^2}\right]$

95. Which of the following is a unit of zero order reaction? [PMT-2000]
(A) mol. l^{-1}. s^{-1} (B) l. mol. l^{-1}. s^{-1}
(C) l^{-1} mol. l^{-1}. s^{-1} (D) l. mol. s

96. For a reaction $2NO(g) + Cl_2 \rightleftharpoons 2NOCl(g)$, when concentration of Cl_2 is doubled, the rate of reaction becomes two-times of the original. When the concentration of NO is doubled the rate becomes four times. What is the order of the reaction ? [PMT-2000]
(A) 1 (B) 2 (C) 3 (D) 4

97. For a first order reaction velocity constant. k = 10 Two third life for it would be : [PMT-2001]
(A) 1100 s (B) 2200 s (C) 3300 s (D) 4400 s

98. The reaction $2NO(g) + O_2(g) \rightleftharpoons 2NO_2(g)$ is of first order. If volume of reaction vessel is reduced to 1/3 the rate of reaction would be : [PMT-2001]
(A) 1/3 times (B) 2/3 times
(C) 3 times (D) 6 times

99. For the functioning of enzymes which of the following statements is not correct ? [PMT-2001]
(A) An optimum temperature is needed
(B) An optimum pH is needed
(C) They are substrate specific
(D) They always increase activation energy

100. Hydrolysis of ethyl acetate is catalysed by aqueous [PMT-2002]
(A) Na_2SO_4 (B) K_2SO_4 (C) H_2SO_4 (D) $BaSO_4$

101. When the concentration of 'A' in the reaction $A+B \rightleftharpoons AB$ is doubled, the rate of reaction will be [PMT-2002]
(A) Doubled (B) Decreased by half
(C) Unchanged (D) Increased by four times

102. Which of the following expressions is correct for the first order reactions (Co refers to initial concentration and $t_{1/2}$ refers to half life time : [PMT-2003]
(A) $t_{1/2} \propto Co^0$ (B) $t_{1/2} \propto Co^{-1}$
(C) $t_{1/2} \propto Co$ (D) $t_{1/2} \propto Co^{1/2}$

103. For the reaction $A + B \rightarrow$ product it is found that if the concentration A is doubled, velocity increase, four times but when concentrations of B is doubled velocity of reaction is not effected. The right velocity equation is : [PMT-2003]
(A) Velocity = $k/4\,[A]^2$ (B) Velocity = $k\,[A]\,[B]$
(C) Velocity = $k\,[A]^2\,[B]^0$ (D) Velocity = $k\,[A]^2\,[B]^2$

104. For an endothermic reaction where ΔH represents the enthalpy of the reaction in kJ/mol, the minimum value for the energy of activation will be [PET-93]
(A) Less than ΔH (B) Zero
(C) More than ΔH (D) Equal to ΔH

105. For the reaction $H_2(g) + Br_2(g) \rightarrow 2HBr(g)$ the experimental data suggest rate = $k[H_2][Br_2]^{1/2}$ The molecularity and order of the reaction are respectively [PET-93]
(A) $2, \frac{3}{2}$ (B) $\frac{3}{2}, \frac{3}{2}$ (C) 1, 1 (D) $1, \frac{1}{2}$

106. The rate of the reaction $CCl_3CHO + NO \rightarrow CHCl_3 + NO + CO$ is given by equation. rate = $k\,[CCl_3CHO]\,[NO]$. If concentrations are expressed in moles/litre the units of k are [PET-93]
(A) $litre^2\ mole^{-2}\ sec^{-1}$ (B) $mole\ litre^{-1}\ sec^{-1}$
(C) $litre\ mole^{-1}\ sec^{-1}$ (D) $second^{-1}$

107. For reaction $2A + B \rightarrow$ Products, the active mass of B is kept constant and that of A is doubled. The rate of reaction will then [PET-93]
(A) Increase 2 times (B) Increase 4 times
(C) Decrease 2 times (D) Decrease 4 times

108. In a reaction $2A + B \rightarrow A_2B$ the reactant A will disappear at [PET-93]
(A) Half the rate B will decease
(B) The same rate that B will decrease
(C) Twice the rate that B will decrease
(D) The same rate that A_2B will form

109. In the reaction $2A + B \rightarrow A_2B$, if the concentration of A is doubled and of B is halved, then the rate of the reaction will [PET-94]
(A) Increase by four times (B) Decrease by two times
(C) Increases by two times (D) Remain the same

110. The units of rate constant of a first order reaction are [PET-94]
(A) mol lit^{-1} sec^{-1} (B) sec^{-1}
(C) mol^{-1} sec (D) litre sec^{-1} mol^{-1}

111. The rate constant (K') of one reaction is double of the rate constant (K") of another reaction. Then the relationship between the corresponding activation energies of the two reactions (E' a and Ea") will be [PET-94]
(A) Ea' > Ea" (B) Ea' = Ea"
(C) Ea' < Ea" (D) Ea' = 4 Ea"

112. An increase in the rate of a reaction for a rise in temperature is due to [PET-95]
(A) The increase in the number of collisions
(B) The decrease in the number of activated molecules
(C) The shortening of the mean free path
(D) The lowering of the activation energy

113. A reaction rate constant is given by $k = 1.2 \times 10^{14} e^{-25000/RT}\ sec^{-1}$. It means [PET-95]

(A) log k versus log T will give a straight line with slope as - 25000
(B) log k versus T will give a straight line with slope as - 25000
(C) log k versus log 1/T will give a straight line with slope as - 25000
(D) log k versus 1/T will give a straight line

114. The decomposition of N_2O_5 is a first order reaction represented by
$N_2O_5 \rightarrow N_2O_4 + \frac{1}{2}O_2$.
After 15 minutes the volume of O_2 produced is 9 ml and at the end of the reaction 35 ml. The rate constant is equal to - [PET-95]
(A) $\frac{1}{15}\ln\frac{35}{44}$ (B) $\frac{1}{15}\ln\frac{44}{26}$
(C) $\frac{1}{15}\ln\frac{44}{35}$ (D) $\frac{1}{15}\ln\frac{35}{26}$

115. The thermal decomposition of a compound is first order. If a sample of the compound decomposes 50% in 120 minutes, in what time will it undergo 90% decomposition ? [PET-96]
(A) Nearly 240 minutes (B) Nearly 480 minutes
(C) Nearly 450 minutes (D) Nearly 400 minutes

116. The order of a reaction with rate equals to $k\,C_A^{3/2}\,C_B^{-1/2}$ is [PET-96]
(A) 2 (B) 1 (C) - 1/2 (D) 3/2

117. A first order reaction is 75% complete after 32 minutes. When was 50% of the reaction completed ? [PET-97]
(A) 16 minutes (B) 8 minutes
(C) 4 minutes (D) 32 minutes

118. The Arrhenius equation expressing the effect of temperature on the rate constant is : [PET-97]
(A) $k = e^{-E_a/RT}$ (B) $k = \frac{E_a}{RT}$
(C) $k = \log_e \frac{E_a}{RT}$ (D) $k = Ae^{-E_a/RT}$

119. The first order rate constant for the decomposition of N_2O_5 is 6.2×10^{-4} second^{-1}. The half life period for this decomposition is : [PET-97]
(A) 1117.7 (B) 111.7 (C) 223.4 (D) 160.9

120. For the first order reaction with the constant k, which expression givens the half-life period ? (Initial concentration = a) [PET-98]
(A) $\frac{\ln 2}{k}$ (B) $\frac{1}{k_a}$ (C) $\frac{0.693}{k_a}$ (D) $\frac{3}{2k_{a^2}}$

121. The rate constant is given by the equation $k = pze^{-E/RT}$. Which factor should register a decrease for the reaction to proceed more rapidly ? [PET-98]
(A) T (B) Z (C) E (D) P

122. For a reaction of the type A + B → products it is observed that doubling the concentration of A causes the reaction rate to be four times as great, but doubling the amount of B does not affect the rate. The rate equation is : [PET-98]
(A) Rate = k [A] [B] (B) Rate = k/4 $[A]^2$
(C) Rate = k $[A]^2$ [B] (D) Rate =k $[A]^2$ $[B]^2$

123 For the following homogeneous reaction, the unit of rate constant is :
$A + B \longrightarrow C$ [PET-99]
(A) sec^{-1} (B) sec^{-1} mole (C) sec^{-1} $mole^{-1}$ (D) sec

124. The temperature coefficient of most of the reactions lies between : [PET-99]
(A) 1 and 3 (B) 2 and 3 (C) 1 and 4 (D) 2 and 4

125. The rate constant of a first order reaction whose half-life is 480 seconds, is : [PET-99]
(A) 2.88×10^{-3} sec^{-1} (B) 1.44×10^{-3} sec^{-1}
(C) 1.44 sec^{-1} (D) 0.72×10^{-3} sec^{-1}

126. The conversion of A → B follows second order kinetics. Doubling the concentration of A will increase the rate of formation of B by a factor : [PET-99]
(A) 1/4 (B) 2 (C) 1/2 (D) 4

127. The given reaction $2FeCl_3 + SnCl_2 \rightarrow 2FeCl_2 + SnCl_4$ is an example of : [PET-99]
(A) First order reaction (B) Second order reaction
(C) Third order reaction (D) None of these

128. If reaction between A and B to give C shows first order kinetics in A and second order in B, the rate equation can be written as : [PET-99]
(A) Rate = k [A] $[B]^{1/2}$ (B) Rate = k $[A]^{1/2}$ [B]
(C) Rate = k [A] $[B]^2$ (D) Rate = k $[A]^2$ [B]

129. Inversion of sucrose is : [PET-2000]
(A) First order reaction (B) Zero order reaction
(C) Second order reaction (D) Third order reaction

130. For the reaction $2HI \rightleftharpoons H_2 + I_2$, the rate of the reaction is proportional to $[HI]^2$ this means that the reaction is : [PET-2000]
(A) Unimolecular (B) Bimolecular
(C) Of first order (D) Of second order

131. For a reaction 2A + B → Products, doubling the initial concentration of both the reactants increases the rate by a factor of 8 and doubling the concentration of B alone doubles the rate. The rate law for the reaction is : [PET-2001]
(A) γ = k [A] $[B]^2$ (B) γ = k $[A]^2$[B]
(C) γ = k [A] [B] (D) γ = k $[A]^2$ $[B]^2$

132. For the first order reaction A → Product that half-life time is 100 seconds. The rate constant of the reaction

is : [PET-2001]
(A) $6.93 \times 10^{-2}\ s^{-1}$ (B) $6.93 \times 10^{-4}\ s^{-1}$
(C) $6.93 \times 10^{-3}\ s$ (D) 6.93 s

133. The time of completion of 90% of a first order reaction is approximately [PET-2002]
(A) 1.1 times that of half-life
(B) 2.2 times that of half-life
(C) 3.3 times that of half-life
(D) 4.4 times that of half-life

134. The rate law for the reaction
$2N_2O_5 \rightarrow 4NO_2 + O_2$ is [PET-2002]
(A) $r = K[N_2O_5]$ (B) $r = K[N_2O_5]^2$
(C) $r = K[N_2O_5]^0$ (D) $r = K[NO_2]^4[O_2]$

135. The half-life period for a first order reaction is 693 seconds. The rate constants for this reaction would be [PET-2002]
(A) 0.1 sec^{-1} (B) 0.01 sec^{-1}
(C) 0.001 sec^{-1} (D) 0.0001 sec^{-1}

136. A graph plotted between log k Vs. 1/T for calculating activation energy is shown by [PET-2002]

(A) 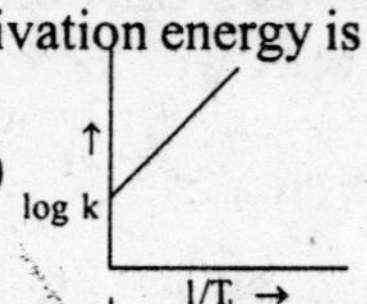

(B) 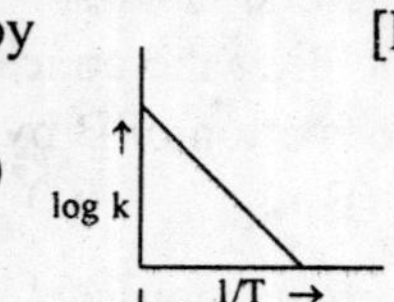

(C) 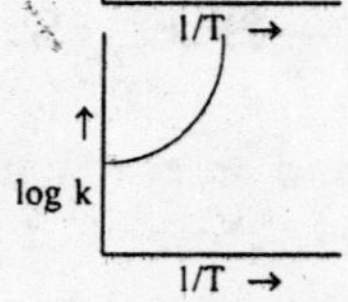

(D) 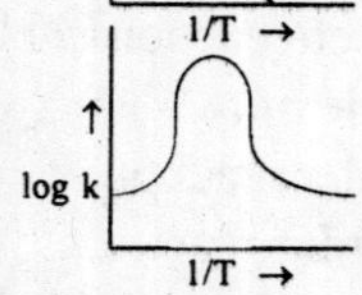

137. A gaseous reaction has a rate = [A] [B] if volume of reaction vessel is reduced to $\frac{1}{4}$ the rate becomes in comparision to original. [PET-2003]
(A) $\frac{1}{8}$ (B) $\frac{1}{16}$ (C) 8 times (D) 16 times

138. In a reaction reactant 'A' decomposes 10% in 1 hour 20% in 2 hour and 30% in 3 hour, the unit of rate constant of this reaction in [PET-2003]
(A) s^{-1} (B) litre $mol^{-1}\ s^{-1}$
(C) mol $litre^{-1}\ s^{-1}$ (D) $litre^2\ mole^{-2}\ s^{-1}$

139. For the reaction : the order of reaction is
$H_2 + Cl_2 \xrightarrow{\text{sunlight}} 2HCl$ [AIIMS-2002]
(A) 0 (B) 1 (C) 2 (D) 3

140. The order of the reaction occurring by following mechanism should be [JIPMER-95]
(i) $A_2 \rightarrow A + A$ (fast)
(ii) $A + B_2 \leftrightarrow AB + B$ (slow)
(iii) $A + B \rightarrow$ (fast)
(A) $1\frac{1}{2}$ (B) 2 (C) $3\frac{1}{2}$ (D) None of these

141. The rate of reaction depends upon [AFMC-95]
(A) Initial concentration of reactants
(B) Temperature (C) Pressure (D) Volume

142. For a first order reaction, the unit of rate constant is:
(A) mol $l^{-1}\ s^{-1}$ (B) $mol^{-1}\ l^{-1}$ (C) s^{-1} (D) mol $l^{-1}s$

143. The order of a reaction with rate equal to $kC_A^{+3/2}\ C_B^{-1/2}$ will be : [PET-2001]

P.E.
(A) 2 (B) 1 (C) – 1/2 (D) 3/2

144. Which respect to the figure given which of the following statement is correct ? [EAMCET-95]

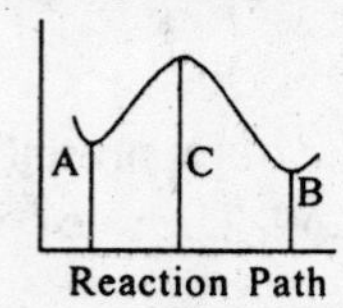

(A) ΔE for the forward reaction is C–B
(B) ΔE for the forward reaction is B–A
(C) $E_{(forward)} > E_{(backward)}$
(D) E (for reverse reaction) = C–A

145. Consider the following reaction :
$N_2(g) + 3H_2(g) \rightarrow 2NH_3(g)$
The rate of this reaction in terms of N_2 at T (K) is $-\frac{d[N_2]}{dt} = 0.02$ mol. $lit^{-1}\ sec^{-1}$. What is the value of $-\frac{d[H_2]}{dt}$ (in units of mol.$lit^{-1}sec^{-1}$) at the same temperature? [EAMCET-2002]
(A) 0.02 (B) 50 (C) 0.06 (D) 0.04

146. A + B → Product. The value of rate constant depends on [BHU-2003]
(A) conc. of A and B (B) temperature
(C) pressure (D) all of these

147. $r = k[NO]^2[O_2]$. Rate constant can be increased by [BHU-2003]
(A) increasing [NO] (B) increasing $[O_2]$
(C) increasing temperature (D) decreasing temperature

148. Rate of a first order reaction depends upon [AFMC-2003]
(A) Time (B) Temperature
(C) Concentration of the reactants
(D) All of these

149. The rate of a first order reaction is 1.5×10^{-2} mol $L^{-1}\ min^{-1}$ at 0.5 M concentration of the reactant. The half-life of the reaction is [CBSE-PMT-2004]
(A) 0.383 min (B) 23.1 min (C) 8.73 min (D) 7.53 min

150. In the first order reaction, X → Y ; the concentration of X changes from 0.1 M to 0.025 M in 40 min. The rate of reaction (mol/L/min) when concentration of X is 0.01 M, would be [IIT-2004]
(A) 3.47×10^{-5} (B) 3.47×10^{-4}
(C) 5.8×10^{-4} (D) 5.8×10^{-6}

Kinetics

1.	(D)	51.	(A) (D)	101.	(A)
2.	(C)	52.	(D)	102.	(A)
3.	(A)	53.	(D)	103.	(C)
4.	(A)	54.	(B)	104.	(D)
5.	(A)	55.	(A)	105.	(A)
6.	(C)	56.	(D)	106.	(C)
7.	(B)	57.	(A)	107.	(B)
8.	(C)	58.	(D)	108.	(C)
9.	(B)	59.	(A)	109.	(C)
10.	(B)	60.	(D)	110.	(B)
11.	(D)	61.	(C)	111.	(C)
12.	(B)	62.	(C)	112.	(A)
13.	(D)	63.	*	113.	(D)
14.	(B)	64.	(B)	114.	(D)
15.	(B)	65.	*	115.	(D)
16.	(B)	66.	*	116.	(B)
17.	(C)	67.	(D)	117.	(A)
18.	(B)	68.	(C)	118.	(D)
19.	(D)	69.	(A)	119.	(A)
20.	(C)	70.	(C)	120.	(A)
21.	(B)	71.	(B)	121.	(C)
22.	(A)	72.	(A)	122.	(B)
23.	(C)	73.	(A)	123.	(C)
24.	(D)	74.	(B)	124.	(B)
25.	(D)	75.	(A)	125.	(B)
26.	(C)	76.	(B)	126.	(D)
27.	(C)	77.	(B)	127.	(C)
28.	(A)	78.	(C)	128.	(C)
29.	(B)	79.	(D)	129.	(A)
30.	(D)	80.	(A)	130.	(D)
31.	(A)	81.	(A)	131.	(B)
32.	(A)	82.	(D)	132.	(C)
33.	(B)	83.	(C)	133.	(C)
34.	(B)	84.	(B)	134.	(A)
35.	(B)	85.	(C)	135.	(C)
36.	(D)	86.	(D)	136.	(B)
37.	(A)	87.	(D)	137.	(D)
38.	(C)	88.	(D)	138.	(A)
39.	(B)	89.	(B)	139.	(A)
40.	(C)	90.	(A)	140.	(A)
41.	(D)	91.	(C)	141.	(A)
42.	(A)	92.	(D)	142.	(C)
43.	(A)	93.	(B)	143.	(B)
44.	(B)	94.	(B)	144.	(B)
45.	(B)	95.	(A)	145.	(C)
46.	(D)	96.	(C)	146.	(B)
47.	*	97.	(A)	147.	(C)
48.	*	98.	(C)	148.	(D)
49.	(B)	99.	(D)	149.	(B)
50.	*	100.	(C)	150.	(B)

47. 1.82×10^{-14} kg hr^{-1}
48. 7/3
50. Infinite temperature
63. IInd order
65. 10^oC
66. Lowers

ELECTROCHEMISTRY

Electrolytic Conduction : Conductor : A substance which allows an electric current to flow through it.

conductors –
- Metallic conductors (remains unchanged as current flows)
- Solutions of salts like NaCl (docomposed by the passage of electric current.)

Insulator : Which doesn't allow electric current to flow.

Electrolytes : Substances whose solutions conduct the electric current and which are decomposed by the passage of electric current are called electrolytes.

Non electrolytes : whose aqueous solution do not conduct electric current.

Conductance depends on :

1. Number of ions per unit volume.
2. Charge on the ions.
3. Speed of the ions at which they move.

Effect (a) on dilution : Conductance decreases since the no. of ions per unit volume decreases.

(b) on heating : conductance increases as viscosity lowers.

(c) on melting : conductance increases as ions are free to move.

Electrolytes –
- **Non electrolytes** e.g. glycerine, alcohols, ethers, esters, urea, hydrocarbons, sugar etc.
- **weak electrolytes** e.g. RCOOH; H_2CO_3, NH_4OH, HCN, Ph-COOH etc
- **strong electrolytes** e.g. halides, hydroxides, acetates, nitrates, chlorates, sulphates of mono & divalent cations. HCl, NH_4Cl, HBr, HI, H_2SO_4, HNO_3 etc.

strong acids, strong bases and salts are strong electrolytes.

Measurement of Electrolytic conductivity : If the solution of an electrolyte is kept between two parallel electrodes of cross section A cm^2 and l cm. apart, then its

$$\text{Resistance, } R = \rho \times \frac{l}{A}$$

$$\text{Resistivity, } \rho = R \times \frac{A}{l}$$

The specific conductance k(kappa) is reciprocal of resistivity, ρ

$k = \frac{1}{\rho} = \frac{1}{R} \times \frac{l}{A}$ unit.: $ohm^{-1} cm^{-1}$; mhos

k = observed conductance $(\frac{1}{R})$ x cell constant

(for any particular cell, the value $\frac{l}{A}$ is constant)

cell constant, $\frac{l}{A} = k \times R$

Electrolytic conductivity is an additive property.

The conductance of a 1 c.c of solution is known as the specific conductance of that solution. It is denoted by κ (kappa)

Equivalent conductivity : It is defined as the conductance due to all the ions produced by 1 gram equivalent of an electrolyte in a given volume of the solution. It is denoted by λ (lambda).

$\Lambda = k \times v$ (k = specific conductance; v = c.c. of soln. containing 1g. equivalent of the electrolyte) unit : $ohm^{-1} cm^2 eq^{-1}$

Molecular conductivity : The conductivity of a solution containing one gram mole of the electrolyte when placed between two sufficiently large electrodes which are 1cm. apart is known as molecular conductivity of the solution (molar conductance is defined as the conductance of all the ions produced by one mole of an electrolyte when dissolved in a certain volume, v c.c.).

$\mu = k \times v$ (v in c. c. of the soln. containing 1gm. mole of the electrolyte). unit : $ohm^{-1} cm^2 mol^{-1}$

Kohlrausch's Law :

At infinite dilution, when dissociation is complete, each ion makes a definite contribution towards equivalent conductance of the electrolyte irrespective of the nature of the other ion with which it is associated.

The equivalent conductance at infinite dilution for any electrolyte

$$\Lambda^o = \lambda_c^o + \lambda_a^o$$

where λ_c^o and λ_a^o is contribution of cation and anion respectively (these contributions are called ionic conductances at infinite dilution).

e.g.in case of $NaCl \rightleftharpoons Na^+ + Cl^-$; $Al_2(SO_4)_3 \rightleftharpoons Al^{3+} + 3SO_4^{2-}$

Application of Kohlrausch's Law :

1. Calculation of equivalent conductance at infinite dilution for weak electrolytes. Equivalent conductances of strong electrolytes at infinite dilution can be obtained graphically by extrapolation method. This method is not applicable for weak electrolytes since they are very feebly ionised. In such cases, Kohlrausch's method enables indirect evaluations.
2. Determination of degree of dissociation of weak electrolytes

$$\alpha = \frac{\Lambda_c}{\Lambda_o}$$

Λ_c = equivalent conductance of the weak electrolyte at the given concentration

Λ_o = equivalent conductance at zero or infinite dilution

3. Determination of solubility of sparingly soluble salts.

Ostwald's dilution law :

Degree of dissociation of a weak electrlyte is proportional to the square root of the dilution (applicable to weak electrolytes like weak acids, CH_3COOH and weak bases, NH_4OH etc.).

e.g. $AB \rightleftharpoons A^+ + B^-$ $\quad k = \frac{[A^+][B^-]}{[AB]}$

Let 1 g. mole of the electrolyte be dissolved in V litres and its degree of dissociation,

$$AB \rightleftharpoons A^+ + B^-$$

Initial conc. $\frac{1}{v}$ $\quad 0 \quad 0$

Eq. conc. $\left(\frac{1-\alpha}{v}\right) \quad \frac{\alpha}{v} \quad \frac{\alpha}{v}$

$$k = \frac{\frac{\alpha}{v} \times \frac{\alpha}{v}}{1-\frac{\alpha}{v}} = \frac{\alpha^2}{(1-\alpha)v} \quad \text{or } k = \frac{C\alpha^2}{1-\alpha},\ \alpha = \sqrt{\frac{k}{C}}$$

for polybasic acids

1. $H_3PO_4 \rightleftharpoons H^+ + H_2PO_4^-$ $\quad k_1$
2. $H_2PO_4^- \rightleftharpoons H^+ + HPO_4^{2-}$ $\quad k_2$
3. $HPO_4^{2-} \rightleftharpoons H^+ + PO_4^{3-}$ $\quad k_3$

$k_1 > k_2 > k_3$ [proton comes from a neutral species in the first step while in the second & third steps from-vely charged species (very difficult)].

Faradays Laws of Electrolysis :

They hold at all temperatues and pressures.

First Law : The amount of any substance that is deposited or liberated at an electrode is directly proportional to the quantity of electricity passed through the electrolyte.

$W \propto Q$

$W = Itz$ since $Q = It$

$z = \frac{M}{nF}$

$\therefore W = \frac{M}{nF} \times It$

(M = molar mass)

W = wt. of the ions liberated
Q = quantity of electricity in coulombs
I = strength of current in amperes
t = time in seconds
z = proportionality constant also called electrochemical equivalent

Electrochemical equivalent of an ion is the weight of the ion deposited by one amphere current in one second. One Faraday of electricity (96, 485 coulombs) will deposit one gram equivalent of the metal.

Second Law : when the same amount of electricity flows through different electrolytes, the amounts of different substances produced at the electrodes are directly proportional to their equivalent weights.

Theoretical importance of second law is that the amount of charge carried by one gram equivalent of any ion is one Faraday. i.e. 96, 500 coulombs.

When electricity is passed through a series of $AgNO_3$ solution and $CuSO_4$ solution

$$W \propto E,\ \frac{W_1}{E_1} = \frac{W_2}{E_2} \text{ or } \frac{W_1}{W_2} = \frac{E_1}{E_2}$$

$$\frac{\text{wt. of Ag deposited}}{\text{wt. of Cu deposited}} = \frac{\text{Eq. wt. of Ag}}{\text{Eq. wt. of Cu}} = \frac{107.88}{31.80}$$

Electrochemical cell : There are two types of electrochemical cells :

1. Electrolytic cell : Electrical energy causes chemical reactions to occur i.e. electrical energy is converted into chemical energy. It is widely used in industrial processes e.g. for the production of Al, Mg etc.

2. Galvanic or Voltaic cell : Here, the chemical energy is converted into electrical energy. Dry cell is a common example of galvanic cell (Daniel cell). It is used in lightning of a torch light bulb; in hearing aids, car lights, transistor radios, camera flash lights etc.

Polarity of the electrodes : In electrolytic cells, negative polarity of the cathode arises from the fact that there is a surplus of electrons at the cathode. The assignment of +ve polarity to the cathode of a galvanic cell originates from the fact that the electrons enter the cell at this electrode.

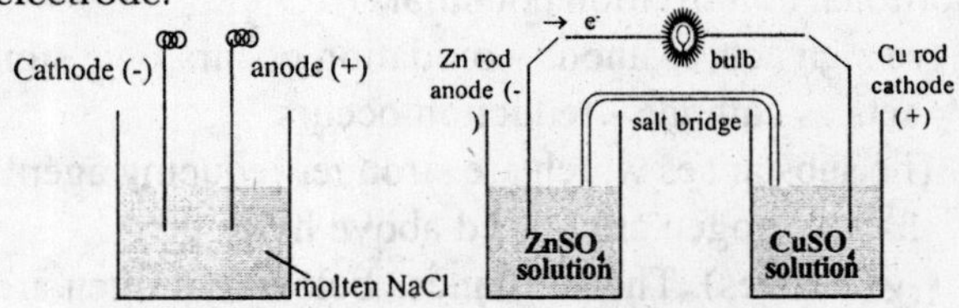

Electrolytic cell

Voltaic cell (galvanic)

Significance of salt bridge :

1. Connects the solutions of two half cells and completes the cell circuit.
2. It prevents transference or diffusion of the solutions from one half cell to the other.
3. It keeps the solutions in two half cells electrically neutral.
4. It prevents the liquid - liquid junction potential i.e. the potential difference which arises between two solutions, when in contact with each other .
5. It minimises the e.m.f. fluctuation of cell.

Salt bridge can be replaced by a porous partition which allows the migration of ions without allowing the solutions to intermix.

Limitations of Salt Bridge:If the electrolyte solution contains Ag^+,Hg_2^{2+}, Ti^+ ions, we can't use KCl in salt bridge because precipitation of AgCl, Hg_2Cl_2,TiCl take place.

Electrode potential : The electrode potential is given a +ve or -ve sign on the basis of reduction or oxidation reactions taking place at the electrodes.

When an electrode is coupled with standard hydrogen electrode, if the electrode reaction involves reduction i.e. +ve ions take electrons from the electrode and deposit on it as atoms, electrode potential is given a positive sign. On the other hand, when the electrode is coupled with S H E and if oxidation reaction occurs at the electrode, the electrode is given a negative sign.

e.g. Let the Zn / Zn^{++} electrode is coupled with S H E

$Zn(s) \rightleftharpoons Zn^{++} + 2e^-$ oxidation reaction, negative electrode

$H_2(g) \rightleftharpoons 2H^+ + 2e^-$

$Cu^{++} + 2e^- \longrightarrow Cu(S)$ copper is a positive electrode

Standard Electrode Potential : The term standard electrode potential is used to designate that potential which is obtained when the concentration or the activity of the ions in the solution is unity and the temp is 25°C. It is denoted by symbol E^o.

Standard electrode potentials of various electrodes have been calculated by coupling the electrode under consideration with S H E as reference electrode whose potential has been arbitrarily fixed as zero.

- Reduction occurs at electrode with higher SRP
- Oxidation occurs at electrode with lower SRP

The metal with LSR potential will displace the metal from its salt solution with HSR. Standard electrode potential is directly proportional to atomic size but inversely proportional to Ionisation potential.

1. -ve sign acts as anode - oxidation occurs, +ve sign acts as cathode - reduction occurs.
2. The substances which are stronger reducing agents than hydrogen are placed above hydrogen (-ve values). The substances below hydrogen are weaker reducing agents.
3. At the top are the active metals. Non metals are at the bottom.
4. The substances which are stronger oxidation agents than H^+ ions are placed below hydrogen in the series.
5. Metals below hydrogen are less reacting and do not evolve hydrogen from water.
6. Metals below hydrogen don't undergo corrosion.
7. Metals with low reduction potentials are extracted by electrolytic method e.g. Na, Al, Li, Mg etc.

Electromotive Force :

The potential difference which causes a current to flow from an electrode of higher potential to an electrode of lower potential is called electromotive force of the cell. (The potential difference between the two electrodes or half cells is known as the emf of the cell or cell potential). It is not affected by the internal resistance of the cell.

$$E^\circ_{cell} = E^\circ_R - E_L \; (E^\circ_{cell} = E^\circ_{cathode} - E^\circ_{anode})$$

E°_{cell} = EMF; E°_R = Reduction potential of cathode;
E°_L = Reduction potential of anode

EMF is expressed in volts. Potentiometers are used to measure emf. Weston standard cell is used to standardise potentiometers.

1. The EMF of a reversible electrode

Nernst equation gives the potential of any electrode or of a reversible cell. The concentration of pure solids and liquids are taken as unity.

$$E = E^\circ - \frac{2.303RT}{nF} \log k$$

(n = no. of electrons transferred in the half cell).

$$E = E^\circ - \frac{0.0591}{n} \log_{10} \frac{[M^{n+}]}{[M]}$$

$$= E^\circ - \frac{0.0591}{n} \log_{10} [M^{n+}] \qquad \text{At Equilibrium, } E = 0$$

$$\therefore E^\circ = \frac{2.303}{nF} \log_{10} k$$

$$= \frac{0.0591}{n} \log_{10} k$$

$$\log_{10} k = \frac{n E^\circ_{cell}}{0.0591}$$

2. Application of EMF for calculation of decrease in free energy of a reaction (chemical affinity)

$$\Delta G^\circ = -nFE^\circ_{cell}$$

(ΔG° = standard free energy change for the cell, E°_{cell} = standard EMF of the cell).

Similarly maximum obtainable work from the cell,

$W = - nFE^\circ_{cell}$

also $\Delta G^\circ = -2.303RT \log_{10} k$

Primary cells : They do not have indefinite life e.g. dry cell or Leblanche cell

Secondary cells : can be recharged e.g. lead strorage battery, Ni-Cd battery

Fuel cells : A compact source of power is called a fuel cell.

Fuel cells are pollution free source of power.

Mainly employed in space craft.

The materials undergoing oxidation at the anode or reduction at the cathode are stored outside the cell. These are not the part of the cell as in dry cell.

e.g. 1. hydrogen-oxygen fuel cell.

$2H_2(g) + 4OH^-(aq) \rightarrow 4H_2O(l) + 4e^-$ - Left electrode

$O_2(g) + 2H_2O(l) + 4e^- \rightarrow 4OH^-(aq)$ - Right electrode

Hydrogen is oxidised at anode and oxygen is reduced at cathode. The overall cell reaction produces water

2. hydrocarbons / air fuel

3. Pt electrode / conc. H_3PO_4

- hydrogen-oxygen cell using conc. NaOH as an electrolyte was used in the Apollo space programme.

Corrosion

Bronze, Copper - green coating
Silver - greyish

Rusting of iron : Rust is $Fe(OH)_3 + Fe_2O_3$.

dissolved oxygen oxidizes Fe^{+2} to Fe^{+3} at the anode. The OH^- ions at the anode combine with Fe^{+3} ions forming $Fe(OH)_3$.

Corrosion is accelerated by NaCl solution (because of the presence of electrolytes, electrical conduction enhances) Cl^- ions enhance rusting.

Prevention of corrosion :

1. Coating by metallic coatings such as Sn, Zn, Ni, Cr, Al etc.

2. Coating with H_3PO_4 or orthophosphates produces insoluble tenacious film of iron phosphate.

3. Applying greese and oil.

Galvanisation : Process of coating iron with zinc. Galvanised iron containers can not be used for the purpose of canning food because Zn will pass into the solution thereby poisoning the food.

Things to remember

Conductance depends upon : (a) number of ions per unit volume (b) charge on the ions (c) speed of the ions at which they move.

Specific conductance : k (kappa) = $\frac{1}{\rho} = \frac{1}{R} \times \frac{l}{A}$

k = observed conductance($\frac{1}{R}$) x cell constant ($\frac{l}{A}$)

unit: $ohm^{-1}\ cm^{-1}$

Equivalent conductance : $\Lambda = k \times v$

(v = vol. containing 1gram equivalent of an electrolyte)

unit : $ohm^{-1}\ cm^{2}\ eq^{-1}$

Molar conductance : $\mu = k \times v$

(v = volume containing 1gram mole of an electrolyte)

Kohlrausch's Law : $\Lambda_m^{\infty} = v_+ \lambda_+^{\infty} + v_- \lambda_-^{\infty}$

Degree of dissociation of weak electrolyte $\propto = \frac{\Lambda_c}{\Lambda_o}$

Faraday First Law : $W \propto Q$; W = It Z (since Q = It);

$z = \frac{M}{nF}$ $\therefore W = \frac{M}{nF} \times It$

Faraday Second Law : $W \propto Q$; $\frac{W_1}{E_1} = \frac{W_2}{E_2}$ or $\frac{W_1}{W_2} = \frac{E_1}{E_2}$

Nernst equation : $E = E^{o} - \frac{2.303}{nF} RT \log k$

At equilibrium $E^{o}_{cell} = \frac{0.0591}{n} \log k$

or $\log k = \frac{n \times E^{o}cell}{0.0591}$

EMF and free energy change are related by

$\Delta G = -nFE^{o}_{cell}$

Rust is $Fe(OH)_3 + Fe_2O_3$

Types of cells : (a) Primary : Cant be recharged e.g. dry cell (Leblonche cell)

(b) Secondary : They can be recharged e.g. lead storage battery; Ni-Cd battery etc.

In galvanic or voltaic cell, the cathode is assigned negative polarity.

Solved Examples

1. 0.5N solution of a salt surrounding two platinum electrodes 0.02m apart and $4 \times 10^{-4}\ m^2$ in area was found to offer a resistance of 25 ohm. Calculate (i) conductance (ii) cell constant (iii) electrolytic conductivity and (iv) the equivalent conductivity of the solution.

Sol. R = 25 ohm

conductance = $\frac{1}{R} = \frac{1}{25} = 0.04$ S

cell constant = $\frac{l}{A} = \frac{0.02}{4 \times 10^{-4}} = 50\ m^{-1}$

Electrolytic conductivity, k = cell constant x conductance

$= 50 \times 0.04 = 2.0\ Sm^{-1} = 0.02\ Scm^{-1}$

Equivalent conductivity = $\frac{1000k}{N} = \frac{1000 \times 0.02}{0.5}$

$= 40\ S\ cm^2\ eq^{-1}$.

2. The resistance of a 0.01N solution of an electrolyte was found to be 210 ohm at 25°C. Calculate the equivalent conductance of the solution at 25°C. Cell constant is 0.88

Sol. we know $\Lambda_{eq} = \frac{1000 \times k}{N}$ or $\Lambda_{eq} = k \times v$

(v = volume of the solution containing 1gram equivalent of the substance)

Specific conductance k = $\frac{\text{cell constant}}{\text{resistance}} = \frac{0.88}{210}\ S\ cm^{-1}$

$\Lambda_{eq} = \frac{1000 \times 0.88}{210 \times 0.01} = 419.05\ S\ cm^2\ equiv.^{-1} (ohm^{-1} cm^2 eq^{-1})$

3. The cell constant of a particular cell is 1.52 and it gives a resistance of 72.18 ohms for a decinormal solution of $ZnSO_4$. Find the equivalent and molecular conductance of $ZnSO_4$ solution.

Sol. Specific conductance,

k = observed conductance ($\frac{1}{R}$) x cell constant

$= \frac{1}{72.18} \times 1.52 = 0.021058$ mhos

$\Lambda_{eq} = \frac{1000 \times k}{N} = \frac{1000 \times 0.021058}{0.10}$

$= 210.58$ mhos or $s\ cm^2\ equiv^{-1}$.

$\Lambda_m = \frac{1000 \times k}{M} = \frac{1000 \times 0.021058}{0.05}$

$= 421.16$ mhos or $s\ cm^2\ mol^{-1}$.

4. If the molar ionic conductances of Al^{3+} is 189 $s\ cm^2\ mol^{-1}$ and SO_4^{2-} is 160 $s\ cm^2\ mol^{-1}$, calculate the molar conductance of $Al_2(SO_4)_3$.

Sol. $Al_2(SO_4)_3 \rightleftharpoons 2Al^{3+} + 3SO_4^{2-}$

$\Lambda^{\infty}_m = \lambda_+^{\infty} v_+ + \lambda_-^{\infty} v_-$

$= (189 \times 2) + (160 \times 3) = 858\ s\ cm^2\ mol^{-1}$

5. The equivalent conductances of sodium acetate, hydrochloric acid and sodium chloride at infinite dilution are 91.0, 426.0 and 126.0 ohm^{-1} respectively at 25°C. Calculate the equivalent conductance at infinite dilution for acetic acid.

Sol. $\Lambda_m^{\infty} CH_3COONa = \lambda^{\infty}_{Na^+} + \lambda^{\infty}_{CH_3COO^-} = 91.0\ ohm^{-1}$

$\Lambda^{\infty}_m HCl = \lambda^{\infty}_{H^+} + \lambda^{\infty}_{Cl^-} = 426.0\ ohm^{-1}$

$\Lambda^{\infty}_m NaCl = \lambda^{\infty}_{Na^+} + \lambda^{\infty}_{Cl^-} = 126.0\ ohm^{-1}$

from the data we can write the equation as

$CH_3COO\,Na + HCl \rightarrow CH_3COOH + NaCl$

in other words, $CH_3COOH \rightarrow CH_3COONa + HCl - NaCl$
$\Lambda^{\infty}_{m} CH_3COONa = 91.0 + 426.0 - 126.0 = 391 ohm^{-1}$.

6. The equivalent conductivity of a normal solution at 18°C is 98.2 and for infinite dilution at the same temperature was 131. Calculate the degree of ionisation of KCl at this dilution.

Sol. we know that $\propto = \frac{\lambda_c}{\lambda_o} = \frac{98.2}{131} = 0.75$ or 75%

7. A current of 0.8 amphere was passed through a voltameter containing copper sulphate solution for 40 minutes. Calculate the amount of copper that would be deposited.

Sol.: we know, $W \propto Q$

$Q = It = 0.8 \times 2400 = 1920$ coulombs

$CuSO_4 \rightleftharpoons Cu^{2+} + SO_4^{2-}$

From the reaction at cathode $Cu^{2+} + 2e^- \rightarrow Cu$ (deposits) one can draw the conclusion that 2 electrons are involved per atom of copper ∴ Eq. weight= $\frac{\text{Atomic wt}}{2} = \frac{63.6}{2} = 31.8$

96,500 coulombs of electricity deposits 31.8 gm of Cu

1920 coulombs of electricity deposits $= \frac{1920}{96,500} \times 31.8$
$= 0.6327$ g.

8. A current was passed through a solution of a salt of a metal 'A' and a solution of zinc sulphate connected in series. After a certain time 1.264 g. of zinc and 0.348 g. of 'A' were deposited. Calculate the equivalent weight of 'A' (Eq. wt. of Zn is 32.7)

Sol. we know that $\frac{W_1}{W_2} = \frac{E_1}{E_2}$

$\frac{\text{Eq. wt. of 'A'}}{\text{Eq. wt. of Zn}} = \frac{\text{wt. of x}}{\text{wt. of Zn}}$

substituting the values, $\frac{\text{Eq. wt. of 'A'}}{32.7} = \frac{0.348}{1.264}$

Eq. wt. of 'A' $= \frac{32.7 \times 0.348}{1.264} = 9.0$

9. Consider the reaction $2Ag^+ + Cd \rightarrow 2Ag + Cd^{2+}$
The standard electrode potentials for $Ag^+ \rightarrow Ag$ and $Cd^{2+} \rightarrow Cd$ couples are 0.80 and -0.40 volt, respectively.
(i) what is the standard potential E° for this reaction ?
(ii) for the electrochemical cell in which this reaction takes place, which electrode is -ve electrode ?

Sol. $2Ag^+ + 2e^- \rightarrow 2Ag$; $E^o_{Ag^+/Ag} = 0.80$ volt
Reduction
$Cd \rightarrow Cd^{2+} + 2e^-$; $E^o_{Cd^{2+}/Cd} = -0.40$ volt
Oxidation $E^o_{Cd/Cd^{2+}} = +0.40$ volt

$E^o = E^o_{Cd/Cd^{2+}} + E^o_{Ag^+/Ag} = 0.40 + 0.80 = 1.20$ volts

(ii) the negative electrode is always the electrode whose reduction potential has smaller value or the electrode where oxidaton occurs. Thus Cd electrode is the negative electrode.

10. Calculate the e.m.f. of the cell, $Mg(s)/Mg^{2+}(0.2M)//Ag^+(1 \times 10^{-3}M)/Ag$. $E^o_{Ag^+/Ag} = 0.8$ volt; $E^o_{Mg^{2+}/Mg} = -2.37$ volt What will be the effect on e.m.f. if concentration of Mg^{2+} ion is decreased to 0.1M ?

Sol. $E^o_{cell} = E^o_{cathode} - E^o_{anode}$
$= 0.80 - (-2.37) = 3.17$ volts

cell reaction $Mg + 2Ag^+ \rightarrow 2Ag + Mg^{2+}$

$E_{cell} = E^o_{cell} - \frac{0.591}{n} \log \frac{[Mg^{2+}]}{[Ag^+]}$

$= 3.17 - \frac{0.591}{2} \log \frac{0.2}{[1 \times 10^{-3}]^2}$

$= 3.17 - 0.1566 = 0.0134$ volts

$Mg^{2+} = 0.1$ M

$E_{cell} = E^o_{cell} = \frac{0.591}{2} \log \frac{0.1}{(1 \times 10^{-3})^2}$

$= (3.17 - 0.1477)$ volt
$= 3.0223$ volt

11. Calculate the electrode potential at a copper electrode dipped in a 0.1M solution of copper sulphate at 25°C. The standard electrode potential of Cu^{2+}/Cu system is 0.34 volt at 298 K.

Sol. we know that $E_{red} = E^o_{red} + \frac{0.591}{n} \log_{10} [\text{ion}]$

$= 0.34 + \frac{0.591}{2} \log_{10} [0.1]$

$= 0.34 + 0.02955 \times (-1) = 0.34 - 0.02955 = 0.31045$ volt.

12. What is the single electrode potential of a half cell for Zinc electrode dipping in 0.01M $ZnSO_4$ solution at 25°C ? The standard electrode potential of Zn/Zn^{2+} system is 0.763 volt at 25°C.

Sol. we know that

$E_{oxi} = E^o_{oxi} - \frac{0.591}{n} \log_{10} [\text{ion}]$

$= 0.763 - \frac{0.591}{2} \log_{10} [0.01]$

$= 0.763 - 0.02955 \times (-2) = 0.763 + 0.0591 = 0.8221$ volt.

13. For the cell reaction $Mg / Mg^{2+}(aq) // Ag^+(aq) / Ag$, calculate the equilibrium constant at 25°C and maximum work that can be obtained by operating the cell.
$E^o_{Mg^{2+}/Mg} = -2.37$ volts and $E^o_{Ag^+/Ag} = +0.80$ volt

Sol. $E^o_{cell} = 0.80 + 2.37 = 3.17$

$\log k_c = \frac{nE^o_{cell}}{0.0591} = \frac{2 \times 3.17}{0.0591} = 107.2758$

$k_c = 1.89 \times 10^{107}$

$-\Delta G$ = maximum work $= nFE^o = 2 \times 96500 \times 3.17$
$= 611810$ J.

14. Calculate the equilibrium constant for the electrochemical reaction in the $Zn/Zn^{2+}(IM)//Ag^+(IM)/Ag$ at 25°C from the given data; $E^o(Zn^{2+}/Zn) = -0.765$ V and $E^o(Ag^+/Ag) = 0.80$ V.

Sol. we know $E = E^o - \frac{2.303}{nF} RT \log_{10} k_{eq}$.

At equilibrium, E = 0

$\therefore E^o = \frac{2.303}{nF} RT \log_{10} k_{eq} = \frac{0.591}{n} RT \log_{10} k_{eq}$.

$E^o_{cell} = E^o_R - E^o_L = 0.80 - (-0.765) = 1.565V$

Substituting the values

$$1.565 = \frac{0.591}{2} RT \log_{10} k_{eq}$$

$$1.565 = 0.0295 \log_{10} k_{eq}.$$

$$\log_{10} k_{eq} = \frac{1.565}{0.0295} = 53.0508$$

$$k_{eq} = \text{Antilog of } 53.0508 = 1.124 \times 10^{53}.$$

15. Calculate the free energy change taking place in hydrogen/oxgen fuel cell in which the following reactions take place :

(i) $O_2 + 4H^+ + 4e^- \rightarrow 2H_2O$ at the cathode; $E^o = 1.229$ V

(ii) $2H_2 \rightarrow 4H^+ + 4e^-$ at the anode

Sol. $E^o_{cell} = E^o_R - E^o_L = 1.229 - 0.000 = 1.229$ V

we know $\Delta G^o = -nFE^o$

Substituting the values $\Delta G^o = -4 \times 96{,}500 \times 1.229$

$= -474.4$ k J

Reduction - Oxidation

Oxidation :

1. Combination with oxygen

e.g. $C + O_2 \rightarrow CO_2$; $S + O_2 \rightarrow SO_2$; $2Mg + O_2 \rightarrow 2MgO$;

2. Removal of hydrogen

e.g. $H_2S + I_2 \rightarrow 2HI + S$; $Zn + HCl \rightarrow ZnCl_2 + H_2$; $4HI + O_2 \rightarrow 2I_2 + 2H_2O$

3. Combination with an electronegative element or removal of electropositive element. e.g. $SnCl_2 + Cl_2 \rightarrow SnCl_4$; $K_2MnO_4 + Cl_2 \rightarrow 2KMnO_4 + 2KCl$; $2FeCl_2 + Cl_2 \rightarrow 2FeCl_3$.

4. When valency of electropositive element increases $2FeCl_2 + Cl_2 \rightarrow 2FeCl_3$; $SnCl_2 + Cl_2 \rightarrow SnCl_4$

5. Electronic concept : Where electrons are lost (according to Franklin) $Sn^{+2} \rightarrow Sn^{+4} + 2e^-$; $Zn \rightarrow Zn^{+2} + 2e^-$; $Fe^{+2} \rightarrow Fe^{+3} + e^-$.

Reduction

1. Combination with hydrogen

e.g. $H_2 + Cl_2 \rightarrow 2HCl$; $H_2 + S \rightarrow H_2S$; $H_2 + Br_2 \rightarrow 2HBr$

2. Removal of oxygen

e.g. $ZnO + C \rightarrow Zn + CO$; $Fe_3O_4 + 4H_2 \rightarrow 3Fe + 4H_2O$; $CuO + H_2 \rightarrow Cu + H_2O$

3. Positive content of a compound increases and negative content decrease

e.g. $\underset{1:2}{2CuCl_2} \rightarrow \underset{1:1}{Cu_2Cl_2} + Cl_2$; $\underset{1:2}{HgCl_2} + SnCl_2 \rightarrow \underset{2:2}{Hg_2Cl_2} + SnCl_4$

4. The valency of electropositive element drcreases $2FeCl_3 + H_2S \rightarrow 2FeCl_2 + 2HCl + S\downarrow$;

$2HgCl_2 + SnCl_2 \rightarrow Hg_2Cl_2 + SnCl_4$

5. Electronic concept: where electrons are gained $Cl_2 + 2e^- \rightarrow 2Cl^-$; $2H^+ + 2e^- \rightarrow H_2$; $Fe^{+3} + e^- \rightarrow Fe^{+2}$

Redox reactions : oxidation and reduction take place simultaneously

e.g. $2HgCl_2 + SnCl_2 \rightarrow Hg_2Cl_2 + SnCl_4$; $HgCl_2$ is reduced by $SnCl_2$ and $SnCl_2$ is oxidised by $HgCl_2$.

$PbS + 4H_2O_2 \rightarrow PbSO_4 + 4H_2O$; Pbs is oxidised; H_2O_2 is reduced.

Oxidising agent : which is capable of accepting electrons. e.g. O_2, HNO_3, halogens, MnO_4^-, $Cr_2O_7^{2-}$

Reducing agent : which is capable of donating electrons. e.g. K, Mg, CO, HBr, HI, $SnCl_2$, NO_2^- etc.

Reducing as well as oxidising agents : O_3, H_2O_2, SO_2 etc.

1. Ozone : $PbS + 4O_3 \rightarrow PbSO_4 + 4O_2$; O_3 is oxidising agent

$H_2O_2 + O_3 \rightarrow H_2O + 2O_2$; O_3 is reducing agent

2. H_2O_2 : $PbS + 4H_2O_2 \rightarrow PbSO_4 + 4H_2O$; H_2O_2 as oxidising agent

$Ag_2O + H_2O_2 \rightarrow 2Ag + H_2O + O_2$; H_2O_2 as reducing agent

3. SO_2 : $2H_2S + SO_2 \rightarrow 2H_2O + 3S$; SO_2 oxidising agent

Valency and oxidation number:

Valency : (1) The number of hydrogen atoms which combine with one atom of the element.

(2) The number of electrons which an atom can lend, borrow or share.

Oxidation number : The formal charge that has been left on a central atom when all the surrounding atoms, groups or ions have been withdrawn.

Rules for determining oxidation number

1. Uncombined or free atoms have always been assigned zero oxidation number. e.g. C^o, H_2^o, Na^o, Cl_2^o etc.

2. Except in ionic hydrides, where the oxidation number is -1, oxidation number of hydrogen is +1.

3. Oxidation number of oxygen is always -2. Exceptions: (a) in peroxides -1 and (b) OF_2 is +2.

4. The algebric sum of oxidation numbers of all atoms in a molecule is always zero.

5. In a polyatomic ion, the algebric sum of oxidation numbers of all atoms is equal to net charge on the ion.

Calculation of oxidation number : e.g.

(i) $K\underline{Mn}O_4$, $1 + x - 2 \times 4 = 0 ; x = 8 - 1 = 7$

(ii) $H_2\underline{S}O_4$, $2 \times 1 + x - 2 \times 4 = 0; \ x = 8 - 2 = 6$

(iii) $K_2\underline{Cr_2}O_7$, $x - 2 \times 4 = -3 ; \ x = -3 + 8 = 5$

(iv) $Ni\ (CO)_4$, 0, $SO_4^{2-} = x - 2 \times -4 = -2; x = 6$

Redox reactions and oxidation number :

Oxidation is a process in which the oxidation number of the element is increased and conversely in reduction the oxidation number of the element decreases.

$Na \rightarrow Na^+ + e^-$ (0 → 1 ∴ Na is oxidised)

$Cl_2 + 2e^- \rightarrow 2Cl^-$ oxidation no. of Cl from O→ -1decreased.

e.g. $Pb^{2+}S^{2-} + 4H_2^{+1}O_2^{-1} \rightarrow Pb^{2+}S^{6+}O_4^{-2} + H_2^{+1}O^{-2}$

Sulphur atom in PbS has oxidation number -2; S in $PbSO_4$ has+6 oxidation. ∴ S is oxidised.

Oxidation number of oxygen in H_2O_2 is -1, which changes to -2 in water, Hence oxygen is reduced. Thus overall PbS is oxidised to $PbSO_4$ and H_2O_2 is reduced to H_2O.

Identification of oxidising and reducing agents on the basis of oxidation number.

e.g. $2FeCl_3 + H_2S \rightarrow 2FeCl_2 + 2HCl + S$; $FeCl_3$ is reduced ∴ it is oxidising agent

S in H_2S is -2→O ∴H_2S is oxidized ∴H_2S is reducing agent.

e.g. $MnO_2 + 4HCl \rightarrow MnCl_2 + 2H_2O + Cl_2$

$Mn^{+4} \rightarrow Mn^{+2}$, Mn is reduced ∴ MnO_2 is oxidising agent.

$Cl^{-1} \rightarrow Cl_2^o$ ∴HCl is reducing agent.

The substances in which the oxidation number of any element increases is supposed to have been oxidized hence it acts as reducing agent, because the substance loses electrons and vice-versa.

Equivalent weight of oxidising agent

The equivalent weight of an oxidising agent is the weight which accepts an electron in a chemical reaction.

Eq. wt. = molecular wt./ no. of electrons gained by one mole of the oxidising agent.

The equivalent weight can also be calculated by dividing its molecular weight by the decrease of oxidation number of any of its atoms.

e.g. $K_2Cr_2O_7$: $Cr_2O_7^{2-} + 14H^+ + 6e^- \rightarrow 2Cr^{3+} + 7H_2O$

Here the oxidation number of chromuum decreases from +12 to +6 ∴ Eq. wt. = mol. wt. /6 = $\frac{294.2}{6}$ = 49.033

e.g. $KMnO_4$: (1) $MnO_4^{-1} + 8H^+ + 5e^- \rightarrow MnO_2 + 4H_2O$ acidic medium (2) $MnO_4^{-1} + 2H_2O + 3e^- \rightarrow MnO_2 + 4OH^-$ alkaline/neutral. In (1) $Mn^{+7} \rightarrow Mn^{+2}$ change in oxidation no is +5

$$\text{Eq.wt. of } KMnO_4 = \frac{\text{mol. wt.}}{5} = \frac{158.04}{5} = 31.6$$

In (2) $Mn^{+7} \rightarrow Mn^{+4}$ Eq. wt. = $\frac{158.04}{3}$ = 52.68

Equivalent weight of reducing agents :

The equivalent weight of a reducing agent is the weight which looses an electron in a chemical reaction

$$\text{Eq. wt.} = \frac{\text{mol wt.}}{\text{no. of electrons lost by one mole}}$$

The eq. wt. can also be calculated by dividing its molecular wt. by the increase in oxidation number of any atom in it.

e.g. $H_2C_2O_4$: $C_2O_4^{2-} \rightarrow 2CO_2 + 2e^-$

oxidation no. of 2 carbon atoms in $C_2O_4^{2-}$ = 6
oxidation no. of 2 carbon atoms in CO_2 = 8 } change in oxidation number = +2

$$\text{Eq. wt.} = \frac{\text{mol. wt.}}{2} = \frac{126}{2} = 63$$

e.g. H_2S : $S^{2-} \rightarrow S^o + 2e^-$ change in oxidation no. is 2

$$\therefore \text{Eq. wt.} = \frac{\text{mol. wt.}}{2} = \frac{34}{2} = 17$$

Balancing redox reactions : Oxidation change method (Johnson 1880)

$SO_3^{2-} + MnO^{4-} \rightarrow Mn^{2+} + SO_4^{2-}$ (acid medium)

a. In the ionic redox equation, select the atom being oxidized in the reducing agent, indicate the no. of electrons lost $S^{4+}O_3^{2-} \rightarrow S^{+6}O_4^{2-}$ - loss of two electrons --- (1)

b. Select the atom in the oxidising agent which is being reduced and indicate the number of electrons gained $MnO^- \rightarrow Mn^{+2}$ +7 to +2, gain of five electrons --- (2)

c. Cross multiply the oxidant and reductant by a suitable coefficient so that the number of electrons gained and lost are multiplied.

multiply (1) by 5 and (2) by 2 and then adding

$$5SO_3^{2-} + 2MnO_4^{-1} \rightarrow 5SO_4^{2-} + 2Mn^{+2}$$

Charges -10 -2 -10 +4

Balance -12 → -6 on LHS

d. Add the equations and balance them chargewise by either adding H^+ ions (acidic medium) or by adding OH^- ions (in alkaline medium). Now balance the equation, atom wise also.

Now add $6H^+$ on LHS and balance the equation by adding $3H_2O$ on RHS.

$$5SO_3^{2-} + 2MnO_4^- + 6H^+ \rightarrow 5SO_4^{2-} + 2Mn^{2+} + 3H_2O$$

Ion electron method :

(1) $P \rightarrow PH_3+H_2PO_2^-$ (alkaline medium)

a. The oxidation and reduction steps are separately written in the form of two half reactions.

b. These half reactions are balanced, either by H^+ or OH^- according to whether the reaction is carried out in acidic or alkaline medium.

c. The charges are balanced on both sides with electrons

d. The half reactions are then multiplied by a suitable coefficient so that the electrons on both sides are equal and therfore cancelled.

a. (i) $P \rightarrow PH_3$ (ii) $P \rightarrow H_2PO_2^-$

b. In this step we shall add OH^- ions as required

(i) $P+3H_2O \rightarrow PH_3+3OH^-$ (ii) $P+2OH^- \rightarrow H_2PO_2^-$

c. Now the electrons are adjusted as required

(i) $P+3e^-+3H_2O \rightarrow PH_3+3OH^-$ (ii) $P+2OH^- \rightarrow H_2PO_2^-+e^-$

d. Multiply (ii) by 3 and add to (i) so that all electrons on both sides are cancelled.

$P+3H_2O+3P+6OH^- \rightarrow PH_3+3OH^-+3H_2PO_2^-$

$4P+3OH^-+3H_2O \rightarrow PH_3+3H_2PO_2^-$

(2) $MnO_4^-+Fe^{2+} \rightarrow Mn^{2+}+Fe^{3+}+H_2O$ (acid medium)

a. (i) $MnO_4^- \rightarrow Mn^{2+}$ (ii) $Fe^{2+} \rightarrow Fe^{3+}$

b. (i) $MnO_4^-+8H^+ \rightarrow Mn^{2+}+4H_2O$ (ii) $Fe^{2+} \rightarrow Fe^{3+}+e^-$

c. (i) $MnO_4^-+8H^++5e^- \rightarrow Mn^{2+}+4H_2O$ (ii) $Fe^{2+} \rightarrow Fe^{3+}+e^-$

d. multiply (ii) by 5 and then adding to (i)

$5Fe^{2+}+MnO_4^-+8H^+ \rightarrow Mn^{2+}+5Fe^{3+}+4H_2O$.

1. Oxidation number of Mn in K_2MnO_4 is [CBSE-91]
(A) +7 (B) -5 (C) +6 (D) -6

2. The charge on a cation M is +2 and on an anion A is -3. The compound formed by these has the formula, [CBSE-91]
(A) MA_2 (B) M_2A (C) M_3A_2 (D) M_2A_3

3. What is the oxidation number of S in $Na_2S_4O_6$? [CBSE-93]
(A) 2 (B) 3 (C) 1½ (D) 2½

4. Which substance is serving as a reducing agent in the following reaction ? [CBSE-94]
$14H^+ + Cr_2O_7^{2-} + 3Ni \rightarrow 2Cr^{3+} + 7H_2O + 3Ni^{2+}$
(A) H_2O (B) Ni (C) H^+ (D) Cr_2O_7

5. The standard reduction potentials 25°C of Li^+ / Li, Ba^{2+} / Ba, Na^+ / Na and Mg^{2+} / Mg are –3.05, –2.73, –2.7 and –2.37 volt respectively. Which one of the following is the strongest oxidising agent ? [CBSE-94]
(A) Na^+ (B) Li^+ (C) Ba^{2+} (D) Mg^{2+}

6. The resistance of a 0.1N solution of acetic acid is 250 ohm, when measured in a cell of cell constant 1.15 cm^{-1}. The equivalent conductance (in ohm^{-1} cm^2, $equiv^{-1}$) of 0.1 N acetic acid is. [CBSE-94]
(A) 4.6 (B) 9.2 (C) 18.4 (D) 2.3

7. The most powerful oxidizing agent from the following is [CBSE-94]
(A) H_2SO_4 (B) H_3BO_3 (C) HPO_3 (D) H_3PO_4

8. The oxidation state of I in $H_4IO_6^-$ is [CBSE-94]
(A) +7 (B) –1 (C) +5 (D) +1

9. An electrochemical cell is set up as follows :
Pt (H_2.1atm)/0.1MHCl/0.1M acetic acid/ (H_2,1atm) Pt.
E.M.F. of this cell will not be zero because [CBSE-95]
(A) the pH of 0.1 M HCl and 0.1 M acetic acid is not the same.
(B) acids used in the two compartments are different.
(C) E.M.F. of a cell depends on the molarities of acids used .
(D) the temperature is constant.

10. A 5 ampere is passed through a solution of zinc sulphate for 40 minutes. Find the amount of zinc deposited at the cathode. [CBSE-96]
(A) 40.65 gm (B) 4.065 gm
(C) 0.4065 gm (D) 65.04 gm

11. Reducton potential for the following half-cell reactions are:

$$Zn = Zn^{2+} + 2e^-; E_0 = 0.76 \text{ V}$$
$$Fe = Fe^{2+} + 2e^-; E_0 = +0.44\text{V}$$

The EMF for the cell reaction $Fe^{2+} + Zn = Zn^{2+} + Fe$ will be [CBSE-96]

(A) -1.20 V (B) +0.32 V (C) -0.32 V (D) +1.20 V

12. Which of the following elements is responsible for oxidation of water to O_2 in biological processes ? [CBSE-97]

(A) Fe (B) Mn (C) Cu (D) Mo

13. E^0 for the cell, Zn | Zn^{2+}(aq) || Cu^{2+}(aq) | Cu is 1.10 V at 25°C, the equilibrium constant for the reaction $Zn^+ + Cu^{2+}(aq) \rightleftharpoons Cu + Zn^{2+}(aq)$ is order of [CBSE-97]

(A) 10^{-28} (B) 10^{-37} (C) 10^{-18} (D) 10^{17}

14. Which of the following is redox reaction ?[CBSE-97]

(A) H_2SO_4 with NaOH

(B) In atmosphere O_3 from O_2 by lightning

(C) Nitrogen oxides from nitrogen and oxygen by lightning

(D) Evaporation of H_2O.

15. The oxidation number of phosphorus in $Ba(H_2PO_2)_2$ is [CBSE-97]

(A) -1 (B) +1 (C) +2 (D) -2

16. Equivalent conductance of NaCl, HCl and C_2H_5COONa at infinite dilution are 126.45, 426.16 and 91 ohm^{-1} cm^2 respectivly. The equivalent conductance of C_2H_5COONa is [CBSE-97]

(A) 201.28 ohm^{-1}/cm^{-2} (B) 390.71/ohm cm^2

(C) 698.28/ohm cm^2 (D) 540.48/ohm cm^2

17. For the cell reaction Cu^{2+}[Cu (aq)]+Zn(s)= Zn^{2+}(aq)]+ Cu(s) of an electrochemical cell, the change in free energy, ΔG, at a given temperature is a function of [CBSE-98]

(A) $\ln(C_1)$ (B) $\ln(C_2/C_1)$ (C) $\ln C_1 + C_2$ (D) $\ln(C_2)$

18. In PO_4^{3-} ion, the formal charge on each oxygen atom and P-O bond order respectively are [CBSE-98]

(A) –0.75, 1.25 (B) –3,1.25

(C) –0.75, 1.0 (D) –0.75, 0.6

19. Oxidation state of Fe in Fe_3O_4 is [CBSE-99]

(A) 3/2 (B) 4/5 (C) 5/4 (D) 8/3

20. The specific conductance of a 0.1 N KCl solution at 23°C is 0.012 $ohm^{-1}cm^{-1}$. The resistance of cell containing the solution at the same temperature was found to be 55 ohm. The cell constant will be [CBSE-99]

(A) 0.142 cm^{-1} (B) 0.616 cm^{-1}

(C) 0.918 cm^{-1} (D) 1.12 cm^{-1}

21. Purification of aluminium, by electrolyte refining, is known is [CBSE-99]

(A) Hall's process (B) Baeyer's process

(C) Hoope's process (D) Serpeck's process

22. At infinite dilution for an aq. solution of $BaCl_2$, molar conductivity for Ba^{2+} and Cl^- ions are Λ_m^∞ = 127.32 s cm^2/ mol, Λ_m^∞ = 76.34 s cm^2/mol respectively. What is Λ_m^∞ for $BaCl_2$ at same dilution ? [CBSE-2000]

(A) 280 s cm^2 mol^{-1} (B) 330.98 s cm^2 mol^{-1}

(C) 90.98 s cm^2 mol^{-1} (D) 203.6 s cm^2 mol^{-1}

23. What is E_0 for the cell involving disproportionation of unstable Cu^+ ion into its constituent oxidized-reduced states, if E_0 Cu^{++}/Cu^+ = 0.15V, E_0 Cu^+/Cu = 0.340V ? [CBSE-2000]

(A) –0.19 V (B) –0.38V (C) +0.38V (D) 0.49 V

24. A, B, C are three elements forming a part of compound in oxidation states of +2, +5 and -2 respectively. What could be the compound ? [CBSE-2000]

(A) $A_2(BC)_2$ (B) $A_2(BC_4)_3$ (C) $A_3(BC_4)_2$ (D) ABC

25. Std. electrode potentials are [CBSE-2001]

Fe^{+2}/Fe, E^0 = -0.44 and Fe^{+3}/Fe^{+2} E^0 = 0.77

Now an iron piece is dipped in aqueous solution of mixture of Fe^{3+} and Fe^{2+} ions then -

(A) Fe^{3+} increases (B) Fe^{3+} decrease

(C) Fe^{+2}/Fe^{3+}remains unchanged (D) Fe^{+2}decrease

26. Most convenient method to protect bottom of ship made of iron is [CBSE-2001]

(A) Coating it with red led oxide

(B) White tin plating (C) Connecting it with Mg block

(D) Connecting it with Pb block

27. In electrolysis of NaCl when Pt electrode is taken then H_2 is liberated at cathode while with Hg cathode it forms sodium amalgam [CBSE-2002]

(A) Hg is more inert than Pt

(B) More voltage is required to reduce H^+ at Hg than at Pt

(C) Na is dissolved in Hg while it does not dissolve in Pt

(D) Conc. of H^+ ions is larger when Pt electrode is taken.

28. The emf of a Daniell cell at 298 K is E_1. [CBSE-2003]

$$Zn \left| \begin{matrix} ZnSO_4 \\ (0.01M) \end{matrix} \right|\left| \begin{matrix} CuSO_4 \\ (1.0M) \end{matrix} \right| Cu$$

When the concentration of $ZnSO_4$ is 1.0M and that of $CuSO_4$ is 0.01 M, the emf changed to E_2. What is the relationship between E_1 and E_2 ?

(A) $E_2 = 0 \neq E_1$ (B) $E_1 > E_2$ (C) $E_1 < E_2$ (D) $E_1 = E_2$

29. On the basis on the information available from the reaction [CBSE-2003]

$4/3\ Al + O_2 \rightarrow 2/3\ Al_2O_3$, $\Delta G = -827$ KJ mol^{-1} of O_2 the minimum emf required to carry out an electrolysis of Al_2O_3 is (F = 96500 C mol^{-1})

(A) 8.56 V (B) 2.14 V (C) 4.28 V (D) 6.42 V

30. In which cell the free energy of a chemical reaction is directly converted into electricity? [AIIMS-94]

(A) Fuel cell (B) Leclanche cell

(C) Concentration cell (D) Lead storage battery

31. Which of the following is not true about e.m.f. of a cell ? [AIIMS-94]
(A) It is maximum voltage obtainable from the cell
(B) It is responsible for the flow of steady current in the cell
(C) Work calculated from it is not the maximum work obtainable from the cell
(D) It is the potential difference between two electrodes when no current is flowing in circuit

32. Which is an oxidizing substance amongst the following ? [AIIMS-94]
(A) CO_2 (B) NO_2 (C) SO_2 (D) SO_3

33. The number of electrons required to deposit 1gm equivalent aluminium (at. wt. = 27) from a solution of aluminium chloride will be [AIIMS-95]
(A) 1 (B) 2 (C) 3 (D) 4

34. In order to protect, iron from corrosion, the iron is coated with [AIIMS-95]
(A) Nickel (B) Copper (C) Zinc (D) Tin

35. By the electrolysis of aqueous solution of $CuSO_4$, the products obtained at both the electrodes are [AIIMS-95]
(A) H_2 at anode and Cu at cathode
(B) $H_2S_2O_8$ at anode and O_2 at cathode
(C) O_2 at anode and H_2 at cathode
(D) O_2 at anode and Cu at cathode

36. How many electrons are delivered at the cathode during electrolysis by a current of 1 A in 60 seconds ? [AIIMS-95]
(A) 6.0×10^{23} (B) 6.0×10^{20}
(C) 3.74×10^{20} (D) 7.48×10^{21}

37. Assertion : Copper liberates hydrogen from a solution of hydrochloric acid. Reason : Hydrogen is below copper in the electrochemical series. [AIIMS-95]
(A) A (B) B (C) C (D) D

38. The equivalent weight of oxygen, when it is converted to oxide is equal to [AIIMS-95]
(A) Molecular weight (B) $\frac{\text{Molecular weight}}{2}$
(C) $\frac{\text{Molecular weight}}{3}$ (D) $\frac{\text{Molecular weight}}{4}$

39. Assertion : Sodium ions are discharged in preference to hydrogen ions at a mercury cathode. Reason : The nature of cathode can affect the order of discharge of cations. [AIIMS-96]
(A) A (B) B (C) C (D) D

40. Assertion : In electrolysis, the quantity of electricity needed for depositing 1 mole of silver is different from that required for 1 mole of copper. Reason : The molecular weights of sliver and copper are different. [AIIMS-96]
(A) A (B) B (C) C (D) D

41. In the reaction :
$Cu(s) + 2Ag^+(aq) \rightarrow Cu^{2+}(aq) + Ag(s)$
The reduction half cell reaction is [AIIMS-97]
(A) $Cu + 2e^- \rightarrow Cu^{2-}$ (B) $Cu - 2e^- \rightarrow Cu^{2+}$
(C) $Ag^+ + e^- \rightarrow Ag$ (D) $Ag - e^- \rightarrow Ag^+$

42. In the reaction : $4Fe + 3O_2 \rightarrow 4Fe^{3+} + 6O_2^{2-}$, which of the following statements is incorrect ? [AIIMS-98]
(A) It is a redox reaction (B) Fe^{3+} is an oxidizing agent
(C) Metallic iron is a reducing agent
(D) Metallic iron is reduced to Fe^{3+}

43. The oxidation numbers of hydrogen in KH, MgH_2 and NaH are respectively [AIIMS-98]
(A) −1, −1 and −1 (B) +1, +1 and +1
(C) +2, +1 and −2 (D) −2, −3 and −1

44. The specific conductance of a N/10 KCl at 25°C is 0.0112 ohm^{-1} cm^{-1}. The resistance of cell containing solution at the same temperature was found to be 55 ohms. The cell constant will be [AIIMS-98]
(A) 0.0616 cm^{-1} (B) 0.616 cm^{-1}
(C) 6.16 cm^{-1} (D) 616 cm^{-1}

45. Which of the following has been universally accepted as a reference electrode at all temperatures and has been assigned a value of zero volt ? [AIIMS-98]
(A) Graphite electrode (B) Copper electrode
(C) Platinum electrode (D) Standard hydrogen electrode

46. A certain current liberated 0.504 g of hydrogen in 2 hours. How many grams of copper can be liberated by the same current flowing for the same time in copper sulphate solution [AIIMS-98]
(A) 12.7 g (B) 16.0 g (C) 31.8 g (D) 63.5 g

47. The specific conductivity of 0.1 N KCl solution at 20°C is 0.0212 ohm^{-1} cm^{-1}. The resistance of cell containing this solution at 20°C is 55 ohm. The cell constant is [AIIMS-99]
(A) 1.166 cm^{-1} (B) 2.173 cm^{-1}
(C) 3.324 cm^{-1} (D) 4.616 cm^{-1}

48. The equivalent weight of phosphoric acid (H_3PO_4) in the reaction $NaOH + H_3PO_4 \rightarrow NaH_2PO_4 + H_2O$ is [AIIMS-99]
(A) 25 (B) 49 (C) 59 (D) 98

49. Which of the following will not undergo hydrolysis in water ? [AIIMS-99]
(A) CCl_4 (B) $SiCl_4$ (C) $GeCl_4$ (D) $SnCl_4$

50. Which of the following statement is true for a electrochemical cell ? [AIIMS-99]
(A) H_2 is cathode and Cu is anode
(B) H_2 is anode and Cu is cathode

(C) Reduction occurs at H_2 electrode
(D) Oxidation occurs at Cu electrode

51. Assertion : Equivalent conductance of all electrolytes decreases with increasing concentration Reason : Lesser number of ions are available per gram equivalent at higher concentration. [AIIMS-99]
(A) A (B) B (C) C (D) D

52. Assertion : Zinc displaces copper from copper sulphate solution. Reason : The E°_{298} of Zn is –0.76 volts and that of Cu is +0.34 volts [AIIMS-99]
(A) A (B) B (C) C (D) D

53. Assertion : Identification of cathode and anode is done by the use of a thermometer. Reason : Higher is the value of reduction potential greater world be its reducing power. [AIIMS-99]
(A) A (B) B (C) C (D) D

54. A gas z is bubbled through a solution containing x^- and y^-. If the reduction potential are in the order $x > y > z$, then [AIIMS-2000]
(A) y will oxidize z and not x
(B) y will oxidize x and not z
(C) y will oxidize both x and z
(D) y will oxidize both x and z

55. Fe^{+3} and Sn^{2+} cannot stay together because : [AIIMS-2001]
(A) Fe^{+3} reduces into Fe^{+2} and Sn^{+2} oxidises into Sn^{+4}
(B) No reaction occurs (C) Precipitation occurs
(D) Fe^{+3} get oxidizes and Sn^{+2} gets reduced

56. Assertion (A) : Stannous chloride is a powerful oxidizing agent which oxidizes mercuric chloride, to metallic mercury. Reason (R) Stannous chloride gives grey precipitate with mercuric chloride, but stannic chloride does not do so. [AIIMS-2002]
(A) A (B) B (C) C (D) D (E) E

57. Which of the following reactions is used to make a fuel cell ? [AIIMS-2003]
(A) $Cd(s) + 2Ni(OH)_3(s) \rightarrow CdO(s) + 2Ni(OH)_2(s) + H_2O(l)$
(B) $2Pb(s) + PbO_2(s) + 2H_2SO_4(aq) \rightarrow 2PbSO_4(s) + 2H_2O(l)$
(C) $2H_2(g) + O_2(g) \rightarrow 2H_2O(l)$
(D) $2Fe(s) + O_2(g) + 4H^+(aq) \rightarrow 2Fe^{2+}(aq) + 2H_2O(l)$

58. Time required to deposit one milli mole of aluminium metal by the passage of 9.65 amperes through aqueous solution of aluminium ion is- [AIIMS-2003]
(A) 30s (B) 10s (C) 30,000s (D) 10,000s

59. The process of covering iron sheets with a layer of zinc is called : [JIPMER-91]
(A) Galvanising (B) Zinc plating
(C) Tempering (D) Dressing

60. Approximate atomic weight of an element is 26.89. If its equivalent is 8.9, the exact atomic weight of the element would be : [JIPMER-92]
(A) 25.89 (B) 8.9 (C) 26.7 (D) 17.8

61. How many coulombs of electricity are required for the oxidation of 1 mol of H_2O to O_2? [JIPMER-93]
(A) $1.93 \times 10^5 C$ (B) $9.65 \times 10^4 C$
(C) $3.86 \times 10^5 C$ (D) $4.825 \times 10^4 C$

62. For the cell reaction, $X(s) + 2Y^+ \rightarrow X^{2+} + 2Y$, K_c has been found to be 10^{12}. The E^0 cell is ? [JIPMER-93]
(A) 0.708 V (B) 1.36 V (C) 0.354 V (D) 1.006 V

63. In a cell $Zn/Zn^{2+} || Cu^{2+}/Cu$ the positive electrode will be [JIPMER-94]
(A) Cu (B) Cu^{2+} (C) Zn (D) Zn^{2+}

64. Which of the following statements about galvanic cell is incorrect ? [JIPMER-95]
(A) Reduction occurs at cathode (B) Anode is positive
(C) Oxidation occurs at the electrode with lower reduction potential. (D) Cathode is positive.

65. What is E° for electrode represented by Pt, O_2 (1atm.)/ $2H^+$(1 M) ? [JIPMER-95]
(A) 0.018 V (B) Unpredictable (C) Zero (D) 0.118 V

66. Weak electrolytes are only partly dissociated in solutions. The extent to which they are dissociated in a given solution is known as the degree of dissociation;. Which of the following statement for this degree of dissociation in aqueous solutions is true ? [JIPMER-96]
(A) It is inversely proportional to the initial concentration of the electrolyte
(B) It is directly proportional to the initial concentration of the electrolyte
(C) It is independent of the initial concentration of the electrolyte
(D) It depends on the equilibrium concentration of the electrolyte

67. Which of the substances Na, Hg, S, Pt and graphite, can be used as electrodes in electrolytic cells having aqueous solutions ? [JIPMER-96]
(A) Hg, Pt and graphite (B) Na, S and Hg
(C) Pt and graphite only (D) Na and S only

68. Two electrodes A^{2+}/A and B^+/B have reduction potentials (E°) 0.62 V and –1.25 V respectively. The cell voltage will be : [JIPMER-96]
(A) 0.63 V (B) –0.63V
(C) 1.87V (D) None of the above

69. The conductivity of an aqueous solution of strong electrolyte [JIPMER-96]
(A) bears no relationship with concentration
(B) decreases with increase in dilution

(C) remains constant at all concentration
(D) increases with dilution

70. When a molten ionic hydride is electrolysed [JIPMER-98]
(A) Hydrogen is liberated at the cathode
(B) Hydrogen is liberated at the anode
(C) There is no reaction
(D) H- ions produced migrate to the cathode

71. Which of the following substances when dissolved in water will give a solution that does not conduct electricity? [JIPMER-98]
(A) Hydrogen chloride (B) Potassium Hydroxide
(C) Sodium acetate (D) Urea

72. The oxidation potential of a hydrogen electrode at pH = 10 and P_{H_2}=1 is [JIPMER-2000]
(A) 0.059V (B) +0.59V (C) 0.00V (D) 0.51V

73. Which solution will show highest resistance during the passage of current [JIPMER-2001]
(A) 0.05 N NaCl (B) 2 N NaCl
(C) 0.1 N NaCl (D) 1 N NaCl

74. The voltage of a cell whose half cells are given below is
$Mg^{2+} + 2e \rightarrow Mg$ (s); E° = – 2.37 V
$Cu^{2+} + 2e \rightarrow Cu$ (s); E° = + 0.34 V [JIPMER-2001]
(A) 1.36 (B) +2.71 (C) -2.03 (D) +2.71

75. In 1 gram of a metal oxide, metal precipitated is 0.68gram. what is the equivalent weight of metal. [JIPMER-2002]
(A) 17 (B) 34 (C) 68 (D) 52

76. M is the molecular mass of $KMnO_4$. The equivalent mass of $KMnO_4$ when it is converted into K_2MnO_4 is [JIPMER-2002]
(A) M (B) M/3 (C) M/15 (D) M/7

77. Which of the following is not a reducing agent? [JIPMER-2002]
(A) SO_2 (B) H_2O_2 (C) CO_2 (D) NO_2

78. The standard reduction potential of Li^+/Li, Ba^{2+}/Ba, Na^+/Na and Mg^{2+}/Mg are –3.05, –2.73, –2.71 and –2.37 volts respectively which one of the following is strongest oxidizing agent ? [JIPMER-2002]
(A) Na^+ (B) Li^+ (C) Ba^{+2} (D) Mg^{2+}

79. If the E^0 = 1.10 volt for the cell reaction $Zn(s) + Cu^{2+}(aq) \rightarrow Cu(s) + Zn^{2+}(aq)$. Then value of ΔG^0 is [BHU-95]
(A) 212.3 kJ (B) 106.05 kJ
(C) –212.3 kJ (D) 424.6 kJ

80. Of $[Zn^{2+}]$ = 0.1 M and E° = –0.76 V then half cell potential at 298 K for the reaction $Zn^{2+}(aq) + 2e^- \rightarrow Zn(s)$ is [BHU-95]
(A) 0.789 V (B) –0.789 V
(C) –0.698 V (D) 0.698 V

81. The standard reduction potentials for the following half-cell reaction are $Zn \rightarrow Zn^{2+} + 2e^+$; E° = –0.76 V ; $Fe \rightarrow Fe^{2+} + 2e^-$; E° = –0.44 V The e.m.f for the cell reaction $Fe^{2+} + Zn \rightarrow Zn^{2+} + Fe$ will be [BHU-95]
(A) –0.32 V (B) +0.32 V (C) +1.20 V (D) –1.20 V

82. What is the amount of chlorine evolved when 2 amperes of current is passed for 3.0 minutes in an aqueous solution of NaCl ? [BHU-98]
(A) 66 g (B) 1.32 g (C) 33 g (D) 99 g

83. Which among the following is the strongest reducing agent ? Given $Fe^{2+} + 2e^- \rightarrow Fe(-0.44\ V)$; Ni^{2+} + 327°C, $e^- \rightarrow Ni\ (-0.25V)$; $Sn^{2+} + 2e^- \rightarrow Sn(-0.14V)$ and , $Fe^{3+} + e^- \rightarrow Fe^{2+}$ (–77 V) ? [BHU-98]
(A) Fe (B) Fe^{2+} (C) Ni (D) Sn

84. Vapour density of a metal chloride is 77. If its equivalent weight is 3, its atomic weight will be [BHU-99]
(A) 3 (B) 6 (C) 9 (D) 12

85. For a redox reaction,
$HNO_4^- + C_2O_4^{2-} + H^+ \rightarrow Mn^{2+} + CO_2 + H_2O$
Correct stoichiometric coefficients of HNO_4^- , $C_2O_4^{2-}$ OH^+ are [BHU-2000]
(A) 2,5,16 (B) 16,5,2 (C) 5,11,2 (D) 2,16,5

86. Faraday constant [BHU-2001]
(A) Depends on equivalent mass
(B) Depends on the number of electrons
(C) Is a numerical constant
(D) Depends upon the current passed

87. Calculate the number of coulombs required to deposit 40.5 g of Al when the electrode reaction is
$Al^{3+} + 3e^- \rightarrow Al$ [BHU-2002]
(A) 4.34 x 10^5 C (B) 434 x 10^5 C
(C) 4.43 x 10^5 C (D) 43.40 x 10^5 C

88. Unit of specific conductivity is [AFMC-91]
(A) mho cm (B) ohm cm^{-1} (C) mho (D) mho cm^{-1}

89. When current is passed through an electrolyte containing ions, [AFMC-92]
(A) All ions move to oppositely charged plates
(B) Ions do not move
(C) All ions moves to respective charged plates
(D) All the above are false

90. The amount of electricity that can deposit 108 gm of silver from $AgNO_3$ solution is [AFMC-93]
(A) Ampere (B) 1 Coulomb
(C) 1 Faraday (D) None of these

91. Use of electrolysis is [AFMC-95]

(A) Electroplating (B) Electrorefining
(C) Both (a) and (b) (D) None of these

92. Which of the following behaves as both oxidizing and reducing agents ? [AFMC-95]
(A) H_2SO_4 (B) SO_2 (C) H_2S (D) HNO_3

93. Which of the following conducts electricity ? [AFMC-95]
(A) Fused NaCl (B) CO_2 (C) Br_2 (D) Si

94. Electroplating is discovered by [AFMC-96]
(A) Faraday (B) Arrhenius
(C) Newton (D) Dalton

95. Which of the following is ferroelectric compound ? [AFMC-97]
(A) $BaTiO_3$ (B) $K_4[Fe(CN)_6]$
(C) Pb_2O_3 (D) None of these

96. Which of the following lose charge at cathode [AFMC-2000]
(A) Ions (B) Cations (C) Anions (D) None of these

97. Laws of electrolysis were given by [AFMC-2001]
(A) Lamarck (B) Ostwald
(C) Faraday (D) Arrhenius

98. For the redox reaction [IIT-92]
$MnO_4^- + C_2O_4^{2-} + H^+ \rightarrow Mn^{2+} + CO_2 + H_2O$
The correct coefficient for the balanced reaction:

	MnO_4^-	$C_2O_4^{2-}$	H^+
(A)	2	5	16
(B)	16	5	2
(C)	5	16	2
(D)	2	16	5

99. A dilute aqueous solution of Na_2SO_4 is electrolysed using platinium electrodes. The products at the anode and cathode are : [IIT-96]
(A) O_2, H_2 (B) $S_2O_8^{2-}$, Na
(C) O_2, Na (D) $S_2O_8^{2-}$, H_2

100. A solution of sodium in liquid ammonia at -33°C conducts electricity. On cooling, the conductivity of this solution .... [IIT-97]

101. When an aqueous solution of sodium fluoride is electrolysed, the gas liberated at the anode is .... [IIT-97]

102. A gas X at 1atm is bubbled through a solution containing a mixture of 1MY⁻ and 1MZ⁻ at 25°C. If the reduction potential is Z > Y > X , then, [IIT-99]
(A) Y will oxidize X and not Z
(B) Y will oxidize Z and not X
(C) Y will oxidize both X and Z
(D) Y will reduce both X and Z

103. The oxidation number of sulphur in S_8, S_2F_2, H_2S respectively, are [IIT-99]
(A) 0, + 1 and −2 (B) +2, + 1 and −2
(C) 0, + 1 and +2 (D) −2, + 1 and −2

104. For the electrochemical cell, $M|M^+||X^-|X$, $E°(M^+/M) = 0.44$ V and $E°(X/X^-) = 0.33$ V. From this data one can deduce that [IIT-2000]
(A) $M + X \rightarrow M^+ + X^-$ is the spontaneous reaction
(B) $M^+ + X^- \rightarrow M + X$ is the spontaneous reaction
(C) $E_{cell} = 0.77$ V (D) $E_{cell} = -0.77$ V

105. Amongst the following, identify the species with an atom in +6 oxidation state [IIT-2000]
(A) MnO_4^- (B) $Cr(CN)_6^{3-}$ (C) NiF_6^{2-} (D) CrO_2Cl_2

106. In the standardization of $Na_2S_2O_3$ using $K_2Cr_2O_7$ by iodometry, the equivalent weight of $K_2Cr_2O_7$ is [IIT-2001]
(A) (molecular weight)/2 (B) (molecular weight)/6
(C) (molecular weight)/3 (D) same as molecular weight

107. The correct order of equivalent conductance at infinite dilution of LiCl, NaCl and KCl is [IIT-2001]
(A) LiCl > NaCl > KCl (B) KCl > NaCl > LiCl
(C) NaCl > KCl > LiCl (D) LiCl > KCl > NaCl

108. Satutrated solution of KNO_3 is used to make 'salt-bridge' because [IIT-2001]
(A) velocity of K^+ is greater than that of NO_3^-
(B) velocity of NO_3^- is greater than that of K^+
(C) velocity of both K^- and NO_3^- are nearly the same
(D) KNO_3 is highly soluble in water

109. Standard electrode potential data are useful for understanding the suitability of an oxidant in a redox titration. Some half cell reactions and their standard potentials are given below :
$MnO_4^-(aq.) + 8H^+(aq.) + 5e^- \rightarrow Mn^{2-}(aq.) + 4H_2O(l)$
E° = 1.51 V
$Cr_2O_7^{2-}(aq.) + 14H^+(aq) + 6e^- \rightarrow 2Cr^{3+}(aq.) + 7H_2O(l)$
E° = 1.38 V
$Fe^{3+}(aq.) + e^- \rightarrow Fe^{2+}(aq.)$ E° = 0.77 V
$Cl_2(g) + 2e^- \rightarrow 2Cl^-(aq.)$ E° = 1.40 V
Identify the only incorrect statement regarding the quantitative estimation of aqueous $Fe(NO_3)_2$ [IIT-2002]
(A) MnO_4^- can be used in aqueous HCl
(B) $Cr_2O_7^{2-}$ can be used in aqueous HCl
(C) MnO_4^- can be used in aqueous H_2SO_4
(D) $Cr_2O_7^{2-}$ can be used in aqueous H_2SO_4

110. In an electrolytic cell flow of e⁻ will be [IIT-2003]
(A) from anode to internal source
(B) from cathode to internal source
(C) from cathode to anode (D) None of these

111. EMF of a cell in terms of reduction potential of its left and right electrodes is [AIEEE-2002]

(A) $E = E_{left} - E_{right}$ (B) $E = E_{left} + E_{right}$
(C) $E = E_{right} - E_{left}$ (D) $E = -(E_{right} + E_{left})$

112. Which of the following is a redox reaction ? [AIEEE-2002]

(A) $NaCl + KNO_3 \rightarrow NaNO_3 + KCl$
(B) $CaC_2O_4 + 2HCl \rightarrow CaCl_2 + H_2C_2O_4$
(C) $Mg(OH)_2 + 2NH_4Cl \rightarrow MgCl_2 + 2NH_4OH$
(D) $Zn + 2AgCN \rightarrow 2Ag + Zn(CN)_2$

113. Conductivity (unit Siemens S) is directly proportional to area of the vessel and the concentration of the solution in it and is inversely proportional to the length of the vessel. Then the unit of the constant of proportionality is [AIEEE-2002]

(A) $Sm\ mol^{-1}$ (B) $Sm^2\ mol^{-1}$
(C) $S^{-2}m^2\ mol$ (D) $S^2m^2\ mol^{-2}$

114. If ϕ denotes reduction potential, then which is true? [AIEEE-2002]

(A) $E^o_{cell} = \phi_{right} - \phi_{left}$ (B) $E^o_{cell} = \phi_{left} + \phi_{right}$
(C) $E^o_{cell} = \phi_{left} - \phi_{right}$ (D) $E^o_{cell} = -(\phi_{left} + \phi_{right})$

115. What will be the emf for the given cell
$Pt|H_2(P_1)|H^+(aq)||H_2(P_2)|Pt$ [AIEEE-2002]

(A) $\frac{RT}{f}\log\frac{P_1}{P_2}$ (B) $\frac{RT}{2f}\log\frac{P_1}{P_2}$
(C) $\frac{RT}{f}\log\frac{P_2}{P_1}$ (D) None of these

116. Which of the following reactions is possible at anode? [AIEEE-2002]

(A) $2Cr^{3+} + 7H_2O \rightarrow Cr_2O_7^{2-} + 14H^+$ (B) $F_2 \rightarrow 2F^-$
(C) $(1/2)O_2 + 2H^+ \rightarrow H_2O$ (D) None of these

117. When the sample of copper with zinc impurity is to be purified by electrolysis, the appropriate electrodes are. [AIEEE-2002]

	Cathode	Anode
(A)	Pure zinc	Pure copper
(B)	Impure sample	Pure copper
(C)	Impure zinc	Impure sample
(D)	Pure copper	Impure sample

118. When during electrolysis of a solution of $AgNO_3$ 9650 coulombs of charge pass through the electroplating bath, the mass of silver deposited on the cathode will be [AIEEE-2003]

(A) 1.08 g. (B) 10.8 g. (C) 21.6 g. (D) 108 g.

119. For a cell reaction involving a two-electron change, the standard emf of the cell is found to be 0.295 V at 25°C. The equilibrium constant of the reaction at 25°C will be [AIEEE-2003]

(A) 1×10^{-10} (B) 29.5×10^{-2} (C) 10 (D) 1×10^{10}

120. Standard reduction electrode potentials of three metals A, B and C are respectively + 0.5 V, –3.0 V and –1.2 V. The reducing powers of these metals are [AIEEE-2003]

(A) B > C > A (B) A > B > C
(C) C > B > A (D) A > C > B

121. Several blocks of magnesium are fixed to the bottom of a ship to [AIEEE-2003]

(A) Keep away the sharks (B) Make the ship lighter
(C) Prevent action of water and salt
(D) Prevent puncturing by under-sea rocks.

122. A silver cup is plated with silver by passing 965 coloumbs of electricity, the amount of silver deposited is : [EAMCET-90]

(A) 9.89 gm (B) 107.87gm
(C) 1.0787 gm (D) 1.002 gm

123. A metallic oxide contains 60% of the metal. The equivalent weight of the metal is ....[EAMCET-90]

124. Three Faradays of electricity was passed through an aqueous solution of Iron (II) bromide. The weight of iron metal (at. wt.= 56) deposited at the cathode in grams is [EAMCET-91]

(A) 56 (B) 84 (C) 112 (D) 168

125. In the raction $C_2O_4^{2-} + MnO_4 + H^+ \rightarrow Mn^{+2} + CO_2$ the reductant is .... [EAMCET-91]

126. When 9.65 coulombs of electricity is passed through a solution of silver nitrate (atomic weight of Ag=107.87g taking as 108.0) the amount of silver deposited is [EAMCET-92]

(A) 10.8mg (B) 5.4mg (C) 16.2mg (D) 21.2mg

127. 0.84 gm of a metal carbonate reacts exactly with 40ml of N/2 H_2SO_4. The equivalent weight of the metal carbonate is [EAMCET-92]

(A) 84gm (B) 64 gm (C) 42 gm (D) 38 gm

128. The oxidation number of chlorine in $KClO_3$ is .... [EAMCET-93]

129. The metal that has highest electrical conductivity is [EAMCET-94]

(A) Al (B) Cu (C) Ag (D) Au

130. The formula of a metal chloride is MCl_3 if it contains 20% of the metal, the atomic weight of the metal is approximately. [EAMCET-94]

(A) 25.5 (B) 11.8 (C) 21.3 (D) 106.5

131. The metal that forms a self protecting film of oxide to prevent corrosion is [EAMCET-94]

(A) Cu (B) Al (C) Pt (D) Au

132. How many grams of Copper will be deposited from a solution of $CuSO_4$ by passing 0.5 Faraday of electricity (At. wt. of Cu = 63.5) [EAMCET-94]

(A) 31.75 (B) 63.5 (C) 15.875 (D) 127

133. The single electrode reactions.
$Cd \rightleftharpoons Cd^{2+} + 2e^- : E^o = 0.4030V$
$Cu^{2+} + 2e^- \rightleftharpoons Cu : E^o = 0.337V$
Can be combined to give the cell reaction
$Cd + Cu^{2+} \rightleftharpoons Cd^{2+} + Cu$ the e.m.f. such cell is [EAMCET-94]
(A) 0.74V (B) 0.1354V (C) 0.66 (D) 0.066V

134. Bleaching action of chlorine in the presence of moisture is an example of [EAMCET-94]
(A) Oxidation (B) Reduction
(C) Displacement (D) Substitution

135. In the electrolysis of HF–KF mixture for the production of fluorine, the role of KF is [EAMCET-94]
(A) To lower the oxidation potential of fluoride
(B) To make HF a conducting solution
(C) To minimize corrosion
(D) To lower the solubility of fluorine

136. In the reaction $NO_2^- + OCl^- \rightarrow NO_3^- + Cl^-$ the oxidation state of chlorine [EAMCET-94]
(A) Does not change (B) Changes from +1 to –1
(C) Changes from –2 to–1 (D) Changes from 0 to –1

137. Which of the following alkali metals has the biggest tendency for the half reaction ?
$M(g) \rightarrow M^+(aq) + e$ [EAMCET-95]
(A) Sodium (B) Lithium (C) Potassium (D) Cesium

138. When a copper wire is dipped in aqueous $AgNO_3$ solution, the solution turns blue. The reason for this is [EAMCET-95]
(A) Oxidation of Silver (B) Reduction of copper
(C) Oxidation of copper (D) Reduction of Silver

139. How many grammes of copper would be deposited if 3.00 amperes of current is passed through a solution of copper sulphate for 4.0 Hours ? (At, Wt, of Cu = 63.54) [EAMCET-95]
(A) 7.11 (B) 14.22 (C) 28.44 (D) 56.88

140. The standard reduction potentials at 298 K for the following half cell reaction are given below :
$Zn^{++}(aq) + 2e^- \rightarrow Zn(s)$ –0.762
$Cr^{+++}(aq) + 3e^- \rightarrow Cr(s)$ –0.740
$2H^+(aq) + 2e^- \rightarrow H_2(g)$ 0.000
$Fe^{+++}(aq) + e^- \rightarrow Fe^{++}(aq)$ 0.770
which one is the strongest reducing agent ? [EAMCET-96]
(A) Zn(s) (B) Cr(s) (C) $H_2(g)$ (D) $Fe^{++}(aq)$

141. When Zn metal is added to $CuSO_4$ solution, Cu is precipitated it is due to [EAMCET-96]
(A) Oxidation of Cu^{2+} (B) Reduction of Cu^{2+}
(C) Hydrolysis of $CuSO_4$ (D) Ionisation of $CuSO_4$

142. In the electrochemical cell $H_2(g)$ 1 atm. / H^+ (1M) //Cu^{2+} (1M) Cu (S) which one of the following statements is true ? [EAMCET-97]
(A) H_2 is cathode : Cu is anode
(B) Oxidation occurs at Cu electrode
(C) Reduction occurs at H_2 electrode
(D) H_2 is anode : Cu is cathode

143. On passing a current through a KCl solution, 19.5 gm. of potassium is deposited. If the same quantity of electricity is passed through a solution of aluminium chloride, the amount of Aluminium deposited is : [EAMCET-97]
(A) 4.5g (B) 9.0 g (C) 13.5 g (D) 27 g

144. The cell reaction of a cell is $Mg(s) + Cu^{2+}(aq) \rightarrow Cu(s) + Mg^{2+}(aq)$ If the standard reduction potentials of Mg & Cu are –2.37 & 0.34 V respectively, the emf of the cell is [EAMCET-98]
(A) 2.03 (B) –2.03 (C) 2.71 (D) –2.71

145. The standard reduction potentials at 298 K for the following half cell reactions are given below. Find out the strongest reducing agent [EAMCET-98(MBBS)]
(A) $Zn^{2+}(aq) + 2e^- \rightarrow Zn(s)$ – 0.762
(B) $Cr^{3+}(aq) + 3e^- \rightarrow Cr(s)$ – 0.74
(C) $2H^+(aq) + 2e^- \rightarrow H_2(g)$ – 0.00
(D) $Fe^{3+}(aq) + e^- \rightarrow Fe^{2+}(ag)$ + 0.77

146. For a cell the cell reaction is $Mg(s) + Cu^{2+}(aq) \rightarrow Cu(s) + Mg^{2+}(aq)$. If the standard reduction potential of Mg and Cu are –2.37 and +0.34 V respectively; the emf of the cell is [EAMCET-98(MBBS)]
(A) 2.03V (B) –2.03V (C) 2.71V (D) –2.71V

147. Chlorine is passed into dilute cold KOH solution. What are the oxidation numbers of products formed ? [EAMCET-99]
(A) –1, +5 (B) –1, +3 (C) +1, +7 (D) +1, –1

148. Standard reduction potentials of $Sn^{2+}|Sn$ and $Zn^{2+}|Zn$ electrodes are –0.14 and –0.74V respectively. What is the E° of the following cell in volts ? $Zn|Zn^{2+} || Sn^{2+}|Sn$ [EAMCET-99(MBBS)]
(A) –0.6 (B) –0.88 (C) 0.88 (D) 0.6

149. What is the approximate quantity of electricity (in coulombs) required to deposit all the silver from 250 ml of 1 M $AgNO_3$ aqueous solution ? (At. wt. of Ag = 108) [EAMCET-2000]
(A) 96500 (B) 24125 (C) 48250 (D) 12062.5

150. Which of the following aqueous solutions conducts electricity ? [EAMCET(MBBS)-2000]
(A) Urea (B) Glucose (C) Sucrose (D) NaCl

151. One ampere of current is passed for 9650 second through molten $AlCl_3$. What is the weight in grams of Al deposited at cathode ? (Atomic weight of Al = 27)

[EAMCET-2001]

(A) 0.9 (B) 9.0 (C) 0.09 (D) 90.0

152. Molten $CuCl_2$ is electrolysed using platinum electrodes. The reaction occurring at anode is _____ [EAMCET-2001]

(A) $2Cl^- \rightarrow Cl_2(g) + 2e^-$ (B) $Cl_2(g) + 2e^- \rightarrow 2Cl^-$
(C) $Cu^{2+} + 2e^- \rightarrow Cu(s)$ (D) $Cu(s) \rightarrow Cu^{2+} + 2e^-$

153. How many moles of acidified $FeSO_4$ can be completely oxidized by one mole of $KMnO_4$? [EAMCET(MBBS)-2001]

(A) 10 (B) 5 (C) 6 (D) 4

154. One Faraday of electricity is passed separately through one litre of one molar aqueous solutions of (i) $AgNO_3$ (ii) $SnCl_4$ and (iii) $CuSO_4$. The number of moles of Ag, Sn and Cu deposited at cathode are respectively : [EAMCET(MBBS)-2001]

(A) 1.0, 0.25, 0.5 (B) 1.0, 0.5, 0.25
(C) 0.5, 1.0, 0.25 (D) 0.25, 0.5, 1.0

155. Aqueous copper sulphate solution is electrolysed using platinum electrodes. The electrode reaction occurring at cathode is : [EAMCET(MBBS)-2001]

(A) $Cu^{2+}(aq) + 2e^- \rightarrow Cu(s)$
(B) $Cu(s) \rightarrow Cu^{2+}(aq) + 2e^-$
(C) $2H_2O(l) \rightarrow O_2(g) + 4H^+(aq) + 4e^-$
(D) $O_2(g) + 4H^+(aq) + 4e^- \rightarrow 2H_2O(l)$

156. Consider the following reaction
$6NaOH + 3Cl_2 \rightarrow 5NaCl + A + 3H_2O$ What is the oxidation number of chlorine in "A" [EAMCET(MBBS)-2001]

(A) +5 (B) 1 (C) +3 (D) +1

157. What is the reduction electrode potential (in volts.) of copper electrode when $[Cu^{2+}] = 0.01$ M in a solution at 25°C? (E° of Cu^{2+}/Cu electrode is +0.34 V) [EAMCET-2002]

(A) 0.3991 (B) 0.2809 (C) 0.3105 (D) 0.3695

158. 0.066 grams of metal was deposited when a current of 2 amperes is passed through a metal ion solution for 100 seconds. What is the electrochemical equivalent (in gram coulomb^{-1}) of the metal ? [EAMCET-2002]

(A) 3.3×10^{-6} (B) 3.3×10^{-4} (C) 0.033 (D) 3.3

159. Consider the following four electrodes :
A = Cu^{2+} (0.0001 M) | Cu(s)
B = Cu^{2+} (0.01 M) | Cu(s)
C = Cu^{2+} (0.01 M) | Cu(s)
D = Cu^{2+} (0.001 M) | Cu(s)
If the standard reduction potential of Cu^{2+} | Cu is + 0.34 V, the reduction potentials (in volts) of the above electrodes follow the order.... [EAMCET(med.)-2002]

(A) A > D > C > B (B) B > C > D > A
(C) C > D > B > A (D) A > B > C > D

160. Aqueous NaCl solution is electrolyzed using platinum electrodes. What is the product formed at cathode ? [EAMCET(med.)-2002]

(A) Na (B) H_2 (C) O_2 (D) Cl_2

161. When X amperes of current is passed through molten $AlCl_3$ for 96.5 seconds, 0.09 grams of aluminium is deposited. What is the value of X ? [EAMCET-2003]

(A) 10 (B) 20 (C) 30 (D) 40

162. Which of the following is used to flow electricity ? [CPMT-96]

(A) Fe (B) Al (C) Cu (D) Ag

163. If the reduction potential is more [CPMT-96]

(A) It is easily oxidized (B) Easily reducing
(C) Act as oxidizing agent (D) Redox nature

164. Which of the following cannot work as oxidizing agent ? [CPMT-96]

(A) O_2 (B) $KMnO_4$ (C) H_2O_2 (D) I_2

165. Which connot displace H_2 from its compound? [CPMT-96]

(A) Hg (B) Zn (C) Cu (D) None of these

166. E_0 of a cell a A + bB → cC + dD is- [CPMT-97]

(A) $-\frac{RT}{nF}\log\frac{[C]^c[D]^d}{[A]^a[B]^b}$ (B) $-\frac{RT}{nF}\log\frac{[a]^A[b]^B}{[c]^C[d]^D}$

(C) $-\frac{RT}{nF}\log\frac{[C]^c[d]^D}{[A]^a[B]^B}$ (D) $-\frac{RT}{nF}\log\frac{[C]^c[d]^D}{[a]^A[B]^b}$

167. Oxidation numbers of Mn in K_2MnO_4 and $MnSO_4$ are respectively [CPMT-97]

(A) +7, + 2 (B) +6, + 2 (C) +5, +2 (D) +2, +6

168. The unit of equivalent conductivity is [CPMT-99]

(A) Ohm cm (B) Ohm^{-1}cm^2 (g equivalent)$^{-1}$
(C) Ohm cm^2 (g-equivalent) (D) S cm^{-2}

169. At infinite dilution, the percentage ionisation for both strong and weak electrolytes is [CPMT-99]

(A) 1% (B) 20% (C) 50% (D) 100%

170. The standard reduction potentials of 4 elements are given below. Which of the following will be the most suitable reducing agent ? I = – 3.04V, II = –1.90 V, III = 0V, IV = 1.90 V [CPMT-99]

(A) I (B) II (C) III (D) IV

171. The values of x and y in the following redox reaction $x\,Cl_2 + 6OH^- \rightarrow ClO_3^- + y\,Cl^- + 3H_2O$ are - [CPMT-2000]

(A) x = 5, y = 3 (B) x = 2, y = 4
(C) x = 3, y = 5 (D) x = 4, y = 2

172. Which is a redox reaction ? [CPMT-2000]

(A) $BaCl_2 + H_2SO_4 \rightarrow BaSO_4 + 2HCl$
(B) $H_2SO_4 + 2NaOH \rightarrow Na_2SO_4 + 2H_2O$
(C) $CH_3COOH + C_2H_5OH \rightarrow CH_3COOC_2H_5 + H_2O$
(D) $2FeCl_3 + SnCl_2 \rightarrow 2FeCl_2 + SnCl_4$

173. Standard electrode potentials of two half cells are given as
$Zn^{2+}/Zn = -0.76$ V ;
$Ag^+/Ag = 0.80$ V
The EMF of the cell $Zn \mid Zn^{2+}$ (1M) $\|$ Ag^+ (1M) $\mid Ag$ will be [CPMT-2000]
(A) 1.56 V (B) 1.20 V (C) 0.04 V (D) -1.56V

174. The equivalent weight of $C_6(COOH)_6$ is [CPMT-2001]
(A) 57 (B) 69 (C) 171 (D) 181

175. The oxidation number of As in $H_2AsO_4^-$ is [CPMT-2001]
(A) +6 (B) +5 (C) −7 (D) +9

176. M^{+3} ion loses 3e. Its oxidation number in water is ? [CPMT-2002]
(A) 0 (B) +3 (C) +6 (D) -3

177. To prevent corrosion of iron by zinc coating is called [CPMT-2002]
(A) galvanisation (B) rusting
(C) protection (D) none of these

178. The standard electrode potential of the half cells are given below: [CPMT-2003]
$Zn^{2+} + 2e^- \rightarrow Zn$; E = −7.62 V
$Fe^{2-} + 2e^- \rightarrow Fe$; E = −7.81 V
The emf of the cell $Fe^{2+} + Zn \rightarrow Zn^{2+} + Fe$ is
(A) 1.54 V (B) −1.54 V (C) −0.19 V (D) +0.19 V

179. The conductivity of strong electrolyte[CPMT-2003]
(A) increases on dilution slightly
(B) decrease on dilution (C) do not change with dilution
(D) depend upon density of electrolytes itself

180. Hydrogen-oxygen fuel cells are used in space craft to supply: [PMT-93]
(A) Power for heat and light (B) Power for pressure
(C) Oxygen (D) Water

181. Prevention of corrosion of Iron metal by Zinc coating is called : [PMT-93]
(A) Galvanisation (B) Cathodic protection
(C) Electrolysis (D) Photo-electrolysis

182. If M is the molecular weight of $KMnO_4$ its equivalent weight will be when it is converted into K_2MnO_4 : [PMT-93]
(A) M (B) M/3 (C) M/5 (D) M/7

183. If NaOH is added to an aqueous solution of Zinc ions, a white precipitate appears and on adding excess NaOH the precipitate dissolves. In this solution Zinc exists in the : [PMT-93]
(A) Cationic Part (B) Anionic Part
(C) Both in Cationic and Anionic Parts
(D) There is no Zinc in the solution

184. When one ampere current flows for 1 sec. through a conductor, this quantity of electricity is known as : [PMT-93]
(A) Faraday (B) Coulomb
(C) E. M. F. (D) Ohm

185. Which of the following statements is correct ?
Galvanic cell cenverts : [PMT-93]
(A) chemical energy into electrical energy
(B) electrical energy into chemical energy
(C) metal from its elemental state to the combined state
(D) electrolyte into individual ions

186. In the following reaction
$3Br_2 + 6CO_3^{2-} + 3H_2O \rightarrow 5Br^- + BrO_3^- + 6HCO_3^-$ [PMT-94]
(A) Bromine is oxidised and carbonate is reduced.
(B) Bromine is reduced and carbonate is oxidised.
(C) Bromine is neither reduced nor oxidised.
(D) Bromine is reduced and oxidised both.

187. When electricity is passed through a solution of aluminium chloride, 13.5 gms of Al are deposited. The number of Faraday must be : [PMT-94]
(A) 2.0 (B) 1.5 (C) 1.0 (D) 0.5

188. Which will not affect the degree of ionisation ? [PMT-94]
(A) Temperature (B) Concentration
(C) Type of solvent (D) Current

189. In the following reaction [PMT-94]
$P_4 + 3NaOH + 3H_2O \rightarrow PH_3 + 3NaH_2PO_2$
(A) Phosphorus is oxidised
(B) Phosphorus is oxidised and reduced
(C) Phosphorus is reduced (D) Sodium is oxidised

190. The electroplating of chromium is undertaken because [PMT-94]
(A) electroysis of chromium is easier
(B) chromium can form alloys with other metals
(C) chromium gives protective and decorative coating to the base metal
(D) of the high reactivity of metallic chromium

191. Oxidation number of P in $Mg_2P_2O_7$ is [PMT-95]
(A) + 3 (B) + 2 (C) + 5 (D) - 3

192. In the reaction [PMT-95]
$3\ Br_2 + 6\ Co_3^{-2} + 3H_2O = 5\ Br^- + BrO_3^- + 6HCO_3^-$
(A) Bromine is oxidised, carbonate is reduced
(B) Both bromine and carbonate are reduced

(C) Bromine is neither reduced nor oxidised
(D) Bromine is reduced as well as oxidised

193. In the cell Zn | Zn^{2+} || Cu^{2+} | Cu the negative electrode is [PMT-95]
(A) Cu (B) Cu^{2+} (C) Zn (D) Zn^{2+}

194. 96500 coulombs of electric current liberates from $CuSO_4$ solution [PMT-95]
(A) 63.5 gm Cu (B) 31.75 gm Cu
(C) 96500 gm Cu (D) 100 gm Cu

195. Electrode potentials of five elements A, B, C, D and E are respectively, -1.36, -0.32. 0. -1.26 and -0.42. The reactivity order of these elements are in the order of [PMT-95]
(A) A D E B and C (B) C B E D and A
(C) B D E A and C (D) C A E D and B

196. Rusting on iron needs [PMT-95]
(A) dry air (B) air and water
(C) distilled water and carbon dioxide
(D) oxygen and carbon dioxide

197. When 96500 coulomb of electricity is passed through a copper sulphate solution, the amount of copper deposited will be [PMT-96]
(A) 0.25 mol (B) 0.50 mol
(C) 1.00 mol (D) 2.00 mol

198. The increase in equivalent conductance of an electrolyte solution with dilution is due to the increase in [PMT-96]
(A) ionic attraction (B) molecular attraction
(C) degree of association of the electrolyte
(D) degree of ionisation of the electrolyte

199. Iron displaces copper from its salt solution, because [PMT-96]
(A) atomic number of iron is less than that of copper
(B) the standard reduction potential of iron is less than that of copper
(C) the standard reduction potential of iron is more than that of copper
(D) the iron salt is more soluble in water than the copper salt

200. Amongst the following electrodes the one with zero electrode potential is [PMT-97]
(A) Calomel electrode (B) Standard hydrogen electrode
(C) Glass electrode (D) Gas electrode

201. Which of the following is correct expression for electrode potantial of cell - [PMT-97]

(A) $E = E^\circ \frac{RT}{nF} \ln \frac{[product]}{[reactant]}$

(B) $E = E^\circ + \frac{RT}{F} \ln \frac{[product]}{[reactant]}$

(C) $E = E^\circ \frac{RT}{nF} \ln \frac{[reactant]}{[product]}$

(D) $E = -\frac{RT}{F} \ln \frac{[product]}{[reactant]}$

202. In the chemical reaction $Cl_2 + H_2S \rightarrow 2HCl + S$. The oxidation number of sulphur changes from : [PMT-99]
(A) 0 to 2 (B) 2 to 0 (C) - 2 to 0 (D) -2 to -1

203. In acidic medium, equivalent weight of potassium permanganate is equal to : [PMT-99]
(A) molecular weight of $KMnO_4$
(B) $\frac{1}{5}$ x molecular weight of $KMnO_4$
(C) $\frac{1}{3}$ x molecular weight of $KMnO_4$
(D) $\frac{1}{2}$ x molecular weight of $KMnO_4$

204. 2.5 F of electricity are passed through a $CuSO_4$ solution. The number of gm equivalent of Cu deposited on anode is : [PMT-2001]
(A) Zero (B) 1.25 (C) 2.5 (D) 5.0

205. The equivalent weight of $KMnO_4$ in alkaline medium will be : [PMT-2001]
(A) 31.60 (B) 52.66 (C) 79.00 (D) 158.00

206. In galvanic cell, the salt bridge is used to [PMT-2002]
(A) Complete the circuit
(B) Reduce the electric resistance in the cell
(C) Separate cathode from anode
(D) Carry salts for the chemical reaction

207. The reference electrode is made by using [PMT-2002]
(A) $ZnCl_2$ (B) $CuSO_4$ (C) $HgCl_2$ (D) Hg_2Cl_2

208. In the acidic solution of copper sulphate and alkaline solution of copper cyanide respectively, same electric current is flown for a definite time which of the following statements is correct for this. [PMT-2003]
(A) Amount of copper deposited in both the solutions is the same.
(B) Amount of copper deposited in alkaline copper cyanide solution is less.
(C) Amount of copper deposited in acidic copper sulphate solution is less. (D) None of these

209. The following four colourless solutions of different salts are taken separately in four beakers. A copper metal piece is dropped in each of the solution. Which of the four solutions is changed into blue colour after some time? [PMT-2003]
(A) Zinc nitrate solution (B) Barium nitrate solution
(C) Silver nitrate solution (D) Sodium nitrate solution

210. What is the emf of the cell, whose half cells are given below [PMT-2003]
$Mg^{2+} + 2e^- \leftrightarrows Mg(s)$ E = –2.37 V
$Cu^{2+} + 2e^- \leftrightarrows Cu(s)$ E = +0.34 V
(A) -2.30 V (B)1.336 V (C) 2.71 V (D) 2.03 V

211. Which of the following reactions is a redox reaction. [PMT-2003]
(A) $2AgNO_3 + BaCl_2 = 2AgCl + Ba(NO_3)_2$
(B) $BaO_2 + H_2SO_4 = BaSO_4 + H_2O_2$
(C) $P_2O_5 + 2H_2O = H_4P_2O_7$
(D) $Cu + 2AgNO_3 = 2Ag + Cu(NO_3)_2$

212. In the reaction [PMT-2003]
$2Fe^{3+}(aq) + Sn^{2+}(aq) \rightarrow 2Fe^{2+}(aq) + A$. A belongs to
(A) $Sn^{4+}(aq)$ (B) $Sn^{2+}(aq)$ (C) $Sn^{3+}(aq)$ (D) Sn^0

213. When lead storage battery is charged [PET-93]
(A) Lead dioxide dissolves
(B) Sulphuric acid is regenerated
(C)The lead electrode becomes coated with lead sulphate
(D) The amount of sulpuric acid decreases

214. Which of the following statements is correct about equivalent weight of $KMnO_4$ [PET-94]
(A) It is one third of its molecular weight in alkaline medium
(B) It is one fifth of its molecular weight in alkaline medium
(C) It is equal to its molecular weight in acidic medium
(D) It is one third of its molecular weight in acidic medium

215. A current of 2.5 amperes was passed through a copper sulphate solution for 6 minutes 26 seconds. The amount of copper deposited is
Atomic weight of copper = 63.5
1Faraday = 96500 Coulmbs [PET-94]
(A) 6.35 gms (B) 0.3175 gms
(C) 0.635 gms (D) 3.175 gms

216. What is the emf of the cell, whose half cells are given below $Mg^{2+} + 2e^- \rightleftharpoons Mg(s)$ E = –2.37 V [PET-94]
$Cu^{2+} + 2e^- \rightleftharpoons Cu(s)$ E = + 0.34 V
(A) - 2.30V (B) 1.336V (C) 2.71V (D) 2.03V

217. The relation between Δ G and E for cell is Δ G = -nFE; the cell reaction will be spontaneous, if [PET-95]
(A) G is negative (B) G is positive
(C) E is negative (D) E is positive

218. Faraday has the dimensions of [PET-95]
(A) Coulombs (B) Coulomb equivalent
(C) Coulomb per equivalent
(D) Coulomb per degree Kelvin

219. Zinc displaces copper from the solution of its salt because [PET-95]
(A) Atomic number of zinc is more than that of copper
(B) Zinc salt is more soluble in water than the copper salt
(C) Gibb's free energy of zinc is less than that of copper
(D) Zinc is placed higher than copper in electrochemical series

220. The unit of cell constant is [PET-96]
(A) $ohm^{-1}cm^{-1}$ (B) ohm cm (C) cm (D) cm^{-1}

221. The chemical reaction taking place at the anode of a cell is [PET-96]
(A) Ionization (B) Reduction
(C) Oxidation (D) Hydrolysis

222. In dry cell the reaction which takes place at the graphite anode is [PET-96]
(A) $Zn^{2+} + 2e^- \rightarrow Zn(s)$ (B) $Zn(s) \rightarrow Zn^{2+} + 2e^-$
(C) $Mn^{2+} + 2e^- \rightarrow Mn(s)$ (D) $Mn(s) \rightarrow Mn^+ + e^- + 1.5V$

223. In Galvanisation, iron surface is coated with [PET-96]
(A) Tin (B) Chromium (C) Zinc (D) Nickel

224. Which one of the following statements is correct ? [PET-97]
(A) The oxidation number of oxygen in KO_2 is zero
(B) The specific conductance of an electrolyte solution decreases with increase in dilution
(C) Sn^{2+} oxidizes Fe^{3+}
(D) Zn / $ZnSO_4$ is a reference electrode

225. What would be the weight of deposited silver on passing 965 coulombs of electricity in a solution of silver nitrate ? [PET-97]
(A) 1.08 gm (B) 2.16 gm (C) 0.54 gm (D) 0.27 gm

226. Mark the false statement : [PET-97]
(A) A salt bridge is used to eliminate liquid junction potential
(B) The Gibb's free energy change, Δ G is related with electromotive force (E), as Δ G = –nFE
(C) Nernst equation for single electrode $E = E^o - RT/nF \ln a_{M^{n+}}$
(D) The efficiency of a hydrogen oxygen fuel cell is 23%

227. In acidic medium one mole of MnO_4^- accepts how many moles of electrons in a redox process ? [PET-98]
(A) 1 (B) 3 (C) 5 (D) 6

228. In acidic medium potassium dichromate acts as an oxidant according to the equation
$Cr_2O_7^{2-} + 14H^+ + 6e^- \rightarrow 2Cr^{3+} + 7H_2O$
What is the equivalent of $K_2Cr_2O_7$ (mol.wt.=M) [PET-98]
(A) M (B) M/2 (C) M/3 (D) M/6

229. In which cell the free energy of a chemical reaction is directly converted into electricity ? [PET-98]
(A) Leclanche cell (B) Concentration cell
(C) Fuel cell (D) Lead storage battery

230. The number of coulombs required for the deposition

of 107.870 g of silver is : [PET-98]
(A) 96.500 (B) 45, 250 (C) 1,93,000 (D) 10,000

231. Prevention of corrosion of iron by Zn coating is called: [PET-99]
(A) Galvanization (B) Cathodic protection
(C) Electrolysis (D) Photoelectrolysis

232. Hydrogen-Oxygen fuel cells are used in space craft to supply: [PET-99]
(A) Power for heat and light (B) Power for pressure
(C) Oxygen (D) Water

233. The reaction is spontaneous if the cell potential is: [PET-99]
(A) Positive (B) Negative (C) Zero (D) Infinite

234. 0.1914 g of an organic acid is dissolved in about 20 ml of water, 25 ml of 0.12 N NaOH is required for the complete neutralization of the acid solution. The equivalent weight of the acid is : [PET-2000]
(A) 65 (B) 64 (C) 63.80 (D) 62.50

235. Incorrect statement regarding rusting is : [PET-2000]
(A) Metallic iron is oxidized to Fe^{3+} ions
(B) Metallic iron is reduced to Fe^{2+} ions
(C) Oxygen gas is reduced to oxide ions
(D) Yellowish - brown product is formed

236. If an iron wire is placed in a solution of copper sulphate : [PET-2000]
(A) Copper will precipitate out
(B) Iron will dissolve and precipitate out thereafter
(C) Iron and copper both will go into solution
(D) No reaction will take place

237. To which of the following, the mass of an element deposited at an electrode is directly proportional ? [PET-2000]
(A) Atomic weight (B) Equivalent weight
(C) Molecular weight (D) Atomic number

238. From the solution of which of the following, one faraday of electricity will liberate one gram atom of metal ? [PET-2000]
(A) NaCl (B) $BaCl_2$ (C) $CuSO_4$ (D) $AlCl_3$

239. The oxidation number of C in CO_2 is : [PET-2001]
(A) −2 (B) +2 (C) −4 (D) +4

240. One faraday of electricity is equal to : [PET-2001]
(A) 96500 Coulomb (B) 10^6 Coulomb
(C) 3.7 x 10^6 Coulomb (D) 6.23 x 10^{23} Coulomb

241. Which of the following compounds will not undergo decomposition on passing electricity through aqueous solution ? [PET-2001]
(A) Sugar (B) Sodium Chloride
(C) Sodium Bromide (D) Sodium Acetate

242. The relationship between standard reduction potential of cell and equilibrium constant is shown by [PET-2002]
(A) $E^\circ_{cell} = \frac{n}{0.059} \log K_C$ (B) $E^\circ_{cell} = \frac{0.059}{n} \log K_C$
(C) $E^\circ_{cell} = 0.059 \log K_C$ (D) $E^\circ_{cell} = \frac{\log K_C}{n}$

243. The oxidation number of S in $H_2S_2O_8$ is[PET-2002]
(A) +2 (B) +4 (C) +6 (D) +7

244. The standard cell potential of $Zn \mid Zn^{2+}(aq) \| Cu^{2+}(aq) \mid Cu$ cell is 1.10 V. The maximum work obtained by this cell will be [PET-2002]
(A) 106.15 kilo joule (B) −212.30 kilo joule
(C) −318.45 kilo joule (D) −424.60 kilo joule

245. In order to prepare one litre normal solution of $KMnO_4$, how many gm of $KMnO_4$ are required if the solution is to be used in acid medium for oxidation [PET-2002]
(A) 158 gm (B) 31.60 gm (C) 62.0 gm (D)790 gm

246. Which of the following is not used to make salt bridge? [PET-2003]
(A) KCl (B) KNO_3 (C) NH_4NO_3 (D) CH_3COOK

247. The electricity needed to liberate one gram equivalent of Cu^{2+} is [PET-2003]
(A) 96500 amperes (B) 1 Faraday
(C) 96500 Faraday (D) 1 Coulomb

248. When lead storage battery is charged [PET-2003]
(A) PbO_2 is dissolved (B) H_2SO_4 is regenerated
(C) Lead is deposited on lead electrode
(D) $PbSO_4$ is deposited on lead electrode.

249. Which of the following reactions occurs at the cathode during the charging of a lead storage battery
(A) $Pb^{2+} + 2e^- \rightarrow Pb$ (B) $Pb^{2+} + SO_4^{2-} \rightarrow PbSO_4$
(C) $Pb \rightarrow Pb^{2+} + 2e^-$
(D) $PbSO_4 + 2H_2O \rightarrow PbO_2 + 4H^+ + SO_4^{2-} + 2e^-$

250. 1 A current is supplied for 30 minutes through NaCl. How much chlorine will be liberated? [BHU-2003]
(A) 0.66g (B) 0.66 mole (C) 0.33g (D) 0.33 mole

251. $KMnO_4 + H_2C_2O_4 \xrightarrow{H^+}$. Change in the oxidation number of Mn is [BHU-2003]
(A) 7 to 2 (B) 7 to 5 (C) 2 to 7 (D) 5 to 7

252. $Zn^{2+} + 2e^- \rightarrow Zn(s)$; $E^0 = 0.76$
$Fe^{3+} + e^- \rightarrow Fe^{2+}$; $E^0 = +0.77$
$Cr^{3+} + 3e^- \rightarrow Cr$; $E^0 = -0.79$
$H^+ + e^- \rightarrow 1/2\, H_2$; $E^0 = 0.00$
Stronger reducing agent is **[BHU-2003]**
(A) Fe^{2+} (B) Zn (C) Cr (D) H_2

253. Which of the following 1 M soltions conducts

more electricity? [AFMC-2003]

(A) Boric acid (B) Nitric acid

(C) Phosphorous acid (D) Sulphuric acid

254.Galvanization includes plating of iron with [AFMC-2003]

(A) Aluminium (B) Zinc

(C) Copper (D) Tin

255. Which of the following is used in electroplating? [AFMC-2003]

(A) AgNO3 (B) AgCl

(C) Sodium argentocyanide (D) None of these

256. The standard e.m.f. of a galvanic cell involving cell reaction with n = 2 is found to be 0.295 V at 25°C. The equilibrium constant of the reaction would be
(Given F = 96500 C mol^{-1}; R = 8.314 J K^{-1} mol^{-1}) [CBSE-PMT-2004]

(A) 2.0×10^{11} (B) 4.0×10^{12} (C) 1.0×10^{2} (D) 1.0×10^{10}

257.The emf of the given cell,
Zn | Zn^{2+} (0.1 M) || Fe^{2+} (0.01 M) | Fe is 0.2905 V at 298 K. K_{eq} of this cell will be [IIT-2004]

(A) not available (B) $10^{0.32/0.059}$

(C) $10^{0.32/0.0295}$ (D) $e^{0.32/0.0295}$

Electrochemistry and Redox Reactions

1.	(C)	51.	(A)	101.	*	151.	(A)	201.	(A)	251.	(A)
2.	(C)	52.	(A)	102.	(A)	152.	(A)	202.	(C)	252.	(C)
3.	(D)	53.	(D)	103.	(A)	153.	(B)	203.	(B)	253.	(D)
4.	(B)	54.	(A)	104.	(B)	154.	(A)	204.	(A)	254.	(B)
5.	(D)	55.	(A)	105.	(D)	155.	(A)	205.	(D)	255.	(A)
6.	(A)	56.	(E)	106.	(C)	156.	(A)	206.	(A)	256.	(D)
7.	(A)	57.	(C)	107.	(A)	157.	(B)	207.	(D)	257.	(C)
8.	(A)	58.	(A)	108.	(C)	158.	(B)	208.	(B)		
9.	(A)	59.	(A)	109.	(B)	159.	(B)	209.	(C)		
10.	(B)	60.	(C)	110.	(B)	160.	(B)	210.	(C)		
11.	(B)	61.	(A)	111.	(C)	161.	(A)	211.	(D)		
12.	(A)	62.	(C)	112.	(D)	162.	(C)	212.	(A)		
13.	(B)	63.	(C)	113.	(B)	163.	(D)	213.	(D)		
14.	(C)	64.	(B)	114.	(A)	164.	(D)	214.	(A)		
15.	(B)	65.	(C)	115.	(D)	165.	(A)	215.	(B)		
16.	(B)	66.	(B)	116.	(A)	166.	(A)	216.	(C)		
17.	(B)	67.	(A)	117.	(D)	167.	(B)	217.	(A)		
18.	(A)	68.	(C)	118.	(B)	168.	(B)	218.	(C)		
19.	(D)	69.	(B)	119.	(D)	169.	(D)	219.	(D)		
20.	(B)	70.	(B)	120.	(A)	170.	(A)	220.	(D)		
21.	(C)	71.	(D)	121.	(C)	171.	(C)	221.	(C)		
22.	(B)	72.	(B)	122.	(C)	172.	(D)	222.	(B)		
23.	(B)	73.	(B)	123.	*	173.	(A)	223.	(C)		
24.	(D)	74.	(B)	124.	(B)	174.	(A)	224.	(B)		
25.	(B)	75.	(A)	125.	*	175.	(C)	225.	(A)		
26.	(C)	76.	(A)	126.	(A)	176.	(C)	226.	(D)		
27.	(B)	77.	(C)	127.	(C)	177.	(A)	227.	(C)		
28.	(B)	78.	(D)	128.	*	178.	(C)	228.	(D)		
29.	(B)	79.	(C)	129.	(C)	179.	(A)	229.	(C)		
30.	(A)	80.	(B)	130.	(A)	180.	(B)	230.	(A)		
31.	(C)	81.	(B)	131.	(A)	181.	(A)	231.	(A)		
32.	(D)	82.	(C)	132.	(C)	182.	(A)	232.	(A)		
33.	(C)	83.	(A)	133.	(A)	183.	(B)	233.	(A)		
34.	(C)	84.	(D)	134.	(A)	184.	(B)	234.	(C)		
35.	(D)	85.	(A)	135.	(B)	185.	(A)	235.	(B)		
36.	(C)	86.	(C)	136.	(B)	186.	(D)	236.	(A)		
37.	(D)	87.	(A)	137.	(B)	187.	(B)	237.	(B)		
38.	(D)	88.	(D)	138.	(C)	188.	(D)	238.	(A)		
39.	(A)	89.	(A)	139.	(B)	189.	(B)	239.	(D)		
40.	(B)	90.	(C)	140.	(A)	190.	(C)	240.	(A)		
41.	(C)	91.	(C)	141.	(B)	191.	(C)	241.	(A)		
42.	(D)	92.	(B)	142.	(D)	192.	(D)	242.	(B)		
43.	(A)	93.	(A)	143.	(A)	193.	(C)	243.	(C)		
44.	(B)	94.	(A)	144.	(C)	194.	(B)	244.	(B)		
45.	(D)	95.	(A)	145.	(A)	195.	(A)	245.	(B)		
46.	(B)	96.	(C)	146.	(C)	196.	(B)	246.	(D)		
47.	(A)	97.	(C)	147.	(D)	197.	(B)	247.	(B)		
48.	(D)	98.	(A)	148.	(D)	198.	(D)	248.	(B)		
49.	(A)	99.	(A)	149.	(B)	199.	(B)	249.	(D)		
50.	(B)	100.	*	150.	(D)	200.	(B)	250.	(A)		

100. decreases

101. oxygen

123. 12

125. $O_2O_4^{2-}$

128. +5

COLLOIDS

Most of the naturally occuring substances such as blood, milk, foods, wood, rubber etc. are colloidal in nature.

Colloid particle size : 1-100 nm (10^{-5} - 10^{-7} cm.)
Suspension : 100 n m (10^{-3} - 10^{-5}cm)
True solution : 1 n m ($<10^{-7}$cm)

Types of colloids :

1. Depending upon the states of aggregation of the dispersed phase and dispersion medium.

Since two gases are completely miscible, they always form the true solution. A colloidal solution of the type gas in gas therefore is not possible.

2. Based on interaction of phases : If the dispersed phase has little affinity for the dispersion medium, they form lyophobic colloids e.g. colloidal gold (suspensoid), $Fe(OH)_3$, As_2S_3, S etc. If the dispersed phase has considerable affinity for the dispersion medium, they form lyophilic colloids e.g. gelatin & water, naturally occouring proteins, polysaccharides, gums and resins (Emulsoid).

Preparation of Colloids :

Breaking down the coarser aggregates into particles of colloidal size is done by dispersion methods while grouping molecules into larger aggregates of colloidal size is done by condensation methods.

Dispersion methods :

(a) Mechanical grinding (b) peptisation (c) Bredig's arc method:

Condensation methods :

(a) oxidation (b) reduction (c) by exchange of solvent.

Colloidal solutions are purified by (1) Dialysis (2) Ultrafiltration (3) Ultra centrifuging.

Colligative properties : Like true solutions, the colloidal solutions also show colligative properties.

Optical properties :

1. Tyndall effect : When a beam of light is passed through a true solution, it can not be seen unless the eye is placed directly in its path. However, when the same beam is passed through a colloidal dispersion it becomes visible as a bright streak. This phenomenon is due to scattering of light from the surface of the colloidal particles. In true solutions, there are no particles of sufficient size to scatter light.

In lyophobic colloids, the tyndall effect is quite well defined. In lyophilic colloids, the tyndall effect is very weak.

a. The projection beam of a cinema projector is visible due to scattering of light by dust and smoke particles present in the theatre.

b. Ultra-microscope is built on the principles of Tyndall effect.

c. Tail of comet : Comets fly with tremendous velocity and leave behind tiny solid particles suspended in air. These particles scatter light and are visible as tyndall cone.

d. Blue colour of sky and see water

e. Twinkling of stars is also due to Tyndall effect.

2. Brownian moment : The Brownian moment is due to the bombardment of colloidal particles by molecules of dispersion medium which are in constant motion. Brownion moment offers a visible proof of the random kinetic motion of molecules in a liquid. Bownian moment is due to the particles in a state of ceaseless erratic and random motion.

Electrical Properties : Colloidal particles almost invariably carry positive or negative charge. All particles of a given sol carry the same charge.

+ vely charged : $Fe(OH)_3$, $Al(OH)_3$, $Cr(OH)_3$, methylene blue sol etc.

- vely charged : As_2S_3, Sb_2S_3, Ag, Cu, Au sols, eosin, congo red etc.

Electrical double layer : The surface of a colloidal particle acquires a positive or negative charge by elective adsorption of ions carrying positive or negative charge rspectively. A charged colloidal particle is surrounded by ions of opposite kind. Ions of one type gets absorbed and forms a fixed layer on colloidal particles. Ions of opposite charge from liquid tend to approach this layer forming a diffused layer. The charges of opposite signs on the fixed and diffused layers lead to a difference in potential. This difference of potential between the fixed and diffused layers is called electrokinetic or zeta potential.

Cataphoresis or Electrophoresis : The migration of colloidal particles under the influence of an electric field is termed cataphoresis. The charged particles of a colloidal solution move relative to the dispersion medium under the influence of an electric field.

Electro-osmosis: The moment of dispersion medium under the influence of electric field when the dispersed particles are prevented from moving is known as Electro-osmosis.

Isoelectric point : At a particular H^+ ion concentration, the colloidal particles become neutral and exhibit no moment in an electric field. This is referred to as Isoelectric point of the sol.

Coagulation of colloidal solutions : The stability of colloidal state rests on the existence of electrical charge on the particles. This charge on account of repulsion, prevents the particles from approaching sufficiently close to one another and to coalesce. However, when the charge is neutralised or lowered to a certain critical value, the precipitation occurs.

This phenomenon of change of colloidal state to suspended state is known as coagulation or flocculation of colloidal solutions. This is achieved by many ways.

By the action of electrolytes : If the electrolytes are persent in larger amounts, the colloidal particles take up the ions carrying charge opposite to that present on themselves. The ion carrying the opposite charge is called the flocculating ion.

Hardy-Schulze rule : The greater the valency of the flocculating ion added, the greater is its power to cause precipitation.

Flocculation Power : The order of coagulation of +vely charged $Fe(OH)_3$ sol is

$$[Fe(CN)_6]^{4-} > [PO_4]^{3-} > [SO_4]^{2-} > Cl^-$$

Coagulation of - vely charged AS_2S_3 sol : $Al^{+3} > Ba^{+2} > Na^+$

Flocculation value : The minimum concentration of an electrolyte required to cause coagulation or flocculation of a sol is called flocculation value. $AlCl_3 < BaSO_4 < NaCl$

The minimum concentration in milli-moles per lit. required to cause precipitation of a sol (in 2 hours) is called flocculation value.

Protection : Lyophilic colloids possess the property of protecting lyophobic colloids from precipitation by the action of electrolytes. When a lyophilic sol is added to a lyophobic sol, it is believed that lyophilic particles form envelops around the lyophobic particles and thus protect the latter from the action of electrolytes. Lyophilic colloids used for such purposes are called protective colloids e.g. gelatin (lyophilic sol) is added to a gold sol (lyophobic sol).

Gold number : It is given by Zsigmondy.

It is the power to measure the protective power of different colloids. It is defined as the weight in milligrams of a protective colloid which prevents the coagulation of 10ml. of a given gold sol on adding 1ml. of a 10% solution of NaCl.

The smaller the gold number of a lyophilic colloid, the greater is its protective power. e.g. gelatin (0.005 - 0.01); Haemoglobin (0.03); potato starch (25), gum arabic (0.15), Egg albumin (0.08 - 0.10).

Emulsions : Emulsions are colloidal systems in which dispersed phase as well as dispersion medium are normally liquids. Particles of dispersed phase in emulsions are generally bigger than those in sols. Emulsions show tyndall effect and Brownian moment.

Types of Emulsions :

(a) oil in water : e.g. milk, vanishing cream

(b) water in oil : e.g. butter, cold creams

Emulsifier : The function of an emulsifier is to lower the interfacial tension between water and oil so as to facilitate the mixing of the two liquids. e.g. soaps of various kinds, proteins, gums and agar-agar act as emulsifiers. Cleaning action of soap is due to emulsification of greasy or oily minerals by soap solution. A large number of pharmaceutical preparations are emulsions.

Gels : Colloidal system containing a liquid dispersed in solid. On standing, a gel may shrink and lose the liquid held by it. It is called syneresis or weeping. Some gels liquify on shaking and reset when allowed to stand. This transformation is known as thixotropy. e.g. table gelly (gel of water in gelatin), soáps, boot polishes, animal tissues, many food stuffs have a gel structure. Curd is a gel of water in casein.

Adsorption : The phenomenon of higher concentration of any molecular species at the surface than in the bulk of a solid (or a liquid) is called adsorption.

The forces involved are molecular forces i.e. van der Waal's forces. In absorption a substance is distributed uniformly throughout the body of a solid or a liquid. The process in which a substance shows absorption and adsorption is called sorption. e.g. H_2/Charcoal.

The decrease of temperature and increase of pressure, both tend to cause increase in the magnitude of adsorption of a gas on a solid.

Curve showing affect of temp on the adsorption - Isobar

Curve showing variation of pressure with temp. - Isostere

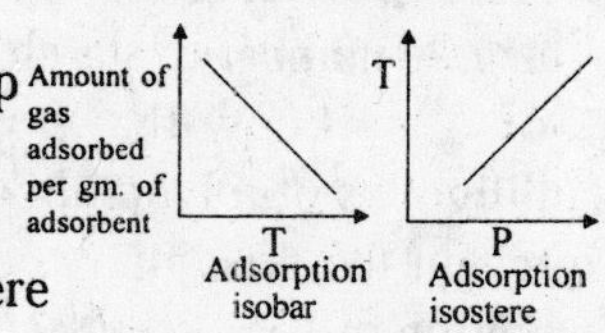

Chemisorption : Adsorption in not necessarily a physical phenomenon always. It may as well be a chemical process involving chemical interaction between the surface atoms of the adsorbent and the atoms of the adsorbate. This type of adsorption is called as chemisorption.

e.g. 1. oxygen is chemisorbed by Tungsten.

2. H_2 as well as C is chemisorbed by Ni.

In each case, a stable surface compound, frequently referred to as surface complex results.

Chemisorption is irreversible.

Micelles : The substances whose molecules aggregate spontaneously in a given solvent yielding stable particles of colloidal dimensions. The aggregates which the colloids form in solution are called micelles. They have both lyophilic and lyophobic groups. Due to their dual nature, they are called amphithetic for e.g. Sodium oleate on dissolution give Na+ and $C_{17}H_{33}COO^-$ (oleate ion).

The number of molecules in the ionic detergent micelle is 10 - 100. Rod shaped micelles are formed at high concentration, while spherical or flattened-spherical micelles are formed near critical micelle concentration.

Micelles are used as drug carriers, cleaners in froath flotation, in petroleum recovery etc.

When a cloth with spot of dirt is soaked with soap solution, soap dissolves dirt by forming micelles.

Things to remember

Colloid particle size: 1- 100 n m.

Blood, milk, foods, rubber wood etc. are colloidal in nature.

Colloidal solutions of Au, Cu, Ag, Pt are prepared by Bredigs Arc method (dispersion method).

Tyndall effect is due to scattering of light.

Brownian moment - a proof for random kinetic motion of molecules.

Hardy Schulze rule : The greater the valency of flocculating ion added, the greater is its power to cause precipitation.

Emulsion : (a) oil in water e.g. milk, vanishing cream

(b) water in oil e.g.: butter, cold cream etc.

Gel : a liquid dispersed in solid e.g. table Jelly, curd etc.

Solved Examples

(1) The coagulation of 100ml of a colloidal solution of gold is completely prevented by adding 0.25 g. of starch to it before adding 10 ml. of 10% NaCl solution. Calculate the gold number of starch.

Sol. : we know that gold number is the number of milligrams of a lyophilic colloid that is required to just prevent the precipitaion of 10 ml. of a gold sol on the addition of 1 ml. of 10% NaCl solution.

10 ml of 10% NaCl solution is added to 100 ml. gold sol. one can say 1 ml. of 10% NaCl solution is added to 10 ml. of gold sol.

100 ml. gold sol required 0.25g. starch = 0.25×10^3 mg.

10 ml. gold sol require = $\frac{10}{100} \times 0.25 \times 10^3 = 25$ mg. starch

∴ gold number of starch is 25.

(2) 5 ml. of 1M NaCl is required for the coagulation of 100 ml. of As_2S_3 sol. Calculate the coagulation power of NaCl.

Sol.: we know that the minimum concentration of an electrolyte in milli-moles per litre required to cause precipititaion is to be calculated.

Resultant volume after the addition of 5 ml. of 1M NaCl solution = 105

concentration in millimoles / litre = $\frac{5}{105} \times 1000 = 47.6$

(3) The coagulation of 10ml. of gold solution is completely prevented by addition of 0.025 g. of starch to it before adding 1ml. of 10% NaCl solution. Find out the gold number of starch.

Sol. : 10 ml of gold sol required 0.025 g. of starch = 0.025×10^3 mg.

∴ gold no. of starch is 25.

1. In Faraday-Tyndall effect the colloidally suspended particles [CBSE-90]
(A) trace out the path of a strong beam of light
(B) show Brownian movement
(C) show electrophoresis (D) coagulate

2. A catalyst is a substance which [CBSE-90]
(A) alters the chemical equilibrium
(B) changes the equilibrium of a reaction so that the concentration of product increases
(C) lowers the activation energy which the reactant molecules must aquire in order to form the intermediate activated complex
(D) None of these

3. For the adsorption of a gas on a solid, the plot of log x/m versus log P is linear with slope equal to [CBSE-94]
(A) k (B) log k (C) n (D) 1/n

4. A colloidal system has particles of what size ? [CBSE-96]
(A) 10^{-4} m to 10^{-10} m (B) 10^{-5} m to 10^{-7} m
(C) 10^{-9} m to 10^{-12} m (D) 10^{-6} m to 10^{-9} m

5. The ability to bring about coagulation of a given colloid depends upon [CBSE-97]
(A) its charge (B) the sign of the charge alone
(C) the magnitude of charge
(D) both magnitude and charge

6. Colloidal solutions can be destroyed by [CBSE-2000]
(A) Separation by animal parchment
(B) Electrophoresis (C) Dialysis
(D) Addition of electrolytes

7. Position of non polar & polar part in miscelle [CBSE-2002]
(A) Polar at outer surface but non polar at inner surface.
(B) Polar at inner surface non polar at outer surface.
(C) Distributed over all the surface.
(D) Are present on the surface only.

8. Assertion : Physical absorption of molecules on surface requires activation energy. Reason : Because the bonds of the absorbed molecules are broken. [AIIMS-94]
(A) A (B) B (C) C (D) D

9. The physical adsorption of gases on the solid surface is due to [AIIMS-98]
(A) Hydrogen bonding (B) Covalent bonding
(C) van der Waal's force (D) All of these

10. The size of colloidal particle is [AIIMS-99]
(A) 10^{-6} to 10^{-9} m (B) 10^{-9} to 10^{-12} m
(C) 10^{-3} to 10^{-9} m (D) 10^{-12} to 10^{-19} m

11. Which amongst the following will be fastest hydrolysed

with alc. KOH ? [AIIMS-2001]

(A) $CH_3COOC_2H_5$ (B) C_2H_5CN

(C) $CH_3CH_2CONHCH_3$ (D) $(CH_3CO)_2O$

12. When in colloidal solution in both sides electrodes are placed then the movement of medium is called [AIIMS-2001]

(A) Electrical osmosis (B) Cataphoresis

(C) Coagulation (D) Brownian movement

13. Freundich equation is : [AIIMS-2001]

(A) $\log \frac{m}{x} = \log K + \frac{1}{n}\log P$

(B) $\log \frac{x}{m} = \log K - \frac{1}{n}\log P$

(C) $\log \frac{x}{m} = \log K + \frac{1}{n}\log P$

(D) $\log \frac{x}{m} = \log K + \frac{1}{P}\log n$

14. The movement of colloidal particles, under applied electric field is known as [AIIMS-2001]

(A) Dialysis (B) Electrophoresis

(C) Electrodialysis (D) None of these

15. The size of colloidal particle is [AIIMS-2002]

(A) 10^{-3} to 10^{-9} m (B) 10^{-6} to 10^{-9} m

(C) 10^{-9} to 10^{-12} m (D) 10^{-12} to 10^{-19} m

16. The process of separating colloidal particles from particles of molecular dimensions is : [JIPMER-92]

(A) Crystallisation (B) Filtration

(C) Dialysis (D) Sublimation

17. Butter is colloid formed when : [JIPMER-92]

(A) Fat is dispersed in solid casein

(B) Water is dispersed in fat

(C) Fat globules are dispersed in water

(D) Suspension of casein in water

18. An example of miscelle is: [JIPMER-93]

(A) Aqueous NaCl solution (B) Sodium stearate

(C) Gold sol (D) Milk

19. The coagulation power of an electrolyte for arsenous sulphide sol is maximum for : [JIPMER-93]

(A) $AlCl_3$ (B) Na_3PO_4 (C) $K_4[Fe(CN)_6]$ (D) HCl

20. The movement of the colloidal particles under the influence of applied electric field is called :[JIPMER-93]

(A) Dialysis (B) Tyndall effect

(C) Osmosis (D) Electrophoresis

21. How many phases are present in a colloidal system ? [JIPMER-94]

(A) 4 (B) 3 (C) 2 (D) 1

22. A random motion in a zig-zag manner of colloidal particles is called [JIPMER-95]

(A) Coagulation (B) Osmosis

(C) Electrophoresis (D) Brownian movement

23. A strong beam of light passed through a colloidal solution is [JIPMER-95]

(A) Scattered (B) Reflected

(C) Refracted (D) Absorbed completely

24. Which of the following represents surfactant molecule? [JIPMER-95]

(A) H_2O (B) C_2H_{36}

(C) $C_{17}H_{35}COONa$ (D) None of these

25. The movement of colloidal particles with reference to dispersion medium when subjected to an electrical potential is called [JIPMER-96]

(A) Electro osmosis (B) Electrophoresis

(C) Tyndall effect (D) Brownian movement

26. Which of the following has the highest protective power on lyophobic colloids [JIPMER-96]

(A) gelatin (B) sodium oleate

(C) gum arabic (D) starch

27. The tyndall effect is associated with colloidal particles is due to [JIPMER-97]

(A) The presence of electrical charge

(B) Scattering of light

(C) Absorption of light (D) Reflection of light

28. Colloidal solutions of gold prepared by different methods have different colours owing to. [JIPMER-98]

(A) The difference in the size of the colloidal particles.

(B) The fact the gold exhibits a variable valency of +1 and +3.

(C) Different concentrations of gold.

(D) Presence of different types of foreign particles depending upon the method of preparation of the colloid.

29. The adsorption of hydrogen by platinum is known as [JIPMER-99]

(A) Hydrogenation (B) Reduction

(C) Occlusion (D) Dehydrogenation

30. Milk is a colloid in which [JIPMER-2002]

(A) Liquid is dispersed in liquid

(B) A solid is dispersed in liquid

(C) A gas is dispersed in liquid

(D) Some sugar is dispersed in water

31. Milk is an example of [BHU-96]

(A) Gel (B)Sol (C) Suspension (D) Emulsion

32. When a strong beam of light is passed through a colloidal solution, the light will [BHU-96]

(A) Give a rainbow (B) Be scattered

(C) Be reflected (D) Pass unchanged

33. Dialysis can separate [BHU-98]

(A) Glucose and fructose (B) Glucose and sucrose

(C) Glucose and NaCl (D) Glucose and proteins

34. Gold number is associated with [BHU-99]
(A) Electrophoresis (B) Purple of Cassius
(C) Protective colloids (D) Amount of pure gold

35. The size of the colloidal particles is [AFMC-91]
(A) Less than 10^{-10} m (B) More than 10^{-10} m
(C) More than 10^{-7} m (D) Between 10^{-10} and 10^{-7} m

36. Crystalloids differ from colloids mainly in the aspect of [AFMC-93]
(A) Electrical behaviour (B) Particle size
(C) Particle nature (D) Solubility

37. Brownian moment is found in [AFMC-96]
(A) Colloidal solution (B) Suspension
(C) Saturated solution (D) Unsaturated solution

38. Colloidal found effective in treating eye disease [AFMC-96]
(A) Colloidal sulphur (B) Colloidal antimony
(C) Colloidal gold (D) Colloidal silver

39. Lyophilic colloids are stable due to [AFMC-98]
(A) Charge on the particle (B) Large size of article
(C) Small size of the particle
(D) Layer of dispersion medium on the particles.

40. Tyndall effect can be observed in a [AFMC-99]
(A) Colloidal solution (B) Solution
(C) Solvent (D) Precipitate

41. The movement of colloidal particles towards the charged electrodes on passing electricity is [AFMC-2000]
(A) Brownian movement (B) Cataphoresis
(C) Tyndall effect (D) None of these

42. Colloidal sol is [AFMC-2000]
(A) True solution (B) Suspension
(C) Heterogeneous sol (D) Homogeneous sol

43. Soap removes grease by [AFMC-2000]
(A) Adsorption (B) Emulsification
(C) Coagulation (D) None of these

44. Physical adsorption is inversely proportional to the [AFMC-2000]
(A) Volume (B) Concentration
(C) Temperature (D) All of these

45. The movement of colloidal particle towards the oppositely charged electrodes on passing electricity is known [AFMC-2000]
(A) Cataphoresis (B) Tyndall effect
(C) Brownian (D) None of these

46. Light scattering takes place in [AFMC-2001]
(A) Solutions of electrolyte (B) Colloidal solution
(C) Electrolysis (D) Electroplating

47. Blue colour of water in sea is due to [AFMC-2001]
(A) Refraction of blue light of impurities in sea water
(B) Refraction of blue sky by water
(C) Scattering of light by water (D) None of these

48. In physical adsorption, the extent of absorbate increases with [IIT-2003]
(A) decrease in temperature
(B) increase in temperature
(C) decrease in pressure (D) All of these

49. The formation of gas at the surface of tungsten due to adsorption is the reaction of order. [AIEEE-2002]
(A) 0 (B) 1 (C) 2 (D) Insufficient data

50. Which one of the following characteristics is not correct for physical adsorption ? [AIEEE-2003]
(A) Adsorption on solids is reversible
(B) Adsorption increases with increase in temperature
(C) Adsorption is spontaneous
(D) Both enthalpy and entropy of adsorption are negative

51. During the cleaning action of soap, ... part of soap dissolves in the dirt and encapsulates of form 'micelle'. [EAMCET(med.)-2002]
(A) Both hydrophobic and hydrophilic
(B) Hydrophilic (C) Hydrophobic (D) Cation

52. Colour of colloidal solution is due to- [CPMT-96]
(A) Diff. size of colloid (B) Due to formation of complex
(C) Due to formation of hydrated crystal
(D) None of these

53. Which of the following is property of colloid ? [CPMT-96]
(A) Scattering of light (B) They show attraction
(C) Dialysis (D) Emulsion

54. The stability of lyophilic colloid is due to - [CPMT-96]
(A) Charge on their particles
(B) A layer of medium of dispersion on their particles
(C) Small size of particles
(D) The large size of their particles

55. The size of colloidal particle is - [CPMT-97]
(A) 10^{-6} to 10^{-9} m (B) 10^{-9} to 10^{-12} m
(C) 10^{-3} to 10^{-9} m (D) 10^{-12} to 10^{-19} m

56. The movement of colloidal particles under electric field is called [CPMT-2001]
(A) Electrodialysis (B) Brownian movement
(C) Electrophoresis (D) Tyndall effect

57. A colloidal system in which gas bubbles are dispersed in a liquid is known as : [PMT-93]
(A) Foam (B) Sol (C) Aerosol (D) Emulsion

58. According to Langmuir adsorption isotherm, the

amount of gas adsorbed at very high pressures : [PMT-93]

(A) reaches a constant limiting value
(B) goes on increasing with pressure
(C) goes on decreasing with pressure
(D) increases first and decreases later with pressure

59. Butter is : [PMT-94]
(A) Fat dispersed in milk. (B) Fat dispersed in water
(C) Water dispersed in fat (D) Water dispersed in oil

60. Which of the following makes the lyophilic sol. unstable ? [PMT-94]

(A) Dialysis (B) Addition of electrolyte
(C) Addition of alcohol
(D) Addition of alcohol and electrolyte both

61. The size of the colloidal particles is in between [PMT-95]

(A) $10^{-7} - 10^{-9}$ cm (B) $10^{-9} - 10^{-11}$ cm
(C) $10^{-5} - 10^{-7}$ cm (D) $10^{-2} - 10^{-3}$ cm

62. The stability of lyophillic colloids is due to which of the following ? [PMT-95]

(A) Charge on their particles
(B) Large size of their particles
(C) Smaller size of their particles
(D) A layer of medium of dispersion on their particles

63. Milk is [PMT-95]
(A) Fat dispersed in water (B) Water dispersed in fat
(C) Water dispersed in oil (D) Fat dispersed in fat

64. When the temperature is lowered and pressure is raised, the adsorption of a gas on a solid - [PMT-97]
(A) Decreases (B) Increases
(C) Remains unaffected (D) Decreases first then increases

65. On adding few drops of dil. HCl to freshly precipitated ferric hydroxide, a red coloured colloid is obtained. This process is called [PMT-97]
(A) Dissolution (B) Protective action
(C) Peptization (D) Dialysis

66. Which of the following is most effective in causing the coagulation of ferric hydroxide sol ? [PMT-2000]
(A) KCl (B) KNO_3 (C) K_2SO_4 (D) $K_3Fe(CN)_6$

67. Blood may be purified by : [PMT-2000]
(A) dialysis (B) electro-osmosis
(C) coagulation (D) filtration

68. Colloids can be purified by : [PMT-2001]
(A) Dialysis (B) Electrolysis
(C) Distillation (D) Chromatography

69. The As_2S_3 sol is charged due to adsorption of : [PMT-2001]

(A) S^{2-} (B) OH^- (C) H^+ (D) As^{3+}

70. Surface tension of lyophilic sols is [PMT-2002]
(A) Lower than that of H_2O (B) More than that of H_2O
(C) Equal to that of H_2O (D) None of the above

71. Milk is a colloid in which a [PMT-2002]
(A) Liquid is dispersed in a liquid
(B) Solid is dispersed in a liquid
(C) Gas is dispersed in a liquid
(D) Sugar is dispersed in a liquid

72. Adsorbed acetic acid on activated charcoal is [PMT-2002]

(A) Adsorber (B) Absorber (C) Adsorbent (D) Adsorbate

73. Brownian movement is due to [PMT-2003]
(A) Attraction force between colloidal particles and the molecules of dispersion medium
(B) Convection currents in liquid
(C) Impact of the molecules of dispersion medium on colloidal particles
(D) Changes in heat energy in liquid state.

74. Colloidal solutions are purified by [PET-93]
(A) Coagulation (B) Dialysis
(C) Electro-osmosis (D) Filtration

75. Which of the following statements is not correct ? [PET-93]

(A) Physical adsorption is due to Vander Waals forces
(B) Chemical adsorption decreases at high temperature and low pressure
(C) Physical adsorption is reversible
(D) Adsorption energy for a chemical adsorption is generally greater than hat of physical adsorption

76. Which of the following is most effective in coagulating a ferric hydroxide sol ? [PET-93]
(A) KCl (B) KNO_3 (C) K_2SO_4 (D) K_3 $[Fe(CN)_6]$

77. In the adsorption of acetic acid on activated charcoal, the acetic acid is an [PET-94]
(A) Adsorber (B) Absorber
(C) Adsorbent (D) Adsorbate

78. The ability of an ion to bring about coagulation of a given colloid depends upon [PET-94]
(A) Its size (B) The magnitude of its charge only
(C) The sign of its charge
(D) Both magnitude and sign of its charges

79. Emulsifier [PET-95]
(A) Accelerates the dispersion
(B) Homogenesis the emulsion (C) Stabilizes the emulsion
(D) Aids the flocculation of emulsion

80. Milk is an example of [PET-96]
(A) Emulsion (B) Suspension (C) Sol (D) Gel

81. Fog is a colloidal system of [PET-96]
(A) A gas in liquid (B) Liquid in gas

(C) Gas in gas (D) Solid in gas

82. In physical adsorption the gas molecules are held on solid surface by [PET-96]
(A) Chemical forces (B) Electrostatic forces
(C) Gravitational forces (D) van der Waal's forces

83. Point out the false statement : [PET-97]
(A) Brownian movement and Tyndall effect is shown by colloidal systems.
(B) Gold number is a measure of the protective power of a lyophilic colloid.
(C) The colloidal solution of a liquid in liquid is called as gel.
(D) Hardy-Schulze rule is related with coagulation.

84. Which one of the following will have the highest coagulation power for a Ferric hydroxide Sol ? [PET-97]
(A) NaCl (B) $BaCl_2$
(C) K_2CrO_4 (D) $K_3[Fe(CN)_6]$

85. In which of the following Tyndall effect is not observed ? [PET-98]
(A) Suspensions (B) Emulsions
(C) Sugar solution (D) Gold sol

86. Which of the following is a lyophilic colloid ? [PET-98]
(A) Milk (B) Gum (C) Fog (D) Blood

87. An arsenious sulphide sol carries a negative charge. The maximum precipitating power for this sol is possessed by : [PET-99]
(A) K_2SO_4 (B) $CaCl_2$ (C) Na_3PO_4 (D) $AlCl_3$

88. Tyndall effect would be observed in a : [PET-99]
(A) Solution (B) Solvent
(C) Colloidal solution (D) Precipitate

89. Adsorption is multilayer in the case of : [PET-99]
(A) Physical adsorption (B) Chemisorption
(C) Both (D) None of both

90. Which is an emulsion among the following ? [PET-2001]
(A) Milk (B) Wood (C) Butter (D) Air

91. Which one of the following is an incorrect statement for physisorption [PET-2002]
(A) It is a reversible process
(B) It requires less heat of adsorption
(C) It requires activation energy
(D) It takes place at low temperature

92. Which of the following substances is not used for preparing lyophilic sols [PET-2002]
(A) Starch (B) Gum (C) Gelatin (D) Metal sulphide

93. At the critical micelle concentration the surfactant molecules [CBSE-98]
(A) decomposes (B) becomes completely soluble
(C) associates (D) dissociates

94. Which is not correct regarding adsorption of a gas on surface of solid ? [CBSE-2001]
(A) On increasing temperature adsorption increases continuously
(B) Enthalpy and entropy changes are -ve
(C) Adsorption is more for some specific substance
(D) Reversible

95. (A) The micelle formed by sodium stearate in water has –COO groups at the surface.
(R) Surface tension of water is reduced by the addition of stearate.
(A) (B) (C) (D)

96. The coagulating power of an electrolyte for arsenious sulphide decreases in order of
(A) $Na^+ > Al^{3+} > Ba^{2+}$ (B) $SO_4^{2-} > PO_4^{3-} > Cl^-$
(C) $PO_4^{-3} > SO_4^{2-} > Cl^-$ (D) $Al^{3+} > Ba^{2+} > Na^+$

97. A colloidal solution is subjected to an electric field. The colloidal particles move towards the anode, coagulation of the same colloidal solution is studied using NaCl, $BaCl_2$ and $AlCl_3$ solution. The coagulating power is
(A) $NaCl > BaCl_2 > AlCl_3$ (B) $AlCl_3 > BaCl_2 > NaCl_3$
(C) $NaCl > AlCl_3 > BaCl_2$ (D) $BaCl_2 > NaCl > AlCl_3$

98. Which of the following is not a Colloid ?
(A) Saline (B) Blood (C) Smoke (D) Rubber latex

99. As_2S_3 is a
(A) Positive colloid (B) Negative colloid
(C) Neutral colloid (D) None of the above

100. Small liquid droplets dispersed in another liquid is called [CPMT-99]
(A) Suspension (B) Emulsion (C) Gel (D) True solution

101. Which one of the following characteristics is not correct for physical adsorption ?
(A) Adsorption on solids is reversible
(B) Adsorption increases with increase in temperature
(C) Adsorption is spontaneous
(D) Both enthalpy and entropy of adsorption are negative

102. Which one of the following salts forms a "micelle" ?
(A) Sodium formate (B) Sodium acetate
(C) Sodium stearate (D) Sodium chloride

103. Which one of the following statements in not correct?
(A) Physical adsorption decreases with increase in the temperature
(B) Physical adsorption is multilayered.
(C) Activation energy of physical adsorption is very high
(D) Enthalpy change of physical adsorption is about 20 $kJ.mol^{-1}$.

104. Emulsifier is an agent which [JIPMER-96]

(A) Homogenises the emulsion (B) Stabilises the emulsion
(C) Accelerates the dispersion (D) Breaks up the emulsion

105. Wood charcoal is used to decolourise sugar because it

(A) adsorbs coloured material
(B) absorbs decolourised material
(C) reduces coloured material (D) none of these

106. Brownian movement is : [PET-2000]

(A) Zig-zag motion of the colloidal particles
(B) Migration of colloidal particles under the influence of electric field
(C) Scattering of light by colloidal particles
(D) None of the above

107. Gold number is related with : [PET-2000]

(A) Colloids (B) Radioactivity
(C) Gas equation (D) Kinetic energy

108. The process of separation of a colloid from crystalloid through semi-permeable membrane is known as: [PMT-99]

(A) dialysis (B) coagulation
(C) ultrafiltration (D) titration

109. Which one of the following equations represents Freundlich adsorption isotherm. [EAMCET-2003]

(A) $\frac{x}{m} = kp$ (B) $\frac{x}{m} = kp^n$
(C) $\log \frac{x}{m} = kp^n$ (D) $\log \frac{x}{m} = kn \log p$

110. Which of the following forms cationic micelles above certain concentration ? [CBSE-PMT-2004]

(A) Sodium dodecyl sulphate (B) Sodium acetate
(C) Urea (D) Cetyltrimethylammonium bromide

Colloids

1.	(B)	51.	(C)	101.	(C)
2.	(C)	52.	(A)	102.	(C)
3.	(C)	53.	(A)	103.	(C)
4.	(D)	54.	(B)	104.	(B)
5.	(D)	55.	(A)	105.	(A)
6.	(D)	56.	(C)	106.	(A)
7.	(A)	57.	(A)	107.	(A)
8.	(D)	58.	(A)	108.	(A)
9.	(C)	59.	(C)	109.	(B)
10.	(A)	60.	(A)	110.	(D)
11.	(D)	61.	(C)		
12.	(A)	62.	(D)		
13.	(C)	63.	(A)		
14.	(B)	64.	(B)		
15.	(B)	65.	(C)		
16.	(C)	66.	(D)		
17.	(C)	67.	(A)		
18.	(B)	68.	(A)		
19.	(A)	69.	(A)		
20.	(D)	70.	(A)		
21.	(C)	71.	(A)		
22.	(D)	72.	(D)		
23.	(A)	73.	(C)		
24.	(C)	74.	(B)		
25.	(B)	75.	(B)		
26.	(A)	76.	(D)		
27.	(B)	77.	(D)		
28.	(A)	78.	(D)		
29.	(C)	79.	(C)		
30.	(A)	80.	(A)		
31.	(D)	81.	(A)		
32.	(B)	82.	(D)		
33.	(D)	83.	(C)		
34.	(C)	84.	(D)		
35.	(D)	85.	(C)		
36.	(B)	86.	(B)		
37.	(A)	87.	(D)		
38.	(D)	88.	(C)		
39.	(C)	89.	(A)		
40.	(A)	90.	(A)		
41.	(B)	91.	(C)		
42.	(D)	92.	(D)		
43.	(B)	93.	(C)		
44.	(C)	94.	(A)		
45.	(A)	95.	(B)		
46.	(B)	96.	(D)		
47.	(C)	97.	(B)		
48.	(A)	98.	(A)		
49.	(A)	99.	(B)		
50.	(C)	100.	(C)		

CATALYSIS

Catalyst : Substance which increases the rate of a chemical reaction without its being permanently changed is called a catalyst.

Types of Catalysis :

1. Positive catalysis : which increases the rate of a reaction e.g. (a) $N_2(g) + 3H_2(g) \xrightarrow{Fe(s)} 2NH_3$
(b) $2H_2 + O_2 \xrightarrow{Pt} 2H_2O$ etc.

2. Negative catalysis : decreases the velocity of a chemical reaction
e.g. (a) oxidation of chloroform $\xrightarrow{C_2H_5OH}$ slows down
(b) Decomposition of $H_2O_2 \xrightarrow{H_3PO_4;\ H_2SO_4}$ retards
(c) Antifreezes like glycerol retards rusting of the machines
(d) $(C_2H_5)_4$ Pb is added to petrol to retard the ignition of petrol

3. Autocatalysis : In certain reactions, one of the products of the reaction acts as a catalyst for that reaction
e.g. (a) $KMnO_4 + C_2H_2O_4 +$ dil. H_2SO_4. The reaction is slow initially but brisk later because $MnSO_4$ (Mn^{+2} ions) formed accelerates the reaction.

4. Induced catalysis : Here, one of the reactions induces the other reaction. e.g. Sodium arsenite is not oxidized by air but Na_2SO_3 is oxidized by air. If air is passed through their mixture, oxidation of Na_2SO_3 induces the oxidation of sodium arsenite.

Promoters : A substance which by itself has no catalytic property but promotes the catalytic activity of a catalyst is called a promoter.
e.g. $N_2 + 3H_2 \xrightarrow[\text{Mo-promoter}]{\text{Fe-catalyst}} 2NH_3$
- oxides of Al, Th & U also promote the catalytic property of Fe.
- Cu & Te promote the catalytic activity of Ni in hydrogenation of oils.

Homogeneous catalysis : The catalyst and the reactants are in the same phase .
e.g. (1) $2SO_2(g) + O_2(g) \xrightarrow{NO(g)} 2SO_3(g)$
(2) Inversion of cane sugar; hydrolysis of ester
(3) $2CO + O_2 \xrightarrow{NO(g)} 2CO_2$
Depending on the participation H^+ (acid) or OH^- (base), the calatylytic pocess is known as acid-base catalysis.

Heterogeneous catalysis: The catalyst is not in the same phase as the reactants e.g.(1) $2SO_2 + O_2 \xrightarrow{Pt} 2SO_3$
(2) $2H_2O_2 \xrightarrow{Pt} 2H_2O + O_2$
(3) $N_2 + 3H_2 \xrightarrow{Fe} 2NH_3$

Enzyme catalysis : e.g.

$$\underset{\text{starch}}{(C_6H_{10}O_5)_n} + nH_2O \xrightarrow{\text{Diastage}} \underset{\text{maltose}}{nC_{12}H_{22}O} \xrightarrow[\text{maltase}]{H_2O} \underset{\text{glucose}}{2C_6H_{12}O_6} \xrightarrow{\text{zymase}} 2C_2H_5OH + CO_2$$

Characteristics of ctalytic reactions :
1. The catalyst remains unchanged in amount and chemical composition at the end of the reaction (it may undergo physical change).
2. Only a small quantity of the catalyst ;10^{-6} to 10^{-8} g/l is needed.
3. The catalyst doesnot alter the state of the equilibrium in a reversible rection.
4. The catalyst does not initiate the reaction.
5. The catalyst is specific in its action
e.g. MnO_2 can catalyze the decomposition of $KClO_3$ but not $KClO_4$.
6. The catalyst can not alter the nature of the porducts of the reactions

Exceptions :(a) 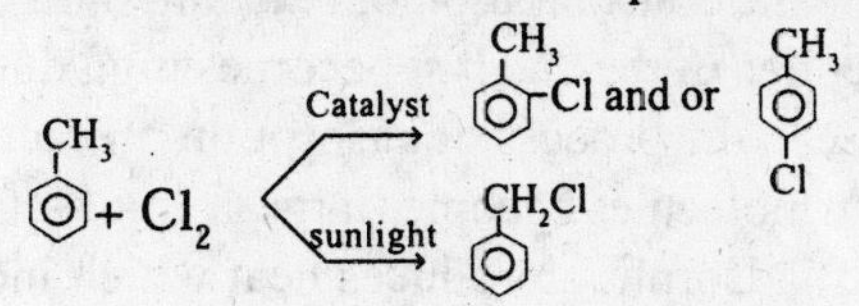

(b) $CO + 3H_2 \xrightarrow{Ni} CH_4 + H_2O$
$CO + 2H_2 \xrightarrow{ZnO + Cr_2O_3} CH_3OH$
$CO + H_2 \xrightarrow{Cu} HCHO$

1. A chemical reaction is catalysed by a catalyst X. Hence X [CBSE-95]
(A) does not affect equilibrium constant of the reaction
(B) decreases rate constant of the reaction.
(C) reduces enthalpy of the reaction.
(D) increases activation energy of the reaction.

2. According to the adsorption theory of catalysis, the speed of the reaction increases because [CBSE-2003]
(A) adsorption lowers the activation energy of the reaction
(B) the concentration of reactant, molecules at the active centres of the catalyst becomes high due to adsorption.
(C) in the process of adsorption, the activation energy of the molecules becomes large.
(D) adsorption produces heat which increases the speed of the reaction

3. Co-enzyme acts as [AIIMS-2000]
(A) A catalyst (B) A promoter for catalyst
(C) Inhibitor for catalyst (D) None of these

4. A catalyst in a chemical reaction [BHU-98]
(A) Does not initiate a reaction
(B) Increases the activation energy of the reaction
(C) Changes the equilibrium constant of a reaction
(D) Does not change the rate of the reaction

5. A catalyst [BHU-2001]
(A) Is used up in reaction
(B) Is not used up in the reaction
(C) Changes in composition during the reaction
(D) Increases equilibrium constant

6. Which is not a characteristic of a catalyst ? [AFMC-92]
(A) It changes the equilibrium constant
(B) It initiates the reaction
(C) It increases the rate of reaction
(D) It increases the average K.E. of the molecules.

7. The turbidity of a polymer solution measures [AFMC-94]
(A) The light absorbed by the solution
(B) The light transmitted by the solution
(C) The light scattered by the solution
(D) None of the above

8. Catalyst only [AFMC-95]
(A) Decreases activation energy
(B) Increases activation energy
(C) Comes to equilibrium (D) All of the above

9. The reactivity of a catalyst depends on [AFMC-2000]
(A) Particle size (B) Solubility
(C) Molecular weight (D) None of these

10. Diethyl ether is prepared by passing ethyl alcohol vapours over a catalyst under high pressure and temperature. The catalyst is [EAMCET-96]
(A) SiO_2 (B) CuO (C) Al_2O_3 (D) Ag_2O

11. Which one of the following reactions is an example of heterogeneous catalysis ? [EAMCET-2001]
(A) $2CO(g) + O_2(g) \xrightarrow{NO(g)} 2CO_2$
(B) $2SO_2(g) + O_2(g) \xrightarrow{NO(g)} 2SO_3$
(C) $2CO(g) + O_2(g) \xrightarrow{Pt(s)} 2CO_2$
(D) $CH_3CHO(g) \xrightarrow{I_2(g)} CH_4 + CO$

12. Which of the following statements is correct ? A catalyst is a substance which [PMT-93]
(A) increases the rate of forward reaction
(B) increases both rate of forward reaction and reverse reaction equally
(C) does not influence the reaction rate
(D) alters the equilibrium constant of a reaction

13. Oxidation of oxalic acid by a acidified $KMnO_4$ is an example of autocatalysis, it is due to presence of [PMT-94]
(A) SO_4^{-2} (B) MnO_4^- (C) Mn^{+2} (D) K^+

14. Which of the following kinds of catalysis can be explained by the adsorption theory ? [AIIMS-95]
(A) Enzymes catalysis (B) Acid base catalysis
(C) Heterogeneous catalysis (D) Homogeneous catalysis

15. When warm oxalic acid is titrated against $KMnO_4$ in acid medium, first the reaction is slow and thereafter fast. This is due to [JIPMER-97]
(A) Promoter (B) Catalysis
(C) Autocatalysis (D) None of the above

16. When a catalyst is added to a system the
(A) Value of equilibrium constant is decreased
(B) The rate of forward reaction is increased and that of backward reaction is decreased
(C) Equilibrium concentrations are unchanged
(D) Equilibrium concentrations are increased.

17. A biological catalyst is essential a/an
(A) Enzyme (B) Amino acid
(C) Carbohydrate (D) Nitrogen molecule

18. In the hydrogenation of oils, the catalyst used is [AFMC-98]
(A) Co (B) Ni (C) Pd (D) Pt

19. A biological catalyst is essentially [AFMC-98]
(A) An amino acid (B) An enzyme
(C) A carbohydrate (D) A nitrogen molecule

20. Organic catalysts differ from inorganic catalysts in the fact that [AFMC-2001]
(A) They are consumed

(B) They are proteinaceous in nature
(C) They are not much effective
(D) They have no difference

21. The role of a catalyst in a reversible reaction is
(A) It increases the rate of forward reaction
(B) It decreases the rate of backward reaction
(C) It alters the equilibrium constant
(D) It allows the equilibrium to be achieved quickly

22. What is the catalyst used during the manufacture of nitric acid by Ostwald's method ?
(A) Al_2O_3 (B) Fe (C) V_2O_5 (D) Platinum guage

23. Catalyst is [CPMT-96]
(A) specific (B) Take part in reaction
(C) do not take part in reaction
(D) None of these

24. Catalyst is used [CPMT-97]
(A) To increase the product
(B) To decrease the product
(C) To minimise the time of reaction
(D) None of these

25. A catalyst increases rate of reaction by [CPMT-99]
(A) Decreasing enthalpy (B) Decreasing internal energy
(C) Decreasing activation energy
(D) Increasing activation energy

26. Which of the following catalyst is used in the polymerisation of $HC \equiv CH$ to C_6H_6 ? [CPMT-99]
(A) $AlCl_3$ (B) $HgSO_4$ (C) $NbCl_5$ (D) HCl

27. Small liquid droplets dispersed in another liquid is called
(A) Suspension (B) Emulsion
(C) Gel (D) True solution

28. Butter is a colloid formed when
(A) water is disperesed in fat
(B) milk is dispersed in fat
(C) fat is dispersed in water (D) milk is dispersed in water

29. A catalyst can affect reversible reaction by [CPMT-2002]
(A) changing equilibrium constant
(B) slowing forward reaction
(C) attaining equilibria in both directions
(D) none of these

30. In chemical adsorption, how many layers are adsorbed?
(A) One (B) Two (C) Multi (D) Zero

31. What is the role of a catalyst in a catalysed reaction ? [PMT-96]
(A) Lowers the activation energy
(B) Increases the activation energy
(C) Affects the free energy change
(D) Affects the enthalpy change

32. Platinised asbestos helps in the formation of SO_3 from SO_2 and O_2. But, if even a small amout of As_2O_3 is present, the platinised asbestos does not help in the formation of SO_2. As_2O_3 acts here as [PMT-97]
(A) A positive catalyst (B) A negative catalyst
(C) An autocatalyst (D) A poison

33. Amongst the following chemical reactions, the one representing homogeneous catalysis is : [PMT-99]
(A) $N_2 (g) + 3H_2 (g) \xrightarrow{Fe} 2NH_3 (g)$
(B) $2SO_2 (g) + O_2 (g) \xrightarrow{2NO} 2SO_3 + 2NO (g)$
(C) $CO (g) + 3H_2 (g) \xrightarrow{Ni} CH_4 (g) + H_2O$
(D) $2SO_2 (g) + O_2 (g) \xrightarrow{V_2O_5} 2SO_3 (g)$

34. A chemical reaction is catalysed by 'X' therefore 'X' [PMT-2003]
(A) Increases the activation energy
(B) Does not effect the equilibrium constant of the reaction
(C) Decreases the enthalpy of the reaction
(D) Decreases the velocity constant of the reaction

35. In which of the following processes shape-selective catalysis is occurring ? [PET-94]
(A) Conversion of alcohol to gasoline
(B) Synthesis of methanol from CO and H_2
(C) Polymerisation of ethylene (D) Synthesis of ammonia

36. In the reaction
$2SO_2 + O_2 \xrightarrow[As_2O_3]{pt} 2SO_3$, As_2O_3 acts as a -
(A) Autocatalyst (B) Poison
(C) Promotor (D) Positive catalyst

37. The catalyst remains unchanged at end of the reaction as regards - [PET-95]
(A) Quantity and physical state
(B) Chemical composition and physical state
(C) Chemical composition, quantity & physical state
(D) Quantity and chemical composition

38. Which one is false in the following statement ? [PET-97]
(A) A catalyst is specific in its action
(B) A very small amount of the catalyst alters the rate of a reaction
(C) The number of free valencies on the surface of the catalyst increases on subdivision
(D) Ni is used as catalyst in the manufacture of ammonia

39. Which of the following kinds of catalysis can be explained by the adsorption theory ? [PET-98]
(A) Homogeneous catalysis (B) Acid-base catalysis
(C) Heterogeneous catalysis (D) Enzyme catalysis

40. Which of the following requires catalyst ?

(A) $S + O_2 \rightarrow SO_2$ (B) $2SO_2 + O_2 \rightarrow 2SO_3$

(C) $C + O_2 \rightarrow CO_2$ (D) All of the above

41. When a catalyst is added to a system the [JIPMER-2001]

(A) value of equilibrium constant is decreased

(B) the rate of forward reaction is increased and that of backward reaction is decreased

(C) equilibrium concentrations are unchanged

(D) equilibrium concentrations are increased

42. The catalyst used in the manufacture of methanol from water gas is [PET-2002]

(A) V_2O_5 (B) Ni + Mo

(C) $ZnO + Cr_2O_3$ (D) Pt + W

Catalysis

1. (A)
2. (A)
3. (B)
4. (A)
5. (B)
6. (A)
7. (C)
8. (A)
9. (A)
10. (C)
11. (C)
12. (B)
13. (C)
14. (C)
15. (C)
16. (C)
17. (A)
18. (B)
19. (B)
20. (B)
21. (D)
22. (D)
23. (C)
24. (C)
25. (C)
26. (A)
27. (B)
28. (A)
29. (C)
30. (A)
31. (A)
32. (D)
33. (B)
34. (B)
35. (A)
36. (B)
37. (D)
38. (D)
39. (C)
40. (B)
41. (C)
42. (C)

Thought Provoking Questions with Hints

1. Atomic volume is not a direct measure of the size of the atom, why ?

Hint: Atoms can pack together in different crystal structures. This means that, for example, atoms that pack in very open structure will have a large atomic volume; slightly larger atoms that pack in a tighter arrangement will have a smaller atomic volume. This is the reverse of the pattern in their sizes.

2. Why is it not strictly correct to consider electro negativity values as physical properties of elements?

Hint: We can't measure electro negativities because they have been invented by chemists particularly Linus Pauling. They are not properties of atoms. However, the values have been conjured from measurements we can make. e.g. ionisation potentials.

3. Give a short explanation of diagonal relationship in the periodic table in terms of changes in nuclear charge and shielding of electrons.

Hint: Moving one place to the right across a period the increased nuclear charge holds the electrons more tightly to the atom. Moving down one place in the group, the extra shell of electron lessens the attraction of the nucleus for the outer electrons. A diagonal move means that these effects tend to compensate for one another.

4. What is the equation for the nuclear reaction that takes place when tritium undergoes beta decay? What is produced in the reaction?

Hint: ${}_{1}^{3}H \rightarrow {}_{2}^{3}He + {}_{-1}^{0}e$. The product is an isotope of helium.

5. The mass spectrum of a sample of hydrogen gas corresponding to a mass to charge ratio (m/e) value of 2. However there are other lines at higher m/e value. What would their m/e values be?

Hint: A sample of hydrogen gas will contain H_2, HD and D_2 molecules which have m/e values of 2, 3, and 4 respectively. As you would expect, the later two peaks are extremely small in intensities.

6. Explain why AlH_3 would be expected to bond with a hydride ion?

Hint: Like $AlCl_3$, there is an empty orbital on the aluminium which can accept a lone pair of electrons. The hydride ion has the necessary pair of electrons.

7. Write down the formula of the molecule made when $LiAlH_4$ reacts with $H_2C{=}CHCOOH$.

Hint: $H_2C{=}CHCH_2OH$. The double bond is not affected during the reduction. The acid group is converted to an alcohol.

8. Suggest a way of making PH_3, starting with PCl_3.

Hint: One possibility is to react phosphorous trichloride with $LiAlH_4$

$3LiAlH_4 + 4PCl_3 \rightarrow 3LiAl + 3AlCl_3 + 4PH_3$

9. What does molecular orbital theory have to say about the presence of lone pairs in a water molecule?

Hint: The lone pairs are not to be found in molecular orbital theory. This goes to show that we should not mistake the results of mathematical theories as a direct picture of realities.

10. Why is $K^{39.098}$ placed after $Ar^{39.948}$ in the periodic table?

Hint: The elements are arranged in the order of increasing atomic number in the modern classifications. Therefore Ar_{18} is placed before K_{19}.

11. Water shows an amphoteric oxide behaviour when it reacts with HCl and NH_3. Explain.

Hint: $\underset{\text{base}}{H_2O\,(l)} + \underset{\text{acid}}{HCl\,(g)} \rightarrow \underset{\text{acid}}{H_3O^+\,(aq.)} + \underset{\text{base}}{Cl^-\,(aq.)}$

$$\underset{\text{acid}}{H_2O\,(l)} + \underset{\text{base}}{NH_3\,(aq.)} \rightleftharpoons \underset{\text{acid}}{NH_4^+\,(aq.)} + \underset{\text{base}}{OH^-\,(aq.)}$$

In the first case, H_2O is a proton acceptor and in the second a proton donor according to Bronsted-Lowry theory of acids and bases.

12. Calculate the electro negativity of fluorine from the following data. $E_{H-H} = 104.2$ k.cal.mol^{-1}, $E_{F-F} = 36.6$ k.cal.mol^{-1}, $E_{H-F} = 134.6$ k.cal.mol^{-1}; $X_H = 2.1$.

Hint: Substituting the values in Pauling's equation,

$X_F - X_H = 0.208\,[E_{H-F} - (E_{F-F} \times E_{H-H})^{½}]^{½}$

we get, $X_F - 2.1 = 0.208\,[134.6 - (104.2 \times 36.6)^{½}]^{½}$

$X_F = 3.4$

13. Calculate the electronegativity of fluorine if its IE and EA are respectively 17.5 and 3.5 eV.

Hint: Substituting the values in Mulliken equation,

X = IE + EA/ 5.6

we get, for $X_F = 17.5 + 3.5 / 5.6 = 3.75$

14. Calculate the electro negativity of an element with atomic no. 14 by Allred–Rochow method if its covalent radius is 1.18 Å.

Hint: First we calculate the Z_{eff} from the Allred–Rochow equation,

$$X = 0.359 \times \frac{Z_{eff}}{r^2} + 0.744$$

$$Z_{eff} = 14 - (0.35{\times}4 + 0.85{\times}8 + 2{\times}1) = 3.80$$

Now substituting the value we get,

$$X = 0.359\frac{3.80}{(1.18)^2} + 0.744 = 1.72$$

15. What would you expect to happen if you heat washing soda crystals?

Hint: The crystals dissolve in their own water of crystallisation and the steam is given off.

16. Why are washing soda crystals used in bath salts together with a little of colouring and perfume?

Hint: The carbonate ions present in these crystals give a deposit of $CaCO_3$ with Ca^{2+} ions often found in hard water.

17. Why is it that lithium salts have a grater degree of covalent characters than the other halides of the group?

Hint: The small size of Li^+ gives it a huge polarising power. This leads to covalency.

18. Suggest a reason why LiF has the lowest solubility of the group 1 metal halides?
Hint: The small size of both the Li^+ and F^- ions leads to a very large lattice energy, which means the crystal is very hard to break.
19. How many electrons does a hydride ion possess? What does this tell about the statement that all noble gas structures are very stable?
Hint: Two; $1s^2$. It is wrong.
20. A student suggested that calcium should be made if CaO is reacted with Al powder. Was the student correct?
ΔG_f^0 (CaO) = –604.2 kJ mol^{-1};
ΔG_f^0 (Al_2O_3) = –1582.4 kJ mol^{-1}.
Hint: The equation is,
$3CaO (s) + 2Al (s) \rightarrow 3Ca (s) + Al_2O_3 (s)$
$\Delta G^0 = \Delta G_f^0(Al_2O_3) - 3\,\Delta G_f^0(CaO) = +230.2$ kJ mol^{-1}
The positive sign tells us that the reaction cannot occur under standard conditions. But at high temperatures, the free energy change is negative and the reaction takes place. This method is used to extract calcium in industry.
21. How are Zinti compounds formed?
Hint: The first group alkali elements form a number of compounds with 13–16 group members in which they are present as complex polynuclear anions.
22. Why is MgO used for the lining of the steel making furnace?
Hint: MgO is a basic flux. It is helpful in the removal of acidic impurities such as SO_2, P_2O_5, SiO_2 etc from the steel if present forming $MgSO_3$, $Mg_3(PO_4)_2$ and $MgSiO_3$ as slag in each case.
23. Why is the white ash obtained on burning Mg in air gives odour of ammonia on treatment with water?
Hint: Mg_3N_2 (magnesium nitride) formed is present in small quantity in the ash. It undergoes hydrolysis to give ammonia.
$Mg_3N_2 + 6H_2O \rightarrow Mg(OH)_2 + 2NH_3$
24. Why do crystalline salts of alkaline earth metals (2nd group) contain more water of crystallisation than the corresponding salts of alkali metals (1st group)?
Hint: As the ionic radii of second group metals are smaller compared to the first group metal ions, their degree of hydration is higher.
25. Apart from lattice energy, which other energy change is important in accounting for solubilities?
Hint: The solvation energies of ions.
26. Beryllium makes complex ion BeF_4^{2-}. Predict the shape of this ion?
Hint: The electronic configuration of Be^{2+} is $1s^2$. The ion can make four bonds by making use of its empty 2s and 2p orbital. As there are no lone pairs and only four bond pairs to the four fluorine ions, the electron repulsion predicts that the BeF_4^{2-} ion should be tetrahedral in shape and it is.
27. At a temperature of 15°C the energy absorbed when water changes from liquid to gas is about 2500 kJ Kg^{-1}. Estimate the energy used in the evaporation of water over land and sea each year. How much energy is released over land each year if around 125×10^{12} tonnes of precipitation takes place? (Total fall of water to the earth's surface is ~500×10^{12} tonnes and about one quarter of this precipitation for water falling is rain, snow etc.)
Hint: $500\times10^{12}\times10^3$ Kg × 2500 kJ Kg^{-1} = 1.25×10^{21} kJ. This energy is absorbed and leads to a lowering of temperature.
$125\times10^{12}\times10^3$ Kg × 2500 kJ Kg^{-1} = 0.31×10^{21} kJ. This energy is released and thereby tends to increase the temperature of the atmosphere.
28. Why I_2 dissolves more in KI solution than in water?
Hint: Being a covalent molecule, I_2 is sparingly soluble in water. In KI solution it forms a polyhalide KI_3, an ionic compound which is more soluble.
29. Why is neon used in warning signal illuminations?
Hint: The neon illuminated signals are visible from long distances. They are visible under the mist, fog etc. conditions also.
30. Why Tl has a greater tendency to give Tl^+ ions in solution rather than Tl^{3+} ions?
Hint: $E^\circ_{Tl^+/Tl}$ is more negative than $E^\circ_{Tl^{3+}/Tl}$ which suggest that Tl has a greater tendency to turn into Tl^+ in solution.
31. How can the thermit reaction not be stopped by water like an ordinary fire?
Hint: In an ordinary fire the reactions are kept going by the material reacting with oxygen in the air. Water not only cools the fire, but also keeps oxygen out. In the thermit reaction, the oxygen involved is in the metal oxide form, so keeping atmospheric reaction out has no effect. Also, at the temperatures of the thermit reactions ~1000°C water will give off hydrogen with aluminium.
32. What type of bonding holds the layers together in graphite? Why will graphite conduct electricity well in a direction parallel to the planes of the hexagon?
Hint: van der Waal's bonding holds the layers together. The delocalised orbitals in graphite lie parallel to the planes. Electrons can move along the planes making use of these orbitals. Conduction is good parallel to the planes. There is no overlap of orbitals perpendicular to the planes, so conduction is poor in this direction.
33. Which theory other than VBT could be used to explain the bonding in SiF_6^{2-} and predict the shape?
Hint: Molecular orbital theory could be used. Octahedral shape.
34. Copper articles corrode to produce green patches on the surface. What is the chemistry of this corrosion?
Hint: The copper present in the articles on exposure to moist air (oxygen) and carbon dioxide present in atmosphere is converted into basic carbonate which is seen on the surface of layers as green patches.

$2Cu + CO_2 + H_2O + O_2 \rightarrow [Cu(OH)_2CuCO_3]$

35. Why is it reasonable to think that the release of chloride ions from $SiCl_4$ is a favourable process?

Hint: Chloride ions, Cl^- are kinetically and energetically stable in water and they can be solvated by water molecules, so they are good leaving groups in the reaction.

36. Why is the shape of PCl_5 trigonal bipyramidal?

Hint: It minimises the repulsion.

37. Triple superphosphate $Ca(H_2PO_4)_2$ commands a higher price than ordinary super phosphate. Why?

Hint: It contains no anhydride $CaSO_4$, so each kilogram is richer in fertilizer.

38. Even reactive metals like Na do not react with liquid oxygen. Why?

Hint: At the temperature of liquid oxygen (less than 90K) the sodium and oxygen have insufficient energy to get over the energy barrier.

39. How do CO and CN^- ions stop haemoglobin to take up oxygen?

Hint: CO and CN^- are able to bond more strongly to the iron atom in haemoglobin than can oxygen. Haemoglobin would become saturated with these species blocking it to take up oxygen leading to death.

40. Why is it necessary to remove sulphur compounds from oil and natural gas?

Hint: Sulphur compounds easily poison the catalyst used in many industrial processes.

41. Why is $SOCl_2$ preferred to PCl_5 for replacement of OH group in organic reactions?

Hint: The side products from $SOCl_2$ are gases and are easily removed. PCl_5 gives non-volatile $POCl_3$.

42. SO_2 and H_2S react together according to

$2H_2S\ (g) + SO_2\ (g) \rightarrow 3S\ (s) + 2H_2O\ (l)$. Both have reducing properties. Which is more powerful reducing agent?

Hint: The oxidation number of sulphur in SO_2 is +4, and in H_2S is –2. In both cases the sulphur is left with an oxidation number of 0. The SO_2 gains electrons, the H_2S losses them; therefore H_2S reducing agent.

43. Explain why the displacement reactions of the halogens are best regarded as redox reactions.

Hint: When, for example chlorine converts iodide ions into iodine, the iodide ions lose electrons and the chlorine atoms gain electrons. i.e. chlorine oxidises the iodide.

44. Is it true that iodine will not displace any other halogen?

Hint: No. It would displace Astatine from solution.

45. Why does ClF_3 have a distorted T–shape?

Hint: Each fluorine will provide one of the electrons in the three bonding pairs. Chlorine has 7 electrons in its valence shells. This makes ten electrons in total. As six of them are used in bonding, four are left. These make the two lone pairs. Lone pair – bond pair repulsion causes distortion of the molecule.

46. Why should you expect chromium (VI) compounds to show oxidising properties?

Hint: Ions in high oxidation states tend to take electrons from another chemical and return to a lower oxidation state. As it loses electrons the other chemical must have been oxidised.

47. Explain how to know the concentration of a solution of chloride in a silver nitrate solution.

Hint: In titrations $AgNO_3$ is mostly used to determine the concentrations of solutions that contain chloride ions. When all the chloride ions are used up there will be no more precipitate made. To know this point, one can use K_2CrO_4 (potassium chromate) solution as indicator. CrO_4^{2-} ions give a red precipitate of silver chromate. (Ag_2CrO_4) with silver ions which takes place after the precipitation of AgCl is completed.

48. Which is the stronger oxidising agent, manganate(VII) ions in acid or alkaline conditions?

Hint: In alkaline solution. It has the more positive redox potential.

49. Solutions of iron (III) salts are often acidic. Why is it so?

Hint: One may assume that $[Fe(H_2O)_6]^{3+}$ ions are present in the solution. The complex ion can act as a proton donor,

$[Fe(H_2O)_6]^{3+} + H_2O\ (l) \rightarrow [Fe(H_2O)_5OH]^{3+}(aq) + H_3O^+(l)$

50. For dichlorobis (ethane–1,2–diamine) cobalt (III) $[Co(NH_2CH_2CH_2NH_2)_2]Cl_2$ three isomers are possible two are optically active while one is not. Explain.

Hint:

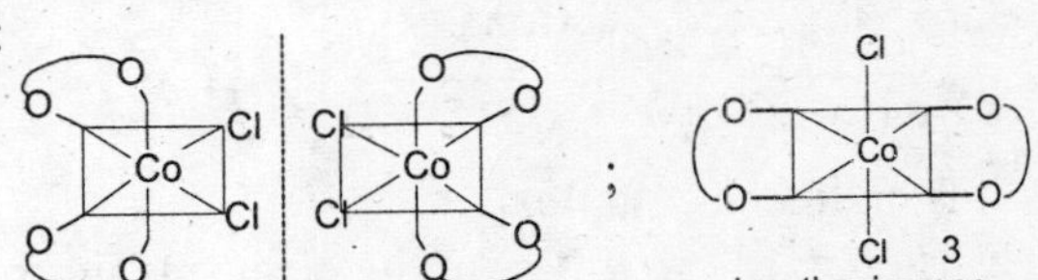

1 optically active 2 ; 3 inactive isomer

The structures 1 and 2 are enantiomers and are optically active. The third one is inactive isomer as it posses both a plane of symmetry and centre of symmetry.

51. What appears to be the oxidation state of iron atoms in $FeFe(CN)^-_6$?

Hint: Each cyanide carries a charge of -1. The two iron atoms take up five negative charges so their average oxidation number is +2.5. In practice there is a mixture of Fe^{2+} and Fe^{3+} ions.

52. Zinc(II) chloride is extremely deliquescent. What might be the reason for the insolubility of zinc(II) fluoride?

Hint: ZnF_2 has very high lattice energy compared to $ZnCl_2$.

53. Dimercury(I) iodide Hg_2I_2, is a greenish colour and is precipitated if iodide ions are added to a solution of dimercury(I) sulphate. Likewise the red mercury(II) iodide HgI_2, is precipitated from a solution of mercury(II) sulphate. However both precipitates dissolve in excess iodide solution. What might be the reason for it?

Hint: A complex ion HgI_4^{2-} is produced. This is a mercury (II) complex. Dimercury(I) complexes do not occur. If a complex can be made in a reaction, it is common for dimercury(I) to change its oxidation state, giving the mercury(II) complex. Infact disproportionation reactions occur.

e.g. $Hg_2^{2+}(aq) + 4I^-(aq) \rightarrow HgI_4^{2-}(aq) + Hg(I)$

54. Why is mercury(II) nitrate an exception to the normal pattern of decomposition by heat?

Hint: Because HgO is easily decomposed by heat, the product of reaction is liquid mercury rather than the oxide.

55. How do you separate ZnS from CdS?

Hint: The solubility product of ZnS is greater than that of CdS. If the acid is present the concentration of free sulphide ions is reduced and becomes too low for ZnS to be precipitated. However CdS will be precipitated. Then add a little acid to the solution before adding the sulphide solution to the mixture. Filter the precipitate of the CdS. Then add little alkali to filtrate to neutralise the acid. Now ZnS will precipitate.

56. How do you account for low spin and high spin complexes with a suitable example?

Hint: d^4, d^5, d^6, d^7 configuration in octahedral complexes exhibits low and high spin.

e.g. (i) $[Fe(H_2O)_6]^{3+}$ in this complex iron is in +3 state i.e. the ion possess 23 electrons. (d^5 configuration)

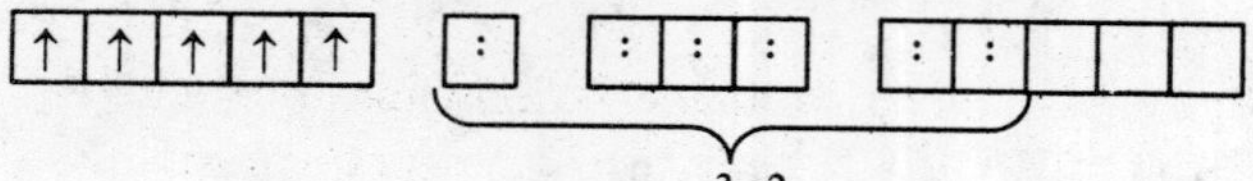

sp^3d^2

There are 5 unpaired electrons.

$\mu_{eff} = \sqrt{n(n+2)}$ BM $= \sqrt{5(5+2)} = \sqrt{35} = 5.92$ BM.

Thus it is high spin complex.

e.g. (ii) $[Fe(CN)_6]^{3-}$ in this case the electrons in iron +3 state can be represented as

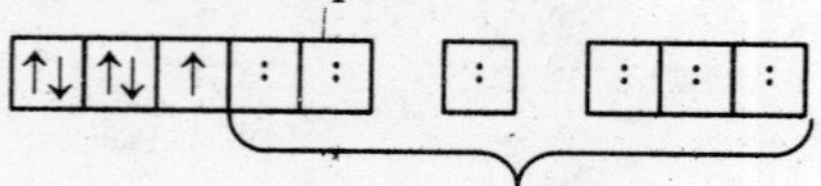

d^2sp^3

$\mu_{eff} = \sqrt{n(n+2)}$ BM $= \sqrt{1(1+2)} = \sqrt{2} = 1.73$ BM

Thus it is low spin complex.

57. Why have the energy levels negative sign?

Hint: Because the zero of potential energy is the world beyond the atom, electrons inside the atoms are lower in potential energy, so they have negative values. We have to put energy in to get the electrons out.

58. What happens to gaps between the energy levels as 'n' increases?

Hint: The gap becomes increasingly small; eventually zero.

59. Why could the mass of a neutron not determined in a mass spectrometer?

Hint: The neutron has no charge. Only charged particles can be deflected and detected in a mass spectrometer.

60. The relative atomic mass of chlorine is quoted as 35.5. A sample of chlorine made in the laboratory is a mixture of Cl_{17}^{35} and Cl_{17}^{37}. What are the percentages of the two isotopes in chlorine?

Hint: We must have $\frac{X}{100} \times 37 + \left[\frac{100-X}{100}\right] \times 35 = 35.5$

X = 25%; therefore, chlorine contains 75% Cl_{17}^{35} and 25% Cl_{17}^{37}.

61. Power stations are about 30% efficient i.e. they convert about 30% of heat energy to electrical energy suppose a nuclear power station uses fuel that generates electrical power at the rate of 1MW g^{-1}, what would be the thermal power generated? How many grams of fuel would have to undergo fission each day? Assuming that 1 tonne of coal can release 25 GJ, how many tonnes of coal would have to be burnt in order to generate 1000 MW of electrical power?

Hint: If 25 GJ i.e. 25000 MJ were released in one second the power would be 25000 MW. So to produce 3 MW would need 0.12 tonnes of coal.

If the power is needed at this level for the whole of one day, we need $= 0.12 \times 24 \times 60 \times 60 = 10368$ tonnes.

62. 236 g U^{236} isotope could provide about 1.6×10^{13} J of heat energy. Because 1 J s^{-1} is the same as one watt, if the 236 g suffered fission in 1 second the power output would be 1.6×10^{13} W. What would be the power output if 1 g of the isotope were used up in one day?

Hint: The power generated is

$$\frac{1.6 \times 10^{13}}{236 \times 24 \times 60 \times 60} = 7.8 \times 10^5 \text{ W which rounds up to 1 MW.}$$

63. Suppose a carbon – 12 atom is formed in a star, which fusion reaction would convert it into an atom of nitrogen – 13?

Hint: $_6C^{12} + {}_1H^1 \rightarrow {}_7N^{13}$

64. It has been estimated that the C^{14} in the atmosphere is responsible for producing 60 atoms of N^{14} and 60 electrons every hour for each gram of carbon. We can quote this disintegration rate as 60 counts hr^{-1} g^{-1}. A sample of a sea shell found near a sea shore was found to have a count of 4 counts $hr^{-1}g^{-1}$. Estimate the age of the shell. ($t_{0.5}$ of is 5730 years).

Hint: Every 5730 years the rate would have halved.

Count rate / counts $hr^{-1}g^{-1}$	60	30	15	7.5	3.75
No. of $t_{0.5}$	0	1	2	3	4

If the count has reduced to 3.75 counts $hr^{-1}g^{-1}$, the lapse of time would be $4 \times 5370 = 21480$. The actual count rate is a little higher than 3.75 counts $hr^{-1}g^{-1}$. So the shell must be less than the calculated age i.e. ~21000 years.

65. What is the maximum distance away from the nucleus that the electron can get? What has happened to the atom then?

Hint: Infinity. It has been ionised.

66. Why could we not discover the ionisation energy of the hydrogen atom by looking at the line in the Balmer series?

Hint: The ionisation energy is the energy difference between the energy levels with $n = 1$ and $n = \infty$. The Balmer series only show transition to the level $n = 2$. We have to look in Lyman series as they contain transitions to the lowest energy level, $n = 1$.

67. Methanoic acid 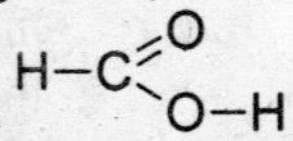 has one carbon–oxygen bond of length 123 pm, and another of 136 pm. But the both carbon oxygen bond in $HCOO^-$ ion have the same length of 127 pm. What do you infer from this?

Hint: In 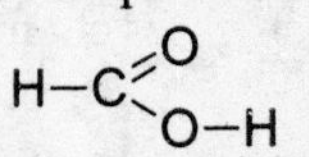 the C=O bond is shorter than the other. In $HCOO^-$, the two bonds have the same length due to resonating structures.

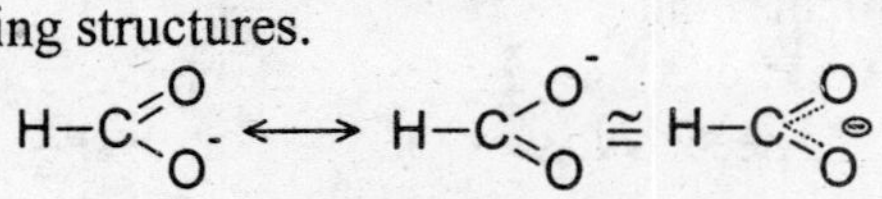

68. What is an important difference between the ions as they appear in the equations for ionisation energies and electron affinities, and the ions in a real ionic compound?

Hint: The key difference is that the equations for the ionisation energies are for atoms and ions in the gaseous state. In practice the ions are found in (solid) crystals. It is the attractions in the crystal that make all the difference in deciding weather an ionic compound exist.

69. Which of the halogens have the highest polarisibility and why?

Hint: It is iodine. It has the largest number of electrons and therefore the most shielding and the greatest polarisability.

70. Whether intermolecular forces exist in NaCl?

Hint: NaCl is an ionic substance. As such it is held together by electrostatic attractions between the ions and not by intermolecular forces. There are no molecules in ionic substances.

71. Which type of intermolecular force would hold H_2, benzene and pentan–3–one together?

Hint: H_2 and benzene are non-polar. These molecules are held together in liquid or solid by dispersion forces. Pentan–3–one has a permanent dipole moment, so the forces are between permanent dipoles.

72. Why does smoke from a fire or a cigarette often have a blue tinge to its colour?

Hint: The smoke is colloidal, so when it is viewed with an angle to the source light, it appears blue due to Tyndall effect.

73. What type of forces hold the layer of carbon atoms in graphite together?

Hint: van der Waal's forces.

74. Why is ZnS not completely ionic in nature?

Hint: The sulphide ion is large and squashy. It is easily polarised by the much smaller zinc ion, and covalence results.

75. Explain what happens to the pressure of an ideal gas if its volume is halved at the same temperature.

Hint: The pressure increases. Halving the volume will double number of collisions of the particles with walls of the container as particles are travelling across the container have a shorter distance to travel before they reach a wall.

76. 100 cc of a solution of HCl were neutralised by 0.12grams of Mg ribbon. What was the concentration of the acid (mol. Wt. of Mg is 24)?

Hint: $Mg\ (s) + 2HCl\ (aq) \rightarrow MgCl_2\ (aq) + H_2O\ (g)$

From this equation we understand that 1 mole of Mg will react with 2 moles of HCl.

No. of moles of Mg used = 0.12 / 24 = 0.005 mole.

No. of moles of HCl used = 0.01 mole, since 1 mol of Mg reacts with two moles of HCl.

Concentration of the acid = 0.1 mole.

77. Why is it important to wash a pipette and especially a burette with distilled water after a titration using alkali?

Hint: If alkali is left in the pipette, it will become encrusted with solid Na_2CO_3 and the jet may become blocked. This can be removed using dil. HNO_3. Alkali left in the burette may block the jet. However in the case of a burette it can led to a deposit of solid sodium carbonate in the socket of the tap. The alkali can also attack the grease that may be used to lubricate the socket.

78. Which of the following reactions represent redox reaction?

(a) $BaCl_2\ (aq) + H_2SO_4(aq) \rightarrow BaSO_4(s) + 2HCl\ (aq)$

(b) $ClO_3^-(aq) + 3Zn(s) + 6H^+(aq) \rightarrow Cl^-(aq) + 3Zn^{2+}(aq) + 3H_2O\ (l)$

(c) $2Cu^{2+}(aq) + 4I^-(aq) \rightarrow 2CuI\ (s) + I_2\ (aq)$

Hint: (a) This is a precipitation reaction not a redox reaction.

(b) Zn atoms are converted into Zn ions. The oxidation number changes from 0 to +2. Zn atoms have lost electrons. They have been oxidised by chlorate (V) ClO_3^- ions. Chlorine in ClO_3^- has +5 oxidation number. As zinc atoms have given up electrons, they are the reducing agents. This is a redox reaction.

(c) The copper(II) ions are converted to copper(I) ions. Some, but not all of the iodide ions which start with an oxidation number of -1 are converted to molecules, which have an oxidation number of 0. Iodide ions have been oxidised. Therefore, in this reaction Cu^{2+} ions are oxidising agents while I^- are reducing agents.

79. Which is likely to be stronger oxidising agent: PbO or PbO_2?

Hint: PbO_2 is a better oxidising agent as Pb is in (IV) oxidation state i.e. higher oxidation state.In PbO, it is (II).

80. In an exothermic reaction how do the strengths of the bonds in the reactants compare with the strengths of the bonds in the products?

Hint: The bonds in the products are (overall) stronger than in the reactants. The stronger the bonds that are made, the more energy is released.

81. 1,3-butadiene has structure that is often shown as

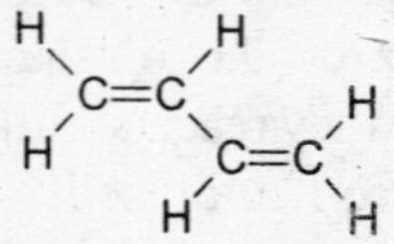

Its heat of hydration is -239 kJmol^{-1}.

The heat of hydrogentation of ethene is -132 kJ mol^{-1}. Could there be delocalised pi bonds in 1,3–butadiene?

Hint: If there were two entirely separate double bonds in 1,3–butadiene we would expect its heat of hydrogenation to be twice that of ethene i.e. -264 kJ mol^{-1}. Infact the value is -239 kJ mol^{-1}. The value is 25 kJ mol^{-1} less than expected from the values, it appears that the pi bond in this molecule is delocalised.

82. Hydrazine N_2H_4 is an endothermic compound. Its heat of formation is 50.6 kJ mol^{-1}. Calculate the enthalpy change for the reaction of burning hydrazine in oxygen: $N_2H_4(l)+O_2 \rightarrow N_2(g) + 2H_2O\ (l)$. Why do you think hydrazine has had some success as a rocket fuel? $\Delta H_f(H_2O) = -285.9$ kJ mol^{-1} : $\Delta H_f\,(N_2H_4) = 50.6$ kJ mol^{-1}.

Hint: $\Delta H_{(reaction)} = -2\ \Delta H_{f}\,(H_2O) - \Delta H_f\,(N_2H_4)$

$= 2\times(-285.9) - (50.6) = -622.4$ kJ mol^{-1}.

The reaction is strongly exothermic. It is an explosively fast reaction. Both the qualities are prime for a rocket fuel.

83. Ammonium nitrate NH_4NO_3 is widely used as a fertilizer. Unfortunately it is also found use as an explosive by terrorists. $\Delta H_f\,(NH_4NO_3) = -364.6$ kJ mol^{-1}. How will you account for the statement?

Hint: We know

$2NH_4NO_3\,(s) \rightarrow 2N_2(g) + 4H_2O(g) + O_2(g)$.

The enthalpy change for the reaction is

$\Delta H_f\,(H_2O) - 2\Delta H_f(NH_4NO_3)$

$= 4\times(-285.9) - 2\times(-364.4) = -414.4$ kJ mol^{-1}.

A good explosive must decompose rapidly and produce large changes in volume by releasing gases. NH_4NO_3 fulfils these requirements. The 7 moles of the gas produced have a volume of over 150000 cc. It is the rapid expansion that forces shock waves through the air which causes the damage.

84. Calculate the lattice energy for KF from the following data

(i) $K(s) + ½\,F_2 \rightarrow KF(s)$ -562.6 kJ mol^{-1}

(ii) $K(s) \rightarrow K\,(g)$ 89.6 kJ mol^{-1}

(iii) $K\,(g) \rightarrow K^+\,(g)$ 419.0 kJ mol^{-1}

(iv) $½\,F_2\,(g) \rightarrow F\,(g)$ 79.1 kJ mol^{-1}

(v) $F\,(g) \rightarrow F^-\,(g)$ -332.6 kJ mol^{-1}

Hint: We know

– (lattice energy) = – (heat of formation) + heats of atomization + ionisation energies + electron affinities.

On substitution we get

= -(-562.6)+89.6+79.1+419-332.6

= 562.6+89.6+79.1+419-332.6 = 817.7

Lattice energy = -817.7 kJ mol^{-1}

85. What relation can be found between ionic radii and the heats of hydration?

Hint: The smaller the ionic radius, the greater is the hydration energy. One of the reason for this is that the water molecules can get closer to the centre of charge, because of which the attractions are increased.

86. Which of the following are spontaneous changes:

(a) a ball rolling downhill

(b) A mixture of nitrogen and oxygen gases changing into separate samples of the two gasses

(c) Instant coffee granules dissolving in water

(d) Petrol burning?

Hint: We know that spontaneous changes take place without the need to do work and non-spontaneous changes take place when work is done.

(a) (c) and (d) do not require us to do work, so they are spontaneous.

87. Why are many metal oxides insoluble in water?

Hint: The oxide ion is small and has a negative charge. Both factors lead to large lattice energies of metal oxides and therefore to insolubility.

88. Why NaI is more soluble in water than NaCl?

Hint: Lattice energy of NaCl is greater than that of NaI by about 82 kJ mol^{-1}. We know that high solubility is favoured by a small lattice energy and a high hydration energy.

89. Would you expect a mixture of (a) Hexane and pentane; (b) propan–1–ol and water will show deviations from Raoult's law?

Hint: (a) Both are non-polar hydrocarbons and therefore no deviation.

(b) propan–1–ol is polar and it can take part in hydrogen bonding. Therefore this mixture shows deviation.

90. Why is benzene added to ethanol and water mixture before distillation?

Hint: Ethanol and water give an azeotropic mixture containing about 95.6 percent ethanol. If benzene is added to this mixture then azeotropic mixture does not form. This enables one to separate ethanol from the mixture completely pure.

91. A solution contains equal concentrations of zinc ions and nickel(II) ions. What would you expect to happen if H_2S is passed through the solution?

Hint: The solubility product of nickel(II) sulphide is 100 times smaller than that of zinc sulphide. Therefore nickel(II) sulphide will be precipitated first. Soon after zinc sulphide would also be precipitated as the solubility products are not very different.

92. Explain why blocks of magnesium can be attached to the hulls of the ships or attached to iron pipes with the aim of preventing rusting.

Hint: Because Mg has a more negative electrode potential than iron. Magnesium acts as an anode. Mg is worn away before the iron.

93. Briefly explain why equal number of electrons must be lost from the cathode as gained at the anode during electrolysis.

Hint: When electricity passes between two points charges are on the move. However the charges are not used up. Therefore the electrons that leave the cathode must enter into anode.

94. When copper(II) sulphate solution is electrolysed using platinum electrodes, the blue colour of the solution gradually fades. Why? After some time when the solution is almost colourless bubbles appear at the cathode. Why do they appear, what are they?

Hint: The blue colour is due to the presence of Cu^{2+} (aq). During the electrolysis, these ions are removed as they change into Cu(s). Hence the colour fades. When the concentration of Cu(II) ions becomes very small, water molecules begin to be discharged. The bubbles are hydrogen.

95. When does electrolysis of Cu(II) sulphate using copper electrodes start?

Hint: When the anode has been worn away so much that it no longer dips properly into the electrolyte.

96. Li^+ has radius of 68 pm and Na^+ has 98 pm radius. Why does the Li^+ travel less easily than Na^+ ion?

Hint: Li^+ ion has a very dense centre of charge and water molecules are attached to it extremely strongly. The ion carries several layers of water molecules around it which greatly increase the effective size of the ion. Hence Li^+ ion does not conduct so well as Na^+ ion.

97. A bottle containing an acid was marked with pH 5. What can we understand from it?

Hint: A weak acid will have a pH in this range. However it is possible that the bottle contains a very dilute solution of a strong acid like H_2SO_4. The pH alone does not tell us whether a solution contains a strong or a weak acid. pH depends upon concentration.

98. Imagine a hydrogen atom and a chlorine atom coming together with a total energy of 10×10^{-19} J. Why is it that the molecule is likely to fall apart almost as soon as it is made?

Hint: The energy available is greater than the energy needed to bring the bond between them. The outcome is that the molecule breaks up and the atoms fly apart as soon as it is made. If the collision is to be successful there has to be a way for the hydrogen chloride molecule to lose energy before it has time to break apart. This can be achieved in the presence of, say nitrogen, where the molecule can give up some of its energy to the nitrogen molecule. It can also happen that gaseous reactions are influenced by collisions of the molecules with the walls of the container.

99. Suggest the best method for measuring the rate of the reaction for $5C_2O_4^{2-}(aq) + 2MnO_4^-(aq) + 16H^+(aq) \rightarrow 10CO_2(g)+2Mn^{2+}(aq)+ 8H_2O$.

Hint: Manganate(VII) ions (MnO_4^-) are purple in solution while Mn^{2+} ions are pale pink. Colorimetric method is the best. It could also be done by measuring the volume of CO_2 given off. This approach has the disadvantage that CO_2 is somewhat soluble in water, so initial measurements of gas volume at the start of the experiment will be erroneous.

100. Two bond energies are sometimes quoted for the bond in the chlorine molecule 242 kJ mol^{-1} and 1150kJmol^{-1}. Which do you think refers to homolysis and to heterolysis?

Hint: The smaller value is for the homolytic fission. Heterolytic fission requires a process very much like ionisation to occur. This requires a great deal of energy.

101. The reaction between hydrogen and bromine is $H_2(g)+Br_2(g) \rightarrow 2HBr(g)$, predict the rate law.

Hint: Rate = $k_1[H_2][Br_2]^{3/2} / k_2[Br_2]+[HBr]$.

The mechanism involves free radicals. Therefore it is wrong to represent it as rate = $k[H_2(g)][Br_2(g)]$.

102. How does the addition of an inert gas at constant volume and constant pressure effect the equilibrium constant?

Hint: The addition of an inert gas at constant volume causes a increase in the pressure of the system. Since K_p is independent of pressure, the degree of advancement of the reaction is not affected. The addition of inert gas at constant pressure causes an increase in the volume of the system. Thus the reaction moves towards larger number of gaseous reactants or products.

103. How is the energy of activation different from the energy of intermediate complex?

Hint: By definition the energy of activation is the difference in the average of reactants and the energy of intermediate complex which is formed while going from reactants to products. The intermediate complex generally lies above the average energy of products.

104. What will happen to the pressure of a gas containing fixed number of molecules at constant volume if the temperature is raised?

Hint: The pressure of the gas is due to the collisions of the molecules with the side of the vessel. An increase in temperature causes increase in average molecular speeds. As the volume is kept constant the molecules will collide more frequently with the sides of the vessel and also impart more change in momentum. Therefore the pressure increases.

105. How many years it would take to spend one Avogadro's number of Rupees at a rate of 10 lakh of rupees in one second?

Hint: Rupees spent in one second = 1000000 i.e. 10^6.
Taking 365 days equivalent to a year
Rs. that can be spent = $10^6 \times 365 \times 24 \times 60 \times 60$
Avogadro's number = 6.02×10^{23}
$\therefore$ Number of years = $(6.02 \times 10^{23}) / (10^6 \times 365 \times 24 \times 60 \times 60)$
= 1.91×10^{10} years.

106. The composition of a sample of Wüstite is $Fe_{0.93}O_{1.00}$. What percentage of the iron is present in the form of Fe(III)?

Hint: The compound is non-stoichiometric where electrical neutrality is achieved by converting appropriate number of Fe^{2+} to Fe^{3+} ions. 7 Fe^{2+} ions are missing out of the expected 100 Fe^{2+} ions. The missing 2×7 positive charge is compensated by the presence of Fe^{3+} ions. Replacement of one Fe^{2+} ion by a Fe^{3+} ion causes the positive charge to increase by 1 unit. Thus the 14 positive charge is compensated by 14 Fe^{3+} ions out of a total of 93 iron ions.

107. What could be inferred from the following graph?

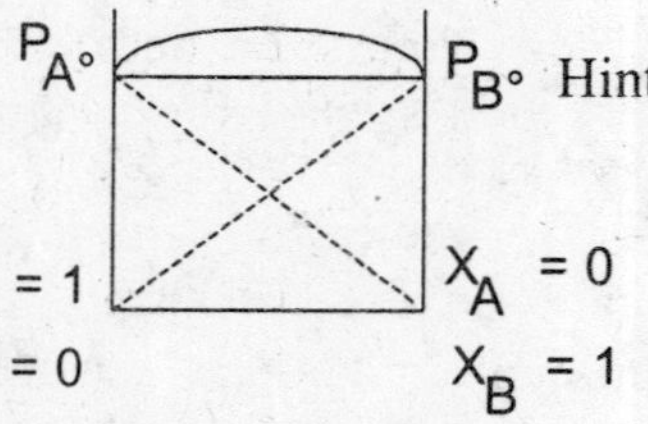

Hint: This shows that there is a positive deviation from Raoult's law.

108. What causes the explosions in houses built on land that were used for rubbish dumps?

Hint: Methane is released when the rubbish decomposes under the ground. It can happen that methane escapes into houses built over the site. The explosion is the rapid burning of methane if, for example a spark starts the reaction.

109. Explain the hybridisation of the carbon atoms in 1,2–butadiene.

Hint: The formula for 1,2–butadiene is

$H_2\underset{1}{C}=\underset{2}{C}=\underset{3}{C}H-\underset{4}{C}H_3$ It is an example of cumulative dienes. The hybridisation of carbon atoms is $sp^2(C_1)$ $sp(C_2)$ $sp^2(C_3)$ and $sp^3(C_4)$. The cumulative double bonds cannot participate in resonance and thereby the delocalisation or displacement of electrons is not possible as is seen in conjugated dienes.

110. Draw the structures of the molecules having $C_2H_2Cl_2$ formula and note which of them show dipole moment?

Hint:

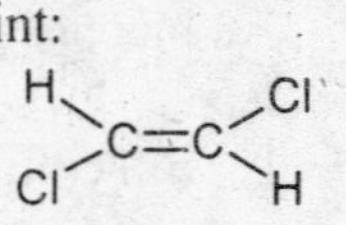
I

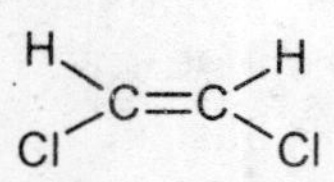
II

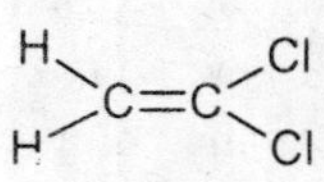
III

The Structure I is trans 1,2–dichloroethene it has no dipole moment. The structure II which is cis 1,2–dichloroethene and the III, 1,1–dichloroethene have dipole moment.

111. If bromine water is used to test for an alkene, the water contains HOBr. What will be made if this molecule adds across the double in ethene?

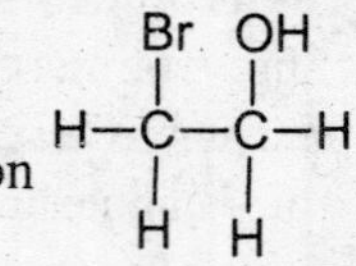

Hint: 2-bromoethan-1-ol(ethylene bromohydrin) is formed. The solution is decolourised.

112. After an ozonolysis experiment the only product obtained is CH_3CHO. What could be the starting compound?

Hint: The Starting alkene could be any of the following as both will deliver acetaldehyde.

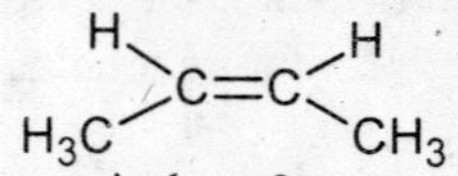
cis-but-2-ene

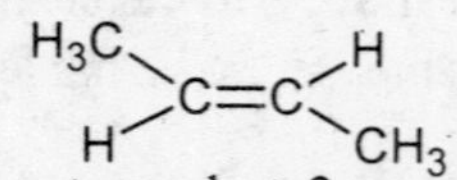
trans-but-2-ene

113. Why is ethyne used in oxy-acetylene welding?

Hint: The exothermic nature of the reaction allows the very high temperature to melt metals to be reached.

114. How do you account for the stability of acylinium ion?

Hint: among the resonance structuresin the acylinium ion (II),every atom is complete with octet configuration while in the structure I the Carbon atom has only six electrons.

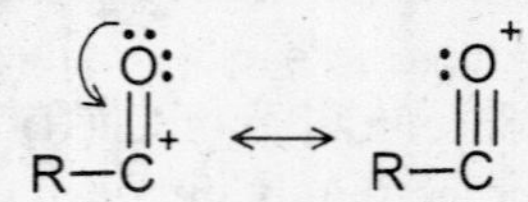
I II

115. How do you account for the oil slick formation on the surface of the ocean (in case of oil tanker accidents)?

Hint: Petroleum which mainly contains paraffins along with other fractions has lower density than water therefore it forms oil slicks on the surface.

116. Predict the structure of alcohols that will be made by the addition of H_2SO_4 followed with reaction by water to but–1–ene and but–2–ene.

Hint:

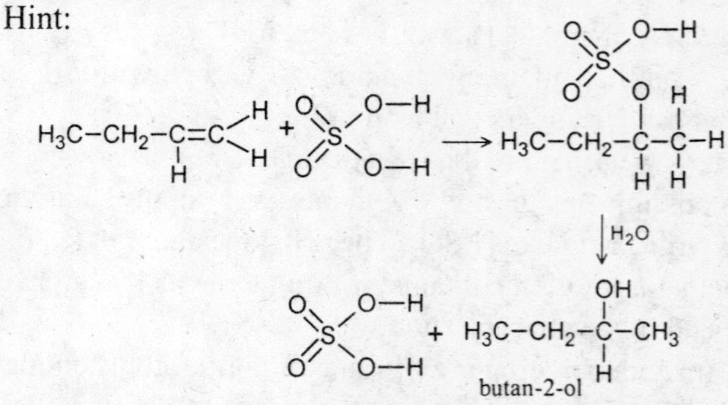
butan-2-ol

With but–2–ene also the product formed is butan–2–ol just in the case of but–1–ene.

117. Why is that if benzene is burned in air, it produces a very smoky flame and large amounts of soot are sent into the air?

Hint: Going to the very large proportion of carbon in benzene, a great deal of oxygen is needed to convert all of it to carbon dioxide.

$C_6H_6(l) + 15O_2(g) \rightarrow 12CO_2(g) + 6H_2O\ (g)$.

In air there is not enough oxygen available close to the flame and some of the carbon remains as soot.

118. Suggest a reason why benzene diazonium ion attacks the para rather than the ortho position on phenol.

Hint: There is steric hindrance at work. The large benzene diazonium ion cannot easily occupy the ortho positions to the phenolic OH group.

119. PCl_5 and $SOCl_2$ can be used to convert ethanol to chloroethane. In which method chloroethane can be separated easily and why?

Hint: $C_2H_5OH + SOCl_2 \rightarrow C_2H_5Cl + SO_2 + HCl$

Sulphur dioxide and hydrogen chloride are gasses and therefore bubble out of the reaction mixture, leaving a reasonably pure product. While with PCl_5, the products $C_2H_5OH + PCl_5 \rightarrow C_2H_5Cl + POCl_3 + HCl$ formed are not volatile as in the former case.

120. What is alcoholometry?

Hint: The method or process by which the percentage of alcohol present in a sample is determined is known as alcoholometry. In this method, the specific gravity of the given sample is measured with the help of a hydrometer. The alcohol percentage is noted from the standard tables coinciding the measured specific gravity (percentage of alcohol in some beverages : Beer 3 to 6, Wine 8 to 10, Whisky 35 to 40, Brandy 40 to 45, Gin 40 to 45, Rum 45 to 55, etc.).

121. How is Pyrene helpful in extinguishing fire?

Hint: CCl_4 is known as pyrene. It is used as fire extinguisher. Its vapour density is very high. The vapours of CCl_4 form a layer around the burning object thus cutting off oxygen supply from the air needed to continue the combustion.

122.Mention the self addition compounds of acetylene.

Hint:(a) 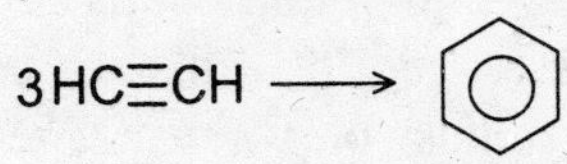

(b) 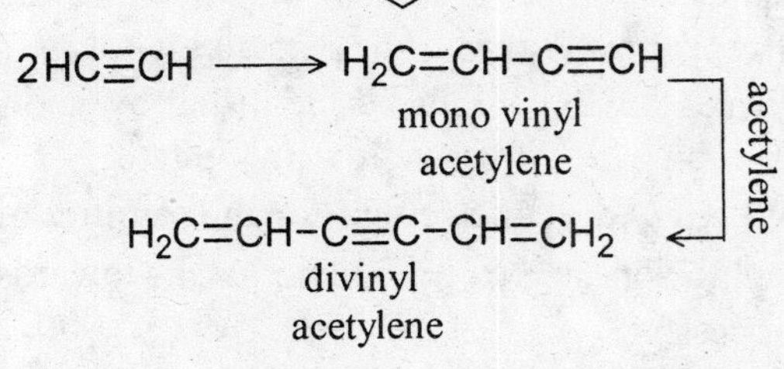

(c) 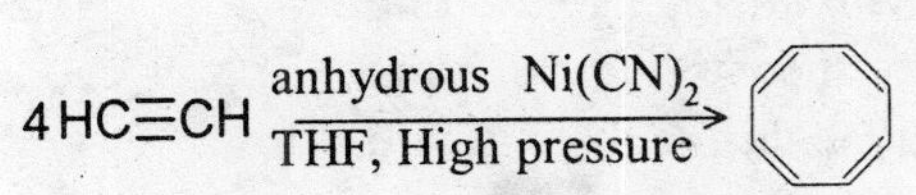

cyclo-octatetraene

123. How do you account for the reaction of HI with 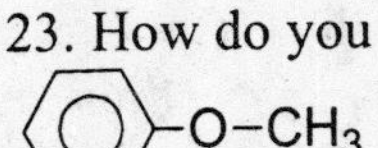 ?

Hint: The possibilities are

(i) 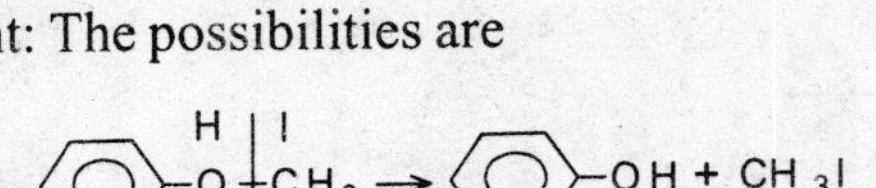

(ii) 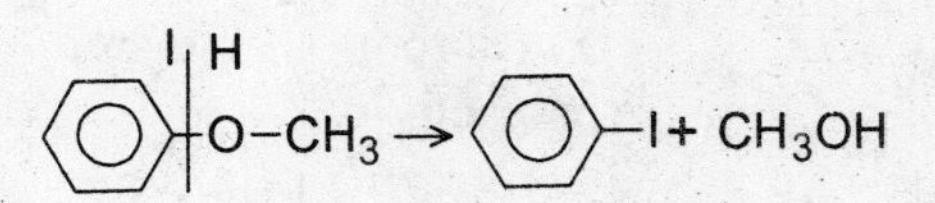

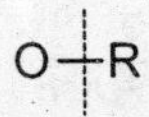 bond cleavage leads to a more stable R^+ (carboaction) which in turn reacts with I^- to give RI.

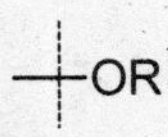 bond cleavage leads to a less stable 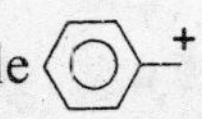 species. Nucleophilic substitution on the benzene ring ($^SN^2$) is not possible while the high energetic species 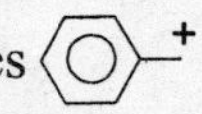 can't form by ($^SN^1$) mechanism.

124. Why benzoic acid is stronger acid than phenol?

Hint: Phenoxide ion shows more resonating structures than the benzoate ion.

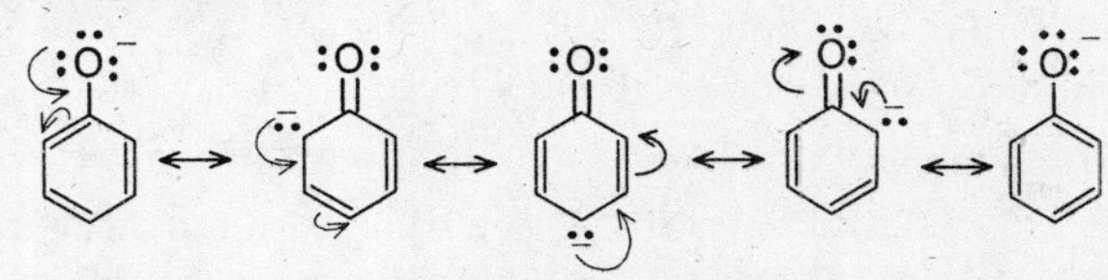

The benzoate ion is more stabilised because the electronegative charge is carried by the oxygen atom.

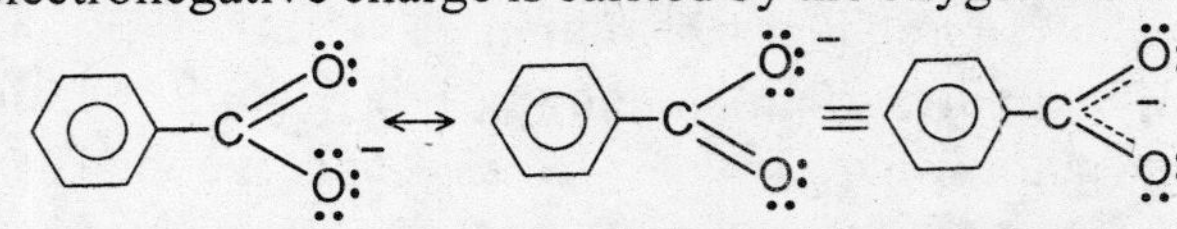

In the resonating structures of phenoxide ions, the negative charge is present on the less electronegative carbon atoms of the ring.

125. Why is it easier to carry out experiments with iodoethane rather than chloroethane?

Hint: C_2H_5Cl is a gas while C_2H_5I is a liquid at room temperature and atmospheric pressure. It is much easier to handle liquids rather than gases in a reaction.

126. Why do alcohols with a long carbon backbone become immiscible with water?

Hint: The miscibility of alcohols with water depends on (i) the attractions between the molecules owing to the hydrogen bonds between them and (ii) the entropy of mixing. If the carbon chain becomes too long. Its presence in water disrupts many hydrogen bonds between the water molecules. This is energetically unfavourable. Eventually the enthalpy change for mixing becomes positive and overcomes the favourable entropy of mixing.

127. What will be the result of using pent–2–ene, (i) by absorbing it in conc. H_2SO_4 followed by dilution with H_2O (ii) by reacting it with diborane followed by H_2O_2?

Hint: (i) The addition is Markownikoff addition. The product formed is $CH_3–CH_2–CH_2–CHOH–CH_3$ (pentan–2–ol).

(ii) It is anti-Markownikoff addition. The product formed is $CH_3–CH_2–CH_2–CH_2–CH_2OH$ (pentan–1–ol).

128. To convert butan–1–ol to 1–chlorobutane using HCl requires heat and a $ZnCl_2$ catalyst. The isomeric 2–methyl–propan–2–ol is converted to 2–chloro–2–methyl

propane by conc. HCl acid at room temperature. Why is there such a difference in reactivity?

Hint: The intermediate formed in the reaction with 2-methyl propan-2-ol is a tertiary carbocation while a primary carbocation is obtained from butan–1–ol. Tertiary carbocation formation is energetically favoured and also more the stability of the carbocation, earlier it forms.

129. What type of alcohol is 1–phenyl ethanol (Ph–CHOH–CH_3). What is the product formed on oxidation with $K_2Cr_2O_7$ solution. What product is expected when this alcohol is subjected to dehydration with conc. H_2SO_4?

Hint: It is a secondary alcohol. Acetophenone, a ketone is formed. On dehydration it forms styrene (phenyl ethene).

130. Phenyl group is known to exert negative inductive effect. But each phenyl ring in diphenyl is more reactive than benzene towards electrophilic substitution. Why?

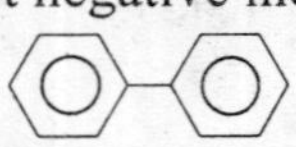

Hint: In diphenyl one of the phenyl groups acts as an electron donor while the other is an electron acceptor. This makes phenyl ring more reactive than benzene towards electrophilic substitution.

131. o–hydroxy benzaldehyde is a liquid at room temperature while p-hydroxy benzaldehyde is a solid with high melting point. Explain.

Hint: In o–hydroxy benzaldehyde intramolecular hydrogen bonding is present while intermolecular hydrogen bonding in the case of p-hydroxy benzaldehyde allows the association of the molecules leading to the solid nature.

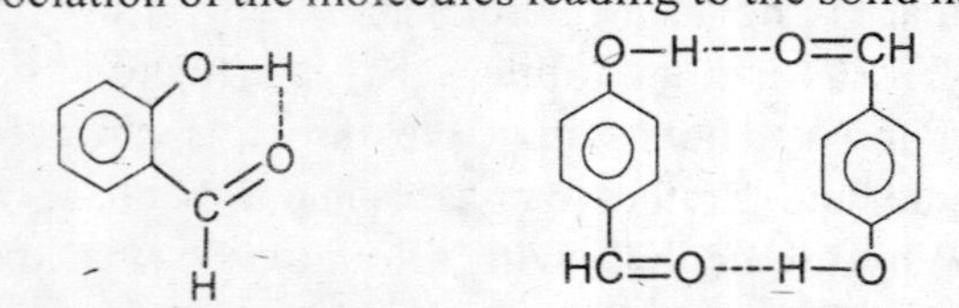

o-hydroxy benzaldehyde p-hydroxy benzaldehyde

132. Why is propanone a liquid at room temperature and pressure while propane is a gas?

Hint: The attractions between dipole moments on different molecules are sufficiently strong to keep the molecules together as a liquid.

133. From which alkene, butanone and pentan–2–one can be obtained?

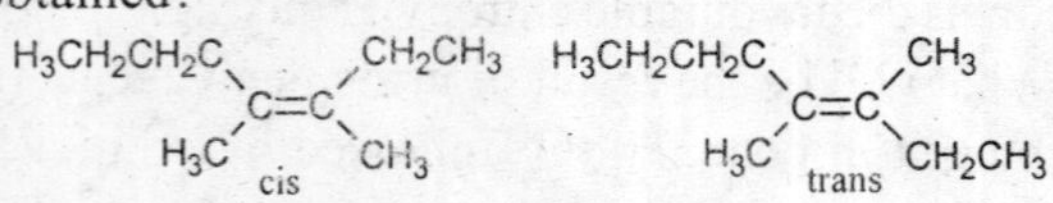

3,4–dimethylhept–3–ene

Hint: both will deliver the same products on ozonolysis.

134. What are crown ethers?

Hint:They are heterocyclic polyethers generally having 4 or more oxygen atoms. They have crown like appearance: The complex of 18–Crown–6 and K^+ $[SCN]^-$ is shown here. There are 18 atoms present in total in the ring. The ring contains 6 oxygen atoms.

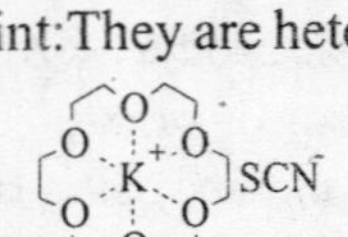

135. The labels of different reagent bottles containing CH_3OH, C_2H_5OH, CH_3COOH and CH_3COCH_3 were destroyed. How would you identify them?

Hint: The following tests are performed in each case.

(i) The one which forms a white precipitate with $NaHSO_3$ is CH_3COOCH_3

(ii) The one which gives brisk effervescence (due to release of CO_2 gas) is CH_3COOH.

Now methanol and ethanol are left. They can be distinguished by iodoform test. C_2H_5OH responds to the test with I_2 and NaOH forming a yellow precipitate.

136. Why does a solution of 2–hydroxypropanoic acid (lactic acid made by the method) $CH_3CHO \rightarrow CH_3CH(OH)CN \rightarrow CH_3CH(OH)COOH$ is optically inactive?

Hint: The cyanide ion can attack the aldehyde from above or below the plane of molecule. The products are mirror images of each other. The resulting solution is optically inactive as it contains equal amount of d and l varieties.

137. What happens when iodine solution is added to propanol?

Hint: Iodine is decolourised. The reaction involves keto–enol tautomerism.

$$H_3C-\overset{\overset{\Large O}{\|}}{C}-CH_3 \rightleftharpoons H_3C-\overset{\overset{\Large OH}{|}}{C}=CH_2$$

138. If the labels of different reagent bottles containing CH_3COCH_3, CH_3CHICH_3, CH_3CH_2CHO and heptane were destroyed, how would you test the samples and identify them?

Hint: (i) CH_3COCH_3 would response iodoform test giving yellow precipitate

(ii) CH_3CH_2CHO would give positive test with schiff's reagent, turning the solution pink. In this case Fehling's solution or Tollen's reagent can also be used for the identification where reddish brown Cu_2O precipitate and silver mirror is produced respectively.

(iii) CH_3CHICH_3 can be identified by adding $AgNO_3$ solution which gives yellow precipitate.

139. How do you account for the stability of α–amino naphthalene and α–nitro naphthalene towards oxidation process?

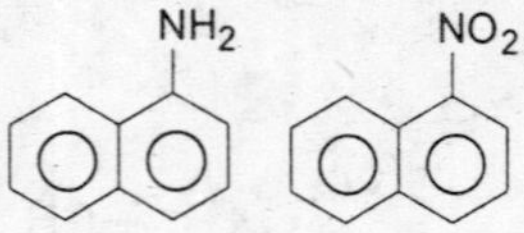

Hint: Oxidising agent is one which accepts the electrons given out during the oxidation process. Therefore the oxidising agent is an electron acceptor one. In 1–nitro naphthalene, the nitro group is an electron withdrawing and thus deactivates the ring i.e. the electron density of the ring is lowered and thereby making it resistant towards oxidation. The amino group being electron repelling group activates the ring and thus making it susceptible to oxidation.

140. In the preparation of a carboxylic acid from nitrile using alkaline hydrolysis. (i) How would you know that

the reaction is complete? (ii) Why very little ammonia is given off when acid is used for hydrolysis?

Hint: (i) No more ammonia is given off.

(ii) ammonia formed in the reaction is converted to NH_4^+ ions by taking H^+ ions from the acid.

141. Why is $CaCl_2$ guard tube used in the preparation of ethanoyl chloride?

Hint: They are used to prevent water vapour entering from the atmosphere and reacting with ethanoyl chloride.

142. How do you account for the b.p. of carboxylic acids and esters?

Hint: Carboxylic acids have higher b.p. due to intermolecular hydrogen bonding (association of molecules takes place). In the case of esters hydrogen bonding formation is not possible.

143. How do you account for the basicity of $CH_3CH_2CH_2NH_2$, $CH_2=CHCH_2NH_2$ and $HC\equiv CCH_2NH_2$?

Hint: In all the cases the nitrogen atom is having a lone pair of electrons showing the basic nature. It can be observed that at β–position with amino group the carbon atom is attached to a single, double and triple bond respectively by sp^3, sp^2 and sp hybradisation. More the s-character in the hybradised orbitals forming the bonds, more the electronegativity and thus increasing the electron – withdrawing capability. Therefore the order is $CH_3CH_2CH_2NH_2 > CH_2=CHCH_2NH_2 > HC\equiv CCH_2NH_2$.

144. How do you account for the acid hydrolysis and alkaline hydrolysis of ester?

Hint: Esters undergo hydrolysis very slowly even with boiling water. In the presence of acids or alkalies, the rate of hydrolysis is higher. As the acid hydrolysis is reversible, it is slower than alkaline hydrolysis which is irreversible.

Acid hydrolysis $RCOOR' + H_2O \overset{H^+}{\rightleftharpoons} RCOOH + R'OH$

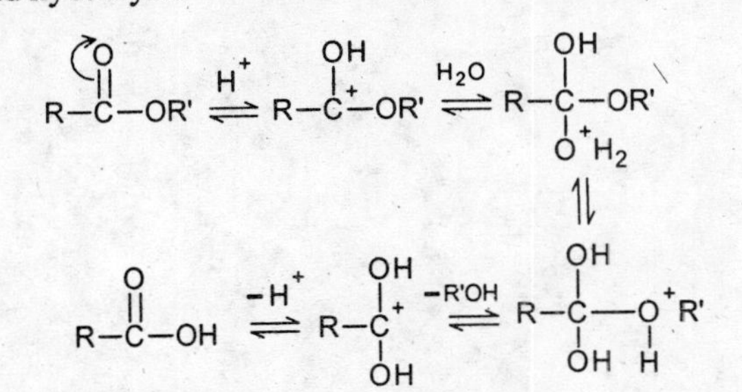

Alkaline hydrolysis

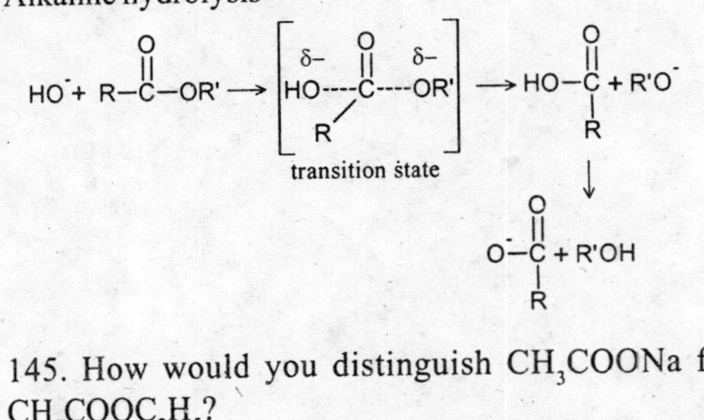

145. How would you distinguish CH_3COONa from $CH_3COOC_2H_5$?

Hint: Sodium acetate is a salt formed by the neutralisation reaction,

$CH_3COOH + NaOH \rightarrow CH_3COONa + H_2O$. It is an ionic solid compound which is soluble in water and undergoes ionisation. Salts of W.A. + S.B. or W.B. + S.A. undergo hydrolysis. Ethyl acetate is an ester formed by esterification,

$CH_3COOH + HOH_5C_2 \rightarrow CH_3COOC_2H_5$.

It is a covalent compound and is a liquid. It undergoes hydrolysis in the presence of acid, base or even water to give acetic acid and ethyl alcohol.

146. Why cyanides are soluble in water while isocynaides are sparingly soluble?

Hint: Alkyl cyanides or nitriles are polar molecules. They form intermolecular hydrogen bonding with water. Thus accounting for higher solubilites. In isonitriles the hydrogen bonding with water molecule is not possible and therefore are insoluble.

147. Why is ethoxyethane more soluble in acids like HCl

Hint: Just as the oxygen atom in water has one of its lone pairs promoted to make the oxonium ion, so a lone pair on the oxygen atom too in an ether can be promoted.Hydrogen bonding can occur between oxygen atom on an ether and hydrogen atoms on surrounding water molecules. Hydrogen bonding cannot occur between oxygen atoms in water molecules and hydrogen atoms bonded to carbon in ether. These bonds are not sufficiently polar to give the hydrogen atoms an appreciable positive charge.

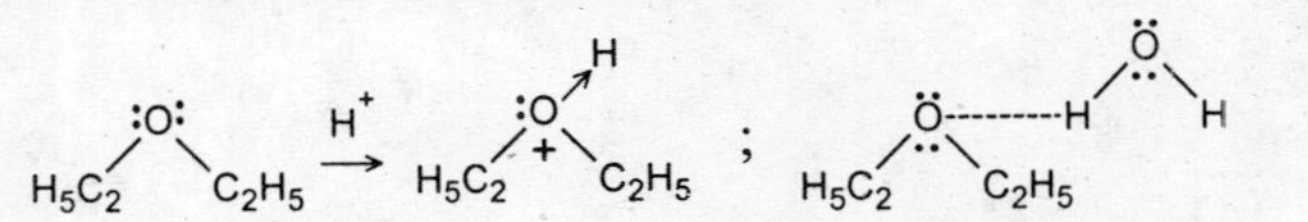

Protonation Hydrogen bonding

148. What would happen if you were to add silver nitrate solution to quaternary ammonium salts?

Hint: The quaternary salt is ionic and when it dissolves, the component ions are free to move through water. The chloride and silver ions give a white precipitate of AgCl.

149. Why are β–keto acids unstable?

Hint: It has been observed that some β–keto acids start loosing CO_2 even at normal temperatures. The same will be brisk at higher temperatures ~100°C.

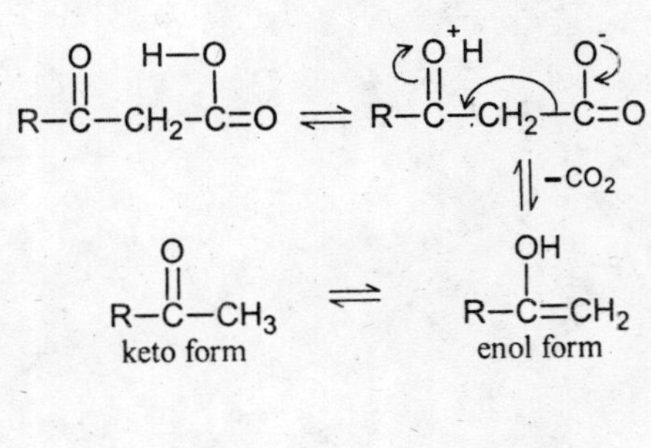

This is due to the possibility of a cyclic transition structure.

150. Why aniline doesn't undergo Friedel-Craft's reaction?

Hint: The lone pair of electrons on nitrogen atom of NH_2 group form a coordinate covalent bond with the Friedel- Craft catalyst i.e. dry $AlCl_3$ and the complex formed is an electron withdrawing thus deactivating the ring towards electrophilic substitution

NOBEL LAUREATES IN CHEMISTRY

Year	Name	Discovery/Contribution
1901	Jacobus H. van 't Hoff	The laws of chemical dynamics and osmotic pressure in solutions.
1902	Emil Fischer	Sugar and purine syntheses.
1903	Svante Arrhenius	Electrolytic theory of dissociation.
1904	Sir William Ramsay	Inert gaseous elements in air, and determination of their place in the periodic system.
1905	Adolf von Baeyer	Organic dyes and hydro-aromatic compounds.
1906	Henri Moissan	Isolation of the element fluorine, and the adoption in the service of science of the electric furnace called after him.
1907	Eduard Buchner	Biochemical researches and discovery of cellfree fermentation.
1908	Ernest Rutherford	Disintegration of the elements, and the chemistry of radioactive substances.
1909	Wilhelm Ostwald	Catalysis and the fundamental principles governing chemical equilibria and rates of reaction.
1910	Otto Wallach	Pioneer work in the field of alicyclic compounds.
1911	Marie Curie	Discovery of the elements radium and polonium.
1912	Victor Grignard	Grignard reagent, which in recent years has greatly advanced the progress of organic chemistry.
	Paul Sabatier	Hydrogenation of organic compounds in presence of finely disintegrated metals.
1913	Alfred Werner	The linkage of atoms in molecules especially in inorganic chemistry.
1914	Theodore W. Richards	Accurate determinations of the atomic weight of a large number of chemical elements.
1915	Richard Willstätter	Plant pigments, especially chlorophyll.
1918	Fritz Haber	Synthesis of ammonia from its elements.
1920	Walther Nernst	Thermochemistry.
1921	Frederick Soddy	Chemistry of radioactive substances, investigations into the origin and nature of isotopes.
1922	Francis W. Aston	Mass Spetrograph of isotopes, in a large number of non-radioactive elements and enunciation of the whole-number rule.
1923	Fritz Pregl	Micro-analysis of organic substances.
1925	Richard Zsigmondy	Heterogenous nature of colloid solutions.
1926	The Svedberg	Disperse systems.
1927	Heinrich Wieland	Constitution of the bile acids and related substances.
1928	Adolf Windaus	Constitution of the sterols and their connection with the vitamins.

NOBEL LAUREATES IN CHEMISTRY

Year	Name	Discovery/Contribution
1929	Arthur Harden and Hans von Euler-Chelpin	Fermentation of sugar and fermentative enzymes.
1930	Hans Fischer	Constitution of haemin and chlorophyll and synthesis of haemin.
1931	Carl Bosch and Friedrich Bergius	Chemical high pressure methods.
1932	Irving Langmuir	Surface chemistry.
1934	Harold C. Urey	Heavy hydrogen.
1935	Frédéric Joliot and Irène Joliot-Curie	Synthesis of new radioactive elements.
1936	Peter Debye	Dipole moments and diffraction of X-rays and electrons in gases.
1937	Norman Haworth and Paul Karrer	Investigations on carbohydrates and vitaminC. Investigations on carotenoids, flavins and vitamins A and B2.
1938	Richard Kuhn	Carotenoids and vitamins.
1939	Adolf Butenandt and Leopold Ruzicka	Sex hormones. Polymethylenes and higher terpenes.
1943	George de Hevesy	Use of isotopes as tracers in the study of chemical processes.
1944	Otto Hahn	Fission of heavy nuclei.
1945	Artturi Virtanen	Agricultural and nutrition chemistry, especially for fodder preservation method.
1946	James B. Sumner and John H. Northrop, Wendell M. Stanley	Crystallisation of enzymes. Preparation of enzymes and virus proteins in a pure form.
1947	Sir Robert Robinson	Plant products of biological importance, the alkaloids.
1948	Arne Tiselius	Electrophoresis and adsorption analysis, especially the complex nature of the serum proteins.
1949	William F. Giauque	Chemical thermodynamics, particularly concerning behaviour of substances at very low temperatures.
1950	Otto Diels and Kurt Alder	Diene synthesis.
1951	Edwin M. McMillan and Glenn T. Seaborg	Chemistry of the transuranium elements.
1952	Archer J.P. Martin and Richard L.M. Synge	Partition chromatography.
1953	Hermann Staudinger	Macromolecular chemistry.
1954	Linus Pauling	Nature of the chemical bond and its application to

NOBEL LAUREATES IN CHEMISTRY

Year	Name	Discovery/Contribution
		the elucidaion of the structure of complex substances.
1955	Vincent du Vigneaud	Biochemically important sulphur compounds, especially for the elucidation of the structure of complex substances.
1956	Sir Cyril Hinshelwood and Nikolay Semenov	Mechanism of chemical reactions.
1957	Lord Todd	Nucleotides and nucleotidic co-enzymes.
1958	Frederick Sanger	Structure of proteins, especially that of Insulin.
1959	Jaroslav Heyrovsky	Polarographic methods of analysis.
1960	Willard F. Libby	Use of Carbon-14 for age determination in archaeology, geology, geophysics and other branches of science.
1961	Melvin Calvin	Carbon dioxide assimilation in plants.
1962	Max F. Perutz and John C. Kendrew	Structure of globular proteins
1963	Karl Ziegler and Giulio Natta	Chemistry and technology of high polymers.
1964	Dorothy Crowfoot Hodgkin	Determination by X-ray techniques of the structures of important biochemical substances.
1965	Robert B. Woodward	For the art of organic synthesis.
1966	Robert S. Mulliken	Chemical bonds and the electronic structure of molecules by the molecular orbital method.
1967	Manfred Eigen, Ronald G.W. Norrish, George Porter	Extremely fast chemical reactions, affected by disturbing the equilibrium by means of very short pulses of energy.
1968	Lars Onsager	The reciprocal relations bearing his name, which are fundamental for the thermodynamics of irreversible processes.
1969	Derek Barton, Odd Hassel	Development of the concept of conformation and its application in chemistry.
1970	Luis Leloir	Sugar nucleotides and their role in the biosynthesis of carbohydrates.
1971	Gerhard Herzberg	Electronic structure and geometry of molecules, particularly free radicals.
1972	Christian Anfinsen,	Ribonuclease, especially concerning the connection between the amino acid sequence and the biologically active confirmation.
	Stanford Moore, William H. Stein	Understanding of he connection between chemical structure and catalytic activity of the active centre of the ribonuclease molecule.

NOBEL LAUREATES IN CHEMISTRY

Year	Name	Discovery/Contribution
1973	Ernst Otto Fischer, Geoffrey Wilkinson	Chemistry of the organometallic, so called sandwich compounds.
1974	Paul J. Flory	Both theoretical and experimental, in the physical chemistry of the macromolecules.
1975	John Cornforth,	Stereochemistry of enzyme-catalyzed reactions .
	Vladimir Prelog	Stereochemistry of organic molecules and reactions.
1976	William Lipscomb	Structure of boranes, illuminating problems of chemical bonding.
1977	Ilya Prigogine	Non-equilibrium thermodynamics, particularly the theory of dissipative structures.
1978	Peter Mitchell	Understanding of biological energy transfer through the formulation of the chemiosmotic theory.
1979	Herbert C. Brown, Georg Wittig	Use of boron-and phosohorus-containing compounds, respectively, into the reagents in organic synthesis.
1980	Paul Berg,	Biochemistry of nucleic acids, with particular regard to recombinant-DNA.
	Walter Gilbert, Frederick Sanger	Determination of base sequences in nucleic acids.
1981	Kenichi Fukui, Roald Hoffmann	Theories, developed independently, concerning the course of chemical reactions.
1982	Aaron Klug	Crystallographic electron microscopy and structural elucidation of biologically important nucleic acid-pro tein complexes.
1983	Henry Taube	Mechanisms of electron transfer reactions, especially in metal complexes.
1984	Bruce Merrifield	Methodology of chemical synthesis on a solid matrix.
1985	Herbert A. Hauptman, Jerome Karle	Direct methods for the determination of crystal structures.
1986	Dudley R. Herschbach, Yuan T. Lee, John C. Polanyi	Dynamics of chemical elementary processes.
1987	Donald J. Cram, Jean-Marie Lehn, Charles J. Pedersen	Development and use of molecules with structure-specific interactions of high selectivity.
1988	Johann Deisenhofer, Robert Huber, Hartmut Michel	Determination of the three-dimensional structure of a photosynthetic reaction centre.
1989	Sidney Altman, Thomas R. Cech	Catalytic properties of R.N.A.

NOBEL LAUREATES IN CHEMISTRY

Year	Name	Discovery/Contribution
1990	Elias James Corey	Development of the theory and methodology of organic synthesis.
1991	Richard R. Ernst	Methodology of high resolution nuclear magnetic resonance (NMR) spectroscopy.
1992	Rudolph A. Marcus	Theory of electron transfer reactions in chemical systems.
1993	Kary B. Mullis, Michael Smith	Polymerase chain reaction (PCR). Oligonucleotide based site-directed mutagenesis and its development for protein studies
1994	George A. Olah	Carbocation chemistry.
1995	Paul J. Crutzen, Mario J. Molina, F. Sherwood Rowland	Atmospheric chemistry particulary concerning the formation and decomposition of ozone.
1996	Robert F. Curl Jr., Sir Harold Kroto, Richard E. Smalley	Fullerenes.
1997	Paul D. Boyer, John E. Walker, Jens C. Skou	For their elucidation of the enzymatic mechanism underlying the synthesis of adenosine triphosphate (A.T.P). Ion-transporting enzyme, Na^+K^+-ATPase.
1998	Walter Kohn, John Pople	Density-functional theory. Computational methods in quantum chemistry.
1999	Ahmed Zewail	Studies of transition states and chemical reactions using femtosecond spectroscopy.
2000	Alan Heeger, Alan G. MacDiarmid, Hideki Shirakawa	Conductive polymers.
2001	William S. Knowles, Ryoji Noyori, K. Barry Sharpless	Catalytic asymmetric synthesis and development of new drugs and material.
2002	John B. Fenn, Koichi Tanaka, Kurt Wüthrich	For developing electro-spray ionisation technique for mass spectrometry. Soft LASER desorption technique for mass spectroscopy 3-D NMR for the identification of biological molecules.
2003	Peter Agre, Roderick MacKinnon	Fundamental chemical knowledge on how cell functions.

Test - taking Tips

1. *Read all directions carefully.*
2. *Read all questions carefully.*
3. *Be sure you know what is being asked.*
4. *Look at all choices before you answer.*
5. *Eliminate answers you know are wrong.*
6. *Paraphrase the questions.*
7. *Think carefully.*
8. *Write Neatly.*
9. *Answer to the point.*
10. *Check your work.*

THINK

If you think you are beaten, you are,
If you think you dare not, you don't,
If you like to win, but you think you can
It is almost certain you won't.

If you think you'll lose. you're lost.
For out of the world we find,
Success begins with a fellow's will
It's all in the state of mind.

If you think you are outclassed, you are,
You've got to think high to rise,
You've got to be sure of yourself before
You can ever win a prize.

Life's battles don't always go
To the stronger or faster man,
But sooner or later the man who wins
Is the man WHO THINKS HE CAN!